HEAR

DO

All of these reso... ...rse that motivates students every step of the way. And ... ...nstructor the tools to monitor student progress and engagement throughout the course. Interested? View details or take an interactive tour: www.wileyplus.com/experience

PRINCIPLES OF HUMAN ANATOMY

Eleventh Edition

Gerard J. Tortora
Bergen Community College

Mark T. Nielsen
University of Utah

WILEY

John Wiley & Sons, Inc.

Executive Editor	Bonnie Roesch
Executive Marketing Manager	Clay Stone
Developmental Editor	Karen Trost
Senior Production Editor	Lisa Wojcik
Senior Media Editor	Linda Muriello
Project Editor	Lorraina Raccuia
Editorial Assistant	Lauren Morris
Senior Designer	Madelyn Lesure
Text Designer	Brian Salisbury/Karin Gerdes Kincheloe
Cover Design	Howard Grossman
Photo Manager	Hilary Newman
Cover Photo	Photograph by Mark Nielsen
Senior Illustration Editors	Claudia Durrell, Anna Melhorn
Illustrators	Kevin Somerville, Imagineering Art, Rich Coombs, John Gibb

This book was typeset by Aptara Corporation and printed and bound by R.R. Donnelley. The cover was printed by Phoenix Color Corporation.

This book is printed on acid free paper. ∞

To order books or for customer service please, call 1-800-CALL WILEY (225-5945).

ISBN-13 978-0-471-78931-4

Printed in the United States of America

10 9 8 7 6 5 4 3 2 1

ABOUT THE AUTHORS

Jerry Tortora is Professor of Biology and former Biology Coordinator at Bergen Community College in Paramus, New Jersey, where he teaches human anatomy and physiology as well as microbiology. He received his bachelor's degree in biology from Fairleigh Dickinson University and his master's degree in science education from Montclair State College. He is a member of many professional organizations, including the Human Anatomy and Physiology Society (HAPS), the American Society of Microbiology (ASM), American Association for the Advancement of Science (AAAS), National Education Association (NEA), and the Metropolitan Association of College and University Biologists (MACUB).

To my mother,
Angelina M. Tortora. Her love,
guidance, faith, support, and example
continue to be the cornerstone of my
personal and professional life.
G.J.T.

Above all, Jerry is devoted to his students and their aspirations. In recognition of this commitment, Jerry was the recipient of MACUB's 1992 President's Memorial Award. In 1996, he received a National Institute for Staff and Organizational Development (NISOD) excellence award from the University of Texas and was selected to represent Bergen Community College in a campaign to increase awareness of the contributions of community colleges to higher education.

Jerry is the author of several best-selling science textbooks and laboratory manuals, a calling that often requires an additional 40 hours per week beyond his teaching responsibilities. Nevertheless, he still makes time for four or five weekly aerobic workouts that include biking and running. He also enjoys attending college basketball and professional hockey games and performances at the Metropolitan Opera House.

Mark Nielsen is a Professor in the Department of Biology at the University of Utah and for the past twenty-one years has taught anatomy and its related subjects to over 16,000 students. In addition to teaching human anatomy in the Department of Biology, he also teaches neuroantomy, embryology, a human dissection course, a teaching human anatomy course, and assits with the comparative vertebrate morphology course. He developed the anatomy course for the physican assistant program at the University of Utah School of Medicine, where he taught for five years, and taught in the cadaver lab at the University of Utah School of Medicine. He developed and continues to help maintain the anatomy and physiology program for the Utah College of Massage Therapy (presently the largest massage school in the United States) and taught his program there for twelve years during its inception and development. His graduate training is in anatomy and his anatomy expertise has a strong basis in dissection. He has prepared and participated in hundreds of dissections of both humans and other vertebrate animals. All his courses incorporate a cadaver-based component in the training with an outstanding exposure to cadaver anatomy. He is a member of the American Association of Anatomists (AAA), the Human Anatomy and Physiology Society (HAPS), and the Anatomical Society of Great Britain and Ireland (ASGBI).

Mark has a passion for teaching anatomy and sharing his knowledge with his students. In addition to the many students he has taught, he has trained and mentored 900 students who have worked in his anatomy laboratory as teaching assistants. His concern for students and his teaching excellence have been acknowledged through numerous awards. He received the prestigous Presidential Teaching Scholar Award at the University of Utah, is a five-time recipient of the University of Utah Student Choice Award for Outstanding Teacher and Mentor, a two time winner of the Outstanding Teacher in the Physician Assistant Program, recipient of the American Massage Therapy Association Jerome Perlinski Teacher of the Year Award, and a two time recipient of Who's Who Among America's Teachers.

To my academic mentors Professors John Legler and Dennis Bramble,

I was lucky to be nurtured academically by two great anatomists, scholars, and teachers. Thank you, I can never thank you enough or repay you for your contributions to my academic career.
M.T.N.

PREFACE

◉ PRINCIPLES OF HUMAN ANATOMY, *eleventh edition,* is designed for introductory courses in human anatomy. The highly successful approach of previous editions–to provide students with an accurate, clearly written, and expertly illustrated presentation of the structure of the human body, to offer insights into the connections between structure and function, and to explore the practical and relevant applications of anatomical knowledge to everyday life and career development—has been enhanced in this edition by innovations designed to increase student motivation and success.

◉ NEW TO THIS EDITION

A very obvious change to this eleventh edition is the addition of a new co-author, Professor Mark Nielsen, of the Univeristy of Utah. Marks's passion for the study of anatomy is evident in his award winning teaching and in the contributions that he has made for the revisions in this text. We have worked with each other in the past in less formal ways, and are pleased to now have the opportunity to fully collaborate on this text revision.

◉ TEXT UPDATES

Every chapter in the eleventh edition of *Principles of Human Anatomy* incorporates a host of improvements to both the text and the art suggested by reviewers, educators, or students. Some noteworthy text changes include the conversion of the chapters on the axial skeleton, the appendicular skeleton, and surface anatomy to the exhibit format used so successfully in many of the other chapters; the transfer of updated versions of most clinical applications and applications to health to a companion clinical supplement with expansion of selected clinical applications into a new text feature called Clinical Connections; and complete rewrites of all chapter introductions.

In addition, every chapter of the text has been revised and updated; see, for example, the revision of the sections on transport across plasma membranes and control of cell destiny in Chapter 2; the revised section on hair growth in Chapter 5; the sections on skeletal muscle structure, nerve and blood supply of skeletal muscle, and development of muscles in Chapter 10; the revised discusssion of the classification of sensory receptors in Chapter 21; and the changes to the discussions of kidney function, incontinence, and hemodialysis in Chapter 25.

Some coverage has been reorganized, such as the section on bone formation in Chapter 6; the move from Chapter 7 to Chapter 6 of the sections on types of bones and bone surface markings; the coverage of joint classification in Chapter 9; and the revised order of the sections on spinal nerves and spinal cord functions in Chapter 18.

New coverage includes new sections on the human body and disease in Chapter 1; transcytosis in Chapter 2; tattooing and body piercing, anti-aging treatments, and sun damage in Chapter 5; factors affecting bone growth, and reticulofibrous and lamellar bone in Chapter 6; knee replacements in Chapter 9; isotonic and isometric contractions in Chapter 10; anatomical and functional organization of the nervous system in Chapter 17; and comparison of the sympathetic and parasympathetic divisions of the autonomic nervous system in Chapter 20.

◉ ART AND DESIGN

This edition boasts a new, larger trim size to allow more flexibility in the layout of the line illustrations and photographs so important to the study of anatomy. The simple design allows the illustrations to be the focal point on each page. In addition, you'll notice the re-design for the updated Clinical Connections within each chapter. These now include illustrations or photos and are boxed, making them more readily apparent and interesting to students. New to this edition is the expanded use of red print used to indicate all callouts of figures and tables. Not only is the reader alerted to refer to the figure or table, but the color print also serves as a place locator for easy return to the narrative.

An outstanding illustration program has always been a signature feature of this text. Beautiful artwork, carefully chosen photographs and photomicrographs, and unique pedagogical enhancements all combine to make the visual appeal and usefulness of the illustration program in *Principles of Human Anatomy* distinctive.

Continuing in this tradition, you will find exciting new three-dimensional illustrations gracing the pages of nearly every chapter in the text. In addition, every orientation diagram in the text has been redrawn to make it easier for students to understand what they are seeing. While there are many new illustrations throughout the text, we paid particular attention to updating and redrawing all of the skeletal bones and muscles [see any figure in chapters 7, 8, 9, or 11]. These new illustrations are among the best that we have ever seen in an anatomy textbook.

◉ CADAVER PHOTOGRAPHS

Mark's talent and expertise have brought about a dramatic increase in the number of cadaver photos in this edition, especially in Chapters 7, 8, 11, and 15, as well as replacement of existing photos with selections from his outstanding collection. These distinctive images were photographed by Mark in his laboratory at the University of Utah. Most of the meticulous dissections are the work of his colleague [and former student] Shawn Miller. Others are dissected by students under Mark's guidance. The matching of these photographs with the line art brings your students that much closer to experiencing an actual cadaver lab.

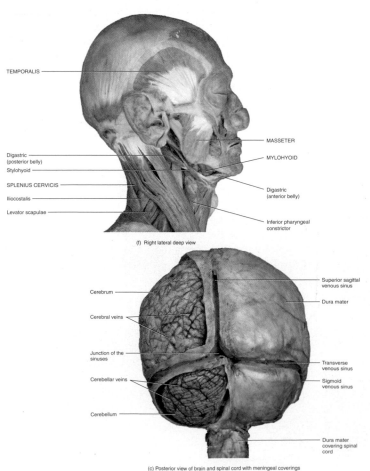

(f) Right lateral deep view

(c) Posterior view of brain and spinal cord with meningeal coverings dissected to show dural sinuses and veins on brain

◉ PHOTOMICROGRAPHS

Mark is also responsible for most of the new photomicrographs included in this edition. Some, showing exploded segments at higher magnification allow students to clearly see specific anatomical details.

(c) Photomicrograph

MP3 DOWNLOADS

An exciting new feature has been added to the illustration program for this edition. MP3 downloads linked to identified illustrations in each chapter give the students the opportunity to hear while they study—as they would in lecture—about the importance and relevance of the structures or concepts that are depicted. These illustrations are identified in each chapter by a a distinctive icon.

◉ COMPLETE TEACHING AND LEARNING PACKAGE

The eleventh edition of principles of human anatomy is accompanied by a host of dynamic resources designed to help you and your students maximize your time and energies. Please contact your Wiley representative for details about these and other resources or visit our website at www.wiley.com/college/sc/totora and click on the text cover to explore the assets more fully.

REAL ANATOMY

◉ *NEW!—REAL ANATOMY*   Mark Nielsen and Shawn Miller, of the University of Utah, led a team of media and anatomical experts in the creation of this powerful new DVD, *Real Anatomy*. Their extensive experience in undergraduate anatomy classrooms and cadaver laboratories as well as their passion for the subject matter shine through this new, user-friendly program with its intuitive interface. The 3-D imaging software allows students to dissect through numerous layers of a three-dimensional real human body to study and learn the anatomical structures of all body systems from multiple perspectives. Histology is viewed via a virtual microscope at varied levels of magnification. Professors can use the program to capture and customize images from a large database of stunning cadaver photographs and clear histology micrographs for presentations, quizzing, or testing.

WileyPLUS is a powerful online tool that provides students and instructors with an integrated suite of teaching and learning resources in one easy-to-use website. With *WileyPLUS*, students will come to class better prepared for your lectures, get immediate feedback and context-sensitive help on assignments and quizzes, and have access to a full range of interactive learning resources, including a complete online version of their text. A description of some of the resources available to students within *WileyPLUS* appears on the front endpapers of this text. Instructors benefit as well with *WileyPLUS*, with all the tools and resources included to prepare and present dynamic lectures as well as assess student progress and learning. New, within Wiley*PLUS*, **Quickstart** is an organizing tool that makes it possible for you to spend less time preparing lectures and grading quizzes and more time teaching and interacting with students. Ask your sales representative to set you up with a test drive, or view a demo online.

◉ VISUAL LIBRARY FOR ANATOMY AND PHYSIOLOGY

4.0 A cross-platform DVD includes all of the illustrations from the textbook in labeled, unlabeled, and unlabeled with leader lines formats. In addition, many illustrations and photographs not included in the text, but which could easily be added to enhance lecture or lab, are included. Search for images by chapter, or by using keywords.

◉ COMPANION WEBSITES

A dynamic website for students, rich with many activities for review and exploration includes self-quizzes for each chapter, Visual Anatomy review exercises, and weblinks. An access code is bundled with each new text. A dedicated companion website for instructors provides many resources for preparing and presenting lectures. Additionally, this website provides a web version of the Visual Library for Anatomy and Physiology, additional critical thinking questions with answers, an editable test bank, a computerized test bank, transparencies on demand, and clicker questions. These websites can be accessed through www.wiley.com/college/tortora.

◉ CLINICAL CONNECTIONS

This manual helps students make the connections between the normal anatomy they are learning and the abnormal situations they might encounter in life or their future careers. Organized by body system, information on diseases and disorders, medical tests and procedures, and medical terminology is added to and enhances the clinical connections within the text. The supplement is available at no extra cost to all students purchasing a new copy of the text.

◉ HUMAN ANATOMY LABORATORY MANUAL by Allen and Harper

This laboratory manual includes multiple activities to enhance student laboratory experience. Illustrations and terminology closely match the text, making this manual the perfect companion. Also available are a Cat Dissection Laboratory Guide and a Fetal Pig Laboratory Guide depending upon your dissection needs. These guides are available to package at no additional cost with the main laboratory manual or as a stand-alone dissection guides.

◉ PHOTOGRAPHIC ATLAS OF THE HUMAN BODY

2e by Tortora Loaded with excellent cadaver photographs and micrographs, the high quality imagery can be used in the classroom, laboratory or for studying and review.

ACKNOWLEDGMENTS

We wish to especially thank several academic colleagues for their helpful contributions to this edition.

- **Kathleen Anderson** of the University of Iowa contributed to the completion of this edition by updating and revising the Self-Quizzes and Critical Thinking Questions at the end of each chapter. In the process she supplied many new, challenging, and diverse test questions.
- **Shawn Miller** of the University of Utah spent many hours in the cadaver lab preparing the exquisite dissections that Mark photographed for use in this text. As you will see as you look through and use this text, his skill is unmatched. We are indebted to him for his major contribution to the quality of the art program we present here. Also, **Nate Mortensen** of the University of Utah, another outstanding dissector, assisted with the dissection and preparation of numerous cadavers.

In addition to the enhancements to the text contributed by those named above, a talented group of academics contributed their insights and expertise in assembling the vast array of ancillary support available for this edition. We wish to acknowledge each and thank them for their work. Special thanks to **Izak Paul**, Mount Royal College for the coordination and review of the instructor's resources included in the **QuickStart** Wiley*Plus* course. Thanks to **Candice Francis**, Palomar College for the creation of the scripts for the MP3 files for selected illustrations; **Richard Connett**, Monroe Community College, **Jon Jackson**, University of North Dakota, and **Peggie Williamson**, Central Texas College, for the Testbank; **Rodney Erickson**, University of Bridgeport, **Virginia Gutierrez-Osborne**, Fresno City College and **Kelly Johnson**, University of Kansas for the Lecture Slides in PowerPoint; **Betsy Brantley**, Lansing Community College for the Student Practice Quizzes; **Frank Baker**, Golden West Coast College, **James Marker**, University of Wisconsin, Green Bay, **Rennee Moore**, Solano Community College, **Julie Rapps** of Carroll College, and **John Wilkins**, Ball State University for the Pre-lecture and Post-lecture Quizzes; **Gary Allen**, Dalhousie University for the creation of Cadaver Lab Videos and Quizzes; and **Robyn O'Kane**, LaGuardia Community College for consulting on and reviewing the exercises in Visual Anatomy. Special thanks to **James Witte** and **Prasanthi Pallapu** of Auburn University and the Institute for Learning Styles Research and **Tom Lancraft** of St. Petersburg College for their collaboration with us in developing questions and tools for students to assess, understand, and apply their learning style preferences.

We are also very grateful to our colleagues who have reviewed the manuscript or participated in focus groups and offered numerous suggestions for improvement: Frank Baker, Golden West Coast College; Leann Blem, Virginia Commonwealth University; Christine Byrd, Western Michigan University; Richard Connett, Monroe Community College; Jonathan R. Day, California State University—Chico; Ruth Ebeling, Biola University; Candice Francis, Palomar College; Roberto Gonzales, Northwest Vista College; Wanda Hargroder, Louisiana State University; Kelly Johnson, University of Kansas; Ray McCoy, The College of

William and Mary; Robin McFarland, Cabrillo College; Michele Monlux, Modesto College; Patricia Munn, Longview Community College; James Rankin, University of Toledo; Montashem Samsam, University of Central Florida; Suzanne Strait-Holman, Marshall University; David Woodman, University of Nebraska; and Kelly A. Young, California State University—Long Branch.

This beautiful textbook would not be possible without the talent and skill of several outstanding medical illustrators. Kevin Sommerville has contributed many illustrations for us over numerous editions. For this edition, many new drawings are the work of his talented hands. We so value the long relationship we have with Kevin. We also welcome two new illustrators to our "team", both former students of Mark's. John Gibb is responsible for all of the new skeletal art and most of the outstanding new muscle illustrations. Richard Coombs contributed several new illustrations for Chapter 1, 16, and 25. And we thank the artists of Imagineering Media Services for all they do to enhance the visuals within this text.

Finally, our hats are off to everyone at Wiley. We enjoy collaborating with this enthusiastic, dedicated, and talented team of publishing professionals. Our thanks to the entire team—Bonnie Roesch, Executive Editor; Karen Trost, Developmental Editor; Lorraina Raccuia, Project Editor; Lauren Morris, Editorial Assistant; Lisa Wojcik, Production Editor; Hilary Newman, Photo Editor; Anna Melhorn and Claudia Durrell, Illustration Coordinators; Madelyn Lesure, Designer; Karin Kincheloe, Page Makeup; and Clay Stone, Executive Marketing Manager.

Last, but not least, thanks to our families and friends who have supported our writing activities in more ways then could ever be mentioned here. Their understanding and encouragement will always be appreciated and will never be taken for granted.

Gerard J. Tortora
Department of Science and Health, S229
Bergen Community College
400 Paramus Road
Paramus, NJ 07652

Mark Nielsen
Department of Biology
University of Utah
257 South 1400 East
Salt Lake City, UT 84112

NOTE TO STUDENTS

Your book has a variety of special features that will make your time studying anatomy a more rewarding experience. These have been developed based on feedback from students—like you—who have used previous editions of the text.

Begin by anticipating what is to be learned from each chapter. A brief, page referenced overview called **Contents at a Glance** begins each chapter. As you start to read each section of the chapter, be sure to take note of the **Objectives** at the beginning of the section to help you focus on what is important as you read. At the end of the section, take time to try and answer the **Checkpoint** questions placed there. If you can, then you are ready to move on. If you have trouble answering the questions, you may want to re-read the section before continuing.

◉ OBJECTIVES
- Describe the components, connective tissue coverings, and branching of a spinal nerve.
- Define a plexus, and identify the distribution of nerves of the cervical, brachial, lumbar, and sacral plexuses.
- Describe the clinical significance of dermatomes.

◉ CHECKPOINT
6. How are the spinal nerves named and numbered? Why are spinal nerves classified as mixed nerves?
7. Describe how a spinal nerve is connected to the spinal cord.
8. Describe the branches and innervations of a typical spinal nerve.
9. Describe the principal plexuses and the regions they supply.

In some cases it is easier to understand the relevance of anatomical structures and the functions they support by considering what happens when they don't work the way they should. The **Clinical Connection** features, which appear throughout the text and within selected Exhibits, present a variety of clinical perspectives related to the text discussion. For example, a discussion of skin cancer appears in the chapter on the integumentary system (Chapter 5), and a Clinical Connection about rotator cuff injury and dislocated and separated shoulder appears within the exhibit on the shoulder joint in Chapter 9, the joints chapter.

Studying the figures (illustrations that include artwork and photographs) in this book is as important as reading the text. To get the most out of the visual parts of this book, use the tools we have added to the figures to help you understand the concepts being presented. Start by reading the **Legend**, which explains what the figure is about. Next, study the **Key Concept Statement**, indicated by a "key" icon, which reveals a basic idea portrayed in the figure. Added to many figures you will also find an **Orientation Diagram** to help you understand the perspective from which you are viewing a particular piece of anatomical art. Finally, at the bottom

● CLINICAL CONNECTION Hypertension

About 50 million Americans have **hypertension**, or persistently high blood pressure. It is the most common disorder affecting the heart and blood vessels and is the major cause of heart failure, kidney disease, and stroke. In May 2003, the Joint National Committee on Prevention, Detection, Evaluation, and Treatment of High Blood Pressure published new guidelines for hypertension because clinical studies have linked what were once considered fairly low blood pressure readings to an increased risk of cardiovascular disease. The new guidelines are as follows:

Category	Systolic (mmHg)	Diastolic (mmHg)
Normal	Less than 120 *and*	Less than 80
Prehypertension	120–139 *or*	80–89
Stage 1 hypertension	140–159 *or*	90–99
Stage 2 hypertension	Greater than 160 *or*	Greater than 100

Using the new guidelines, the normal classification was previously considered optimal; prehypertension now includes many more individuals previously classified as normal or high-normal; stage 1 hypertension is the same as in previous guidelines; and stage 2 hypertension now combines the previous stage 2 and stage 3 categories since treatment options are the same for the former stages 2 and 3.

Types and Causes of Hypertension

Between 90% and 95% of all cases of hypertension are **primary hypertension**, a persistently elevated blood pressure that cannot be attributed to any identifiable cause. The remaining 5–10% of cases are **secondary hypertension**, which has an identifiable underlying cause. Several disorders cause secondary hypertension:

- *Obstruction of renal blood flow* or disorders that damage renal tissue may cause the kidneys to release excessive amounts of renin into the blood. The resulting high level of angiotensin II causes vasoconstriction, thus increasing systemic vascular resistance.
- *Hypersecretion of aldosterone* resulting, for instance, from a tumor of the adrenal cortex stimulates excess reabsorption of salt and water by the kidneys, which increases the volume of body fluids.
- *Hypersecretion of epinephrine and norepinephrine* by a **pheochromocytoma** (fē-ō-krō′-mō-sī-TŌ-ma), a tumor of the adrenal medulla. Epinephrine and norepinephrine increase heart rate and contractility and increase systemic vascular resistance.

Damaging Effects of Untreated Hypertension

High blood pressure is known as the "silent killer" because it can cause considerable damage to the blood vessels, heart, brain, and kidneys before it causes pain or other noticeable symptoms. It is a major risk factor for the number-one (heart disease) and number-three (stroke) causes of death in the United States. In blood vessels, hypertension causes thickening of the tunica media, accelerates development of atherosclerosis and coronary artery disease, and increases systemic vascular resistance. In the heart, hypertension increases the afterload, which forces the ventricles to work harder to eject blood.

The normal response to an increased workload due to vigorous and regular exercise is hypertrophy of the myocardium, especially in the of each figure you will find a **Figure Question**, accompanied by a "question mark" icon. If you try to answer these questions as you go along, they will serve as self-checks to help you understand the material. Often it will be possible to answer a question by examining the figure itself. Other questions will encourage you to integrate the knowledge you've gained by carefully reading the text associated with the figure. Still other questions may prompt you to think critically about the topic at hand or predict a consequence in advance of its description in the text. You will find the answer to each figure question at the end of the chapter in which the figure appears. Selected figures include **Functions** boxes, brief summaries of the functions of the anatomical structure of system shown.

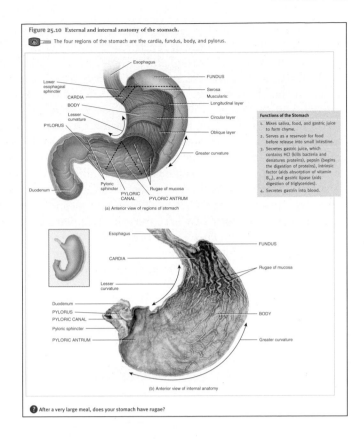

Figure 25.10 External and internal anatomy of the stomach.

The four regions of the stomach are the cardia, fundus, body, and pylorus.

Functions of the Stomach
1. Mixes saliva, food, and gastric juice to form chyme.
2. Serves as a reservoir for food before release into small intestine.
3. Secretes gastric juice, which contains HCl (kills bacteria and denatures proteins), pepsin (begins the digestion of proteins), intrinsic factor (aids absorption of vitamin B$_{12}$), and gastric lipase (aids digestion of triglycerides).
4. Secretes gastrin into blood.

(a) Anterior view of regions of stomach

(b) Anterior view of internal anatomy

After a very large meal, does your stomach have rugae?

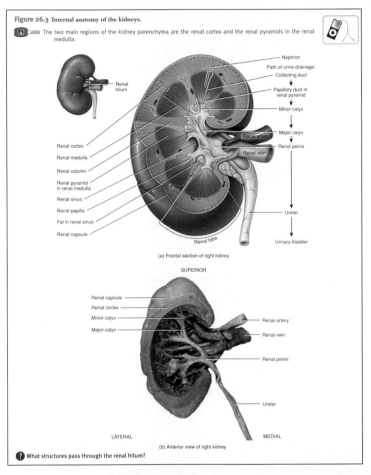

Figure 26.3 Internal anatomy of the kidneys.

The two main regions of the kidney parenchyma are the renal cortex and the renal pyramids in the renal medulla.

(a) Frontal section of right kidney

(b) Anterior view of right kidney

What structures pass through the renal hilum?

In each chapter you will find that several illustrations are marked with an icon that looks like an MP3 player. This is an indication that a download which, narrates and discusses the important elements of that particular illustration is available for your study. You can access these downloads on the student companion website.

Learning the complex anatomy and all of the terminology involved for each body system can be a daunting task. For many topics—including the bones, joints, skeletal muscles, surface anatomy, blood vessels, lymph nodes, and nerves—we have created special **Exhibits** that organize the material into manageable segments. Each Exhibit consists of an objective, an overview, a tabular summary of the relevant anatomy, an associated group of illustrations or photographs, and checkpoint questions. Some Exhibits also contain a relevant Cinical Connection.

At the end of each chapter are other resources that you will find useful. The **Key Medical Terms** section includes selected terms dealing with both normal and pathological conditions. The **Study Outline** is a concise statement of important topics discussed in the chapter. Page numbers are listed next to key concepts so you can refer easily to specific passages in the text for clarification or amplification. The **Self-quiz Questions** are designed to help you evaluate your understanding of the chapter contents. **Critical Thinking Questions** are word problems that allow you to apply the concepts you have studied in the chapter to specific situations. Answers to the Self-Quiz Questions and suggested answers to the Critical Thinking Quesitons (some of which have no one right answer) appear in an appendix at the end of the book so you can check your progress.

At times, you may require extra help to learn specific anatomical features of the various body systems. One way to do this is

through the use of **Mnemonics**, aids to help memory. Mnemonics are included throughout the text, some displayed in figures, tables, or exhibits, and some included within the text discussion. We encourage you to not only use the mnemonics provided, but also to create your own to help you learn the multitude of terms involved in your study of human anatomy.

Throughout the text we have included **Pronunciations** and, sometimes, **Word Roots**, for many terms that may be new to you. These appear in parentheses immediately following the new words, and the pronunciations are repeated in the glossary at the back of the book. Look at the words carefully and say them out loud several times. Learning to pronounce a new word will help you remember it and make it a useful part of your medical vocabulary. Take a few minutes now to read the following pronunciation key, so it will be familiar as you encounter new words. The key is repeated at the beginning of the Glossary, page G-1.

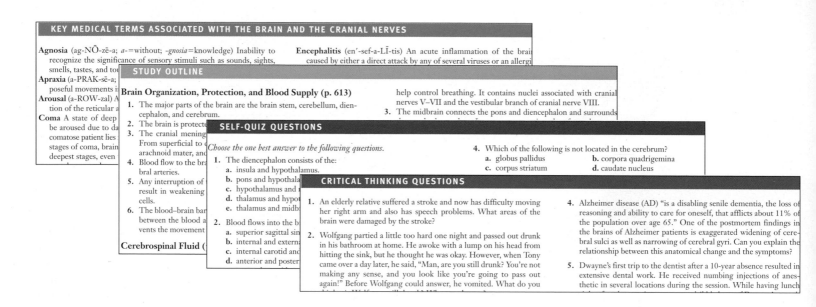

◉ PRONUNCIATION KEY

1. The most strongly accented syllable appears in capital letters, for example, bilateral (bī-LAT-er-al) and diagnosis (dī-ag-NŌ-sis).

2. If there is a secondary accent, it is noted by a prime ('), for example, physiology (fiz'-ē-OL-o-jē). Any additional secondary accents are also noted by a prime, for example, decarboxylation (dē-kar-bok'-si-LĀ-shun).

3. Vowels marked by a line above the letter are pronounced with the long sound, as in the following common words:

ā as in *māke*   *ī* as in *īvy*   *ū* as in *cūte*

ē as in *bē*   *ō* as in *pōle*

4. Vowels not marked by a line above the letter are pronounced with the short sound, as in the following words:

a as in *above* or *at*   *i* as in *sip*   *u* as in *bud*

e as in *bet*   *o* as in *not*

5. Other vowel sounds are indicated as follows:

oy as in *oil*   *oo* as in *root*

Consonant sounds are pronounced as in the following words:

b as in *bat*   *m* as in *mother*
ch as in *chair*   *n* as in *no*
d as in *dog*   *p* as in *pick*
f as in *father*   *r* as in *rib*
g as in *get*   *s* as in *so*
h as in *hat*   *t* as in *tea*
j as in *jump*   *v* as in *very*
k as in *can*   *w* as in *welcome*
ks as in *tax*   *z* as in *zero*
kw as in *quit*   *zh* as in *lesion*
l as in *let*

To provide more assistance in learning the language of anatomy, a full **Glossary** of terms with phonetic pronunciations appears at the end of the book. The basic building blocks of medical terminology—**Combining Forms, Word Roots, Prefixes, and Suffixes**—are listed inside the back cover, as is a listing of **Eponyms**, traditional terms that include reference to a person's name, along with the current terminology.

BRIEF TABLE OF CONTENTS

CONTENTS

- **CLINICAL CONNECTION**
Prostate Cancer **857**   Breast Cancer **876**
Sexually Transmitted Diseases **880**

1 | AN INTRODUCTION TO THE HUMAN BODY

INTRODUCTION You are about to begin a study of the human body to learn how it is organized and how it functions. In order to understand what happens to the body when it is injured, diseased, or placed under stress, you must know how it is organized and how its different parts normally work. Much of what you study in this chapter will help you visualize the body as anatomists do, and you will learn a basic anatomical vocabulary that will help you talk about the body in a way that is understood by both scientists and health-care professionals. •

ANATOMY DEFINED

◉ **OBJECTIVE**

• Define anatomy and physiology, and name several subdisciplines of anatomy.

Anatomy (a-NAT-ō-mē; *ana-*=up;=*-tomy*=process of cutting) is primarily the study of *structure* and the relationships among structures. It was first studied by **dissection** (dis-SEK-shun; *dis-*=apart; *-section*=act of cutting), the careful cutting apart of body structures to study their relationships. Today, a variety of imaging techniques also contribute to the advancement of anatomical knowledge. We will describe and compare some common imaging techniques in Table 1.3 on pages 20–22. The anatomy of the human body can be studied at various levels of structural organization, ranging from microscopic to macroscopic. These levels and the different methods used to study them provide the basis for the subdisciplines of anatomy, several of which are described in Table 1.1.

Anatomy deals mostly with structures of the body. A related discipline, **physiology** (fiz′-ē-OL-ō-jē; *physio-*=nature; *-logy*=study of), deals with *functions* of body parts—that is, how they work. Because function cannot be completely separated from structure, you will learn how each structure of the body often reflects its functions. Some of the structure–function relationships are very obvious; for example, the bones of the skull are tightly joined to protect the brain. The bones of the fingers, by contrast, are more loosely joined to permit various movements. The external ear is shaped in such a way as to collect and localize sound waves, which facilitates hearing. Other relationships are not as obvious; for example, the passageways that carry air into the lungs branch extensively when they reach the lungs. Tiny air sacs—about 300 million—cluster at the ends of the large number of airway branches. Similarly, the vessels carrying blood into the lungs branch extensively to form tiny tubes that surround the small air sacs. Because of these anatomical features, the total surface area within the lungs is about the size of a handball court. This large surface area is the key to the lungs' ability to carry out their efficient exchange of oxygen and carbon dioxide between the air and blood.

◉ **CHECKPOINT**

1. Which subdisciplines of anatomy would be used when dissecting a cadaver?
2. Give several examples of how the structure and function of the human body are related.

TABLE 1.1

Selected Subdisciplines of Anatomy

SUBDISCIPLINE	STUDY OF
Embryology (em′-brē-OL-ō-jē; *embry-*=embryo; *-logy*=study of)	In humans, the first eight weeks of development after the egg is fertilized.
Developmental biology	The complete developmental history of an individual from fertilization to death.
Cell biology	Cellular structure and function.
Histology (hiss′-TOL-ō-jē; *hist-*=tissue)	Microscopic structure of tissues.
Surface anatomy	Surface markings of the body to understand the relations of deep or internal anatomy through visualization and palpation (gentle touch).
Cross-sectional anatomy	Internal structure and relationships of the body through the use of cross-sections.
Gross anatomy	Structures that can be examined without using a microscope.
Systemic anatomy	Structure of specific systems of the body such as the nervous or respiratory systems.
Regional anatomy	Specific regions of the body such as the head or chest.
Radiographic anatomy (rā-dē-ō-GRAF-ik; *radio-*=ray; *-graphic*=to write)	Body structures that can be visualized with x-rays.
Pathological anatomy (path′-ō-LOJ-i-kal; *path-*=disease)	Structural changes (from gross to microscopic) associated with disease.

Several **noninvasive diagnostic techniques** are commonly used by health-care professionals and students to assess certain aspects of body structure and function. In **inspection**, the first diagnostic technique, the examiner observes the body for any changes that deviate from normal (Figure A). In **palpation** (pal-PĀ-shun; *palpa-*=to touch) the examiner feels body surfaces with the hands (Figure B). An example is palpating a blood vessel (called an artery) to find the pulse and measure the heart rate. In **auscultation** (aus-cul-TĀ-shun; *ausculta-*=to listen to) the examiner listens to body sounds to evaluate the functioning of certain organs, often using a stethoscope to amplify the sounds (Figure C). An example is auscultation of the lungs during breathing to check for crackling sounds associated with abnormal fluid accumulation in the lungs. In **percussion** (pur-KUSH-un; *percus-*=to beat) the examiner taps on the body surface with the fingertips and listens to the resulting echo (Figure D). For example, percussion may reveal the abnormal presence of fluid in the lungs or air in the intestines. It is also used to reveal the size, consistency, and position of an underlying structure. •

(A) Inspection of oral (mouth) cavity

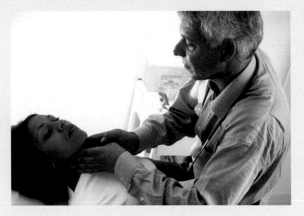

(B) Palpation of lymph nodes in neck

(C) Auscultation of lungs

(D) Percussion of lungs

LEVELS OF BODY ORGANIZATION

◉ **OBJECTIVES**

- Describe the levels of structural organization that make up the human body.
- List the 11 systems of the human body, the organs present in each, and their general functions.

The levels of organization of a language—letters of the alphabet, words, sentences, paragraphs, and so on—can be compared to the levels of organization of the human body. Your exploration of the human body will extend from some of the smallest body structures and their functions to the largest structure—an entire person. From the smallest to the largest size of their components, six levels of organization will help you to understand anatomy: the chemical, cellular, tissue, organ, system, and organismal levels of organization (Figure 1.1).

❶ The **chemical level**, which can be compared to the letters of the alphabet, includes **atoms**, the smallest units of matter that participate in chemical reactions, and **molecules**, two or more atoms joined together. Certain atoms, such as carbon (C), hydrogen (H), oxygen (O), nitrogen (N), phosphorus (P), and calcium (Ca), are essential for life. Two familiar molecules found in the body are deoxyribonucleic acid (DNA), the genetic material passed from one generation to the next, and glucose, commonly known as blood sugar.

❷ At the **cellular level**, molecules combine to form **cells**, the basic structural and functional units of an organism. Just as words are the small building blocks of language, cells are the smallest living units in the human body. Among the many kinds of cells in your body are muscle cells, nerve cells, and blood cells. Figure 1.1 shows a smooth muscle cell, one of three types of muscle cells in the body. The cellular level of organization is the focus of Chapter 2.

❸ The next level of structural organization is the **tissue level**. **Tissues** are groups of cells and the materials surrounding them that work together to perform a particular function. There are just four basic types of tissue in your body: *epithelial tissue, connective tissue, muscular tissue,* and *nervous tissue.* Chapter 3 describes the tissue level of organization. Shown in Figure 1.1 is smooth muscle tissue, which consists of tightly packed smooth muscle cells.

❹ At the **organ level**, different types of tissues are joined together. Similar to the relationship between sentences and paragraphs, **organs** are structures that are composed of two or more different types of tissues; they have specific functions and usually have recognizable shapes. Examples of organs are the stomach, heart, liver, lungs, and brain. Figure 1.1 shows how several tissues make up the stomach. The stomach's outer covering is a *serous membrane*, a layer of epithelial tissue and connective tissue that reduces friction when the stomach moves and rubs against other organs.

Underneath are the *smooth muscle tissue layers*, which contract to churn and mix food and push it on to the next digestive organ, the small intestine. The innermost lining is an *epithelial tissue layer* that produces fluid and chemicals responsible for digestion in the stomach.

⑤ The next level of structural organization in the body is the **system level**, also called the **organ-system level**. A **system** (or chapter in our analogy) consists of related organs (paragraphs) with a common function. An example is the digestive system, which breaks down and absorbs food. Its organs include the mouth, salivary glands, pharynx (throat), esophagus (food tube), stomach, small intestine, large intestine, liver, gallbladder, and pancreas. Sometimes an organ is part of more than one system. The pancreas, for example, is part of both the digestive system and the hormone-producing endocrine system.

⑥ The largest organizational level is the **organismal level**. An **organism**, any living individual, can be compared to a book in our analogy. All the parts of the human body functioning together constitute the total organism.

In the following chapters, you will study the anatomy and some physiology of the body systems. Table 1.2 on pages 6–7 introduces the components and functions of these systems in the order they are discussed in the book.

◉ CHECKPOINT

3. Define the following terms: atom, molecule, cell, tissue, organ, system, and organism.
4. Referring to Table 1.2, which body systems help eliminate wastes?

Figure 1.1 **Levels of structural organization in the human body.**

🔑 The levels of structural organization are chemical, cellular, tissue, organ, system, and organismal.

1 CHEMICAL LEVEL

Atoms (C, H, O, N, P)

Molecule (DNA)

2 CELLULAR LEVEL

Smooth muscle cell

3 TISSUE LEVEL

Smooth muscle tissue

4 ORGAN LEVEL

Serous membrane

Smooth muscle tissue layers

Stomach   Epithelial tissue

5 SYSTEM LEVEL

Esophagus
Liver
Stomach
Pancreas
Gallbladder
Small intestine
Large intestine

Digestive system

6 ORGANISMAL LEVEL

❓ **Which level of structural organization is composed of two or more different types of tissues that work together to perform a specific function?**

TABLE 1.2

The Eleven Systems of the Human Body

Hair

Skin and associated glands

Fingernails (and toenails)

Bone

Cartilage

Joint

Skeletal muscle

INTEGUMENTARY SYSTEM (CHAPTER 5)

Components: Skin, and structures associated with it, such as hair, nails, sweat glands, and oil glands.

Functions: Protects the body; helps regulate body temperature; eliminates some wastes; helps make vitamin D; and detects sensations such as touch, pain, warmth, and cold.

SKELETAL SYSTEM (CHAPTERS 6–9)

Components: Bones and joints of the body and their associated cartilages.

Functions: Supports and protects the body; provides a surface area for muscle attachments; aids body movements; houses cells that produce blood cells; stores minerals and lipids (fats).

MUSCULAR SYSTEM (CHAPTERS 10, 11)

Components: Muscles composed of skeletal muscle tissue, so-named because it is usually attached to bones.

Functions: Produces body movements, such as walking; stabilizes body position (posture); generates heat.

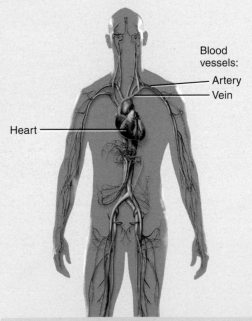

Blood vessels:

Artery

Vein

Heart

Tonsil

Thymus

Thoracic duct

Spleen

Lymph node

Lymphatic vessel

Brain

Spinal cord

Nerve

CARDIOVASCULAR SYSTEM (CHAPTERS 13–15)

Components: Blood, heart, and blood vessels.

Functions: Heart pumps blood through blood vessels; blood carries oxygen and nutrients to cells and carbon dioxide and wastes away from cells and helps regulate acid–base balance, temperature, and water content of body fluids; blood components help defend against disease and repair damaged blood vessels.

LYMPHATIC SYSTEM AND IMMUNITY (CHAPTER 16)

Components: Lymphatic fluid (lymph) and vessels; also includes spleen, thymus, lymph nodes, and tonsils.

Functions: Returns proteins and fluid to blood; carries lipids from gastrointestinal tract to blood; includes structures where lymphocytes that protect against disease-causing microbes mature and proliferate.

NERVOUS SYSTEM (CHAPTERS 17–22)

Components: Brain, spinal cord, nerves, and special sense organs, such as the eyes and ears.

Functions: Generates action potentials (nerve impulses) to regulate body activities; detects changes in the body's internal and external environments, interprets the changes, and responds by causing muscular contractions or glandular secretions.

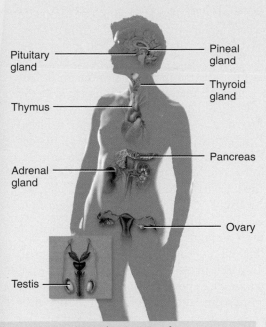

Pituitary gland

Pineal gland

Thyroid gland

Thymus

Pancreas

Adrenal gland

Ovary

Testis

Larynx (voice box)

Pharynx (throat)

Trachea (windpipe)

Bronchus

Lung

Mouth

Pharynx

Salivary gland

Esophagus

Liver

Gallbladder (posterior and inferior to liver)

Stomach

Pancreas (posterior to stomach)

Large intestine

Small intestine

Anus

ENDOCRINE SYSTEM (CHAPTER 23)

Components: Hormone-producing glands (pineal gland, hypothalamus, pituitary gland, thymus, thyroid gland, parathyroid glands, adrenal glands, pancreas, ovaries, and testes) and hormone-producing cells in several other organs.

Functions: Regulates body activities by releasing hormones, which are chemical messengers transported in blood from an endocrine gland or tissue to a target organ.

RESPIRATORY SYSTEM (CHAPTER 24)

Components: Lungs and air passageways such as the pharynx (throat), larynx (voice box), trachea (windpipe), and bronchial tubes leading into and out of the lungs.

Functions: Transfers oxygen from inhaled air to blood and carbon dioxide from blood to exhaled air; helps regulate acid–base balance of body fluids; air flowing out of lungs through vocal cords produces sounds.

DIGESTIVE SYSTEM (CHAPTER 25)

Components: Organs of gastrointestinal tract—a long tube that includes the mouth, pharynx (throat), esophagus, stomach, small and large intestines, and anus; also includes accessory organs that assist in digestive processes, such as the salivary glands, liver, gallbladder, and pancreas.

Functions: Achieves physical and chemical breakdown of food; absorbs nutrients; eliminates solid wastes.

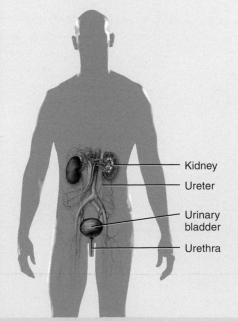

Kidney

Ureter

Urinary bladder

Urethra

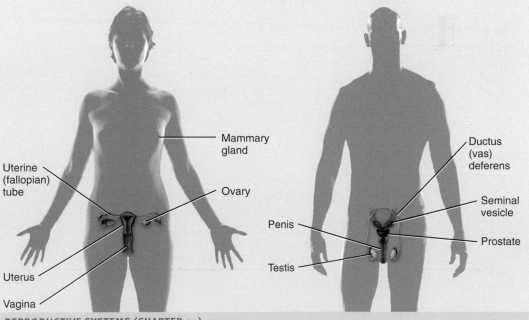

Mammary gland

Uterine (fallopian) tube

Ovary

Uterus

Vagina

Penis

Testis

Ductus (vas) deferens

Seminal vesicle

Prostate

URINARY SYSTEM (CHAPTER 26)

Components: Kidneys, ureters, urinary bladder, and urethra.

Functions: Produces, stores, and eliminates urine; eliminates wastes and regulates volume and chemical composition of blood; helps maintain the acid–base balance of body fluids; maintains body's mineral balance; helps regulate production of red blood cells.

REPRODUCTIVE SYSTEMS (CHAPTER 27)

Components: Gonads (testes in males and ovaries in females) and associated organs (uterine tubes, uterus, and vagina in females and epididymis, ductus deferens, and penis in males). Also, various accessory glands (mammary and vestibular glands in females and prostate and seminal vesicles for males).

Functions: Gonads produce gametes (sperm or oocytes) that unite to form a new organism; gonads also release hormones that regulate reproduction and other body processes; associated organs transport and store gametes.

BASIC ANATOMICAL TERMINOLOGY

◉ **OBJECTIVES**

- Describe the orientation of the human body in the anatomical position.
- Relate the common names to the corresponding anatomical descriptive terms for various regions of the human body.
- Define the anatomical planes, sections, and the directional terms used to describe the human body.

Scientists and health-care professionals use a common language of special terms when referring to body structures and their functions. The language of anatomy has precisely defined meanings that allow us to communicate without using unneeded or ambiguous words. For example, is it correct to say, "The wrist is above the fingers?" This might be true if your upper limbs (described shortly) are at your sides. But if you hold your hands up above your head, your fingers would be above your wrists. To prevent this kind of confusion, anatomists use a standard anatomical position and a special vocabulary for relating body parts to one another.

Anatomical Position

In the study of anatomy, it is important to have a reference point for descriptive purposes. The **anatomical position** is the reference point from which anatomical structures are described. In the anatomical position, the subject stands erect facing the observer, with the head level and the eyes facing directly forward. The feet are flat on the floor and directed forward, and the upper limbs are at the sides with the palms turned forward (Figure 1.2). Once the body is in the anatomical position, it is easier to visualize and understand how it is organized into various regions.

In the anatomical position, the body is upright. Two terms describe a reclining body. If the body is lying face down, it is in the **prone** position. If the body is lying face up, it is in the **supine** position.

Regional Names

The human body is divided into several major regions that can be identified externally. These are the head, neck, trunk, upper limbs, and lower limbs (Figure 1.2). The head consists of the skull and face. The skull encloses and protects the brain, while the face is the front portion of the head that includes the eyes, nose, mouth, forehead, cheeks, and chin. The neck supports the head and attaches it to the trunk. The trunk consists of the chest, abdomen, and pelvis. Each upper limb (extremity) is attached to the trunk and consists of the shoulder, armpit, arm (portion of the limb from the shoulder to the elbow), forearm (portion of the limb from the elbow to the wrist), wrist, and hand. Each lower limb (extremity) is also attached to the trunk and consists of the buttock, thigh (portion of the limb from the buttock to the knee), leg (portion of the limb from the knee to the ankle), ankle, and foot. The *groin* is the area on the front surface of the body marked by a crease on each side, where the trunk attaches to the thighs.

Figure 1.2 shows the common names of major parts of the body. The corresponding anatomical descriptive form (adjective) for each part appears in parentheses next to the common name. For example, if you receive a tetanus shot in your *buttock*, it is a *gluteal* injection. Why does the descriptive form of a body part look different from its common name? The reason is that the descriptive form is based on a Greek or Latin word or "root" for the same part or area. The Latin word for armpit is *axilla* (ak-SIL-a), for example, and thus one of the nerves passing within the armpit is named the axillary nerve. Understanding the word roots of anatomical terms can help you in your learning process. You will learn more about the word roots of anatomical terms as you read this book.

Figure 1.2 The anatomical position. The common names and corresponding anatomical descriptive forms (in parentheses) are indicated for specific body regions. For example, the head is the cephalic region of the body.

In the anatomical position, the subject stands erect facing the observer with the head level and the eyes facing forward. The feet are flat on the floor and directed forward, and the upper limbs are at the sides with the palms facing forward.

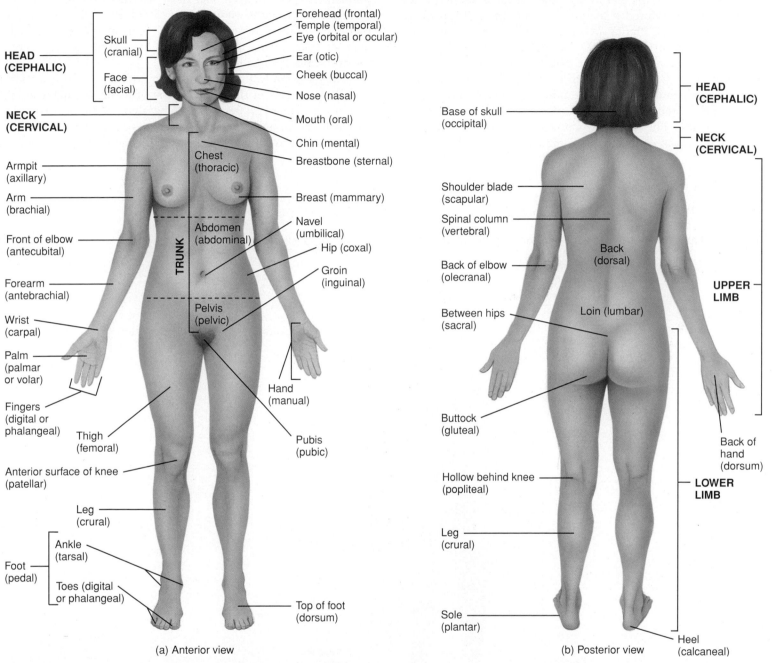

HEAD (CEPHALIC)
Skull (cranial)
Face (facial)
NECK (CERVICAL)

Forehead (frontal)
Temple (temporal)
Eye (orbital or ocular)
Ear (otic)
Cheek (buccal)
Nose (nasal)
Mouth (oral)
Chin (mental)
Breastbone (sternal)

Armpit (axillary)
Arm (brachial)
Front of elbow (antecubital)
Forearm (antebrachial)
Wrist (carpal)
Palm (palmar or volar)
Fingers (digital or phalangeal)
Thigh (femoral)
Anterior surface of knee (patellar)
Leg (crural)
Ankle (tarsal)
Foot (pedal)
Toes (digital or phalangeal)

Chest (thoracic)
Abdomen (abdominal)
TRUNK
Pelvis (pelvic)

Breast (mammary)
Navel (umbilical)
Hip (coxal)
Groin (inguinal)
Hand (manual)
Pubis (pubic)
Top of foot (dorsum)

(a) Anterior view

Base of skull (occipital)
Shoulder blade (scapular)
Spinal column (vertebral)
Back of elbow (olecranal)
Between hips (sacral)
Buttock (gluteal)
Hollow behind knee (popliteal)
Leg (crural)
Sole (plantar)

HEAD (CEPHALIC)
NECK (CERVICAL)
UPPER LIMB
Back (dorsal)
Loin (lumbar)
Back of hand (dorsum)
LOWER LIMB
Heel (calcaneal)

(b) Posterior view

? What is the usefulness of defining one standard anatomical position?

Planes and Sections

As you have just seen, by making reference to various body regions in the anatomical position, it is possible to study the body from the outside. It is also possible to study the internal structure of the body by slicing the body in different ways and examining the sections. The terms that follow describe the different planes and sections encountered in anatomical studies, as well as in many medical procedures. **Planes** are imaginary flat surfaces that pass through the body (Figure 1.3). A **sagittal plane** (SAJ-i-tal; *sagitta-*=arrow) is a vertical plane that divides the body or organ into right and left sides. More specifically, when such a plane passes through the midline of the body or organ and divides it into *equal* right and left sides, it is called a **midsagittal plane**, or a **median plane**. The *midline* is an imaginary vertical line that divides the body into equal left and right sides. If the sagittal plane does not pass through the midline but instead divides the body or organ into *unequal* right and left sides, it is called a **parasagittal plane** (*para-*=beside, near). A **frontal**, or **coronal**, **plane** (kō-RŌ-nal; *corona*=crown) divides the body or organ into front and back portions. A **transverse plane** divides the body or organ into upper and lower portions. A transverse plane may also be termed a *cross-sectional plane* or *horizontal plane*. Sagittal, frontal, and transverse planes are all at right angles to one another. An **oblique plane**, by contrast, passes through the body or organ at an angle between a transverse plane and a sagittal plane or between a transverse plane and a frontal plane.

When you study a body region, you often view it in section. **Sections** are cuts of the body or one of its organs made along one of the planes just described. A section produces a flat two-dimensional surface of the original three-dimensional structure. It is important to know the plane of the section so you can under-

Figure 1.3 Planes through the human body.

Frontal, transverse, sagittal, and oblique planes divide the body in specific ways.

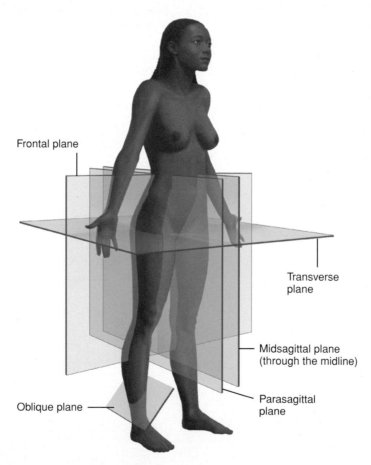

Frontal plane

Transverse plane

Midsagittal plane (through the midline)

Parasagittal plane

Oblique plane

Right anterolateral view

? **Which plane divides the heart into anterior and posterior portions?**

stand the anatomical relationship of one part to another. Figure 1.4 indicates how three different sections—a *transverse (cross) section*, a *frontal section*, and a *midsagittal section*—provide different views of the brain.

◉ **CHECKPOINT**

5. Describe the anatomical position and explain why it is used.

6. Locate each region shown in Figure 1.2 on your own body, and then identify it by its common name and the corresponding anatomical descriptive form.
7. Which of the planes that divide the body are vertical?
8. What is the difference between a plane and a section?

Figure 1.4 **Planes and sections through different parts of the brain.** The diagrams (left) show the planes, and the photographs (right) show the resulting sections. **Note:** The arrows in the diagrams indicate the direction from which each section is viewed. This aid is used throughout the book to indicate viewing perspective.

Planes divide the body in various ways to define sections.

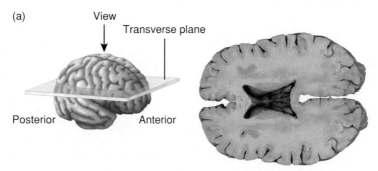

(a)

View

Transverse plane

Posterior   Anterior

Transverse section

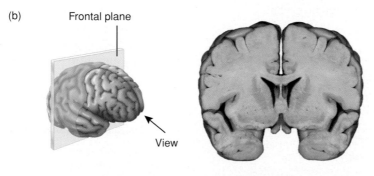

(b)

Frontal plane

View

Frontal section

(c)

Midsagittal plane

View

Midsagittal section

? Which plane divides the brain into unequal right and left portions?

EXHIBIT 1.1 Directional Terms

OBJECTIVE

- Define each directional term used to describe the human body.

Overview

To help improve communication when discussing the basic parts of the body and the relationships those parts have to one another, anatomists and health-care professionals use specific **directional terms**, words that describe the position of one body part relative to another. Most of the directional terms used to describe the relationship of one part of the body to another can be grouped into pairs that have opposite meanings. For example, **superior** means toward the upper part of the body, and **inferior** means toward the lower part of the body. It is important to understand that directional terms have relative meanings; they make sense only when used to describe the position of one structure relative to another. For example, your knee is superior to your ankle, even though both are located in the inferior half of the body. Study the directional terms below and the example of how each is used. As you read the examples, look at Figure 1.5 to see the location of each structure.

CHECKPOINT

Which directional terms can be used to specify the relationships between (1) the elbow and the shoulder, (2) the left and right shoulders, (3) the sternum and the humerus, and (4) the heart and the diaphragm?

DIRECTIONAL TERM	DEFINITION	EXAMPLE OF USE
Superior (soo′-PĒR-ē-or) (cephalic or cranial)	Toward the head, or the upper part of a structure.	The heart is superior to the liver.
Inferior (in′-FĒR-ē-or) (caudal)	Away from the head, or the lower part of a structure.	The stomach is inferior to the lungs.
Anterior (an-TĒR-ē-or) (ventral)*	Nearer to or at the front of the body.	The sternum (breastbone) is anterior to the heart.
Posterior (pos-TĒR-ē-or) (dorsal)*	Nearer to or at the back of the body.	The esophagus is posterior to the trachea (windpipe).
Medial (MĒ-dē-al)	Nearer to the midline.†	The ulna is medial to the radius.
Lateral (LAT-er-al)	Farther from the midline.	The lungs are lateral to the heart.
Intermediate (in′-ter-MĒ-dē-at)	Between two structures.	The transverse colon is intermediate between the ascending and descending colons.
Ipsilateral (ip-si-LAT-er-al)	On the same side of the body as another structure.	The gallbladder and ascending colon are ipsilateral.
Contralateral (CON-tra-lat-er-al)	On the opposite side of the body from another structure.	The ascending and descending colons are contralateral.
Proximal (PROK-si-mal)	Nearer to the attachment of a limb to the trunk; nearer to the origination of a structure.	The humerus is proximal to the radius.
Distal (DIS-tal)	Farther from the attachment of a limb to the trunk; farther from the origination of a structure.	The phalanges are distal to the carpals.
Superficial (soo′-per-FISH-al) (external)	Toward or on the surface of the body.	The ribs are superficial to the lungs.
Deep (internal)	Away from the surface of the body.	The ribs are deep to the skin of the chest and back.

*Note that the terms *anterior* and *ventral* mean the same thing in humans. However, in four-legged animals *ventral* refers to the belly side and is therefore *inferior*. Similarly, the terms *posterior* and *dorsal* mean the same thing in humans, but in four-legged animals *dorsal* refers to the back side and is therefore *superior*.

†Recall that the midline is an imaginary vertical line that divides the body into equal right and left sides.

Figure 1.5 Directional terms.

Directional terms precisely locate various parts of the body relative to one another.

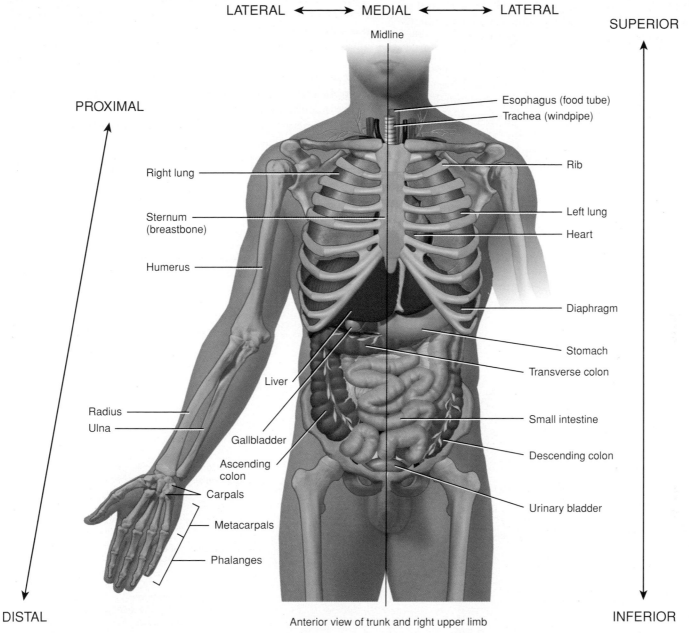

LATERAL ←——→ MEDIAL ←——→ LATERAL

SUPERIOR

Midline

PROXIMAL

Esophagus (food tube)
Trachea (windpipe)

Right lung

Rib

Sternum
(breastbone)

Left lung

Heart

Humerus

Diaphragm

Stomach
Transverse colon

Liver

Radius

Small intestine

Ulna

Descending colon

Gallbladder

Ascending
colon

Carpals

Urinary bladder

Metacarpals

Phalanges

DISTAL

Anterior view of trunk and right upper limb

INFERIOR

Is the radius proximal to the humerus? Is the esophagus anterior to the trachea? Are the ribs superficial to the lungs?
Is the urinary bladder medial to the ascending colon? Is the sternum lateral to the descending colon?

EXHIBIT 1.1   **13**

BODY CAVITIES

⊙ OBJECTIVE

- Describe the principal body cavities, the organs they contain, and their associated linings.

Body cavities are spaces within the body that help protect, separate, and support internal organs. Bones, muscles, and ligaments separate the various body cavities from one another. Here we discuss several of the larger body cavities (Figure 1.6).

The cranial bones form the **cranial cavity**, which contains the brain. The bones of the vertebral column (backbone) form the **vertebral (spinal) canal**, which contains the spinal cord. Three layers of protective tissue, the **meninges** (me-NIN-jēz), line the cranial cavity and the vertebral canal.

The major body cavities of the trunk are the thoracic and abdominopelvic cavities. The **thoracic cavity** (thor-AS-ik; *thorac-*=chest), or chest cavity (Figure 1.6 and Figure 1.7 on pages 16–17), is formed by the ribs, the muscles of the chest, the sternum (breastbone), and the thoracic (chest) portion of the vertebral column (backbone). Within the thoracic cavity are the **pericardial cavity** (per-i-KAR-dē-al; *peri-*=around; *-cardial*=heart), a space containing a small amount of lubricating fluid that surrounds the heart, and two **pleural cavities** (PLOOR-al; *pleur-*=rib or side). Each pleural cavity surrounds one lung and also contains a small amount of lubricating fluid. These cavities will be described in more detail shortly. The central portion of the thoracic cavity is an anatomic region called the **mediastinum** (mē-dē-as-TĪ-num; *media-*=middle; *-stinum*=partition). The mediastinum is between the medial walls of the two pleural cavities and extends from the sternum to the vertebral column, and from the first rib to the diaphragm (Figure 1.7a–d). The mediastinum contains all thoracic organs except the lungs themselves. Among the structures in the mediastinum are the heart, esophagus, trachea, thymus, and several large blood vessels. The **diaphragm** (DĪ-a-fram=partition or wall) is a dome-shaped muscle that separates the thoracic cavity from the abdominopelvic cavity.

Figure 1.6 Body cavities. The dashed lines in (a) and (b) indicate the border between the abdominal and pelvic cavities.

🔑 The major body cavities of the trunk are the thoracic and abdominopelvic cavities.

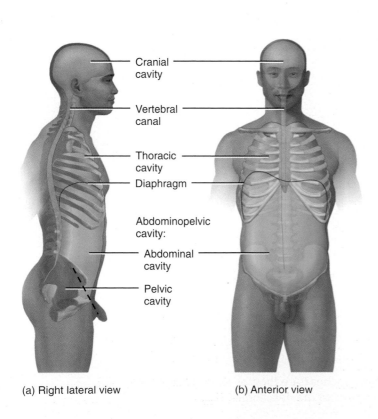

(a) Right lateral view (b) Anterior view

CAVITY	COMMENTS
Cranial cavity	Formed by cranial bones and contains brain.
Vertebral canal	Formed by vertebral column and contains spinal cord and the beginnings of spinal nerves.
Thoracic cavity*	Chest cavity; contains pleural and pericardial cavities and mediastinum.
Pleural cavity	Each surrounds a lung; the serous membrane of each pleural cavity is the pleura.
Pericardial cavity	Surrounds the heart; the serous membrane of the pericardial cavity is the pericardium.
Mediastinum	Anatomic region in the central portion of the thoracic cavity between the medial walls of pleural cavities; extends from sternum to vertebral column and from first rib to diaphragm; contains all the structures of the thoracic cavity other than the lungs, including, for example, the heart, thymus, esophagus, trachea, and several large blood vessels.
Abdominopelvic cavity	Subdivided into abdominal and pelvic cavities.
Abdominal cavity	Contains stomach, spleen, liver, gallbladder, small intestine, and most of large intestine; the serous membrane of the abdominal cavity is the peritoneum.
Pelvic cavity	Contains urinary bladder, portions of large intestine, and internal organs of reproduction.

* See Figure 1.7 for details of the thoracic cavity.

The **abdominopelvic cavity** (ab-dom-i-nō-PEL-vik; see Figure 1.6), extends from the diaphragm to the groin and is encircled by the abdominal muscle wall and the bones and muscles of the pelvis. As the name suggests, the abdominopelvic cavity is divided into two portions, even though no wall separates them. The superior portion, the **abdominal cavity** (*abdomin-*=belly), contains the stomach, spleen, liver, gallbladder, small intestine, and most of the large intestine. The inferior portion, the **pelvic cavity** (*pelv-*=basin), contains the urinary bladder, portions of the large intestine, and internal organs of the reproductive system. Organs inside the thoracic and abdominopelvic cavities are termed the **viscera** (VIS-er-a).

Thoracic and Abdominal Cavity Membranes

A thin, slippery **serous membrane** covers the viscera within the thoracic and abdominal cavities and also lines the walls of the thorax and abdomen. The parts of a serous membrane are (1) the *parietal layer* (pa-RĪ-e-tal=wall of a cavity), which lines the walls of the cavities, and (2) the *visceral layer*, which covers and adheres to the viscera within the cavities. A small amount of lubricating fluid in the cavity between the two layers reduces friction, allowing the viscera to slide somewhat during movements— for example, when the lungs inflate and deflate during breathing.

The serous membrane associated with the lungs is called the **pleura** (PLOO-ra) (Figure 1.7a, c, f). The *visceral pleura* clings to the surface of the lungs; the *parietal pleura* lines the chest wall and covers the superior surface of the diaphragm. In between is the *pleural cavity*, filled with a small volume of lubricating fluid. The serous membrane of the heart is the **pericardium** (see Figure 1.7a–c, e). The *visceral pericardium* covers the surface of the heart, and the *parietal pericardium* lines the fibrous pericardium that surrounds the heart. Between them is the *pericardial cavity*, which contains a small amount of lubricating fluid. The **peritoneum** (per′-i-tō-NĒ-um) is the serous membrane of the abdominal

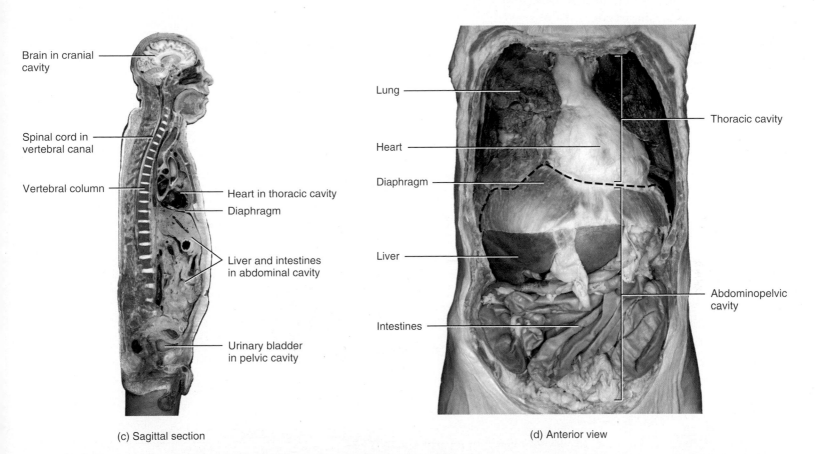

Brain in cranial cavity

Spinal cord in vertebral canal

Vertebral column

Heart in thoracic cavity

Diaphragm

Liver and intestines in abdominal cavity

Urinary bladder in pelvic cavity

(c) Sagittal section

Lung

Heart

Diaphragm

Liver

Intestines

Thoracic cavity

Abdominopelvic cavity

(d) Anterior view

? In which cavities are the following organs located: urinary bladder, stomach, heart, small intestine, lungs, internal female reproductive organs, thymus, spleen, liver? Use the following symbols for your response: T = thoracic cavity, A = abdominal cavity, or P = pelvic cavity.

cavity (see Figure 25.3a on page 785). The *visceral peritoneum* covers the abdominal viscera, and the *parietal peritoneum* lines the abdominal wall and covers the inferior surface of the diaphragm. Between them is the *peritoneal cavity*, which contains a small amount of lubricating fluid. Most abdominal organs are surrounded by the peritoneal cavity. However, some are located between the parietal peritoneum and the posterior abdominal wall. Such organs are said to be *retroperitoneal* (re-trō-per′-i-tō-NĒ-al; *retro-*=behind). The kidneys, adrenal glands, pancreas, duodenum of the small intestine, ascending and descending colons of the large intestine, and the abdominal aorta and inferior vena cava are retroperitoneal.

In addition to the body cavities just described, you will learn about other body cavities in later chapters. These include the *oral* (*mouth*) *cavity*, which contains the tongue and teeth; the *nasal cavity* in the nose; the *orbital cavities*, which contain the eyeballs; the *middle ear cavities*, which contain small bones in the middle ear; and *synovial cavities*, which are found in freely movable joints and contain synovial fluid. A summary of the major body cavities and their membranes is presented in Figure 1.6.

⦿ **CHECKPOINT**

9. What structures separate the various body cavities from one another?
10. What structures are found in the mediastinum?

Figure 1.7 The thoracic cavity. The dashed lines in (a), (b), and (d) indicate the borders of the mediastinum. Notice that the pericardial cavity surrounds the heart, and that the pleural cavities surround the lungs. **Note:** When transverse sections, such as those shown in (b) and (c), are viewed inferiorly (from below) the anterior aspect of the body appears on the top of the illustration and the left side of the body appears on the right side of the illustration.

🔑 The mediastinum is an anatomical region that is medial to the lungs and extends from the sternum to the vertebral column and from the first rib to the diaphragm.

(a) Anterior view

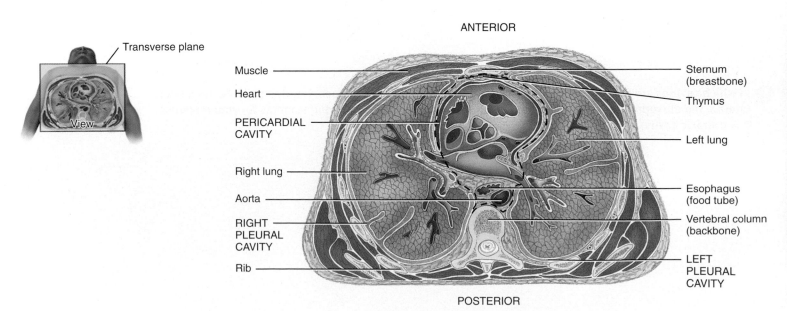

(b) Inferior view of transverse section of thoracic cavity

ANTERIOR

Parietal pericardium

Parietal pleura

RIGHT PLEURAL CAVITY

Skin

Visceral pleura

Right lung

Rib

Sternum

PERICARDIAL CAVITY

Visceral pericardium

Heart

Esophagus

Aorta

Vertebral column (backbone)

Spinal cord

LATERAL

MEDIAL

POSTERIOR

(c) Inferior view of transverse section of thoracic cavity

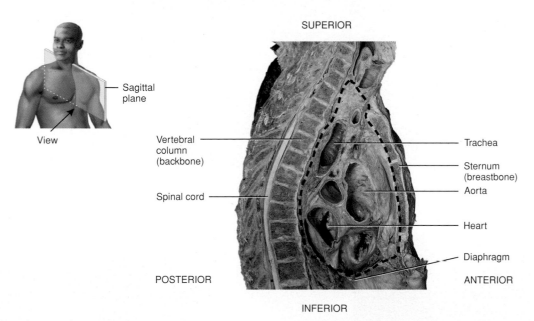

Sagittal plane

View

SUPERIOR

Vertebral column (backbone)

Spinal cord

Trachea

Sternum (breastbone)

Aorta

Heart

Diaphragm

POSTERIOR

ANTERIOR

INFERIOR

(d) Sagittal section of thoracic cavity

Lung

Pleural cavity

Visceral pleura

Parietal pleura

(e) Layers of the pleura

Parietal pericardium

Pericardial cavity

Visceral pericardium

Fibrous pericardium

Heart

(f) Layers of the pericardium

? **Which of the following structures are contained in the mediastinum: right lung, heart, esophagus, spinal cord, trachea, rib, thymus, left pleural cavity?**

ABDOMINOPELVIC REGIONS AND QUADRANTS

○ **OBJECTIVE**

• Name and describe the abdominopelvic regions and the abdominopelvic quadrants.

To describe the location of the many abdominal and pelvic organs more easily, anatomists and clinicians use two methods of dividing the abdominopelvic cavity into smaller areas. In the first method, two transverse and two vertical lines, aligned like a tick-tack-toe grid, partition this cavity into nine **abdominopelvic regions** (Figure 1.8a, b). The top horizontal line, the *subcostal*

Figure 1.8 Abdominopelvic regions. (a) The nine regions in surface view. (b) The nine regions with the greater omentum removed. The greater omentum is a serous membrane that contains fatty tissue that covers some of the abdominal organs. (c) Organs or parts of organs in the nine regions. The internal reproductive organs in the pelvic cavity are shown in Figures 27.1 on page 847 and 27.11 on pages 860–861.

🔑 The nine-region designation is used for anatomical studies.

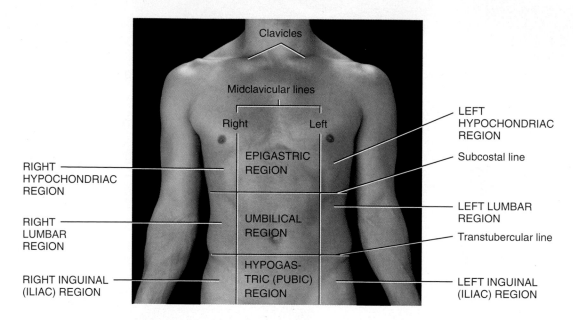

(a) Anterior view showing abdominopelvic regions

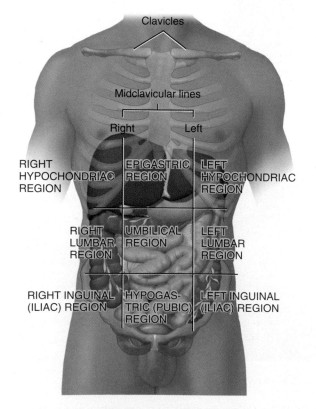

(b) Anterior view showing location of abdominopelvic regions

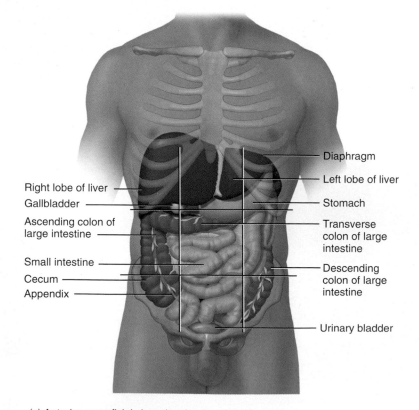

(c) Anterior superficial view showing location of abdominopelvic organs

❓ **In which abdominopelvic region is each of the following found: left lobe of the liver, transverse colon, urinary bladder, spleen?**

line, is drawn just inferior to the rib cage, across the inferior portion of the stomach; the bottom horizontal line, the *transtubercular line*, is drawn just inferior to the tops of the hip bones. Two vertical lines, the left and right *midclavicular lines*, are drawn through the midpoints of the clavicles (collar bones), just medial to the nipples. The four lines divide the abdominopelvic cavity into a larger middle section and smaller left and right sections. The names of the nine abdominopelvic regions are **right hypochondriac, epigastric, left hypochondriac, right lumbar, umbilical, left lumbar, right inguinal (iliac), hypogastric (pubic)**, and **left inguinal (iliac)**. Note which organs and parts of organs are in the different regions by carefully examining Figure 1.8c. Although some parts of the body in these illustrations are most likely unfamiliar to you at this point, they will be discussed in detail in later chapters.

The second method is simpler and divides the abdominopelvic cavity into **quadrants** (KWOD-rantz; *quad-*=one-fourth), as shown in Figure 1.9. In this method, a transverse line and a midsagittal line are passed through the **umbilicus** (um-BIL-i-kus or um-bi-LĪ-kus; *umbilic-*=navel) or *belly button*. The names of the abdominopelvic quadrants are **right upper quadrant (RUQ), left upper quadrant (LUQ), right lower quadrant (RLQ)**, and **left lower quadrant (LLQ)**. The nine-region division is more widely used for anatomical studies to determine organ location; quadrants are more commonly used by clinicians for describing the site of abdominopelvic pain, tumor, injury, or other abnormality.

Figure 1.9 Quadrants of the abdominopelvic cavity. The two lines intersect at right angles at the umbilicus (navel).

🔑 The quadrant designation is used to locate the site of pain, a tumor, or some other abnormality.

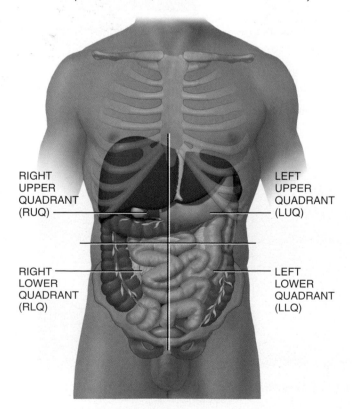

RIGHT UPPER QUADRANT (RUQ)

LEFT UPPER QUADRANT (LUQ)

RIGHT LOWER QUADRANT (RLQ)

LEFT LOWER QUADRANT (LLQ)

Anterior view

❓ In which quadrant would the pain from appendicitis (inflammation of the appendix) be felt?

CHECKPOINT
11. Locate the abdominopelvic regions and the abdominopelvic quadrants on yourself, and list some of the organs found in each.

• CLINICAL CONNECTION | Autopsy

An **autopsy** (AW-top-sē; *auto-*=self; *-opsy*=to see) is a postmortem (after-death) examination of the body and dissection of its internal organs to confirm or determine the cause of death. An autopsy can uncover the existence of diseases not detected during life, determine the extent of injuries, and explain how those injuries may have contributed to a person's death. It also may support the accuracy of diagnostic tests, establish the beneficial and adverse effects of drugs, reveal the effects of environmental influences on the body, provide more information about a disease, assist in the accumulation of statistical data, and educate health-care students. Moreover, an autopsy can reveal conditions that may affect offspring or siblings (such as congenital heart defects). An autopsy may be legally required, such as in the course of a criminal investigation, or may be used to resolve disputes between beneficiaries and insurance companies about the cause of death. •

THE HUMAN BODY AND DISEASE

OBJECTIVE
• Distinguish between a symptom and a sign of a disease.

A **disorder** is any abnormality of structure and/or function. **Disease** is a more specific term for an illness characterized by a recognizable set of symptoms and signs in which body structures and functions are altered in characteristic ways. A person with a disease may experience **symptoms**, *subjective* changes in body functions that are not apparent to an observer. Examples of symptoms are headache, nausea, and anxiety. *Objective* changes that a clinician can observe and measure are called **signs**. Signs of disease can be either anatomical, such as swelling or a rash, or physiological, such as fever, high blood pressure, or paralysis. A *local disease* affects one part or a limited region of the body; a *systemic disease* affects either the entire body or several parts of it.

The science that deals with why, when, and where diseases occur and how they are transmitted among individuals in a community is known as **epidemiology** (ep′-i-dē-mē-OL-ō-jē: *epi*= upon; *-demi*=people). **Pharmacology** (far′-ma-KOL-ō-jē; *pharmac*=drug) is the science that deals with the effects and uses of drugs in the treatment of disease.

Diagnosis (dī′-ag-NŌ-sis; *dia*=through; *-gnosis*=knowledge) is the science and skill of distinguishing one disorder or disease from another. The patient's symptoms and signs, his or her medical history, a physical exam, and laboratory tests provide the basis for making a diagnosis. Taking a *medical history* consists of collecting information about events that might be related to a patient's illness. These include the chief complaint (primary reason for seeking medical attention), history of present illness, past medical problems, family medical problems, social history, and review of symptoms and signs. A *physical examination* is an orderly evaluation of the body and its functions. This process includes the noninvasive techniques of inspection, palpation, auscultation, and percussion that you learned about earlier in the chapter, along with measurement of vital signs (temperature, pulse, respiratory rate, and blood pressure), and sometimes laboratory tests.

CHECKPOINT
12. What factors are considered when making a diagnosis?

MEDICAL IMAGING

⦿ OBJECTIVE

• Describe the principles and importance of medical imaging procedures in the evaluation of organ functions and the diagnosis of disease.

Medical imaging refers to techniques and processes used to create images of the human body. Various kinds of medical imaging allow visualization of structures inside our bodies and are increasingly helpful for precise diagnosis of a wide range of anatomical and physiological disorders. The grandparent of all medical im-

---

TABLE 1.3

Common Medical Imaging Procedures

RADIOGRAPHY

Procedure: A single barrage of x-rays passes through the body, producing an image of interior structures on x-ray-sensitive film. The resulting two-dimensional image is a *radiograph* (RĀ-dē-ō-graf´), commonly called an *x-ray*.

Comments: Radiographs are relatively inexpensive, quick, and simple to perform, and usually provide sufficient information for diagnosis. X-rays do not easily pass through dense structures so bones appear white. Hollow structures, such as the lungs, appear black. Structures of intermediate density, such as skin, fat, and

muscle, appear as varying shades of gray. At low doses, x-rays are useful for examining soft tissues such as the breast **(mammography)** and for determining bone density **(bone densitometry)**.

It is necessary to use a substance called a contrast medium to make hollow or fluid-filled structures visible in radiographs. X-rays make structures that contain contrast media appear white. The medium may be introduced by injection, orally, or rectally, depending on the structure to be imaged.

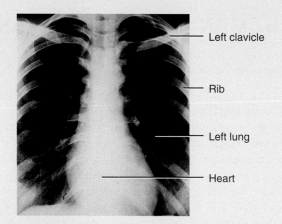

Left clavicle

Rib

Left lung

Heart

Radiograph of the thorax in anterior view

Mammogram of a female breast showing a cancerous tumor (white mass with uneven border)

Bone densitometry scan of the lumbar spine in anterior view

Contrast x-rays are used to image blood vessels **(angiography)**, the urinary system **(intravenous urography)**, and the gastrointestinal tract **(barium contrast x-ray)**.

Angiogram of an adult human heart showing a blockage in a coronary artery (arrow)

Intravenous urogram showing a kidney stone (arrow) in the right ureter

Barium contrast x-ray showing a cancer of the ascending colon (arrow)

aging techniques is conventional radiography (x-rays), in medical use since the late 1940s. The newer imaging technologies not only contribute to diagnosis of disease, but they also are advancing our understanding of normal anatomy and physiology. Table 1.3 describes some commonly used medical imaging techniques.

MAGNETIC RESONANCE IMAGING (MRI)

Procedure: The body is exposed to a high-energy magnetic field, which causes protons (small positive particles within atoms, such as hydrogen) in body fluids and tissues to arrange themselves in relation to the field. Then a pulse of radio waves "reads" these ion patterns, and a color-coded image is assembled on a video monitor. The result is a two- or three-dimensional blueprint of cellular chemistry.

Comments: Relatively safe, but can't be used on patients with metal in their bodies. Shows fine details for soft tissues but not for bones. Most useful for differentiating between normal and abnormal tissues. Used to detect tumors and artery-clogging fatty plaques, reveal brain abnormalities, measure blood flow, and detect a variety of musculoskeletal, liver, and kidney disorders.

Magnetic resonance image of the brain in sagittal section

ULTRASOUND SCANNING

Procedure: High-frequency sound waves produced by a handheld wand reflect off body tissues and are detected by the same instrument. The image, which may be still or moving, is called a *sonogram* (SON-ō-gram) and is shown on a video monitor.

Comments: Safe, noninvasive, painless, and uses no dyes. Most commonly used to visualize the fetus during pregnancy. Also used to observe the size, location, and actions of organs and blood flow through blood vessels **(doppler ultrasound)**.

COMPUTED TOMOGRAPHY (CT) [formerly called computerized axial tomography (CAT) scanning]

Procedure: Computer-assisted radiography in which an x-ray beam traces an arc at multiple angles around a section of the body. The resulting transverse section of the body, called a *CT scan*, is shown on a video monitor.

Comments: Visualizes soft tissues and organs with much more detail than conventional radiographs. Differing tissue densities show up as various shades of gray. Multiple scans can be assembled to build three-dimensional views of structures. In recent years, whole-body CT scanning has emerged. Typically, such scans actually target the torso. Whole-body CT scanning appears to provide the most benefit in screening for lung cancers, coronary artery disease, and kidney cancers.

ANTERIOR

POSTERIOR

Computed tomography scan of the thorax in inferior view

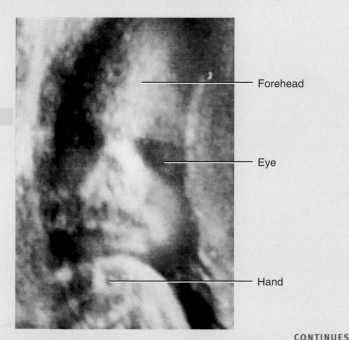

Sonogram of a fetus (Courtesy of Andrew Joseph Tortora and Damaris Soler)

CONTINUES

TABLE 1.3 CONTINUED

Common Medical Imaging Procedures

POSITRON EMISSION TOMOGRAPHY (PET)

Procedure: A substance that emits positrons (positively charged particles) is injected into the body, where it is taken up by tissues. The collision of positrons with negatively charged electrons in body tissues produces gamma rays (similar to x-rays) that are detected by gamma cameras positioned around the subject. A computer receives signals from the gamma cameras and constructs a *PET scan* image, displayed in color on a video monitor. The PET scan shows where the injected substance is being used in the body. In the PET scan image shown here, the black and blue colors indicate minimal activity; the red, orange, yellow, and white colors indicate areas of increasingly greater activity.

Comments: Used to study the physiology of body structures, such as metabolism in the brain or heart.

ANTERIOR

POSTERIOR

Positron emission tomography scan of a transverse section of the brain (darkened area at upper left indicates where a stroke has occurred)

RADIONUCLIDE SCANNING

Procedure: A *radionuclide* (radioactive substance) is introduced intravenously into the body and carried by the blood to the tissue to be imaged. Gamma rays emitted by the radionuclide are detected by a gamma camera outside the subject and fed into a computer. The computer constructs a *radionuclide image* and displays it in color on a video monitor. Areas of intense color take up a lot of the radionuclide and represent high tissue activity; areas of less intense color take up smaller amounts of the radionuclide and represent low tissue activity. **Single-photo-emission computerized tomography (SPECT) scanning** is a specialized type of radionuclide scanning that is especially useful for studying the brain, heart, lungs, and liver.

Comments: Used to study activity of a tissue or organ, such as searching for malignant tumors in body tissues or scars that may interfere with heart muscle function.

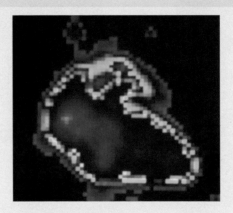

Radionuclide (nuclear) scan of a normal human heart

Single-photon-emission computerized tomography (SPECT) scan of a transverse section of the brain (green area at lower left indicates a migraine attack)

ENDOSCOPY

Procedure: The visual examination of the inside of body organs or cavities using a lighted instrument with lenses called an *endoscope*. The image is viewed through an eyepiece on the endoscope or projected onto a monitor.

Comments: Examples of endoscopy include colonoscopy, laparoscopy, and arthroscopy. *Colonoscopy* is used to examine the interior of the colon, which is part of the large intestine. *Laparoscopy* is used to examine the organs within the abdominopelvic cavity. *Arthroscopy* is used to examine the interior of a joint, usually the knee.

Interior view of the colon as shown by colonoscopy

◉ **CHECKPOINT**

13. Which forms of medical imaging use radiation? Which do not?
14. Which imaging technique best reveals the physiology of a structure?
15. Which imaging technique would best show a broken bone?

MEASURING THE HUMAN BODY

◉ **OBJECTIVE**

• Explain why measurements are important for evaluating the human body.

An important aspect of describing the body and understanding how it works is *measurement*—what the dimensions of an organ are, how much it weighs, how long it takes for a physiological event to occur. Such measurements also have clinical importance,

for example, in determining how much of a given medication should be administered. As you will see, measurements involving time, weight, temperature, size, and volume are a routine part of your studies in a medical science program.

Whenever you come across a measurement in this text, the measurement will be given in metric units. The metric system is the standard used in the sciences. To help you compare the metric unit to a familiar unit, the approximate U.S. equivalent will also be given in parentheses directly after the metric unit. For example, you might be told that the length of a particular part of the body is 2.54 cm (1 in.).

To help you understand the correlation between the metric system and the U.S. system of measurement, there are five tables located in Appendix A for your reference.

◉ **CHECKPOINT**

16. Can you provide any other examples of measurements that are useful in evaluating the human body?

Anatomy Defined (p. 3)

1. Anatomy is the science of body structures and the relationships among structures; physiology is the science of body functions.
2. Some subdisciplines of anatomy are embryology, developmental biology, cell biology, histology, surface anatomy, cross-sectional anatomy, gross anatomy, systemic anatomy, regional anatomy, radiographic anatomy, and pathological anatomy (see Table 1.1 on page 3).

Levels of Body Organization (p. 4)

1. The human body consists of six levels of structural organization: chemical, cellular, tissue, organ, system, and organismal.
2. Cells are the basic structural and functional living units of an organism and the smallest living units in the human body.
3. Tissues are groups of cells and the materials surrounding them that work together to perform a particular function.
4. Organs are composed of two or more different types of tissues; they have specific functions and usually have recognizable shapes.
5. Systems consist of related organs that have a common function.
6. An organism is any living individual.
7. Table 1.2 on pages 6–7 introduces the 11 systems of the human organism: the integumentary, skeletal, muscular, nervous, endocrine, cardiovascular, lymphatic, respiratory, digestive, urinary, and reproductive systems.

Basic Anatomical Terminology (p. 8)

Anatomical Position (p. 8)

1. Descriptions of any region of the body assume the body is in the anatomical position, in which the subject stands erect facing the observer, with the head level and the eyes facing directly forward. The feet are flat on the floor and directed forward, and the upper limbs are at the sides, with the palms turned forward.
2. A body lying face down is prone, and a body lying face up is supine.

Regional Names (p. 8)

1. Regional names are terms given to specific regions of the body. The principal regions are the head, neck, trunk, upper limbs, and lower limbs.
2. Within the regions, specific body parts have common names and are specified by corresponding anatomical terms. Examples are chest (thoracic), nose (nasal), and wrist (carpal).

Planes and Sections (p. 10)

1. Planes are imaginary flat surfaces that are used to divide the body or organs into definite areas. A midsagittal plane divides the body or organ into equal right and left sides; a parasagittal plane divides the body or organ into unequal right and left sides; a frontal plane divides the body or organ into anterior and posterior portions; a transverse plane divides the body or organ into superior and inferior portions; and an oblique plane passes through the body or organ at an angle between a transverse plane and a sagittal plane or a transverse plane and a frontal plane.
2. Sections are cuts of the body or one of its organs made along a plane. Sections are named according to the plane along which the cut is made and include transverse, frontal, and sagittal.

Directional Terms (p. 12)

1. Directional terms indicate the relationship of one part of the body to another.
2. Exhibit 1.1 on pages 12–13 summarizes commonly used directional terms.

Body Cavities (p. 14)

1. Spaces in the body that help protect, separate, and support internal organs are called body cavities.
2. The cranial cavity contains the brain and the vertebral canal contains the spinal cord. The meninges are protective tissues that line the cranial cavity and vertebral canal.
3. The diaphragm separates the thoracic cavity from the abdominopelvic cavity. Viscera are organs within the thoracic and abdominopelvic cavities. A serous membrane lines the wall of the cavity and adheres to the viscera.
4. Within the thoracic cavity is a pericardial cavity, which surrounds the heart, and two pleural cavities, each of which surrounds one lung.
5. The central portion of the thoracic cavity is an anatomical region called the mediastinum. It is located between the medial wall of each pleural cavity and extends from the sternum to the vertebral column and from the first rib to the diaphragm. It contains all thoracic viscera except the lungs.
6. The abdominopelvic cavity is divided into a superior abdominal cavity and an inferior pelvic cavity.
7. Viscera of the abdominal cavity include the stomach, spleen, liver, gallbladder, small intestine, and most of the large intestine.

8. Viscera of the pelvic cavity include the urinary bladder, portions of the large intestine, and internal organs of the reproductive system.

Thoracic and Abdominal Cavity Membranes (p. 15)

1. Serous membranes line the walls of the thoracic and abdominal cavities and cover the organs within them. They include the pleura, associated with the lungs; the pericardium, associated with the heart; and the peritoneum, associated with the abdominal cavity.
2. Figure 1.6 on pages 14–15 summarizes body cavities and their membranes.

Abdominopelvic Regions and Quadrants (p. 18)

1. To describe the location of organs easily, the abdominopelvic cavity may be divided into nine regions by drawing four imaginary lines (left midclavicular, right midclavicular, subcostal, and transtubercular).
2. The names of the abdominopelvic regions are right hypochondriac, epigastric, left hypochondriac, right lumbar, umbilical, left lumbar, right iliac (inguinal), hypogastric (pubic), and left iliac (inguinal).
3. To locate the site of an abdominopelvic abnormality in clinical studies, the abdominopelvic cavity may be divided into quadrants by passing imaginary horizontal and vertical lines through the umbilicus.
4. The names of the abdominopelvic quadrants are right upper quadrant (RUQ), left upper quadrant (LUQ), right lower quadrant (RLQ), and left lower quadrant (LLQ).

The Human Body and Disease (p. 19)

1. A disorder is a general term for any abnormality of structure and/or function. A disease is an illness with a definite set of symptoms and signs.
2. Symptoms are subjective changes in body functions that are not apparent to an observer. Signs are objective changes that can be observed and measured.

Medical Imaging (p. 20)

1. Medical imaging refers to techniques and processes used to create images by the human body. They allow visualization of internal structures to diagnose abnormal anatomy and deviations from normal physiology.
2. Table 1.3 on pages 20–22 summarizes several medical imaging techniques.

Measuring the Human Body (p. 23)

1. Measurements involving time, weight, temperature, size, and volume are used in clinical situations.
2. Measurements in this book are given in metric units followed by U.S. equivalents in parentheses.

SELF-QUIZ QUESTIONS

Choose the one best answer to the following questions.

1. Which word describes the location of the lungs with reference to the liver?
   a. anterior   b. inferior   c. distal
   d. medial   e. superior

2. Which is located most inferiorly?
   a. abdominal cavity   b. pleural cavity   c. mediastinum
   d. cranial cavity   e. pelvic cavity

3. The spleen, tonsils, and thymus are all organs in which system?
   a. nervous   b. lymphatic and immune
   c. cardiovascular   d. digestive   e. endocrine

4. A transverse plane is:
   a. the same as a horizontal plane.
   b. perpendicular to a sagittal plane.
   c. one that divides the body into right and left sides.
   d. Both a and b are correct.
   e. Only c is correct.

5. Which of the following statements about the function of the respiratory system is *not* true?
   a. It supplies oxygen.
   b. It eliminates carbon dioxide.
   c. It helps regulate acid–base balance in the body.
   d. It filters blood.
   e. It includes the lungs.

6. The body in the anatomical position:
   a. is standing with upper limbs parallel to the floor.
   b. is prone.
   c. has the palms turned anteriorly.
   d. has the head turned to the right.
   e. Both a and c are correct.

Complete the following.

7. The four basic types of tissues in the body are _____, _____, _____, and _____.

8. The serous membrane lining the abdominal cavity is the _____.

9. The abdominal region directly superior to the umbilical region is the _____.

10. A _____ plane divides the body into unequal right and left sections.

11. Body structures known as _____ are composed of two or more different tissues, have specific functions, and usually have recognizable shapes.

12. Because the liver and ascending colon are both located on the right side of the abdomen, they could be described as _____-lateral.

13. In terms of position, the spleen is _____ to the diaphragm and _____ to the stomach.

Are the following statements true or false? If the statement is False, *find and fix the mistake to make the statement* True.

14. The joints of the body are part of the muscular system.

15. "Tissue" is a higher level of structural organization than "organ."

16. The parietal layer of a serous membrane adheres to the organs of the thoracic and abdominal cavities.

17. The diaphragm separates the abdominal and pelvic cavities.

18. The mediastinum is part of the thoracic cavity.

Matching

19. Match the following common and anatomical terms:
    ____ (a) front of elbow   (1) gluteal
    ____ (b) arm   (2) inguinal
    ____ (c) shoulder   (3) cervical
    ____ (d) buttock   (4) plantar
    ____ (e) groin   (5) antecubital
    ____ (f) thigh   (6) brachial
    ____ (g) neck   (7) acromial
    ____ (h) sole   (8) femoral

20. Match the following:

____ **(a)** superficial
____ **(b)** deep
____ **(c)** anterior (ventral)
____ **(d)** posterior (dorsal)
____ **(e)** medial
____ **(f)** lateral
____ **(g)** proximal
____ **(h)** distal

(1) toward the front of the body
(2) point on a limb nearer to the trunk
(3) toward the surface of the body
(4) nearer the midsagittal plane
(5) at the back of the body
(6) point on a limb farther from the trunk
(7) farther from the midsagittal plane
(8) away from the surface of the body

CRITICAL THINKING QUESTIONS

1. To make an incision into the body to expose the kidney, the surgeon does not need to penetrate the abdominal cavity. How is this possible?

2. Eight-year-old Taylor was going for the record for the longest upside-down hang from the monkey bars. She didn't make it and she may have done extensive damage to her entire upper limb. The emergency room technician would like an x-ray film of her arm in the anatomical position. Use the proper anatomical terms to describe the position of Taylor's arm, forearm, and hand in the x-ray.

3. An alien landed in your backyard, abducted your cat, and flew off. Being an observant student of anatomy, you later described the alien's appearance to the FBI as follows: "It had 2 caudal extensions, 6 bilateral extremities, 4 axillae, and 1 oral orifice in place of an umbilicus." What did the alien look like?

4. Your anatomy professor displays an MRI scan that shows a parasagittal section of the torso taken through the midpoint of the left mammary gland. Name five organs you would expect to see in this image. (You may consult a general photo of the human body.)

5. Mikhail has been diagnosed with a ruptured appendix, which has allowed bacteria from his intestinal tract to infect his peritoneum. The doctors are very concerned. Why do they consider this condition (peritonitis) to be so dangerous?

❓ ANSWERS TO FIGURE QUESTIONS

1.1 Organs are composed of two or more different types of tissues that work together to perform a specific function.

1.2 Having one standard anatomical position allows directional terms to be clearly defined, so that any body part can be described in relation to any other part.

1.3 The frontal plane divides the heart into anterior and posterior portions.

1.4 The parasagittal plane divides the brain into unequal right and left portions.

1.5 No, the radius is distal to the humerus; No, the esophagus is posterior to the trachea; Yes, the ribs are superficial to the lungs; Yes, the urinary bladder is medial to the ascending colon; No, the sternum is medial to the descending colon.

1.6 Urinary bladder = P, stomach = A, heart = T, small intestine = A, lungs = T, internal female reproductive organs = P, thymus = T, spleen = A, liver = A.

1.7 Structures in the mediastinum include the heart, esophagus, trachea, and thymus.

1.8 The liver is mostly in the epigastric region; the transverse colon is in the umbilical region; the urinary bladder is in the hypogastric region; the spleen is in the left hypochondriac region.

1.9 The pain associated with appendicitis would be felt in the right lower quadrant (RLQ).

2 | CELLS

INTRODUCTION **Cells** are the basic, living, structural and functional units of the body. There are about 200 different types of cells in the body. Cells are composed of characteristic parts, whose coordinated functioning enables each cell type to fulfill a unique biochemical or structural role. **Cell biology** is the study of cellular structure and function. As you study the various parts of a cell and their relationships to one another, you will learn that cell structure and function are interdependent and inseparable. Cells perform numerous chemical reactions to create life processes. One way that they do this is by *compartmentalization*, the isolation of specific kinds of chemical reactions within specialized structures inside the cell. The isolated reactions are coordinated to maintain life in a cell, tissue, organ, system, and organism. •

A GENERALIZED CELL

◉ **OBJECTIVE**

- Name and describe the three principal parts of a cell.

Figure 2.1 is a cell that is a composite of many different cell types in the body. Most cells have many of the features shown in this diagram, but few have all of them. For ease of study, we can divide a cell into three principal parts: plasma membrane, cytoplasm, and nucleus.

- The **plasma membrane** forms the cell's flexible outer surface, separating the cell's internal environment (inside the cell) from the external environment (outside the cell). This selective barrier regulates the flow of materials into and out of a cell, helping to establish and maintain the appropriate environment for normal cellular activities. The plasma membrane also plays a key role in communication among cells and between cells and their external environment.

- The **cytoplasm** (SĪ-tō-plazm; -*plasm*=formed or molded) is all the cellular contents between the plasma membrane and the nucleus. This compartment has two components: the cytosol and organelles. **Cytosol** (SĪ-tō-sol), the fluid portion of cytoplasm, contains water, dissolved solutes, and suspended particles. Surrounded by cytosol are several different types of **organelles** (or-ga-NELZ=little organs). Each type of organelle has a characteristic shape and specific functions. Examples include the cytoskeleton, ribosomes, endoplasmic reticulum, Golgi complex, lysosomes, peroxisomes, and mitochondria.

- The **nucleus** (NOO-klē-us=nut or kernel) is a large organelle that houses most of a cell's DNA. Within the nucleus, each **chromosome** (*chromo-*=colored), a single molecule of DNA associated with several proteins, contains thousands of hereditary units called **genes** that control most aspects of cellular structure and function.

◉ **CHECKPOINT**

1. What are the three main parts of a cell and the general features of each of these parts?

Figure 2.1 Typical structures found in body cells.

The cell is the basic, living, structural and functional unit of the body.

Flagellum

Cilium

Secretory vesicle

Cytoskeleton:
  Microtubule
  Microfilament
  Intermediate filament

Microvilli

Centrosome:
  Pericentriolar material
  Centrioles

PLASMA MEMBRANE

Lysosome

Smooth endoplasmic reticulum

Peroxisome

Mitochondrion

Microtubule

NUCLEUS:
  Chromatin
  Nuclear pore
  Nuclear envelope
  Nucleolus

Glycogen granules

CYTOPLASM (cytosol plus organelles except the nucleus)

Rough endoplasmic reticulum

Ribosome

Golgi complex

Microfilament

Sectional view

❓ **What are the three main parts of a cell?**

THE PLASMA MEMBRANE

◉ **OBJECTIVES**

- Describe the structure and functions of the plasma membrane.
- Describe the processes that transport substances across the plasma membrane.

The **plasma membrane** is a flexible yet sturdy barrier that surrounds and contains the cytoplasm of a cell. The *fluid mosaic model* describes its structure. According to this model, the molecular arrangement of the plasma membrane resembles an ever-moving sea of lipids that contains a "mosaic" of many different proteins (Figure 2.2). The proteins may float freely like icebergs or be moored at specific locations. The lipids act as a barrier to the entry or exit of various substances, while some of the proteins in the plasma membrane act as "gatekeepers" that regulate the passage of other molecules and ions (charged particles) into and out of the cell.

Structure of the Membrane

The basic structural framework of the plasma membrane is the **lipid bilayer**, two back-to-back layers made up of three types of lipid molecules—phospholipids, cholesterol, and glycolipids (Figure 2.2). About 75% of the membrane lipids are *phospholipids*, lipids that contain phosphorus. About 20% of plasma membrane lipids are *cholesterol* molecules, which are interspersed among the other lipids in both layers of the membrane. *Glycolipids*, which are lipids attached to carbohydrates (*glyco-*=carbohydrate), account for the other 5%.

Membrane proteins are divided into two categories—integral and peripheral—according to whether or not they are firmly embedded in the membrane (Figure 2.2). **Integral proteins** extend into or through the lipid bilayer and are firmly embedded in it. Most integral proteins are **transmembrane proteins**, which means that they span the entire lipid bilayer and protrude into both the cytosol and extracellular fluid. **Peripheral proteins**, by contrast, are not as firmly embedded in the membrane and are attached to membrane lipids or integral proteins at the inner or outer surface of the membrane.

Many membrane proteins are **glycoproteins**, proteins with carbohydrate groups attached to the ends that protrude into the extracellular fluid. The carbohydrate portions of glycolipids and glycoproteins form an extensive sugary coat called the **glycocalyx** (glī-kō-KĀL-iks). The composition of the glycocalyx acts like a molecular "signature" that enables cells to recognize one another. For example, a white blood cell's ability to detect a "foreign" glycocalyx is one basis of the immune response that helps us destroy invading organisms. In addition, the glycocalyx enables cells to adhere to one another in some tissues, and it protects cells from being digested by enzymes in the extracellular fluid. The chemical properties of the glycocalyx attract a film of fluid to the surface of many cells. This action makes red blood cells slippery as they flow through narrow blood vessels and protects cells that line the airways and the gastrointestinal tract from drying out.

Functions of Membrane Proteins

Generally, the types of lipids in cellular membranes vary only slightly from one membrane to another. In contrast, the membranes

Figure 2.2 Structure of the plasma membrane.

The basic framework of the plasma membrane is the lipid bilayer.

? What is the glycocalyx?

of different cells and various intracellular organelles have re-markably different assortments of proteins that determine many of the membrane's functions.

- Some integral membrane proteins form **ion channels**, *pores* or holes through which specific ions, such as potassium ions (K^+) can flow to get into or out of the cell. Most ion channels are *selective*; they allow only a single type of ion to pass through.
- Other integral proteins act as **carriers** or **transporters**, selectively moving a polar substance or ion from one side of the membrane to the other.
- Integral proteins called **receptors** serve as cellular recognition sites. Each type of receptor recognizes and binds a specific type of molecule. For instance, insulin receptors bind the hormone insulin. A specific molecule that binds to a receptor is called a **ligand** (LĪ-gand; *liga-*=tied) of that receptor.
- Some integral proteins are **enzymes** that catalyze specific chemical reactions at the inside or outside surface of the cell.
- Integral proteins may also serve as **linkers**, which anchor proteins in the plasma membranes of neighboring cells to one another or to protein filaments inside and outside the cell. Peripheral proteins also serve as enzymes and linkers.
- Membrane glycoproteins and glycolipids often serve as **cell-identity markers**. They may enable a cell to recognize other cells of the same kind during tissue formation or to recognize and respond to potentially dangerous foreign cells. The ABO blood-type markers are one example of cell identity markers. When you receive a blood transfusion, the blood type must be compatible with your own.

In addition, peripheral proteins help support the plasma membrane, anchor integral proteins, and participate in mechanical activities such as moving materials and organelles within cells, changing cell shape in dividing and muscle cells, and attaching cells to one another.

Membrane Permeability

The term *permeable* means that a structure, such as a membrane, permits the passage of substances through it, while *impermeable* means that a structure does not permit the passage of substances through it. Although plasma membranes are not completely permeable to any substance, they do permit some substances to pass more readily than others. This property of membranes is called **selective permeability**.

The lipid bilayer portion of the membrane is permeable to some molecules such as oxygen, carbon dioxide, and steroids, but is impermeable to ions and molecules such as glucose. It is also permeable to water. Transmembrane proteins that act as channels and transporters increase the plasma membrane's permeability to a variety of small- and medium-sized charged substances (including ions) that cannot cross the lipid bilayer without help. These proteins are very selective—each one helps only a specific molecule or ion to cross the membrane. Macromolecules, such as proteins, cannot pass through the plasma membrane except by endocytosis and exocytosis (discussed later in this chapter).

◉ CHECKPOINT
2. What is the composition of the lipid bilayer?
3. Distinguish integral from peripheral proteins.
4. What are the major functions of membrane proteins?

Transport Across the Plasma Membrane

Before discussing how materials move into and out of cells, we will describe the locations of the various fluids through which the substances move. Fluid within cells is called **intracellular** (*intra-*=within, inside) **fluid (ICF)** or **cytosol**. Fluid outside body cells is called **extracellular** (*extra-*=outside) **fluid (ECF)** and is found in several locations: (1) The ECF filling the microscopic spaces between the cells of tissues is called **interstitial** (in-ter-STISH-al) **fluid** (*inter-*=between; *stit*=to place or set), or *intercellular fluid*. (2) The ECF in blood vessels is termed **plasma** (Figure 2.3); in lymphatic vessels it is called **lymph**. Among the substances in extracellular fluid are gas and nutrient molecules and ions—all needed for the maintenance of life.

Extracellular fluid circulates through the blood and lymphatic vessels (circulatory vessels) and moves into and out of the spaces between the tissue cells and the circulatory vessels. Thus, it is in constant motion throughout the body. Essentially, all body cells are surrounded by the same fluid environment. The movement of substances across a plasma membrane and across membranes within cells is essential to the life of the cell and the organism. Certain substances—oxygen, for example—must move into the cell to support life, and waste materials or harmful substances must be moved out. Plasma membranes regulate the movements of such materials.

Substances generally move across cellular membranes via transport processes that can be classified as passive or active, depending on whether they require cellular energy. In *passive processes*, a substance moves down its concentration gradient or electrical gradient to cross the membrane using only its own kinetic energy. There is no input of energy from the cell. In *active processes*, cellular energy is used to drive the substance "uphill"

Figure 2.3 Body fluids. Intracellular fluid (ICF) is the fluid within cells. Extracellular fluid (ECF) is found outside cells, in blood vessels as plasma, lymphatic vessels as lymph, and between tissue cells as interstitial fluid.

🔒 ⚙ Plasma membranes regulate fluid movements from one compartment to another.

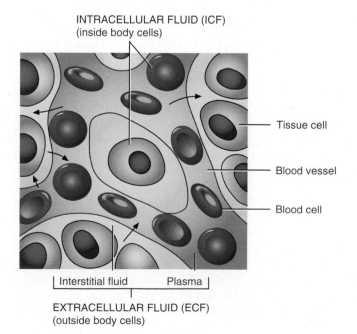

INTRACELLULAR FLUID (ICF)
(inside body cells)

— Tissue cell

— Blood vessel

— Blood cell

| Interstitial fluid | Plasma |

EXTRACELLULAR FLUID (ECF)
(outside body cells)

❓ What is another name for intracellular fluid?

against its concentration or electrical gradient. The cellular energy used is usually in the form of adenosine triphosphate (ATP).

Some materials cross cellular membranes simply by moving through the lipid bilayer or membrane channels using their own *kinetic energy* (energy of motion). Kinetic energy is intrinsic to the particles that are moving. Examples of processes that rely on kinetic energy include diffusion and osmosis. Other substances must bind to specific transporter proteins to cross a cellular membrane, as in facilitated diffusion and active transport. Still other substances pass through cellular membranes within small, spherical sacs called *vesicles* that bud off from an existing membrane. Examples include endocytosis, in which vesicles detach from the plasma membrane while bringing materials into a cell, and exocytosis, the merging of vesicles with the plasma membrane to release materials from a cell. Thus, materials may cross plasma membranes by using kinetic energy, transporter proteins, or vesicles.

Kinetic Energy Transport

DIFFUSION Diffusion (di-FŪ-zhun; *diffus-*=spreading) is a passive process in which there is a *net* (greater) movement of a substance from a region of higher concentration to a region of lower concentration—that is, the substance moves from an area where there is more of it to an area where there is less of it. The substance moves because of its kinetic energy. A good example of diffusion in the body is the movement of oxygen from the blood into the body cells and the movement of carbon dioxide from the cells back into the blood. This movement ensures that cells receive adequate amounts of oxygen and eliminate carbon dioxide as part of their normal metabolism.

OSMOSIS Osmosis (oz-MŌ-sis; *osmo-*=a pushing) is the net movement of water molecules through a selectively permeable membrane from an area of higher water concentration (lower concentration of *solutes*, dissolved substances) to an area of lower water concentration (higher solute concentration). It is also a passive process. The water molecules pass through pores (holes) in integral proteins in the membrane and between neighboring phospholipid molecules due to their kinetic energy, and the net movement continues until equilibrium is reached. Water moves between various compartments of the body by osmosis.

Transport by Transporter Proteins

FACILITATED DIFFUSION Facilitated diffusion is a passive process that is accomplished with the assistance of transmembrane proteins functioning as carriers. In this process, some large molecules and molecules that are insoluble in lipids can still pass through the plasma membrane. Among these are various sugars, especially glucose. In facilitated diffusion, glucose binds to a specific carrier protein on one side of the plasma membrane, the carrier undergoes a change in shape, and, as a result, glucose is released on the opposite side of the membrane.

ACTIVE TRANSPORT The process by which substances, usually ions, are transported across plasma membranes with the expenditure of energy by the cell, typically from an area of lower concentration to an area of higher concentration, is called **active transport**. In active transport the substance being moved makes contact with a specific site on a transporter protein. Then the ATP splits, and the energy from the breakdown of ATP causes a change in the shape of the transporter protein that expels the substance on the opposite side of the membrane. Active transport is considered an active process because energy is required for transporter proteins to move substances across the membrane against a concentration

gradient. Active transport is vitally important in maintaining ion concentrations in both body cells and extracellular fluids. For example, before a nerve cell can conduct a nerve impulse, the concentration of potassium ions (K^+) must be considerably higher inside the nerve cell than outside and the concentration of sodium ions (Na^+) must be higher outside than inside. Active transport makes this possible.

Transport in Vesicles

A **vesicle** (VES-i-kul; *vescula*=little blister or bladder), as noted earlier, is a small, spherical, membranous sac formed by budding off from an existing membrane. Vesicles transport substances from one structure to another within cells, take in substances from extracellular fluid, or release substances into extracellular fluid. In **endocytosis** (*endo-*=within), materials move into a cell in a vesicle formed from the plasma membrane. In **exocytosis** (*exo-*=out), materials move out of a cell by the fusion of vesicles, formed inside a cell, with the plasma membrane. Both endocytosis and exocytosis require cellular energy supplied by the breakdown of ATP. Thus transport in vesicles is an active process.

ENDOCYTOSIS Here we consider three types of endocytosis: receptor-mediated endocytosis, phagocytosis, and bulk-phase endocytosis.

In **receptor-mediated endocytosis**, a highly selective type of endocytosis, cells take up specific ligands. (Recall that ligands are molecules that bind to specific receptors.) A vesicle forms after a receptor protein in the plasma membrane recognizes and binds to a particular particle in the extracellular fluid. For instance, cells take up cholesterol contained in low-density lipoproteins (LDLs), transferrin (an iron-transporting protein in the blood), some vitamins, antibodies, and certain hormones by receptor-mediated endocytosis. Receptor-mediated endocytosis of LDLs (and other ligands) occurs as follows (Figure 2.4):

❶ **Binding.** On the extracellular side of the plasma membrane, an LDL particle that contains cholesterol binds to a specific receptor in the plasma membrane to form a receptor-LDL complex. The receptors are integral membrane proteins that are concentrated in regions of the plasma membrane called *clathrin-coated pits*. Here, a protein called *clathrin* attaches to the membrane on its cytoplasmic side. Many clathrin molecules come together, forming a basketlike structure around the receptor-LDL complexes that causes the membrane to invaginate (fold inward).

❷ **Vesicle formation.** The invaginated edges of the membrane around the clathrin-coated pit fuse and a small piece of the membrane pinches off. The resulting vesicle, known as a *clathrin-coated vesicle*, contains the receptor-LDL complexes.

❸ **Uncoating.** Almost immediately after it is formed, the clathrin-coated vesicle loses its clathrin coat to become an *uncoated vesicle*. Clathrin molecules either return to the inner surface of the plasma membrane or help form coats on other vesicles inside the cell.

❹ **Fusion with endosome.** The uncoated vesicle quickly fuses with a vesicle known as an *endosome*. Within an endosome, the LDL particles separate from their receptors.

❺ **Recycling of receptors to plasma membrane.** Most of the receptors accumulate in elongated protrusions of the endosome. These pinch off, forming transport vesicles that return the receptors to the plasma membrane. An LDL receptor is returned to the plasma membrane about 10 minutes after it enters a cell.

Figure 2.4 Receptor-mediated endocytosis of a low-density lipoprotein (LDL) particle.

🔑 Receptor-mediated endocytosis imports materials that are needed by cells.

❓ **What are several other examples of ligands that can undergo receptor-mediated endocytosis?**

Phagocytosis (fag-ō-sī-TŌ-sis; *phago-*=to eat) is a form of endocytosis in which the cell engulfs large solid particles, such as worn-out cells, whole bacteria, or viruses (Figure 2.5). Only a few body cells, termed **phagocytes**, are able to carry out phagocytosis. Two main types of phagocytes are *macrophages*, located in many body tissues, and *neutrophils*, a type of white blood cell. Phagocytosis begins when the particle binds to a plasma membrane receptor on the phagocyte, causing it to extend **pseudopods** (SOO-dō-pods; *pseudo-*=false; *-pods*=feet), projections of its plasma membrane and cytoplasm. Pseudopods surround the particle outside

Figure 2.5 Phagocytosis. Pseudopods surround a particle and the membranes fuse to form a phagosome.

🔑 Phagocytosis is a vital defense mechanism that helps protect the body from disease.

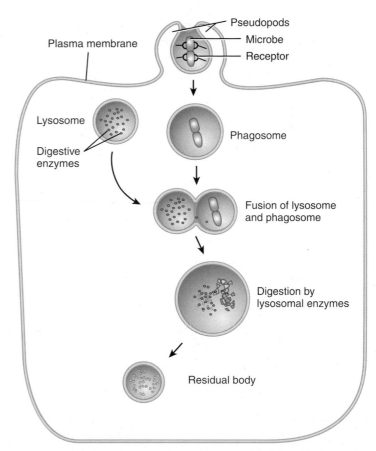

(a) Diagram of the process

6️⃣ **Degradation in lysosomes.** Other transport vesicles, which contain the LDL particles, bud off the endosome and soon fuse with a *lysosome*. Lysosomes contain many digestive enzymes. Certain enzymes break down the large protein and lipid molecules of the LDL particle into amino acids, fatty acids, and cholesterol. These smaller molecules then leave the lysosome. The cell uses cholesterol for rebuilding its membranes and for synthesis of steroids, such as estrogen. Fatty acids and amino acids can be used for ATP production or to build other molecules needed by the cell.

TEM about 3700x

(b) White blood cell engulfs microbe

TEM about 3700x

(c) White blood cell destroys microbe

❓ **What triggers pseudopod formation?**

the cell, and the membranes fuse to form a vesicle called a *phagosome*, which enters the cytoplasm. The phagosome fuses with one or more lysosomes, and lysosomal enzymes break down the ingested material. In most cases, any undigested materials in the phagosome remain indefinitely in a vesicle called a *residual body*. The process of phagocytosis is a vital defense mechanism that helps protect the body from disease. Through phagocytosis, macrophages dispose of invading microbes and billions of aged, worn-out red blood cells every day; neutrophils also help rid the body of invading microbes. Pus is a mixture of dead neutrophils, macrophages, and tissue cells and fluid in an infected wound.

Most body cells carry out **bulk-phase endocytosis**, also called **pinocytosis** (pi-nō-sī-TŌ-sis; *pino-*=to drink), a form of endocytosis in which tiny droplets of extracellular fluid are taken up. No receptor proteins are involved; all solutes dissolved in the extracellular fluid are brought into the cell. During bulk-phase endocytosis, the plasma membrane folds inward and forms a vesicle containing a droplet of extracellular fluid. The vesicle detaches or "pinches off" from the plasma membrane and enters the cytosol. Within the cell, the vesicle fuses with a lysosome, where enzymes degrade the engulfed solutes. The resulting smaller molecules, such as amino acids and fatty acids, leave the lysosome to be used elsewhere in the cell. Bulk-phase endocytosis occurs in most cells, especially absorptive cells in the intestines and kidneys.

EXOCYTOSIS In contrast with endocytosis, which brings materials into a cell, exocytosis releases materials from a cell. Just remember that *endo* means "in," and *exo* means "out." All cells carry out exocytosis, but it is especially important in two types of cells: (1) secretory cells that liberate digestive enzymes, hormones, mucus, or other secretions; and (2) nerve cells that release substances called *neurotransmitters*. In some cases, wastes are also released by exocytosis. During exocytosis, membrane-enclosed vesicles called *secretory vesicles* form inside the cell, fuse with the plasma membrane, and release their contents into the extracellular fluid.

Segments of the plasma membrane lost through endocytosis are recovered or recycled by exocytosis. The balance between endocytosis and exocytosis keeps the surface area of a cell's plasma membrane relatively constant. Membrane exchange is quite extensive in certain cells. In your pancreas, for example, the cells that secrete digestive enzymes can recycle an amount of plasma membrane equal to the cell's entire surface area in 90 minutes.

TRANSCYTOSIS Transport in vesicles may also be used to successively move a substance into, across, and out of a cell. In this active process, called *transcytosis*, vesicles undergo endocytosis on one side of a cell, move across the cell, and then undergo exocytosis on the opposite side. As the vesicles fuse with the plasma membrane, the vesicular contents are released into the extracellular fluid. Transcytosis occurs most often across the endothelial cells that line blood vessels and is a means for materials to move between blood plasma and interstitial fluid. For instance, when a woman is pregnant, some of her antibodies cross the placenta into the fetal circulation via transcytosis.

Table 2.1 summarizes the processes by which materials move into and out of cells.

◉ **CHECKPOINT**

5. Distinguish a passive process from an active process.
6. Why are simple diffusion, facilitated diffusion, osmosis, and active transport important to the body?
7. Describe each type of transport in vesicles. How is each important to the body?

TABLE 2.1

Transport of Materials Into and Out of Cells

TRANSPORT PROCESS	DESCRIPTION
Kinetic energy transport Diffusion	A passive process in which a substance moves from an area of higher to lower concentration.
Osmosis	A passive process that involves the movement of water molecules across a selectively permeable membrane from an area of higher water concentration to an area of lower water concentration.
Transport by transporter proteins Facilitated diffusion	Passive movement of a substance down its concentration gradient via transmembrane proteins that act as transporters.
Active transport	An active process in which cell expends energy to move a substance across the membrane against its concentration gradient through transmembrane proteins that act as transporters.
Transport in vesicles	An active process that involves the movement of substances into or out of a cell in vesicles that bud from the plasma membrane.
Endocytosis	Movement of substances into a cell in vesicles.
Receptor-mediated endocytosis	Ligand–receptor complexes trigger infolding of a clathrin-coated pit that forms a vesicle containing ligands.
Phagocytosis	"Cell eating"; movement of a solid particle into a cell after pseudopods engulf it to form a phagosome.
Bulk-phase endocytosis	"Cell drinking"; movement of extracellular fluid into a cell by infolding of plasma membrane to form a vesicle.
Exocytosis	Movement of substances out of a cell in secretory vesicles that fuse with the plasma membrane and release their contents into the extracellular fluid.
Transcytosis	Movement of a substance through a cell as a result of endocytosis on one side and exocytosis on the opposite side.

CYTOPLASM

◉ **OBJECTIVE**

• Describe the structure and function of cytoplasm, cytosol, and organelles.

Cytosol

Cytoplasm consists of all the cellular contents within the plasma membrane except for the nucleus, and has two components:

(1) cytosol and (2) organelles, tiny structures that perform different functions in the cell.

The **cytosol (intracellular fluid)** is the fluid portion of the cytoplasm that surrounds organelles (see Figure 2.1) and constitutes about 55% of total cell volume. Although it varies in composition and consistency from one part of a cell to another, cytosol is 75–90% water plus various dissolved and suspended components. Among these are different types of ions, glucose, amino acids, fatty acids, proteins, lipids, ATP, and waste products. Also present in some cells are various organic molecules that aggregate into masses for storage. These aggregations may appear and disappear at different times in the life of a cell. Examples include *lipid droplets* that contain triglycerides, and clusters of glycogen molecules called *glycogen granules* (see Figure 2.1).

The cytosol is the site of many chemical reactions required for a cell's existence. For example, enzymes in cytosol catalyze numerous chemical reactions. As a result of these reactions, energy is released and captured to drive cellular activities. In addition, these reactions provide the building blocks for maintaining cell structure, function, and growth.

Organelles

As noted earlier, **organelles** are specialized structures within the cell that have characteristic shapes; they perform specific functions in cellular growth, maintenance, and reproduction. Despite the many chemical reactions going on in a cell at any given time, there is little interference among reactions because they are confined to different organelles. Each type of organelle has its own set of enzymes that carry out specific reactions, and serves as a functional compartment for specific biochemical processes. The numbers and types of organelles vary in different cells, depending on the cell's function. Even though the nucleus is an organelle, it is discussed in a separate section because of its special importance in directing the life of a cell.

The Cytoskeleton

The **cytoskeleton** is a network of protein filaments that extends throughout the cytosol (see Figure 2.1). Three types of filamentous proteins contribute to the cytoskeleton's structure, as well as the structure of other organelles. In the order of their increasing diameter, these structures are microfilaments, intermediate filaments, and microtubules.

MICROFILAMENTS These are the thinnest elements of the cytoskeleton. Microfilaments are concentrated at the periphery (near the plasma membrane) of a cell (Figure 2.6a). They are composed of the protein *actin* and have two general functions: movement and mechanical support. With respect to movement, microfilaments are involved in muscle contraction, cell division, and cell locomotion. Cell locomotion occurs during the migration of embryonic cells during development, the invasion of tissues by white blood cells to fight infection, or the migration of skin cells during wound healing.

Microfilaments provide much of the mechanical support that is responsible for the basic strength and shapes of cells. They anchor the cytoskeleton to integral proteins in the plasma membrane. Microfilaments also provide mechanical support for cell extensions called **microvilli** (*micro-*=small; *-villi*=tufts of hair; singular is *microvillus*), which are nonmotile, microscopic fingerlike projections of the plasma membrane. Within a microvillus is a core of parallel microfilaments that supports it and attaches it to other parts of the cytoskeleton (Figure 2.6a). Microvilli increase the surface area of the plasma membrane and are abundant on the surfaces of cells involved in absorption, such as the epithelial cells that line the small intestine. Some microfilaments extend beyond the plasma membrane and help cells attach to one another or to extracellular material.

INTERMEDIATE FILAMENTS As their name suggests, intermediate filaments are thicker than microfilaments but thinner than microtubules (Figure 2.6b). Several different proteins can compose intermediate filaments, which are exceptionally strong. They are found in parts of cells subject to mechanical stress and help anchor organelles such as the nucleus and attach cells to one another.

MICROTUBULES The largest of the cytoskeletal components, microtubules are long, unbranched hollow tubes composed mainly of a protein called *tubulin*. The centrosome (discussed shortly) serves as the initiation site for the assembly of microtubules. The microtubules grow outward from the centrosome toward the periphery of the cell (Figure 2.6c). Microtubules help determine cell shape and function in the intracellular transport of organelles, such as secretory vesicles, and the migration of chromosomes during cell division. They also participate in the movement of specialized cell projections such as cilia and flagella.

Figure 2.6 Cytoskeleton.

The cytoskeleton is a network of three kinds of protein filaments that extend throughout the cytoplasm: microfilaments, intermediate filaments, and microtubules.

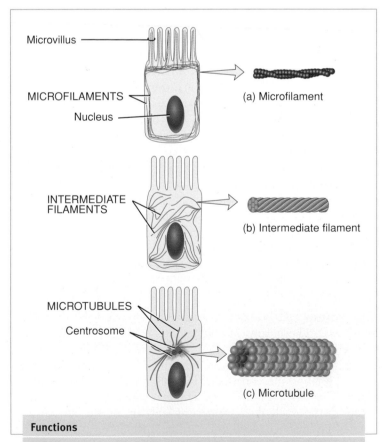

(a) Microfilament

(b) Intermediate filament

(c) Microtubule

Functions

1. Serves as a scaffold that helps to determine a cell's shape and to organize the cellular contents.
2. Aids movement of organelles within the cell, of chromosomes during cell division, and of whole cells such as phagocytes.

❓ **Which cytoskeletal component helps form the structure of centrioles, cilia, and flagella?**

Centrosome

The **centrosome**, located near the nucleus, consists of two components: a pair of centrioles and pericentriolar material (Figure 2.7a). The two **centrioles** are cylindrical structures, each composed of nine clusters of three microtubules (triplets) arranged in a circular pattern (Figure 2.7b). The long axis of one centriole is at a right angle to the long axis of the other (Figure 2.7a). Surrounding the centrioles is **pericentriolar material** (per′-ē-sen′-trē-O-lar), which contains hundreds of ring-shaped complexes composed of the protein tubulin. These tubulin complexes are the organizing centers for growth of the mitotic spindle, which plays a critical role in cell division, and for microtubule formation in nondividing cells. During cell division, centrosomes replicate so that succeeding generations of cells have the capacity for cell division.

Cilia and Flagella

Microtubules are the dominant components of cilia and flagella, which are motile projections of the cell surface (Figure 2.1). **Cilia** (SIL-ē-a=eyelashes; singular is *cilium*) are numerous, short, hairlike projections that extend from the surface of the cell (see Figure 2.1). Each cilium contains a core of 20 microtubules surrounded by plasma membrane (Figure 2.8a). These 20 microtubules are arranged with one pair in the center surrounded by nine clusters of two fused microtubules (doublets). Each cilium is anchored to a *basal body* just below the surface of the plasma membrane. A basal body is similar in structure to a centriole and functions in initiating the assembly of cilia and flagella.

The coordinated movement of many cilia on the surface of a cell causes the steady movement of fluid along the cell's surface. Many cells of the respiratory tract, for example, have hundreds of cilia that help sweep foreign particles trapped in mucus away from the lungs. The extremely thick mucus secretions that are produced in people who suffer from cystic fibrosis interfere with ciliary action and the normal functions of the respiratory tract. The movement of cilia is also paralyzed by nicotine in cigarette smoke. For this reason, smokers cough often to remove foreign particles from their airways. Because cells that line the uterine (fallopian) tubes also have cilia that sweep oocytes (egg cells) toward the uterus, females who smoke have an increased risk of ectopic (outside the uterus) pregnancy.

Flagella (fla-JEL-a=whip; singular is *flagellum*) are similar in structure to cilia but are typically much longer (see Figure 2.1). Flagella usually move an entire cell. The only example of a flagellum in the human body is a sperm cell's tail, which propels the sperm toward the oocyte in the uterine tube.

Ribosomes

Ribosomes (RĪ-bō-sōms; -*somes*=bodies) are sites of protein synthesis. These tiny organelles are packages of **ribosomal RNA (rRNA)** and many ribosomal proteins. Ribosomes are so named because of their high content of *ribonucleic* acid. Structurally, a ribosome consists of two subunits, one about half the size of the other (Figure 2.9). The two subunits are made separately in the nucleolus, a spherical body inside the nucleus (see page 41). Once produced, they exit the nucleus and join together in the cytosol where they become functional.

Figure 2.7 Centrosome.

 Located near the nucleus, the centrosome consists of a pair of centrioles and pericentriolar material.

(a) Details of a centrosome

(b) Arrangement of microtubules in centrosome

(c) Centrioles

Function

The pericentriolar material of the centrosome contains tubulins that build microtubules in nondividing cells and form the mitotic spindle during cell division.

? If you observed that a cell did not have a centrosome, what could you predict about its capacity for cell division?

Figure 2.8 Cilia and flagella.

A cilium contains a core of microtubules with one pair in the center surrounded by nine clusters of doublet microtubules.

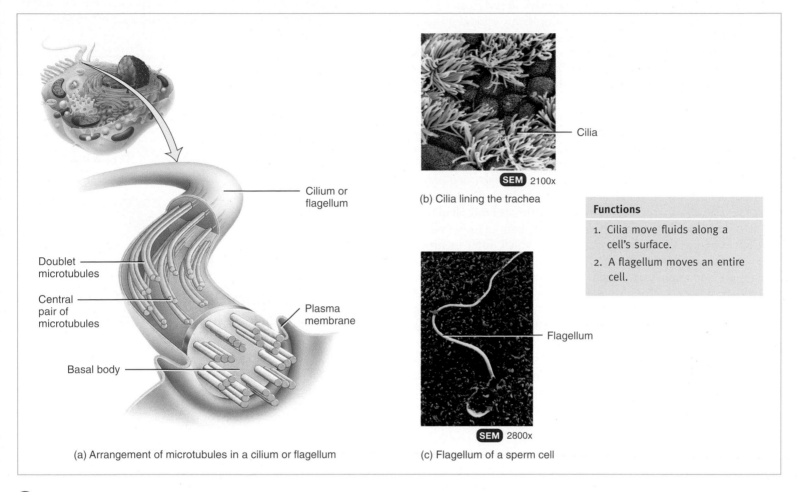

Cilium or flagellum

Doublet microtubules

Central pair of microtubules

Plasma membrane

Basal body

(a) Arrangement of microtubules in a cilium or flagellum

Cilia

SEM 2100x

(b) Cilia lining the trachea

Functions

1. Cilia move fluids along a cell's surface.
2. A flagellum moves an entire cell.

Flagellum

SEM 2800x

(c) Flagellum of a sperm cell

? What is the functional difference between cilia and flagella?

Figure 2.9 Ribosomes.

Ribosomes are the sites of protein synthesis.

Functions

1. Ribosomes associated with endoplasmic reticulum synthesize proteins destined for insertion in the plasma membrane or secretion from the cell.
2. Free ribosomes synthesize proteins used in the cytosol.

Large subunit   Small subunit   Complete functional ribosome

Details of ribosomal subunits

? Where are subunits of ribosomes synthesized and assembled?

Some ribosomes, called *free ribosomes*, are unattached to any structure in the cytoplasm. Free ribosomes primarily synthesize proteins used *inside* the cell. Other ribosomes, called *membrane-bound ribosomes*, attach to the nuclear membrane and to an extensively folded membrane called the endoplasmic reticulum (see Figure 2.10b). These ribosomes synthesize proteins destined for specific organelles, for insertion in the plasma membrane, or for export from the cell. Ribosomes are also located within mitochondria, where they synthesize mitochondrial proteins.

Endoplasmic Reticulum

The **endoplasmic reticulum** (en-dō-PLAS-mik re-TIK-ū-lum; *-plasmic*=cytoplasm; *reticulum*=network), or **ER**, is a network of membranes in the form of flattened sacs or tubules (Figure 2.10). The ER extends from the nuclear envelope (membrane around the nucleus), to which it is connected, throughout the cytoplasm. The ER is so extensive that it constitutes more than half of the membranous surfaces within the cytoplasm of most cells.

Cells contain two distinct forms of ER, which differ in structure and function. **Rough ER** is continuous with the nuclear membrane and usually is folded into a series of flattened sacs. The outer surface of rough ER is studded with ribosomes, the sites of protein synthesis. Proteins synthesized by ribosomes attached to rough ER enter spaces within the ER for processing and sorting. In some cases, enzymes attach the proteins to carbohydrates to form glycoproteins. In other cases, enzymes attach the proteins to phospholipids, also synthesized by rough ER. These molecules may be incorporated into the membranes of organelles, inserted into the plasma membrane or secreted via exocytosis. Thus rough ER produces secretory proteins, membrane proteins, and many organellar proteins.

Smooth ER extends from the rough ER to form a network of membrane tubules (Figure 2.10). Unlike rough ER, smooth ER does not have ribosomes on the outer surfaces of its membrane. However, smooth ER contains unique enzymes that make it functionally more diverse than rough ER. Because it lacks ribosomes, smooth ER does not synthesize proteins, but it does synthesize fatty acids and steroids, such as estrogens and testosterone. In liver cells, enzymes of the smooth ER help release glucose into the bloodstream and inactivate or detoxify lipid-soluble drugs or potentially harmful substances, such as alcohol, pesticides, and *carcinogens* (cancer-causing agents). In liver, kidney, and intestinal cells a smooth ER enzyme removes the phosphate group from glucose-6-phosphate, which allows the "free" glucose to enter the bloodstream. In muscle cells, the calcium ions (Ca^{2+}) that trigger contraction are released from the sarcoplasmic reticulum, a form of smooth ER.

Figure 2.10 Endoplasmic reticulum.

 The endoplasmic reticulum is a network of membrane-enclosed sacs or tubules that extend throughout the cytoplasm and connect to the nuclear envelope.

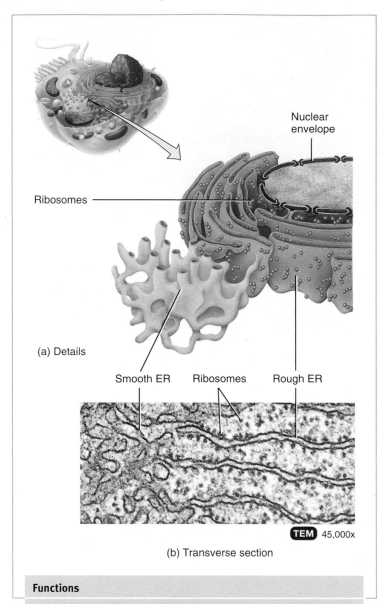

(a) Details

Nuclear envelope

Ribosomes

Smooth ER   Ribosomes   Rough ER

TEM 45,000x

(b) Transverse section

Functions

1. Rough ER synthesizes glycoproteins and phospholipids that are transferred into cellular organelles, inserted into the plasma membrane, or secreted during exocytosis.

2. Smooth ER synthesizes fatty acids and steroids, such as estrogens and testosterone; inactivates or detoxifies drugs and other potentially harmful substances; removes the phosphate group from glucose-6-phosphate; and stores and releases calcium ions that trigger contraction in muscle cells.

? **What are the structural and functional differences between rough and smooth ER?**

Golgi Complex

Most of the proteins synthesized by ribosomes attached to rough ER are ultimately transported to other regions of the cell. The first step in the transport pathway is through an organelle called the **Golgi complex** (GOL-jē). It consists of 3 to 20 **cisternae** (sis-TER-nē=cavities; singular is *cisterna*), small, flattened membranous sacs with bulging edges that resemble a stack of pita bread (Figure 2.11). The cisternae are often curved, giving the Golgi complex a cuplike shape. Most cells have several Golgi complexes, and Golgi complexes are more extensive in protein-secreting cells, a clue to the organelle's role in the cell.

The cisternae at the opposite ends of a Golgi complex differ from each other in size, shape, and enzymatic activity. The convex **entry** or *cis* **face** is a cisterna that faces the rough ER. The concave **exit** or ***trans* face** is a cisterna that faces the plasma membrane. Sacs between the entry and exit faces are called **medial cisternae.** Transport vesicles (described shortly) from the ER merge to form the entry face. From the entry face, the cisternae are thought to mature, in turn becoming medial and then exit cisternae.

Different enzymes in the entry, medial, and exit cisternae of the Golgi complex permit each of these areas to modify, sort, and package proteins for transport to different destinations. The entry face receives and modifies proteins produced by the rough ER. The medial cisternae add carbohydrates to proteins to form glycoproteins and lipids to proteins to form lipoproteins. The exit face modifies the molecules further and then sorts and packages them for transport to their destinations.

Figure 2.11 Golgi complex.

The opposite faces of a Golgi complex differ in size, shape, content, and enzymatic activity.

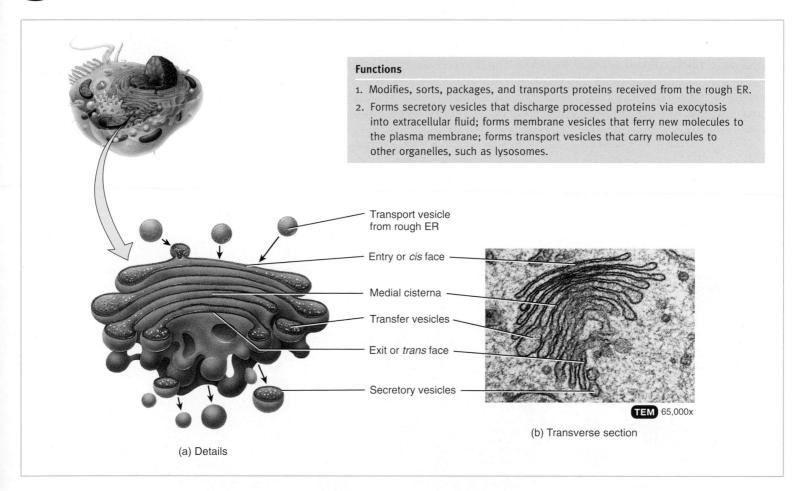

Functions

1. Modifies, sorts, packages, and transports proteins received from the rough ER.
2. Forms secretory vesicles that discharge processed proteins via exocytosis into extracellular fluid; forms membrane vesicles that ferry new molecules to the plasma membrane; forms transport vesicles that carry molecules to other organelles, such as lysosomes.

Transport vesicle from rough ER

Entry or *cis* face

Medial cisterna

Transfer vesicles

Exit or *trans* face

Secretory vesicles

TEM 65,000x

(b) Transverse section

(a) Details

? How do the entry and exit faces differ in function?

Proteins arriving at, passing through, and exiting the Golgi complex do so through maturation of the cisternae and exchanges that occur via transfer vesicles (Figure 2.12):

❶ Proteins synthesized by ribosomes on the rough ER are surrounded by a piece of the ER membrane, which eventually buds from the membrane surface to form **transport vesicles**.

❷ Transport vesicles move toward the entry face of the Golgi complex.

❸ Fusion of several transport vesicles creates the entry face of the Golgi complex and releases proteins into its lumen (space).

❹ The proteins move from the entry face into one or more medial cisternae. Enzymes in the medial cisternae modify the proteins to form glycoproteins, glycolipids, and lipoproteins. **Transfer vesicles** that bud from the edges of the cisternae move specific enzymes back toward the entry face and move some partially modified proteins toward the exit face.

❺ The products of the medial cisternae move into the lumen of the exit face.

❻ Within the exit face cisterna, the products are further modified and are sorted and packaged.

❼ Some of the processed proteins leave the exit face and are stored in **secretory vesicles**. These vesicles deliver the proteins to the plasma membrane, where they are discharged by exocytosis into the extracellular fluid. For example, certain pancreatic cells release the hormone insulin in this way.

❽ Other processed proteins leave the exit face in **membrane vesicles** that deliver their contents to the plasma membrane for incorporation into the membrane. In doing so, the Golgi complex adds new segments of plasma membrane as existing segments are lost and modifies the number and distribution of membrane molecules.

❾ Finally, some processed proteins leave the exit face in transport vesicles that will carry the proteins to another cellular destination. For instance, transport vesicles carry digestive enzymes to lysosomes; the structure and functions of these important organelles are discussed next.

Lysosomes

Lysosomes (LĪ-sō-sōms; *lyso-*=dissolving; *-somes*=bodies) are membrane-enclosed vesicles that form from the Golgi complex (Figure 2.13). Inside can be as many as 60 kinds of powerful digestive enzymes that are capable of breaking down a wide variety of molecules. Lysosomes fuse with vesicles formed during endocytosis and the lysosomal enzymes break down the contents of the vesicles. Proteins in the lysosomal membrane allow the final products of digestion, such as sugars, fatty acids, and amino acids, to be transported into the cytosol. In a similar way, lysosomes in phagocytes can break down and destroy microbes, such as bacteria and viruses.

Lysosomal enzymes also recycle the cell's own structures. A lysosome can engulf another organelle, digest it, and return the digested components to the cytosol for reuse. In this way, old

Figure 2.12 Processing and packaging of synthesized proteins by the Golgi complex.

🔑 All proteins exported from the cell are processed in the Golgi complex.

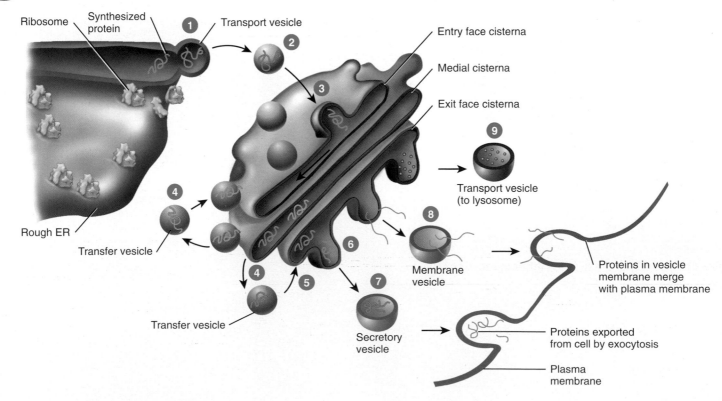

❓ What are the three general destinations for proteins that leave the Golgi complex?

organelles are continually replaced. The process by which entire worn-out organelles are digested is called **autophagy** (aw-TOF-a-jē; *auto-*=self; *-phagy*=eating). In autophagy, the organelle to be digested is enclosed by a membrane derived from the ER to create a vesicle called an **autophagosome** (aw-tō-FĀ-gō-som); the vesicle then fuses with a lysosome. In this way, a human liver cell, for example, recycles about half its cytoplasmic contents every week. Autophagy also contributes to cellular differentiation, control of growth, tissue remodeling, adaptation to adverse environments, and cell defense. Lysosomal enzymes may also destroy the entire cell that contains them, a process known as **autolysis** (aw-TOL-i-sis). Autolysis occurs in some pathological conditions and also is responsible for the tissue deterioration that occurs immediately after death.

Although most of the digestive processes involving lysosomal enzymes occur within a cell, there are some instances in which the enzymes operate in extracellular digestion. One example is the release of lysosomal enzymes during fertilization. Enzymes from lysosomes in the head of a sperm cell help the sperm cell penetrate the surface of the ovum.

Peroxisomes

Another group of organelles similar in structure to lysosomes, but smaller, are called **peroxisomes** (pe-ROKS-i-sōms; *peroxi-*=peroxide; *-somes*=bodies; see Figure 2.1). Peroxisomes, also called *microbodies*, contain several enzymes called *oxidases* that can oxidize (remove hydrogen atoms) from various organic substances. For example, amino acids and fatty acids are oxidized in peroxisomes as part of normal metabolism. In addition, enzymes in peroxisomes oxidize toxic substances, such as alcohol. Thus, peroxisomes are very abundant in the liver, where detoxification of alcohol and other damaging substances takes place. A byproduct of the oxidation reactions is hydrogen peroxide (H_2O_2), a potentially toxic compound. However, peroxisomes also contain an enzyme called *catalase*, which decomposes H_2O_2. Because the generation and degradation of H_2O_2 occur within the same organelle, peroxisomes protect other parts of the cell from the toxic effects of H_2O_2. Like mitochondria, peroxisomes self replicate. One way that new peroxisomes may form is by budding off from preexisting ones by enlarging and dividing.

Proteasomes

Although lysosomes degrade proteins delivered to them in vesicles, cytosolic proteins also require disposal at certain times in the life of a cell. Continuous destruction of unneeded, damaged, or faulty proteins is the function of tiny barrel-shaped structures called **proteasomes** (PRŌ-tē-a-sōms=protein bodies). For example, proteins that are part of metabolic pathways are degraded after they have accomplished their function. Such protein destruction halts the pathway once the appropriate response has been achieved. A typical body cell contains many thousands of proteasomes, in both the cytosol and the nucleus. They are far too small to discern under the light microscope and do not show up well in electron micrographs. Proteasomes were so named because they contain numerous *proteases*, enzymes that cut proteins into small peptides. Once the enzymes of a proteasome have chopped up a protein into smaller chunks, other enzymes then break down the peptides into amino acids, which can be recycled into new proteins.

Mitochondria

Because they generate most of the ATP through aerobic (oxygen-requiring) respiration, **mitochondria** (mī-tō-KON-drē-a; *mito-*=thread; *-chondria*=granules; singular is *mitochondrion*) are referred to as the "powerhouses" of the cell. A cell may have as few as a hundred or as many as several thousand mitochondria, depending on the activity of the cell. Active cells, such as those found in the muscles, liver, and kidneys, which use ATP at a high rate, have a large number of mitochondria. Mitochondria are usually located within the cell where oxygen enters the cell or where the ATP is used, for example, among the contractile proteins in muscle cells.

Figure 2.13 Lysosomes.

Lysosomes contain several kinds of powerful digestive enzymes.

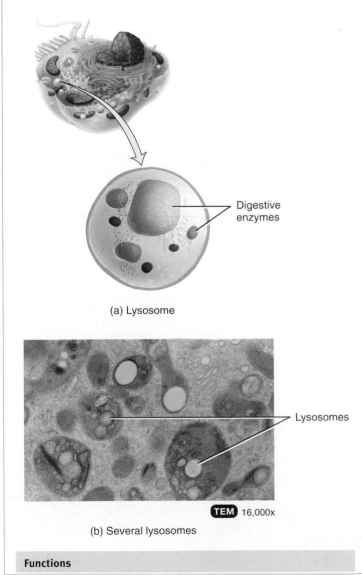

(a) Lysosome

Digestive enzymes

TEM 16,000x

(b) Several lysosomes

Lysosomes

Functions

1. Digest substances that enter a cell via endocytosis and transport final products of digestion into cytosol.
2. Carry out autophagy, the digestion of worn-out organelles.
3. Carry out autolysis, the digestion of entire cell.
4. Carry out extracellular digestion.

? What is the name of the process by which worn-out organelles are digested by lysosomes?

A mitochondrion consists of an **outer mitochondrial membrane** and an **inner mitochondrial membrane** with a small fluid-filled space between them (Figure 2.14). Both membranes are similar in structure to the plasma membrane. The inner mitochondrial membrane contains a series of folds called **cristae**

Figure 2.14 Mitochondria.

Within mitochondria, chemical reactions of aerobic cellular respiration generate ATP.

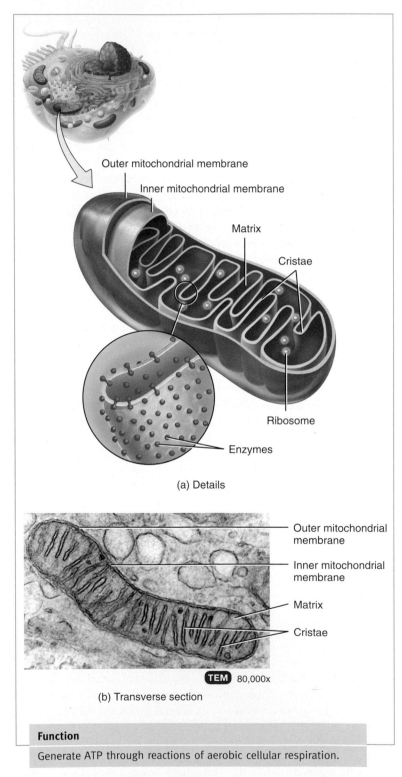

Outer mitochondrial membrane

Inner mitochondrial membrane

Matrix

Cristae

Ribosome

Enzymes

(a) Details

Outer mitochondrial membrane

Inner mitochondrial membrane

Matrix

Cristae

TEM 80,000x

(b) Transverse section

Function

Generate ATP through reactions of aerobic cellular respiration.

? How do the cristae of a mitochondrion contribute to its ATP-producing function?

(KRIS-tē=ridges). The central fluid-filled cavity of a mitochondrion, enclosed by the inner mitochondrial membrane, is the **matrix**. The elaborate folds of the cristae provide an enormous surface area for the chemical reactions that are part of the aerobic phase of *cellular respiration*, the reactions that produce most of a cell's ATP. The enzymes that catalyze these reactions are located on the cristae and in the matrix of the mitochondria.

Like peroxisomes, mitochondria self-replicate, a process that occurs during times of increased cellular energy demand or before cell division. Synthesis of some of the proteins needed for mitochondrial functions occurs on the ribosomes that are present in the mitochondrial matrix. Mitochondria even have their own DNA, in the form of multiple copies of a circular DNA molecule that contains 37 genes. These mitochondrial genes help build mitochondrial components.

Although the nucleus of each somatic cell contains genes from both your mother and father, mitochondrial genes are inherited only from your mother. The head of a sperm (the part that penetrates and fertilizes an oocyte) normally lacks most organelles, such as mitochondria, ribosomes, endoplasmic reticulum, and the Golgi complex, and any sperm mitochondria that do enter the oocyte are soon destroyed.

⊙ CHECKPOINT

8. What does cytoplasm have that cytosol lacks?
9. Which organelles are surrounded by a membrane and which are not?
10. Which organelles contribute to synthesizing protein hormones and packaging them into secretory vesicles?
11. What happens on the cristae and in the matrix of mitochondria?

NUCLEUS

⊙ OBJECTIVE

• Describe the structure and functions of the nucleus.

The **nucleus** is a spherical or oval-shaped structure that usually is the most prominent feature of a cell (Figure 2.15). Most cells have a single nucleus, although some, such as mature red blood cells, have none. In contrast, skeletal muscle cells and a few other types of cells have multiple nuclei. A double membrane called the **nuclear envelope** separates the nucleus from the cytoplasm. Both layers of the nuclear envelope are lipid bilayers similar to the plasma membrane. The outer membrane of the nuclear envelope is continuous with rough ER and resembles it in structure. Many openings called **nuclear pores** extend through the nuclear envelope. Each nuclear pore consists of a circular arrangement of proteins surrounding a large central opening that is about 10 times wider than the pore of a channel protein in the plasma membrane.

Nuclear pores control the movement of substances between the nucleus and the cytoplasm. Small molecules and ions move through the pores passively by diffusion. Most large molecules, such as RNAs and proteins, cannot pass through the nuclear pores by diffusion. Instead, their passage involves an active transport process in which the molecules are recognized and selectively transported through the nuclear pore into or out of the nucleus. For example, proteins needed for nuclear functions move from the cytosol into the nucleus; newly formed RNA molecules move from the nucleus into the cytosol in this manner.

Inside the nucleus are one or more spherical bodies called **nucleoli** (noo-KLĒ-ō-lī; singular is *nucleolus*) that function in producing ribosomes. Each nucleolus is simply a cluster of protein, DNA, and RNA, that is not enclosed by a membrane. Nucleoli are the sites of rRNA synthesis and the assembly of rRNA and proteins into ribosomal subunits. Nucleoli are quite prominent in cells that synthesize large amounts of protein, such as muscle and liver cells. Nucleoli disperse and disappear during cell division and reorganize once new cells are formed.

Figure 2.15 Nucleus.

The nucleus contains most of a cell's genes, which are located on chromosomes.

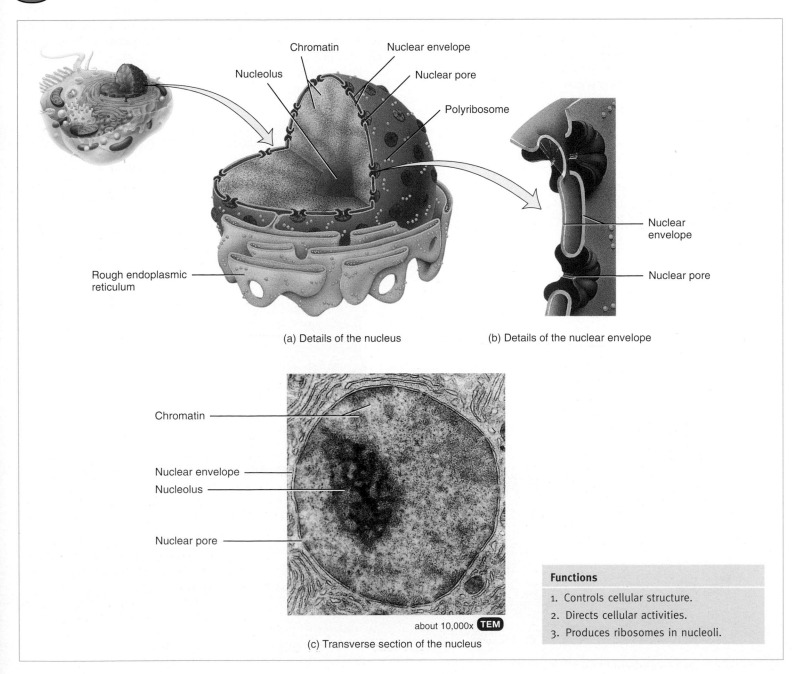

(a) Details of the nucleus

(b) Details of the nuclear envelope

about 10,000x **TEM**

(c) Transverse section of the nucleus

Functions

1. Controls cellular structure.
2. Directs cellular activities.
3. Produces ribosomes in nucleoli.

? What is chromatin?

Within the nucleus are most of the cell's hereditary units, called **genes**, which control cellular structure and direct cellular activities. Genes are arranged along **chromosomes** (*chromo*=colored). Human somatic (body) cells have 46 chromosomes, 23 inherited from each parent. Each chromosome is a long molecule of DNA that is coiled together with several proteins (Figure 2.16). This complex of DNA, proteins, and some RNA is called **chromatin**. The total genetic information carried in a cell or an organism is its **genome**.

In cells that are not dividing, the chromatin appears as a diffuse, granular mass. Electron micrographs reveal that chromatin has a beads-on-a-string structure. Each bead is a **nucleosome** and consists of double-stranded DNA wrapped twice around a core of eight proteins called **histones**, which help organize the coiling and folding of DNA. The string between the beads is **linker DNA**, which holds adjacent nucleosomes together. In cells that are not dividing, another histone promotes coiling of nucleosomes into a larger-diameter **chromatin fiber**, which then folds into large loops. Just before cell division takes place, however, the DNA replicates (duplicates) and the loops condense even more, forming a pair of **chromatids**. As you will see shortly, during cell division a pair of chromatids constitutes a chromosome.

The main parts of a cell, their descriptions, and their functions are summarized in Table 2.2.

Figure 2.16 **Packing of DNA into a chromosome in a dividing cell.** When packing is complete, two identical DNAs and their histones form a pair of chromatids, held together by a centromere.

A chromosome is a highly coiled and folded DNA molecule that is combined with protein molecules.

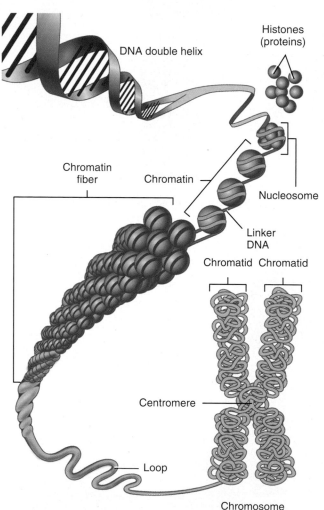

What are the components of a nucleosome?

Human Genome Project: logo with DNA chromosome

In the last decade of the twentieth century, the genomes of humans, mice, fruit flies, and more than 50 microbes were sequenced. As a result, research in the field of **genomics**, the study of the relationships between the genome and the biological functions of an organism, has flourished. The Human Genome Project began in June 1990 as an effort to sequence all of the nearly 3.2 billion nucleotides of our genome, and was completed in April 2003. Scientists now know that the total number of genes in the human genome is about 30,000, far fewer than the 100,000 previously predicted to exist. Information regarding the human genome and how it is affected by the environment seeks to identify and discover the functions of the specific genes that play a role in genetic diseases. Genomic medicine also aims to design new drugs and to provide screening tests to enable physicians to provide more effective counseling and treatment for disorders with significant genetic components such as hypertension (high blood pressure), obesity, diabetes, and cancer. •

◉ CHECKPOINT
12. How is DNA packed into the nucleus?

TABLE 2.2

Cell Parts and Their Functions

PART	DESCRIPTION	FUNCTIONS
Plasma membrane	Fluid-mosaic lipid bilayer (phospholipids, cholesterol, and glycolipids) studded with proteins; surrounds cytoplasm.	Protects cellular contents; makes contact with other cells; contains channels, transporters, receptors, enzymes, cell-identity markers, and linker proteins; mediates the entry and exit of substances.
Cytoplasm	Cellular contents between the plasma membrane and nucleus—cytosol and organelles.	Site of all intracellular activities except those occurring in the nucleus.
Cytosol	Composed of water, solutes, suspended particles, lipid droplets, and glycogen granules.	Medium in which many of cell's metabolic reactions occur.
Organelles	Specialized structures with characteristic shapes.	Each organelle has specific functions.
Cytoskeleton	Network of three types of protein filaments: microfilaments, intermediate filaments, and microtubules.	Maintains shape and general organization of cellular contents; responsible for cellular movements.
Centrosome	A pair of centrioles plus pericentriolar material.	The pericentriolar material contains tubulins, which are used for growth of the mitotic spindle and microtubule formation.
Cilia and flagella	Motile cell surface projections that contain 20 microtubules and a basal body.	Cilia move fluids over a cell's surface; flagella move an entire cell.
Ribosome	Composed of two subunits containing ribosomal RNA and proteins; may be free in cytosol or attached to rough ER.	Protein synthesis.
Endoplasmic reticulum (ER)	Membranous network of flattened sacs or tubules. Rough ER is covered by ribosomes and is attached to the nuclear envelope; smooth ER lacks ribosomes.	Rough ER synthesizes glycoproteins and phospholipids that are transferred to cellular organelles, inserted into the plasma membrane, or secreted during exocytosis. Smooth ER synthesizes fatty acids and steroids; inactivates or detoxifies drugs; removes the phosphate group from glucose-6-phosphate; and stores and releases calcium ions in muscle cells.
Golgi complex	Consists of 3–20 flattened membranous sacs called cisternae; structurally and functionally divided into entry (cis) face, medial cisternae, and exit (trans) face.	Entry (cis) face accepts proteins from rough ER; medial cisternae form glycoproteins, glycolipids, and lipoproteins; exit (trans) face modifies the molecules further, then sorts and packages them for transport to their destinations.
Lysosome	Vesicle formed from Golgi complex; contains digestive enzymes.	Fuses with and digests contents of endosomes, pinocytic vesicles, and phagosomes and transports final products of digestion into cytosol; digests worn-out organelles (autophagy), entire cells (autolysis), and extracellular materials.
Peroxisome	Vesicle containing oxidases (oxidative enzymes) and catalase (decomposes hydrogen peroxide); new peroxisomes bud from preexisting ones.	Oxidizes amino acids and fatty acids; detoxifies harmful substances, such as alcohol; produces hydrogen peroxide.
Proteasome	Tiny barrel-shaped structure that contains proteases (proteolytic enzymes).	Degrades unneeded, damaged, or faulty proteins by cutting them into small peptides.
Mitochondrion	Consists of an outer and an inner mitochondrial membrane, cristae, and matrix; new mitochondria form from preexisting ones.	Site of aerobic cellular respiration reactions that produce most of a cell's ATP.
Nucleus	Consists of a nuclear envelope with pores, nucleoli, and chromosomes, which exist as a tangled mass of chromatin in interphase cells.	Nuclear pores control the movement of substances between the nucleus and cytoplasm, nucleoli produce ribosomes, and chromosomes consist of genes that control cellular structure and direct cellular functions.

Flagellum — Cilium

Intermediate filament
Centrosome

Lysosome
Smooth ER
Peroxisome

Microtubule

NUCLEUS
CYTOPLASM
PLASMA MEMBRANE
Ribosome on rough ER
Golgi complex
Mitochondrion
Microfilament

CELL DIVISION

◉ **OBJECTIVE**

• Discuss the stages, events, and significance of somatic and reproductive cell division.

Most cells of the human body undergo **cell division**, the process by which cells reproduce themselves. The two types of cell division—somatic cell division and reproductive cell division—accomplish different goals for the organism.

A **somatic cell** (*soma*=body) is any cell of the body other than a germ cell. A **germ cell** is a gamete (sperm or oocyte) or any precursor cell destined to become a gamete. In **somatic cell division**, a cell undergoes a nuclear division called **mitosis** (mī-TŌ-sis; *mitos*=thread) and a cytoplasmic division called **cytokinesis** (sī-tō-ki-NĒ-sis; *cyto*=cell; *kinesis*=movement) to produce two identical cells, each with the same number and kind of chromosomes as the original cell. Somatic cell division replaces dead or injured cells and adds new ones during tissue growth.

Reproductive cell division is the mechanism that produces gametes, the cells needed to form the next generation of sexually reproducing organisms. This process consists of a special two-step division called **meiosis**, in which the number of chromosomes in the nucleus is reduced by half.

Somatic Cell Division

The **cell cycle** is an orderly sequence of events by which a somatic cell duplicates its contents and divides in two. Human cells, such as those in the brain, stomach, and kidneys, contain 23 pairs of chromosomes, for a total of 46. One member of each pair is inherited from each parent. The two chromosomes that make up each pair are called **homologous chromosomes** (hō-MOL-ō-gus; *homo*=same) or **homologs**; they contain similar genes arranged in the same (or almost the same) order. When examined under a light microscope, homologous chromosomes generally look very similar. The exception to this rule is one pair of chromosomes called the **sex chromosomes**, designated X and Y. In females the homologous pair of sex chromosomes consists of two large X chromosomes; in males the pair consists of an X and a much smaller Y chromosome. Because somatic cells contain two sets of chromosomes, they are called **diploid cells** (DIP-loid; *dipl-*=double; *-oid*=form), symbolized **2n**.

When a cell reproduces, it must replicate (duplicate) all its chromosomes to pass its genes to the next generation of cells. The cell cycle consists of two major periods: interphase, when a cell is not dividing, and the mitotic (M) phase, when a cell is dividing (Figure 2.17).

Interphase

During **interphase** the cell replicates its DNA through a process that will be described shortly. It also produces additional organelles and cytosolic components in anticipation of cell division. Interphase is a state of high metabolic activity; it is during this time that the cell does most of its growing. Interphase consists of three phases: G₁, S, and G₂ (Figure 2.17). The phase designated S stands for *synthesis* of DNA. Because the G phases are periods when there is no activity related to DNA

Figure 2.17 The cell cycle. Not illustrated is cytokinesis, division of the cytoplasm, which usually occurs during late anaphase of the mitotic phase.

🔑 In a complete cell cycle, a cell duplicates its contents and divides into two identical cells.

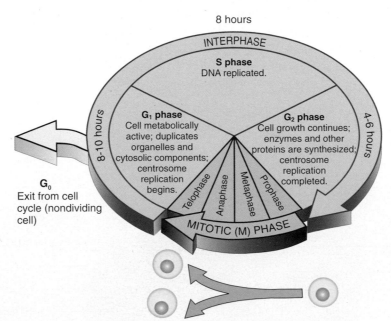

❓ **In which phase of the cell cycle does DNA replication occur?**

duplication; they are thought of as *gaps* or interruptions in DNA duplication.

The **G₁ phase** is the interval between the mitotic phase and the S phase. During G₁ the cell is metabolically active; it replicates most of its organelles and cytosolic components but not its DNA. Replication of centrosomes also begins in the G₁ phase. Virtually all the cellular activities described in this chapter happen during G₁. For a cell with a total cell cycle time of 24 hours, G₁ lasts 8 to 10 hours. However, the duration of this phase is quite variable. It is very short in many embryonic cells or cancer cells. Cells that remain in G₁ for a very long time, perhaps destined never to divide again, are said to be in the **G₀ phase**. Most nerve cells are in the G₀ phase. Once a cell enters the S phase, however, it is committed to go through cell division.

The **S phase**, the interval between G₁ and G₂, lasts about 8 hours. During the S phase, DNA replication occurs. As a result, the two identical cells formed during cell division will have the same genetic material. The **G₂ phase** is the interval between the S phase and the mitotic phase. It lasts 4 to 6 hours. During G₂ cell growth continues, enzymes and other proteins are synthesized in preparation for cell division, and replication of centrosomes is completed. When DNA replicates during the S phase, its helical structure partially uncoils, and the two strands separate at the points where hydrogen bonds connect base pairs (Figure 2.18). Each exposed base of the old DNA strand then pairs with the complementary base of a newly synthesized nucleotide. A new DNA strand takes shape as chemical bonds form between neighboring nucleotides. The uncoiling and complementary base pairing continues until each of the two original DNA strands is joined with a newly formed complementary DNA strand. The original DNA molecule has become two identical DNA molecules.

Figure 2.18 Replication of DNA. The two strands of the double helix separate by breaking the hydrogen bonds (shown as dotted lines) between nucleotides. New, complementary nucleotides attach at the proper sites, and a new strand of DNA is synthesized alongside each of the original strands. Arrows indicate hydrogen bonds forming again between pairs of bases.

🔑 Replication doubles the amount of DNA.

Key:
- Ⓐ = Adenine
- Ⓖ = Guanine
- Ⓣ = Thymine
- Ⓒ = Cytosine

Hydrogen bond

Old strand   New strand   New strand   Old strand

❓ **Why is it crucial that DNA replication occurs prior to cytokinesis in somatic cell division?**

A microscopic view of a cell during interphase shows a clearly defined nuclear envelope, a nucleolus, and a tangled mass of chromatin (Figure 2.19a). Once a cell completes its activities during the G_1, S, and G_2 phases of interphase, the mitotic phase begins.

Mitotic Phase

The **mitotic (M) phase** of the cell cycle consists of a nuclear division, or mitosis, and a cytoplasmic division, or cytokinesis, to form two identical cells. The events that occur during mitosis and cytokinesis are plainly visible under a microscope because chromatin condenses into discrete chromosomes.

NUCLEAR DIVISION: MITOSIS Mitosis, as noted earlier, is the distribution of two sets of chromosomes into two separate nuclei. The process results in the *exact* partitioning of genetic information. For convenience, biologists divide the process into four stages: prophase, metaphase, anaphase, and telophase. However, mitosis is a continuous process; one stage merges seamlessly into the next.

1. **Prophase.** During early prophase, the chromatin fibers condense and shorten into chromosomes that are visible under the light microscope (Figure 2.19b). The condensation process may prevent entangling of the long DNA strands as they move during mitosis. Because longitudinal DNA replication took place during the S phase of interphase, each prophase chromosome consists of a pair of identical strands called **chromatids**. A constricted region called a **centromere** holds the chromatid pair together. At the outside of each centromere is a protein complex known as the **kinetochore** (ki-NET-ō-kor). Later in prophase, tubulins in the pericentriolar material of the centrosomes start to form the **mitotic spindle**, a football-shaped assembly of microtubules that attach to the kinetochore (Figure 2.19b). As the microtubules lengthen, they push the centrosomes to the poles (opposite ends) of the cell so that the spindle extends from pole to pole. The mitotic spindle is responsible for the separation of chromatids to opposite poles of the cell. Then, the nucleolus disappears and the nuclear envelope breaks down.
2. **Metaphase.** During metaphase, the microtubules of the mitotic spindle align the centromeres of the chromatid pairs at the exact center of the mitotic spindle (Figure 2.19c). This midpoint region is called the **metaphase plate**.
3. **Anaphase.** During anaphase, the centromeres split, separating the two members of each chromatid pair, which move toward opposite poles of the cell (Figure 2.19d). Once separated, the chromatids are termed chromosomes. As the chromosomes are pulled by the microtubules of the mitotic spindle during anaphase, they appear V-shaped because the centromeres lead the way, dragging the trailing arms of the chromosomes toward the pole.
4. **Telophase.** The final stage of mitosis, telophase, begins after chromosomal movement stops (Figure 2.19e). The identical sets of chromosomes, now at opposite poles of the cell, uncoil and revert to the threadlike chromatin form. A nuclear envelope forms around each chromatin mass, nucleoli reappear in the identical nuclei, and the mitotic spindle breaks up.

CYTOPLASMIC DIVISION: CYTOKINESIS As noted earlier, division of a cell's cytoplasm and organelles into two identical cells is called **cytokinesis** (sī=-tō-ki-NĒ-sis; -*kinesi*=motion). This process usually begins in late anaphase with the formation of a **cleavage furrow**, a slight indentation of the plasma membrane, and is completed after telophase. The cleavage furrow usually appears midway between the centrosomes and extends around the periphery of the cell (Figures 2.19d and e). Actin microfilaments that lie just inside the plasma membrane form a *contractile ring* that pulls the plasma membrane progressively inward. The ring constricts the center of the cell, like tightening a belt around the waist, and ultimately pinches it in two. Because the plane of the cleavage furrow is always perpendicular to the mitotic spindle, the two sets of chromosomes end up in separate cells. When cytokinesis is complete, interphase begins (Figure 2.19f).

The sequence of events can be summarized as

$$G_1 \rightarrow S \text{ phase} \rightarrow G_2 \text{ phase} \rightarrow \text{mitosis} \rightarrow \text{cytokinesis}$$

Table 2.3 summarizes the events of the cell cycle in somatic cells.

TABLE 2.3

Events of the Somatic Cell Cycle

PHASE	ACTIVITY
Interphase	Period between cell divisions; chromosomes not visible under light microscope.
G_1 phase	Metabolically active cell duplicates organelles and cytosolic components; replication of chromosomes begins. (Cells that remain in the G_1 phase for a very long time, and possibly never divide again, are said to be in the G_0 state.)
S phase	Replication of DNA and centrosomes.
G_2 phase	Cell growth, enzyme and protein synthesis continues; replication of centrosomes complete.
Mitotic phase	Parent cell produces identical cells with identical chromosomes; chromosomes visible under light microscope.
Mitosis	Nuclear division; distribution of two sets of chromosomes into separate nuclei.
Prophase	Chromatin fibers condense into paired chromatids; nucleolus and nuclear envelope disappear; each centrosome moves to an opposite pole of the cell.
Metaphase	Centromeres of chromatid pairs line up at metaphase plate.
Anaphase	Centromeres split; identical sets of chromosomes move to opposite poles of cell.
Telophase	Nuclear envelopes and nucleoli reappear; chromosomes resume chromatin form; mitotic spindle disappears.
Cytokinesis	Cytoplasmic division; contractile ring forms cleavage furrow around center of cell, dividing cytoplasm into separate and equal portions.

Figure 2.19 Cell division: mitosis and cytokinesis. Begin the sequence at at the top of the figure and read clockwise until you complete the process.

In somatic cell division, a single diploid cell divides to produce two identical diploid cells.

1

Centrosome:
  Centrioles
  Pericentriolar material
Nucleolus
Nuclear envelope
Chromatin
Plasma membrane
Cytosol

LM all at 700x

(a) INTERPHASE

6

2

Kinetochore

Centromere

Chromosome
(two chromatids
joined at
centromere)

Mitotic spindle
(microtubules)

Fragments of
nuclear envelope

Early   Late

(b) PROPHASE

(f) IDENTICAL CELLS IN INTERPHASE

Metaphase plate

5

3

(c) METAPHASE

Cleavage furrow

(e) TELOPHASE

4

Chromosome

Cleavage furrow

Late   Early

(d) ANAPHASE

? When does cytokinesis begin?

Control of Cell Destiny

A cell has three possible destinies—to remain alive and functioning without dividing, to grow and divide, or to die. Homeostasis is maintained when there is a balance between cell proliferation and cell death. The signals that tell a cell when to exist in the G_0 phase, when to divide, and when to die have been the subjects of intense and fruitful research in recent years.

Within a cell, there are enzymes called **cyclin-dependent protein kinases (Cdks)** that can transfer a phosphate group from ATP to a protein to activate the protein; other enzymes can remove the phosphate group from the protein to deactivate it. The activation and deactivation of Cdks at the appropriate time is crucial in the initiation and regulation of DNA replication, mitosis, and cytokinesis.

Switching the Cdks on and off is the responsibility of cellular proteins called **cyclins** (SĪK-lins), so named because their levels rise and fall during the cell cycle. The joining of a specific cyclin and Cdk molecule triggers various events that control cell division.

The activation of specific cyclin–Cdk complexes is responsible for progression of a cell from G_1 to S to G_2 to mitosis in a specific order. If any step in the sequence is delayed, all subsequent steps are delayed in order to maintain the normal sequence. The levels of cyclins in the cell are very important in determining the timing and sequence of events in cell division. For example, the level of the cyclin that helps drive a cell from G_2 to mitosis rises throughout the G_1, S, and G_2 phases and into mitosis. The high level triggers mitosis, but toward the end of mitosis, the level declines rapidly and mitosis ends. Destruction of this cyclin, as well as others in the cell, is by proteasomes.

Cellular death is also regulated. Throughout the lifetime of an organism, certain cells undergo **apoptosis** (ap'-op-TŌ-sis or ap'-ō-TŌ-sis=a falling off), an orderly, genetically programmed death. In apoptosis, a triggering agent from either outside or inside the cell causes "cell-suicide" genes to produce enzymes that damage the cell in several ways, including disruption of its cytoskeleton and nucleus. As a result, the cell shrinks and pulls away from neighboring cells. Although the plasma membrane remains intact, the DNA within the nucleus fragments and the cytoplasm shrinks. Phagocytes in the vicinity then ingest the dying cell. This function of phagocytes involves a receptor protein in the plasma membrane of the phagocyte that binds to a lipid in the plasma membrane of the suicidal cell. Apoptosis removes unneeded cells during fetal development, such as the webbing between digits. It continues to occur after birth to regulate the number of cells in a tissue and eliminate potentially dangerous cells such as cancer cells.

Apoptosis is a normal type of cell death; in contrast, **necrosis** (ne-KRŌ-sis=death) is a pathological type of cell death that results from tissue injury. In necrosis, many adjacent cells swell, burst, and spill their cytoplasm into the interstitial fluid. The cellular debris usually stimulates an inflammatory response by the immune system, a process that does not occur in apoptosis.

Reproductive Cell Division

In the process called *sexual reproduction*, each new organism is the result of the union of two different gametes (fertilization), one produced by each parent. If gametes had the same number of chromosomes as somatic cells, the number of chromosomes would double at fertilization. **Meiosis** (mī-Ō-sis; *mei-*=lessening; *-osis*=condition of), the reproductive cell division that occurs in the gonads (ovaries and testes), produces gametes with half the number of chromosomes. As a result, gametes contain a single set of 23 chromosomes and thus are **haploid (*n*) cells** (HAP-loyd; *hapl-*=single). Fertilization restores the diploid number of chromosomes.

Meiosis

Unlike mitosis, which is complete after a single round, meiosis occurs in two successive stages: **meiosis I** and **meiosis II**. During the interphase that precedes meiosis I, the chromosomes of the diploid cell start to replicate. As a result of replication, each chromosome consists of two sister (genetically identical) chromatids, which are attached at their centromeres. This replication of chromosomes is similar to the one that precedes mitosis in somatic cell division.

MEIOSIS I Meiosis I, which begins once chromosomal replication is complete, consists of four phases: prophase I, metaphase I, anaphase I, and telophase I (Figure 2.20a). Prophase I is an extended phase in which the chromosomes shorten and thicken, the nuclear envelope and nucleoli disappear, and the mitotic spindle forms. Two events that are not seen in mitotic prophase occur during prophase I of meiosis. First, the two sister chromatids of each pair of homologous chromosomes pair off, an event called **synapsis** (Figure 2.20b). The resulting four chromatids form a structure called a **tetrad**. Second, parts of the chromatids of two homologous chromosomes may be exchanged with one another. Such an exchange between parts of nonsister (genetically different) chromatids is termed **crossing-over** (Figure 2.20b). This process permits an exchange of genes between the chromatids. Crossing-over results in *genetic recombination*—that is, the formation of new combinations of genes—and accounts for part of the great genetic variation among humans and other organisms that form gametes via meiosis.

In metaphase I, the tetrads formed by the homologous pairs of chromosomes line up along the metaphase plate of the cell, with homologous chromosomes side by side (Figure 2.20a). During anaphase I, the members of each homologous pair of chromosomes separate as they are pulled to opposite poles of the cell by the microtubules attached to the centromeres. The paired chromatids, held by a centromere, remain together. (Recall that during mitotic anaphase, the centromeres split and the sister chromatids separate.) Telophase I and cytokinesis of meiosis are similar to telophase and cytokinesis of mitosis. The net effect of meiosis I is that each resulting cell contains the haploid number of chromosomes because it contains only one member of each pair of the homologous chromosomes present in the starting cell.

MEIOSIS II The second stage of meiosis, meiosis II, also consists of four phases: prophase II, metaphase II, anaphase II, and telophase II (Figure 2.20a). These phases are similar to those that occur during mitosis; the centromeres split, and the sister chromatids separate and move toward opposite poles of the cell.

In summary, meiosis I begins with a diploid starting cell and ends with two cells, each with the haploid number of chromosomes. During meiosis II, each of the two haploid cells formed during meiosis I divides; the net result is four haploid gametes that are genetically different from the original diploid starting cell.

Figure 2.21 on page 50 compares the events of meiosis and mitosis.

Figure 2.20 Meiosis, reproductive cell division. Details of events are discussed in the text.

In reproductive cell division, a single diploid starting cell undergoes meiosis I and meiosis II to produce four haploid gametes that are genetically different from the starting cell that produced them.

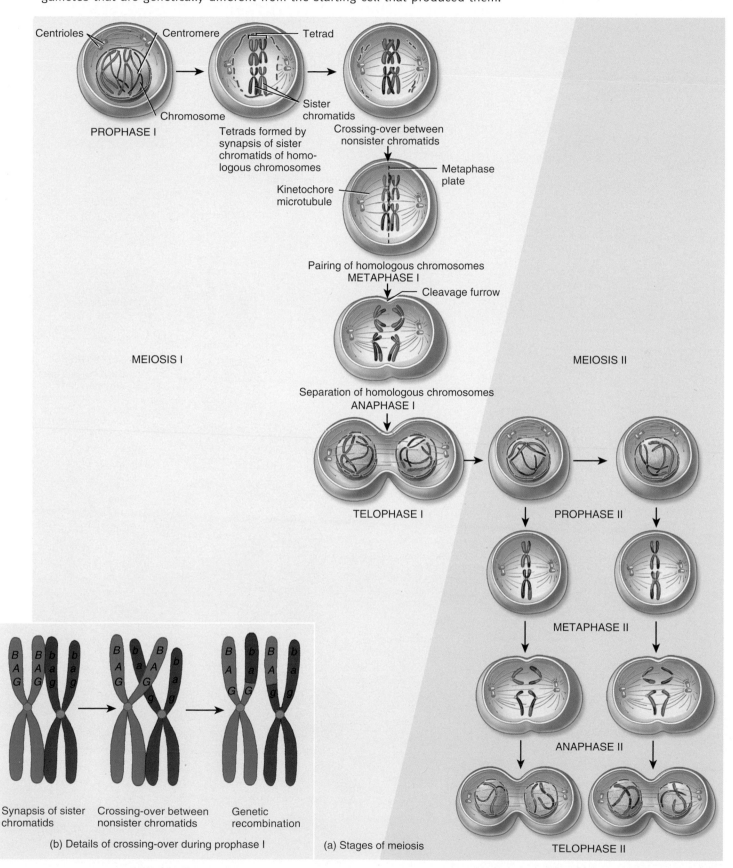

PROPHASE I

Centrioles

Centromere

Tetrad

Chromosome

Sister chromatids

Tetrads formed by synapsis of sister chromatids of homologous chromosomes

Crossing-over between nonsister chromatids

Kinetochore microtubule

Metaphase plate

Pairing of homologous chromosomes
METAPHASE I

Cleavage furrow

MEIOSIS I

Separation of homologous chromosomes
ANAPHASE I

TELOPHASE I

MEIOSIS II

PROPHASE II

METAPHASE II

ANAPHASE II

TELOPHASE II

B A G / *B A G* / *b a g* / *b a g*

Synapsis of sister chromatids

Crossing-over between nonsister chromatids

Genetic recombination

(b) Details of crossing-over during prophase I

(a) Stages of meiosis

How does crossing-over affect the genetic content of the four haploid gametes?

Figure 2.21 Comparison between mitosis (left) and meiosis (right) in which the parent cell has two pairs of homologous chromosomes.

The phases of meiosis II and mitosis are similar.

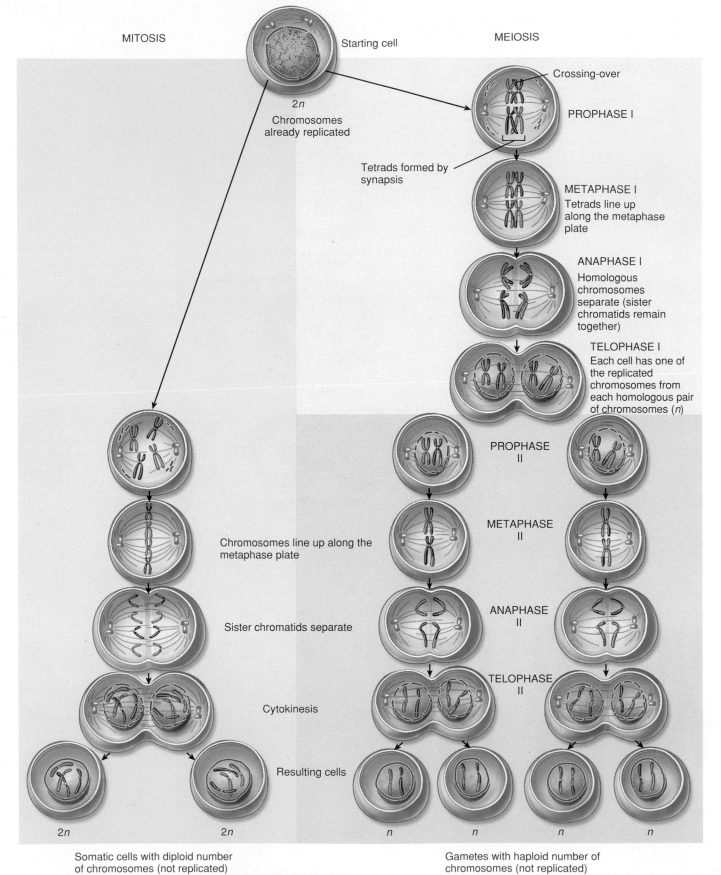

MITOSIS

MEIOSIS

Starting cell

2n
Chromosomes already replicated

Crossing-over

PROPHASE I

Tetrads formed by synapsis

METAPHASE I
Tetrads line up along the metaphase plate

ANAPHASE I
Homologous chromosomes separate (sister chromatids remain together)

TELOPHASE I
Each cell has one of the replicated chromosomes from each homologous pair of chromosomes (n)

PROPHASE II

Chromosomes line up along the metaphase plate

METAPHASE II

Sister chromatids separate

ANAPHASE II

Cytokinesis

TELOPHASE II

Resulting cells

2n   2n   n   n   n   n

Somatic cells with diploid number of chromosomes (not replicated)

Gametes with haploid number of chromosomes (not replicated)

? How does anaphase I of meiosis differ from anaphase of mitosis?

13. Distinguish between somatic and reproductive cell division. Why is each important?
14. Define interphase. When does DNA replicate?
15. What are the major events of each stage of the mitotic phase?
16. How are apoptosis and necrosis similar and different?
17. How are haploid (*n*) and diploid (*2n*) cells different?
18. What are homologous chromosomes?

CELLULAR DIVERSITY

● OBJECTIVE

• Describe how cells differ in size and shape.

The body of an average human adult is composed of nearly 100 trillion cells. All of these cells can be classified into about 200 different cell types. Cells vary considerably in size. The sizes of cells are measured in units called *micrometers.* One micrometer (μm) is equal to 1 one-millionth of a meter, or 10^{-6} (1/25,000 of an inch). High-powered microscopes are needed to see the smallest cells of the body. The largest cell, a single oocyte, has a diameter of about 140 μm and is barely visible to the unaided eye. A red blood cell has a diameter of 8 μm. To better visualize this, an average hair from the top of your head is approximately 100 μm in diameter.

The shapes of cells also vary considerably (Figure 2.22). They may be round, oval, flat, cube-shaped, column-shaped, elongated, star-shaped, cylindrical, or disc-shaped. A cell's shape is related to its function in the body. For example, a sperm cell has a long whiplike tail (flagellum) that it uses for locomotion. The disc shape of a red blood cell gives it a large surface area that enhances its ability to pass oxygen to other cells. The long, spindle shape of a relaxed smooth muscle cell shortens as it contracts. This change in shape allows groups of smooth muscle cells to narrow or widen the passage for blood flowing through blood vessels. In this way, they regulate blood flow through various tissues. Recall that some cells contain microvilli, which greatly increase their surface area. Microvilli are common in the epithelial cells that line the small intestine, where the large surface area speeds the absorption of digested food. Nerve cells have long extensions that permit them to conduct nerve impulses over great distances. As you will see in the next several chapters, cellular diversity also permits organization of cells into more complex tissues and organs.

● CHECKPOINT

19. How is cell shape related to function?

Figure 2.22 Diverse shapes and sizes of human cells. The relative difference in size between the smallest and largest cells of the human body is actually much greater than shown here.

🔑 The nearly 100 trillion cells in an average adult can be classified into about 200 different cell types.

Sperm cell

Smooth muscle cell

Nerve cell

Red blood cell

Epithelial cell

❓ **Why are sperm the only body cells that need a flagellum?**

• CLINICAL CONNECTION Cancer

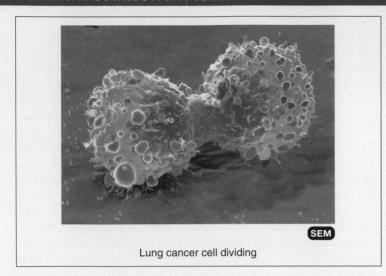

Lung cancer cell dividing

Cancer is a group of diseases characterized by uncontrolled or abnormal cell proliferation. When cells in a part of the body divide without control, the excess tissue that develops is called a **tumor** or **neoplasm** (NĒ-ō-plazm; *neo*-=new). The study of tumors is called **oncology** (on-KOL-ō-jē; *onco*-=swelling or mass). Tumors may be cancerous and often fatal, or they may be harmless. A cancerous neoplasm is called a **malignant tumor** or **malignancy**. One property of most malignant tumors is their ability to undergo **metastasis** (me-TAS-ta-sis), the spread of cancerous cells to other parts of the body.

A **benign tumor** is a neoplasm that does not metastasize. An example is a wart. Most benign tumors may be removed surgically if they interfere with normal body function or become disfiguring. Some benign tumors can be inoperable and perhaps fatal.

Growth and Spread of Cancer

Cells of malignant tumors duplicate rapidly and continuously. As malignant cells invade surrounding tissues, they often trigger **angiogenesis**, the growth of new networks of blood vessels. Proteins that stimulate angiogenesis in tumors are called **tumor angiogenesis factors (TAFs)**. The formation of new blood vessels can occur either by overproduction of TAFs or by the lack of naturally occurring angiogenesis inhibitors. As the cancer grows, it begins to compete with normal tissues for space and nutrients. Eventually, the normal tissue decreases in size and dies. Some malignant cells may detach from the initial (primary) tumor and invade a body cavity or enter the blood or lymph, then circulate to and invade other body tissues, establishing secondary tumors. Malignant cells resist the antitumor defenses of the body. The pain associated with cancer develops when the tumor presses on nerves or blocks a passageway in an organ so that secretions build up pressure, or as a result of dying tissue or organs.

Causes of Cancer

Several factors may trigger a normal cell to lose control and become cancerous. One cause is environmental agents: substances in the air we breathe, the water we drink, and the food we eat. A chemical agent or radiation that produces cancer is called a **carcinogen** (car-SIN-ō-jen). Carcinogens induce **mutations**, permanent changes in the DNA base sequence of a gene. The World Health Organization estimates that carcinogens are associated with 60–90% of all human cancers. Examples of carcinogens are hydrocarbons found in cigarette tar, radon gas from the earth, and ultraviolet (UV) radiation in sunlight.

Intensive research efforts are now directed toward studying cancer-causing genes, or **oncogenes** (ON-kō-jēnz). When inappropriately activated, these genes have the ability to transform a normal cell into a cancerous cell. Most oncogenes derive from normal genes called **proto-oncogenes** that regulate growth and development. The proto-oncogene undergoes some change that causes it either to be expressed inappropriately or to make its products in excessive amounts or at the wrong time. Some oncogenes cause excessive production of growth factors, chemicals that stimulate cell growth. Others may trigger changes in a cell-surface receptor, causing it to send signals as though it were being activated by a growth factor. As a result, the growth pattern of the cell becomes abnormal.

Proto-oncogenes in every cell carry out normal cellular functions until a malignant change occurs. It appears that some proto-oncogenes are activated to oncogenes by mutations in which the DNA of the proto-oncogene is altered. Other proto-oncogenes are activated by a rearrangement of the chromosomes so that segments of DNA are exchanged. Rearrangement activates proto-oncogenes by placing them near genes that enhance their activity.

Damage to genes called **tumor-suppressor genes**, which produce proteins that normally inhibit cell division, causes some types of cancer. Loss or alteration of a tumor-suppressor gene called *p53* on chromosome 17 is the most common genetic change leading to a wide variety of tumors, including breast and colon cancers. The normal p53 protein arrests cells in the G_1 phase, which prevents cell division. Normal p53 protein also assists in repair of damaged DNA and induces apoptosis in the cells where DNA repair was not successful. For this reason, the p53 gene is nicknamed "the guardian angel of the genome."

Some cancers have a viral origin. Viruses are tiny packages of nucleic acids, either RNA or DNA, that can reproduce only while inside the cells they infect. Some viruses, termed **oncogenic viruses**, cause cancer by stimulating abnormal proliferation of cells. For instance, the *human papillomavirus (HPV)* causes virtually all cervical cancers in women. The virus produces a protein that causes proteasomes to destroy p53, a protein that normally suppresses unregulated cell division. In the absence of this suppressor protein, cells proliferate without control.

Recent studies suggest that certain cancers may be linked to a cell having abnormal numbers of chromosomes. As a result, the cell could potentially have extra copies of oncogenes or too few copies of tumor-suppressor genes, which in either case could lead to uncontrolled cell proliferation. There is also some evidence suggesting that cancer may be caused by normal stem cells that develop into cancerous stem cells capable of forming malignant tumors.

Carcinogenesis: A Multistep Process

Carcinogenesis (kar′-si-nō-JEN-e-sis), the process by which cancer develops, is a multistep process in which as many as 10 distinct mutations may have to accumulate in a cell before it becomes cancerous. The progression of genetic changes leading to cancer is best understood for colon (colorectal) cancer. Such cancers, as well as lung and breast cancer, take years or decades to develop. In colon cancer, the tumor begins as an area of increased cell proliferation that results from one mutation. This growth then progresses to abnormal, but noncancerous, growths called *adenomas*. After two or three additional mutations, a mutation of the tumor-suppressor gene p53 occurs and a carcinoma develops. The fact that so many mutations are needed for a cancer to develop indicates that cell growth is normally controlled with many sets of checks and balances. A compromised immune system is also a significant component in carcinogenesis.

Treatment of Cancer

Many cancers are removed surgically. However, when cancer is widely distributed throughout the body or exists in organs such as the brain whose functioning would be greatly harmed by surgery, chemotherapy and radiation therapy may be used instead. Sometimes surgery, chemotherapy, and radiation therapy are used in combination. Chemotherapy involves administering drugs that cause death of cancerous cells. Radiation therapy breaks chromosomes, thus blocking cell division. Because cancerous cells divide rapidly, they are more vulnerable to the destructive effects of chemotherapy and radiation therapy than are normal cells. Unfortunately for the patients, hair follicle cells, red bone marrow cells, and cells lining the gastrointestinal tract also are rapidly dividing. Hence, the side effects of chemotherapy and radiation therapy include hair loss due to death of hair follicle cells, vomiting and nausea due to death of cells lining the stomach and intestines, and susceptibility to infection due to slowed production of white blood cells in red bone marrow.

Treating cancer is difficult because it is not a single disease and because the cells in a single tumor population rarely behave all in the same way. Although most cancers are thought to derive from a single abnormal cell, by the time a tumor reaches a clinically detectable size, it may contain a diverse population of abnormal cells. For example, some cancerous cells metastasize readily and others do not. Some are sensitive to chemotherapy drugs and some are drug-resistant. Because of differences in drug resistance, a single chemotherapeutic agent may destroy susceptible cells but permit resistant cells to proliferate.

Another potential treatment for cancer that is currently under development is *virotherapy,* the use of viruses to kill cancer cells. The viruses employed in this strategy are designed so that they specifically target cancer cells without affecting the healthy cells of the body. For example, proteins (such as antibodies) that specifically bind to receptors found only in cancer cells are attached to viruses. Once inside the body, the viruses bind to cancer cells and then infect them. The cancer cells are eventually killed once the viruses cause cellular lysis.

Researchers also are investigating the role of *metastasis regulatory genes* that control the ability of cancer cells to undergo metastasis. Scientists hope to develop therapeutic drugs that can manipulate these genes and, therefore, block metastasis of cancer cells.

AGING AND CELLS

◉ **OBJECTIVE**

• Describe the cellular changes that occur with aging.

Aging is a normal process accompanied by a progressive alteration of the body's homeostatic adaptive responses to maintain normal conditions. It produces observable changes in structure and function and increases vulnerability to environmental stress and disease. The specialized branch of medicine that deals with the medical problems and care of elderly persons is **geriatrics** (jer-ē-AT-riks; *ger-*=old age; *-iatrics*=medicine). **Gerontology** (jer-on-TOL-ō-jē) is the scientific study of the process and problems associated with aging.

Although many millions of new cells normally are produced each minute, several kinds of cells in the body—heart muscle cells, skeletal muscle cells, and nerve cells—do not divide because they are arrested permanently in the G_0 phase (see page 45). Experiments have shown that many other cell types have only a limited capability to divide. Normal cells grown outside the body divide only a certain number of times and then stop. These observations suggest that cessation of mitosis is a normal, genetically programmed event. According to this view, "aging genes" are part of the genetic blueprint at birth. These genes have an important function in normal cells but their activities slow over time. They bring about aging by slowing down or halting processes vital to life.

Another aspect of aging involves **telomeres** (TĒ-lō-merz), specific DNA sequences found only at the tips of each chromosome. These pieces of DNA protect the tips of chromosomes from erosion and from sticking to one another. However, in most normal body cells each cycle of cell division shortens the telomeres. Eventually, after many cycles of cell division, the telomeres can be completely gone and even some of the functional chromosomal material may be lost. These observations suggest that erosion of DNA from the tips of our chromosomes contributes greatly to aging and death of cells. It has recently been learned that individuals who experience high levels of stress have significantly shorter telomere length.

Glucose, the most abundant sugar in the body, plays a role in the aging process. It is haphazardly added to proteins inside and outside cells, forming irreversible cross-links between adjacent protein molecules. With advancing age, more cross-links form, which contributes to the stiffening and loss of elasticity that occur in aging tissues.

Free radicals produce oxidative damage in lipids, proteins, or nucleic acids by "stealing" an electron to accompany their unpaired electrons. Some effects are wrinkled skin, stiff joints, and hardened arteries. Normal cellular metabolism—for example, aerobic cellular respiration in mitochondria—produces some free radicals. Others are present in air pollution, radiation, and certain foods we eat. Naturally occurring enzymes in peroxisomes and in the cytosol normally dispose of free radicals. Certain dietary substances, such as vitamin E, vitamin C, beta carotene, zinc, and selenium, are antioxidants that inhibit free radical formation.

Some theories of aging explain the process at the cellular level, while others concentrate on regulatory mechanisms operating within the entire organism. For example, the immune system may start to attack the body's own cells. This *autoimmune response* might be caused by changes in cell-identity markers at the surface of cells (certain plasma membrane glycoproteins and glycolipids) that cause antibodies to attach to and mark the cell for destruction. As changes in the proteins on the plasma membrane of cells increase, the autoimmune response intensifies, producing the well-known signs of aging. In the chapters that follow, we will discuss the effects of aging on each body system in sections similar to this one.

◉ **CHECKPOINT**

20. What is one reason that some tissues become stiffer as they age?

KEY MEDICAL TERMS ASSOCIATED WITH CELLS

Note to the Student *Most chapters in this text are followed by a glossary of key medical terms that include both normal and pathological conditions. You should familiarize yourself with the terms because they will play an essential role in your medical vocabulary.*

Anaplasia (an-a-PLĀ-zē-a; *an-*=not; *-plasia*=to shape) The loss of tissue differentiation and function that is characteristic of most malignancies.

Atrophy (AT-rō-fē; *a-*=without; *-trophy*=nourishment) A decrease in the size of cells, with a subsequent decrease in the size of the affected tissue or organ; wasting away.

Dysplasia (dis-PLĀ-zē-a; *dys-*=abnormal) Alteration in the size, shape, and organization of cells due to chronic irritation or inflammation; may progress to neoplasia (tumor formation, usually malignant) or revert to normal if the irritation is removed.

Hyperplasia (hī-per-PLĀ-zē-a; *hyper-*=over) Increase in the number of cells of a tissue due to an increase in the frequency of cell division.

Hypertrophy (hī-PER-trō-fē) Increase in the size of cells without cell division.

Karyotype (KAR-ē-ō-tīp; *karyo-*=nucleus) The chromosomal characteristics of an individual presented as a systematic arrangement of pairs of metaphase chromosomes arrayed in descending order of size and according to the position of the centromere; useful in judging whether chromosomes are normal in number and structure.

Metaplasia (met-a-PLĀ-zē-a; *meta-*=change) The transformation of one type of cell into another.

Progeny (PROJ-e-nē; *pro-*=forward; *-geny*=production) Offspring or descendants.

Tumor marker A substance introduced into circulation by tumor tissue that indicates the presence of a tumor as well as the specific type. Tumor markers may be used to screen, diagnose, assess prognosis, evaluate a response to treatment, and monitor for recurrence of cancer.

STUDY OUTLINE

Introduction (p. 26)
1. A cell is the basic, living, structural and functional unit of the body.
2. Cell biology is the scientific study of cellular structure and function.

A Generalized Cell (p. 27)
1. Figure 2.1 on page 27 provides an overview of the typical structures in body cells.
2. The principal parts of a cell are (1) the plasma membrane; (2) the cytoplasm, the cellular contents between the plasma membrane and nucleus; and (3) the nucleus.

The Plasma Membrane (p. 28)
Structure of the Membrane (p. 28)
1. The plasma membrane surrounds and contains the cytoplasm of a cell.
2. The membrane is composed of proteins and lipids that are held together by forces other than chemical bonds.
3. According to the fluid mosaic model, the membrane is a mosaic of proteins floating like icebergs in a lipid bilayer sea.
4. The lipid bilayer consists of two back-to-back layers of phospholipids, cholesterol, and glycolipids.
5. Integral proteins extend into or through the lipid bilayer; peripheral proteins associate with membrane lipids or integral proteins at the inner or outer surface of the membrane.
6. Many integral proteins are glycoproteins, with sugar groups attached to the ends that face the extracellular fluid. Together with glycolipids, the glycoproteins form a glycocalyx on the extracellular surface of cells.

Functions of Membrane Proteins (p. 28)
1. Membrane proteins have a variety of functions. Integral proteins are channels and transporters that help specific solutes cross the membrane; receptors that serve as cellular recognition sites; enzymes that catalyze specific chemical reactions; and linkers that anchor proteins in the plasma membranes to protein filaments inside and outside the cell. Peripheral proteins serve as enzymes and linkers; support the plasma membrane; anchor integral proteins; and participate in mechanical activities. Membrane glycoproteins function as cell-identity markers.

Membrane Permeability (p. 29)
1. The membrane's selective permeability permits some substances to pass more readily than others. The lipid bilayer is permeable to most nonpolar, uncharged molecules. It is impermeable to ions and charged or polar molecules other than water and urea. Channels and transporters increase the plasma membrane's permeability to small- and medium-sized polar and charged substances, including ions, that cannot cross the lipid bilayer.

Transport Across the Plasma Membrane (p. 29)
1. In passive processes, a substance moves down its concentration gradient across the membrane using its own kinetic energy of motion. In active processes, cellular energy is used to drive the substance "uphill" against its concentration gradient.
2. Substances cross plasma membranes by using kinetic energy, by binding to specific transporter proteins, and by utilizing vesicles.
3. In diffusion, molecules or ions move from an area of higher concentration to an area of lower concentration until equilibrium is reached.
4. Osmosis is the net movement of water through a selectively permeable membrane from an area of higher water concentration to an area of lower water concentration.
5. In facilitated diffusion, a solute such as glucose binds to a specific transporter on one side of the membrane and is released on the other side after the transporter undergoes a change in shape.
6. Active transport is the movement of a substance (usually ions) across a cell membrane from lower to higher concentration using energy derived from ATP and a carrier protein.
7. Transport in vesicles includes both endocytosis and exocytosis.
8. Receptor-mediated endocytosis is the selective uptake of large molecules and particles (ligands) that bind to specific receptors in membrane areas called clathrin-coated pits.
9. Phagocytosis is the ingestion of solid particles. It is an important process used by some white blood cells to destroy bacteria that enter the body.
10. Bulk-phase endocytosis is the ingestion of extracellular fluid. In this process, the fluid becomes surrounded by a pinocytic vesicle.
11. Exocytosis involves movement of secretory or waste products out of a cell by fusion of vesicles with the plasma membrane.

Cytoplasm (p. 32)
1. Cytoplasm is all the cellular contents between the plasma membrane and the nucleus.
2. Cytoplasm consists of cytosol and organelles.

Cytosol (p. 32)
1. Cytosol (intracellular fluid) is the fluid portion of cytoplasm.
2. Cytosol contains mostly water, plus ions, glucose, amino acids, fatty acids, proteins, lipids, ATP, and waste products.
3. Cytosol is the site of many chemical reactions required for a cell's existence.

Organelles (p. 33)

1. Organelles are specialized structures with characteristic shapes and specific functions.
2. The cytoskeleton is a network of several kinds of protein filaments that extend throughout the cytoplasm. Components of the cytoskeleton are microfilaments, intermediate filaments, and microtubules. The cytoskeleton provides a structural framework for the cell and is responsible for cell movements.
3. The centrosome consists of pericentriolar material and a pair of centrioles. The pericentriolar material organizes microtubules in nondividing cells and the mitotic spindle in dividing cells.
4. Cilia and flagella, motile projections of the cell surface, are formed by basal bodies. Cilia move fluid along the cell surface; flagella move an entire cell.
5. Ribosomes consist of two subunits made in the nucleus that are composed of ribosomal RNA and ribosomal proteins. They serve as sites of protein synthesis.
6. Endoplasmic reticulum (ER) is a network of membranes that form flattened sacs or tubules; it extends from the nuclear envelope throughout the cytoplasm.
7. Rough ER is studded with ribosomes that synthesize proteins; the proteins then enter the space within the ER for processing and sorting. Rough ER produces secretory proteins, membrane proteins, and organelle proteins; forms glycoproteins; synthesizes phospholipids; and attaches proteins to phospholipids.
8. Smooth ER lacks ribosomes. It synthesizes fatty acids and steroids; inactivates or detoxifies drugs and other potentially harmful substances; removes phosphate from glucose-6-phosphate; and releases calcium ions that trigger contraction in muscle cells.
9. The Golgi complex consists of flattened sacs called cisternae. The entry, medial, and exit regions of the Golgi complex contain different enzymes that permit each to modify, sort, and package proteins for transport in secretory vesicles, membrane vesicles, or transport vesicles to different cellular destinations.
10. Lysosomes are membrane-enclosed vesicles that contain digestive enzymes. Endosomes, phagosomes, and pinocytic vesicles deliver materials to lysosomes for degradation. Lysosomes function in digestion of worn-out organelles (autophagy), digestion of a host cell (autolysis), and extracellular digestion.
11. Peroxisomes contain oxidases that oxidize amino acids, fatty acids, and toxic substances; the hydrogen peroxide produced in the process is destroyed by catalase.
12. The proteases contained in proteasomes continually degrade unneeded, damaged, or faulty proteins by cutting them into small peptides.
13. Mitochondria consist of a smooth outer membrane, an inner membrane containing cristae, and a fluid-filled cavity called the matrix. These so-called "powerhouses" of the cell produce most of a cell's ATP.

Nucleus (p. 40)

1. The nucleus consists of a double nuclear envelope; nuclear pores, which control the movement of substances between the nucleus and cytoplasm; nucleoli, which produce ribosomes; and genes arranged on chromosomes, which control cellular structure and direct cellular activities.
2. Human somatic cells have 46 chromosomes, 23 inherited from each parent. The total genetic information carried in a cell or an organism is its genome.

Cell Division (p. 44)

1. Cell division is the process by which cells reproduce themselves. It consists of nuclear division (mitosis or meiosis) and cytoplasmic division (cytokinesis).
2. Cell division that replaces cells or adds new ones is called somatic cell division and involves mitosis and cytokinesis.

3. Cell division that results in the production of gametes (sperm and ova) is called reproductive cell division and consists of meiosis and cytokinesis.

Somatic Cell Division (p. 44)

1. The cell cycle, an orderly sequence of events in which a somatic cell duplicates its contents and divides in two, consists of interphase and a mitotic phase.
2. Human somatic cells contain 23 pairs of homologous chromosomes and are thus diploid (2n).
3. Before the mitotic phase, the DNA molecules, or chromosomes, replicate themselves so that identical sets of chromosomes can be passed on to the next generation of cells.
4. A cell between divisions that is carrying on every life process except division is said to be in interphase, which consists of three phases: G_1, S, and G_2.
5. During the G_1 phase, the cell replicates its organelles and cytosolic components, and centrosome replication begins; during the S phase, DNA replication occurs; during the G_2 phase, enzymes and other proteins are synthesized and centrosome replication is completed.
6. Mitosis is the splitting of the chromosomes and the distribution of two identical sets of chromosomes into separate and equal nuclei; it consists of prophase, metaphase, anaphase, and telophase.
7. In cytokinesis, which usually begins in late anaphase and ends once mitosis is complete, a cleavage furrow forms at the cell's metaphase plate and progresses inward, pinching in through the cell to form two separate portions of cytoplasm.

Control of Cell Destiny (p. 48)

1. A cell can either remain alive and functioning without dividing, grow and divide, or die. The control of cell division depends on specific cyclin-dependent protein kinases and cyclins.
2. Apoptosis is normal, programmed cell death. It first occurs during embryological development and continues throughout the lifetime of an organism.
3. Certain genes regulate both cell division and apoptosis. Abnormalities in these genes are associated with a wide variety of diseases and disorders.

Reproductive Cell Division (p. 48)

1. In sexual reproduction, each new organism is the result of the union of two different gametes, one from each parent.
2. Gametes contain a single set of chromosomes (23) and thus are haploid (n).
3. Meiosis is the process that produces haploid gametes; it consists of two successive nuclear divisions called meiosis I and meiosis II.
4. During meiosis I, homologous chromosomes undergo synapsis (pairing) and crossing-over; the net result is two haploid cells that are genetically unlike each other and unlike the starting diploid parent cell that produced them.
5. During meiosis II, two haploid cells divide to form four haploid cells.

Cellular Diversity (p. 51)

1. The almost 200 different types of cells in the body vary considerably in size and shape.
2. The sizes of cells are measured in micrometers. One micrometer (μm) equals 10^{-6} (1/25,000 of an inch). Cells in the body range from 8 μm to 140 μm in size.
3. A cell's shape is related to its function.

Aging and Cells (p. 53)

1. Aging is a normal process accompanied by progressive alteration of the body's homeostatic adaptive responses.
2. Many theories of aging have been proposed, including genetically programmed cessation of cell division, buildup of free radicals, and an intensified autoimmune response.

Choose the one best answer to the following questions.

1. The fluid portion of the cytoplasm is called the:
   a. cytosol   **b.** interstitial fluid
   c. plasma   **d.** cisternae
   e. cristae

2. Respiratory gases move between the lungs and blood by means of:
   a. simple diffusion   **b.** osmosis
   c. active transport   **d.** phagocytosis
   e. bulk-phase endocytosis

3. Which means of active transport does the human immunodeficiency virus (HIV) use to enter white blood cells?
   a. osmosis
   b. facilitated diffusion
   c. receptor-mediated endocytosis
   d. filtration
   e. exocytosis

4. The glycocalyx is a surface coat on cells that:
   a. aids the movement of red blood cells through small blood vessels.
   b. consists of carbohydrate portions of membrane glycolipids and glycoproteins.
   c. serves as a recognition signal for other body cells.
   d. facilitates the adherence of cells to each other in some tissues.
   e. All of the above are correct.

5. Cytokinesis is:
   a. a passive form of transmembrane transport.
   b. how macrophages engulf bacteria.
   c. the separation of chromatid pairs during metaphase.
   d. the division of the cytoplasm during somatic cell division.
   e. the development of flagella on sperm.

6. As a result of somatic cell division, each of the two cells produced has:
   a. the haploid (*n*) number of chromosomes.
   b. twice as many chromosomes as the original cell.
   c. the diploid (*2n*) number of chromosomes.
   d. exactly the same number of chromosomes as the original cell.
   e. *two* of the above.

7. The process by which worn-out cells or whole bacteria are engulfed and destroyed is called:
   a. exocytosis   **b.** hemolysis
   c. phagocytosis   **d.** autolysis
   e. autophagy

8. Division of the nucleus during somatic cell division is called:
   a. apoptosis   **b.** mitosis
   c. meiosis   **d.** metastasis
   e. metaphase

Complete the following.

9. Solvents move down a pressure gradient across a selectively permeable membrane by a passive process called _____.

10. Reproductive cell division consists of a special two-step process of division called _____.

11. Clathrin-coated pits participate in a membrane transport process called _____.

12. The plasma membrane is composed of two main chemical components: a bilayer of _____ and protein.

13. Specific molecules that bind to receptors are called _____.

14. The extracellular fluid filling the microscopic spaces between the cells of tissues is called _____.

15. The net movement of water molecules through a selectively permeable membrane from an area of higher water concentration to an area of lower water concentration is called _____.

Are the following statements true or false? If the statement is False, *find and fix the mistake to make the statement* True.

16. Cholesterol molecules make up the largest percentage of lipid molecules in the plasma membrane.

17. Peripheral proteins span the entire lipid bilayer.

18. Facilitated diffusion is considered an active form of membrane transport because it requires the assistance of a transporter protein.

Matching

19. Match the following:
    ____ **(a)** directs cellular activities by means of genes located here
    ____ **(b)** sites of protein synthesis; may occur attached to ER or scattered freely in cytoplasm
    ____ **(c)** system of membranous channels providing pathways for transport within the cell and surface areas for chemical reactions; may be rough or smooth
    ____ **(d)** stacks of cisternae; involved in packaging and secretion of glycolipids, glycoproteins, and lipoproteins
    ____ **(e)** may release enzymes that lead to autolysis of the cell
    ____ **(f)** similar to lysosomes, but smaller; contain the enzyme catalase
    ____ **(g)** cristae-containing structures, called "powerhouses of the cell" because ATP production occurs within them
    ____ **(h)** part of cytoskeleton; peripherally located in cytoplasm; involved with cell migration and contraction
    ____ **(i)** part of cytoskeleton; give shape to cell; found in flagella, cilia, centrioles, and mitotic spindle
    ____ **(j)** help organize mitotic spindle used in cell division
    ____ **(k)** long, hairlike structures that help move an entire cell
    ____ **(l)** short, hairlike structures that move particles over cell surface

    (1) centrosomes
    (2) cilia
    (3) endoplasmic reticulum
    (4) flagella
    (5) lysosomes
    (6) Golgi complex
    (7) microfilaments
    (8) mitochondria
    (9) nucleus
    (10) microtubules
    (11) ribosomes
    (12) peroxisomes

20. Match the following:
    ____ **(a)** time during mitosis when chromosomes move toward opposite poles of the cell
    ____ **(b)** the result of reproductive cell division
    ____ **(c)** time during mitosis when chromosomes uncoil into chromatin, the mitotic spindle breaks up, and the nuclear envelope appears
    ____ **(d)** spherical bodies that function in producing ribosomes
    ____ **(e)** time during interphase when a cell is duplicating organelles and cytosolic components, but not DNA
    ____ **(f)** cells that contain 23 pairs of chromosomes
    ____ **(g)** members of a chromosome pair, one inherited from each parent
    ____ **(h)** time during mitosis when the nuclear envelope disappears, chromatin thickens into distinct chromosomes, and mitotic spindle forms
    ____ **(i)** proteins that form the core around which DNA wraps in chromatin
    ____ **(j)** time during interphase when DNA replication occurs
    ____ **(k)** the genetically programmed death of certain cells
    ____ **(l)** time during mitosis when chromatids line up on the metaphase plate of the mitotic spindle

    (1) S phase
    (2) anaphase
    (3) metaphase
    (4) prophase
    (5) telophase
    (6) homologs
    (7) gametes
    (8) diploid cells
    (9) apoptosis
    (10) G_1 phase
    (11) histones
    (12) nucleoli

CRITICAL THINKING QUESTIONS

1. Jim had some blood work done recently. When his white blood cells were counted, the test indicated that 95% of them were neutrophils; the normal percentage of neutrophils in a white blood cell count is 75%. What does the increased percentage of neutrophils in Jim's blood suggest?

2. You may inherit your brown eyes from either your mother or your father but some traits can be passed down only from mother to child. "Maternal inheritance" is due to genetic material that is not located in the nucleus. Can you suggest an explanation?

3. In the "old days," intravenous solutions did not come "prepackaged" and had to be mixed at the hospital. Maureen wasn't very good at arithmetic and misplaced her decimal point when calculating how much glucose to add to the intravenous solution for her patient. Instead of making a 0.9% saline solution, she made a 9.0% saline solution. Using your knowledge of osmosis, predict what would happen to her patient's blood cells if she injected this solution into her patient's bloodstream.

4. A child was brought to the emergency room after eating rat poison containing arsenic. Arsenic kills rats by blocking the function of the mitochondria. What effect would the poison have on the child's body functions?

5. Imagine that researchers have discovered a new chemotherapy agent that disrupts microtubules in cancerous cells but leaves normal cells unaffected. What effect would this agent have on the cancer cells?

6. "Gene therapy" relies on delivering healthy genes into cells containing defective genes. One of the current processes undergoing research is to attach healthy genes to a virus that can "carry" the gene into the cells. With your knowledge of transport processes, provide an explanation of how this can work, and speculate on potential obstacles.

? ANSWERS TO FIGURE QUESTIONS

2.1 The three main parts of the cell are the plasma membrane, cytoplasm, and nucleus.

2.2 The glycocalyx is a coat on the extracellular surface of the plasma membrane that is composed of the carbohydrate portions of membrane glycolipids and glycoproteins.

2.3 Another name for intracellular fluid is cytosol.

2.4 Cholesterol, iron, vitamins, and hormones are examples of ligands that can undergo receptor-mediated endocytosis.

2.5 The binding of particles to a plasma membrane receptor triggers pseudopod formation.

2.6 Microtubules help form the structure of centrioles, cilia, and flagella.

2.7 A cell without a centrosome probably would not be able to undergo cell division.

2.8 Cilia move fluids across cell surfaces; flagella move entire cells.

2.9 Large and small ribosomal subunits are synthesized in the nucleolus in the nucleus and assembled in the cytoplasm.

2.10 Rough ER has attached ribosomes; smooth ER does not. Rough ER synthesizes proteins that will be exported from the cell; smooth ER is associated with lipid synthesis and other metabolic reactions.

2.11 The entry face receives and modifies proteins from rough ER. The exit face modifies, sorts, packages, and transports molecules to other destinations.

2.12 Some proteins are discharged from the cell by exocytosis, some are incorporated into the plasma membrane, and some occupy storage vesicles that become lysosomes.

2.13 Digestion of worn-out organelles by lysosomes is called autophagy.

2.14 Mitochondrial cristae increase the surface area available for chemical reactions and contain the enzymes needed for ATP production.

2.15 Chromatin is a complex of DNA, proteins, and some RNA.

2.16 A nucleosome is a double-stranded molecule of DNA wrapped twice around a core of eight histones (proteins).

2.17 DNA replication occurs during the S phase of the cell cycle.

2.18 DNA replication occurs before cytokinesis so that each of the new cells will have a complete set of genes.

2.19 Cytokinesis usually starts in late anaphase.

2.20 The result of crossing-over is that the cells that follow are genetically unlike each other and genetically unlike the starting cell that produced them.

2.21 During anaphase I of meiosis, the paired chromatids are held together by a centromere and do not separate. During mitotic anaphase the centromeres split and sister chromatids separate.

2.22 Sperm, which use the flagella for locomotion, are the only body cells required to move considerable distances.

3 | TISSUES

INTRODUCTION As you learned in Chapter 2, a cell is a complex collection of compartments, each of which carries out a host of biochemical reactions that make life possible. However, a cell seldom functions as an isolated unit in the body. Instead, cells usually work together in groups called tissues. A **tissue** is a group of cells that usually have a common embryonic origin and function together to carry out specialized activities. The structure and properties of a specific tissue are influenced by factors such as the nature of the extracellular material that surrounds the tissue cells and the connections between the cells that compose the tissue. Tissues may be hard, semisolid, or even liquid in their consistency, a range exemplified by bone, fat, and blood. In addition, tissues vary tremendously with respect to the kinds of cells present, how the cells are arranged, and the types of fibers present, if any.

Think about taking a shower. As you wash your body, your skin feels soft and pliable, but the bones beneath your skin are hard and inflexible. What makes these structures so different? Why doesn't the water that splashes on your body penetrate through this surface? The water you drink is absorbed from your digestive system into your blood. What differences between the skin and digestive system lining are responsible for this difference in water permeability?

Occasionally, people break one of their bones. While these breaks can be very painful, they are usually repaired relatively quickly, leaving little or no evidence of the once painful damage. On the other hand, cartilage damage in a joint can lead to prolonged pain and misery. Why these differences?

The answers to these questions deal with understanding the nature of different types of tissues. Tissues form an important level in the hierarchy of body design introduced in Chapter 1. Cells form tissues, and tissues make up the organs that work together to maintain stability in the internal fluid environment. As you study and understand the structure of tissues in this chapter, you will gain more insight into their functional roles in maintaining normal body conditions.

Histology (hiss-TOL-ō-jē; *hist-*=tissue; *-ology*=study of) is the science that deals with the study of tissues. A **pathologist** (pa-THOL-ō-gist; *patho-*=disease) is a physician who specializes in laboratory studies of cells and tissues to help other physicians make accurate diagnoses. One of the principal functions of a pathologist is to examine tissues for any changes that might indicate disease. •

TYPES OF TISSUES AND THEIR ORIGINS

◉ OBJECTIVE
• Name the four basic types of tissues that make up the human body, and state the characteristics of each.

Body tissues can be classified into four basic types according to structure and their function.

1. **Epithelial tissue** covers body surfaces, lines hollow organs, body cavities, and ducts; it also forms glands. This tissue allows the body to interact with both its internal and external environments.
2. **Connective tissue** protects and supports the body and its organs. Various types of connective tissue bind organs together, store energy reserves as fat, and help provide immunity to disease-causing organisms.
3. **Muscular tissue** is composed of cells specialized for contraction and generation of force. In the process, muscular tissue generates heat that warms the body.
4. **Nervous tissue** detects changes in a variety of conditions inside and outside the body and responds by generating nerve impulses that activate muscular contractions and glandular secretions.

Epithelial tissue and all connective tissues, except cartilage, bone, and blood, are more general in nature and have a wide distribution in the body. These tissues are components of most body organs and have a wide range of structure and function. We will look at epithelial tissue and connective tissue in some detail in this chapter. The general features of bone tissue and blood will be introduced here, but their detailed discussion is presented in Chapters 6 and 13, respectively. Similarly, the structure and function of muscular tissue and nervous tissue are introduced here and examined in detail in Chapters 10 and 17, respectively.

Liver biopsy

A **biopsy** (BĪ-op-sē; *bio-*=life; *-opsy*=to view) is the removal of a sample of living tissue for microscopic examination. This procedure is used to help diagnose many disorders, especially cancer, and to discover the cause of unexplained infections and inflammations. Both normal and potentially diseased tissues are removed for purposes of comparison. Once the tissue sample is removed, either surgically or through a needle and syringe, it may be preserved, stained to highlight special properties, or cut into thin sections for microscopic observation. Sometimes a biopsy is conducted while a patient is anesthetized during surgery to help a physician determine the most appropriate treatment. For example, if a biopsy of thyroid tissue reveals malignant cells, the surgeon can immediately proceed with the most appropriate procedure. •

◉ CHECKPOINT
1. Define a tissue.
2. What are the four basic types of human tissue?

CELL JUNCTIONS

◉ **OBJECTIVE**

• Describe the structure and functions of the five main types of cell junctions.

Before looking more specifically at the types of tissue, we will first examine how cells are held together to form tissues. Most epithelial cells and some muscle and nerve cells are tightly joined into functional units. **Cell junctions** are contact points between the plasma membranes of tissue cells. Here we consider the five most important types of cell junctions: tight junctions,

adherens junctions, desmosomes, hemidesmosomes, and gap junctions (Figure 3.1).

Tight Junctions

Tight junctions consist of weblike strands of transmembrane proteins that fuse together the outer surfaces of adjacent plasma membranes to seal off the passageways between adjacent cells (Figure 3.1a). Cells of epithelial tissues that line the stomach, intestines, and urinary bladder have many tight junctions. They retard the passage of substances between cells and thus prevent the

Figure 3.1 Cell junctions.

🔋⬚ Most epithelial cells and some muscle and nerve cells contain cell junctions.

Adjacent plasma membranes
Connexons (composed of connexins)
Gap between cells

(e) Gap junction

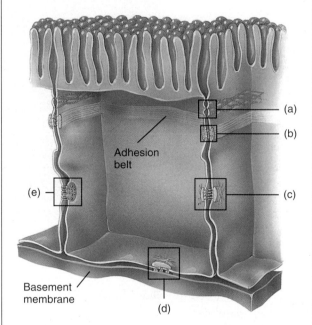

Adhesion belt
Basement membrane

(a)
(b)
(c)
(d)
(e)

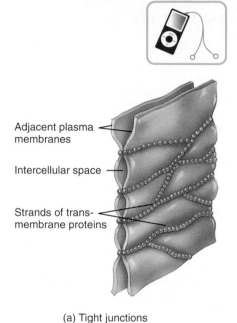

Adjacent plasma membranes
Intercellular space
Strands of trans-membrane proteins

(a) Tight junctions

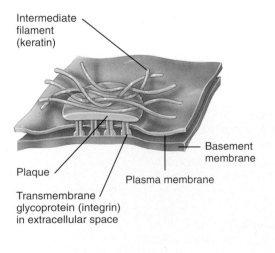

Intermediate filament (keratin)
Basement membrane
Plaque
Plasma membrane
Transmembrane glycoprotein (integrin) in extracellular space

(d) Hemidesmosome

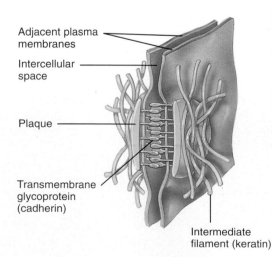

Adjacent plasma membranes
Intercellular space
Plaque
Transmembrane glycoprotein (cadherin)
Intermediate filament (keratin)

(c) Desmosome

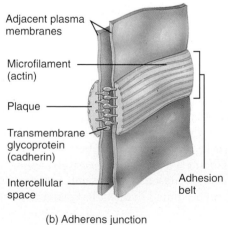

Adjacent plasma membranes
Microfilament (actin)
Plaque
Transmembrane glycoprotein (cadherin)
Intercellular space
Adhesion belt

(b) Adherens junction

❓ **Which type of cell junction functions in communication between adjacent cells?**

contents of these organs from leaking into the blood or surrounding tissues.

Adherens Junctions

Adherens junctions (ad-HER-ens) contain *plaque*, a dense layer of proteins on the inside of the plasma membrane that attaches to both membrane proteins and to microfilaments of the cytoskeleton (Figure 3.1b). Transmembrane glycoproteins called **cadherins** join the cells. Each cadherin inserts into the plaque from the opposite side of the plasma membrane, partially crosses the intercellular space (the space between the cells), and connects to cadherins of an adjacent cell. In epithelial cells, adherens junctions often form extensive zones called *adhesion belts* because they encircle the cell similar to the way a belt encircles your waist. Adherens junctions help epithelial surfaces resist separation during various contractile activities, as when food moves through the intestines.

Desmosomes

Like adherens junctions, **desmosomes** (DEZ-mō-sōms; *desmo-* =band) contain plaque and have transmembrane glycoproteins (cadherins) that extend into the intercellular space between adjacent cell membranes and attach cells to one another (Figure 3.1c). However, unlike adherens junctions, the plaque of desmosomes does not attach to microfilaments. Instead, a desmosome plaque attaches to elements of the cytoskeleton known as intermediate filaments, which consist of the protein keratin. The intermediate filaments extend from desmosomes on one side of the cell across the cytosol to desmosomes on the opposite side of the cell. This structural arrangement contributes to the stability of the cells and tissue. These spot-weld-like junctions are common among the cells that make up the epidermis (the outermost layer of the skin) and among cardiac muscle cells in the heart. Desmosomes prevent epidermal cells from separating under tension and cardiac muscle cells from pulling apart during contraction.

Hemidesmosomes

Hemidesmosomes (*hemi-*=half) resemble desmosomes but they do not link adjacent cells. The name arises from the fact that they look like half of a desmosome (Figure 3.1d). However, the transmembrane glycoproteins in hemidesmosomes are *integrins* rather than cadherins. On the inside of the plasma membrane, integrins attach to intermediate filaments made of the protein keratin. On the outside of the plasma membrane, the integrins attach to the protein *laminin*, which is present in the basement membrane (discussed shortly). Thus, hemidesmosomes anchor cells not to each other but to the basement membrane.

Gap Junctions

At **gap junctions**, membrane proteins called *connexins* form tiny fluid-filled tunnels called *connexons* that connect neighboring cells (Figure 3.1e). The plasma membranes of gap junctions are not fused together as in tight junctions but are separated by a very narrow intercellular gap (space). Through the connexons, ions and small molecules can diffuse from the cytosol of one cell to another, but the passage of large molecules such as vital intracellular proteins is prevented. The transfer of nutrients, and perhaps wastes, takes place through gap junctions in avascular tissues such as the lens and cornea of the eye. Gap junctions allow the cells in a tissue to communicate with one another. In a developing embryo, some of the chemical and electrical signals that regulate growth and cell differentiation travel via gap junctions. Gap junctions also enable nerve or muscle impulses to spread rapidly among cells, a process that is crucial for the normal operation of some parts of the nervous system and for the contraction of muscle in the heart, gastrointestinal tract, and uterus.

◉ CHECKPOINT

3. Which type of cell junctions allows cellular communication?
4. Which types of cell junctions are found in epithelial tissues?

COMPARISON OF EPITHELIAL AND CONNECTIVE TISSUE

Before examining epithelial tissue and connective tissue in more detail, let's compare these two widely distributed tissues. Major structural differences between epithelial tissue and connective tissue are immediately obvious under a light microscope. The first obvious difference is the cell number. In epithelial tissue, many cells are tightly packed together, while in connective tissue a large amount of extracellular material separates widely scattered cells. The second obvious difference is that epithelial tissue has no blood vessels, while most connective tissues have significant networks of blood vessels. While these key structural distinctions account for some of the major functional differences between these tissue types, they also lead to a common bond. Because epithelial tissues lack blood vessels, they are always found in close association with blood-vessel-rich connective tissues to enable them to make the exchanges with blood necessary for their survival and function.

EPITHELIAL TISSUE

◉ OBJECTIVES

• Describe the general features of epithelial tissue.
• List the location, structure, and function of each different type of epithelial tissue.

An **epithelial tissue** (ep-i-THĒ-lē-al), or **epithelium** (plural is *epithelia*), consists of cells arranged in continuous sheets, in either single or multiple layers. Because the cells are closely packed and are held tightly together by many cell junctions, there is little intercellular space between adjacent plasma membranes. Epithelial tissue forms coverings and linings throughout the body. It is never covered by another tissue, so it always has a free surface. Epithelial tissues have three major functions: (1) selective barriers that limit or aid the transfer of substances into and out of the body; (2) secretory surfaces that release products produced by the cells onto their free surfaces; and (3) protective surfaces that resist the abrasive influences of the environment.

The various surfaces of epithelial cells often differ in structure and have specialized functions. The **apical (free) surface** of an epithelial cell faces the body surface, a body cavity, the lumen (interior space) of an internal organ, or a tubular duct that receives cell secretions (Figure 3.2). Apical surfaces may contain cilia or microvilli. The **lateral surfaces** of an epithelial cell, which face the adjacent cells on either side, may contain tight junctions, adherens junctions, desmosomes, and/or gap junctions. The **basal surface** of an epithelial cell is opposite the apical surface. The basal surfaces of the deepest layer of epithelial cells adhere to extracellular materials such as the basement membrane. Hemidesmosomes in the basal surfaces of the deepest layer of epithelial cells anchor the epithelium to the basement membrane (described next). In discussing epithelia with multiple layers the term *apical layer* refers to the most superficial layer of cells, and the *basal layer* is the deepest layer of cells.

The **basement membrane** is a thin extracellular layer that commonly consists of two layers, the basal lamina and reticular lamina. The *basal lamina* (*lamina*=thin layer) is closer to—and secreted by—the epithelial cells. It contains proteins such as laminin and collagen (described shortly), as well as glycoproteins and proteoglycans (also described shortly). As you have already learned, the laminin molecules in the basal lamina adhere to integrins in hemidesmosomes and thus attach epithelial cells to the basement membrane (see Figure 3.1d). The *reticular lamina* is closer to the underlying connective tissue and contains proteins such as collagen produced by connective tissue cells called *fibroblasts* (see Figure 3.6). In addition to attaching to and supporting the overlying epithelial tissue, basement membranes have other functions. They form a surface along which epithelial cells migrate during growth or wound healing, restrict passage of larger molecules between epithelium and connective tissue, and participate in filtration of blood in the kidneys.

Epithelial tissue has its own nerve supply, but as mentioned previously it is **avascular** (*a-*=without; *vascular*=vessel), relying on the blood vessels of the adjacent connective tissue to bring nutrients and remove wastes. Exchange of substances between epithelium and connective tissue occurs by diffusion.

Because epithelial tissue forms boundaries between the body's organs, or between the body and the external environment, it is repeatedly subjected to physical stress and injury. A high rate of cell division allows epithelial tissue to constantly renew and repair itself by sloughing off dead or injured cells and replacing them with new ones. Epithelial tissue plays many different roles in the body; the most important are protection, filtration, secretion, absorption, and excretion. In addition, epithelial tissue combines with nervous tissue to form special organs for smell, hearing, vision, and touch.

Classification of Epithelial Tissue

Types of epithelial tissue are classified according to two characteristics: the arrangement of cells into layers and the shapes of the cells (Figure 3.3).

1. **Arrangement of cells in layers.** Epithelial cells are arranged in one or more layers depending on function:
   a. *Simple (unilaminar) epithelium* is a single layer of cells that functions in diffusion, osmosis, filtration, secretion, or absorption. **Secretion** is the production and release of substances such as mucus, sweat, or enzymes. **Absorption** is the intake of fluids or other substances such as digested food from the intestinal tract.
   b. *Pseudostratified epithelium* (*pseudo-*=false) appears to have multiple layers of cells because the cell nuclei lie at different levels and not all cells reach the apical surface, but it is actually a simple epithelium because all its cells rest on the basement membrane. Cells that do extend to the apical surface may contain cilia; others (goblet cells) secrete mucus.
   c. *Stratified (multilaminar) epithelium* (*stratum*=layer) consists of two or more layers of cells that protect underlying tissues in locations where there is considerable wear and tear.
2. **Cell shapes.** Epithelial cells vary in shape depending on their function:
   a. *Squamous* cells (SKWĀ-mus=flat) are arranged like floor tiles and are thin, which allows for the rapid passage of substances.

Figure 3.2 Surfaces of epithelial cells and the structure and location of the basement membrane.

The basement membrane is found between epithelium and connective tissue.

Apical (free) surface
Lateral surfaces
Epithelium
Basal surface
Basal lamina
Reticular lamina
Basement membrane
Connective tissue
Nerve
Blood vessel

? **What are the functions of the basement membrane?**

Figure 3.3 Cell shapes and arrangement of layers for covering and lining epithelium.

Cell shapes and arrangement of layers are the bases for classifying covering and lining epithelium.

? **Which cell shape is best adapted for the rapid movement of substances from one cell to another?**

b. *Cuboidal* cells are as tall as they are wide and are shaped like cubes or hexagons. They may have microvilli at their apical surface and function in either secretion or absorption.

c. *Columnar* cells are much taller than they are wide, like columns, and protect underlying tissues. Their apical surfaces may have cilia or microvilli, and they often are specialized for secretion and absorption.

d. *Transitional* cells change shape, from flat to cuboidal and back, as organs such as the urinary bladder stretch (distend) to a larger size and then collapse to a smaller size.

When we combine the two characteristics (arrangements of layers and cell shapes), we come up with the following types of epithelia:

I. Simple epithelium
  A. Simple squamous epithelium
  B. Simple cuboidal epithelium
  C. Simple columnar epithelium (nonciliated and ciliated)
  D. Pseudostratified columnar epithelium (nonciliated and ciliated)
II. Stratified epithelium
  A. Stratified squamous epithelium (keratinized, when surface cells are dead and become hardened and nonkeratinized, when surface cells remain alive)*
  B. Stratified cuboidal epithelium*
  C. Stratified columnar epithelium*
  D. Transitional epithelium

---

*This classification is based on the shape of the cells in the apical layer.

Epithelial tissue may be further divided into two types. (1) **Covering and lining epithelium** forms the outer covering of the skin and some internal organs. It also forms the inner lining of blood vessels, ducts, and body cavities, and the interior of the respiratory, digestive, urinary, and reproductive systems. (2) **Glandular epithelium** makes up the secreting portion of glands such as the thyroid gland, adrenal glands, and sweat glands.

Each of these epithelia is described in the following sections and illustrated in Table 3.1 on pages 64–68 and Table 3.2 on page 70. The illustration of each type consists of a photomicrograph, a corresponding diagram, and an inset that identifies a major location of the tissue in the body. Descriptions, locations, and functions of the tissues accompany each illustration.

Covering and Lining Epithelium

Simple Epithelium

SIMPLE SQUAMOUS EPITHELIUM   This tissue consists of a single layer of flat cells that resembles a tiled floor when viewed from the apical surface (Table 3.1A). The nucleus of each cell is a flattened oval or sphere and is centrally located. Simple squamous epithelium is present at sites where the processes of filtration (such as blood filtration in the kidneys) or diffusion (such as diffusion of oxygen into blood vessels of the lungs) occur. It is not found in body areas that are subject to mechanical stress (wear and tear).

Simple squamous epithelium occurs in two common locations in the body. The simple squamous epithelium that lines the cardiovascular and lymphatic system, that is, the heart, blood vessels,

TABLE 3.1

Epithelial Tissues: Covering and Lining Epithelia

SIMPLE EPITHELIUM

A. Simple squamous epithelium

Description: Single layer of flat cells; centrally located nucleus.

Location: Lines heart, blood vessels, lymphatic vessels, air sacs of lungs, glomerular (Bowman's) capsule of kidneys, and inner surface of the tympanic membrane (eardrum); forms epithelial layer of serous membranes, such as the peritoneum, pericardium, and pleura.

Function: Filtration, diffusion, osmosis, and secretion in serous membranes.

Peritoneum

Nucleus of simple squamous cell

Cytoplasm

Plasma membrane

LM 243x

Surface view of simple squamous epithelium of mesothelial lining of peritoneum

Flat nucleus of simp squamous cell

Connective tissue

Muscular tissue

LM 630x

Small intestine

Sectional view of simple squamous epithelium of small intestine

Simple squamous cell

Basement membrane

Connective tissue

Simple squamous epithelium

and lymphatic vessel linings, is known as **endothelium** (*endo-* =within; *-thelium*=covering). The simple squamous epithelium that forms the epithelial layer of serous membranes, such as the peritoneum, pleura, or pericardium, is called **mesothelium** (*meso-*=middle).

SIMPLE CUBOIDAL EPITHELIUM The cuboidal shape of the cells in simple cuboidal epithelium (Table 3.1B) is obvious when

the tissue is sectioned and viewed from the side. Cell nuclei are usually round and centrally located. Simple cuboidal epithelium is found in organs such as the thyroid gland and kidneys and performs the functions of secretion and absorption.

SIMPLE COLUMNAR EPITHELIUM When viewed from the side, the cells of simple columnar epithelium appear taller than they are wide, with oval nuclei near the base. Simple columnar epithe-

SIMPLE EPITHELIUM

B. Simple cuboidal epithelium

Description: Single layer of cube-shaped cells; centrally located nucleus.

Location: Covers surface of ovary, lines anterior surface of capsule of the lens of the eye, forms the pigmented epithelium at the posterior surface of the retina of the eye, lines kidney tubules and smaller ducts of many glands, and makes up the secreting portion of some glands such as the thyroid gland and the ducts of some glands such as the pancreas.

Function: Secretion and absorption.

Sectional view of simple cuboidal epithelium of intralobular duct of pancreas

Simple cuboidal epithelium

C. Nonciliated simple columnar epithelium

Description: Single layer of nonciliated column-like cells with nuclei near base of cells; contains goblet cells and cells with microvilli in some locations.

Location: Lines the gastrointestinal tract (from the stomach to the anus), ducts of many glands, and gallbladder.

Function: Secretion and absorption.

Sectional view of nonciliated simple columnar epithelium of lining of jejunum of small intestine

Nonciliated simple columnar epithelium

CONTINUES

lium exists in two forms: nonciliated simple columnar epithelium and ciliated simple columnar epithelium.

Nonciliated simple columnar epithelium contains two types of cells—columnar epithelial cells with microvilli at their apical surface, and goblet cells (Table 3.1C). **Microvilli**, fingerlike cytoplasmic projections, increase the surface area of the plasma membrane (see Figure 2.1 on page 27), thus increasing the rate of absorption by the cell. **Goblet cells** are modified columnar ep-

ithelial cells that secrete mucus, a slightly sticky fluid, at their apical surfaces. Before it is released, mucus accumulates in the upper portion of the cell, causing it to bulge out and making the whole cell resemble a goblet or wine glass. Secreted mucus serves as a lubricant for the linings of the digestive, respiratory, and reproductive tracts, and most of the urinary tract. Mucus also helps prevent destruction of the stomach lining by acidic gastric juice secreted by the stomach.

TABLE 3.1 CONTINUED

Epithelial Tissues: Covering and Lining Epithelia

SIMPLE EPITHELIUM

D. Ciliated simple columnar epithelium

Description: Single layer of ciliated column-like cells with nuclei near base; contains goblet cells in some locations.

Location: Lines some bronchioles (small tubes) of respiratory tract, uterine (fallopian) tubes, uterus, some paranasal sinuses, central canal of spinal cord, and ventricles of the brain.

Function: Moves mucus and other substances by ciliary action.

Sectional view of ciliated simple columnar epithelium of uterine tube

Ciliated simple columnar epithelium

E. Pseudostratified columnar epithelium

Description: Not a true stratified tissue; nuclei of cells are at different levels; all cells are attached to basement membrane, but not all reach the apical surface.

Location: Pseudostratified ciliated columnar epithelium lines the airways of most of upper respiratory tract; pseudostratified nonciliated columnar epithelium lines larger ducts of many glands, epididymis, and part of male urethra.

Function: Secretion and movement of mucus by ciliary action.

Sectional view of pseudostratified ciliated columnar epithelium of trachea

Pseudostratified ciliated columnar epithelium

Ciliated simple columnar epithelium contains columnar epithelial cells with cilia at the apical surface (Table 3.1D). In certain parts of the smaller airways of the upper respiratory tract, goblet cells are interspersed among ciliated columnar epithelia. Mucus secreted by the goblet cells forms a film over the airway surface that traps inhaled foreign particles. The cilia beat in unison, moving the mucus and any foreign particles toward the throat, where it can be coughed up and swallowed or spit out. Coughing and sneezing speed up the movement of cilia and mucus. Cilia also help move oocytes expelled from the ovaries through the uterine (fallopian) tubes into the uterus.

STRATIFIED EPITHELIUM

F. Stratified squamous epithelium

Description: Several layers of cells; cuboidal to columnar shape in deep layers; squamous cells form the apical layer and several layers deep to it; cells from the basal layer replace surface cells as they are lost.

Location: Keratinized variety forms superficial layer of skin; nonkeratinized variety lines wet surfaces, such as lining of the mouth, esophagus, part of epiglottis, part of pharynx, and vagina, and covers the tongue.

Function: Protection.

Vagina

Stratified squamous epithelium

Connective tissue

LM 400x

Sectional view of stratified squamous epithelium of vagina

Flattened squamous cell at apical surface

Basement membrane

Connective tissue

Stratified squamous epithelium

G. Stratified cuboidal epithelium

Description: Two or more layers of cells in which the cells in the apical layer are cube-shaped.

Location: Ducts of adult sweat glands and esophageal glands and part of male urethra.

Function: Protection and limited secretion and absorption.

Esophagus

Nucleus of stratified cuboidal cell

Lumen of duct

Stratified cuboidal epithelium

Connective tissue

380x

Sectional view of stratified cuboidal epithelium of the duct of an esophageal gland

Apical surface

Basement membrane

Connective tissue

Stratified cuboidal epithelium

CONTINUES

Pseudostratified Columnar Epithelium

As noted earlier, pseudostratified columnar epithelium appears to have several layers because the nuclei of the cells are at various depths (Table 3.1E), but all the cells are attached to the basement membrane in a single layer and some cells do not extend to the apical surface. In *pseudostratified ciliated columnar epithelium*, the cells that extend to the surface either secrete mucus (goblet cells) or bear cilia. The secreted mucus traps foreign particles and the cilia sweep away mucus for eventual elimination from the body. *Pseudostratified nonciliated columnar epithelium* contains cells without cilia and lacks goblet cells.

TABLE 3.1 CONTINUED

Epithelial Tissues: Covering and Lining Epithelia

STRATIFIED EPITHELIUM

H. Stratified columnar epithelium

Description: Several layers of irregularly shaped cells; only the apical layer has columnar cells.

Location: Lines part of urethra, large excretory ducts of some glands, such as esophageal glands, small areas in anal mucous membrane, and part of the conjunctiva of the eye.

Function: Protection and secretion.

Esophagus

Stratified columnar epithelium

Lumen of duct

Nucleus of stratified columnar cell

Connective tissue

LM 300x

Sectional view of stratified columnar epithelium of the duct of an esophageal gland

Apical surface

Basement membrane

Connective tissue

Stratified columnar epithelium

I. Transitional epithelium

Description: Appearance is variable (transitional); shape of cells in apical layer ranges from squamous (when stretched) to cuboidal (when relaxed).

Location: Lines urinary bladder and portions of ureters and urethra.

Function: Permits distension.

Urinary bladder

Lumen of urinary bladder

Nucleus of transitional cell

Transitional epithelium

Connective tissue

LM 400x

Sectional view of transitional epithelium of urinary bladder in relaxed state

Apical surface

Basement membrane

Connective tissue

Relaxed transitional epithelium

Stratified Epithelium

In contrast to simple epithelium, stratified epithelium has two or more layers of cells. Because of this, it is more durable and can better protect underlying tissues. Some cells of stratified epithelia also produce secretions. The name of the specific kind of stratified epithelium depends on the shape of the cells in the apical layer.

STRATIFIED SQUAMOUS EPITHELIUM   Cells in the apical layer of this type of epithelium are flat; those of the deep layers vary in shape from cuboidal to columnar (Table 3.1F). The basal (deepest) cells continually undergo cell division. As new cells grow, the cells of the basal layer are pushed upward toward the apical layer. As they move farther from the deeper layers and from their blood supply in the underlying connective tissue, they become dehydrated, shrunken, and harder, and then die. At the apical layer, after the dead cells lose their cell junctions they are sloughed off, but they are replaced continuously as new cells emerge from the basal layer.

Using your fingernails, roughly scratch the back of your hand. Notice that the affected skin reddens somewhat but does not bleed (unless of course you interpreted a rough scratch as having more force than we intended). Don't do this, but if you applied the same force of scratch to the tissue on the roof of your mouth, you would get a much different result: The roof of your mouth would start to bleed. What accounts for these different results?

Stratified squamous epithelium exists in both keratinized and nonkeratinized forms. In *keratinized stratified squamous epithelium*, the apical layer of cells and several cell layers deep to it are partially dehydrated and contain **keratin**, a tough, fibrous intracellular protein that helps protect the skin and underlying tissues from heat, microbes, and chemicals. The keratin fills the cells and the organelles die. Keratinized stratified squamous epithelium forms the superficial layer of the skin. *Nonkeratinized stratified squamous epithelium*, which is found, for example, lining the mouth and esophagus, does not contain large amounts of keratin in the apical layer and several layers deep to it and remains moist. The organelles are not replaced in this epithelium. Both types of stratified squamous epithelium form the first line of defense against microbes.

STRATIFIED CUBOIDAL EPITHELIUM   This is a fairly rare type of epithelium in which cells in the apical layer are cuboidal (Table 3.1G). Stratified cuboidal epithelium mainly serves a protective function, but it also has a limited role in secretion and absorption.

STRATIFIED COLUMNAR EPITHELIUM   Like stratified cuboidal epithelium, stratified columnar epithelium also is uncommon. Usually the basal layers consist of shortened, irregularly shaped cells; only the apical layer has cells that are columnar in shape (Table 3.1H). This type of epithelium functions in protection and secretion.

TRANSITIONAL EPITHELIUM   Transitional epithelium, a type of stratified epithelium, is present only in the urinary system and has a variable appearance. In its relaxed or unstretched state (Table 3.1I), transitional epithelium looks like stratified cuboidal epithelium, except that the cells in the apical layer tend to be large and rounded. As the tissue is stretched, its cells become flatter, giving the appearance of stratified squamous epithelium. Because of its elasticity, transitional epithelium is ideal for lining hollow structures that are subjected to expansion from within, such as the uri-

nary bladder. It allows the urinary bladder to stretch to hold a variable amount of fluid without rupturing, while still serving as a multilayered protective lining.

CLINICAL CONNECTION | Papanicolaou Test

A **Papanicolaou test** (pa-pa-NI-kō-lō), also called a **Pap test** or **Pap smear,** involves collection and microscopic examination of epithelial cells that have been scraped off the apical layer of a tissue. A very common type of Pap test involves examining the cells from the nonkeratinized stratified squamous epithelium of the vagina and cervix (inferior portion) of the uterus. This type of Pap test is performed mainly to detect early changes in the cells of the female reproductive system that may indicate a precancerous condition or cancer. In performing a Pap smear, a physician collects cells, which are then smeared on a microscope slide. The slides are sent to a laboratory for analysis. An annual Pap test is recommended for all women as part of a routine pelvic exam. They are also recommended for all females over age 18 and for sexually active females.   •

Glandular Epithelium

The function of glandular epithelium is secretion and is accomplished by glandular cells that often lie in clusters deep to the covering and lining epithelium. A **gland** may consist of a single cell or a group of cells that secrete substances into ducts (tubes), onto a surface, or into the blood. All glands of the body are classified as either endocrine or exocrine.

The secretions made by **endocrine glands** (*endo-*=inside; *-crine*=secretion; Table 3.2A) enter the interstitial fluid and then diffuse directly into the bloodstream without flowing through a duct. These secretions, called *hormones*, regulate many metabolic and physiological activities to maintain homeostasis. The pituitary, thyroid, and adrenal glands are examples of endocrine glands. Endocrine glands will be described in detail in Chapter 23.

Exocrine glands (*exo-*=outside; Table 3.2B) secrete their products into ducts that empty onto the surface of a covering and lining epithelium such as the skin surface or the lumen of a hollow organ. The secretions of exocrine glands include mucus, sweat, oil, earwax, saliva, and digestive enzymes. Examples of exocrine glands include sudoriferous (sweat) glands, which produce sweat to help lower body temperature, and salivary glands, which secrete saliva. Saliva contains mucus and digestive enzymes among other substances. As you will learn later in the text, some

TABLE 3.2

Epithelial Tissue: Glandular Epithelium

A. Endocrine glands

Description: Secretory products (hormones) diffuse into blood after passing through interstitial fluid.

Location: Examples include pituitary gland at base of brain, pineal gland in brain, thyroid and parathyroid glands near larynx (voice box), adrenal glands superior to kidneys, pancreas near stomach, ovaries in pelvic cavity, testes in scrotum, and thymus in thoracic cavity.

Function: Produce hormones that regulate various body activities.

Sectional view of endocrine gland (thyroid gland)

B. Exocrine glands

Description: Secretory products released into ducts.

Location: Sweat, oil, and earwax glands of the skin; digestive glands such as salivary glands, which secrete into mouth cavity, and pancreas, which secretes into the small intestine.

Function: Produce substances such as sweat, oil, earwax, saliva, or digestive enzymes.

Sectional view of the secretory portion of an exocrine gland (sweat gland)

glands of the body, such as the pancreas, ovaries, and testes, are mixed glands that contain both endocrine and exocrine tissue.

Structural Classification of Exocrine Glands

Exocrine glands are classified as unicellular or multicellular. As the name implies, **unicellular glands** are single-celled glands. Goblet cells are important unicellular exocrine glands that secrete mucus directly onto the apical surface of a lining epithelium. Most glands are **multicellular glands**, composed of many cells that form a distinctive microscopic structure or macroscopic organ. Examples include sudoriferous, sebaceous (oil), and salivary glands.

Multicellular glands are categorized according to two criteria: (1) whether their ducts are branched or unbranched and (2) the shape of the secretory portions of the gland (Figure 3.4). If the duct of the gland does not branch, it is a **simple gland**. If the duct branches, it is a **compound gland**. Glands with tubular secretory parts are **tubular glands**; those with rounded secretory portions are **acinar glands** (AS-i-nar; *acin-*=berry), also called *alveolar glands*. **Tubuloacinar glands** have both tubular and more rounded secretory parts.

Combinations of these features are the criteria for the following structural classification scheme for multicellular exocrine glands:

I. Simple glands
   A. **Simple tubular.** Tubular secretory part is straight and attaches to a single unbranched duct. Example: glands in the large intestine.
   B. **Simple branched tubular.** Tubular secretory part is branched and attaches to a single unbranched duct. Example: gastric glands.
   C. **Simple coiled tubular.** Tubular secretory part is coiled and attaches to a single unbranched duct. Example: sweat glands.
   D. **Simple acinar.** Secretory portion is rounded and attaches to a single unbranched duct. Example: glands of the penile urethra.
   E. **Simple branched acinar.** Rounded secretory part is branched and attaches to a single unbranched duct. Example: sebaceous glands.

II. Compound glands
   A. **Compound tubular.** Secretory portion is tubular and attaches to a branched duct. Example: bulbourethral (Cowper's) glands.
   B. **Compound acinar.** Secretory portion is rounded and attaches to a branched duct. Example: mammary glands.
   C. **Compound tubuloacinar.** Secretory portion is both tubular and rounded and attaches to a branched duct. Example: acinar glands of the pancreas.

Figure 3.4 Multicellular exocrine glands. Pink represents the secretory portion; lavender represents the duct.

Structural classification of multicellular exocrine glands is based on the branching pattern of the duct and the shape of the secreting portion.

Simple tubular   Simple branched tubular   Simple coiled tubular   Simple acinar   Simple branched acinar

Compound tubular   Compound acinar   Compound tubuloacinar

? How do simple multicellular glands differ from compound ones?

Functional Classification of Exocrine Glands

The functional classification of exocrine glands is based on how their secretions are released. Each of these secretory processes begins with the endoplasmic reticulum and Golgi complex working together to form intracellular secretory vesicles that contain the secretory product. Secretions of **merocrine glands** (MER-ō-krin; *mero-*=a part) are synthesized on ribosomes attached to rough ER; processed, sorted, and packaged by the Golgi complex; and released from the cell in secretory vesicles via exocytosis (Figure 3.5a). Most exocrine glands of the body are merocrine

glands. Examples include the salivary glands and pancreas. **Apocrine glands** (AP-ō-krin; *apo-*=from) accumulate their secretory product at the apical surface of the secreting cell. Then, that portion of the cell pinches off from the rest of the cell to release the secretion (Figure 3.5b). The remaining part of the cell repairs itself and repeats the process. Electron microscopy has confirmed that this is the mechanism of secretion in the mammary glands. Recent evidence reveals that the sweat glands of the skin, named apocrine sweat glands after this mode of secretion, actually undergo merocrine secretion. The cells of **holocrine glands** (HŌ-lō-krin; *holo-*=entire) accumulate a secretory product in their cytosol. As the secretory cell matures, it ruptures and becomes the secretory product (Figure 3.5c). The sloughed off cell is replaced by a new cell. One example of a holocrine gland is a sebaceous (oil) gland of the skin.

⦿ CHECKPOINT

5. Describe the various layering arrangements and cell shapes of epithelium.
6. What characteristics are common to all epithelial tissues?
7. How is the structure of the following kinds of epithelium related to their functions: simple squamous, simple cuboidal, simple columnar (nonciliated and ciliated), pseudostratified columnar (ciliated and nonciliated), stratified squamous (keratinized and nonkeratinized), stratified cuboidal, stratified columnar, and transitional?
8. Where are endothelium and mesothelium located?
9. What is the difference between endocrine glands and exocrine glands? Name and give examples of the three functional classes of exocrine glands.

Figure 3.5 Functional classification of multicellular exocrine glands.

 The functional classification of exocrine glands is based on whether a secretion is a product of a cell or consists of an entire or a partial glandular cell.

Secretion
Secretory vesicle
Golgi complex
Rough ER
Nucleus

(a) Merocrine secretion

Pinched off portion of cell is secretion

(b) Apocrine secretion

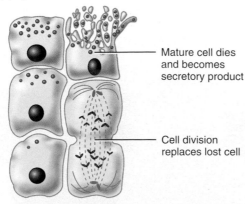

Mature cell dies and becomes secretory product

Cell division replaces lost cell

(c) Holocrine secretion

❓ What class of glands are sebaceous (oil) glands? Salivary glands?

CONNECTIVE TISSUE

⦿ OBJECTIVES

• Describe the general features of connective tissue.
• Describe the structure, location, and function of the various types of connective tissues.

Connective tissue is one of the most abundant and widely distributed tissues in the body. In its various forms, connective tissue has a variety of functions. It binds together, supports, and strengthens other body tissues; protects and insulates internal organs; compartmentalizes structures such as skeletal muscles; serves as the major transport system within the body (blood, a fluid connective tissue); is the primary location of stored energy reserves (adipose or fat tissue); and is the main source of immune responses.

General Features of Connective Tissue

Connective tissue consists of two basic elements: extracellular matrix and cells. A connective tissue's **extracellular matrix** is the material located between its widely spaced cells. The extracellular matrix consists of *protein fibers (fibrillar extracellular matrix) and ground substance (interfibrillar extracellular matrix)*, the material between the cells and the fibers. The extracellular fibers are secreted by the connective tissue cells and the fibers control the surrounding watery environment via specific proteoglycan molecules. The structure of the extracellular matrix determines much of the tissue's qualities. For instance, in cartilage, the extracellular matrix is firm but pliable. The extracellular matrix of bone, by contrast, is hard and inflexible.

In contrast to epithelia, connective tissues do not usually occur on body surfaces. Also, unlike epithelia, connective tissues usually are highly vascular; that is, they have a rich blood supply. Exceptions include cartilage, which is avascular, and tendons, with a scanty blood supply. Except for cartilage, connective tissues, like epithelia, are supplied with nerves.

Connective Tissue Cells

Mesodermal embryonic cells called *mesenchymal cells* give rise to the cells of connective tissue. Each major type of connective tissue contains an immature class of cells with a name ending in *-blast*, which means "to bud or sprout." These immature cells are called *fibroblasts* in loose and dense connective tissue (described shortly), *chondroblasts* in cartilage, and *osteoblasts* in bone. Blast cells retain the capacity for cell division and secrete the extracellular matrix that is characteristic of the tissue. In cartilage and bone, once the extracellular matrix is produced, the immature cells differentiate into mature cells with names ending in *-cyte*, namely chondrocytes and osteocytes. Mature cells have reduced capacities for cell division and extracellular matrix formation and are mostly involved in maintaining the matrix.

The types of connective tissue cells vary according to the type of tissue and include the following (Figure 3.6):

1. **Fibroblasts** (FĪ-brō-blasts; *fibro-*=fibers) are large, flat cells with branching processes. They are present in all the general connective tissues, and usually are the most numerous. Fibroblasts migrate through the connective tissue, secreting the fibers and certain components of the ground substance of the extracellular matrix.

2. **Macrophages** (MAK-rō-fā-jez; *macro-*=large; *-phage*= eaters) develop from monocytes, a type of white blood cell.

Macrophages have an irregular shape with short branching projections and are capable of engulfing bacteria and cellular debris by phagocytosis. *Fixed macrophages* reside in a particular tissue; examples include alveolar macrophages in the lungs or splenic macrophages in the spleen. *Wandering macrophages* have the ability to move throughout the tissue and gather at sites of infection or inflammation to carry on phagocytosis.

3. **Plasma cells** are small cells that develop from a type of white blood cell called a *B lymphocyte*. Plasma cells secrete antibodies, proteins that attack or neutralize foreign substances in the body. Thus, plasma cells are an important part of the body's immune response. Although they are found in many places in the body, most plasma cells reside in connective tissues, especially in the gastrointestinal and respiratory tracts. They are also abundant in the salivary glands, lymph nodes, spleen, and red bone marrow.

4. **Mast cells** are abundant alongside the blood vessels that supply connective tissue. They produce histamine, a chemical that dilates small blood vessels as part of the inflammatory response, the body's reaction to injury or infection. In addition, researchers have recently discovered that mast cells can bind to, ingest, and kill bacteria.

5. **Adipocytes**, also called fat cells or *adipose* cells, are connective tissue cells that store triglycerides (fats). They are found deep to the skin and around organs such as the heart and kidneys.

6. **White blood cells** are not found in significant numbers in normal connective tissue. However, in response to certain conditions they migrate from blood into connective tissues. For example, *neutrophils* gather at sites of infection, and *eosinophils* migrate to sites of parasitic invasions and allergic responses.

Figure 3.6 **Representative cells and fibers present in connective tissue.**

Fibroblasts are usually the most numerous connective tissue cells.

? **What is the function of fibroblasts?**

Connective Tissue Extracellular Matrix

Each type of connective tissue has unique properties, based on the specific extracellular materials between the cells. The extracellular matrix consists of two major components: (1) the ground substance, or interfibrillar extracellular matrix, and (2) the fibers, or fibrillar extracellular matrix.

Ground Substance

As noted earlier, the **ground substance (interfibrillar extracellular matrix)** (*inter*=between; *fibrillar*=fibers) is the component of a connective tissue between the cells and fibers. The ground substance may be fluid, semifluid, gelatinous, or calcified. It supports cells, binds them together, stores water, and provides a medium for exchange of substances between the blood and cells. It plays an active role in how tissues develop, migrate, proliferate, and change shape, and in how they carry out their metabolic functions.

Ground substance contains water and an assortment of large organic molecules, many of which are complex combinations of polysaccharides and proteins. The polysaccharides include hyaluronic acid, chondroitin sulfate, dermatan sulfate, and keratan sulfate. Collectively, they are referred to as **glycosaminoglycans** (glī-kos-a-mē-nō-GLĪ-kans) or **GAGs**. Except for hyaluronic acid, the GAGs are associated with proteins called **proteoglycans** (prō-tē-ō-GLĪ-kans). The proteoglycans form a core protein and the GAGs project from the protein like the bristles of a brush. One of the most important properties of GAGs is that they trap water, making the ground substance more jellylike.

Hyaluronic acid (hī-a-loo-RON-ik) is a viscous, slippery substance that binds cells together, lubricates joints, and helps maintain the shape of the eyeballs. White blood cells, sperm cells, and some bacteria produce *hyaluronidase*, an enzyme that breaks apart hyaluronic acid, thus causing the ground substance of connective tissue to become more liquid. The ability to produce hyaluronidase helps white blood cells move more easily through connective tissues to reach sites of infection and aids penetration of an oocyte by a sperm cell during fertilization. It also accounts for the rapid spread of bacteria through connective tissues. **Chondroitin sulfate** (kon-DROY-tin) provides support and adhesiveness in cartilage, bone, skin, and blood vessels. The skin, tendons, blood vessels, and heart valves contain **dermatan sulfate**; bone, cartilage, and the cornea of the eye contain **keratan sulfate**. Also present in the ground substance are **adhesion proteins**, which are responsible for linking components of the ground substance to one another and to the surfaces of cells. The main adhesion protein of connective tissue is **fibronectin**, which binds to both collagen fibers (discussed shortly) and ground substance, linking them together. Fibronectin also attaches cells to the ground substance.

Fibers

Three types of **fibers (fibrillar extracellular matrix)** are embedded in the ground substance between the cells: collagen fibers, elastic fibers, and reticular fibers. They function to strengthen and support connective tissues.

Collagen fibers (*colla*=glue) are very strong and resist pulling forces, but they are not stiff, which allows tissue flexibility. The properties of different types of collagen fibers vary from tissue to tissue. For example, the collagen fibers found in cartilage and bone form different associations with surrounding molecules. As a result of these associations, the collagen fibers in cartilage attract more water molecules than those in bone, which gives cartilage a more cushioning effect. Collagen fibers often occur in parallel bundles (see Figure 3.6). The bundle arrangement adds great tensile strength to the tissue. Chemically, collagen fibers consist of the protein *collagen*, which is the most abundant protein in your body, representing about 25% of the total protein. Collagen fibers are found in most types of connective tissues, especially bone, cartilage, tendons (which attach muscle to bone), and ligaments (which attach bone to bone).

Elastic fibers, which are smaller in diameter than collagen fibers, branch and join together to form a network within a tissue. An elastic fiber consists of molecules of the protein *elastin* surrounded by a glycoprotein named *fibrillin*, which adds strength and stability. Because of their unique molecular structure, elastic fibers are strong but can be stretched up to 150% of their relaxed length without breaking. Equally important, elastic fibers have the ability to return to their original shape after being stretched, a property called *elasticity*. Elastic fibers are plentiful in skin, blood vessel walls, and lung tissue.

Reticular fibers (*reticul-*=net), consisting of *collagen* arranged in fine bundles with a coating of glycoprotein, provide support in the walls of blood vessels and form a network around the cells in some tissues, such as areolar connective tissue, adipose tissue, and smooth muscle tissue. Produced by fibroblasts, reticular fibers are much thinner than collagen fibers and form branching networks. Like collagen fibers, reticular fibers provide support and strength. Reticular fibers are plentiful in reticular connective tissue, which forms the **stroma** (=bed or covering) or supporting framework of many soft organs, such as the spleen and lymph nodes. These fibers also help form the basement membrane.

Classification of Connective Tissues

Because of the diversity of cells and extracellular matrix and the differences in their relative proportions, the classification of connective tissues is not always clear-cut and multiple classifications exist. We offer the following classification scheme:

I. Embryonic connective tissue
   A. Mesenchyme
   B. Mucous connective tissue
II. Mature connective tissue
   A. Loose connective tissue
      1. Areolar connective tissue
      2. Adipose tissue
      3. Reticular connective tissue
   B. Dense connective tissue
      1. Dense regular connective tissue
      2. Dense irregular connective tissue
      3. Elastic connective tissue
   C. Cartilage
      1. Hyaline cartilage
      2. Fibrocartilage
      3. Elastic cartilage
   D. Bone tissue
   E. Liquid connective tissue
      1. Blood tissue
      2. Lymph

Note that our classification scheme has two major subclasses of connective tissue: embryonic and mature. **Embryonic connective tissue** is present primarily in the *embryo*, the developing human from fertilization through the first two months of preg-

TABLE 3.3

Embryonic Connective Tissues

A. Mesenchyme

Description: Consists of irregularly shaped mesenchymal cells embedded in a semifluid ground substance that contains reticular fibers.

Location: Under skin and along developing bones of embryo; some mesenchymal cells are found in adult connective tissue, especially along blood vessels.

Function: Forms almost all other types of connective tissue.

Embryo

LM 300x

Sectional view of mesenchyme of a developing embryo

Ground substance

Nucleus of mesenchymal cell

Reticular fiber

Mesenchyme

B. Mucous connective tissue

Description: Consists of widely scattered fibroblasts embedded in a viscous, jellylike ground substance that contains fine collagen fibers.

Location: Umbilical cord of fetus.

Function: Support.

Fetus

Umbilical cord

LM 200x

Sectional view of mucous connective tissue of the umbilical cord

Epithelial surface cell of umbilical cord

Collagen fiber

Nucleus of fibroblast

Ground substance

Mucous connective tissue

nancy, and in the *fetus*, the developing human from the third month of pregnancy to birth.

One example of embryonic connective tissue found almost exclusively in the embryo is **mesenchyme** (MEZ-en-kīm), the tissue from which most connective tissues eventually arise (Table 3.3A). Mesenchyme is composed of irregularly shaped cells, a semifluid ground substance, and delicate reticular fibers. Another kind of embryonic tissue is **mucous connective tissue (Whar-**

ton's jelly), found mainly in the umbilical cord of the fetus. Mucous connective tissue is a form of mesenchyme that contains widely scattered fibroblasts, a more viscous jellylike ground substance, and collagen fibers (Table 3.3B).

The second major subclass of connective tissue, **mature connective tissue**, is present in the newborn. Its cells arise primarily from mesenchyme. In the next section we explore the numerous types of mature connective tissue.

Types of Mature Connective Tissue

The five types of mature connective tissue are (1) loose connective tissue, (2) dense connective tissue, (3) cartilage, (4) bone tissue, and (5) liquid connective tissue (blood tissue and lymph). We now examine each in detail.

Loose Connective Tissue

The fibers of **loose connective tissue** are loosely intertwined between cells. The types of loose connective tissue are areolar connective tissue, adipose tissue, and reticular connective tissue.

AREOLAR CONNECTIVE TISSUE One of the most widely distributed connective tissues in the body is areolar connective tissue (a-RĒ-ō-lar; *areol-*=a small space). It contains several types of cells, including fibroblasts, macrophages, plasma cells, mast cells, and adipocytes (Table 3.4A). All three types of fibers—collagen, elastic, and reticular—are arranged randomly throughout the tissue. The ground substance contains hyaluronic acid, chondroitin sulfate, dermatan sulfate, and keratan sulfate. Combined with adipose tissue, areolar connective tissue forms the *subcutaneous layer*; the layer of tissue that attaches the skin to underlying tissues and organs.

ADIPOSE TISSUE Adipose tissue is a loose connective tissue in which the cells, called **adipocytes** (*adipo-*=fat), are specialized for storage of triglycerides (fats) (Table 3.4B). Adipocytes are derived from fibroblasts. Because the cell fills up with a single, large triglyceride droplet, the cytoplasm and nucleus are pushed to the periphery of the cell. Adipose tissue is found wherever areolar connective tissue is located. Adipose tissue is a good insulator and can therefore reduce heat loss through the skin. It is a major energy reserve and generally supports and protects various organs. As a person gains weight, the amount of adipose tissue increases and new blood vessels form. Thus, an obese person has many more blood vessels than does a lean person, a situation that can cause high blood pressure, since the heart has to work harder.

Most adipose tissue in adults is *white adipose tissue*, the type just described. Another type, called *brown adipose tissue (BAT)*, obtains its darker color from a very rich blood supply, along with numerous pigmented mitochondria that participate in aerobic cellular respiration. Although BAT is widespread in the fetus and infant, in adults only small amounts are present. BAT generates considerable heat and probably helps to maintain body temperature in the newborn. The heat generated by the many mitochondria is carried away to other body tissues by the extensive blood supply.

RETICULAR CONNECTIVE TISSUE Reticular connective tissue consists of fine interlacing reticular fibers (a thin form of collagen fiber) and reticular cells (Table 3.4C). Reticular connective tissue forms the stroma (supporting framework) of the liver, spleen, and lymph nodes and helps bind together smooth muscle cells. Additionally, reticular fibers in the spleen filter blood and remove worn-out blood cells, and reticular fibers in lymph nodes filter lymph and remove bacteria.

Dense Connective Tissue

Dense connective tissue contains more numerous, thicker, and denser fibers (packed more closely) but considerably fewer cells than loose connective tissue. There are three types: dense regular connective tissue, dense irregular connective tissue, and elastic connective tissue.

CLINICAL CONNECTION | Liposuction

A surgical procedure, called **liposuction** (*lip-*=fat) or **suction lipectomy** (*-ectomy*=to cut out), involves suctioning out small amounts of adipose tissue from various areas of the body. After an incision is made in the skin, the fat is removed through a stainless steel tube, called a cannula, with the assistance of a unit that suctions out the fat. The technique can be used as a body-contouring procedure in regions such as the thighs, buttocks, arms, breasts, and abdomen, and to transfer fat to another area of the body. Postsurgical complications that may develop include fat that may enter blood vessels broken during the procedure and obstruct blood flow, infection, loss of feeling in the area, fluid depletion, injury to internal structures, and severe postoperative pain. •

Abdominal Liposuction

DENSE REGULAR CONNECTIVE TISSUE In dense regular connective tissue, bundles of collagen fibers are *regularly* arranged in parallel patterns that provide the tissue with great strength (Table 3.4D). The tissue withstands pulling along the length of the fibers. Fibroblasts, which produce the fibers and ground substance, appear in rows between the fibers. The tissue is silvery white and tough, yet somewhat pliable. Examples include tendons and most ligaments.

DENSE IRREGULAR CONNECTIVE TISSUE Dense irregular connective tissue contains tightly packed, woven meshes of collagen fibers that are usually *irregularly* arranged (Table 3.4E). Found in parts of the body where pulling forces are exerted in various directions, this tissue often occurs in sheets, such as in the dermis of the skin, which is deep to the epidermis, or the fibrous pericardium around the heart. Heart valves, the perichondrium (the tissue surrounding cartilage), and the periosteum (the tissue surrounding bone) are dense irregular connective tissues, although they have a fairly orderly arrangement of collagen fibers.

ELASTIC CONNECTIVE TISSUE Branching elastic fibers predominate in elastic connective tissue (Table 3.4F), giving the unstained tissue a yellowish color. Fibroblasts are present in the spaces between the fibers. Elastic connective tissue is quite strong and can recoil to its original shape after being stretched. Elasticity is important to the normal functioning of lung tissue, which recoils as you exhale, and elastic arteries, which recoil between heartbeats to maintain blood flow.

Cartilage

Cartilage consists of a dense network of collagen fibers or elastic fibers firmly embedded in chondroitin sulfate, a gel-like component of the ground substance. Cartilage can endure considerably more stress than loose and dense connective tissues. The strength

TABLE 3.4

Mature Connective Tissues

LOOSE CONNECTIVE TISSUE

A. Areolar connective tissue

Description: Consists of fibers (collagen, elastic, and reticular) and several kinds of cells (fibroblasts, macrophages, plasma cells, adipocytes, and mast cells) embedded in a semifluid ground substance.

Location: Subcutaneous layer deep to skin; papillary (superficial) region of dermis of skin; lamina propria of mucous membranes; and around blood vessels, nerves, and body organs.

Function: Strength, elasticity, and support.

LM 400x

Sectional view of subcutaneous areolar connective tissue

Areolar connective tissue

B. Adipose tissue

Description: Consists of adipocytes, cells specialized to store triglycerides (fats) as a large centrally located droplet; nucleus and cytoplasm are peripherally located.

Location: Subcutaneous layer deep to skin, around heart and kidneys, yellow bone marrow, and padding around joints and behind eyeball in eye socket.

Function: Reduces heat loss through skin, serves as an energy reserve, supports, and protects. In newborns, brown adipose tissue generates considerable heat that helps maintain proper body temperature.

LM 630x

LM 200x

Sectional view of adipose tissue showing adipocytes of white fat and details of an adipocyte

Adipose tissue

of cartilage is due to its collagen fibers, and its *resilience* (ability to assume its original shape after deformation) is due to chondroitin sulfate.

Like other connective tissues, cartilage has few cells and large quantities of extracellular matrix. It differs from other connective tissues, however, in not having nerves or blood vessels in its extracellular matrix. Interestingly, cartilage does not have a blood supply because it secretes an *antiangiogenesis* (an′-tē-an′-gē-ō-JEN-e-sis) *factor* (*anti-*=against; *angio-*=vessel; *genesis*=production), a substance that prevents blood vessel growth. Because of this property, antiagiogenesis factor is being studied as a possible cancer treatment to stop cancer cells from promoting new vessel growth, which supports their rapid rate of cell division and expansion.

The cells of mature cartilage, called **chondrocytes** (KON-drō-sīts; *chondro-*=cartilage), occur singly or in groups within spaces called **lacunae** (la-KOO-nē=little lakes; singular *lacuna*) in the extracellular matrix. A covering of dense irregular

---

TABLE 3.4 CONTINUED

Mature Connective Tissues

LOOSE CONNECTIVE TISSUE

C. Reticular connective tissue

Description: A network of interlacing reticular fibers and reticular cells.

Location: Stroma (supporting framework) of liver, spleen, lymph nodes; red bone marrow, which gives rise to blood cells; reticular lamina of the basement membrane; and around blood vessels and muscles.

Function: Forms stroma of organs; binds together smooth muscle tissue cells; filters and removes worn-out blood cells in the spleen and microbes in lymph nodes.

Lymph node

LM 400x

Sectional view of reticular connective tissue of a lymph node

Nucleus of reticular cell

Reticular fiber

Reticular connective tissue

DENSE CONNECTIVE TISSUE

D. Dense regular connective tissue

Description: Extracellular matrix looks shiny white; consists mainly of collagen fibers regularly arranged in bundles; fibroblasts present in rows between bundles.

Location: Forms tendons (attach muscle to bone), most ligaments (attach bone to bone), and aponeuroses (sheetlike tendons that attach muscle to muscle or muscle to bone).

Function: Provides strong attachment between various structures.

Tendon

Skeletal muscle

LM 200x

Sectional view of dense regular connective tissue of a tendon

Nucleus of fibroblast

Collagen fiber

Dense regular connective tissue

FIBROCARTILAGE Chondrocytes are scattered among clearly visible thick bundles of collagen fibers within the extracellular matrix of fibrocartilage (Table 3.4H). Fibrocartilage lacks a perichondrium. With a combination of strength and rigidity, this tissue is the strongest of the three types of cartilage. One location of fibrocartilage is the intervertebral discs, the discs between vertebrae (backbones).

ELASTIC CARTILAGE The chondrocytes of elastic cartilage are located within a threadlike network of elastic fibers within the extracellular matrix (Table 3.4I). A perichondrium is present. Elastic cartilage provides strength and elasticity and maintains the shape of certain structures, such as the external ear.

CARTILAGE

I. Elastic cartilage

Description: Consists of chondrocytes located in a threadlike network of elastic fibers within the extracellular matrix.

Location: Lid on top of larynx (epiglottis), part of external ear (auricle), and auditory (eustachian) tubes.

Function: Gives support and maintains shape.

Auricle of ear

Perichondrium

Nucleus of chondrocyte

Elastic fiber in ground substance

Lacuna containing chondrocyte

LM 400x

Sectional view of elastic cartilage of auricle of ear

Elastic cartilage

CONTINUES

TABLE 3.4 CONTINUED

Mature Connective Tissues

BONE TISSUE

J. Compact bone

Description: Compact bone tissue consists of osteons (haversian systems) that contain lamellae, lacunae, osteocytes, canaliculi, and central (haversian) canals. By contrast, spongy bone tissue (see Figure 6.4a,b on pages 146–147) consists of thin columns called trabeculae; spaces between trabeculae are filled with red bone marrow.

Location: Both compact and spongy bone tissue make up the various parts of bones of the body.

Function: Support, protection, storage; houses blood-forming tissue; serves as levers that act with muscle tissue to enable movement.

Sectional view of several osteons (haversian systems) of femur (thigh bone) and details of one osteon

Details of an osteocyte

LIQUID CONNECTIVE TISSUE

K. Blood

Description: Consists of blood plasma and formed elements: red blood cells (erythrocytes), white blood cells (leukocytes), and platelets (thrombocytes).

Location: Within blood vessels (arteries, arterioles, capillaries, venules, and veins) and within the chambers of the heart.

Function: Red blood cells transport oxygen and some carbon dioxide; white blood cells carry on phagocytosis and are involved in allergic reactions and immune system responses; platelets are essential for the clotting of blood.

Blood in blood vessels

Blood smear

Bone Tissue

Cartilage, joints, and bones make up the skeletal system. The skeletal system supports soft tissues, protects delicate structures, and works with skeletal muscles to generate movement. Bones store calcium and phosphorus; house red bone marrow, which produces blood cells; and contain yellow bone marrow, a storage site for triglycerides. Bones are organs composed of several different connective tissues, including **bone** or **osseous tissue** (OS-ē-us), the periosteum, red and yellow bone marrow, and the endosteum (a membrane that lines a space within bone that stores yellow bone marrow). Bone tissue is classified as either compact or spongy, depending on how its extracellular matrix and cells are organized.

The basic unit of **compact bone** is an **osteon** or **haversian system** (Table 3.4J). Each osteon has four parts:

1. The **lamellae** (la-MEL-lē=little plates; singular is lamella) are concentric rings of extracellular matrix that consist of mineral salts (mostly calcium and phosphates), which give bone its hardness, and collagen fibers, which give bone its strength. The lamellae are responsible for the compact nature of this type of bone tissue.

2. **Lacunae** (la-KOO-nē; singular is *lacuna* pronounced la-KOO-na) are small spaces between lamellae that contain mature bone cells called **osteocytes**.

3. Projecting from the lacunae are **canaliculi** (kan-a-LIK-ū-lī=little canals), networks of minute canals containing the processes of osteocytes. Canaliculi provide routes for nutrients to reach osteocytes and for wastes to leave them.

4. A **central (haversian) canal** contains blood vessels and nerves.

Spongy bone lacks osteons. Rather, it consists of columns of bone called **trabeculae** (tra-BEK-ū-lē=little beams), which contain lamellae, osteocytes, lacunae, and canaliculi. Spaces between trabeculae are filled with red bone marrow. Chapter 6 presents the histology of bone tissue in more detail.

Liquid Connective Tissue

BLOOD TISSUE **Blood tissue** (or simply blood) is a connective tissue with a liquid extracellular matrix called **blood plasma**, a pale yellow fluid that consists mostly of water with a wide variety of dissolved substances—nutrients, wastes, enzymes, plasma proteins, hormones, respiratory gases, and ions (Table 3.4K). Suspended in the blood plasma are formed elements—red blood cells (erythrocytes), white blood cells (leukocytes), and platelets (thrombocytes). **Red blood cells** transport oxygen to body cells and remove some carbon dioxide from them. **White blood cells** are involved in phagocytosis, immunity, and allergic reactions. Platelets participate in blood clotting. The details of blood are considered in Chapter 13.

LYMPH **Lymph** is the extracellular fluid that flows in lymphatic vessels. It is a connective tissue that consists of several types of cells in a clear liquid extracellular matrix similar to blood plasma but with much less protein. The composition of lymph varies from one part of the body to another. For example, lymph leaving lymph nodes includes many lymphocytes, a type of white blood cell, in contrast to lymph from the small intestine, which has a high content of newly absorbed dietary lipids. The details of lymph are considered in Chapter 16.

◉ CHECKPOINT

10. In what ways do connective tissues differ from epithelia?

11. What are the features of the cells, ground substance, and fibers that make up connective tissue?

12. How are connective tissues classified? List the various types.

13. Describe how the structures of the following connective tissues are related to their functions: areolar connective tissue, adipose tissue, reticular connective tissue, dense regular connective tissue, dense irregular connective tissue, elastic connective tissue, hyaline cartilage, fibrocartilage, elastic cartilage, bone tissue, blood tissue, and lymph.

MEMBRANES

◉ OBJECTIVES

- Define a membrane.
- Describe the classification of membranes.

Membranes are flat sheets of pliable tissue that cover or line a part of the body. The combination of an epithelial layer and an underlying connective tissue layer constitutes an **epithelial membrane**. The principal epithelial membranes of the body are mucous membranes, serous membranes, and the cutaneous membrane, or skin. Another kind of membrane, a **synovial membrane**, lines joints and contains connective tissue but no epithelium.

Epithelial Membranes

Mucous Membranes

A **mucous membrane** or **mucosa** lines a body cavity that opens directly to the exterior. Mucous membranes line the entire digestive, respiratory, and reproductive tracts, and much of the urinary tract. They consist of a lining layer of epithelium and an underlying layer of connective tissue (Figure 3.7a on page 84).

The epithelial layer of a mucous membrane is an important feature of the body's defense mechanisms because it is a barrier that microbes and other pathogens have difficulty penetrating. Usually, tight junctions connect the cells, so materials cannot leak in between them. Goblet cells and other cells of the epithelial layer of a mucous membrane secrete mucus, and this slippery fluid prevents the cavities from drying out. It also traps particles in the respiratory passageways and lubricates food as it moves through the gastrointestinal tract. In addition, the epithelial layer secretes some of the enzymes needed for digestion and is the site of food and fluid absorption in the gastrointestinal tract. The epithelia of mucous membranes vary greatly in different parts of the body. For example, the mucous membrane of the small intestine is nonciliated simple columnar epithelium (see Table 3.1C), and the large airways to the lungs consist of pseudostratified ciliated columnar epithelium (see Table 3.1E).

The connective tissue layer of a mucous membrane is areolar connective tissue and is called the **lamina propria** (LAM-i-na PRŌ-prē-a; *propria*=one's own), so named because it belongs to the mucous membrane. The lamina propria supports the epithelium, binds it to the underlying structures, allows some flexibility of the membrane, and affords some protection for underlying structures. It also holds blood vessels in place and is the vascular source for the overlying epithelium. Oxygen and nutrients diffuse from the lamina propria to the epithelium covering it; carbon dioxide and wastes diffuse in the opposite direction.

Figure 3.7 Membranes.

A membrane is a flat sheet of pliable tissue that covers or lines a part of the body.

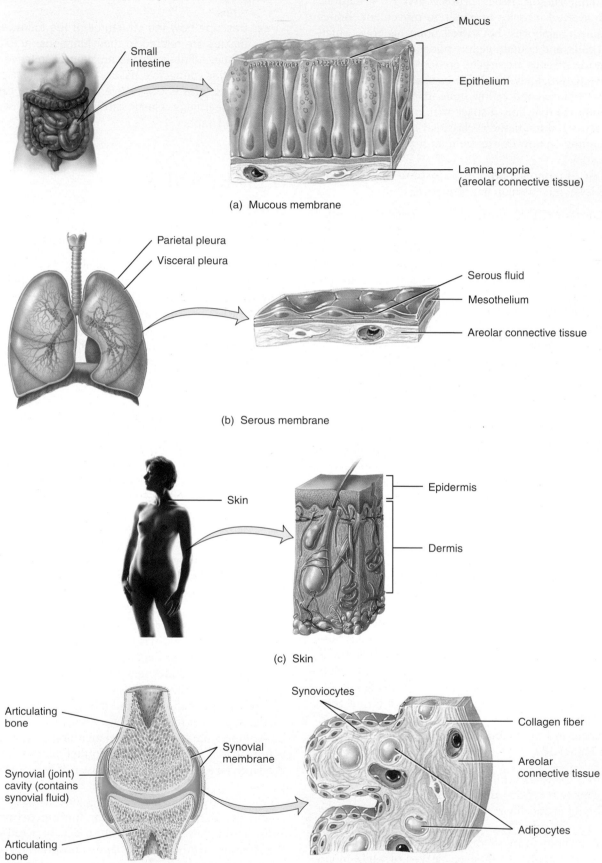

Small intestine

Mucus

Epithelium

Lamina propria (areolar connective tissue)

(a) Mucous membrane

Parietal pleura

Visceral pleura

Serous fluid

Mesothelium

Areolar connective tissue

(b) Serous membrane

Skin

Epidermis

Dermis

(c) Skin

Articulating bone

Synoviocytes

Collagen fiber

Synovial membrane

Synovial (joint) cavity (contains synovial fluid)

Areolar connective tissue

Articulating bone

Adipocytes

(d) Synovial membrane

? What is an epithelial membrane?

Serous Membranes

A **serous membrane** (*serous*=watery) or **serosa** lines a body cavity that does not open directly to the exterior (thoracic or abdominal cavities), and it covers the organs that are within the cavity. Serous membranes consist of areolar connective tissue covered by mesothelium (simple squamous epithelium) (Figure 3.7b). Serous membranes have two layers: The layer attached to and lining the cavity wall is called the **parietal layer** (pa-RĪ-e-tal; *pariet-*=wall); the layer that covers and adheres to the organs within the cavity is the **visceral layer** (*viscer-*=body organ) (see Figure 1.7a on page 16). The mesothelium of a serous membrane secretes **serous fluid**, a watery lubricant that allows organs to glide easily over one another or to slide against the walls of cavities.

The serous membrane lining the thoracic cavity and covering the lungs is the **pleura**. The serous membrane lining the heart cavity and covering the heart is the **pericardium**. The serous membrane lining the abdominal cavity and covering the abdominal organs is the **peritoneum**.

Cutaneous Membrane

The **cutaneous membrane** or **skin** covers the entire surface of the body and consists of a superficial portion called the *epidermis* and a deeper portion called the *dermis* (Figure 3.7c). The epidermis consists of keratinized stratified squamous epithelium, which protects underlying tissues. The dermis consists of dense irregular connective tissue and areolar connective tissue. Details of the cutaneous membrane are presented in Chapter 5.

Synovial Membranes

Synovial membranes (sin-Ō-vē-al; *syn-*=together, referring here to a place where bones come together; *ova*=egg, because of their resemblance to the slimy white of an uncooked egg) line the cavities of freely movable joints. Like serous membranes, synovial membranes line structures that do not open to the exterior. Un-like mucous, serous, and cutaneous membranes, they lack an epithelium and are therefore not epithelial membranes. Synovial membranes are composed of a discontinuous layer of cells called **synoviocytes** (si-NŌ-vē-ō-sīts), which are closer to the synovial cavity (space between the bones) and a layer of connective tissue (areolar and adipose) deep to the synoviocytes (Figure 3.7d). Synoviocytes secrete some of the components of synovial fluid. **Synovial fluid** lubricates and nourishes the cartilage covering the bones at movable joints and contains macrophages that remove microbes and debris from the joint cavity.

◉ CHECKPOINT

14. Define the following kinds of membranes: mucous, serous, cutaneous, and synovial.
15. Where is each type of membrane located in the body? What are their functions?

MUSCULAR TISSUE

◉ OBJECTIVES

- Describe the general features of muscular tissue.
- Contrast the structure, location, and mode of control of skeletal, cardiac, and smooth muscle tissue.

Muscular tissue consists of elongated cells called *muscle fibers*, or *myocytes*, that can use ATP to generate force. As a result, muscular tissue produces body movements, maintains posture, and generates heat. It also provides protection. Based on its location and certain structural and functional features, muscular tissue is classified into three types: skeletal, cardiac, and smooth (Table 3.5).

Skeletal muscle tissue is named for its location—it is usually attached to the bones of the skeleton (Table 3.5A). Another distinctive feature is its *striations*, alternating light and dark bands within the fibers that are visible under a light microscope. Skeletal muscle is considered *voluntary* because it can be made to con-

---

TABLE 3.5

Muscular Tissues

A. Skeletal muscle tissue

Description: Long, cylindrical, striated fibers with many peripherally located nuclei; voluntary control.

Location: Usually attached to bones by tendons.

Function: Motion, posture, heat production, and protection.

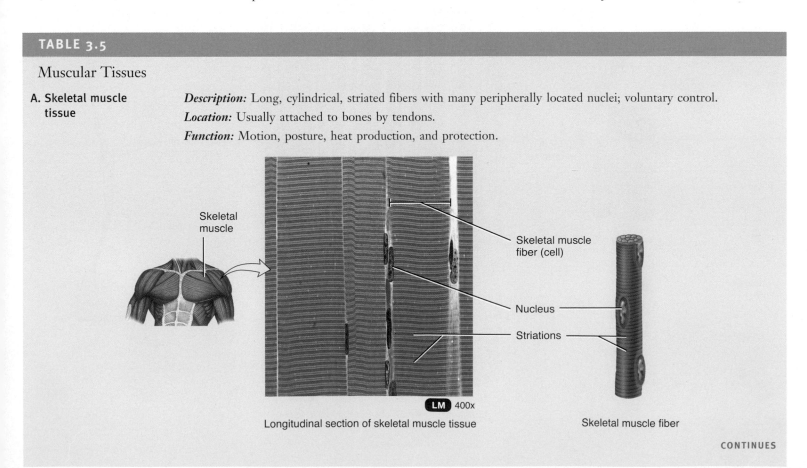

Skeletal muscle

Skeletal muscle fiber (cell)

Nucleus

Striations

LM 400x

Longitudinal section of skeletal muscle tissue

Skeletal muscle fiber

CONTINUES

TABLE 3.5 CONTINUED

Muscular Tissues

B. Cardiac muscle tissue

Description: Branched striated fibers with one or two centrally located nuclei; contains intercalated discs; involuntary control.

Location: Heart wall.

Function: Pumps blood to all parts of the body.

Heart

LM 500x

Longitudinal section of cardiac muscle tissue

Cardiac muscle fibers

C. Smooth muscle tissue

Description: Spindle-shaped (thickest in middle and tapering at both ends), nonstriated fibers with one centrally located nucleus; involuntary control.

Location: Iris of the eyes, walls of hollow internal structures such as blood vessels, airways to the lungs, stomach, intestines, gallbladder, urinary bladder, and uterus.

Function: Motion (constriction of blood vessels and airways, propulsion of foods through gastrointestinal tract, contraction of urinary bladder and gallbladder).

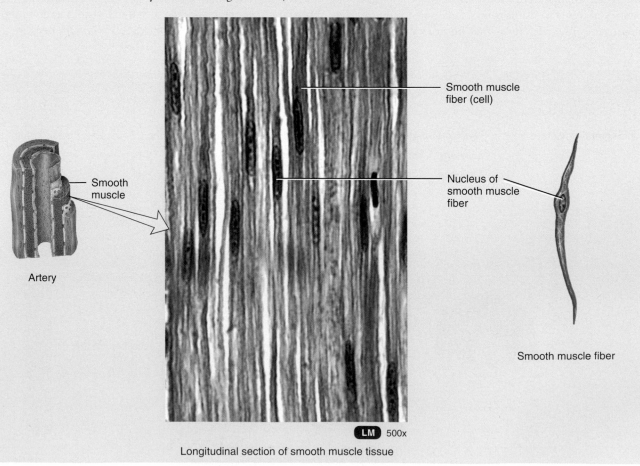

Artery

LM 500x

Longitudinal section of smooth muscle tissue

Smooth muscle fiber

tract or relax by conscious control. Skeletal muscle fibers can vary greatly in length, ranging from a few centimeters in short muscles to up to 30–40 cm (about 12–16 in.) in your longest muscles. Roughly cylindrical in shape, a muscle fiber is a multinucleated cell with the nuclei located at the cell's periphery.

Cardiac muscle tissue forms most of the wall of the heart (Table 3.5B). Like skeletal muscle, it is striated. However, unlike skeletal muscle tissue, it is *involuntary;* its contraction is not consciously controlled. Cardiac muscle fibers are branched and usually have only one centrally located nucleus; an occasional cell has two nuclei. They attach end to end by transverse thickenings of the plasma membrane called **intercalated discs** (*intercalat-*=to insert between), which contain both desmosomes and gap junctions. Intercalated discs are unique to cardiac muscle tissue. The desmosomes strengthen the tissue and hold the fibers together during their vigorous contractions. The gap junctions provide a route for quick conduction of muscle action potentials throughout the heart.

Smooth muscle tissue is located in the walls of hollow internal structures such as blood vessels, airways to the lungs, the stomach, intestines, gallbladder, and urinary bladder (Table 3.5C). Its contraction helps constrict or narrow the lumen of blood vessels, physically break down and move food along the gastrointestinal tract, move fluids through the body, and eliminate wastes. Smooth muscle fibers are usually *involuntary*, and they are nonstriated (lack striations), hence the term *smooth*. A smooth muscle fiber is a small spindle-shaped cell that is thickest in the middle and tapers at each end. It contains a single, centrally located nucleus. Gap junctions connect many individual fibers in some smooth muscle tissues, for example, in the wall of the intestines. Such muscle tissues can produce powerful contractions as many muscle fibers contract in unison. In other locations, such as the

iris of the eye, smooth muscle fibers contract individually, like skeletal muscle fibers, because gap junctions are absent.

Chapter 10 provides a more detailed discussion of muscular tissue.

◉ **CHECKPOINT**

16. Which types of muscular tissue are striated? Which are smooth?
17. Which types of muscular tissue have gap junctions?

NERVOUS TISSUE

◉ **OBJECTIVE**

• Describe the structural features and functions of nervous tissue.

Despite the tremendous complexity of the nervous system, the nervous tissue that makes up this system is a highly specialized cellular tissue consisting of two principal types of cells—neurons and neuroglia, or glial cells. Little extracellular material is present. **Neurons** (*neuro*=nerve) comprise a broad category of unusually shaped cell types with one common structural feature: Long cytoplasmic processes (extensions) known as dendrites and axons project from the main **cell body** where the nucleus and other typical organelles are located (Table 3.6). These processes can vary in number (from two to thousands) and length (from fractions of a millimeter to a meter). Neurons form complex networks within the central nervous system (the brain and spinal cord), somewhat comparable to the circuit boards of sophisticated computers. About 98% of nervous tissue is located in the central nervous system. Outside the central nervous system, long neuronal processes come together to form peripheral nerves that

TABLE 3.6

Nervous Tissue

Description: Consists of neurons (nerve cells) and neuroglia. Neurons consist of a cell body and processes extending from the cell body (one to multiple dendrites and a single axon). Neuroglia do not generate or conduct nerve impulses but have other important supporting functions.

Location: Nervous system.

Function: Exhibits sensitivity to various types of stimuli, converts stimuli into nerve impulses (action potentials), and conducts nerve impulses to other neurons, muscle fibers, or glands.

Spinal cord

Dendrite
Nucleus of neuroglial cell
Nucleus in cell body
Axon

LM 400x

Neuron of spinal cord

travel between the central nervous system and the various tissues of the body. Functionally, neurons are highly excitable cells with the ability to initiate and propagate **action potentials (nerve impulses)** when excited by physical or chemical agents in their environment. **Dendrites**, which are tapering, highly branched, and usually short cell processes, receive information from other sources, most commonly from other neurons, and electrically transmit it toward the cell body. The **axon** (*axo-*=axis), a single, thin, cylindrical process that can be very long, carries action potentials away from the cell body to other cells. When an action potential reaches the end of the axon, the neuron releases a **neurotransmitter**, a locally acting chemical messenger that influences another cell (or cells), which can be another neuron, a muscle cell, or a gland cell. Because of this unique combination of structure and function, nervous tissue plays important roles in communicating information from one part of the body to another and in regulating muscular and glandular tissues.

Neuroglia (noo-RŌG-lē-a; *-glia*=glue) do not generate or conduct nerve impulses. Instead, they support neurons both physically and metabolically. Without neuroglia, the neurons could not survive and function. Neuroglia, which outnumber the neurons approximately five to one, come in a variety of shapes, each specialized to accomplish specific functions. The detailed structure and function of neurons and neuroglia are considered in Chapter 17.

◉ **CHECKPOINT**

18. What are the functions of the dendrites, cell body, and axon of a neuron?

AGING AND TISSUES

◉ **OBJECTIVE**

• Describe the effects of aging on tissues.

Generally, tissues heal faster and leave less obvious scars in the young than in the aged. In fact, surgery performed on fetuses leaves no scars at all. The younger body is generally in a better nutritional state, its tissues have a better blood supply, and its cells have a higher metabolic rate. Thus, its cells can synthesize needed materials and divide more quickly. The extracellular components of tissues also change with age. Glucose, the most abundant sugar in the body, plays a role in the aging process. As the body ages, glucose is haphazardly added to proteins inside and outside cells, forming irreversible cross-links between adjacent protein molecules. With advancing age, more cross-links form, which contributes to the stiffening and loss of elasticity that occur in aging tissues. Collagen fibers, responsible for the strength of tendons, increase in number and change in quality with aging. Changes in the collagen of arterial walls affect the flexibility of arteries as much as the fatty deposits associated with atherosclerosis (see page 473). Elastin, another extracellular component, is responsible for the elasticity of blood vessels and skin. It thickens, fragments, and acquires a greater affinity for calcium with age—changes that may also be associated with the development of atherosclerosis.

◉ **CHECKPOINT**

19. What common changes occur in epithelial and connective tissues with aging?

KEY MEDICAL TERMS ASSOCIATED WITH TISSUES

Atrophy (AT-rō-fē; *a-*=without; *-trophy*=nourishment) A decrease in the size of cells, with a subsequent decrease in the size of the affected tissue or organ.

Hypertrophy (hī-PER-trō-fē; *hyper-*=above or excessive) Increase in the size of a tissue because its cells enlarge without undergoing cell division.

Tissue rejection An immune response of the body directed at foreign proteins in a transplanted tissue or organ. Immunosuppressive drugs, such as cyclosporine, have largely overcome tissue rejection in heart-, kidney-, and liver-transplant patients.

Tissue transplantation The replacement of a diseased or injured tissue or organ. The most successful transplants involve use of a person's own tissues or those from an identical twin.

Xenotransplantation (zen'-ō-trans-plan-TĀ-shun; *xeno-*=strange, foreign) The replacement of a diseased or injured tissue or organ with cells or tissues from an animal. Porcine (from pigs) and bovine (from cows) heart valves are used for some heart-valve replacement surgeries.

STUDY OUTLINE

Types of Tissues and Their Origins (p. 59)

1. A tissue is a group of similar cells, usually with a similar embryological origin, that is specialized for a particular function.
2. The various tissues of the body are classified into four basic types: epithelial, connective, muscular, and nervous tissue.

Cell Junctions (p. 60)

1. Cell junctions are points of contact between adjacent plasma membranes.
2. Tight junctions form fluid-tight seals between cells; adherens junctions, desmosomes, and hemidesmosomes anchor cells to one another or to the basement membrane; and gap junctions permit electrical and chemical signals to pass between cells.

Comparison of Epithelial and Connective Tissue (p. 61)

1. Epithelial tissues have many cells tightly packed together and are avascular.
2. Connective tissues have relatively few cells with lots of extracellular material.

Epithelial Tissue (p. 61)

1. An epithelium consists mostly of cells with little extracellular material between adjacent plasma membranes. The apical, lateral, and basal surfaces of epithelial cells are modified in various ways to carry out specific functions. Epithelium is arranged in sheets and attached to a basement membrane. Although it is avascular, it has a nerve supply. Epithelia are derived from all three primary germ layers and have a high capacity for renewal.

2. Epithelial layers can be simple (one layer) or stratified (several layers). The cell shapes may be squamous (flat), cuboidal (cubelike), columnar (rectangular), or transitional (variable).

3. The subtypes of epithelia include covering and lining epithelia and glandular epithelia.

4. Simple squamous epithelium consists of a single layer of flat cells (Table 3.1A). It is found in parts of the body where filtration or diffusion are priority processes. One type, called endothelium, lines the heart and blood vessels. Another type, called mesothelium, forms the serous membranes that line the thoracic and abdominopelvic cavities and cover the organs within them.

5. Simple cuboidal epithelium consists of a single layer of cube-shaped cells that function in secretion and absorption (Table 3.1B). It is found covering the ovaries, in the kidneys and eyes, and lining some glandular ducts.

6. Nonciliated simple columnar epithelium, a single layer of nonciliated rectangular cells (Table 3.1C), lines most of the gastrointestinal tract. Specialized cells containing microvilli perform absorption. Goblet cells secrete mucus.

7. Ciliated simple columnar epithelium consists of a single layer of ciliated rectangular cells (Table 3.1D). It is found in a few portions of the upper respiratory tract, where it moves foreign particles trapped in mucus out of the respiratory tract.

8. Pseudostratified columnar epithelium has only one layer but gives the appearance of many (Table 3.1E). A ciliated variety contains goblet cells and lines most of the upper respiratory tract; a nonciliated variety has no goblet cells and lines ducts of many glands, the epididymis, and part of the male urethra.

9. Stratified squamous epithelium consists of several layers of cells. Cells of the apical layer and several layers deep to it are flat (Table 3.1F). A nonkeratinized variety lines the mouth. A keratinized variety forms the epidermis, the most superficial layer of the skin.

10. Stratified cuboidal epithelium consists of several layers of cells. Cells at the apical layer are cube-shaped (Table 3.1G). It is found in adult sweat glands and a portion of the male urethra.

11. Stratified columnar epithelium consists of several layers of cells. Cells of the apical layer have a columnar shape (Table 3.1H). It is found in a portion of the male urethra and large excretory ducts of some glands.

12. Transitional epithelium consists of several layers of cells whose appearance varies with the degree of stretching (Table 3.1I). It lines the urinary bladder.

13. A gland is a single cell or a group of epithelial cells adapted for secretion.

14. Endocrine glands secrete hormones into interstitial fluid and then into the blood (Table 3.2A).

15. Exocrine glands (mucous, sweat, oil, and digestive glands) secrete into ducts or directly onto a free surface (Table 3.2B).

16. The structural classification of exocrine glands includes unicellular and multicellular glands.

17. The functional classification of exocrine glands includes holocrine, apocrine, and merocrine glands.

Connective Tissue (p. 72)

1. Connective tissue is one of the most abundant body tissues.

2. Connective tissue consists of relatively few cells and an abundant extracellular matrix of ground substance and fibers. It does not usually occur on free surfaces, has a nerve supply (except for cartilage), and is highly vascular (except for cartilage, tendons, and ligaments).

3. Cells in connective tissue are derived from mesenchymal cells.

4. Cell types include fibroblasts (secrete extracellular matrix), macrophages (perform phagocytosis), plasma cells (secrete antibodies), mast cells (produce histamine), adipocytes (store fat), and white blood cells (migrate from blood in response to infections).

5. The ground substance and fibers make up the extracellular matrix.

6. The ground substance supports and binds cells together, provides a medium for the exchange of materials, stores water, and is active in influencing cell functions.

7. Substances found in the ground substance include water and polysaccharides such as hyaluronic acid, chondroitin sulfate, dermatan sulfate, and keratan sulfate (glycosaminoglycans). Also present are proteoglycans and adhesion proteins.

8. The fibers in the extracellular matrix provide strength and support and are of three types: (a) Collagen fibers (composed of collagen) are found in large amounts in bone, tendons, and ligaments. (b) Elastic fibers (composed of elastin, fibrillin, and other glycoproteins) are found in skin, blood vessel walls, and lungs. (c) Reticular fibers (composed of collagen and glycoprotein) are found around fat cells, nerve fibers, and skeletal and smooth muscle cells.

9. The two major subclasses of connective tissue are embryonic connective tissue (found in the embryo and fetus) and mature connective tissue (present in the newborn).

10. The embryonic connective tissues are mesenchyme, which forms almost all other connective tissues (Table 3.3A), and mucous connective tissue, found in the umbilical cord of the fetus, where it gives support (Table 3.3B).

11. Mature connective tissue differentiates from mesenchyme, and is subdivided into several types: loose or dense connective tissue, cartilage, bone tissue, and liquid connective tissue.

12. Loose connective tissue includes areolar connective tissue, adipose tissue, and reticular connective tissue.

13. Areolar connective tissue consists of the three types of fibers (collagen, elastic, and reticular), several types of cells, and a semifluid ground substance (Table 3.4A). It is found in the subcutaneous layer, in mucous membranes, and around blood vessels, nerves, and body organs.

14. Adipose tissue consists of adipocytes, which store triglycerides (Table 3.4B). It is found in the subcutaneous layer, around organs, and in yellow bone marrow. Brown adipose tissue (BAT) generates heat.

15. Reticular connective tissue consists of reticular fibers and reticular cells and is found in the liver, spleen, and lymph nodes (Table 3.4C).

16. Dense connective tissue includes dense regular connective tissue, dense irregular connective tissue, and elastic connective tissue.

17. Dense regular connective tissue consists of parallel bundles of collagen fibers and fibroblasts (Table 3.4D). It forms tendons, most ligaments, and aponeuroses.

18. Dense irregular connective tissue usually consists of randomly arranged collagen fibers and a few fibroblasts (Table 3.4E). It is found in fasciae, the dermis of skin, and membrane capsules around organs.

19. Elastic connective tissue consists of branching elastic fibers and fibroblasts (Table 3.4F). It is found in the walls of large arteries, lungs, trachea, and bronchial tubes.

20. Cartilage contains chondrocytes and has a rubbery extracellular matrix (chondroitin sulfate) containing collagen and elastic fibers.

21. Hyaline cartilage, which consists of a gel-like ground substance and appears bluish white in the body, is found in the embryonic skeleton, at the ends of bones, in the nose, and in respiratory structures (Table 3.4G). It is flexible, allows movement, and provides support, and is usually surrounded by a perichondrium.

22. Fibrocartilage is found in the pubic symphysis, intervertebral discs, and menisci (cartilage pads) of the knee joint (Table 3.4H). It contains chondrocytes scattered among clearly visible bundles of collagen fibers.

23. Elastic cartilage maintains the shape of organs such as the epiglottis of the larynx, auditory (eustachian) tubes, and external ear (Table 3.4I). It contains chondrocytes located within a threadlike network of elastic fibers, and has a perichondrium.

24. Bone or osseous tissue consists of an extracellular matrix of mineral salts and collagen fibers that contribute to the hardness of bone, and osteocytes that are located in lacunae (Table 3.4J). It supports and protects the body, provides a surface area for muscle attachment, helps the body move, stores minerals, and houses blood-forming tissue.

25. Blood tissue is liquid connective tissue that consists of blood plasma and formed elements—red blood cells, white blood cells, and platelets (Table 3.4K). Its cells transport oxygen and carbon

dioxide, carry on phagocytosis, participate in allergic reactions, provide immunity, and bring about blood clotting.

26. Lymph, the extracellular fluid that flows in lymphatic vessels, is also a liquid connective tissue. It is a clear fluid similar to blood plasma but with less protein.

Membranes (p. 83)

1. An epithelial membrane consists of an epithelial layer overlying a connective tissue layer. Examples are mucous, serous, and cutaneous membranes.
2. Mucous membranes line cavities that open to the exterior, such as the gastrointestinal tract.
3. Serous membranes line closed cavities (pleura, pericardium, peritoneum) and cover the organs in the cavities. These membranes consist of parietal and visceral layers.
4. Synovial membranes line joint cavities, bursae, and tendon sheaths and consist of areolar connective tissue instead of epithelium.

Muscular Tissue (p. 85)

1. Muscular tissue consists of cells called muscle fibers or myocytes that are specialized for contraction. It provides motion, maintenance of posture, heat production, and protection.
2. Skeletal muscle tissue is attached to bones and is striated and voluntary (Table 3.5A).

3. The action of cardiac muscle tissue, which forms most of the heart wall and is striated, is involuntary (Table 3.5B).
4. Smooth muscle tissue is found in the walls of hollow internal structures (blood vessels and viscera) and is nonstriated and involuntary (Table 3.5C).

Nervous Tissue (p. 87)

1. The nervous system is composed of neurons (nerve cells) and neuroglia (protective and supporting cells) (Table 3.6).
2. Neurons respond to stimuli by converting the stimuli into electrical signals called action potentials (nerve impulses), and conducting nerve impulses to other cells.
3. Most neurons consist of a cell body and two types of processes, dendrites and axons.

Aging and Tissues (p. 88)

1. Tissues heal faster and leave less obvious scars in the young than in the aged; surgery performed on fetuses leaves no scars.
2. The extracellular components of tissues, such as collagen and elastic fibers, also change with age.

SELF-QUIZ QUESTIONS

Choose the one best answer to the following questions.

1. A group of similar cells that has the same embryological origin and operates together to perform a specialized activity is called a(n):
   a. organ **b.** tissue **c.** system **d.** organelle **e.** organism

2. Which of the following is an avascular tissue?
   a. bone
   b. fibrocartilage
   c. stratified squamous epithelium
   d. Both **b** and **c** are correct.
   e. All three types of tissue are avascular.

3. The two layers of a basement membrane are the:
   a. apical and basal layers.
   b. parietal and visceral layers.
   c. lamina propria and epithelial layers.
   d. avascular and vascular layers.
   e. basal lamina and reticular lamina.

4. Mesenchyme is:
   a. part of the extracellular matrix of connective tissues.
   b. the embryonic connective tissue from which almost all other connective tissues arise.
   c. a mucus-producing cell.
   d. a type of fiber found in connective tissues.
   e. the covering of bone and cartilage.

5. The connective tissue best designed to resist pulling forces in various directions is:
   a. fibrocartilage.
   b. bone.
   c. dense regular connective tissue.
   d. dense irregular connective tissue.
   e. areolar connective tissue.

6. Supporting cells in nervous tissue are called:
   a. neuroglia **b.** neurons **c.** cadherins
   d. mesenchyme **e.** synoviocytes

7. Which statement best describes connective tissue?
   a. usually contains a large amount of extracellular matrix
   b. always arranged in a single layer of cells

   c. primarily concerned with secretion
   d. cells always very closely packed together
   e. is avascular

8. Which type of muscular tissue contains an alternating pattern of myofilaments and is under conscious control?
   a. skeletal muscle tissue **b.** cardiac muscle tissue
   c. smooth muscle tissue **d.** *Two* of the above.
   e. *All* of the above.

Complete the following.

9. The muscular tissue that forms most of the wall of the heart is _____ muscle tissue.

10. A gland that secretes its product into a duct is referred to as a(n) _____ gland.

11. Membrane proteins called connexins form tunnels called connexons in the type of cell junction known as a(n) _____.

12. The three types of fibers that strengthen and support the extracellular matrix of connective tissues are _____ fibers, _____ fibers, and _____ fibers.

13. Hollow organs belonging to systems that do not open to the outside of the body are lined with a _____ membrane.

14. The connective tissue layer of a mucous membrane is called the _____.

Are the following statements true or false? If the statement is False, *find and fix the mistake to make the statement* True.

15. Blood and lymph are types of connective tissues.

16. Tight junctions are common between epithelial cells lining the stomach or urinary bladder where they prevent fluid from passing between the lining cells.

17. A biopsy is the removal of a sample of dead cells for microscopic examination.

18. Desmosomes connect lateral surfaces of adjacent cells to one another, and hemidesmosomes connect the basal surface of a cell to a basement membrane.

Matching

19. Match the following:

___ **(a)** a mature bone cell; resides in a space called a lacuna
___ **(b)** large phagocytic cell derived from a monocyte; engulfs bacteria and cleans up debris; wandering or fixed
___ **(c)** derived from a fibroblast; specialized for triglyceride storage
___ **(d)** a mucus-producing cell
___ **(e)** large, branched cell; secretes the extracellular matrix of connective tissues
___ **(f)** located along the walls of blood vessels; secrete histamine,which dilates blood vessels during inflammation
___ **(g)** cartilage cell; resides in a space called a lacuna
___ **(h)** derived from a B lymphocyte; produces antibodies as part of an immune response

(1) fibroblast
(2) mast cell
(3) plasma cell
(4) osteocyte
(5) adipocyte
(6) chondrocyte
(7) macrophage
(8) goblet cell

20. Match the following:

___ **(a)** lines inner surface of the stomach and intestine
___ **(b)** avascular connective tissue that occurs as three main types
___ **(c)** lines urinary bladder, ureters, and urethra
___ **(d)** forms fasciae and dermis of skin
___ **(e)** has cells that are specialized for triglyceride storage
___ **(f)** all cells attached to the basement membrane, not all of which reach the surface of tissue
___ **(g)** subcutaneous tissue containing several kinds of cells and all three fiber types
___ **(h)** lines the mouth; present on the outer surface of the skin
___ **(i)** lines air sacs of the lungs; suited for diffusion of gases
___ **(j)** strong tissue found in the lungs; can recoil back to its original shape after being stretched

(1) dense irregular connective tissue
(2) adipose tissue
(3) simple columnar epithelium
(4) stratified squamous epithelium
(5) cartilage
(6) areolar connective tissue
(7) transitional epithelium
(8) pseudostratified columnar epithelium
(9) elastic connective tissue
(10) simple squamous epithelium

CRITICAL THINKING QUESTIONS

1. Mike has had a series of respiratory tract infections this winter. His doctor has just prescribed a mucus-thinning drug. Using your knowledge of the structure of the mucous membrane lining the respiratory tract, how do you think this type of drug will help Mike get better?

2. As part of a routine pelvic exam for all women, a biopsy of cervical and vaginal tissue is taken (*Pap smear*). Describe the kinds of cells that you would see in the normal smear. How would the cells appear in *cervical dysplasia*?

3. Janelle has been an anorexic for several years. As a result of her chronically low daily caloric intake, her adipocytes are storing little or no triglycerides. What structural problems might Janelle suffer as a result?

4. In order to show-off to his fifth-grade classmates, Jonathan stuck a common pin and sewing needle into his fingertip. His classmates were amazed that there was no blood. What type of tissue did Jonathan pierce, and why was there no visible bleeding?

5. Sometime during your daily activities you come in contact with a colony of bacteria that gets onto your skin. You have no cuts in the area, and your skin is in good condition. However, these microbes manage to penetrate your tissues and get into your blood. What specific structural barriers would the bacteria have to overcome to accomplish this?

? ANSWERS TO FIGURE QUESTIONS

3.1 Gap junctions allow cellular communication via passage of electrical and chemical signals between adjacent cells.

3.2 The basement membrane provides a physical support for the epithelium.

3.3 Because they are so thin, squamous cells allow substances to move most rapidly from one cell to another.

3.4 Simple multicellular exocrine glands have a nonbranched duct; compound multicellular exocrine glands have a branched duct.

3.5 Sebaceous (oil) glands are holocrine glands, and salivary glands are merocrine glands.

3.6 Fibroblasts secrete the fibers and ground substance of the extracellular matrix.

3.7 An epithelial membrane is a membrane that consists of an epithelial layer and an underlying layer of connective tissue.

4 DEVELOPMENT

INTRODUCTION   Before we examine the first body system in Chapter 5 (The Integumentary System), we will take a look at how body systems develop. Knowledge of their origins will enhance your understanding of the different systems of the human body. Later in the book, you will learn more about development in the context of the various body systems.

As you learned in Chapter 2, **sexual reproduction** is the process by which organisms produce offspring by making sex cells called **gametes** (GAM-ēts=spouses). Male gametes are called **sperm (spermatozoa)** and female gametes are called **secondary oocytes.** The organs that produce gametes are called *gonads*, the testes in the male and the ovaries in the female. The details of sperm formation in the testes and secondary oocyte formation in the ovaries are discussed in Chapter 27. Once sperm have developed and are deposited in the female reproductive tract and a secondary oocyte has been released from the ovary, fertilization can occur.

Pregnancy is a sequence of events that begins with fertilization, proceeds to implantation, embryonic development, and fetal development, and normally ends with birth about 38 weeks later, or 40 weeks after the last menstrual period.

Developmental biology is the study of the sequence of events from the fertilization of a secondary oocyte by a sperm cell to the formation of an adult organism. From fertilization through the eighth week of development, a stage called the **embryonic period**, the developing human is called an **embryo** (*em-*=into; *-bryo*=grow). **Embryology** (embrē-OL-ō-jē) is the study of development from the fertilized egg through the eighth week. The **fetal period** begins at week nine and continues until birth. During this time, the developing human is called a **fetus** (FĒ-tus=offspring).

Prenatal development (prē-NĀ-tal; *pre-*=before; *natal*=birth) is the time from fertilization to birth and includes both the embryonic and fetal periods. Prenatal development is divided into periods of three calendar months each, called **trimesters**.

1. The **first trimester** is the most critical stage of development during which of all the major organ-systems begin to form. Because of this, it is also the period when the developing organism is most vulnerable to the effects of drugs, radiation, and microbes.

2. The **second trimester** is characterized by the nearly complete development of organ systems. By the end of this stage, the fetus assumes distinctively human features.

3. The **third trimester** represents a period of rapid fetal growth. During the early stages of this period, most of the organ systems become fully functional.

Among the female reproductive organs involved in the major events of the embryonic and fetal periods are the *ovaries*, paired organs in the superior portion of the pelvic cavity on either side of the uterus (Figure 4.1). During the reproductive years of a female, they produce secondary oocytes and discharge them into the peritoneal cavity each month, a process called *ovulation*. The *uterine (fallopian) tubes*, which extend laterally from the uterus, provide an exit route from the peritoneal cavity for the secondary oocyte. The uterine tube is also where the secondary oocyte can come in contact with sperm. Additionally, the uterine tubes also transport a fertilized (or unfertilized) oocyte to the uterus. The *uterus*, an inverted pear-shaped organ in the pelvic and lower abdominal cavity, consists of a superior portion (fundus), middle portion (body), and inferior portion (cervix). The interior of the body of the uterus is called the *uterine cavity*, and the interior of the cervix is known as the *cervical canal*. After puberty, in preparation of implantation of the fertilized ovum, the uterus forms a vascular, glandular lining called the *endometrium* during each menstrual cycle. Part of the endometrium is shed during menstruation. Deep to the endometrium is a muscle layer called the *myometrium*. The largest mass of smooth muscle in the human body, the myometrium is responsible for the strong contractions that expel the endometrium during menstruation and help push the fetus out of the uterus during labor. So the uterus is the site of implantation of a fertilized ovum, development of the fetus during pregnancy, and labor. The cervix of the uterus connects to the *vagina*, a canal that opens to the exterior, which serves as the receptacle for the penis during sexual intercourse, the outlet for the menstrual flow, and the passageway for childbirth. •

Figure 4.1 Uterus and associated structures. The left side of the uterine tube, uterus, and vagina have been sectioned.

 The ovaries produce female gametes called secondary oocytes.

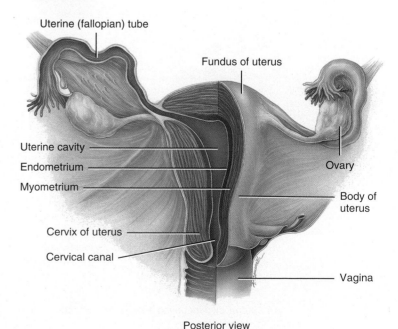

View

Uterine (fallopian) tube

Fundus of uterus

Uterine cavity

Endometrium

Myometrium

Ovary

Body of uterus

Cervix of uterus

Cervical canal

Vagina

Posterior view

? **What are the functions of the uterine tubes?**

EMBRYONIC PERIOD

⦿ **OBJECTIVE**

- Explain the major developmental events that occur during the embryonic period.

First Week of Development

The first week of development is characterized by several significant events including fertilization, cleavage of the zygote, blastocyst formation, and implantation.

Fertilization

During **fertilization** (fer-til-i-ZĀ-shun; *fertil-*=fruitful) the genetic material from a haploid sperm cell and a haploid secondary oocyte merges into a single diploid nucleus. Of the 200 million sperm introduced into the vagina, fewer than 2 million (1%) reach the cervix and only about 200 (0.0001%) reach the secondary oocyte. Fertilization normally occurs in the uterine (fallopian) tube within 12 to 24 hours after ovulation. Sperm can remain viable for about 48 hours after deposition in the vagina, although a secondary oocyte is viable for only about 24 hours after ovulation. Thus, pregnancy is *most likely* to occur if intercourse takes place during a 3-day window—from 2 days before ovulation to 1 day after ovulation.

Sperm swim from the vagina into the cervical canal propelled by the whiplike movements of their tails (flagella). The passage of sperm through the rest of the uterus and into the uterine tube results mainly from contractions of the walls of these organs. Sperm that reach the vicinity of the oocyte within minutes after ejaculation *are not capable* of fertilizing it until about seven hours later.

During this time in the female reproductive tract, mostly in the uterine tube, sperm undergo **capacitation** (ka-pas'-i-TĀ-shun; *capacit-*=capable of), a series of functional changes that cause the sperm's tail to beat even more vigorously and prepare its plasma membrane to fuse with the oocyte's plasma membrane. During capacitation, sperm are acted upon by secretions in the female reproductive tract that result in the removal of cholesterol, glycoproteins, and proteins from the plasma membrane around the head of the sperm cell.

For fertilization to occur, a sperm cell first must penetrate the **corona radiata** (kō-RŌ-na=crown; rā-dē-A-ta=to shine), the cells that surround the secondary oocyte, and the **zona pellucida** (ZŌ-na=zone; pe-LOO-si-da=allowing passage of light), the clear glycoprotein layer between the corona radiata and the oocyte's plasma membrane (Figure 4.2a). The *acrosome*, a helmet-like structure that covers the head of a sperm, contains several enzymes (see Figure 27.6 on page 852). Enzymes and strong tail movements by the sperm help it penetrate the cells of the corona radiata and come in contact with the zona pellucida. One of the glycoproteins in the zona pellucida acts as a sperm receptor, binding to specific membrane proteins on the acrosome to trigger the **acrosomal reaction**, the release of the contents of the acrosome. The acrosomal enzymes digest a path through the zona pellucida as the lashing sperm tail pushes the sperm cell onward. Although many sperm undergo acrosomal reactions, only the first sperm cell to penetrate the entire zona pellucida and reach the oocyte's plasma membrane undergoes **syngamy** (*syn-*=coming together; *-gamy*=marriage), fusion with the oocyte. This penetration of the zona pellucida results in a reaction that changes the properties of the zona pellucida and makes it impermeable to the other sperms, blocking **polyspermy**, fertilization by more than one sperm cell.

Figure 4.2 Selected structures and events in fertilization. (a) A sperm cell penetrating the corona radiata and zona pellucida around a secondary oocyte. (b) A sperm cell in contact with a secondary oocyte. (c) Male and female pronuclei.

🔑 ▭ During fertilization, genetic material from a sperm cell and a secondary oocyte merge to form a single diploid nucleus.

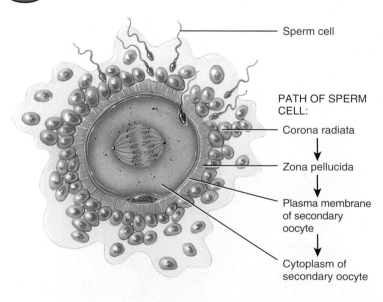

PATH OF SPERM CELL:

Corona radiata
↓
Zona pellucida
↓
Plasma membrane of secondary oocyte
↓
Cytoplasm of secondary oocyte

(a) Sperm cell penetrating a secondary oocyte

SEM 1100x

(b) Sperm cell in contact with a secondary oocyte

LM 250x

(c) Male and female pronuclei

❓ **What is capacitation?**

Once a sperm cell enters a secondary oocyte, the oocyte completes meiosis II (see Figure 2.20 on page 49). It divides into a larger ovum (mature egg) and a smaller second polar body, which contains DNA but no cytoplasm. The polar body eventually fragments and disintegrates (see Figure 27.15 on page 866). The nucleus in the head of the sperm develops into the **male pronucleus**, and the nucleus in the fertilized ovum develops into the **female pronucleus** (Figure 4.2c). After the male and female pronuclei form, they fuse, producing a single diploid nucleus that contains 23 chromosomes from each pronucleus. Fertilization has taken place. Thus, the fusion of the haploid (*n*) pronuclei restores the diploid number (*2n*) of 46 chromosomes. The fertilized ovum now is called a **zygote** (ZĪ-gōt; *zygon*=yolk). The process of fertilization takes approximately 24 hours.

Dizygotic (fraternal) twins are produced from the independent release of two secondary oocytes and the subsequent fertilization of each by different sperm. They are the same age and in the uterus at the same time, but genetically they are as dissimilar as any other siblings. Dizygotic twins may or may not be the same sex. Because **monozygotic (identical) twins** develop from a single fertilized ovum, they contain exactly the same genetic material and are always the same sex. Monozygotic twins arise from separation of the developing zygote into two embryos, which occurs within 8 days of fertilization 99% of the time. Separations that occur later than 8 days are likely to produce **conjoined twins**, a situation in which the twins are joined together and share some body structures.

Cleavage of the Zygote

After fertilization, the zygote undergoes mitotic cell divisions called **cleavage** (KLĒV-ij) that initially increase the number of cells without increasing the overall size of the cell mass (Figure 4.3). The first division of the zygote begins about 24 hours after fertilization and is completed about 6 hours later. Each succeeding division takes slightly less time. By the second day after fertilization, the second cleavage is completed and there are four cells (Figure 4.3b). By the end of the third day, there are 16 cells. The progressively smaller cells produced by cleavage are called **blastomeres** (BLAS-tō-mērz; *blasto-*=germ or sprout; *-meres*=parts). Successive cleavages eventually become more rapid and produce a solid sphere of cells called the **morula** (MOR-ū-la; *morula*=mulberry). The morula is still surrounded by the zona pellucida and is about the same size as the original zygote (Figure 4.3c).

Blastocyst Formation

By the end of the fourth day, the number of cells in the morula increases as it continues to move through the uterine tube toward the uterine cavity. When the morula enters the uterine cavity on day 4 or 5, a glycogen-rich secretion from the glands of the endometrium passes into the uterine cavity and enters the morula through the zona pellucida. This secretion, called **uterine milk**, along with nutrients stored in the cytoplasm of the blastomeres of the morula, provides nourishment for the developing morula. At the 32-cell stage, the fluid enters the morula, collects between the blastomeres, and reorganizes them around a large fluid-filled cavity called the **blastocyst cavity** (BLAS-tō-sist; *blasto-*=germ or sprout; *-cyst*=bag) or *blastocele* (Figure 4.3e). With the formation of this cavity, the developing mass is then called the **blastocyst**. Though it now has hundreds of cells, the blastocyst is still about the same size as the original zygote.

During the formation of the blastocyst two distinct cell populations arise: the embryoblast and trophoblast (Figure 4.3e). The **embryoblast**, or **inner cell mass**, is located internally and eventually develops into the embryo. The **trophoblast** (TRŌF-ō-blast; *tropho-*=develop or nourish) is the outer superficial layer of cells that forms the sphere-like wall of the blastocyst. It will ultimately develop into the outer chorionic sac that surrounds the fetus and the fetal portion of the placenta, the site of exchange of nutrients and wastes between the mother and fetus. On about the fifth day after fertilization, the blastocyst "hatches" from the zona pellucida by digesting a hole in it with an enzyme, and then squeezing through the hole. This shedding of the zona pellucida is necessary in order to permit the next step, implantation (attachment) into the vascular, glandular endometrial lining of the uterus.

Figure 4.3 Cleavage and the formation of the morula and blastocyst.

Cleavage refers to the early, rapid mitotic divisions of a zygote.

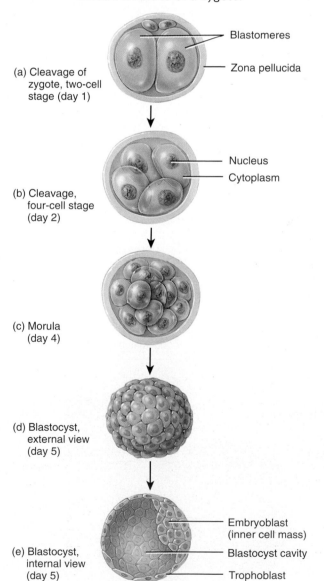

(a) Cleavage of zygote, two-cell stage (day 1) — Blastomeres, Zona pellucida

(b) Cleavage, four-cell stage (day 2) — Nucleus, Cytoplasm

(c) Morula (day 4)

(d) Blastocyst, external view (day 5)

(e) Blastocyst, internal view (day 5) — Embryoblast (inner cell mass), Blastocyst cavity, Trophoblast

? **What is the histological difference between a morula and a blastocyst?**

Implantation

The blastocyst remains free within the uterine cavity for about 2 days before it attaches to the uterine wall. During this period it receives nourishment from the secretions of the endometrium. About 6 days after fertilization, the blastocyst loosely attaches to the endometrium in a process called **implantation** (Figure 4.4). As the blastocyst implants, usually in either the posterior portion of the fundus or the body of the uterus, it orients with the embryoblast toward the endometrium (Figure 4.4b). About seven days after fertilization, the blastocyst attaches to the endometrium more firmly, endometrial glands in the vicinity enlarge, and the endometrium becomes more vascular by forming new blood vessels.

Following implantation, the endometrium is known as the **decidua** (dē-SID-ū-a=falling off). The decidua separates from the endometrium after the fetus is delivered much as it does in normal menstruation. Different regions of the decidua are named based on their positions relative to the site of the implanted blastocyst (Figure 4.4c). The **decidua basalis** is the portion of the endometrium between the embryo and the stratum basalis of the uterus; it provides large amounts of glycogen and lipids for the developing embryo and fetus and later becomes the maternal part of the placenta. The **decidua capsularis** is the portion of the endometrium located between the embryo and the uterine cavity. The **decidua parietalis** (par-rī-e-TAL-is) is the remaining modified endometrium that lines the noninvolved areas of the rest of the uterus. As the embryo and later the fetus enlarges, the decidua capsularis bulges into the uterine cavity and fuses with the decidua parietalis, thereby obliterating the uterine cavity. By about 27 weeks, the decidua capsularis degenerates and disappears.

The major events associated with the first week of development are summarized in Figure 4.5.

Figure 4.4 Relation of a blastocyst to the endometrium of the uterus at the time of implantation.

Implantation, the attachment of a blastocyst to the endometrium, occurs about 6 days after fertilization.

(a) External view of blastocyst, about 6 days after fertilization

(b) Frontal section through endometrium of uterus and blastocyst, about 6 days after fertilization

(c) Regions of the decidua

❓ How does the blastocyst merge with and burrow into the endometrium?

• CLINICAL **CONNECTION** | Stem Cell Research and Therapeutic Cloning

Stem cells are unspecialized cells that have the ability to divide for indefinite periods and give rise to specialized cells. A stem cell that has the potential to form an entire organism is known as a *totipotent stem cell* (tō -TIP-ō -tent; *totus-*=whole; *-potentia*=power). The cells of a zygote and an embryo through the morula stage of development are totipotent stem cells. Embryoblast cells of a blastocyst, by contrast, can give rise to many (but not all) different types of cells. They are called *pluripotent stem cells* (ploo-RIP-ō-tent; *plur-*=several). Later, pluripotent stem cells can undergo further specialization into cells that have a specific function. These cells are called *multipotent stem cells* (mul-TIP-ō-tent). Their primary role is to maintain or repair a tissue. Examples include keratinocytes that produce new skin cells, myeloid and lymphoid stem cells that develop into blood cells, and spermatogonia that give rise to sperm. Pluripotent stem cells currently used in research are derived from (1) the embryoblast of embryos in the blastocyst stage that were destined to be used for infertility treatments but were not needed and from (2) nonliving fetuses terminated during the first three months of pregnancy.

On October 13, 2001, researchers reported cloning of the first human embryo to grow cells to treat human diseases. **Therapeutic cloning** is envisioned as a procedure in which the genetic material of a patient with a particular disease is used to create pluripotent stem cells to treat the disease. Using the principles of therapeutic cloning, scientists hope to make an embryo clone of a patient, remove the pluripotent stem cells from the

Human embryonic stem cell

embryo, and then use them to grow tissues to treat particular diseases. For example, they might be used to generate cells and tissues for transplantation to treat conditions such as cancer, Parkinson and Alzheimer disease, spinal cord injury, diabetes, heart disease, stroke, burns, birth defects, osteoarthritis, and rheumatoid arthritis. Presumably, the tissues would not be rejected since they would contain the patient's own genetic material. Stem cells may also be used in the development of new drugs and may increase understanding of cancer and birth defects.

Scientists are also investigating the potential clinical applications of *adult stem cells*—stem cells that remain in the body throughout adulthood. Recent experiments suggest that the ovaries of adult mice contain stem cells that can develop into new ova (eggs). If these same types of stem cells are found in the ovaries of adult women, scientists could potentially harvest some of them from a woman about to undergo a sterilizing medical treatment (such as chemotherapy), store them, and then return the stem cells to her ovaries after the medical treatment is completed in order to restore fertility. Studies have also suggested that stem cells in human adult red bone marrow have the ability to differentiate into cells of the liver, kidney, heart, lung, skeletal muscle, skin, and organs of the gastrointestinal tract. In theory, adult stem cells from red bone marrow could be harvested from a patient and then used to repair other tissues and organs in that patients's body without having to use stem cells from embryos.

Figure 4.5 Summary of events associated with the first week of development.

 Fertilization usually occurs in the uterine tube.

Frontal plane

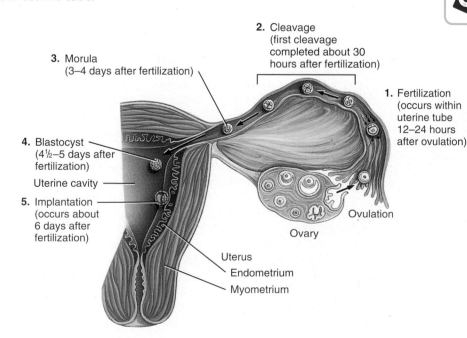

2. Cleavage (first cleavage completed about 30 hours after fertilization)

3. Morula (3–4 days after fertilization)

1. Fertilization (occurs within uterine tube 12–24 hours after ovulation)

4. Blastocyst (4½–5 days after fertilization)

Uterine cavity

5. Implantation (occurs about 6 days after fertilization)

Ovulation

Ovary

Uterus

Endometrium

Myometrium

Frontal section through uterus, uterine tube, and ovary

? **In which portion of the uterus does implantation usually occur?**

◉ CHECKPOINT

1. Where does fertilization normally occur?
2. What is a morula, and how is it formed?
3. Describe the layers of a blastocyst, and their eventual fates.
4. When, where, and how does implantation occur?
5. On what basis are the three regions of the decidua named?

Second Week of Development

Development of the Trophoblast

About 8 days after fertilization, the trophoblast develops into two layers in the region of contact between the blastocyst and endometrium. These layers are a **syncytiotrophoblast** (sin-sīt-ē-ō-TRŌF-ō-blast) that contains no distinct cell boundaries, and a **cytotrophoblast** (sī-tō-TRŌF-ō-blast) between the embryoblast and syncytiotrophoblast (Figure 4.6a). The two layers of trophoblast become part of the chorion (one of the fetal membranes) as they undergo further growth (see Figure 4.11a on page 105). During implantation, the syncytiotrophoblast secretes enzymes that digest and liquefy the endometrial cells so the blastocyst can penetrate the uterine lining. Eventually, the blastocyst becomes buried in the endometrium and inner one-third of the myometrium (muscle layer of the uterus). Another secretion of the trophoblast is human chorionic gonadotropin (hCG), the key hormone in the maintenance of pregnancy and the hormone that is detected by a home pregnancy test. Its presence in the maternal urine or blood is an indication that an implanted embryo is present in the uterus. Human chorionic gonadotropin mimics a maternal hormone, luetenizing hormone (LH), which maintains the endocrine secretions of progesterone and estrogens from the ovary. In the typical menstrual cycle, LH production declines, resulting in uterine contractions and the expulsion of the endometrium from the uterus. If this were to happen in a pregnant cycle the embryo would be aborted. By producing hCG, the embryo ensures the maintenance of a pregnant cycle. Peak secretion of hCG occurs about the ninth week of pregnancy, at which time the placenta is well developed and can now produce the progesterone and estrogens on its own.

• CLINICAL **CONNECTION** Early Pregnancy Tests

Home pregnancy test

Early pregnancy tests detect the tiny amounts of human chorionic gonadotropin (hCG) in the urine that begin to be excreted about 8 days after fertilization. The test kits can detect pregnancy as early as the first day of a missed menstrual period—that is, at about 14 days after fertilization. Chemicals in the kits produce a color change if a reaction occurs between hCG in the urine and hCG antibodies included in the kit.

Several of the test kits available at pharmacies are as sensitive and accurate as test methods used in many hospitals. Still, false-negative and false-positive results can occur. A false-negative result (the test is negative, but the woman is pregnant) may be due to testing too soon or to an ectopic pregnancy. A false-positive result (the test is positive, but the woman is not pregnant) may be due to excess protein or blood in the urine or to hCG produced from a rare type of uterine cancer. Thiazide diuretics, hormones, steroids, and thyroid drugs may also affect the outcome of an early pregnancy test.

Development of the Bilaminar Embryonic Disc

Cells of the embryoblast also differentiate into two layers around 8 days after fertilization: a **hypoblast (primitive endoderm)** and **epiblast (primitive ectoderm)** (Figure 4.6a). Cells of the hypoblast and epiblast together form a flat disc referred to as the **bilaminar embryonic disc** (bī-LAM-in-ar=two-layered). During further development, the cells of the hypoblast continue as a single squamous (flat) layer of cells, while the epiblast cells become stratified. Soon a small cavity appears within the layers of epiblast cells and eventually enlarges to form the **amniotic cavity** (am-nē-OT-ik; *amnio-*=lamb).

Development of the Amnion

As the amniotic cavity enlarges, a single layer of squamous cells forms a dome-like roof above the epiblast cells called the amnion (AM-nē-on) (Figure 4.6a). Thus, the amnion forms the roof of the amniotic cavity and the epiblast forms the floor. Initially, the amnion overlies only the bilaminar embryonic disc. However, as the embryonic disc increases in size and begins to fold, the amnion eventually surrounds the entire embryo (see Figure 4.11a inset), creating the amniotic cavity that becomes filled with amniotic fluid. Most amniotic fluid is initially derived from maternal blood. Later, the fetus contributes to the fluid by excreting urine into the amniotic cavity. Amniotic fluid serves as a shock absorber for the fetus, helps regulate fetal body temperature, helps prevent the fetus from drying out, and prevents adhesions between the skin of the fetus and surrounding tissues. The amnion usually ruptures just before birth; it and its fluid constitute the "bag of waters." Embryonic cells are normally sloughed off into amniotic fluid. They can be examined in a procedure called *amniocentesis*, which involves withdrawing some of the amniotic fluid that bathes the developing fetus and analyzing the fetal cells and dissolved substances (see page 112).

Development of the Yolk Sac

Also on the eighth day after fertilization, cells of the hypoblast migrate and cover the inner surface of the blastocyst wall (Figure 4.6a). The migrating cells form a thin membrane called the **exocoelomic membrane** (ek-sō-sē-LŌ-mik; *exo-*=outside; *-koilos*=space). Together with the hypoblast, the exocoelomic membrane forms the wall of the **yolk sac**, the former blastocyst cavity during earlier development (Figure 4.6b). As a result, the bilaminar embryonic disc is now positioned between the amniotic cavity and yolk sac.

Since human embryos receive their nutrients from the endometrium, the yolk sac is relatively empty, small, and decreases in size as development progresses (see Figure 4.11a inset). Nevertheless, the yolk sac has several important functions in humans. It supplies nutrients to the embryo during the second and third weeks of development, is the source of blood cells from the third through sixth weeks, contains the first cells (primordial germ cells) that will eventually migrate into the developing gonads and differentiate into gametes, forms part of the gut (gastrointestinal tract), and helps prevent drying out of the embryo.

Figure 4.6 **Figure 4.6** **Principal events of the second week of development.**

About 8 days after fertilization, the trophoblast develops into a syncytiotrophoblast and a cytotrophoblast; the embryoblast (inner cell mass) develops into a hypoblast and epiblast (bilaminar embryonic disc).

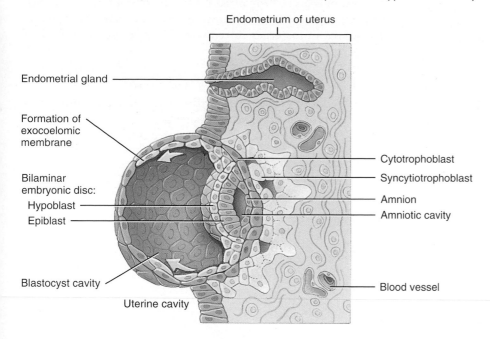

Endometrium of uterus

Endometrial gland

Formation of exocoelomic membrane

Bilaminar embryonic disc:
Hypoblast
Epiblast

Blastocyst cavity

Uterine cavity

Cytotrophoblast
Syncytiotrophoblast
Amnion
Amniotic cavity

Blood vessel

(a) Frontal section through endometrium of uterus showing blastocyst, about 8 days after fertilization

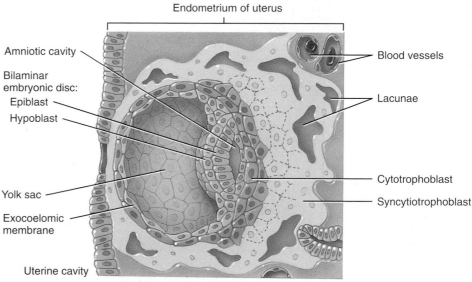

Endometrium of uterus

Amniotic cavity

Bilaminar embryonic disc:
Epiblast
Hypoblast

Yolk sac

Exocoelomic membrane

Uterine cavity

Blood vessels

Lacunae

Cytotrophoblast
Syncytiotrophoblast

(b) Frontal section through endometrium of uterus showing blastocyst, about 9 days after fertilization

Endometrium of uterus

Lacunae

Yolk sac

Lacunar network

Uterine cavity

Maternal sinusoids

Chorion:
Extraembryonic mesoderm

Syncytiotrophoblast
Cytotrophoblast

Amnion
Amniotic cavity

Bilaminar embryonic disc:
Epiblast
Hypoblast

Endometrial gland (right) and sinusoid (left) emptying into lacunar network

(c) Frontal section through endometrium of uterus showing blastocyst, about 12 days after fertilization

❓ **How is the bilaminar embryonic disc connected to the trophoblast?**

Development of Sinusoids

On the ninth day after fertilization, the blastocyst becomes completely embedded in the endometrium. As the syncytiotrophoblast expands into the endometrium and around the yolk sac, small spaces called **lacunae** (la-KOO-nē=little lakes) develop within it (Figure 4.6b).

By the twelfth day of development, the lacunae fuse to form larger, interconnecting spaces called **lacunar networks** (Figure 4.6c). Endometrial capillaries (microscopic maternal blood vessels) around the developing embryo expand and are referred to as **maternal sinusoids**. As the syncytiotrophoblast erodes some of the sinusoids and endometrial glands, maternal blood and glandular secretions enter the lacunar networks, which serve as both a rich source of materials for embryonic nutrition and a disposal site for the embryo's wastes.

Development of the Extraembryonic Coelom

About the twelfth day after fertilization, the **extraembryonic mesoderm** develops. These mesodermal cells are derived from the yolk sac and form a connective tissue (mesenchyme) around

Figure 4.7 Gastrulation.

Gastrulation involves the rearrangement and migration of cells from the epiblast.

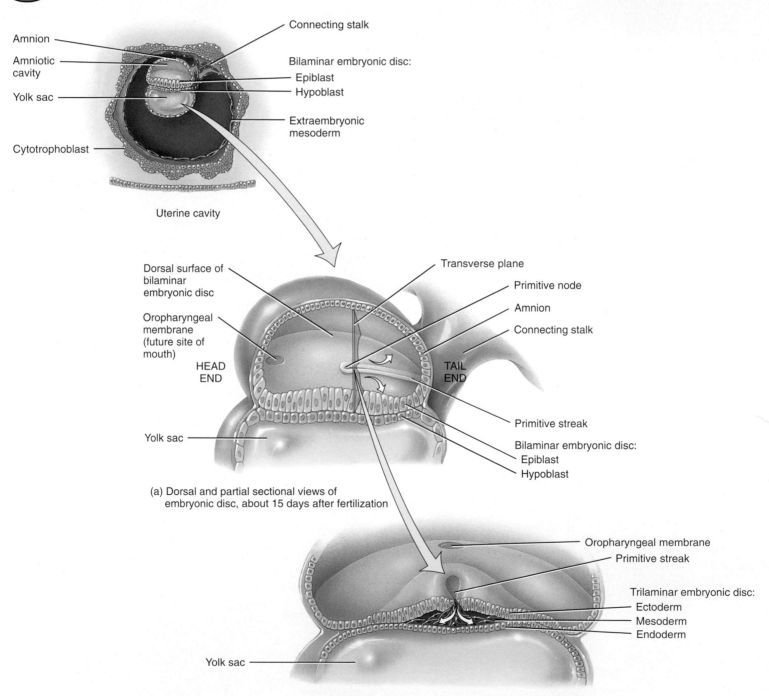

Amnion

Amniotic cavity

Yolk sac

Cytotrophoblast

Connecting stalk

Bilaminar embryonic disc:
Epiblast
Hypoblast

Extraembryonic mesoderm

Uterine cavity

Dorsal surface of bilaminar embryonic disc

Oropharyngeal membrane (future site of mouth)

HEAD END

Yolk sac

Transverse plane

Primitive node

Amnion

Connecting stalk

TAIL END

Primitive streak

Bilaminar embryonic disc:
Epiblast
Hypoblast

(a) Dorsal and partial sectional views of embryonic disc, about 15 days after fertilization

Oropharyngeal membrane

Primitive streak

Trilaminar embryonic disc:
Ectoderm
Mesoderm
Endoderm

Yolk sac

(b) Transverse section of trilaminar embryonic disc, about 16 days after fertilization

? **What is the significance of gastrulation?**

the amnion and yolk sac (Figure 4.6c). Soon a number of large cavities develop in the extraembryonic mesoderm, which then fuse to form a single, larger cavity called the **extraembryonic coelom** (SĒ-lōm).

Development of the Chorion

The extraembryonic mesoderm and the two layers of the trophoblast (the cytotrophoblast and the syncytiotrophoblast) together form the **chorion** (KOR-ē-on=membrane) (Figure 4.6c). It surrounds the embryo and, later, the fetus (see Figure 4.11a inset). Eventually the chorion becomes the principal embryonic part of the placenta, the structure for exchange of materials between mother and fetus. The chorion also protects the embryo and fetus from the immune responses of the mother in two ways: (1) It secretes proteins that block antibody production by the mother. (2) It promotes the production of T lymphocytes that suppress the normal immune response in the uterus. Finally, the chorion produces human chorionic gonadotropin (hCG), an important hormone of pregnancy.

The inner layer of the chorion eventually fuses with the amnion. With the development of the chorion, the extraembryonic coelom is now referred to as the **chorionic cavity.** By the end of the second week of development, the bilaminar embryonic disc becomes connected to the trophoblast by a band of extraembryonic mesoderm called the **connecting (body) stalk** (see Figure 4.7, inset), the future umbilical cord.

◉ CHECKPOINT

6. What are the functions of the trophoblast?
7. How is the bilaminar embryonic disc formed?
8. Describe the formation of the amnion, yolk sac, and chorion and explain their functions.
9. Why are sinusoids important during embryonic development?

Third Week of Development

The third embryonic week begins a six-week period of rapid development and differentiation. During the third week, the three primary germ layers are established, which lays the groundwork for organ development in weeks four through eight.

Gastrulation

The first major event of the third week of development, **gastrulation** (gas-troo-LĀ-shun), occurs about 15 days after fertilization. In this process, the bilaminar (two-layered) embryonic disc, consisting of epiblast and hypoblast, is transformed into a **trilaminar** (three-layered) **embryonic disc** consisting of three primary germ layers: the ectoderm, mesoderm, and endoderm. The primary germ layers are the major embryonic tissues from which the various tissues and organs of the body develop.

Gastrulation involves the well-coordinated and important rearrangement and migration of cells from the epiblast to set the stage for the important interactions of the newly positioned cells. The first evidence of gastrulation is the formation of the **primitive streak**, a faint groove on the dorsal surface of the epiblast that elongates from the posterior to the anterior part of the embryo (Figure 4.7a). The primitive streak clearly establishes the head and tail ends of the embryo, as well as its right and left sides. At the head end of the primitive streak a small group of epiblast cells forms a rounded structure called the **primitive node**.

Following formation of the primitive streak, cells of the epiblast move inward below the primitive streak and detach from the epiblast (Figure 4.7b). Once the cells begin to migrate beneath the overlying epiblast, some of them displace the original hypoblast by pushing it outward from beneath the epiblast and completely replace it with a new layer of cells. This new layer of cells forming the roof of the yolk sac is the **endoderm** (*endo-* = inside; *-derm*=skin). Other migrating cells remain between the epiblast and newly formed endoderm to form the **mesoderm** (*meso-*=middle). Cell placement and differentiation of cells within this layer are crucial to inducing further development of the final body plan. Cells remaining in the epiblast then form the **ectoderm** (*ecto-*=outside). Table 4.1 provides details about the fates of these primary germ layers; coverage is also included in later chapters in the context of the various body systems.

TABLE 4.1		
Structures Produced by the Three Primary Germ Layers		
ENDODERM	**MESODERM**	**ECTODERM**
Epithelial lining of gastrointestinal tract (except the oral cavity and anal canal) and the eqithelium of its glands.	All skeletal and cardiac muscle tissue and most smooth muscle tissue.	All nervous tissue.
Epithelial lining of urinary bladder, gallbladder, and liver.	Cartilage, bone, and other connective tissues.	Epidermis of skin.
Epithelial lining of pharynx, auditory (eustachian) tubes, tonsils, larynx, trachea, bronchi, and lungs.	Blood, red bone marrow, and lymphatic tissue.	Hair follicles, arrector pili muscles, nails, epithelium of skin glands (sebaceous and sudoriferous), and mammary glands.
Epithelium of thyroid gland, parathyroid glands, pancreas, and thymus.	Endothelium of blood vessels and lymphatic vessels.	Lens, cornea, and internal eye muscles.
Epithelial lining of prostate and bulbourethral (Cowper's) glands, vagina, vestibule, urethra, and associated glands such as the greater (Bartholin's) vestibular and lesser vestibular glands.	Dermis of skin.	Internal and external ear.
	Fibrous tunic and vascular tunic of eye.	Neuroepithelium of sense organs.
	Middle ear.	Epithelium of oral cavity, nasal cavity, paranasal sinuses, salivary glands, and anal canal.
	Mesothelium of thoracic, abdominal, and pelvic cavities.	Epithelium of pineal gland, pituitary gland, and adrenal medullae.
	Epithelium of kidneys and ureters.	
	Epithelium of adrenal cortex.	
	Epithelium of gonads and genital ducts.	

About 16 days after fertilization, mesodermal cells from the primitive node migrate toward the head end of the embryo and form a hollow tube of cells in the midline called the **notochordal process** (nō-tō-KOR-dal) (Figure 4.8). By days 22–24, the notochordal process becomes a solid cylinder of cells called the **notochord** (nō-tō-KORD; *noto-*=back; *-chord*=cord) (see Figure 4.9a). This structure plays an extremely important role in **induction** (in-DUK-shun), the process by which one tissue (*inducing tissue*) stimulates the specialization of an adjacent tissue (*responding tissue*). An inducing tissue usually produces a chemical substance that influences the responding tissue. The notochord induces certain neighboring mesodermal cells to develop into parts of vertebrae (back bones), induces the overlying ectoderm to fold inward to form the nervous system, and contributes to the formation of intervertebral discs between vertebrae (see Figure 7.15 on pages 193–194).

Also during the third week of development, two faint depressions appear on the dorsal surface of the embryo. The structure closer to the head end is called the **oropharyngeal membrane** (or-ō-fa-RIN-jē-al; *oro-*=mouth; *-pharyngeal*=pertaining to the pharynx) (Figure 4.8a, b). It breaks down during the fourth week to connect the mouth cavity to the pharynx (throat) and the remainder of the gastrointestinal tract. The structure closer to the tail end is called the **cloacal membrane** (klō-Ā-kul=sewer), which degenerates in the seventh week to form the openings of the anus and urinary and reproductive tracts. These two membranes are regions in the trilaminar embryo where the mesoderm is not present between the outer ectoderm and inner endoderm.

When the cloacal membrane appears, the wall of the yolk sac forms a small vascularized outpouching called the **allantois** (a-LAN-tō-is; *allant-*=sausage) that extends into the connecting stalk (Figure 4.8b). In most other mammals, the allantois is used for gas exchange and waste removal. Because the placenta performs these functions, the allantois is not a prominent structure in humans (see Figure 4.11a, inset). Nevertheless, it does func-

tion in early formation of blood and blood vessels and it is associated with the development of the urinary bladder.

Neurulation

As mentioned previously, in addition to inducing mesodermal cells to develop into parts of vertebrae, the notochord also induces ectodermal cells over it to form the **neural plate** (Figure 4.9a). (Also see Figure 19.26 on page 653). By the end of the third week the neural plate begins a process of **invagination** (in-vaj-in-NA-shun), an infolding of the ectoderm neural plate cells into the underlying mesoderm. The lateral edges of the neural plate become more elevated and form the **neural folds** (Figure 4.9b). The depressed region between the folds is called the **neural groove** (Figure 4.9c). As the fold deepens, the neural folds approach each other and fuse, thus converting the neural plate into a **neural tube** that is pushed beneath the surface ectoderm into the underlying mesoderm (Figure 4.9d). Neural tube cells then develop into the brain and spinal cord. This embryonic process by which the neural plate, neural folds, and neural tube form is called **neurulation** (noor-oo-LĀ-shun).

As the neural tube forms, some of the ectodermal cells migrate to form several layers of cells called the **neural crest** (see Figure 19.26 on page 653). Neural crest cells give rise to some of the peripheral nerves, the meninges (coverings) of the brain and spinal cord, and most of the skeletal and connective tissue components of the head.

At about four weeks after fertilization, the head end of the neural tube develops into three enlarged areas called **primary brain vesicles** (see Figure 19.27 on page 654). The parts of the brain that develop from the various brain vesicles are described on page 653 in Chapter 19.

Development of Somites

By about the seventeenth day after fertilization, the mesoderm adjacent to the notochord and neural tube forms paired longitu-

Figure 4.8 Development of the notochordal process.

🔑 The notochordal process develops from the primitive node and later becomes the notochord.

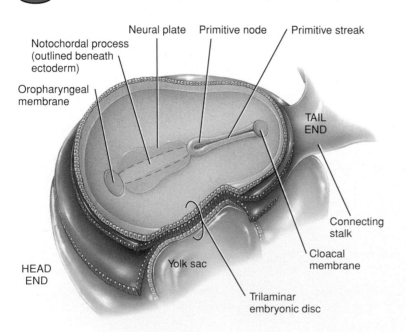

(a) Dorsal and partial sectional views of trilaminar embryonic disc, about 16 days after fertilization

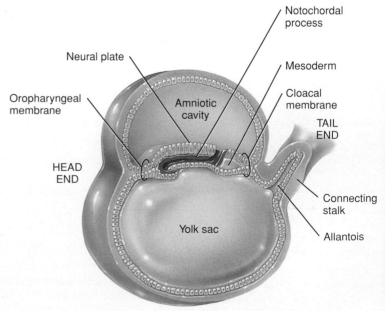

(b) Sagittal section of trilaminar embryonic disc, about 16 days after fertilization

❓ **What is the significance of the notochord?**

dinal columns of **paraxial mesoderm** (par-AK-sē-al; *para-*=near) (Figure 4.9b). The mesoderm lateral to the paraxial mesoderm forms paired cylindrical masses called **intermediate mesoderm**. The mesoderm lateral to the intermediate mesoderm consists of a pair of flattened sheets called **lateral plate mesoderm**. The paraxial mesoderm soon segments into a series of paired, cube-shaped structures called **somites** (SŌ-mīts=little bodies) (Figure 4.9c, d). By the end of the fifth week, 42–44 pairs of somites are present. The number of somites that develop over a given period can be correlated to the approximate age of the embryo.

Each somite differentiates into three regions: a **myotome**, a **dermatome**, and a **sclerotome** (see Figure 10.10 on page 304). The myotomes develop into the skeletal muscles of the neck, trunk, and limbs; the dermatomes form connective tissue, including the dermis of the skin; and the sclerotomes give rise to the vertebrae and ribs.

Development of the Intraembryonic Coelom

In the third week of development, small spaces appear in the lateral plate mesoderm. These spaces soon merge to form a larger cavity called the **intraembryonic coelom** (SĒ-lom=cavity). This cavity splits the lateral plate mesoderm into two parts called the splanchnic mesoderm, which is adjacent to the endoderm and yolk sac, and somatic mesoderm, which is adjacent to the ectoderm and amnion (Figure 4.9d). **Splanchnic mesoderm** (SPLANGK-nik=visceral) forms the heart and the visceral layer

Figure 4.9 Neurulation and the development of somites.

Neurulation is the process by which the neural plate, neural folds, and neural tube form.

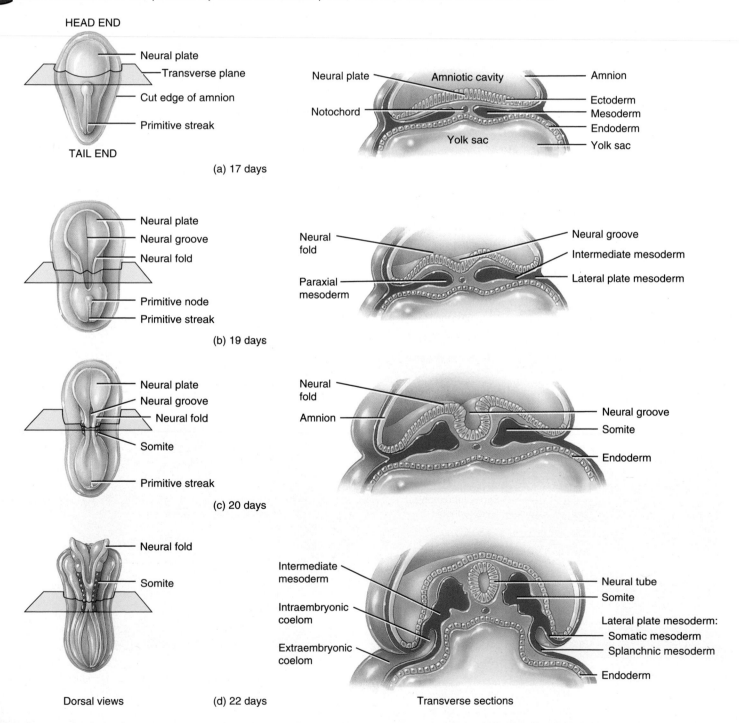

HEAD END

Neural plate
Transverse plane
Cut edge of amnion
Primitive streak

TAIL END

(a) 17 days

Neural plate
Notochord
Amniotic cavity
Amnion
Ectoderm
Mesoderm
Endoderm
Yolk sac
Yolk sac

Neural plate
Neural groove
Neural fold
Primitive node
Primitive streak

(b) 19 days

Neural fold
Paraxial mesoderm
Neural groove
Intermediate mesoderm
Lateral plate mesoderm

Neural plate
Neural groove
Neural fold
Somite
Primitive streak

(c) 20 days

Neural fold
Amnion
Neural groove
Somite
Endoderm

Neural fold
Somite

Intermediate mesoderm
Intraembryonic coelom
Extraembryonic coelom

Neural tube
Somite
Lateral plate mesoderm:
Somatic mesoderm
Splanchnic mesoderm
Endoderm

Dorsal views (d) 22 days Transverse sections

Which structures develop from the neural tube and somites?

of the serous pericardium, blood vessels, the smooth muscle and connective tissues of the respiratory and digestive organs, and the visceral layer of the serous membrane of the pleurae and peritoneum. **Somatic mesoderm** (sō-MAT-ik; *soma-*=body) gives rise to the bones, ligaments, and connective tissue of the limbs and the parietal layer of the serous membrane of the pericardium, pleurae, and peritoneum. During the second month of development, the intraembryonic coelom divides into the pericardial, pleural, and peritoneal cavities.

Development of the Cardiovascular System

At the beginning of the third week, **angiogenesis** (an-jē-ō-JEN-e-sis; *angio-*=vessel; *-genesis*=production), the formation of blood vessels, begins in the extraembryonic mesoderm of the yolk sac, connecting stalk, and chorion. This early development is necessary because there is insufficient yolk in the yolk sac and ovum to provide adequate nutrition for the rapidly developing embryo. Angiogenesis is initiated when mesodermal cells differentiate into **hemangioblasts**. These then develop into cells called **angioblasts**, which aggregate to form isolated masses of cells referred to as **blood islands** (see Figure 15.18 on page 535). As the

blood islands grow and fuse, they soon form an extensive system of blood vessels throughout the embryo.

About 3 weeks after fertilization, blood cells and blood plasma begin to develop *outside* the embryo from hemangioblasts in the blood vessels in the walls of the yolk sac, allantois, and chorion. These then develop into pluripotent stem cells that form blood cells. Blood formation begins *within* the embryo at about the fifth week in the liver and the twelfth week in the spleen, red bone marrow, and thymus.

The heart forms from splanchnic mesoderm in the head end of the embryo on day 18 or 19 after fertilization. This region of mesodermal cells is called the **cardiogenic area** (kar-dē-ō-JEN-ik; *cardio-*=heart; *-genic*=producing). In response to induction signals from the underlying endoderm, these mesodermal cells ultimately form a pair of **endocardial tubes** (see Figure 14.12 on page 477). The tubes then fuse to form a single **primitive heart tube**. By the end of the third week, the primitive heart tube bends on itself, becomes S-shaped, and begins to beat. It then joins blood vessels in other parts of the embryo, connecting stalk, chorion, and yolk sac to form a primitive cardiovascular system.

Development of the Chorionic Villi and Placenta

As the embryonic tissue invades the uterine wall, maternal uterine vessels are eroded and maternal blood fills spaces, called **lacunae**, within the invading tissue. By the end of the second week of development, **chorionic villi** (ko-rē-ON-ik VIL-ī) begin to develop. These fingerlike projections consist of chorion (syncytiotrophoblast surrounded by cytotrophoblast) that project into the endometrial wall of the uterus (Figure 4.10a). By the end of the third week, blood capillaries develop in the chorionic villi (Figure 4.10b). Blood vessels in the chorionic villi connect to the embryonic heart by way of the umbilical arteries and umbilical vein through the connecting (body) stalk, which will eventually become the umbilical cord (Figure 4.10c). The fetal blood capillaries within the chorionic villi project into the lacunae, which unite to form the **intervillous spaces** that bathe the chorionic villi with maternal blood. As a result, maternal and fetal blood

---

Figure 4.10 Development of chorionic villi.

🔑 Blood vessels in chorionic villi connect to the embryonic heart via the umbilical arteries and umbilical vein.

Lacunar networks
Extraembryonic mesoderm
Chorion:
  Syncytiotrophoblast
  Cytotrophoblast
Chorionic villi
Connecting stalk
Amniotic cavity
Maternal sinusoid
Yolk sac
Extraembryonic coelom

(a) Frontal section through uterus showing blastocyst, about 13 days after fertilization

Maternal sinusoid
Chorion:
  Syncytiotrophoblast
  Cytotrophoblast
Intervillous space
Connecting stalk
Extraembryonic mesoderm
Blood capillary

(b) Details of two chorionic villi, about 21 days after fertilization

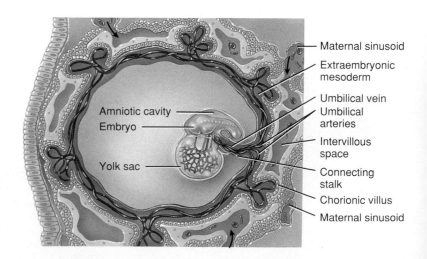

Maternal sinusoid
Extraembryonic mesoderm
Umbilical vein
Umbilical arteries
Amniotic cavity
Embryo
Intervillous space
Connecting stalk
Yolk sac
Chorionic villus
Maternal sinusoid

(c) Frontal section through uterus showing an embryo and its vascular supply, about 21 days after fertilization

❓ **Why is development of chorionic villi important?**

vessels are in close proximity. Note, however, that maternal and fetal blood vessels do not join, and the blood they carry does not normally mix. Instead, oxygen and nutrients in the blood of the mother's intervillous spaces, which are the spaces between chorionic villi, diffuse across the cell membranes into the capillaries of the villi. Waste products such as carbon dioxide diffuse in the opposite direction.

Placentation (plas-en-TĀ-shun) is the process of forming the **placenta** (pla-SEN-ta=flat cake), the site of exchange of nutrients and wastes between the mother and fetus. The placenta also produces hormones needed to sustain the pregnancy. The placenta is unique because it develops from two separate individuals, the mother and the fetus.

By the beginning of the twelfth week, the placenta has two distinct parts: (1) the fetal portion formed by the chorionic villi of the chorion and (2) the maternal portion formed by the maternal blood in the intervillous spaces and the decidua basalis of the endometrium (Figure 4.11a). When fully developed, the placenta is a rounded disc-like structure (Figure 4.11b). Functionally, the placenta allows oxygen and nutrients to diffuse from maternal blood into fetal blood while carbon dioxide and wastes diffuse from fetal blood into maternal blood. The placenta also is a protective barrier because most microorganisms cannot pass through it. However, certain viruses, such as those that cause AIDS, German measles, chickenpox, measles, encephalitis, and poliomyelitis, can cross the placenta. Many drugs, alcohol, and some substances that can cause birth defects also pass freely through the placenta. The placenta stores nutrients such as carbohydrates, proteins, calcium, and iron, which are released into fetal circulation as required.

Figure 4.11 Placenta and umbilical cord.

The placenta is formed by the chorionic villi of the embryo and the decidua basalis of the endometrium of the mother.

(a) Details of placenta and umbilical cord

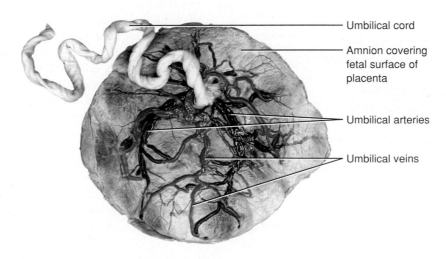

(b) Fetal aspect of placenta

? **What is the function of the placenta?**

The actual connection between the placenta and embryo, and later the fetus, is through the **umbilical cord** (um-BIL-i-kul=navel), which develops from the connecting stalk and is usually about 2 cm (1 in.) wide and about 50–60 cm (20–24 in.) in length. The umbilical cord consists of two umbilical arteries that carry deoxygenated fetal blood to the placenta, one umbilical vein that carries oxygen and nutrients acquired from the mother's intervillous spaces into the fetus, and supporting mucous connective tissue called **Wharton's jelly** derived from the allantois. A layer of amnion surrounds the entire umbilical cord and gives it a shiny appearance (Figure 4.11b). In some cases, the umbilical vein is used to transfuse blood into a fetus or to introduce drugs for various medical treatments.

In about 1 in 200 newborns, only one of the two umbilical arteries is present in the umbilical cord. It may be due to failure of the artery to develop or degeneration of the vessel early in development. Nearly 20% of infants with this condition develop cardiovascular defects.

After the birth of the baby, the placenta detaches from the uterus and is therefore termed the **afterbirth**. At this time, the umbilical cord is tied off and then severed. The small portion (about an inch) of the cord that remains attached to the infant begins to wither and falls off, usually within 12 to 15 days after birth. The area where the cord was attached becomes covered by a thin layer of skin, and scar tissue forms. The scar is the **umbilicus** (navel).

Pharmaceutical companies use human placentas as a source of hormones, drugs, and blood; portions of placentas are also used for burn coverage. The placental and umbilical cord veins can also be used in blood vessel grafts, and cord blood can be frozen to provide a future source of pluripotent stem cells, for example, to repopulate red bone marrow following radiotherapy for cancer.

In some cases, the entire placenta or part of it may become implanted in the inferior portion of the uterus, near or covering the opening of the uterus into the vagina. This condition is called **placenta previa** (PRĒ-vē-a=before or in front of). Although placenta previa may lead to spontaneous abortion, it also occurs in approximately 1 in 250 live births. It is dangerous to the fetus because it may cause premature birth and intrauterine hypoxia due to maternal bleeding. Maternal mortality is increased due to hemorrhage and infection. The most important symptom is sudden, painless, bright-red vaginal bleeding in the third trimester. Cesarean section is the preferred method of delivery in placenta previa.

◉ CHECKPOINT

10. What is the significance of gastrulation?
11. How do the three primary germ layers form? Why are they important?
12. What is the function of the notochord?
13. Describe how neurulation occurs. Why is it important?
14. What are the functions of somites?
15. How does the cardiovascular system develop?
16. How does the placenta form and what is its importance?

Fourth Week of Development

The fourth through eighth weeks are very significant in embryonic development because all major organs appear during this time. The term **organogenesis** (or′-ga-nō-JEN-e-sis) refers to the formation of body organs and systems. By the end of the

eighth week, all the major body systems have begun to develop, although their functions for the most part are minimal. Organogenesis requires the presence of blood vessels to supply developing organs with oxygen and other nutrients. However, recent studies suggest that blood vessels play a significant role in organogenesis even before blood begins to flow within them. The endothelial cells of blood vessels apparently provide some type of developmental signal, either a secreted substance or a direct cell-to-cell interaction, that is necessary for organogenesis.

During the fourth week of development, the embryo undergoes dramatic changes in shape and nearly triples its size. It is essentially converted from a flat, two-dimensional trilaminar embryonic disc to a three-dimensional cylinder, via a process called **embryonic folding** (Figure 4.12). The cylinder consists of endoderm in the center (gut lining), ectoderm on the outside (epidermis) with some internal ectoderm from neurulation, and mesoderm in between. The main force responsible for embryonic folding is the different rates of growth of various parts of the embryo, especially the rapid longitudinal growth of the nervous system (neural tube). Folding in the median plane produces a **head fold** and a **tail fold**, while folding in the horizontal plane results in the two **lateral folds**. Overall, due to the foldings, the embryo curves into a C-shape.

The head fold brings the developing heart and mouth into their eventual adult positions. The tail fold brings the developing anus into its eventual adult position. The lateral folds form as the lateral margins of the trilaminar embryonic disc bend ventrally. As they move toward the midline, the lateral folds incorporate the dorsal part of the yolk sac into the embryo as the **primitive gut**, the forerunner of the gastrointestinal tract (Figure 4.12b). The primitive gut differentiates into an anterior **foregut**, an intermediate **midgut**, and a posterior **hindgut** (Figure 4.12c). The fates of the foregut, midgut, and hindgut are described on page 814. Recall that the oropharyngeal membrane is located in the head end of the embryo (see Figure 4.8). It separates the future pharyngeal (throat) region of the foregut from the **stomodeum** (stō-mō-DĒ-um; *stomo-*=mouth), the future oral (mouth) cavity. Because of head folding, the oropharyngeal membrane moves downward and the foregut and stomodeum move closer to their final positions. When the oropharyngeal membrane ruptures during the fourth week, the pharyngeal region of the pharynx and stomodeum are brought into contact with each other.

In a developing embryo, the last part of the hindgut expands into a cavity called the **cloaca** (see Figure 26.13a on page 841). On the outside of an embryo is a small cavity in the tail region called the **proctodeum** (prok′-tō-DĒ-um; *procto-*=anus) (Figure 4.12c). Separating the cloaca from the proctodeum is the **cloacal membrane** (see Figure 4.8). During embryonic development, the cloaca divides into a ventral urogenital sinus and a dorsal anorectal canal. As a result of tail folding, the cloacal membrane moves downward and the urogenital sinus, anorectal canal, and proctodeum move closer to their final positions. When the cloacal membrane ruptures during the seventh week of development, the urogenital and anal openings are created.

Along with embryonic folding, development of somites and the neural tube (previously described) occurs during the fourth week. In addition, eventually five pairs of **pharyngeal** or **branchial arches** (*branch*=gill) develop on each side of the future head and neck regions (Figure 4.13 on page 108). These five paired structures begin their development on the twenty-second day after fertilization and form swellings on the surface of the embryo. Each pharyngeal arch consists of an outer covering of

Figure 4.12 Embryonic folding.

Embryonic folding converts the two-dimensional trilaminar embryonic disc into a three-dimensional cylinder.

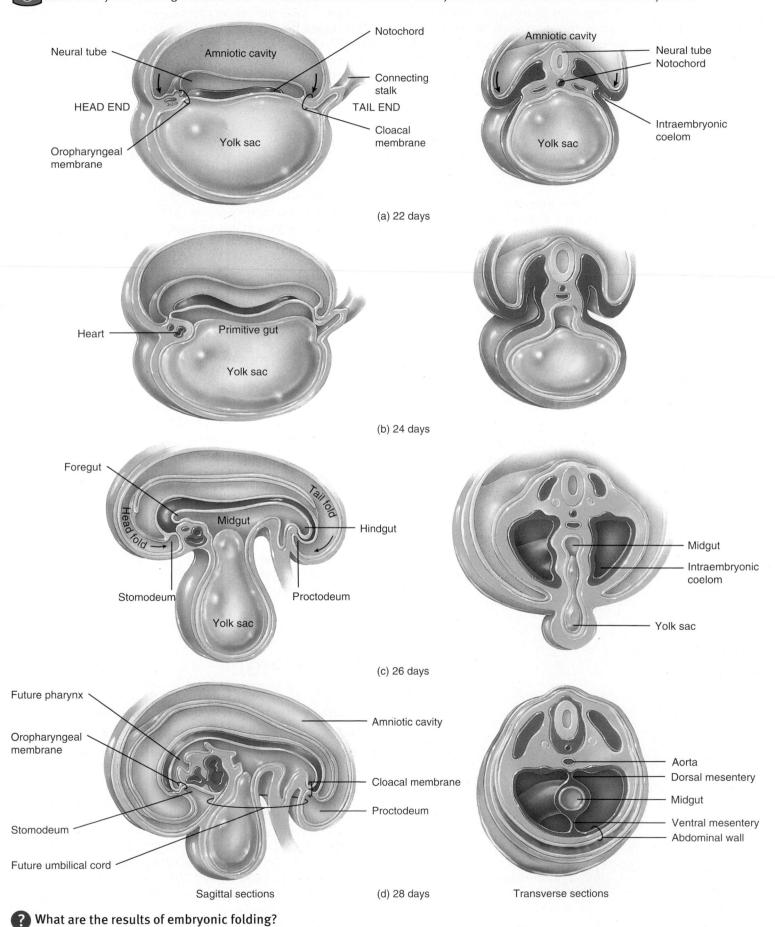

(a) 22 days

(b) 24 days

(c) 26 days

(d) 28 days

Sagittal sections   Transverse sections

? What are the results of embryonic folding?

ectoderm, an inner covering of endoderm, and mesoderm in between. The pharyngeal arches also contain an artery, a cranial nerve, cartilage, and muscle tissue. Also on the outside of the embryo are a series of grooves between the pharyngeal arches called **pharyngeal clefts**, which separate the pharyngeal arches (Figure 4.13a). The clefts meet corresponding balloonlike outgrowths in the pharyngeal lining called **pharyngeal pouches**. Where the pharyngeal cleft and pouch meet to separate the arches, the outer ectoderm of the cleft contacts the inner endoderm of the pouch and there is no mesoderm between (Figure 4.13b).

Just as the somite gives rise to specified structures in the body wall, each pharyngeal arch, cleft, and pouch gives rise to specified structures in the head and neck. Each pharyngeal arch is a developmental unit and includes a skeletal component, muscle, nerve, and blood vessels.

In the human embryo, there are four obvious pharyngeal arches and and two less distinct arches. Each of these arches develops into a specific and unique component of the head and neck region. For example, the first pharyngeal arch is often called the *mandibular arch* because it forms the jaws (the *mandible* is the lower jawbone).

The first sign of a developing ear is a thickened area of ectoderm, the **otic placode** (future internal ear), which can be distinguished about 22 days after fertilization (see Figure 4.13a). A

thickened area of ectoderm called the **lens placode** (see Figure 4.13a), which will become the eye, also appears at this time.

By the middle of the fourth week, the upper limbs begin their development as outgrowths of mesoderm covered by ectoderm called **upper limb buds** (see Figure 8.13a on page 245). By the end of the fourth week, the **lower limb buds** develop. The heart also forms a distinct projection on the ventral surface of the embryo called the **heart prominence**, just caudal to the pharyngeal arches (see Figure 8.13b on page 246), and a **tail** becomes visible (see Figure 8.13a on page 245).

Fifth Through Eighth Weeks of Development

During the fifth week of development, there is a very rapid development of the brain and thus growth of the head is considerable. By the end of the sixth week, the head grows even larger relative to the trunk, and the limbs show substantial development (see Figure 8.13b on page 246). In addition, the neck and trunk begin to straighten, and the heart is now four-chambered. By the seventh week, the various regions of the limbs become distinct and the beginnings of digits appear (see Figure 8.13c on page 246). At the start of the eighth week, the final week of the embryonic period, the digits of the hands are short and webbed and the tail is still visible, but shorter. In addition, the eyes are open and the auricles of the ears are visible (see Figure 8.13b on page 246). By the end of the eighth week, all regions of limbs are apparent; the digits are distinct and no longer webbed due to removal of cells via apoptosis. Also, the eyelids come together and may fuse, the tail disappears, and the external genitals begin to differentiate. The embryo now has clearly human characteristics.

> ◉ **CHECKPOINT**
> 17. How does embryonic folding occur?
> 18. How does the primitive gut form and what is its significance?
> 19. What is the origin of the structures of the head and neck?
> 20. What are limb buds?
> 21. What changes occur in the limbs during the second half of the embryonic period?

Figure 4.13 Development of pharyngeal arches, pharyngeal clefts, and pharyngeal pouches.

🔑 The five pairs of pharyngeal pouches consist of ectoderm, mesoderm, and endoderm and contain blood vessels, cranial nerves, cartilage, and muscular tissue.

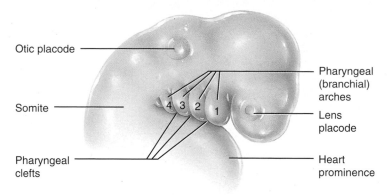

(a) External view, about 28-day embryo

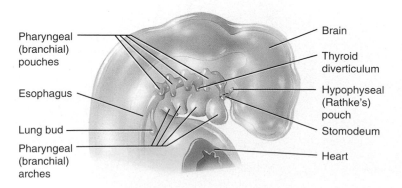

(b) Sagittal section, about 28-day embryo

❓ Why are pharyngeal arches, clefts, and pouches important?

FETAL PERIOD

◉ **OBJECTIVE**
• Describe the major events of the fetal period.

During the fetal period, tissues and organs that developed during the embryonic period grow and differentiate. Very few new structures appear during the fetal period, but the rate of body growth is remarkable, especially during the second half of intrauterine life. For example, during the last two-and-one-half months of intrauterine life, the weight of the fetus doubles. At the beginning of the fetal period, the head is half the length of the body. By the end of the fetal period, the head size is only one-quarter the length of the body. During the same period, the limbs also increase in size from one-eighth to one-half the fetal length. During the fetal period, the fetus is less vulnerable to the damaging effects of drugs, radiation, and microbes than it was as an embryo.

A summary of the major developmental events of the embryonic and fetal period is presented in Table 4.2 and illustrated in Figure 4.14 on page 110.

TABLE 4.2

Summary of Changes During Embryonic and Fetal Development

TIME	APPROXIMATE SIZE AND WEIGHT	REPRESENTATIVE CHANGES
Embryonic period		
1–4 weeks	0.6 cm (3/16 in.)	Primary germ layers and notochord develop. Neurulation occurs. Primary brain vesicles, somites, and intraembryonic coelom develop. Blood vessel formation begins and blood forms in yolk sac, allantois, and chorion. Heart forms and begins to beat. Chorionic villi develop and placental formation begins. The embryo folds. The primitive gut, pharyngeal arches, and limb buds develop. Eyes and ears begin to develop, tail forms, and body systems begin to form.
5–8 weeks	3 cm (1.25 in.) 1 g (1/30 oz)	Limbs become distinct and digits appear. Heart becomes four-chambered. Eyes are far apart and eyelids are fused. Nose develops and is flat. Face is more humanlike. Bone formation begins. Blood cells start to form in liver. External genitals begin to differentiate. Tail disappears. Major blood vessels form. Many internal organs continue to develop.
Fetal period		
9–12 weeks	7.5 cm (3 in.) 30 g (1 oz)	Head constitutes about half the length of the fetal body, and fetal length nearly doubles. Brain continues to enlarge. Face is broad, with eyes fully developed, closed, and widely separated. Nose develops a bridge. External ears develop and are low set. Bone formation continues. Upper limbs almost reach final relative length but lower limbs are not quite as well developed. Heartbeat can be detected. Gender is distinguishable from external genitals. Urine secreted by fetus is added to amniotic fluid. Red bone marrow, thymus, and spleen participate in blood cell formation. Fetus begins to move, but its movements cannot be felt yet by the mother. Body systems continue to develop.
13–16 weeks	18 cm (6.5–7 in.) 100 g (4 oz)	Head is relatively smaller than rest of body. Eyes move medially to their final positions, and ears move to their final positions on the sides of the head. Lower limbs lengthen. Fetus appears even more humanlike. Rapid development of body systems occurs.
17–20 weeks	25–30 cm (10–12 in.) 200–450 g (0.5–1 lb)	Head is more proportionate to rest of body. Eyebrows and head hair are visible. Growth slows but lower limbs continue to lengthen. Vernix caseosa (fatty secretions of oil glands and dead epithelial cells) and lanugo (delicate fetal hair) cover fetus. Brown fat forms and is the site of heat production. Fetal movements are commonly felt by mother (quickening).
21–25 weeks	27–35 cm (11–14 in.) 550–800 g (1.25–1.5 lb)	Head becomes even more proportionate to rest of body. Weight gain is substantial, and skin is pink and wrinkled.
26–29 weeks	32–42 cm (13–17 in.) 110–1350 g (2.5–3 lb)	Head and body are more proportionate and eyes are open. Toenails are visible. Body fat is 3.5% of total body mass and additional subcutaneous fat smoothes out some wrinkles. Testes begin to descend toward scrotum at 28 to 32 weeks. Red bone marrow is major site of blood cell production. Many fetuses born prematurely during this period survive if given intensive care because lungs can provide adequate ventilation, and central nervous system is developed enough to control breathing and body temperature.
30–34 weeks	41–45 cm (16.5–18 in.) 2000–2300 g (4.5–5 lb)	Skin is pink and smooth. Fetus assumes upside-down position. Body fat is 8% of total body mass. Fetuses 33 weeks and older usually survive if born prematurely.
35–38 weeks	50 cm (20 in.) 3200–3400 g (7–7.5 lb)	By 38 weeks circumference of fetal abdomen is greater than that of head. Skin is usually bluish-pink, and growth slows as birth approaches. Body fat is 16% of total body mass. Testes are usually in scrotum in full-term male infants. Even after birth, an infant is not completely developed; an additional year is required, especially for complete development of the nervous system.

4   8   12   16   20   24   28   32   36   (weeks)

Figure 4.14 Summary of representative developmental events of the embryonic and fetal periods.
The embryos and fetuses are not shown at their actual sizes.

6 ⚷ Development during the fetal period is mostly concerned with the growth and differentiation of tissues and organs formed during the embryonic period.

(a) 20-day embryo

- Neural plate
- Neural groove
- Cut edge of amnion
- Somite
- Yolk sac
- Primitive streak

(b) 24-day embryo

- Developing brain
- Heart prominence
- Developing spinal cord
- Somite

(c) 32-day embryo

- Pharyngeal arches
- Lens placode
- Heart prominence
- Upper limb bud
- Tail
- Lower limb bud

(d) 44-day embryo

- Otic placode
- Developing nose
- Upper limb
- Lower limb
- Umbilical cord

(e) 52-day embryo

- Ear
- Eye
- Nose
- Upper limb
- Umbilical cord
- Lower limb

(f) Ten-week fetus

- Ear
- Eye
- Nose
- Upper limb
- Yolk sac
- Rib
- Umbilical cord
- Placenta
- Lower limb

(g) Thirteen-week fetus

- Ear
- Eye
- Nose
- Mouth
- Upper limb
- Umbilical cord
- Lower limb

(h) Twenty-six-week fetus

- Ear
- Eye
- Nose
- Mouth
- Upper limb
- Lower limb

? How much weight is gained by a fetus during the last two-and-a-half months of intrauterine life?

Throughout the text we will discuss the developmental biology of the various body systems in their respective chapters. The following list of these sections is presented for your reference.

◉ CHECKPOINT
22. What are the general developmental trends during the fetal period?
23. Using Table 4.2 as a guide, select any one body structure in weeks 9 through 12 and trace its development through the remainder of the fetal period.

MATERNAL CHANGES DURING PREGNANCY

◉ OBJECTIVE
- Describe the effects of pregnancy on various body systems.

Near the end of the third month of pregnancy, the uterus occupies most of the pelvic cavity. As the fetus continues to grow, the uterus extends higher into the abdominal cavity. Toward the end of a full-term pregnancy, the uterus fills almost the entire abdominal cavity. It pushes the maternal intestines, liver, and stomach superiorly, elevates the diaphragm, and widens the thoracic cavity. Pressure on the stomach may force the stomach contents superiorly into the esophagus, resulting in heartburn. In the pelvic cavity, compression of the ureters and urinary bladder occurs.

Besides the anatomical changes associated with pregnancy, pregnancy-induced physiological changes also occur, including weight gain due to the fetus, amniotic fluid, the placenta, uterine enlargement, and increased total body water; increased storage of nutrients; marked breast enlargement in preparation for milk secretion and ejection; and lower back pain due to lordosis (hollow back).

Several changes occur in the maternal cardiovascular system. For example, heart rate increases and blood volume increases, mostly during the second half of pregnancy. These increases are necessary to meet the additional demands of the fetus for nutri-

ents and oxygen. The total volume of air inhaled and exhaled per minute also increases to meet the added oxygen demands of the fetus.

With regard to the digestive activities and movements of the gastrointestinal tract, pregnant women experience an increase in appetite. A general decrease in gastrointestinal tract movement can cause constipation, and produce nausea, vomiting, and heartburn.

Pressure on the urinary bladder by the enlarging uterus can produce urinary symptoms, such as increased frequency and urgency of urination, and incontinence (inability to retain urine). An increase in blood flow through the kidneys allows faster elimination of the extra wastes produced by the fetus.

Changes in the skin during pregnancy are more apparent in some women than in others. Included are increased pigmentation around the eyes and cheekbones in a masklike pattern and in the circular area around the nipples of the breasts. Some females also exhibit a dark, vertical line along the lower abdomen. Stretch marks over the abdomen can occur as the uterus enlarges, and hair loss increases.

Changes in the reproductive system include swelling of the external genitals, as well as increased blood supply, and increased pliability of the vagina. The uterus increases from its nonpregnant mass of 60–80 g to 900–1200 g at term due to growth of muscle fibers in the myometrium.

◉ CHECKPOINT
24. What structural and functional changes occur in the mother during pregnancy?

LABOR

◉ OBJECTIVE
- Explain the events associated with the three stages of labor.

Obstetrics (ob-STET-riks; *obstetrix*=midwife) is the branch of medicine that deals with the management of pregnancy, labor, and the **neonatal period**, the first 28 days after birth. **Labor** is the process by which the fetus is expelled from the uterus through the vagina. **Parturition** (par-toor-ISH-un; *parturit-*=childbirth) also means giving birth.

Uterine contractions occur in waves that start at the top of the uterus and move downward, eventually expelling the fetus. **True labor** begins when uterine contractions occur at regular intervals, usually producing pain. As the interval between contractions shortens, the contractions intensify. Another symptom of true labor in some women is localization of pain in the back that is intensified by walking. The reliable indicator of true labor is dilation (expansion) of the cervix and the "show," a discharge of a blood-containing mucus that appears in the cervical canal during labor. In **false labor,** pain is felt in the abdomen at irregular intervals, but it does not intensify and walking does not alter it significantly. There is no "show" and no cervical dilation.

True labor can be divided into three stages (Figure 4.15):

❶ **Stage of dilation.** The time from the onset of labor to the complete dilation of the cervix is the *stage of dilation.* This stage, which typically lasts 6–12 hours, features regular contractions of the uterus, usually a rupturing of the amniotic sac, and complete dilation (to 10 cm) of the cervix. If the amniotic sac does not rupture spontaneously, it is ruptured intentionally.

❷ **Stage of expulsion.** The time (10 minutes to several hours) from complete cervical dilation to delivery of the baby is the *stage of expulsion.*

❸ **Placental stage.** The time (5–30 minutes or more) after delivery until the placenta or "afterbirth" is expelled by powerful uterine contractions is the *placental stage.* These contractions also constrict blood vessels that were torn during delivery, thereby reducing the likelihood of hemorrhage.

As a rule, labor lasts longer with first babies, typically about 14 hours. For women who have previously given birth, the average duration of labor is about 8 hours—although the time varies enormously among births.

Delivery of a physiologically immature baby carries certain risks. A **premature infant** or "preemie" is generally considered a baby who weighs less than 2500 g (5.5 lb) at birth. Poor prenatal care, drug abuse, history of a previous premature delivery, and mother's age below 16 or above 35 increase the chance of premature delivery. The body of a premature infant is not yet ready to sustain some critical functions, and thus its survival is uncertain without medical intervention.

About 7% of pregnant women do not deliver by 2 weeks after their due date. Such cases carry an increased risk of brain damage to the fetus, and even fetal death, due to inadequate supplies of oxygen and nutrients from an aging placenta. Post-term deliveries may be facilitated by inducing labor, initiated by administration of oxytocin (Pitocin®), or by surgical delivery (cesarean section).

◉ **CHECKPOINT**

25. What is the difference between false labor and true labor?
26. What happens during the stage of dilation, the stage of expulsion, and the placental stage of true labor?

Figure 4.15 Stages of true labor.

❻ ⌇ The term *parturition* refers to birth.

- Urinary bladder
- Vagina
- Ruptured amniotic sac
- Rectum

❶ Stage of dilation

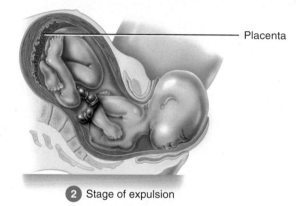

- Placenta

❷ Stage of expulsion

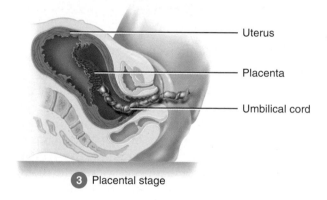

- Uterus
- Placenta
- Umbilical cord

❸ Placental stage

? **What event marks the beginning of the stage of expulsion?**

KEY MEDICAL TERMS ASSOCIATED WITH DEVELOPMENT

Amniocentesis (am′-nē-ō-sen-TĒ-sis; *amnio-*=amnion; *-centesis*=puncture to remove fluid) A prenatal diagnostic test in which some of the amniotic fluid that bathes the developing fetus is withdrawn and the fetal cells and dissolved substances are analyzed.

Chorionic villi sampling (CVS) (ko-rē-ON-ik VIL-ī) A prenatal diagnostic test in which a catheter is guided through the vagina and cervix of the uterus and then advanced to the chorionic villi and about 30 milligrams of tissue are suctioned out and prepared for chromosomal analysis.

Fetal alcohol syndrome (FAS) A specific pattern of fetal malformation due to intrauterine exposure to alcohol. FAS is one of the most common causes of mental retardation and the most common preventable cause of birth defects in the United States.

Fetal surgery Surgical procedure to repair and correct certain fetal disorders prior to childbirth; also called **in-utero surgery**.

Lethal gene (LĒ-thal JĒN; *lethum*=death) A gene that, when expressed, results in death either in the embryonic state or shortly after birth.

Maternal alpha-fetoprotein (AFP) test A prenatal diagnostic test in which the mother's blood is analyzed for the presence of AFP, a protein synthesized in the fetus that passes into the maternal circulation.

Mutation (mū-TĀ-shun) Any change in the sequence of bases in a DNA molecule resulting in a permanent alteration of an inherited trait.

Premature infant Generally refers to a baby who weighs less than 2500 g (5.5 lb) at birth.

Puerperal fever (pū-ER-per-al; *puer*=child) An infectious disease of childbirth, also called **puerperal sepsis** and **childbed fever**. The disease, which results from an infection originating in the birth canal, affects the endometrium.

Teratogen (TER-a-tō-jēn; *terato-*=monster; *-gen*=creating) Any agent or influence that causes developmental defects in the embryo. Examples include alcohol, pesticides, industrial chemicals, antibiotics, thalidomide, LSD, and cocaine.

STUDY OUTLINE

Embryonic Period (p. 94)

1. Pregnancy is a sequence of events that begins with fertilization, and proceeds to implantation, embryonic development, and fetal development. It normally ends in birth.
2. During fertilization a sperm cell penetrates a secondary oocyte and their pronuclei unite. Penetration of the zona pellucida is facilitated by enzymes in the sperm's acrosome. The resulting cell is a zygote.
3. Normally, only one sperm cell fertilizes a secondary oocyte.
4. Early rapid cell division of a zygote is called cleavage, and the cells produced by cleavage are called blastomeres. The solid sphere of cells produced by cleavage is a morula.
5. The morula develops into a blastocyst, a hollow ball of cells differentiated into a trophoblast and an embryoblast (inner cell mass).
6. The attachment of a blastocyst to the endometrium is termed implantation; it occurs as a result of enzymatic degradation of the endometrium.
7. The trophoblast develops into the syncytiotrophoblast and cytotrophoblast, both of which become part of the chorion.
8. The embryoblast differentiates into hypoblast and epiblast, the bilaminar (two-layered) embryonic disc.
9. The amnion is a thin protective membrane that develops from the cytotrophoblast.
10. The exocoelomic membrane and hypoblast form the yolk sac, which transfers nutrients to the embryo, forms blood cells, produces primordial germ cells, and forms part of the gut.
11. Erosion of sinusoids and endometrial glands provides blood and secretions, which enter lacunar networks to supply nutrition to and remove wastes from the embryo.
12. The extraembryonic coelom forms within extraembryonic mesoderm.
13. The extraembryonic mesoderm and trophoblast form the chorion, the principal embryonic part of the placenta.
14. The third week of development is characterized by gastrulation, the conversion of the bilaminar disc into a trilaminar (three-layered) embryonic disc consisting of ectoderm, mesoderm, and endoderm.
15. The first evidence of gastrulation is formation of the primitive streak and then the primitive node, notochordal process, and notochord.
16. The three primary germ layers form all tissues and organs of the developing organism. Table 4.1 on page 101 summarizes the structures that develop from the primary germ layers.
17. Also during the third week, the oropharyngeal and cloacal membranes form. The wall of the yolk sac forms a small vascularized outpouching called the allantois, which functions in blood formation and development of the urinary bladder.
18. The process by which the neural plate, neural folds, and neural tube form is called neurulation. The brain and spinal cord develop from the neural tube.
19. Paraxial mesoderm segments to form somites from which skeletal muscles of the neck, trunk, and limbs develop. Somites also form connective tissues and vertebrae.
20. Blood vessel formation, called angiogenesis, begins in mesodermal cells called angioblasts.
21. The heart develops from mesodermal cells called the cardiogenic area. By the end of the third week, the primitive heart beats and circulates blood.
22. Chorionic villi, projections of the chorion, connect to the embryonic heart so that maternal and fetal blood vessels are brought into close proximity so that nutrients and wastes can be exchanged between maternal and fetal blood.
23. Placentation refers to formation of the placenta, the site of exchange of nutrients and wastes between the mother and fetus. The placenta also functions as a protective barrier, stores nutrients, and produces several hormones to maintain pregnancy.
24. The actual connection between the placenta and embryo (and later the fetus) is the umbilical cord.
25. Organogenesis refers to the formation of body organs and systems and occurs during the fourth week of development.
26. Conversion of the flat, two-dimensional trilaminar embryonic disc to the three-dimensional cylinder occurs by a process called embryonic folding.
27. Embryonic folding brings various organs into their final adult positions and helps form the gastrointestinal tract.
28. Pharyngeal arches, clefts, and pouches give rise to the structures of the head and neck.
29. By the end of the fourth week, upper and lower limb buds develop and by the end of the eighth week the embryo has clearly human features.

Fetal Period (p. 108)

1. The fetal period is primarily concerned with the growth and differentiation of tissues and organs that developed during the embryonic period.
2. The rate of body growth is remarkable, especially during the ninth and sixteenth weeks.
3. The principal changes associated with embryonic and fetal growth are summarized in Table 4.2 on page 109.

Maternal Changes During Pregnancy (p. 111)

1. During pregnancy, several anatomical and physiological changes occur in the mother.
2. The uterus nearly fills the abdominal cavity toward the end of full-term pregnancy, pushing the viscera out of their normal position.
3. Physiological changes include weight gain, increased skin pigmentation in certain areas, and various alterations in the cardiovascular, respiratory, digestive, urinary, and reproductive systems.

Labor (p. 111)

1. Labor is the process by which the fetus is expelled from the uterus through the vagina to the outside.
2. True labor involves dilation of the cervix, expulsion of the fetus, and delivery of the placenta.

SELF-QUIZ QUESTIONS

Choose the one best answer to the following questions.

1. The embryonic period includes:
   a. the first and second trimesters.
   b. weeks 9–38 of development.
   c. the time between ovulation and fertilization.
   d. the first 8 weeks of development.
   e. the full 38 weeks of development.

2. Approximately how long after fertilization does implantation of an embryo usually occur?
   a. 3 weeks   b. about 6 days
   c. 1 day   d. about 3 days
   e. 30 minutes

3. Dizygotic (fraternal) twins result from:
   a. two secondary oocytes and two sperm.
   b. two secondary oocytes and one sperm.
   c. one secondary oocyte and two sperm.
   d. one secondary oocyte and one sperm.
   e. one secondary oocyte and three sperm.

4. A series of functional changes that cause a sperm's tail to beat more vigorously and prepare its plasma membrane to fuse with the oocyte's plasma membrane is called:
   a. fertilization.   b. implantation.
   c. capacitation.   d. syngamy.
   e. cleavage.

5. Chemicals from the notochord induce the development of:
   a. the heart.
   b. parts of the vertebral column and the neural plate.
   c. openings to the urinary and reproductive tracts.
   d. serous membranes.
   e. the maternal portion of the placenta.

6. Oxygen and nutrients from the maternal blood must pass through which of the following structures before entering the fetal blood?
   a. umbilical vein   b. umbilical artery
   c. decidua capsularis   d. intervillous spaces
   e. allantois

Complete the following.

7. Fusion of a sperm with a secondary oocyte is called _____.

8. The primary germ layers are the _____, the _____, and the _____.

9. The primary germ layers form by a process called _____, which is the major event in the third week of development.

10. _____, which is derived initially from maternal blood, serves as a shock absorber for the fetus and helps regulate fetal body temperature.

11. The process of blood vessel formation is called _____.

12. Following implantation, the endometrium of the uterus becomes known as the _____.

13. The trophoblast secretes a hormone called _____, which maintains the corpus luteum.

Are the following statements true or false? If the statement is False, *find and fix the mistake to make the statement* True.

14. Organ systems are almost completely developed by the end of the first trimester.

15. The chorionic villi form the maternal portion of the placenta.

16. Development of the placenta occurs during the third month of pregnancy.

17. The primary germ layers, which give rise to all tissues and organs, are present at the time of implantation.

18. The umbilical cord contains two umbilical arteries that carry oxygenated blood to the fetus, and one umbilical vein that carries fetal blood to the placenta.

19. The most common site for an ectopic pregnancy is the vagina.

Matching

20. Match the following (answers are used more than once):
    ____ (a) epithelial lining of all gastrointestinal, respiratory, and genitourinary tracts except near openings to the exterior of the body
    ____ (b) epidermis of skin, epithelial lining of entrances to the body (such as mouth, nose, and anus), hair, nails
    ____ (c) all of the skeletal system (bone, cartilage, and other connective tissues)
    ____ (d) muscle (skeletal, cardiac, and most smooth)
    ____ (e) blood and all blood and lymphatic vessels
    ____ (f) entire nervous system
    ____ (g) thyroid, parathyroid, thymus, and pancreas

    (1) develop(s) from ectoderm
    (2) develop(s) from endoderm
    (3) develop(s) from mesoderm

21. Match the following:
    ____ (a) induces the formation of the neural plate
    ____ (b) forms the heart
    ____ (c) develops into the brain and spinal cord
    ____ (d) the initial result of gastrulation; establishes head and tail ends of the embryo
    ____ (e) forms bones and ligaments of the limbs
    ____ (f) degenerates to form the openings of the anus and the urinary and reproductive tracts
    ____ (g) forms the meninges (coverings) of the brain and spinal cord

    (1) primitive streak
    (2) notochord
    (3) cloacal membrane
    (4) neural tube
    (5) neural crest
    (6) splanchnic mesoderm
    (7) somatic mesoderm

22. Match the following:
    ____ (a) the structure that implants into the uterine wall
    ____ (b) the result of the fusion of the haploid pronuclei
    ____ (c) the layer of the trophoblast that secretes enzymes that liquefy endometrial cells during implantation and eventually develops into the chorion
    ____ (d) the form of the female gamete that is fertilized by a sperm cell
    ____ (e) the portion of the blastocyst that eventually develops into the embryo
    ____ (f) with the decidua basalis, forms the placenta
    ____ (g) the amnion eventually develops from this structure
    ____ (h) the form of the female gamete that is the result of meiosis II following fertilization
    ____ (i) with the exocoelomic membrane, forms the wall of the yolk sac
    ____ (j) the solid sphere of blastomeres that results from cleavage of the zygote

    (1) secondary oocyte
    (2) ovum
    (3) zygote
    (4) morula
    (5) blastocyst
    (6) inner cell mass
    (7) syncytiotrophoblast
    (8) chorionic villi
    (9) hypoblast
    (10) epiblast

CRITICAL THINKING QUESTIONS

1. Fetal alcohol syndrome (FAS) is one of the most common causes of mental retardation and the most common preventable cause of birth defects in the United States. How is the alcohol that the woman drinks able to affect her developing fetus?

2. Some disorders of the nervous system may cause particular skin problems. For example, a person with a type of nervous system tumor may show coffee-colored spots on the skin. How are the structures of the nervous system and the skin related?

3. Josefina, in the last 2 weeks of her first pregnancy, anxiously called her doctor to ask if she should leave for the hospital. She was experiencing irregular "labor" pains that were thankfully eased by walking. She had no other signs to report. The doctor told Josefina to stay home—it wasn't time yet. Why did the doctor tell Josefina to stay home? How did he know it "wasn't time"?

4. Infection of the mother by certain viruses is known to result in severe birth defects in the child. If there is no mixing of maternal and fetal blood in the placenta, how can a viral infection in the mother cause problems in the child? At what point during the pregnancy do you think the greatest risk to the child would occur?

5. Larry's wife Elena is eight months pregnant. He has been fussing at her lately about the increase in their weekly grocery bill "even though there are still only two of us," and he has complained that they have been "going through toilet paper and antacids like crazy lately." When he wakes up in the morning, he discovers that Elena has stolen his extra pillow to prop herself up higher. What are some of the specific anatomical and physiological changes Elena has been experiencing that might explain these lifestyle changes?

? ANSWERS TO FIGURE QUESTIONS

4.1 The uterine tubes provide a route for sperm to reach a secondary oocyte and transport a fertilized (or unfertilized) oocyte to the uterus.

4.2 Capacitation is the series of changes in sperm after they have been deposited in the female reproductive tract that enable them to fertilize a secondary oocyte.

4.3 A morula is a solid ball of cells; a blastocyst consists of a rim of cells (trophoblast) surrounding a cavity (blastocyst cavity) and an embryoblast (inner cell mass).

4.4 The blastocyst secretes digestive enzymes that eat away the endometrial lining at the site of implantation.

4.5 Implantation usually occurs in the posterior portion of the fundus or body of the uterus.

4.6 The bilaminar embryonic disc is attached to the trophoblast by the connecting stalk.

4.7 Gastrulation converts a two-dimensional bilaminar embryonic disc into a two-dimensional trilaminar embryonic disc.

4.8 The notochord induces mesodermal cells to develop into parts of vertebrae and forms a portion of the intervertebral discs.

4.9 The neural tube forms the brain and spinal cord; somites develop into skeletal muscles, connective tissue, and the vertebrae.

4.10 Chorionic villi help to bring the fetal and maternal blood vessels into close proximity allowing for the more efficient movement of oxygen and nutrients from maternal into fetal blood and carbon dioxide and wastes from fetal into maternal blood.

4.11 The placenta exchanges materials between the fetus and the mother, protects the fetus against many microbes, and stores nutrients.

4.12 Because of embryonic folding, the embryo curves into a C-shape, various organs are brought into their eventual adult positions, and the primitive gut is formed.

4.13 Pharyngeal arches, clefts, and pouches give rise to structures of the head and neck.

4.14 During the last two-and-a-half months of development, fetal weight doubles.

4.15 Complete dilation of the cervix marks the onset of the stage of expulsion.

5 | THE INTEGUMENTARY SYSTEM

INTRODUCTION When you meet someone for the first time, before you exchange a single word or even shake hands, you form a first impression from the person's appearance. Aside from clothing, your first impression is based largely on the most visible part of the body, the skin (integument) and its associated structures. For example, you note the distribution, color, length, and condition of hair, which gives clues about health and age. You might also note freckles, moles, and other pigmentation differences or observe blemishes, pimples, and scars. In addition to these physical features, the skin might reveal other things below the surface. When people are cold, they develop goose bumps and their hairs stand on end. If they are hot, they sweat. The skin can also inform you about a person's history of recent and past sun exposure or reveal the red blush of embarrassment. Most of us are so conscious of the importance of first impressions that we expend considerable time, money, and effort in our daily grooming rituals to "put our best face forward."

Why are there different colors and textures of skin? What is a freckle, a blackhead, a blister, a callus? Why is the skin thicker in some areas of the body than in others? What are fingerprints? How is hair related to the skin and how does hair grow? Why is hair curly or straight, and why does it turn gray? These are just a few of the questions you will be able to answer as you learn about the structure and function of this important body system.

Recall from Chapter 1 that a system (organ system) consists of a group of organs working together to perform specific activities. The **integumentary** (in-teg-ū-MEN-tar-ē; *in*=inward; *tegere*=to cover; *inte-*=whole; *-gument*= body covering) **system** is composed of organs such as the skin, hair, oil and sweat glands, nails, and receptors. The integumentary system helps maintain a constant body temperature, protects the body, and provides sensory information about the surrounding environment. Of all the body's organs, none is more easily inspected or more exposed to infection, disease, and injury. Although the skin's location makes it vulnerable to damage from trauma, sunlight, microbes, and pollutants in the environment, it has protective features that ward off such damage. As noted previously, skin reflects our emotions (frowning, blushing) and some aspects of normal physiology (such as sweating). Changes in skin color may indicate homeostatic imbalances, such as the bluish skin color characteristic of hypoxia (oxygen deficiency at the tissue level). Abnormal skin eruptions or rashes such as chickenpox, cold sores, or measles may reveal systemic infections or diseases of internal organs. **Dermatology** (der′-ma-TOL-ō-jē; *dermato-*=skin; *-logy*=study of) is the medical specialty that deals with the diagnosis and treatment of integumentary system disorders. •

STRUCTURE OF THE SKIN

◉ OBJECTIVES

- Describe the layers of the epidermis and the cells that compose them.
- Compare the composition of the papillary and reticular regions of the dermis.
- Explain the basis for different skin colors.

The **skin**, **cutaneous membrane**, or **integument**, which covers the external surface of the body, is the largest organ of the body in both surface area and weight. In adults, the skin covers an area of about 2 square meters (22 square feet) and weighs 4.5–5 kg (10–11 lb), about 16% of total body weight. It ranges in thickness from 0.5 mm (.02 in.) on the eyelids to 4.0 mm (.16 in.) on the heels. However, over most of the body it is 1–2 mm (.04–.08 in.) thick. Structurally, the skin consists of two main parts (Figure 5.1 on pages 118–119). The superficial, thinner portion, which is composed of *epithelial tissue*, is the **epidermis** (ep-i-DERM-is; *epi-*=above; *dermis*=skin). The deeper, thicker, *connective tissue* portion is the **dermis**. Deep to the dermis, but not part of the skin, is the **subcutaneous** (*sub*=below; *cutis*=skin) **(subQ) layer**, or **hypodermis** (*hypo-*=below), which consists of areolar and adipose tissues.

Epidermis

The **epidermis** is composed of keratinized stratified squamous epithelium. It contains four principal types of cells: keratinocytes, melanocytes, Langerhans cells, and Merkel cells (Figure 5.2 on page 120). About 90% of epidermal cells are **keratinocytes** (ker-a-TIN-ō-sīts; *keratino-*=hornlike, after the tough nature of the horns on animals; *-cytes*=cells), which are arranged in four or five layers and produce the protein **keratin** (Figure 5.2a). Recall from Chapter 3 that keratin is a tough, fibrous intracellular protein that helps protect the skin and underlying tissues from heat, microbes, and chemicals. Keratinocytes also produce lamellar granules, which release a water-repellent sealant that decreases water entry and loss and inhibits the passage of foreign materials.

About 8% of the epidermal cells are **melanocytes** (MEL-a-nō-sīts; *melano-*=black), which develop from the neural crest of a developing embryo and produce the pigment melanin (Figure 5.2b). Melanocytes have a cuboidal cell body with long armlike processes that project between the neighboring keratinocytes. By means of these processes, one melanocyte can contact approximately 30 neighboring keratinocytes. Their long, slender projections transfer melanin granules to the neighboring keratinocytes. **Melanin** is a yellow-red or brown-black pigment that contributes to skin color and absorbs damaging ultraviolet (UV) light. Once inside keratinocytes, the melanin granules cluster to form a protective veil over the nucleus, on the side toward the skin surface. In this way, they shield the nuclear DNA from damage by UV light. Although their melanin granules effectively protect keratinocytes, melanocytes themselves are particularly susceptible to damage by UV light.

Langerhans cells (LANG-er-hans), or **intraepidermal macrophage cells**, arise from red bone marrow and migrate to the epidermis (Figure 5.2c), where they constitute a small fraction of the epidermal cells. Like melanocytes, Langerhans cells have long armlike processes and situate themselves among many surrounding keratinocytes. They participate in immune responses mounted against microbes that invade the skin. Langerhans cells and other cells of the immune system recognize a foreign microbe or substance so that it can be destroyed. Langerhans cells are easily damaged by UV light.

Figure 5.1 Components of the integumentary system. The skin consists of a superficial, thin epidermis and a deep, thicker dermis. Deep to the skin is the subcutaneous layer, which attaches the dermis to the fascia of the body.

The integumentary system includes organs such as the skin and hairs and its accessory structures—nails and skin glands—along with associated muscles and nerves.

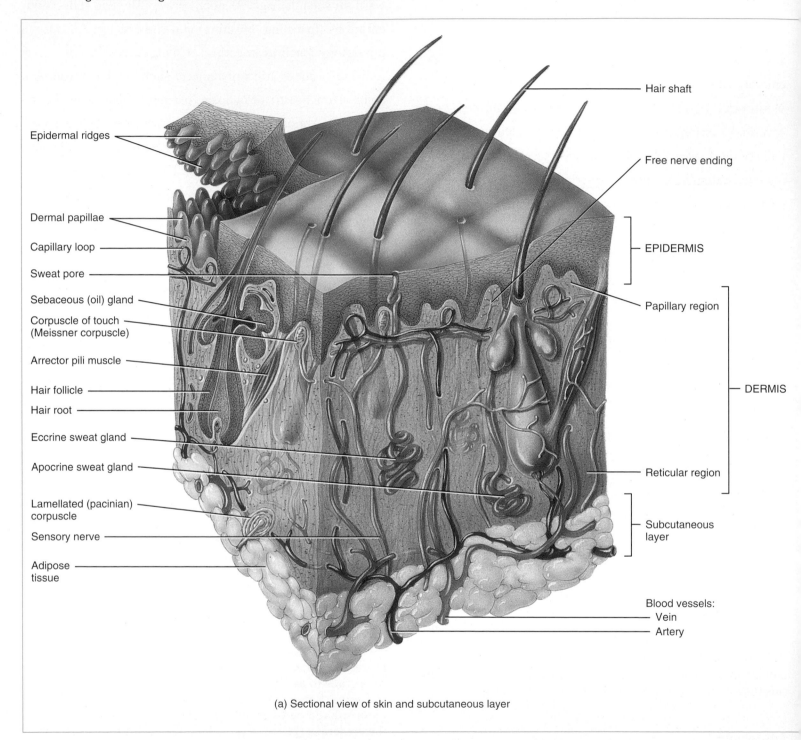

(a) Sectional view of skin and subcutaneous layer

Functions

1. Regulates body temperature.
2. Stores blood.
3. Protects body from external environment.
4. Detects cutaneous sensations.
5. Excretes and absorbs substances.
6. Synthesizes vitamin D.

EPIDERMIS

Papillary region

DERMIS

Reticular region

Sebaceous (oil) gland

Hair root

Hair follicle

LM 60x

(b) Sectional view of skin

Sweat pores

Epidermal ridges

(c) Epidermal ridges and sweat pores

 What types of tissues make up the epidermis and the dermis?

Figure 5.2 Types of cells in the epidermis. Besides keratinocytes, the epidermis contains melanocytes, which produce the pigment melanin; Langerhans cells, which participate in immune responses; and Merkel cells, which function in the sensation of touch.

Most of the epidermis consists of keratinocytes, which produce the protein keratin (which protects underlying tissues) and lamellar granules (which contain a waterproof sealant).

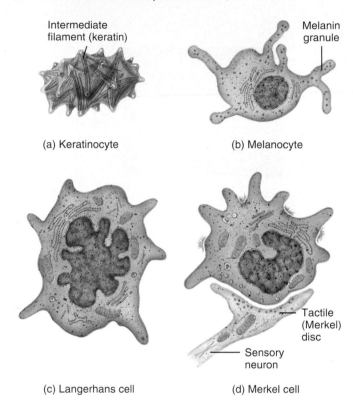

Intermediate filament (keratin)

Melanin granule

(a) Keratinocyte

(b) Melanocyte

Tactile (Merkel) disc

Sensory neuron

(c) Langerhans cell

(d) Merkel cell

? What is the function of melanin?

The **Merkel cells**, or **tactile epithelial cells**, are the least numerous of the epidermal cells. They are located in the deepest layer of the epidermis, where they contact the flattened process of a sensory neuron (nerve cell), a structure called a **tactile (Merkel) disc** (Figure 5.2d). Merkel cells and tactile discs detect different aspects of touch sensations.

Several distinct layers of keratinocytes in various stages of development form the epidermis (Figure 5.3). In most regions of the body the epidermis has four strata or layers—stratum basale, stratum spinosum, stratum granulosum, and a thin stratum corneum. This is called **thin skin**. Where exposure to friction is greatest, such as in the fingertips, palms, and soles, the epidermis has five layers—stratum basale, stratum spinosum, stratum granulosum, stratum lucidum, and a thick stratum corneum. This is called **thick skin**. The details of thin and thick skin are discussed later in the chapter.

Stratum Basale

The deepest layer of the epidermis, the **stratum basale** (ba-SA-lē; *basal-*=base), is composed of a single row of cuboidal or columnar keratinocytes, some of which are *stem cells* that undergo cell division to continually produce new keratinocytes. The nuclei of these keratinocytes are large, and their cytoplasm contains many ribosomes, a small Golgi complex, a few mitochondria, and some rough endoplasmic reticulum. The cytoskeleton within keratinocytes of the stratum basale includes scattered intermediate filaments called *tonofilaments*. Tonofilaments are composed of a protein that forms keratin in the more superficial epidermal layers. Tonofilaments attach to desmosomes, which bind cells of the stratum basale to each other, to the cells of the adjacent stratum spinosum, and to hemidesmosomes, which bind the keratinocytes to the basement membrane positioned between the epidermis and the dermis. The tough protein keratin, which is produced by the keratinocytes, also protects deeper layers from injury. Melanocytes and Merkel cells (with their associated tactile discs) are scattered among the keratinocytes of the basal layer. The stratum basale is sometimes referred to as the **stratum germinativum** (jer-mi-na-TĒ-vum; *germ-*=sprout) to indicate its role in forming new cells.

Stratum Spinosum

Superficial to the stratum basale is the **stratum spinosum** (spi-NŌ-sum; *spinos-*=thornlike). Other than an occasional Merkel cell, this stratum consists of numerous keratinocytes arranged in 8–10 layers. Cells in the more superficial layers become somewhat flattened. The keratinocytes in the stratum spinosum, which are produced by the stem cells in the basal layer, have the same organelles as cells of the stratum basale and some retain their ability to divide. Although they are rounded and larger in living tissue, cells of the stratum spinosum shrink and pull apart when prepared for microscopic examination so that they appear to be covered with thornlike spines (thus, the name) (Figure 5.3a). At each spinelike projection, bundles of tonofilaments insert into desmosomes, which tightly join the cells to one another. This arrangement provides both strength and flexibility to the skin. Langerhans cells and projections of melanocytes are also present in this layer.

Stratum Granulosum

At about the middle of the epidermis, the **stratum granulosum** (gran-ū-LŌ-sum; *granulos-*=little grains) consists of three to five layers of flattened keratinocytes that are undergoing apoptosis. (Recall from Chapter 2 that apoptosis is an orderly, genetically programmed cell death in which the nucleus fragments before the cells die.) The nuclei and other organelles of these cells begin to degenerate, and tonofilaments become more apparent because the integrity of these cells has become compromised as they have moved farther from their source of nutrition (the dermal blood vessels). A distinctive feature of cells in this layer is the presence of darkly staining granules of a protein called **keratohyalin** (ker′-a-tō-HĪ-a-lin), which converts the tonofilaments into keratin. Also present in the keratinocytes are membrane-enclosed **lamellar granules**, which release a lipid-rich secretion. This secretion fills the spaces between cells of the stratum granulosum, stratum lucidum, and stratum corneum. The lipid-rich secretion acts as a water-repellent sealant, retarding loss of body fluids and entry of foreign materials. As their nuclei break down during apoptosis, the keratinocytes of the stratum granulosum can no longer carry on vital metabolic reactions, and they die. Thus, the stratum granulosum marks the transition between the deeper, metabolically active strata and the dead cells of the more superficial strata.

Figure 5.3 Layers of the epidermis.

The epidermis consists of keratinized stratified squamous epithelium.

(a) Four principal cell types in epidermis

(b) Photomicrograph of a portion of the skin

LM 240x

? Which epidermal layer includes stem cells that continually undergo cell division?

Stratum Lucidum

The **stratum lucidum** (LOO-si-dum; *lucid-*=clear) is present in the thick skin of areas such as the fingertips, palms, and soles. It consists of three to five layers of clear, flat, dead keratinocytes that contain large amounts of keratin and thickened plasma membranes.

Stratum Corneum

The **stratum corneum** (COR-nē-um; *corne-*=horn or horny) consists on average of 25 to 30 layers of flattened dead keratinocytes, but can range in thickness from a few cells in thin skin to 50 or more cell layers in thick skin. The cells are extremely thin, flat, plasma-membrane enclosed packages of keratin and are called **corneocytes** or **squames** (*squame*=scale), which no longer contain a nucleus or any internal organelles. They are the final product of the differentiation process of the keratinocytes. The corneocytes within each layer overlap one another like the scales on the skin of a snake. Neighboring layers of corneocytes also form strong connections with one another. The plasma membranes of adjacent corneocytes are arranged in complex, wavy folds that fit together like pieces of a jigsaw puzzle to hold the layers together. In this outer stratum of the epidermis, often referred to as the *cornified layer*, cells are continuously shed and replaced by cells from the deeper strata. Its multiple layers of dead cells help to pro-

tect deeper layers from injury and microbial invasion. Constant exposure of skin to friction stimulates the formation of a **callus**, an abnormal thickening of the stratum corneum.

Keratinization and Growth of the Epidermis

Newly formed cells in the stratum basale are pushed slowly to the surface. As the cells move from one epidermal layer to the next, they accumulate more and more keratin in a process called **keratinization** (ker'-a-tin-i-ZĀ-shun). Then they undergo apoptosis. Eventually the keratinized cells slough off and are replaced by underlying cells that in turn become keratinized. This differentiation process accounts for the different characteristic features of the keratinocytes as they mature into corneocytes. The whole process by which cells form in the stratum basale, rise to the surface, become keratinized, and slough off takes about four weeks in an average epidermis of 0.1 mm (.04 in.) thickness. The rate of cell division in the stratum basale increases when the outer layers of the epidermis are stripped away, as occurs in abrasions and burns. The mechanisms that regulate this remarkable growth are not well understood, but hormonelike proteins such as **epidermal growth factor (EGF)** play a role. An excessive amount of keratinized cells shed from the skin of the scalp is called **dandruff**.

Table 5.1 presents a summary of the distinctive features of the epidermal strata.

TABLE 5.1	
Summary of Epidermal Strata	
STRATUM	**DESCRIPTION**
Stratum basale	Deepest layer, composed of a single row of cuboidal or columnar keratinocytes that contain scattered tonofilaments (intermediate filaments); stem cells undergo cell division to produce new keratinocytes; melanocytes, Langerhans cells, and Merkel cells associated with tactile discs are scattered among the keratinocytes.
Stratum spinosum	Eight to 10 rows of many-sided keratinocytes with bundles of tonofilaments; includes armlike processes of melanocytes and Langerhans cells.
Stratum granulosum	Three to five rows of flattened keratinocytes, in which organelles are beginning to degenerate; cells contain the protein keratohyalin, which converts tonofilaments into keratin, and lamellar granules, which release a lipid-rich, water-repellent secretion.
Stratum lucidum	Present only in skin of fingertips, palms, and soles; consists of three to five rows of clear, flat, dead keratinocytes with large amounts of keratin.
Stratum corneum	Twenty-five to 30 rows of dead, flat keratinocytes that contain mostly keratin.

■ CHECKPOINT

1. Which tissue forms the epidermis? Which tissue forms the dermis?
2. List the distinctive features of the epidermal layers from deepest to most superficial.

Dermis

The second, deeper part of the skin, the **dermis**, is composed of a strong connective tissue containing collagen and elastic fibers. This woven network of fibers has great tensile strength (resists pulling or stretching forces). The dermis also has the ability to stretch and recoil easily. Leather, which we use for belts, shoes, baseball gloves, and basketballs, is the dried and treated dermis of other animals. The few cells present in the dermis include predominantly fibroblasts, with some macrophages, and a few adipocytes near its boundary with the subcutaneous layer. Blood vessels, nerves, glands, and hair follicles (epithelial invaginations of the epidermis) are embedded in the dermal layer. The dermis is essential to the survival of the epidermis, and these adjacent layers form many important structural and functional relations. Based on its tissue structure, the dermis can be divided into a superficial papillary region and a deeper reticular region.

The **papillary region** makes up about one-fifth of the thickness of the total layer (see Figure 5.1). It consists of areolar connective tissue containing thin collagen fibers and fine elastic fibers. Its surface area is greatly increased by small, fingerlike structures that project into the undersurface of the epidermis called **dermal papillae** (pa-PIL-ē=nipples) that greatly increase its surface contact with the epidermis. The dermal papillae can vary greatly in size and number throughout different parts of the dermis. In the thin skin covering most of the body, the dermal papillae are relatively few in number, small, and irregularly scattered. By contrast, in the thick skin of the palmar surfaces of the hands and plantar surfaces of the feet, the papillae are relatively numerous, tall, and arranged in patterned rows. These nipple-shaped structures project into the epidermis and some contain **capillary loops** (blood capillaries). Other dermal papillae contain touch receptors, in addition to the capillaries, called **corpuscles of touch**, or **Meissner corpuscles**. These structures are sensitive nerve endings that detect touch. Also present in the dermal papillae are **free nerve endings**, dendrites that lack any apparent structural specialization. Different free nerve endings initiate signals that produce sensations of warmth, coolness, pain, tickling, and itching.

The **reticular region** (*reticul-*=netlike), which is attached to the subcutaneous layer, consists of dense irregular connective tissue containing fibroblasts, bundles of thick collagen, and some coarse elastic fibers (see Figure 5.1). The collagen fibers in the reticular region are arranged in a netlike manner and have a more regular arrangement than those in the papillary region. The more regular orientation of the large collagen fibers parallels the local tensile forces to help the skin resist stretching. A few adipose cells, hair follicles, nerves, sebaceous (oil) glands, and sudoriferous (sweat) glands occupy the spaces between fibers.

Both of the layers of the dermis contain dense horizontal networks of small blood vessels. These vascular networks arise from vessels in the underlying skeletal muscles that send branches into the subcutaneous layer and then to the dermis. These dermal vascular networks feed the extensive capillary loops in the dermal papillae.

The combination of collagen and elastic fibers in the reticular region provides the skin with strength, **extensibility** (ability to stretch), and **elasticity** (ability to return to original shape after stretching). The extensibility of skin can be readily seen around joints and in pregnancy and obesity. Because of the collagenous, vascular structure of the dermis, **striae** (STRĪ-ē=streaks) or **stretch marks**, a form of intrinsic scarring, can result from the internal damage to this layer that occurs when the skin is stretched too much. When the the skin is overstretched, the mechanical demands placed on this layer exceed the strength of the collagen bonding and the stretch available from both the elastic fibers and the creases in the skin. As a result, the lateral bonding of adjacent collagen fibers is disrupted and small dermal blood vessels rupture. Stretch marks intially appear as reddish streaks at these sites. Later, after a poorly vascularized scar tissue forms at these sites of dermal breakdown, the stretch marks remain visible as silvery white streaks. Stretch marks often occur in the abdominal skin during pregnancy, on the skin of weight-lifters where the skin is stretched by a rapid increase in muscle mass, and in the stretched skin accompanying gross obesity.

In certain regions of the body, collagen fibers tend to orient more in one direction than another. **Lines of cleavage (tension lines)** in the skin indicate the predominant direction of underlying collagen fibers (Figure 5.4a, b). The lines are especially evident on the palmar surfaces of the fingers, where they are arranged parallel to the long axis of the digits. Knowledge of lines of cleavage is especially important to plastic surgeons. For example, a surgical incision running parallel to the collagen fibers will heal with only a fine scar. A surgical incision made across the rows of fibers disrupts the collagen, and the wound tends to gape open and heal in a broad, thick scar.

The surfaces of the palms, fingers, soles, and toes are marked by series of ridges and grooves. They appear either as straight lines or as a pattern of loops and whorls, as on the tips of the digits. These **epidermal ridges** are produced during the third month of fetal development as the epidermis projects downward into the dermis between the dermal papillae of the papillary region (see Figure 5.1a, c). The epidermal ridges increase the surface area of the epidermis and thus function to increase the grip of the hand or foot by increasing friction. Because the ducts of sweat glands open on the tops of the epidermal ridges as sweat pores, the sweat and ridges form **fingerprints** (or **footprints**) when a smooth object is touched. The epidermal ridge pattern is genetically determined and is unique for each individual. Normally, the ridge pattern does not change during life, except to enlarge, and thus can serve as the basis for identification through fingerprints or footprints. The study of the pattern of epidermal ridges that is concerned with the identification and classification of fingerprints is referred to as **dermatoglyphics** (der-ma-tō-GLIF-iks; *glyphe*=carved work).

In addition to forming epidermal ridges, the complex papillary surface of the dermis has other functional properties. The dermal papillae greatly increase the contact area between the dermis and epidermis. This increased dermal contact area, with its extensive network of small blood vessels, serves as an important source of

TABLE 5.2

Summary of Papillary and Reticular Regions of the Dermis

REGION	DESCRIPTION
Papillary region	The superficial portion of the dermis (about one-fifth); consists of areolar connective tissue with thin collagen fibers and fine elastic fibers; contains dermal papillae that house capillaries, corpuscles of touch, and free nerve endings.
Reticular region	The deeper portion of the dermis (about four-fifths); consists of dense irregular connective tissue with bundles of thick collagen and some coarse elastic fibers. Spaces between fibers contain some adipose cells, hair follicles, nerves, sebaceous glands, and sudoriferous glands.

nutrition for the overlying epidermis. Molecules diffuse from the small blood capillaries in the dermal papilla to the cells of the stratum basale, allowing the basal epithelial stem cells to divide and the keratinocytes to grow and develop. As keratinocytes push toward the surface and away from the dermal blood source, they are no longer able to obtain the nutrition they require, leading to the eventual breakdown of their organelles.

The dermal papillae fit together with the complementary epidermal ridge to form an extremely strong junction between the two layers. This jigsaw puzzle-like connection strengthens the skin against forces that attempt to separate the epidermis from the dermis.

Table 5.2 summarizes the structural features of the papillary and reticular regions of the dermis.

CHECKPOINT

3. Compare the structure and functions of the epidermis and dermis.
4. Compare the composition of the papillary and reticular regions of the dermis.
5. How are epidermal ridges formed?

The Structural Basis of Skin Color

Melanin, hemoglobin, and carotene are three pigments that impart a wide variety of colors to skin. The amount of **melanin** causes the skin's color to vary from pale yellow to reddish-brown to black. The difference between the two forms of melanin, *pheomelanin* (yellow to red) and *eumelanin* (brown to black), is most apparent in the hair. Melanocytes, the melanin-producing cells, are most plentiful in the epidermis of the penis, nipples of the breasts, area just around the nipples (areolae), face, and limbs. They are also present in mucous membranes. Because the *number* of melanocytes is about the same in all people, differences in skin color are due mainly to the *amount of pigment* the melanocytes produce and transfer to keratinocytes. In some people, melanin accumulates in patches called *freckles*. As a person gets older, *age (liver) spots* may develop. These flat blemishes look like freckles and range in color from light brown to black. Like freckles, age spots are accumulations of melanin. A round, flat, or raised area that usually develops in childhood or adolescence and represents a benign localized overgrowth of melanocytes is called a **nevus** (NĒ-vus), or a **mole**.

Figure 5.4 Lines of cleavage.

Lines of cleavage in the skin indicate the predominant direction of underlying collagen fibers in the reticular region.

(a) Anterior view   (b) Posterior view

? Why are lines of cleavage clinically important?

Melanocytes synthesize melanin from the amino acid *tyrosine* in the presence of an enzyme called *tyrosinase*. Synthesis occurs in an organelle called a **melanosome**. Exposure to ultraviolet (UV) light increases the enzymatic activity within melanosomes and thus increases melanin production. Both the amount and darkness of melanin increase upon UV exposure, which gives the skin a tanned appearance and helps protect the body against further UV radiation. A tan is lost when the melanin-containing keratinocytes are shed from the stratum corneum. Melanin absorbs UV radiation, prevents damage to DNA in epidermal cells, and neutralizes free radicals that form in the skin following damage by UV radiation. Thus, within limits, melanin serves a protective function. As you will see later, however, repeatedly exposing the skin to UV light may cause skin cancer.

Dark-skinned individuals have large amounts of melanin in the epidermis. Consequently, the epidermis has a dark pigmentation and skin color ranges from yellow to reddish-brown to black. Light-skinned individuals have little melanin in the epidermis. Thus, the epidermis appears translucent and skin color ranges from pink to red depending on the oxygen content of the blood moving through capillaries in the dermis. The red color is due to **hemoglobin**, the oxygen-carrying pigment in red blood cells.

Carotene (KAR-ō-tēn; *carot*=carrot) is a yellow-orange pigment that gives egg yolk and carrots their color. This precursor of vitamin A, which is used to synthesize pigments needed for vision, is stored in the stratum corneum and fatty areas of the dermis and subcutaneous layer in response to excessive dietary intake. In fact, so much carotene may be deposited in the skin after eating large amounts of carotene-rich foods that the skin actually turns orange, which is especially apparent in light-skinned individuals. Decreasing carotene intake alleviates the problem.

Albinism (AL-bin-izm; *albin-*=white) is the inherited inability of an individual to produce melanin. Most **albinos** (al-BĪ-nōs), people affected by albinism, have melanocytes that are unable to synthesize tyrosinase. Melanin is missing from their hair, eyes, and skin.

In another condition, called **vitiligo** (vit-i-LĪ-gō), the partial or complete loss of melanocytes from patches of skin produces irregular white spots. The loss of melanocytes may be related to an immune system malfunction in which antibodies attack the melanocytes.

Tattooing and Body Piercing

Tattooing is a permanent coloration of the skin in which a foreign pigment is deposited with a needle into the dermis. It is believed that the practice originated in Ancient Egypt between 4000 and 2000 B.C. Today, tattooing is performed in one form or another by nearly all peoples of the world, and it is estimated that about one in five U.S. college students has one or more. Tattoos are created by injecting ink with a needle that punctures the epidermis and moves between 50 to 3000 times per minute and deposits the ink in the dermis. Since the dermis is stable (unlike the epidermis which is shed about every four weeks), tattoos are permanent. However, they can fade over time due to exposure to sunlight, improper healing, picking scabs, and flushing away ink particles by the lymphatic system. Tattoos can be removed by lasers that use concentrated beams of light. In the procedure, which requires a series of treatments, the tattoo inks and pigments selectively absorb the high-intensity laser light without destroying normal surrounding skin tissue. The laser causes the tattoo to dissolve into small ink particles that are eventually removed by the immune system. Laser removal of tattoos involves a considerable investment in time and money and can be quite painful.

Body piercing, the insertion of jewelry through an artificial opening, is also an ancient practice employed by Egyptian pharaohs and Roman soldiers, and a current tradition among many Americans. Today it is estimated that about one in three U.S. college students has had a body piercing. For most piercing locations, the practitioner cleans the skin with an antiseptic, retracts the skin with forceps, and pushes a needle through the skin. Then the jewelry is connected to the needle and pushed through the skin. Total healing can take up to a year. Among the sites that are pierced are the ears, nose, eyebrows, lips, tongue, nipples, navel, and genitals. Potential complications of body piercing are infections, allergic reactions, and anatomical damage (such as nerve damage or cartilage deformation). In addition, body-piercing jewelry may interfere with certain medical procedures such as masks used for resuscitation, airway management procedures, urinary catheterization, radiographs, and delivery of a baby.

◉ CHECKPOINT

6. What are the three pigments in the skin and how do they contribute to skin color?
7. How do albinism and vitiligo differ?

Subcutaneous Layer or Hypodermis

The **subcutaneous** (*under the skin*) **layer**, which is deep to the dermis, is also referred to as the **hypodermis** (*hypo*=below) (see Figure 5.1). This connective tissue layer, which is not part of the skin, differs from region to region in the body. In some areas it is a thin layer comprised of the loose connective tissue called *areolar tissue*, while in other regions it is a thick, tough layer of fibrous bands of collagen accompanied by adipose tissue. Fibers that extend from the dermis anchor the skin to the subcutaneous layer, which in turn attaches to the underlying *fascia*, connective tissue that surrounds muscles and bones. The subcutaneous layer serves as a storage depot for fat and contains large blood vessels that supply and drain the capillaries of the skin. The amount of fat deposited in subcutaneous regions of adipose tissue varies greatly among different individuals, ranging from a very thin layer with minimal fat deposits in a lean person to a layer four to six inches thick in an obese individual. This region (and sometimes the dermis) also contains nerve endings called **lamellated (pacinian) corpuscles** (pa-SIN-ē-an) that are sensitive to pressure (see Figure 5.1).

The subcutaneous layer has multiple functions. It functions as a loose binding tissue that unites the upper layers of the skin to deeper structures, while at the same time allowing the skin to move freely over these deeper structures. In certain regions, such as the skin of the soles and palms, the subcutaneous layer forms tough fat pads, composed of fibrous bands of collagen and adipose tissue, which absorb shock and protect underlying muscle and bone. As noted previously, the subcutaneous layer is also the principal site of energy storage in the body. Adipose tissue is an active metabolic tissue with numerous nerve endings and rich vascular networks that help regulate and mobilize the energy stores. Finally, the subcutaneous layer serves as a layer of insulation that helps retard heat loss from the body.

◉ CHECKPOINT

8. How does the subcutaneous layer vary from one body region to another?
9. State three functions of the subcutaneous layer.

• CLINICAL **CONNECTION** Skin Cancer

Excessive exposure to the sun causes virtually all of the 1 million cases of **skin cancer** diagnosed in the United States annually. There are three common forms of skin cancer. **Basal cell carcinomas** account for about 78% of all skin cancers. The tumors arise from cells in the stratum basale of the epidermis and rarely metastasize. **Squamous cell carcinomas**, which account for about 20% of all skin cancers, arise from squamous cells of the epidermis, and they have a variable tendency to metastasize. Most arise from preexisting lesions of damaged tissue in sun-exposed skin. Basal and squamous cell carcinomas are together known as *non-melanoma skin cancer* and are 50% more common in males than in females. **Malignant melanomas** arise from melanocytes and account for about 2% of all skin cancers. The American Academy of Dermatology estimates that the lifetime risk of developing melanoma is currently 1 in 75, double the risk only 15 years ago. In part, this increase is due to depletion of the ozone layer, which absorbs some UV light high in the atmosphere. But the main reason for the increase is that more people are spending more time in the sun. Malignant melanomas metastasize rapidly and can kill a person within months of diagnosis.

The key to successful treatment of malignant melanoma is early detection. The early warning signs of malignant melanoma are identified by the acronym ABCD (Figures A, B). A is for asymmetry; malignant melanomas tend to lack symmetry. B is for border; malignant melanomas have notched, indented, scalloped, or indistinct borders. C is for color; malignant melanomas have uneven coloration and may contain several colors. D is for diameter; ordinary moles tend to be smaller than 6 mm (0.25 in.), about the size of a pencil eraser. Once a malignant melanoma has the characteristics of A, B, and C, it is usually larger than 6 mm.

(A) Normal nevus (mole)   (B) Malignant melanoma

Among the risk factors for skin cancer are the following:

1. **Skin type.** Individuals with light-colored skin who never tan but always burn are at high risk.
2. **Sun exposure.** People who live in areas with many days of sunlight per year and at high altitudes (where ultraviolet light is more intense) have a higher risk of developing skin cancer. Likewise, people who engage in outdoor occupations and those who have suffered three or more severe sunburns have a higher risk.
3. **Family history.** Skin cancer rates are higher in some families than in others.
4. **Age.** Older people are more prone to skin cancer owing to longer total exposure to sunlight.
5. **Immunological status.** Individuals who are immunosuppressed have a higher incidence of skin cancer. •

ACCESSORY STRUCTURES OF THE SKIN

◉ OBJECTIVE

• Compare the structure, distribution, and functions of hair, skin glands, and nails.

Accessory structures of the skin—hair, skin glands, and nails—develop from the embryonic epidermis. Among their important functions are protection of the body by hair and nails, and regulation of body temperature by sweat glands.

Hair

The body has a dense covering of **hairs**, or *pili* (PI-lē), not nearly as numerous as the hairs on the skin of some other mammals. The distribution of hairs varies from approximately 600 per cm^2 (3,900 per $in.^2$) on facial skin to 60 per cm^2 (390 per $in.^2$) on the skin over the rest of the body. When you look at a square centimeter of your skin, it is difficult to notice 600 or even 60 hairs on this small patch of skin, because hairs can range in length from a fraction of a millimeter (.04 in.) to over a meter (1.09 yards), with widths from 0.005 mm (the diameter of some of the smallest cells in the body) to 0.6 mm. So, while some hairs are very obvious to the unaided eye, others are only visible with the use of a microscope.

Hairs are absent from the palms, palmar surfaces of the fingers, soles, and plantar surfaces of the toes. In adults, hair is usually most heavily distributed across the scalp, over the brows of the eyes, and around the external genitalia. Genetic and hormonal influences largely determine the thickness and pattern of hair distribution.

Although the protection it offers is limited, hair on the head guards the scalp from injury and the sun's rays. It also decreases heat loss from the scalp. Eyebrows and eyelashes protect the eyes from foreign particles, as does hair in the nostrils and in the external ear canal. Touch receptors associated with hair follicles (hair root plexuses) are activated by the slightest movement of a hair.

Anatomy of a Hair

Each hair is composed of columns of dead, keratinized epidermal cells bonded together by extracellular proteins. The **shaft** is the superficial portion of the hair, most of which projects from the surface of the skin (Figure 5.5a). The transverse-sectional shape of the shafts of hairs varies with different ethnic groups. Straight hair has round shafts; wavy hair has oval shafts; and curly hair has kidney-shaped shafts.

The **root** is the portion of the hair deep to the shaft that penetrates into the dermis, and sometimes into the subcutaneous layer. The shaft and root both consist of three concentric layers (Figure 5.5c, d). The inner *medulla* is composed of two or three rows of irregularly shaped cells containing pigment granules and air spaces. The middle *cortex* forms the major part of the shaft and consists of elongated cells that contain pigment granules in dark hair but mostly air in gray or white hair. The *cuticle* of the hair, the outermost layer, consists of a single layer of thin, flat cells that are the most heavily keratinized. Cuticle cells are arranged like shingles on the side of a house, with their free edges pointing toward the free end of the hair (Figure 5.5b).

Figure 5.5 Hair.

Hairs are growths of epidermis composed of dead, keratinized epithelial cells.

Hair shaft

Epidermal cells

SEM 70x

(b) Several hair shafts showing the shinglelike cuticle cells

Hair shaft

Hair root

Sebaceous gland

Arrector pili muscle

Hair root plexus

Eccrine sweat gland

Bulb

Papilla of the hair

Apocrine sweat gland

Blood vessels

(a) Hair and surrounding structures

Hair root:
Medulla
Cortex
Cuticle of the hair

Hair follicle:
Internal root sheath
External root sheath
} Epithelial root sheath

Dermal root sheath

Hair matrix

Melanocyte

Papilla of the hair

Blood vessels

Bulb

(c) Frontal section of hair root

Hair root:
Cuticle of the hair
Cortex
Medulla

Hair follicle:
Internal root sheath
External root sheath
Dermal root sheath

(d) Transverse section of hair root

CONTINUES

Surrounding the root of the hair is the **hair follicle**, which is made up of an external root sheath and an internal root sheath, together referred to as an **epithelial root sheath** (Figure 5.5c, d). The *external root sheath* is a downward continuation of the epidermis. Near the surface of the skin, it contains all the epidermal layers. At the base of the hair follicle, the external root sheath contains only the stratum basale. The *internal root sheath* is pro-duced by the hair matrix (described shortly) and forms a cellular tubular sheath of epithelium between the external root sheath and the hair. The dense dermis surrounding the hair follicle is called the **dermal root sheath**.

The base of each hair follicle and its surrounding dermal root sheath is an onion-shaped structure, the **bulb** (Figure 5.5c). This structure houses a nipple-shaped indentation, the **papilla of the**

FIGURE 5.5 CONTINUED ▶

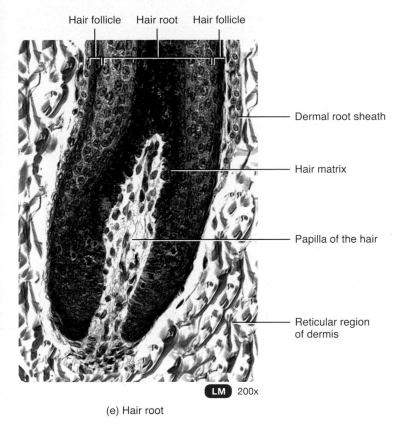

Hair follicle   Hair root   Hair follicle

Dermal root sheath

Hair matrix

Papilla of the hair

Reticular region of dermis

LM 200x

(e) Hair root

? Why does it hurt when you pluck a hair but not when you have a haircut?

hair, which contains areolar connective tissue and many blood vessels that nourish the growing hair follicle. The bulb also contains a germinal layer of cells called the **hair matrix**. Hair matrix cells arise from the stratum basale, the site of cell division. Hence, hair matrix cells are responsible for the growth of existing hairs, and they produce new hairs when old hairs are shed. This replacement process occurs within the same follicle. Hair matrix cells also give rise to the cells of the internal root sheath.

Sebaceous (oil) glands (discussed shortly) and a bundle of smooth muscle cells are also associated with hairs (Figure 5.5a). The smooth muscle is called **arrector pili** (a-REK-tor PI-le; *arrect-*=to raise). It extends from the superficial dermis of the skin to the connective tissue sheath around the hair follicle. In its normal position, hair emerges at an angle to the surface of the skin. Under physiological or emotional stress, such as cold or fright, autonomic nerve endings stimulate the arrector pili muscles to contract, which pulls the hair shafts perpendicular to the skin surface. This action causes "goosebumps" or "gooseflesh" because the skin around the shaft forms slight elevations. Why do we activate the arrector pili muscles and erect the body's hairs? As mentioned previously, the arrector pili muscles contract when a person is cold or experiences emotions associated with fear or anger. What is it about cold or fear that induces the contraction of these muscles? It is difficult to see the significance of this function in humans, because of our reduced hair coverage. However, in animals with a lot of hair, this mechanism can be very important for survival. By erecting the hairs on their bodies, most mammals can trap dead (unmoving) air between the hairs. Dead air

space (as used in the double-pane insulated glass windows that trap a thin layer of dead air space), the most effective form of insulation known, helps reduce heat loss from the body when it is exposed to cold. Hair erection can also make mammals look bigger and more formidable, which might help them intimidate predators. Humans still exhibit hair erection, but it has lost its effectiveness in reducing heat loss or intimidating enemies because it has become so sparse.

Surrounding each hair follicle are dendrites of neurons, called **hair root plexuses**, that are sensitive to touch (Figure 5.5a). The hair root plexuses generate nerve impulses if the hair shaft is moved, as when an insect bumps into it when crawling across your skin.

Each hair follicle goes through a growth cycle, which consists of a growth stage, a regression stage, and a resting stage. During the **growth (anagen) stage**, cells of the hair matrix divide. As new cells from the hair matrix are added to the base of the hair root, existing cells of the hair root are pushed upward and the hair grows longer. While the cells of the hair are being pushed upward, they become keratinized and die. Following the growth stage is the **regression (catagen) stage**, when cells of the hair matrix stop dividing, the hair follicle atrophies (shrinks), and the hair stops growing. After the regression stage, the hair follicle enters a **resting (telogen) stage**. Following the resting stage, a new growth cycle begins. The old hair root falls out or is pushed out of the hair follicle, and a new hair begins to grow in its place. Scalp hair is in the growth stage for 2 to 6 years, the regression stage for 2 to 3 weeks, and the resting stage for about 3 months. At any time, about 85% of scalp hairs are in the growth stage. Visible hair is dead, but until the hair is pushed out of its follicle by a new hair, portions of its root within the scalp are alive.

Hair Growth

Normal hair loss in an adult scalp is about 70–100 hairs per day. Both the rate of growth and the replacement cycle can be altered by illness, diet, high fever, surgery, blood loss, severe emotional stress, and gender. Rapid weight-loss diets that severely restrict calories or protein increase hair loss. An increase in the rate of hair shedding can also occur with certain drugs, after radiation therapy for cancer, and for 3–4 months after childbirth. **Alopecia** (al-o-PE-she-a), the partial or complete lack of hair, may result from genetic factors, aging, endocrine disorders, chemotherapy for cancer, or skin disease.

Types of Hairs

Hair follicles develop at about twelve weeks after fertilization as downgrowths of the stratum basale of the epidermis into the dermis (see Figure 5.8e). Usually by the fifth month of development, the follicles produce very fine, nonpigmented hairs called **lanugo** (la-NOO-gō=wool or down) that cover the body of the fetus. This hair is shed before birth, except in the scalp, eyebrows, and eyelashes. A few months after birth, slightly thicker hairs replace these downy hairs. Over the remainder of the body of an infant, a new growth of short, fine hair occurs. These hairs are known as **vellus hairs** (VEL-us=fleece), commonly called "peach fuzz." In response to hormones (androgens) secreted at puberty, coarse pigmented and frequently curly hair develops in the axillae (armpits) and pubic region. In males, these hairs also appear on the face and other parts of the body. The hairs that develop at puberty, together with those of the head, eyebrows, and eyelashes, are called **terminal hairs**. About 95% of body hair on males is terminal hair (5% vellus hair), but only about 35% of body hair on females is terminal (65% vellus hair).

Hair Color

Like the surface epidermis, hair color is the result of melanocyte activity in the stratum basale at the base of the hair follicle. During the active phase of the growth cycle, large melanocytes in the central portion of the bulb distribute melanosomes to the keratinocytes of the medulla and cortex of the developing hair (Figure 5.5c). These melanocytes form melanosomes containing either the black-brown or the red-yellow form of the melanin pigment. The resulting hair color arises from the ratio of the different pigment-containing melanosomes produced. With age, the melanocyte number and activity level at the base of the follicle declines, which leads to a decrease in the pigmentation of the hair. In addition to this decrease in the natural melanin pigmentation, the number of air spaces within the medulla of the hair increase. The air spaces change the way the hair reflects light, resulting in metallic (gray) and white reflections of light.

Skin Glands

Recall from Chapter 3 that glands are single or groups of epithelial cells that secrete a substance. Several kinds of exocrine glands are associated with the skin: sebaceous (oil) glands, sudoriferous (sweat) glands, ceruminous glands, and mammary glands. Mammary glands, which are specialized sudoriferous glands that secrete milk, are discussed in Chapter 27 along with the female reproductive system.

Sebaceous Glands

Sebaceous glands (se-BĀ-shus; *sebace-*=greasy), or **oil glands**, are simple, branched acinar glands. With few exceptions, they are connected to hair follicles (Figures 5.6a and see Figures 5.1 and 5.5a). The secreting portion of a sebaceous gland lies in the dermis, typically situated in the angle the arrector pili muscle forms with the outer wall of a hair follicle, and usually opens into the neck of the hair follicle. In other locations, such as the lips, glans penis, and labia minora, sebaceous glands open directly onto the surface of the skin. Sebaceous glands, which vary in size and shape, are found in the skin over all regions of the body except the palms and soles. Because of their relationship to hairs, sebaceous glands are most numerous where hairs are most numerous. They are small in most areas of the trunk and limbs, but large in the skin of the breasts, face, neck, and upper chest.

The sac-like base of a sebaceous gland has a lining of cuboidal cells that resemble basal epithelial cells. Like the basal epithelial cells, these cells produce cells that are pushed away from the lining as newer generations of cells are produced. The lumen of the base gradually fills with cells as multiple generations of cells pile on top of each other from the base up. These cells differentiate by developing large lipid-filled vesicles in their cytoplasm. The cells eventually become so distended that they rupture and fill the duct of the gland with an oily secretion called **sebum** (SĒ-bum; *sebum*=oily substance), which is derived from the released

Figure 5.6 Histology of skin glands.

🔑 Sebaceous glands are simple branched acinar glands; eccrine and apocrine sweat glands are simple coiled tubular glands.

(a) Sebaceous gland

Collagen fibers in dermis
Hair
Hair follicle
Sebaceous gland
Arrector pili muscle

LM 50x

(b) Eccrine sweat gland

Duct
Secretory portion

LM 200x

Location of skin glands

Sebaceous gland
Eccrine sweat gland
Apocrine sweat gland

(c) Apocrine sweat gland

Collagen fibers in dermis
Duct
Secretory portion
Adipose cell

LM 100x

❓ **What is the main function of eccrine sweat glands?**

lipid-filled vesicles along with the cellular debris. Sebum is a mixture of fats, cholesterol, proteins, and inorganic salts. Sebum coats the surface of hairs and helps keep them from drying and becoming brittle. Sebum also prevents excessive evaporation of water from the skin, keeps the skin soft and pliable, and inhibits the growth of certain bacteria. During childhood, sebaceous glands are relatively small and inactive. At puberty, androgens from the testes, ovaries, and adrenal glands stimulate sebaceous glands to grow in size and increase their production of sebum. **Acne** is an inflammation of sebaceous glands that usually begins at puberty when the sebaceous glands are stimulated by androgens.

Sudoriferous Glands

There are three to four million **sweat glands**, or s**udoriferous glands** (sū-dor-IF-er-us; *sudor-*=sweat; *-ferous*=bearing). The cells of sweat glands release their secretions by exocytosis and empty them onto the skin surface through pores or into hair follicles. Depending on their structure, location, and type of secretion, the glands are classified as either eccrine or apocrine.

Eccrine sweat glands (EK-krin; *eccrine*=secreting outwardly), also known as **merocrine sweat glands**, are simple, coiled tubular glands (Figure 5.6b) and are much more common than apocrine sweat glands. They are distributed throughout the skin of most parts of the body, except for the margins of the lips, nail beds of the fingers and toes, glans penis, glans clitoris, labia minora, and eardrums. Eccrine sweat glands are most numerous in the skin of the forehead, palms, and soles; their density can be as high as 450 per square centimeter (3000 per square inch) in the palms. The secretory portion of eccrine sweat glands is located mostly in the deep dermis (sometimes in the upper subcutaneous layer). The excretory duct projects through the dermis and epidermis and ends as a pore at the surface of the epidermis (see also Figure 5.1 and 5.4c). On the palms and soles, they open all along the apex of the epidermal ridges. This fact makes fingerprint detection possible.

The sweat produced by eccrine sweat glands (about 600 mL per day) consists of water, ions (mostly Na^+ and Cl^-), urea, uric acid, ammonia, amino acids, glucose, and lactic acid. The main function of eccrine gland sweat is to help regulate body temperature through evaporation. As sweat evaporates, large quantities of heat energy leave the body surface. This response, which helps regulate body temperature, is called *thermoregulatory sweating*. Sweat, or perspiration, usually occurs first on the forehead and scalp, extends to the face and the rest of the body, and occurs last on the palms and soles. Eccrine sweat also plays a small role in

eliminating wastes such as urea, uric acid, and ammonia. Eccrine sweat glands also release sweat in response to an emotional stress such as fear or embarrassment. This type of sweating is referred to as **emotional sweating** or a *cold sweat*. In contrast to thermoregulatory sweating, emotional sweating first occurs on the palms, soles, and axillae and then spreads to other areas of the body. Sweat that evaporates from the skin before it is perceived as moisture is referred to as **insensible perspiration** (*in*=not). Sweat that is excreted in larger amounts and is sensed as moisture on the skin is called **sensible perspiration**.

Apocrine sweat glands (AP-ō-krin; *apo*=separated from) are also simple, coiled tubular glands. These large glands extend not only into the lower portion of the dermis but also into the subcutaneous layer. Of the three basic skin glands, apocrine sweat glands are the least numerous. They are found mainly in the skin of the axilla (armpit), groin, areolae (pigmented areas around the nipples) of the breasts, and bearded regions of the face in adult males (Figure 5.6c). These glands were once thought to release their secretions in an apocrine manner—by pinching off a portion of the cell. We now know, however, that their secretion is released by exocytosis, which is characteristic of the release of secretions by merocrine glands. Nevertheless, the term *apocrine* is still used. The secretory portion of these sweat glands is located mostly in the subcutaneous layer, and the excretory duct opens into hair follicles (see also Figure 5.1). The secretory portion of the gland has a simple cuboidal lining of secretory cells whose free surfaces project into the lumen of the gland. Surrounding the secretory portion of the gland is a vascular connective tissue containing numerous nerve endings. Surrounding the upper portions of the coils and the duct of the gland is a thin coat of **myoepitheliocytes** (*myo*=muscle). These are muscle-like contractile epithelial cells that help squeeze out the secretions from the gland into the hair follicle. Their secretory product is slightly viscous compared to eccrine secretions and contains the same components as eccrine sweat plus lipids and proteins. At first this substance is odorless, but as it spreads onto the hairs, bacteria decompose the proteins, creating a strong musky odor that is called body odor. In women, cells of apocrine sweat glands enlarge about the time of ovulation and shrink during menstruation. Eccrine sweat glands start to function soon after birth, but apocrine sweat glands do not begin to function until puberty. Apocrine sweat glands are stimulated during emotional stress and sexual excitement; these secretions are commonly known as a "cold sweat." In contrast to eccrine sweat glands, apocrine sweat glands are not active during thermoregulatory sweating.

Table 5.3 presents a comparison of eccrine and apocrine sweat glands.

TABLE 5.3		
Comparison of Eccrine and Apocrine Sweat Glands		
FEATURE	**ECCRINE SWEAT GLANDS**	**APOCRINE SWEAT GLANDS**
Distribution	Throughout skin of most regions of the body, especially in skin of forehead, palms, and soles.	Skin of the axilla, groin, areolae, bearded regions of the face, clitoris, and labia minora.
Location of secretory portion	Mostly in deep dermis.	Mostly in subcutaneous layer.
Termination of excretory duct	Surface of epidermis.	Hair follicle.
Secretion	Less viscous; consists of water, ions (Na^+, Cl^-), urea, uric acid, ammonia, amino acids, glucose, and lactic acid.	More viscous; consists of the same components as eccrine sweat glands plus lipids and proteins.
Functions	Regulation of body temperature, waste removal, and stimulated during emotional stress.	Stimulated during emotional stress and sexual excitement.
Onset of function	Soon after birth.	Puberty.

Ceruminous Glands

Modified sweat glands in the external ear, called **ceruminous glands** (se-ROO-mi-nus; *cer-*=wax), produce a waxy lubricating secretion. The secretory portions of ceruminous glands lie in the subcutaneous layer, deep to sebaceous glands. Their excretory ducts open either directly onto the surface of the external auditory canal (ear canal) or into ducts of sebaceous glands. The combined secretion of the ceruminous and sebaceous glands is called **cerumen**, or earwax. Cerumen in the external auditory canal provides a sticky barrier that prevents the entrance of foreign bodies, such as insects. Cerumen also waterproofs the canal and keeps bacteria and fungi from entering cells.

Nails

Nails are plates of tightly packed, hard, dead keratinized epidermal cells. The cells form a clear, solid covering over the dorsal surfaces of the distal portions of the digits. Each nail consists of a nail body, a free edge, and a nail root (Figure 5.7). The **nail body (plate)** is the visible portion of the nail. It is comparable to the stratum corneum of the general epidermis, with the exception that its flattened, keratinized cells fill with a harder type of keratin and do not shed. Below the nail body is a layer of epithelium and a deeper layer of dermis. Most of the nail body appears pink because of blood flowing through underlying capillaries. The **free edge** is the part of the nail body that may extend past the distal

Figure 5.7 Nails. Shown is a fingernail.

Nail cells arise by transformation of superficial cells of the nail matrix.

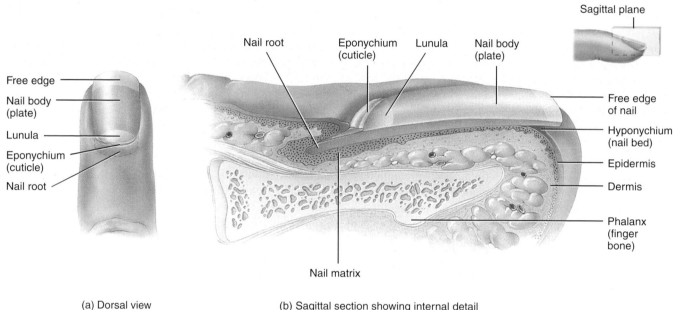

(a) Dorsal view

(b) Sagittal section showing internal detail

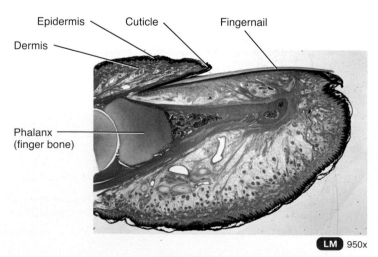

(c) Photomicrograph of fingertip

? Why are nails so hard?

end of the digit. The free edge is white because there are no underlying capillaries. The **nail root** is the part of the nail that is buried in a fold of skin. The whitish, crescent-shaped area of the proximal end of the nail body is called the **lunula** (LOO-noo-la = little moon). It appears whitish because the thickened stratum basale in this area does not allow the vascular tissue underneath to show through. Beneath the free edge is a thickened region of stratum corneum called the **hyponychium** (hī-pō-NIK-ē-um; *hypo-* = below; *-onych* = nail), or **nail bed**, which secures the nail to the fingertip. The **eponychium** (ep-ō-NIK-ē-um; *ep-* = above) or **cuticle** is a narrow band of epidermis that extends from and adheres to the margin (base and lateral border) of the nail wall. It occupies the proximal border of the nail and consists of stratum corneum.

The proximal portion of epithelium deep to the nail root is known as the **nail matrix**. The nail matrix cells divide mitotically to produce growth. Nail growth occurs by the transformation of superficial nail matrix cells into nail cells. In the process, the harder outer layer is pushed forward over the stratum basale. The growth rate of nails is determined by the rate at which nail matrix cells divide, which is influenced by factors such as your age, health, and nutritional status. Nail growth also varies according to the season, the time of day, and environmental temperature. The average growth in the length of fingernails is about 1 mm (.04 in.) per week. The growth rate is somewhat slower in toenails. The longer the digit, the faster the nail grows.

Nails have a variety of functions:

1. They protect the distal end of the digits.
2. They provide support and counterpressure to the palmar surface of the fingers to enhance touch perception and manipulation.
3. They allow us to grasp and manipulate small objects, and they can be used to scratch and groom the body in various ways.

⊙ **CHECKPOINT**
   10. Describe the structure of a hair.
   11. What produces "goosebumps"?
   12. Contrast the locations and functions of sebaceous (oil) glands, sudoriferous (sweat) glands, and ceruminous glands.
   13. Describe the principal parts of a nail.

TYPES OF SKIN

⊙ **OBJECTIVE**
   • Compare structural and functional differences in thin and thick skin.

Although the skin over the entire body is similar in structure, there are quite a few local variations related to thickness of the epidermis, strength, flexibility, degree of keratinization, distribution and type of hair, density and types of glands, pigmentation, vascularity (blood supply), and innervation (nerve supply). Two major types of skin are recognized on the basis of certain structural and functional properties: **thin (hairy) skin** and **thick (hairless) skin**.

Table 5.4 presents a comparison of the features of thin and thick skin.

⊙ **CHECKPOINT**
   14. What criteria are used to distinguish thin and thick skin?

TABLE 5.4		
Comparison of Thin and Thick Skin		
FEATURE	**THIN SKIN**	**THICK SKIN**
Distribution	All parts of the body except areas such as palms, and palmar surface of digits, and soles.	Areas such as palms, palmar surface of digits, and soles.
Epidermal thickness	0.10–0.15 mm (0.004–0.006 in.).	0.6–4.5 mm (0.024–0.18 in.).
Epidermal strata	Stratum lucidum essentially lacking; thinner strata spinosum and corneum.	Thick strata lucidum, spinosum, and corneum.
Epidermal ridges	Lacking due to poorly developed and fewer and less well organized dermal papillae.	Present due to well-developed and more numerous dermal papillae organized in parallel rows.
Hair follicles and arrector pili muscles	Present.	Absent.
Sebaceous glands	Present.	Absent.
Sudoriferous glands	Fewer.	More numerous.
Sensory receptors	Sparser.	Denser.

FUNCTIONS OF THE SKIN

◉ OBJECTIVE

- Describe how the skin contributes to the regulation of body temperature, storage of blood, protection, sensation, excretion and absorption, and synthesis of vitamin D.

Now that you have a basic understanding of the structure of the skin, you can better appreciate its many functions. Among the numerous functions of the integumentary system (mainly the skin) are the following:

1. **Regulation of body temperature (thermoregulation).** In response to high environmental temperature or strenuous exercise, the evaporation of sweat from the skin surface helps lower an elevated body temperature to normal. In response to low environmental temperature, production of sweat is decreased, which helps conserve heat.

2. **Blood reservoir.** The dermis houses an extensive network of blood vessels that carry 8–10% of the total blood flow in a resting adult. For this reason, the skin acts as a *blood reservoir*. During moderate exercise, the flow of blood through the skin increases, which increases the amount of heat radiated from the body. During very strenuous exercise, however, skin blood vessels constrict (narrow) somewhat, diverting more blood to contracting skeletal muscles and to the heart. Because of this shunting of blood away from the skin, however, less heat is lost from the skin, and body temperature tends to rise.

3. **Protection.** The skin covers the body and provides protection in various ways. Keratin in the skin protects underlying tissues from microbes, abrasion, heat, and chemicals and the tightly interlocked keratinocytes resist invasion by microbes. Lipids released by lamellar granules inhibit evaporation of water from the skin surface, thus protecting the body from dehydration; they also retard entry of water across the skin surface during showers and swims. The oily sebum from the sebaceous glands protects skin and hairs from drying out and contains bactericidal chemicals that kill surface bacteria. The pigment melanin provides some protection against the damaging effects of ultraviolet light. Protective functions that are immunological in nature are carried out by epidermal Langerhans cells, which alert the immune system to the presence of potentially harmful microbial invaders, and by macrophages in the dermis, which phagocytize bacteria and viruses that manage to penetrate the skin surface.

4. **Cutaneous sensations.** *Cutaneous sensations* are sensations that arise in the skin. These include tactile sensations—touch, pressure, vibration, and tickling—as well as thermal sensations such as warmth and coolness. Another cutaneous sensation, pain, usually is an indication of impending or actual tissue damage. Some of the wide variety of abundantly distributed nerve endings and receptors in the skin include the tactile discs of the epidermis, the corpuscles of touch in the dermis, and hair root plexuses around each hair follicle. Chapter 21 provides more details on the topic of cutaneous sensations.

5. **Excretion and absorption.** The skin plays minor roles in *excretion*, the elimination of substances from the body, and *absorption*, the passage of materials from the external environment into body cells. Besides removing water and heat (by evaporation), sweat also is the vehicle for excretion of small amounts of salts, carbon dioxide, and two organic products of protein breakdown—ammonia and urea. The absorption of water-soluble substances through the skin is negligible, but certain lipid-soluble materials do penetrate the skin. These include fat-soluble vitamins (A, D, E, and K) and oxygen and carbon dioxide gases. Toxic materials that can be absorbed through the skin include organic solvents such as acetone (in some nail polish removers) and carbon tetrachloride (dry-cleaning fluid); salts of heavy metals such as lead, mercury, and arsenic; and the toxins in poison ivy and poison oak. The skin can also absorb, through patches placed on it, birth control hormones, drugs to relieve chronic pain, nitroglycerin for heart patients, medication for motion sickness, and nicotine to help people stop smoking.

6. **Synthesis of vitamin D.** Synthesis of vitamin D requires activation of a precursor molecule in the skin by ultraviolet rays in sunlight. Enzymes in the liver and kidneys then modify the activated molecule, finally producing *calcitriol*, the most active form of vitamin D. Calcitriol is a hormone that aids in the absorption of calcium in foods from the gastrointestinal tract into the blood. Only a small amount of exposure to ultraviolet light (about 10 to 15 minutes at least twice a week) is required for vitamin D synthesis. People who avoid sun exposure and individuals who live in colder, northern climates may experience vitamin D deficiency if it is not included in their diet as supplements.

BLOOD SUPPLY OF THE INTEGUMENTARY SYSTEM

◉ OBJECTIVE

- Describe the blood supply of the integumentary system.

Although the epidermis is avascular, the dermis is well supplied with blood (see Figure 5.1). The arteries supplying the dermis are generally derived from branches of arteries supplying skeletal muscles in a particular region. Some arteries supply the skin directly. One plexus (network) of arteries, the **cutaneous plexus**, is located at the junction of the dermis and subcutaneous layer and sends branches that supply the sebaceous (oil) and sudoriferous (sweat) glands, the deep portions of hair follicles, and adipose tissue. The **papillary plexus**, formed at the level of the papillary region, sends branches that supply the capillary loops in the dermal papillae, sebaceous (oil) glands, and the superficial portion of hair follicles. The arterial plexuses are accompanied by venous plexuses that drain blood from the dermis into larger subcutaneous veins.

Nutrients and oxygen diffuse to the avascular epidermis from blood vessels in the dermal papillae. The epidermal cells of the stratum basale are closest to these blood vessels and receive most of the nutrients and oxygen. These cells are the most active metabolically and continuously undergo cell division to produce new keratinocytes. As the new keratinocytes are pushed further from the blood supply by continuing cell division, the epidermal strata above the stratum basale receive fewer nutrients and the cells become less active and eventually die.

◉ CHECKPOINT

15. How do the distributions of the cutaneous plexus and the papillary plexus differ?

• CLINICAL CONNECTION | Burns

A **burn** is tissue damage caused by excessive heat, electricity, radioactivity, or corrosive chemicals that break down (denature) the proteins in the skin cells. Burns destroy some of the skin's important contributions to homeostasis—protection against microbial invasion and desiccation, and regulation of body temperature.

Burns are graded according to their severity. A *first-degree burn* involves only the epidermis (Figure A). It is characterized by mild pain and erythema (redness) but no blisters. Skin functions remain intact. The pain and damage caused by a first-degree burn may be lessened by immediately flushing it with cold water. Generally, a first-degree burn will heal in about 3–6 days and may be accompanied by flaking or peeling. One example of a first-degree burn is a mild sunburn.

A *second-degree burn* destroys a portion of the epidermis and possibly parts of the dermis (Figure B). Some skin functions are lost. In a second-degree burn, redness, blister formation, edema, and pain result. (Blister formation is caused by separation of the epidermis from the dermis due to the accumulation of tissue fluid between the layers.) Associated structures, such as hair follicles, sebaceous glands, and sweat glands, usually are not injured. If there is no infection, second-degree burns heal without skin grafting in about 3–4 weeks, but scarring may result. First- and second-degree burns are collectively referred to as *partial-thickness burns*.

A *third-degree burn,* or *full-thickness burn,* destroys a portion of the epidermis, the underlying dermis, and associated structures (Figure C). Most skin functions are lost. Such burns vary in appearance from marble-white to mahogany colored to charred, dry wounds. There is marked edema, and the burned region is numb because sensory nerve endings have been destroyed. Regeneration occurs slowly, and much granulation tissue forms before being covered by epithelium. Skin grafting may be required to promote healing and to minimize scarring.

The injury to the skin tissues directly in contact with the damaging agent is the *local effect* of a burn. Generally, however, the *systemic effects* of a major burn are a greater threat to life. The systemic effects of a burn may include (1) a large loss of water, plasma, and plasma proteins, which causes shock; (2) bacterial infection; (3) reduced circulation of blood; (4) decreased production of urine; and (5) diminished immune responses.

The seriousness of a burn is determined by its depth and extent of area involved, as well as the person's age and general health. According to the American Burn Association's classification of burn injury, a major burn includes third-degree burns over 10% of body surface area; second-degree burns over 25% of body surface area; or any third-degree burns on the face, hands, feet, or *perineum* (per-i-NĒ-um, which includes the anal and urogenital regions). When the burn area exceeds 70%, more than half the victims die.

Many people who have been burned in fires also inhale smoke. If the smoke is unusually hot or dense or if inhalation is prolonged, serious problems can develop. The hot smoke can damage the trachea (windpipe), causing its lining to swell. As the swelling narrows the trachea, airflow into the lungs is obstructed. Further, small airways inside the lungs can also narrow, producing wheezing and shortness of breath. A person who has inhaled smoke is given oxygen through a face mask, and a tube may be inserted into the trachea to assist breathing. •

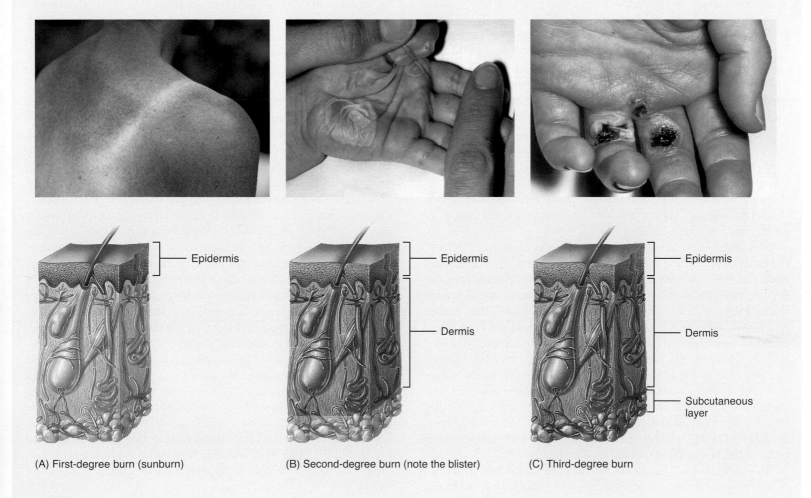

(A) First-degree burn (sunburn)   (B) Second-degree burn (note the blister)   (C) Third-degree burn

DEVELOPMENT OF THE INTEGUMENTARY SYSTEM

◉ **OBJECTIVE**

• Describe the development of the epidermis, its accessory structures, and the dermis.

The *epidermis* is derived from the **ectoderm**, which covers the surface of the embryo. Initially, at about the fourth week after fertilization, the epidermis consists of only a single layer of ectodermal cells (Figure 5.8a). At the beginning of the seventh week the single layer, called the **basal layer**, divides and forms a superficial protective layer of flattened cells called the **periderm** (Figure 5.8b). The peridermal cells are continuously sloughed off and, by the fifth month of development, secretions from sebaceous glands mix with them and with hairs to form a fatty substance called **vernix caseosa** (VER-niks KĀ-sē-ō-sa; *vernix*=varnish, *caseosa*=cheese). This substance covers and protects the skin of the fetus from the constant exposure to the amniotic fluid in which it is bathed. In addition, the slippery vernix caseosa facilitates the birth of the fetus and protects the skin from being damaged by the nails.

By about 11 weeks, the basal layer forms an intermediate layer of cells (Figure 5.8c). Proliferation of the basal cells eventually forms all layers of the epidermis, which are present at birth (Figure 5.8d). *Epidermal ridges* form along with the epidermal layers (Figure 5.8c). By about the eleventh week, cells from the neural crest (see Figure 19.26b on page 653) migrate into the dermis and differentiate into **melanoblasts** (Figure 5.8c). These cells soon enter the epidermis and differentiate into **melanocytes**. Later in the first trimester of pregnancy, *Langerhans cells*, which arise from red bone marrow, invade the epidermis. *Merkel cells*, also derived from the neural crest cells, appear in the epidermis between 8 and 12 weeks.

The *dermis* arises from **lateral plate mesoderm** (see Figure 4.9b on page 103) and the **dermatome of somites** (see Figure 10.10 on page 304). The mesoderm from these two sources is located deep to the surface ectoderm. The mesoderm gives rise to a loosely organized embryonic connective tissue called **mesenchyme** (MEZ-en-kīm; see Figure 5.8a). By 11 weeks, the mesenchymal cells differentiate into fibroblasts and begin to form collagen and elastic fibers. As the epidermal ridges form, parts of the superficial dermis project into the epidermis and develop into the *dermal papillae*, which contain capillary loops, corpuscles of touch, and free nerve endings (Figure 5.8c).

Hair follicles develop at about twelve weeks as downgrowths of the basal layer of the epidermis into the deeper dermis called **hair buds** (Figure 5.8d). As the hair buds penetrate deeper into the dermis, their distal ends become club-shaped and are called **hair bulbs** (Figure 5.8f). Invaginations of the hair bulbs, called *papillae of the hair*, fill with mesoderm in which blood vessels and nerve endings develop (Figure 5.8f). Cells in the center of a hair bulb develop into the *hair matrix*, which forms the *hair*. The peripheral cells of the hair bulb form the *epithelial root sheath* (Figure 5.8g). Mesenchyme in the surrounding dermis develops into the *dermal root sheath* and *arrector pili muscle* (Figure 5.8g). By the fifth month, the hair follicles produce lanugo (delicate fetal hair; see page 127). It is produced first on the head and then on other parts of the body, and is usually shed prior to birth.

Most *sebaceous (oil) glands* develop as outgrowths from the sides of hair follicles at about four months and remain connected to the follicles (Figure 5.8e). Most *sudoriferous (sweat)* glands are derived from downgrowths (buds) of the stratum basale of the epidermis into the dermis (Figure 5.8d). As the buds penetrate into the dermis, the proximal portion forms the duct of the sweat gland and the distal portion coils and forms the secretory portion of the gland (Figure 5.8g). Sweat glands appear at about five months on the palms and soles and a little later in other regions.

Nails are developed at about 10 weeks. Initially they consist of a thick layer of epithelium called the **primary nail field**. The nail itself is keratinized epithelium and grows distally from its base. It is not until the ninth month that the nails actually reach the tips of the digits.

◉ **CHECKPOINT**

16. Which structures develop as downgrowths of the basal layer of epidermal cells?

AGING AND THE INTEGUMENTARY SYSTEM

◉ **OBJECTIVE**

• Describe the effects of aging on the integumentary system.

The pronounced effects of skin aging do not become noticeable until people reach their late forties. Most of the age-related changes occur in the dermis. Collagen fibers in the dermis begin to decrease in number, stiffen, break apart, and disorganize into a shapeless, matted tangle. Elastic fibers lose some of their elasticity, thicken into clumps, and fray, an effect that is greatly accelerated in the skin of smokers. Fibroblasts, which produce both collagen and elastic fibers, decrease in number. As a result, the skin forms the characteristic crevices and furrows known as *wrinkles*.

With further aging, Langerhans cells dwindle in number and dermal macrophages become less-efficient phagocytes, thus decreasing the skin's immune responsiveness. Decreased size of sebaceous glands leads to dry and broken skin that is more susceptible to infection. The production of sweat diminishes, which probably contributes to the increased incidence of heat stroke in the elderly. There is a decrease in the number of functioning melanocytes, resulting in gray hair and atypical skin pigmentation. Hair loss increases with aging as hair follicles stop producing hairs. About 25% of males begin to show signs of hair loss by age 30 and about two-thirds have significant hair loss by age 60. Both males and females develop pattern baldness. An increase in the size of some melanocytes produces pigmented blotching (age spots). Walls of blood vessels in the dermis become thicker and less permeable, and subcutaneous adipose tissue is lost. Aged skin (especially the dermis) is thinner than young skin, and the migration of cells from the basal layer to the epidermal surface slows considerably. With the onset of old age, skin heals poorly and becomes more susceptible to pathological conditions such as skin cancer and pressure ulcers. **Rosacea** (rō-ZĀ-shē-a=rosy) is a skin condition that affects mostly fair-skinned adults between ages 30–60. It is characterized by redness, tiny pimples, and noticeable blood vessels, usually in the central area of the face.

Growth of nails and hair slows during the second and third decades of life. The nails also may become more brittle with age, often due to dehydration or repeated use of cuticle remover or nail polish.

Figure 5.8 Development of the integumentary system.

🔒 ▭ The epidermis develops from ectoderm and the dermis develops from mesoderm.

(a) Fourth week
- Ectoderm
- Mesenchyme

(b) Seventh week
- Periderm
- Basal layer

(c) Eleven weeks
- Intermediate layer
- Epidermal ridge
- Basal layer
- Dermal papilla
- Melanoblast
- Developing collagen and elastic fibers

(d) Twelve weeks
- Basal layer
- Bud of developing sudoriferous gland
- Hair bud

(e) Fourteen weeks
- Developing sudoriferous gland
- Developing sebaceous gland
- Hair bulb

(f) Sixteen weeks
- Developing sebaceous gland
- Hair shaft
- Papilla of the hair

(g) Eighteen weeks
- Hair shaft
- Sweat pore
- Duct of sudoriferous gland
- Arrector pili muscle
- Epithelial root sheath
- Dermal root sheath
- Secretory portion of sudoriferous gland
- Sebaceous gland
- Bulb
- Papilla of the hair
- Blood vessels

(h) At birth
- Epidermis
- Dermis
- Stratum corneum
- Stratum lucidum
- Stratum granulosum
- Stratum spinosum
- Stratum basale
- Melanocyte

❓ What is the composition of vernix caseosa?

Several cosmetic anti-aging treatments are available to diminish the effects of aging or sun-damaged skin. These include the following:

1. **Topical products** that bleach the skin to tone down blotches and blemishes (hydroquinone) or decrease fine wrinkles and roughness (retinoic acid)
2. **Microdermabrasion** (mī-krō-DER-ma-brā-zhun; *mikros-*=small; *derm*=skin; *-abrasio*=to wear away), a process that uses tiny crystals under pressure to remove and vacuum the skin's surface cells to improve skin texture and reduce blemishes
3. **Chemical peel**, the application of a mild acid (such as glycolic acid) to the skin to remove surface cells to improve skin texture and reduce blemishes
4. **Laser resurfacing**, use of a laser to clear up blood vessels near the skin surface, even out blotches and blemishes, and decrease fine wrinkles
5. **Dermal fillers**, injections of collagen from cows, hyaluronic acid, or calcium hydroxylapatite that plumps up the skin to smooth out wrinkles and fill in furrows, such as those around the nose and mouth and between the eyebrows

6. **Fat transplantation,** a process in which fat from one part of the body is injected into another location such as around the eyes
7. **Botulinum toxin** or **Botox®**, a diluted version of the toxin that inhibits nerve impulse transmission to skeletal muscles, which is injected into the skin to paralyze muscles that cause the skin to wrinkle
8. **Radio frequency nonsurgical facelift**, a procedure that uses radio frequency emissions to tighten the deeper layers of the skin of the jowls, neck, and sagging eyebrows and eyelids
9. **Facelift, browlift,** or **necklift**, invasive surgery in which loose skin and fat are removed surgically and the underlying connective tissue and muscle are tightened
10. **Thread lift,** a noninvasive surgical facelift in which threads that have tiny barbs are inserted into sagging facial tissues to gently lift and reposition them into a more vertical and youthful position

◉ **CHECKPOINT**

17. What factors contribute to the susceptibility of aging skin to infection?

KEY MEDICAL TERMS ASSOCIATED WITH THE INTEGUMENTARY SYSTEM

Abrasion (a-BRĀ-shun; *ab-*=away; *-rasion*=scraped) An area where skin has been scraped away.

Athlete's foot A superficial fungus infection of the skin of the foot.

Blister A collection of serous fluid within the epidermis or between the epidermis and dermis, due to short-term but severe friction. The term *bulla* (BUL-a) refers to a large blister.

Callus (KAL-lus=hard skin) An area of hardened and thickened skin that is usually seen in palms and soles and is due to persistent pressure and friction.

Cold sore A lesion, usually in the oral mucous membrane, caused by Type I herpes simplex virus (HSV) transmitted by oral or respiratory routes. The virus remains dormant until triggered by factors such as ultraviolet light, hormonal changes, and emotional stress. Also called a *fever blister*.

Comedo (KOM-ē-dō; *comedo*=to eat up) A collection of sebaceous material and dead cells in the hair follicle and excretory duct of the sebaceous (oil) gland. Usually found over the face, chest, and back, and more commonly during adolescence. Also called *blackhead*.

Contact dermatitis (der-ma-TĪ-tis; *dermat-*=skin; *-itis*=inflammation of) Inflammation of the skin characterized by redness, itching, and swelling and caused by exposure of the skin to chemicals that bring about an allergic reaction, such as poison ivy toxin.

Contusion (kon-TOO-shun; *contundere*=to bruise) Condition in which tissue deep to the skin is damaged, but the epidermis is not broken.

Corn A painful conical thickening of the stratum corneum of the epidermis found principally over toe joints and between the toes, often caused by friction or pressure.

Cyst (SIST; *cyst*=sac containing fluid) A sac with a distinct connective tissue wall, containing a fluid or other material.

Eczema (EK-ze-ma, *ekzeo-*=to boil over) An inflammation of the skin characterized by patches of red, blistering, dry, extremely itchy skin. It occurs mostly in skin creases in the wrists, backs of the kness, and front of elbows. It typically begins in infancy, and many children outgrow the condition. The cause is unknown but is linked to genetics and allergies.

Frostbite Local destruction of skin and subcutaneous tissue on exposed surfaces as a result of extreme cold. In mild cases, the skin is blue and swollen and there is slight pain. In severe cases there is considerable swelling, some bleeding, no pain, and blistering. If untreated, gangrene may develop. Frostbite is treated by rapid rewarming.

Furuncle (FŪ-rung-kul) A boil; an abscess resulting from infection of a hair follicle. Usually caused by the bacterium, *Staphylococcus aureus.*

Hemangioma (he-man′-jē-Ō-ma; *hem-*=blood; *-angi-*=blood vessel; *-oma*=tumor) Localized benign tumor of the skin and subcutaneous layer that results from an abnormal increase in blood vessels.

Hives Condition of the skin marked by reddened elevated patches that are often itchy. Most commonly caused by infections, physical trauma, medications, emotional stress, food additives, and certain food allergies. Also called *urticaria* (ūr-ti-KAR-ē-a).

Intradermal (in-tra-DER-mal; *intra-*=within) Within the skin. Also called *intracutaneous.*

Keloid (KĒ-loid; *kelis*=tumor) An elevated, irregular darkened area of excess scar tissue caused by collagen formation during healing. It extends beyond the original injury and is tender and frequently painful. It occurs in the dermis and underlying subcutaneous tissue, usually after trauma, surgery, a burn, or severe acne; more common in people of African descent.

Keratosis (ker-a-TŌ-sis; *kera-*=horn) Formation of a hardened growth of epidermal tissue, such as a *solar keratosis*, a premalignant lesion of the sun-exposed skin of the face and hands.

Laceration (las-er-Ā-shun; *lacer-*=torn) An irregular tear of the skin.

Papule (PAP-ūl; *papula*=pimple) A small, round skin elevation less than 1 cm in diameter. One example is a pimple.

Pruritus (p-RI-tus; *pruri-*=to itch) Itching, one of the most common dermatological disorders. It may be caused by skin disorders (infections), systemic disorders (cancer, kidney failure), psychogenic factors (emotional stress), or allergic reactions.

Psoriasis (sō-RĪ-a-sis) A common and chronic skin disorder in which keratinocytes divide and move more quickly than normal from the stratum basale to the stratum corneum. As a result, the surface cells never get a chance to cycle into the later keratinizing stages. The surface cells are shed immaturely and on the scalp are called dandruff.

Topical In reference to a medication, applied to the skin surface rather than ingested or injected.

Wart Mass produced by uncontrolled growth of epithelial skin cells; caused by a papilloma virus. Most warts are noncancerous.

STUDY OUTLINE

Structure of the Skin (p. 117)

1. The integumentary system consists of organs such as the skin and hair, and other structures such as nails.
2. The skin is the largest organ of the body in surface area and weight. The principal parts of the skin are the epidermis (superficial) and dermis (deep).
3. The subcutaneous layer (hypodermis) is deep to the dermis and not part of the skin. It anchors the dermis to underlying tissues and organs, and it contains lamellated (pacinian) corpuscles.
4. The types of cells in the epidermis are keratinocytes, melanocytes, Langerhans cells, and Merkel cells.
5. The epidermal layers, from deep to superficial, are the stratum basale, stratum spinosum, stratum granulosum, stratum lucidum (in thick skin only), and stratum corneum (see Table 5.1 on page 122). Stem cells in the stratum basale undergo continuous cell division, producing keratinocytes for the other layers.
6. The dermis consists of papillary and reticular regions. The papillary region is composed of areolar connective tissue containing thin collagen and fine elastic fibers, dermal papillae, and Meissner corpuscles. The reticular region is composed of dense irregular connective tissue containing bundles of thick collagen and some coarse elastic fibers, adipose tissue, hair follicles, nerves, sebaceous (oil) glands, and ducts of sudoriferous (sweat) glands.
7. Epidermal ridges provide the basis for fingerprints and footprints.
8. The color of skin is due to melanin, carotene, and hemoglobin.
9. In tattooing, a pigment is deposited with a needle in the dermis. Body piercing is the insertion of jewelry through an artificial opening.

Accessory Structures of the Skin (p. 125)

1. Accessory structures of the skin—hair, skin glands, and nails—develop from the embryonic epidermis.
2. A hair consists of a shaft, most of which is superficial to the surface, a root that penetrates the dermis and sometimes the subcutaneous layer, and a hair follicle.
3. Associated with each hair follicle is a sebaceous (oil) gland, an arrector pili muscle, and a hair root plexus.
4. New hairs develop from division of hair matrix cells in the bulb; hair replacement and growth occur in a cyclic pattern consisting of growth, regression, and resting stages.
5. Hairs offer a limited amount of protection—from the sun, heat loss, and entry of foreign particles into the eyes, nose, and ears. They also function in sensing light touch.
6. Lanugo hairs of the fetus are shed before birth. Most body hair on males is terminal (coarse, pigmented); most body hair on females is vellus (fine).
7. Sebaceous (oil) glands are usually connected to hair follicles; they are absent from the palms and soles. Sebaceous glands produce sebum, which moistens hairs and waterproofs the skin. Clogged sebaceous glands may produce acne.

8. There are two types of sudoriferous (sweat) glands: eccrine and apocrine. Eccrine sweat glands have an extensive distribution; their ducts terminate at pores at the surface of the epidermis. Eccrine sweat glands are involved in thermoregulation and waste removal and are stimulated during emotional stress. Apocrine sweat glands are limited to the skin of the axillae, groin, and areolae; their ducts open into hair follicles. Apocrine sweat glands are stimulated during emotional stress and sexual excitement.
9. Ceruminous glands are modified sudoriferous glands that secrete cerumen (ear wax). They are found in the external auditory canal (ear canal).
10. Nails are hard, dead, keratinized epidermal cells over the dorsal surfaces of the distal portions of the digits.
11. The principal parts of a nail are the nail body, free edge, nail root, lunula, eponychium, and nail matrix. Cell division of the matrix cells produces new nails.

Types of Skin (p. 131)

1. Thin skin covers all parts of the body except for the palms, palmar surfaces of the digits, and the soles.
2. Thick skin covers the palms, palmar surfaces of the digits, and soles.
3. See Table 5.4 on page 131 for a comparison of thin and thick skin.

Functions of the Skin (p. 132)

1. Skin functions include body temperature regulation (thermoregulation), blood storage, protection, sensation, excretion and absorption, and synthesis of vitamin D.
2. The skin participates in thermoregulation by liberating sweat at its surface and functions as a blood reservoir.
3. The skin provides physical, chemical, and biological barriers that help protect the body.
4. Cutaneous sensations include touch, hot and cold, and pain.

Blood Supply of the Integumentary System (p. 132)

1. The epidermis is avascular.
2. The dermis is supplied by the cutaneous and papillary plexuses.

Development of the Integumentary System (p. 134)

1. The epidermis develops from the embryonic ectoderm, and the accessory structures of the skin (hair, nails, and skin glands) are epidermal derivatives.
2. The dermis is derived from mesodermal cells.

Aging and the Integumentary System (p. 134)

1. Most effects of aging begin to occur when people reach their late forties.
2. Among the effects of aging are wrinkling, loss of subcutaneous fat, atrophy of sebaceous glands, and decrease in the number of melanocytes and Langerhans cells.

SELF-QUIZ QUESTIONS

Choose the one best answer to the following questions.

1. Which of the following statements about the function of skin is not true?
   a. It helps regulate body temperature.
   b. It protects against bacterial invasion and dehydration.
   c. It absorbs water and salts.
   d. It participates in the synthesis of vitamin D.
   e. It detects stimuli related to temperature and pain.

2. The layer of the skin from which new epidermal cells are derived is the:
   a. stratum corneum
   b. stratum basale
   c. stratum lucidum
   d. stratum granulosum
   e. stratum spinosum

3. The substance that prevents excessive evaporation of water from the skin, keeps the skin soft and pliable, and inhibits the growth of bacteria is:
   a. sebum
   b. keratin
   c. cerumen
   d. sweat
   e. carotene

4. The subcutaneous layer consists mostly of:
   a. melanin
   b. simple squamous epithelial tissue
   c. keratin
   d. areolar and adipose tissue
   e. smooth muscle

5. Tyrosinase is an enzyme necessary for the production of:
   a. melanin
   b. vitamin D
   c. sweat
   d. carotene
   e. sebum

6. Lanugo hairs:
   a. a water-repelling protein in the epidermis.
   b. a type of skin pigment.
   c. the cuticle of a nail.
   d. nonpigmented hairs covering the body of the fetus.
   e. hairs that appear at puberty in response to hormone changes.

7. What word does the letter *C* stand for in the ABCD rule for detecting malignant melanoma?
   a. *cancer*
   b. *corners* are asymmetrical and uneven
   c. *colors* in large variety
   d. *crust* forms on a mole
   e. *circular* shape

Complete the following.

8. _____ glands, which secrete earwax, are located in the lining of the external auditory meatus.

9. The two principal parts of the skin are the superficial _____, and the deeper _____.

10. Sweat glands are also known as _____ glands.

11. 98% of all skin cancers are either _____ or _____.

12. According to the American Burn Association's classification of burn injury, a *major* burn includes third-degree burns over _____% of the body surface area; or second-degree burns over _____% of the body surface area.

Are the following statements true or false? If the statement is False, *find and fix the mistake to make the statement* True.

13. Hair follicles and sebaceous glands are found only in areas covered by thick skin.

14. The scientific name for the cuticle of a nail is the lunula.

15. The dermis receives its blood supply from two arterial networks or plexuses: the cutaneous plexus and the epidermal plexus.

16. Eccrine sweat glands are more common than apocrine sweat glands.

17. The dermis consists of two regions; the papillary region is superficial, and the reticular region is deep.

18. Racial differences in skin color are mainly due to the number of melanocytes in the epidermis.

19. Skin color is considered to be a risk factor associated with skin cancer.

20. Increased incidence of heat stroke in the elderly probably results from increased sweat production.

Matching

21. Match the following:
    ____ (a) assist in immune responses; easily damaged by UV light
    ____ (b) produce pigment that shields cell nuclei from UV light
    ____ (c) located in subcutaneous tissue; sensitive to pressure
    ____ (d) most abundant epidermal cells
    ____ (e) touch receptor cells in the stratum basale layer
    ____ (f) tactile receptors located in dermal papillae

    (1) melanocytes
    (2) keratinocytes
    (3) Merkel cells
    (4) Langerhans cells
    (5) Meissner corpuscles (corpuscles of touch)
    (6) lamellated (pacinian) corpuscles

22. Match the following:
    ____ (a) helps control the rate of cell division in the stratum basale
    ____ (b) lipid-rich secretion of keratinocytes that fills spaces between the layers of the epidermis and that acts as a water-repellent sealant
    ____ (c) reddish brown/black pigment that neutralizes free radicals in the skin following damage by UV radiation
    ____ (d) a yellow-orange pigment found in the stratum corneum, dermis, and subcutaneous layer
    ____ (e) tough, fibrous protective protein in the epidermis
    ____ (f) fatty substance that helps prevent water evaporation from skin and inhibits bacterial growth

    (1) melanin
    (2) keratin
    (3) lamellar granules
    (4) epidermal growth factor
    (5) carotene
    (6) sebum

23. Match the following:
    ____ (a) a painful conical thickening of the stratum corneum principally found over toe joints and between the toes
    ____ (b) an inflammation of the skin characterized by patches of red, blistering, dry, and extremely itchy skin
    ____ (c) an elevated, irregular darkened area of excess scar tissue caused by collagen formation during healing
    ____ (d) a hardened growth of epidermal tissue
    ____ (e) a mole or birthmark

    (1) keloid
    (2) eczema
    (3) nevus
    (4) corn
    (5) keratosis

CRITICAL THINKING QUESTIONS

1. Your 65-year-old aunt has spent every sunny day at the beach for as long as you can remember. Her skin looks a lot like a comfortable lounge chair—brown and wrinkled. Recently her dermatologist removed a suspicious growth from the skin of her face. What would you suspect is the problem and its likely cause?

2. Lindsay is partying with his friends on a cold winter night. The outside temperature is below freezing, but they are not wearing their coats when they step outside to smoke. They say they don't need their coats because they are drinking alcoholic beverages, which they say keeps them warm enough. You have learned that alcohol dilates the blood vessels in the skin. Do you think Lindsay and his friends have chosen an effective way to control their body temperature? Why or why not?

3. Courtney and her college roomate made travel plans to go to Cancun for spring break. Neither one of them ever bothered to wear sunscreen, thinking that they would get a better suntan without it. When the women reached their forties, they were both diagnosed with malignant melanoma. Why has the risk of developing melanoma doubled over the last 15 years? What is the prognosis for someone diagnosed with malignant melanoma?

4. Your nephew has been learning about cells in science class and now refuses to take a bath. He asks, "If all cells have a semipermeable membrane, and my skin is made out of cells, then won't I swell up and pop when I take a bath?" Explain this dilemma to your nephew before he starts attracting flies.

5. People always say, "It's not the heat; it's the humidity," when complaining about summer weather. With regard to the integumentary system, why do you think people feel hotter when it's 95 degrees Fahrenheit and 95% humidity than when it's 95 degrees and 30% humidity? (Humidity refers to the amount of water vapor in the air.)

? ANSWERS TO FIGURE QUESTIONS

5.1 The epidermis is composed of epithelial tissue, and the dermis is made up of connective tissue.

5.2 Melanin protects the DNA of the nucleus of keratinocytes from being damaged by UV light.

5.3 The stratum basale is the layer of the epidermis with stem cells that continually undergo cell division.

5.4 Lines of cleavage are clinically important because surgical incisions parallel to them leave only fine scars, but a surgical incision made across several layers of lines of cleavage will tend to heal in a broad, thick scar.

5.5 Plucking a hair stimulates hair root plexuses in the dermis, some of which are sensitive to pain. Because the cells of a hair shaft are already dead and the hair shaft lacks nerves, cutting hair is not painful.

5.6 The main function of eccrine sweat glands is to help regulate body temperature through evaporation.

5.7 Nails are hard because they are composed of tightly packed, dead, keratinized epidermal cells that do not shed and are filled with a harder type of keratin than in the epidermis.

5.8 Vernix caseosa consists of secretions from sebaceous glands, sloughed off peridermal cells, and hairs.

6 | BONE TISSUE

INTRODUCTION A **bone** is an organ made up of several different tissues working together: bone (osseous) tissue, cartilage, dense connective tissues, epithelium, adipose tissue, and nervous tissue. The entire framework of bones and their cartilages constitute the **skeletal system**. Bone tissue is a complex and dynamic living tissue. It continually engages in a process called *remodeling*—building new bone tissue and breaking down old bone tissue. For example, in the early days of space exploration, young, healthy men in prime physical shape returned from their space flights, only to alarm the physicians who were monitoring their physical health. Examinations of the astronauts revealed that they had lost up to 20% of their total bone density during their extended stay in space. The zero-gravity (weightless) environment of space, coupled with the fact that the men traveled in small capsules that greatly limited their movement for these extended periods of time, placed minimal strain on their bones. Contrast these early astronauts with athletes who subject their bones to great forces, which place significant strain on the bone tissue. These athletes show an increase in overall bone density. How is bone capable of changing in response to the different mechanical demands placed on it? Why do high activity levels that strain bone tissue greatly improve bone health? This chapter will survey the various components of bones so you can understand how bones form, how they age, and how exercise affects their density and strength. The study of bone structure and the treatment of bone disorders is referred to as **osteology** (os-tē-OL-ō-jē; *osteo-*=bone; *-logy*=study of). •

FUNCTIONS OF BONE AND THE SKELETAL SYSTEM

◉ **OBJECTIVE**

• Describe the six main functions of the skeletal system.

Bone tissue makes up about 18% of the weight of the human body. The skeletal system performs several basic functions:

1. **Support.** The skeleton serves as the structural framework for the body by supporting soft tissues and providing attachment points for the tendons of most skeletal muscles.
2. **Protection.** The skeleton protects the most important internal organs from injury. For example, cranial bones protect the brain, vertebrae (backbones) protect the spinal cord, and the rib cage protects the heart and lungs.
3. **Assistance in movement.** Most skeletal muscles attach to bones; when they contract, they pull on bone to produce movement.
4. **Mineral storage and release.** Bone tissue stores several minerals, especially calcium and phosphorus, which contribute to the strength of bone. Bone tissue stores about 99% of total body calcium. On demand, bone releases minerals into the blood to maintain critical mineral balances and to distribute the minerals to other parts of the body.
5. **Blood cell production.** Within certain bones, a connective tissue called **red bone marrow** produces red blood cells, white blood cells, and platelets, a process called **hemopoiesis** (hēm-ō -poy-Ē-sis; *hemo-*=blood; *poiesis-*=making). Red bone marrow consists of developing blood cells, adipocytes, fibroblasts, and macrophages within a network of reticular fibers. It is present in developing bones of the fetus and in some adult bones, such as the hip bones (pelvic bones), ribs, sternum (breastbone), vertebrae (backbones), skull, and ends of the humerus (arm bone) and femur (thigh bone). In a newborn, all bone marrow is red and is involved in hemopoiesis. With increasing age, much of the bone marrow changes from red to yellow.
6. **Triglyceride storage. Yellow bone marrow** consists mainly of adipose cells, which store triglycerides. The stored triglycerides are a potential chemical energy reserve.

◉ **CHECKPOINT**

1. What types of tissues make up the skeletal system?
2. How do red and yellow bone marrow differ in composition and function?

TYPES OF BONES

◉ **OBJECTIVE**

• Classify bones on the basis of shape and location.

A bone can serve as a library of information about many other aspects of the body's anatomy. Bones serve as an enduring record of an individual's life because bones do not deteriorate following death like soft tissues do. For example, age, size, stature, gender, health, and race can all be determined from an examination of bones. Such information can be useful to forensic specialists seeking to identify the skeletal remains of a victim or anthropologists who study ancient skeletal remains. The shape of a given bone also reveals a great deal of information about its functional role in the body, such as the physical strength of the bone and the type of forces the bone experienced as muscles moved it. Each little bump, groove, hole, projection, and ridge on a bone has a story to tell.

The 206 individual bones in an adult skeleton come in a variety of sizes and shapes. They range in size from the small ear bones in the skull to the nearly two-foot-long femur (thigh bone) of an adult. The various shapes of the bones are designed to accomplish particular functions. Anatomists recognize five types of bones in the skeleton on the basis of shape: long, short, flat, irregular, and sesamoid (Figure 6.1).

Long bones have greater length than width and consist of a shaft and a variable number of extremities (ends). They are slightly curved for strength. A curved bone absorbs the stress of the body's weight at several different points so that it is evenly distributed. If such bones were straight, the weight of the body would be unevenly distributed and the bone would fracture easily. Long bones consist mostly of *compact bone tissue*, which is dense and has smaller spaces, but they also contain considerable amounts of *spongy bone tissue*, which has larger spaces (see Figure 6.4 on page 146–147). Long bones include the humerus (arm bone), ulna and radius (forearm bones), femur (thigh bone), tibia and fibula (leg bones), phalanges (finger and toe bones).

Short bones are somewhat cube-shaped and nearly equal in length and width. They consist of spongy bone except at the surface, where there is a thin layer of compact bone. Examples of short bones are the carpal (wrist) bones (except for the pisiform, which is a sesamoid bone) and tarsal (ankle) bones (except for the calcaneus, which is an irregular bone).

Flat bones are generally thin and composed of two nearly parallel plates of compact bone enclosing a layer of spongy bone. The layers of compact bone are called external and internal *tables*,

Figure 6.1 Types of bones based on shape. The bones are not drawn to scale.

🔑 The shapes of bones largely determine their functions.

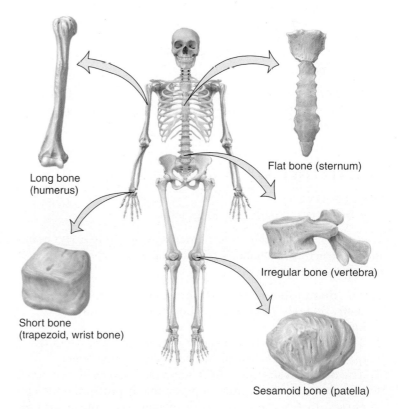

Long bone (humerus)

Flat bone (sternum)

Irregular bone (vertebra)

Short bone (trapezoid, wrist bone)

Sesamoid bone (patella)

❓ **Which type of bone primarily protects and provides a large surface area for muscle attachment?**

and the spongy bone sandwiched between them is referred to as **diploe** (DIP-lō-ē) (see Figure 6.6). Flat bones afford considerable protection and provide extensive areas for muscle attachment. Flat bones include the cranial (skull) bones, which protect the brain; the sternum (breastbone) and ribs, which protect organs in the thorax; and the scapulae (shoulder blades).

Irregular bones have complex shapes and cannot be grouped into any of the three categories just described. They also vary in the amount of spongy and compact bone present. Such bones include the vertebrae (backbones), certain facial bones, and the aforementioned calcaneus (heel bone).

Sesamoid bones (meaning shaped like a sesame seed) develop in certain tendons where there is considerable friction, compression, and physical stress. They are not always completely ossified and measure only a few millimeters in diameter except for the two patellae (kneecaps), the largest of the sesamoid bones. Sesamoid bones vary in number from person to person except for the patellae, which are located in the quadriceps femoris tendon (see Figure 11.24a, b on page 400–401) and are normally present in all individuals. Functionally, sesamoid bones protect tendons from excessive wear and tear, and they often change the direction of pull of a tendon, which improves the mechanical advantage at a joint.

In the upper limbs, sesamoid bones usually occur only in the joints of the palmar surface of the hands. Two frequently encountered sesamoid bones are in the tendons of the adductor pollicis and flexor pollicis brevis muscles at the metacarpophalangeal joint of the thumb (see Figure 8.6a on page 226). In the lower limbs, aside from the patellae, two constant sesamoid bones occur on the plantar surface of the foot in the tendons of the flexor hallucis brevis muscle at the metatarsophalangeal joint of the great (big) toe (see Figure 8.12b on page 242).

An additional type of bone is not included in this classification by shape, but instead is classified by location. **Sutural** (SOŌ-chur-al; *sutura*=seam) or **Wormian bones** (named after a Danish anatomist, O. Worm, 1588–1654) are small bones located within the sutures (joints) of certain cranial bones (see Figure 7.5a on page 172). The number of sutural bones varies greatly from person to person.

◉ **CHECKPOINT**

3. Give examples of long, short, flat, irregular, and sesamoid bones.

ANATOMY OF A BONE

◉ **OBJECTIVE**

• Describe the parts of a long bone.

Using a long bone such as the humerus as an example, let's examine the structure of a bone. A typical long bone consists of the following parts (Figure 6.2):

1. The **diaphysis** (dī-AF-i-sis=growing between) is the bone's shaft, or body—the long, cylindrical, main portion of the bone.
2. The **epiphyses** (e-PIF-i-sēz=growing own; singular is *epiphysis*) are the proximal and distal ends of the bone.
3. The **metaphyses** (me-TAF-i-sēz; *meta*-=between; singular is *metaphysis*) are the regions between the diaphysis and the epiphyses. In a *growing bone*, each metaphysis contains an **epiphyseal plate** (ep-i-FIZ-ē-al), a layer of hyaline cartilage that allows the diaphysis of the bone to grow in length (described later in the chapter). When bone growth in length stops

between the ages of about 18 to 21, the cartilage in the epiphyseal plate is replaced by bone and the resulting bony structure is known as the **epiphyseal line**.

4. The **articular cartilage** is a thin layer of hyaline cartilage covering the part of the epiphysis where the bone forms an articulation (joint) with another bone. Articular cartilage reduces friction and absorbs shock at freely movable joints. Because articular cartilage lacks a perichondrium and lacks blood vessels, repair of damage is limited.
5. The **periosteum** (per-ē-OS-tē-um; *peri*-=around) is a tough sheath that surrounds the bone surface wherever it is not covered by articular cartilage. It is composed of an *outer fibrous layer* of dense irregular connective tissue and an *inner osteogenic layer* that consists of cells. Some of the cells enable bone to grow in thickness, but not in length. The periosteum also protects the bone, assists in fracture repair, helps nourish bone tissue, and serves as an attachment point for ligaments and tendons. The periosteum is attached to the underlying bone by **perforating (Sharpey's) fibers**, thick bundles of collagen that extend from the periosteum into the bone extracellular matrix.
6. The **medullary cavity** (MED-ū-lar-ē; *medulla*-=marrow, pith), or **marrow cavity**, is a hollow, cylindrical space within the diaphysis that contains fatty yellow bone marrow and numerous blood vessels in adults. This cavity minimizes the weight of the bone by reducing the dense bony material where it is least needed. The long bones' tubular design provides maximum strength with minimum weight.
7. The **endosteum** (end-OS-te-um; *endo*-=within) is a thin membrane that lines the medullary cavity. It contains a single layer of bone-forming cells and a small amount of connective tissue.

Using the design of the long bone as a model, the structure of the other bone types is not difficult to understand. Each of the other bone types—short, flat, irregular, and sesamoid bones—have designs that closely resemble the epiphysis of a long bone. In fact, long bones are the only bones with a diaphysis and a medullary cavity. The other bone types have outer plates of compact bone covering an inner core of spongy bone. In certain flat bones of the skull, the spongy bone is sandwiched between two plates of compact bone and houses red bone marrow.

◉ **CHECKPOINT**

4. Diagram the parts of a long bone, and list the functions of each part.

BONE SURFACE MARKINGS

◉ **OBJECTIVE**

• Describe the principal surface markings on bones and the functions of each.

In addition to having a variety of shapes, bones have different surface textures and contours. A closer look at the surface of a bone reveals a great deal of information. The surface of a bone is marked by a variety of bumps, grooves, concavities (indentations), projections, and holes. These characteristic features of bones are called **surface markings** or **osseous landmarks**, and they are structural features adapted for specific functions. Most are not present at birth but develop later in response to certain forces and are most prominent during adult life. Various bumps, ridges, or rough areas indicate where soft tissues, such as

Figure 6.2 **Parts of a long bone.** The spongy bone tissue of the epiphysis and metaphysis contains red bone marrow, and the medullary cavity of the diaphysis contains yellow bone marrow (in adults).

 A long bone is covered by articular cartilage at its proximal and distal epiphyses and by periosteum around the diaphysis.

(a) Partially sectioned humerus (arm bone)

(b) Sectioned femur (thigh bone)

Functions of Bone Tissue

1. Supports soft tissue and provides attachment for skeletal muscles.
2. Protects internal organs.
3. Assists in movement along with skeletal muscles.
4. Stores and releases minerals.
5. Contains red bone marrow, which produces blood cells.
6. Contains yellow bone marrow, which stores triglycerides (fats).

? **What is the functional significance of the periosteum?**

tendons and ligaments, attach to the bone. (*Tendons* attach skeletal muscle to bone; *ligaments* attach one bone to another bone.) Groove-like impressions or holes indicate locations where nerves and blood vessels pass through the bones. Smooth surfaces indicate areas of movement between neighboring bones. In general, a smooth-surface depression on one bone accommodates a smooth-surfaced projection from another bone to form the joint surfaces so that bones can fit together. In response to tension on

a bone surface where tendons, ligaments, aponeuroses, and fasciae pull on the periosteum of bone, new bone is deposited, resulting in raised or roughened areas.

There are two major types of surface markings: (1) *depressions and openings*, which form joints or allow the passage of soft tissues (such as blood vessels and nerves), and (2) *processes*, projections or outgrowths that either help form joints or serve as attachment points for connective tissue (such as ligaments and tendons).

Table 6.1 describes the various surface markings and provides examples of each.

◉ **CHECKPOINT**

5. List and describe several bone surface markings, and give an example of each. Check your list against Table 6.1.

HISTOLOGY OF BONE TISSUE

◉ **OBJECTIVE**

• Describe the histological features of bone tissue.

We will now examine the structure of bone at the microscopic level. Like other connective tissues, **bone**, or **osseous tissue** (OS-ē-us), contains an abundant extracellular matrix that surrounds widely separated cells. The extracellular matrix is about 25% water, 25% collagen fibers, and 50% crystallized mineral salts. The most abundant mineral salt is calcium phosphate $[Ca_3(PO_4)_2]$. It combines with another mineral salt, calcium hydroxide $[Ca(OH)_2]$, to form crystals of **hydroxyapatite** $[Ca_{10}(PO_4)_6(OH)_2]$. As the crystals form, they combine with still other mineral salts, such as calcium carbonate $(CaCO_3)$, and ions such as magnesium, fluoride, potassium, and sulfate. As these mineral salts are deposited in the framework formed by the collagen fibers of the extracellular matrix, they crystallize and the tissue hardens. This process, called **calcification** (kal-si-fi-KĀ-shun), is initiated by bone-building cells called *osteoblasts*, to be described shortly.

It was once thought that calcification simply occurred when enough mineral salts were present to form crystals. Now, however, we know that the process occurs only in the presence of collagen fibers. Mineral salts begin to crystallize in the microscopic spaces between collagen fibers. After the spaces are filled, mineral crystals accumulate around the collagen fibers. The combination of crystallized salts and collagen fibers is responsible for the hardness that is characteristic of bone.

Although a bone's *hardness* depends on the crystallized inorganic mineral salts, a bone's *flexibility* depends on its collagen fibers. Like reinforcing metal rods in concrete, collagen fibers and other organic molecules provide *tensile strength*, which is

TABLE 6.1

Bone Surface Markings

MARKING	DESCRIPTION	EXAMPLE
DEPRESSIONS AND OPENINGS: Sites allowing the passage of soft tissue (nerves, blood vessels, ligaments, tendons) or formation of joints		
Fissure (FISH-ur)	Narrow slit between adjacent parts of bones through which blood vessels or nerves pass.	Superior orbital fissure of the sphenoid bone (Figure 7.6f and 7.10b on pages 177 and 186).
Foramen (fō-RĀ-men=hole; plural is *foramina*)	Opening through which blood vessels, nerves, or ligaments pass.	Optic foramen of the sphenoid bone (Figure 7.6f and 7.10b on pages 177 and 186).
Fossa (FOS-a=trench; plural is *fossae*, FOS-ē)	Shallow depression (*fossa*=trench).	Coronoid fossa of the humerus (Figure 8.4a on page 220).
Sulcus (SUL-kus=groove; plural is *sulci*, SUL-sī)	Furrow along a bone surface that accommodates a blood vessel, nerve, or tendon.	Intertubercular sulcus of the humerus (Figure 8.4a on page 220).
Meatus (mē-Ā-tus= passageway; plural is *meati*, me-Ā-tī)	Tubelike opening.	External and internal auditory meatus of the temporal bone (Figure 7.2a on page 168).
PROCESSES: Projections or outgrowths on bone that form joints or attachment points for connective tissue, such as ligaments and tendons.		
Processes that form joints:		
Condyle (KON-dil; *condylus*=knuckle)	Large, round protuberance at the end of a bone.	Lateral condyle of the femur (Figure 8.10a on page 236).
Facet (FAS-et or fa-SET)	Smooth flat articular surface.	Superior articular facet of a vertebra (Figure 7.16a on page 195).
Head	Rounded articular projection supported on the neck (constricted portion) of a bone.	Head of the femur (Figure 8.10a on page 236).
Processes that form attachment points for connective tissue:		
Crest	Prominent ridge or elongated projection.	Iliac crest of the hip bone (Figure 8.8a on page 230).
Epicondyle (*epi-*=above)	Projection above a condyle.	Medial epicondyle of the femur (Figure 8.10a on page 236).
Line	Long, narrow ridge or border (less prominent than a crest).	Linea aspera of the femur (Figure 8.10b on page 236).
Spinous process	Sharp, slender projection.	Spinous process of a vertebra (Figure 7.16a on page 195).
Trochanter (tro-KAN-ter)	Very large projection.	Greater trochanter of the femur (Figure 8.10a on page 236).
Tubercle (TOO-ber-kul; *tuber*=knob)	Small, rounded projection.	Greater tubercle of the humerus (Figure 8.4a on page 220).
Tuberosity	Large, rounded, usually roughened projection.	Ischial tuberosity of the hip bone (Figure 8.8b on page 230).

resistance to being stretched or torn apart. Soaking a bone in an acidic solution, such as vinegar, dissolves its mineral salts, causing the bone to become rubbery and flexible. As you will see shortly, bone cells called *osteoclasts* secrete enzymes and acids that break down the extracellular matrix of bone.

Four types of cells are present in bone tissue: osteogenic cells, osteoblasts, osteocytes, and osteoclasts (Figure 6.3).

1. **Osteogenic cells** (os-tē-ō-JEN-ik; *-genic*=producing) are unspecialized stem cells derived from mesenchyme, the tissue from which almost all connective tissues are formed. They are the only bone cells to undergo cell division; the resulting cells develop into osteblasts. Osteogenic cells are found along the inner portion of the periosteum, in the endosteum, and in the canals within bone that contain blood vessels.

2. **Osteoblasts** (OS-tē-ō-blasts; *-blasts*=buds or sprouts) are bone-building cells. They synthesize and secrete collagen fibers and other organic components needed to build the extracellular matrix of bone tissue, and they initiate calcification. As osteoblasts surround themselves with extracellular matrix, they become trapped in their secretions and become osteocytes. (Note: Cells with the suffix *blast* in bone or any other connective tissue secrete extracellular matrix.)

3. **Osteocytes** (OS-tē-ō-sīts; *-cytes*=cells), mature bone cells, are the main cells in bone tissue and maintain its daily metabolism, such as the exchange of nutrients and wastes with the blood. Like osteoblasts, osteocytes do not undergo cell division. (Note: Cells with the suffix *cyte* in bone or any other tissue maintain the tissue.)

4. **Osteoclasts** (OS-tē-ō-clasts; *-clast*=break) are huge cells derived from the fusion of as many as 50 monocytes (a type of white blood cell) and are concentrated in the endosteum. On the side of the cell that faces the bone surface, the osteoclast's plasma membrane is deeply folded into a *ruffled border*. Here

the cell releases powerful lysosomal enzymes and acids that digest the protein and mineral components of the underlying extracellular matrix of bone. This breakdown of the extracellular matrix of bone, termed **resorption** (rē-SORP-shun), is part of the normal development, growth, maintenance, and repair of bone. (Note: Cells with the suffix *clast* in bone break down extracellular matrix.) As you will see later in response to certain hormones, osteoclasts help regulate blood calcium level. They are also target cells for drug therapy used to treat osteoporosis.

As with other tissues, anatomists recognize various categories of bone tissue. Two distinct arrangements of the mineralized collagen extracellular matrix constitute the two main types of bone tissue: *reticulofibrous (woven) bone* and *lamellar bone.*

Reticulofibrous bone, also called **woven bone** or **primary bone**, is found at the initial sites of bone development during fetal life and at sites of bone repair in mature bones. It consists of a mixed array of fine and coarse mineralized collagen fibers in an irregular arrangement, giving the appearance of a woven fabric. Reticulofibrous bone is a rapidly deposited tissue characterized by highly active osteoblasts and large (70–200 μm) vascular canals.

The most common type of bone tissue, comprising almost the entire mature skeleton, is **lamellar bone**. Unlike reticulofibrous bone, lamellar bone consists of mineralized collagen fibers arranged in highly ordered plates, or *lamellae (lamina*=thin plate), which are each approximately 7 μm thick. The arrangement of the lamellar plates of mineralized extracellular matrix varies within the tissue, but they are always closely associated with bone cells and blood vessels. Based on the arrangement of lamellae and the overall porosity (the volume occupied by "pores" or empty spaces compared to the total volume of the tissue), lamellar bone is placed into two subcategories: *compact bone* and *spongy bone* (see Figure 6.2). Overall, about 80% of the skeleton is compact bone and 20% is spongy bone.

Figure 6.3 **Types of cells in bone tissue.**

Osteogenic cells undergo cell division and develop into osteoblasts, which secrete bone extracellular matrix.

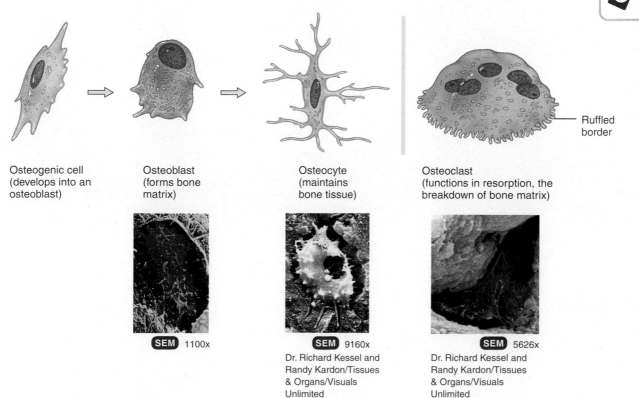

Osteogenic cell
(develops into an
osteoblast)

Osteoblast
(forms bone
matrix)

Osteocyte
(maintains
bone tissue)

Osteoclast
(functions in resorption, the
breakdown of bone matrix)

Ruffled
border

SEM 1100x

SEM 9160x
Dr. Richard Kessel and
Randy Kardon/Tissues
& Organs/Visuals
Unlimited

SEM 5626x
Dr. Richard Kessel and
Randy Kardon/Tissues
& Organs/Visuals
Unlimited

? **Why is bone resorption important?**

Compact Bone Tissue

Compact (cortical) bone is the type of bone tissue observed at the surface of a bone, but it also can extend deeper into the bone tissue and makes up the bulk of the diaphyses of long bones (see Figure 6.2). Compact bone tissue contains small spaces; in fact, to the unaided eye, compact bone looks like a dense, solid material. However, when viewed under a microscope compact bone is quite porous, with an abundance of microscopic spaces and canals. It provides protection and support and resists the stresses produced by weight and movement.

The surface layers of the compact bone form as thin concentric layers of bone called **circumferential lamellae**. These thin layers of superficial bone are deposited by osteoblasts that reside in the periosteum. The structural arrangement of the circumferential lamellae helps the bone resist torsional (twisting) strains applied to the bone (Figure 6.4a).

Beneath the circumferential lamellae, which form the surface of the compact bone, the lamellar structure changes. In this deeper compact bone, the **osteon**, or **haversian system** (ha-VER-shan), is the basic structural unit. Resembling the growth rings of a tree, the osteon is a series of **concentric lamellae**, circular plates of mineralized extracellular matrix of increasing diameter, surrounding a small network of blood vessels, lymphatics, and nerves located in a hollow **central canal** (Figure 6.4a). These tube-like units of bone generally form a series of parallel cylinders that, in long bones, tend to run parallel to the long axis of the bone. Between the concentric lamellae are small spaces called **lacunae** (la-KOO-nē = little lakes; singular is *lacuna*), which contain osteocytes. Radiating in all directions from the lacunae are tiny **canaliculi** (kan-a-LIK-ū-lī = small channels), which are filled with extracellular fluid. Inside the canaliculi are slender fingerlike processes of osteocytes (see inset at right of Figure 6.4a). Neighboring osteocytes communicate via gap junctions. The canaliculi connect lacunae with one another and with the central canals, forming an intricate, miniature system of interconnected canals throughout the bone. This system provides many routes for nutrients and oxygen to reach the osteocytes and for the removal of wastes.

Figure 6.4 Histology of compact and spongy bone. (a) Sections through the diaphysis of a long bone, from the surrounding periosteum on the right, to compact bone in the middle, to spongy bone and the medullary cavity on the left. The inset at the upper right shows an osteocyte in a lacuna. (b and c) Details of spongy bone. See part (d) for a photomicrograph of compact bone tissue and Figure 6.13a for a scanning electron micrograph of spongy bone tissue.

Bone tissue is arranged in concentric lamellae around a central (haversian) canal in compact bone, and in irregularly arranged lamellae in the trabeculae in spongy bone.

(a) Osteons (haversian systems) in compact bone and trabeculae in spongy bone

Osteons in compact bone tissue are aligned in the same direction along lines of stress. In the shaft, for example, they are parallel to the long axis of the bone. As a result, the shaft of a long bone resists bending or fracturing even when considerable force is applied from either end. Compact bone tissue tends to be thickest in those parts of a bone where stresses are applied in relatively few directions. The lines of stress in a bone are not static. They change as a person learns to walk and in response to repeated strenuous physical activity, such as occurs when a person undertakes weight training. The lines of stress in a bone also can change because of fractures or physical deformity. Thus, the organization of osteons is not static but changes over time in response to the physical demands placed on the skeleton.

The areas between neighboring osteons contain **interstitial lamellae**, which also have lacunae with osteocytes and canaliculi. Interstitial lamellae are fragments of older osteons that have been partially destroyed during bone rebuilding or growth.

Blood vessels, lymphatic vessels, and nerves from the periosteum penetrate the compact bone through transverse **perforating** or **Volkmann's canals** (FOLK-mans). The vessels and nerves of the perforating canals connect with those of the medullary cavity, periosteum, and central canals.

Spongy Bone Tissue

In contrast to compact bone tissue, **spongy (trabecular) bone tissue** does not contain osteons (Figure 6.4b, c). It is always located in the *interior* of a bone, protected by a covering of compact bone. It consists of lamellae that are arranged in an irregular lattice of thin columns of bone called **trabeculae** (tra-BEK-ū-lē =little beams; singular is *trabecula*). Between these small beams of bone are spaces that are visible to the unaided eye. These macroscopic spaces between the trabeculae are filled with red bone marrow in some bones, which produces blood cells, and yellow

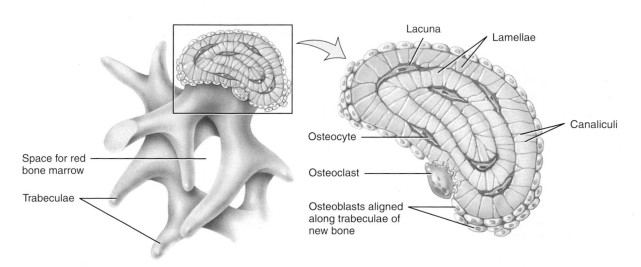

Lacuna

Lamellae

Osteocyte

Osteoclast

Osteoblasts aligned along trabeculae of new bone

Canaliculi

Space for red bone marrow

Trabeculae

(b) Enlarged aspect of spongy bone trabeculae

(c) Details of a section of a trabecula

LM 100x

LM 400x

Canaliculi

Central (haversian) canal

Lacuna

Lamellae

(d) Sectional view of several osteons (haversian systems) of femur (thigh bone) and details of one osteon

? As people age, some central (haversian) canals may become blocked. What effect would this have on the osteocytes?

bone marrow (adipose tissue) in other bones. Both these types of bone marrow contain numerous small blood vessels. Within each trabecula are osteocytes that lie in lacunae. Radiating from the lacunae are canaliculi. Osteocytes receive nourishment directly from the blood circulating through the blood vessels in the spaces between trabeculae.

Spongy bone tissue makes up most of the interior bone tissue of short, flat, sesamoid, and irregularly shaped bones; all but the paper-thin compact bone covering of the epiphyses of long bones; and a narrow rim around the medullary cavity of the diaphysis of long bones. Spongy bone is always covered by a layer of compact bone for protection.

At first glance, the trabeculae of spongy bone tissue may appear to be less organized than the osteons of compact bone tissue. However, the trabeculae of spongy bone tissue are precisely oriented along lines of stress, a characteristic that helps bones resist stresses and transfer force without breaking. Spongy bone tissue tends to be located where bones are not heavily stressed or where stresses are applied from many directions.

Spongy bone tissue is different from compact bone tissue in two respects. First, spongy bone tissue is light, which reduces the overall weight of a bone. This reduction in weight allows the

• CLINICAL **CONNECTION** | Bone Scan

Hot spot (white area) in a cancerous vertebra

A **bone scan** is a diagnostic procedure that takes advantage of the fact that bone is living tissue. A small amount of a radioactive tracer compound that is readily absorbed by bone is injected intravenously. The degree of uptake of the tracer is related to the amount of blood flow to the bone. A scanning device (gamma camera) measures the radiation emitted from the bones, and the information is translated into a photograph that can be read like an x-ray on a monitor. Normal bone tissue is identified by a consistent gray color throughout because of its uniform uptake of the radioactive tracer. Darker or lighter areas may indicate bone abnormalities. Darker areas called "hot spots" are areas of increased metabolism that absorb more of the radioactive tracer due to increased blood flow. Hot spots may indicate bone cancer, abnormal healing of fractures, or abnormal bone growth. Lighter areas called "cold spots" are areas of decreased metabolism that absorb less of the radioactive tracer due to decreased blood flow. Cold spots may indicate problems such as degenerative bone disease, decalcified bone, fractures, bone infections, Paget's disease, and rheumatoid arthritis. A bone scan detects abnormalities 3 to 6 months sooner than standard x-ray procedures and exposes the patient to less radiation. A bone scan is the standard test for bone density screening, particularly for osteoporosis in females. •

bone to move more readily when pulled by a skeletal muscle. Second, the trabeculae of spongy bone tissue support and protect the red bone marrow. Spongy bone in the hip bones, ribs, sternum (breastbone), vertebrae, and the proximal ends of the humerus and femur is the only site where red bone marrow is stored and, thus, the site where hemopoiesis (blood cell production) occurs in adults.

◉ CHECKPOINT

6. Why is bone considered a connective tissue?
7. Describe the four types of cells in bone tissue.
8. What is the composition of the extracellular matrix of bone tissue?
9. Distinguish between spongy and compact bone tissue in terms of microscopic appearance, location, and function.

BLOOD AND NERVE SUPPLY OF BONE

◉ OBJECTIVE

• Describe the blood and nerve supply of bone.

Bone is richly supplied with blood. Blood vessels, which are especially abundant in portions of bone containing red bone marrow, pass into bones from the periosteum. We will consider the blood supply of a long bone such as the mature tibia (shin bone) shown in Figure 6.5.

Periosteal arteries (per′-ē-OS-tē-al) accompanied by nerves enter the diaphysis through numerous perforating (Volkmann's) canals and supply the periosteum and outer part of the compact bone (see Figure 6.4a). Near the center of the diaphysis, a large **nutrient artery** enters the compact bone through a hole called the **nutrient foramen** (Figure 6.5). On entering the medullary cavity, the nutrient artery divides into proximal and distal branches that supply both the inner part of compact bone tissue of the diaphysis and the spongy bone tissue and red marrow as far as the epiphyseal plates (or lines). Some bones, like the tibia, have only one nutrient artery; others like the femur (thigh bone) have several. The ends of long bones are supplied by the metaphyseal and epiphyseal arteries, which arise from arteries that supply the associated joint. The **metaphyseal arteries** (met-a-FIZ-ē-al) enter the metaphyses of a long bone and, together with the nutrient artery, supply the red bone marrow and bone tissue of the metaphyses. The **epiphyseal arteries** (ep-i-FIZ-ē-al) enter the epiphyses of a long bone and supply the red bone marrow and bone tissue of the epiphyses.

Veins that carry blood away from long bones are evident in three places: (1) One or two **nutrient veins** accompany the nutrient artery and exit in the diaphysis; (2) numerous **epiphyseal veins** and **metaphyseal veins** accompany their respective arteries and exit in the epiphyses; and (3) many small **periosteal veins** accompany their respective arteries and exit the periosteum. The periosteum surrounding the bone has numerous lymphatic capillaries and lymph vessels, but there is no evidence of any lymphatic vessels within the bone tissue.

Nerves accompany the blood vessels that supply bones. The periosteum is rich in sensory nerves, some of which carry pain sensations. These nerves are especially sensitive to tearing or tension, which explains the severe pain resulting from a fracture or a bone tumor. For the same reason there is some pain associated with a bone marrow needle biopsy. In this procedure, a needle is inserted into the middle of the bone to withdraw a sample of red bone marrow for microscopic examination for conditions such as

Figure 6.5 Blood supply of a mature long bone, the tibia (shinbone).

Bone is richly supplied with blood vessels.

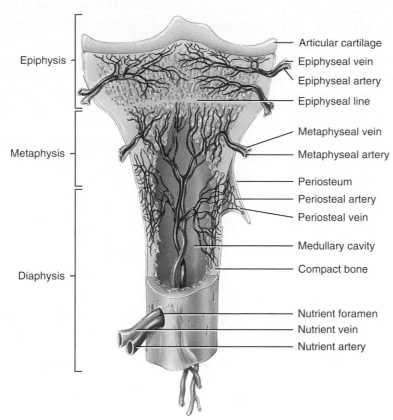

? Where do periosteal arteries enter bone tissue?

leukemias, metastatic neoplasms, lymphoma, Hodgkin's disease, and aplastic anemia. As the needle penetrates the periosteum, pain is felt. Once it passes through, there is little pain.

◉ CHECKPOINT

10. Explain the location and roles of the nutrient arteries, nutrient foramina, epiphyseal arteries, and periosteal arteries.

BONE FORMATION

◉ OBJECTIVES

- Explain the importance of bone formation during different phases of a person's lifetime.
- Describe the process of fracture repair.

The process by which bone forms is called **ossification** (os'-i-fi-KĀ-shun; *ossi-*=bone; *-fication*=making). Bone formation occurs in four principal situations: (1) the initial formation of bones in an embryo and fetus, (2) the growth of bones during infancy, childhood, and adolescence until their adult sizes are reached, (3) the remodeling of bone (replacement of old bone by new bone tissue throughout life), and (4) the repair of fractures (breaks in bones) throughout life.

Initial Bone Formation in an Embryo and Fetus

We will first consider the initial formation of bone in an embryo and fetus. The embryonic "skeleton" is at first composed of mesenchyme shaped like bones which become the sites where subsequent cartilage formation and then ossification occurs. This

begins during the sixth week of embryonic development and follows one of two patterns.

The two methods of bone formation, which both involve the replacement of a preexisting connective tissue with bone, do not lead to differences in the structure of mature bones, but are simply different methods of bone development. In the first type of ossification, called **intramembranous ossification** (in'-tra-MEM-bra-nus; *intra-*=within; *membran*=membrane), bone forms directly within mesenchyme arranged in sheetlike layers that resemble membranes. In the second type, **endochondral ossification** (en'-dō-KON-dral; *endo-*=within; *chondral*=cartilage), bone forms within hyaline cartilage that develops from mesenchyme.

Intramembranous Ossification

Intramembranous ossification is the simpler of the two methods of bone formation. The flat bones of the skull and mandible (lower jawbone) are formed in this way. Also, the "soft spots" that help the fetal skull pass through the birth canal later harden as they undergo intramembranous ossification, which occurs as follows (Figure 6.6 on page 150):

① **Development of the ossification center.** At the site where the bone will develop, specific chemical messages cause the mesenchymal cells to cluster together and differentiate, first into osteogenic cells and then into osteoblasts. The site of such a cluster is called an **ossification center**. Osteoblasts secrete the organic extracellular matrix of bone until they are surrounded by it.

❷ Calcification. Next, the secretion of extracellular matrix stops and the cells, now called osteocytes, lie in lacunae and extend their narrow cytoplasmic processes into canaliculi that radiate in all directions. Within a few days, calcium and other mineral salts are deposited and the extracellular matrix hardens or calcifies (calcification).

❸ Formation of trabeculae. As the bone extracellular matrix forms, it develops into trabeculae that fuse with one another to form spongy bone. Blood vessels grow into the spaces between the trabeculae. Connective tissue that is associated with the blood vessels in the trabeculae differentiates into red bone marrow.

❹ Development of the periosteum. In conjunction with the formation of trabeculae, the mesenchyme at the periphery of the bone condenses and develops into the periosteum. Eventually, a thin layer of compact bone replaces the surface layers of the spongy bone, but spongy bone remains in the center. Much of the newly formed bone is remodeled (destroyed and reformed) as the bone is transformed into its adult size and shape.

Endochondral Ossification

The replacement of cartilage by bone is called **endochondral ossification**. Although most bones of the body are formed in this way, the process is best observed in a long bone. It proceeds as follows (Figure 6.7):

❶ Development of the cartilage model. At the site where the bone is going to form, specific chemical messages cause the mesenchymal cells to crowd together in the shape of the future bone, and then develop into chondroblasts. The chondroblasts secrete cartilage extracellular matrix, producing a **cartilage model** consisting of hyaline cartilage. A covering called the **perichondrium** (per-i-KON-drē-um) develops around the cartilage model.

❷ Growth of the cartilage model. Once chondroblasts become deeply buried in the cartilage extracellular matrix, they are called chondrocytes. The cartilage model grows in length by continual cell division of chondrocytes accompanied by further secretion of the cartilage extracellular matrix. This type of growth is termed **interstitial growth**

Figure 6.6 Intramembranous ossification. Illustrations ❶ and ❷ show a smaller field of vision at higher magnification than illustrations ❸ and ❹. Refer to this figure as you read the corresponding numbered paragraphs in the text.

🔑 Intramembranous ossification involves the formation of bone within mesenchyme arranged in sheet-like layers that resemble membranes.

❶ Development of ossification center

❷ Calcification

❸ Formation of trabeculae

❹ Development of the periosteum

❓ **Which bones of the body develop by intramembranous ossification?**

(growth from within) and results in an increase in length. In contrast, growth of the cartilage in thickness is due mainly to the addition of more extracellular matrix material to the periphery of the model by new chondroblasts that develop from the perichondrium. This growth pattern, in which extracellular matrix is deposited on the cartilage surface, is called **appositional growth** (a-pō-ZISH-i-nal), meaning growth of the outer surface.

As the cartilage model continues to grow, chondrocytes in its midregion hypertrophy (increase in size) and the surrounding cartilage extracellular matrix begins to calcify. Other chondrocytes within the calcifying cartilage die because nutrients can no longer diffuse quickly enough through the extracellular matrix. As these chon-

drocytes die, lacunae form and eventually merge into small cavities.

❸ **Development of the primary ossification center.** Primary ossification proceeds *inward* from the external surface of the bone. A nutrient artery penetrates the perichondrium and the calcifying cartilage model through a nutrient foramen in the midregion of the cartilage model, stimulating osteogenic cells in the perichondrium to differentiate into osteoblasts. Once the perichondrium starts to form bone, it is known as the **periosteum**. Near the middle of the model, periosteal capillaries grow into the disintegrating calcified cartilage, inducing growth of a **primary ossification center**, a region where bone tissue will replace most of the cartilage. Osteoblasts then begin to deposit bone extracellular matrix

Figure 6.7 Endochondral ossification.

During endochondral ossification, bone gradually replaces a cartilage model.

1 Development of cartilage model

2 Growth of cartilage model

3 Development of primary ossification center

4 Development of the medullary cavity

5 Development of secondary ossification center

6 Formation of articular cartilage and epiphyseal plate

❓ If radiographs of an 18-year-old basketball star player show clear epiphyseal plates but no epiphyseal lines, is she likely to grow taller?

over the remnants of calcified cartilage, forming spongy bone trabeculae.

④ **Development of the medullary (marrow) cavity.** As the primary ossification center grows toward the ends of the bone, osteoclasts break down some of the newly formed spongy bone trabeculae. This activity leaves a cavity, the medullary (marrow) cavity, in the diaphysis (shaft). Eventually, most of the wall of the diaphysis is replaced by compact bone.

⑤ **Development of the secondary ossification centers.** When branches of the epiphyseal artery enter the epiphyses, **secondary ossification centers** develop, usually around the time of birth. Bone formation is similar to that in primary ossification centers. One difference, however, is that spongy bone remains in the interior of the epiphyses (no medullary cavities are formed here). In contrast to primary ossification, secondary ossification proceeds *outward* from the center of the epiphysis toward the outer surface of the bone.

⑥ **Formation of articular cartilage and the epiphyseal plate.** The hyaline cartilage that covers the epiphyses becomes the articular cartilage. Prior to adulthood, hyaline cartilage remains between the diaphysis and epiphysis as the epiphyseal (growth) plate, which is responsible for the lengthwise growth of long bones and is discussed next.

◉ CHECKPOINT

11. What are the major events of intramembranous ossification and endochondral ossification and how are they different?

Bone Growth During Infancy, Childhood, and Adolescence

During infancy, childhood, and adolescence, bones throughout the body grow in thickness by appositional growth, and long bones lengthen by the addition of bone material on the diaphyseal side of the epiphyseal plate by interstitial growth.

Growth in Length

The growth in length of a long bone involves (1) interstitial growth of cartilage on the epiphyseal side of the epiphyseal plate and (2) replacement of cartilage with bone by endochondral ossification on the diaphyseal side of the epiphyseal plate.

To understand how a bone grows in length, you need to know some of the details of the structure of the epiphyseal plate (Figure 6.8). The **epiphyseal (growth) plate** (ep′-i-FIZ-ē-al) is a layer of hyaline cartilage in the metaphysis of a growing bone that consists of four zones (Figure 6.8b):

1. **Zone of resting cartilage.** This layer is nearest the epiphysis and consists of small, scattered chondrocytes. The term *resting* is used because the cells do not function in bone growth. Rather, they anchor the epiphyseal plate to the epiphysis of the bone.

2. **Zone of proliferating cartilage.** Slightly larger chondrocytes in this zone are arranged like stacks of coins. These chondrocytes undergo interstitial growth as they divide and secrete extracellular matrix. The chondrocytes in this zone divide to replace those that die at the diaphyseal side of the epiphyseal plate.

3. **Zone of hypertrophic cartilage** (hī-per-TRŌ-fik). This layer consists of large, maturing chondrocytes arranged in columns.

4. **Zone of calcified cartilage.** The final zone of the epiphyseal plate is only a few cells thick and consists mostly of chondrocytes

Figure 6.8 **The epiphyseal plate is a layer of hyaline cartilage in the metaphysis of a growing bone.** The epiphyseal plate appears as a dark band between whiter calcified areas in the radiograph (x-ray) shown in part (a).

⑥ 🔑 The epiphyseal plate allows the diaphysis of a bone to increase in length.

(a) Radiograph showing the epiphyseal plate of the femur of a 3-year-old

Femur

EPIPHYSEAL PLATE

Tibia

Diaphyseal side

Developing bone of diaphysis

Zone of calcified cartilage

Zone of hypertrophic cartilage

Zone of proliferating cartilage

Zone of resting cartilage

Epiphyseal side **LM** 400x

(b) Histology of the epiphyseal plate

❓ **What activities of the epiphyseal plate account for the lengthwise growth of the diaphysis?**

that are dead because the extracellular matrix around them has calcified. Osteoclasts dissolve the calcified cartilage, and osteoblasts and capillaries from the diaphysis invade the area. The osteoblasts lay down bone extracellular matrix, replacing the calcified cartilage by the process of endochondral ossification. As a result, the zone of calcified cartilage becomes "new diaphysis" that is firmly cemented to the rest of the diaphysis of the bone.

The activity of the epiphyseal plate is the only way that the diaphysis can increase in length. As a bone grows, chondrocytes proliferate on the epiphyseal side of the plate. New chondrocytes cover older ones, which are then destroyed by calcification. Thus, the cartilage is replaced by bone on the diaphyseal side of the plate. In this way the thickness of the epiphyseal plate remains relatively constant, but the bone on the diaphyseal side increases in length.

When adolescence comes to an end at about age 18 in females and 21 in males, the epiphyseal plates close; that is, the epiphyseal cartilage cells stop dividing and bone replaces all the cartilage. The epiphyseal plate fades, leaving a bony structure called the **epiphyseal line**. With the appearance of the epiphyseal line, bone growth in length stops. If a bone fracture damages the epiphyseal plate, the fractured bone may be shorter than normal once adult stature is reached. This is because damage to cartilage, which is avascular, accelerates closure of the epiphyseal plate, thus inhibiting lengthwise growth of the bone.

Growth in Thickness

Unlike cartilage, which can thicken by both interstitial and appositional growth, bone can grow in thickness (diameter) only by **appositional growth** (Figure 6.9):

❶ At the bone surface, periosteal cells differentiate into osteoblasts, which secrete the collagen fibers and other organic molecules that form bone extracellular matrix. The osteoblasts become surrounded by extracellular matrix and develop into osteocytes. This process forms bone ridges on either side of a periosteal blood vessel. The ridges slowly enlarge and create a groove for the periosteal blood vessel.

❷ Eventually, the ridges fold together and fuse, and the groove becomes a tunnel that encloses the blood vessel. The former periosteum now becomes the endosteum that lines the tunnel.

❸ Osteoblasts in the endosteum deposit bone extracellular matrix, forming new concentric lamellae. The formation of additional concentric lamellae proceeds inward toward the periosteal blood vessel. In this way, the tunnel fills in, and a new osteon is created.

❹ As an osteon is forming, osteoblasts under the periosteum deposit new circumferential lamellae, further increasing the thickness of the bone. As additional periosteal blood vessels become enclosed as in step ❶, the growth process continues.

Recall that as new bone tissue is being deposited on the outer surface of bone, the bone tissue lining the medullary cavity is destroyed by osteoclasts in the endosteum. In this way, the medullary cavity enlarges as the bone increases in thickness.

▣ CHECKPOINT

12. Describe the zones of the epiphyseal plate and their functions.
13. Explain how bone grows in length and thickness.
14. What is the significance of the epiphyseal line?

Figure 6.9 Bone growth in thickness.

As new bone tissue is deposited on the outer surface of bone by osteoblasts, the bone tissue lining the medullary cavity is destroyed by osteoclasts in the endosteum.

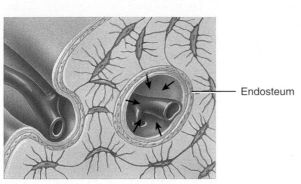

❶ Ridges in periosteum create groove for periosteal blood vessel.

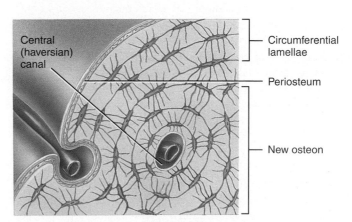

❷ Periosteal ridges fuse, forming an endosteum-lined tunnel.

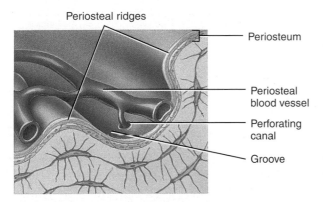

❸ Osteoblasts in endosteum build new concentric lamellae inward toward center of tunnel, forming a new osteon.

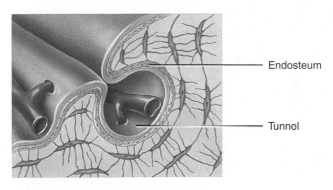

❹ Bone grows outward as osteoblasts in periosteum build new circumferential lamellae. Osteon formation repeats as new periosteal ridges fold over blood vessels.

❓ **How does the medullary cavity enlarge during growth in thickness?**

Remodeling of Bone

Like skin, bone forms before birth but continually renews itself thereafter. **Bone remodeling** is the ongoing replacement of old bone tissue by new bone tissue. It involves **bone resorption**, the removal of minerals and collagen fibers from bone by osteoclasts, and **bone deposition**, the addition of minerals and collagen fibers to bone by osteoblasts. Thus, bone resorption results in the destruction of bone extracellular matrix, while bone deposition results in the formation of bone extracellular matrix. At any given time, about 5% of the total bone mass in the body is being remodeled. The renewal rate for compact bone tissue is about 4% per year and for spongy bone tissue it is about 20% per year. Remodeling also takes place at different rates in different regions of the body. The distal portion of the femur is replaced about every four months. By contrast, bone in certain areas of the shaft of the femur will not be replaced completely during an individual's life. Even after bones have reached their adult shapes and sizes, old bone is continually destroyed and new bone is formed in its place. Remodeling also removes injured bone, replacing it with new bone tissue. Remodeling may be triggered by factors such as exercise, sedentary lifestyle, and changes in diet.

Remodeling has several other benefits. Since the strength of bone is related to the degree to which it is stressed, if newly formed bone is subjected to heavy loads, it will grow thicker and therefore be stronger than the old bone. Also, the shape of a bone can be altered for proper support based on the stress patterns experienced during the remodeling process. Finally, new bone is more resistant to fracture than old bone.

A delicate balance exists between the actions of osteoclasts and osteoblasts. Should too much new tissue be formed, the bones become abnormally thick and heavy. If too much mineral material is deposited in the bone, the surplus may form thick bumps, called *spurs*, on the bone that interfere with movement at joints. Excessive loss of calcium or tissue weakens the bones, and they may break, as occurs in osteoporosis, or they may become too flexible, as in rickets and osteomalacia. Abnormal acceleration of the remodeling process results in a condition called *Paget's disease*, in which the newly formed bone, especially that of the pelvis, limbs, lower vertebrae, and skull, becomes hard and brittle and fractures easily.

◉ CHECKPOINT

15. What is bone remodeling? Why is it important?

● CLINICAL **CONNECTION** Osteoporosis

SEM 30x   SEM 30x

(A) Normal bone   (B) Osteoporotic bone

Comparison of spongy bone tissue from (A) a normal young adult and (B) a person with osteoporosis.

Osteoporosis (os′-tē-ō-pō-RŌ-sis; *por*=passageway;-*osis*=condition) is literally a condition of porous bones that affects 10 million people a year in the United States. In addition, 18 million people have low bone mass (osteopenia), which puts them at risk for osteoporosis. The basic problem is that bone resorption (destruction) outpaces bone deposition (formation). In large part this is due to depletion of calcium from the body—more calcium is lost in urine, feces, and sweat than is absorbed from the diet. Bone mass becomes so depleted that bones fracture, often spontaneously, under the mechanical stresses of everyday living. For example, a hip fracture might result from simply sitting down too quickly. In the United States, osteoporosis causes more than one and a half million fractures a year, mainly in the hip, wrist, and vertebrae. Osteoporosis afflicts the entire skeletal system. In addition to fractures, osteoporosis causes shrinkage of vertebrae, height loss, hunched backs, and bone pain.

Osteoporosis primarily affects middle-aged and elderly people, 80 percent of them women. Older women suffer from osteoporosis more often than men for two reasons: (1) Women's bones are less massive than men's bones, and (2) production of estrogens in women declines dramatically at menopause, while production of the main androgen,

testosterone, in older men wanes gradually and only slightly. Estrogens and testosterone stimulate osteoblast activity and synthesis of bone matrix. Besides gender, risk factors for developing osteoporosis include a family history of the disease, European or Asian ancestry, thin or small body build, an inactive lifestyle, cigarette smoking, a diet low in calcium and vitamin D, more than two alcoholic drinks a day, and the use of certain medications.

Osteoporosis is diagnosed by taking a family history and undergoing a *bone mineral density* (BMD) test. BMD tests are performed like x rays and measure bone density. They can also be used to confirm a diagnosis of osteoporosis, determine the rate of bone loss, and monitor the effects of treatment.

Treatment options for osteoporosis are varied. With regard to nutrition, a diet high in calcium is important to reduce the risk of fractures. Vitamin D is necessary for the body to utilize calcium. In terms of exercise, regularly performing weight-bearing exercises has been shown to maintain and build bone mass. Among these exercises are walking, jogging, hiking, climbing stairs, playing tennis, and dancing. Resistance exercises, such as weight lifting, build bone strength and muscle mass.

Medications used to treat osteoporosis are generally of two types: (1) antiroabsorptive drugs slow down the progression of bone loss and (2) bone-building drugs promote increasing bone mass. Among the anteroabsorptive drugs are (1) *biphosphonates*, which inhibit osteoclasts (Fosamax®, Actonel®, Boniva®, and calcitonin); (2) selective estrogen receptor modulators, which mimic the effects of estrogens without unwanted side effects (Raloxifene®, Evista®); and (3) estrogen replacement therapy (ERT), which replaces estrogens lost during and after menopause (Premarin®) and hormone replacement therapy, which replaces estrogens and progesterone lost during and after menopause (Prempro®). ERT helps maintain and increase bone mass after menopause. Women on ERT have a slight increased risk of stroke and blood clots. HRT also helps maintain and increase bone mass and women on HRT have an increased risk of heart disease, breast cancer, stroke, blood clots, and dementia.

Among the bone-forming drugs is parathyroid hormone (PTH), which stimulates osteoblasts to produce new bone (Forteo®). Others are under development. ●

FRACTURES

A **fracture** is any break in a bone. Fractures are named according to their severity, the shape or position of the fracture line, or even the physician who first described them. Among the common types of fractures are the following (Figure 6.10):

- **Open (compound) fracture:** The broken ends of the bone protrude through the skin (Figure 6.10a). Conversely, a closed (simple) fracture does not break the skin.

- **Comminuted fracture** (KOM-i-noo-ted; *com-*=together; *-minuted*=crumbled): The bone is spintered, crushed, or broken into pieces at the site of impact, and smaller bone fragments lie between the two main fragments (Figure 6.10b). This is the most difficult fracture to treat.

- **Greenstick fracture:** A partial fracture in which one side of the bone is broken and the other side bends; occurs only in children, whose bones are not yet fully ossified and contain more organic material than inorganic material (Figure 6.10c).

Figure 6.10 Types of bone fractures. Illustrations are shown on the left and radiographs (x-rays) are shown on the right.

A fracture is any break in a bone.

(a) Open fracture

(b) Comminuted fracture

(c) Greenstick fracture

(d) Impacted fracture

(e) Pott's fracture

(f) Colles' fracture

? What is a fibrocartilaginous callus?

- **Impacted fracture:** One end of the fractured bone is forcefully driven into the interior of the other (Figure 6.10d).
- **Pott's fracture:** A fracture of the distal end of the lateral leg bone (fibula), with serious injury of the distal tibial articulation (Figure 6.10e).
- **Colles' fracture** (KOL-ez): A fracture of the distal end of the lateral forearm bone (radius) in which the distal fragment is displaced posteriorly (Figure 6.10f).

In some cases, a bone may fracture without visibly breaking. A **stress fracture** is a series of microscopic fissures in bone that forms without any evidence of injury to other tissues. In healthy adults, stress fractures result from repeated, strenuous activities such as running, jumping, or aerobic dancing. Stress fractures are quite painful and also result from disease processes that disrupt normal bone calcification, such as osteoporosis. About 25% of stress fractures involve the tibia. Although standard x-ray images often fail to reveal the presence of stress fractures, they show up clearly in a bone scan.

The repair of bone fracture involves the following steps (Figure 6.11):

❶ **Formation of fracture hematoma.** Blood vessels crossing the fracture line are broken. As blood leaks from the torn ends of the vessels, it forms a mass of blood (usually clotted) around the site of the fracture. This clot, called a **fracture hematoma** (hē-ma-TŌ-ma; *hemat-*=blood; *-oma*=tumor), usually forms 6 to 8 hours after the injury. Because the circulation of blood stops at the site where the fracture hematoma forms, nearby bone cells die. Swelling and inflammation occur in response to dead bone cells, producing additional cellular debris. Phagocytes (neutrophils and macrophages) and osteoclasts begin to remove the dead or damaged tissue in and around the fracture hematoma. This stage may last up to several weeks.

❷ **Fibrocartilaginous callus formation.** Fibroblasts from the periosteum invade the fracture site and produce collagen fibers. In addition, cells from the periosteum develop into chondroblasts and begin to produce fibrocartilage in this region. These events lead to the development of a **fibrocartilaginous callus** (fī-brō-kar-ti-LAJ-i-nus), a mass of repair tissue consisting of collagen fibers and cartilage that bridges the broken ends of the bone. Formation of the fibrocartilaginous callus takes about 3 weeks.

❸ **Bony callus formation.** In areas closer to well-vascularized healthy bone tissue, osteogenic cells develop into osteoblasts,

Figure 6.11 Steps involved in repair of a bone fracture.

Bone heals more rapidly than cartilage because its blood supply is more plentiful.

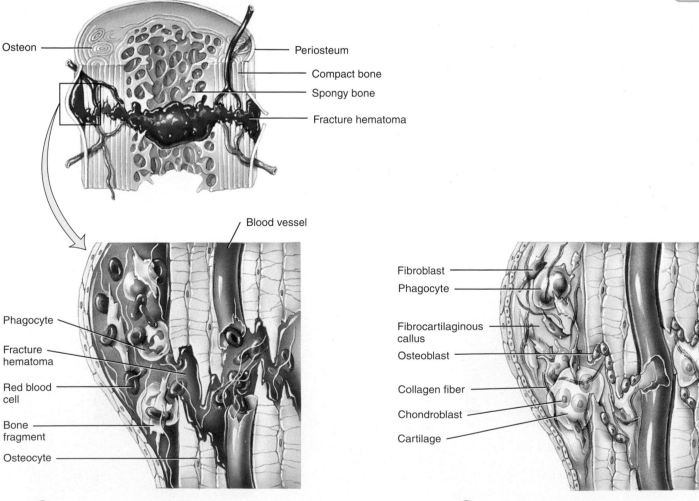

❶ Formation of fracture hematoma

❷ Fibrocartilaginous callus fomation

which begin to produce spongy bone trabeculae. The trabeculae join living and dead portions of the original bone fragments. In time, the fibrocartilage is converted to spongy bone, and the callus is then referred to as a **bony callus**. The bony callus lasts about 3 to 4 months.

④ **Bone remodeling.** The final phase of fracture repair is **bone remodeling** of the callus. Dead portions of the original fragments of broken bone are gradually resorbed by osteoclasts. Compact bone replaces spongy bone around the periphery of the fracture. Sometimes, the repair process is so thorough that the fracture line is undetectable, even in a radiograph (x-ray). However, a thickened area on the surface of the bone remains as evidence of a healed fracture, so that a healed bone may eventually be stronger than it was before the break.

Although bone has a generous blood supply, healing sometimes takes months. The calcium and phosphorus needed to strengthen and harden new bone are deposited only gradually, and bone cells generally grow and reproduce slowly. The temporary disruption in their blood supply also helps explain the slowness of healing of severely fractured bones.

◉ CHECKPOINT

16. Define a fracture and outline the four steps involved in fracture repair.

FACTORS AFFECTING BONE GROWTH

◉ OBJECTIVE

- Explain why minerals, vitamins, and hormones are important in bone growth.

Bone growth in the young, bone remodeling in the adult, and the repair of fractured bone depend on several factors and include the following:

1. **Minerals.** Large amounts of calcium and phosphorus are needed while bones are growing, as are smaller amounts of magnesium, fluoride, and manganese. These minerals are also necessary during bone remodeling.

2. **Vitamins.** Vitamin A stimulates activity of osteoblasts. Vitamin C is needed for synthesis of collagen, the main bone protein. As you will soon learn, vitamin D helps build bone by increasing the absorption of calcium from foods in the gastrointestinal tract into the blood. Vitamins K and B_{12} are also needed for synthesis of bone proteins.

3. **Hormones.** During childhood, the hormones most important to bone growth are the insulinlike growth factors (IGFs), which are produced by the liver and bone tissue. IGFs stimulate osteoblasts, promote cell division at the epiphyseal plate and in the periosteum, and enhance synthesis of the proteins needed to build new bone. IGFs are produced in response to the secretion of human growth hormone (hGH) from the anterior lobe of the pituitary gland. Thyroid hormones (T_3 and T_4) from the thyroid gland also promote bone growth by stimulating osteoblasts. In addition, the hormone insulin from the pancreas promotes bone growth by increasing the synthesis of bone proteins.

At puberty, the secretion of hormones known as sex hormones causes a dramatic effect on bone growth. The **sex hormones** include estrogens (produced by the ovaries) and androgens such as testosterone (produced by the testes). Although females have much higher levels of estrogens and males have higher levels of androgens, females also have low levels of androgens, and males have low levels of estrogens. The adrenal glands of both sexes produce androgens, and other tissues, such as adipose tissue, can convert androgens to estrogens. These hormones are responsible for increased osteoblast activity and synthesis of bone extracellular matrix and the sudden "growth spurt" that occurs during the teenage years. Estrogens also promote changes in the skeleton that are typical of females, such as widening of the pelvis. Ultimately sex hormones, especially estrogens in both sexes, shut down growth at epiphyseal plates, causing elongation of the

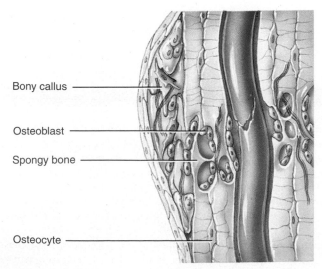

Bony callus

Osteoblast

Spongy bone

Osteocyte

③ Bony callus formation

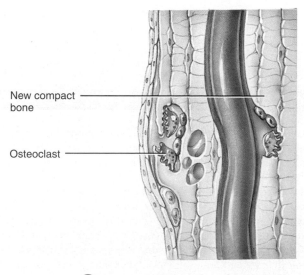

New compact bone

Osteoclast

④ Bone remodeling

❓ **Why does it sometimes take months for a fracture to heal?**

bones to cease. Lengthwise growth of bones typically ends earlier in females than in males due to their higher levels of estrogens.

During adulthood, sex hormones contribute to bone remodeling by slowing resorption of old bone and promoting deposition of new bone. One way that estrogens slow resorption is by promoting apoptosis (programmed death) of osteoclasts. Parathyroid hormone, calcitriol (the active form of vitamin D), and calcitonin are other hormones that can affect bone remodeling.

◉ **CHECKPOINT**
17. What factors affect bone growth?

EXERCISE AND BONE TISSUE

◉ **OBJECTIVE**
• Describe how exercise and mechanical stress affect bone tissue.

Within limits, bone tissue has the ability to alter its strength in response to changes in mechanical stress. When placed under stress, bone tissue becomes stronger through increased deposition of mineral salts and production of collagen fibers by osteoblasts. Without mechanical stress, bone does not remodel normally because bone resorption occurs more quickly than bone formation.

The main mechanical stresses on bone are those that result from the pull of skeletal muscles and the pull of gravity. If a person is bedridden or has a fractured bone in a cast, the strength of the unstressed bones diminishes because of the loss of bone minerals and decreased numbers of collagen fibers. Astronauts subjected to the microgravity of space also lose bone mass. In both cases, bone loss can be dramatic—as much as 1% per week. In contrast, the bones of athletes, which are repetitively and highly stressed, become notably thicker and stronger than those of astronauts or nonathletes. Weight-bearing activities, such as walking or moderate weight lifting, help build and retain bone mass. Adolescents and young adults should engage in regular weight-bearing exercise prior to the closure of the epiphyseal plates to help build total mass prior to its inevitable reduction with aging. However, people of all ages can and should strengthen their bones by engaging in weight-bearing exercise.

◉ **CHECKPOINT**
18. How do mechanical stresses strengthen bone tissue?

AGING AND BONE TISSUE

◉ **OBJECTIVE**
• Describe the effects of aging on bone tissue.

From birth through adolescence, more bone tissue is produced than is lost during bone remodeling. In young adults the rates of bone deposition and resorption are about the same. As the level of sex steroids diminishes during middle age, especially in women after menopause, a decrease in bone mass occurs because bone resorption outpaces bone deposition. In old age, loss of bone through resorption by osteoclasts occurs more rapidly than bone gain by osteoblasts. Because women's bones generally are smaller and less massive than men's bones to begin with, loss of bone mass in old age typically has a greater adverse effect in women. These factors contribute to a higher incidence of osteoporosis in females.

There are two principal effects of aging on bone tissue: loss of bone mass and brittleness. Loss of bone mass results from **demineralization** (dē-min'-er-al-i-ZĀ-shun), the loss of calcium and other minerals from bone extracellular matrix. This loss usually begins after age 30 in females, accelerates greatly around age 45 as levels of estrogens decrease, and continues until as much as 30% of the calcium in bones is lost by age 70. Once bone loss begins in females, about 8% of bone mass is lost every 10 years. In males, calcium loss typically does not begin until after age 60, and about 3% of bone mass is lost every 10 years. The loss of calcium from bones is one of the problems in osteoporosis.

The second principal effect of aging on the skeletal system, brittleness, results from a decreased rate of protein synthesis. Recall that the organic part of bone matrix, mainly collagen fibers, gives bone its tensile strength. The loss of tensile strength causes the bones to become very brittle and susceptible to fractures. In some elderly people, collagen fiber synthesis slows, in part, due to diminished production of human growth hormone. In addition to increasing the susceptibility to fractures, loss of bone mass also leads to deformity, pain, loss of height, and loss of teeth.

◉ **CHECKPOINT**
19. What is demineralization, and how does it affect the functioning of bone?
20. What changes occur in the organic part of bone extracellular matrix with aging?

KEY MEDICAL TERMS ASSOCIATED WITH BONE TISSUE

Osteoarthritis (os'-tē-ō-ar-THRĪ-tis; *arthr*=joint) The degeneration of articular cartilage such that the bony ends touch; the resulting friction of bone against bone worsens the condition. Usually associated with the elderly.

Osteogenic sarcoma (os'-tē-ō-JEN-ik sar-KŌ-ma; *sarcoma*=connective tissue tumor) Bone cancer that primarily affects osteoblasts and occurs most often in teenagers during their growth spurt; the most common sites are the metaphyses of the femur, tibia, and humerus. Metastases occur most often in lungs; treatment consists of multidrug chemotherapy and removal of the malignant growth, or amputation of the limb.

Osteomyelitis (os'-tē-ō-mī-el-Ī-tis) An infection of bone characterized by high fever, sweating, chills, pain, and nausea, pus formation,

edema, and warmth over the affected bone and rigid overlying muscles. It is often caused by bacteria, usually *Staphylococcus aureus*. The bacteria may reach the bone from outside the body (through open fractures, penetrating wounds, or orthopedic surgical procedures); from other sites of infection in the body (abscessed teeth, burn infections, urinary tract infections, or upper respiratory infections) via the blood; and from adjacent soft tissue infections (as occurs in diabetes mellitus).

Osteopenia (os'-tē-ō-PĒ-nē-a; *-penia*=poverty) Reduced bone mass due to a decrease in the rate of bone synthesis to a level insufficient to compensate for normal bone resorption; any decrease in bone mass below normal. An example is osteoporosis.

Introduction (p. 140)

1. A bone is made up of several different tissues: bone, or osseous tissue, cartilage, dense connective tissues, epithelium, various blood-forming tissues, adipose tissue, and nervous tissue.
2. The entire framework of bones and their cartilages constitute the skeletal system.

Functions of Bone and the Skeletal System (p. 141)

1. The skeletal system functions in support, protection, movement, mineral storage and release, blood cell production, and triglyceride storage.

Types of Bones (p. 141)

1. On the basis of shape, bones are classified as long, short, flat, irregular, or sesamoid. Sesamoid bones develop in tendons or ligaments.
2. Sutural bones are found within the sutures of certain cranial bones.

Anatomy of a Bone (p. 142)

1. Parts of a typical long bone are the diaphysis (shaft), proximal and distal epiphyses (ends), metaphyses, articular cartilage, periosteum, medullary (marrow) cavity, and endosteum.

Bone Surface Markings (p. 142)

1. Surface markings are structural features visible on the surfaces of bones.
2. Each marking—whether a depression, an opening, or a process—is structured for a specific function, such as joint formation, muscle attachment, or passage of nerves and blood vessels (see Table 6.1 on page 144).

Histology of Bone Tissue (p. 144)

1. Bone tissue consists of widely separated cells surrounded by large amounts of extracellular matrix.
2. The four principal types of cells in bone tissue are osteogenic cells, osteoblasts, osteocytes, and osteoclasts.
3. The extracellular matrix of bone contains abundant mineral salts (mostly hydroxyapatite) and collagen fibers.
4. Compact bone tissue consists of osteons (haversian systems) with little space between them.
5. Compact bone tissue lies over spongy bone tissue in the epiphyses and makes up most of the bone tissue of the diaphysis. Functionally, compact bone tissue protects, supports, and resists stress.
6. Spongy bone tissue does not contain osteons. It consists of trabeculae surrounding spaces filled with red bone marrow.
7. Spongy bone tissue forms most of the structure of short, flat, and irregular bones, and the epiphyses of long bones. Functionally, spongy bone tissue trabeculae offer resistance along lines of stress, support and protect red bone marrow, and make bones lighter for easier movement.

Blood and Nerve Supply of Bone (p. 148)

1. Long bones are supplied by periosteal, nutrient, metaphyseal, and epiphyseal arteries; veins accompany the arteries.
2. Nerves accompany blood vessels in bone; the periosteum is rich in sensory neurons.

Bone Formation (p. 149)

1. The process by which bone forms is called **ossification** (os′-i-fi-KĀ-shun; *ossi-*=bone; *-fication*=making) or **osteogenesis** (os′-tē-ō-

JEN-e-sis). Bone formation occurs in four principal situations: (1) the initial formation of bones in an embryo and fetus, (2) the growth of bones during infancy, childhood, and adolescence until their adult sizes are reached, (3) the remodeling of bone (replacement of old bone by new bone tissue throughout life), and (4) the repair of fractures (breaks in bones) throughout life.
2. Bone development begins during the sixth or seventh week of embryonic development. The two types of ossification, intramembranous and endochondral, involve the replacement of a preexisting connective tissue with bone.
3. Intramembranous ossification refers to bone formation directly within mesenchyme arranged in sheet-like layers that resemble membranes.
4. Endochondral ossification refers to bone formation within hyaline cartilage that develops from mesenchyme. The primary ossification center of a long bone is in the diaphysis. Cartilage degenerates, leaving cavities that merge to form the medullary cavity. Osteoblasts lay down bone. Next, ossification occurs in the epiphyses, where bone replaces cartilage, except for the epiphyseal (growth) plate.
5. The epiphyseal plate consists of four zones: resting cartilage, proliferating cartilage, hypertrophic cartilage, and calcified cartilage.
6. Because of the activity at the epiphyseal plate, the diaphysis of a bone increases in length.
7. Bone grows in width as a result of the addition of new bone tissue by periosteal osteoblasts around the outer surface of the bone (appositional growth).
8. Bone remodeling is the replacement of old bone tissue by new bone tissue.
9. Old bone tissue is constantly destroyed by osteoclasts, whereas new bone is constructed by osteoblasts.

Fractures (p. 155)

1. A fracture is any break in a bone.
2. Fracture repair involves formation of a fracture hematoma, a fibrocartilaginous callus, and a bony callus, followed by bone remodeling.
3. Types of fractures include closed (simple), open (compound), comminuted, greenstick, impacted, stress, Pott's, and Colles'.

Factors Affecting Bone Growth (p. 157)

1. Normal growth depends on minerals (calcium, phosphorus, magnesium), vitamins (A, C, D, K, and B_{12}), hormones (human growth hormone, insulinlike growth factors, insulin, thyroid hormones, sex hormones, and parathyroid hormone), and weight-bearing exercise.

Exercise and Bone Tissue (p. 158)

1. Mechanical stress increases bone strength by increasing deposition of mineral salts and production of collagen fibers.
2. Removal of mechanical stress weakens bone through demineralization and collagen fiber reduction.

Aging and Bone Tissue (p. 158)

1. The principal effect of aging is a loss of calcium from bones, which may result in osteoporosis.
2. Another effect is a decreased production of extracellular matrix proteins (mostly collagen fibers), which makes bones more brittle and thus more susceptible to fracture.

SELF-QUIZ QUESTIONS

Choose the one best answer to the following questions:

1. Which of the following is *not* considered a function of bone tissue and the skeletal system?
   a. attachment site of muscles, tendons, and ligaments
   b. storage of calcium and phosphorus
   c. formation of blood cells
   d. support and protection of soft organs and tissues
   e. synthesis of vitamin D

2. Yellow bone marrow consists mainly of:
   a. adipose cells   b. hemopoietic tissue
   c. calcium salts   d. collagen
   e. elastic cartilage

3. Haversian canals contain:
   a. high concentrations of calcium salts
   b. blood vessels
   c. osteocytes
   d. articular cartilage
   e. the epiphyseal plate

4. The thin columns of bone that form an irregular lattice in spongy bone are called:
   a. osteons   b. canaliculi
   c. lacunae   d. Volkmann's canals
   e. trabeculae

5. Periosteal arteries enter a bone through:
   a. haversian canals   b. lacunae
   c. canaliculi   d. Volkmann's canals
   e. trabeculae

6. Increased osteoblast cell activity would have what effect on bone tissue?
   a. reduced bone mass   b. osteopenia
   c. osteomyelitis   d. increased bone mass
   e. *Two* of the above

7. Which bone cells are bone-building cells?
   a. osteoclasts   b. osteogenic cells
   c. osteocytes   d. osteoblasts
   e. osteons

Complete the following:

8. The periosteum is made of _____ tissue.

9. Most bones of the body are formed from cartilage by the process of _____ ossification.

10. The process by which mineral salts crystallize around collagen fibers in bone is called _____.

11. Osteocytes are located in spaces called _____.

12. The medullary cavity in the long bone of an adult contains _____.

13. The cells responsible for the resorption (destruction) of bone tissue are _____.

14. The process by which a bone grows in thickness is called _____ growth.

15. _____ disease is characterized by an abnormal acceleration of the remodeling process resulting in deformed, hard, brittle, easily fractured bones, especially in the bones of the limbs, pelvis, lower vertebrae, and skull.

16. _____ refers to bone cancer that primarily affects osteoblasts and occurs most often in teenagers during their growth spurt.

Are the following statements true or false? If the statement is False, *find and fix the mistake to make the statement* True.

17. The flat bones of the skull develop by intramembranous ossification.

18. Canaliculi are microscopic canals running longitudinally through bone.

19. Bones do not have a nerve supply.

20. A fracture hematoma is a mass of fibrocartilage that bridges the broken ends of the bones.

21. The organic extracellular matrix of bone tissue is secreted by osteoblasts.

22. Osteoporosis is a disease in which bone deposition outpaces bone resorption.

23. Osteoartbritis is an infection of bone often caused by bacteria, usually *Staphylococcus aureus*.

Matching

24. Match the following.
   ____ (a) thin layer of hyaline cartilage at end of a long bone
   ____ (b) region of mature bone where diaphysis joins epiphysis
   ____ (c) covering over bone to which ligaments and tendons attach
   ____ (d) a layer of hyaline cartilage in the metaphysis of a growing bone
   ____ (e) lining of the medullary cavity; contains osteoprogenitor cells and osteoblasts

   (1) articular cartilage
   (2) endosteum
   (3) fibrous periosteum
   (4) metaphysis
   (5) epiphyseal plate

CRITICAL THINKING QUESTIONS

1. Lynda, a petite 55-year-old couch potato who smokes heavily, wants to lose 50 pounds before her next class reunion. Her diet consists mostly of diet soda and crackers. Explain the effects of her age and lifestyle on her bone composition.

2. Hannah and her anthropology classmates were studying some bones from their museum collection. Scott noticed some thick bumps on some of the bones, and noted that these were *spurs*. In terms of bone remodeling, can you offer an explanation as to how the spurs were formed?

3. Aunt Edith is 95 years old today. She comments that she's been getting shorter every year and soon she'll fade out all together. What's happening to Aunt Edith?

4. Astronaut John Glenn exercised every day while in space, yet he and the other astronauts experienced weakness upon their return to earth. Why?

5. Chantal was concerned that her new baby was a "cone head" when she saw him for the first time. Later on, after carefully following the advice to always lay the baby on his back for sleeping, she became concerned that the back of his head was getting flat. If bone is hard, why does the baby's head keep changing shape?

? ANSWERS TO FIGURE QUESTIONS

6.1 Flat bones protect and provide a large surface area for muscle attachment.

6.2 The periosteum is essential for growth in bone thickness, bone repair, and bone nutrition. It also serves as a point of attachment for ligaments and tendons.

6.3 Bone resorption is necessary for the development, maintenance, and repair of bone.

6.4 The central (haversian) canals are the main blood supply to the osteocytes of an osteon (haversian system), so their blockage would lead to death of the osteocytes.

6.5 Periosteal arteries enter bone tissue through perforating (Volkmann's) canals.

6.6 Flat bones of the skull and mandible (lower jawbone) develop by intramembranous ossification.

6.7 Yes, she probably will grow taller. The absence of epiphyseal lines, indications of growth zones that have ceased to function, indicates that the bone is still lengthening.

6.8 The lengthwise growth of the diaphysis is caused by cell divisions in the zone of proliferating cartilage and replacement of the zone of calcified cartilage with bone (new diaphysis).

6.9 The medullary cavity enlarges by activity of the osteoclasts in the endosteum.

6.10 A fibrocartilaginous callus is a mass of fibrocartilaginous repair tissue that bridges the ends of broken bones.

6.11 Healing of bone fractures can take months because calcium and phosphorus deposition is a slow process, and bone cells generally grow and reproduce slowly.

7 | THE SKELETAL SYSTEM: THE AXIAL SKELETON

INTRODUCTION Without bones, you would be unable to perform movements such as walking or grasping. The slightest blow to your head or chest could damage your brain or heart. It would even be impossible for you to chew food. Every day your skeleton is put through a wide range of activities. Think about the movements and forces your skeleton experiences as you go about your daily life. As you walk, run, climb stairs, or lift things, your skeleton must support the weight of the body and the various loads that the body carries. In addition to the support it provides, the skeleton must also let you move freely from place to place. The wonderful capacities of the skeleton become even more impressive when you consider the loads and strains it encounters during various exercise and sports activities, such as gymnastics meets, figure-skating competitions, or basketball games. In a basketball game, for example, a player runs from one end of the court to the other hundreds of times, often making sharp turns and abrupt starts and stops. In addition to this rigorous running, the player also jumps explosively in an attempt to shoot, rebound, and defend other players. All this running and jumping occurs on the firm hardwood surface of the basketball court or the blacktop of the playground. During the course of the game the skeleton must be able to withstand powerful movements, move freely and gracefully, support the body against a wide range of gravitational and muscular loads, and absorb significant amounts of impact. To make this possible, the skeleton must be strong and light in weight with bones in a variety of shapes and sizes. Because the skeletal system forms the framework of the body, a familiarity with the names, shapes, and positions of individual bones will help you locate other organ systems. For example, the radial artery, the site where pulse is usually taken, is named for its proximity to the radius, the lateral bone of the forearm. The frontal lobe of the brain lies deep to the frontal (forehead) bone. The tibialis anterior muscle is located near the anterior surface of the tibia (shinbone). The ulnar nerve is named for its proximity to the ulna, the medial bone of the forearm. Parts of certain bones also outline the lungs, heart, and abdominal and pelvic organs.

Movements such as throwing a ball, biking, and walking require an interaction between bones and muscles. To understand how muscles produce different movements, you will learn where the muscles attach to individual bones and the types of joints acted on by the contracting muscles. The bones, muscles, and joints together form an integrated system called the **musculoskeletal system**. The branch of medical science that is concerned with the prevention or correction of disorders of the musculoskeletal system is called **orthopedics** (or′-thō-PĒ-diks; *ortho-*=correct; *pedi*=child). •

DIVISIONS OF THE SKELETAL SYSTEM

◉ OBJECTIVE
• Describe how the skeleton is divided into axial and appendicular divisions.

The adult human skeleton consists of 206 named bones, most of which are paired, with one member of each pair on the right and left sides of the body. The skeletons of infants and children have more than 206 bones because some of their bones fuse later in life. Examples are the hip bones and some bones of the vertebral column (backbone).

Bones of the adult skeleton are grouped into two principal divisions: the **axial skeleton** and the **appendicular skeleton**

(*appendic-*=to hang onto). Table 7.1 presents the 80 bones of the axial skeleton and the 126 bones of the appendicular skeleton. Figure 7.1 on pages 164–165 shows how both divisions join to form the complete skeleton (the bones of the axial skeleton are shown in gray). The axial skeleton consists of the bones that lie around the longitudinal *axis* of the human body, an imaginary vertical line that runs through the body's center of gravity from the head to the space between the feet: skull bones, auditory ossicles (ear bones), hyoid bone (see Figure 7.4a), ribs, sternum (breastbone), and bones of the vertebral column. The appendicular skeleton consists of the bones of the **upper** and **lower limbs**. The limbs are made up of bones that form the **limb girdles** and bones that form the **free limbs**. These are discussed in detail in Chapter 8.

We will organize our study of the skeletal system around the two divisions of the skeleton, with emphasis on how the many

TABLE 7.1

The Bones of the Adult Skeletal System

DIVISION OF THE SKELETON	STRUCTURE		NUMBER OF BONES
Axial skeleton	**Skull**		
	Cranium		8
	Face		14
	Hyoid		1
	Auditory ossicles		6
	Vertebral column		26
	Thorax		
	Sternum		1
	Ribs		24
		Subtotal =	80

DIVISION OF THE SKELETON	STRUCTURE		NUMBER OF BONES
Appendicular skeleton	**Upper limbs**		
	Pectoral (shoulder) girdles		
	Clavicle		2
	Scapula		2
	Free upper limbs		
	Humerus		2
	Ulna		2
	Radius		2
	Carpals		16
	Metacarpals		10
	Phalanges		28
	Lower limb		
	Pelvic (hip) girdle		
	Hip, pelvic, or coxal bone		2
	Free lower limbs		
	Femur		2
	Patella		2
	Fibula		2
	Tibia		2
	Tarsals		14
	Metatarsals		10
	Phalanges		28
		Subtotal =	126
Axial & Appendicular		Total =	206

Figure 7.1 Divisions of the skeletal system. The axial skeleton is indicated in gray, the appendicular skeleton in gold. Note the position of the hyoid bone in Figure 7.3a.

The adult human skeleton consists of 206 bones grouped into axial and appendicular divisions.

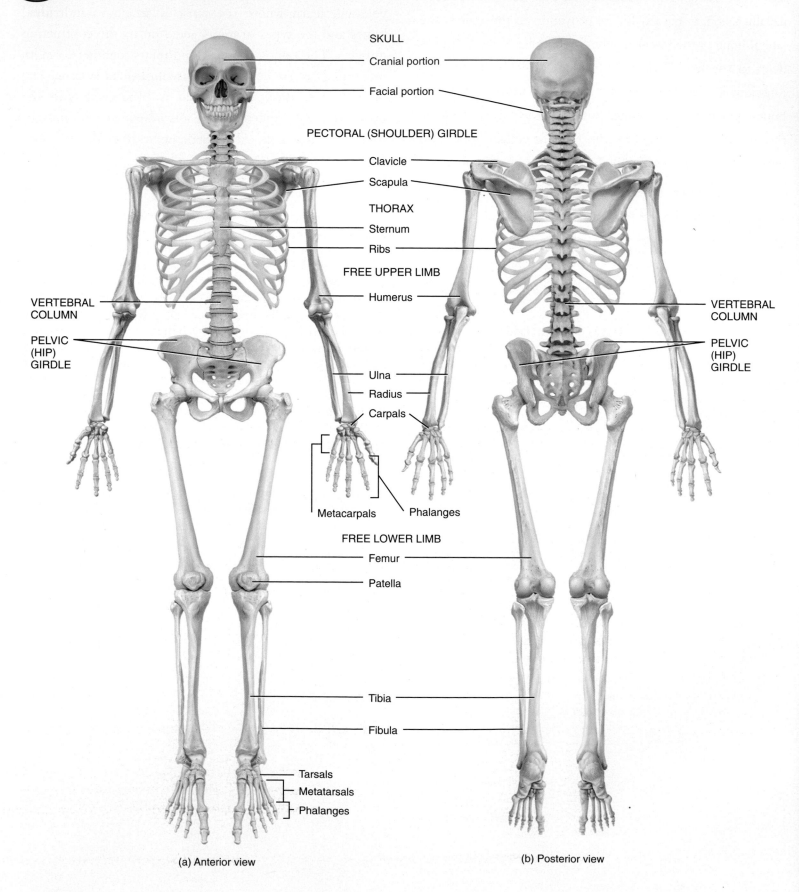

SKULL
Cranial portion
Facial portion

PECTORAL (SHOULDER) GIRDLE

Clavicle
Scapula

THORAX

Sternum
Ribs

FREE UPPER LIMB

Humerus

VERTEBRAL COLUMN

PELVIC (HIP) GIRDLE

VERTEBRAL COLUMN

PELVIC (HIP) GIRDLE

Ulna
Radius
Carpals

Metacarpals   Phalanges

FREE LOWER LIMB

Femur
Patella

Tibia

Fibula

Tarsals
Metatarsals
Phalanges

(a) Anterior view

(b) Posterior view

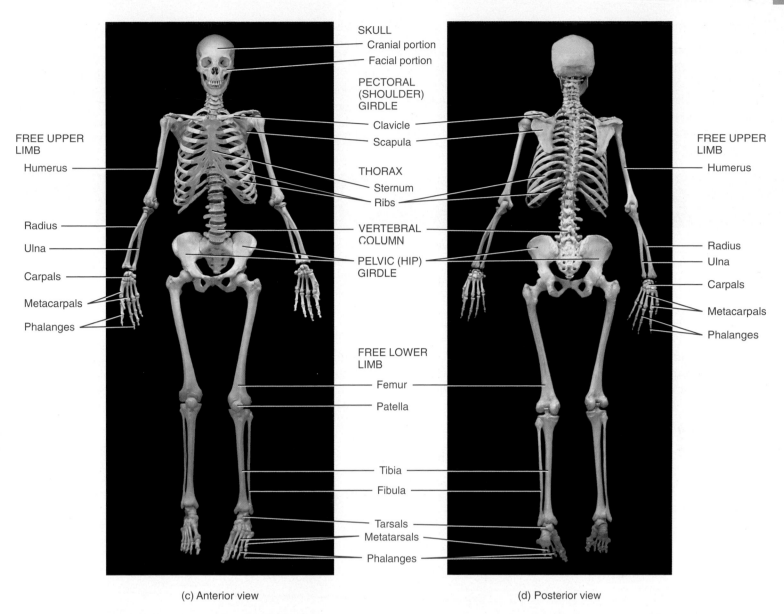

(c) Anterior view

(d) Posterior view

? Which of the following structures are part of the axial skeleton, and which are part of the appendicular skeleton? Skull, clavicle, vertebral column, shoulder girdle, humerus, pelvic girdle, and femur.

bones of the body are interrelated. In this chapter we focus on the axial skeleton, looking first at the skull and then at the bones of the vertebral column and the chest.

◉ CHECKPOINT

1. List the bones that make up the axial and appendicular divisions of the human skeleton.

SKULL (CRANIUM)

◉ OBJECTIVES

- Name the cranial and facial bones and indicate the number of each.
- Describe the following unique features of the skull: sutures, paranasal sinuses, and fontanels.

The **skull** or **cranium**, which contains 22 bones, rests on the superior end of the vertebral column (backbone). It includes two sets of bones: cranial bones and facial bones (Table 7.2). The **cranial bones** (*crani-*=skull) form the cranial cavity, which encloses and protects the brain. The eight cranial bones are the frontal bone, two parietal bones, two temporal bones, the occipital bone, the sphenoid bone, and the ethmoid bone. Fourteen **facial bones**

form the face: two nasal bones, two maxillae (or maxillas), two zygomatic bones, the mandible, two lacrimal bones, two palatine bones, two inferior nasal conchae, and the vomer. Figures 7.2 through 7.8 in Exhibits 7.1 through 7.9 illustrate the bones of the skull from different viewing directions.

TABLE 7.2

Summary of Bones of the Adult Skull

CRANIAL BONES	FACIAL BONES
Frontal (1)	Nasal (2)
Parietal (2)	Maxillae (2)
Temporal (2)	Zygomatic (2)
Occipital (1)	Mandible (1)
Sphenoid (1)	Lacrimal (2)
Ethmoid (1)	Palatine (2)
Inferior nasal conchae (2)	
Vomer (1)	

The numbers in parentheses indicate how many of each bone are present.

General Features and Functions

Besides forming the large cranial cavity, the skull also forms several smaller cavities, including the nasal cavity and orbits (eye sockets), which open to the exterior. Certain skull bones also contain cavities called paranasal sinuses that are lined with mucous membranes and open into the nasal cavity. Also within the skull are small cavities that house the structures involved in hearing and equilibrium.

Other than the auditory ossicles, which are involved in hearing and are located within the temporal bones, the mandible is the only bone of the skull that has a large range of movement. Most of the skull bones are held together by sutures, which are immovable joints in most cases in adults. Sutures are especially noticeable on the outer surface of the skull (see Figure 7.3 on page 168).

The skull has numerous surface markings, such as foramina and fissures through which blood vessels and nerves pass. You will

---

EXHIBIT 7.1 Cranial Bones—Frontal Bone

■ OBJECTIVE
• Identify the location and surface features of the frontal bone.

Description

The **frontal bone** forms the forehead (the anterior part of the cranium), the roofs of the *orbits* (eye sockets), and most of the anterior part of the cranial floor (Figure 7.2). In most individuals it is an unpaired bone. Soon after birth the left and right sides of the frontal bone are united by the *metopic suture*, which usually disappears by age 6–8.

Surface Features

If you examine the anterior view of the skull in Figure 7.2, you will note the *frontal squama*, a thick, scalelike plate of bone that forms the forehead. It gradually slopes inferiorly from the

Figure 7.2 Anterior view of the skull.

🔑 The skull consists of cranial bones and facial bones.

FRONTAL BONE

Frontal squama

PARIETAL BONE

Squamous suture

SPHENOID BONE

Orbit

ETHMOID BONE

LACRIMAL BONE

Zygomaticofacial foramen

ZYGOMATIC BONE

Perpendicular plate of ethmoid bone

INFERIOR NASAL CONCHA

VOMER

Mental foramen

Coronal suture

Supraorbital foramen

Supraorbital margin

Optic foramen (canal)

Superior orbital fissure

TEMPORAL BONE

NASAL BONE

Inferior orbital fissure

Middle nasal concha

Infraorbital foramen

MAXILLA

Alveolar process of maxilla

Alveolar process of mandible

MANDIBLE

(a) Anterior view

learn the names of important skull bone surface markings as the various bones are described.

The cranial bones have other functions in addition to protecting the brain. Their inner surfaces attach to membranes (meninges) that stabilize the positions of the brain, blood vessels, and nerves. The outer surfaces of cranial bones provide large areas of attachment for muscles that move various parts of the head. The bones also provide attachment for some muscles that are in-volved in producing facial expressions. Besides forming the framework of the face, the facial bones protect and provide support for the entrances to the digestive and respiratory systems. Together, the cranial and facial bones protect and support the delicate special sense organs for vision, taste, smell, hearing, and equilibrium (balance). The inner ear contains sensory structures important for hearing and others that monitor the position and movement of the head, factors important in the sense of balance.

coronal suture (its joint with the paired parietal bones) on top of the skull, then angles abruptly and becomes almost vertical above the orbits. At the superior border of the orbits the frontal bone thickens, forming the *supraorbital margin* (*supra-*=above; *-orbital*= circular). From this margin the frontal bone extends as a horizontal plate of bone posteriorly to form the roof of the orbit and part of the floor of the cranial cavity. Within the supraorbital margin, slightly medial to its midpoint, is a hole called the *supraorbital foramen* through which the supraorbital nerve and artery pass. Sometimes this foramen is incomplete and is called the *supraorbital notch*. Near the midline, within the vertical portion of the frontal squama, the bone is hollow. These hollow spaces are the paranasal sinuses called the *frontal sinuses*. Paranasal sinuses, mucous membrane–lined cavities within certain skull bones, will be discussed later in the chapter.

◉ **CHECKPOINT**
What structures pass through the supraorbital foramen?

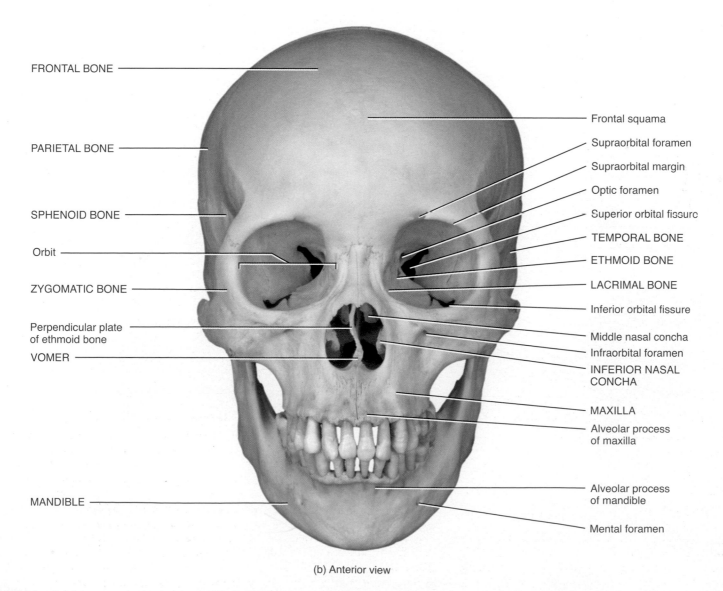

FRONTAL BONE

Frontal squama

PARIETAL BONE

Supraorbital foramen

Supraorbital margin

Optic foramen

SPHENOID BONE

Superior orbital fissure

TEMPORAL BONE

Orbit

ETHMOID BONE

ZYGOMATIC BONE

LACRIMAL BONE

Inferior orbital fissure

Perpendicular plate
of ethmoid bone

Middle nasal concha

VOMER

Infraorbital foramen

INFERIOR NASAL
CONCHA

MAXILLA

Alveolar process
of maxilla

Alveolar process
of mandible

MANDIBLE

Mental foramen

(b) Anterior view

❓ **Which of the bones shown here are cranial bones?**

EXHIBIT 7.1 **167**

EXHIBIT 7.2 Cranial Bones—Parietal Bones

● OBJECTIVE

- Identify the location and surface features of the parietal bones.

Description

The two **parietal bones** (pa-RĪ-e-tal; *pariet-*=wall) are a pair of large, quadrilateral (four-sided) bones that form the greater portion of the sides and roof of the cranial cavity (Figure 7.3). Each

bone articulates with five other bones. The inferior border forms a beveled articular surface, while the anterior, posterior, and superior borders form deeply denticulate articular surfaces.

Figure 7.3 Right lateral view of the skull.

🔑 The zygomatic arch is formed by the zygomatic process of the temporal bone and the temporal process of the zygomatic bone.

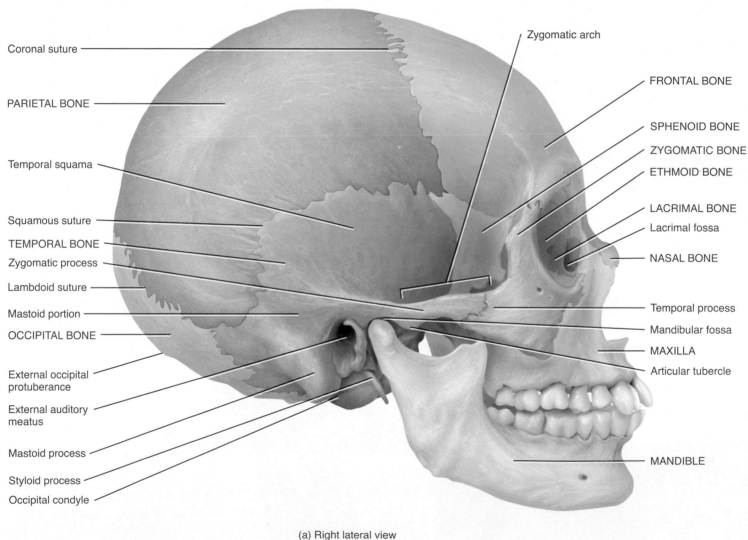

Coronal suture

PARIETAL BONE

Temporal squama

Squamous suture

TEMPORAL BONE

Zygomatic process

Lambdoid suture

Mastoid portion

OCCIPITAL BONE

External occipital protuberance

External auditory meatus

Mastoid process

Styloid process

Occipital condyle

Zygomatic arch

FRONTAL BONE

SPHENOID BONE

ZYGOMATIC BONE

ETHMOID BONE

LACRIMAL BONE

Lacrimal fossa

NASAL BONE

Temporal process

Mandibular fossa

MAXILLA

Articular tubercle

MANDIBLE

(a) Right lateral view

Surface Features

The external surface of each of these bones is slightly convex (curved outward, like the outside of a sphere), while the internal surface is concave. The internal surfaces of the parietal bones contain many protrusions and depressions that accommodate the blood vessels supplying the dura mater, the superficial membrane (meninx) covering the brain.

☉ **CHECKPOINT**
How do the parietal bones relate to the cranial cavity?

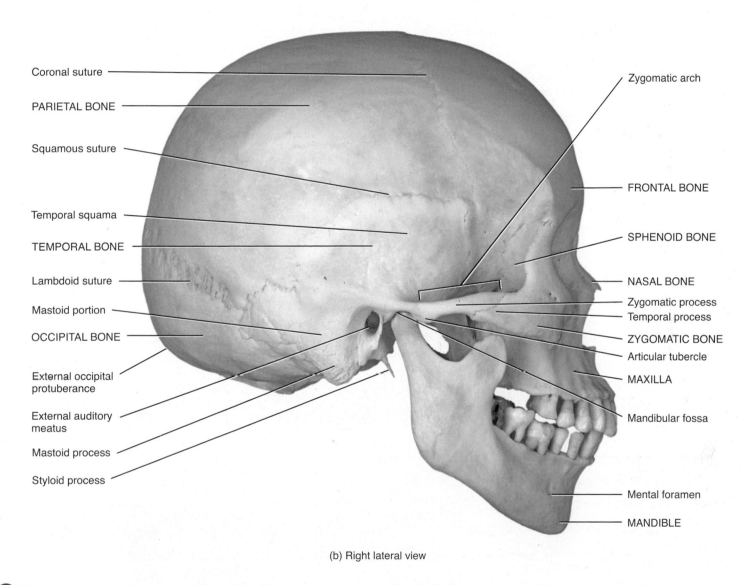

(b) Right lateral view

Coronal suture
PARIETAL BONE
Squamous suture
Temporal squama
TEMPORAL BONE
Lambdoid suture
Mastoid portion
OCCIPITAL BONE
External occipital protuberance
External auditory meatus
Mastoid process
Styloid process

Zygomatic arch
FRONTAL BONE
SPHENOID BONE
NASAL BONE
Zygomatic process
Temporal process
ZYGOMATIC BONE
Articular tubercle
MAXILLA
Mandibular fossa
Mental foramen
MANDIBLE

❓ **What are the major bones on either side of the squamous suture, the lambdoid suture, and the coronal suture?**

EXHIBIT 7.2   **169**

EXHIBIT 7.3 Cranial Bones—Temporal Bones

⦿ **OBJECTIVE**

• Identify the location and surface features of the temporal bones.

Description

The two **temporal bones** (*tempor-*=temple) form the inferior lateral aspects of the cranium and part of the cranial floor. The terms *temporal* and *temple* are derived from the Latin word *tempus*, meaning "time," in reference to the graying of hair in the temple area, a sign of time's passing.)

Surface Features

In the lateral view of the skull (see Figure 7.3), note the *temporal squama*, the thin, flat portion of the temporal bone that forms the anterior and superior part of the *temple* (the region of the cranium around the ear). Projecting anteriorly from the inferior portion of the temporal squama is the *zygomatic process*, which articulates (forms a joint) with the temporal process of the zygomatic (cheek) bone. Together, the zygomatic process of the temporal bone and the temporal process of the zygomatic bone form the *zygomatic arch*. You can easily palpate this horizontal arch of bone immediately anterior to the ear.

On the inferoposterior surface of the zygomatic process of the temporal bone is a socket called the *mandibular fossa*. Anterior to the mandibular fossa is a rounded elevation, the *articular tubercle* (see Figure 7.3). The mandibular fossa and articular tubercle articulate with the mandible (lower jawbone) to form the *temporomandibular joint (TMJ)*.

Located posterior and inferior on the temporal bone is the *mastoid portion* (*mastid*=breast-shaped) (see Figure 7.3). The mastoid portion is located posterior and inferior to the *external auditory meatus* (*meatus*=passageway), or ear canal, which directs

Figure 7.4 Medial view of sagittal section of the skull. Although the hyoid bone is not part of the skull, it is included in the illustration for reference.

🔑 The cranial bones are the frontal, parietal, temporal, occipital, sphenoid, and ethmoid bones. The facial bones are the nasal bone, maxillae, zygomatic bones, lacrimal bones, palatine bones, mandible, and vomer.

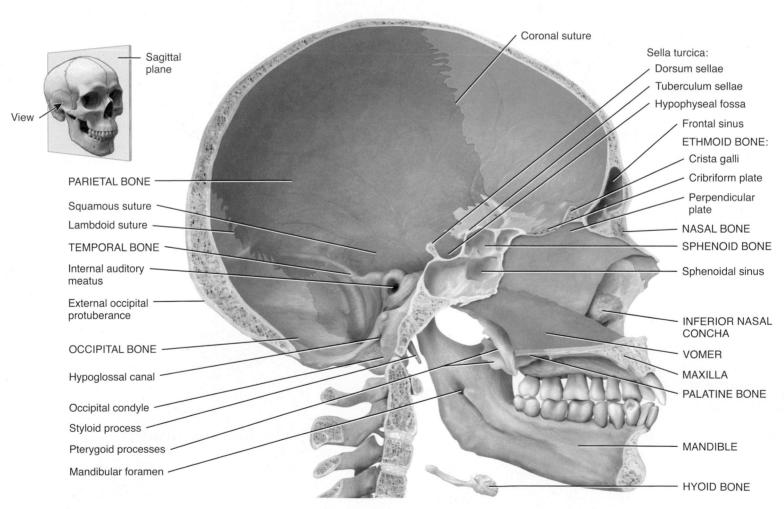

(a) Medial view of sagittal section

sound waves into the ear. In the adult, this portion of the bone contains several *mastoid "air cells."* These tiny air-filled compartments are separated from the brain by thin bony partitions. **Mastoiditis** (inflammation of the mastoid air cells) can spread an infection to the middle ear and to the brain.

The *mastoid process* is a rounded projection of the mastoid portion of the temporal bone posterior and inferior to the external auditory meatus. It serves as a point of attachment for several neck muscles (see Figure 7.3). The *internal auditory meatus* (Figure 7.4) is the opening through which the facial (VII) and vestibulocochlear (VIII) cranial nerves pass. The *styloid process* (*styl-*=stake or pole) projects inferiorly from the inferior surface of the temporal bone and serves as a point of attachment for muscles and ligaments of the tongue and neck (see Figure 7.3). Between the styloid process and the mastoid process is the *stylomastoid foramen*, through which the facial (VII) nerve and stylomastoid artery pass (see Figure 7.6).

At the floor of the cranial cavity (see Figure 7.6a) is the *petrous portion* (*petrous*=rock) of the temporal bone. This portion is pyramidal (having the shape of a pyramid) and located at the base of the skull between the sphenoid and occipital bones. The petrous portion houses the internal ear and the middle ear, structures involved in hearing and equilibrium. It also contains the *carotid foramen*, through which the carotid artery passes (see Figure 7.6a and 7.6c). Posterior to the carotid foramen and anterior to the occipital bone is the *jugular foramen*, a passageway for the jugular vein, formed by adjacent notches in the temporal and occipital bones.

◉ **CHECKPOINT**
What structures form the zygomatic arch?

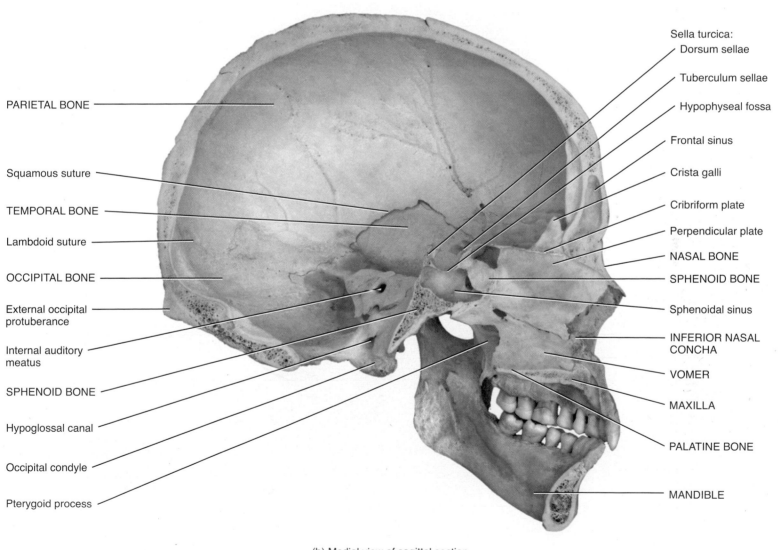

PARIETAL BONE

Squamous suture

TEMPORAL BONE

Lambdoid suture

OCCIPITAL BONE

External occipital protuberance

Internal auditory meatus

SPHENOID BONE

Hypoglossal canal

Occipital condyle

Pterygoid process

Sella turcica:
Dorsum sellae

Tuberculum sellae

Hypophyseal fossa

Frontal sinus

Crista galli

Cribriform plate

Perpendicular plate

NASAL BONE

SPHENOID BONE

Sphenoidal sinus

INFERIOR NASAL CONCHA

VOMER

MAXILLA

PALATINE BONE

MANDIBLE

(b) Medial view of sagittal section

❓ **With which bones does the temporal bone articulate?**

EXHIBIT 7.3   **171**

EXHIBIT 7.4   Cranial Bones—Occipital Bone

◉ OBJECTIVE

- Identify the location and surface features of the occipital bone.

Description

The **occipital bone** (ok-SIP-i-tal; *occipit-*=back of head) forms the posterior part and most of the base of the cranium. When viewed from behind, it appears as a platelike bone with a somewhat triangular shape. Its inferior portion is a thick, blocklike region that surrounds the junction of the brain and spinal cord (Figure 7.5; also see Figure 7.3).

Surface Features

The *foramen magnum* (=large hole) is in the inferior part of the bone. Within this foramen, the medulla oblongata (inferior part of the brain) connects with the spinal cord. The vertebral and spinal arteries also pass through this foramen. The *occipital condyles* are oval processes with convex surfaces, one on either side of the foramen magnum (see Figure 7.6c). They articulate with depressions on the first cervical vertebra (atlas) to form the *atlanto-occipital joints*. Superior to each occipital condyle on the inferior surface of the skull is the *hypoglossal canal* (*hypo-*=under;

Figure 7.5 Posterior view of the skull. The sutures are exaggerated for emphasis.

The occipital bone forms most of the posterior and inferior portions of the cranium.

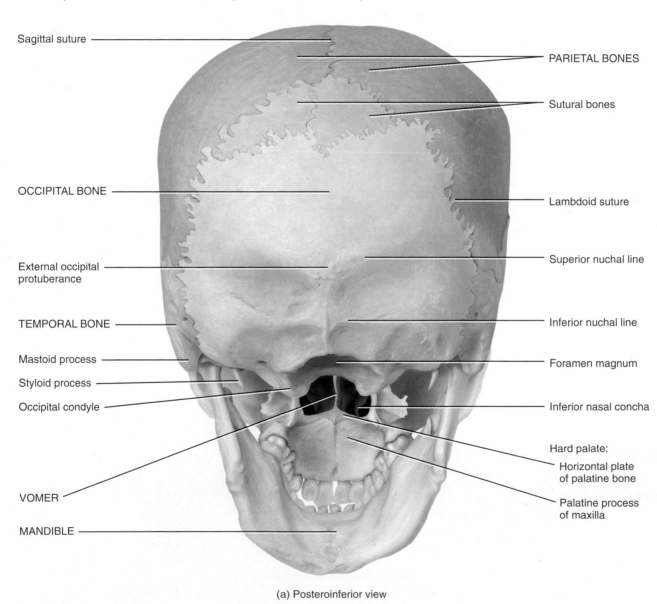

(a) Posteroinferior view

-glossal=tongue), through which the hypoglossal (XII) nerve and a branch of the ascending pharyngeal artery pass (see Figure 7.4).

The *external occipital protuberance* is the most prominent midline projection on the posterior surface of the bone just superior to the foramen magnum. You may be able to feel this structure as a definite bump, the most prominent protrusion on the back of your head, just above your neck (see Figure 7.5a). A large fibrous, elastic ligament, the *ligamentum nuchae* (*nucha-*=nape of neck), which helps support the head, extends from the external occipital protuberance to the seventh cervical vertebra. Extending laterally from the protuberance are two curved lines, the *superior nuchal*

lines, and below these are two *inferior nuchal lines*, which are areas of muscle attachment (see Figure 7.5a). It is possible to view the parts of the occipital bone, as well as surrounding structures, in the inferior view of the skull in Figure 7.6c.

◉ **CHECKPOINT**
What structures pass through the hypoglossal canal?

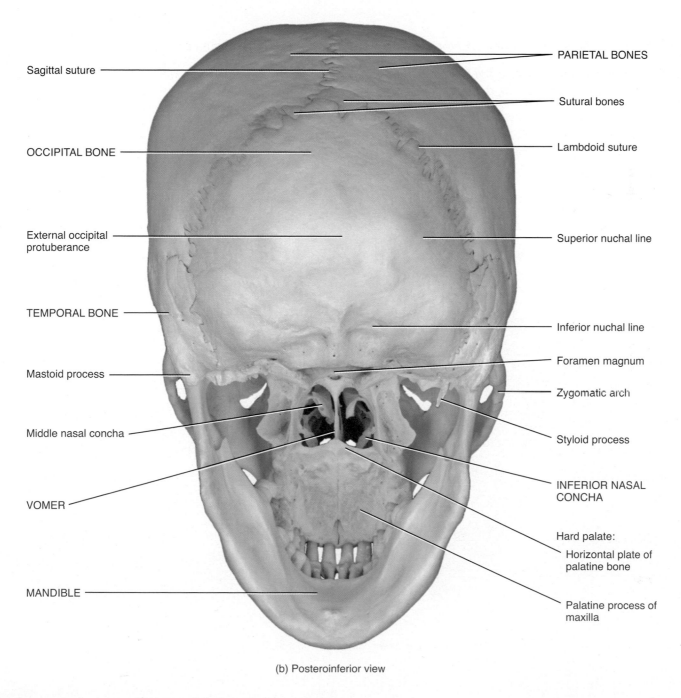

(b) Posteroinferior view

? **Which bones form the posterior, lateral portion of the cranium?**

EXHIBIT 7.4   **173**

EXHIBIT 7.5 Cranial Bones—Sphenoid Bone

OBJECTIVE

- Identify the location and surface features of the sphenoid bone.

Description

The **sphenoid bone** (SFĒ-noyd=wedge-shaped) lies at the middle part of the base of the skull (Figure 7.6 a–d). This bone is called the keystone of the cranial floor because it articulates with all the other cranial bones, holding them together. Viewing the floor of the cranium superiorly (Figure 7.6a, b), note the sphenoid articulations: anteriorly with the frontal bone, laterally with

the temporal bones, anterolaterally with the parietal bones, and posteriorly with the occipital bone. It lies posterior and slightly superior to the nasal cavity and forms part of the floor, sidewalls, and rear wall of the orbit (see Figure 7.10).

Surface Features

The shape of the sphenoid resembles a butterfly with outstretched wings (Figure 7.6e, f). The *body* of the sphenoid is the hollowed, cubelike medial portion between the ethmoid and occipital bones. The hollow of the body is the *sphenoidal sinus*, which drains into the nasal cavity (see Figure 7.11c). The *sella turcica*

Figure 7.6 Sphenoid bone. (See Tortora, *A Photographic Atlas of the Human Body*, 2e, Figure 3.9.)

The sphenoid bone is called the keystone of the cranial floor because it articulates with all other cranial bones, holding them together.

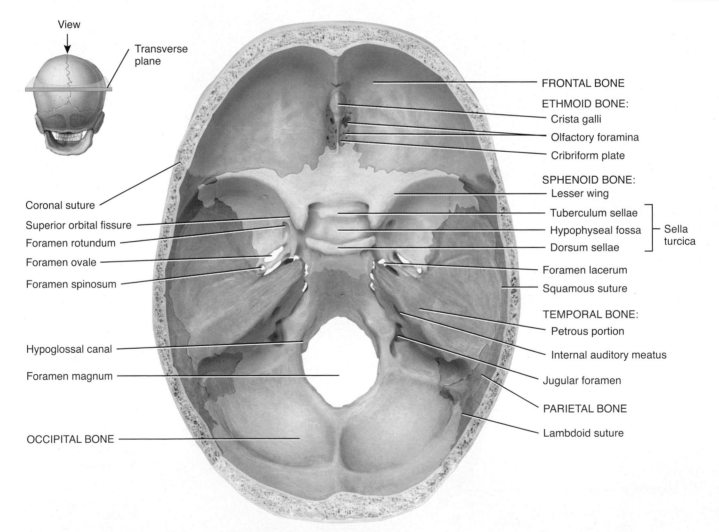

(a) Superior view of sphenoid bone in floor of cranium

(SEL-a TUR-si-ka; *sella*=saddle; *turcica*=Turkish) is a bony saddle-shaped structure on the superior surface of the body of the sphenoid (Figure 7.6a, b). The anterior part of the sella turcica, which forms the horn of the saddle, is a ridge called the *tuberculum sellae*. The seat of the saddle is a depression, the *hypophyseal fossa* (hī-pō-FIZ-ē-al), which contains the pituitary gland. The posterior part of the sella turcica, which forms the back of the saddle, is another ridge called the *dorsum sellae*.

The *greater wings* of the sphenoid project laterally from the body, forming the anterolateral floor of the cranium (Figure 7.6a, b, e, f). The greater wings also form part of the lateral wall of the skull just anterior to the temporal bone and can be viewed exter-nally. The *lesser wings*, which are smaller than the greater wings, form a ridge of bone anterior and superior to the greater wings. They form part of the floor of the cranium and the posterior part of the orbit of the eye.

Between the body and lesser wing just anterior to the sella tur-cica is the *optic foramen* or *canal* (*optic*=eye), through which the optic (II) nerve and ophthalmic artery pass into the orbit (Figure 7.6e, f). Lateral to the body between the greater and lesser wings is an elongated, triangular slit called the *superior orbital fissure*. This fissure may also be seen in the anterior view of the orbit in Figure 7.10. Blood vessels and cranial nerves pass through this fissure.

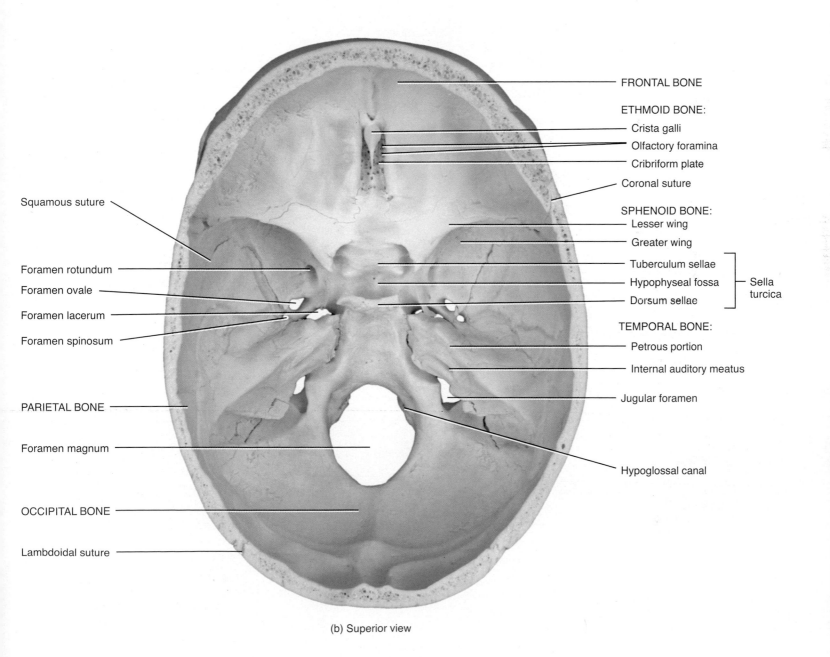

(b) Superior view

⬤ EXHIBIT 7.5 CONTINUES ▶

EXHIBIT 7.5 **175**

EXHIBIT 7.5   Cranial Bones—Sphenoid Bone CONTINUED

The *pterygoid processes* (TER-i-goyd=winglike) extend from the inferior part of the sphenoid bone (Figure 7.6c, d). These structures project inferiorly from the points where the body and greater wings unite and they form the lateral posterior region of the nasal cavity. Some of the muscles that move the mandible attach to the pterygoid processes. At the base of the lateral pterygoid process in the greater wing is the *foramen ovale* (=oval), an opening for the mandibular branch of the trigeminal (V) nerve (Figure 7.6a–d). Another foramen, the *foramen spinosum* (=resembling a spine, because of its proximity to the sharp spine of the sphenoid), lies at the posterior angle of the sphenoid and transmits the middle meningeal blood vessels. The *foramen lacerum* (=lacerated) is bounded anteriorly by the sphenoid bone and posteriorly by the temporal and occipital bones. The

● **FIGURE 7.6 CONTINUED** ▶

(c) Inferior view

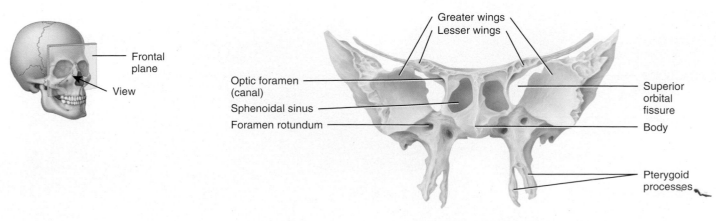

(e) Anterior view of sphenoid bone

foramen is covered in part by a layer of fibrocartilage in living subjects and is a joint uniting the three bones. It transmits a branch of the ascending pharyngeal artery. Another foramen associated with the sphenoid bone is the *foramen rotundum* (=round) located at the junction of the anterior and medial parts of the sphenoid bone. Through it passes the maxillary branch of the trigeminal (V) nerve.

◉ CHECKPOINT
Why is the sphenoid bone called the keystone of the cranial floor?

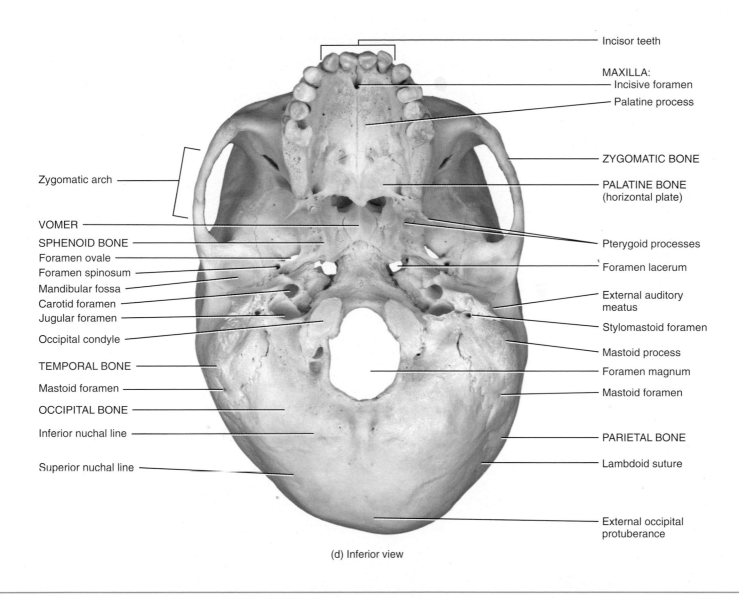

Incisor teeth

MAXILLA:
Incisive foramen
Palatine process

ZYGOMATIC BONE

PALATINE BONE
(horizontal plate)

Zygomatic arch

VOMER
SPHENOID BONE
Foramen ovale
Foramen spinosum
Mandibular fossa
Carotid foramen
Jugular foramen
Occipital condyle

TEMPORAL BONE

Mastoid foramen

OCCIPITAL BONE

Inferior nuchal line

Superior nuchal line

Pterygoid processes

Foramen lacerum

External auditory meatus

Stylomastoid foramen

Mastoid process
Foramen magnum
Mastoid foramen

PARIETAL BONE

Lambdoid suture

External occipital protuberance

(d) Inferior view

Superior orbital fissure

Foramen rotundum

Greater wing
Lesser wing
Optic foramen
Sphenoidal sinus

Body

Pterygoid processes

(f) Anterior view

❓ **Name the bones that articulate with the sphenoid bone, starting at the crista galli of the ethmoid bone and going in a clockwise direction.**

EXHIBIT 7.5 **177**

EXHIBIT 7.6

Cranial Bones—Ethmoid Bone

◉ OBJECTIVE

- Identify the location and surface features of the ethmoid bone.

Description

The unpaired **ethmoid bone** (ETH-moyd=like a sieve) is a light, spongelike bone located on the midline in the anterior part of the cranial floor between the two orbits (Figure 7.7). It is anterior to the sphenoid and posterior to the nasal bones. The ethmoid bone forms (1) part of the anterior portion of the cranial floor; (2) the thin, medial wall of the orbits; (3) the superior portion of the nasal septum, a partition that divides the nasal cavity into right and left sides; and (4) most of the superior sidewalls of the nasal cavity. The ethmoid is a major superior supporting structure of the nasal cavity.

Figure 7.7 Ethmoid bone. (See Tortora, *A Photographic Atlas of the Human Body*, 2e, Figure 3.10.)

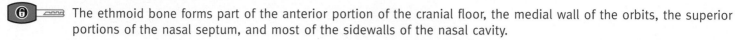

The ethmoid bone forms part of the anterior portion of the cranial floor, the medial wall of the orbits, the superior portions of the nasal septum, and most of the sidewalls of the nasal cavity.

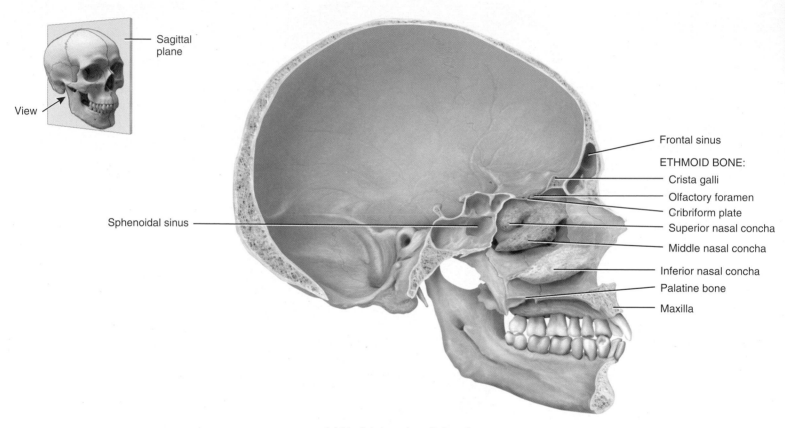

(a) Medial view of sagittal section

(b) Superior view

(c) Anterior view

Surface Features

The *lateral masses* of the ethmoid bone compose most of the wall between the nasal cavity and the orbits. They contain 3 to 18 air spaces called *ethmoidal cells*. The ethmoidal cells together form the *ethmoidal sinuses* (see Figure 7.11a, b). The midline *perpendicular plate* forms the superior portion of the nasal septum (see Figure 7.9). The horizontal *cribriform plate* (*cribri-*=sieve) lies in the anterior floor of the cranium and forms the roof of the nasal cavity. The cribriform plate contains the *olfactory foramina* (*olfact-*=smell), through which the olfactory (I) nerves pass. Projecting superiorly from the cribriform plate is a triangular process called the *crista galli* (*crista*=crest; *galli*=cock), which serves as a point of attachment for the membrane that separates the two hemispheres (sides) of the brain.

(d) Anterior view of position of ethmoid bone in skull

Right orbit

Crista galli

Perpendicular plate

Middle nasal concha

Inferior nasal concha

Vomer

Ethmoidal cells of ethmoid sinus

Crista galli

Olfactory foramina

Cribriform plate

Perpendicular plate

Lateral mass

(e) Superolateral view of right side

■ EXHIBIT 7.6 CONTINUES ▶

EXHIBIT 7.6   **179**

EXHIBIT 7.6 Cranial Bones—Ethmoid Bone CONTINUED

The lateral masses of the ethmoid bone contain two thin, scroll-shaped projections lateral to the nasal septum. These are called the *superior nasal concha* (KONG-ka=shell) or *turbinate* and the *middle nasal concha (turbinate)*. The plural form is *conchae* (KONG-kē). A third pair of conchae, the inferior nasal conchae, are separate bones (discussed shortly). The conchae increase the vascular and mucous membrane surface area in the nasal cavity, which warms, moistens, and humidifies inhaled air before it passes into the lungs. The conchae also cause inhaled air to swirl; the result is that many inhaled particles become trapped in the mucus

FIGURE 7.7 CONTINUED

Frontal plane

View

Superior meatus
Superior nasal concha
Perpendicular plate
Middle nasal concha
Middle meatus
Inferior nasal concha
Inferior meatus
Vomer
Maxilla

Brain
Crista galli
Orbital fat
Ethmoidal cells of ethmoidal sinus
Maxillary sinus
Nasal cavity
Hard palate
Oral cavity
Tongue
Mandible

(f) Frontal section through the head

that lines the nasal cavity. This action of the conchae helps cleanse inhaled air before it passes into the rest of the respiratory passageways. The superior nasal conchae are near the olfactory foramina of the cribriform plate where the sensory receptors for olfaction (smell) terminate in the mucous membrane of the superior nasal conchae. Thus, they increase the surface area for the sense of smell.

◉ CHECKPOINT

The ethmoid bone forms which other cranial structures?

(g) Frontal section through the head

❓ **What part of the ethmoid bone forms the superior part of the nasal septum? The medial walls of the orbits?**

EXHIBIT 7.6 **181**

◉ OBJECTIVE

- Identify the location of the nasal, lacrimal, palatine, inferior nasal conchae, and vomer.

Nasal Bones

Description

The paired **nasal bones** are small, flattened, rectangular-shaped bones that form the bridge of the nose (see Figure 7.2). These small bones protect the upper entry to the nasal cavity and provide attachment for a couple of thin muscles of facial expression. For those of you who wear glasses, they are the bones that form the resting place for the bridge of the glasses. The major structural portion of the nose consists of cartilage.

Lacrimal Bones

Description

The paired **lacrimal bones** (LAK-ri-mal; *lacrim-*=teardrops) are thin and roughly resemble a fingernail in size and shape (see Figures 7.2, 7.3, and 7.10). These bones, the smallest bones of the face, are posterior and lateral to the nasal bones and form a part of the medial wall of each orbit.

Surface Features

The lacrimal bones each contain a *lacrimal fossa*, a vertical tunnel formed with the maxilla, that houses the lacrimal sac, a structure that gathers tears and passes them into the nasal cavity (see Figure 7.10). The medial surface of the bones is covered with mucous membrane and forms part of the upper wall of the nasal cavity.

Palatine Bones

Description

The two L-shaped **palatine bones** (PAL-a-tīn) form the posterior portion of the hard palate. In addition to the roof of the mouth they form the posterior portion of the floor and walls of the nasal cavity. They also form a small portion of the floors of the orbits bordering the optic foramen (canal) (see Figures 7.6c, e).

Surface Features

The posterior portion of the hard palate, which separates the nasal cavity from the oral cavity, is formed by the *horizontal plates* of the palatine bones (see Figure 7.6c). The perpendicular plates of the palatine bones contribute to the walls of the nasal cavity and orbits.

Inferior Nasal Conchae

Description

The two **inferior nasal conchae** are inferior to the middle nasal conchae of the ethmoid bone (see Figures 7.2 and 7.7a). These scroll-like bones form a part of the inferior lateral wall of the nasal cavity and project into the nasal cavity. The inferior nasal conchae are separate bones; they are not part of the ethmoid bone. All three pairs of nasal conchae (superior, middle, and inferior) increase the mucosal covered surface area of the nasal cavity and help swirl and filter air before it passes into the lungs. However, only the superior nasal conchae of the ethmoid bone are involved in the sense of smell.

Vomer

Description

The unpaired **vomer** (VŌ-mer=plowshare) is a roughly triangular bone on the floor of the nasal cavity that articulates superiorly with the perpendicular plate of the ethmoid bone and inferiorly with both the maxillae and palatine bones along the midline (see Figures 7.2, 7.5, and 7.9). It is one of the components of the nasal septum, the partition that divides the nasal cavity into right and left sides.

● CLINICAL **CONNECTION** | Black Eye

A **black eye** is a bruising around the eye, commonly due to an injury to the face, rather than an eye injury. In response to trauma, blood and other fluids accumulate in the space around the eye, causing the swelling and dark discoloration. One cause might be a blow to the sharp ridge just superior to the supraorbital margin that fractures the frontal bone, resulting in bleeding. Another is a blow to the nose. Certain surgical procedures (face lift, eyelid surgery, jaw surgery, or nasal surgery) can also result in black eyes. ●

◉ CHECKPOINT

Which bones form the hard palate?

EXHIBIT 7.8 Facial Bones—Maxillae and Zygomatic Bones

OBJECTIVE

- Identify the location and surface features of the maxillae and zygomatic bones.

Maxillae

Description

The paired **maxillae** (mak-SIL-ē=jawbones; singular is *maxilla*) unite to form the upper jawbone. They articulate with every bone of the face except the mandible (lower jawbone) (see Figures 7.2 and 7.7). The maxillae form part of the floors of the orbits, part of the lateral walls and floor of the nasal cavity, and most of the hard palate. The hard palate is the bony roof of the mouth, and is formed by the palatine processes of the maxillae and horizontal plates of the palatine bones. The hard palate separates the nasal cavity from the oral cavity.

Surface Features

Each maxilla has a central body, which is hollowed out internally to form the mucosal-lined *maxillary sinus*. Like the previously mentioned paranasal sinuses, the mucosa of the maxillary sinus is continuous with the mucosa of the nasal cavity (see Figure 7.11). The *alveolar process* (al-VĒ-ō-lar; *alveol-*=small cavity) is an arch that contains the *alveoli* (sockets) for the maxillary (upper) teeth (see Figure 7.2a). The *palatine process* is a horizontal projection of the maxilla that forms the anterior three-quarters of the hard palate (see Figure 7.6c, e). The union and fusion of the maxillary bones is normally completed before birth.

The *infraorbital foramen* (*infra-*=below; *orbital*=orbit), which can be seen in the anterior view of the skull in Figure 7.2, is an opening in the maxilla inferior to the orbit. Through it passes the infraorbital nerve and blood vessels and a branch of the maxillary division of the trigeminal (V) nerve. Another prominent foramen in the maxilla is the *incisive foramen* (=incisor teeth) just posterior to the incisor teeth (see Figure 7.6c, e). It transmits branches of the greater palatine blood vessels and nasopalatine nerve. A final structure associated with the maxilla and sphenoid bone is the *inferior orbital fissure*, which is located between the greater wing of the sphenoid and the posterior aspect of the maxilla (see Figure 7.10).

Zygomatic Bones

Description

The two **zygomatic bones** (*zygo-*=yokelike), commonly called cheekbones, form the prominences of the cheeks and part of the

• CLINICAL **CONNECTION** | Cleft Palate and Cleft Lip

Usually the palatine processes of the maxillary bones unite during weeks 10 to 12 of embryonic development. Failure to do so can result in one type of **cleft palate**. The condition may also involve incomplete fusion of the horizontal plates of the palatine bones (see Figure 7.6c, d). Another form of this condition, called **cleft lip**, involves a split in the upper lip. Cleft lip and cleft palate often occur together. Depending on the extent and position of the cleft, speech and swallowing may be affected. In addition, children with cleft palate tend to have many ear infections that can lead to hearing loss. Facial and oral surgeons recommend closure of cleft lip during the first few weeks following birth, and surgical results are excellent. Repair of cleft palate typically is done between 12 and 18 months of age, ideally before the child begins to talk. Speech therapy may be needed, because the palate is important for pronouncing consonants, and orthodontic therapy may be needed to align the teeth. Again, results are usually excellent. Folic acid (one of the B vitamins) supplementation during pregnancy decreases the incidence of cleft palate and cleft lip. •

lateral wall and floor of each orbit (see Figure 7.10). They articulate with the frontal, maxilla, sphenoid, and temporal bones.

Surface Features

The *temporal process* of the zygomatic bone projects posteriorly and articulates with the zygomatic process of the temporal bone to form the *zygomatic arch* (see Figure 7.3a, b). Posteriorly the bone forms a concave temporal surface that contributes to the temporal region of the skull and houses the tendon of the strong jaw-closing muscle called the *temporalis*. A foramen associated with the zygomatic bone is the *zygomaticofacial foramen* near the center of the bone (Figure 7.2a). It transmits the zygomaticofacial nerve and vessels.

CHECKPOINT

Which structures pass through the zygomaticofacial foramen?

EXHIBIT 7.8 **183**

EXHIBIT 7.9 Facial Bones—Mandible

- Identify the location and surface features of the mandible.

Description

The **mandible** (*mand-*=to chew), or lower jawbone, is the largest, strongest facial bone (Figure 7.8). This large arch-shaped bone is the only movable skull bone (other than the auditory ossicles).

Surface Features

In the lateral view, you can see that the mandible consists of a curved, horizontal portion, the *body*, and two perpendicular portions, the *rami* (RĀ-mī=branches). The *angle* of the mandible is the area where each *ramus* (singular form) meets the body. Each ramus has a posterior *condylar process* (KON-di-lar) that articulates with the mandibular fossa and articular tubercle of the temporal bone (see Figures 7.3a and 7.6c) to form the **temporomandibular joint (TMJ)**. It also has an anterior *coronoid process* (KOR-ō-noyd) to which the temporalis muscle attaches. The depression between the coronoid and condylar processes is called the *mandibular notch*. The *alveolar process* is an arch containing the *alveoli* (sockets) for the mandibular (lower) teeth.

The *mental foramen* (*ment-*=chin) is approximately inferior to the second premolar tooth. It is near this foramen that dentists reach the mental nerve when injecting anesthetics. Another fora-men associated with the mandible is the *mandibular foramen* on the medial surface of each ramus, another site often used by dentists to inject anesthetics. The mandibular foramen is the beginning of the *mandibular canal*, which runs obliquely in the ramus and anteriorly to the body. The inferior alveolar nerves and blood vessels pass through the canal and are distributed to the mandibular teeth.

● **CLINICAL CONNECTION** | Temporomandibular Joint Syndrome

One problem associated with the temporomandibular joint is **temporomandibular joint (TMJ) syndrome**. It is characterized by dull pain around the ear, tenderness of the jaw muscles, a clicking or popping noise when opening or closing the mouth, limited or abnormal opening of the mouth, headache, tooth sensitivity, and abnormal wearing of the teeth. TMJ syndrome can be caused by improperly aligned teeth, grinding or clenching the teeth, trauma to the head and neck, or arthritis. Treatments include application of moist heat or ice, limiting the diet to soft foods, administration of pain relievers such as aspirin, muscle retraining, use of a splint or bite plate to reduce clenching and teeth grinding (especially when worn at night), adjustment or reshaping of the teeth (orthodontic treatment), and surgery. •

■ **CHECKPOINT**
What structures form the temporomandibular joint?

Figure 7.8 Mandible.

⑥ ⚿ The mandible is the largest and strongest facial bone.

(a) Right lateral view

(b) Right lateral view

❓ **What is the functional feature of the mandible that distinguishes it from all the other skull bones?**

Nasal Septum

The inside of the nose, called the nasal cavity, is divided into right and left sides by a vertical partition called the **nasal septum**. The three components of the nasal septum are the vomer, the perpendicular plate of the ethmoid bone (which together form the bony nasal septum), and the septal cartilage (Figure 7.9). The inferior border of the perpendicular plate of the ethmoid bone joins the superoposterior border of the vomer to form the more posterior bony part of the septum. The septal cartilage, which is hyaline cartilage, articulates with the anterior margins of the two bones to form the anterior portion of the septum. The term "broken nose" usually refers to damage to the septal cartilage rather than the nasal bones.

Orbits

Seven bones of the skull join to form each **orbit** (eye socket), which contains the eyeball and associated structures (Figure 7.10 on page 186). The three cranial bones of the orbit are the frontal, sphenoid, and ethmoid; the four facial bones are the palatine, zygomatic, lacrimal, and maxilla. Each pyramid-shaped orbit has four regions that converge posteriorly:

1. Parts of the frontal and sphenoid bones comprise the *roof* of the orbit.
2. Parts of the zygomatic and sphenoid bones form the *lateral wall* of the orbit.

3. Parts of the maxilla, zygomatic, and palatine bones make up the *floor* of the orbit.
4. Parts of the maxilla, lacrimal, ethmoid, and sphenoid bones form the *medial wall* of the orbit.

Associated with each orbit are five openings:

1. The *optic foramen (canal)* is at the junction of the roof and medial wall.
2. The *superior orbital fissure* is at the superior lateral angle of the apex.
3. The *inferior orbital fissure* is at the junction of the lateral wall and floor.
4. The *supraorbital foramen* is on the medial side of the supraorbital margin of the frontal bone.
5. The *lacrimal fossa* is in the lacrimal bone.

Foramina

We mentioned most of the **foramina** (openings for blood vessels, nerves, or ligaments) of the skull in the descriptions of the cranial and facial bones that they penetrate. As preparation for studying other systems of the body, especially the nervous and cardiovascular systems, these foramina and the structures passing through them are listed in Table 7.3 on page 187. For your convenience and for future reference, the foramina are listed alphabetically.

Figure 7.9 **Nasal septum.** (See Tortora, *A Photographic Atlas of the Human Body*, 2e, Figure 3.4.)

 The structures that form the nasal septum are the perpendicular plate of the ethmoid bone, the vomer, and septal cartilage.

Sagittal section

 What is the function of the nasal septum?

Figure 7.10 Details of the orbit (eye socket).

🔑 The orbit is a pyramid-shaped structure that contains the eyeball and associated structures.

FRONTAL BONE

Supraorbital margin

Supraorbital foramen

SPHENOID BONE

Optic foramen (canal)

Superior orbital fissure

NASAL BONE

LACRIMAL BONE

ETHMOID BONE

Lacrimal fossa

MAXILLA

Zygomaticofacial foramen

ZYGOMATIC BONE

Infraorbital foramen

Inferior orbital fissure

(a) Anterior view showing the bones of the right orbit

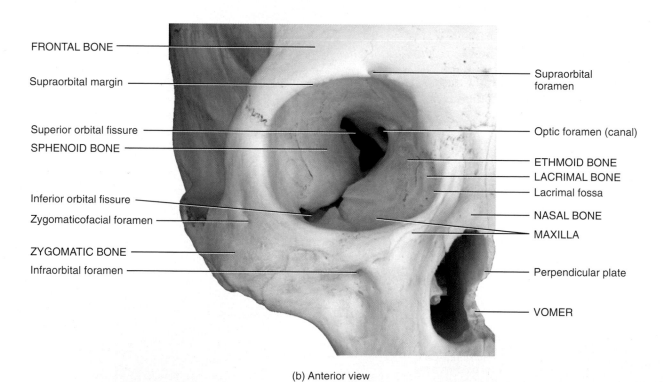

FRONTAL BONE

Supraorbital margin

Supraorbital foramen

Superior orbital fissure

SPHENOID BONE

Optic foramen (canal)

ETHMOID BONE

LACRIMAL BONE

Lacrimal fossa

Inferior orbital fissure

Zygomaticofacial foramen

NASAL BONE

MAXILLA

ZYGOMATIC BONE

Infraorbital foramen

Perpendicular plate

VOMER

(b) Anterior view

❓ **Which seven bones form the orbit and which one is not visible in the figure?**

TABLE 7.3

Principal Foramina of the Skull

FORAMEN	LOCATION	STRUCTURES PASSING THROUGH
Carotid (relating to carotid artery in neck)	Petrous portion of temporal bone (Figure 7.6c).	Internal carotid artery and sympathetic nerves for eyes.
Hypoglossal (*hypo-*=under; *glossus*=tongue)	Superior to base of occipital condyles (Figure 7.6a).	Cranial nerve XII (hypoglossal) and branch of ascending pharyngeal artery.
Incisive (*incisive*=pertaining to incisor teeth)	Posterior to incisor teeth in maxilla (Figure 7.6c).	Branches of greater palatine blood vessels and nasopalatine nerve.
Infraorbital (*infra*=below)	Inferior to orbit in maxilla (Figure 7.10a).	Infraorbital nerve and blood vessels and a branch of the maxillary division of cranial nerve V (trigeminal).
Jugular (*jugular*=throat)	Posterior to carotid canal between petrous portion of temporal bone and occipital bone (Figure 7.6a).	Internal jugular vein, cranial nerves IX (glossopharyngeal), X (vagus), and XI (accessory).
Lacerum (*lacerum*=lacerated)	Bounded anteriorly by sphenoid bone, posteriorly by petrous portion of temporal bone, and medially by the sphenoid bone and occipital bone (Figure 7.6a).	Branch of ascending pharyngeal artery in palatine bones.
Magnum (*magnum*=large)	Occipital bone (Figure 7.6c).	Medulla oblongata and its membranes (meninges), cranial nerve XI (accessory), and vertebral and spinal arteries.
Mandibular (*mand*=to chew)	Medial surface of ramus of mandible (Figure 7.8).	Inferior alveolar nerve and blood vessels.
Mastoid (=breast-shaped)	Posterior border of mastoid process of temporal bone (Figure 7.6c).	Emissary vein to transverse sinus and branch of occipital artery to dura mater.
Mental (*ment-*=chin)	Inferior to second premolar tooth in mandible (Figure 7.8).	Mental nerve and vessels.
Olfactory (*olfact*=to smell)	Cribriform plate of ethmoid bone (Figure 7.6a).	Cranial nerve I (olfactory).
Optic (canal) (=eye)	Between superior and inferior portions of small wing of sphenoid bone (Figure 7.10).	Cranial nerve II (optic) and ophthalmic artery.
Ovale (*ovale*=oval)	Greater wing of sphenoid bone (Figure 7.6a).	Mandibular branch of cranial nerve V (trigeminal).
Rotundum (=round)	Junction of anterior and medial parts of sphenoid bone (Figure 7.6a).	Maxillary branch of cranial nerve V (trigeminal).
Spinosum (=resembling a spine)	Posterior angle of sphenoid bone (Figure 7.6a).	Middle meningeal blood vessels.
Stylomastoid (*stylo*=stake or pole)	Between styloid and mastoid processes of temporal bone (Figure 7.6c).	Cranial nerve VII (facial) and stylomastoid artery.
Supraorbital (*supra-*=above)	Supraorbital margin of orbit in frontal bone (Figure 7.10a).	Supraorbital nerve and artery.

Unique Features of the Skull

The skull exhibits several unique features not seen in other bones of the body. These include sutures, paranasal sinuses, and fontanels.

Sutures

A **suture** (SOO-chur=seam) is an immovable joint, in most cases in an adult skull, that holds most skull bones together. Sutures in the skulls of infants and children often are movable. In older individuals, many of the sutures in the skull fuse and become completely immovable. The names of many sutures reflect the bones they unite. For example, the frontozygomatic suture is between the frontal bone and the zygomatic bone. Similarly, the sphenoparietal suture is between the sphenoid bone and the parietal bone. In other cases, however, the names of sutures are not so obvious. Of the many sutures found in the skull, we will identify only four prominent ones:

1. The **coronal suture** (kō-RŌ-nal; *coron-*=crown) unites the frontal bone and both parietal bones (see Figure 7.3).
2. The **sagittal suture** (SAJ-i-tal; *sagitt-*=arrow) unites the two parietal bones on the superior midline of the skull (see Figure 7.5). The sagittal suture is so named because in the infant, before the bones of the skull are firmly united, the suture and the fontanels (soft spots) associated with it resemble an arrow.
3. The **lambdoid suture** (LAM-doyd) unites the two parietal bones to the occipital bone. This suture is so named because of its resemblance to the Greek letter lambda (λ), as can be seen in Figure 7.5. Sutural bones may occur within the sagittal and lambdoid sutures.
4. The **squamous sutures** (SKWĀ-mus; *squam-*=flat) unite the parietal and temporal bones on the lateral aspects of the skull (see Figure 7.3).

Paranasal Sinuses

The **paranasal sinuses** (*para-*=beside) are cavities within certain cranial and facial bones near the nasal cavity. They are most evident in a sagittal section of the skull (Figure 7.11). The paranasal sinuses are lined with mucous membranes that are continuous with the lining of the nasal cavity. Secretions produced by the mucous membrane of the paranasal sinuses drain into the nasal cavity. Paranasal sinuses are not present at birth, but develop during the prepubescent years as outgrowths of the nasal mucosa. (*Prepubescent* means before becoming capable of reproduction, typically the early to mid-teenage years.) Skull bones containing the paranasal sinuses are the frontal, sphenoid, ethmoid, and maxillary. The sinuses lighten the mass of the skull and increase the surface area of the nasal mucosa, thus increasing the production of mucus to help moisten and cleanse inhaled air. In addition, the paranasal sinuses serve as resonating (echo) chambers within the skull that intensify and prolong sounds, thereby enhancing the quality of the voice. The influence of the sinuses on your voice becomes obvious when you have a cold; the passageways through which sound travels into and out of the sinuses become blocked by excess mucus production, changing the quality of your voice.

Fontanels

The skeleton of a newly formed embryo consists of cartilage or mesenchyme arranged in sheetlike layers that resemble membranes shaped like bones. Gradually, ossification occurs—bone replaces the cartilage and mesenchyme. At birth, mesenchyme-filled spaces called **fontanels** (fon-ta-NELZ=little fountains) are present between the cranial bones (Figure 7.12 on page 190). Commonly called "soft spots," fontanels are areas of unossified mesenchyme. Eventually, they will be replaced with bone by in-

Figure 7.11 Paranasal sinuses. (a and b) Location of the paranasal sinuses. (c) Openings of the paranasal sinuses into the nasal cavity. Portions of the conchae have been sectioned. (d) Dissection into maxillary sinus. (See Tortora, *A Photographic Atlas of the Human Body*, 2e, Figure 3.4.)

Paranasal sinuses are mucous membrane–lined spaces in the frontal, sphenoid, ethmoid, and maxillary bones that connect to the nasal cavity.

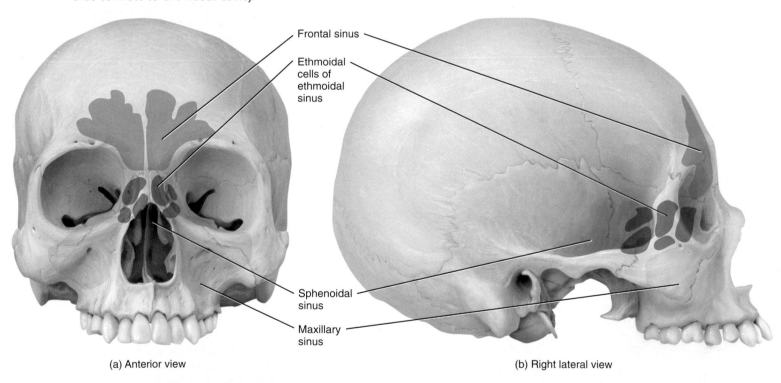

(a) Anterior view

(b) Right lateral view

tramembranous ossification and become sutures. Functionally, the fontanels provide some flexibility to the fetal skull. They allow the skull to change shape as it passes through the birth canal and permit rapid growth of the brain during infancy. Although an infant may have many fontanels at birth, the form and location of six are fairly constant:

1. The unpaired **anterior fontanel**, located at the midline between the two parietal bones and the frontal bones, is roughly diamond-shaped and is the largest fontanel. It usually closes 18 to 24 months after birth.
2. The unpaired **posterior fontanel** is located at the midline between the two parietal bones and the occipital bone. Because it is much smaller than the anterior fontanel, it generally closes about 2 months after birth.

3. The paired **anterolateral fontanels**, located laterally between the frontal, parietal, temporal, and sphenoid bones, are small and irregular in shape. Normally, they close about 3 months after birth.
4. The paired **posterolateral fontanels**, located laterally between the parietal, occipital, and temporal bones, are irregularly shaped. They begin to close 1 to 2 months after birth, but closure is generally not complete until 12 months.

The amount of closure in fontanels helps a physician gauge the degree of brain development. In addition, the anterior fontanel serves as a landmark for withdrawal of blood for analysis from the superior sagittal sinus (a large vein on the midline surface of the brain). See Figure 15.11a on pages 516–517.

Sagittal plane

View

Sphenoidal sinus
Hypophyseal fossa
Nasal conchae (cut):
 Superior
 Middle
 Inferior

Orifice of auditory (eustachian) tube

Frontal sinus
Opening of sphenoidal sinus
Opening of posterior ethmoidal cells
Opening of frontal sinus
Opening of maxillary sinus
Nasal cavity
Opening of nasolacrimal duct
Hard palate:
 Maxilla
 Palatine bone

(c) Sagittal section

Brain

Eyeball

Opening of maxillary sinus

Maxillary sinus

Maxilla (cut)

Tongue

(d) Dissection into maxillary sinus

? What are the functions of the paranasal sinuses?

Figure 7.12 Fontanels at birth. (See Tortora, *A Photographic Atlas of the Human Body*, 2e, Figure 3.12.)

Fontanels are mesenchyme-filled spaces between cranial bones that are present at birth.

Parietal bone

Future squamous suture

POSTERIOR FONTANEL

POSTEROLATERAL FONTANEL

Occipital bone

ANTERIOR FONTANEL

Future coronal suture

Frontal bone

ANTEROLATERAL FONTANEL

Sphenoid bone

Temporal bone

(a) Right lateral view

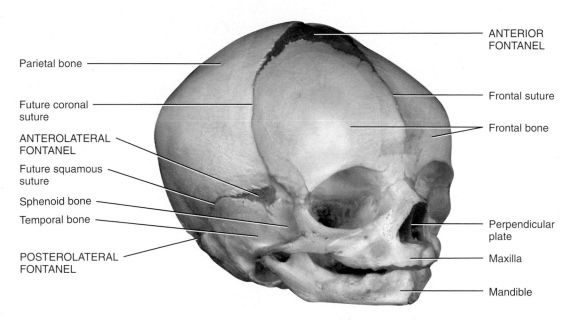

Parietal bone

Future coronal suture

ANTEROLATERAL FONTANEL

Future squamous suture

Sphenoid bone

Temporal bone

POSTEROLATERAL FONTANEL

ANTERIOR FONTANEL

Frontal suture

Frontal bone

Perpendicular plate

Maxilla

Mandible

(b) Right anterolateral view

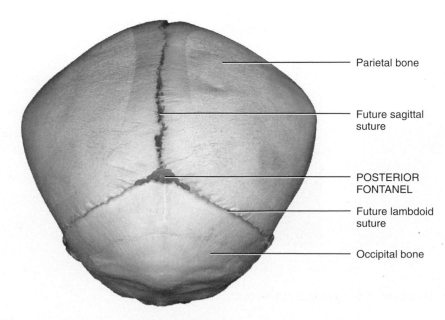

Parietal bone

Future sagittal suture

POSTERIOR FONTANEL

Future lambdoid suture

Occipital bone

(c) Posterior view

? Which fontanel is bordered by four different skull bones?

Cranial Fossae

The internal shape and appearance of the cranium resembles the outer surface of the brain. As the cranial bones form around the developing brain, they create a mold that approximates the brain's shape. If the *calvarium*, the roof of the cranium, is removed, the floor of the cranium displays three distinct levels that correspond to the major contours of the inferior surface of the brain. The three distinct regions of the floor of the cranium are called **cranial fossae** (Figure 7.13). Each cranial fossa is at a different level, creating the appearance of stairs that step down from anterior to posterior inside the skull. A specific structural region of the brain fits snugly into each cranial fossa. From anterior to posterior, they are named the anterior cranial fossa, middle cranial fossa, and posterior cranial fossa. The highest level, the *anterior cranial fossa*, is formed largely by the portion of the frontal bone that constitutes the roof of the orbits and nasal cavity, the crista galli and cribriform plate of the ethmoid bone, and the lesser wings and part of the body of the sphenoid bone. This fossa houses the frontal lobes of the cerebral hemispheres of the brain. The rough surface of the frontal bone can lead to tearing of the frontal lobes of the cerebral

hemispheres during head trauma. The *middle cranial fossa* is inferior and posterior to the anterior cranial fossa. It is shaped like a butterfly with a small median portion and two expanded lateral portions. The median portion is formed by part of the body of the sphenoid bone, and the lateral portions are formed by the greater wings of the sphenoid bone, temporal squama, and parietal bone. The middle cranial fossa cradles the temporal lobes of the cerebral hemispheres. The last fossa, at the most inferior level, is the *posterior cranial fossa*, the largest of the fossae. It is formed mostly by the occipital bone and the petrous and mastoid portions of the temporal bone. It is a very deep fossa that accommodates the cerebellum, pons, and medulla oblongata of the brain.

⊙ **CHECKPOINT**

2. Describe the general features of the skull.
3. What bones constitute the orbit?
4. What structures make up the nasal septum?
5. Define the following: foramen, suture, paranasal sinus, and fontanel.
6. Name the cranial fossae from inferior to superior.

Figure 7.13 Cranial fossae.

 Cranial fossae are levels in the cranial floor that contain depressions for brain convolutions, grooves for blood vessels, and numerous foramina.

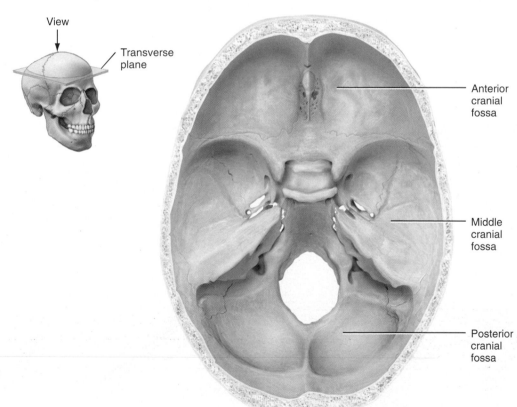

Superior view of floor of cranium

 Which cranial fossa is the largest?

HYOID BONE

⊙ **OBJECTIVE**

• Describe the relationship of the hyoid bone to the skull.

The single **hyoid bone** (=U-shaped) is a unique component of the axial skeleton because it does not articulate with any other bone. Rather, it is suspended from the styloid processes of the temporal bones by ligaments and muscles (see Figure 11.7a on page 329). Located in the anterior neck between the mandible and larynx (Figure 7.14a), the hyoid bone supports the tongue, providing attachment sites for some tongue muscles and for muscles of the neck and pharynx. The hyoid bone also helps to keep the larynx (voice box) open at all times. The hyoid bone consists of a horizontal, rectangular *body* and paired projections called the *lesser horns* and the *greater horns* (Figure 7.14b, c). Muscles and ligaments attach to these paired projections.

The hyoid bone, as well as cartilages of the larynx and trachea, are often fractured during strangulation. As a result, they are carefully examined at autopsy when strangulation is a suspected cause of death.

⊙ **CHECKPOINT**

7. What are the functions of the hyoid bone?

Figure 7.14 Hyoid bone.

 The hyoid bone supports the tongue, providing attachment sites for muscles of the tongue, neck, and pharynx.

Mandible

Hyoid bone

Larynx

(a) Position of hyoid

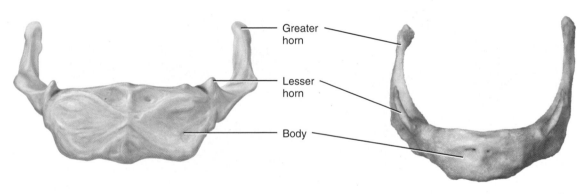

Greater horn
Lesser horn
Body

(b) Anterior view

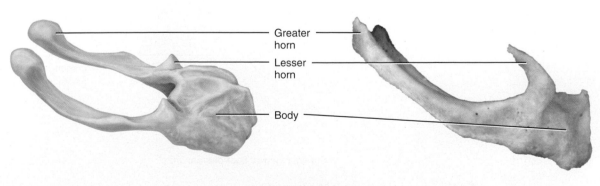

Greater horn
Lesser horn
Body

(c) Right lateral view

? In what way is the hyoid bone different from all other bones of the axial skeleton?

VERTEBRAL COLUMN

◉ **OBJECTIVE**

- Identify the regions and normal curves of the vertebral column and describe its structural and functional features.

The **vertebral column**, also called the *spine*, *backbone*, or *spinal column*, makes up about two-fifths of the total height of the body and is composed of a series of bones called **vertebrae** (VER-te-brē; singular is *vertebra*). The vertebral column consists of bone and connective tissue that surrounds and protects the nervous tissue of the spinal cord. The length of the column is about 71 cm (28 in.) in an average adult male and about 61 cm (24 in.) in an average adult female. The vertebral column functions as a strong, flexible rod with elements that can move forward, move backward, move sideways, and rotate. It encloses and protects the spinal cord, supports the head, and serves as a point of attachment for the ribs, pelvic girdle, and muscles of the back and upper limbs.

The total number of vertebrae during early development is 33. Then, several vertebrae in the sacral and coccygeal regions fuse. As a result, the adult vertebral column typically contains 26 vertebrae (Figure 7.15a). These are distributed as follows:

7 **cervical vertebrae** (*cervic-*=neck) in the neck region.

12 **thoracic vertebrae** (*thorax*=chest) posterior to the thoracic cavity.

5 **lumbar vertebrae** (*lumb-*=loin) supporting the lower back.

1 **sacrum** (SĀ-krum=sacred bone) consisting of five fused **sacral vertebrae**.

1 **coccyx** (KOK-siks=cuckoo, because the shape resembles the bill of a cuckoo bird), consisting of four fused **coccygeal vertebrae** (kok-SIJ-ē-al).

The cervical, thoracic, and lumbar vertebrae are movable, but the sacrum and coccyx are not. We will discuss each of these regions in detail shortly.

Normal Curves of the Vertebral Column

When viewed from anterior or posterior, a normal adult vertebral column appears straight. But when viewed from the side, it shows four slight bends called **normal curves** (Figure 7.15b). Relative to the front of the body, the *cervical* and *lumbar curves* are convex (bulging out), and the *thoracic* and *sacral curves* are concave (cupping in). The curves of the vertebral column increase its strength, help maintain balance in the upright position, absorb shocks during walking, and help protect the vertebrae from fracture.

Figure 7.15 Vertebral column. The numbers in parentheses in (a) indicate the number of vertebrae in each region. In (d), the relative size of the disc has been enlarged for emphasis. (See Tortora, *A Photographic Atlas of the Human Body*, 2e, Figure 3.15.)

🔑 ⚊ The adult vertebral column typically contains 26 vertebrae.

(a) Anterior view showing regions of the vertebral column

Cervical vertebrae (7)

Thoracic vertebrae (12)

Lumbar vertebrae (5)

Intervertebral disc

Sacrum (1)

Coccyx (1)

POSTERIOR

ANTERIOR

Cervical curve (formed by 7 cervical vertebrae)

Thoracic curve (formed by 12 thoracic vertebrae)

Intervertebral disc

Intervertebral foramen

Lumbar curve (formed by 5 lumbar vertebrae)

Sacrum

Sacral curve (formed by 5 fused sacral vertebrae)

Coccyx

(b) Right lateral view showing four normal curves

CONTINUES ▶

FIGURE 7.15 CONTINUED ▶

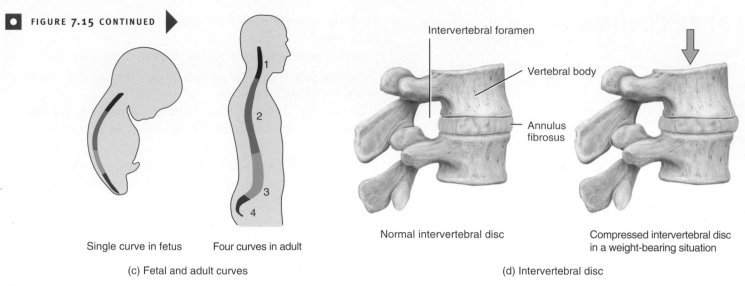

Single curve in fetus   Four curves in adult

(c) Fetal and adult curves

Normal intervertebral disc

Compressed intervertebral disc in a weight-bearing situation

(d) Intervertebral disc

? **Which curves of the adult vertebral column are concave (relative to the anterior side of the body)?**

In the fetus, there is only a single anteriorly concave curve (Figure 7.15c). At about the third month after birth, when an infant begins to hold its head erect, the cervical curve develops. Later, when the child sits up, stands, and walks, the lumbar curve develops. The thoracic and sacral curves are called *primary curves* because they form first during fetal development. The cervical and lumbar curves are known as *secondary curves* because they begin to form later, several months after birth. All curves are fully developed by age 10. However, secondary curves may be progressively lost in old age.

Intervertebral Discs

Between the bodies of adjacent vertebrae from the second cervical vertebra to the sacrum are **intervertebral discs** (Figure 7.15d). They account for 25% of the height of the vertebral

---

• CLINICAL CONNECTION | **Abnormal Curves of the Vertebral Column**

Various conditions may exaggerate the normal curves of the vertebral column, or the column may acquire a lateral bend, resulting in **abnormal curves** of the vertebral column.

Scoliosis (skō-lē-Ō-sis; *scolio-*=crooked) is a lateral bending of the vertebral column, usually in the thoracic region (Figure A). This is the most common of the abnormal curves. It may result from congenitally (present at birth) malformed vertebrae, chronic sciatica, paralysis of muscles on one side of the vertebral column, poor posture, or one leg being shorter than the other.

Kyphosis (kī-FŌ-sis; *kyphos-*=hump) is an increase in the thoracic curve of the vertebral column (Figure B). In tuberculosis of the spine, vertebral bodies may partially collapse, causing an acute angular bending of the vertebral column. In the elderly, degeneration of the intervertebral discs leads to kyphosis. Kyphosis may also be caused by rickets and poor posture. It is also common in females with advanced osteoporosis. The term *round-shouldered* is an expression for mild kyphosis.

Lordosis (lor-DŌ-sis; *lord-*=bent backward), sometimes called *hollow* or *sway back*, is an increase in the lumbar curve of the vertebral column (Figure C). It may result from increased weight of the abdomen as in pregnancy, or extreme obesity, poor posture, rickets, osteoporosis, or tuberculosis of the spine. •

(A) Scoliosis

(B) Kyphosis

(C) Lordosis

column. Each disc has an outer fibrous ring consisting of fibro-cartilage called the *annulus fibrosus* (*annulus*=ringlike) and an inner soft, pulpy, highly elastic substance called the *nucleus pulposus* (*pulposus*=pulplike). The discs form strong joints, permit various movements of the vertebral column, and absorb vertical shock. During the course of a day the discs compress so that we are a bit shorter at night. While we are sleeping there is less compression so that we are taller when we awake in the morning. Under compression, they flatten and broaden, with age, the nucleus pulposus hardens and becomes less elastic. Narrowing of the discs and compression of vertebrae results in a decrease in height with age.

Since intervertebral discs are avascular, the annulus fibrosus and nucleus pulposus rely on blood vessels from the bodies of vertebrae to obtain oxygen and nutrients and remove wastes. Certain stretching exercises, for example yoga, decompress discs and increase blood circulation, both of which speed up the uptake of oxygen and nutrients by discs and the removal of wastes.

Parts of a Typical Vertebra

Even though vertebrae in different regions of the spinal column vary in size, shape, and detail, they are similar enough that we can discuss the structures (and the functions) of a typical vertebra (Figure 7.16). Vertebrae typically consist of a body, a vertebral arch, and several processes.

Body

The **body** is the largest portion of a vertebra and forms the anterior blocklike mass of the bone. All the adjacent vertebral bodies lined up in a row create the columnlike axis of the skeleton. The vertebral bodies serve as the main weight-bearing component of the vertebral column. The cartilaginous intervertebral discs lie between adjacent vertebral bodies and the superior and inferior surfaces are roughened for the attachment of the discs. The anterior and posterior surfaces contain nutrient foramina, openings for blood vessels that deliver nutrients and oxygen and remove carbon dioxide and wastes from bone tissue.

Vertebral Arch

The **vertebral arch** extends posteriorly from the body of the vertebra and together with the body of the vertebra surrounds the spinal cord. The vertebral arch is variable and complex and consists of numerous bony projections that provide attachment surfaces and lever arms for muscles. Two short, thick processes, the *pedicles* (PED-i-kuls=little feet), form the vertebral arch. The pedicles project posteriorly from the body to unite with the laminae. The *laminae* (LAM-i-nē=thin layers) are the flat parts that join to form the posterior portion of the vertebral arch. The *vertebral foramen* lies between the vertebral arch and body and contains the spinal cord and its meningeal coverings, adipose tissue, areolar connective tissue, and blood vessels. Collectively, the vertebral foramina of all vertebrae form the *vertebral (spinal) canal*. The pedicles exhibit superior and inferior indentations called *vertebral notches*. When the vertebral notches are stacked on top of one another, they form an opening between adjoining vertebrae on both sides of the column. Each opening, called an *intervertebral foramen*, permits the passage of a single spinal nerve.

Processes

Seven **processes** arise from the vertebral arch. At the point where a lamina and pedicle join, a *transverse process* extends laterally on each side. A single *spinous process (spine)* projects posteriorly from the junction of the laminae. These three processes serve as points of attachment for muscles. The remaining four processes form joints with other vertebrae above or below. The

Figure 7.16 **Structure of a typical vertebra, as illustrated by a thoracic vertebra.** In (b), only one spinal nerve has been included, and it has been extended beyond the intervertebral foramen for clarity.

A vertebra consists of a body, a vertebral arch, and several processes.

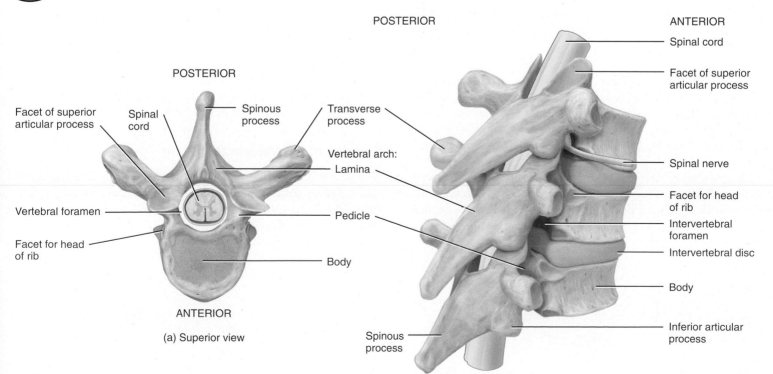

(a) Superior view

(b) Right posterolateral view of articulated vertebrae

CONTINUES ▶

FIGURE 7.16 CONTINUED

(c) Superior view

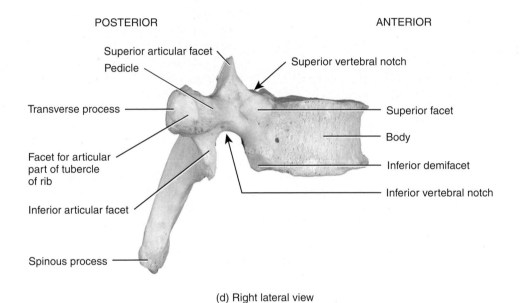

(d) Right lateral view

? What are the functions of the vertebral and intervertebral foramina?

two *superior articular processes* of a vertebra articulate (form joints) with the two inferior articular processes of the vertebra immediately superior to them. In turn, the two *inferior articular processes* of that vertebra articulate with the two superior articular processes of the vertebra immediately inferior to them, and so on.

The articulating surfaces of the articular processes are referred to as *facets* (FAS-et or fa-SET=little faces), and are covered with hyaline cartilage. The articulations formed between the bodies and articular facets of successive vertebrae are called *intervertebral joints*.

● CHECKPOINT

8. What are the functions of the vertebral column?

9. When do the secondary vertebral curves develop?

Regions of the Vertebral Column

In Exhibits 7.10 through 7.15, we turn now to the five regions of the vertebral column, beginning superiorly and moving inferiorly. These regions are cervical, thoracic, lumbar, sacral, and coccygeal. Note that vertebrae in each region are numbered in sequence, from superior to inferior.

EXHIBIT 7.10　Vertebral Regions—Cervical Vertebrae

• Identify the location and surface features of the cervical vertebrae.

Description

The **cervical vertebrae** (C1–C7) are the most variable of the vertebrae. This variability is due to the fact that the general form of the first two vertebrae in this region is modified significantly from the remaining five cervical vertebrae. The cervical vertebrae form a delicate column of bones that vary considerably in the range of mobility at their joint surfaces. The bodies of cervical vertebrae are smaller than those of thoracic vertebrae (Figure 7.17a). The vertebral arches, however, are larger. A unique character that identifies the cervical vertebrae is a foramen in their transverse process.

Surface Features

All cervical vertebrae have three foramina: one vertebral foramen and two transverse foramina (Figure 7.17d). The *vertebral foramina* of cervical vertebrae are the largest in the spinal column because they house the cervical enlargement of the spinal cord. Each cervical transverse process contains a *transverse foramen*, a

distinguishing feature of cervical vertebrae, through which the vertebral artery and its accompanying vein and nerve fibers pass. The spinous processes of C2 through C6 are often *bifid*—that is, split into two parts (Figure 7.17a, d).

The first two cervical vertebrae differ considerably from the others. Named after the mythological Atlas, who supported the world on his shoulders, the first cervical vertebra (C1), the **atlas,** supports the head (Figure 7.17a, b). The atlas is a ring of bone with *anterior* and *posterior arches* and large *lateral masses*. It lacks a body and a spinous process. The superior surfaces of the lateral masses, called *superior articular facets* (FAS-ets or fa-SETS=little faces), are concave and articulate with the occipital condyles of the occipital bone to form the paired *atlanto-occipital joints*. These articulations permit the movement seen when moving the head to signify "yes." The inferior surfaces of the lateral masses, the *inferior articular facets*, articulate with the second cervical vertebra. The transverse processes and transverse foramina of the atlas are quite large.

The second cervical vertebra (C2), called the **axis** (Figure 7.17a, c, d), does have a body. Projecting from its body is a peglike process called the *dens* (=tooth) or *odontoid process*. The dens projects superiorly through the anterior portion of the vertebral foramen of the atlas. The dens makes a pivot on which the atlas and head rotate. This arrangement permits side-to-side movement of the head, as when you move your head to signify "no." The

Figure 7.17　Cervical vertebrae.

The cervical vertebrae are found in the neck region.

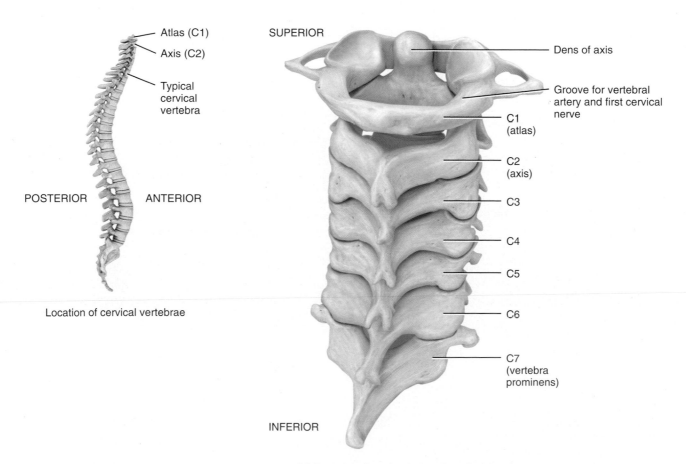

(a) Posterior view of articulated cervical vertebrae

■ EXHIBIT 7.10 CONTINUES ▶

EXHIBIT 7.10　**197**

● **FIGURE 7.17 CONTINUED** ▶

POSTERIOR

Posterior arch

Groove for vertebral artery and first cervical spinal nerve

Superior articular facet

Vertebral foramen

Transverse foramen

Lateral mass

Transverse process

Articular surface for dens of axis

Anterior arch

ANTERIOR

(b) Superior view of atlas (C1)

POSTERIOR

Spinous process

Lamina

Inferior articular process

Vertebral foramen

Dens

Transverse process

Superior articular facet

ANTERIOR

Body

(c) Superior view of axis (C2)

articulation formed between the anterior arch of the atlas and dens of the axis, and between their articular facets, is called the *atlanto-axial joint*. In some instances of trauma, the dens of the axis may be driven into the medulla oblongata of the brain. This type of injury is the usual cause of death from whiplash injuries.

The third through sixth cervical vertebrae (C3–C6), represented by the vertebra in Figure 7.17e, correspond to the structural pattern of the typical cervical vertebra previously described. The seventh cervical vertebra (C7), called the *vertebra prominens*, is somewhat different (Figure 7.17a). It has a single large spinous process that may be seen and felt at the base of the neck.

The superior and inferior articular facets of the last 5 cervical vertebrae contact each other in the transverse plane. This articular orientation provides the neck region with a varied range of mobility, which includes lateral bending, forward and backward bending, and rotational movements as the articular surfaces slide over one another.

● CHECKPOINT
How do the atlas and axis differ from the other cervical vertebrae?

Dens
Articular surface for anterior arch of atlas
Superior articular facet
Lamina
Spinous process
Transverse foramen
Body
Transverse process
Inferior articular facet
POSTERIOR
ANTERIOR

(d) Right lateral view

POSTERIOR
Bifid spinous process
Lamina
Vertebral foramen
Superior articular facet
Pedicle
Transverse process
Transverse foramen
Body
ANTERIOR

(c) Superior view of a typical cervical vertebra

❓ **Which bones permit the movement of the head to signify "no"?**

EXHIBIT 7.10   **199**

EXHIBIT 7.11   Vertebral Regions—Thoracic Vertebrae

- Identify the location and surface features of the thoracic vertebrae.

Description

Thoracic vertebrae (T1–T12; Figure 7.18) are considerably larger and stronger than cervical vertebrae and become progressively larger from superior to inferior. In addition, the spinous processes on T1 through T10 are long, laterally flattened, and directed inferiorly and they overlap the process on the vertebra directly inferior. In contrast, the spinous processes on T11 and T12 are shorter, broader, and directed more posteriorly. Compared to cervical vertebrae, thoracic vertebrae also have longer and larger transverse processes. They are easily identified by their *costal facets*, which are articular surfaces for the ribs.

Surface Features

The most distinguishing feature of thoracic vertebrae is that they articulate with the ribs. Except for T11 and T12, the transverse processes have facets that articulate with the *tubercles* of the ribs. The bodies of thoracic vertebrae also have either facets or demifacets (half facets) that articulate with the *heads* of the ribs. The articulations between the thoracic vertebrae and ribs are called *vertebrocostal joints* and are distinguishing features of thoracic vertebrae. As you can see in Figure 7.18, T1 has a superior facet and an inferior demifacet, one on each side of the vertebral body.

Figure 7.18 Thoracic vertebrae.

🔒 ⚞⚟ The thoracic vertebrae are found in the chest region and articulate with the ribs.

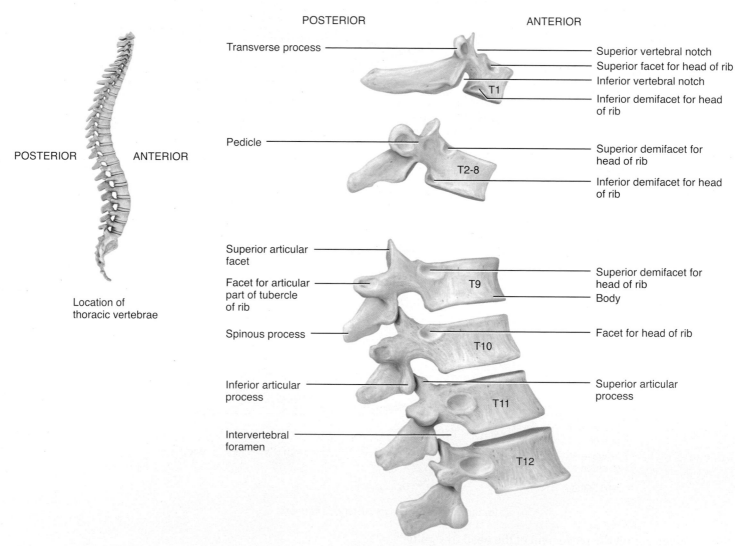

(a) Right lateral view of several articulated thoracic vertebrae

T2–T8 have a superior and inferior demifacet, one on each side of the vertebral body. T9 has a superior demifacet on each side of the vertebral body, and T10–T12 have a facet on each side of the vertebral body. Movements of the thoracic region are limited by thin intervertebral discs and by the attachment of the ribs to the sternum.

The superior and inferior articular processes form flat articular surfaces positioned in the frontal plane. The surface of the superior articular facet faces posterior and the surface of the inferior articular facet faces anterior. These arterior and posterior facing surfaces contact each other and limit forward and backward bending of this region of the vertebral column, while allowing for small amounts of rotational movement and significant lateral bending.

● CHECKPOINT

Describe several distinguishing features of thoracic vertebrae.

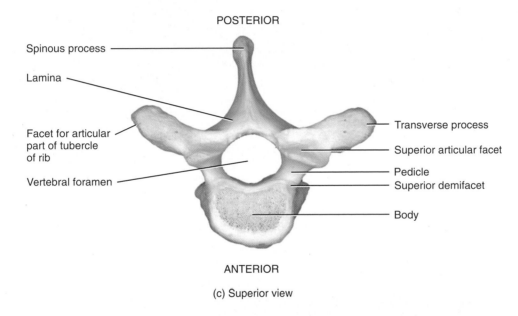

POSTERIOR

Spinous process

Lamina

Facet for articular part of tubercle of rib

Vertebral foramen

Transverse process

Superior articular facet

Pedicle

Superior demifacet

Body

ANTERIOR

(c) Superior view

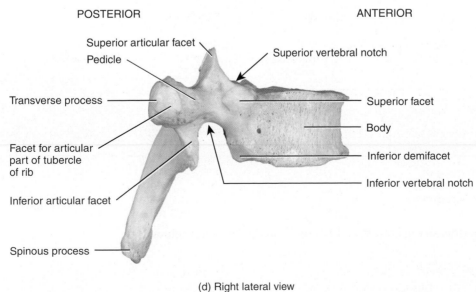

POSTERIOR

ANTERIOR

Superior articular facet

Pedicle

Transverse process

Facet for articular part of tubercle of rib

Inferior articular facet

Spinous process

Superior vertebral notch

Superior facet

Body

Inferior demifacet

Inferior vertebral notch

(d) Right lateral view

? Which parts of thoracic vertebrae articulate with the ribs?

EXHIBIT 7.11 **201**

EXHIBIT 7.12   Vertebral Regions—Lumbar Vertebrae

☉ **OBJECTIVE**

• Identify the location and surface features of the lumbar vertebrae.

Description

The **lumbar vertebrae** (L1–L5) are the largest and strongest of the unfused vertebrae in the vertebral column (Figure 7.19) because the amount of body weight supported by the vertebrae increases toward the inferior end of the backbone. They are readily identifiable by their large size, lack of transverse foramina, and absence of costal articular facets. They have large blocklike bodies with kidney-shaped articular surfaces.

Surface Features

Their various projections are short and thick. The superior articular processes are directed medially and the inferior articular

Figure 7.19 Lumbar vertebrae.

🔑 Lumbar vertebrae are found in the lower back.

Location of lumbar vertebrae

(a) Right lateral view of articulated lumbar vertebrae

(b) Superior view

(c) Right lateral view

❓ **Why are the lumbar vertebrae the largest and strongest in the vertebral column?**

processes are directed laterally. These surfaces contact each other in the sagittal plane allowing this region of the vertebral column to bend anteriorly and posteriorly, while limiting rotational movements. The transverse processes are flattened lateral projections that are taller than they are wide, serving as strong lever arms for muscles that bend the vertebral column. The spinous processes are quadrilateral in shape, thick and broad, and project nearly straight posteriorly. The spinous processes are well-adapted for the attachment of the large back muscles.

A summary of the major structural differences among cervical, thoracic, and lumbar vertebrae is presented in Table 7.4.

◉ **CHECKPOINT**
Why are the lumbar vertebrae larger and stronger then either thoracic or cervical vertebrae?

TABLE 7.4

Comparison of Major Structural Features of Cervical, Thoracic, and Lumbar Vertebrae

CHARACTERISTIC	CERVICAL	THORACIC	LUMBAR
Overall structure			
Size	Small	Larger	Largest
Foramina	One vertebral and two transverse	One vertebral	One vertebral
Spinous processes	Slender and often bifid (C2–C6)	Long and fairly thick (most project inferiorly)	Short and blunt (project posteriorly rather than inferiorly)
Transverse processes	Small	Fairly large	Large and blunt
Articular facets for ribs	Absent	Present	Absent
Direction of articular facets			
Superior	Posterosuperior	Posterolateral	Medial
Inferior	Anteroinferior	Anteromedial	Lateral
Size of intervertebral discs	Thick relative to size of vertebral bodies	Thin relative to vertebral bodies	Massive

EXHIBIT 7.12 **203**

EXHIBIT 7.13 Vertebral Regions—Sacral and Coccygeal Vertebrae

◉ OBJECTIVE

- Identify the location and surface features of the sacral and coccygeal vertebrae.

Sacrum

Description

The **sacrum** is a triangular bone formed by the union of five sacral vertebrae (S1–S5), indicated in Figure 7.20a. The sacral vertebrae begin to fuse in individuals between 16 and 18 years of age, and the process is usually completed by age 30. The sacrum serves as a strong foundation for the pelvic girdle. It is positioned at the posterior portion of the pelvic cavity medial to the two hip bones. The female sacrum is shorter, wider, and more curved between S2 and S3 than the male sacrum (see Table 8.1 on page 235).

Surface Features

The concave anterior side of the sacrum faces the pelvic cavity. It is smooth and contains four *transverse lines (ridges)* that mark the joining of the sacral vertebral bodies (Figure 7.20a). At the ends of these lines are four pairs of *anterior sacral foramina*. The lateral portion of the sacrum forms a broad expanded region called the *sacral ala* (ĀL-a=wing; plural is *alae*, ĀL-ē), which is formed by the fused costal (rib) processes of the upper sacral vertebrae.

The convex, posterior surface of the sacrum contains a *median sacral crest*, which is the fused spinous processes of the upper sacral vertebrae; a *lateral sacral crest*, which is the fused transverse processes of the sacral vertebrae; and four pairs of *posterior sacral foramina* (Figure 7.20b). These foramina, along with the anterior sacral foramina, connect with the sacral canal to allow passage of nerves and blood vessels. The *sacral canal* is a continuation of the vertebral canal. The laminae of the fifth sacral vertebra, and sometimes the fourth, fail to meet. This leaves an inferior opening to the vertebral canal called the *sacral hiatus* (hī-Ā-tus=opening). On either side of the sacral hiatus is a *sacral cornu* (KOR-noo=horn; plural is *cornua*, KOR-noo-a), the inferior articular processes of the fifth sacral vertebra. They are connected by ligaments to the coccyx.

The narrow inferior portion of the sacrum is known as the *apex*. The broad superior portion of the sacrum is called the *base*. The anteriorly projecting border of the base, called the *sacral promontory* (PROM-on-tō-rē), is one of the points used for measurements of the pelvis. On both lateral surfaces the sacrum has a large ear-shaped *auricular surface* that articulates with the ilium of each hip bone to form the *sacroiliac joint*. Posterior to the auricular surface is a roughened surface, the *sacral tuberosity*, that

Figure 7.20 Sacrum and coccyx. The sacrum and coccyx are detached in (c) and (d).

🔑 ⌐ The sacrum is formed by the union of five sacral vertebrae, and the coccyx is formed by the union of usually four coccygeal vertebrae.

ANTERIOR

Location of sacrum and coccyx

Base of sacrum
Sacral ala
Anterior sacral promontory
Anterior sacral foramen
Transverse line
Apex of sacrum

S1
S2
S3
S4
S5
Co1
Co2
Co3
Co4

Superior articular process

Sacrum
Sacral hiatus
Coccyx

(a) Anterior view

Superior articular facet

Sacral canal
Sacral tuberosity
Auricular surface
Lateral sacral crest
Posterior sacral foramen
Median sacral crest
Sacral cornu
Coccygeal cornu
Transverse process

Co1
Co2
Co3
Co4

(b) Posterior view

contains depressions for the attachment of ligaments. The sacral tuberosity is another surface of the sacrum that unites with the hip bones to form the sacroiliac joints. The *superior articular processes* of the sacrum articulate with the fifth lumbar vertebra, and the base of the sacrum articulates with the body of the fifth lumbar vertebra, to form the *lumbosacral joint*.

Coccyx

Description

The **coccyx** is the terminal end of the vertebral column, indicated in Figure 7.20 as Co1–Co4. In humans it represents a short remnant of longer tail vertebrae found in many other vertebrate species. It is a triangular bone that forms from the fusion of three to five vertebral segments, most commonly representing four fused vertebrae. The first segment is the most characteristic of the vertebral plan, with succeeding segments becoming highly reduced. The coccygeal vertebrae fuse when a person is between 20 and 30 years of age.

Surface Features

The dorsal surface of the body of the coccyx contains two long *coccygeal cornua*. They are connected by ligaments to the sacral

cornua. The coccygeal cornua are the pedicles and superior articular processes of the first coccygeal vertebra. They are on the lateral surfaces of the coccyx, a series of *transverse processes*, the first pair being the largest. The coccyx articulates superiorly with the apex of the sacrum. In females, the coccyx points inferiorly; in males, it points anteriorly (see Table 8.1 on page 235).

⊙ **CHECKPOINT**
How many vertebrae fuse to form the sacrum and the coccyx?

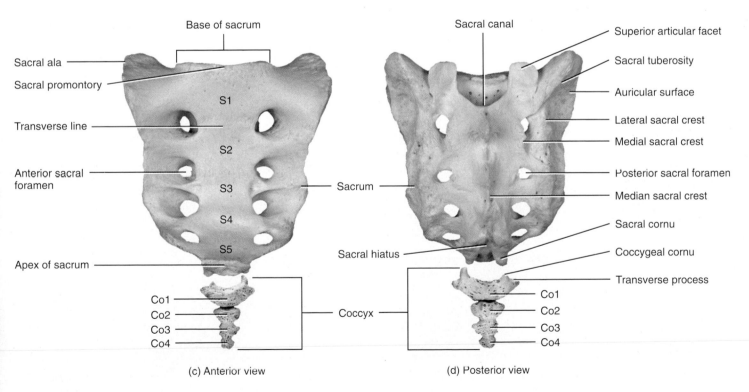

(c) Anterior view   (d) Posterior view

? How many foramina pierce the sacrum, and what is their function?

EXHIBIT 7.13   **205**

THORAX

⊙ **OBJECTIVE**

Identify the bones of the thorax.

The term **thorax** refers to the entire chest. The skeletal part of the thorax, the **thoracic cage**, is a bony enclosure formed by the sternum, ribs and their costal cartilages, and the bodies of the thoracic vertebrae (see Figure 7.21). The thoracic cage is narrower at its superior end and broader at its inferior end and is flattened from front to back. It encloses and protects the organs in the thoracic and superior abdominal cavities, provides support for the upper limbs, and plays a role in breathing.

EXHIBIT 7.14	Thoracic Bones—Sternum

⊙ **OBJECTIVE**

• Identify the location and surface features of the sternum.

Description

The **sternum**, or breastbone, is a flat, narrow bone located in the center of the anterior thoracic wall that measures about 15 cm (6 in.) in length and consists of three parts (Figure 7.21). The superior part is the **manubrium** (ma-NOO-brē-um=handle-like); the middle and largest part is the **body**; and the inferior, smallest part is the **xiphoid process** (ZĪ-foyd=sword-shaped). The segments of the sternum typically fuse by age 25 and the points of fusion are marked by transverse ridges.

Surface Features

The junction of the manubrium and body forms the *sternal angle*. The manubrium has a depression on its superior surface, the *suprasternal notch*. Lateral to the suprasternal notch are *clavicular notches* that articulate with the medial ends of the clavicles to form

Figure 7.21 Skeleton of the thorax.

🔑 The bones of the thorax enclose and protect organs in the thoracic cavity and upper abdominal cavity.

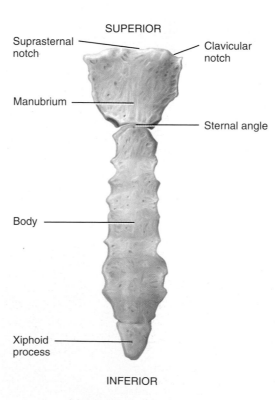

(a) Anterior view of sternum

(b) Anterior view

◉ CHECKPOINT
10. What bones form the skeleton of the thorax?
11. What are the functions of the bones of the thorax?

the *sternoclavicular joints*. The manubrium also articulates with the costal cartilages of the first and second ribs to form the *sternocostal joints*. The body of the sternum articulates directly or indirectly with the costal cartilages of the second through tenth ribs. The xiphoid process consists of hyaline cartilage during infancy and childhood and does not ossify completely until about age 40. No ribs are attached to it, but the xiphoid process provides attachment for some abdominal muscles. Incorrect positioning of the hands of a rescuer during cardiopulmonary resuscitation (CPR) may fracture the xiphoid process, driving it into internal organs.

◉ CHECKPOINT
Why is the xiphoid process clinically significant?

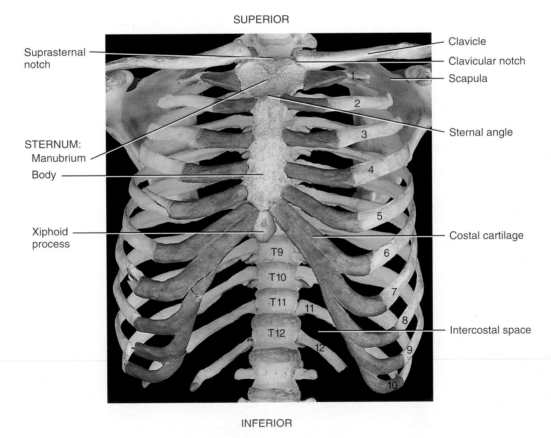

(c) Anterior view of skeleton of thorax

❓ **With which ribs does the body of the sternum articulate?**

EXHIBIT 7.14   **207**

EXHIBIT 7.15 Thoracic Bones—Ribs

◉ OBJECTIVE
- Identify the location and surface features of the ribs.

Description

Twelve pairs of **ribs** numbered 1–12 from superior to inferior, give structural support to the sides of the thoracic cavity (see Figure 7.21c). The ribs increase in length from the first through seventh, then decrease in length to the twelfth rib. Each articulates posteriorly with its corresponding thoracic vertebra. Although we describe only the 12 thoracic ribs, there are in reality ribs accompanying each vertebra. The cervical, lumbar, and sacral ribs fuse to their corresponding vertebrae to become the major contributor to what is described as the transverse processes of the cervical and lumbar vertebrae and the alae of the sacrum.

Surface Features

The first through seventh pairs of ribs have a direct anterior attachment to the sternum by a strip of hyaline cartilage called *costal cartilage* (*cost-*=rib). The costal cartilages contribute to the elasticity of the thoracic cage and prevent various blows to the chest from fracturing the sternum and/or ribs. The ribs that have costal cartilages and attach directly to the sternum are called *true (vertebrosternal) ribs*. The remaining five pairs of ribs are termed *false*

ribs because their costal cartilages either attach indirectly to the sternum or do not attach to the sternum at all. The cartilages of the eighth, ninth, and tenth pairs of ribs attach to one another and then to the cartilages of the seventh pair of ribs. These false ribs are called *vertebrochondral ribs*. The eleventh and twelfth pairs of ribs are false ribs designated as *floating (vertebral) ribs* because the costal cartilage at their anterior ends does not attach to the sternum at all. These ribs attach only posteriorly to the thoracic vertebrae. Inflammation of one or more costal cartilages, called *costochondritis*, is characterized by local tenderness and pain in the anterior chest wall that may radiate. The symptoms mime the chest pain associated with a heart attack (angina pectoris).

Figure 7.22a shows the parts of a typical (third through ninth) rib. The *head* is a projection at the posterior end of the rib that contains a pair of articular *facets* (superior and inferior). The facet of the head fits into either a facet on the body of one vertebra or into the demifacets of two adjoining vertebrae to form *vertebrocostal joints*. The *neck* is a constricted portion just lateral to the head. A knoblike structure on the posterior surface where the neck joins the body is called a *tubercle* (TOO-ber-kul). The *nonarticular part* of the tubercle attaches to a ligament (lateral costotransverse ligament), which in turn attaches the transverse process of a vertebra to the nonarticular part of the corresponding rib. The *articular part* of the tubercle articulates with the facet of a transverse process of the inferior of the two vertebrae to

Figure 7.22 The structure of ribs. Each rib has a head, a neck, and a body. The facets and the articular part of the tubercle of a rib articulate with a thoracic vertebra. (See Tortora, *A Photographic Atlas of the Human Body*, 2e, Figure 3.21.)

⬡🔑 Each rib articulates posteriorly with its corresponding thoracic vertebra.

(a) Posterior view of left rib

(b) Posterior view of left ribs articulated with thoracic vertebrae and sternum

View

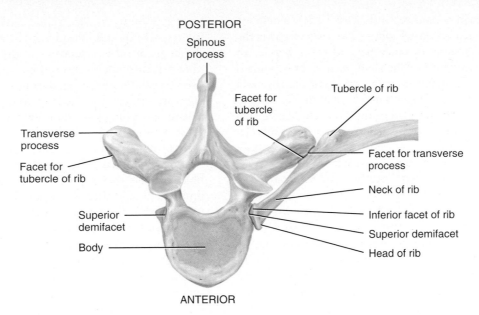

POSTERIOR

Spinous
process

Tubercle of rib

Facet for
tubercle
of rib

Transverse
process

Facet for transverse
process

Facet for
tubercle of rib

Neck of rib

Superior
demifacet

Inferior facet of rib

Superior demifacet

Body

Head of rib

ANTERIOR

(c) Superior view of left rib articulated with thoracic vertebra

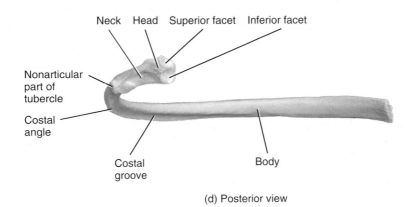

Neck   Head   Superior facet   Inferior facet

Nonarticular
part of
tubercle

Costal
angle

Costal
groove

Body

(d) Posterior view

POSTERIOR

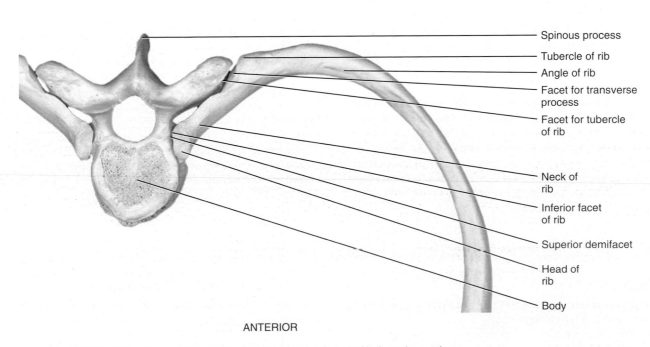

Spinous process

Tubercle of rib

Angle of rib

Facet for transverse
process

Facet for tubercle
of rib

Neck of
rib

Inferior facet
of rib

Superior demifacet

Head of
rib

Body

ANTERIOR

(e) Superior view of left rib and part of right rib articulated with thoracic vertebra

 How does a rib articulate with a thoracic vertebra?

■ EXHIBIT **7.15** CONTINUES ▶

EXHIBIT 7.15   **209**

EXHIBIT 7.15 Thoracic Bones—Ribs CONTINUED

which the head of the rib is connected (Figure 7.22c). These articulations also form vertebrocostal joints. The *body (shaft)* is the main part of the rib. A short distance beyond the tubercle, an abrupt change in the curvature of the shaft occurs. This point is called the *costal angle*. The inner inferior surface of the rib has a *costal groove* that protects blood vessels and a small nerve.

In summary, the posterior portion of the rib is connected to a thoracic vertebra by its head and the articular part of a tubercle. The facet of the head fits into a facet on the body of one vertebra or into the demifacets of two adjoining vertebrae. The articular part of the tubercle articulates with the facet of the transverse process of the vertebra.

If you examine Figure 7.21c, you will notice that the first rib is the shortest, broadest, and most sharply curved. The first rib is an important landmark because of its close relationship to the nerves of the *brachial plexus* (the entire nerve supply of the shoulder and upper limb), two major blood vessels, the subclavian artery and vein, and two skeletal muscles, the anterior and medial scalene muscles. The superior surface of the first rib has two shallow grooves, one for the subclavian vein and one for the subclavian artery and inferior trunk of the brachial plexus. The second rib is thinner, less curved, and considerably longer than the first. Unlike the paired facets of the typical ribs, the tenth rib has a single articular facet on its head. The eleventh and twelfth ribs also have single articular facets on their heads, but no necks, tubercles, or costal angles.

Spaces between ribs, called *intercostal spaces*, are occupied by intercostal muscles, blood vessels, and nerves. Surgical access to the lungs or other structures in the thoracic cavity is commonly obtained through an intercostal space. Special rib retractors are used to create a wide separation between ribs. The costal cartilages are sufficiently elastic in younger individuals to permit considerable bending without breaking.

Structures passing between the thoracic cavity and the neck pass through an opening called the **superior thoracic aperture**. Among these structures are the trachea, esophagus, nerves, and blood vessels that supply and drain the head, neck, and upper limbs. The aperture is bordered by the first thoracic vertebra (posteriorly), the first pair of ribs and their cartilages, and the superior border of the manubrium of the sternum. Structures passing between the thoracic cavity and abdominal cavity pass through the **inferior thoracic aperture**. This large opening, which is closed by the diaphragm, allows passage of structures such as the esophagus, nerves, and large blood vessels. The inferior thoracic aperture is bordered by the twelfth thoracic vertebra (posteriorly), the eleventh and twelfth pair of ribs, the costal cartilages of ribs 7 through 10, and the joint between the body and xiphoid process of the sternum (anteriorly).

● **CHECKPOINT**
How are ribs classified?

KEY MEDICAL TERMS ASSOCIATED WITH THE AXIAL SKELETON

Chiropractic (kī-rō-PRAK-tik; *cheir*=hand; *praktikos*=efficient). A holistic health care discipline that focuses on nerves, muscles, and bones.

Chiropractor A health care professional who is concerned with the diagnosis, treatment, and prevention of mechanical disorders of the musculoskeletal system and the effects of these disorders on the nervous system and health in general. Treatment involves using the hands to apply specific force to adjust joints of the body (manual adjustment), especially the vertebral column. Chiropractors may also use massage, heat therapy, ultrasound, electric stimulation, and acupuncture. Chiropractors often provide information about diet, exercise, changes in lifesytle, and stress management. Chiropractors do not prescribe drugs or perform surgery.

Craniostenosis (krā′-nē-ō-sten-Ō-sis; *cranio-*=skull; *-stenosis*=narrowing) Premature closure of one or more cranial sutures during the first 18 to 20 months of life, resulting in a distorted skull that can restrict brain growth and development. Premature closure of the sagittal suture produces a long narrow skull; premature closure of the coronal suture results in a broad skull. Surgery is necessary to prevent brain damage.

Craniotomy (krā-nē-OT-ō-mē; *cranio-*=skull; *-tome*=cutting) Surgical procedure in which part of the cranium is removed to extract a blood clot, a brain tumor, or a sample of brain tissue for biopsy.

Laminectomy (lam′-i-NEK-tō-mē; *lamina-*=layer) Surgical procedure involving removal of a vertebral lamina to access the vertebral canal and relieve the symptoms of a herniated disc.

Lumbar spine stenosis (*sten-*=narrowed) Narrowing of the spinal canal in the lumbar part of the vertebral column, due to hypertrophy of surrounding bone or soft tissues. It may be caused by arthritic changes in the intervertebral discs and is a common cause of back and leg pain.

Spinal fusion (FYŪ-zhun) Surgical procedure in which two or more vertebrae of the vertebral column are stabilized with a bone graft or synthetic device to treat a fracture of a vertebra or following removal of a herniated disc.

Whiplash injury Injury to the neck region due to severe hyperextension (backward tilting) of the head followed by severe hyperflexion (forward tilting) of the head, usually associated with a rear-end automobile collision. Symptoms are related to stretching and tearing of ligaments and muscles, vertebral fractures, and herniated vertebral discs.

STUDY OUTLINE

Introduction (p. 162)

1. The bones of the skeleton come in a variety of shapes and sizes and must be strong yet relatively light weight.
2. Bones protect soft body parts and make movement possible; they also serve as landmarks for locating parts of other body systems.
3. The musculoskeletal system is composed of the bones, joints, and muscles working together.

Divisions of the Skeletal System (p. 163)

1. The axial skeleton consists of bones arranged along the longitudinal axis. The parts of the axial skeleton are the skull, auditory ossicles (ear bones), hyoid bone, vertebral column, sternum, and ribs.
2. The appendicular skeleton consists of the bones of the upper and lower limbs. The limbs are made up of limb girdles and free limbs.

Skull or Cranium (p. 165)

1. The 22 bones of the skull include cranial bones and facial bones.
2. The eight cranial bones include the frontal, parietal (2), occipital, temporal (2), sphenoid, and ethmoid.
3. The 14 facial bones are the nasal (2), lacrimal (2), palatine (2), inferior nasal conchae (2), vomer, maxillae (2), zygomatic (2), and mandible.
4. The nasal septum consists of the vomer, perpendicular plate of the ethmoid, and septal cartilage. The nasal septum divides the nasal cavity into left and right sides.
5. Seven skull bones form each of the orbits (eye sockets).
6. The foramina of the skull bones provide passages for nerves and blood vessels (Table 7.3 on page 187).
7. Sutures are immovable joints in adults that connect most bones of the skull. Examples are the coronal, sagittal, lambdoid, and squamous sutures.
8. Paranasal sinuses are cavities in bones of the skull lined with mucous membranes that communicate with the nasal cavity. The frontal, sphenoid, and ethmoid bones and the maxillae contain paranasal sinuses.
9. Fontanels are mesenchyme-filled spaces between the cranial bones of fetuses and infants. The major fontanels are the anterior, posterior, anterolaterals (2), and posterolaterals (2). After birth, the fontanels fill in with bone and become sutures.
10. The three cranial fossae are regions within the cranium that contain depressions for brain convolutions, grooves for cranial blood vessels, and numerous foramina.

Hyoid Bone (p. 192)

1. The hyoid bone is a U-shaped bone that does not articulate with any other bone.
2. It supports the tongue and provides attachment for some tongue muscles and for some muscles of the throat and neck.

Vertebral Column (p. 193)

1. The vertebral column, sternum, and ribs constitute the skeleton of the body's trunk.
2. The 26 bones of the adult vertebral column are the cervical vertebrae (7), the thoracic vertebrae (12), the lumbar vertebrae (5), the sacrum (5 fused vertebrae), and the coccyx (usually 4 fused vertebrae).
3. The adult vertebral column contains four normal curves (cervical, thoracic, lumbar, and sacral) that give strength, support, and balance.
4. Between the bodies of adjacent vertebrae are the cartilaginous intervertebral discs. They form strong joints that permit various movements and absorb vertical shock.
5. The vertebrae are similar in structure, each usually consisting of a body, vertebral arch, and seven processes. Vertebrae in the different regions of the column vary in size, shape, and detail.

Thorax (p. 206)

1. The thoracic skeleton consists of the sternum, ribs and costal cartilages, and thoracic vertebrae.
2. The thoracic cage protects vital organs in the chest area and upper abdomen.

SELF-QUIZ QUESTIONS

Choose the one best answer to the following questions.

1. A foramen is:
   a. a cavity within a bone.
   b. a depression.
   c. a hole for blood vessels and nerves.
   d. a ridge.
   e. a site for muscle attachment.

2. Normal skulls have two of each of the following *except*:
   a. parietal bones.
   b. nasal bones.
   c. temporal bones.
   d. sphenoid bones.
   e. maxillae.

3. Tears pass into the nasal cavity through a tunnel formed in part by the:
   a. nasal bone.
   b. lacrimal bone.
   c. vomer.
   d. Both a and b are correct.
   e. a, b, and c are all correct.

4. Which of the following bones is correctly matched with its process?
   a. zygomatic bone–temporal process
   b. maxilla–pterygoid process
   c. vomer–mandibular process
   d. temporal–occipital process
   e. sphenoid–coronoid process

5. All of the following articulate with the maxilla except the:
   a. nasal bone.
   b. frontal bone.
   c. palatine bone.
   d. mandible.
   e. zygomatic bone.

6. Which of the following is *not* a component of the orbit?
   a. temporal
   b. ethmoid
   c. zygomatic
   d. maxilla
   e. lacrimal

7. The nasal septum is formed by parts of the:
   a. ethmoid bone.
   b. nasal bone.
   c. vomer.
   d. palatine bone.
   e. Both a and c are correct.

8. Which of the following does *not* contain a paranasal sinus?
   a. zygomatic
   b. sphenoid
   c. ethmoid
   d. frontal
   e. maxilla

9. The skeleton of the thorax:
   a. is formed by 12 pairs of ribs and costal cartilages, the sternum, and 12 thoracic vertebrae.
   b. protects the internal chest organs, as well as the liver.
   c. is narrower at its superior end.
   d. aids in supporting the bones of the shoulder girdle.
   e. is described by all of the above.

10. The meninges (coverings) of the brain attach to the crista galli, which is part of the:
    a. frontal bone.
    b. temporal bone.
    c. sphenoid bone.
    d. parietal bone.
    e. ethmoid bone.

Complete the following.

11. The first cervical vertebra is the _____, and the second cervical vertebra is the _____.

12. According to shape classification, phalanges are _____ bones, carpals are _____ bones, and ribs are _____ bones.

13. Bones located in sutures are called sutural or _____ bones.

14. The only nonarticulating bone of the axial skeleton is the _____.

Are the following statements true or false? If the statement is False, *find and fix the mistake to make the statement* True.

15. The opening in the occipital bone through which the medulla oblongata connects with the spinal cord is the foramen magnum.

16. The sagittal suture joins the parietal bones.

17. Sesamoid bones protect tendons from wear and tear and may improve the mechanical advantage at a joint by altering the direction of pull of a tendon.

18. The sacral promontory projects anteriorly from the apex of the sacrum.

19. The tubercle of a rib articulates with demifacets on the bodies of adjacent vertebrae.

Matching

20. Match the following (answers may be used more than once):

____ **(a)** mandibular fossa	**(1)**	zygomatic
____ **(b)** optic foramen	**(2)**	frontal
____ **(c)** superior nasal concha	**(3)**	axis
____ **(d)** superior and inferior nuchal lines	**(4)**	ethmoid
____ **(e)** horizontal plate	**(5)**	maxilla
____ **(f)** pterygoid process	**(6)**	temporal
____ **(g)** coronoid process	**(7)**	sternum
____ **(h)** mastoid process	**(8)**	occipital
____ **(i)** temporal process	**(9)**	palatine
____ **(j)** supraorbital margin	**(10)**	mandible
____ **(k)** olfactory foramina	**(11)**	sphenoid
____ **(l)** sella turcica		
____ **(m)** mental foramen		
____ **(n)** external auditory meatus		
____ **(o)** perpendicular plate		
____ **(p)** palatine process		
____ **(q)** xiphoid process		
____ **(r)** odontoid process		
____ **(s)** horizontal plate		

CRITICAL THINKING QUESTIONS

1. Four-year-old Pattee finds a soft spot on the baby's skull and announces that the baby needs to go back because "it's not finished yet." Explain the presence of soft spots in the infant.

2. Dave has a tumor on his pituitary gland that requires surgery. The operation will require exceptional surgical skill both to reach the gland and to avoid serious, possibly fatal, complications. Describe the location of the pituitary gland and the associated skeletal (and other) anatomical features that make such an operation so delicate.

3. The ad reads "New Postureperfect mattress! Keeps spine perfectly straight—just like when you were born! A straight spine means a great sleep!" Would you buy a mattress from this company?

4. Trina slipped on the ice, fell backward, and hit her head on the ice. Why might a serious blow to the occipital bone be fatal?

5. John had complaints of severe headaches and a yellow-green nasal discharge. His physician ordered x-rays of his head. John was surprised to see several fuzzy-looking holes on the x-ray. Why does John have holes in his head?

? ANSWERS TO FIGURE QUESTIONS

7.1 The skull and vertebral column are part of the axial skeleton. The clavicle, shoulder girdle, humerus, pelvic girdle, and femur are part of the appendicular skeleton.

7.2 The frontal, parietal, sphenoid, ethmoid, and temporal bones are cranial bones.

7.3 The parietal and temporal bones are on either side of the squamous suture. The parietal and occipital bones are on either side of the lambdoid suture. The parietal and frontal bones surround the coronal suture.

7.4 The temporal bone articulates with the parietal, sphenoid, zygomatic, and occipital bones.

7.5 The parietal bones form the posterior, lateral portion of the cranium.

7.6 From the crista galli of the ethmoid bone, the sphenoid articulates with the frontal, parietal, temporal, occipital, temporal, parietal, and frontal bones, ending again at the crista galli of the ethmoid bone.

7.7 The perpendicular plate of the ethmoid bone forms the superior part of the nasal septum, and the lateral masses compose most of the medial walls of the orbits.

7.8 The mandible is the only movable skull bone, other than the auditory ossicles.

7.9 The nasal septum divides the nasal cavity into right and left sides.

7.10 Bones forming the orbit are the frontal, sphenoid, zygomatic, maxilla, lacrimal, ethmoid, and palatine. The palatine bone is not visible in the figure.

7.11 The paranasal sinuses produce mucus and serve as resonating chambers for vocalization.

7.12 The anterolateral fontanel is bordered by four different skull bones.

7.13 The posterior cranial fossa is the largest.

7.14 The hyoid bone does not articulate with any other bone.

7.15 The thoracic and sacral curves of the vertebral column are concave relative to the anterior of the body.

7.16 The vertebral foramina enclose the spinal cord; the intervertebral foramina provide spaces for spinal nerves to exit the vertebral column.

7.17 The atlas moving on the axis permits movement of the head to signify "no."

7.18 The facets and demifacets on the bodies of the thoracic vertebrae articulate with the heads of the ribs, and the facets on the transverse processes of these vertebrae articulate with the tubercles of the ribs.

7.19 The lumbar vertebrae are the largest and strongest in the body because the amount of weight supported by vertebrae increases toward the inferior end of the vertebral column.

7.20 There are four pairs of sacral foramina, for a total of eight. Each anterior sacral foramen joins a posterior sacral foramen at the intervertebral foramen. Nerves and blood vessels pass through these tunnels in the bone.

7.21 The body of the sternum articulates directly or indirectly with ribs 3–10.

7.22 The facet on the head of a rib fits into a facet on the body of a vertebra, and the articular part of the tubercle of a rib articulates with the facet of the transverse process of a vertebra.

8

THE SKELETAL SYSTEM: THE APPENDICULAR SKELETON

INTRODUCTION As noted in Chapter 7, the two main divisions of the skeletal system are the axial skeleton and the appendicular skeleton. The axial skeleton is the skeletal axis, or core of the body, and helps protect the internal organs. The primary function of the appendicular skeleton, the focus of this chapter, is movement. The appendicular skeleton consists of the upper and lower limbs. Each limb is composed of limb girdles and free limbs. The upper limbs consist of the pectoral girdles and free upper limbs; the lower limbs consist of the pelvic girdles and free lower limbs. The girdles attach the free limbs to the axial skeleton. As you progress through this chapter, you will see how the bones of the appendicular skeleton are connected with one another and with skeletal muscles, making possible a wide array of movements. This arrangement permits us to walk, write, use a computer, dance, swim, and play a musical instrument. •

SKELETON OF THE UPPER LIMB

◉ **OBJECTIVE**
• Identify the bones of the skeleton of the upper limb.

Each upper limb skeleton consists of 32 bones, which form two distinct regions—(1) the pectoral girdle and (2) the free upper limb. The **pectoral** (PEK-tō-ral) or **shoulder girdles** attach the bones of the free upper limbs to the axial skeleton (Figure 8.1a, b). Each of the two pectoral girdles consists of a clavicle and a scapula. The clavicle is the anterior bone and the scapula is the posterior bone. The pectoral girdles do not articulate with the vertebral column and are held in position and supported by a group of large muscles that extend from the vertebral column and rib cage to the scapula. Each **free upper limb (extremity)**

Figure 8.1 **Skeleton of the right upper limb—pectoral (shoulder) girdle and free upper limb.**
(See Tortora, *A Photographic Atlas of the Human Body*, 2e, Figure 3.1.)

🔑 ▭▭ Each upper limb skeleton consists of 32 bones.

Pectoral girdle:
Clavicle
Scapula

(a) Anterior view of pectoral girdle

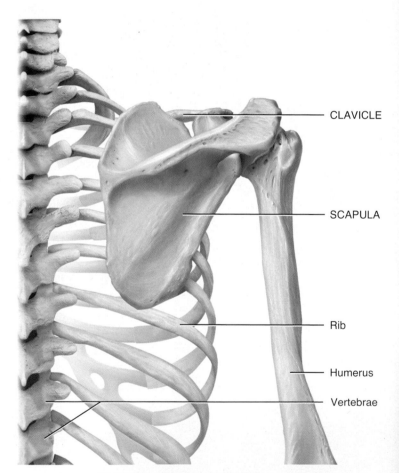

(b) Posterior view of pectoral girdle

has 30 bones in three locations—(1) the humerus in the arm; (2) the ulna and radius in the forearm; and (3) the 8 carpals in the carpus (wrist), the 5 metacarpals in the metacarpus (palm), and the 14 phalanges in the digits (fingers and thumb) of the hand (Figure 8.1c).

⊙ **CHECKPOINT**
1. What is the function of the pectoral girdle?

(c) Anterior view of free upper limb

? Which bones make up the pectoral girdle and free upper limb?

EXHIBIT 8.1 Pectoral Girdle—Clavicle

◉ OBJECTIVE

• Describe the location and surface features of the clavicle.

Description

Each slender, S-shaped **clavicle** (KLAV-i-kul=key), or *collarbone*, lies horizontally across the anterior part of the thorax superior to the first rib (see Figure 8.1a). The medial half of the clavicle is convex anteriorly, and the lateral half is concave anteriorly. It is typically smoother and straighter in females while being rougher and more curved in males. The clavicle forms the anterior strut (a rod or bar that resists compression) of the pectoral girdle that props the shoulder joint away from the rib cage. It is subcutaneous in location and easily palpable throughout its length.

Surface Features

The medial end of the clavicle, called the *sternal end* (Figure 8.2), is rounded and articulates with the manubrium of the sternum to form the *sternoclavicular joint* (see Figure 8.1a). The broad, flat, lateral end, the *acromial end* (a-KRŌ-mē-al) (Figure 8.2), articulates with the acromion of the scapula at the *acromioclavicular joint* (see Figure 8.1a).

The *conoid tubercle* (KŌ-noyd=conelike) on the inferior surface of the lateral end of the bone is a point of attachment for the conoid ligament. As its name implies, the *impression for the costoclavicular ligament* on the inferior surface of the sternal end is a point of attachment for the costoclavicular ligament.

● CLINICAL CONNECTION Fractured Clavicle

The clavicle transmits mechanical force from the free upper limb to the trunk. If the force transmitted to the clavicle is excessive, as in falling on your outstretched arm, a **fractured clavicle** may result. A fractured clavicle may also result from a blow to the superior part of the anterior thorax. In fact, the clavicle is one of the most frequently broken bones in the body. Because the junction of the clavicle's two curves is its weakest point, the clavicular midregion is the most frequent fracture site. Even in the absence of fracture, compression of the clavicle as a result of automobile accidents involving the use of shoulder-harness seatbelts often causes damage to the median nerve, which lies between the clavicle and the second rib. A fractured clavicle is usually treated with a regular sling to keep the arm from moving outward. •

Fractured clavicle

◉ CHECKPOINT

Which joints are formed by the articulation of the clavicle with other bones?

Figure 8.2 Right clavicle.

The clavicle articulates medially with the manubrium of the sternum and laterally with the acromion of the scapula.

EXHIBIT 8.1 **217**

EXHIBIT 8.2   Pectoral Girdle—Scapula

● OBJECTIVE

- Describe the location and surface features of the scapula.

Description

Each **scapula** (SCAP-ū-la; plural is *scapulae*), or *shoulder blade*, is a large, triangular, flat bone with a ridge on its posterior surface. The scapula occupies the superior part of the posterior thorax between the levels of the second and seventh ribs a few finger breadths lateral to the vertebral column (see Figure 8.1a, b).

Surface Features

A prominent ridge called the *spine* runs diagonally across the posterior surface of the scapula (Figure 8.3). The lateral end of the spine projects as a flattened, expanded process called the *acromion* (a-KRŌ-mē-on; *acro-*=peak; *omos*=shoulder), easily felt as the high point, or peak, of the shoulder. Tailors measure the length of the free upper limb from the acromion. The acromion articulates with the acromial end of the clavicle to form the *acromioclavicular joint*. Inferior to the acromion is a shallow depression, the *glenoid cavity*, that accepts the head of the humerus (arm bone) to form the *glenohumeral joint* (see Figure 8.1a).

Figure 8.3 Right scapula (shoulder blade).

🔑 The glenoid cavity of the scapula articulates with the head of the humerus to form the shoulder joint.

Scapula

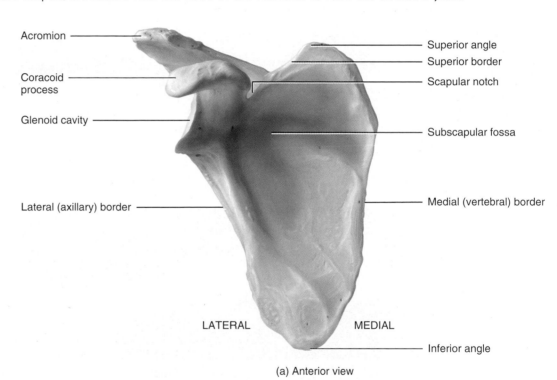

Acromion
Coracoid process
Glenoid cavity
Lateral (axillary) border
Superior angle
Superior border
Scapular notch
Subscapular fossa
Medial (vertebral) border
LATERAL   MEDIAL
Inferior angle

(a) Anterior view

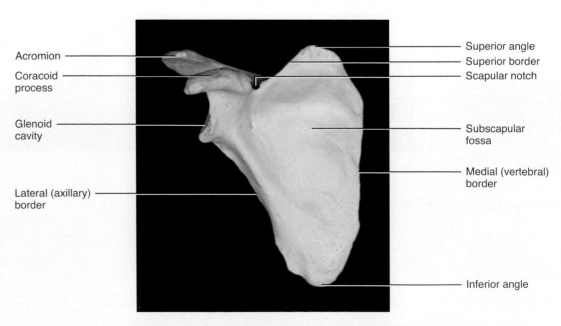

Acromion
Coracoid process
Glenoid cavity
Lateral (axillary) border
Superior angle
Superior border
Scapular notch
Subscapular fossa
Medial (vertebral) border
Inferior angle

Anterior view

The thin edge of the scapula closer to the vertebral column is called the *medial (vertebral) border*. It lies about 5 cm (2 in.) from the vertebral column. The thick edge of the scapula closer to the arm is called the *lateral (axillary) border*. The medial and lateral borders join at the *inferior angle*. The superior edge of the scapula, called the *superior border*, joins the vertebral border at the *superior angle*. The *scapular notch* is a prominent indentation along the superior border through which the suprascapular nerve passes.

At the lateral end of the superior border of the scapula is a projection of the anterior surface called the *coracoid process* (KOR-a-koyd=like a crow's beak), to which the tendons of muscles attach. Superior and inferior to the spine are two fossae: the *supraspinous fossa* (sū-pra-SPĪ-nus) and the *infraspinous fossa* (in-fra-SPĪ-nus). The supraspinous fossa and infraspinous fossa serve as surfaces of attachment for the supraspinatus and infraspinatus muscles of the shoulder, respectively. On the anterior surface is a slightly hollowed-out area called the *subscapular fossa*, also an attachment surface for the subscapularis muscle.

◉ CHECKPOINT

Which joints are formed by articulations of the scapula with other bones?

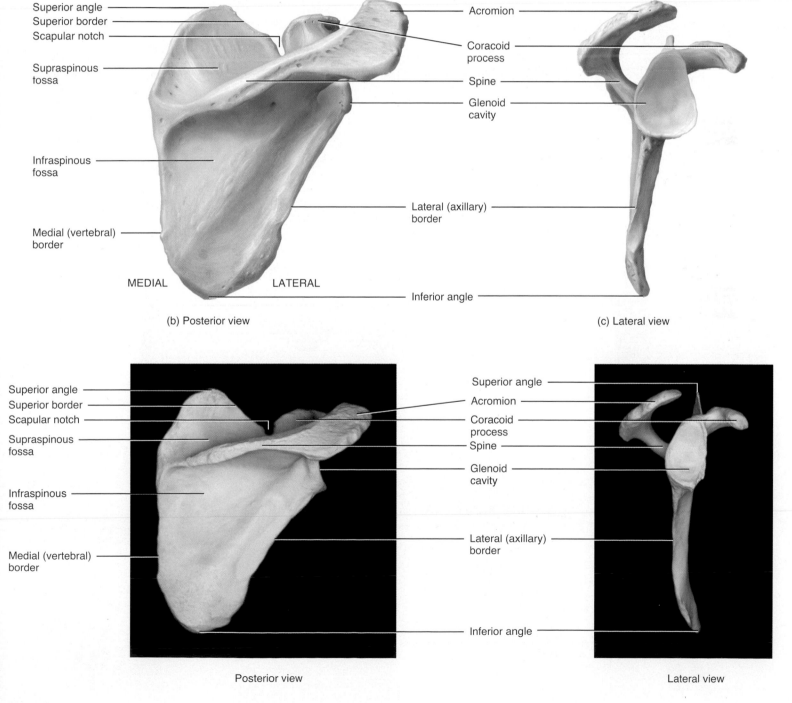

Superior angle
Superior border
Scapular notch
Supraspinous fossa
Infraspinous fossa
Medial (vertebral) border
MEDIAL   LATERAL

Acromion
Coracoid process
Spine
Glenoid cavity
Lateral (axillary) border
Inferior angle

(b) Posterior view

(c) Lateral view

Superior angle
Superior border
Scapular notch
Supraspinous fossa
Infraspinous fossa
Medial (vertebral) border

Superior angle
Acromion
Coracoid process
Spine
Glenoid cavity
Lateral (axillary) border
Inferior angle

Posterior view

Lateral view

❓ Which part of the scapula forms the high point of the shoulder?

EXHIBIT 8.2   **219**

EXHIBIT 8.3   Skeleton of the Arm—Humerus

☉ **OBJECTIVE**

• Identify the location and surface landmarks of the humerus.

Description

The **humerus** (HŪ -mer-us), or arm bone, is the longest and largest bone of the free upper limb (Figure 8.4). It has a ball-like proximal end with two prominent projections of bone at the base of the ball, a cylindrical, tubular shaft that makes up the majority of its length, and an expanded flattened distal end. It articulates proximally with the scapula and distally at the elbow with both the ulna and the radius.

Surface Features

The proximal end of the humerus features a rounded *head* that articulates with the glenoid cavity of the scapula to form the *glenohumeral joint.* Distal to the head is the *anatomical neck,* which is visible as an oblique groove. The *greater tubercle* is a lateral projection distal to the anatomical neck. It is the most laterally palpable bony landmark of the shoulder region. The *lesser tubercle* projects anteriorly. Between both tubercles runs an *intertubercular sulcus.* The *surgical neck* is a constriction in the

Humerus

Figure 8.4 Right humerus in relation to the scapula, ulna, and radius.

 The humerus is the longest and largest bone of the free upper limb.

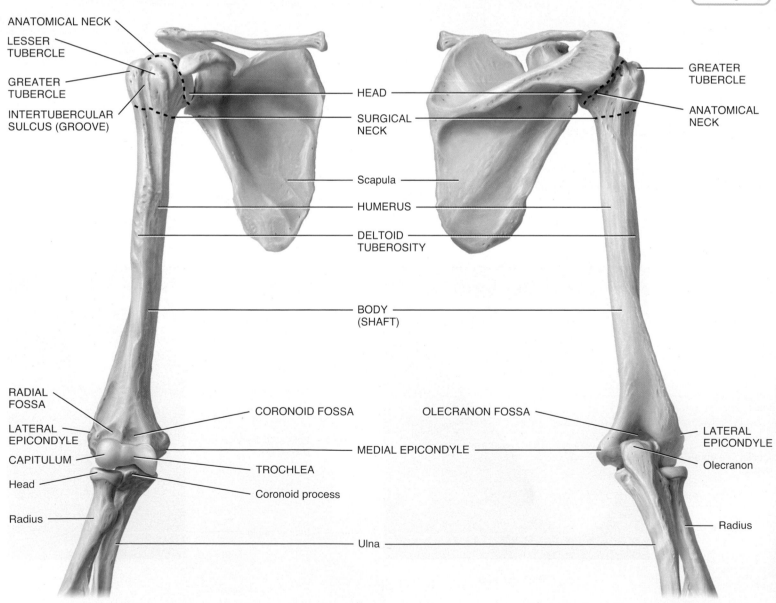

(a) Anterior view

(b) Posterior view

humerus just distal to the tubercles, where the head tapers to the shaft; it is so named because fractures often occur here.

The *body (shaft)* of the humerus is roughly cylindrical at its proximal end, but it gradually becomes triangular until it is flattened and broad at its distal end. Laterally, at the middle portion of the shaft, there is a roughened, V-shaped area called the *deltoid tuberosity*. This area serves as a point of attachment for the tendon of the deltoid muscle.

Several prominent features are evident at the distal end of the humerus. The *capitulum* (ka-PIT-ū-lum; *capit-*=head) is a rounded knob on the lateral aspect of the bone that articulates with the head of the radius. The *radial fossa* is an anterior depression above the capitulum that receives the head of the radius when the forearm is flexed (bent). The *trochlea* (TRŌK-lē-a =pulley), located medial to the capitulum, is a spool-shaped surface that articulates with the ulna. The *coronoid fossa* (KOR-ō-

noyd=crown-shaped) is an anterior depression that receives the coronoid process of the ulna when the forearm is flexed. The *olecranon fossa* (ō-LEK-ra-non=elbow) is the large posterior depression that receives the olecranon of the ulna when the forearm is extended (straightened). The *medial epicondyle* and *lateral epicondyle* are rough projections on either side of the distal end of the humerus to which the tendons of most muscles of the forearm are attached. The ulnar nerve may be palpated by rolling a finger over the skin surface above the posterior surface of the medial epicondyle. This nerve is the one that makes you feel a very severe pain when you hit your elbow.

⦿ **CHECKPOINT**
Distinguish between the anatomical neck and the surgical neck of the humerus.

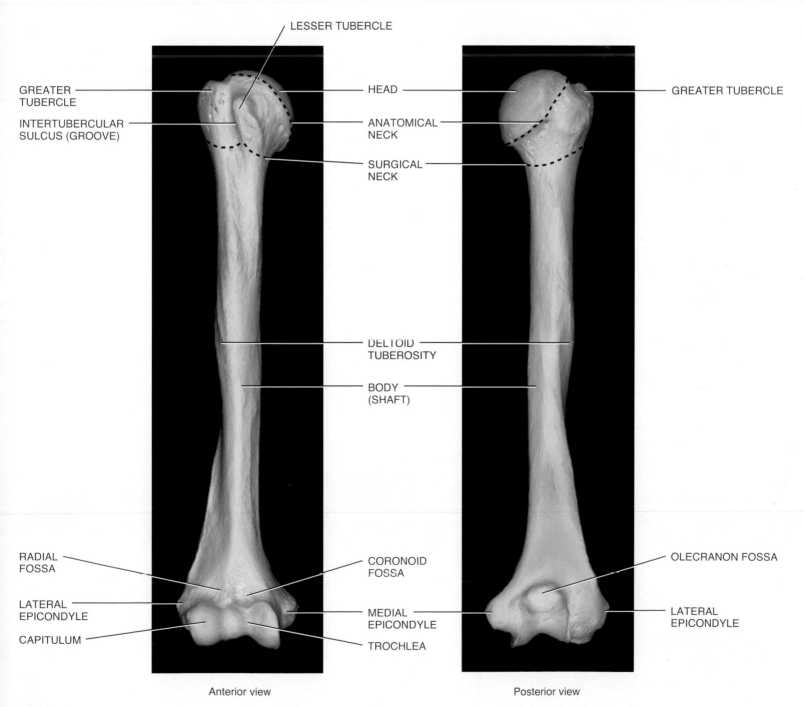

LESSER TUBERCLE

GREATER TUBERCLE

INTERTUBERCULAR SULCUS (GROOVE)

HEAD

ANATOMICAL NECK

SURGICAL NECK

GREATER TUBERCLE

DELTOID TUBEROSITY

BODY (SHAFT)

RADIAL FOSSA

LATERAL EPICONDYLE

CAPITULUM

CORONOID FOSSA

MEDIAL EPICONDYLE

TROCHLEA

OLECRANON FOSSA

LATERAL EPICONDYLE

Anterior view

Posterior view

? **Which parts of the humerus articulate with the radius at the elbow? with the ulna at the elbow?**

EXHIBIT 8.3 **221**

EXHIBIT 8.4 Skeleton of the Forearm—Ulna and Radius

◉ **OBJECTIVE**

• Identify the location and surface landmarks of the ulna and radius.

Ulna

Description

The **ulna** is located on the medial aspect (the little-finger side) of the forearm and is longer than the radius (Figure 8.5a, b). It is sometimes convenient to use an aid to help remember informa-tion that may be unfamiliar. Such an aid is called a *mnemonic device* (nē-MON-ik=memory). One such mnemonic to help you remember the location of the ulna in relation to the hand is "p.u." (the *p*inky is on the *u*lna side). The ulna is thick and notched at its proximal end, and its wide triangular shaft tapers to become more narrow and cylindrical distally.

Surface Features

At the proximal end of the ulna (Figure 8.5b) is the *olecranon*, which forms the prominence of the elbow. The *coronoid process*

Figure 8.5 Right ulna and radius.

🔑 ⚿ In the forearm, the longer ulna is on the medial side, and the shorter radius is on the lateral side.

(a) Anterior view

(b) Posterior view

(*corone*=crow) (Figure 8.5a, c, d) is an anterior projection distal to a large notch, the *trochlear notch*. This notch, on the anterior side of the olecranon, receives the trochlea of the humerus to form part of the elbow joint (Figure 8.5c, d). The *radial notch* is a depression that is lateral and inferior to the trochlear notch and articulates with the head of the radius. Just inferior to the coronoid process is the *ulnar tuberosity*. The distal end of the ulna consists of a *head* that is separated from the wrist by a disc of fibrocartilage. A *styloid process* is located on the posterior side of the ulna's distal end.

Radius

Description

The **radius**, the shorter of the two bones, is located on the lateral aspect (thumb side) of the forearm (Figure 8.5a, b). In contrast to the ulna, the radius is narrow at its proximal end and widens at its distal end.

Radius   Ulna

Anterior view

Ulna   Radius

Posterior view

● EXHIBIT **8.4** CONTINUES ▶

EXHIBIT 8.4   Skeleton of the Forearm—Ulna and Radius CONTINUED

Surface Features

The proximal end of the radius has a disc-shaped *head* that articulates with the capitulum of the humerus and the radial notch of the ulna. Inferior to the head is the constricted *neck*. A roughened area inferior to the neck on the medial side, called the *radial tuberosity*, is a point of attachment for the tendon of the biceps brachii muscle. The shaft of the radius widens distally to form a *styloid process* on the lateral side. Fracture of the distal end of the radius is the most common fracture in adults older than 50 years.

The ulna and radius articulate with the humerus at the *elbow joint*. The articulation occurs in two places: where the head of the radius articulates with the capitulum of the humerus (Figure 8.5c), and where the trochlear notch of the ulna receives the trochlea of the humerus (Figure 8.5c, d).

The ulna and the radius connect with one another at three sites. First, a broad, flat, fibrous connective tissue called the

⬤ **FIGURE 8.5 CONTINUED** ▶

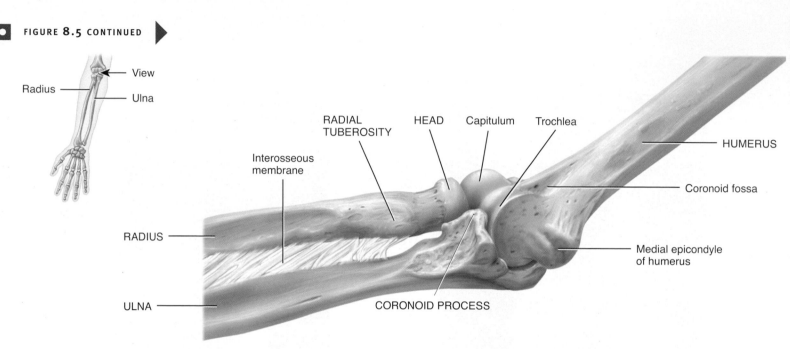

(c) Medial view in relation to humerus

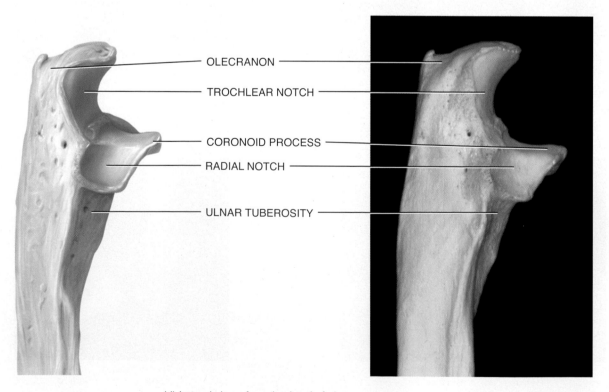

(d) Lateral view of proximal end of ulna

interosseous membrane (in'-ter-OS-ē-us; inter-=between, osse-=bone) joins the shafts of the two bones (Figure 8.5a, b). This membrane also provides a site of attachment for some tendons of deep skeletal muscles of the forearm. The ulna and radius articulate at their proximal and distal ends. Proximally, the head of the radius articulates with the ulna's *radial notch*, a depression that is lateral and inferior to the trochlear notch (Figure 8.5d). This articulation is the *proximal radioulnar joint*. Distally, the head of the ulna articulates with the *ulnar notch* of the radius

(Figure 8.5e). This articulation is the *distal radioulnar joint*. Finally, the distal end of the radius articulates with three bones of the wrist—the lunate, the scaphoid, and the triquetrum—to form the *radiocarpal (wrist) joint*.

⊙ **CHECKPOINT**
How many joints are formed between the ulna and radius, and what are their names?

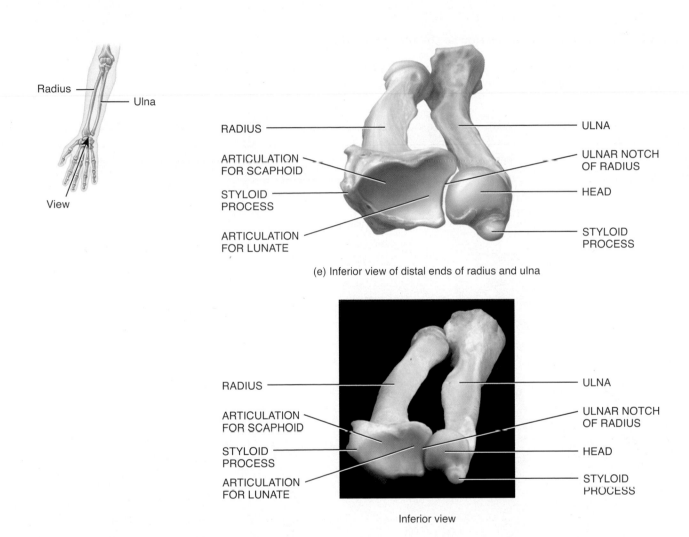

(e) Inferior view of distal ends of radius and ulna

Inferior view

❓ **What part of the ulna is called the "elbow"?**

EXHIBIT 8.4 **225**

EXHIBIT 8.5   Skeleton of the Hand

☉ OBJECTIVE

• Identify the location and surface landmarks of the bones of the hand.

Carpals

Description

The **carpus** (wrist) is the proximal region of the hand and consists of eight small bones, the **carpals**, joined to one another by ligaments (Figure 8.6). Articulations between carpal bones are called *intercarpal joints*. The carpals are arranged in two transverse rows of four bones each. Their names reflect their shapes.

Surface Features

The carpals in the proximal row, from lateral to medial, are the **scaphoid** (SKAF-oid=boatlike), **lunate** (LOO-nāt=moon-shaped), **triquetrum** (trī-KWĒ-trum=three-cornered), and **pisiform** (PIS-i-form=pea-shaped). The proximal row of carpal bones articulates with the distal end of the ulna and radius to

Figure 8.6 **Right wrist and hand in relation to the ulna and radius.**

🔑 The skeleton of the hand consists of the proximal carpals, the intermediate metacarpals, and the distal phalanges.

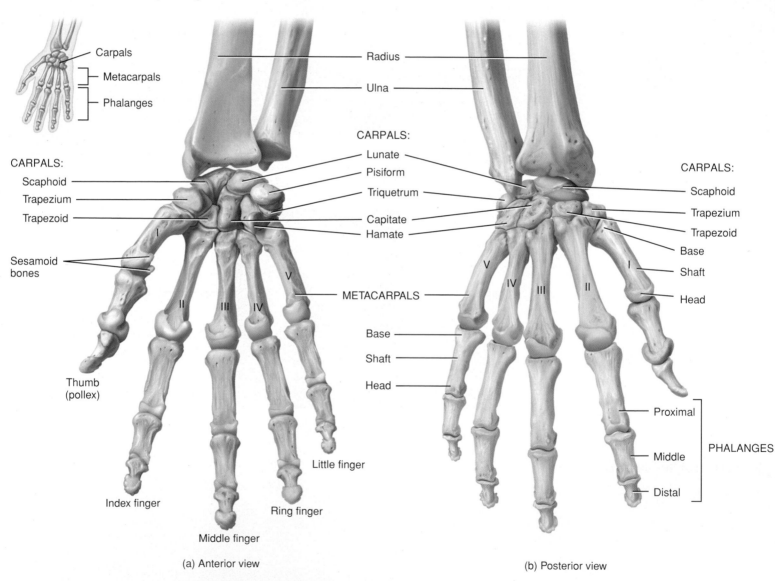

(a) Anterior view   (b) Posterior view

MNEMONIC for carpal bones*:

Stop **L**etting **T**hose **P**eople **T**ouch **T**he **C**adaver's **H**and.

Scaphoid **L**unate **T**riquetrum **P**isiform	**T**rapezium **T**rapezoid **C**apitate **H**amate
Proximal row	Distal row
Lateral ───────────→ Medial	Lateral ───────────→ Medial

* *Edward Tanner, University of Alabama, SOM*

form the *wrist joint*. The carpals in the distal row, from lateral to medial, are the **trapezium** (tra-PĒ-zē-um=four-sided figure with no two sides parallel), **trapezoid** (TRAP-e-zoid=four-sided figure with two sides parallel), **capitate** (KAP-i-tāt=head-shaped), and **hamate** (HAM-āt=hooked).

The capitate is the largest carpal bone; its rounded projection, the head, articulates with the lunate. The hamate is named for a large hook-shaped projection on its anterior surface. In about 70% of carpal fractures, only the scaphoid is broken. This is because the force of a fall on an outstretched hand is transmitted from the capitate through the scaphoid to the radius.

The anterior concave space formed by the pisiform and hamate (on the ulnar side), and the scaphoid and trapezium (on the radial side), with the roof-like covering of the *flexor retinaculum* (fibrous bands of deep fascia), is the **carpal tunnel**. The long flexor tendons of the digits and thumb and the median nerve pass through the carpal tunnel. Narrowing of the carpal tunnel may give rise to a condition called *carpal tunnel syndrome* (described in Chapter 11).

A mnemonic for learning the names of the carpal bones is shown in Figure 8.6. The first letters of the carpal bones from lateral to medial (proximal row, then distal row) correspond to the first letter of each word in the mnemonic.

Metacarpals

Description

The **metacarpus** (*meta-*=beyond), or palm, is the intermediate region of the hand and consists of five bones called **metacarpals**.

Surface Features

Each metacarpal bone consists of a proximal *base*, an intermediate *shaft*, and a distal *head* (Figure 8.6b). The metacarpal bones are numbered I to V (or 1–5), starting with the thumb, from lateral to medial. The bases articulate with the distal row of carpal bones to form the *carpometacarpal joints*. The heads articulate with the proximal phalanges to form the *metacarpophalangeal (MP) joints*. The heads of the metacarpals are commonly called "knuckles" and are readily visible in a clenched fist.

Phalanges

Description

The **phalanges** (fa-LAN-jēz; *phalan-*=a battle line), or bones of the digits, make up the distal region of the hand. There are 14 phalanges in the five digits of each hand and, like the metacarpals, the digits are numbered I to V (or 1–5), beginning with the thumb (pollex), from lateral to medial. A single bone of a digit is referred to as a *phalanx* (FĀ-lanks).

Surface Features

Each phalanx consists of a proximal *base*, an intermediate *shaft*, and a distal *head* (Figure 8.6b). The thumb (*pollex*) has two phalanges, and there are three phalanges in each of the other four digits. In order from the thumb, these other four digits are commonly referred to as the index finger, middle finger, ring finger, and little finger. The first row of phalanges, the *proximal row*, articulates with the metacarpal bones and second row of phalanges. The second row of phalanges, the *middle row*, articulates with the proximal row and the third row. The third row of phalanges, the *distal row*, articulates with the middle row. The thumb has no middle phalanx. Joints between phalanges are called *interphalangeal (IP) joints*.

◉ CHECKPOINT
Which is the most frequently fractured wrist bone?

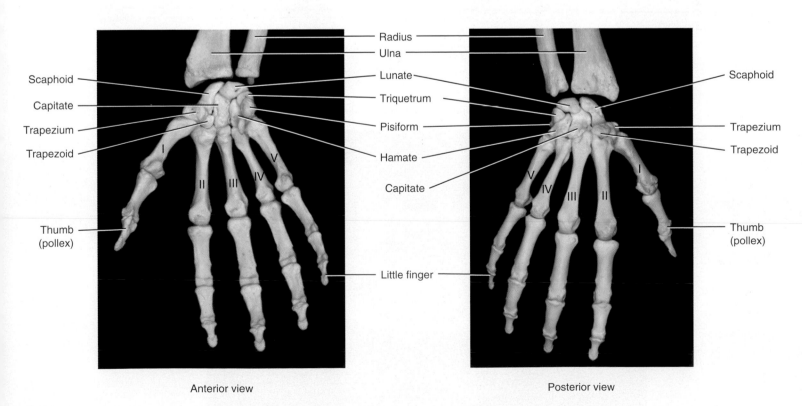

Anterior view

Posterior view

❓ **Which is the most frequently fractured wrist bone?**

EXHIBIT 8.5   **227**

SKELETON OF THE LOWER LIMB

⦿ **OBJECTIVE**

• Identify the bones of the skeleton of the lower limb.

Each lower limb skeleton consists of 31 bones, which form two distinct regions: (1) the pelvic girdle and (2) the free lower limb. The **pelvic (hip) girdle** consists of the two **hip bones**, also called **coxal (pelvic) bones**, or **os coxae**, (KOK-sal;

cox-=hip) (Figure 8.7a). The hip bones unite anteriorly at a joint called the **pubic symphysis** (PŪ-bik SIM-fi-sis). They unite posteriorly with the sacrum at the *sacroiliac joints.* The complete ring composed of the hip bones, pubic symphysis, and sacrum forms a deep, basinlike structure called the **bony pelvis** (*pelv-*=basin). The plural is *pelves* (PEL-vēz) or *pelvises.* Functionally, the bony pelvis provides a strong and stable support for the vertebral column and pelvic and lower abdominal organs. The pelvic girdle of the bony pelvis

Pelvic (hip) girdle

Figure 8.7 Skeleton of the right lower limb—pelvic (hip) girdle and free lower limb.

🔑 ⚏ Each free lower limb skeleton consists of 30 bones.

Hip bone

Sacrum

Coccyx

Pubic symphysis

Sacroiliac joint

Sacral promontory

Pelvic brim

Acetabulum

Obturator foramen

(a) Anterosuperior view of pelvic girdle

Hip bone

Sacrum

Pubic symphysis

Sacroiliac joint

Sacral promontory

Pelvic brim

Acetabulum

Obturator foramen

Anterosuperior view

also connects the bones of the free lower limbs to the axial skeleton.

Each of the two hip bones of a newborn consists of three bones separated by cartilage: a superior *ilium*, an inferior and anterior *pubis*, and an inferior and posterior *ischium*. By age 23, the three separate bones fuse together (see Figure 8.8a). Although the hip bones function as single bones, anatomists commonly discuss them as though they still consisted of three bones.

Each **free lower limb (extremity)** (Figure 8.7b) has 30 bones in four locations—(1) the femur in the thigh; (2) the patella (kneecap); (3) the tibia and fibula in the leg; and (4) the 7 tarsals in the tarsus (ankle), the 5 metatarsals in the metatarsus, and the 14 phalanges (bones of the digits) in the foot.

⊙ **CHECKPOINT**

2. What is the function of the pelvic girdle?

- Hip bone
- Sacrum
- FEMUR
- PATELLA
- TIBIA
- FIBULA
- TARSALS
- METATARSALS
- PHALANGES

(b) Anterior view of free lower limb

❓ **Which bones make up the pelvic girdle and free part of the lower limb?**

EXHIBIT 8.6 Bones of the Pelvic Girdle

⊙ **OBJECTIVE**

• Identify the locations and surface features of the three components of the hip bone.

Ilium

Description

The **ilium** (IL-ē-um=flank) is the largest of the three components of the hip bone (Figure 8.8b, c). It is thick near the hip joint and expands into a large curved plate of bone superiorly.

Figure 8.8 Right hip bone. The lines of fusion of the ilium, ischium, and pubis depicted in (a) are not always visible in an adult.

🔑 ▭ The acetabulum is the socket for the head of the femur where the three parts of the hip bone converge and ossify.

(a) Lateral view showing parts of hip bone

(b) Detailed lateral view

Lateral view

Surface Features

A superior *ala* (=wing) and an inferior *body*, which enters into the formation of the *acetabulum*, the socket for the head of the femur, comprise the ilium. Its superior border, the *iliac crest*, ends anteriorly in a blunt *anterior superior iliac spine*. Bruising of the anterior superior iliac spine and associated soft tissues, such as occurs in body contact sports, is called a **hip pointer**. Below this spine is the *anterior inferior iliac spine*. Posteriorly, the iliac crest ends in a sharp *posterior superior iliac spine*. Below this spine is the *posterior inferior iliac spine*. The spines serve as points of attachment for the tendons of the muscles and ligaments of the trunk, hip, and thighs. Below the posterior inferior iliac spine is the *greater sciatic notch* (sī-AT-ik), through which the sciatic nerve, the largest nerve in the body, passes.

(c) Detailed medial view

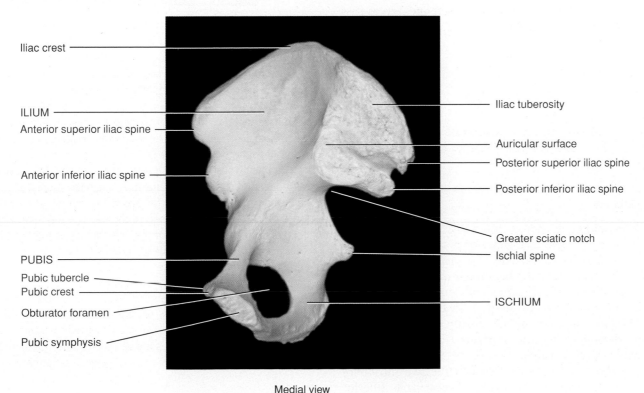

Medial view

? Which part of the hip bone articulates with the femur? With the sacrum?

EXHIBIT **8.6** CONTINUES ▶

EXHIBIT 8.6 Bones of the Pelvic Girdle CONTINUED

The medial surface of the ilium contains the *iliac fossa*, a concavity where the tendon of the iliacus muscle attaches. Posterior to this fossa are the *iliac tuberosity*, a point of attachment for the sacroiliac ligament, and the *auricular surface* (*auric-*=ear-shaped), which articulates with the sacrum to form the *sacroiliac joint* (see Figure 8.7a). Projecting anteriorly and inferiorly from the auricular surface is a ridge called the *arcuate line* (AR-kū-āt; *arc-*=bow).

The other conspicuous markings of the ilium are three arched lines on its lateral surface called the *posterior gluteal line* (*glut-*=buttock), the *anterior gluteal line*, and the *inferior gluteal line*. The gluteal muscles attach to the ilium between these lines.

Ischium

Description

The **ischium** (IS-kē-um=hip) (Figure 8.8b, c), the inferior and posterior portion of the hip bone, is situated between the body of the ilium and the inferior ramus of the pubis. The ischium is a sideways-arched or U-shaped structure, with its concave, notched margin contributing to the posterior two-thirds of the obturator foramen (the large hole on the anterior surface of the hip bone) (Figure 8.8b, c).

Surface Features

The ischium is comprised of a superior *body* and an inferior *ramus* (*ram-*=branch; plural is *rami*). The ramus joins the pubis. Features of the ischium include the prominent *ischial spine*, a *lesser sciatic notch* below the spine, and a rough and thickened *ischial tuberosity*. This prominent tuberosity may hurt someone's thigh when you sit on his or her lap. Together, the ramus and the pubis surround the *obturator foramen* (OB-too-rā-tōr; *obtur-*=closed up), the largest foramen in the skeleton. The foramen is so-named because, even though blood vessels and nerves pass through it, it is nearly completely closed by the fibrous *obturator membrane*.

Pubis

Description

The **pubis** is the inferior, anterior portion of the hip bone and, like the ischium, has the form of a sideways arch or a U-shape (Figure 8.8b, c).

Surface Features

A *superior ramus*, an *inferior ramus*, and a *body* between the rami comprise the pubis. The anterior, superior border of the body is the *pubic crest*, and at its lateral end is a projection called the *pubic tubercle*. This tubercle is the beginning of a raised line, the *pectineal line* (pek-TIN-ē-al), which extends superiorly and laterally along the superior ramus to merge with the arcuate line of the ilium. This line, as you will see shortly, is one important landmark for distinguishing the superior (false) pnd inferior (true) portions of the bony pelvis.

The *pubic symphysis* is the joint between the two hip bones (see Figure 8.7a). It consists of a disc of fibrocartilage. Inferior to this joint, the inferior rami of the two pubic bones converge to form the *pubic arch*. In the later stages of pregnancy, the hor-

mone relaxin (produced by the ovaries and placenta) increases the flexibility of the pubic symphysis to ease delivery of the baby. Weakening of the joint, together with an already compromised center of gravity due to an enlarged uterus, also alters the gait during pregnancy.

The *acetabulum* (as-e-TAB-ū-lum=vinegar cup) is a deep fossa formed by the ilium, ischium, and pubis. It functions as the socket that accepts the rounded head of the femur. Together, the acetabulum and the femoral head form the *hip (coxal) joint*. On the inferior side of the acetabulum is a deep indentation, the *acetabular notch*. It forms a foramen through which blood vessels and nerves pass, and it serves as a point of attachment for a ligament of the femur (that is, the ligament of the head of the femur).

● **CLINICAL CONNECTION** | Hip Fracture

Fractured femur

Although any region of the hip girdle may fracture, the term **hip fracture** most commonly applies to a break in the bones associated with the hip joint—the head, neck, or trochanteric regions of the femur, or the bones that form the acetabulum. In the United States, 300,000–500,000 people sustain hip fractures each year. The incidence of hip fractures is increasing, due in part to longer life spans. Decreases in bone mass due to osteoporosis (which occurs more often in females), along with an increased tendency to fall, predispose elderly people to hip fractures.

Hip fractures often require surgical treatment, the goal of which is to repair and stabilize the fracture, increase mobility, and decrease pain. Sometimes the repair is accomplished by using surgical pins, screws, nails, and plates to secure the head of the femur. In severe hip fractures, the femoral head or the acetabulum of the hip bone may be replaced by *prostheses* (artificial devices). The procedure of replacing either the femoral head or the acetabulum is *hemiarthroplasty* (hem-ē-AR-thrō-plas-tē; *hemi-*=one half; *arthro*=joint; *-plasty*=molding). Replacement of both the femoral head and acetabulum is *total hip arthroplasty*. The acetabular prosthesis is made of plastic, and the femoral prosthesis is metal; both are designed to withstand a high degree of stress. The prostheses are attached to healthy portions of bone with acrylic cement and screws (see Figure (C) on page 283). ●

◉ **CHECKPOINT**
Why is the obturator foramen so named?

FALSE AND TRUE PELVES

◉ **OBJECTIVE**
- Distinguish between the false and true pelves and explain why they are important clinically.

The bony pelvis is divided into superior and inferior portions by a boundary called the *pelvic brim* (Figure 8.9a). You can trace the pelvic brim by following the landmarks around parts of the hip bones to form the outline of an oblique plane. Beginning posteriorly at the *sacral promontory*, trace laterally and inferiorly along the *arcuate lines* of the ilium. Continue inferiorly along the *pectineal lines* of the pubis. Finally, trace anteriorly along the *pubic crest* to the superior portion of the *pubic symphysis*. Together, these points form an oblique plane that is higher in the back than in the front. The circumference of this plane is the pelvic brim.

The portion of the bony pelvis superior to the pelvic brim is the **false (greater) pelvis** (Figure 8.9b). It is bordered by the lumbar vertebrae posteriorly, the upper portions of the hip bones laterally, and the abdominal wall anteriorly. The space

Figure 8.9 True and false pelves. Shown here is the female pelvis. For simplicity, in part (a) the landmarks of the pelvic brim are shown only on the left side of the body, and the outline of the pelvic brim is shown only on the right side. The entire pelvic brim is shown in Table 8.1 on page 235. (See Tortora, *A Photographic Atlas of the Human Body, 2e,* Figure 3.27.)

 The true and false pelves are separated by the pelvic brim.

(a) Anteroposterior view of pelvic girdle

(b) Midsagittal section indicating locations of true and false pelves

❓ **What is the significance of the pelvic axis?**

enclosed by the false pelvis is part of the lower abdomen; it does not contain pelvic organs, except for the urinary bladder (when it is full) and the uterus, ovaries, and uterine tubes of the female, which project into this region and are covered by the peritoneal serous membrane. This becomes more evident during pregnancy. The portion of the bony pelvis inferior to the pelvic brim is the **true (lesser) pelvis** (Figure 8.9b). It is bounded by the sacrum and coccyx posteriorly, inferior portions of the ilium and ischium laterally, and the pubic bones anteriorly. The true pelvis surrounds the pelvic cavity (see Figure 1.6 on page 14). The superior opening of the true pelvis, bordered by the pelvic brim, is called the *pelvic inlet*; the inferior opening of the true pelvis is the *pelvic outlet*. The *pelvic axis* is an imaginary line that curves through the true pelvis from the central point of the plane of the pelvic inlet to the central point of the plane of the pelvic outlet. During childbirth the pelvic axis is the route taken by the baby's head as it descends through the pelvis.

Pelvimetry is the measurement of the size of the inlet and outlet of the birth canal, which may be done by ultrasonography or physical examination. Measurement of the pelvic cavity in pregnant females is important because the fetus must pass through the narrower opening of the pelvis at birth.

◉ CHECKPOINT

3. Why are the false and true pelves important clinically?

COMPARISON OF FEMALE AND MALE PELVES

◉ OBJECTIVE

• Describe the principal differences between female and male pelves.

In the following discussion it is assumed that the male and female are comparable in age and physical stature. Generally, the bones of a male are larger and heavier than those of a female and have larger surface markings. Sex-related differences in the features of bones are readily apparent when comparing the female and male pelves. Most of the structural differences in the pelves are adaptations to the requirements of pregnancy and childbirth. The female's pelvis is wider and shallower than the male's. Consequently, there is more space in the true pelvis of the female, especially in the pelvic inlet and pelvic outlet, which accommodate the passage of the infant's head at birth. Other significant differences between the pelves of females and males are listed and illustrated in Table 8.1.

◉ CHECKPOINT

4. Why are differences between female and male pelves important?

COMPARISON OF PECTORAL AND PELVIC GIRDLES

◉ OBJECTIVE

• Compare the principal difference between the pectoral and pelvic girdles.

Now that we have have studied the structures of the pectoral and pelvic girdles, we can note some of their significant differences. The pectoral girdle does not directly articulate with the vertebral column, but the pelvic girdle does so via the sacroiliac joint. The sockets (glenoid fossae) for the free upper limbs in the pectoral girdle are shallow and maximize movement, in contrast to the sockets (acetabula) for the free lower limbs in the pelvic girdle, which are deep and allow less movement. Overall, the structure of the pectoral girdle offers more mobility than strength, and that of the pelvic girdle offers more strength than mobility. This difference can be traced back to early evolutionary modifications in the two limbs. The pectoral limb became more mobile because of its modifications as the steering limb, while the pelvic limb became more stable and fixed to the axial skeleton because of its role as the locomotor limb.

◉ CHECKPOINT

5. Why are differences between the pectoral and pelvic girdle important?

TABLE 8.1

Comparison of Female and Male Pelves

POINT OF COMPARISON	FEMALE	MALE
General structure	Light and thin.	Heavy and thick.
False (greater) pelvis	Shallow.	Deep.
Pelvic brim (inlet)	Larger and more oval.	Smaller and heart-shaped.
Acetabulum	Small and faces anteriorly.	Large and faces laterally.
Obturator foramen	Oval.	Round.
Pubic arch	Greater than 90° angle.	Less than 90° angle.

Anterior views

Iliac crest	Less curved.	More curved.
Ilium	Less vertical.	More vertical.
Greater sciatic notch	Wide.	Narrow.
Coccyx	More movable and more curved anteriorly.	Less movable and less curved anteriorly.
Sacrum	Shorter, wider (see anterior views), and more curved anteriorly.	Longer, narrower (see anterior views), and less curved anteriorly.

Right lateral views

Pelvic outlet	Wider.	Narrower.
Ischial tuberosity	Shorter, farther apart, and more laterally projecting.	Longer, closer together, and more medially projecting.

Inferior views

EXHIBIT 8.7   Skeleton of the Thigh—Femur and Patella

■ **OBJECTIVE**

• Identify the location and surface features of the femur and patella.

Femur

Description

The **femur**, or thigh bone, is the longest, heaviest, and strongest bone in the body (Figure 8.10). Its proximal end articulates with the acetabulum of the hip bone. Its distal end articulates with the tibia and patella. The *body (shaft)* of the femur angles medi-

ally and, as a result, the knee joints are closer to the midline. The angle (angle of convergence) is greater in females because the female pelvis is broader.

Surface Features

The proximal end of the femur consists of a rounded *head* that articulates with the acetabulum of the hip bone to form the *hip (coxal) joint.* The head contains a small centered depression (pit) called the *fovea capitis* (FŌ-vē-a CAP-i-tis; *fovea*=pit; *capitis*=of the head) (Figure 8.10c). The ligament of the head of the femur connects the fovea capitis of the femur to the

Figure 8.10 Right femur and patella.

🔑 ⚊ The patella articulates with the lateral and medial condyles of the femur.

(a) Anterior view   (b) Posterior view

acetabulum of the hip bone. The *neck* of the femur is a constricted region distal to the head. The *greater trochanter* (trō-KAN-ter) and *lesser trochanter* are projections that serve as points of attachment for the tendons of some of the thigh and buttock muscles. The greater trochanter is the prominence felt and seen anterior to the hollow on the side of the hip. It is a landmark commonly used to locate the site for intramuscular injections into the lateral surface of the thigh. The lesser trochanter is inferior and medial to the greater trochanter. Between the anterior surfaces of the trochanters is a narrow *intertrochanteric line* (Figure 8.10a). A ridge called the *intertrochanteric crest* appears between the posterior surfaces of the trochanters (Figure 8.10b).

Inferior to the intertrochanteric crest on the posterior surface of the body of the femur is a vertical ridge called the *gluteal tuberosity*. It blends into another vertical ridge called the *linea aspera* (LIN-ē-a AS-per-a; *asper*=rough). Both ridges serve as attachment points for the tendons of several thigh muscles.

The expanded distal end of the femur includes the *medial condyle* (condyle=knuckle) and the *lateral condyle*. These articulate with the medial and lateral condyles of the tibia. Superior to the condyles are the *medial epicondyle* and the *lateral epicondyle*, to which ligaments of the knee joint attach. A depressed area between the condyles on the posterior surface is called the *intercondylar fossa* (in-ter-KON-di-lar). The *patellar surface* is located between the condyles on the anterior surface.

GREATER TROCHANTER
INTERTROCHANTERIC LINE
LATERAL EPICONDYLE
LATERAL CONDYLE

HEAD
NECK
LESSER TROCHANTER
BODY (SHAFT)
LINEA ASPERA
MEDIAL EPICONDYLE
MEDIAL CONDYLE

GREATER TROCHANTER
INTERTROCHANTERIC CREST
GLUTEAL TUBEROSITY
INTERCONDYLAR FOSSA
LATERAL EPICONDYLE
LATERAL CONDYLE

Anterior view

Posterior view

Femur
View

HEAD
FOVEA CAPITIS
GREATER TROCHANTER
NECK
INTERTROCHANTERIC CREST
LESSER TROCHANTER

(c) Medial view of proximal end of femur

EXHIBIT 8.7 CONTINUES

EXHIBIT 8.7 Skeleton of the Thigh—Femur and Patella CONTINUED

Patella

Description

The **patella** (=little dish), or kneecap, is a small, triangular bone located anterior to the knee joint (Figure 8.10d,e). It is a sesamoid bone that develops in the tendon of the quadriceps femoris muscle. The functions of the patella are to increase the leverage of the tendon of the quadriceps femoris muscle, to maintain the position of the tendon when the knee is bent (flexed), and to protect the knee joint.

Surface Features

The broad proximal end of the patella is called the *base*. The pointed distal end is the *apex*. The posterior surface contains two *articular facets*, one for the medial condyle of the femur and the other for the lateral condyle of the femur. The patellar ligament attaches the patella to the tibial tuberosity. The *patellofemoral joint*, between the posterior surface of the patella and the patellar surface of the femur, is the intermediate component of the *tibiofemoral (knee) joint*.

• CLINICAL CONNECTION | Patellofemoral Stress Syndrome

Patellofemoral stress syndrome ("runner's knee") is one of the most common problems runners experience. During normal flexion and extension of the knee, the patella tracks (glides) superiorly and inferiorly in the groove between the femoral condyles. In patellofemoral stress syndrome, normal tracking does not occur; instead, the patella tracks laterally as well as superiorly and inferiorly, and the increased pressure on the joint causes aching or tenderness around or under the patella. The pain typically occurs after a person has been sitting for awhile, especially after exercise. It is worsened by squatting or walking down stairs. One cause of runner's knee is constantly walking, running, or jogging on the same side of the road. Because roads slope down on the sides, the knee that is closer to the center of the road endures greater mechanical stress because it does not fully extend during a stride. Other predisposing factors include having knock-knees, running on hills, and running long distances. •

◉ CHECKPOINT
Why is the greater trochanter important clinically?

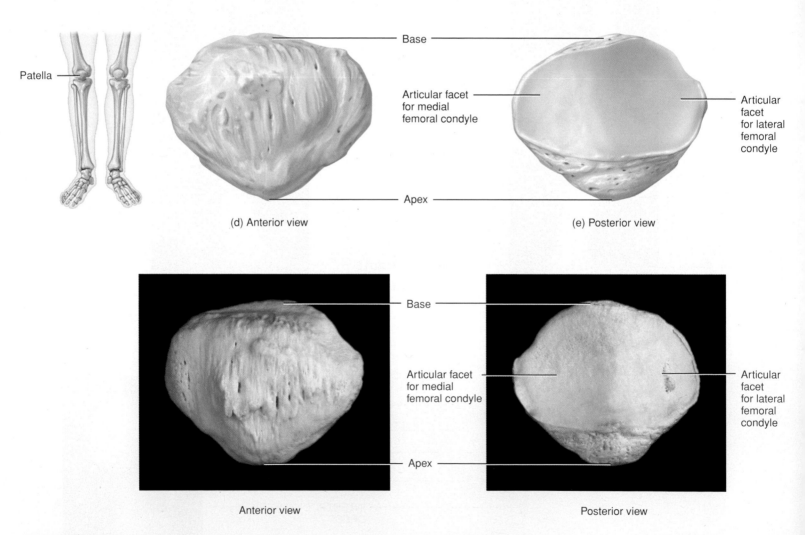

(d) Anterior view

(e) Posterior view

Anterior view

Posterior view

❓ Why is the angle of convergence of the femurs greater in females than males?

EXHIBIT 8.8 Skeleton of the Leg—Tibia and Fibula

⊙ **OBJECTIVE**
- Identify the location and surface features of the tibia and fibula.

Tibia

Description

The **tibia**, or shin bone, is the larger, medial, weight-bearing bone of the leg (Figure 8.11) and is triangular in transverse section. The term *tibia* means flute, as the tibial bones of birds were used in ancient times to make musical instruments. The tibia is the second longest bone of the body, only exceeded in length by the femur. The tibia articulates at its proximal end with the femur and fibula and at its distal end with the fibula and the talus bone of the ankle. The tibia and fibula, like the ulna and radius, are connected by an interosseous membrane.

Figure 8.11 Right tibia and fibula.

🔑 The tibia articulates with the femur and fibula proximally, and with the fibula and talus distally.

MNEMONIC for location of tibia and fibula:
The fibu**LA** is **LA**teral.

(a) Anterior view

(b) Posterior view

⊙ **EXHIBIT 8.8 CONTINUES** ▶

EXHIBIT 8.8   Skeleton of the Leg—Tibia and Fibula   CONTINUED

Surface Features

The proximal end of the tibia is expanded into a *lateral condyle* and a *medial condyle* (Figure 8.11a, b). These articulate with the condyles of the femur to form the lateral and medial *tibiofemoral (knee) joints*. The inferior surface of the lateral condyle articulates with the head of the fibula. The slightly concave condyles are separated by an upward projection called the *intercondylar eminence* (Figure 8.11b). The *tibial tuberosity* on the anterior surface is a point of attachment for the patellar ligament. Inferior to and continuous with the tibial tuberosity is a sharp ridge that can be felt below the skin and is known as the *anterior border (crest)* or shin.

The medial surface of the distal end of the tibia forms the *medial malleolus* (mal-LĒ-ō-lus=hammer). This structure articulates with the talus of the ankle and forms the prominence that can be felt on the medial surface of the ankle. The *fibular notch* (Figure 8.11c) articulates with the distal end of the fibula to form the *distal tibiofibular joint*. Of all the long bones of the body, the tibia is the most frequently fractured and is also the most frequent site of an open (compound) fracture.

● **FIGURE 8.11 CONTINUED** ▶

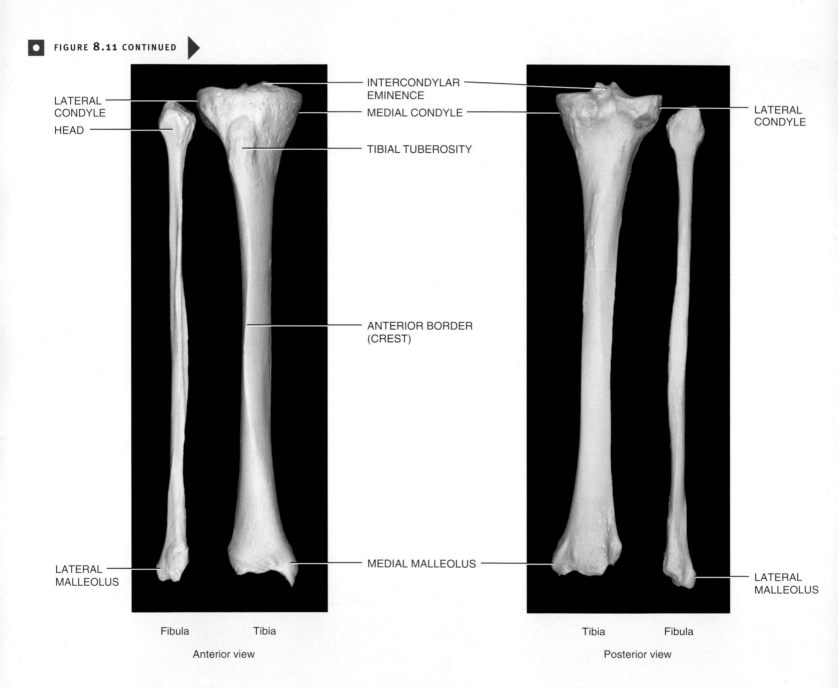

Fibula   Tibia

Anterior view

Tibia   Fibula

Posterior view

Fibula

Description

The **fibula**, one in each leg, is a slender, splint-like bone that is slightly expanded at both ends and is the developmental counterpart of the ulna in the free upper limb. The fibula is parallel and lateral to the tibia, but is considerably smaller. Unlike the tibia, the fibula does not articulate with the femur and is non-weight-bearing, but it does help stabilize the ankle joint.

Surface Features

The *head* of the fibula, the proximal end, articulates with the inferior surface of the lateral condyle of the tibia below the level of the knee joint to form the *proximal tibiofibular joint*. The distal end has a projection called the *lateral malleolus* that articulates with the talus of the ankle. This forms the prominence on the lateral surface of the ankle. As noted previously, the fibula also articulates with the tibia at the fibular notch to form the distal tibiofibular joint.

(c) Lateral view of distal end of tibia

Lateral view

◉ CHECKPOINT

Which structures form the medial and lateral prominences of the ankle?

❓ Which leg bone bears the weight of the body?

EXHIBIT 8.8 **241**

EXHIBIT 8.9 Skeleton of the Foot

OBJECTIVE

- Identify the location and surface features of the bones of the foot.

Tarsals

Description

The **tarsus** (ankle) is the proximal region of the foot and consists of seven **tarsal bones** (Figure 8.12a, b). The tarsal bones are much greater in size than the small carpal bones, with the two most proximal bones—the talus and calcaneus—being significantly larger than their more distal counterparts. Joints between tarsal bones are called *intertarsal joints*.

Surface Features

The tarsals include the **talus** (TĀ-lus=ankle bone) and **calcaneus** (kal-KĀ-nē-us=heel), located in the posterior part of the foot. The calcaneus is the largest and strongest tarsal bone. The anterior tarsal bones are the **navicular** (=like a little boat), three **cuneiform bones** (=wedge-shaped) called the **third (lateral)**,

Figure 8.12 Right foot and arches of the foot.

The skeleton of the foot consists of the proximal tarsals, the intermediate metatarsals, and the distal phalanges.

Superior view

Inferior view

Tarsals
Metatarsals
Phalanges

LATERAL   POSTERIOR   MEDIAL   POSTERIOR   LATERAL

TARSALS:
Calcaneus

Cuboid

Base

Shaft

Head

TARSALS:
Talus
Navicular
Third (lateral) cuneiform
Second (intermediate) cuneiform
First (medial) cuneiform

METATARSALS:

Sesamoid bones

PHALANGES:
Proximal
Middle
Distal
Great toe (hallux)

TARSALS:
Calcaneus

Cuboid

(a) Superior view   (b) Inferior view

MNEMONIC for tarsal bones:

Tall	Centers	Never	Take	Shots	From	Corners.
Talus	Calcaneus	Navicular	Third cuneiform	Second cuneiform	First cuneiform	Cuboid

second (intermediate), and first (medial) cuneiforms, and the cuboid (=cube-shaped). The talus, the most superior tarsal bone, is the only bone of the foot that articulates with the fibula and tibia. It articulates on one side with the medial malleolus of the tibia and on the other side with the lateral malleolus of the fibula. These articulations form the *talocrural (ankle) joint*. During walking, the talus transmits about half the weight of the body to the calcaneus. The remainder is transmitted to the other tarsal bones.

Metatarsals

Description

The **metatarsus** is the intermediate region of the foot and consists of five **metatarsal bones** numbered I to V (or 1–5) from the medial to lateral position (Figure 8.12a, b). They are convex dorsally and concave on their plantar surfaces.

Surface Features

Like the metacarpals of the palm, each metatarsal consists of a proximal *base*, an intermediate *shaft*, and a distal *head*. The metatarsals articulate proximally with the first, second, and third cuneiform bones and with the cuboid to form the *tarsometatarsal joints*. Distally, they articulate with the proximal row of phalanges to form the *metatarsophalangeal joints*. The first metatarsal is thicker than the others because it bears more weight.

Phalanges

Description

The **phalanges** comprise the distal component of the foot and resemble those of the hand both in number and arrangement (Figure 8.12a, b). The toes are numbered I to V (or 1–5) beginning with the great toe (hallux), from medial to lateral.

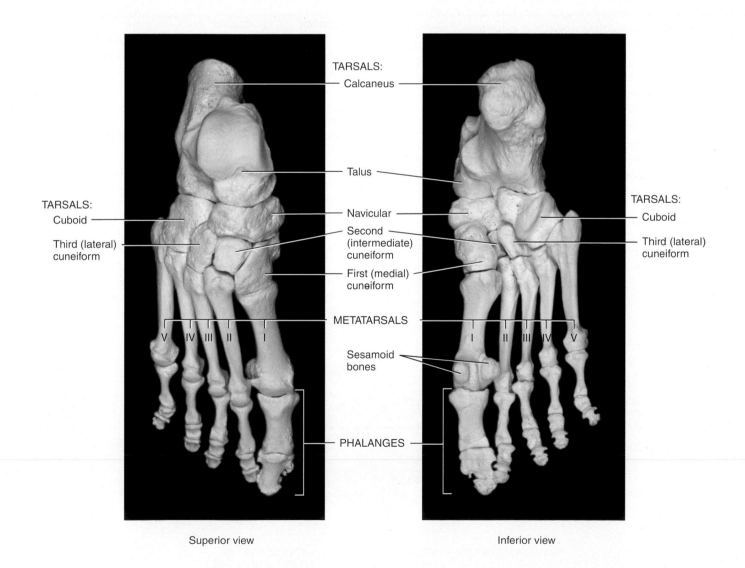

Superior view   Inferior view

EXHIBIT **8.9** CONTINUES ▶

EXHIBIT 8.9 Skeleton of the Foot CONTINUED

Surface Features

Each *phalanx* (singular) consists of a proximal *base*, an intermediate *shaft*, and a distal *head*. The great or big toe (*hallux*) has two large, heavy phalanges called proximal and distal phalanges. The other four toes each have three phalanges—proximal, middle, and distal. Joints between phalanges of the foot, like those of the hand, are called *interphalangeal joints*.

Arches of the Foot

The bones of the foot are arranged in two **arches** (Figure 8.12c). The arches enable the foot to support the weight of the body, provide an ideal distribution of body weight over the soft and hard tissues of the foot, and provide leverage when walking. The arches are not rigid; they yield as weight is applied and spring back when the weight is lifted, thus helping to absorb shocks. Usually, the arches are fully developed by the time children reach age 12 or 13.

The **longitudinal arch** has two parts, both of which consist of tarsal and metatarsal bones arranged to form an arch from the anterior to the posterior part of the foot. The *medial part* of the longitudinal arch, which originates at the calcaneus, rises to the talus and descends through the navicular, the three cuneiforms, and the heads of the three medial metatarsals. The *lateral part* of the longitudinal arch also begins at the calcaneus. It rises at the cuboid and descends to the heads of the two lateral metatarsals. The medial portion of the longitudinal arch is so high that the medial portion of the foot between the ball and heel does not touch the ground when you walk on a hard surface.

The **transverse arch** is found between the medial and lateral aspects of the foot and is formed by the navicular, three cuneiforms, and the bases of the five metatarsals.

As noted earlier, one function of the arches is to distribute body weight over the soft and hard tissues of the body. Normally, the ball of the foot carries about 40 percent of the weight and the heel carries about 60 percent. The ball of the foot is the padded portion of the sole superficial to the heads of the metatarsals. When a person wears high-heeled shoes, however, the distribution of weight changes so that the ball of the foot may carry up to 80 percent and the heel 20 percent. As a result, the fat pads at the ball of the foot are damaged, joint pain develops, and structural changes in bones may occur.

● CLINICAL **CONNECTION** Flatfoot and Clawfoot

The bones composing the arches are held in position by ligaments and tendons. If these ligaments and tendons are weakened, the height of the medial longitudinal arch may decrease or "fall." The result is **flatfoot**, the causes of which include excessive weight, postural abnormalities, weakened supporting tissues, and genetic predisposition. A custom-designed arch support often is prescribed to treat flatfoot.

Clawfoot is a condition in which the medial longitudinal arch is abnormally elevated. It is often caused by muscle deformities, such as may occur in diabetics whose neurological lesions lead to atrophy of muscles of the foot. ●

◉ CHECKPOINT

What are the functions of the arches of the foot?

◉ FIGURE **8.12** CONTINUED ▶

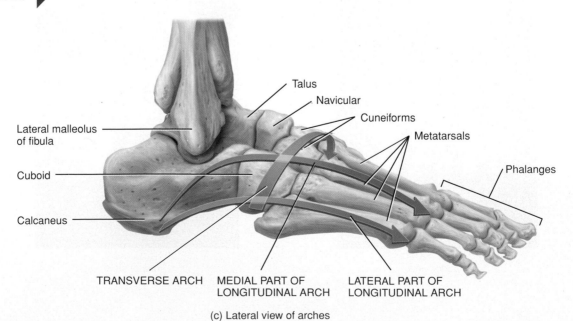

(c) Lateral view of arches

Talus
Navicular
Cuneiforms
Metatarsals
Phalanges
Lateral malleolus of fibula
Cuboid
Calcaneus
TRANSVERSE ARCH
MEDIAL PART OF LONGITUDINAL ARCH
LATERAL PART OF LONGITUDINAL ARCH

? **Which tarsal bone articulates with the tibia and fibula?**

DEVELOPMENT OF THE SKELETAL SYSTEM

◉ **OBJECTIVE**

• Describe the development of the skeletal system.

Most skeletal tissue arises from *mesenchymal cells*, connective tissue cells derived from **mesoderm**. However, much of the skeleton of the skull arises from neural crest cells of ectodermal origin. In either case, the cells condense and form models of bone in areas where the bones themselves will ultimately form. In some cases, the bones form directly within the mesenchyme (intramembranous ossification; see Figure 6.5 on page 149). In other cases, the bones form within hyaline cartilage that develops from mesenchyme (endochondral ossification; see Figure 6.6 on page 150).

The *skull* begins development during the fourth week after fertilization. It develops from tissue around the developing brain and consists of two major portions: **neurocranium** (mesoderma in origin), which forms the bones of the skull, and **viscerocranium** (ectodermal in origin), which forms the bones of the face (Figure 8.13a). The neurocranium is divided into two parts called the **cartilaginous neurocranium** (mesodermal in origin) and **membranous neurocranium** (neural crest in origin). The cartilaginous neurocranium consists of hyaline cartilage developed from mesenchyme at the base of the developing skull. It later undergoes endochondral ossification to form the *bones at the base of the skull*. The membranous neurocranium consists of neural crest mesenchyme and later undergoes intramembranous ossification to form the *flat bones that make up the roof and sides of the skull*. During fetal life and infancy the flat bones are separated by membrane-filled spaces called *fontanels* (see Figure 7.19 on page 202). The viscerocranium, like the neurocranium, is divided into two parts: **cartilaginous viscerocranium** and **membranous viscerocranium**. The cartilaginous viscerocranium is derived from the cartilage of the first two pharyngeal (branchial) arches (see Figure 4.13 on page 108). Endochondral ossification of these cartilages forms the *ear bones* and *hyoid bone*. The membranous viscerocranium is derived from neural crest mesenchyme in the first pharyngeal arch and, following intramembranous ossification, forms the facial bones.

The *vertebrae* and *ribs* are derived from the sclerotomes of somites (see Figure 10.10 on page 304). Mesenchymal cells from these regions surround the notochord (see Figure 10.10) at about four weeks after fertilization. The notochord induces the mesenchymal cells to form five distinct centers of bone formation at each segmental level—a *vertebral body*, *two costal (rib) centers*, *and two vertebral arch centers*. Between the vertebral bodies, the notochord induces mesenchymal cells to form the *nucleus pulposus* of an intervertebral disc and surrounding mesenchymal cells form the *annulus fibrosus* of an intervertebral disc. As development continues, the two vertebral arch centers unite with one another and with the vertebral body and the *vertebral arch* then surrounds the spinal cord (failure of the vertebral arch to develop properly results in a condition called spina bifida). In the thoracic region,

Figure 8.13 Development of the skeletal system. Bones that develop from the cartilaginous neurocranium are indicated in light blue; from the cartilaginous viscerocranium in dark blue; from the membranous neurocranium in dark red; and from the membranous viscerocranium in light red.

 After the free limb buds develop, endochondral ossification of the free limb bones begins by the end of the eighth embryonic week.

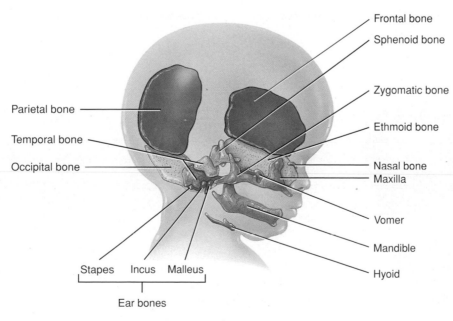

(a) Development of the skull

CONTINUES ▶

the costal centers enlarge and grow anteriorly to form the *ribs*. In the cervical, lumbar, and sacral regions the costal centers do not enlarge to the same degree as they do in the thorax, and they also fuse with the body and vertebral arch. The *sternum* develops from mesoderm in the anterior body wall.

The *skeleton of the free limbs* is derived from the somatic region of the lateral plate mesoderm (see Figure 4.9d on page 103). During the middle of the fourth week after fertilization, the free upper limbs appear as small elevations at the sides of the trunk called **free upper limb buds** (Figure 8.13b). About 2 days later, the **free lower limb buds** appear. The free limb buds consist of **mesenchyme** covered by **ectoderm**. At this point, a mesenchymal skeleton exists in the free limbs; the masses of mesoderm surrounding the developing bones, which will become the skeletal muscles of the free limbs, have migrated into the free limbs from the myotome of the somite.

By the sixth week, the free limb buds develop a constriction around the middle portion. The constriction produces flattened distal segments of the free upper buds called **hand plates** and distal segments of the lower buds called **foot plates** (Figure 8.13c). These plates represent the beginnings of the hands and feet, respectively. At this stage of free limb development, a cartilaginous skeleton formed from mesenchyme is present. By the seventh week (Figure 8.13d), the *arm, forearm,* and *hand* are evident in the free upper limb bud, and the *thigh, leg,* and *foot* appear in the free lower limb bud. By the eighth week (Figure 8.13e), as the shoulder, elbow, and wrist areas become apparent, the free upper limb bud is appropriately called the free upper limb, and the lower limb bud is now the free lower limb.

Endochondral ossification of the free limb bones begins by the end of the eighth week after fertilization. By the twelfth week, primary ossification centers are present in most of the free limb bones. Most secondary ossification centers appear after birth.

◉ **CHECKPOINT**

6. When and how do the free limbs develop?

☐ **FIGURE 8.13 CONTINUED** ▶

(b) Four-week embryo showing development of free limb buds

(c) Six-week embryo showing development of hand and foot plates

(d) Seven-week embryo showing development of arm, forearm, and hand in free upper limb bud and thigh, leg, and foot in free lower limb bud

(e) Eight-week embryo in which free limb buds have developed into free upper and lower limbs

❓ **Which of the three basic embryonic tissues—ectoderm, mesoderm, and endoderm—gives rise to the skeletal system?**

Clubfoot or **talipes equinovarus** (*pes*=foot; *equino-*=horse) An inherited deformity in which the foot is twisted inferiorly and medially, and the angle of the arch is increased; occurs in 1 of every 1000 births.

Genu valgum (JĒ-noo VAL-gum; *genu-*=knee; *valgum*=bent outward) A deformity in which the knees are abnormally close together and the space between the ankles is increased due to a lateral angulation of the tibia. Also called **knock-knee**.

Genu varum (JĒ-noo VAR-um; *varum*=bent toward the midline) A deformity in which the knees are abnormally separated and the free lower limbs are bowed medially. Also called **bowleg**.

Hallux valgus (HAL-uks VAL-gus; *hallux*=great toe) Angulation of the great toe away from the midline of the body, typically caused by wearing tightly fitting shoes. When the great toe angles toward the next toe, there is a bony protrusion at the base of the great toe. Also called a **bunion**.

STUDY OUTLINE

Skeleton of the Upper Limb (p. 214)

1. Each pectoral (shoulder) girdle consists of a clavicle and scapula (see Exhibits 8.1 and 8.2).
2. Each pectoral girdle attaches the free upper limb to the axial skeleton.
3. Each of the two free upper limbs (extremities) contains 30 bones.
4. The bones of each of the upper limb include the humerus, ulna, radius, carpals, metacarpals, and phalanges (see Exhibits 8.3–8.5).

Skeleton of the Lower Limb (p. 228)

1. The pelvic (hip) girdle consists of two hip bones.
2. Each hip bone consists of three fused bones: the ilium, pubis, and ischium (see Exhibit 8.6).
3. The hip bones, sacrum, and pubic symphysis form the bony pelvis. It supports the vertebral column and pelvic viscera and attaches the free lower limbs to the axial skeleton.
4. Each of the two free lower limbs (extremities) contains 30 bones.
5. The bones of each free lower limb include the femur, the patella, the tibia, the fibula, the tarsals, the metatarsals, and the phalanges (see Exhibits 8.7–8.9).
6. The bones of the foot are arranged in two arches, the longitudinal arch and the transverse arch, to provide support and leverage.

False and True Pelves (p. 233)

1. The true pelvis is separated from the false pelvis by the pelvic brim.

Comparison of Pectoral and Pelvic Girdles (p. 234)

1. The pectoral girdle does not articulate directly with the vertebral column; the pelvic girdle does via the sacroiliac joint.
2. The glenoid fossae are shallow and maximize movement; the acetabula are deep and allow less movement.
3. The pectoral girdle affords more mobility than strength; the pelvic girdle offers more strength than mobility.

Comparison of Female and Male Pelves (p. 234)

1. Bones of males are generally larger and heavier than bones of females, with more prominent markings for muscle attachment.
2. The female pelvis is adapted for pregnancy and childbirth. Gender-related differences in pelvic structure are listed and illustrated in Table 8.1 on page 235.

Development of the Skeletal System (p. 245)

1. Bone forms from mesoderm by intramembranous or endochondral ossification.
2. Bones of the free limbs develop from free limb buds, which consist of mesoderm and ectoderm.

SELF-QUIZ QUESTIONS

Choose the one best answer to the following questions.

1. The olecranon is at the proximal end of the:
   a. humerus.   **b.** radius.   **c.** ulna.
   d. tibia.   **e.** scapula.

2. The tibia articulates distally with the:
   a. femur.   **b.** fibula.   **c.** talus.
   d. cuboid.   **e.** Both b and c are correct.

3. Which structures are on the posterior surface of the free upper limb?
   a. radial fossa and radial notch
   b. trochlea and capitulum
   c. coronoid process and coronoid fossa
   d. olecranon process and olecranon fossa
   e. lesser tubercle and intertubercular sulcus

4. The bones of the pectoral girdle:
   a. articulate with the sternum anteriorly and the vertebrae posteriorly.
   b. include both clavicles, both scapulae, and the manubrium of the sternum.
   c. are considered to be part of the axial skeleton.
   d. articulate with the head of the humerus, forming the shoulder joint.
   e. are described by none of the above.

5. The anatomical name for the socket into which the humerus fits is the:
   a. acetabulum.   **b.** coronoid fossa.   **c.** glenoid cavity.
   d. supraspinous fossa.   **e.** iliac fossa.

6. All of the following are tarsal bones *except* the:
   a. cuboid.   **b.** triquetrum.   **c.** navicular.
   d. first cuneiform.   **e.** third cuneiform.

7. Which of the following articulates with the sacrum?
   a. ilium   **b.** ischium
   c. pubis   **d.** Both a and b are correct.
   e. All three bones articulate with the sacrum.

8. The greater trochanter is a large bony prominence located:
   a. on the proximal part of the humerus.
   b. on the proximal part of the femur.
   c. near the tibial tuberosity.
   d. on the ilium.
   e. on the posterior surface of the scapula.

9. Which of the following statements regarding the male pelvis is *not* true?
   a. The bones are heavier and thicker than in the female.
   b. The male pelvis is narrow and deep.
   c. The coccyx is less curved anteriorly.
   d. The pelvic outlet is narrower than in the female.
   e. None of the above statements is true.

10. Which of the following is not a carpal bone?
    a. hamate   b. cuboid   c. pisiform
    d. trapezium   e. scaphoid

11. What bone of the antebrachium is located directly proximal to the pisiform carpal bone?
    a. humerus   b. radius   c. hamate
    d. ulna   e. proximal phalanx

12. The most lateral projection of the scapula is the
    a. spine.   b. acromion process.   c. glenoid fossa.
    d. inferior angle.   e. scapular notch.

Complete the following.

13. The greater sciatic notch is an indentation seen on the _____.

14. The sesamoid bone that forms in the tendon of the quadriceps muscle is the _____.

15. The lesser trochanter is on the _____ surface of the _____.

16. The portion of the pelvis superior to the pelvic brim is the _____ pelvis.

17. The _____ of the _____ fits into a depression called the acetabulum.

18. The talus bone is located _____ to the navicular bone.

19. On the distal humerus, the _____ is located medial to the _____.

Are the following statements true or false? If the statement is False, *find and fix the mistake to make the statement* True.

20. The metatarsals are proximal to the tarsals.

21. The hamate is a carpal bone with a hook-shaped projection on its anterior surface.

22. The acromial extremity is located at the lateral end of the clavicle.

23. The capitulum of the humerus articulates with the styloid process of the radius.

24. The medial end of the clavicle is located immediately lateral to the manubrium.

Matching

25. Match the following bony landmarks and bones. Answers may be used more than once.

    ____ **(a)** fibular notch   **(1)** radius
    ____ **(b)** acromion   **(2)** femur
    ____ **(c)** coracoid process   **(3)** tibia
    ____ **(d)** lateral malleolus   **(4)** humerus
    ____ **(e)** deltoid tuberosity   **(5)** scapula
    ____ **(f)** radial tuberosity   **(6)** ischium
    ____ **(g)** anterior inferior iliac spine   **(7)** ulna
    ____ **(h)** linea aspera   **(8)** fibula
    ____ **(i)** ischial tuberosity   **(9)** ilium
    ____ **(j)** lateral styloid process **(10)** clavicle
    ____ **(k)** medial malleolus
    ____ **(l)** conoid tubercle
    ____ **(m)** glenoid cavity
    ____ **(n)** coronoid fossa
    ____ **(o)** obturator foramen
    ____ **(p)** gluteal tuberosity
    ____ **(q)** trochlear notch

Matching

26. Match the following bony landmarks and bones (answers may be used more than once):

    ____ **(a)** lateral to lunate and   **(1)** capitate
            proximal to trapezoid   **(2)** triquetrum
    ____ **(b)** medial to trapezoid   **(3)** cuboid
            and lateral to hamate   **(4)** scaphoid
    ____ **(c)** distal to calcaneus and   **(5)** navicular
            lateral to third cuneiform   **(6)** middle phalanx
    ____ **(d)** distal to calcaneus and
            proximal to all three
            cuneiform bones
    ____ **(e)** proximal to metacarpal V
            and posterior to pisiform
    ____ **(f)** distal to metatarsal I and
            proximal to distal phalanx
            of great toe

CRITICAL THINKING QUESTIONS

1. The *Local News* reported that farmer Bob Ramsey caught his hand in a piece of machinery on Tuesday. He lost the lateral two fingers of his left hand. Science reporter Kent Clark, a high school junior, reports that Farmer Ramsey has 3 remaining phalanges. Is Kent correct or is he anatomy-challenged?

2. Rose had flat feet as a child and was told to take up ballet dancing to correct the condition. Now Rose has hallux valgus and trouble with her talocrural joint, but at least her flat feet are cured. Explain how Rose's problems relate to each other.

3. On the coast of Alaska, archaeologists have unearthed an ancient burial site that contains the skeletal remains of some Native Americans alongside some very large kayaks (a type of paddled boat). The humerus bones show unusually pronounced deltoid tuberosities. Why/how might these projections have grown so large?

4. Amy was visiting her Grandmother Amelia in the hospital. Grandmother said she had a broken hip but when Amy peeked at the chart, it said she had a fractured femur. Explain the discrepancy. Meanwhile, Grandpa Jeremiah is complaining that his "sacroiliac is acting up," making it painful for him to stand or sit for long or to walk very far. Does Grandpa really know what he's talking about? Explain.

5. Derrick, the high school gym teacher, jogs on the same banked high school track at the same time and in the same direction every day. Recently he has been experiencing an ache in one knee, especially after he sits and reads the newspaper following his jog. Nothing else in his routine has changed. What could be the problem with his knee?

? ANSWERS TO FIGURE QUESTIONS

8.1 The pectoral girdle is made up of the clavicle and scapula. The free upper limb includes the humerus, ulna, radius, carpus, metacarpus, and phalanges.

8.2 The weakest part of the clavicle is its midregion at the junction of the two curves.

8.3 The acromion forms the high point of the shoulder.

8.4 The radius articulates at the elbow with the capitulum and radial fossa of the humerus. The ulna articulates at the elbow with the trochlea, coronoid fossa, and olecranon fossa of the humerus.

8.5 The olecranon is the "elbow" part of the ulna.

8.6 The scaphoid is the most frequently fractured wrist bone.

8.7 The pelvic girdle is made up of the hip bones. The free lower limb includes the femur, patella, tibia, fibula, tarsus, metatarsus, and phalanges.

8.8 The femur articulates with the acetabulum of the hip bone; the sacrum articulates with the auricular surface of the hip bone.

8.9 The pelvic axis is the course taken by a baby's head as it descends through the pelvis during childbirth.

8.10 The angle of convergence of the femurs is greater in females than males because the female pelvis is broader.

8.11 The tibia is the weight-bearing bone of the leg.

8.12 The talus is the only tarsal bone that articulates with the tibia and fibula.

8.13 Most of the skeletal system arises from embryonic mesoderm.

9 JOINTS

INTRODUCTION Have you ever looked beneath the hood of a car and examined the engine and its many moving parts? An engine is a complex machine that consists of numerous parts, many of which rub against each other and generate considerable *friction* (resistance between rubbing parts). These various friction-generating parts produce the necessary forces required to power the movement of the car, and they have a limited lifetime because of constant wear and tear. Most people are thrilled if they get 10 to 15 years of use from their car's engine, but our bodies, which endure similar wear and tear, must last a lifetime.

The human skeleton needs to move but bones are too rigid to bend without being damaged. Fortunately, flexible connective tissues hold bones together while still permitting, in most cases, some degree of movement. A **joint**, also called an **articulation** (ar-tik′-ū-LĀ-shun) or **arthrosis** (ar-THRŌ-sis), is a point of contact between two bones, between bone and cartilage, or between bone and teeth. When we say one bone *articulates* with another bone, we mean that the bones form a joint. Think for a moment about the amazing motion and the complexity of the coordinated movements that occur as the bones of the body move against one another, movements that are far more complex than those of almost any machine. These joint actions are repeated daily and carry us from childhood, into adolescence, and through our adult lives. Joints have a phenomenal capacity to function efficiently into the latter years of life, exhibiting life spans that are many times greater than those of most machines. How does the structure of a joint make this possible? Why do joints sometimes fail and cause our movements to become painful? How can we prolong the efficient function of our joints? These are just some of the questions that you will be able to answer as you learn about the structure and function of joints. The scientific study of joints is termed **arthrology** (ar-THROL-ō-jē; *arthr-*=joint; *-ology*=study of). The study of motion of the human body is called **kinesiology** (ki-nē-sē-OL-ē-jē; *kinesi-*=movement). •

JOINT CLASSIFICATIONS

◉ OBJECTIVE
• Describe the structural basis for the classifications of joints.

Bones never contact one another directly. There is always some other connective tissue between the bony contact surfaces. Joints have a wide variety of structures and just as many functions. The most familiar joints are junctions between neighboring bones where movements of the skeleton occur, but there are many joints that permit little to no movement. Why are some joints designed to produce a wide array of movement and others little to no movement at all? Keep reading and you will soon find out.

Over the years anatomists have used a variety of schemes to classify the various joints of the body, some based on joint structure, others using joint movements. Present-day classification and the current anatomical nomenclature are based on joint structure.

The structure of joints includes the connecting tissues between the neighboring bones. At the simplest level, there are two basic ways bones connect with one another to form joints. Bones are either held together by solid masses of connective tissue, or they are connected by a connective tissue capsule that surrounds a lubricated cavity.

Joints formed by a solid mass of connective tissue between the neighboring bones are called **synarthroses** (sin-ar-THRŌ-sēz; *syn*=together). There are two types of synarthroses—**fibrous joints** (FI-brus), which have connective tissue masses of dense irregular connective tissue, and **cartilaginous joints** (kar-ti-LAJ-i-nus), which use some form of cartilage as the connecting tissue between the bones.

The capsular joints, with their lubricated cavity called a synovial cavity, are referred to as **synovial joints** (si-NŌ-vē-al; *syn*=together; *ovial*=egg, because the synovial tissues resembled fresh egg whites) or **diarthroses** (di-ar–THRŌ-sēz;=movable joints). These joints are divided into various types based on the structure of the joint surfaces, but all share the common design of a joint capsule surrounding a lubricated synovial cavity.

The different structural relations of the binding tissues, the types of connective tissues involved, and the structure of the opposing bone surfaces become the defining factors for joint classification (see Table 9.1 on page 258).

Ligaments

Throughout this chapter on joints you will often encounter the term ligament. **Ligaments** (*ligare*=to bind) are dense irregular or dense regular connective tissue structures that bind one bone to another bone (see Figure 9.15). They come in a variety of forms and are integral parts of joints. They can serve as intrinsic binding structures constituting the substance of the joint itself (for example, sutural ligaments or periodontal ligaments), or as extrinsic supporting bands that stabilize joints while limiting their range of motion (for example, the anterior cruciate ligament). As you study the various joints of the body in the sections that follow, you will learn about a variety of ligaments and their structure, location, and function.

Ligaments are noncontractile tissues (unlike muscles they cannot contract or shorten) that resist stretching. They are typically damaged when forces that exceed their strength stretch them beyond their limits. Damage to these connective tissue structures is referred to as a sprain.

A **sprain** is the forcible wrenching or twisting of a joint that stretches or tears its ligaments but does not dislocate the bones. It occurs when the ligaments are stressed beyond their normal capacity. Sprains also may damage surrounding blood vessels, muscles, tendons, or nerves. Severe sprains may be so painful that the joint cannot be moved. There is considerable swelling, which results from hemorrhage of ruptured blood vessels. The ankle joint is most often sprained; the lower back is another frequent location. A **strain** is a stretched or partially torn muscle. It often occurs when a muscle contracts suddenly and powerfully—for example, in the leg muscles of sprinters when they spring from the blocks at the starting line. Like muscle injuries, ligament injuries can be graded. Grade I sprains range from a stretch without a tear to a 20% tear of the ligament's collagen fibers. A grade II sprain involves tearing 20 to 75% of the ligament. Grade III sprains involve damage of more than 75% of the ligament's collagen to complete tearing of the ligament. Severe ligament injuries should be taken care of immediately, as they can lead to major joint instability and malfunction.

◉ CHECKPOINT
1. On what basis are joints classified?

FIBROUS JOINTS

◉ **OBJECTIVE**

• Describe the structure and functions of the three types of fibrous joints.

Fibrous joints are synarthroses in which the neighboring bones are joined together by a solid mass of dense irregular connective tissue. The adjoining connective tissue can vary from small fibrous strands of connective tissue, to large thick bands, to extensive membranous sheets. Within this classification of joints

Figure 9.1 Fibrous joints.

🔑 ▭ At a fibrous joint the bones are held together by dense irregular connective tissue.

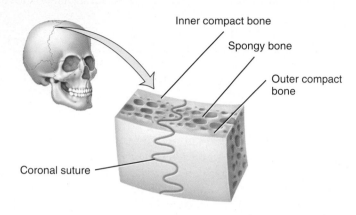

Inner compact bone
Spongy bone
Outer compact bone
Coronal suture

(a) Suture between skull bones

Fibula
Tibia
Anterior tibiofibular ligament

Syndesmosis between tibia and fibula

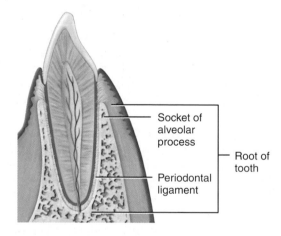

Socket of alveolar process
Root of tooth
Periodontal ligament

Syndesmosis between tooth and socket of alveolar process (gomphosis)

(b) Syndesmosis

Fibula
Interosseous membrane
Tibia

(c) Interosseous membrane between tibia and fibula

❓ **What are the different types of fibrous joints and where are they found in the body?**

there are three types: sutures, syndesmoses, and interosseous membranes.

Sutures

A **suture** (SOO-chur; *sutur-*=seam) is a fibrous joint composed of a thin layer of dense irregular connective tissue. Sutures are found only between bones of the skull. An example is the coronal suture between the frontal and parietal bones (Figure 9.1a on page 252). The irregular, interlocking edges of sutures give them added strength and decrease their chance of fracturing. Sutures are joints that form as the numerous bones of the skull come in contact during development. They are immovable to slightly movable and play important roles in shock absorption in the skull.

Some sutures, although present during growth of the skull, are replaced by bone in the adult. Such a suture is called a **synostosis** (sin′-os-TŌ-sis; *os-*=bone), or bony joint—a joint in which there is a complete fusion of two separate bones into one. For example, the frontal bone grows in halves that grow together across a suture line. Usually they are completely fused by age 6 and the suture becomes obscure. If the suture persists beyond age 6, it is called a **metopic suture** (me-TŌ-pik; *metopon*=forehead).

Syndesmoses

A **syndesmosis** (sin′-dez-MŌ-sis; *syndesmo-*=band or ligament) is a fibrous joint in which there is a greater distance between the articulating surfaces and more dense irregular connective tissue than in a suture. The dense irregular connective tissue is typically arranged as a bundle (ligament) and the joint permits limited movement. One example of a syndesmosis is the distal tibiofibular joint, where the anterior tibiofibular ligament connects the tibia and fibula (Figure 9.1b, left). Another example of a syndesmosis is called a **gomphosis** (gom-FŌ-sis; *gompho-*=bolt or nail) or *dentoalveolar joint*, in which a cone-shaped peg fits into a socket. The only examples of gomphoses in the human body are the articulations between the roots of the teeth and their sockets (alveoli) in the maxillae and mandible (Figure 9.1b, right). The dense irregular connective tissue between a tooth and its socket is the thin periodontal ligament (membrane). A gomphosis permits no movement. Inflammation and degeneration of the gums, periodontal ligament, and bone is called *periodontal disease*.

Interosseous Membranes

The final category of fibrous joint is the interosseous membrane. An **interosseous membrane** is a substantial sheet of dense irregular connective tissue that binds neighboring long bones and permits slight movement. There are two principal interosseous membrane joints in the human body. One occurs between the radius and ulna in the forearm (see Figure 8.5a, b on page 222) and the other occurs between the tibia and fibula in the leg (Figure 9.1c). These strong connective tissue sheets not only help hold these adjacent long bones together, they also play an important role in defining the range of motion between the neighboring bones and providing an increased attachment surface for muscles that produce movements of the digits of the hand and foot.

◉ **CHECKPOINT**

2. What are the similarities in the three types of fibrous joints? How do they differ?

CARTILAGINOUS JOINTS

◉ **OBJECTIVE**

• Describe the structure and functions of the two types of cartilaginous joints.

Like a fibrous joint, a **cartilaginous joint** is a synarthrosis. Here there is a solid connective tissue joint that allows little or no movement. The articulating bones are tightly connected, either by hyaline cartilage or by fibrocartilage (see Table 3.4H on page 80). The two types of cartilaginous joints are synchondroses and symphyses.

Synchondroses

A **synchondrosis** (sin′-kon-DRŌ-sis; *chondro-*=cartilage; plural is *synchondroses*) is a cartilaginous joint in which the connecting material is hyaline cartilage. An example of a synchondrosis is the epiphyseal (growth) plate that connects the epiphysis and diaphysis of a growing bone (Figure 9.2a). A synchondrosis is immovable. A photomicrograph of the epiphyseal plate is shown in Figure 6.8b on page 152. When bones stop growing in length, bone replaces the hyaline cartilage, and the synchondrosis becomes a synostosis, or bony joint.

In an x-ray of a young person's skeleton, the synchondroses are easily seen as thin dark areas between the white-appearing bone tissue (see Figure 6.8a on page 152). This is how a medical professional can view an x-ray and determine that a person still has some degree of growth remaining. Breaks in a bone that extend into the epiphyseal plate and damage the cartilage of the

Figure 9.2 Cartilaginous joints.

🔓 ⚊ At a cartilaginous joint the bones are held together by cartilage.

Epiphyseal plates

Epiphysis

Epiphysis

Diaphysis

(a) Synchondrosis

Hip bones

Pubic symphysis

(b) Symphysis

❓ What is the structural difference between a synchondrosis and a symphysis?

synchondrosis can affect further growth of the bone, leading to shortened development of the length of the bone.

Symphyses

A **symphysis** (SIM-fi-sis=growing together; plural is *symphyses*) is a cartilaginous joint in which the ends of the articulating bones are covered with hyaline cartilage, but the bones are connected by a broad, flat disc of fibrocartilage. All symphyses occur in the midline of the body. The pubic symphysis between the anterior surfaces of the hip bones is one example of a symphysis (Figure 9.2b) and is slightly movable. This type of joint is also found at the junction of the manubrium and body of the sternum (see Figure 7.21a on page 206) and at the intervertebral joints between the bodies of vertebrae (see Figure 7.15a on page 194). A portion of the intervertebral disc is made up of fibrocartilage. The structure of the intervertebral disc helps to provide the limited range of motion in the vertebral column while serving as an important shock-absorbing pad between the vertebral bodies.

☑ CHECKPOINT

3. How do symphyses differ from synchondroses? Where is each found in the body?

SYNOVIAL JOINTS

☑ OBJECTIVES

• Describe the structure of synovial joints.
• Outline the structure and function of bursae and tendon sheaths.
• List the six types of synovial joints.

Structure of Synovial Joints

Synovial joints (si-NŌ-vē-al) have certain characteristics that distinguish them from other joints. The unique characteristic of a synovial joint is the presence of a space called a **synovial (joint) cavity** that is surrounded by a connective tissue capsule that connects the articulating bones (Figure 9.3). Synovial joints range from slightly movable to the most mobile joints of the body. For example, the synovial joints between a few of the carpal bones have very limited movement, while the shoulder joint can move freely in all directions. The bone surfaces within the capsule of a synovial joint are covered by a layer of hyaline cartilage called **articular cartilage**. The smooth cartilage covers the articulating surfaces of the bones but does not bind them together. The lubricated articular cartilage reduces friction between bones in the joint during movement and helps to absorb shock.

Articular Capsule

Not to be confused with articular cartilage, a sleevelike **articular (joint) capsule** surrounds a synovial joint, encloses the synovial cavity, and unites the articulating bones. The articular capsule is composed of two layers, an outer fibrous membrane and an inner synovial membrane (Figure 9.3). The **fibrous membrane** usually consists of dense irregular connective tissue (mostly collagen fibers) that attaches to the periosteum of the articulating bones (it is literally a thickened continuation of the periosteum between the two bones). The flexibility of the fibrous membrane permits considerable movement at a joint while its great tensile strength (resistance to stretching) helps prevent the bones from dislocating. The fibers of some fibrous membranes are thicker and

arranged in parallel bundles that are highly adapted for resisting tensile strains. Such fiber bundles, called *ligaments* are often designated by individual names. The strength of ligaments is one of the principal mechanical factors that holds bones close together in a synovial joint. The inner layer of the articular capsule, the **synovial membrane**, is composed of areolar connective tissue with elastic fibers. At many synovial joints, the synovial membrane includes accumulations of adipose tissue that varies in thickness. The thicker regions of adipose tissue are called **articular fat pads**. An example is the infrapatellar fat pad in the knee (see Figure 9.15a, c).

A "double-jointed" person does not really have extra joints. Individuals who are double-jointed have greater flexibility in their articular capsules and ligaments; the resulting increase in range of motion allows them to entertain fellow partygoers with activities such as touching their thumbs to their wrists and putting their ankles or elbows behind their necks. Unfortunately, such flexible joints are less structurally stable and are more easily dislocated.

Synovial Fluid

The synovial membrane secretes **synovial fluid** (*ov-*=egg), which forms a thin film over the surfaces within the articular capsule. This viscous, clear or pale yellow fluid was named for its similarity in appearance and consistency to uncooked egg white. Synovial fluid consists of hyaluronic acid secreted by fibroblast-like cells in the synovial membrane and interstitial fluid filtered from blood plasma. Its functions include reducing friction by lubricating the joint, absorbing shocks, and supplying oxygen and nutrients to and removing carbon dioxide and metabolic wastes from the chondrocytes within articular cartilage. (Recall that cartilage is an avascular tissue, so it does not have blood vessels to perform the latter function.) Synovial fluid also contains phagocytic cells that remove microbes and the debris that results from normal wear and tear in the joint. When a synovial joint is immobile for a time, the fluid becomes quite viscous (gel-like), but as joint movement increases, the fluid becomes less viscous. One of the benefits of warming up before exercise is that it stimulates the production and secretion of synovial fluid; more fluid means less stress on the joints during exercise.

We are all familiar with the cracking sounds heard as certain joints move, or the popping sounds that arise when a person pulls on his fingers to crack his knuckles. According to one theory, when the synovial cavity expands, the pressure inside the synovial cavity decreases, creating a partial vacuum. The suction draws carbon dioxide and oxygen out of blood vessels in the synovial membrane, forming bubbles in the fluid. When the fingers are flexed (bent) the volume of the cavity decreases and the pressure increases; this bursts the bubbles and creates the cracking or popping sound as the gases are driven back into solution.

Accessory Ligaments, Articular Discs, and Labra

Many synovial joints also contain **accessory ligaments** called extracapsular ligaments and intracapsular ligaments. *Extracapsular ligaments* lie outside the articular capsule, such as the fibular and tibial collateral ligaments of the knee joint (see Figure 9.15). *Intracapsular ligaments* occur within the articular capsule but are excluded from the synovial cavity by folds of the synovial membrane. The anterior and posterior cruciate ligaments of the knee joint are intracapsular ligaments (see Figure 9.15d, g).

A few synovial joints in the body contain **articular discs**, fibrocartilage structures not covered by synovial membrane that divide the synovial cavity into two smaller cavities (see Figure 9.11c). As you will see later, separate movements can occur in the two cavities, such as those of the temporomandibular joint (TMJ) (see pages 268–269). Peripherally articular discs attach to the fibrous membrane. In some joints, incomplete discs, **menisci** (me-NIS-sī or me-NIS-kī; the singular is *meniscus*) partially divide the joint (see Figure 9.15c, d, g). These crescent-shaped discs are prominent features of the knee joint. Like the articular discs, menisci bind strongly to the inside of the fibrous membrane. During development, the synovial membrane covers the meniscus, as the meniscus develops between the fibrous membrane and synovial membrane. As the joint becomes functional, the synovial membrane is worn off the surface of the meniscus where considerable friction occurs. The synovial membrane,

then, stops at the base of the meniscus and does not continue onto the meniscal articular surface. The functions of the menisci are not completely understood, but are known to include the following: (1) shock absorption; (2) a better fit between articulating bony surfaces; (3) providing adaptable surfaces for combined movements; (4) weight distribution over a greater contact surface; and (5) distribution of synovial lubricant across the articular surfaces of the joint.

A **labrum** (the plural is *labra*), prominent in the ball-and-socket joints of the shoulder and hip (see Figure 9.12c, d and 9.14c, d), is the fibrocartilaginous lip that extends from the edge of the joint socket. The labrum helps deepen the joint socket and increases the area of contact over the ball-like opposite surface.

Figure 9.3 Structure of a typical synovial joint. Note the two layers of the articular capsule—the fibrous membrane and the synovial membrane. Synovial fluid lubricates the joint cavity between the synovial membrane and the articular cartilage.

The distinguishing feature of a synovial joint is the synovial cavity between the articulating bones.

(a) Frontal section

(b) Sagittal section of right elbow joint

? **What structure of synovial joints allows efficient movement between neighboring bones?**

Monitor: view of joint as seen via the arthroscope

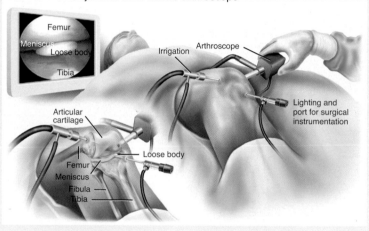

Nerve and Blood Supply

The nerves that supply a joint are the same as those that supply the skeletal muscles that move the joint. Synovial joints contain many nerve endings that are distributed to the articular capsule and associated ligaments. Some of the nerve endings convey information about pain from the joint to the spinal cord and brain for processing. Other nerve endings are responsive to the degree of movement and stretch at a joint. This information is also relayed to the spinal cord and brain, which may respond by sending impulses through different nerves to the muscles to adjust body movements.

Although many of the components of synovial joints are avascular, arteries in the vicinity send out numerous branches that penetrate the ligaments and articular capsule to deliver oxygen and nutrients. Veins remove carbon dioxide and wastes from the joints. The arterial branches from several different arteries typically join together around a joint before penetrating the articular capsule. The chondrocytes in the articular cartilage of a synovial joint receive oxygen and nutrients from synovial fluid, which is derived from capillary blood; all other joint tissues are supplied directly by capillaries. Carbon dioxide and wastes pass from chondrocytes of articular cartilage into synovial fluid and then into lymphatic capillaries in the surrounding tissues; carbon dioxide and wastes from all other joint structures pass directly into capillaries that drain into veins.

Bursae and Tendon Sheaths

The various movements of the body create friction between moving parts. Saclike structures called **bursae** (BER-sē=purses; singular is *bursa*) are strategically situated to alleviate friction around some joints, such as the shoulder and knee joints (see Figures 9.12 a–c and 9.15c, h). Bursae are not strictly parts of synovial joints, but they do resemble joint capsules because their walls consist of a thin dense connective tissue outer fibrous membrane lined by a synovial membrane. They are also filled with a small amount of fluid similar to synovial fluid. Bursae occur in areas of the body where adjacent tissues create considerable friction with one another. They are located between the skin and bone, tendons and bones, muscles and bones, and ligaments and bones. The lubricated bursal sacs reduce friction during the movement of these adjacent body parts. An acute or chronic inflammation of a bursa, called **bursitis**, is usually caused by irritation from repeated, excessive exertion of a joint. The condition may also be caused by trauma, an acute or chronic infection (including syphilis and tuberculosis), or by rheumatoid arthritis. Symptoms include pain, swelling, tenderness, and limited movement.

Structures called *tendon sheaths* also reduce friction at joints. **Tendon (synovial) sheaths** are tubelike bursae that wrap around tendons that experience considerable friction on all sides. This occurs where tendons pass through fibro-osseous tunnels (tunnels formed by connective tissue bands and bone), such as the tendon of the biceps brachii muscle at the shoulder joint (see Figure 9.12c) and the tendons of wrist, ankle, fingers, and toes.

Types of Synovial Joints

Although all synovial joints are similar in structure, the shapes of the articulating surfaces vary to allow various types of movements. Accordingly, synovial joints are divided into six types: planar, hinge, pivot, condyloid, saddle, and ball-and-socket.

Planar Joints

The articulating surfaces of bones in a **planar joint** are flat or slightly curved (Figure 9.4a). Planar joints primarily permit side-to-side and back-and-forth gliding movements between the flat surfaces of bones (described shortly). Many planar joints are *biaxial*, meaning that they permit movement around two axes. An axis is a straight line around which a bone rotates (revolves). Some examples of planar joints are the intercarpal joints (between carpal bones at the wrist); intertarsal joints (between tarsal bones at the ankle); sternoclavicular joints (between the manubrium of the sternum and the clavicle); acromioclavicular joints (between the acromion of the scapula and the clavicle); sternocostal joints (between the sternum and ends of the costal cartilages at the tips of the second through seventh pairs of ribs); and vertebrocostal joints (between the heads and tubercles of ribs and transverse processes of thoracic vertebrae).

Hinge Joints

In a **hinge joint**, the convex surface of one bone fits into the concave surface of another bone (Figure 9.4b). As the name implies, hinge joints produce an angular, opening-and-closing motion like that of a hinged door. Hinge joints permit only flexion (bending of joints) and extension (straightening of joints). In most joint movements, one bone remains in a fixed position while the other moves around an axis. Hinge joints are said to be *monaxial (uniaxial)* because they typically allow motion around a single axis. Examples of hinge joints are the knee (actually a modified hinge joint), elbow, ankle, and interphalangeal joints.

Figure 9.4 Types of synovial joints. For each type, a drawing of the actual joint and a simplified diagram are shown.

Synovial joints are classified into six principal types on the basis of the shapes of the articulating bone surfaces.

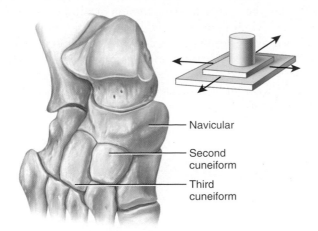

(a) Planar joint between navicular and second and third cuneiforms of tarsus in foot

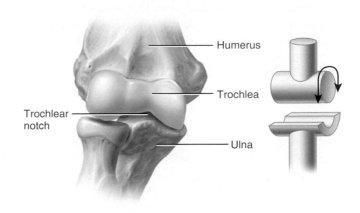

(b) Hinge joint between trochlea of humerus and trochlear notch of ulna at the elbow

(c) Pivot joint between head of radius and radial notch of ulna

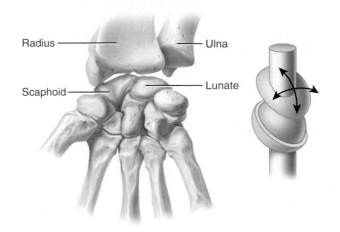

(d) Condyloid joint between radius and scaphoid and lunate bones of carpus (wrist)

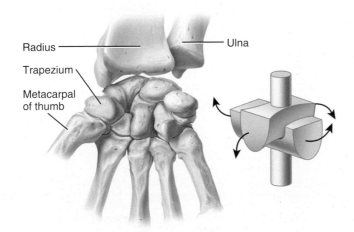

(e) Saddle joint between trapezium of carpus (wrist) and metacarpal of thumb

(f) Ball-and-socket joint between head of femur and acetabulum of hip bone

? Which of the joints shown in this figure are biaxial?

Pivot Joints

In a **pivot joint**, the rounded or pointed surface of one bone articulates with a ring formed partly by another bone and partly by a ligament (Figure 9.4c). A pivot joint is monaxial because it allows rotation (revolving) around its own longitudinal axis only. Examples of pivot joints are the atlanto-axial joint, in which the atlas rotates around the axis and permits the head to turn from side-to-side as in signifying "no" (see Figure 9.9a), and the radioulnar joints that enable the palms to turn anteriorly and posteriorly (see Figure 9.10h).

Condyloid Joints

In a **condyloid joint** (KON-di-loyd; *condyl-*=knuckle), also called an *ellipsoidal joint*, the convex oval-shaped projection of one bone fits into the oval-shaped depression of another bone (Figure 9.4d). A condyloid joint is *biaxial* because it permits movement around two axes (flexion–extension and abduction–adduction). These are described shortly. Notice that you can move your index finger both forward and backward and from side to side. Examples are the wrist and the metacarpophalangeal joints of the second through fifth digits.

TABLE 9.1

Summary of Structural Classification of Joints

STRUCTURAL CLASSIFICATION	DESCRIPTION	FUNCTION (MOVEMENT)	EXAMPLE
Fibrous No synovial cavity; articulating bones held together by dense irregular connective tissue.			
Suture	Articulating bones united by a thin layer of dense irregular connective tissue, found between bones of the skull. With age, some sutures are replaced by a synostosis, in which separate cranial bones fuse into a single bone.	Immovable to slightly movable, helps absorb shock in the skull.	Coronal suture.
Syndesmosis	Articulating bones united by more dense irregular connective tissue.	Immovable to slightly movable.	Distal tibiofibular joint.
Interosseous membrane	Articulating bones united by a substantial sheet of dense irregular connective tissue.	Slightly movable.	Between the tibia and fibula.
Cartilaginous No synovial cavity; articulating bones united by hyaline cartilage or fibrocartilage.			
Synchondrosis	Connecting material is hyaline cartilage.	Immovable.	Epiphyseal plate between the diaphysis and epiphysis of a long bone.
Symphysis	Connecting material is a broad, flat disc of fibrocartilage.	Slightly movable and helps absorb shock in the vertebral column.	Pubic symphysis and intervertebral joints.
Synovial Characterized by a synovial (joint) cavity, articular cartilage, and an articular (joint) capsule; may contain accessory ligaments, articular discs, and bursae.			
Planar	Articulated surfaces are flat or slightly curved.	Many biaxial; bones permit back-and-forth and side-to-side movements	Intercarpal, intertarsal, sternocostal (between sternum and the second–seventh pairs of ribs), and vertebrocostal joints.
Hinge	Convex surface fits into a concave surface.	Monaxial; flexion–extension.	Knee (modified hinge) elbow, ankle, and interphalangeal joints.
Pivot	Rounded or pointed surface fits into a ring formed partly by bone and partly by a ligament.	Monaxial; rotation.	Atlanto-axial and radioulnar joints.
Condyloid	Oval-shaped projection fits into an oval-shaped depression.	Biaxial; flexion–extension and abduction–adduction.	Radiocarpal and metacarpophalangeal joints.
Saddle	Articular surface of one bone is saddle-shaped, and the articular surface of the other bone "sits" in the saddle.	Triaxial; flexion–extension, abduction–adduction, and rotation.	Carpometarcarpal joint between trapezium and thumb.
Ball-and-socket	Ball-like surface fits into a cuplike depression.	Triaxial; flexion–extension, abduction–adduction, circumduction, and rotation.	Shoulder and hip joints.

Saddle Joints

In a **saddle joint**, the articular surface of one bone is saddle-shaped, and the articular surface of the other bone fits into the "saddle" as a sitting rider would (Figure 9.4e). A saddle joint is a modified condyloid joint in which the movement is somewhat freer. Saddle joints are *triaxial*, permitting movements around three axes (flexion–extension, abduction–adduction, and rotation). These are also described shortly. An example of a saddle joint is the carpometacarpal joint between the trapezium of the carpus and metacarpal of the thumb.

Ball-and-Socket Joints

A **ball-and-socket joint** consists of the ball-like surface of one bone fitting into a cuplike depression of another bone (Figure 9.4f). Such joints are *triaxial* (flexion–extension, abduction–adduction, and rotation). Examples of ball-and-socket joints are the shoulder and hip joints. At the shoulder joint the head of the humerus fits into the glenoid cavity of the scapula. At the hip joint the head of the femur fits into the acetabulum of the hip bone.

Table 9.1 on page 258 summarizes the categories of joints and their functions (movements).

◉ **CHECKPOINT**

4. Name the six types of synovial joints and describe the structure of each.
5. What are the functions of articular cartilage, synovial fluid, and articular discs?
6. What types of sensations are perceived at joints, and from what sources do joint components receive their nourishment?
7. In what ways are bursae similar to joint capsules? How do they differ?
8. Which types of joints are nonaxial, monaxial, biaxial, and multiaxial?

TYPES OF MOVEMENTS AT SYNOVIAL JOINTS

◉ **OBJECTIVE**

• Describe the types of movements that can occur at synovial joints.

Anatomists, physical therapists, and kinesiologists (professionals who treat injury or disease by movements of various kinds) use specific terminology to designate movements that can occur at synovial joints. These precise terms may indicate the form of motion, the direction of movement, or the relationship of one body part to another during movement. Movements at synovial joints are grouped into four main categories: (1) gliding (translational movements), (2) angular movements, (3) rotation, and (4) special movements. This last category includes movements that occur only at certain joints.

Gliding

Gliding is a simple movement in which relatively flat bone surfaces move back-and-forth and from side-to-side with respect to one another (Figure 9.5). There is no significant alteration of the angle between the bones. Gliding movements are limited in range due to the structure of the articular capsule and associated ligaments and bones. The intercarpal and intertarsal joints are examples of articulations where gliding movements occur.

Figure 9.5 Gliding movements at synovial joints.

 Gliding motion consists of side-to-side and back-and-forth movements.

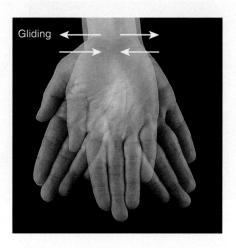

Intercarpal joints

❓ **Which types of synovial joints exhibit gliding movements?**

Angular Movements

In **angular movements**, there is an increase or a decrease in the angle between articulating bones. The major angular movements are flexion, extension, lateral flexion, hyperextension, abduction, adduction, and circumduction. These movements are discussed with respect to the body in the anatomical position (see Figure 1.2 on page 9).

Flexion, Extension, Lateral Flexion, and Hyperextension

Flexion and extension are opposite movements. In **flexion** (FLEK-shun; *flex-*=to bend) there is a decrease in the angle between articulating bones; in **extension** (eks-TEN-shun; *exten-*=to stretch out) there is an increase in the angle between articulating bones, often to restore a part of the body to the anatomical position after it has been flexed (Figure 9.6). Both movements usually occur along the sagittal plane. All of the following are examples of flexion (as you have probably already guessed, extension is simply the reverse of these movements):

• Bending the head toward the chest at the atlanto-occipital joint between the atlas (the first vertebra) and the occipital bone of the skull, and at the cervical intervertebral joints between the cervical vertebrae (Figure 9.6a)
• Bending the trunk forward at the intervertebral joints, as in doing a crunch with your abdominal muscles
• Moving the humerus forward at the shoulder joint, as in swinging the arms forward while walking (Figure 9.6b)
• Moving the forearm toward the arm at the elbow joint, that is, bending the elbow (Figure 9.6c)
• Moving the palm toward the forearm at the wrist or radio-carpal joint, as in shooting a basketball (Figure 9.6d)
• Bending of the digits of the hand or feet at the interphalangeal joints, as in clenching your fingers to make a fist
• Moving the thigh forward at the hip joint, as in climbing stairs (Figure 9.6e)
• Moving the leg toward the thigh at the knee or tibiofemoral joint, as occurs when bending the knee (Figure 9.6f)

Figure 9.6 Angular movements at synovial joints—flexion, extension, hyperextension, and lateral flexion.

In angular movements, there is an increase or decrease in the angle between articulating bones.

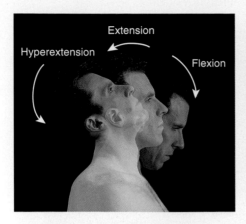

(a) Atlanto-occipital and cervical intervertebral joints

(b) Shoulder joint

(c) Elbow joint

(d) Wrist joint

(e) Hip joint

(f) Knee joint

(g) Intervertebral joints

? **What two examples of flexion do not occur along the sagittal plane?**

Although flexion and extension usually occur along the sagittal plane, there are a few exceptions. For example, flexion of the thumb involves movement of the thumb medially across the palm at the carpometacarpal, metacarpophalangeal, and interphalangeal joints of the thumb, as in touching the thumb to the opposite side of the palm (see Figure 11.20g on page 378). Another example is movement of the trunk sideways to the right or left at the waist as in a sidebend. This movement, which occurs along the frontal plane and involves the intervertebral joints, is called **lateral flexion** (Figure 9.6g).

Continuation of extension beyond the anatomical position is called **hyperextension** (*hyper-*=beyond or excessive). Examples of hyperextension include:

- Bending the head backward at the atlanto-occipital and cervical intervertebral joints (Figure 9.6a)
- Bending the trunk backward at the intervertebral joints as in a backbend
- Moving the humerus backward at the shoulder joint, as in swinging the arms backward while walking (Figure 9.6b)
- Moving the palm backward at the wrist joint (Figure 9.6d)
- Moving the femur backward at the hip joint, as in walking (Figure 9.6e)

Hyperextension of hinge joints, such as the elbow, interphalangeal, and knee joints, is usually prevented by the arrangement of ligaments and the anatomical alignment of the bones.

Abduction, Adduction, and Circumduction

Abduction (ab-DUK-shun; *ab-*=away; *-duct*=to lead) is the movement of a bone away from the midline; **adduction** (ad-DUK-shun; *ad-*=toward) is the movement of a bone toward the midline. Both movements usually occur along the frontal plane. Examples of abduction include moving the humerus laterally at the shoulder joint (Figure 9.7a), moving the palm laterally at the wrist joint (Figure 9.7b), and moving the femur laterally at the hip

Figure 9.7 Angular movements at synovial joints—abduction and adduction.

Abduction and adduction usually occur along the frontal plane.

(a) Shoulder joint

(b) Wrist joint

(c) Hip joint

(d) Metacarpophalangeal joints of the fingers (not the thumb)

? In what way is considering adduction as "adding your limb to your trunk" an effective learning device?

joint (Figure 9.7c). The movement that returns each of these body parts to the anatomical position is adduction (Figure 9.7a–c).

The midline of the body is not used as a point of reference for abduction and adduction of the digits. In abduction of the fingers

(but not the thumb), an imaginary line is drawn through the longitudinal axis of the middle (longest) finger, and the fingers move away (spread out) from the middle finger (Figure 9.7d). In abduction of the thumb, the thumb moves away from the palm in the sagittal plane (see Figure 11.20g on page 378). Abduction of the toes is relative to an imaginary line drawn through the second toe. Adduction of the fingers and toes involves returning them to the anatomical position. Adduction of the thumb moves the thumb toward the palm in the sagittal plane (see Figure 11.20g on page 378).

Circumduction (ser-kum-DUK-shun; *circ-*=circle) is movement of the distal end of a body part in a circle (Figure 9.8). Circumduction is not an isolated movement by itself but rather a continuous sequence of flexion, abduction, extension, and adduction. Examples of circumduction are moving the humerus in a circle at the shoulder joint (Figure 9.8a), moving the hand in a circle at the wrist joint, moving the thumb in a circle at the carpometacarpal joint, moving the fingers in a circle at the metacarpophalangeal joints (between the metacarpals and phalanges), and moving the femur in a circle at the hip joint (Figure 9.8b). Although both the shoulder and hip joints permit circumduction, flexion, abduction, extension, and adduction are more limited in the hip joints than in the shoulder joints due to the tension on certain ligaments and muscles and the greater depth of the acetabulum in the hip joint (see Exhibits 9.2 and 9.4).

Figure 9.8 Angular movements at synovial joints—circumduction.

Circumduction is the movement of the distal end of a body part in a circle.

(a) Shoulder joint

(b) Hip joint

 ? Which movements in continuous sequence produce circumduction?

Rotation

In **rotation** (rō-TĀ-shun; *rota-*=revolve), a bone revolves around its own longitudinal axis. One example is turning the head from side-to-side at the atlanto-axial joint (between the atlas and the axis), as when you shake your head "no" (Figure 9.9a). Another is turning the trunk from side-to-side at the intervertebral joints while keeping the hips and free lower limbs in the anatomical position. In the free limbs, rotation is defined relative to the midline, and specific qualifying terms are used. If the anterior surface of a bone of the free limb is turned toward the midline, the movement is called *medial (internal) rotation.* You can medially rotate the humerus at the shoulder joint as follows: Starting in the anatomical position, flex your elbow and then move your palm across the chest (Figure 9.9b). In order to laterally rotate the humerus, move in the opposite direction (Figure 9.9b). You can medially rotate the femur at the hip joint as follows: Lie on your back, bend your knee, and then move your leg and foot laterally from the midline. Although you are moving your leg and foot laterally, the femur is rotating medially (Figure 9.9c). Medial rotation of the leg at the knee joint can be produced by sitting on a chair, bending your knee, raising your free lower limb off the floor, and turning your toes medially. If the anterior surface of the bone of a free limb is turned away from the midline, the movement is called *lateral (external) rotation* (see Figure 9.9b, c).

Special Movements

Special movements occur only at certain joints. They include elevation, depression, protraction, retraction, inversion, eversion, dorsiflexion, plantar flexion, supination, pronation, and opposition (Figure 9.10):

- **Elevation** (el-e-VĀ-shun=to lift up) is an upward movement of a part of the body, such as closing the mouth at the temporomandibular joint (between the mandible and temporal bone) to elevate the mandible (Figure 9.10a) or shrugging the shoulders at the acromioclavicular joint to elevate the scapula. Its opposing movement is depression. Other bones that may be elevated (or depressed) include the hyoid, clavicle, and ribs.
- **Depression** (dē-PRESH-un=to press down) is a downward movement of a part of the body, such as opening the mouth to depress the mandible (Figure 9.10b) or returning shrugged shoulders to the anatomical position to depress the scapula.
- **Protraction** (prō-TRAK-shun=to draw forth) is a movement of a part of the body anteriorly in the transverse plane. Its opposing movement is retraction. You can protract your mandible at the temporomandibular joint by thrusting it outward (Figure 9.10c) or protract your clavicles at the acromioclavicular and sternoclavicular joints by crossing your arms.

Figure 9.9 Rotation at synovial joints.

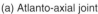 In rotation, a bone revolves around its own longitudinal axis.

(a) Atlanto-axial joint

(b) Shoulder joint

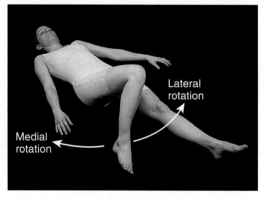

(c) Hip joint

? How do medial and lateral rotation differ?

- **Retraction** (rē-TRAK-shun=to draw back) is a movement of a protracted part of the body back to the anatomical position (Figure 9.10d).
- **Inversion** (in-VER-zhun=to turn inward) is movement of the sole medially at the intertarsal joints (between the tarsals) (Figure 9.10e). Its opposing movement is eversion.
- **Eversion** (ē-VER-zhun=to turn outward) is a movement of the sole laterally at the intertarsal joints (Figure 9.10f).
- **Dorsiflexion** (dor′-si-FLEK-shun) refers to bending of the foot at the ankle or talocrural joint (between the tibia, fibula, and talus) in the direction of the dorsum (superior surface) (Figure 9.10g). Dorsiflexion occurs when you stand on your heels. Its opposing movement is plantar flexion.
- **Plantar flexion** involves bending of the foot at the ankle joint in the direction of the plantar or inferior surface (Figure 9.10g), as when you elevate your body by standing on your toes.
- **Supination** (soo-pi-NĀ-shun) is a movement of the forearm at the proximal and distal radioulnar joints that turns the palm anteriorly (Figure 9.10h). This position of the palms is one of the defining features of the anatomical position. Its opposing movement is pronation.

- **Pronation** (prō-NĀ-shun) is a movement of the forearm at the proximal and distal radioulnar joints in which the distal end of the radius crosses over the distal end of the ulna and the palm is turned posteriorly (Figure 9.10h).
- **Opposition** (op-ō-ZISH-un) is the movement of the thumb at the carpometacarpal joint (between the trapezium and metacarpal of the thumb) in which the thumb moves across the palm to touch the tips of the fingers on the same hand (Figure 9.10i). This is the distinctive digital movement that gives humans and other primates the ability to grasp and manipulate objects very precisely.

A summary of the movements that occur at synovial joints is presented in Table 9.2 (see page 264).

⊙ **CHECKPOINT**

9. What are the four major categories of movements that occur at synovial joints?
10. Alone or with a partner, demonstrate each movement listed in Table 9.2.

Figure 9.10 Special movements at synovial joints.

🔑 Special movements occur only at certain synovial joints.

(a) Temporomandibular joint (b) Elevation / Depression

(c) Temporomandibular joint (d) Protraction / Retraction

(e) Intertarsal joints (f) Inversion / Eversion

(g) Ankle joint Dorsiflexion / Plantar flexion

Palm posterior / Palm anterior

Pronation / Supination

(h) Radioulnar joint

Opposition

(i) Opposition

❓ **What movement of the shoulder girdle is involved in bringing the arms forward until the elbows touch?**

TABLE 9.2

Summary of Movements at Synovial Joints

MOVEMENT	DESCRIPTION	MOVEMENT	DESCRIPTION
Gliding	Movement of relatively flat bone surfaces back-and-forth and side-to-side over one another; little change in the angle between bones.	Rotation	Movement of a bone around its longitudinal axis; in the limbs, it may be medial (toward midline) or lateral (away from midline).
Angular	Increase or decrease in the angle between bones.	Special	Occurs at specific joints.
Flexion	Decrease in the angle between articulating bones, usually in the sagittal plane.	Elevation	Superior movement of a body part.
Lateral flexion	Movement of the trunk in the frontal plane.	Depression	Inferior movement of a body part.
		Protraction	Anterior movement of a body part in the transverse plane.
Extension	Increase in the angle between articulating bones, usually in the sagittal plane.	Retraction	Posterior movement of a body part in the transverse plane.
Hyperextension	Extension beyond the anatomical position.	Inversion	Medial movement of the sole.
Abduction	Movement of a bone away from the midline, usually in the frontal plane.	Eversion	Lateral movement of the sole.
		Dorsiflexion	Bending the foot in the direction of the dorsum (superior surface).
Adduction	Movement of a bone toward the midline, usually in the frontal plane.	Plantar flexion	Bending the foot in the direction of the plantar surface (sole).
Circumduction	Flexion, abduction, extension, and adduction in succession, in which the distal end of a body part moves in a circle.	Supination	Movement of the forearm that turns the palm anteriorly.
		Pronation	Movement of the forearm that turns the palm posteriorly.
		Opposition	Movement of the thumb across the palm to touch fingertips on the same hand.

FACTORS AFFECTING CONTACT AND RANGE OF MOTION AT SYNOVIAL JOINTS

◉ OBJECTIVE

- Describe six factors that influence the type of movement and range of motion possible at a synovial joint.

The articular surfaces of synovial joints contact one another and determine the type and possible range of motion. **Range of motion (ROM)** refers to the range, measured in degrees of a circle, through which the bones of a joint can be moved. The following factors contribute to keeping the articular surfaces in contact and affect range of motion:

1. **Structure or shape of the articulating bones.** The structure or shape of the articulating bones determines how closely they can fit together. The articular surfaces of some bones have a complementary relationship. This spatial relationship is very obvious at the hip joint, where the head of the femur articulates with the acetabulum of the hip bone. An interlocking fit allows rotational movement.
2. **Strength and tension (tautness) of the joint ligaments.** The different components of a fibrous capsule are tense or taut only when the joint is in certain positions. Tense ligaments not only restrict the range of motion but also direct the movement of the articulating bones with respect to each other. In the knee joint, for example, the anterior cruciate ligament is taut and the posterior cruciate ligament is loose when the knee is straightened, and the reverse occurs when the knee is bent.
3. **Arrangement and tension of the muscles.** Muscle tension reinforces the restraint placed on a joint by its ligaments, and thus restricts movement. A good example of the effect of muscle tension on a joint is seen at the hip joint. When the thigh is fixed with the knee extended, the movement is restricted by the tension of the hamstring muscles on the posterior surface of the thigh. But if the knee is flexed, the tension on the hamstring muscles is lessened, and the thigh can be raised farther.
4. **Contact of soft parts.** The point at which one body surface contacts another may limit mobility. For example, if you bend your arm at the elbow, it can move no further after the anterior surface of the forearm meets with and presses against the biceps brachii muscle of the arm. Joint movement may also be restricted by the presence of adipose tissue.
5. **Hormones.** Joint flexibility may also be affected by hormones. For example, relaxin, a hormone produced by the placenta and ovaries, increases the flexibility of the fibrocartilage of the pubic symphysis and loosens the ligaments between the sacrum and hip bone toward the end of pregnancy. These changes permit expansion of the pelvic outlet, which assists in delivery of the baby.
6. **Disuse.** Movement at a joint may be restricted if a joint has not been used for an extended period. For example, if an elbow joint is immobilized by a cast, range of motion at the joint may be limited for a time after the cast is removed. Disuse may also result in decreased amounts of synovial fluid, diminished flexibility of ligaments and tendons, and *muscular atrophy*, a reduction in size or wasting of a muscle.

◉ CHECKPOINT

11. How do the strength and tension of ligaments determine range of motion?

● CLINICAL **CONNECTION** | Rheumatism and Arthritis

Rheumatism (ROO-ma-tizm) is any painful disorder of the supporting structures of the body—bones, ligaments, tendons, or muscles—that is not caused by infection or injury. Arthritis is a form of rheumatism in which the joints are swollen, stiff, and painful. It afflicts about 45 million people in the United States, and is the leading cause of physical disability among adults over age 65.

Osteoarthritis

Osteoarthritis (OA) (os´-tē-ō-ar-THRĪ-tis) is a degenerative joint disease in which joint cartilage is gradually lost. It results from a combination of aging, obesity, irritation of the joints, muscle weakness, and wear and abrasion. Commonly known as "wear-and-tear" arthritis, osteoarthritis is the most common type of arthritis.

Osteoarthritis is a progressive disorder of synovial joints, particularly weight-bearing joints. Articular cartilage deteriorates and new bone forms in the subchondral areas and at the margins of the joint. The cartilage slowly degenerates, and as the bone ends become exposed, spurs (small bumps) of new osseous tissue are deposited on them in a misguided effort by the body to protect against the friction. These spurs decrease the space of the joint cavity and restrict joint movement. Unlike rheumatoid arthritis (described next), osteoarthritis affects mainly the articular cartilage, although the synovial membrane often becomes inflamed late in the disease. There are two major distinctions between osteoarthritis and rheumatoid arthritis: Osteoarthritis first afflicts the larger joints (knees, hips) and is due to wear and tear, and rheumatoid arthritis first strikes smaller joints and is an active attack of the cartilage. Osteoarthritis is the most common reason for hip- and knee-replacement surgery.

Rheumatoid Arthritis

Rheumatoid arthritis (RA) is an autoimmune disease in which the immune system of the body attacks its own tissues—in this case, its own cartilage and joint linings. RA is characterized by inflammation of the joint, which causes swelling, pain, and loss of function (Figures A and B). Usually, this form of arthritis occurs bilaterally: If one wrist is affected, the other is also likely to be affected, although often not to the same degree.

The primary symptom of RA is inflammation of the synovial membrane. If untreated, the membrane thickens, and synovial fluid accumulates. The resulting pressure causes pain and tenderness. The membrane then produces an abnormal granulation tissue, called *pannus,* that adheres to the surface of the articular cartilage and sometimes erodes the cartilage completely. When the cartilage is destroyed, fibrous tissue joins the exposed bone ends. The fibrous tissue ossifies and fuses the joint so that it becomes immovable—the ultimate crippling effect of rheumatoid arthritis. The growth of the granulation tissue causes the distortion of the fingers that characterizes hands of RA sufferers.

Gouty Arthritis

Uric acid (a substance that gives urine its name) is a waste product produced during the metabolism of nucleic acid (DNA and RNA) subunits. A person who suffers from **gout** (GOWT) either produces excessive amounts of uric acid or is not able to excrete as much as normal. The result is a buildup of uric acid in the blood. This excess acid then reacts with sodium to form a salt called sodium urate. Crystals of this salt accumulate in soft tissues such as the kidneys and in the cartilage of the ears and joints.

In **gouty arthritis**, sodium urate crystals are deposited in the soft tissues of the joints. Gout most often affects the joints of the feet, especially at the base of the big toe. The crystals irritate and erode the cartilage, causing inflammation, swelling, and acute pain. Eventually, the crystals destroy all joint tissues. If the disorder is untreated, the ends of the articulating bones fuse, and the joint becomes immovable. Treatment consists of pain relief (ibuprofen, naproxen, colchicine, and cortisone) followed by administration of allopurinol to keep uric acid levels low so that crystals do not form. ●

(A Gamma ray photograph of swollen joints (bright spots) due to RA

(B) Photograph of an individual with severe RA

SELECTED JOINTS OF THE BODY

In Chapters 7 and 8 we discussed the major bones and their markings. In this chapter we have examined how joints are classified according to their structure and function, and we have introduced the movements that occur at joints. Table 9.3 (selected joints of the axial skeleton) and Table 9.4 (selected joints of the appendicular skeleton) will help you integrate the information you have learned in all three chapters. These tables list some of the major joints of the body according to their articular components (the bones that enter into their formation), their structural and functional classification, and the type(s) of movement that occur(s) at each joint.

Next we examine in detail six selected joints of the body in a series of exhibits. Each exhibit considers a specific synovial joint and contains (1) a definition—a description of the type of joint and the bones that form the joint; (2) the anatomical components—a description of the major connecting ligaments, articular disc (if present), articular capsule, and other distinguishing features of the joint; and (3) the joint's possible movements. Each exhibit also refers you to a figure that illustrates the joint. The joints described are the temporomandibular joint (TMJ), shoulder (humeroscapular or glenohumeral) joint, elbow joint, hip (coxal) joint, knee (tibiofemoral) joint, and ankle (talocrural) joint. Because these joints are described in detail in Exhibits 9.1 through 9.6 (Figures 9.11 through 9.16), they are not included in Tables 9.3 and 9.4.

TABLE 9.3

Selected Joints of the Axial Skeleton

JOINT	ARTICULAR COMPONENTS	STRUCTURAL CLASSIFICATION	MOVEMENTS
Suture	Between skull bones.	Fibrous.	None to slight shock absorbing motion.
Atlanto-occipital	Between superior articular facets of atlas and occipital condyles of occipital bone.	Synovial (condyloid).	Flexion and extension of head and slight lateral flexion of head to either side.
Atlanto-axial	(1) Between dens of axis and anterior arch of atlas and (2) between lateral masses of atlas and axis.	Synovial (pivot) between dens and anterior arch, and synovial (planar) between lateral masses.	Rotation of head.
Intervertebral	(1) Between vertebral bodies and (2) between vertebral arches.	Cartilaginous (symphysis) between vertebral bodies, and synovial (planar) between vertebral arches.	Flexion, extension, lateral flexion, and rotation of vertebral column.
Vertebrocostal	(1) Between facets of heads of ribs and facets of bodies of adjacent thoracic vertebrae and intervertebral discs between them and (2) between articular part of tubercles of ribs and facets of transverse processes of thoracic vertebrae.	Synovial (planar).	Slight gliding.
Sternocostal	Between sternum and first seven pairs of ribs.	Cartilaginous (synchondrosis) between sternum and first pair of ribs, and synovial (planar) between sternum and second through seventh pairs of ribs.	None between sternum and first pair of ribs; slight gliding between sternum and second through seventh pairs of ribs.
Lumbosacral	(1) Between body of fifth lumbar vertebra and base of sacrum and (2) between inferior articular facets of fifth lumbar vertebra and superior articular facets of first vertebra of sacrum.	Cartilaginous (symphysis) between body and base, and synovial (planar) between articular facets.	Flexion, extension, lateral flexion, and rotation of vertebral column.

TABLE 9.4

Selected Joints of the Appendicular Skeleton

JOINT	ARTICULAR COMPONENTS	STRUCTURAL CLASSIFICATION	MOVEMENTS
Sternoclavicular	Between sternal end of clavicle, manubrium of sternum, and first costal cartilage.	Synovial (planar and pivot).	Gliding, with limited movements in nearly every direction.
Acromioclavicular	Between acromion of scapula and acromial end of clavicle.	Synovial (planar).	Gliding and rotation of scapula on clavicle.
Radioulnar	Proximal radioulnar joint between head of radius and radial notch of ulna; distal radioulnar joint between ulnar notch of radius and head of ulna.	Synovial (pivot).	Rotation of forearm.
Wrist (radiocarpal)	Between distal end of radius and scaphoid, lunate, and triquetrum of carpus.	Synovial (condyloid).	Flexion, extension, abduction, adduction, circumduction, and slight hyperextension of wrist.
Intercarpal	Between proximal row of carpal bones, distal row of carpal bones, and between both rows of carpal bones (midcarpal joints).	Synovial (planar), except for hamate, scaphoid, and lunate (midcarpal) joint, which is synovial (saddle).	Gliding plus flexion, extension, abduction, adduction, and slight rotation at midcarpal joints.
Carpometacarpal	Carpometacarpal joint of thumb between trapezium of carpus and first metacarpal; carpometacarpal joints of remaining digits formed between carpus and second through fifth metacarpals.	Synovial (saddle) at thumb and synovial (planar) at remaining digits.	Flexion, extension, abduction, adduction, and circumduction at thumb, and gliding at remaining digits.
Metacarpophalangeal and metatarsophalangeal	Between heads of metacarpals (or metatarsals) and bases of proximal phalanges.	Synovial (condyloid).	Flexion, extension, abduction, adduction, and circumduction of phalanges.
Interphalangeal	Between heads of phalanges and bases of more distal phalanges.	Synovial (hinge).	Flexion and extension of phalanges.
Sacroiliac	Between auricular surfaces of sacrum and ilia of hip bones.	Synovial (planar).	Slight gliding (even more so during pregnancy).
Pubic symphysis	Between anterior surfaces of hip bones.	Cartilaginous (symphysis).	Slight movements (even more so during pregnancy).
Tibiofibular	Proximal tibiofibular joint between lateral condyle of tibia and head of fibula; distal tibiofibular joint between distal end of fibula and fibular notch of tibia.	Synovial (planar) at proximal joint, and fibrous (syndesmosis) at distal joint.	Slight gliding at proximal joint, and slight rotation of fibula during dorsiflexion of foot.
Intertarsal	Subtalar joint between talus and calcaneus of tarsus; talocalcaneonavicular joint between talus and calcaneus and navicular of tarsus; calcaneocuboid joint between calcaneus and cuboid of tarsus.	Synovial (planar) at subtalar and calcaneocuboid joints, and synovial (saddle) at talocalcaneonavicular joint.	Inversion and eversion of foot.
Tarsometatarsal	Between three cuneiforms of tarsus and bases of five metatarsal bones.	Synovial (planar).	Slight gliding.

EXHIBIT 9.1 Temporomandibular Joint

◉ OBJECTIVE

- Describe the anatomical components of the temporo-mandibular joint and explain the movements that can occur at this joint.

Description

The **temporomandibular joint (TMJ)** is a combined hinge and planar joint formed by the condylar process of the mandible and the mandibular fossa and articular tubercle of the temporal bone. The temporomandibular joint is the only movable joint between skull bones (with the exception of the ear ossicles); all other skull joints are sutures and therefore immovable.

Anatomical Components

1. **Articular disc (meniscus).** Fibrocartilage disc that separates the joint cavity into superior and inferior compartments, each with a synovial membrane (Figure 9.11c).
2. **Articular capsule.** Thin, fairly loose envelope around the circumference of the joint (Figure 9.11a, b).
3. **Lateral ligament.** Two short bands on the lateral surface of the articular capsule that extend inferiorly and posteriorly from the inferior border and tubercle of the zygomatic process of the temporal bone to the lateral and posterior aspect of the neck of the mandible. The lateral ligament is covered by the parotid gland and helps strengthen the joint laterally and prevent displacement of the mandible (Figure 9.11a).
4. **Sphenomandibular ligament.** Thin band that extends inferiorly and anteriorly from the spine of the sphenoid bone to the ramus of the mandible (Figure 9.11b). It does not contribute significantly to the strength of the joint.

Figure 9.11 Right temporomandibular joint (TMJ).

 The TMJ is the only movable joint between skull bones.

(a) Right lateral view

(b) Left medial view

(c) Sagittal section viewed from right

5. Stylomandibular ligament. Thickened band of deep cervical fascia that extends from the styloid process of the temporal bone to the inferior and posterior border of the ramus of the mandible. This ligament separates the parotid gland from the submandibular gland and limits movement of the mandible at the TMJ (Figure 9.11a, b).

Movements

In the temporomandibular joint, only the mandible moves because the temporal bone is firmly anchored to other bones of the skull by sutures. Accordingly, the mandible may function in depression (jaw opening) and elevation (jaw closing), which occurs in the inferior compartment, and protraction, retraction, lateral displacement, and slight rotation, which occur in the superior compartment (see Figure 9.10a–d).

◉ **CHECKPOINT**

What distinguishes the temporomandibular joint from the other joints of the skull?

Brain

Mandibular fossa of temporal bone

External auditory meatus

Superior compartment

ARTICULAR DISC

Inferior compartment

Articular tubercle of temporal bone

Condylar process of mandible

SYNOVIAL CAVITY

(d) Sagittal section

? **Which ligament prevents displacement of the mandible?**

EXHIBIT 9.1 **269**

EXHIBIT 9.2 Shoulder Joint

● **OBJECTIVE**

• Describe the anatomical components of the shoulder joint and the movements that can occur at this joint.

Description

The **shoulder joint** is a ball-and-socket joint formed by the head of the humerus and the glenoid cavity of the scapula. It also is referred to as the *humeroscapular* or *glenohumeral joint*.

Anatomical Components

1. **Articular capsule.** Thin, loose sac that completely envelops the joint and extends from the glenoid cavity to the anatomical neck of the humerus. The inferior part of the capsule is its weakest area (Figure 9.12).
2. **Coracohumeral ligament.** Strong, broad ligament that strengthens the superior part of the articular capsule and extends from the coracoid process of the scapula to the greater tubercle of the humerus (Figure 9.12a, b). The ligament

strengthens the superior part of the articular capsule, and reinforces the anterior aspect of the articular capsule.
3. **Glenohumeral ligaments.** Three thickenings of the articular capsule over the anterior surface of the joint that extend from the glenoid cavity to the lesser tubercle and anatomical neck of the humerus. These ligaments are often indistinct or absent and provide only minimal strength (Figure 9.12a, b). The ligaments play a role in joint stabilization when the humerus approaches or exceeds it limits of motion.
4. **Transverse humeral ligament.** Narrow sheet extending from the greater tubercle to the lesser tubercle of the humerus (Figure 9.12a). The ligament functions as a retinaculum to hold the long head of the biceps brachii muscle.
5. **Glenoid labrum.** Narrow rim of fibrocartilage around the edge of the glenoid cavity that slightly deepens and enlarges the glenoid cavity (Figure 9.12b, c, d).
6. **Bursae.** Four bursae (see page 256) are associated with the shoulder joint. They are the *subscapular bursa* (Figure 9.12a), *subdeltoid bursa*, *subacromial bursa* (Figure 9.12a–c), and *subcoracoid bursa*.

Figure 9.12 Right shoulder (humeroscapular or glenohumeral) joint.

⑥ ▭ Most of the stability of the shoulder joints results from the arrangement of the rotator cuff muscles.

Clavicle

Acromioclavicular ligament

Coracoacromial ligament

Acromion of scapula

SUBACROMIAL BURSA

CORACOHUMERAL LIGAMENT

GLENOHUMERAL LIGAMENTS

TRANSVERSE HUMERAL LIGAMENT

Tendon of subscapularis muscle

Humerus

Coracoclavicular ligament:
Conoid ligament
Trapezoid ligament

Superior transverse scapular ligament

Coracoid process of scapula

SUBSCAPULAR BURSA

ARTICULAR CAPSULE

Scapula

Tendon of biceps brachii muscle (long head)

(a) Anterior view

Movements

The shoulder joint allows flexion, extension, hyperextension abduction, adduction, medial rotation, lateral rotation, and circumduction of the arm (see Figures 9.5–9.8). It has more freedom of movement than any other joint of the body. This freedom results from the looseness of the articular capsule and shallowness of the glenoid cavity in relation to the large size of the head of the humerus.

Although the ligaments of the shoulder joint strengthen it to some extent, most of the strength results from the muscles that surround the joint, especially the *rotator cuff muscles*. These muscles (supraspinatus, infraspinatus, teres minor, and subscapularis) join the scapula to the humerus (see also Figure 11.17 on pages 358–362). The tendons of the rotator cuff muscles encircle the joint (except for the inferior portion) and fuse with the articular capsule. The rotator cuff muscles work as a group to hold the head of the humerus in the glenoid cavity.

◉ **CHECKPOINT**
What is the function of the rotator cuff?

● CLINICAL **CONNECTION** | **Rotator Cuff Injury and Dislocated and Separated Shoulder**

Rotator cuff injury is a strain or tear in the rotator cuff muscles and is a common injury among baseball pitchers, volleyball players, racket sports players, swimmers and violinists, due to shoulder movements that involve vigorous circumduction. It also occurs as a result of wear and tear, aging, trauma, poor posture, improper lifting, and repetitive motions in certain jobs, such as placing items on a shelf above your head. Most often, there is tearing of the supraspinatus muscle tendon of the rotator cuff. This tendon is especially predisposed to wear-and-tear because of its location between the head of the humerus and acromion of the scapula, which compresses the tendon during shoulder movements.

The joint most commonly dislocated in adults is the shoulder joint because its socket is quite shallow and the bones are held together by supporting muscles. Usually in a **dislocated shoulder,** the head of the humerus becomes displaced inferiorly, where the articular capsule is least protected. Dislocations of the mandible, elbow, fingers, knee, or hip are less common.

A **separated shoulder** refers to an injury of the acromioclavicular joint, a joint formed by the acromion of the scapula and the acromial end of the clavicle. This condition is usually the result of forceful trauma to the joint, as when the shoulder strikes the ground in a fall. •

SUPERIOR

Acromion of scapula

SUBACROMIAL BURSA

Tendon of biceps brachii muscle (long head)

Tendon of infraspinatus muscle

GLENOID CAVITY (covered with articular cartilage)

ARTICULAR CAPSULE

Tendon of teres minor muscle

POSTERIOR

Coracoacromial ligament

Tendon of supraspinatus muscle

CORACOHUMERAL LIGAMENT

CORACOID PROCESS OF SCAPULA

Tendon of subscapularis muscle

GLENOHUMERAL LIGAMENTS

GLENOID LABRUM

ANTERIOR

View

(b) Lateral view (opened) with synovial membrane removed

◘ **EXHIBIT 9.2 CONTINUES** ▶

EXHIBIT 9.2 **271**

EXHIBIT 9.2 Shoulder Joint CONTINUED

● **FIGURE 9.12 CONTINUED** ▶

Frontal plane

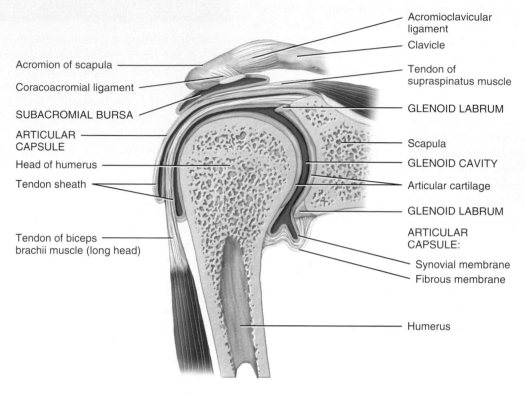

Acromion of scapula

Coracoacromial ligament

SUBACROMIAL BURSA

ARTICULAR CAPSULE

Head of humerus

Tendon sheath

Tendon of biceps brachii muscle (long head)

Acromioclavicular ligament

Clavicle

Tendon of supraspinatus muscle

GLENOID LABRUM

Scapula

GLENOID CAVITY

Articular cartilage

GLENOID LABRUM

ARTICULAR CAPSULE:

Synovial membrane

Fibrous membrane

Humerus

(c) Frontal section

LATERAL

MEDIAL

Acromion of scapula

GLENOID LABRUM

ARTICULAR CARTILAGE

Glenoid cavity

Head of humerus

ARTICULAR CAPSULE

Supraspinatus muscle

Scapula

Subscapularis muscle

Teres major muscle

Latissimus dorsi muscle

(d) Frontal section

? **Why does the shoulder joint have more freedom of movement than any other joint of the body?**

EXHIBIT 9.3 Elbow Joint

OBJECTIVE

- Describe the anatomical components of the elbow joint and the movements that can occur at this joint.

Description

The **elbow joint** is a hinge joint formed by the trochlea and capitulum of the humerus, the trochlear notch of the ulna, and the head of the radius.

Anatomical Components

1. **Articular capsule.** The anterior part of the articular capsule covers the anterior part of the elbow joint, from the radial and coronoid fossae of the humerus to the coronoid process of the ulna and the annular ligament of the radius. The posterior part extends from the capitulum, olecranon fossa, and lateral epicondyle of the humerus to the annular ligament of the radius, the olecranon of the ulna, and the ulna posterior to the radial notch (Figure 9.13a, b).

2. **Ulnar collateral ligament.** Thick, triangular ligament that extends from the medial epicondyle of the humerus to the coronoid process and olecranon of the ulna (Figure 9.13a). Part of this ligament deepens the socket for the trochlea of the humerus.

3. **Radial collateral ligament.** Strong, triangular ligament that extends from the lateral epicondyle of the humerus to the annular ligament of the radius and the radial notch of the ulna (Figure 9.13b). The annular ligament of the radius holds the head of the radius in the radial notch of the ulna.

Movements

The elbow joint allows flexion and extension of the forearm (see Figure 9.6c).

CHECKPOINT

At the elbow joint, which ligaments connect (a) the humerus and the ulna, and (b) the humerus and the radius?

Figure 9.13 Right elbow joint.

The elbow joint is formed by parts of three bones: humerus, ulna, and radius.

(a) Medial aspect

(b) Lateral aspect

? **Which movements are possible at a hinge joint?**

EXHIBIT 9.3 **273**

EXHIBIT 9.4 | Hip Joint

OBJECTIVE

- Describe the anatomical components of the hip joint and the movements that can occur at this joint.

Description

The **hip joint** *(coxal joint)* is a ball-and-socket joint formed by the head of the femur and the acetabulum of the hip bone.

Anatomical Components

1. **Articular capsule.** Very dense and strong capsule that extends from the rim of the acetabulum to the neck of the femur (Figure 9.14c, d). One of the strongest structures of the body, the capsule consists of circular and longitudinal fibers. The circular fibers, called the *zona orbicularis*, form a collar around the neck of the femur. Accessory ligaments known as the iliofemoral ligament, pubofemoral ligament, and ischiofemoral ligament reinforce the longitudinal fibers of the articular capsule.

2. **Iliofemoral ligament.** Thickened portion of the articular capsule that extends from the anterior inferior iliac spine of the hip bone to the intertrochanteric line of the femur (Figure 9.14a, b). The ligament is said to be the body's strongest ligament and prevents hyperextension of the femur at the hip joint.

3. **Pubofemoral ligament.** Thickened portion of the articular capsule that extends from the pubic part of the rim of the acetabulum to the neck of the femur (Figure 9.14a). The ligament prevents overabduction of the femur at the hip joint and strengthens the articular capsule.

4. **Ischiofemoral ligament.** Thickened portion of the articular capsule that extends from the ischial wall of the acetabulum to the neck of the femur (Figure 9.14b). The ligament slackens during adduction and tenses during abduction, and strengthens the articular capsule.

5. **Ligament of the head of the femur.** Flat, triangular band (primarily a synovial fold) that extends from the fossa of the acetabulum to the fovea capitis of the head of the femur (Figure 9.14c, d). The ligament usually contains a small artery to the head of the femur.

Figure 9.14 Right hip (coxal) joint.

The articular capsule of the hip joint is one of the strongest structures in the body.

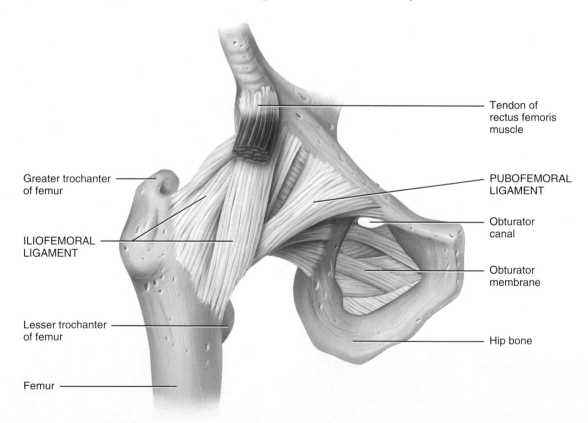

(a) Anterior view

6. **Acetabular labrum.** Fibrocartilage rim attached to the margin of the acetabulum that enhances the depth of the acetabulum. Because the diameter of the acetabular rim is smaller than that of the head of the femur, dislocation of the femur is rare (Figure 9.14c, d).

7. **Transverse ligament of the acetabulum.** Strong ligament that crosses over the acetabular notch. It supports part of the acetabular labrum and is connected with the ligament of the head of the femur and the articular capsule (Figure 9.14c).

Movements

The hip joint allows flexion, extension, abduction, adduction, circumduction, medial rotation, and lateral rotation of thigh (see Figures 9.6–9.8). The extreme stability of the hip joint is related to the very strong articular capsule and its accessory ligaments, the manner in which the femur fits into the acetabulum, and the muscles surrounding the joint. Although the shoulder and hip joints are both ball-and-socket joints, the movements at the hip joints do not have as wide a range of motion. Flexion is limited by the anterior surface of the thigh coming into contact with the anterior abdominal wall when the knee is flexed and by tension of the hamstring muscles when the knee is extended. Extension is limited by tension of the iliofemoral, pubofemoral, and ischiofemoral ligaments. Abduction is limited by the tension of the pubofemoral ligament, and adduction is limited by contact with the opposite limb and tension in the ligament of the head of the femur. Medial rotation is limited by the tension in the ischiofemoral ligament, and lateral rotation is limited by tension in the iliofemoral and pubofemoral ligaments.

◉ **CHECKPOINT**
What factors limit the degree of flexion and abduction at the hip joint?

(b) Posterior view

▣ EXHIBIT **9.4** CONTINUES ▶

EXHIBIT 9.4 **275**

EXHIBIT 9.4 Hip Joint CONTINUED

● FIGURE 9.14 CONTINUED ▶

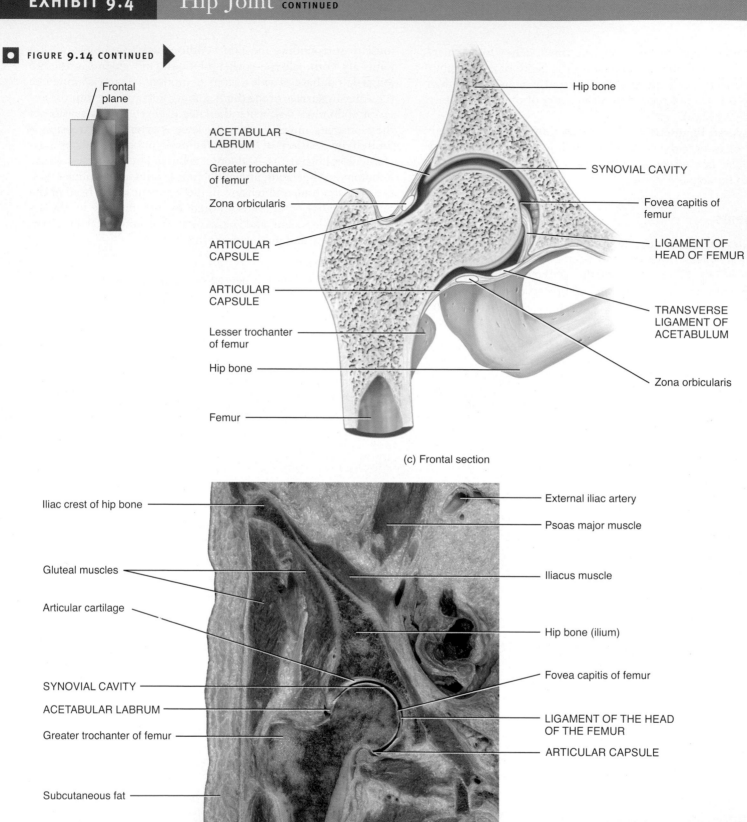

Frontal plane

ACETABULAR LABRUM

Greater trochanter of femur

Zona orbicularis

ARTICULAR CAPSULE

ARTICULAR CAPSULE

Lesser trochanter of femur

Hip bone

Femur

Hip bone

SYNOVIAL CAVITY

Fovea capitis of femur

LIGAMENT OF HEAD OF FEMUR

TRANSVERSE LIGAMENT OF ACETABULUM

Zona orbicularis

(c) Frontal section

Iliac crest of hip bone

Gluteal muscles

Articular cartilage

SYNOVIAL CAVITY

ACETABULAR LABRUM

Greater trochanter of femur

Subcutaneous fat

Vastus lateralis muscle

Femur

External iliac artery

Psoas major muscle

Iliacus muscle

Hip bone (ilium)

Fovea capitis of femur

LIGAMENT OF THE HEAD OF THE FEMUR

ARTICULAR CAPSULE

Adductor muscles

LATERAL

MEDIAL

(d) Frontal section

❓ **Which ligaments limit the degree of extension that is possible at the hip joint?**

EXHIBIT 9.5 | Knee Joint

OBJECTIVE

- Describe the main anatomical components of the knee joint and explain the movements that can occur at this joint.

Description

The **knee joint** (*tibiofemoral joint*) is the largest and most complex joint of the body. It is a modified hinge joint that actually consists of three joints within a single synovial cavity:

1. Laterally is a tibiofemoral joint, between the lateral condyle of the femur, lateral meniscus, and lateral condyle of the tibia, which is weight-bearing.
2. Medially is a second tibiofemoral joint, between the medial condyle of the femur, medial meniscus, and medial condyle of the tibia.
3. An intermediate patellofemoral joint is between the patella and the patellar surface of the femur.

Anatomical Components

1. **Articular capsule.** No complete, independent capsule unites the bones of the knee joint. The ligamentous sheath surrounding the joint consists mostly of muscle tendons or their expansions (Figure 9.15a, b, f, h). There are, however, some capsular fibers connecting the articulating bones.
2. **Medial and lateral patellar retinacula.** Fused tendons of insertion of the quadriceps femoris muscle and the fascia lata (deep fascia of thigh) that strengthen the anterior surface of the joint (Figure 9.15a).
3. **Patellar ligament.** Continuation of the common tendon of insertion of the quadriceps femoris muscle that extends from the patella to the tibial tuberosity. This ligament also strengthens the anterior surface of the joint. The posterior surface of the ligament is separated from the synovial membrane of the joint by an *infrapatellar fat pad* (Figure 9.15a, c, e, i).
4. **Oblique popliteal ligament.** Broad, flat ligament that extends from the intercondylar fossa of the femur to the head of the tibia and lateral condyle of the femur to the medial condyle of the tibia (Figure 9.15b). The ligament and tendon strengthen the posterior surface of the joint.
5. **Arcuate popliteal ligament.** Extends from the lateral condyle of the femur to the styloid process of the head of the fibula. It strengthens the lower lateral part of the posterior surface of the joint (Figure 9.15b).
6. **Tibial collateral ligament.** Broad, flat ligament on the medial surface of the joint that extends from the medial condyle of the femur to the medial condyle of the tibia (Figure 9.15a, b, d, f–h). Tendons of the sartorius, gracilis, and semitendinosus muscles, all of which strengthen the medial aspect of the joint, cross the ligament. Because the tibial collateral ligament is firmly attached to the medial meniscus, tearing of the ligament frequently results in tearing of the meniscus and damage to the anterior cruciate ligament, described under 8a.
7. **Fibular collateral ligament.** Strong, rounded ligament on the lateral surface of the joint that extends from the lateral condyle of the femur to the lateral side of the head of the fibula (Figure 9.15a, b, d, f–i). It strengthens the lateral aspect of the joint. The ligament is covered by the tendon of the biceps femoris muscle. The tendon of the popliteal muscle is deep to the ligament.
8. **Intracapsular ligaments.** Ligaments within the capsule that connect the tibia and femur. The anterior and posterior cruciate ligaments (KROO-shē-āt=like a cross) are named based on their origins relative to the intercondylar area of the tibia. From their origins, they cross on their way to their destinations on the femur.

   a. **Anterior cruciate ligament (ACL).** Extends posteriorly and laterally from a point anterior to the intercondylar area of the tibia to the posterior part of the medial surface of the lateral condyle of the femur (Figure 9.15d, g). The ACL limits hyperextension of the knee (which normally does not occur at this joint) and prevents the anterior sliding of the tibia on the femur. This ligament is stretched or torn in about 70% of all serious knee injuries.

   b. **Posterior cruciate ligament (PCL).** Extends anteriorly and medially from a depression on the *posterior* intercondylar area of the tibia and lateral meniscus to the anterior part of the lateral surface of the medial condyle of the femur (Figure 9.15d, g). The PCL prevents the posterior sliding of the tibia (and anterior sliding of the femur) when the knee is flexed. This is very important when walking down stairs or a steep incline.

9. **Articular discs (menisci).** Two fibrocartilage discs between the tibial and femoral condyles help compensate for the irregular shapes of the bones and circulate synovial fluid.

   a. **Medial meniscus.** Semicircular piece of fibrocartilage (C-shaped). Its anterior end is attached to the anterior intercondylar fossa of the tibia, anterior to the anterior cruciate ligament. Its posterior end is attached to the posterior intercondylar fossa of the tibia between the attachments of the posterior cruciate ligament and lateral meniscus (Figure 9.15d, e, g, h).

   b. **Lateral meniscus.** Nearly circular piece of fibrocartilage (approaches an incomplete O in shape) (Figure 9.15c, d, f, g, i). Its anterior end is attached anteriorly to the intercondylar eminence of the tibia, and laterally and posteriorly to the anterior cruciate ligament. Its posterior end is attached posteriorly to the intercondylar eminence of the tibia, and anteriorly to the posterior end of the medial meniscus. The anterior surfaces of the medial and lateral menisci are connected to each other by the *transverse ligament of the knee* (Figure 9.15d) and to the margins of the head of the tibia by the *coronary ligaments* (not illustrated).

10. The more important bursae of the knee include the following:

    a. *Prepatellar bursa* between the patella and skin (Figure 9.15c).

    b. *Infrapatellar bursa* between superior part of tibia and patellar ligament (Figure 9.15a, c, e).

    c. *Suprapatellar bursa* between inferior part of femur and deep surface of quadriceps femoris muscle (Figure 9.15a, c, e, h).

EXHIBIT 9.5 CONTINUES

EXHIBIT 9.5 **277**

EXHIBIT 9.5 Knee Joint CONTINUED

Movements

The knee joint allows flexion, extension, slight medial rotation, and lateral rotation of leg in the flexed position (see Figures 9.6f and 9.9c).

◉ **CHECKPOINT**

What are the opposing functions of the anterior and posterior cruciate ligaments?

Figure 9.15 Right knee (tibiofemoral) joint.

The knee joint is the largest and most complex joint in the body.

(a) Anterior superficial view

(b) Posterior deep view

(c) Sagittal section

(d) Anterior deep view

The knee joint is the joint most vulnerable to damage because it is a mobile, weight-bearing joint and its stability depends almost entirely on its associated ligaments and muscles. Further, there is no correspondence of the articulating bones. A **swollen knee** may occur immediately or hours after an injury. The initial swelling is due to escape of blood from damaged blood vessels adjacent to areas involving rupture of the anterior cruciate ligament, damage to synovial membranes, torn menisci, fractures, or collateral ligament sprains. Delayed swelling is due to excessive production of synovial fluid, a condition commonly re-ferred to as "water on the knee." A common type of knee injury in foot-ball is **rupture of the tibial collateral ligaments**, often associated with tearing of the anterior cruciate ligament and medial meniscus (torn cartilage). Usually, a hard blow to the lateral side of the knee while the foot is fixed on the ground causes the damage. A **dislocated knee** refers to the displacement of the tibia relative to the femur. The most common type is dislocation anteriorly, resulting from hyperextension of the knee. A frequent consequence of a dislocated knee is damage to the popliteal artery. •

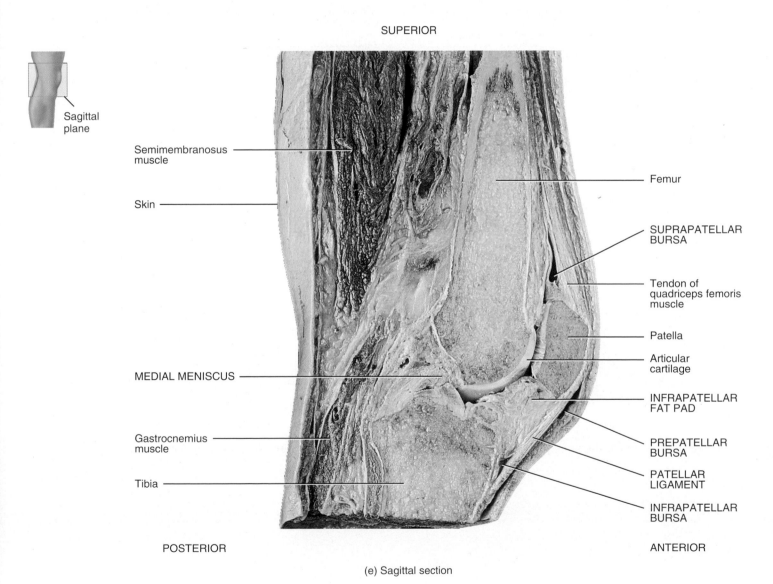

(e) Sagittal section

EXHIBIT **9.5** CONTINUES ▶

EXHIBIT 9.5 **279**

EXHIBIT 9.5 Knee Joint CONTINUED

FIGURE 9.15 CONTINUED ▶

Tendon of quadriceps femoris muscle

Femur

Patella

FIBULAR COLLATERAL LIGAMENT

LATERAL MENISCUS

ARTICULAR CAPSULE

Patellar ligament

Fibula

TIBIAL COLLATERAL LIGAMENT

Tibia

LATERAL — MEDIAL

(f) Anterior view

Articular cartilage

Femur

POSTERIOR CRUCIATE LIGAMENT (PCL)

ANTERIOR CRUCIATE LIGAMENT (ACL)

TIBIAL COLLATERAL LIGAMENT

FIBULAR COLLATERAL LIGAMENT

MEDIAL MENISCUS

LATERAL MENISCUS

OBLIQUE POPLITEAL LIGAMENT (CUT)

Posterior ligament of head of fibula

Tibia

Fibula

MEDIAL — LATERAL

(g) Posterior view

Tendon of quadriceps femoris muscle

Femur

SUPRAPATELLAR BURSA

Patella

ARTICULAR CAPSULE

PATELLAR LIGAMENT

MEDIAL MENISCUS

Tibia

TIBIAL COLLATERAL LIGAMENT

(h) Medial view

Femur

Tendon of quadriceps femoris muscle

SYNOVIAL MEMBRANE

Patella

Articular cartilage

LATERAL MENISCUS

ARTICULAR CAPSULE (cut)

FIBULAR COLLATERAL LIGAMENT

PATELLAR LIGAMENT

Fibula

Periosteum

(i) Lateral view

? What movement occurs at the knee joint when the quadriceps femoris (anterior thigh) muscles contract?

EXHIBIT 9.6   Ankle Joint

OBJECTIVE

- Describe the anatomical components of the ankle joint and explain the movements that can occur at this joint.

Description

The **ankle joint** (*talocrural joint*) is a hinge joint formed by (1) the distal end of the tibia and its medial malleolus with the talus and (2) the lateral malleolus of the fibula with the talus. It is a strong and stable joint due to the shapes of the articulating bones, the strength of its ligaments, and the tendons that surround it.

Anatomically, the ankle is the region that extends from the distal region of the leg to the proximal region of the foot and contains the ankle joint. In this transition region, there is a change in orientation from a vertical position of the bones and muscles and associated structures in the leg to a horizontal position of the structures in the foot. As a result, there is a turning anteriorly of the tendons, blood vessels, and nerves in the ankle as they enter the foot.

As the structures pass from the leg into the foot at the ankle from a vertical to a horizontal orientation, they are anchored by thickenings of deep fascia called **retinacula** (ret-i-NAK-yoo-la; *retineo*=to hold back). Two principal retinacula are the superior and inferior extensor retinacula (see Figure 11.24 on pages 400–405).

Anatomical Components

1. **Articular capsule.** Completely surrounds the joint and is attached superiorly to the tibia and fibula and inferiorly to the talus. The capsule is thin (and weak) anteriorly and posteriorly to permit dorsiflexion and plantar flexion.
2. **Deltoid (medial) ligament.** Strong, flat, triangular ligament that extends from the medial malleolus to the talus, navicular, and calcaneus of the tarsus. It is divisible into superficial and deep parts. The superficial components from anterior to posterior are the *tibionavicular ligament, tibiocalcaneal ligament*, and *posterior tibiotalar ligament*. The deep component is the *anterior tibiotalar ligament* (Figure 9.16a). The deltoid ligament strengthens the medial aspect of the ankle joint.

Figure 9.16 Right ankle (talocrural) joint.

The strength and stability of the ankle joint are due to the shapes of the articulating bone, the strength of its ligaments, and the tendons that surround it.

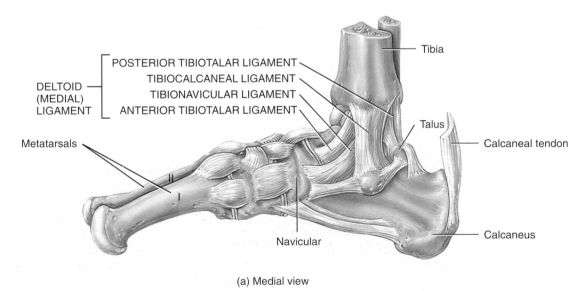

(a) Medial view

EXHIBIT 9.6 CONTINUES

EXHIBIT 9.6   **281**

EXHIBIT 9.6 Ankle Joint CONTINUED

3. **Lateral ligament.** Not as strong as the deltoid ligament. It extends from the lateral malleolus to the talus and calcaneus and is divisible into three components: *anterior talofibular ligament, posterior talofibular ligament*, and *calcaneofibular ligament* (Figure 9.16b). The lateral ligament strengthens the lateral aspect of the ankle joint.

Movements

The ankle joint permits dorsiflexion and plantar flexion (see Figure 9.10g).

◉ **CHECKPOINT**

How is the ankle defined anatomically?

● **CLINICAL CONNECTION** | Ankle Sprains and Fractures

The ankle is the most frequently injured major joint in the body. **Ankle sprains** are the most common ankle injuries and often occur in sports that involve running and jumping. *Sprains of the lateral ankle* occur more frequently than those of the medial ankle and are usually caused by excessive inversion (supination) of the foot with plantar flexion of the ankle. As a result, the weaker lateral ligament is partially torn and there is considerable pain and local swelling. Less common *sprains of the medial ankle* occur as a result of excessive eversion (pronation). In the process, the deltoid ligament may be torn but usually the ligament does not tear due to its great strength; instead it may pull off the tip of the medial malleolus. Ankle sprains are treated with RICE: rest, ice, compression, and elevation. Severe sprains may require cast immobilization or surgery.

A **fracture** of the distal end of the leg that involves both the medial and lateral malleoli is called a *Pott's fracture*. Such a fracture may result in dislocation of the talus. Sometimes only lateral malleolar fractures occur; even less frequently, only medial malleolar fractures occur. •

◉ **FIGURE 9.16 CONTINUED** ▶

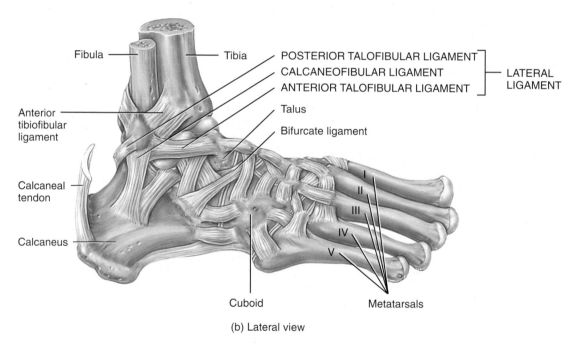

(b) Lateral view

❓ How is the ankle defined anatomically?

AGING AND JOINTS

◉ **OBJECTIVE**
• Explain the effects of aging on joints.

Aging usually results in decreased production of synovial fluid in joints. In addition, the articular cartilage becomes thinner with age, and ligaments shorten and lose some of their flexibility. The effects of aging on joints, which vary considerably from one person to another, are influenced by genetic factors and by wear and tear. Although degenerative changes in joints may begin in those as young as 20 years of age, most changes do not occur until much later. By age 80, almost everyone develops some type of degeneration in the knees, elbows, hips, and shoulders. It is also common for elderly individuals to develop degenerative changes in the vertebral column, resulting in a hunched-over posture and pressure on nerve roots. One type of arthritis, called osteoarthritis (see Clinical Connection on page 265), is at least partially age-related. Nearly everyone over age 70 has evidence of some osteoarthritic changes. Stretching and aerobic exercises that attempt to maintain full range of motion are very important in minimizing the effects of aging. They help to maintain the effective functioning of ligaments, tendons, muscles, synovial fluid, and articular cartilage.

◉ **CHECKPOINT**
12. Which joints show evidence of degeneration in nearly all individuals as aging progresses?

• CLINICAL **CONNECTION** | Arthroplasty

Joints that have been severely damaged by diseases such as arthritis, or by injury, may be replaced surgically with artificial joints in a procedure referred to as **arthroplasty** (AR-thrō-plas′-tē; *arthr-*=joint; *-plasty*=plastic repair of). Although most joints in the body can be repaired by arthroplasty, the ones most commonly replaced are the hips, knees, and shoulders. About 400,000 hip replacements and 300,000 knee replacements are performed annually in the United States. During the procedure, the ends of the damaged bones are removed and metal, ceramic, or plastic components are fixed in place. The goals of arthroplasty are to relieve pain and increase range of motion.

Partial hip replacements involve only the femur. **Total hip replacements** involve both the acetabulum and head of the femur (Figure A–C). The damaged portions of the acetabulum and the head of the femur are replaced by prefabricated prostheses (artificial devices). The acetabulum is shaped to accept the new socket, the head of the femur is removed, and the center of the femur is shaped to fit the femoral component. The acetabular component consists of a plastic such as polyethylene, and the femoral component is composed of a metal such as cobalt-chrome, titanium alloys, or stainless steel. These materials are designed to withstand a high degree of stress and to prevent a response by the immune system. Once the appropriate acetabular and femoral components are selected, they are attached to the healthy portion of bone with acrylic cement, which forms an interlocking mechanical bond.

Knee replacements are actually a resurfacing of cartilage and, like hip replacements, may be partial or total. In a **partial knee replacement**, also called a **unicompartmental knee replacement**, only one side of the knee joint is replaced. Once the damaged cartilage is removed from the distal end of the femur, the femur is reshaped and a metal femoral component is cemented in place. Then the damaged cartilage from the proximal end of the tibia is removed, along with the meniscus. The tibia is reshaped and fitted with a plastic tibial component that is cemented into place.

In a **total knee replacement** (Figure D, E), the damaged cartilage is removed from the distal end of the femur, proximal end of the tibia, and the back surface of the patella (If the back surface of the patella is not badly damaged, it may be left intact). The femur is reshaped and fitted with a metal femoral component and cemented in place. The tibia is reshaped and fitted with a plastic tibial component that is cemented in place. If the back surface of the patella is badly damaged, it is replaced with a plastic implant.

Researchers are continually seeking to improve the strength of the cement and devise ways to stimulate bone growth around the implanted area. Potential complications of arthroplasty include infection, blood clots, loosening or dislocation of the replacement components, and nerve injury. •

(A) Preparation for total hip replacement

(B) Artificial hip joint

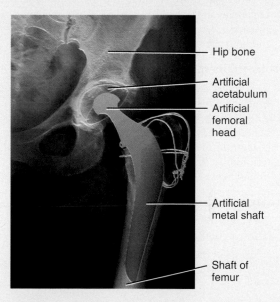

(C) Radiograph of an artificial hip joint

(D) Components of a total knee replacement

(E) Radiograph of a total knee replacement

KEY MEDICAL TERMS ASSOCIATED WITH JOINTS

Arthralgia (ar-THRAL-jē-a; *arthr-*=joint; *-algia*=pain) Pain in a joint.
Bursectomy (bur-SEK-tō-mē; *-ectomy*=removal of) Removal of a bursa.
Chondritis (kon-DRĪ-tis; *chondr-*=cartilage) Inflammation of cartilage.

Subluxation (sub-luks-Ā-shun) A partial or incomplete dislocation.
Synovitis (sin′-ō-VĪ-tis) Inflammation of a synovial membrane in a joint.

STUDY OUTLINE

Introduction (p. 250)

1. A joint (articulation or arthrosis) is a point of contact between two bones, between bone and cartilage, or between bone and teeth.
2. The scientific study of joints is termed arthrology and the study of the motion of the human body is called kinesiology.

Joint Classifications (p. 251)

1. Joints are connective tissue junctions between bones that have varying functions and range from immovable to highly movable junctions.
2. Joint classification is based on the presence or absence of a synovial cavity (diarthroses or synovial joints) versus the presence of a solid mass of connective tissue (synarthroses). Solid connective tissue joints can have either cartilage (cartilaginous joints) or dense connective tissue (fibrous joints) as the joining tissue.
3. Ligaments are dense irregular or dense regular connective tissue structures that bind one bone to another bone. They stabilize joints and limit the range of motion at a joint.

Fibrous Joints (p. 252)

1. The bones of fibrous joints are held together by dense irregular connective tissue.
2. These joints include immovable sutures (found between skull bones), immovable to slightly movable syndesmoses (such as roots of teeth in the sockets in the mandible and maxilla and the distal tibiofibular joint), and slightly movable interossious membranes (found between the radius and ulna in the forearm).

Cartilaginous Joints (p. 253)

1. The bones of cartilaginous joints are held together by cartilage.
2. These joints include immovable synchondroses united by hyaline cartilage (epiphyseal plates between diaphyses and epiphyses) and slightly movable symphyses united by fibrocartilage (pubic symphysis).

Synovial Joints (p. 254)

1. Synovial joints, or diarthroses, contain a space between bones called the synovial cavity.
2. Other characteristics of synovial joints are the presence of articular cartilage covering the adjacent surfaces of bone and an articular capsule, made up of an outer fibrous membrane and an inner synovial membrane.
3. The synovial membrane secretes synovial fluid, which forms a thin, viscous film over the articular cartilages and other surfaces within the articular capsule.
4. Many synovial joints also contain accessory ligaments (extracapsular and intracapsular) and articular discs or menisci. A labrum is a fibrocartilaginous lip that extends from the edge of the joint socket and helps deepen the socket.
5. Synovial joints contain an extensive nerve and blood supply. The nerves convey information about pain, joint movements, and the degree of stretch at a joint. Blood vessels penetrate the articular capsule and ligaments.
6. Bursae are saclike structures, similar in structure to joint capsules, that alleviate friction as soft tissues such as muscles, tendons, and the skin rub against each other or neighboring bones. They are common around joints such as the shoulder and knee joints.
7. Tendon sheaths are tubelike bursae that wrap around tendons where there is considerable friction, such as in the wrist, fingers, ankles, and toes.
8. Types of synovial joints are planar, hinge, pivot, condyloid, saddle, and ball-and-socket.
9. In a planar joint the articulating surfaces are flat, and the bones glide back-and-forth and side-to-side (many are biaxial); examples are joints between carpals and tarsals.
10. In a hinge joint, the convex surface of one bone fits into the concave surface of another, and the motion is angular around one axis (monaxial); examples are the elbow, knee (modified hinge joint), and ankle joints.
11. In a pivot joint, a round or pointed surface of one bone fits into a ring formed by another bone and a ligament, and movement is rotational (monaxial); examples are the atlanto-axial and radioulnar joints.
12. In a condyloid joint, an oval projection of one bone fits into an oval cavity of another, and motion is angular around two axes (biaxial); examples include the wrist joint and metacarpophalangeal joints of the second through fifth digits.
13. In a saddle joint, the articular surface of one bone is shaped like a saddle and the other bone fits into the "saddle" like a sitting rider; motion is angular around three axes (triaxial). An example is the carpometacarpal joint between the trapezium and the metacarpal of the thumb.
14. In a ball-and-socket joint, the ball-shaped surface of one bone fits into the cuplike depression of another; motion is around three axes (triaxial). Examples include the shoulder and hip joints.
15. Table 9.1 on page 258 summarizes the categories of joints and their functions (movements).

Types of Movements at Synovial Joints (p. 259)

1. In a gliding movement, the nearly flat surfaces of bones move back-and-forth and side-to-side.
2. In angular movements, a change in the angle between bones occurs. Examples are flexion–extension, lateral flexion, hyperextension, and abduction–adduction. Circumduction refers to flexion, abduction, extension, and adduction in succession.
3. In rotation, a bone moves around its own longitudinal axis.
4. Special movements occur at specific synovial joints. Examples are elevation–depression, protraction–retraction, inversion–eversion, dorsiflexion–plantar flexion, supination–pronation, and opposition.
5. Table 9.2 on page 264 summarizes the various types of movements at synovial joints.

Factors Affecting Contact and Range of Motion at Synovial Joints (p. 264)

1. The ways that articular surfaces of synovial joints contact one another determines the type of movement possible.
2. Factors that contribute to keeping the surfaces in contact and affect range of motion are structure or shape of the articulating bones, strength and tension of the ligaments, arrangement and tension of the muscles, apposition of soft parts, hormones, and disuse.

Selected Joints of the Body (p. 266)

1. A summary of selected joints of the body, including articular components, structural and functional classifications, and movements, is presented in Tables 9.3 and 9.4 on pages 266 and 267.
2. The temporomandibular joint (TMJ) is between the condyle of the mandible and mandibular fossa and articular tubercle of the temporal bone (Exhibit 9.1 on pages 268–269).
3. The shoulder (humeroscapular or glenohumeral) joint is between the head of the humerus and glenoid cavity of the scapula (Exhibit 9.2 on pages 270–272).
4. The elbow joint is between the trochlea of the humerus, the trochlear notch of the ulna, and the head of the radius (Exhibit 9.3 on page 273).
5. The hip (coxal) joint is between the head of the femur and acetabulum of the hip bone (Exhibit 9.4 on pages 274–276).

6. The knee (tibiofemoral) joint is between the patella and patellar surface of the femur; the lateral condyle of the femur, the lateral meniscus, and the lateral condyle of the tibia; and the medial condyle of the femur, the medial meniscus, and the medial condyle of the tibia (Exhibit 9.5 on pages 277–280).
7. The ankle (talocrural) joint is formed by the distal end of the tibia and its medial malleolus with the talus and the lateral malleolus of the fibula (Exhibit 9.6 on pages 281–282).

Aging and Joints (p. 282)

1. With aging, a decrease in synovial fluid, thinning of articular cartilage, and decreased flexibility of ligaments occur.
2. Most individuals experience some degeneration in the knee, elbow, hip, and shoulder joints due to the aging process.

SELF-QUIZ QUESTIONS

Choose the one best answer to the following questions.

1. Choose the pair of terms that is most closely associated or matched:
   a. wrist joint–pronation
   b. intertarsal joints–inversion
   c. elbow joint–hyperextension
   d. ankle joint–eversion
   e. interphalangeal joint–circumduction

2. Which of the following is a ball-and-socket joint?
   a. the temporomandibular joint
   b. the knee joint
   c. the shoulder joint
   d. the elbow joint
   e. Both b and c are correct.

3. Which of the following structures is not associated with the knee joint?
   a. glenoid labrum
   b. patellar ligament
   c. infrapatellar bursa
   d. cruciate ligaments
   e. tibial collateral ligament

4. The lambdoid suture is an example of a(n):
   a. synarthrosis.
   b. amphiarthrosis.
   c. diarthrosis.
   d. fibrous joint.
   e. Both a and d are correct.

5. Pads of fibrocartilage that extend into the space between articulating bones in the knee are called:
   a. ligaments.
   b. bursae.
   c. articular capsules.
   d. menisci.
   e. gomphoses.

6. The joint at which the atlas rotates around the axis is an example of a:
   a. hinge joint.
   b. ball-and-socket joint.
   c. gliding joint.
   d. saddle joint.
   e. pivot joint.

7. The inner layer of the articular capsule is the:
   a. synovial membrane.
   b. fibrous capsule.
   c. bursa.
   d. meniscus.
   e. articular cartilage.

8. A broad flat disc of fibrocartilage joins the bones at the:
   a. knee joint.
   b. shoulder joint.
   c. pubic symphysis.
   d. coronal suture.
   e. atlanto-occipital joint.

9. Which department in the hospital would someone report to who was experiencing severe joint pain (arthralgia)?
   a. Rheumatology
   b. Dermatology
   c. Neurology
   d. Urology
   e. Oncology

10. Which of the following disorders is an autoimmune disorder in which cartilage and joint linings of the skeleton are attacked by the body's own immune system?
    a. osteoarthritis
    b. rheumatoid arthritis
    c. gout
    d. chondritis
    e. arthralgia

Complete the following:

11. The epiphyseal plate is an example of the structural joint classification known as a _____ because _____ joins the epiphysis and diaphysis of the growing bone.

12. When the fibers of the fibrous capsule of an articular capsule are arranged in parallel bundles, the structure is called a(n) _____.

13. The temporomandibular joint is functionally classified as a(n) _____.

14. The ends of bones at synovial joints are covered with a protective layer of _____.

15. Sac-like structures that reduce friction between body parts at joints are called _____.

16. _____ of the L4 vertebra indicates that there is a partial or incomplete disclocation of L4.

17. A common type of knee injury seen in football, usually caused by a hard blow to the lateral side of the knee while the foot is fixed on the ground, involves rupture of the _____ and tearing of the _____ and the _____, and is also known as the "unhappy triad."

Are the following statements true or false? If the statement is False, *find and fix the mistake to make the statement* True.

18. Synovial fluid becomes more viscous as joint activity increases.

19. Tipping the head backward at the atlanto-occipital joint is an example of hyperextension.

20. Raising the arm to point straight ahead is an example of abduction.

21. The nerves that supply a joint are different from those that supply the skeletal muscles that move that joint.

22. A shoulder dislocation is an injury to the acromioclavicular joint.

23. Osteoarthritis is a common autoimmune disease seen in most older persons.

24. Gout refers to a disease in which a large bump is noticeable on the anterior neck.

Matching

25. Match the following:

____ **(a)** distal tibiofibular joint
____ **(b)** pubic symphysis
____ **(c)** coronal suture
____ **(d)** tooth in alveolar socket
____ **(e)** atlanto-axial joint
____ **(f)** intercarpal joints
____ **(g)** elbow joint
____ **(h)** carpometacarpal joint of thumb
____ **(i)** epiphyseal plate

(1) synovial, saddle
(2) gomphosis
(3) syndesmosis
(4) synovial, pivot
(5) cartilaginous (fibrocartilage)
(6) cartilaginous (hyaline cartilage)
(7) fibrous joint, synarthrosis
(8) synovial, gliding
(9) synovial, hinge

26. Match the following:

____ **(a)** movement of a body part anteriorly, horizontally to the ground
____ **(b)** horizontal movement of an anteriorly projected body part back into the anatomical position
____ **(c)** movement of the sole medially at the intertarsal joints
____ **(d)** movement of the sole laterally
____ **(e)** action that occurs when you stand on your tiptoes
____ **(f)** position of foot when the heel is on the floor and rest of the foot is raised
____ **(g)** movement of the forearm to turn the palm anteriorly
____ **(h)** movement of the forearm to turn the palm posteriorly

(1) pronation
(2) plantar flexion
(3) eversion
(4) retraction
(5) inversion
(6) protraction
(7) dorsiflexion
(8) supination

CRITICAL THINKING QUESTIONS

1. Burt and Al have been golf partners for 50 years. Burt's golf game improved by 5 points this spring and he credits the hip replacement he had last year. Al's knee has been bothering him for years but when he asked his orthopedist about a new knee joint he was told "it's not that simple." Al told Burt "one joint's like any other joint" and wants a second opinion. What will the new doctor say?

2. After your second Human Anatomy exam, you dropped to one knee, raised your arm over your head with one hand clenched into a fist, pumped your arm up and down, bent your head back, looked straight up, and yelled "*Yes!*" Use the proper terms to describe the movements at the various joints.

3. Lars was just getting the hang of bodysurfing during his first trip to the shore when he got caught in the break of a wave. While he was being rolled in the surf, he felt his shoulder "pop." When Lars finally got back to his towel he was out of breath, in pain, and his arm was hanging at an odd angle. What's the prognosis for the rest of Lars's vacation at the shore?

4. Arthur was in a serious motorcycle accident in which he broke his tibia, fibula, and patella. His knee was twisted sideways in the process. His bones have since healed, but he still experiences considerable pain in the knee joint and says it feels like "there's stuff in there." He also has trouble with stability on the joint while walking. To what specific anatomical problems would you attribute Arthur's joint difficulties?

5. Chuck has gone to the chiropractor because his back hurts. The chiropractor tells him, "You know, your pelvis is out of alignment." This news came shortly after his orthopedist told him he had slightly bowed legs. On top of all this, he has severe osteoarthritis in his ankle joint. Do you think all of these things could be related? Why or why not?

? ANSWERS TO FIGURE QUESTIONS

9.1 The fibrous joints are sutures found in the skull, syndesmoses found in a variety of places in the skeleton (such as the distal tibiofibular joint), and interosseous membranes found between the radius and ulna and between the tibia and fibula.

9.2 The structural difference between a synchondrosis and a symphysis is the type of cartilage that holds the joint together: hyaline cartilage in a synchondrosis and fibrocartilage in a symphysis.

9.3 Synovial joints are capsular joints containing a thin coat of lubricating fluid to reduce friction at the articulating cartilage surfaces.

9.4 Many planar and condyloid joints are biaxial joints.

9.5 Gliding movements occur at planar joints.

9.6 Two examples of flexion that do not occur in the sagittal plane are flexion of the thumb and lateral flexion of the trunk.

9.7 When you adduct your arm or leg, you bring it closer to the midline of the body, thus "adding" it to the trunk.

9.8 Circumduction involves flexion, abduction, extension, and adduction in continuous sequence.

9.9 The anterior surface of a bone or limb rotates toward the midline in medial rotation, and away from the midline in lateral rotation.

9.10 Bringing the arms forward until the elbows touch is an example of protraction.

9.11 The lateral ligament prevents displacement of the mandible.

9.12 The shoulder joint is the most freely movable joint in the body because of the looseness of its articular capsule and the shallowness of the glenoid cavity in relation to the size of the head of the humerus.

9.13 A hinge joint permits flexion and extension.

9.14 Tension in three ligaments—iliofemoral, pubofemoral, and ischiofemoral—limits the degree of extension at the hip joint.

9.15 Contraction of the quadriceps femoris muscle causes extension at the knee joint.

9.16 Anatomically, the ankle is the region between the distal aspect of the leg and the proximal aspect of the foot.

10 | MUSCULAR TISSUE

INTRODUCTION Machines are part of our daily lives, from simple can openers to complex computers, cars, and copy machines. But what exactly is a machine? According to *Webster's* dictionary, a machine is "an assemblage of parts that transmit forces, motion, and energy one to another in a predetermined manner." So even the most complex machines are composed of simpler parts such as levers, fulcra, latches, notches, receptors, energy sources, wires, and cables. These simple structures combine to create the more complex structure we define as a machine. Now think about your body, as a machine, and its systems and organs as parts. We can more easily comprehend the structure and function of the machinery we call the human body when we better understand the operation of its parts.

The major "machinery" of the human body is muscular tissue, which weighs in at 40–50% of total body mass. Most of the work generated by the body, such as pumping blood through the blood vessels, regulating flow in those vessels, moving food through the gastrointestinal tract, moving urine out of the urinary bladder, generating heat, creating the words we use in daily conversation, maintaining our posture, and producing the movements of our skeletons, is a result of the activity of muscles. In this chapter we will explore the organs we call muscles, from the level of the cell to entire muscles, some of which are quite large. We will also learn how the simple parts of this muscle machinery work together to produce powerful contractile forces that generate most of the activity of the human body. The scientific study of muscles is known as **myology** (mī-OL-ō-jē; *myo-*=muscle;*-logy*=study of). •

OVERVIEW OF MUSCULAR TISSUE

◉ OBJECTIVE

• Compare the three types of muscular tissue with their functions and special properties.

Types of Muscular Tissue

There are three types of muscular tissue: skeletal, cardiac, and smooth (see the comparison in Table 3.5 on pages 85–86). Although the three types of muscular tissue share some properties, they differ from one another in their microscopic anatomy, location, and control by the nervous and endocrine systems.

Skeletal muscle tissue is so named because the function of most skeletal muscles is to move bones of the skeleton. (A few skeletal muscles attach to structures other than bone, such as the skin or even other skeletal muscles.) Skeletal muscle tissue is termed *striated* because alternating light and dark bands (*striations*) are visible when the tissue is examined under a microscope (see Figure 10.4). Skeletal muscle tissue works primarily in a *voluntary* manner; that is, its activity can be consciously (voluntarily) controlled.

Cardiac muscle tissue is found only in the heart, where it forms most of the heart wall. Like skeletal muscle, cardiac muscle is *striated*, but its action is *involuntary*—its alternating contraction and relaxation cannot be consciously controlled. The heart beats because it has a pacemaker that initiates each contraction; this built-in (intrinsic) rhythm is called **autorhythmicity**. Several hormones and neurotransmitters adjust heart rate by speeding or slowing the pacemaker.

Smooth muscle tissue is located in the walls of hollow internal structures, such as blood vessels, airways, and most organs in the abdominopelvic cavity. It is also attached to hair follicles in the skin. Smooth muscle tissue gets its name from the fact that, under a microscope, it appears *nonstriated* or *smooth*. The action of smooth muscle is usually *involuntary*, and, like cardiac muscle, some smooth muscle tissue has autorhythmicity. Both cardiac muscle and smooth muscle are regulated by neurons that are part of the autonomic (involuntary) division of the nervous system and by hormones released by endocrine glands.

Functions of Muscular Tissue

Through sustained contraction or alternating contraction and relaxation, muscular tissue has four key functions: producing body movements, stabilizing body positions, storing and moving substances within the body, and generating heat:

1. **Producing body movements.** Total body movements such as walking and running, and localized movements such as grasping a pencil or nodding the head as a result of muscular contractions, rely on the integrated functioning of skeletal muscles, bones, and joints.
2. **Stabilizing body positions.** Skeletal muscle contractions stabilize joints and help maintain body positions, such as standing or sitting. Postural muscles contract continuously when a person is awake; for example, sustained contractions in neck muscles hold the head upright.
3. **Storing and moving substances within the body.** Sustained contractions of ringlike bands of smooth muscles called *sphincters* may prevent outflow of the contents of a hollow organ. Temporary storage of food in the stomach or urine in the urinary bladder is possible because smooth muscle sphincters close off the outlets of these organs. Cardiac muscle contractions pump blood through the body's blood vessels. Contraction and relaxation of smooth muscle in the walls of blood vessels help adjust their diameter and thus regulate the rate of blood flow. Smooth muscle contractions also move food and substances such as bile and enzymes through the gastrointestinal tract, push gametes (sperm and oocytes) through the reproductive systems, and propel urine through the urinary system. Skeletal muscle contractions promote the flow of lymph and aid the return of blood to the heart.
4. **Producing heat.** As muscular tissue contracts, it also produces heat. Much of the heat released by muscle is used to maintain normal body temperature. Involuntary contractions of skeletal muscles, known as shivering, can dramatically increase the rate of heat production.

Properties of Muscular Tissue

Muscular tissue has four special properties that enable it to function and contribute to homeostasis:

1. **Electrical excitability**, a property of both muscle cells and neurons, is the ability to respond to certain stimuli by producing electrical signals—for example, **action potentials (impulses)**. The action potentials travel along the plasma membrane due to the presence of specific ion channels. The stimuli that trigger action potentials may be electrical signals arising in the muscle tissue itself, as in the heart's pacemaker, or chemical stimuli, such as neurotransmitters released by neurons, hormones distributed by the blood, or even local changes in pH.
2. **Contractility** is the ability of muscular tissue to contract forcefully when stimulated by an action potential. When muscle contracts, it generates tension (force of contraction) while pulling on its attachment points. In some muscle contractions, the muscle develops tension but does not shorten. An example is holding a book in an outstretched hand. In other muscle contractions, the tension generated is great enough to overcome the load (resistance) of the object being moved so the muscle shortens and movement occurs. An example is lifting a book off a table.
3. **Extensibility** is the ability of muscular tissue to stretch without being damaged. Extensibility allows a muscle to contract forcefully even if it is already stretched. Normally, smooth muscle is subject to the greatest amount of stretching. For example, each time the stomach fills with food, the muscle in the wall is stretched. Cardiac muscle also is stretched each time the heart fills with blood.
4. **Elasticity** is the ability of muscular tissue to return to its original length and shape after contraction or extension.

Skeletal muscle is the focus of much of this chapter. Cardiac muscle and smooth muscle are described briefly here. Cardiac muscle is discussed in more detail in Chapter 14 (the heart), and smooth muscle is included in Chapter 20 (the autonomic nervous system), as well as in discussions of the various organs containing smooth muscle.

◉ CHECKPOINT

1. What features distinguish the three types of muscular tissue?
2. Summarize the functions of muscular tissue.
3. Describe the properties of muscular tissue.

SKELETAL MUSCLE TISSUE

 OBJECTIVES

- Explain the importance of connective tissue components, blood vessels, and nerves to skeletal muscles.
- Describe the microscopic anatomy of a skeletal muscle fiber.
- Explain how a skeletal muscle fiber contracts and relaxes.

Each skeletal muscle is a separate organ composed of hundreds to thousands of skeletal muscle cells, also called **muscle fibers** because of their elongated shapes. Connective tissues surround muscle fibers and whole muscles, and carry blood vessels and nerves into the skeletal muscles to exert their effects on individual muscle fibers (Figure 10.1). To understand skeletal muscle contraction, you first need to learn about its gross and microscopic anatomy.

Figure 10.1 Organization of skeletal muscle and its connective tissue coverings.

A skeletal muscle consists of individual muscle fibers (cells) bundled into fasciculi and surrounded by three connective tissue layers.

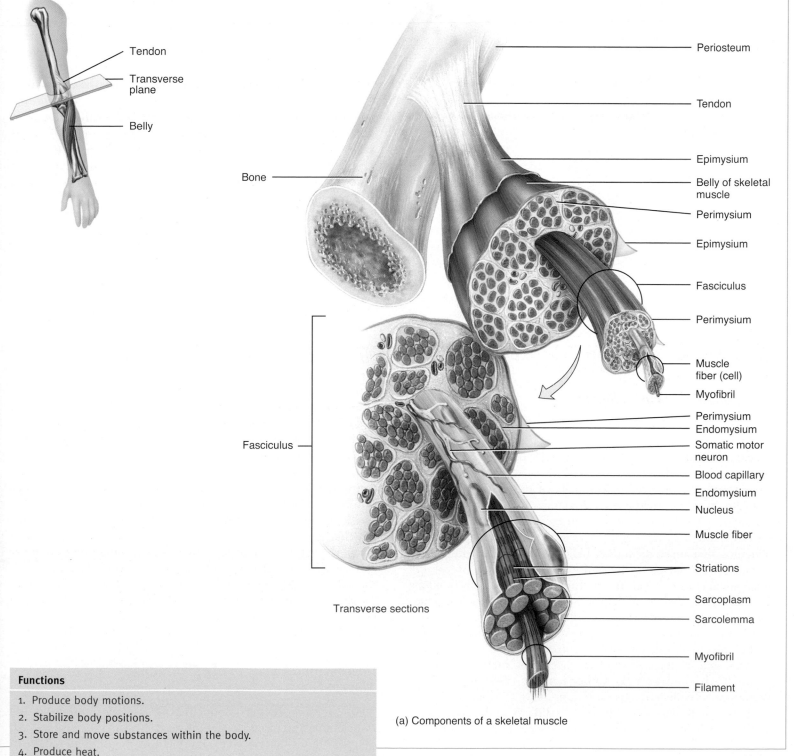

(a) Components of a skeletal muscle

Functions

1. Produce body motions.
2. Stabilize body positions.
3. Store and move substances within the body.
4. Produce heat.

CONTINUES ▶

Epidermis

Dermis

Subcutaneous layer

Fascia

Periosteum

Tibia

Medullary cavity

Interosseous membrane

Fibula

Tendon

Retinaculum

Muscle (cut)

Tendon

(b) Step dissection of leg

? **Which connective tissue coat surrounds groups of muscle fibers, separating them into fasciculi?**

Structure of a Skeletal Muscle

A typical skeletal muscle consists of a muscle belly connected by tendons to the skeleton (Figure 10.1a). The reddish or meatlike appearance that we associate with muscular tissue arises from the large population of well-vascularized muscle cells in the **muscle belly (body)** of the organ. The belly of the muscle can be an elongate, thick, rounded mass, a triangular shape, a thick rectangular mass, or a thin, flat sheet of muscular tissue. In contrast, the **tendons**, tough, glistening white dense regular connective tissue structures that attach the muscle belly to the bones, are minimally vascular, lack muscle cells, and consist primarily of parallel arrangements of collagen fibers. Like the muscle belly, tendons display a great variety of shapes: Some are long ropelike structures, while others are flat sheets called **aponeuroses** (ap'-ō-noo-RŌ-sez; *apo-*=from, *neu-*=a sinew). An example of an aponeurosis is the epicranial aponeurosis on top of the skull between the occipital and frontal bellies of the occipitofrontalis muscle (see Figure 11.4c on page 321). Other tendons are short bands or extensions of connective tissue that are so short that they make the muscle body appear as if it attaches directly to the bone. With this basic understanding of the parts of a skeletal muscle, let's further explore the relationship between the muscle body and its tendons.

Because muscles generate tremendous forces when they contract, it is important to understand how the muscle belly connects to the tendons and how the tendons connect to the bones. What makes these connections so strong that they do not pull apart when you lift something heavy? Another look at Figure 10.1 will help you find the answer.

The bulk of the muscle belly consists of striated skeletal muscle fibers. Notice, however, that there is more to the muscle belly than just muscle fibers. Surrounding each muscle fiber is a thin wrapping of loose connective tissue called the **endomysium** (en'-dō-MĪZ-ē-um; *endo-*=within). This surrounding connective tissue helps to bind the muscle fibers together, yet it is loose enough to allow them to move freely over one another. In addition, the endomysium carries the small blood vessels that supply the fibers with nutrients. Next, notice that groups of muscle fibers form bundles wrapped in a thicker layer of connective tissue. This muscle fiber bundle is a **fasciculus** (fa-SIK-ū-lus=little bundle; plural is *fasciculi*), also called *fascicle*, and its dense connective tissue covering is called the **perimysium** (per'-i-MĪZ-ē-um; *peri-*=around). The perimysium also allows a certain degree of freedom of motion between neighboring fasciculi and transmits blood vessels. Around the periphery of the muscle is a somewhat thicker covering of dense connective tissue called the **epimysium** (ep'-i-MĪZ-ē-um; *epi-*=upon), which binds all the fasciculi together to form the muscle belly. So, the muscle belly is not just a large mass of individual muscle fibers, but groups of fibers wrapped in connective tissue. It is important to realize that, even though we give the connective tissue three separate names, it is a continuous interconnected network of collagenous connective tissue wrapped around groups of muscle cells. But what does all this have to do with the connections among muscle, tendons, and bones?

To clarify this, it is important to note that the muscle fibers in the belly of the muscle eventually taper to round blunt ends, but the connective tissues associated with the fibers continue beyond the blunt ends. This continuing mass of collagenous connective tissue takes on a glistening whitish appearance, resulting from highly ordered collagen fibers, reduced numbers of blood vessels, and an absence of muscle cells. As a result of this design, the tendons are a continuous mass of connective tissue that runs through the muscle as the endomysium, perimysium, and epimysium and emerges from the belly of the muscle as the tendons of origin and insertion at either end. It is this design that makes your muscles so incredibly strong. At its junction with the bone, the surface tissue of the tendon is continuous with the periosteum, while its deeper collagen fibers enter the bone to blend with the collagen of the osseous extracellular matrix. This continuous network of connective tissue is essential to the function of the musculoskeletal system.

The various skeletal muscles of the body are further grouped together and protected by large connective tissue sheets, called **fascia** (FASH-ē-a=bandage), which wrap around groups of muscles much like a sock encircles your foot. For example, underneath the skin and subcutaneous layer in the free lower limbs a thin, tough, glistening sheet of dense connective tissue called the *fascia of the free lower limbs* surrounds all the muscles (Figure 10.1b). The free upper limbs and trunk have similar sheets of fascia.

Microscopic Anatomy of a Skeletal Muscle Fiber

The most important components of a skeletal muscle are the muscle fibers themselves. Mature muscle fibers range from 10 to 100 μm* in diameter. Typical muscle fiber length is about 10 cm

---

*One micrometer (μm) is 10^{-6} meter (1/25,000 in.).

(4 in.) in humans, although some are up to 30 cm (12 in.) long in the free lower limbs. During embryonic development, each skeletal muscle fiber arises from the fusion of a hundred or more small mesodermal cells called *myoblasts* (Figure 10.2a). Hence, each mature skeletal muscle fiber is a single cell with a hundred or more nuclei. Once fusion has occurred, the muscle fiber loses its ability to undergo cell division. Thus, most skeletal muscle fibers arise before birth, and most of these cells last a lifetime.

Figure 10.2 Microscopic organization of skeletal muscle. (a) During embryonic development, many myoblasts fuse to form one skeletal muscle fiber. Once fusion has occurred, a skeletal muscle fiber loses the ability to undergo cell division, but satellite cells retain this ability. (b and c) The sarcolemma of the muscle fiber encloses sarcoplasm and myofibrils, which are striated. Sarcoplasmic reticulum (SR) wraps around each myofibril. Thousands of transverse tubules, filled with interstitial fluid, invaginate from the sarcolemma toward the center of the muscle fiber. A triad is a transverse tubule and the two terminal cisterns of the sarcoplasmic reticulum on either side of it. A photomicrograph of skeletal muscle tissue is shown in Table 3.5A on page 85.

🔑 The contractile elements of muscle fibers are the myofibrils, which contain overlapping thick and thin filaments.

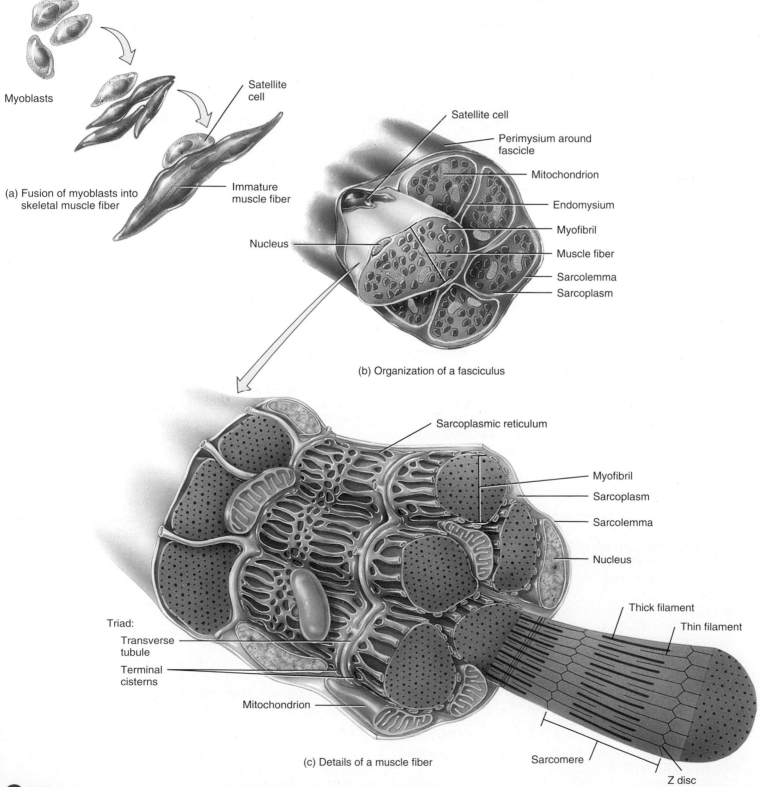

(a) Fusion of myoblasts into skeletal muscle fiber

(b) Organization of a fasciculus

(c) Details of a muscle fiber

❓ **Which structure shown here releases calcium ions to trigger muscle contraction?**

The dramatic muscle growth that occurs after birth occurs by enlargement of existing muscle fibers, called **hypertrophy** (hī-PER-trō-fē; *hyper-*=above or excessive; *-trophy*=nourishment), rather than by **hyperplasia** (hī-per-PLĀ-zē-a; *-plasis*=molding), an increase in the number of fibers. During childhood, human growth hormone and other hormones stimulate an increase in the size of skeletal muscle fibers. The hormone testosterone (from the testes in males and in small amounts from other tissues such as the ovaries in females) promotes further enlargement of muscle fibers. A few myoblasts do persist in mature skeletal muscle as *satellite cells* (Figure 10.2a, b). Satellite cells retain the capacity to fuse with one another or with damaged muscle fibers to regenerate functional muscle fibers. However, the number of new skeletal muscle fibers that can be formed by satellite cells is not enough to compensate for significant skeletal muscle damage or degeneration. In such cases, skeletal muscle tissue undergoes **fibrosis**, the replacement of muscle fibers by fibrous scar tissue. Thus, regeneration of skeletal muscle tissue is limited.

Sarcolemma, T Tubules, and Sarcoplasm

The multiple nuclei of a skeletal muscle fiber are located just beneath the **sarcolemma** (*sarc-*=flesh; *-lemma*=sheath), the plasma membrane of a muscle fiber (Figure 10.2b, c). Thousands of tiny invaginations of the sarcolemma, called **transverse tubules (T tubules)** tunnel in from the surface toward the center of each muscle fiber. Transverse tubules are open to the outside of the fiber and thus are filled with interstitial fluid. Muscle action potentials propagate along the sarcolemma and through the transverse tubules, quickly spreading throughout the muscle fiber. This arrangement ensures that all the superficial and deep parts of the muscle fiber become excited by an action potential almost simultaneously.

The sarcolemma surrounds the **sarcoplasm**, the cytoplasm of a muscle fiber. Sarcoplasm includes a substantial amount of glycogen, which can be split into glucose that is used for synthesis of ATP. In addition, the sarcoplasm contains a red-colored protein called **myoglobin** (MĪ-ō-glōb-in). This protein, found only in muscle, binds oxygen molecules that diffuse into muscle fibers from interstitial fluid. Myoglobin releases oxygen when mitochondria need it for ATP production. The mitochondria lie in rows throughout the muscle fiber, strategically close to the muscle proteins that use ATP during contraction.

Myofibrils and Sarcoplasmic Reticulum

At high magnification, the sarcoplasm appears stuffed with little threads. These small structures are the contractile elements of skeletal muscle, the **myofibrils** (Figure 10.2b, c). Myofibrils are about 2 μm in diameter and extend the entire length of the muscle fiber. Their prominent striations make the whole muscle fiber look striped.

A fluid-filled system of membranous sacs called the **sarcoplasmic reticulum** (sar'-kō-PLAZ-mik re-tik'-ū-lum) or **SR** encircles each myofibril (Figure 10.2c). This elaborate system is similar to smooth endoplasmic reticulum in nonmuscle cells. Dilated end sacs of the sarcoplasmic reticulum called **terminal cisterns** (=reservoirs) butt against the transverse tubules from both sides. One transverse tubule and the two terminal cisterns on either side of it form a **triad** (*tri-*=three). In a relaxed muscle fiber, the sarcoplasmic reticulum stores calcium ions (Ca^{2+}). Release of Ca^{2+} from the terminal cisterns into the sarcoplasm triggers muscle contraction.

Muscular Atrophy and Hypertrophy

Muscular atrophy (A-trō-fē; *a-*=without, *-trophy*=nourishment) is a wasting away of muscles. Individual muscle fibers decrease in size as a result of progressive loss of myofibrils. Atrophy that occurs because muscles are not used is termed *disuse atrophy*. Bedridden individuals and people with casts experience disuse atrophy because the flow of nerve impulses (nerve action potentials) to inactive skeletal muscle is greatly reduced. The condition is reversible. If instead the nerve supply to a muscle is disrupted or cut, the muscle undergoes *denervation atrophy*. Over a period of from 6 months to 2 years, the muscle shrinks to about one-fourth its original size, and the muscle fibers are irreversibly replaced by fibrous connective tissue.

As noted previously, **muscular hypertrophy** is an increase in the size of muscle fibers owing to increased production of myofibrils, mitochondria, sarcoplasmic reticulum, and other organelles. It results from very forceful, repetitive muscular activity, such as strength training. Because hypertrophied muscles contain more myofibrils, they are capable of more forceful contractions.

Filaments and the Sarcomere

Within myofibrils are smaller structures called **filaments** (Figure 10.2c). *Thin filaments* are about 8 nm[†] in diameter and 1–2 μm long, while *thick filaments* are 16 nm in diameter and 1–2 μm long. Both thin and thick filaments are directly involved in the contractile process. Overall, there are two thin filaments for every thick filament in the regions of filament overlap. The filaments inside a myofibril do not extend the entire length of a muscle fiber. Instead, they are arranged in compartments called **sarcomeres** (*-mere*=part), the basic functional units of a myofibril (Figure 10.3a). Narrow, plate-shaped regions of dense protein material called **Z discs** separate one sarcomere from the next. Thus, a sarcomere extends from one Z disc to the next Z disc.

The thick and thin filaments overlap one another to a greater or lesser extent, depending on whether the muscle is contracted, relaxed, or stretched. The pattern of their overlap, consisting of a variety of zones and bands (Figure 10.3b), creates the striations that can be seen both in single myofibrils and in whole muscle fibers. The darker middle part of the sarcomere is the **A band**, which extends the entire length of the thick filaments (Figure 10.3b). Toward each end of the A band is a *zone of overlap*, where the thick and thin filaments lie side by side. The **I band** is a lighter, less dense area that contains thin filaments but no thick filaments (Figure 10.3b). A Z disc passes through the center of each I band. A narrow **H zone** in the center of each A band contains thick filaments but no thin filaments. Supporting proteins that hold the thick filaments together at the center of the H zone form the **M line**, so named because it is at the *middle* of the sarcomere. Figure 10.4 shows the relations of the zones, bands, and lines as seen in a transmission electron micrograph.

Muscle Proteins

Myofibrils are built from three kinds of proteins: (1) contractile proteins, which generate force during contraction; (2) regulatory proteins, which help switch the contraction process on and off; and (3) structural proteins, which keep the thick and thin filaments in the proper alignment, give the myofibril elasticity and extensibility, and link the myofibrils to the sarcolemma and extracellular matrix.

[†]One nanometer (nm) is 10^{-9} meter (0.001μm).

Figure 10.3 **The arrangement of filaments within a sarcomere.** A sarcomere extends from one Z disc to the next.

Myofibrils contain two types of contractile filaments: thick filaments and thin filaments.

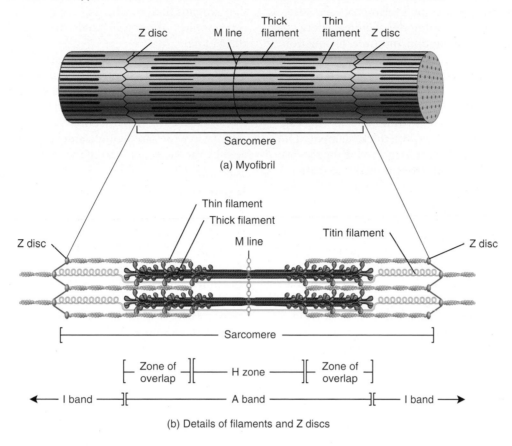

Among the following, which is smallest: muscle fiber, thick filament, or myofibril? Which is largest?

Figure 10.4 **Transmission electron micrograph showing the characteristic zones and bands of a sarcomere.**

The striations of skeletal muscle are alternating darker A bands and lighter I bands.

What is the ratio between thick and thin filaments in skeletal muscle?

The two *contractile proteins* in muscle are myosin and actin, which are the main components of thick and thin filaments, respectively. **Myosin** functions as a motor protein in all three types of muscle tissue. *Motor proteins* push or pull various cellular structures to achieve movement by converting the chemical energy in ATP into mechanical energy of motion or force production. About 300 molecules of myosin form a single thick filament in sleletal muscular tissue. Each myosin molecule is shaped like two golf clubs twisted together (Figure 10.5a). The *myosin tail* (twisted golf club handles) points toward the M line in the center of the sarcomere. Tails of neighboring myosin molecules lie parallel to one another, forming the shaft of the thick filament. The two projections of each myosin molecule (golf club heads) are called *myosin heads*. The heads project outward from the shaft in a spiraling fashion, each extending toward one of the six thin filaments that surround each thick filament.

Thin filaments extend from anchoring points within the Z discs (see Figure 10.3b). Their main component is the protein **actin**. Individual actin molecules join to form an actin filament that is twisted into a helix (Figure 10.5b). On each actin molecule is a *myosin-binding site*, where a myosin head can attach. Smaller amounts of two *regulatory proteins*—**tropomyosin** and **troponin**—are also part of the thin filament. In relaxed muscle, myosin is blocked from binding to actin because strands of tropomyosin cover the myosin-binding site on actin. The tropomyosin strand, in turn, is held in place by troponin molecules.

Besides contractile and regulatory proteins, muscle contains about a dozen *structural proteins*, which contribute to the alignment, stability, elasticity, and extensibility of myofibrils. Several key structural proteins are titin, myomesin, and dystrophin. **Titin** (*titan*=gigantic) is the third most plentiful protein in skeletal muscle (after actin and myosin). This molecule's name reflects its huge size. With a molecular weight of about 3 million daltons, titin is 50 times larger than an average-sized protein. Each titin molecule spans half a sarcomere, from a Z disc to an M line (see Figure 10.3b), a distance of 1 to 1.2 μm in relaxed muscle. Titin anchors a thick filament to both a Z disc and the M line, thereby helping stabilize the position of the thick filament. The part of the titin molecule that extends from the Z disc to the beginning of the thick filament is very elastic. Because it can stretch to at least four times its resting length and then spring back unharmed, titin accounts for much of the elasticity and extensibility of myofibrils. Titin probably helps the sarcomeres return to their resting length after a muscle has contracted or been stretched, may

help prevent overextension of sarcomeres, and maintains the central location of the A bands.

Molecules of the protein **myomesin** form the M line. The M line proteins bind to titin and connect adjacent thick filaments to one another. Myomesin holds the thick filaments in register at the M line.

Dystrophin is a cytoskeletal protein that links thin filaments of the sarcomere to integral membrane proteins of the sarcolemma. In turn, the membrane proteins attach to proteins in the connective tissue extracellular matrix that surrounds muscle fibers. Hence, dystrophin and its associated proteins are thought to reinforce the sarcolemma and help transmit the tension generated by the sarcomeres to the tendons.

Nerve and Blood Supply

Skeletal muscles are well supplied with nerves and blood vessels (see Figure 11.18c on page 365). Most of the skeletal muscles in the body, especially those of the limbs and head, receive one main nerve that supplies them with both motor and sensory functions. Other muscles, such as the sheetlike muscles of the body wall, receive multiple nerves. Nerves typically enter the muscle with the main blood vessels of the muscle as a unit called a **neuromuscular bundle**. These neuromuscular bundles enter the muscle body near the stable attachment or tendon of origin and then spread through the muscle via the perimysial and endomysial connective tissues. The motor fibers initiate the contractile function of muscle cells, while the sensory fibers provide feedback to the nervous system to regulate motor function. The neurons (nerve cells) that stimulate skeletal muscle fibers to contract are called **somatic motor neurons**. A somatic motor neuron has a threadlike extension, called an **axon**, which travels from the neuron cell body in the brain or spinal cord to a group of skeletal muscle fibers in a muscle of the body. A muscle fiber contracts in response to one or more action potentials propagating along its sarcolemma and through its transverse tubule system. Muscle action potentials arise at the **neuromuscular junction (NMJ)**, the synapse between a somatic motor neuron and a skeletal muscle fiber (Figure 10.6a). A **synapse** is a region where communication occurs between two neurons, or between a neuron and a target cell—for example, between a somatic motor neuron and a muscle fiber. At most synapses a small gap, called the **synaptic cleft**, separates the two cells. Because the cells do not physically touch, the action potential from one cell cannot

Figure 10.5 Structure of thick and thin filaments. (a) The thick filament contains about 300 myosin molecules, one of which is pictured. The myosin tails form the shaft of the thick filament, and the myosin heads project outward toward the surrounding thin filaments. (b) Thin filaments contain actin, troponin, and tropomyosin.

Contractile proteins (myosin and actin) generate force during contraction, and regulatory proteins (troponin and tropomyosin) help switch contraction on and off.

(a) A thick filament and a myosin molecule

(b) Portion of a thin filament

? Which proteins connect to the Z disc? Which proteins are present in the A band? In the I band?

"jump the gap" to directly excite the next cell. Instead, the first cell communicates with the second indirectly, by releasing a chemical called a **neurotransmitter**.

At the NMJ, the somatic motor neuron axon terminal divides into a cluster of **synaptic end bulbs** (Figure 10.6b, c). Suspended in the cytosol within each synaptic end bulb are hundreds of membrane-enclosed sacs called **synaptic vesicles**. Inside each synaptic vesicle are thousands of molecules of **acetylcholine** (as′-ē-til-KŌ-lēn), abbreviated **ACh**, the neurotransmitter released at the NMJ. The region of the sarcolemma opposite the synaptic end bulbs is called the **motor end plate**. It contains 30–40 million acetylcholine receptors, which are integral transmembrane proteins that bind specifically to ACh. These receptors are abundant in *junctional folds*, deep grooves in the motor end plate that provide a large surface area for ACh. A neuromuscular junction thus includes all the synaptic end bulbs on one side of the synaptic cleft, plus the motor end plate of the muscle fiber on the other side.

Even though each skeletal muscle fiber has only a single neuromuscular junction, the axon of a somatic motor neuron branches out and forms neuromuscular junctions with many different muscle fibers. A somatic motor neuron plus all the skeletal muscle fibers it stimulates is called a **motor unit**. A single motor neuron makes contact with an average of 150 skeletal muscle fibers, and all muscle fibers in one motor unit contract in unison. Muscles that control precise movements consist of many small motor units. For example, muscles of the larynx (voice box) that control voice production have as few as two or three muscle fibers per motor unit, and muscles controlling eye movements may have 10–20 muscle fibers per motor unit. In contrast, some motor units in skeletal muscles responsible for large-scale and powerful movements, such as the biceps brachii muscle in the arm and the gastrocnemius muscle in the calf of the leg, may have 2000–3000 muscle fibers each. The total strength of a muscle contraction depends, in part, on how

Figure 10.6 **Structure of the neuromuscular junction (NMJ), the synapse between a somatic motor neuron and a skeletal muscle fiber.**

Synaptic end bulbs at the tips of axon terminals contain synaptic vesicles filled with acetylcholine.

(a) Neuromuscular junction

(b) Enlarged view of the neuromuscular junction

(c) Binding of acetylcholine to ACh receptors in the motor end plate

CONTINUES ▶

◻ FIGURE 10.6 CONTINUED ▶

Blood capillary

Axon terminal

Axon collateral (branch)

Synaptic end bulbs

Somatic motor neuron

Axon collateral (branch)

Skeletal muscle fiber

Synaptic end bulbs

SEM 1650x

(d) Neuromuscular junction

? What part of the sarcolemma contains acetylcholine receptors?

large its motor units are and how many motor units are activated at the same time.

Generally, an artery and one or two veins accompany each nerve that penetrates a skeletal muscle. Microscopic blood vessels called *capillaries* are plentiful in muscle tissue; each muscle fiber is in close contact with one or more capillaries (see Figure 10.1a). Capillary blood brings oxygen and nutrients to the muscle fibers and removes heat and the waste products of muscle metabolism. Especially during contraction, a muscle fiber synthesizes and uses considerable ATP (adenosine triphosphate); these reactions require oxygen, glucose, fatty acids, and other substances that are supplied in the blood.

Contraction and Relaxation of Skeletal Muscle Fibers

Arrival of the nerve impulse (nerve action potential) at the synaptic end bulbs causes synaptic vesicles to release ACh (Figure 10.6c ❶). The ACh diffuses across the synaptic cleft between the somatic motor neuron and the motor end plate. Binding of ACh to its receptor in the sarcolemma (Figure 10.6c ❷) allows small cations, most importantly Na^+, to flow across the membrane. The inflow of Na^+ triggers a muscle action potential (Figure 10.6c ❸). The muscle action potential propagates along the sarcolemma through the transverse tubule system and to the sarcoplasmic reticulum, where it causes the release of its stored calcium ions (Ca^{2+}) into the sarcoplasm of the muscle fiber. In the presence of Ca^{2+} and ATP, the skeletal muscle shortens because the thick and thin filaments slide past each other. In this so-called **sliding filament mechanism**, the myosin heads attach to and "walk along" the thin filaments at both ends of a sarcomere, progressively pulling the thin filaments toward the M line (Figure 10.7). When the myosin heads attach to actin during contraction, they are referred to as **crossbridges**. As a result, the thin filaments slide inward and meet at the center of a sarcomere. They may even move so far inward that their ends overlap (Figure 10.7c). As the thin filaments slide inward, the Z discs come closer together, and the sarcomere shortens. Note that the lengths of the individual thick and thin filaments do not change; it is the amount of overlap between them that changes. Shortening of the individual sarcomeres causes shortening of the whole muscle fiber, and ultimately shortening of the entire muscle.

The effect of ACh binding lasts only briefly because the neurotransmitter is rapidly broken down by an enzyme in the synaptic cleft called **acetylcholinesterase (AChE)** (Figure 10.6c ❹).

AChE breaks down ACh into products that cannot activate the ACh receptor. When action potentials cease being transmitted by the motor neuron, ACh release stops and AChE rapidly breaks down the ACh already present in the synaptic cleft. This ends the generation of muscle action potentials, and the Ca^{2+} moves from the sarcoplasm of the muscle fiber back into the membrane of the sarcoplasmic reticulum.

• CLINICAL CONNECTION | **Electromyography**

Electromyography (e-lek′-trō-mī-OG-ra-fē; *electro-*=electricity; *myo-*=muscle; *-graph*=to write) or **EMG** is a test that measures the electrical activity (muscle action potentials) in resting and contracting muscles. Normally, resting muscle produces no electrical activity; a slight contraction produces some electrical activity; and a more forceful contraction produces increased electrical activity. The electrical activity of the muscle is detected by a recording needle placed on or in a muscle to be tested and displayed as waves on an oscilloscope and heard through a loudspeaker. EMG helps to determine if muscle weakness or paralysis is due to a malfunction of the muscle itself or the nerves supplying the muscle. EMG is also used to diagnose certain muscle disorders, such as muscular dystrophy and to understand which muscles function in various stages of complex movements. •

Resting muscle

Contracting muscle

Oscilloscope

Recording needle

Figure 10.7 Sliding filament mechanism of muscle contraction, as it occurs in two adjacent sarcomeres.

During muscle contractions, thin filaments move toward the M line of each sarcomere.

(a) Relaxed muscle

(b) Partially contracted muscle

(c) Maximally contracted muscle

? What happens to the I band and H zone during contraction? Do the lengths of the thick and thin filaments change?

Because skeletal muscle fibers often are very long cells, the NMJ usually is located near the midpoint of the fiber. Muscle action potentials arise at the NMJ and then propagate toward both ends of the fiber. This arrangement permits nearly simultaneous activation (and thus contraction) of all parts of the fiber.

Several plant products and drugs selectively block certain events at the NMJ. Botulinum toxin, produced by the bacterium *Clostridium botulinum*, blocks exocytosis of synaptic vesicles at the NMJ. As a result, ACh is not released, and muscle contraction does not occur. The bacteria proliferate in improperly canned foods and their toxin is one of the most lethal chemicals known. A tiny amount can cause death by paralyzing skeletal muscles. Breathing stops due to paralysis of respiratory muscles, including the diaphragm. Yet, it is also the first bacterial toxin to be used as a medicine (Botox®). Injections of Botox® into the affected muscles can help patients who have strabismus (crossed eyes), or blepharospasm (uncontrollable blinking), or spasms of the vocal cords that interfere with speech. It is also used as a cosmetic treatment to relax muscles that cause facial wrinkles and to alleviate chronic back pain due to muscle spasms in the lumbar region.

The plant derivative *curare*, a poison used by South American Indians on arrows and blowgun darts, causes muscle paralysis by binding to and blocking ACh receptors. Curare-like drugs are often used during surgery to relax skeletal muscles. A family of chemicals called *anticholinesterase agents* have the property of slowing the enzymatic activity of acetylcholinesterase, thus slowing removal of ACh from the synaptic cleft. At low doses, these agents can strengthen weak muscle contractions. One example is neostigmine, which is used to treat patients with myasthenia gravis. Neostigmine is also used as an antidote for curare poisoning and to terminate the effects of curare after surgery.

Muscle Tone

Even while at rest, a skeletal muscle exhibits **muscle tone** (*tonos*=tension), a small amount of tautness or tension in the muscle due to weak, involuntary contractions of its motor units. Recall that skeletal muscle contracts only after it is activated by acetylcholine released by nerve impulses in its somatic motor neurons. Hence, muscle tone is established by neurons in the brain and spinal cord that excite the muscle's somatic motor neurons. When the somatic motor neurons serving a skeletal muscle are damaged or cut, the muscle becomes **flaccid** (FLAK-sid or FLAS-sid=flabby), a state of limpness in which muscle tone is lost. To sustain muscle tone, small groups of motor units are alternately active and inactive in a constantly shifting pattern. Muscle tone keeps skeletal muscles firm, but it does not result in a force strong enough to produce movement. For example, when the muscles in the back of the neck are in normal tonic contraction, they keep the head upright and prevent it from slumping forward on the chest. Muscle tone also is important in smooth muscle tissues, such as those found in the gastrointestinal tract, where the walls of the digestive organs maintain a steady pressure on their contents. The tone of smooth muscle fibers in the walls of blood vessels plays a crucial role in maintaining blood pressure.

Isotonic and Isometric Contractions

Muscle contractions are classified as either isotonic or isometric. In an **isotonic contraction** (*iso-*=equal; *-tonic*=tension), the tension (force of contraction) developed by the muscle remains almost constant while the muscle changes its length. Isotonic contractions are used for body movements and for moving objects. The two types of isotonic contractions are concentric and eccentric. In a **concentric isotonic contraction,** if the tension generated is great enough to overcome the resistance of the object to be moved, the muscle shortens and pulls on another structure, such as a tendon, to produce movement and to reduce the angle at a joint. Picking up a book from a table involves concentric isotonic contractions of the biceps brachii muscle in the arm. By contrast, as you lower the book to place it back on the table, the previously shortened biceps lengthens in a controlled manner while it continues to contract. When the length of a muscle increases during a contraction, the contraction is an **eccentric isotonic contraction**. During an eccentric contraction, the tension exerted by the myosin crossbridges resists movement of a load (the book, in this case) and slows the lengthening process. For reasons that are not well understood, repeated eccentric isotonic contractions (for example, walking downhill) produce more muscle damage and more delayed-onset muscle soreness than do concentric isotonic contractions.

In an **isometric contraction** (-*metro*=measure or length), the tension generated is not enough to exceed the resistance of the object to be moved and the muscle does not change its length. An example would be holding a book steady using an outstretched arm. These contractions are important for maintaining posture and for supporting objects in a fixed position. Although isometric contractions do not result in body movement, energy is still expended. The book pulls the arm downward, stretching the shoulder and arm muscles. The isometric contraction of the shoulder and arm muscles counteracts the stretch. Isometric contractions are important because they stabilize some joints as others are moved. Most activities include both isotonic and isometric contractions.

⦿ CHECKPOINT

4. Describe the types of connective tissue that cover skeletal muscles.
5. Describe the components of a sarcomere. How do thin and thick filaments differ structurally?
6. What proteins compose skeletal muscle tissue and what are their functions?
7. Describe the nerve supply to a skeletal muscle fiber.
8. Why is a rich blood supply so important to muscle contraction?
9. Why is muscle tone important?
10. Define each of the following terms: concentric isotonic contraction, eccentric isotonic contraction, and isometric contraction.

TYPES OF SKELETAL MUSCLE FIBERS

⦿ OBJECTIVE

• Compare the structure and function of the three types of skeletal muscle fibers.

Among other differences, skeletal muscle fibers vary in their content of myoglobin, the red protein that binds oxygen in muscle fibers. Those with a high myoglobin content are called **red muscle fibers**, while those that have a low myoglobin content are called **white muscle fibers**. Red muscle fibers also contain more mitochondria and are supplied by more blood capillaries than white muscle fibers.

Skeletal muscle fibers also contract and relax at different speeds. A fiber is categorized as either slow or fast depending on how rapidly the ATPase in its myosin heads hydrolyzes ATP. In addition, skeletal muscle fibers vary in which metabolic reactions

The term **muscular dystrophy** (DIS-trō-fē; *dys-*=difficult; *-trophy*=nourishment) refers to a group of inherited muscle-destroying diseases that cause progressive degeneration of skeletal muscle fibers. The most common form of muscular dystrophy is *Duchenne muscular dystrophy* (doo-SHĀN) (*DMD*). Because the mutated gene is on the X chromosome, of which males have only one, DMD strikes boys almost exclusively. Worldwide, about 1 in every 3500 male babies—21,000 in all—are born with DMD each year. The disorder usually becomes apparent between the ages of 2 and 5, when parents notice the child falls often and has difficulty running, jumping, and hopping. By age 12 most boys with DMD are unable to walk. Respiratory or cardiac failure usually causes death between the ages of 20 and 30.

In DMD, the gene that codes for the protein dystrophin is mutated, and little or no dystrophin is present. Without the reinforcing effect of dystrophin, the sarcolemma tears easily during muscle contraction. Because their plasma membrane is damaged, muscle fibers slowly rupture and die. The dystrophin gene was discovered in 1987, and by 1990 the first attempts were made to treat DMD patients with gene therapy. The muscles of three boys with DMD were injected with myoblasts bearing functional dystrophin genes, but only a few muscle fibers gained the ability to produce dystrophin. Similar clinical trials with additional patients have also failed. An alternate approach to the problem is to find a way to induce muscle fibers to produce the protein utrophin, which is similar to dystrophin. Experiments with dystrophin-deficient mice suggest this approach may work. •

LM 25x

Variations in skeletal muscle fiber size and increased amounts of connective tissue (red) in Duchenne muscular dystrophy

they use to generate ATP and in how quickly they fatigue. Based on these structural and functional characteristics, skeletal muscle fibers are classified into three main types: (1) slow oxidative fibers, (2) fast oxidative-glycolytic fibers, and (3) fast glycolytic fibers.

Slow Oxidative Fibers

Slow oxidative (SO) or **type I fibers** are smallest in diameter and thus are the least powerful type of muscle fibers. They appear dark red because they contain large amounts of myoglobin and many blood capillaries. Because they have many large mitochondria, SO fibers generate ATP mainly by aerobic (oxygen-requiring) cellular respiration, which is why they are called oxidative fibers. These fibers are said to be "slow" because they use ATP at a slow rate. As a result, SO fibers have a slow speed of contraction. However, slow fibers are very resistant to fatigue and are capable of prolonged, sustained contractions for many hours. These fibers are adapted for maintaining posture and for aerobic, endurance-type activities such as running a marathon.

Fast Oxidative-Glycolytic Fibers

Fast oxidative-glycolytic (FOG) or **type IIa fibers** are intermediate in diameter between the other two types of fibers. Like slow oxidative fibers, they contain large amounts of myoglobin and many blood capillaries. Thus, they have a dark red appearance. FOG fibers can generate considerable ATP by aerobic cellular respiration, which gives them a moderately high resistance to fatigue. Because their intracellular glycogen level is high, they also generate ATP by anaerobic (oxygen-free) glycolysis. FOG fibers are "fast" because they use ATP at a fast rate, which makes their speed of contraction faster than SO fibers. FOG fibers contribute to activities such as walking and sprinting.

Fast Glycolytic Fibers

Fast glycolytic (FG) or **type IIb fibers** are largest in diameter and contain the most myofibrils. Hence, they can generate the most powerful contractions. FG fibers have low myoglobin content, relatively few blood capillaries, few mitochondria, and appear white in color. They contain large amounts of glycogen and generate ATP mainly by anaerobic (nonoxygen-requiring) cellular respiration (glycolysis). Due to their large size and their ability to use ATP at a fast rate, FG fibers contract strongly and quickly. These fast-twitch fibers are adapted for intense anaerobic movements of short duration, such as weight lifting or throwing a ball, but they fatigue quickly. Strength training programs that engage a person in activities requiring great strength for short times produce increases in the size, strength, and glycogen content of fast glycolytic fibers.

The FG fibers of a weight-lifter may be 50% larger than those of a sedentary person or endurance athlete. The increase in size is due to increased synthesis of muscle proteins. The overall result is muscle enlargement due to hypertrophy of the FG fibers.

Most skeletal muscles are a mixture of all three types of skeletal muscle fibers. The proportions vary somewhat, depending on the action of the muscle, the person's training regimen, and genetic factors. For example, the continually active postural muscles of the neck, back, and legs have a high proportion of SO fibers. Muscles of the shoulders and arms, in contrast, are not constantly active but are used briefly now and then to produce large amounts of tension, such as in lifting and throwing. These muscles have a high proportion of FG fibers. Leg muscles, which not only support the body but are also used for walking and running, have large numbers of both SO and FOG fibers.

The skeletal muscle fibers of any given motor unit are all of the same type. However, the different motor units in a muscle are recruited in a specific order, depending on need. For example, if weak contractions suffice to perform a task, only SO motor units are activated. If more force is needed, the motor units of FOG fibers are also recruited. Finally, if maximal force is required, motor units of FG fibers are also called into action. Activation of various motor units is controlled by the brain and spinal cord.

Table 10.1 summarizes the characteristics of the three types of skeletal muscle fibers.

◉ **CHECKPOINT**

11. What is the basis for classifying skeletal muscle fibers into three types?

TABLE 10.1

Characteristics of the Three Types of Skeletal Muscle Fibers

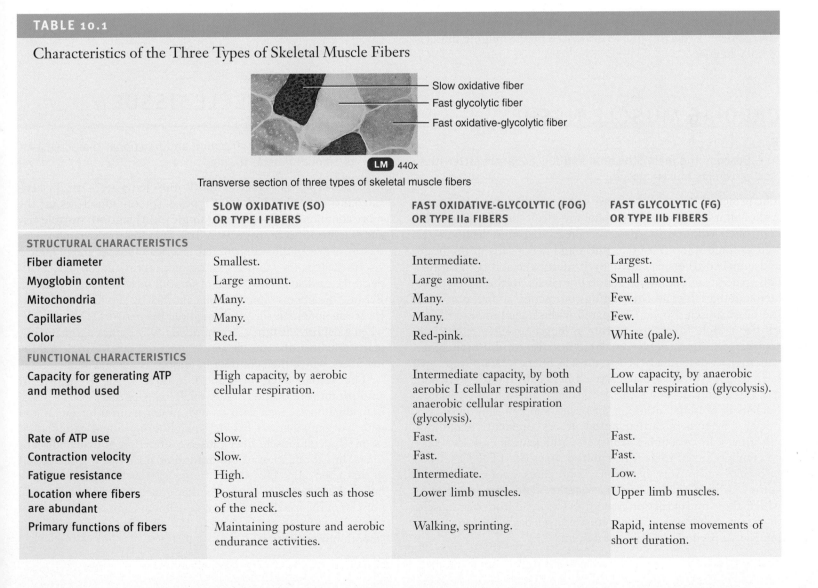

Transverse section of three types of skeletal muscle fibers

	SLOW OXIDATIVE (SO) OR TYPE I FIBERS	FAST OXIDATIVE-GLYCOLYTIC (FOG) OR TYPE IIa FIBERS	FAST GLYCOLYTIC (FG) OR TYPE IIb FIBERS
STRUCTURAL CHARACTERISTICS			
Fiber diameter	Smallest.	Intermediate.	Largest.
Myoglobin content	Large amount.	Large amount.	Small amount.
Mitochondria	Many.	Many.	Few.
Capillaries	Many.	Many.	Few.
Color	Red.	Red-pink.	White (pale).
FUNCTIONAL CHARACTERISTICS			
Capacity for generating ATP and method used	High capacity, by aerobic cellular respiration.	Intermediate capacity, by both aerobic l cellular respiration and anaerobic cellular respiration (glycolysis).	Low capacity, by anaerobic cellular respiration (glycolysis).
Rate of ATP use	Slow.	Fast.	Fast.
Contraction velocity	Slow.	Fast.	Fast.
Fatigue resistance	High.	Intermediate.	Low.
Location where fibers are abundant	Postural muscles such as those of the neck.	Lower limb muscles.	Upper limb muscles.
Primary functions of fibers	Maintaining posture and aerobic endurance activities.	Walking, sprinting.	Rapid, intense movements of short duration.

EXERCISE AND SKELETAL MUSCLE TISSUE

◉ OBJECTIVE

- Describe the effects of exercise on different types of skeletal muscle fibers.

The relative ratio of fast glycolytic (FG) and slow oxidative (SO) fibers in each muscle is genetically determined and helps account for individual differences in physical performance. For example, people with a higher proportion of fast glycolytic (FG) fibers often excel in activities that require periods of intense activity, such as weight lifting or sprinting. People with higher percentages of slow oxidative (SO) fibers are better at activities that require endurance, such as long-distance running.

Although the total number of skeletal muscle fibers usually does not increase, the characteristics of those present can change to some extent. Various types of exercises can induce changes in the fibers in a skeletal muscle. Endurance-type (aerobic) exercises, such as running or swimming, cause a gradual transformation of some fast glycolytic (FG) fibers into fast oxidative-glycolytic (FOG) fibers. The transformed muscle fibers show slight increases in diameter, number of mitochondria, blood supply, and strength. Endurance exercises also result in cardiovascular and respiratory changes that cause skeletal muscles to receive better supplies of oxygen and nutrients, but do not increase muscle mass. By contrast, exercises that require great strength for short periods produce an increase in the size and strength of fast glycolytic (FG) fibers. The increase in size is due to increased synthesis of thick and thin filaments. The overall result is muscle enlargement (hypertrophy), as evidenced by the bulging muscles of body builders.

CARDIAC MUSCLE TISSUE

◉ OBJECTIVE

- Describe the main structural and functional characteristics of cardiac muscle tissue.

The principal tissue in the heart wall is **cardiac muscle tissue**. Although it is striated like skeletal muscle, its activity cannot be controlled voluntarily. Also, certain cardiac muscle fibers (as well as some smooth muscle fibers and nerve cells in the brain and spinal cord) display **autorhythmicity**, the ability to repeatedly generate spontaneous action potentials. In the heart, these action potentials cause alternating contraction and relaxation of the heart muscle fibers. Compared with skeletal muscle fibers, cardiac muscle fibers are shorter in length and less circular in transverse section (Figure 10.8a). They also exhibit branching, which gives individual cardiac muscle fibers a "stair-step" appearance. A typical cardiac muscle fiber is 50–100 μm long and has a diameter of about 14 μm. Usually one centrally located nucleus is present, although an occasional cell may have two nuclei. The ends of cardiac muscle fibers connect to neighboring fibers by irregular transverse thickenings of the sarcolemma referred to as **intercalated discs** (in-TER-kā-lāt-ed; *intercalat-*=to insert between). The discs contain **desmosomes**, which hold the fibers together, and **gap junctions**, which allow muscle action potentials to spread from one muscle fiber to its neighbors. Cardiac muscle tissue has an endomysium, but lacks a perimysium and epimysium.

Mitochondria are larger and more numerous in cardiac muscle fibers than in skeletal muscle fibers. Cardiac muscle fibers have the same arrangement of actin and myosin, and the same bands, zones, and Z discs, as skeletal muscle fibers (Figure 10.8b). The transverse (T) tubules of cardiac muscle are wider but less abundant than those of skeletal muscle; there is one T tubule per sarcomere, located at the Z disc. The sarcoplasmic reticulum of cardiac muscle fibers is somewhat smaller than the SR of skeletal muscle fibers.

Under normal resting conditions, cardiac muscle tissue contracts and relaxes about 75 times per minute. This continuous, rhythmic activity is a major functional difference between cardiac and skeletal muscle tissue. Another difference is the source of stimulation. Skeletal muscle tissue contracts only when stimulated by acetylcholine released by an action potential in a motor neuron. In contrast, cardiac muscle tissue can contract without extrinsic (outside) nervous or hormonal stimulation. Its source of stimulation is a conducting network of specialized cardiac muscle fibers within the heart. Stimulation from the body's nervous or endocrine system merely causes the conducting fibers to increase or decrease their rate of discharge. Cardiac muscle tissue remains contracted 10 to 15 times longer than skeletal muscle tissue, allowing time for the heart chambers to relax and fill with blood between beats. This permits the heart rate to increase significantly but prevents the heart from undergoing *tetanus* (sustained contraction), which would stop blood flow. Like skeletal muscle, cardiac muscle fibers can undergo hypertrophy in response to increased workload. Hence, many athletes have enlarged hearts.

◉ CHECKPOINT

12. How do cardiac and skeletal muscle tissue differ structurally and functionally?

SMOOTH MUSCLE TISSUE

◉ OBJECTIVE

- Describe the main structural and functional characteristics of smooth muscle tissue.

Like cardiac muscle tissue, **smooth muscle tissue** is usually activated involuntarily. Of the two types of smooth muscle tissue, the more common type is **visceral (single-unit) smooth muscle tissue** (Figure 10.9a on page 302). It is found in wraparound sheets that form part of the walls of small arteries and veins and of hollow viscera such as the stomach, intestines, uterus, and urinary bladder. Like cardiac muscle, visceral smooth muscle is autorhythmic. Because the fibers connect to one another by gap junctions, muscle action potentials spread throughout the network. For example, when a neurotransmitter, hormone, or autorhythmic signal stimulates one fiber, the muscle action potential spreads to neighboring fibers, which then contract in unison, as a single unit.

The second kind of smooth muscle tissue is **multiunit smooth muscle tissue** (Figure 10.9b on page 302). It consists of individual fibers, each with its own motor neuron terminals and with few gap junctions between neighboring fibers. Stimulation of one visceral muscle fiber causes contraction of many adjacent fibers, but stimulation of one multiunit fiber causes contraction of that fiber only. The walls of large arteries, the airways to the lungs, the arrector pili muscles that attach to hair follicles, the muscles of the iris that adjust pupil diameter, and the ciliary body that adjusts focus of the lens in the eye all contain multiunit smooth muscle tissue.

Figure 10.8 **Histology of cardiac muscle.** A photomicrograph of cardiac muscle tissue is shown in Table 3.5B on page 86.

Cardiac muscle fibers display autorhythmicity, the ability to repeatedly generate spontaneous action potentials.

(a) Cardiac muscle fibers

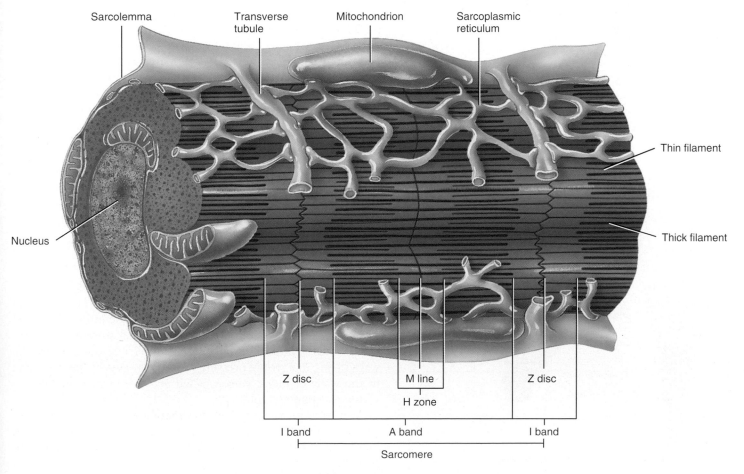

(b) Arrangement of components in a cardiac muscle fiber

? **What are the functions of intercalated discs in cardiac muscle fibers?**

Figure 10.9 **Histology of smooth muscle tissue.** In (a), one autonomic motor neuron synapses with several visceral smooth muscle fibers, and action potentials spread to neighboring fibers through gap junctions. In (b), three autonomic motor neurons synapse with individual multiunit smooth muscle fibers. Stimulation of one multiunit fiber causes contraction of that fiber only. A photomicrograph of smooth muscle tissue is shown in Table 3.5C on page 86.

Smooth muscle fibers have thick and thin filaments but no transverse tubules and little sarcoplasmic reticulum.

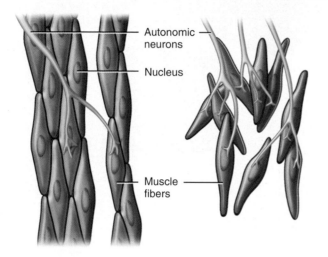

(a) Visceral (single-unit) smooth muscle tissue

(b) Multiunit smooth muscle tissue

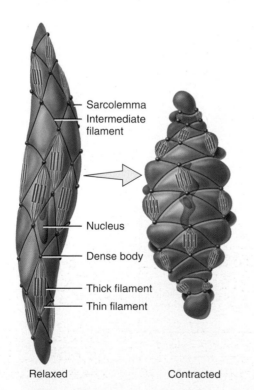

Relaxed   Contracted

(c) Microscopic anatomy of a relaxed and contracted smooth muscle fiber

❓ **Which type of smooth muscle is more like cardiac muscle than skeletal muscle, with respect to both its structure and function?**

Smooth muscle fibers are considerably smaller than skeletal muscle fibers. A single relaxed smooth muscle fiber is 30–200 mm long, thickest in the middle (3–8 mm), and tapered at each end (Figure 10.9c). Within each fiber is a single, oval, centrally located nucleus. The sarcoplasm of smooth muscle fibers contains both thick filaments and thin filaments, in ratios between about 1:10 and 1:15, but they are not arranged in orderly sarcomeres as in striated muscle. Smooth muscle fibers also contain **intermediate filaments**. These filaments contain the protein desmin and appear to have a structural rather than a contractile role. Because the various filaments have no regular pattern of overlap, smooth muscle fibers do not exhibit striations—thus the name *smooth*. Smooth muscle fibers also lack transverse tubules and have little sarcoplasmic reticulum for storage of Ca^{2+}. Smooth muscle tissue has an endomysium, but lacks a perimysium and epimysium.

In smooth muscle fibers, intermediate filaments attach to structures called **dense bodies**, which are functionally similar to Z discs in striated muscle fibers. Some dense bodies are dispersed throughout the sarcoplasm; others are attached to the sarcolemma. Bundles of intermediate filaments stretch from one dense body to another (Figure 10.9c). During contraction, the sliding filament mechanism involving thick and thin filaments generates tension that is transmitted to intermediate filaments. These in turn pull on the dense bodies attached to the sarcolemma, causing a shortening of the muscle fiber (Figure 10.9c). When a smooth muscle fiber contracts, it turns like a corkscrew; it rotates in the opposite direction as it relaxes.

Although the principles of contraction are similar in all three types of muscle tissue, smooth muscle tissue exhibits some important physiological differences. Compared with contraction in a skeletal muscle fiber, contraction in a smooth muscle fiber starts more slowly and lasts much longer. In addition, smooth muscle can both shorten and stretch to a greater extent than other muscle types.

As in striated muscle, an increase in the concentration of Ca^{2+} in the cytosol of smooth muscle initiates contraction. There is far less sarcoplasmic reticulum (the reservoir for Ca^{2+} in striated muscle) in smooth muscle than in skeletal muscle. Calcium ions flow into smooth muscle sarcoplasm from both the interstitial fluid and sarcoplasmic reticulum, but because there are no transverse tubules in smooth muscle fibers, it takes longer for Ca^{2+} to reach the filaments in the center of the fiber and trigger the contractile process. This accounts, in part, for the slow onset and prolonged contraction of smooth muscle.

Calcium ions also move out of the muscle fiber slowly, which delays relaxation. The prolonged presence of Ca^{2+} in the cytosol provides for **smooth muscle tone**, a state of continuous partial contraction. Smooth muscle tissue can thus sustain long-term tone, which is important in the gastrointestinal tract, where the walls maintain a steady pressure on the contents of the tract, and in the walls of blood vessels called *arterioles*, which maintain a steady pressure on blood.

Most smooth muscle fibers contract or relax in response to action potentials from the autonomic nervous system. In addition, many smooth muscle fibers contract or relax in response to stretching, hormones, or local changes in pH, oxygen and carbon dioxide levels, temperature, and ion concentrations.

Unlike striated muscle fibers, smooth muscle fibers can stretch considerably and still maintain their contractile function. When smooth muscle fibers are stretched, they initially contract, developing increased tension. Within a minute or so, the tension decreases. This phenomenon, called the **stress–relaxation response**, allows smooth muscle to undergo great changes in length while still retaining the ability to contract effectively. Even though smooth muscle in the walls of blood vessels and hollow organs such as the stomach, intestines, and urinary bladder can stretch, the pressure on the contents within them changes very little. After the organ empties, however, the smooth muscle rebounds, and the wall of the organ retains its firmness.

Smooth muscle tissue, like skeletal and cardiac muscle tissue, can undergo hypertrophy. In addition, certain smooth muscle fibers, such as those in the uterus, retain their capacity for division and thus can grow by hyperplasia. Also, new smooth muscle fibers can arise from cells called *pericytes*, stem cells found in asso-

ciation with blood capillaries and small veins. Smooth muscle fibers can also proliferate in certain pathological conditions, such as occur in the development of atherosclerosis (see page 000). Although smooth muscle tissue has considerably greater powers of regeneration than skeletal muscle or cardiac muscle, such powers are limited when compared with tissues such as epithelium.

Table 10.2 summarizes the major characteristics of the three types of muscular tissue.

◉ **CHECKPOINT**

13. How do visceral and multiunit smooth muscle differ?
14. Compare the properties of skeletal and smooth muscle.

TABLE 10.2

Summary of the Major Features of the Three Types of Muscular Tissue

CHARACTERISTIC	SKELETAL MUSCLE	CARDIAC MUSCLE	SMOOTH MUSCLE
Microscopic appearance and features	Long cylindrical fiber with many peripherally located nuclei; striated.	Branched cylindrical fiber with one centrally located nucleus; intercalated discs join neighboring fibers; striated.	Fiber is thickest in middle, tapered at each end, and has one centrally positioned nucleus; not striated.
Location	Most commonly attached by tendons to bones.	Heart.	Walls of hollow viscera, airways, blood vessels, iris and ciliary body of eye, arrector pili muscles of hair follicles.
Fiber diameter	Very large (10–100 μm).	Large (10–20 μm).	Small (3–8 μm).
Connective tissue components	Endomysium, perimysium, and epimysium.	Endomysium.	Endomysium.
Fiber length	100 μm–30 cm.	50–100 μm.	30–200 μm.
Contractile proteins organized into sarcomeres	Yes.	Yes.	No.
Sarcoplasmic reticulum	Abundant.	Some.	Very little.
Transverse tubules present	Yes, aligned with each A–I band junction.	Yes, aligned with each Z disc.	No.
Junctions between fibers	None.	Intercalated discs contain gap junctions and desmosomes.	Gap junctions in visceral smooth muscle; none in multiunit smooth muscle.
Autorhythmicity	No.	Yes.	Yes, in visceral smooth muscle.
Source of Ca²⁺ for contraction	Sarcoplasmic reticulum.	Sarcoplasmic reticulum and interstitial fluid.	Sarcoplasmic reticulum and interstitial fluid.
Speed of contraction	Fast.	Moderate.	Slow.
Nervous control	Voluntary (somatic nervous system).	Involuntary (autonomic nervous system).	Involuntary (autonomic nervous system).
Capacity for regeneration	Limited, via satellite cells.	Limited, under certain conditions.	Considerable, via pericytes (compared with other muscular tissues, but limited compared with epithelium).

DEVELOPMENT OF MUSCLES

■ OBJECTIVES

• Describe the development of muscles.

Except for the muscles of the iris of the eyes, all muscles of the body are derived from **mesoderm**. Recall from Chapter 4 that by about the seventeenth day after fertilization, the mesoderm adjacent to the notochord and neural tube forms paired longitudinal columns of paraxial mesoderm (see Figure 4.9b on page 103). The paraxial mesoderm soon undergoes segmentation. Initially the segments form **somitomeres** (sō-MĪT-ō-mērz), small bulges in the paraxial mesoderm. Posterior to the developing head, the somitomeres undergo further development into paired, cube-shaped structures called **somites** (Figure 10.10a). The first pair

of somites appears on the twentieth day of embryonic development. By the end of the fifth week, 42 to 44 pairs of somites are formed. The number of somites that develop over a given period can be correlated to the approximate age of the embryo. In the head region the somitomeres never become somites. Rather, the cells of the somitomeres migrate into the developing pharyngeal arches around the anterior end of the pharynx (throat).

As a result, all the skeletal muscles develop from the *paraxial mesoderm*. The somites give rise to all the muscles of the trunk wall and the limbs. The migrating somitomeres of the pharyngeal arches form the skeletal muscles of the head region.

As you learned in Chapter 4, the cells of a somite differentiate into three regions: (1) a **myotome**, which forms the skeletal muscles of the trunk and limbs; (2) a **dermatome**, which forms the connective tissues, including the dermis of the skin; and (3) a **sclerotome**, which gives rise to the vertebrae (Figure 10.10b).

Figure 10.10 Location and structure of somites, key structures in the development of the muscular system.

🔑 Most muscles are derived from mesoderm.

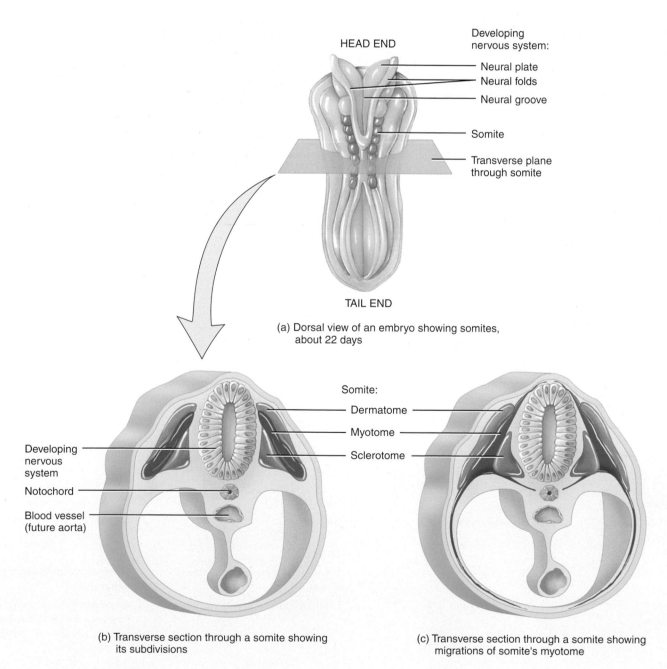

HEAD END

Developing nervous system:

Neural plate
Neural folds
Neural groove

Somite

Transverse plane through somite

TAIL END

(a) Dorsal view of an embryo showing somites, about 22 days

Somite:

Dermatome
Myotome
Sclerotome

Developing nervous system
Notochord
Blood vessel (future aorta)

(b) Transverse section through a somite showing its subdivisions

(c) Transverse section through a somite showing migrations of somite's myotome

❓ **Which part of a somite differentiates into skeletal muscle?**

Cardiac muscle develops from **mesodermal cells** that migrate to and envelop the developing heart while it is still in the form of endocardial tubes (see in Figure 14.12b on page 477).

Smooth muscle develops from lateral plate **mesodermal cells** that migrate to and envelop the developing gastrointestinal tract and viscera.

◉ CHECKPOINT

15. From which type of embryonic tissue do most muscles develop?

AGING AND MUSCULAR TISSUE

◉ OBJECTIVE

- Explain the effects of aging on skeletal muscle.

With aging, humans undergo a slow, progressive loss of skeletal muscle mass that is replaced largely by fibrous connective tissue and adipose tissue. In part, this decline is due to decreased levels of physical activity. Accompanying the loss of muscle mass is a decrease in maximal strength, a slowing of muscle reflexes, and a loss of flexibility. Muscle strength at age 85 is about half that at age 25. In some muscles, a selective loss of muscle fibers of a given type may occur. With aging, the relative number of slow oxidative fibers appears to increase. This could be due to either atrophy of the other fiber types or their conversion into slow oxidative fibers. Whether this is an effect of aging itself or mainly reflects the more limited physical activity of older people is still an unresolved question. Nevertheless, aerobic activities and strength training programs are effective in older people and can slow or even reverse the age-associated decline in muscular performance.

◉ CHECKPOINT

16. Why does muscle strength decrease with aging?

KEY MEDICAL TERMS ASSOCIATED WITH MUSCULAR TISSUE

Muscle strain Tearing of a muscle because of forceful impact, accompanied by bleeding and severe pain. Also known as a Charley horse or pulled muscle. It often occurs in contact sports and typically affects the quadriceps femoris muscle on the anterior surface of the thigh. The condition is treated by RICE therapy: rest (R), ice immediately after the injury (I), compression via a supportive wrap (C), and elevation of the limb (E).

Myalgia (mī-AL-jē-a; *-algia*=painful condition) Pain in or associated with muscles.

Myoma (mī-Ō-ma; *-oma*=tumor) A tumor consisting of muscular tissue.

Myomalacia (mī-Ō-ma-LĀ-shē-a; *-malacia*=soft) Pathological softening of muscular tissue.

Myositis (mī′-ō-SĪ-tis; *-itis*=inflammation of) Inflammation of muscle fibers (cells).

Myotonia (mī′-ō-TŌ-nē-a; *-tonia*=tension) Increased muscular excitability and contractility, with decreased power of relaxation; tonic spasm of the muscle.

Volkmann's contracture (FŌLK-manz kon-TRAK-tur; *contra-*=against) Permanent shortening (contracture) of a muscle due to replacement of destroyed muscle fibers by fibrous connective tissue, which lacks extensibility. Typically occurs in forearm flexor muscles. Destruction of muscle fibers may occur from interference with circulation caused by a tight bandage, a piece of elastic, or a cast.

STUDY OUTLINE

Introduction (p. 287)

1. Muscles constitute 40–50% of total body weight.
2. The prime function of muscle is changing chemical energy into mechanical energy to perform work.

Overview of Muscular Tissue (p. 288)

1. The three types of muscular tissue are skeletal, cardiac, and smooth. Skeletal muscle tissue is primarily attached to bones; it is striated and under voluntary control. Cardiac muscle tissue forms the wall of the heart; it is striated and involuntary. Smooth muscle tissue is located primarily in internal organs; it is nonstriated (smooth) and involuntary.
2. Through contraction and relaxation, muscular tissue performs four important functions: producing body movements, stabilizing body positions, storing and moving substances within the body, and producing heat.
3. Four special properties of muscular tissues are electrical excitability, the property of responding to stimuli by producing action potentials; contractility, the ability to generate tension to do work; extensibility, the ability to be extended (stretched); and elasticity, the ability to return to original shape after contraction or extension.

Skeletal Muscle Tissue (p. 289)

1. Connective tissues surrounding skeletal muscles are the epimysium, covering the entire muscle; perimysium, covering fascicles; and the endomysium, covering muscle fibers. Fascia covers all the muscles of a region and separates muscle from the layers of the skin.
2. Tendons and aponeuroses are extensions of the connective tissue within the muscle belly beyond the muscle fibers that attach the muscle to bone or to other muscle.
3. Each skeletal muscle fiber has 100 or more nuclei because it arises from fusion of many myoblasts. Satellite cells are myoblasts that persist after birth. The sarcolemma is a muscle fiber's plasma membrane; it surrounds the sarcoplasm. T tubules are invaginations of the sarcolemma.
4. Each fiber contains myofibrils, the contractile elements of skeletal muscle. Sarcoplasmic reticulum surrounds each myofibril. Within a myofibril are thin and thick filaments arranged in compartments called sarcomeres.
5. Muscular atrophy is a wasting away of muscles, muscular hypertrophy is an increase in the size of muscles.
6. The overlapping of thick and thin filaments produces striations; darker A bands alternate with lighter I bands.
7. Myofibrils are built from three types of proteins: contractile, regulatory, and structural. The contractile proteins are myosin (thick filament) and actin (thin filament). Regulatory proteins are tropomyosin and troponin (both are part of the thin filament). Structural proteins include titin (links Z disc to M line and stabilizes thick filament).

8. Projecting myosin heads contain actin-binding and ATP-binding sites and are the motor proteins that power muscle contraction.

9. Skeletal muscles are well supplied with nerves and blood vessels. Generally, an artery and one or two veins accompany each nerve that penetrates a skeletal muscle. Blood capillaries bring in oxygen and nutrients and remove heat and waste products of muscle metabolism.

10. Somatic motor neurons provide the nerve impulses that stimulate skeletal muscle to contract.

11. The neuromuscular junction (NMJ) is the synapse between a somatic motor neuron and a skeletal muscle fiber. The NMJ includes the axon terminals and synaptic end bulbs of a motor neuron, plus the adjacent motor end plate of the muscle fiber sarcolemma.

12. When an action potential reaches the end bulbs of a somatic motor neuron, it triggers release of ACh. ACh diffuses across the synaptic cleft and binds to ACh receptors, initiating a muscle action potential. Acetylcholinesterase then quickly destroys ACh.

13. Muscle contraction occurs because myosin heads attach to and "walk" along the thin filaments at both ends of a sarcomere, progressively pulling the thin filaments toward the center of a sarcomere. As the thin filaments slide inward, the Z discs come closer together, and the sarcomere shortens.

14. In a concentric isotonic contraction, the muscle shortens to produce movement and to reduce the angle at a joint. During an eccentric isotonic contraction, the muscle lengthens.

15. Isometric contractions, in which tension is generated without muscle changing its length, are important because they stabilize some joints as others are moved.

Types of Skeletal Muscle Fibers (p. 298)

1. On the basis of their structure and function, skeletal muscle fibers are classified as slow oxidative (SO) or type I, fast oxidative-glycolytic (FOG) or type IIa, and fast glycolytic (FG) or type IIb fibers.

2. Most skeletal muscles contain a mixture of all three fiber types. Their proportions vary with the typical action of the muscle.

3. The motor units of a muscle are recruited in the following order: first SO fibers, then FOG fibers, and finally FG fibers.

4. Table 10.1 on page 299 summarizes the three types of skeletal muscle fibers.

Exercise and Skeletal Muscle Tissue (p. 300)

1. Various types of exercises can induce changes in the fibers in a skeletal muscle. Endurance-type (aerobic) exercises cause a gradual transformation of some fast glycolytic (FG) fibers into fast oxidative-glycolytic (FOG) fibers.

2. Exercises that require great strength for short periods produce an increase in the size and strength of fast-glycolytic (FG) fibers. The increase in size is due to increased synthesis of thick and thin filaments.

Cardiac Muscle Tissue (p. 300)

1. This muscle tissue is found only in the heart. It is striated and involuntary.

2. Cardiac muscle fibers are branching cylinders and usually contain a single centrally located nucleus.

3. Compared to skeletal muscle tissue, cardiac muscle tissue has more mitochondria, smaller sarcoplasmic reticulum, and wider transverse tubules located at Z discs rather than at A–I band junctions.

4. Cardiac muscle fibers branch and are connected end to end via desmosomes.

5. Intercalated discs provide strength and aid in conduction of muscle action potentials by way of gap junctions located in the discs.

6. Unlike skeletal muscle tissue, cardiac muscle tissue contracts continuously and rhythmically.

7. Cardiac muscle tissue can contract without extrinsic stimulation and can remain contracted longer than skeletal muscle tissue.

Smooth Muscle Tissue (p. 300)

1. Smooth muscle tissue is nonstriated and involuntary.

2. Smooth muscle fibers contain intermediate filaments and dense bodies that function as Z discs.

3. Visceral (single-unit) smooth muscle is found in the walls of viscera and small blood vessels. The fibers are arranged in a network.

4. Multiunit smooth muscle is found in large blood vessels, arrector pili muscles, and the iris of the eye. The fibers operate independently rather than in unison.

5. The duration of contraction and relaxation of smooth muscle is longer than in skeletal muscle.

6. Smooth muscle fibers contract in response to nerve impulses, hormones, and local factors.

7. Smooth muscle fibers can stretch considerably without developing tension.

8. Table 10.2 on page 303 summarizes the principal characteristics of the three types of muscle tissue.

Development of Muscles (p. 304)

1. With few exceptions, muscles develop from mesoderm.

2. Skeletal muscles develop from the paraxial mesoderm, via somites and migrating somitomeres. Cardiac muscle and smooth muscle develop from mesodermal cells that migrate during the development process to the heart and to the gastrointestinal tract and viscera, respectively.

Aging and Muscular Tissue (p. 305)

1. With aging, there is a slow, progressive loss of skeletal muscle mass, which is replaced by fibrous connective tissue and fat.

2. Aging also results in a decrease in muscle strength, slower muscle reflexes, and loss of flexibility, which can be compensated for to an extent by increased physical activity.

SELF-QUIZ QUESTIONS

Choose the one best answer to the following questions.

1. The specialized region of the sarcolemma at the neuromuscular junction that contains neurotransmitter receptors is called the:
   a. synapse.   b. synaptic cleft.
   c. motor end plate. d. cisterna.
   e. motor unit.

2. The ability of muscular tissue to respond to a stimulus by producing action potentials is referred to as:
   a. contractility.   b. electrical excitability.
   c. elasticity.   d. extensibility.
   e. conductivity.

3. In Duchenne muscular dystrophy the gene that codes for the protein dystrophin is mutated. The function of dystrophin in muscle cells is to:
   a. contract in response to increases in calcium ion levels.
   b. block the myosin binding sites on actin molecules.
   c. block the actin binding sites on myosin molecules.
   d. stabilize the position of thin filaments.
   e. provide energy to fuel the shortening of the sarcomere.

4. The most fatigue-resistant skeletal muscle fibers are the:
   a. slow oxidative fibers.
   b. fast oxidative-glycolytic fibers.
   c. fast glycolytic fibers.
   d. satellite cells.
   e. All types of fibers are equally fatigue-resistant.

5. The skeletal muscle fibers with the largest diameters that generate the most powerful contractions due to their high myofibril content are the:
   a. slow oxidative fibers.   b. fast oxidative-glycolytic fibers.
   c. fast glycolytic fibers.   d. satellite cells.
   e. All types of fibers produce equally strong contractions.

6. A motor unit is defined as a:
   a. nerve and a muscle.
   b. single motor neuron and a single muscle fiber.
   c. motor neuron and all the muscle fibers it stimulates.
   d. single muscle fiber and the nerves that innervate it.
   e. muscle and the motor and sensory nerves that innervate it.

7. Arrange the following from largest to smallest: (1) myofibril, (2) filament, (3) muscle fiber, (4) fascicle.
   a. 1, 3, 2, 4   b. 3, 4, 2, 1
   c. 3, 1, 4, 2   d. 4, 3, 1, 2
   e. 4, 2, 3, 1

Complete the following.

8. The _____ separates muscle from skin.

9. Connective tissue that separates one muscle fiber from another is the _____.

10. Plate-shaped regions of dense material separating one sarcomere from the next are called _____.

11. The two regulatory proteins in thin filaments are _____ and _____.

12. To describe a sarcomere during contraction, complete each statement with one of the following: L (lengthens), S (shortens), or U (is unchanged in length).
    a. The sarcomere _____.
    b. Each thick myofilament (A band) _____.
    c. Each thin myofilament _____.
    d. The I band _____.
    e. The H zone _____.

13. The muscle tissues that exhibit autorhythmicity are _____ and _____.

14. _____ is a stretched or partially torn muscle; _____ is the forcible wrenching or twisting of a joint that stretches or tears its ligaments.

Are the following statements true or false? If the statement is False, *find and fix the mistake to make the statement* True.

15. Enlargement of existing muscle fibers is called hyperplasia.

16. In adult life, growth of skeletal muscle is due to an increase in the number of muscle cells rather than an increase in the size of existing cells.

17. Duchenne muscular dystrophy (DMD) typically strikes girls in their late teenage years.

18. Stimulation of one visceral (single-unit) smooth muscle fiber causes contraction of several adjacent fibers; stimulation of a multiunit fiber causes contraction of that fiber only.

19. Thick filaments are made mostly of actin.

Matching

20. Match the following:
    _____ (a) involuntary muscle found in blood vessels and the intestines
    _____ (b) involuntary striated muscle
    _____ (c) striated voluntary muscle attached to bones
    _____ (d) spindle-shaped cells
    _____ (e) intercalated discs
    _____ (f) no transverse tubules
    _____ (g) several nuclei, peripherally arranged

    (1) cardiac muscle
    (2) skeletal muscle
    (3) smooth muscle

21. Match the following:
    _____ (a) functional units of myofibrils
    _____ (b) red protein that binds oxygen in the sarcoplasm
    _____ (c) releases acetylcholine into the synaptic cleft at the neuromuscular junction
    _____ (d) invaginations of the plasma membrane of muscle cells
    _____ (e) where the acetylcholine receptors are located at the neuromuscular junction
    _____ (f) stores and releases calcium ions during muscle relaxation and contraction

    (1) sarcolemma
    (2) T tubules
    (3) myoglobin
    (4) sarcoplasmic reticulum
    (5) sarcomeres
    (6) motor neuron

CRITICAL THINKING QUESTIONS

1. A marathon runner and a weight-lifter were discussing muscle types while working out. Both were convinced that their own muscle types were the best. How does the leg muscle of the runner compare to the arm of the lifter?

2. Bill tore some ligaments in his knee while skiing. He was in a toe-to-thigh cast for 6 weeks. When the cast was removed, the newly healed leg was noticeably thinner than the uncasted leg. What happened to his leg?

3. The newspaper reported several cases of botulism poisoning following a potluck fund-raising dinner for the local clinic. The cause appeared to be a potato salad that contained the toxin of the soil bacteria *Clostridium botulinum*. This toxin blocks the release of acetylcholine. What would you expect the major effect of botulism poisoning to be, and why?

4. Research is under way to grow new cardiac muscle cells for ailing hearts. Skeletal muscle transplants have been tried, but they don't work as well as transplants of cardiac muscle. Both skeletal and cardiac muscle are striated, so why does cardiac muscle support rhythmic contraction while skeletal muscle does not?

5. Your study partner says that the contractility, extensibility, and elasticity of muscle tissue can be explained by the "stretchiness" of the contractile proteins in muscle cells. "After all," he says, "why else would they be called 'contractile' proteins?" How should you respond to your study partner? Explain.

? ANSWERS TO FIGURE QUESTIONS

10.1 Perimysium bundles groups of muscle fibers into fasciculi.

10.2 The sarcoplasmic reticulum releases calcium ions to trigger muscle contraction.

10.3 Size, from smallest to largest: thick filament, myofibril, muscle fiber.

10.4 There are two thin filaments for each thick filament in skeletal muscle.

10.5 Actin and titin anchor into the Z disc. A bands contain myosin, actin, troponin, tropomysin, and titin; I bands contain actin, troponin, tropomysin, and titin.

10.6 The part of the sarcolemma that contains acetylcholine receptors is the motor end plate.

10.7 During muscle contraction, the I bands and H zones of the sarcomere disappear. The lengths of the thin and thick filaments do not change.

10.8 The intercalated discs contain desmosomes that hold the cardiac muscle fibers together and gap junctions that enable action potentials to be spread from one muscle fiber to another.

10.9 Visceral smooth muscle and cardiac muscle are similar in that both contain gap junctions, which allow action potentials to spread from one cell to its neighbors.

10.10 As its name suggests, the myotome of a somite differentiates into skeletal muscle.

11 THE MUSCULAR SYSTEM

INTRODUCTION When you look at the world around you, you notice immediately that humans, like all animals, have the ability to move. Probably some of your first observations of movement involve walking, jogging, or running, activities that transport our bodies from one location to another. This type of movement is easy to recognize, and its survival value is unquestionable. But we also move in other ways. Think, for example, about grasping something with your hands and placing it in your mouth, protecting yourself by kicking at something with your lower limb, or throwing something at your roommate to wake him up in time for class. All of these activities occur without moving from one location to another, yet they are movements nonetheless. Reflect for a moment on the wide variety of movements that your roommate makes when he finally stumbles out of bed to get dressed for the day. These range from the simple movements of putting on clothing, to the intricate movements of buttoning shirts and tying shoelaces. Then there is the variety of intricate movements required to eat a meal, movements such as grasping, manipulating, cutting, chewing, and swallowing food. Communication also involves movement, whether it is writing a note on a piece of paper, typing on a keyboard, signaling pleasure and happiness with a smile, or using your voice as well as your throwing arm to wake up your roommate.

Together, the voluntarily controlled muscles of your body comprise the muscular system. The function of most muscles is to produce the movements we have described. A few muscles function mainly to stabilize bones so that other skeletal muscles can execute a movement more effectively. This chapter presents many of the major muscles in the body. For each muscle, we'll identify the attachment sites, actions, and *innervation*—the nerve or nerves that stimulate a muscle to contract—of each muscle described. Developing a working knowledge of these key aspects of skeletal muscle anatomy will help you understand how normal movements occur. This knowledge is especially crucial for professionals, such as those in the allied health and physical rehabilitation fields, whose patients' normal patterns of movement and physical mobility have been disrupted by physical trauma, surgery, or muscular paralysis. •

HOW SKELETAL MUSCLES PRODUCE MOVEMENTS

◉ **OBJECTIVES**

• Describe the relationship between bones and skeletal muscles in producing body movements.
• Define lever and fulcrum, and compare the three types of levers based on location of the fulcrum, effort, and load.
• Identify the types of fascicle arrangements in a skeletal muscle, and relate the arrangements to strength of contraction and range of motion.
• Explain how the prime mover, antagonist, synergist, and fixator in a muscle group work together to produce movement.

Muscle Attachment Sites: Origin and Insertion

Those skeletal muscles that produce movements do so by exerting force on tendons, which in turn pull on bones or other structures (such as skin). Most muscles cross at least one joint and are usually attached to articulating bones that form the joint (Figure 11.1a).

When a skeletal muscle contracts, it pulls one of the articulating bones toward the other. The two articulating bones usually do not move equally in response to contraction. One bone remains stationary or near its original position, either because other muscles stabilize that bone by contracting and pulling it in the opposite direction or because its structure makes it less movable. Ordinarily, the attachment of a muscle's tendon to the stationary bone is called the **origin**; the attachment of the muscle's other tendon to the movable bone is called the **insertion**. A good analogy is a spring on a door. In this example, the part of the spring attached to the frame is the origin; the part attached to the door represents the insertion. A useful rule of thumb is that the origin is usually proximal and the insertion distal, especially in the limbs; the insertion is usually pulled toward the origin. The fleshy portion of the muscle between the tendons is called the **belly (body)**, the coiled middle portion of the spring in our example. The **actions** of a muscle are the main movements that occur when the muscle contracts. In our spring example, this would be the closing of the door.

Muscles that move a body part often do not cover the moving part. Figure 11.1b shows that although one of the functions of the biceps brachii muscle is to move the forearm, the belly of the

Figure 11.1 Relationship of skeletal muscles to bones. (a) Muscles are attached to bones by tendons at their origin and insertion. (b) Skeletal muscles produce movements by pulling on bones. Bones serve as levers, and joints act as fulcrums for the levers. Here the lever-fulcrum principle is illustrated by the movement of the forearm. Note where the load (resistance) and effort are applied in this example.

🔑 In the limbs, the origin of a muscle is usually proximal and the insertion is usually distal.

(a) Origin and insertion of a skeletal muscle

(b) Movement of the forearm lifting a weight

❓ **Where is the belly of the muscle that extends the forearm located?**

muscle lies over the humerus, not over the forearm. You will also see that muscles that cross two joints, such as the rectus femoris and sartorius of the thigh, have more complex actions than muscles that cross only one joint.

Lever Systems

A **lever** is a rigid structure that can move around a fixed point called a **fulcrum**, symbolized by ▲. A lever is acted on at two different points by two different forces: the **effort** (E), which causes movement, and the **load** L or **resistance**, which opposes movement. The effort is the force exerted by muscular contraction; the load is typically the weight of the body part that is moved or some resistance that the moving body part is trying to overcome (such as the weight of a book you might be picking up). Motion occurs when the effort applied to the bone at the insertion exceeds the load. Consider the biceps brachii flexing the forearm at the elbow as an object is lifted (Figure 11.1b). When the forearm is raised, the elbow is the fulcrum. The weight of the forearm plus the weight of the object in the hand is the load. The force of contraction of the biceps brachii pulling the forearm up is the effort.

The relative distance between the fulcrum and load and the point at which the effort is applied determine if a given lever operates at a mechanical advantage or a mechanical disadvantage. For example, if the load is closer to the fulcrum and the effort far-

ther from the fulcrum, then only a relatively small effort is required to move a large load over a small distance. This is called a **mechanical advantage**. If, instead, the load is farther from the fulcrum and the effort is applied closer to the fulcrum, then a relatively large effort is required to move a small load (but at greater speed). This is called a **mechanical disadvantage**. Compare chewing something hard (the load) with your front teeth and the teeth in the back of your mouth. It is much easier to crush the hard food item with the back teeth because they are closer to the fulcrum (the jaw or temporomandibular joint) than are the front teeth. Here is one more example you can try. Straighten out a paper clip. Now get a pair of scissors and try to cut the paper clip with the tip of the scissors (mechanical disadvantage) versus near the pivot point of the scissors (mechanical advantage).

Levers are categorized into three types according to the positions of the fulcrum, the effort, and the load:

1. The fulcrum is between the effort and the load in **first-class levers** (Figure 11.2a). (Think E*F*L.) Scissors and seesaws are examples of first-class levers. A first-class lever can produce either a mechanical advantage or mechanical disadvantage depending on whether the effort or the load is closer to the fulcrum. (Think of an adult and a child on a seesaw.) As we've seen in the preceding examples, if the effort (child) is farther from the fulcrum than the load (adult), a heavy load can be moved, but not very far or fast. If the effort is closer to the ful-

Figure 11.2 Types of levers.

Levers are divided into three types based on the placement of the fulcrum, effort, and load (resistance).

Key:
E = Effort
F = Fulcrum
L = Load

(a) First-class lever

(b) Second-class lever

(c) Third-class lever

? Which type of lever produces the most force?

crum than the load, only a lighter load can be moved, but it moves far and fast. There are few first-class levers in the body. One example is the lever formed by the head resting on the vertebral column (Figure 11.2a). When the head is raised, the contraction of the posterior neck muscles provides the effort (E), the joint between the atlas and the occipital bone (atlanto-occipital joint) forms the fulcrum ⓐ, and the weight of the anterior portion of the skull is the load ⓛ.

2. The load is between the fulcrum and the effort in **second-class levers** (Figure 11.2b). (Think FLE.) Second-class levers operate like a wheelbarrow. They always produce a mechanical advantage because the load is always closer to the fulcrum than the effort. This arrangement sacrifices speed and range of motion for force; this type of lever produces the most force. This class of lever is uncommon in the human body. An example is standing up on your toes. The fulcrum ⓐ is the ball of the foot. The load is the weight of the body. The effort (E) is the contraction of the muscles of the calf, which raise the heel off the ground.

3. The effort is between the fulcrum and the load in **third-class levers** (Figure 11.2c). (Think FEL.) These levers operate like a pair of forceps and are the most common levers in the body. Third-class levers always produce a mechanical disadvantage because the effort is always closer to the fulcrum than the load. In the body, this arrangement favors speed and range of motion over force. The elbow joint, the biceps brachii muscle, and the bones of the arm and forearm are one example of a third-class lever (Figure 11.2c). As we have seen, in flexing the forearm at the elbow, the elbow joint is the fulcrum ⓐ, the contraction of the biceps brachii muscle provides the effort (E), and the weight of the hand and forearm is the load ⓛ.

Effects of Fascicle Arrangement

Recall from Chapter 10 that the skeletal muscle fibers (cells) within a muscle are arranged in bundles known as **fascicles**. Within a fascicle, all muscle fibers are parallel to one another. The fascicles, however, may form one of five patterns with respect to the tendons: parallel, fusiform (spindle-shaped, narrow toward the ends and wide in the middle), circular, triangular, or pennate (shaped like a feather) (Table 11.1).

Fascicular arrangement affects a muscle's power and range of motion. As a muscle fiber contracts, it shortens to about 70% of its resting length. Thus, the longer the fibers in a muscle, the greater the range of motion it can produce. However, the power of a muscle depends not on length but on its total cross-sectional area, because a short fiber can contract as forcefully as a long one. So the more fibers per unit of cross-sectional area a muscle has, the more power it can produce. Fascicular arrangement often represents a compromise between power and range of motion. Pennate muscles, for instance, have a large number of short-fibered fascicles distributed over their tendons, giving them greater power but a smaller range of motion. In contrast, parallel muscles have comparatively few fascicles with long fibers that extend the length of the muscle, so they have a greater range of motion but less power.

Muscle Actions

In the study of muscles, it is common to describe the actions that individual muscles produce at their associated joints. It is important to recognize what movements muscles are capable of producing at the joints they cross, but it is equally important to realize that, in reality, muscles do not work in isolation. Sometimes the movements attributed to muscles are possible only for a certain range of the joint's movement, or occur in combination with the actions of other muscles. As you study the muscles in this chapter, we will introduce you to the primary actions they generate at the joints they cross. With selected muscles we will explore a broader view of their functional roles in the body.

In Chapter 10 we learned that the connective tissues surrounding the contractile components within the muscle belly emerge from either end of the muscle as the tendons, to blend with the periosteum and attach to the bone. When these contractile components work, they generate tension within the muscle as the sliding filaments of the sarcomere attempt to shorten (see page 296). This results in two potential types of isotonic contraction—concentric and eccentric. Recall that an *isotonic contraction* is one in which enough muscle fibers are contracting to shorten the muscle against the load (see page 297). This is in contrast to an *isometric contraction*, where the number of fibers contracting and generating a force are equal to the opposite force of the load and the muscle does not change length. During a **concentric contraction** the muscle shortens as it produces a constant tension and overcomes the load it is moving. In an **eccentric contraction**, the muscle produces a constant tension but lengthens as it gives in to the load it is moving. As a result, the same muscle is the active controller of two opposite movements at a joint. Imagine the following situation. You pick up a book from a table by flexing the elbow joint to lift the book. You then hold the book steady in front of you as you look at it. Next, you slowly lower the book back to the table by extending the elbow joint. The biceps brachii muscle controls this full range of activity. Isotonic concentric contraction of the biceps brachii overcomes the weight of the book and raises it off the table as the elbow joint is flexed. Isometric contraction of the biceps brachii holds the book steady in front of you with a flexed (bent) elbow as you look at it. Finally, isotonic eccentric contraction of the biceps brachii slowly gives way to the load and the book is lowered back down to the table as the elbow joint extends.

Throughout this chapter the actions described for muscles will be the actions produced by concentric (shortening) contractions of the muscles. However, realize that joint actions are not always simply the result of opposing muscles. As we just explained, a single muscle can control opposite joint movements through its concentric and eccentric contractions.

Coordination Among Muscles

Movements often are the result of several skeletal muscles acting as a group. Most skeletal muscles are arranged in opposing (antagonistic) pairs at joints—that is, flexors–extensors, abductors–adductors, and so on. Within opposing pairs, one muscle, called the **prime mover** or **agonist** (=leader), contracts to cause an action while the other muscle, the **antagonist** (*anti-*=against), stretches and yields to the effects of the prime mover. In the process of flexing the forearm at the elbow, for instance, the biceps brachii is the prime mover, and the triceps brachii is the antagonist (see Figure 11.1a). The antagonist and prime mover are usually located on opposite sides of the bone or joint, as is the case in this example.

With an opposing pair of muscles, the roles of the prime mover and antagonist can switch for different movements. For example, while extending the forearm at the elbow against resistance (for example, pushing against some resistance with the back of the hand as you straighten the elbow), the triceps brachii becomes the prime mover, and the biceps brachii is the antagonist. If a prime mover and its antagonist contract at the same time with equal force, there will be no movement.

TABLE 11.1

Arrangement of Fascicles

PARALLEL	FUSIFORM
Fascicles parallel to longitudinal axis of muscle; terminate at either end in flat tendons. *Example:* Stylohyoid muscle (see Figure 11.8a)	Fascicles nearly parallel to longitudinal axis of muscle; terminate in flat tendons; muscle tapers toward tendons, where diameter is less than at belly. *Example:* Digastric muscle (see Figure 11.8a)

CIRCULAR	TRIANGULAR
Fascicles in concentric circular arrangements form sphincter muscles that enclose an orifice (opening). *Example:* Orbicularis oculi muscle (see Figure 11.4a)	Fascicles spread over broad area converge at thick central tendon; gives muscle a triangular appearance. *Example:* Pectoralis major muscle (see Figure 11.3a)

PENNATE

Short fascicles in relation to total muscle length; tendon extends nearly entire length of muscle.

Unipennate	**Bipennate**	**Multipennate**
Fascicles are arranged on only one side of tendon.	Fascicles are arranged on both sides of centrally positioned tendons.	Fascicles attach obliquely from many directions to several tendons.
Example: Extensor digitorum longus muscle (see Figure 11.24b)	*Example:* Rectus femoris muscle (see Figure 11.3a)	*Example:* Deltoid muscle (see Figure 11.17b)

Sometimes a prime mover crosses other joints before it reaches the joint at which its primary action occurs. The biceps brachii, for example, spans both the shoulder and elbow joints, with primary action on the forearm. To prevent unwanted movements at intermediate joints or to otherwise aid the movement of the prime mover, muscles called **synergists** (SIN-er-gists; *syn-=* together; *-ergon*=work) contract and stabilize the intermediate joints. As an example, muscles that flex the fingers (prime movers) cross the intercarpal and radiocarpal joints (intermediate joints). If movement at these intermediate joints was unrestrained, you would not be able to flex your fingers without flexing the wrist at the same time. Synergistic contraction of the wrist extensor muscles stabilizes the wrist joint and prevents unwanted movement, while the flexor muscles of the fingers contract to bring about the primary action, efficient flexion of the fingers. Synergists are usually located close to the prime mover.

Some muscles in a group also act as **fixators**, stabilizing the origin of the prime mover so that the prime mover can act more efficiently. Fixators steady the proximal end of a limb while movements occur at the distal end. For example, the scapula is a freely movable bone that serves as the origin for several muscles that move the arm. When the arm muscles contract, the scapula must be held steady. In abduction of the arm, the deltoid muscle serves as the prime mover, and fixators (pectoralis minor, trapezius, subclavius, serratus anterior muscles, and others) hold the scapula firmly against the back of the chest (see Figure 11.16a-b). The insertion of the deltoid muscle pulls on the humerus to abduct the arm. Under different conditions—that is, for different movements—and at different times, many muscles may act as prime movers, antagonists, synergists, or fixators.

Structure and Function of Muscle Groups

Muscles arise from common masses of muscle tissue in the developing embryo and fetus. The three regions of the body—the free limbs, the trunk, and the head—each have distinct patterns of muscle development.

As developing muscle tissue migrates into the embryonic free limbs it forms two principal masses, an anterior mass of muscle and a posterior mass of muscle. These developing masses of muscle are separated by the developing bones and connective tissue of the free limb. As joints form between the developing bones of the free limb, the muscle masses differentiate into multiple muscles that are enveloped in fascia and separated by the bones, creating anterior and posterior compartments of muscle in the different regions of the free limb. Therefore, a limb **muscle compartment** is a group of skeletal muscles that arose from a common developmental origin. As the muscles of a compartment develop, the nerves and blood vessels develop along with them. Because of this, the muscles of a compartment share common blood and nerve supply. Also, because the muscles of a compartment are grouped on the same side of joints, they tend to have common actions. In the free upper limbs, for example, anterior compartment muscles are flexors of the joints they cross, and posterior compartment muscles are extensors of the joints they cross.

Muscles of the head also arise as functional groups from the embryonic pharyngeal arches and some of the cranial somites. For example, the muscles of mastication arise from the muscle tissue of the first pharyngeal arch. The muscles of facial expression arise from the second pharyngeal arch. Each arch is supplied by a unique cranial nerve, therefore all the muscles of an arch, or functional group, are innervated by one nerve. For example, the fifth cranial nerve (the trigeminal nerve) innervates all the muscles of mastication (first pharyngeal arch muscles) and the seventh cranial nerve (the facial nerve) innervates all the muscles of facial expression (second pharyngeal arch muscles).

◉ **CHECKPOINT**

1. Using the terms origin, insertion, and belly, describe how skeletal muscles produce body movements by pulling on bones.
2. List the three types of levers, and give an example of a first-, second-, and third-class lever found in the body.
3. Describe the various arrangements of fascicles.
4. Explain the difference between a concentric and an eccentric contraction.
5. Define the roles of the prime mover (agonist), antagonist, synergist, and fixator in producing various movements of the free upper limb.

HOW SKELETAL MUSCLES ARE NAMED

◉ **OBJECTIVE**
• Explain seven features used in naming skeletal muscles.

The names of most of the skeletal muscles contain combinations of the word roots of their distinctive features. This works two ways. You can learn the names of muscles by remembering the terms that refer to muscle features, such as the pattern of the muscle's fascicles; the size, shape, action, number of origins, and location of the muscle; and the sites of origin and insertion of the muscle. Knowing the names of a muscle will then give you clues about its features. Study Table 11.2 to become familiar with the terms used in muscle names.

◉ **CHECKPOINT**

6. Select 10 muscles in Figure 11.3 and identify the features on which their names are based. (Hint: *Use the prefix, suffix, and root of each muscle's name as a guide.*)

PRINCIPAL SKELETAL MUSCLES

Exhibits 11.1–11.22 will assist you in learning the names of the principal skeletal muscles in various regions of the body. The muscles in the exhibits are divided into groups according to the part of the body on which they act. As you study groups of muscles in the exhibits, refer to Figure 11.3 on pages 316–317 to see how each group is related to the others.

The exhibits contain the following elements:

• **Objective.** This statement describes what you should learn from the exhibit.
• **Overview.** These paragraphs provide a general introduction to the muscles under consideration and emphasize how the muscles are organized within various regions. The discussion also highlights any distinguishing features of the muscles.
• **Muscle names.** The word roots indicate how the muscles are named. As noted previously, once you have mastered the naming of the muscles, you can more easily understand their actions.
• **Origins, insertions, and actions.** You are also given the origin, insertion, and actions of each muscle.
• **Innervation.** This section lists the nerve or nerves that cause contraction of each muscle. In general, cranial nerves, which arise from the lower parts of the brain, serve muscles in the head region. Spinal nerves, which arise from the spinal cord within the vertebral column, innervate muscles in the rest of the body. Cranial nerves are designated by both a name and a Roman numeral—for example, the facial (VII) nerve. Spinal nerves are numbered in groups according to the part of the spinal cord from which they arise: C=cervical (neck region), T=thoracic (chest region), L=lumbar (lower-back region), and S=sacral (buttocks region). An example is T1, the first thoracic spinal nerve.
• **Relating muscles to movements.** These exercises will help you organize the muscles in the body region under consideration according to the actions they produce.
• **Questions.** These knowledge checkpoints relate specifically to information in each exhibit, and take the form of review, critical thinking, and/or application questions.
• **Clinical Connections.** Selected exhibits include clinical applications, which explore the clinical, professional, or everyday relevance of a particular muscle or its function through descriptions of disorders or clinical procedures.
• **Figures.** The figures in the exhibits may present superficial and deep, anterior and posterior, or medial and lateral views to show each muscle's position as clearly as possible. The muscle names in all capital letters are specifically referred to in the tabular part of the exhibit.

TABLE 11.2

Characteristics Used to Name Muscles

NAME	MEANING	EXAMPLE	FIGURE
DIRECTION: Orientation of muscle fascicles relative to the body's midline.			
Rectus	Parallel to midline	Rectus abdominis	11.12b
Transverse	Perpendicular to midline	Transversus abdominis	11.12b
Oblique	Diagonal to midline	External oblique	11.12a
SIZE: Relative size of the muscle.			
Maximus	Largest	Gluteus maximus	11.22c
Minimus	Smallest	Gluteus minimus	11.22d
Longus	Long	Adductor longus	11.22e
Brevis	Short	Adductor brevis	11.22e
Latissimus	Widest	Latissimus dorsi	11.17b
Longissimus	Longest	Longissimus capitis	11.21a
Magnus	Large	Adductor magnus	11.22b
Major	Larger	Pectoralis major	11.12a
Minor	Smaller	Pectoralis minor	11.16a
Vastus	Huge	Vastus lateralis	11.22a
SHAPE: Relative shape of the muscle.			
Deltoid	Triangular	Deltoid	11.17b
Trapezius	Trapezoid	Trapezius	11.3b
Serratus	Saw-toothed	Serratus anterior	11.16b
Rhomboid	Diamond-shaped	Rhomboid major	11.16d
Orbicularis	Circular	Orbicularis oculi	11.4a
Pectinate	Comblike	Pectineus	11.22a
Piriformis	Pear-shaped	Piriformis	11.22d
Platys	Flat	Platysma	11.4c
Quadratus	Square, four-sided	Quadratus femoris	11.22d
Gracilis	Slender	Gracilis	11.22a
ACTION: Principal action of the muscle.			
Flexor	Decreases a joint angle	Flexor carpi radialis	11.19a
Extensor	Increases a joint angle	Extensor carpi ulnaris	11.19e
Abductor	Moves a bone away from the midline	Abductor pollicis longus	11.19d
Adductor	Moves a bone closer to the midline	Adductor longus	11.22a
Levator	Raises or elevates a body part	Levator scapulae	11.16d
Depressor	Lowers or depresses a body part	Depressor labii inferioris	11.4b
Supinator	Turns palm anteriorly	Supinator	11.19d
Pronator	Turns palm posteriorly	Pronator teres	11.19a
Sphincter	Decreases the size of an opening	External anal sphincter	11.14
Tensor	Makes a body part rigid	Tensor fasciae latae	11.22a,c
Rotator	Rotates a bone around its longitudinal axis	Rotatore	11.21a,b
NUMBER OF ORIGINS: Number of tendons of origin.			
Biceps	Two origins	Biceps brachii	11.18a
Triceps	Three origins	Triceps brachii	11.18b
Quadriceps	Four origins	Quadriceps femoris	11.22a
LOCATION: Structure near which a muscle is found. *Example:* **Temporalis, a muscle near the temporal bone.**			11.4d
ORIGIN AND INSERTION: Sites where muscle originates and inserts. *Example:* **Sternocleidomastoid, originating on the sternum and clavicle and Inserting on mastoid process of temporal bone.**			11.3a

Figure 11.3 Principal superficial skeletal muscles.

Most movements require several skeletal muscles acting in groups rather than individually.

Occipitofrontalis (frontal belly)

Nasalis

Orbicularis oris

Depressor anguli oris

Omohyoid

Sternohyoid

Latissimus dorsi

Rectus abdominis

External oblique

Brachioradialis

Extensor carpi radialis longus

Tensor fasciae latae

Iliacus

Psoas major

Extensor pollicis longus

Pectineus

Adductor longus

Sartorius

Adductor magnus

Gracilis

Vastus lateralis

Rectus femoris

Vastus medialis

Tendon of quadriceps femoris

Patella

Gastrocnemius

Soleus

Tibia

Flexor digitorum longus

Epicranial aponeurosis

Temporalis

Orbicularis oculi

Masseter

Platysma (cut)

Sternocleidomastoid

Scalenes

Trapezius

Deltoid

Pectoralis major

Serratus anterior

Biceps brachii

Brachialis

Triceps brachii

Extensor carpi radialis longus

Brachioradialis

Flexor carpi radialis

Palmaris longus

Flexor digitorum superficialis

Flexor carpi ulnaris

Abductor pollicis longus

Thenar muscles

Hypothenar muscles

Iliotibial tract

Patellar ligament

Tibialis anterior

Fibularis longus

Tibia

Calcaneal (Achilles) tendon

(a) Anterior view

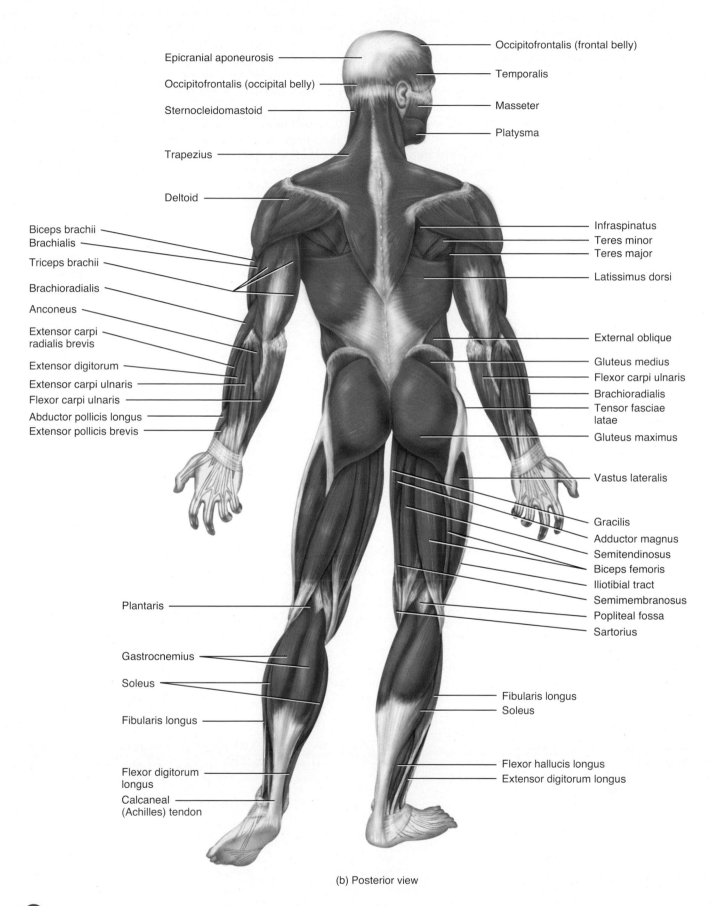

Epicranial aponeurosis

Occipitofrontalis (occipital belly)

Sternocleidomastoid

Trapezius

Deltoid

Biceps brachii
Brachialis

Triceps brachii

Brachioradialis

Anconeus

Extensor carpi radialis brevis

Extensor digitorum

Extensor carpi ulnaris

Flexor carpi ulnaris

Abductor pollicis longus

Extensor pollicis brevis

Plantaris

Gastrocnemius

Soleus

Fibularis longus

Flexor digitorum longus

Calcaneal (Achilles) tendon

Occipitofrontalis (frontal belly)

Temporalis

Masseter

Platysma

Infraspinatus
Teres minor
Teres major

Latissimus dorsi

External oblique

Gluteus medius

Flexor carpi ulnaris

Brachioradialis

Tensor fasciae latae

Gluteus maximus

Vastus lateralis

Gracilis

Adductor magnus

Semitendinosus

Biceps femoris

Iliotibial tract

Semimembranosus

Popliteal fossa

Sartorius

Fibularis longus
Soleus

Flexor hallucis longus
Extensor digitorum longus

(b) Posterior view

? **Give an example of a muscle named for each of the following characteristics: direction of fibers, shape, action, size, origin and insertion, location, and number of tendons of origin.**

EXHIBIT 11.1   Muscles of the Head That Produce Facial Expressions

⦿ **OBJECTIVE**

• Describe the origin, insertion, action, and innervation of the muscles of facial expression.

The muscles of facial expression, which provide us with the ability to express a wide variety of emotions, lie within the subcutaneous layer (Figure 11.4 on pages 320–323). They usually originate in the fascia or bones of the skull and insert into the skin. Because of their insertions, the muscles of facial expression move the skin rather than a joint when they contract.

Among the noteworthy muscles in this group are those surrounding the orifices (openings) of the head such as the eyes, nose, and mouth. These muscles function as *sphincters* (SFINGK-ters), which close the orifices, and *dilators*, which dilate or open the orifices. For example, the **orbicularis oculi** muscle closes the eye, and the levator palpebrae superioris muscle opens it (discussed in Exhibit 11.2). The **occipitofrontalis** is an unusual muscle in this group because it is made up of two parts: an anterior part called the **frontal belly**, which is superficial to the frontal bone, and a posterior part called the **occipital belly**, which is superficial to the occipital bone. The two muscular portions are held together by a strong aponeurosis (sheetlike tendon), the **epicranial aponeurosis** (ep-i-KRĀ-nē-al ap-ō-noo-RŌ-sis), also called the **galea aponeurotica** (GĀ-lē-a ap-ō-noo′-RŌ-ti-ka), that covers the superior and lateral surfaces of the skull. The **buccinator** muscle forms the major muscular portion of the cheek. The duct of the parotid gland (a salivary gland) passes through the buccinator muscle to reach the oral cavity. The buc-

MUSCLE	ORIGIN	INSERTION	ACTION	INNERVATION
SCALP MUSCLES				
Occipitofrontalis (ok-sip′-i-tō-frun-TĀ-lis)				
Frontal belly	Epicranial aponeurosis.	Skin superior to supraorbital margin.	Draws scalp anteriorly, raises eyebrows, and wrinkles skin of forehead horizontally as in a look of surprise.	Facial (VII) nerve.
Occipital belly (*occipit-*=back of the head)	Occipital bone and mastoid process of temporal bone.	Epicranial aponeurosis.	Draws scalp posteriorly.	Facial (VII) nerve.
MOUTH MUSCLES				
Orbicularis oris (or-bi′-kū-LAR-is OR-is; *orb-*=circular; *oris*=of the mouth)	Muscle fibers surrounding opening of mouth.	Skin at corner of mouth.	Closes and protrudes lips, as in kissing; compresses lips against teeth; and shapes lips during speech.	Facial (VII) nerve.
Zygomaticus major (zī-gō-MA-ti-kus; *Zygomatic*=cheek bone; *major*=greater)	Zygomatic bone.	Skin at angle of mouth and orbicularis oris.	Draws angle of mouth superiorly and laterally, as in smiling.	Facial (VII) nerve.
Zygomaticus minor (*minor*=lesser)	Zygomatic bone.	Upper lip.	Raises (elevates) upper lip, exposing maxillary (upper) teeth.	Facial (VII) nerve.
Levator labii superioris (le-VĀ-tor LĀ-bē-ī soo-per′-ē-OR- is; *levator*=raises or elevates; *labii*=lip; *superioris*=upper)	Superior to infraorbital foramen of maxilla.	Skin at angle of mouth and orbicularis oris.	Raises upper lip.	Facial (VII) nerve.
Depressor labii inferioris (de-PRE-sor LĀ-bē-ī; *depressor*=depresses or lowers; *inferioris*=lower)	Mandible.	Skin of lower lip.	Depresses (lowers) lower lip.	Facial (VII) nerve.
Depressor anguli oris (*angul*=angle or corner)	Mandible.	Angle of mouth.	Draws angle of mouth laterally and inferiorly, as in opening mouth.	Facial (VII) nerve.
Levator anguli oris	Inferior to infraorbital foramen.	Skin of lower lip and orbicularis oris.	Draws angle of mouth laterally and superiorly.	Facial (VII) nerve.

cinator muscle is so named because it compresses the cheeks (*bucc-*=cheek) during blowing—for example, when a musician plays a wind instrument such as a trumpet. It functions in whistling, blowing, and sucking and assists in chewing.

Relating Muscles to Movements

Arrange the muscles in this exhibit into two groups: (1) those that act on the mouth and (2) those that act on the eyes.

☉ **CHECKPOINT**

Why do the muscles of facial expression move the skin rather than a joint?

MUSCLE	ORIGIN	INSERTION	ACTION	INNERVATION
MOUTH MUSCLES (Continued)				
Buccinator (BUK-si-nā′-tor; *bucc-*=cheek)	Alveolar processes of maxilla and mandible and pterygomandibular raphe (fibrous band extending from the pterygoid process of the sphenoid bone to the mandible).	Orbicularis oris.	Presses cheeks against teeth and lips, as in whistling, blowing, and sucking; draws corner of mouth laterally; and assists in mastication (chewing) by keeping food between the teeth (and not between teeth and cheeks).	Facial (VII) nerve.
Risorius (ri-ZOR-ē-us; *risor*=laughter)	Fascia over parotid (salivary) gland.	Skin at angle of mouth.	Draws angle of mouth laterally, as in grimacing.	Facial (VII) nerve.
Mentalis (men-TĀ-lis; *ment-*=the chin)	Mandible.	Skin of chin.	Elevates and protrudes lower lip and pulls skin of chin up, as in pouting.	Facial (VII) nerve.
NECK MUSCLE				
Platysma (pla-TIZ-ma; *platy*=flat, broad)	Fascia over deltoid and pectoralis major muscles.	Mandible, muscle around angle of mouth, and skin of lower face.	Draws outer part of lower lip inferiorly and posteriorly as in pouting; depresses mandible.	Facial (VII) nerve.
ORBIT AND EYEBROW MUSCLES				
Orbicularis oculi (or-bi′-kū-LAR-is OK-ū-lī; *oculi*=of the eye)	Medial wall of orbit.	Circular path around orbit.	Closes eye.	Facial (VII) nerve.
Corrugator supercilii (KOR-a-gā′-tor soo′-per-SI-lē-ī; *corrugat*=wrinkle; *supercilii*=of the eyebrow)	Medial end of superciliary arch of frontal bone.	Skin of eyebrow.	Draws eyebrow inferiorly and wrinkles skin of forehead vertically as in frowning.	Facial (VII) nerve.

◘ **EXHIBIT 11.1 CONTINUES** ▶

EXHIBIT 11.1 **319**

Figure 11.4 Muscles of the head that produce facial expressions.

When they contract, muscles of facial expression move the skin rather than a joint.

Epicranial aponeurosis

OCCIPITOFRONTALIS (FRONTAL BELLY)

TEMPORALIS

ORBICULARIS OCULI

LEVATOR LABII SUPERIORIS

ZYGOMATICUS MINOR

ZYGOMATICUS MAJOR

RISORIUS

PLATYSMA

DEPRESSOR ANGULI ORIS

Thyroid cartilage (Adam's apple)

Frontal bone

CORRUGATOR SUPERCILII

Levator palpebrae superioris

Lacrimal gland

Zygomatic bone

Nasalis

Nasal cartilage

Maxilla

BUCCINATOR

MASSETER

ORBICULARIS ORIS

DEPRESSOR LABII INFERIORIS

Mandible

MENTALIS

Omohyoid

Sternohyoid

Sternocleidomastoid

(a) Anterior superficial view   (b) Anterior deep view

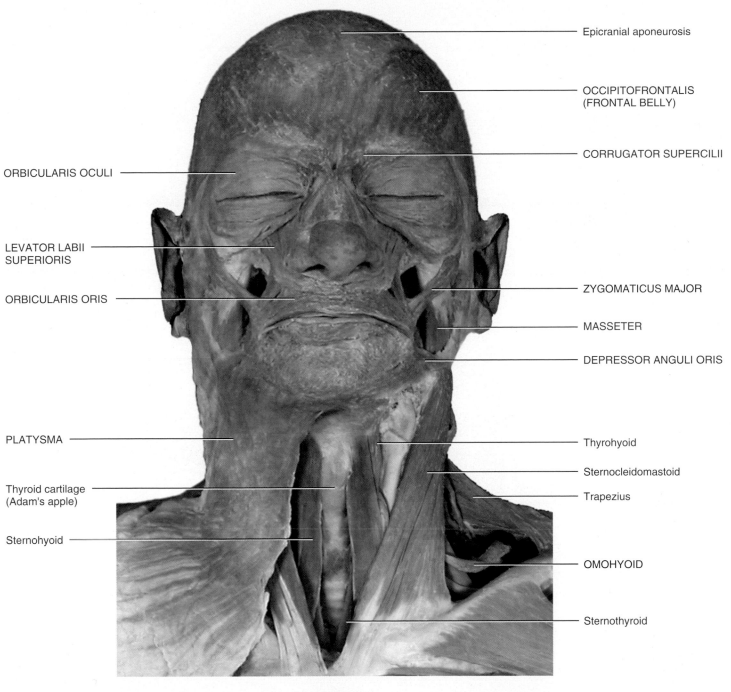

Epicranial aponeurosis

OCCIPITOFRONTALIS
(FRONTAL BELLY)

CORRUGATOR SUPERCILII

ORBICULARIS OCULI

LEVATOR LABII
SUPERIORIS

ORBICULARIS ORIS

ZYGOMATICUS MAJOR

MASSETER

DEPRESSOR ANGULI ORIS

PLATYSMA

Thyrohyoid

Sternocleidomastoid

Thyroid cartilage
(Adam's apple)

Trapezius

Sternohyoid

OMOHYOID

Sternothyroid

(c) Anterior view

EXHIBIT **11.1** CONTINUES ▶

EXHIBIT 11.1   **321**

Epicranial aponeurosis

TEMPORALIS

OCCIPITOFRONTALIS
(OCCIPITAL BELLY)

Zygomatic arch

Posterior auricular

Mandible

MASSETER

Splenius capitis

Sternocleidomastoid

Splenius cervicis

Trapezius

Levator scapulae

Middle scalene

OCCIPITO-
FRONTALIS
(FRONTAL
BELLY)

ORBICULARIS
OCULI

ZYGOMATICUS
MINOR

Nasalis

LEVATOR LABII
SUPERIORIS

ZYGOMATICUS
MAJOR

LEVATOR
ANGULI ORIS

BUCCINATOR

ORBICULARIS
ORIS

RISORIUS

DEPRESSOR
LABII INFERIORIS

MENTALIS

DEPRESSOR
ANGULI ORIS

PLATYSMA

(d) Right lateral superficial view

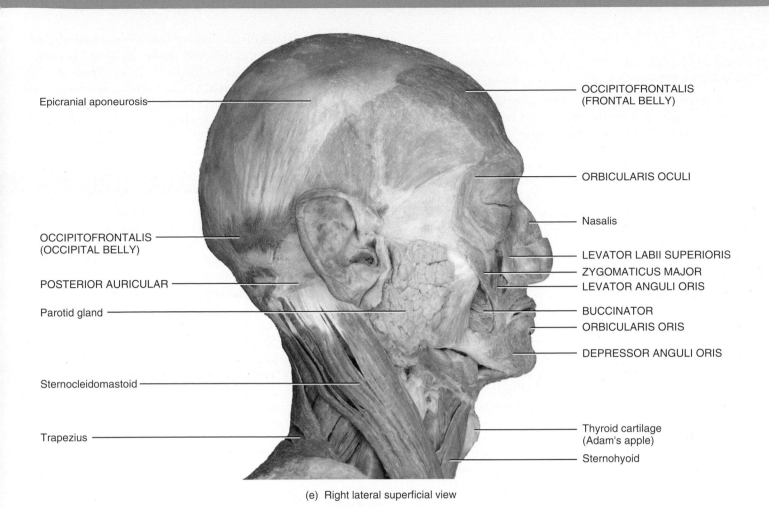

Epicranial aponeurosis

OCCIPITOFRONTALIS (FRONTAL BELLY)

ORBICULARIS OCULI

Nasalis

OCCIPITOFRONTALIS (OCCIPITAL BELLY)

LEVATOR LABII SUPERIORIS

ZYGOMATICUS MAJOR

POSTERIOR AURICULAR

LEVATOR ANGULI ORIS

Parotid gland

BUCCINATOR

ORBICULARIS ORIS

DEPRESSOR ANGULI ORIS

Sternocleidomastoid

Thyroid cartilage (Adam's apple)

Trapezius

Sternohyoid

(e) Right lateral superficial view

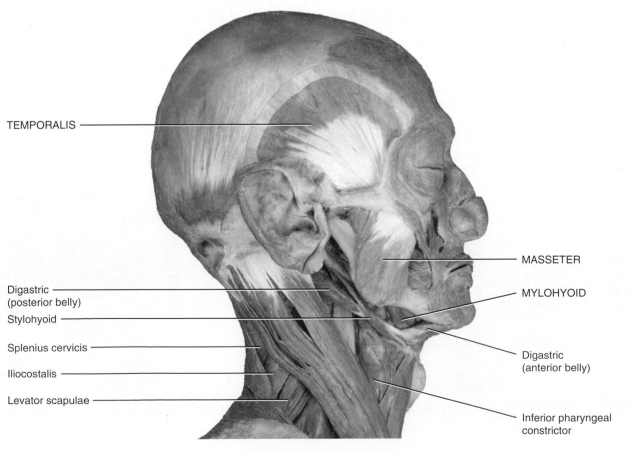

TEMPORALIS

MASSETER

Digastric (posterior belly)

MYLOHYOID

Stylohyoid

Splenius cervicis

Iliocostalis

Digastric (anterior belly)

Levator scapulae

Inferior pharyngeal constrictor

(f) Right lateral deep view

? **Which muscles of facial expression cause frowning, smiling, pouting, and squinting?**

EXHIBIT 11.1 **323**

◉ OBJECTIVE

- Describe the origin, insertion, action, and innervation of the extrinsic eye muscles.

Muscles that move the eyeballs are called **extrinsic eye muscles** because they originate outside the eyeballs (in the orbit) and insert on the outer surface of the sclera ("white of the eye"). (Figure 11.5). The extrinsic eye muscles are some of the fastest contracting and most precisely controlled skeletal muscles in the body.

Three pairs of extrinsic eye muscles control movements of the eyeballs: (1) superior and inferior recti, (2) lateral and medial recti, and (3) superior and inferior oblique. The four recti muscles (superior, inferior, lateral, and medial) arise from a tendinous ring in the orbit and insert into the sclera of the eye. As their names imply, the **superior** and **inferior recti** move the eyeballs superiorly and inferiorly; the **lateral** and **medial recti** move the eyeballs laterally and medially.

The actions of the oblique muscles cannot be deduced from their names. The **superior oblique** muscle originates posteriorly near the tendinous ring, then passes anteriorly, and ends in a round tendon. The tendon extends through a pulleylike loop of fibrocartilaginous tissue called the *trochlea* (=pulley) in the anterior and medial part of the roof of the orbit. Finally, the tendon turns and inserts on the posterolateral aspect of the eyeballs. Accordingly, the superior oblique muscle moves the eyeballs inferiorly and laterally. The **inferior oblique** muscle originates on the maxilla at the anteromedial aspect of the floor of the orbit. It then passes posteriorly and laterally and inserts on the posterolat-

eral aspect of the eyeballs. Because of this arrangement, the inferior oblique muscle moves the eyeballs superiorly and laterally.

Developmentally related to the extrinsic eye muscles is the **levator palpebrae superioris**. Unlike the recti and oblique muscles, it does not move the eyeballs. Rather, it raises the upper eyelids that is, opens the eyes. It is therefore an antagonist to the orbicularis oculi, which closes the eyes.

● CLINICAL CONNECTION Strabismus

Strabismus (stra-BIZ-mus; *strabismos*=squinting) is a condition in which the two eyeballs are not properly aligned. This can be hereditary or it can be due to birth injuries, poor attachments of the muscles, problems with the brain's control center, or localized disease. Strabismus can be constant or intermittent. In strabismus, each eye sends an image to a different area of the brain and because the brain usually ignores the messages sent by one of the eyes, the ignored eye becomes weaker, hence "lazy eye" or *amblyopia*, develops. *External strabismus* results when a lesion in the oculomotor (III) nerve causes the eyeball to move laterally when at rest, and results in an inability to move the eyeball medially and inferiorly. A lesion in the abducens (VI) nerve results in *internal strabismus*, a condition in which the eyeball moves medially when at rest and cannot move laterally.

Treatment options for strabismus depend on the specific type of problem and include surgery, visual therapy (retraining the brain's control center), and orthoptics (eye muscle training to straighten the eyes). •

Figure 11.5 Muscles of the head that move the eyeballs (extrinsic eye muscles) and upper eyelid.

🔑 The extrinsic muscles of the eyeball are among the fastest contracting and most precisely controlled skeletal muscles in the body.

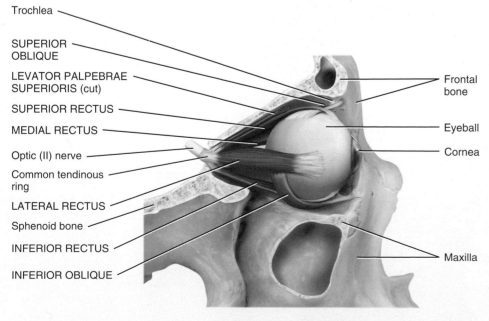

(a) Lateral view of right eyeball

(b) Movements of right eyeball in response to contraction of extrinsic muscles

Relating Muscles to Movements

Arrange the muscles in this exhibit according to their actions on the eyeballs: (1) elevation, (2) depression, (3) abduction, (4) adduction, (5) medial rotation, and (6) lateral rotation. The same muscle may be mentioned more than once.

◉ CHECKPOINT

Which muscles contract and relax in each eye as you gaze to your left without moving your head?

MUSCLE	ORIGIN	INSERTION	ACTION	INNERVATION
Superior rectus (*rectus*=fascicles parallel to midline)	Common tendinous ring (attached to orbit around optic foramen).	Superior and central part of eyeballs.	Moves eyeballs superiorly (elevation) and medially (adduction), and rotates them medially.	Oculomotor (III) nerve.
Inferior rectus	Same as above.	Inferior and central part of eyeballs.	Moves eyeballs inferiorly (depression) and medially (adduction), and rotates them medially.	Oculomotor (III) nerve.
Lateral rectus	Same as above.	Lateral side of eyeballs.	Moves eyeballs laterally (abduction).	Abducens (VI) nerve.
Medial rectus	Same as above.	Medial side of eyeballs.	Moves eyeballs medially (adduction).	Oculomotor (III) nerve.
Superior oblique (*oblique*=fascicles diagonal to midline)	Sphenoid bone, superior and medial to the tendinous ring in the orbit.	Eyeball between superior and lateral recti. The muscle inserts into the superior and lateral surfaces of the eyeballs via a tendon that passes through the trochlea.	Moves eyeballs inferiorly (depression) and laterally (abduction), and rotates them medially.	Trochlear (IV) nerve.
Inferior oblique	Maxilla in floor of orbit.	Eyeballs between inferior and lateral recti.	Moves eyeballs superiorly (elevation) and laterally (abduction) and rotates them laterally.	Oculomotor (III) nerve.
Levator palpebrae superioris (le-VĀ-tor PAL-pe-brē soo'-per'-ē-OR-is; *palpebrae*=eyelids)	Roof of orbit (lesser wing of sphenoid bone).	Skin and tarsal plate of upper eyelid.	Elevates upper eyelids (opens eyes).	Oculomotor (III) nerve.

Frontal bone (cut)

SUPERIOR OBLIQUE

SUPERIOR RECTUS

LEVATOR PALPEBRA SUPERIORIS

MEDIAL RECTUS

LATERAL RECTUS

INFERIOR RECTUS

INFERIOR OBLIQUE

Zygomatic bone (cut)

(c) Right lateral view

? How does the inferior oblique muscle move the eyeball superiorly and laterally ?

EXHIBIT 11.2 **325**

⊙ **OBJECTIVE**

- Describe the origin, insertion, action, and innervation of the muscles that move the mandible.

The muscles that move the mandible (lower jaw bone) at the temporomandibular joint (TMJ) are known as the muscles of mastication (chewing) (Figure 11.6). Of the four pairs of muscles involved in mastication, three are powerful closers of the jaw and account for the strength of the bite: **masseter, temporalis,** and **medial pterygoid**. Of these, the masseter is the strongest muscle of mastication. The medial and **lateral pterygoid** muscles assist in mastication by moving the mandible from side to side to help grind food. Additionally, these muscles protract (protrude) the mandible. These four muscles all develop from the first pharyngeal arch.

Relating Muscles to Movements

Arrange the muscles in this exhibit according to their actions on the mandible: (1) elevation, (2) depression, (3) retraction, (4) protraction, and (5) side-to-side movement. The same muscle may be mentioned more than once.

⊙ **CHECKPOINT**
What would happen if you lost tone in the masseter and temporalis muscles?

Figure 11.6 Muscles that move the mandible (lower jaw bone) and assist in mastication (chewing) and speech.

🔑 The muscles that move the mandible are also known as muscles of mastication.

Parietal bone

TEMPORALIS

Occipital bone

Zygomatic arch (cut)

Temporomandibular joint (TMJ)

MEDIAL PTERYGOID (deep to mandible)

Ramus of mandible (cut and transparent)

Frontal bone

Nasal bone

Zygomatic bone (cut)

LATERAL PTERYGOID

Maxilla

Buccinator

Orbicularis oris

Body of mandible

(a) Right lateral superficial view

MUSCLE	ORIGIN	INSERTION	ACTION	INNERVATION
Masseter (MA-se-ter=a chewer) (see Figure 11.4d)	Maxilla and zygomatic arch.	Angle and ramus of mandible.	Elevates mandible, as in closing mouth.	Mandibular division of trigeminal (V) nerve.
Temporalis (tem'-pō-RĀ-lis; *tempor-*=time or temples)	Temporal bone.	Coronoid process and ramus of mandible.	Elevates and retracts mandible.	Mandibular division of trigeminal (V) nerve.
Medial pterygoid (TER-i-goyd; *medial*= closer to midline; *pterygoid*=like a wing)	Medial surface of lateral portion of pterygoid process of sphenoid bone; maxilla.	Angle and ramus of mandible.	Elevates and protracts (protrudes) mandible and moves mandible from side to side.	Mandibular division of trigeminal (V) nerve.
Lateral pterygoid (TER-i-goyd; *lateral*= farther from midline)	Greater wing and lateral surface of lateral portion of pterygoid process of sphenoid bone.	Condyle of mandible; temporomandibular joint (TMJ).	Protracts mandible, depresses mandible as in opening mouth, and moves mandible from side to side.	Mandibular division of trigeminal (V) nerve.

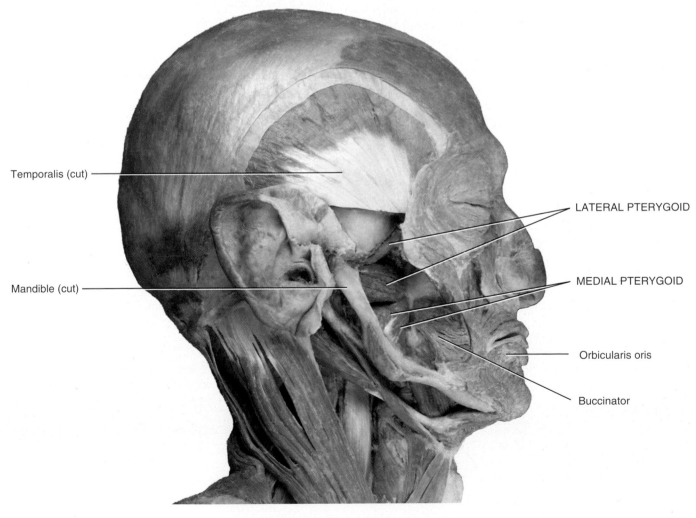

Temporalis (cut)

Mandible (cut)

LATERAL PTERYGOID

MEDIAL PTERYGOID

Orbicularis oris

Buccinator

(b) Right lateral deep view

? Which is the strongest muscle of mastication?

EXHIBIT 11.3 **327**

◉ OBJECTIVE

- Describe the origin, insertion, action, and innervation of the extrinsic muscles of the tongue.

The tongue is a highly mobile structure that is vital to digestive functions such as mastication, detection of taste, and deglutition (swallowing). It is also important in speech. The tongue's mobility is greatly aided by its attachment to the mandible, styloid process of the temporal bone, and hyoid bone.

The tongue is divided into lateral halves by a median fibrous septum. The septum extends throughout the length of the tongue. Inferiorly, the septum attaches to the hyoid bone. Muscles of the tongue are of two principal types: extrinsic and intrinsic. **Extrinsic tongue muscles** originate outside the tongue and insert into it (Figure 11.7). They move the entire tongue in various directions, such as anteriorly, posteriorly, and laterally. **Intrinsic tongue muscles** originate and insert within the tongue. These muscles alter the shape of the tongue rather than moving the entire tongue. The extrinsic and intrinsic muscles of the tongue insert into both lateral halves of the tongue.

When you study the extrinsic tongue muscles, you will notice that all of their names end in *glossus*, meaning tongue. You will also notice that the actions of the muscles are obvious, considering the positions of the mandible, styloid process, hyoid bone, and soft palate, which serve as origins for these muscles. For example, the **genioglossus** (origin: the mandible) pulls the tongue downward and forward, the **styloglossus** (origin: the styloid process) pulls the tongue upward and backward, the **hyoglossus** (origin: the hyoid bone) pulls the tongue downward and flattens it, and the **palatoglossus** (origin: the soft palate) raises the back portion of the tongue.

● CLINICAL CONNECTION | **Intubation During Anesthesia**

When general anesthesia is administered during surgery, a total relaxation of the muscles results. Once the various types of drugs for anesthesia have been given (especially the paralytic agents), the patient's airway must be protected and the lungs ventilated because the muscles involved with respiration are among those paralyzed. Paralysis of the genioglossus muscle causes the tongue to fall posteriorly, which may obstruct the airway to the lungs. To avoid this, the mandible is either manually thrust forward and held in place (known as the "sniffing position"), or a tube is inserted from the lips through the laryngopharynx (inferior portion of the throat) into the trachea (endotracheal intubation). People can also be intubated nasally (through the nose). •

Relating Muscles to Movements

Arrange the muscles in this exhibit according to the following actions on the tongue: (1) depression, (2) elevation, (3) protraction, and (4) retraction. The same muscle may be mentioned more than once.

◉ CHECKPOINT

When your physician says, "Open your mouth, stick out your tongue, and say *ahh*," to examine the inside of your mouth for possible signs of infection, which muscles do you contract?

MUSCLE	ORIGIN	INSERTION	ACTION	INNERVATION
Genioglossus (jē′-nē-ō-GLOS-us; *genio-*=the chin; *glossus*=tongue)	Mandible.	Undersurface of tongue and hyoid bone.	Depresses tongue and thrusts it anteriorly (protraction).	Hypoglossal (XII) nerve.
Styloglossus (stī′-lō-GLOS-us; *stylo*=stake or pole; styloid process of temporal bone)	Styloid process of temporal bone.	Side and undersurface of tongue.	Elevates tongue and draws it posteriorly (retraction).	Hypoglossal (XII) nerve.
Palatoglossus (pal′-a-tō-GLOS-us; *palato-*=the roof of the mouth or palate)	Anterior surface of soft palate.	Side of tongue.	Elevates posterior portion of tongue and draws soft palate down on tongue.	Pharyngeal plexus, which contains axons from the vagus (X) nerve.
Hyoglossus *hypo*=below; (hī′-ō-GLOS-us)	Greater horn and body of hyoid bone.	Side of tongue.	Depresses tongue and draws down its sides.	Hypoglossal (XII) nerve.

Figure 11.7 Muscles of the head that move the tongue and assist in mastication (chewing) and speech—extrinsic tongue muscles.

The extrinsic and intrinsic muscles of the tongue are arranged in both lateral halves of the tongue.

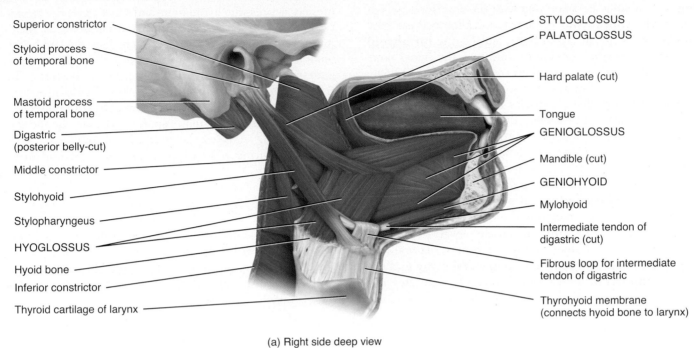

Superior constrictor
Styloid process of temporal bone
Mastoid process of temporal bone
Digastric (posterior belly-cut)
Middle constrictor
Stylohyoid
Stylopharyngeus
HYOGLOSSUS
Hyoid bone
Inferior constrictor
Thyroid cartilage of larynx

STYLOGLOSSUS
PALATOGLOSSUS
Hard palate (cut)
Tongue
GENIOGLOSSUS
Mandible (cut)
GENIOHYOID
Mylohyoid
Intermediate tendon of digastric (cut)
Fibrous loop for intermediate tendon of digastric
Thyrohyoid membrane (connects hyoid bone to larynx)

(a) Right side deep view

Superior constrictor
Styloid process of temporal bone
Mastoid process of temporal bone
Digastric (posterior belly, cut)
STYLOGLOSSUS
Stylohyoid
HYOGLOSSUS
Inferior pharyngeal constrictor
Cricothyroid

PALATOGLOSSUS
Tongue
GENIOGLOSSUS
GENIOHYOID
Mylohyoid (cut)
Digastric (anterior belly, cut)
Thyrohyoid membrane (connects hyoid bone to larynx)
Thyroid cartilage
Trachea

(b) Right lateral deep view

? What are the functions of the tongue?

EXHIBIT 11.4 **329**

EXHIBIT 11.5

Muscles of the Anterior Neck That Assist in Deglutition (Swallowing) and Speech

⊙ OBJECTIVE

- Describe the origin, insertion, action, and innervation of the muscles of the anterior neck.

Two groups of muscles are associated with the anterior aspect of the neck: (1) the **suprahyoid muscles**, so called because they are located superior to the hyoid bone, and (2) the **infrahyoid muscles**, named for their position inferior to the hyoid bone (Figure 11.8). Both groups of muscles stabilize the hyoid bone, allowing it to serve as a firm base on which the tongue can move.

As a group, the suprahyoid muscles elevate the hyoid bone, floor of the oral cavity, and tongue during swallowing. As its name suggests, the **digastric** muscle has two bellies, anterior and posterior, united by an intermediate tendon that is held in position by a fibrous loop. This muscle elevates the hyoid bone and larynx (voice box) during swallowing and speech and depresses the mandible. The **stylohyoid** muscle elevates and draws the hyoid bone posteriorly, thus elongating the floor of the oral cavity during swallowing. The **mylohyoid** muscle elevates the hyoid bone and helps press the tongue against the roof of the oral cavity during swallowing to move food from the oral cavity into the throat. The **geniohyoid** muscle (see Figure 11.7) elevates and draws the hyoid bone anteriorly to shorten the floor of the oral

cavity and to widen the throat to receive food that is being swallowed. It also depresses the mandible.

The infrahyoid muscles are sometimes called "strap" muscles because of their ribbonlike appearance. Most of the infrahyoid muscles depress the hyoid bone and some move the larynx during swallowing and speech. The **omohyoid** muscle, like the digastric muscle, is composed of two bellies connected by an intermediate tendon. In this case, however, the two bellies are referred to as *superior* and *inferior*, rather than anterior and posterior. Together, the omohyoid, **sternohyoid**, and **thyrohyoid** muscles depress the hyoid bone. In addition, the **sternothyroid** muscle depresses the thyroid cartilage (Adam's apple) of the larynx, and the thyrohyoid muscle elevates the thyroid cartilage. The actions are necessary during phonation to produce low and high tones, respectively.

Relating Muscles to Movements

Arrange the muscles in this exhibit according to the following actions on the hyoid bone: (1) elevating it, (2) drawing it anteriorly, (3) drawing it posteriorly, and (4) depressing it; and on the thyroid cartilage: (1) elevating it and (2) depressing it. The same muscle may be mentioned more than once.

MUSCLE	ORIGIN	INSERTION	ACTION	INNERVATION
SUPRAHYOID MUSCLES				
Digastric (dī′-GAS-trik; *di-*=two; *gastr-*=belly)	Anterior belly from inner side of inferior border of mandible; posterior belly from temporal bone.	Body of hyoid bone via an intermediate tendon.	Elevates hyoid bone and depresses mandible, as in opening the mouth.	Anterior belly: mandibular division of trigeminal (V) nerve. Posterior belly: facial (VII) nerve.
Stylohyoid (stī′-lō-HĪ-oid; *stylo-*=stake or pole, styloid process of temporal bone; *hyo-*=U-shaped, pertaining to hyoid bone)	Styloid process of temporal bone.	Body of hyoid bone.	Elevates hyoid bone and draws it posteriorly.	Facial (VII) nerve.
Mylohyoid (mī′-lō-HĪ-oid); *mylo-*=mill)	Inner surface of mandible.	Body of hyoid bone.	Elevates hyoid bone and floor of mouth and depresses mandible.	Mandibular division of trigeminal (V) nerve.
Geniohyoid (jē′-nē-ō-HĪ-oid; *genio-*=chin) (see Figure 11.7)	Inner surface of mandible.	Body of hyoid bone.	Elevates hyoid bone, draws hyoid bone and tongue anteriorly, and depresses mandible.	First cervical spinal nerve.
INFRAHYOID MUSCLES				
Omohyoid (ō-mō-HĪ-oid; *omo-*=relationship to the shoulder)	Superior border of scapula and superior transverse ligament.	Body of hyoid bone.	Depresses hyoid bone.	Branches of spinal nerves C1–C3.
Sternohyoid (ster′-nō-HĪ-oid; *sterno-*=sternum)	Medial end of clavicle and manubrium of sternum.	Body of hyoid bone.	Depresses hyoid bone.	Branches of spinal nerves C1–C3.
Sternothyroid (ster′-nō-THĪ-roid; *thyro-*=thyroid gland)	Manubrium of sternum.	Thyroid cartilage of larynx.	Depresses thyroid cartilage of larynx.	Branches of spinal nerves C1–C3.
Thyrohyoid (thī′-rō-HĪ-oid)	Thyroid cartilage of larynx.	Greater horn of hyoid bone.	Elevates thyroid cartilage and depresses hyoid bone.	Branches of spinal nerves C1–C2 and descending hypoglossal (XII) nerve.

Figure 11.8 Muscles of the anterior neck that assist in deglutition (swallowing) and speech.

🔓 ▭ The suprahyoid muscles elevate the hyoid bone, the floor of the oral cavity, and the tongue during swallowing.

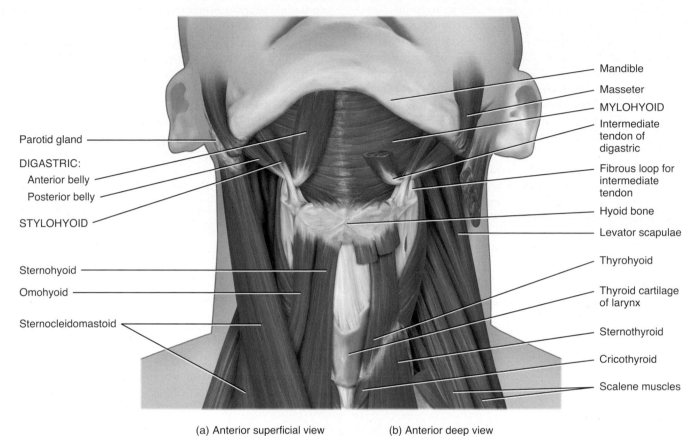

(a) Anterior superficial view   (b) Anterior deep view

Anterior superficial view   (c)   Anterior deep view

❓ **What is the combined action of the suprahyoid and infrahyoid muscles?**

EXHIBIT 11.5   **331**

EXHIBIT 11.6   Muscles of the Neck That Move the Head

OBJECTIVE

• Describe the origin, insertion action, and innervation of the muscles that move the head.

The head is attached to the vertebral column at the atlanto-occipital joints formed by the atlas and occipital bone. Balance and movement of the head on the vertebral column involves the action of several neck muscles. For example, acting together (bilaterally), contraction of the two **sternocleidomastoid** muscles flexes the cervical portion of the vertebral column and extends the head. Acting singly (unilaterally), each sternocleidomastioid muscle laterally flexes and rotates the head. Bilateral contraction of the **semispinalis capitis, splenius capitis,** and **longissimus capitis** muscles extends the head (Figure 11.9a). However, when these same muscles contract unilaterally, their actions are quite different, involving primarily rotation of the head.

The sternocleidomastoid muscle is an important landmark that divides the neck into two major triangles: anterior and posterior (Figure 11.9b). The triangles are important anatomically and surgically because of the structures that lie within their boundaries.

The **anterior triangle** is bordered superiorly by the mandible, medially by the cervical midline, and laterally by the anterior border of the sternocleidomastoid muscle. It has its apex at the sternum (Figure 11.9b). The anterior triangle is subdivided into three paired triangles: *submandibular, carotid,* and *muscular.* An unpaired *submental triangle* is formed by the upper part of the combined right and left anterior triangles. The anterior triangle contains submental, submandibular, and deep cervical lymph nodes; the submandibular salivary gland and a portion of the parotid salivary gland; the facial artery and vein; carotid arteries and internal jugular vein; the thyroid gland and infrahyoid muscles; and the following cranial nerves: glossopharyngeal (IX), vagus (X), accessory (XI), and hypoglossal (XII).

The **posterior triangle** is bordered interiorly by the clavicle, anteriorly by the posterior border of the sternocleidomastoid muscle, and posteriorly by the anterior border of the trapezius muscle (Figure 11.9b). The posterior triangle is subdivided into two triangles, *occipital* and *supraclavicular (omoclavicular),* by the inferior belly of the omohyoid muscle. The posterior triangle contains part of the subclavian artery, external jugular vein, cervical lymph nodes, brachial plexus, and the accessory (XI) nerve.

Figure 11.9 **Muscles of the neck that move the head.**

The sternocleidomastoid muscle divides the neck into two principal triangles: anterior and posterior.

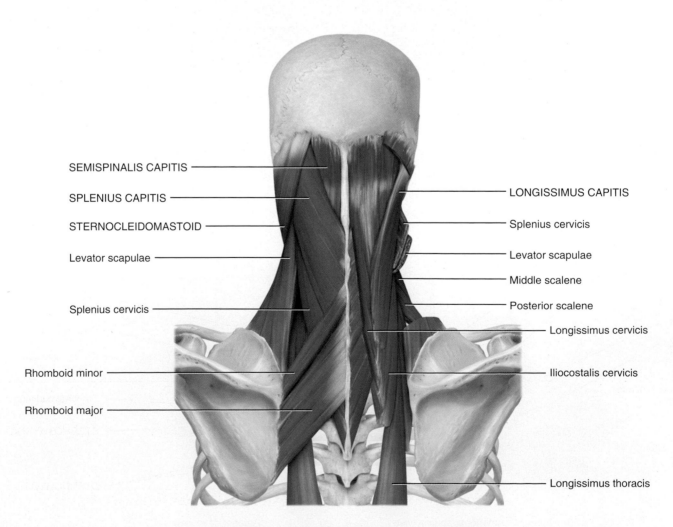

SEMISPINALIS CAPITIS

SPLENIUS CAPITIS

STERNOCLEIDOMASTOID

Levator scapulae

Splenius cervicis

Rhomboid minor

Rhomboid major

LONGISSIMUS CAPITIS

Splenius cervicis

Levator scapulae

Middle scalene

Posterior scalene

Longissimus cervicis

Iliocostalis cervicis

Longissimus thoracis

(a) Posterior superficial view (left) and deep view (right)

Relating Muscles to Movements

Arrange the muscles in this exhibit according to the following actions on the head: (1) flexion, (2) lateral flexion, (3) extension, (4) rotation to side opposite contracting muscle, and (5) rotation to same side as contracting muscle. The same muscle may be mentioned more than once.

⊙ CHECKPOINT
What muscles do you contract to signify "yes" and "no"?

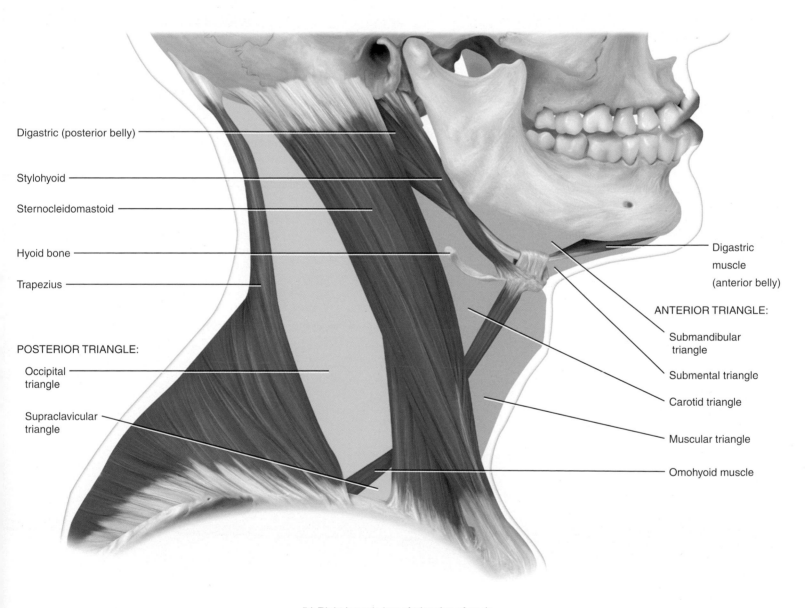

Digastric (posterior belly)

Stylohyoid

Sternocleidomastoid

Hyoid bone

Trapezius

POSTERIOR TRIANGLE:

Occipital triangle

Supraclavicular triangle

Digastric muscle (anterior belly)

ANTERIOR TRIANGLE:

Submandibular triangle

Submental triangle

Carotid triangle

Muscular triangle

Omohyoid muscle

(b) Right lateral view of triangles of neck

❓ **Why are triangles important?**

⊙ EXHIBIT 11.6 CONTINUES ▶

EXHIBIT 11.6   **333**

MUSCLE	ORIGIN	INSERTION	ACTION	INNERVATION
Sternocleidomastoid (ster'-nō-klī'-dō-MAS-toid; *sterno-*=breastbone; *cleido-*=clavicle; *mastoid*=mastoid process of temporal bone)	Sternum and clavicle.	Mastoid process of temporal bone.	Acting together (bilaterally), flex cervical portion of vertebral column; flex head, and elevate sternum during forced inhalation; acting singly (unilaterally), laterally rotate and flex head to opposite side of contracting muscle. The posterior fibers of the muscle can assist in extension of head.	Accessory (XI) nerve.
Semispinalis capitis (se'-mē-spi-NĀ-lis KAP-i-tis; *semi-*=half; *spine*=spinous process; *capit-*=head)	Transverse processes of first six or seven thoracic vertebrae and seventh cervical vertebra, and articular processes of fourth, fifth, and sixth cervical vertebrae.	Occipital bone between superior and inferior nuchal lines.	Acting together, extend head; acting singly, rotate head to side opposite contracting muscle.	Cervical spinal nerves.
Splenius capitis (SPLĒ-nē-us KAP-i-tis; *splenion-*=bandage)	Ligamentum nuchae and spinous processes of seventh cervical vertebra and first three or four thoracic vertebrae.	Occipital bone and mastoid process of temporal bone.	Acting together, extend head; acting singly, rotate head to side opposite contracting muscle.	Cervical spinal nerves.
Longissimus capitis (lon-JIS-i-mus KAP-i-tis; *longissimus*=longest)	Transverse processes of upper four thoracic vertebrae and articular processes of last four cervical vertebrae.	Mastoid process of temporal bone.	Acting together, extend head; acting singly, laterally flex and rotate head to same side as contracting muscle.	Cervical spinal nerves.

OBJECTIVE

- Describe the origin, insertion, action and innervation of the muscles of the larynx.

The muscles of the larynx (voice box), like those of the eyeballs and tongue, are grouped into **extrinsic muscles of the larynx** and **intrinsic muscles of the larynx.** The extrinsic muscles of the larynx, which are associated with the anterior aspect of the neck, are called **infrahyoid muscles** because they lie inferior to the hyoid bone. Refer to Exhibit 11.5 and Figure 11.8 for a description of these muscles. In this Exhibit, we discuss the intrinsic muscles (Figure 11.10).

The extrinsic muscles move the larynx as a whole; the intrinsic muscles of the larynx only move parts of the larynx. Based on their actions, the intrinsic muscles may be grouped into three functional sets. The first set includes the **cricothyroid** and **thyroarytenoid** muscles, which regulate the tension of the vocal folds (true vocal cords). The second set varies the size of the rima glottidis (space between the vocal folds). Those muscles include the **lateral cricoarytenoid,** which brings the vocal folds together (adduction), thus closing the rima glottidis, and the **posterior cricoarytenoid,** which moves the vocal folds apart (abduction), thus opening the rima glottidis. The **transverse arytenoid** closes the posterior portion of the rima glottidis. The last intrinsic muscle functions as a sphincter to control the size of the inlet of the larynx, which is the opening anteriorly from the pharynx (throat) into the larynx. This muscle is the **oblique arytenoid.**

Relating Muscles to Movements

Arrange the intrinsic muscles of the larynx in this exhibit according to the following actions on the vocal cords: (1) increasing tension of the vocal cords; (2) moving the vocal cords apart, and (3) moving the vocal cords together. The same muscle may be mentioned more than once.

CHECKPOINT

How are the intrinsic muscles of the larynx grouped functionally?

MUSCLE	ORIGIN	INSERTION	ACTION	INNERVATION
EXTRINSIC MUSCLES OF THE LARYNX (INFRAHYOID MUSCLES)				
Omohyoid				
Sternohyoid	See Exhibit 11.5			
Sternothyroid				
Thyrohyoid				
INTRINSIC MUSCLES OF THE LARYNX				
Cricothyroid (kri-kō-THĪ-roid; *crico-*=cricoid cartilage of larynx)	Anterior and lateral portion of cricoid cartilage of larynx.	Anterior border of thyroid cartilage of larynx and posterior part of inferior border of thyroid cartilage of larynx.	Elongates and places tension on vocal folds.	Recurrent laryngeal branch of vagus (X) nerve.
Thyroarytenoid (thĪ-rō-ar′-i-TĒ-noid; *-arytaina*=shaped like a jug)	Inferior portion of thyroid cartilage of larynx and middle of cricothyroid ligament.	Base and anterior surface of arytenoid cartilage of larynx.	Shortens and relaxes vocal folds.	Recurrent laryngeal branch of vagus (X) nerve.
Lateral Cricoarytenoid (kri′-kō-ar′-i-TĒ-noid)	Superior border of cricoid cartilage of larynx.	Anterior surface of arytenoid cartilage of larynx.	Brings vocal folds together (adduction), thus closing the rima glottidis.	Recurrent laryngeal branch of vagus (X) nerve.
Posterior Cricoarytenoid	Posterior surface of cricoid cartilage of larynx.	Posterior surface of arytenoid cartilage of larynx.	Moves the vocal folds apart (abduction), thus opening the rima glottidis.	Recurrent laryngeal branch of vagus (X) nerve.
Transverse and Oblique Arytenoid (ar′-i-TĒ-noid)	Posterior surface and lateral border of one arytenoid cartilage of larynx.	Corresponding parts of opposite arytenoid cartilage of larynx.	The transverse arytenoid closes the posterior portion of the rima glottidis; the oblique arytenoid regulates the size of the inlet of the larynx.	Recurrent laryngeal branch of vagus (X) nerve.

EXHIBIT 11.7 CONTINUES ▶

EXHIBIT 11.7   **335**

Figure 11.10 Muscles of the larynx (voice box) that assist in speech.

Intrinsic muscles of the larynx adjust the tension of the vocal folds and open or close the rima glottidis.

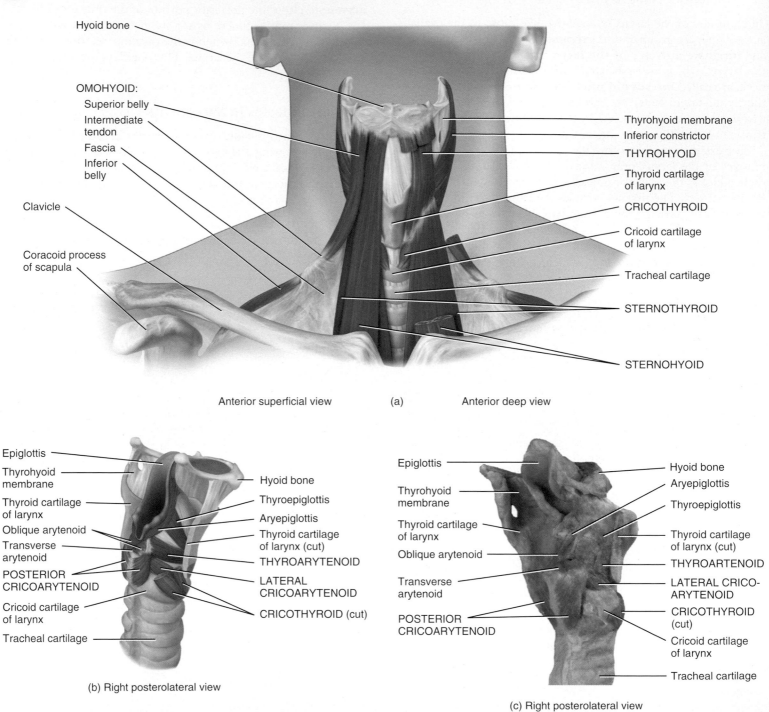

Anterior superficial view   (a)   Anterior deep view

(b) Right posterolateral view

(c) Right posterolateral view

How do the extrinsic and intrinsic muscles of the larynx differ functionally?

EXHIBIT 11.8

Muscles of the Pharynx (Throat) That Assist in Deglutition (Swallowing) and Speech

OBJECTIVE

• Describe the origin, insertion, action, and innervation of the muscles of the pharynx.

The pharynx (throat) is a somewhat funnel-shaped tube posterior to the nasal and oral cavities. It is a common chamber for the respiratory and digestive systems, opening anteriorly into the larynx and posteriorly into the esophagus.

The muscles of the pharynx are arranged in two layers, an outer circular layer and an inner longitudinal layer (Figure 11.11). The **circular layer** is composed of three constrictor muscles, each overlapping the muscle above it, an arrangement that resembles stacked flowerpots. Their names are the inferior, middle, and superior constrictor muscles. The **longitudinal layer** is composed of three muscles that descend from the styloid process of the temporal bone, auditory (eustachian) tube, and soft palate. Their names are the stylopharyngeus, salpingopharyngeus, and palatopharyngeus, respectively.

The **inferior constrictor** muscle is the thickest of the constrictor muscles. Its inferior fibers are continuous with the musculature of the esophagus, whereas its superior fibers overlap the middle constrictor. The **middle constrictor** is fan-shaped and smaller than the inferior constrictor, and it overlaps the superior constric-

tor. The **superior constrictor** is quadrilateral and thinner than the other constrictors. As a group, the constrictor muscles constrict the pharynx during swallowing. The sequential contraction of these muscles moves food and drink from the mouth into the esophagus.

The **stylopharyngeus** muscle is a long, thin muscle that enters the pharynx between the middle and superior constrictors and inserts along with the palatopharyngeus on the thyroid cartilage. The **salpingopharyngeus** muscle is a thin muscle that descends in the lateral wall of the pharynx and also inserts along with the palatopharyngeus muscle. The **palatopharyngeus** muscle also descends in the lateral wall of the pharynx and inserts along with the stylopharyngeus. All three muscles of the longitudinal layer elevate the pharynx and larynx during swallowing and speaking. Elevation of the pharynx widens it to receive food and liquids, and elevation of the larynx causes a structure called the epiglottis to close over the rima glottidis (space between the vocal folds) and seal the respiratory passageway. Food and drink are further kept out of the respiratory tract by the suprahyoid muscles of the larynx, which elevate the hyoid bone. Additionally, the respiratory passageway is sealed by the action of the intrinsic muscles of the larynx, which bring the vocal folds together to close off the rima glottidis. After swallowing, the infrahyoid muscles of the larynx depress the hyoid bone and larynx.

MUSCLE	ORIGIN	INSERTION	ACTION	INNERVATION
CIRCULAR LAYER				
Inferior constrictor (*inferior-*=below; *constrictor*=decreases diameter of a lumen)	Cricoid and thyroid cartilages of larynx.	Posterior median raphe (slender band of collagen fibers) of pharynx.	Constricts inferior portion of pharynx to propel food and drink into esophagus.	Pharyngeal plexus, branches of vagus (X) nerve.
Middle constrictor	Greater and lesser horns of hyoid bone and stylohyoid ligament.	Posterior median raphe of pharynx.	Constricts middle portion of pharynx to propel food and drink into esophagus.	Pharyngeal plexus, branches of vagus (X) nerve.
Superior constrictor (*superior*=above)	Pterygoid process of sphenoid, pterygomandibular raphe, and medial surface of mandible.	Posterior median raphe of pharynx.	Constricts superior portion of pharynx to propel food and drink into esophagus.	Pharyngeal plexus, branches of vagus (X) nerve.
LONGITUDINAL LAYER				
Stylopharyngeus (stī-lō-far-IN-jē-us; *stylo-*=stake or pole; styloid process of temporal bone; *pharyngo-*=pharynx)	Styloid process of temporal bone.	Thyroid cartilage with the palatopharyngeus.	Elevates larynx and pharynx.	Glossopharyngeal (IX) nerve.
Salpingopharyngeus (sal-pin′-gō-far-IN-jē-us; *salping-*=pertaining to the auditory or uterine tube)	Inferior portion of auditory (eustachian) tube.	Thyroid cartilage with the palatopharyngeus.	Elevates larynx and pharynx and opens orifice of auditory (eustachian) tube.	Pharyngeal plexus, branches of vagus (X) nerve.
Palatopharyngeus (pal′-a-tō-far-IN-jē-us; *palato-*=palate)	Soft palate.	Thyroid cartilage with the stylopharyngeus.	Elevates larynx and pharynx and helps close nasopharynx during swallowing.	Pharyngeal plexus, branches of vagus (X) nerve.

EXHIBIT 11.8 CONTINUES

EXHIBIT 11.8   **337**

Figure 11.11 Muscles of the pharynx (throat) that assist in deglutition (swallowing) and speech.

The muscles of the pharynx assist in swallowing and in closing off the respiratory passageway.

(a) Right lateral view

External auditory canal (meatus)
Styloid process of temporal bone
Digastric (posterior belly, cut)
SUPERIOR CONSTRICTOR
Styloglossus (cut)
STYLOPHARYNGEUS
MIDDLE CONSTRICTOR
INFERIOR CONSTRICTOR
Tracheal cartilage
Esophagus

Maxilla
Buccinator
Orbicularis oris
Mandible (cut)
Digastric (anterior belly)
Mylohyoid
Hyoglossus
Stylohyoid (cut)
Hyoid bone
Thyrohyoid membrane
Thyroid cartilage of larynx
Cricothyroid
Cricoid cartilage of larynx

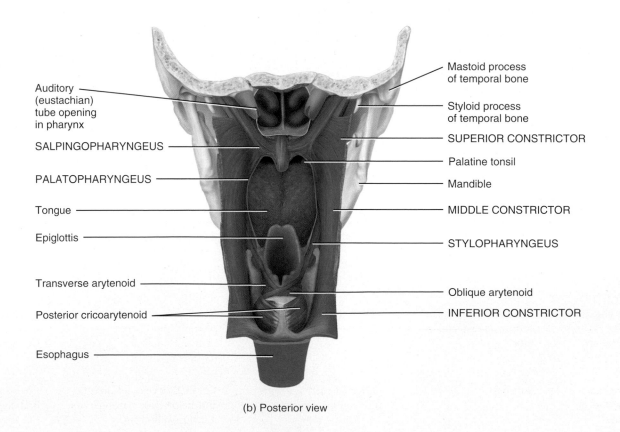

(b) Posterior view

Auditory (eustachian) tube opening in pharynx
SALPINGOPHARYNGEUS
PALATOPHARYNGEUS
Tongue
Epiglottis
Transverse arytenoid
Posterior cricoarytenoid
Esophagus

Mastoid process of temporal bone
Styloid process of temporal bone
SUPERIOR CONSTRICTOR
Palatine tonsil
Mandible
MIDDLE CONSTRICTOR
STYLOPHARYNGEUS
Oblique arytenoid
INFERIOR CONSTRICTOR

Relating Muscles to Movements

Arrange the muscles in this exhibit according to the following actions on the pharynx: (1) constriction and (2) elevation, and according to the following action on the larynx: elevation. The same muscle may be mentioned more than once.

◉ **CHECKPOINT**
What happens when the pharynx and larynx are elevated?

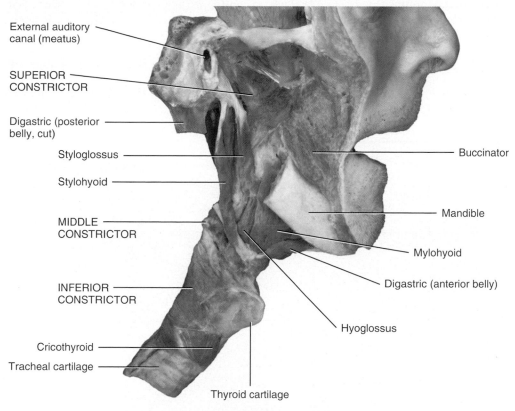

External auditory
canal (meatus)

SUPERIOR
CONSTRICTOR

Digastric (posterior
belly, cut)

Styloglossus

Stylohyoid

MIDDLE
CONSTRICTOR

INFERIOR
CONSTRICTOR

Cricothyroid

Tracheal cartilage

Thyroid cartilage

Buccinator

Mandible

Mylohyoid

Digastric (anterior belly)

Hyoglossus

(c) Right lateral deep view

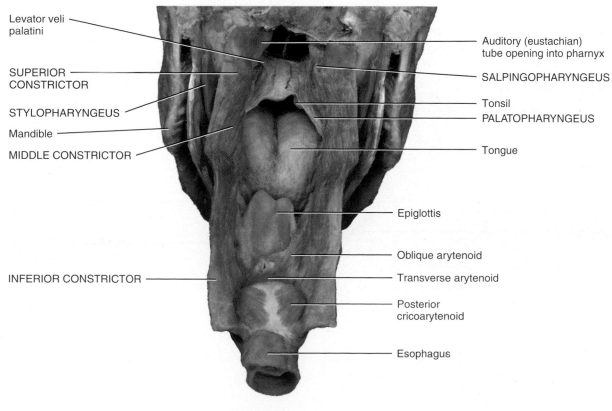

Levator veli
palatini

SUPERIOR
CONSTRICTOR

STYLOPHARYNGEUS

Mandible

MIDDLE CONSTRICTOR

INFERIOR CONSTRICTOR

Auditory (eustachian)
tube opening into pharnyx

SALPINGOPHARYNGEUS

Tonsil
PALATOPHARYNGEUS

Tongue

Epiglottis

Oblique arytenoid

Transverse arytenoid

Posterior
cricoarytenoid

Esophagus

(d) Posterior view

? **What are the antagonistic roles of the longitudinal muscles of the pharynx and the infrahyoid muscles?**

EXHIBIT 11.8   **339**

● OBJECTIVE

• Describe the origin, insertion, action, and innervation of the muscles that act on the abdominal wall.

The anterolateral abdominal wall is composed of skin, fascia, and four pairs of muscles: the external oblique, internal oblique, transversus abdominis, and rectus abdominis (Figure 11.12). The first three muscles named are arranged from superficial to deep.

The **external oblique** is the superficial muscle. Its fascicles extend inferiorly and medially. The **internal oblique** is the intermediate flat muscle. Its fascicles extend at right angles to those of the external oblique. The **transversus abdominis** is the deep muscle, with most of its fascicles directed transversely around the abdominal wall. Together, the external oblique, internal oblique, and transversus abdominis form three layers of muscle around the abdomen. In each layer, the muscle fascicles extend in a different

Figure 11.12 Muscles of the abdomen that protect abdominal viscera and move the vertebral column (backbone).

🔑 The anterolateral abdominal muscles protect the abdominal viscera, move the vertebral column, and assist in forced expiration, defecation, urination, and childbirth.

(a) Anterior superficial view   (b) Anterior deep view

direction. This is a structural arrangement that affords considerable protection to the abdominal viscera, especially when the muscles have good tone.

The **rectus abdominis** muscle is a long muscle that extends the entire length of the anterior abdominal wall, originating at the pubic crest and pubic symphysis and inserting on the cartilages of ribs 5–7 and the xiphoid process of the sternum. The anterior surface of the muscle is interrupted by three transverse fibrous bands of tissue called **tendinous intersections**, believed to be remnants of septa that separated myotomes during embryological development (see Figure 10.10 on page 304).

As a group, the muscles of the anterolateral abdominal wall help contain and protect the abdominal viscera; flex, laterally flex, and rotate the vertebral column at the intervertebral joints; compress the abdomen during forced exhalation; and produce the force required for defecation, urination, and childbirth.

The aponeuroses (sheathlike tendons) of the external oblique, internal oblique, and transversus abdominis muscles form the **rectus sheaths,** which enclose the rectus abdominis muscles. The sheaths meet at the midline to form the **linea alba** (=white line), a tough, fibrous band that extends from the xiphoid process of the sternum to the pubic symphysis. In the latter stages of pregnancy, the linea alba stretches to increase the distance between the rectus abdominis muscles. The inferior free border of the external oblique aponeurosis forms the **inguinal ligament**, which runs from the anterior superior iliac spine to the pubic tubercle (see Figure 11.22a). Just superior to the medial end of the inguinal ligament is a triangular slit in the aponeurosis referred to as the **superficial inguinal ring**, the outer opening of the **inguinal canal** (see Figure 27.2 on page 848). The **inguinal canal** contains the spermatic cord and ilioinguinal nerve in males, and the round ligament of the uterus and ilioinguinal nerve in females.

The posterior abdominal wall is formed by the lumbar vertebrae, parts of the ilia of the hip bones, psoas major and iliacus muscles (described in Exhibit 11.19), and quadratus lumborum muscle. The anterolateral abdominal wall can contract and distend; the posterior abdominal wall is bulky and stable by comparison.

• CLINICAL CONNECTION | Inguinal Hernia

A **hernia** is a protrusion of an organ through a structure that normally contains it, which creates a lump that can be seen or felt through the skin's surface. The inguinal region is a weak area in the abdominal wall. It is often the site an **inguinal hernia**, a rupture or separation of a portion of the inguinal area of the abdominal wall resulting in the protrusion of a part of the small intestine. A hernia is much more common in males than in females because the inguinal canals in males are larger to accommodate the spermatic cord and ilioinguinal nerve. Treatment of hernias most often involves surgery. The organ that protrudes is "tucked" back into the abdominal cavity and the defect in the abdominal muscles is repaired. In addition, a mesh is often applied to reinforce the area of weakness. •

Relating Muscles to Movements

Arrange the muscles in this exhibit according to the following actions on the vertebral column: (1) flexion, (2) lateral flexion, (3) extension, and (4) rotation. The same muscle may be mentioned more than once.

⊙ CHECKPOINT

Which muscles do you contract when you "suck in your gut," thereby compressing the anterior abdominal wall?

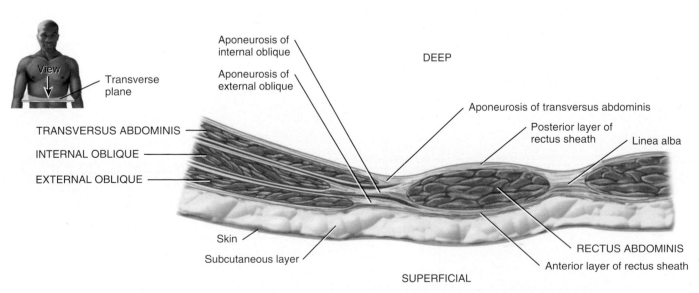

Transverse plane

Aponeurosis of internal oblique

Aponeurosis of external oblique

DEEP

Aponeurosis of transversus abdominis

Posterior layer of rectus sheath

Linea alba

TRANSVERSUS ABDOMINIS

INTERNAL OBLIQUE

EXTERNAL OBLIQUE

Skin

Subcutaneous layer

RECTUS ABDOMINIS

Anterior layer of rectus sheath

SUPERFICIAL

(c) Transverse section of anterior abdominal wall superior to umbilicus (navel)

EXHIBIT 11.9 CONTINUES ▶

EXHIBIT 11.9   **341**

MUSCLE	ORIGIN	INSERTION	ACTION	INNERVATION
Rectus abdominis (REK-tus ab-DOM-in-is; *rectus-*=fascicles parallel to midline; *abdomin*=abdomen)	Pubic crest and pubic Symphysis.	Cartilage of fifth to seventh ribs and xiphoid process.	Flexes vertebral column* (especially lumbar portion), and compresses abdomen to aid in defecation, urination, forced exhalation, and childbirth.	Thoracic spinal nerves T7–T12.
External oblique (ō-BLĒK; *external*=closer to surface; *oblique*=fascicles diagonal to midline)	Ribs 5–12.	Iliac crest and linea alba.	Acting together (bilaterally), compress abdomen and flex vertebral column; acting singly (unilaterally), laterally flex vertebral column, especially lumbar portion, and rotate vertebral column.	Thoracic spinal nerves T7–T12 and the iliohypogastric nerve.
Internal oblique (ō-BLĒK; *internal*=farther from surface)	Iliac crest, inguinal ligament, and thoracolumbar fascia.	Cartilage of ribs 7–10 and linea alba.	Acting together, compress abdomen and flex vertebral column; acting singly, laterally flex vertebral column, especially lumbar portion, and rotate vertebral column.	Thoracic spinal nerves T8–T12, iliohypogastric nerve, and ilioinguinal nerve.
Transversus abdominis (tranz-VER-sus ab-DOM-in-is; *transverse*=fascicles perpendicular to midline)	Iliac crest, inguinal ligament, lumbar fascia, and cartilages of ribs 5–10.	Xiphoid process, linea alba, and pubis.	Compresses abdomen.	Thoracic spinal nerves T8–T12, iliohypogastric nerve, and ilioinguinal nerve.
Quadratus lumborum (kwod-RĀ-tus lum-BOR-um; *quad-*=four; *lumbo-*=lumbar region) (see Figure 11.13)	Iliac crest and iliolumbar ligament.	Inferior border of twelfth rib and first four lumbar vertebrae.	Acting together, pull twelfth ribs interiorly during forced exhalation, fix twelfth ribs to prevent their elevation during deep inhalation, and help extend lumbar portion of vertebral column; acting singly, laterally flex vertebral column, especially lumbar portion.	Thoracic spinal nerve T12 and lumbar spinal nerves L1–L3 or L1–L4.

*The attachments of a muscle (origin and insertion) may be reversed, depending on the action. For example, when the rectus abdominis muscles act from below, they flex the vertebral column. However, when they act from above, they flex the pelvis on the vertebral column.

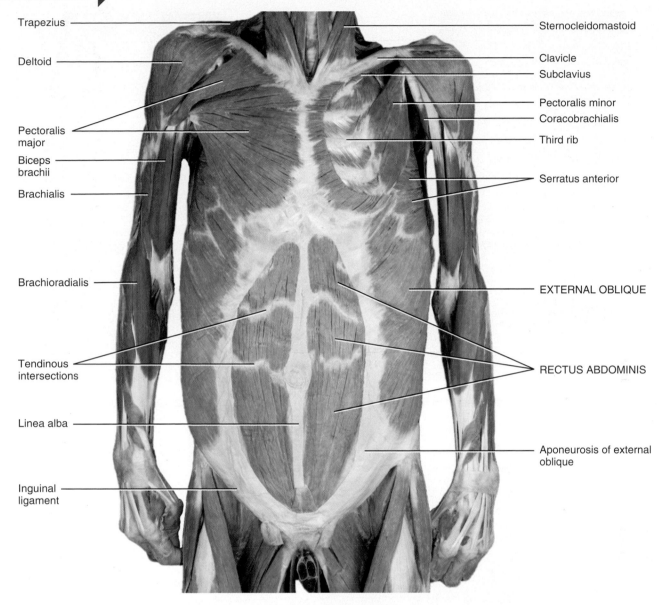

Trapezius

Deltoid

Pectoralis major

Biceps brachii

Brachialis

Brachioradialis

Tendinous intersections

Linea alba

Inguinal ligament

Sternocleidomastoid

Clavicle

Subclavius

Pectoralis minor

Coracobrachialis

Third rib

Serratus anterior

EXTERNAL OBLIQUE

RECTUS ABDOMINIS

Aponeurosis of external oblique

(d) Anterior view

● FIGURE **11.12** CONTINUED ▶

External intercostals

Seventh rib

EXTERNAL OBLIQUE

RECTUS ABDOMINIS

Aponeurosis of external oblique

Inguinal ligament

Linea alba

(e) Right anterolateral superficial view

Seventh rib

Internal intercostals

Aponeurosis of internal oblque

INTERNAL OBLIQUE

Tendinous intersections

(f) Right anterolateral deep view

Seventh rib

Innermost Intercostals

TRANSVERSUS ABDOMINIS

Aponeurosis of transversus abdominis

Iliac crest

(g) Right anterolateral deeper view

? Which abdominal muscle aids in urination?

EXHIBIT 11.10

Muscles of the Thorax (Chest) That Assist in Breathing

● OBJECTIVE

- Describe the origin, insertion, action, and innervation of the muscles used in breathing.

The muscles described here alter the size of the thoracic cavity so that breathing can occur. Inhalation (breathing in) occurs when the thoracic cavity increases in size, and exhalation (breathing out) occurs when the thoracic cavity decreases in size.

The dome-shaped **diaphragm** is the most important muscle that powers breathing. It also separates the thoracic and abdominal cavities. The diaphragm has a convex superior surface that forms the floor of the thoracic cavity (Figure 11.13a) and concave, inferior surface that forms the roof of the abdominal cavity (Figure 11.13d). The **peripheral muscular portion** of the diaphragm originates on the xiphoid process of the sternum, the inferior six ribs and their costal cartilages, and the lumbar vertebrae and their intervertebral discs and the twelfth rib (Figure 11.13d). From their various origins, the fibers of the muscular portion converge and insert into the **central tendon,** a strong aponeurosis located near the center of the muscle (Figure 11.13c, d). The central tendon fuses with the inferior surface of the pericardium (covering of the heart) and the pleurae (coverings of the lungs).

The diaphragm has three major openings through which various structures pass between the thorax and abdomen. These structures include the aorta, along with the thoracic duct and azygous vein, which pass through the **aortic hiatus**; the esophagus with accompanying vagus (X) nerves, which pass through the **esophageal hiatus**; and the inferior vena cava, which passes through the **caval opening (foramen for the vena cava)**. In a condition called a hiatus hernia, the stomach protrudes superiorly through the esophageal hiatus.

Movements of the diaphragm also help return venous blood passing through abdominal veins to the heart. Together with the anterolateral abdominal muscles, the diaphragm helps to increase intra-abdominal pressure to evacuate the pelvic contents during defecation, urination, and childbirth. This mechanism is further assisted when you take a deep breath and close the rima glottidis (the space between vocal folds). The trapped air in the respiratory system prevents the diaphragm from elevating. The increase in intra-abdominal pressure also helps support the vertebral column and prevents flexion during weight lifting. This greatly assists the back muscles in lifting a heavy weight.

Other muscles involved in breathing, called **intercostal muscles,** span the intercostal spaces, the spaces between ribs. These muscles are arranged in three layers. The 11 pairs of **external intercostal muscles** occupy the superficial layer, and their fibers run in an oblique direction interiorly and anteriorly from the rib above to the rib below. They elevate the ribs during inhalation to help expand the thoracic cavity. The 11 pairs of **internal intercostal muscles** occupy the intermediate layer of the intercostal spaces. The fibers of these muscles run at right angles to the external intercostals, in an oblique direction interiorly and posteriorly from the inferior border of the rib above to the superior border of the rib below. They draw adjacent ribs together during forced exhalation to help decrease the size of the thoracic cavity. The deepest muscle layer is made up of the paired **innermost intercostal muscles.** These poorly developed muscles extend in the same direction as the internal intercostals and may have the same role.

As you will see in Chapter 24, the diaphragm and external intercostal muscles are used during quiet inhalation and exhalation. However, during deep, forceful inhalation (during exercise or playing a wind instrument), the sternocleidomastoid, scalene, and pectoralis minor muscles are also used; during deep, forceful exhalation, the external oblique, internal oblique, transversus abdominis, rectus abdominis, and internal intercostals are also used.

MUSCLE	ORIGIN	INSERTION	ACTION	INNERVATION
Diaphragm (DI-a-fram; *dia-*=across; *-praghm*=wall)	Xiphoid process of the sternum, costal cartilages and adjacent portions of the inferior six ribs, lumbar vertebrae and their intervertebral discs.	Central tendon.	Contraction of the diaphragm causes it to flatten and increases the vertical dimension of the thoracic cavity, resulting in inhalation; relaxation of the diaphragm causes it to move superiorly and decreases the vertical dimension of the thoracic cavity, resulting in exhalation.	Phrenic nerve, which contains axons from cervical spinal nerves C3–C5.
External intercostals (in'-ter-KOS-tals; *external*=closer to surface; *inter-*=between; *costa-*=rib)	Interior border of rib above.	Superior border of rib below.	Contraction elevates the ribs and increases the anteroposterior and lateral dimensions of the thoracic cavity, resulting in inhalation; relaxation depresses the ribs and decreases the anteroposterior and lateral dimensions of the thoracic cavity, resulting in exhalation.	Thoracic spinal nerves T2–T12.
Internal intercostals (in'-ter-KOS-tals; *internal*=farther from surface)	Superior border of rib below.	Inferior border of rib above.	Contraction draws adjacent ribs together to further decrease the anteroposterior and lateral dimensions of the thoracic cavity during forced exhalation.	Thoracic spinal nerves T2–T12.

■ EXHIBIT 11.10 CONTINUES ▶

EXHIBIT 11.10   **345**

Relating Muscles to Movements

Arrange the muscles in this exhibit according to the following actions: (1) increase in vertical length, (2) increase in lateral and anteroposterior dimensions, and (3) decrease in lateral and anteroposterior dimensions of the thorax.

◉ CHECKPOINT

What are the names of the three openings in the diaphragm, and which structures pass through each?

Figure 11.13 Muscles of the thorax (chest) that assist in breathing.

🔑 Openings in the diaphragm permit the passage of the aorta, esophagus, and inferior vena cava.

(a) Anterior superficial view   (b) Anterior deep view

(c) Superior view of diaphragm

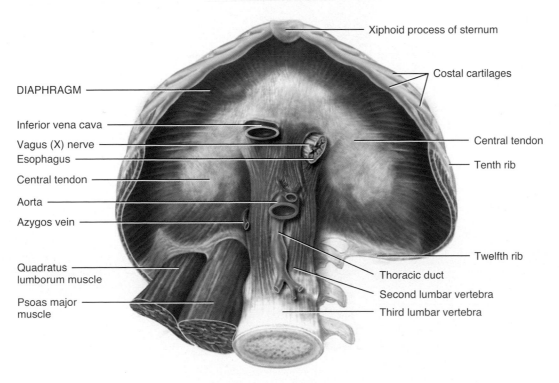

(d) Inferior view of diaphragm

EXHIBIT **11.10** CONTINUES

EXHIBIT 11.10   **347**

● FIGURE **11.13** CONTINUED ▶

Sternum

Pericardium covering central tendon

Pleura (cut)

Ribs (cut)

Inferior vena cava

Esophagus

DIAPHRAGM

Aorta

Body of T8

Serratus anterior

Spinal cord

Latissimus dorsi muscle

Erector spinae muscle

(e) Superior view

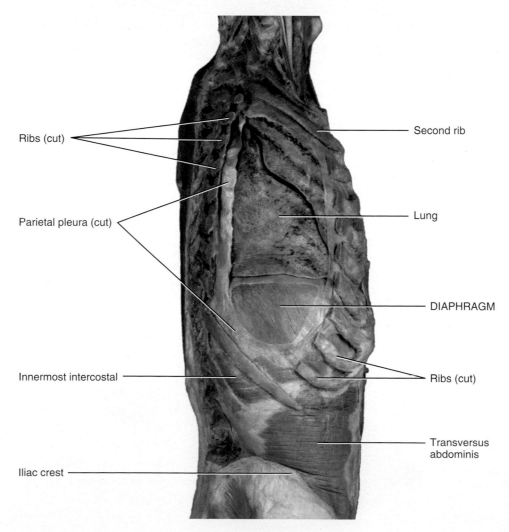

Ribs (cut)

Second rib

Parietal pleura (cut)

Lung

DIAPHRAGM

Innermost intercostal

Ribs (cut)

Transversus abdominis

Iliac crest

(f) Right lateral view

❓ **Which muscle associated with breathing is innervated by the phrenic nerve?**

EXHIBIT 11.11

Muscles of the Pelvic Floor That Support the Pelvic Viscera, Assist in Resisting Intra-abdominal Pressure, and Function as Sphincters

OBJECTIVE

- Describe the origin, insertion, action, and innervation of the muscles of the pelvic floor.

The muscles of the pelvic floor are the levator ani and ischiococcygeus. Together with the fascia covering their internal and external surfaces, these muscles are referred to as the **pelvic diaphragm,** which stretches from the pubis anteriorly to the coccyx posteriorly, and from one lateral wall of the pelvis to the other. This arrangement gives the pelvic diaphragm the appearance of a funnel suspended from its attachments. The pelvic diaphragm separates the pelvic cavity above from the perineum below (see Exhibit 11.12). The anal canal and urethra pierce the pelvic diaphragm in both sexes, and the vagina also goes through it in females.

The two components of the **levator ani** muscle are the **pubococcygeus** and **iliococcygeus.** Figure 11.14 shows these muscles in the female and Figure 11.15 in Exhibit 11.12 illustrates them in the male. The levator ani is the largest and most impor-

tant muscle of the pelvic floor. It supports the pelvic viscera and resists the inferior thrust that accompanies increases in intra-abdominal pressure during functions such as forced exhalation, coughing, vomiting, urination, and defecation. The muscle also functions as a sphincter at the anorectal junction, urethra, and vagina. In addition to assisting the levator ani, the **ischiococcygeus** pulls the coccyx anteriorly after it has been pushed posteriorly during defecation or childbirth.

Relating Muscles to Movements

Arrange the muscles in this exhibit according to the following actions: (1) supporting and maintaining the position of the pelvic viscera; (2) resisting an increase in intra-abdominal pressure; (3) constriction of the anus, urethra, and vagina. The same muscle may be mentioned more than once.

CHECKPOINT

Which muscles are strengthened by Kegel exercises?

MUSCLE	ORIGIN	INSERTION	ACTION	INNERVATION
Levator ani (le-VĀ-tor Ā-nē; *levator*=raises; *ani*=anus)	This muscle is divisible into two parts, the pubococcygeus muscle and the iliococcygeus muscle.			
Pubococcygeus (pū′-bō-kok-SIJ-e-us; *pubo-*=pubis; *-coccygeus*=coccyx)	Pubis and ischial spine.	Coccyx, urethra, anal canal, perineal body of the perineum (a wedge-shaped mass of fibrous tissue in the center of the perineum), and anococcygeal raphe (narrow fibrous band that extends from anus to coccyx).	Supports and maintains position of pelvic viscera; resists increase in intra-abdominal pressure during forced exhalation, coughing, vomiting, urination, and defecation; constricts anus, urethra, and vagina.	Sacral spinal nerves S2–S4.
Iliococcygeus (il′-ē-ō-kok-SIJ-ē-us; *ilio-*=ilium)	Ischial spine.	Coccyx.	As above.	Sacral spinal nerves S2–S4.
Ischiococcygeus (Is′-kē-ō-kok-SIJ-ē-us; *ischio-*=hip)	Ischial spine.	Lower sacrum and upper coccyx.	Supports and maintains position of pelvic viscera; resists increase in intra-abdominal pressure during forced exhalation, coughing, vomiting, urination, and defecation; pulls coccyx anteriorly following defecation or childbirth.	Sacral spinal nerves S4–S5.

EXHIBIT 11.11 CONTINUES ▶

EXHIBIT 11.11   **349**

EXHIBIT 11.11

Muscles of the Pelvic Floor That Support the Pelvic
Viscera, Assist in Resisting Intra-abdominal Pressure, and
Function as Sphincters CONTINUED

• CLINICAL **CONNECTION** Injury of Levator Ani and Urinary Stress Incontinence

During childbirth, the levator ani muscle supports the head of the fetus, and the muscle may be injured during a difficult childbirth or traumatized during an *episiotomy* (a cut made with surgical scissors to prevent or direct tearing of the perineum during the birth of a baby). The consequence of such injuries may be **urinary stress incontinence**, that is, the leakage of urine whenever intra-abdominal pressure is increased—for example, during coughing. One way to treat urinary stress incontinence is to strengthen and tighten the muscles that support the pelvic viscera. This is accomplished by *Kegel exercises*, the alternate contraction and relaxation of muscles of the pelvic floor. To find the correct muscles, the person imagines that she is urinating and then contracts the muscles as if stopping in midstream. The muscles should be held for a count of three, then relaxed for a count of three. This should be done 5–10 times each hour—sitting, standing, and lying down. Kegel exercises are also encouraged during pregnancy to strengthen the muscles for delivery. •

Figure 11.14 Muscles of the pelvic floor that support the pelvic viscera, assist in resisting intra-abdominal pressure, and function as sphincters.

 The pelvic diaphragm supports the pelvic viscera.

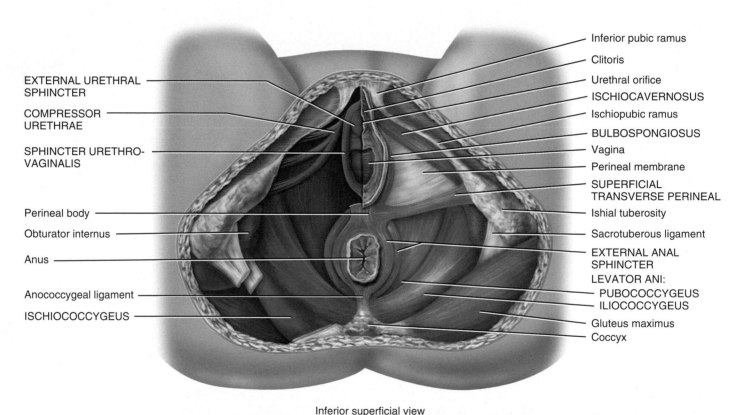

Inferior superficial view

? What are the borders of the pelvic diaphragm?

EXHIBIT 11.12 Muscles of the Perineum

OBJECTIVE

- Describe the origin, insertion, action, and innervation of the muscles of the perineum.

The **perineum** is the region of the trunk inferior to the pelvic diaphragm. It is a diamond-shaped area that extends from the pubic symphysis anteriorly, to the coccyx posteriorly, and to the ischial tuberosities laterally. The female and the male perineums may be compared in Figures 11.14 and 11.15, respectively. A transverse line drawn between the ischial tuberosities divides the perineum into an anterior **urogenital triangle** that contains the external genitals and a posterior anal triangle that contains the anus (see Figure 27.22a on page 874). Several perineal muscles insert into the perineal body of the perineum (described on page 349). Clinically, the perineum is very important to physicians who care for women during pregnancy and treat disorders related to the female genital tract, urogenital organs, and the anorectal region.

The muscles of the perineum are arranged in two layers; **superficial** and **deep**. The muscles of the superficial layer are the **superficial transverse perineal muscle**, the **bulbospongiosus**, and the **ischiocavernosus** (Figures 11.14 and 11.15). The deep muscles of the male perineum are the **deep trans-**verse perineal muscle and **external urethral sphincter** (Figure 11.15). The deep muscles of the female perineum are the **compressor urethrae, sphincter urethrovaginalis,** and **external urethral sphincter** (Figure 11.14). The deep muscles of the perineum assist in urination and ejaculation in males and urination and compression of the vagina in females. The **external anal sphincter** closely adheres to the skin around the margin of the anus and keeps the anal canal and anus closed except during defecation.

Relating Muscles to Movements

Arrange the muscles in this exhibit according to the following actions; (1) expulsion of urine and semen, (2) erection of the clitoris and penis, (3) closure of the anal orifice, and (4) constriction of the vaginal orifice. The same muscle may be mentioned more than once.

CHECKPOINT
What are the borders and contents of the urogenital triangle and the anal triangle?

MUSCLE	ORIGIN	INSERTION	ACTION	INNERVATION
SUPERFICIAL PERINEAL MUSCLES				
Superficial transverse perineal (per-i-NĒ-al; *superficial*=closer to surface; *transverse*=across; *perineus*=perineum)	Ischial tuberosity.	Perineal body of perineum.	Stabilizes perineal body of perineum and supports pelvic floor.	Perineal branch of the pudendal nerve of the sacral plexus.
Bulbospongiosus (bul'-bō-spon'-jē-O-sus; *bulb-*=a bulb; *spongio-*=sponge)	Perineal body of perineum.	Perineal membrane of deep muscles of perineum, corpus spongiosum of penis, and deep fascia on dorsum of penis in male; pubic arch and root and dorsum of clitoris in female.	Helps expel urine during urination, helps propel semen along urethra, assists in erection of the penis in male; constricts vaginal orifice and assists in erection of clitoris in female.	Perineal branch of the pudendal nerve of the sacral plexus.
Ischiocavernosus (is'-kē-ō-ka'-ver-NO-sus; *ischio-*=the hip)	Ischial tuberosity and ischial and pubic rami.	Corpus cavernosum of penis in male and clitoris in female and pubic symphysis.	Maintains erection of penis in male and clitoris in female by decreasing urine drainage.	Perineal branch of the pudendal nerve of the sacral plexus.
DEEP PERINEAL MUSCLES				
Deep transverse perineal (per-i-NĒ-al; *deep*=farther from surface)	Ischial rami.	Perineal body of perineum.	Helps expel last drops of urine and semen in male.	Perineal branch of the pudendal nerve of the sacral plexus.
External urethral sphincter (ū-RĒ-thral SFINGK-ter)	Ischial and pubic rami.	Median raphe in male and vaginal wall in female.	Helps expel last drops of urine and semen in male and urine in female.	Sacral spinal nerve S4 and the inferior rectal branch of the pudendal nerve.
Compressor urethrae (ū-RĒ-thrē)	Ischiopubic ramus.	Blends with partner on other side anterior to urethra.	Serves as an accessory sphincter of the urethra.	Perineal branch of the pudendal nerve of the sacral plexus.
Sphincter urethrovaginalis (ū-rē-thrō-vaj-i-NAL-is)	Perineal body.	Blends with partner on other side anterior to urethra.	Serves as an accessory sphincter of the urethra and facilitates closing of the vagina.	Perineal branch of the pudendal nerve of the sacral plexus.
External anal sphincter (Ā-nal)	Anococcygeal ligament.	Perineal body of perineum.	Keeps anal canal and anus closed.	Sacral spinal nerve S4 and the inferior rectal branch of the pudendal nerve.

EXHIBIT 11.12 CONTINUES

EXHIBIT 11.12 **351**

EXHIBIT 11.12   Muscles of the Perineum CONTINUED

Figure 11.15 Muscles of the perineum.

The urogenital diaphragm assists in urination in females and males, ejaculation in males, and helps strengthen the pelvic floor.

(a) Inferior superficial view

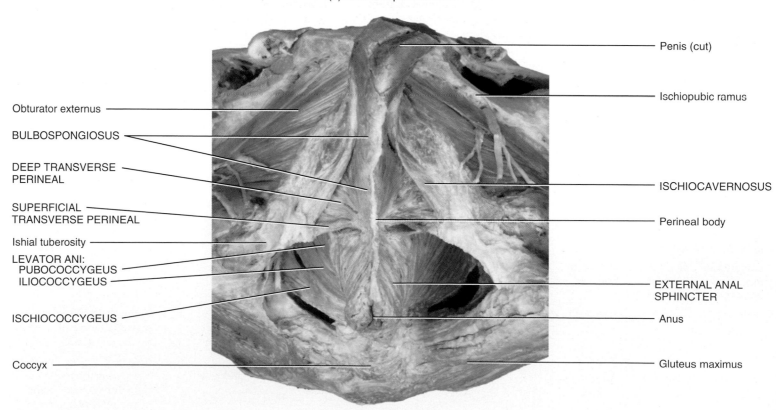

(b) Inferior superficial view

? What are the borders of the perineum?

EXHIBIT 11.13

Muscles of the Thorax (Chest) That Move the Pectoral (Shoulder) Girdle (Clavicle and Scapula)

● **OBJECTIVE**

• Describe the origin, insertion, action, and innervation of the muscles that move the pectoral girdle.

The main action of the muscles that move the pectoral girdle is to stabilize the scapula so it can function as a steady origin for most of the muscles that move the humerus. Because scapular movements usually accompany humeral movements in the same direction, the muscles also move the scapula to increase the range of motion of the humerus. For example, it would not be possible to raise the arm above the head if the scapula did not move with the humerus. During abduction, the scapula follows the humerus by rotating upward.

Muscles that move the pectoral girdle can be classified into two groups based on their location in the thorax: **anterior** and **posterior thoracic muscles** (Figure 11.16). The anterior thoracic muscles are the subclavius, pectoralis minor, and serratus anterior. The **subclavius** is a small, cylindrical muscle under the clavicle that extends from the clavicle to the first rib. It steadies the clavicle during movements of the pectoral girdle. The **pectoralis minor** is a thin, flat, triangular muscle that is deep to the pectoralis major. Besides its role in movements of the scapula, the pectoralis minor muscle assists in forced inhalation. The **serratus anterior** is a large, flat, fan-shaped muscle between the ribs and scapula. It is so named because of the saw-toothed appearance of its origins on the ribs.

The posterior thoracic muscles are the trapezius, levator scapulae, rhomboid major, and rhomboid minor. The **trapezius** is a large, flat, triangular sheet of muscle extending from the skull and vertebral column medially to the pectoral girdle laterally. It is the most superficial back muscle and covers the posterior neck region and superior portion of the trunk. The two trapezius muscles form a trapezoid (diamond-shaped quadrangle)—hence its name. The **levator scapulae** is a narrow, elongated muscle in the posterior

MUSCLE	ORIGIN	INSERTION	ACTION	INNERVATION
ANTERIOR THORACIC MUSCLES				
Subclavius (sub-KLĀ-vē-us, *sub-*=under; *clavius*=clavicle)	First rib.	Clavicle.	Depresses and moves clavicle anteriorly and helps stabilize pectoral girdle.	Subclavian nerve.
Pectoralis minor (pek′-tō-RĀ-lis; *pector-*=the breast, chest, thorax; *minor*=lesser)	Second through fifth, third through fifth, or second through fourth ribs.	Coracoid process of scapula.	Abducts scapula and rotates it downward; elevates third through fifth ribs during forced inhalation when scapula is fixed.	Medial pectoral nerve.
Serratus anterior (ser-Ā-tus; *serratus*=saw-toothed; *anterior*=front)	Superior eight or nine ribs.	Vertebral border and inferior angle of scapula.	Abducts scapula and rotates it upward; elevates ribs when scapula is stabilized; known as "boxer's muscle" because it is important in horizontal arm movements such as punching and pushing.	Long thoracic nerve.
POSTERIOR THORACIC MUSCLES				
Trapezius (tra-PĒ-zē-us; *trapezi-*=trapezoid;-shaped)	Superior nuchal line of occipital bone, ligamentum nuchae, and spines of C7–T12.	Clavicle and acromion and spine of scapula.	Superior fibers upward rotate scapula; middle fibers adduct scapula; inferior fibers depress and upward rotate scapula; superior and inferior fibers together rotate scapula upward; stabilizes scapula.	Accessory (XI) nerve and cervical spinal nerves C3–C5.
Levator scapulae (le-VĀ-tor SKA-pu-le; *levator*=raises; *scapulae*=of the scapula)	Transverse processes of C1–C4.	Superior vertebral border of scapula.	Elevates scapula and rotates it downward.	Dorsal scapular nerve and cervical spinal nerves C3–C5.
Rhomboid major (rom-BOYD; *rhomboid*=rhomboid or diamond-shaped)	Spines of T2–T5.	Vertebral border of scapula inferior to spine.	Elevates and adducts scapula and rotates it downward; stabilizes scapula.	Dorsal scapular nerve.
Rhomboid minor (rom-BOYD)	Spines of C7–T1.	Vertebral border of scapula superior to spine.	Elevates and adducts scapula and rotates it downward; stabilizes scapula.	Dorsal scapular nerve.

● **EXHIBIT 11.13 CONTINUES** ▶

EXHIBIT 11.13 **353**

portion of the neck. It is deep to the sternocleidomastoid and trapezoid muscles. As its name suggests, one of its actions is to elevate the scapula. The **rhomboid major** and **rhomboid minor** lie deep to the trapezius and are not always distinct from each other. They appear as parallel bands that pass inferiorly and laterally from the vertebrae to the scapula. Their names are based on their shape—that is, a rhomboid (an oblique parallelogram). The rhomboid major is about two times wider than the rhomboid minor. Both muscles are used when forcibly lowering the raised upper limbs, as in driving a stake with a sledgehammer.

To understand the actions of muscles that move the scapula, it is first helpful to review the various movements of the scapula:

- **Elevation:** superior movement of the scapula, such as shrugging the shoulders or lifting a weight over the head.
- **Depression:** inferior movement of the scapula, as in pulling down on a rope attached to a pulley.
- **Abduction (protraction):** movement of the scapula laterally and anteriorly, as in doing a "push-up" or punching.
- **Adduction (retraction):** movement of the scapula medially and posteriorly as in pulling the oars in a rowboat.
- **Upward rotation:** movement of the inferior angle of the scapula laterally so that the glenoid cavity is moved upward. This movement is required to move the humerus past the horizontal as in raising the arms in a "jumping jack".

Figure 11.16 **Muscles of the thorax (chest) that move the pectoral (shoulder) girdle (clavicle and scapula).**

Muscles that move the pectoral girdle originate on the axial skeleton and insert on the clavicle or scapula.

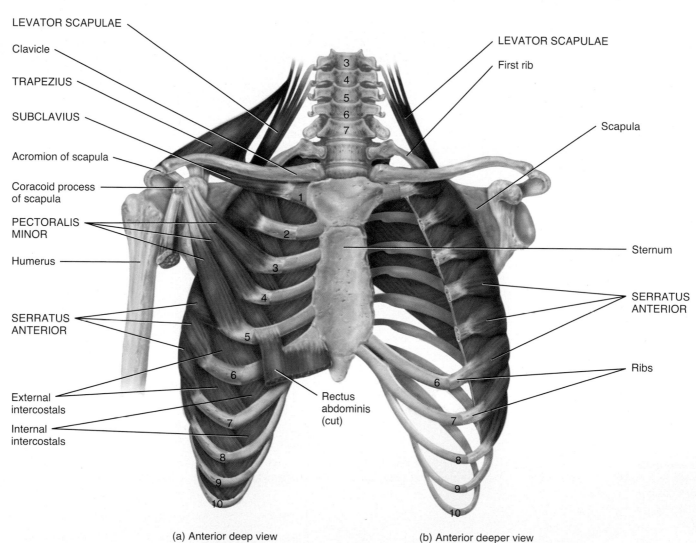

(a) Anterior deep view   (b) Anterior deeper view

- **Downward rotation:** movement of the inferior angle of the scapula medially so that the glenoid cavity is moved downward. This movement is seen when a gymnast on parallel bars supports the weight of the body on the hands.

Relating Muscles to Movements

Arrange the muscles in this exhibit according to the following actions on the scapula: (1) depression, (2) elevation, (3) abduction, (4) adduction, (5) upward rotation, and (6) downward rotation. The same muscle may be mentioned more than once.

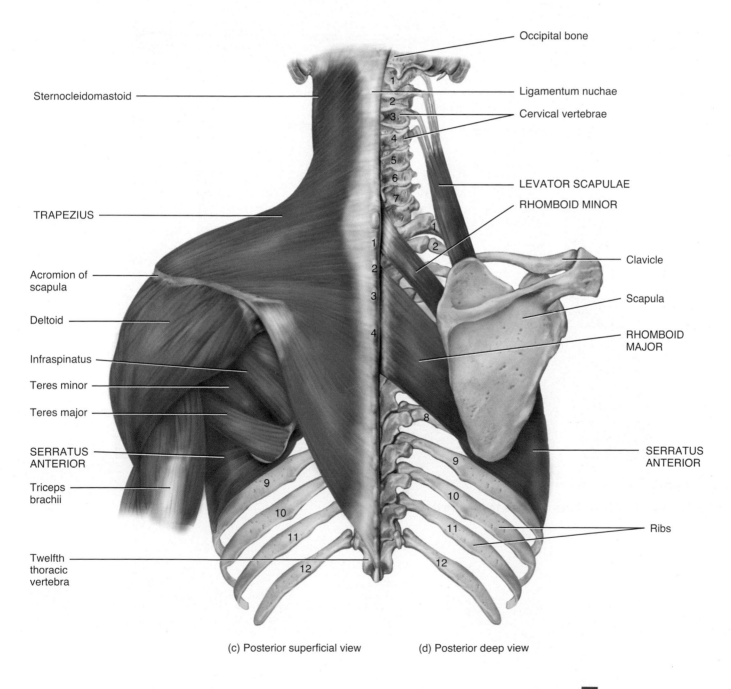

(c) Posterior superficial view   (d) Posterior deep view

⊙ CHECKPOINT
What muscles in this exhibit are used to raise your shoulders, lower your shoulders, join your hands behind your back, and join your hands in front of your chest?

⊙ EXHIBIT 11.13 CONTINUES ▶

EXHIBIT 11.13   **355**

● **FIGURE 11.16 CONTINUED** ▶

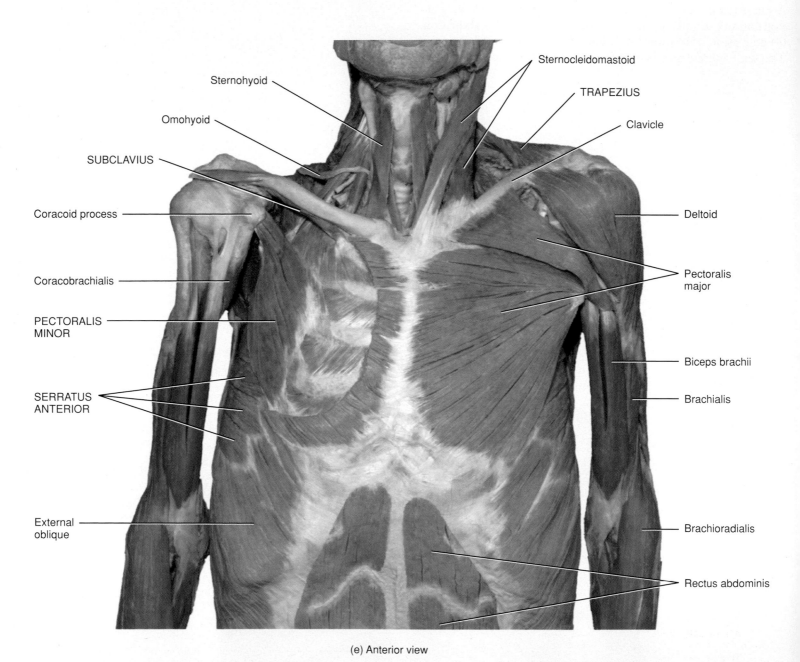

(e) Anterior view

? What is the main action of the muscles that move the pectoral girdle?

EXHIBIT 11.14

Muscles of the Thorax (Chest) and Shoulder That Move the Humerus (Arm Bone)

● **OBJECTIVE**

- Describe the origin, insertion, action, and innervation of the muscles that move the humerus.

Of the nine muscles that cross the shoulder joint, all except the pectoralis major and latissimus dorsi originate on the scapula. The pectoralis major and latissimus dorsi thus are called **axial muscles**, because they originate on the axial skeleton. The remaining seven muscles, the **scapular muscles**, arise from the scapula (Figure 11.17).

Of the two axial muscles that move the humerus, the **pectoralis major** is a large, thick, fan-shaped muscle that covers the superior part of the thorax. It has two origins: a smaller clavicular

head and a larger sternocostal head. The **latissimus dorsi** is a broad, triangular muscle located on the interior part of the back. It is commonly called the "swimmer's muscle" because its many actions are used while swimming: consequently, many competitive swimmers have well-developed "lats."

Among the scapular muscles, the **deltoid** is a thick, powerful shoulder muscle that covers the shoulder joint and forms the rounded contour of the shoulder. This muscle is a frequent site of intramuscular injections. As you study the deltoid, note that its fascicles originate from three different points and that each group of fascicles moves the humerus differently. The **subscapularis** is a large triangular muscle that fills the subscapular fossa of the scapula and forms part of the posterior wall of the

MUSCLE	ORIGIN	INSERTION	ACTION	INNERVATION
AXIAL MUSCLES THAT MOVE THE HUMERUS				
Pectoralis major (pek′-tō-RĀ-lis; *pector-* =chest; *major*= larger) (see also Figure 11.12b)	Clavicle (clavicular head), sternum, and costal cartilages of second to sixth ribs and sometimes first to seventh ribs (sternocostal head).	Greater tubercle and lateral lip of the intertubercular sulcus of humerus.	As a whole, adducts and medially rotates arm at shoulder joint; clavicular head flexes arm, and sternocostal head extends the flexed arm to side of trunk.	Medial and lateral pectoral nerves.
Latissimus dorsi (la-TIS-i-mus DOR-sī; *latissimus*= widest; *dorsi*= of the back)	Spines of T7–L5, crests of sacrum and ilium, inferior four ribs.	Intertubercular sulcus of humerus.	Extends, adducts, and medially rotates arm at shoulder joint; draws arm inferiorly and posteriorly.	Thoracodorsal nerve.
SCAPULAR MUSCLES THAT MOVE THE HUMERUS				
Deltoid (DEL-toyd= triangularly shaped)	Acromial extremity of clavicle (anterior fibers), acromion of scapula (lateral fibers), and spine of scapula (posterior fibers).	Deltoid tuberosity of humerus.	Lateral fibers abduct arm at shoulder joint; anterior fibers flex and medially rotate arm at shoulder joint; posterior fibers extend and laterally rotate arm at shoulder joint.	Axillary nerve.
Subscapularis (sub-scap′-ū-LĀ-ris; *sub-*=below; *scapularis*= scapula)	Subscapular fossa of scapula.	Lesser tubercle of humerus.	Medially rotates arm at shoulder joint.	Upper and lower subscapular nerve.
Supraspinatus (soo-pra-spī-NĀ-tus; *supra-*=above; *spina-*= spine [of the scapula])	Supraspinous fossa of scapula.	Greater tubercle of humerus.	Assists deltoid muscle in abducting arm at shoulder joint.	Suprascapular nerve.
Infraspinatus (in′-fra-spī-NĀ-tus; *infra-*=below)	Infraspinous fossa of scapula.	Greater tubercle of humerus.	Laterally rotates arm at shoulder joint.	Suprascapular nerve.
Teres major (TE-rēz; *teres*=long and round)	Inferior angle of scapula.	Medial lip of intertubercular sulcus of humerus.	Extends arm at shoulder joint and assists in adduction and medial rotation of arm at shoulder joint.	Lower subscapular nerve.
Teres minor (TE-rēz)	Inferior lateral border of scapula.	Greater tubercle of humerus.	Laterally rotates and extends arm at shoulder joint.	Axillary nerve.
Coracobrachialis (kor′-a-kō-brā-kē-Ā-lis; *coraco-*=coracoid process [of the scapula]; *brachi-*=arm)	Coracoid process of scapula.	Middle of medial surface of shaft of humerus.	Flexes and adducts arm at shoulder joint.	Musculocutaneous nerve.

● **EXHIBIT 11.14 CONTINUES** ▶

EXHIBIT 11.14   **357**

axilla. The **supraspinatus**, a rounded muscle named for its location in the supraspinous fossa of the scapula, lies deep to the trapezius. The **infraspinatus** is a triangular muscle, also named for its location in the infraspinous fossa of the scapula. The **teres major** is a thick, flattened muscle inferior to the teres minor that also helps form part of the posterior wall of the axilla. The **teres minor** is a cylindrical, elongated muscle, often inseparable from the infraspinatus, which lies along its superior border. The **coracobrachialis** is an elongated, narrow muscle in the arm.

Four deep muscles of the shoulder—subscapularis, supraspinatus, infraspinatus, and teres minor—strengthen and stabilize the shoulder joint. These muscles join the scapula to the humerus. Their flat tendons fuse together to form the **rotator (musculotendinous) cuff**, a nearly complete circle of tendons around the shoulder joint, like the cuff on a shirtsleeve. The supraspinatus muscle is especially subject to wear and tear because of its location between the head of the humerus and acromion of the scapula, which compress its tendon during shoulder movements, especially abduction of the arm.

Figure 11.17 Muscles of the thorax (chest) and shoulder that move the humerus (arm bone).

The strength and stability of the shoulder joint are provided by the tendons that form the rotator cuff.

(a) Anterior deep view (the intact pectoralis major muscle is shown in Figure 11.12a)

One of the most common causes of shoulder pain and dysfunction in athletes is known as **impingement syndrome**, which is sometimes confused with another common complaint, *compartment syndrome*. The repetitive movement of the arm over the head that is common in baseball, overhead racquet sports, lifting weights over the head, spiking a volleyball, and swimming puts these athletes at risk. Impingement syndrome may also be caused used by a direct blow or stretch injury.

Continual pinching of the supraspinatus tendon as a result of overhead motions causes it to become inflamed and results in pain. If movement is continued despite the pain, the tendon may degenerate near the attachment to the humerus and ultimately may tear away from the bone (rotator cuff injury). Treatment consists of resting the injured tendons, strengthening the shoulder through exercise, and surgery if the injury is particularly severe. ●

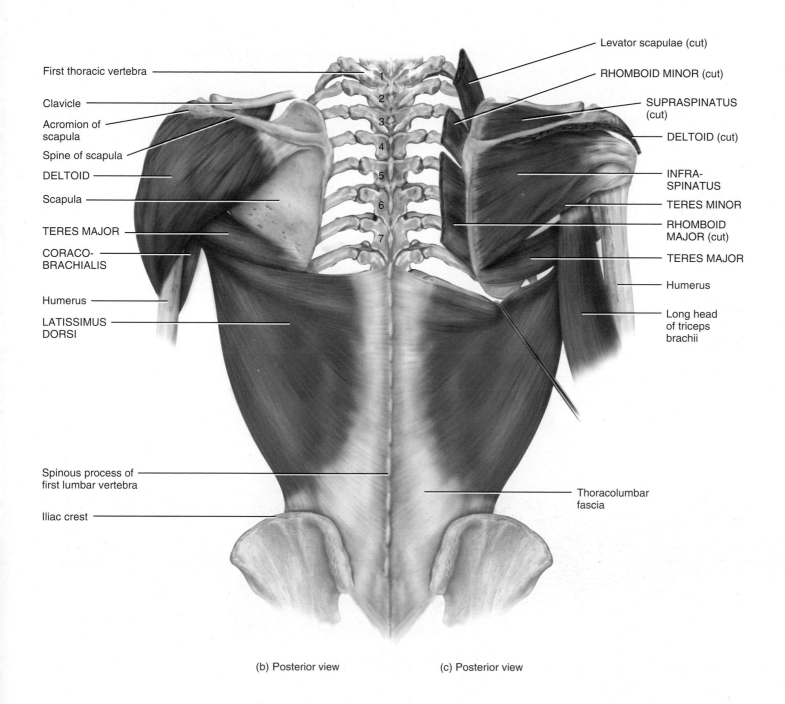

(b) Posterior view

(c) Posterior view

Relating Muscles to Movements

Arrange the muscles in this exhibit according to the following actions on the humerus at the shoulder joint: (1) flexion, (2) extension, (3) abduction, (4) adduction, (5) medial rotation, and (6) lateral rotation. The same muscle may be mentioned more than once.

◉ **CHECKPOINT**

Why are the two muscles that cross the shoulder joint called axial muscles, and the seven others called scapular muscles?

◉ **FIGURE 11.17 CONTINUED** ▶

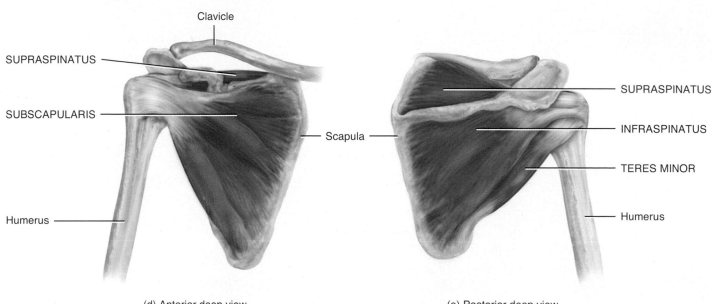

(d) Anterior deep view

(e) Posterior deep view

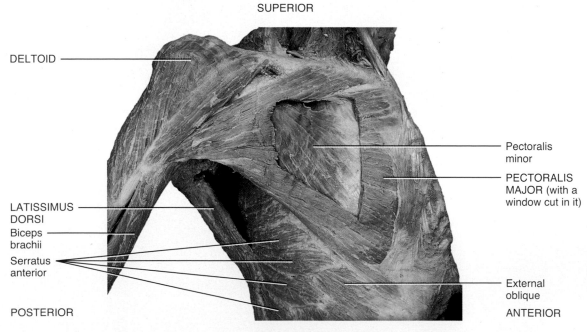

(f) Right lateral superficial view

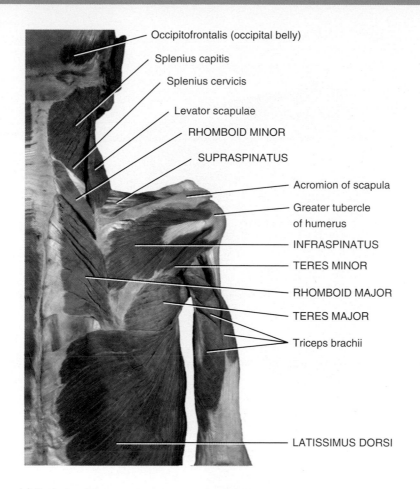

Occipitofrontalis (occipital belly)

Splenius capitis

Splenius cervicis

Levator scapulae

RHOMBOID MINOR

SUPRASPINATUS

Acromion of scapula

Greater tubercle of humerus

INFRASPINATUS

TERES MINOR

RHOMBOID MAJOR

TERES MAJOR

Triceps brachii

LATISSIMUS DORSI

(g) Posterior view

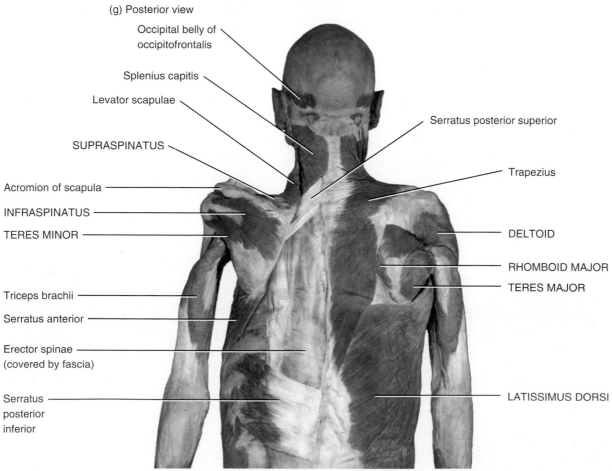

Occipital belly of occipitofrontalis

Splenius capitis

Levator scapulae

SUPRASPINATUS

Acromion of scapula

INFRASPINATUS

TERES MINOR

Triceps brachii

Serratus anterior

Erector spinae (covered by fascia)

Serratus posterior inferior

Serratus posterior superior

Trapezius

DELTOID

RHOMBOID MAJOR

TERES MAJOR

LATISSIMUS DORSI

(h) Posterior view

EXHIBIT **11.14** CONTINUES ▶

EXHIBIT 11.14 **361**

FIGURE 11.17 CONTINUED ▶

(i) Anterior view

❓ Which tendons make up the rotator cuff?

EXHIBIT 11.15

Muscles of the Arm That Move the Radius and Ulna (Forearm Bones)

● OBJECTIVE

- Describe the origin, insertion, action, and innervation of the muscles that move the radius and ulna.

Most of the muscles that move the radius and ulna (forearm bones) cause flexion and extension at the elbow, which is a hinge joint. The biceps brachii, brachialis, and brachioradialis muscles are the flexor muscles. The extensor muscles are the triceps brachii and the anconeus (Figure 11.18).

The **biceps brachii** is the large muscle located on the anterior surface of the arm. As indicated by its name, it has two heads of origin (long and short), both from the scapula. The muscle spans both the shoulder and elbow joints. In addition to its role in flexing the forearm at the elbow joint, it also supinates the forearm at the radioulnar joints and flexes the arm at the shoulder joint. The **brachialis** is deep to the biceps brachii muscle. It is the most powerful flexor of the forearm at the elbow joint. For this reason, it is called the "workhorse" of the elbow flexors. The

MUSCLE	ORIGIN	INSERTION	ACTION	INNERVATION
FOREARM FLEXORS				
Biceps brachii (BĪ-ceps BRA-ke-ī; *biceps*=two heads of origin; *brachii*=arm)	Long head originates from tubercle above glenoid cavity of scapula (supraglenoid tubercle); short head originates from coracoid process of scapula.	Radial tuberosity of radius and bicipital aponeurosis.*	Flexes forearm at elbow joint, supinates forearm at radioulnar joints, and flexes arm at shoulder joint.	Musculocutaneous nerve.
Brachialis (brā-kē-Ā-lis)	Distal, anterior surface of humerus.	Ulnar tuberosity and coronoid process of ulna.	Flexes forearm at elbow joint.	Musculocutaneous and radial nerves.
Brachioradialis (brā′-kē-ō-rā-dē-Ā-lis: *radi*=radius)	Lateral border of distal end of humerus.	Superior to styloid process of radius.	Flexes forearm at elbow joint; supinates and pronates forearm at radioulnar joints to neutral position.	Radial nerve.
FOREARM EXTENSORS				
Triceps brachii (TRĪ-ceps BRA-kē-ī; *triceps*=three heads of origin)	Long head: infraglenoid tubercle, a projection inferior to glenoid cavity of scapula. Lateral head: lateral and posterior surface of humerus superior to radial groove. Medial head: entire posterior surface of humerus inferior to a groove for the radial nerve.	Olecranon of ulna.	Extends forearm at elbow joint and extends arm at shoulder joint.	Radial nerve.
Anconeus (an-KŌ-nē-us; *ancon*= the elbow)	Lateral epicondyle of humerus.	Olecranon and superior portion of shaft of ulna.	Extends forearm at elbow joint.	Radial nerve.
FOREARM PRONATORS				
Pronator teres (PRŌ-nā-tor TE-rēz; *pronator*=turns palm posteriorly)	Medial epicondyle of humerus and coronoid process of ulna.	Midlateral surface of radius.	Pronates forearm at radioulnar joints and weakly flexes forearm at elbow joint.	Median nerve.
Pronator quadratus (PRŌ-nā-tor kwod-RĀ-tus; *quadratus*=square, four-sided)	Distal portion of shaft of ulna.	Distal portion of shaft of radius.	Pronates forearm at radioulnar joints.	Median nerve.
FOREARM SUPINATOR				
Supinator (SOO-pi-nā-tor; *supinator* =turns palm anteriorly)	Lateral epicondyle of humerus and ridge near radial notch of ulna (supinator crest).	Lateral surface of proximal one-third of radius.	Supinates forearm at radioulnar joints.	Deep radial nerve.

*The **bicipital aponeurosis** is a broad aponeurosis from the tendon of insertion of the biceps brachii muscle that descends medially across the brachial artery and fuses with deep fascia over the forearm flexor muscles.

● **EXHIBIT 11.15** CONTINUES ▶

EXHIBIT 11.15 **363**

brachioradialis flexes the forearm at the elbow joint, especially when a quick movement is required or when a weight is lifted slowly during flexion of the forearm.

The **triceps brachii** is the large muscle located on the posterior surface of the arm. It is the more powerful of the extensors of the forearm at the elbow joint. As its name implies, it has three heads of origin, one from the scapula (long head) and two from the humerus (lateral and medial heads). The long head crosses the shoulder joint; the other heads do not. The **anconeus** is a small muscle located on the lateral part of the posterior aspect of the elbow that assists the triceps brachii in extending the forearm at the elbow joint.

Some muscles that move the radius and ulna are involved in pronation and supination at the radioulnar joints. The pronators, as suggested by their names, are the **pronator teres** and **pronator quadratus** muscles. The supinator of the forearm is aptly named the **supinator** muscle. You use the powerful action of the supinator when you twist a corkscrew or turn a screw with a screwdriver.

Figure 11.18 **Muscles of the arm that move the radius and ulna (forearm bones).**

The anterior arm muscles flex the forearm, and the posterior arm muscles extend it.

(a) Anterior view

(b) Posterior view

In the limbs, functionally related skeletal muscles and their associated blood vessels and nerves are grouped together by fascia into regions called **compartments**. In the arm, the biceps brachii, brachialis, and coracobrachialis muscles comprise the **anterior (flexor) compartment.** The triceps brachii muscle forms the **posterior (extensor) compartment.**

(c) Superior view of transverse section of arm

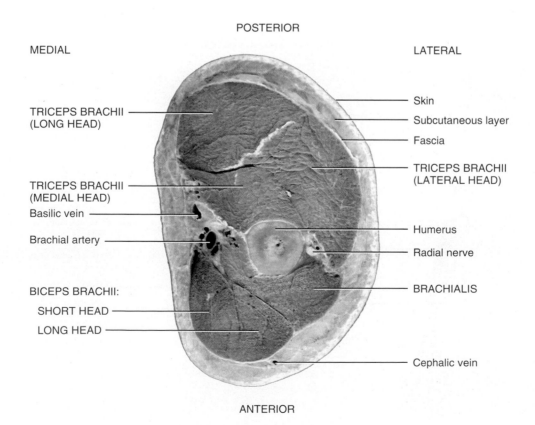

(d) Superior view of transverse section of arm

● **EXHIBIT 11.15 CONTINUES** ▶

EXHIBIT 11.15   **365**

Relating Muscles to Movements

Arrange the muscles in this exhibit according to the following actions on the elbow joint: (1) flexion and (2) extension; the following actions on the forearm at the radioulnar joints: (1) supination and (2) pronation; and the following actions on the humerus at the shoulder joint: (1) flexion and (2) extension. The same muscle may be mentioned more than once.

● CHECKPOINT

Flex your arm. Which group of muscles is contracting? Which group of muscles must relax so that you can flex your arm?

● FIGURE **11.18** CONTINUED ▶

(e) Anterior deep view

(f) Anterior deep view

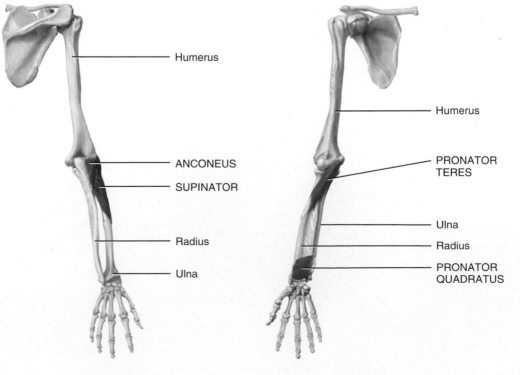

(g) Posterior deep view

(h) Anterior deep view

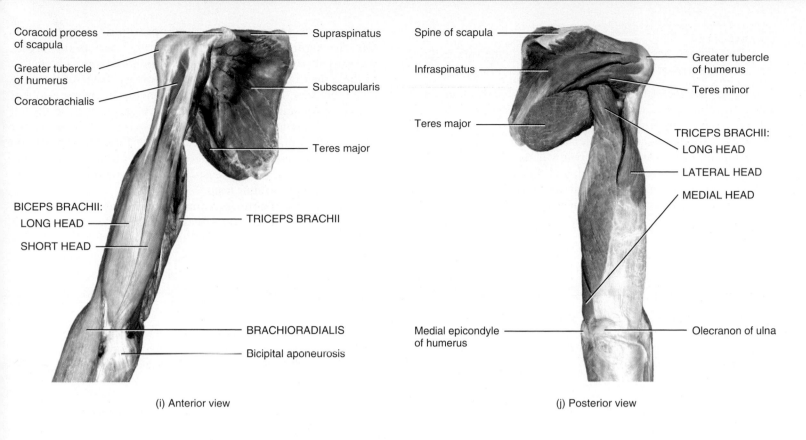

Coracoid process of scapula

Greater tubercle of humerus

Coracobrachialis

BICEPS BRACHII:

LONG HEAD

SHORT HEAD

Supraspinatus

Subscapularis

Teres major

TRICEPS BRACHII

BRACHIORADIALIS

Bicipital aponeurosis

(i) Anterior view

Spine of scapula

Infraspinatus

Teres major

Greater tubercle of humerus

Teres minor

TRICEPS BRACHII:

LONG HEAD

LATERAL HEAD

MEDIAL HEAD

Medial epicondyle of humerus

Olecranon of ulna

(j) Posterior view

Greater tubercle of humerus

Coracobrachialis

BICEPS BRACHII:

SHORT HEAD (cut)

LONG HEAD (cut)

BRACHIALIS

BRACHIORADIALIS

Clavical (cut)

Scapula (cut)

TRICEPS BRACHII

PRONATOR TERES

Flexor carpi radialis

(k) Anterior deep view

? **Which muscles are the most powerful flexor and the most powerful extensor of the forearm?**

EXHIBIT 11.15   **367**

OBJECTIVE

- Describe the origin, insertion, action, and innervation of the muscles that move the wrist, hand, thumb, and fingers.

Muscles of the forearm that move the wrist, hand, thumb, and fingers are many and varied (Figure 11.19 on pages 370–373). Those in this group that act on the digits are known as **extrinsic muscles of the hand** (*ex=* outside) because they originate *outside* the hand and insert within it. As you will see, the names for the muscles that move the wrist, hand, and digits give some indication of their origin, insertion, or action. Based on location and function, the muscles of the forearm are divided into two groups: (1) anterior compartment muscles and (2) posterior compartment muscles. The **anterior (flexor) compartment** muscles of the forearm originate on the humerus, typically insert on the carpals, metacarpals, and phalanges, and function as flexors. The bellies of these muscles form the bulk of the forearm. One of the muscles in the superficial anterior compartment, the palmaris longus muscle, is missing in about 10% of individuals (usually in the left forearm) and is commonly used for tendon repair. The **posterior (extensor) compartment** muscles of the forearm originate on the humerus, insert on the metacarpals and phalanges, and function as extensors. Within each compartment, the muscles are grouped as superficial or deep.

The **superficial anterior compartment** muscles are arranged in the following order from lateral to medial: **flexor carpi radi-**

MUSCLE	ORIGIN	INSERTION	ACTION	INNERVATION
SUPERFICIAL ANTERIOR (FLEXOR) COMPARTMENT OF THE FOREARM				
Flexor carpi radialis (FLEK-sor KAR-pē rā′-dē-A-lis; *flexor*=decreases angle at joint; *carpi*=of the wrist; *radi-*=radius)	Medial epicondyle of humerus.	Second and third metacarpals.	Flexes and abducts hand (radial deviation) at wrist joint.	Median nerve.
Palmaris longus (pal-MA-ris LON-gus; *palma* = palm; *longus*=long)	Medial epicondyle of humerus.	Flexor retinaculum and palmar aponeurosis (deep fascia in center of palm).	Weakly flexes hand at wrist joint.	Median nerve.
Flexor carpi ulnaris (FLEK-sor KAR-pē (ūl-NAR-is; *ulnar-*=ulna)	Medial epicondyle of humerus and superior posterior border of ulna.	Pisiform, hamate, and base of fifth metacarpal.	Flexes and adducts hand (ulnar deviation) at wrist joint.	Ulnar nerve.
Flexor digitorum superficialis (FLEK-sor di-ji-TOR-um soo′-per-fish′-ē-A-lis; *digit*=finger or toe; *superficialis*=closer to surface)	Medial epicondyle of humerus, coronoid process of ulna, and a ridge along lateral margin of anterior surface (anterior oblique line) of radius.	Middle phalanx of each finger.*	Flexes middle phalanx of each finger at proximal interphalangeal joint, proximal phalanx of each finger at metacarpophalangeal joint, and hand at wrist joint.	Median nerve.
DEEP ANTERIOR (FLEXOR) COMPARTMENT OF THE FOREARM				
Flexor pollicis longus (FLEK-sor POL-li-sis LON-gus; *pollic-*=thumb)	Anterior surface of radius and interosseous membrane (sheet of fibrous tissue that holds shafts of ulna and radius together).	Base of distal phalanx of thumb.	Flexes distal phalanx of thumb at interphalangeal joint.	Median nerve.
Flexor digitorum profundus (FLEK-sor di′-ji-TOR-um prō-FUN-dus; *profundus*=deep)	Anterior medial surface of body of ulna.	Base of distal phalanx of each finger.	Flexes distal and middle phalanges of each finger at interphalangeal joints, proximal phalanx of each finger at metacarpophalangeal joint, and hand at wrist joint.	Median and ulnar nerves.
SUPERFICIAL POSTERIOR (EXTENSOR) COMPARTMENT OF THE FOREARM				
Extensor carpi radialis longus (eks-TEN-sor KAR-pē rā′-dē-Ā-lis LON-gus; *tensor*=increases angle at joint)	Lateral supracondylar ridge of humerus.	Second metacarpal.	Extends and abducts hand at wrist joint.	Radial nerve.
Extensor carpi radialis brevis (eks-TEN-sor KAR-pē rā-dē-Ā-lis BREV-is; *brevis*=short)	Lateral epicondyle of humerus.	Third metacarpal.	Extends and abducts hand at wrist joint.	Radial nerve.

alis, palmaris longus and flexor carpi ulnaris (the ulnar nerve and artery are just lateral to the tendon of this muscle at the wrist). The **flexor digitorum superficialis** muscle is deep to the other three muscles and is the largest superficial muscle in the forearm.

The **deep anterior compartment** muscles are arranged in the following order from lateral to medial: **flexor pollicis longus** (the only flexor of the distal phalanx of the thumb) and **flexor digitorum profundus** (ends in four tendons that insert into the distal phalanges of the fingers).

The **superficial posterior compartment** muscles are arranged in the following order from lateral to medial: **extensor carpi radialis longus, extensor carpi radialis brevis, extensor digitorum** (occupies most of the posterior surface of the forearm and divides into four tendons that insert into the middle and distal phalanges of the fingers), **extensor digiti minimi** (a slender muscle usually connected to the extensor digitorum), and the **extensor carpi ulnaris.**

The **deep posterior compartment** muscles are arranged in the following order from lateral to medial: **abductor pollicis longus, extensor pollicis brevis, extensor pollicis longus,** and **extensor indicis.**

The tendons of the muscles of the forearm that attach to the wrist or continue into the hand, along with blood vessels and nerves, are held close to bones by strong fasciae. The tendons are also surrounded by tendon sheaths. At the wrist, the deep fascia is thickened into fibrous bands called **retinacula**

MUSCLE	ORIGIN	INSERTION	ACTION	INNERVATION
Extensor digitorum (eks-TEN-sor di'-ji-TOR-um)	Lateral epicondyle of humerus.	Distal and middle phalanges of each finger.	Extends distal and middle phalanges of each finger at interphalangeal joints, proximal phalanx of each finger at metacarpophalangeal joint, and hand at wrist joint.	Radial nerve.
Extensor digiti minimi (eks-TEN-sor DIJ-i-tē MIN-i-mē; *digit*=finger or toe; *minimi*=smallest)	Lateral epicondyle of humerus.	Tendon of extensor digitorum on fifth phalanx.	Extends proximal phalanx of little finger at metacarpophalangeal joint and hand at wrist joint.	Deep radial nerve.
Extensor carpi ulnaris (eks-TEN-sor KAR-pē ul-NAR-is)	Lateral epicondyle of humerus and posterior border of ulna.	Fifth metacarpal.	Extends and adducts hand at wrist joint.	Deep radial nerve.
DEEP POSTERIOR (EXTENSOR) COMPARTMENT OF THE FOREARM				
Abductor pollicis longus (ab-DUK-tor POL-li-sis LON-gus; *abductor*=moves part away from midline)	Posterior surface of middle of radius and ulna and interosseous membrane.	First metacarpal.	Abducts and extends thumb at carpometacarpal joint and abducts hand at wrist joint.	Deep radial nerve.
Extensor pollicis brevis (eks-TEN-sor POL-li-sis BREV-is)	Posterior surface of middle of radius and interosseous membrane.	Base of proximal phalanx of thumb.	Extends proximal phalanx of thumb at metacarpophalangeal joint, first metacarpal of thumb at carpometacarpal joint, and hand at wrist joint.	Deep radial nerve.
Extensor pollicis longus (eks-TEN-sor POL-li-sis LON-gus)	Posterior surface of middle of ulna and interosseous membrane.	Base of distal phalanx of thumb.	Extends distal phalanx of thumb at interphalangeal joint, first metacarpal of thumb at carpometacarpal joint, and abducts hand at wrist joint.	Deep radial nerve.
Extensor indicis (eks-TEN-sor IN-di-kis: *dicis*=index)	Posterior surface of ulna.	Tendon of extensor digitorum of index finger.	Extends distal and middle phalanges of index finger at interphalangeal joints, proximal phalanx of index finger at metacarpophalangeal joint, and hand at wrist joint.	Deep radial nerve.

*Reminder: The thumb or pollex is the first digit and has two phalanges: proximal and distal. The remaining digits, the fingers, are numbered II-V (2–5), and each has three phalanges: proximal, middle, and distal.

 EXHIBIT 11.16 CONTINUES ▶

EXHIBIT 11.16 **369**

(*retinacul*=holdfast). The **flexor retinaculum** is located over the palmar surface of the carpal bones. The long flexor tendons of the digits and wrist and the median nerve pass deep to the flexor retinaculum. The **extensor retinaculum** is located over the dorsal surface of the carpal bones. The extensor tendons of the wrist and digits pass deep to it.

Relating Muscles to Movements

Arrange the muscles in this exhibit according to the following actions on the wrist joint: (1) flexion, (2) extension, (3) abduction, and (4) adduction; the following actions on the fingers at the metacarpophalangeal joints: (1) flexion and (2) extension; the following actions on the fingers at the interphalangeal joints:

Figure 11.19 Muscles of the forearm that move the wrist, hand, thumb and fingers.

The anterior compartment muscles function as flexors, and the posterior compartment muscles function as extensors.

(a) Anterior superficial view

(b) Anterior intermediate view

(c) Anterior deep view

(1) flexion and (2) extension; the following actions on the thumb at the carpometacarpal, metacarpophalangeal, and interphalangeal joints: (1) extension and (2) abduction; and the following action on the thumb at the interphalangeal joint: flexion. The same muscle may be mentioned more than once.

● CHECKPOINT
Which muscles and actions of the wrist, hand, and digits are used when writing?

Triceps brachii

Humerus
BRACHIORADIALIS
EXTENSOR CARPI RADIALIS LONGUS
Medial epicondyle of humerus
Lateral epicondyle of humerus

Olecranon of ulna
ANCONEUS
Supinator
EXTENSOR CARPI ULNARIS
EXTENSOR DIGITORUM
Ulna
EXTENSOR CARPI RADIALIS BREVIS

FLEXOR CARPI ULNARIS

ABDUCTOR POLLICIS LONGUS

EXTENSOR POLLICIS LONGUS

EXTENSOR INDICIS
EXTENSOR DIGITI MINIMI
EXTENSOR POLLICIS BREVIS

Tendon of extensor carpi ulnaris
Extensor retinaculum

Tendon of extensor digiti minimi

Tendon of extensor indicis

Dorsal interossei

Tendons of extensor digitorum

Extensor expansion

(e) Posterior deep view

(d) Posterior superficial view

● EXHIBIT 11.16 CONTINUES ▶

EXHIBIT 11.16   **371**

● FIGURE **11.19** CONTINUED ▶

Medial epicondyle of humerus

PRONATOR TERES

PALMARIS LONGUS

FLEXOR CARPI ULNARIS

FLEXOR CARPI RADIALIS

BRACHIORADIALIS

FLEXOR DIGITORUM SUPERFICIALIS

FLEXOR POLLICIS LONGUS

PRONATOR QUADRATUS

Flexor retinaculum

Palmaris brevis
Thenar muscles
Hypothenar muscles
Palmar aponeurosis

(f) Anterior superficial view

Brachialis
Biceps brachii
BRACHIORADIALIS

EXTENSOR CARPI RADIALIS LONGUS

EXTENSOR DIGITORUM

EXTENSOR CARPI ULNARIS

EXTENSOR DIGITI MINIMI

ABDUCTOR POLLICIS LONGUS

EXTENSOR POLLICIS BREVIS

EXTENSOR POLLICIS LONGUS

Extensor retinaculum

Dorsal interossei

(g) Posterior superficial view

FLEXOR DIGITORUM PROFUNDUS

FLEXOR POLLICIS LONGUS

FLEXOR DIGITORUM SUPERFICIALIS (cut)

(h) Anterior deep view

SUPINATOR

Interosseous membrane

Radius

Ulna

PRONATOR QUADRATUS

(i) Anterior deeper view

? **What structures pass through the flexor retinaculum?**

EXHIBIT 11.16 **373**

OBJECTIVE

• Describe the origin, insertion, action, and innervation of the intrinsic muscles of the hand.

Several of the muscles discussed in Exhibit 11.16 move the digits in various ways and are known as extrinsic muscles of the hand. They produce the powerful but crude movements of the digits. The **intrinsic muscles of the hand** (Figure 11.20 on pages 376–378) in the palm produce the weak but intricate and precise movements of the digits that characterize the human hand. The muscles in this group are so named because their origins and insertions are *within* the hand.

The intrinsic muscles of the hand are divided into three groups: (1) **thenar**, (2) **hypothenar**, and (3) **intermediate**. The thenar muscles include the abductor pollicis brevis, opponens pollicis, and flexor pollicis brevis. The **abductor pollicis brevis** is a thin, short, relatively broad superficial muscle on the lateral side of the thenar eminence. The **opponens pollicis** is a small, triangular muscle that is deep to the abductor pollicis brevis muscle. The **flexor pollicis brevis** is a short, wide muscle that is medial to the abductor pollicis brevis muscle. The three thenar muscles plus the adductor pollicis form the **thenar eminence**, the lateral rounded contour on the palm that is also called the *ball of the thumb*. The **abductor pollicis** also acts on the thumb. The muscle is fan-shaped and has two heads (oblique and transverse) separated by a gap through which the radial artery passes.

The three hypothenar muscles act on the little finger and form the **hypothenar eminence**, the medial rounded contour on the palm that is also called the ball of the little finger. The hypothenar muscles are the abductor digiti minimi, flexor digiti minimi brevis, and opponens digiti minimi. The **abductor digiti minimi** is a short, wide muscle and is the most superficial of the hypothenar muscles. It is a powerful muscle that plays an important role in grasping an object with outspread fingers. The **flexor digiti minimi brevis** muscle is also short and wide and is lateral to the abductor digiti minimi muscle. The **opponens digiti minimi** muscle is triangular and deep to the other two hypothenar muscles.

The 11 intermediate (midpalmar) muscles act on all the digits. The intermediate muscles include the lumbricals, palmar interossei, and dorsal interossei. The **lumbricals**, as their name indicates, are worm-shaped. They originate from and insert into the tendons of other muscles (flexor digitorum profundus and extensor digitorum). The **palmar interossei** are the smallest and more anterior of the interossei muscles. The **dorsal interossei** are the posterior interossei muscles. Both sets of interossei muscles are located between the metacarpals and are important in abduction, adduction, flexion, and extension of the fingers, and in movements in skilled activities such as writing, typing, and playing a piano.

The functional importance of the hand is readily apparent when you consider that certain hand injuries can result in permanent disability. Most of the dexterity of the hand depends on movements of the thumb. The general activities of the hand are free motion, power grip (forcible movement of the fingers and thumb against the palm, as in squeezing), precision handling (a change in position of a handled object that requires exact control of finger and thumb positions as in winding a watch or threading a needle), and pinch (compression between the thumb and index finger or between the thumb and first two fingers).

Movements of the thumb are very important in the precise activities of the hand, and they are defined in different planes from comparable movements of other digits because the thumb is positioned at a right angle to the other digits. The five principal movements of the thumb are illustrated in Figure 11.20g on page 378) and include *flexion* (movement of the thumb medially across the palm), *extension* (movement of the thumb laterally away from the palm), *abduction* (movement of the thumb in an anteroposterior plane away from the palm), *adduction* (movement of the thumb in an anteroposterior plane toward the palm), and *opposition* (movement of the thumb across the palm so that the tip of the thumb meets the tip of a finger). Opposition is the single most distinctive digital movement that gives humans and other primates the ability to grasp and manipulate objects precisely.

• CLINICAL **CONNECTION** Carpal Tunnel Syndrome

The carpal tunnel is a narrow passageway formed anteriorly by the flexor retinaculum and posteriorly by the carpal bones. Through this tunnel pass the median nerve, the most superficial structure, and the long flexor tendons for the digits (Figure 11.20f on page 377). Structures within the carpal tunnel, especially the median nerve, are vulnerable to compression, and the resulting condition is called **carpal tunnel syndrome**. Compression of the median nerve leads to sensory changes over the lateral side of the hand and muscle weakness in the thenar eminence. This results in pain, numbness, and tingling of the fingers. The condition may be caused by inflammation of the digital tendon sheaths, fluid retention, excessive exercise, infection, trauma, and/or repetitive activities that involve flexion of the wrist, such as keyboarding, cutting hair, and playing a piano. Treatment may involve the use of nonsteroidal anti-inflammatory drugs (such as ibuprofen or aspirin), wearing a wrist splint, corticosteroid injections, or surgery to cut the flexor retinaculum and release presssure on the median nerve. •

Relating Muscles to Movements

Arrange the muscles in this exhibit according to the following actions on the thumb at the carpometacarpal and metacarpophalangeal joints: (1) abduction, (2) adduction, (3) flexion, and (4) opposition; and the following actions on the fingers at the metacarpophalangeal and interphalangeal joints: (1) abduction, (2) adduction, (3) flexion, and (4) extension. The same muscle may be mentioned more than once.

CHECKPOINT

How do the actions of the extrinsic and intrinsic muscles of the hand differ?

MUSCLE	ORIGIN	INSERTION	ACTION	INNERVATION
THENAR (LATERAL ASPECT OF PALM)				
Abductor pollicis brevis (ab-DUK-tor POL-li-sis BREV-is; *abductor*=moves part away from middle; *pollic-*=the thumb; *brevis*=short)	Flexor retinaculum, scaphoid, and trapezium.	Lateral side of proximal phalanx of thumb.	Abducts thumb at carpometacarpal joint.	Median nerve.
Opponens pollicis (op-PŌ-nenz POL-li-sis; *opponens*=opposes)	Flexor retinaculum and trapezium.	Lateral side of first metacarpal (thumb).	Moves thumb across palm to meet any finger (opposition) at the carpometacarpal joint.	Median nerve.
Flexor pollicis brevis (FLEK-sor POL-li-sis BREV-is; *flexor*=decreases angle at joint)	Flexor retinaculum, trapezium, capitate, and trapezoid.	Lateral side of proximal phalanx of thumb.	Flexes thumb at carpometacarpal and metacarpophalangeal joints.	Median and ulnar nerves.
Adductor pollicis (ad-DUK-tor POL-li-sis; *adductor*=moves part toward midline)	Oblique head: capitate and second and third metacarpals; transverse head: third metacarpal.	Medial side of proximal phalanx of thumb by a tendon containing a sesamoid bone.	Adducts thumb at carpometacarpal and metacarpophalangeal joints.	Ulnar nerve.
HYPOTHENAR (MEDIAL ASPECT OF PALM)				
Abductor digiti minimi (ab-DUK-tor DIJ-i-tē MIN-i-mē; *digit*=finger or toe; *minimi*=little)	Pisiform and tendon of flexor carpi ulnaris.	Medial side of proximal phalanx of little finger.	Abducts and flexes little finger at metacarpophalangeal joint.	Ulnar nerve.
Flexor digiti minimi brevis (FLEK-sor DIJ-i-tē MIN-i-mē BREV-is)	Flexor retinaculum and hamate.	Medial side of proximal phalanx of little finger.	Flexes little finger at carpometacarpal and metacarpophalangeal joints.	Ulnar nerve.
Opponens digiti minimi (op-PŌ-nenz DIJ-i-tē MIN-i-mē)	Flexor retinaculum and hamate.	Medial side of fifth metacarpal (little finger).	Moves little finger across palm to meet thumb (opposition) at the carpometacarpal joint.	Ulnar nerve.
INTERMEDIATE (MIDPALMAR)				
Lumbricals (LUM-bri-kals; *lumbric*=earthworm) (four muscles)	Lateral sides of tendons and flexor digitorum profundus of each finger.	Lateral sides of tendons of extensor digitorum on proximal phalanges of each finger.	Flex each finger at metacarpophalangeal joints and extend each finger at interphalangeal joints.	Median and ulnar nerves.
Palmar interossei (in′-ter-OS-ē-ī; *palmar*=palm; *inter-*=between; *ossei*=bones) (three muscles)	Sides of shafts of metacarpals of all digits (except the middle one).	Sides of bases of proximal phalanges of all digits (except the middle one).	Adduct each finger at metacarpophalangeal joints; flex each finger at metacarpophalangeal joints and extend each finger at interphalangeal joints.	Ulnar nerve.
Dorsal interossei (in′-ter-OS-ē-ī; *dorsal*=back surface) (four muscles)	Adjacent sides of metacarpals.	Proximal phalanx of each finger.	Abduct fingers 2–4 at metacarpophalangeal joints; flex fingers 2–4 at metacarpophalangeal joints; and extend each finger at interphalangeal joints.	Ulnar nerve.

● EXHIBIT **11.17** CONTINUES ▶

EXHIBIT 11.17 **375**

Figure 11.20 Muscles of the palm that move the digits—intrinsic muscles of the hand.

The intrinsic muscles of the hand produce the intricate and precise movements of the digits that characterize the human hand.

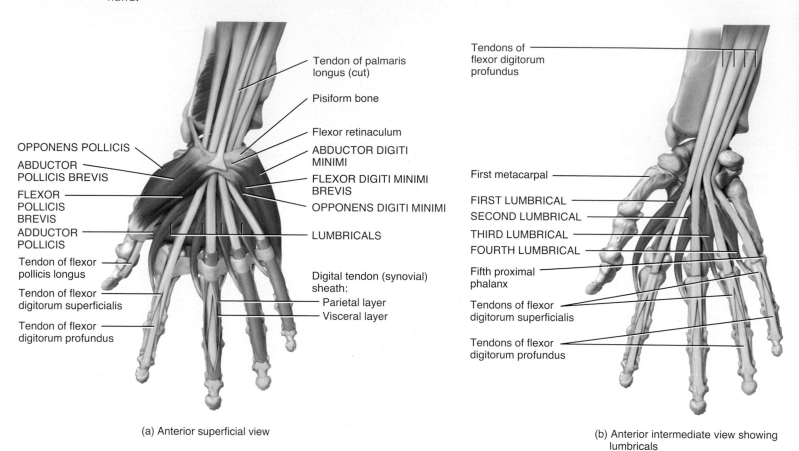

(a) Anterior superficial view

(b) Anterior intermediate view showing lumbricals

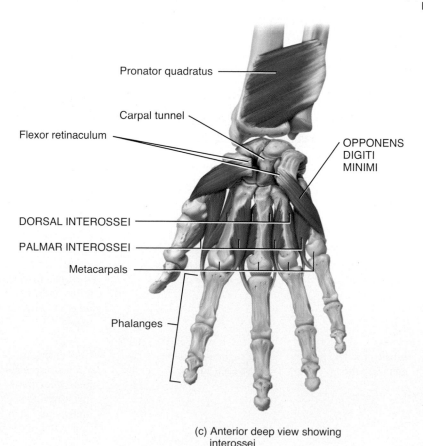

(c) Anterior deep view showing interossei

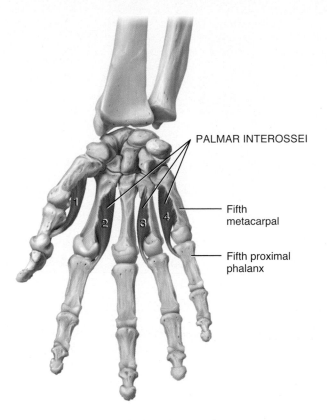

PALMAR INTEROSSEI

1

2 3 4

Fifth
metacarpal

Fifth proximal
phalanx

(d) Anterior deep view of palmar interossei

DORSAL INTEROSSEI

1 2 3 4

Fifth
metacarpal

Fifth
proximal
phalanx

(e) Anterior deep view of dorsal interossei

LATERAL

Transverse
plane

View

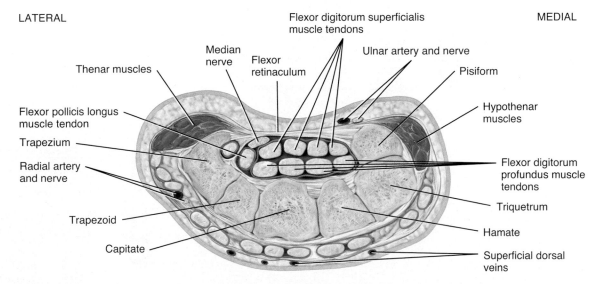

Flexor digitorum superficialis
muscle tendons

MEDIAL

Median
nerve

Flexor
retinaculum

Ulnar artery and nerve

Thenar muscles

Pisiform

Flexor pollicis longus
muscle tendon

Hypothenar
muscles

Trapezium

Radial artery
and nerve

Flexor digitorum
profundus muscle
tendons

Trapezoid

Triquetrum

Capitate

Hamate

Superficial dorsal
veins

(f) Inferior view of transverse section

⬛ **EXHIBIT 11.17 CONTINUES** ▶

EXHIBIT 11.17   **377**

■ FIGURE 11.20 CONTINUED ▶

Flexion   Extension   Abduction   Adduction   Opposition

(g) Movements of the thumb

Flexor retinaculum

ABDUCTOR POLLICIS
BREVIS

ABDUCTOR DIGITI
MINIMI

FLEXOR DIGITI
MINIMI BREVIS

FLEXOR POLLICIS
BREVIS

Tendons of flexor
digitorum superficialis

ADDUCTOR POLLICIS

Tendon of flexor
pollicis longus

LUMBRICALS

(h) Anterior superficial view

OPPONENS
POLLICIS

ABDUCTOR
DIGITI MINIMI

FLEXOR DIGITI
MINIMI BREVIS

OPPONENS
DIGITI MINIMI

ADDUCTOR
POLLICIS

PALMAR
INTEROSSEI

DORSAL
INTEROSSEI

Tendons of
flexors (cut)

(i) Anterior deep view

❓ Muscles of the thenar eminence act on which digit?

⊙ **OBJECTIVE**

- Describe the origin, insertion, action, and innervation of the muscles that move the vertebral column.

The muscles that move the vertebral column (backbone) are quite complex because they have multiple origins and insertions and there is considerable overlap among them. One way to group the muscles is on the basis of the general direction of the muscle bundles and their approximate lengths. For example, the splenius muscles arise from the midline and extend laterally and superiorly to their insertions (Figure 11.21a on page 382). The erector spinae muscle group (consisting of the iliocostalis, longissimus, and spinalis muscles) arises from either the midline or more laterally but usually runs almost longitudinally, with neither a significant lateral nor medial direction as it is traced superiorly. The muscles of the transversospinalis group (semispinalis, multifidus, rotatores) arise laterally but extend toward the midline as they are traced superiorly. Deep to these three muscle groups are small segmental muscles that extend between spinous processes or transverse processes of vertebrae. Note in Exhibit 11.9 that the rectus abdominis, external oblique, internal oblique, and quadratus lumborum muscles also play a role in moving the vertebral column.

The bandage-like **splenius** muscles are attached to the sides and back of the neck. The two muscles in this group are named on the basis of their superior attachments (insertions): **splenius capitis** (head region) and **splenius cervicis** (cervical region). They extend the head and laterally flex and rotate the head.

The **erector spinae** is the largest muscle mass of the back, forming a prominent bulge on either side of the vertebral column. It is the chief extensor of the vertebral column. It is also important in controlling flexion, lateral flexion, and rotation of the vertebral column and in maintaining the lumbar curve, because the main mass of the muscle is in the lumbar region. As noted above, it consists of three groups: iliocostalis (laterally placed), longissimus (intermedially placed), and spinalis (medially placed). These groups, in turn, consist of a series of overlapping muscles, and the muscles within the groups are named according to the regions of the body with which they are associated. The **iliocostalis group**, consists of three muscles: the **iliocostalis cervicis** (cervical region), **iliocostalis thoracis** (thoracic region), and **iliocostalis lumborum** (lumbar region). The **longissimus group** resembles a herringbone and consists of three muscles: the **longissimus capitis** (head region), **longissimus cervicis** (cervical region), and **longissimus thoracis** (thoracic region). The **spinalis group**, also consists of three muscles: the **spinalis capitis**, **spinalis cervicis**, and **spinalis thoracis**.

The **transversospinales** are so named because their fibers run from the transverse processes to the spinous processes of the vertebrae. The semispinalis muscles in this group are also named according to the region of the body with which they are associated: **semispinalis capitis** (head region), **semispinalis cervicis** (cervical region), and **semispinalis thoracis** (thoracic region). These muscles extend the vertebral column and rotate the head. The **multifidus** muscle in this group, as its name implies, is segmented into several bundles. It extends and laterally flexes the vertebral column. This muscle is large and thick in the lumbar region and is important in maintaining the lumbar curve. The

rotatores muscles of this group are short and are found along the entire length of the vertebral column. These small muscles contribute little to vertebral movement, but play important roles in monitoring the position of the vertebral column and providing proprioceptive feedback to the stronger vertebral muscles.

Within the segmental muscle group (Figure 11.21b on page 383), the **interspinales** and **intertransversarii** muscles unite the spinous and transverse processes of consecutive vertebrae. They function primarily in stabilizing the vertebral column during its movements, and providing proprioceptive feedback.

Within the **scalene** group (Figure 11.21c on page 383), the **anterior scalene** muscle is anterior to the middle scalene muscle, the **middle scalene** muscle is intermediate in placement and is the longest and largest of the scalene muscles, and the **posterior scalene** muscle is posterior to the middle scalene muscle and is the smallest of the scalene muscles. These muscles flex, laterally flex, and rotate the head and assist in deep inhalation.

⊙ CLINICAL **CONNECTION** | **Back Injuries and Heavy Lifting**

The four factors associated with increased risk of back injury are amount of force, repetition, posture, and stress applied to the backbone. Poor physical condition, poor posture, lack of exercise, and excessive body weight contribute to the number and severity of sprains and strains. Back pain caused by a muscle strain or ligament sprain will normally heal within a short time and may never cause further problems. However, if ligaments and muscles are weak, discs in the lower back can become weakened and may herniate (rupture) with excessive lifting or a sudden fall, causing considerable pain.

Full flexion at the waist, as in touching your toes, overstretches the erector spinae muscles. Muscles that are overstretched cannot contract effectively. Straightening up from such a position is therefore initiated by the hamstring muscles on the back of the thigh and the gluteus maximus muscles of the buttocks. The erector spinae muscles join in as the degree of flexion decreases. Improperly lifting a heavy weight, however, can strain the erector spinae muscles. The result can be painful muscle spasms, tearing of tendons and ligaments of the lower back, and herniating of intervertebral discs. The lumbar muscles are adapted for maintaining posture, not for lifting. This is why it is important to bend at the knees and use the powerful extensor muscles of the thighs and buttocks while lifting a heavy load. •

Relating Muscles to Movements

Arrange the muscles in this exhibit according to the following actions on the head at the atlanto-occipital and intervertebral joints: (1) extension, (2) lateral flexion, (3) rotation to same side as contracting muscle, and (4) rotation to opposite side as contracting muscle; and arrange the muscles according to the following actions on the vertebral column at the intervertebral joints: (1) flexion, (2) extension, (3) lateral flexion, (4) rotation, and (5) stabilization. The same muscle may be mentioned more than once.

⊙ **CHECKPOINT**

What is the largest muscle group of the back?

⊙ **EXHIBIT 11.18 CONTINUES** ▶

EXHIBIT 11.18 **379**

MUSCLE	ORIGIN	INSERTION	ACTION	INNERVATION
SPLENIUS (SPLĒ-nē-us)				
Splenius capitis (KAP-i-tis; *splenium*= bandage; *capit-*=head)	Ligamentum nuchae and spinous processes of C7–T4.	Occipital bone and mastoid process of temporal bone.	Acting together (bilaterally), extend head; acting singly. (unilaterally), laterally flex and/or rotate head to same side as contracting muscle.	Middle cervical spinal nerves.
Splenius cervicis (SER-vi-kis; *cervic-*=neck)	Spinous processes of T3–T6.	Transverse processes of C1–C2 or C1–C4.	Acting together, extend head; acting singly, later ally flex and/or rotate head to same side as contracting muscle.	Inferior cervical spinal nerves.
ERECTOR SPINAE (e-REK-tor SPI-nē) Consists of iliocostalis muscles (lateral), longissimus muscles (intermediate), and spinalis muscles (medial).				
ILIOCOSTALIS GROUP (Lateral)				
Iliocostalis cervicis (il′-ē-ō-kos-TĀL-is SER-vi-kis; *ilio-*=flank; *costa-*=rib)	Ribs 1–6.	Transverse processes of C4–C6.	Acting together, muscles of each region (cervical, thoracic, and lumbar) extend and maintain erect posture of vertebral column of their respective regions; acting singly, laterally flex vertebral column of their respective regions to the same side as the contracting muscle.	Cervical and thoracic spinal nerves.
Iliocostalis thoracis (il′-ē-ō-kos-TĀL-is thō-RĀ-sis; *thorac-*=chest)	Ribs 7–12.	Ribs 1–6.		Thoracic spinal nerves.
Iliocostalis lumborum (il′-ē-ō-kos-TĀL-is lum-BOR-um)	Iliac crest.	Ribs 7–12.		Lumbar spinal nerves.
LONGISSIMUS GROUP (Intermediate)				
Longissimus capitis (lon-JIS-i-mus KAP-i-tis; *longissimus*=longest)	Articular processes of C4–C7 and transverse processes of T1–T4.	Mastoid process of temporal bone.	Acting together, both longissimus capitis muscles extend head; acting singly, rotate head to same side as contracting muscle.	Middle and inferior cervical spinal nerves.
Longissimus cervicis (lon-JIS-i-mus SER-vi-kis)	Transverse processes of T4–T5.	Transverse processes of C2–C6.	Acting together, longissimus cervicis and both longissimus thoracis muscles extend vertebral column of their respective regions; acting singly, laterally flex vertebral column of their respective regions.	Cervical and superior thoracic spinal nerves.
Longissimus thoracis (lon-JIS-i-mus thō-RĀ-sis)	Transverse processes of lumbar vertebrae.	Transverse processes of all thoracic and superior lumbar vertebrae and ninth and tenth ribs.		Thoracic and lumbar spinal nerves.
SPINALIS GROUP (Medial)				
Spinalis capitis (spi-NĀ-lis KAP-i-tis; *spinal-*=vertebral column)	Arises with semispinalis capitis.	Occipital bone.	Acting together, muscles of each region (cervical, thoracic, and lumbar) extend vertebral column of their respective regions.	Cervical and superior thoracic spinal nerves.
Spinalis cervicis (spi-NĀ-lis SER-vi-kis)	Ligamentum nuchae and spinous process of C7.	Spinous process of axis.		Inferior cervical and thoracic spinal nerves.
Spinalis thoracis (spi-NĀ-lis thō-RĀ-sis)	Spinous processes of T10–L2.	Spinous processes of superior thoracic vertebrae.		Thoracic spinal nerves.

MUSCLE	ORIGIN	INSERTION	ACTION	INNERVATION
TRANSVERSOSPINALES (trans-ver-sò-spi-NĀ-lēz)				
Semispinalis capitis (sem′-ē-spi-NĀ-lis KAP-i-tis; *semi-*= partially or one half)	Articular processes of C4–C6 and transverse processes of C7–T7.	Occipital bone.	Acting together, extend head; acting singly, rotate head to side opposite contracting muscle.	Cervical and thoracic spinal nerves.
Semispinalis cervicis (sem′-ē -spi-NĀ-lis SER-vi-kis)	Transverse processes of T1–T5.	Spinous processes of C1–C5.	Acting together, both semispinalis cervicis and both semispinalis thoracis muscles extend vertebral column of their respective regions; acting singly, rotate head to side opposite contracting muscle.	Cervical and. thoracic spinal nerves.
Semispinalis thoracis (sem′-e-spi-NĀ-lis thō -RĀ-sis)	Transverse processes of T6–T10.	Spinous processes of C6–T4.		Thoracic spinal nerves.
Multifidus (mul-TIF-i-dus; *multi*=many; *bifid-*=segmented)	Sacrum, ilium, transverse processes of lumbar, thoracic, and C4–C7.	Spinous process of a more superior vertebra.	Acting together, extend vertebral column; acting singly, weakly laterally flex vertebral column and weakly rotate vertebral column to side opposite contracting muscle.	Cervical, thoracic, and lumbar spinal nerves.
Rotatores (rō′-ta-TŌ-rēz; singular is **rotatore**; *rotatore*=to rotate)	Transverse processes of all vertebrae.	Spinous process of vertebra superior to the one of origin.	Acting together, weakly extend vertebral column; acting singly, weakly rotate vertebral column to side opposite contracting muscle.	Cervical, thoracic, and lumbar spinal nerves.
SEGMENTAL (seg-MEN-tal)				
Interspinales (in-ter-sp-ī NĀ-lēz; *inter-*=between)	Superior surface of all spinous processes.	Inferior surface of spinous process of vertebra superior to the one of origin.	Acting together, weakly extend, vertebral column; acting singly, stabilize vertebral column during movement.	Cervical, thoracic, and lumbar spinal nerves.
Intertransversarii (in′-ter-trans-vers-AR-ē-ī; singular is *intertransversarius*)	Transverse processes of all vertebrae.	Transverse process of vertebra superior to the one of origin.	Acting together, weakly extend vertebral column; acting singly, weakly laterally flex vertebral column and stabilize it during movements.	Cervical, thoracic, and lumbar spinal nerves.
SCALENES (SKĀ-LĒNZ)				
Anterior scalene (SKĀ-lēn; *anterior*=front; *scalene*=uneven)	Transverse processes of third through sixth cervical vertebrae.	First rib.	Acting together, right and left anterior scalene and middle scalene muscles flex cervical vertebrae and elevate first ribs during deep inhalation; acting singly, laterally flex and rotate cervical vertebrae.	Cervical spinal nerves C5–C6.
Middle scalene (SKĀ-lēn)	Transverse processes of inferior six cervical vertebrae.	First rib.		Cervical spinal nerves C3–C8.
Posterior scalene (SKĀ-lēn)	Transverse processes of fourth through sixth cervical vertebrae.	Second rib.	Acting together, right and left posterior scalene flex cervical vertebrae and elevate second ribs during deep inhalation; acting singly, laterally flex and rotate cervical vertebrae.	Cervical spinal nerves C6–C8.

■ **EXHIBIT 11.18 CONTINUES** ▶

EXHIBIT 11.18 **381**

Figure 11.21 Muscles of the neck and back that move the vertebral column (backbone).

⑥ 🔑 The erector spinae group (iliocostalis, longissimus, and spinalis muscles) is the largest muscular mass of the body and is the chief extensor of the vertebral column.

SPINALIS CAPITIS
LONGISSIMUS CAPITIS
SPINALIS CERVICIS
LONGISSIMUS CERVICIS
ILIOCOSTALIS THORACIS
SPINALIS THORACIS
ILIOCOSTALIS LUMBORUM

SEMISPINALIS CAPITIS
Ligamentum nuchae
SPLENIUS CAPITIS
SPLENIUS CERVICIS
ILIOCOSTALIS CERVICIS
SEMISPINALIS CERVICIS
LONGISSIMUS THORACIS
SEMISPINALIS THORACIS
INTERTRANSVERSARIUS
ROTATORE
MULTIFIDUS

(a) Posterior view

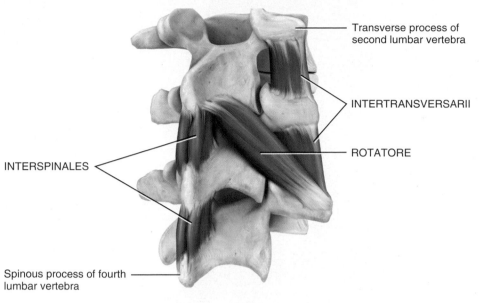

Transverse process of
second lumbar vertebra

INTERTRANSVERSARII

ROTATORE

INTERSPINALES

Spinous process of fourth
lumbar vertebra

(b) Posterolateral view

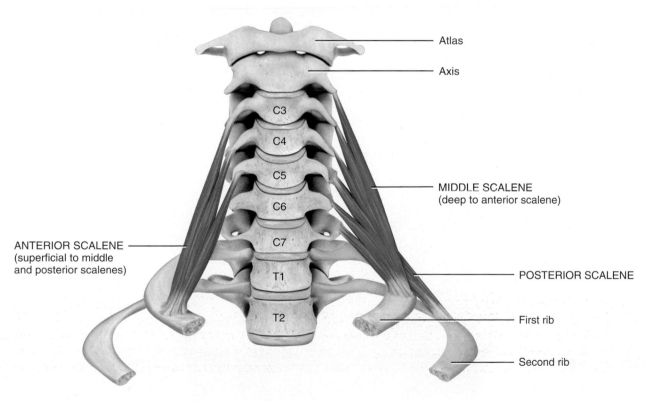

Atlas

Axis

C3

C4

C5

C6

MIDDLE SCALENE
(deep to anterior scalene)

C7

ANTERIOR SCALENE
(superficial to middle
and posterior scalenes)

T1

POSTERIOR SCALENE

T2

First rib

Second rib

(c) Anterior view

● EXHIBIT 11.18 CONTINUES ▶

EXHIBIT 11.18   **383**

● **FIGURE 11.21 CONTINUED** ▶

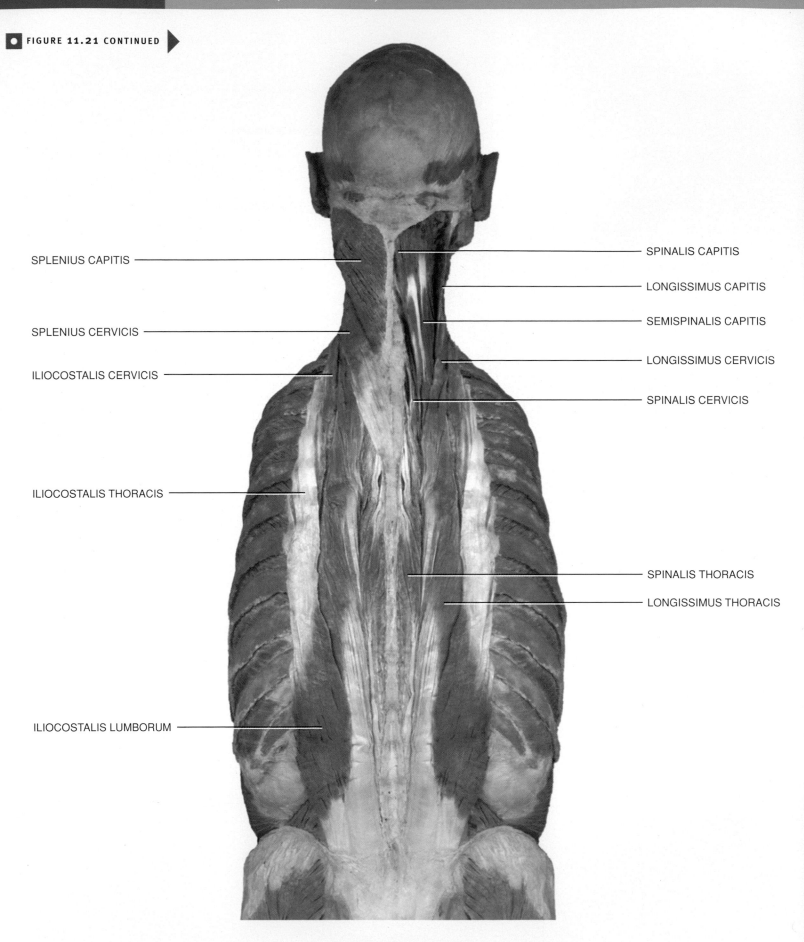

SPLENIUS CAPITIS

SPLENIUS CERVICIS

ILIOCOSTALIS CERVICIS

ILIOCOSTALIS THORACIS

ILIOCOSTALIS LUMBORUM

SPINALIS CAPITIS

LONGISSIMUS CAPITIS

SEMISPINALIS CAPITIS

LONGISSIMUS CERVICIS

SPINALIS CERVICIS

SPINALIS THORACIS

LONGISSIMUS THORACIS

(d) Posterior superficial view

❓ Which muscles originate at the midline and extend laterally and upward to their insertion?

EXHIBIT 11.19

Muscles of the Gluteal Region That Move the Femur (Thigh Bone)

◉ OBJECTIVE

• Describe the origin, insertion, action, and innervation of the muscles that move the femur.

As you will see, muscles of the lower limbs are larger and more powerful than those of the upper limbs because of differences in function. While upper limb muscles are characterized by versatility of movement, lower limb muscles function in stability, locomotion, and maintenance of posture. In addition, muscles of the lower limbs often cross two joints and act equally on both.

The majority of muscles that move the femur originate on the pelvic girdle and insert on the femur (Figure 11.22 on pages 388–394). **The psoas major** and **iliacus** muscles share a common insertion (lesser trochanter of femur) and are collectively known as the **iliopsoas** (il′-ē-ō-SŌ-as) muscle. There are three gluteal muscles: gluteus maximus, gluteus medius, and gluteus minimus. The **gluteus maximus** is the largest and heaviest of the three muscles and is one of the largest muscles in the body. It is the chief extensor of the femur. The **gluteus medius** is mostly deep to the gluteus maximus and is a powerful abductor of the femur at the hip joint. It is a common site for an intramuscular injection. The **gluteus minimus** is the smallest of the gluteal muscles and lies deep to the gluteus medius.

The **tensor fasciae latae** muscle is located on the lateral surface of the thigh. The *fascia lata* is a layer of deep fascia, composed of dense connective tissue, that encircles the entire thigh. It is well developed laterally where, together with the tendons of the tensor fasciae latae and gluteus maximus muscles, it forms a structure called the **iliotibial tract**. The tract inserts into the lateral condyle of the tibia.

The **piriformis**, **obturator internus**, **obturator externus**, **superior gemellus**, **inferior gemellus**, and **quadratus femoris** muscles are all deep to the gluteus maximus muscle and function as lateral rotators of the femur at the hip joint.

Three muscles on the medial aspect of the thigh are the **adductor longus**, **adductor brevis**, and **adductor magnus**. They originate on the pubic bone and insert on the femur. These three muscles adduct the thigh and are unique in their ability to both medially and laterally rotate the thigh. When the foot is on the ground these muscles medially rotate the thigh, but when the foot is off the ground, they are lateral rotators of the thigh. This results from their oblique orientation from their anterior origin to their posterior insertion. In adduction, the adductor longus

flexes the thigh and the adductor magnus extends the thigh. The **pectineus** muscle also adducts and flexes the femur at the hip joint.

Technically, the adductor muscles and pectineus muscles are components of the medial compartment of the thigh and could be included in Exhibit 11.20. However, they are included here because they act on the femur.

At the junction between the trunk and lower limb is a space called the **femoral triangle**. The base is formed superiorly by the inguinal ligament, medially by the lateral border of the adductor longus muscle, and laterally by the medial border of the sartorius muscle. The apex is formed by the crossing of the adductor longus by the sartorius muscle (Figure 11.22a). The contents of the femoral triangle, from lateral to medial, are the femoral nerve and its branches, the femoral artery and several of its branches, the femoral vein and its proximal tributaries, and the deep inguinal lymph nodes.

● CLINICAL **CONNECTION** | Groin Pull

The five major muscles of the inner thigh function to move the legs medially. This muscle group is important in activities such as sprinting, hurdling, and horseback riding. A rupture or tear of one or more of these muscles can cause a **groin pull**. Groin pulls most often occur during sprinting or twisting, or from kicking a solid, perhaps stationary object. Symptoms of a groin pull may be sudden, or may not surface until the day after the injury, and include sharp pain in the inguinal region, swelling, bruising, or inability to contract the muscles. As with most strain injuries, treatment involves RICE therapy, which stands for *Rest*, *Ice*, *Compression*, and *Elevation*. Ice should be applied immediately, and the injured part should be elevated and rested. An elastic bandage should be applied, if possible, to compress the injured tissue. •

Relating Muscles to Movements

Arrange the muscles in this exhibit according to the following actions on the thigh at the hip joint: (1) flexion, (2) extension, (3) abduction, (4) adduction, (5) medial rotation, and (6) lateral rotation. The same muscle may be mentioned more than once.

◉ CHECKPOINT

What is the origin of most muscles that move the femur?

◉ **EXHIBIT 11.19** CONTINUES ▶

EXHIBIT 11.19 **385**

MUSCLE	ORIGIN	INSERTION	ACTION	INNERVATION
Iliopsoas				
Psoas major (SŌ-as; *psoa=* a muscle of the loin)	Transverse processes and bodies of lumbar vertebrae.	With iliacus into lesser trochanter of femur.	Psoas major and iliacus muscles acting together flex thigh at hip joint, rotate thigh laterally, and flex trunk on the hip as in sitting up from the supine position.	Lumbar spinal nerves L2–L3.
Iliacus (il'-ē -A-cus; *iliac-=*ilium)	Iliac fossa and sacrum.	With psoas major into lesser trochanter of femur.		Femoral nerve.
Gluteus maximus (GLOO-tē-us MAK-si-mus; *glute-=*rump or buttock; *maximus=* largest)	Iliac crest, sacrum, coccyx, and aponeurosis of sacrospinalis.	Iliotibial tract of fascia lata and lateral part of linea aspera (gluteal tuberosity) under greater trochanter of femur.	Extends thigh at hip joint and laterally rotates thigh.	Inferior gluteal nerve.
Gluteus medius (GLOO-tē-us MĒ-de-us; *medi-=*middle)	Ilium.	Greater trochanter of femur.	Abducts thigh at hip joint and medially rotates thigh.	Superior gluteal nerve.
Gluteus minimus (GLOO-tē-us MIN-i-mus; *minim-=*smallest)	Ilium.	Greater trochanter of femur.	Abducts thigh at hip joint and medially rotates thigh.	Superior gluteal nerve.
Tensor fasciae latae (TEN-sor FA-shē -ē LĀ-tē ; *tensor=* makes tense; *fasciae=* of the band; *lat-=*wide)	Iliac crest.	Tibia by way of the iliotibial tract.	Flexes and abducts thigh at hip joint.	Superior gluteal nerve.
Piriformis (pir-i-FOR-mis; *piri-=*pear; *form-=*shape)	Anterior sacrum.	Superior border of greater trochanter of femur.	Laterally rotates and abducts thigh at hip joint.	Sacral spinal nerves S1 or S2, mainly S1.
Obturator internus (OB-too-rā'-tor in-TER-nus; *obturator=*obturator foramen; *intern-=*inside)	Inner surface of obturator foramen, pubis, and ischium.	Medial surface of greater trochanter of femur.	Laterally rotates and abducts thigh at hip joint.	Nerve to obturator internus.

MUSCLE	ORIGIN	INSERTION	ACTION	INNERVATION
Obturator externus (OB-too-ra'-tor ex-TER-nus; *extern-*=outside)	Outer surface of obturator membrane.	Deep depression inferior to greater trochanter (trochanteric fossa) of femur.	Laterally rotates and abducts thigh at hip joint.	Obturator nerve.
Superior gemellus (jem-EL-lus; *superior=* above; *gemell-*=twins)	Ischial spine.	Medial surface of greater trochanter of femur.	Laterally rotates and abducts thigh at hip joint.	Nerve to obturator internus.
Inferior gemellus (jem-EL-lus; *inferior*=below)	Ischial tuberosity.	Medial surface of greater trochanter of femur.	Laterally rotates and abducts thigh at hip joint.	Nerve to quadratus femoris.
Quadratus femoris (kwod-RĀ-tus FEM-or-is; *quad-*=square, four-sided; *femoris*=femur)	Ischial tuberosity.	Elevation superior to mid-portion of intertrochanteric crest (quadrate tubercle) on posterior femur.	Laterally rotates and stabilizes hip joint.	Nerve to quadratus femoris.
Adductor longus (LONG-us; *adductor*=moves part closer to midline; *longus*=long)	Pubic crest and pubic symphysis.	Linea aspera of femur.	Adducts and flexes thigh at hip joint and rotates thigh.*	Obturator nerve.
Adductor brevis (BREV-is; *brevis*=short)	Inferior ramus of pubis.	Superior half of linea aspera of femur.	Adducts and flexes thigh at hip joint and rotates thigh.*	Obturator nerve.
Adductor magnus (MAG-nus; *magnus*=large)	Inferior ramus of pubis and ischium to ischial tuberosity.	Linea aspera of femur.	Adducts thigh at hip joint and rotates thigh; anterior part flexes thigh at hip joint, and posterior part extends thigh at hip joint.*	Obturator and sciatic nerves.
Pectineus (pek-TIN-e-us; *pectin-*=a comb)	Superior ramus of pubis.	Pectineal line of femur, between lesser trochanter and linea aspera.	Flexes and adducts thigh at hip joint.	Femoral nerve.

*All adductors are unique muscles that cross the thigh joint obliquely from an anterior origin to a posterior insertion. As a result they laterally rotate the hip joint when the foot is off the ground, but medially rotate the hip joint when the foot is on the ground.

EXHIBIT **11.19** CONTINUES ▶

EXHIBIT 11.19   **387**

Figure 11.22 **Muscles of the gluteal region that move the femur (thigh bone).**

Most muscles that move the femur originate on the pelvic (hip) girdle and insert on the femur.

Twelfth rib

Quadratus lumborum

Iliac crest

ILIACUS

Anterior superior iliac spine

Femoral triangle

TENSOR FASCIAE LATAE

SARTORIUS

QUADRICEPS FEMORIS:

   RECTUS FEMORIS (cut)

VASTUS LATERALIS

VASTUS INTERMEDIUS

VASTUS MEDIALIS

RECTUS FEMORIS (cut)

Iliotibial tract

Section of fascia lata (cut)

Tendon of quadriceps femoris

Patellar ligament

Psoas minor

PSOAS MAJOR

Sacrum

Inguinal ligament

Pubic tubercle

PECTINEUS

ADDUCTOR LONGUS

GRACILIS

ADDUCTOR MAGNUS

Patella

(a) Anterior superficial view (the femoral triangle is indicated by a dashed line)

TENSOR FASCIAE LATAE (cut)

SARTORIUS (cut)

RECTUS FEMORIS (cut)

Iliofemoral ligament
of hip joint

Inguinal ligament

PECTINEUS (cut)

Pubis

OBTURATOR EXTERNUS

ADDUCTOR LONGUS (cut)

PECTINEUS (cut)

ADDUCTOR BREVIS

ADDUCTOR MAGNUS

ADDUCTOR LONGUS (cut)

GRACILIS

Femur

SARTORIUS (cut)

Patella

(b) Anterior deep view (femur rotated laterally)

■ EXHIBIT **11.19** CONTINUES ▶

EXHIBIT 11.19   **389**

● **FIGURE 11.22 CONTINUED** ▶

GLUTEUS MEDIUS

GLUTEUS MAXIMUS

Tensor fasciae latae

GRACILIS

ADDUCTOR MAGNUS

Illiotibial tract

VASTUS LATERALIS

Semitendinosus

BICEPS FEMORIS:

LONG HEAD

SHORT HEAD

Semimembranosus

SARTORIUS

Gastrocnemius

(c) Posterior superficial view

Iliac crest

GLUTEUS MAXIMUS (cut)

Sacrum

Coccyx

OBTURATOR INTERNUS

Ischial tuberosity

Sciatic nerve

GRACILIS

SARTORIUS

GLUTEUS MEDIUS (cut)

GLUTEUS MINIMUS

PIRIFORMIS

SUPERIOR GEMELLUS

Greater trochanter

INFERIOR GEMELLUS

OBTURATOR EXTERNUS

QUADRATUS FEMORIS

GLUTEUS MAXIMUS (cut)

Femur

ADDUCTOR MAGNUS

HAMSTRINGS:

SEMITENDINOSUS

BICEPS FEMORIS

SEMIMEMBRANOSUS

Vastus lateralis

Femur in floor of popliteal fossa

Plantaris

Gastrocnemius

Tendon of biceps femoris

(d) Posterior superficial and deep view

EXHIBIT **11.19** CONTINUES ▶

EXHIBIT 11.19   **391**

FIGURE 11.22 CONTINUED ▶

(e) Anterior deep view

(f) Anterior views

(g) Posterior deep view

Anterior superior iliac spine

ILIACUS

PSOAS MAJOR

TENSOR FASCIAE LATAE

RECTUS FEMORIS (cut)

VASTUS INTERMEDIUS

VASTUS LATERALIS

Tendon of quadriceps femoris

Patella

Pubic symphysis

PECTINEUS

ADDUCTOR LONGUS

GRACILIS

SARTORIUS

VASTUS MEDIALIS

(h) Anterior superficial view

ILIACUS

SARTORIUS (cut)

RECTUS FEMORIS (cut)

TENSOR FASCIAE LATAE

PECTINEUS (cut)

Femur

ADDUCTOR LONGUS (cut)

VASTUS LATERALIS

Tendon of quadriceps femoris

Patella

PSOAS MAJOR

PECTINEUS (cut)

OBTURATOR EXTERNUS

ADDUCTOR LONGUS (cut)

ADDUCTOR BREVIS

GRACILIS

ADDUCTOR MAGNUS

SARTORIUS (cut)

Medial condyle of femur

(i) Anterior deep view

EXHIBIT **11.19** CONTINUES ▶

EXHIBIT 11.19   **393**

FIGURE **11.22** CONTINUED

GLUTEUS MEDIUS

PIRIFORMIS

SUPERIOR GEMELLUS

OBTURATOR INTERNUS

INFERIOR GEMELLUS

QUADRATUS FEMORIS

ADDUCTOR MINIMUS

ADDUCTOR MAGNUS

BICEPS FEMORIS

Popliteal fossa

SARTORIUS

Sacrum

GLUTEUS MAXIMUS

GRACILIS

VASTUS LATERALIS
(covered by iliotibial tract)

SEMITENDINOSUS

SEMIMEMBRANOSUS

Gastrocnemius

(j) Posterior view

? What are the principal differences between the muscles of the free upper and lower limbs?

OBJECTIVE

- Describe the origin, insertion, action, and innervation of the muscles that act on the femur, tibia, and fibula.

Deep fascia separate the muscles that act on the femur (thigh bone) and tibia and fibula (leg bones) into medial, anterior, and posterior compartments (Figure 11.23 on page 397). The muscles of the **medial (adductor) compartment of the thigh** adduct the femur at the hip joint. (See the adductor magnus, adductor longus, adductor brevis, and pectineus, which are components of the medial compartment, in Exhibit 11.19.) The **gracilis**, the other muscle in the medial compartment, not only adducts the thigh, but also flexes the leg at the knee joint. For this reason, it is discussed here. The gracilis is a long, straplike muscle on the medial aspect of the thigh and knee.

The muscles of the **anterior (extensor) compartment of the thigh** extend the leg (and flex the thigh). This compartment contains the quadriceps femoris and sartorius muscles. The **quadriceps femoris** muscle is the largest muscle in the body, covering most of the anterior surface and sides of the thigh. The muscle is actually a composite muscle, usually described as four separate muscles: (1) **rectus femoris**, on the anterior aspect of the thigh; (2) **vastus lateralis**, on the lateral aspect of the thigh; (3) **vastus medialis**, on the medial aspect of the thigh; and (4) **vastus intermedius**, located deep to the rectus femoris between the vastus lateralis and vastus medialis. The common tendon for the four muscles is known as the **quadriceps tendon**, which inserts into the patella. The tendon continues below the patella as the **patellar ligament**, which attaches to the tibial tuberosity. The quadriceps femoris muscle is the great extensor muscle of the leg. The **sartorius** is a long, narrow muscle that forms a band across the thigh from the ilium of the hip bone to the medial side of the tibia. The various movements it produces (flexion of the leg at the knee joint and flexion, abduction, and lateral rotation at the hip joint) help effect the cross-legged sitting position in which the heel of one limb is placed on the knee of the opposite limb. It is known as the tailor's muscle because tailors often assume this cross-legged sitting position. (Because the major action of the sartorius muscle is to move the thigh rather than the leg, it could have been included in Exhibit 11.19.)

The muscles of the **posterior (flexor) compartment of the thigh** flex the leg (and extend the thigh). This compartment is composed of three muscles collectively called the **hamstrings:** (1) **biceps femoris**, (2) **semitendinosus**, and (3) **semimembranosus**. The hamstrings are so named because their tendons are long and stringlike in the popliteal area. Because the hamstrings span two joints (hip and knee), they are both extensors of the thigh and flexors of the leg. The **popliteal fossa** is a diamond-shaped space on the posterior aspect of the knee bordered laterally by the tendons of the biceps femoris muscle and medially by the tendons of the semitendinosus and semimembranosus muscles.

• CLINICAL **CONNECTION** | Pulled Hamstrings

A strain or partial tear of the proximal hamstring muscles is referred to as **pulled hamstrings** or **hamstring strains**. Like pulled groins (see Exhibit 11.19), they are common sports injuries in individuals who run very hard and/or are required to perform quick starts and stops. Sometimes the violent muscular exertion required to perform a feat tears away a part of the tendinous origins of the hamstrings, especially the biceps femoris, from the ischial tuberosity. This is usually accompanied by a contusion (bruising), tearing of some of the muscle fibers, and rupture of blood vessels, producing a hematoma (collection of blood) and sharp pain. Adequate training with good balance between the quadriceps femoris and hamstrings and stretching exercises before running or competing are important in preventing this injury.   •

Relating Muscles to Movements

Arrange the muscles in this exhibit according to the following actions on the thigh at the hip joint: (1) abduction, (2) adduction, (3) lateral rotation, (4) flexion, and (5) extension; and according to the following actions on the leg at the knee joint: (1) flexion and (2) extension. The same muscle may be mentioned more than once.

CHECKPOINT

Which muscles are part of the medial, anterior, and posterior compartments of the thigh?

EXHIBIT 11.20 CONTINUES ▶

EXHIBIT 11.20   **395**

MUSCLE	ORIGIN	INSERTION	ACTION	INNERVATION
MEDIAL (ADDUCTOR) COMPARTMENT OF THE THIGH				
Adductor magnus (MAG-nus)	See Exhibit 11.19			
Adductor longus (LONG-us)				
Adductor brevis (BREV-is)				
Pectineus (pek-TIN-ē-us)				
Gracilis (GRAS-i-lis; *gracilis*=slender)	Body and inferior ramus of pubis.	Medial surface of body of tibia.	Adducts thigh at hip joint, medially rotates thigh, and flexes leg at knee joint.	Obturator nerve.
ANTERIOR (EXTENSOR) COMPARTMENT OF THE THIGH				
Quadriceps femoris (KWOD-ri-ceps FEM-or-is: *quadriceps*=four heads [of origin]; *femoris*=femur)				
Rectus femoris (REK-tus FEM-or-is; *rectus*=fascicles parallel to midline)	Anterior inferior iliac spine.	Patella via quadriceps tendon and then tibial tuberosity via patellar ligament.	All four heads extend leg at knee joint; rectus femoris muscle acting alone also flexes thigh at hip joint.	Femoral nerve.
Vastus lateralis (VAS-tus lat'-e-RA-lis; *vast*=huge; *lateralis*=lateral)	Greater trochanter and linea aspera of femur.			
Vastus medialis (VAS-tus mē-dē-Ā-lis; *medialis*=medial)	Linea aspera of femur.			
Vastus intermedius (VAS-tus in'-ter-MĒ-dē-us; *intermedius*=middle)	Anterior and lateral surfaces of body of femur.			
Sartorius (sar-TOR-ē-us; *sartor*=tailor; longest muscle in body)	Anterior superior iliac spine.	Medial surface of body of tibia.	Weakly flexes leg at knee joint; weakly flexes, abducts, and laterally rotates thigh at hip joint.	Femoral nerve.
POSTERIOR (FLEXOR) COMPARTMENT OF THE THIGH				
Hamstrings A collective designation for three separate muscles.				
Biceps femoris (BĪ-ceps FEM-or-is; *biceps*=two heads of origin)	Long head arises from ischial tuberosity; short head arises from linea aspera of femur.	Head of fibula and lateral condyle of tibia.	Flexes leg at knee joint and extends thigh at hip joint.	Tibial and common fibular nerves from the sciatic nerve.
Semitendinosus (sem'-ē-ten-di-NŌ-sus; *semi-*=half; *tendo*=tendon)	Ischial tuberosity.	Proximal part of medial surface of shaft of tibia.	Flexes leg at knee joint and extends thigh at hip joint.	Tibial nerve from the sciatic nerve.
Semimembranosus (sem'-ē-mem-bra=NŌ-sus; *membran-*=membrane)	Ischial tuberosity.	Medial condyle of tibia.	Flexes leg at knee joint and extends thigh at hip joint.	Tibial nerve from the sciatic nerve.

Figure 11.23 Muscles of the thigh that move the femur (thigh bone) and tibia and fibula (leg bones).

Muscles that act on the leg originate in the hip and thigh and are separated into compartments by deep fascia.

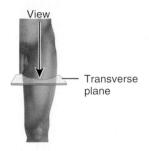

View

Transverse plane

MEDIAL

LATERAL

POSTERIOR

BICEPS FEMORIS

SEMITENDINOSUS

SEMIMEMBRANOSUS

POSTERIOR COMPARTMENT

Sciatic nerve

Gluteus maximus

Skin

Subcutaneous layer

Fascia

Femur

ADDUCTOR MAGNUS

ADDUCTOR BREVIS

GRACILIS

ADDUCTOR LONGUS

MEDIAL COMPARTMENT

VASTUS LATERALIS

VASTUS INTERMEDIUS

VASTUS MEDIALIS

RECTUS FEMORIS

SARTORIUS

ANTERIOR COMPARTMENT

Fascia

Femoral artery

Femoral vein

ANTERIOR

(a) Superior view of transverse section of thigh

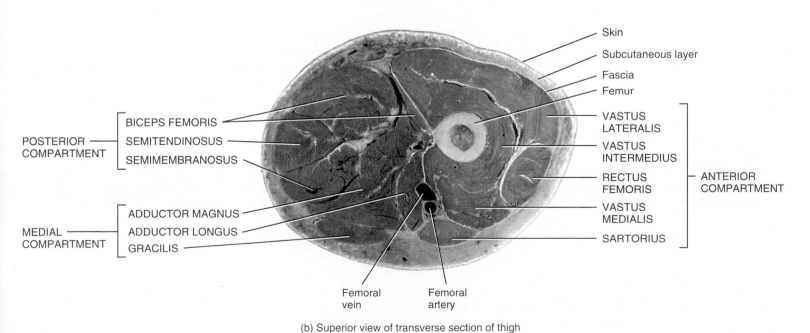

Skin

Subcutaneous layer

Fascia

Femur

BICEPS FEMORIS

SEMITENDINOSUS

SEMIMEMBRANOSUS

POSTERIOR COMPARTMENT

VASTUS LATERALIS

VASTUS INTERMEDIUS

RECTUS FEMORIS

VASTUS MEDIALIS

SARTORIUS

ANTERIOR COMPARTMENT

ADDUCTOR MAGNUS

ADDUCTOR LONGUS

GRACILIS

MEDIAL COMPARTMENT

Femoral vein

Femoral artery

(b) Superior view of transverse section of thigh

? Which muscles constitute the quadriceps femoris and hamstring muscles?

EXHIBIT 11.20 CONTINUES ▶

EXHIBIT 11.20   **397**

EXHIBIT 11.21 Muscles of the Leg That Move the Foot and Toes

OBJECTIVE

- Describe the origin, insertion, action, and innervation of the muscles that move the foot and toes.

Muscles that move the foot and toes are located in the leg (Figure 11.24 on pages 400–404). The muscles of the leg, like those of the thigh, are divided by deep fascia into three compartments: anterior, lateral, and posterior. **The anterior compartment of the leg** consists of muscles that dorsiflex the foot. In a situation analogous to the wrist, the tendons of the muscles of the anterior compartment are held firmly to the ankle by thickenings of deep fascia called the **superior extensor retinaculum** (*transverse ligament of the ankle*) and **inferior extensor retinaculum** (*cruciate ligament of the ankle*).

Within the anterior compartment, the **tibialis anterior** is a long, thick muscle against the lateral surface of the tibia, where it is easy to palpate (feel). The **extensor hallucis longus** is a thin muscle between and partly deep to the tibialis anterior and **extensor digitorum longus muscles**. This featherlike muscle is lateral to the tibialis anterior muscle, where it can also be palpated easily. The **fibularis (peroneus) tertius** muscle is part of the extensor digitorum longus, with which it shares a common origin.

The **lateral (fibular) compartment of the leg** contains two muscles that plantar flex and evert the foot: the **fibularis (peroneus) longus** and **fibularis (peroneus) brevis**.

The **posterior compartment of the leg** consists of muscles in superficial and deep groups. The superficial muscles share a common tendon of insertion, the **calcaneal (Achilles) tendon**, the strongest tendon of the body. It inserts into the calcaneal bone of the ankle. The superficial and most of the deep muscles plantar flex the foot at the ankle joint. The superficial muscles of the posterior compartment are the gastrocnemius, soleus, and plantaris—the so-called calf muscles. The large size of these muscles is directly related to the characteristic upright stance of humans. The **gastrocnemius** is the most superficial muscle and forms the prominence of the calf. The **soleus**, which lies deep to the gastrocnemius, is broad and flat. It derives its name from its resemblance to a flat fish (sole). The **plantaris** is a small muscle that may be absent; conversely, sometimes there are two of them in each leg. It runs obliquely between the gastrocnemius and soleus muscles.

The deep muscles of the posterior compartment are the popliteus, tibialis posterior, flexor digitorum longus, and flexor hallucis longus. The **popliteus** is a triangular muscle that forms the floor of the popliteal fossa. The **tibialis posterior** is the deepest muscle in the posterior compartment. It lies between the flexor digitorum longus and flexor hallucis longus muscles. The **flexor digitorum longus** is smaller than the **flexor hallucis longus**, even though the former flexes four toes, and the latter flexes only the great toe at the interphalangeal joint.

• CLINICAL CONNECTION Shin Splint Syndrome

Shin **splint syndrome**, or simply **shin splints**, refers to pain or soreness along the tibia, specifically the medial, distal two-thirds. It may be caused by tendinitis of the anterior compartment muscles, especially the tibialis anterior muscle, inflammation of the periosteum (periostitis) around the tibia, or stress fractures of the tibia. The tendonitis usually occurs when poorly conditioned runners run on hard or banked surfaces with poorly supportive running shoes. The condition may also occur with vigorous activity of the legs following a period of relative inactivity or running in cold weather without proper warmup. The muscles in the anterior compartment (mainly the tibialis anterior) can be strengthened to balance the stronger posterior compartment muscles. •

MUSCLE	ORIGIN	INSERTION	ACTION	INNERVATION
ANTERIOR COMPARTMENT OF THE LEG				
Tibialis anterior (tib′-ē-Â-lis=tibia; *anterior*=front)	Lateral condyle and body of tibia and interosseous membrane (sheet of fibrous tissue that holds shafts of tibia and fibula together).	First metatarsal and first (medial) cuneiform.	Dorsiflexes foot at ankle joint and inverts foot at intertarsal joints.	Deep fibular (peroneal) nerve.
Extensor hallucis longus (HAL-ū-sis LON-gus; *extensor*=increases angle at joint; *halluc-*=hallux or great toe; *longus*=long)	Anterior surface of fibula and interosseous membrane.	Distal phalanx of great toe.	Dorsiflexes foot at ankle joint and extends proximal phalanx of great toe at metatarsophalangeal joint.	Deep fibular (peroneal) nerve.
Extensor digitorum longus (di′-ji-TOR-um LON-gus)	Lateral condyle of tibia, anterior surface of fibula, and interosseous membrane.	Middle and distal phalanges of toes 2–5.*	Dorsiflexes foot at ankle joint and extends distal and middle phalanges of each toe at interphalangeal joints and proximal phalanx of each toe at metatarsophalangeal joint.	Deep fibular (peroneal) nerve.
Fibularis (Peroneus) tertius (fib-ū-LÃ-ris TER-shus; *peron-*=fibula; *tertius*=third)	Distal third of fibula and interosseous membrane.	Base of fifth metatarsal.	Dorsiflexes foot at ankle joint and everts foot at intertarsal joints.	Deep fibular (peroneal) nerve.

Relating Muscles to Movements

Arrange the muscles in this exhibit according to the following actions on the foot at the ankle joint: (1) dorsiflexion and (2) plantar flexion; according to the following actions on the foot at the intertarsal joints: (1) inversion and (2) eversion; and according to the following actions on the toes at the metatarsophalangeal and interphalangeal joints: (1) flexion and (2) extension. The same muscle may be mentioned more than once.

◉ CHECKPOINT
What are the superior extensor retinaculum and inferior extensor retinaculum?

MUSCLE	ORIGIN	INSERTION	ACTION	INNERVATION
LATERAL (FIBULAR) COMPARTMENT OF THE LEG				
Fibularis (Peroneus) longus (LON-gus)	Head and body of fibula and lateral condyle of tibia.	First metatarsal and first cuneiform.	Plantar flexes foot at ankle joint and everts foot at intertarsal joints.	Superficial fibular (peroneal) nerve.
Fibularis (Peroneus) brevis (BREV-is; *brevis*=short)	Body of fibula.	Base of fifth metatarsal.	Plantar flexes foot at ankle joint and everts foot at intertarsal joints.	Superficial fibular (peroneal) nerve.
SUPERFICIAL POSTERIOR COMPARTMENT OF THE LEG				
Gastrocnemius (gas'-trok-NĒ-mē-us; *gastro-*=belly; *cnem-*=leg)	Lateral and medial condyles of femur and capsule of knee.	Calcaneus by way of calcaneal (Achilles) tendon	Plantar flexes foot at ankle joint and flexes leg at knee joint.	Tibial nerve.
Soleus (SŌ-lē-us; *sole*=a type of flat fish)	Head of fibula and medial border of tibia.	Calcaneus by way of calcaneal (Achilles) tendon.	Plantar flexes foot at ankle joint.	Tibial nerve.
Plantaris (plan-TĀR-is; *plantar-*=sole)	Femur superior to lateral condyle.	Calcaneus by way of calcaneal (Achilles) tendon.	Plantar flexes foot at ankle joint and flexes leg at knee joint.	Tibial nerve.
DEEP POSTERIOR COMPARTMENT OF THE LEG				
Popliteus (pop-LIT-ē-us; *poplit-*=the back of the knee)	Lateral condyle of femur.	Proximal tibia.	Flexes leg at knee joint and medially rotates tibia to unlock the extended knee.	Tibial nerve.
Tibialis posterior (tib'-ē-Ā-lis; *posterior*=back)	Tibia, fibula, and interosseous membrane.	Second, third, and fourth metatarsals; navicular; all three cuneiforms; and cuboid.	Plantar flexes foot at ankle joint and inverts foot at intertarsal joints.	Tibial nerve.
Flexor digitorum longus (di'-ji-TOR-um LON-gus; *digit*=finger or toe)	Posterior surface of tibia.	Distal phalanges of toes 2–5.	Plantar flexes foot at ankle joint; flexes distal and middle phalanges of each toe at interphalangeal joints and proximal phalanx of each toe at metatarsophalangeal joint.	Tibial nerve.
Flexor hallucis longus (HAL-ū-sis; *flexor*=decreases angle at joint)	Inferior two-thirds of fibula.	Distal phalanx of great toe.	Plantar flexes foot at ankle joint; flexes distal phalanx of great toe at interphalangeal joint and proximal phalanx of great toe at metatarsophalangeal joint.	Tibial nerve.

*Reminder: The great toe or hallux is the first toe and has two phalanges: proximal and distal. The remaining toes are numbered II–V (2–5), and each has three phalanges: proximal, middle, and distal.

◖ EXHIBIT 11.21 CONTINUES ▶

EXHIBIT 11.21 **399**

EXHIBIT 11.21 Muscles of the Leg That Move the Foot and Toes CONTINUED

Figure 11.24 Muscles of the leg that move the foot and toes.

The superficial muscles of the posterior compartment share a common tendon of insertion, the calcaneal (Achilles) tendon, which inserts into the calcaneal bone of the ankle.

Quadriceps femoris

Tendon of quadriceps femoris

Iliotibial tract

Biceps femoris

Patella

PLANTARIS

Head of fibula

Patellar ligament

Tibia

TIBIALIS ANTERIOR

GASTROCNEMIUS

FIBULARIS LONGUS

SOLEUS

EXTENSOR DIGITORUM LONGUS

FLEXOR DIGITORUM LONGUS

FIBULARIS BREVIS

FIBULARIS TERTIUS

EXTENSOR HALLUCIS LONGUS

Calcaneal (Achilles) tendon

Fibula

EXTENSOR HALLUCIS BREVIS

EXTENSOR DIGITORUM BREVIS

Metatarsals

Superior extensor retinaculum

Inferior extensor retinaculum

(a) Anterior superficial view

(b) Right lateral superficial view

Gracilis

Sartorius

Biceps femoris

Semitendinosus

Semimembranosus

Femur

Popliteal fossa

PLANTARIS

GASTROCNEMIUS (cut)

Tendon of biceps femoris (cut)

Tibia

POPLITEUS

GASTROCNEMIUS

SOLEUS (cut)

Fibula

TIBIALIS POSTERIOR

SOLEUS

FIBULARIS LONGUS

FLEXOR DIGITORUM LONGUS

FLEXOR HALLUCIS LONGUS

FIBULARIS BREVIS

Tibia

Fibula

Tendon of tibialis posterior

Calcaneal (Achilles) tendon (cut)

(c) Posterior superficial view

(d) Posterior deep view

EXHIBIT **11.21** CONTINUES

EXHIBIT 11.21 **401**

EXHIBIT 11.21 Muscles of the Leg That Move the Foot and Toes CONTINUED

● FIGURE 11.24 CONTINUED ▶

Tibia

TIBIALIS ANTERIOR

Fibia

EXTENSOR HALLUCIS LONGUS

FIBULARIS TERTIUS

EXTENSOR DIGITORUM LONGUS

FIBULARIS LONGUS

FIBULARIS BREVIS

Anterior views

Right lateral view

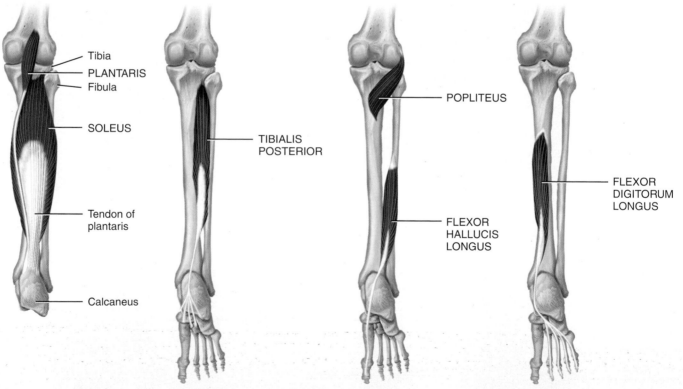

Tibia
PLANTARIS
Fibula

SOLEUS

Tendon of plantaris

Calcaneus

TIBIALIS POSTERIOR

POPLITEUS

FLEXOR HALLUCIS LONGUS

FLEXOR DIGITORUM LONGUS

Posterior deep views

(e) Anterior, right lateral, and posterior views

Iliotibial tract

Patella

Patellar
ligament

Tibia

FIBULARIS
LONGUS

TIBIALIS
ANTERIOR

EXTENSOR
DIGITORUM
LONGUS

FIBULARIS
BREVIS

Tendon of
extensor hallucis
longus

FIBULARIS
TERTIUS

(f) Anterior superficial view

Patella

Lateral condyle
of femur

Patellar ligament

GASTROCNEMIUS

SOLEUS

FIBULARIS
LONGUS

EXTENSOR
DIGITORUM LONGUS

TIBIALIS ANTERIOR

FIBULARIS
BREVIS

FIBULARIS
TERTIUS

Calcaneal
(Achilles) tendon

EXTENSOR HALLUCIS
LONGUS

Lateral malleolus
of fibula

(g) Lateral superficial view

EXHIBIT 11.21 CONTINUES ▶

EXHIBIT 11.21 **403**

EXHIBIT 11.21 Muscles of the Leg That Move the Foot and Toes CONTINUED

○ FIGURE 11.24 CONTINUED ▶

Femur

GASTROCNEMIUS (CUT)

PLANTARIS

POPLITEUS

SOLEUS (cut)

Fibula

GASTROCNEMIUS

TIBIALIS POSTERIOR

FLEXOR DIGITORUM
LONGUS

SOLEUS

FIBULARIS LONGUS

FLEXOR HALLUCIS
LONGUS

FIBULARIS BREVIS

Calcaneal (Achilles)
tendon (cut)

(h) Posterior superficial view

(i) Posterior deep view

? What structures firmly hold the tendons of the anterior compartment muscles to the ankle?

EXHIBIT 11.22 Intrinsic Muscles of the Foot That Move the Toes

- Describe the origin, insertion, action, and innervation of the intrinsic muscles of the foot.

The muscles in this exhibit are termed **intrinsic muscles of the foot** because they originate and insert *within* the foot. The muscles of the hand are specialized for precise and intricate movements, but those of the foot are limited to support and locomotion. The deep fascia of the foot forms the **plantar aponeurosis (fascia)** that extends from the calcaneus bone to the phalanges of the toes. The aponeurosis supports the longitudinal arch of the foot and encloses the flexor tendons of the foot.

The intrinsic muscles of the foot are divided into two groups: **dorsal** and **plantar** (Figure 11.25 on pages 407–409). There is only one dorsal muscle, the **extensor digitorum brevis** a four-part muscle deep to the tendons of the extensor digitorum longus muscle, which extends toes 2–5 at the metatarsophalangeal joints.

The plantar muscles are arranged in four layers. The most superficial layer is called the first layer. Three muscles are in the first layer. The **abductor hallucis**, which lies along the medial border of the sole, is comparable to the abductor pollicis brevis in the hand, and abducts the great toe at the metatarsophalangeal joint. The **flexor digitorum brevis**, which lies in the middle of the sole, flexes toes 2–5 at the interphalangeal and metatarsophalangeal joints. The **abductor digiti minimi**, which lies along the lateral border of the sole, is comparable to the same muscle in the hand, and abducts the little toe.

The second layer consists of the **quadratus plantae**, a rectangular muscle that arises by two heads and flexes toes 2–5 at the metatarsophalangeal joints, and the **lumbricals**, four small muscles that are similar to the lumbricals in the hands. They flex the proximal phalanges and extend the distal phalanges of toes 2–5.

Three muscles comprise the third layer. The **flexor hallucis brevis**, which lies adjacent to the plantar surface of the metatarsal of the great toe, is comparable to the same muscle in the hand, and flexes the great toe. The **adductor hallucis**, which has an oblique and transverse head like the adductor pollicis in the hand, adducts the great toe. The **flexor digiti minimi brevis**, which lies superficial to the metatarsal of the little toe, is comparable to the same muscle in the hand, and flexes the little toe.

The fourth layer is the deepest and consists of two muscle groups. The **dorsal interossei** are four muscles that abduct toes 2–4, flex the proximal phalanges, and extend the distal phalanges. The three **plantar interossei** abduct toes 3–5, flex the proximal phalanges, and extend the distal phalanges. The interossei of the feet are similar to those of the hand. However, their actions are relative to the midline of the second digit rather than the third digit as in the hand.

● CLINICAL **CONNECTION** | Plantar Fasciitis

Plantar fasciitis (fas-ē-Ī-tis) or **painful heel syndrome** is an inflammatory reaction due to chronic irritation of the plantar aponeurosis (fascia) at its origin on the calcaneus (heel bone). The aponeurosis becomes less elastic with age. This condition is also related to weight-bearing activities (walking, jogging, lifting heavy objects), improperly constructed or fitting shoes, excess weight (puts pressure on the feet), and poor biomechanics (flat feet, high arches, and abnormalities in gait may cause uneven distribution of weight on the feet). Plantar fasciitis is the most common cause of heel pain in runners and arises in response to the repeated impact of running. Treatments include ice, deep heat, stretching exercises, weight loss, prosthetics (such as shoe inserts or heel lifts), steroid injections, and surgery. ●

Relating Muscles to Movements

Arrange the muscles in this exhibit according to the following actions on the great toe at the metatarsophalangeal joint: (1) flexion, (2) extension, (3) abduction, and (4) adduction; and according to the following actions on toes 2-5 at the metatarsophalangeal and interphalangeal joints: (1) flexion, (2) extension, (3) abduction, and (4) adduction. The same muscle may be mentioned more than once.

⊙ CHECKPOINT

How do the intrinsic muscles of the hand and foot differ in function?

◢ EXHIBIT **11.22** CONTINUES ▶

EXHIBIT 11.22 **405**

EXHIBIT 11.22 Intrinsic Muscles of the Foot That Move the Toes CONTINUED

MUSCLE	ORIGIN	INSERTION	ACTION	INNERVATION
DORSAL				
Extensor digitorum brevis (*extensor*=increases angle at joint; *digit*=finger or toe; *brevis*=short) (see Figure 11.24a, b)	Calcaneus and inferior extensor retinaculum.	Tendons of extensor digitorum longus on toes 2–4 and proximal phalanx of great toe.*	Extensor hallucis brevis* extends great toe at metatarsophalangeal joint and extensor digitorum brevis extends toes 2–4 at interphalangeal joints.	Deep fibular (peroneal) nerve.
PLANTAR				
First layer (most superficial)				
Abductor hallucis (*abductor*=moves part away from midline; *hallucis*=hallux or great toe)	Calcaneus, plantar aponeurosis, and flexor retinaculum.	Medial side of proximal phalanx of great toe with the tendon of the flexor hallucis brevis.	Abducts and flexes great toe at metatarsophalangeal joint.	Medial plantar nerve.
Flexor digitorum brevis (*flexor*=decreases angle at joint)	Calcaneus and plantar aponeurosis.	Sides of middle phalanx of toes 2–5.	Flexes toes 2–5 at proximal interphalangeal and metatarsophalangeal joints.	Medial plantar nerve.
Abductor digiti minimi (*minimi*=little)	Calcaneus and plantar aponeurosis.	Lateral side of proximal phalanx of little toe with the tendon of the flexor digiti minimi brevis.	Abducts and flexes little toe at metatarsophalangeal joint.	Lateral plantar nerve.
Second layer				
Quadratus plantae (kwod-RĀ-tus; *quad-*= square, four-sided; *planta*=the sole)	Calcaneus.	Tendon of flexor digitorum longus.	Assists flexor digitorum longus to only flex toes 2–5 at interphalangeal and metatarsophalangeal joints.	Lateral plantar nerve.
Lumbricals (LUM-bri-kals; *lumbric-*= earthworm)	Tendons of flexor digitorum longus.	Tendons of extensor digitorum longus on proximal phalanges of toes 2–5.	Extend toes 2–5 at interphalangeal joints and flex toes 2–5 at metatarsophalangeal joints.	Medial and lateral plantar nerves.
Third layer				
Flexor hallucis brevis	Cuboid and third (lateral) cuneiform.	Medial and lateral sides of proximal phalanx of great toe via a tendon containing a sesamoid bone.	Flexes great toe at metatarsophalangeal joint.	Medial plantar nerve.
Adductor hallucis	Metatarsals 2–4, ligaments of 3–5 metatarsophalangeal joints, and tendon of peroneus longus.	Lateral side of proximal phalanx of great toe.	Adducts and flexes great toe at metatarsophalangeal joint.	Lateral plantar nerve.
Flexor digiti minimi brevis	Metatarsal 5 and tendon of peroneus longus.	Lateral side of proximal phalanx of little toe.	Flexes little toe at metatarsophalangeal joint.	Lateral plantar nerve.
Fourth layer (deepest)				
Dorsal interossei (in-ter-OS-ē-ī) (not illustrated)	Adjacent side of all metatarsals.	Proximal phalanges: both sides of toe 2 and lateral side of toes 3 and 4.	Abduct and flex toes 2–4 at metatarsophalangeal joints and extend toes at interphalangeal joints.	Lateral plantar nerve.
Plantar interossei	Metatarsals 3–5.	Medial side of proximal phalanges of toes 3–5.	Adduct and flex proximal metatarsophalangeal joints and extend toes at interphalangeal joints.	Lateral plantar nerve.

*The tendon that inserts into the proximal phalanx of the great toe, together with its belly, is often described as a separate muscle, the extensor hallucis brevis.

Figure 11.25 Intrinsic muscles of the foot that move the toes.

The muscles of the hand are specialized for precise and intricate movements; those of the foot are limited to support and movement.

Tendon of flexor hallucis longus

Tendons of flexor digitorum brevis (cut)

ADDUCTOR HALLUCIS

LUMBRICALS

FLEXOR HALLUCIS BREVIS

PLANTAR INTEROSSEI

FLEXOR DIGITI MINIMI BREVIS

FLEXOR DIGITORUM BREVIS

ABDUCTOR HALLUCIS

ABDUCTOR DIGITI MINIMI

Calcaneus

Tendon of flexor hallucis longus

Tendons of flexor digitorum longus

FLEXOR HALLUCIS BREVIS

Navicular

QUADRATUS PLANTAE

Tendon of tibialis posterior

Tendon of flexor hallucis longus

(a) Plantar superficial view

(b) Plantar intermediate view

EXHIBIT 11.22 Intrinsic Muscles of the Foot That Move the Toes CONTINUED

● FIGURE 11.25 CONTINUED ▶

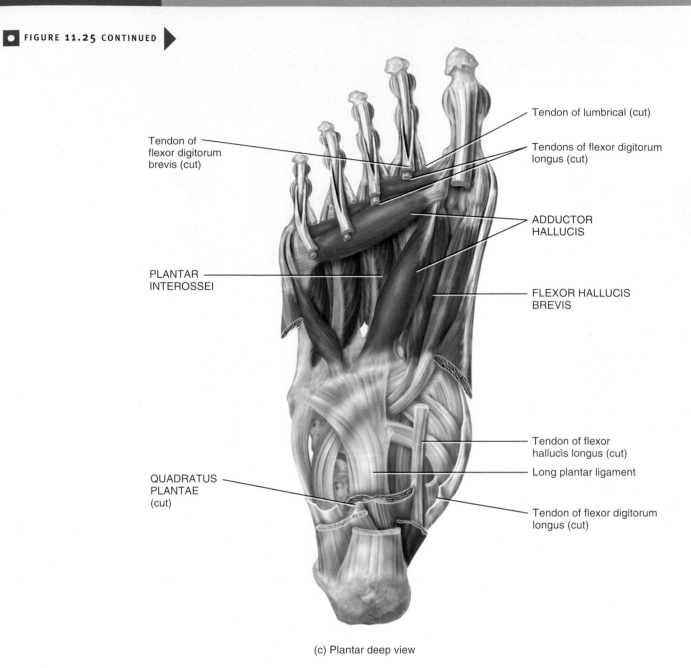

Tendon of
flexor digitorum
brevis (cut)

Tendon of lumbrical (cut)

Tendons of flexor digitorum
longus (cut)

ADDUCTOR
HALLUCIS

PLANTAR
INTEROSSEI

FLEXOR HALLUCIS
BREVIS

Tendon of flexor
hallucis longus (cut)

Long plantar ligament

QUADRATUS
PLANTAE
(cut)

Tendon of flexor digitorum
longus (cut)

(c) Plantar deep view

Phalanges

Metatarsals

PLANTAR
INTEROSSEI

Tarsals

(d) Plantar view

Phalanges

DORSAL
INTEROSSEI

Metatarsals

Tendon of
fibularis longus

Tarsals

Tendon of
fibularis brevis

(e) Plantar view

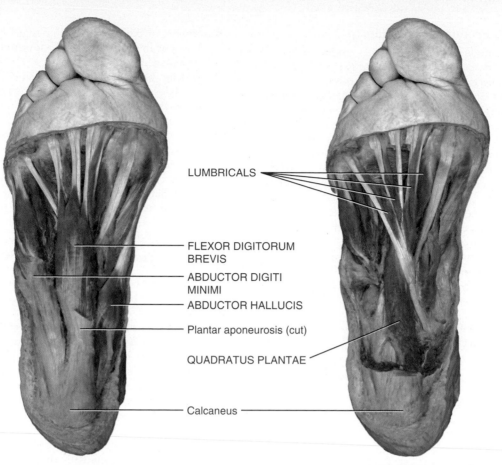

LUMBRICALS

FLEXOR DIGITORUM
BREVIS

ABDUCTOR DIGITI
MINIMI

ABDUCTOR HALLUCIS

Plantar aponeurosis (cut)

QUADRATUS PLANTAE

Calcaneus

(f) Plantar view, first muscle layer

(g) Plantar view, second muscle layer

ADDUCTOR HALLUCIS:
TRANSVERSE HEAD
OBLIQUE HEAD

DORSAL INTEROSSEI

PLANTAR INTEROSSEI

FLEXOR HALLUCIS
BREVIS

FLEXOR DIGITI
MINIMI BREVIS

Tendon of fibularis longus

Tendon of tibialis posterior

Long plantar ligament

Tendon of flexor hallucis
longus (cut)

Tendon of flexor digitorum
longus (cut)

Calcaneus

(h) Plantar view, third muscle layer

(i) Plantar view, fourth muscle layer

? What structure supports the longitudinal arch and encloses the flexor tendons of the foot?

EXHIBIT 11.22   409

How Skeletal Muscles Produce Movements (p. 310)

1. Skeletal muscles that produce movement do so by pulling on bones.
2. The attachment to the more stationary bone is the origin; the attachment to the more movable bone is the insertion.
3. Bones serve as levers, and joints serve as fulcrums. Two different forces act on the lever: load (resistance) and effort.
4. Levers are categorized into three types—first-class, second-class, and third-class (most common)—according to the positions of the fulcrum, the effort, and the load on the lever.
5. Fascicular arrangements include parallel, fusiform, circular, triangular, and pennate. Fascicular arrangement affects a muscle's power and range of motion.
6. Muscles control joints by contracting both concentrically (shortening contractions) and eccentrically (lengthening contractions) to regulate fine muscle control at a joint.
7. A prime mover produces the desired action; an antagonist produces an opposite action. Synergists assist a prime mover by reducing unnecessary movement. Fixators stabilize the origin of a prime mover so that it can act more efficiently.

How Skeletal Muscles Are Named (p. 314)

1. Distinctive features of different skeletal muscles include direction of muscle fascicles; size, shape, action, number of origins (or heads), and location of the muscle; and sites of origin and insertion of the muscle.
2. Most skeletal muscles are named based on combinations of these features.

Principal Skeletal Muscles (p. 314)

1. Muscles of facial expression move the skin rather than a joint when they contract, and they permit us to express a wide variety of emotions.
2. The extrinsic muscles that move the eyeballs are among the fastest contracting and most precisely controlled skeletal muscles in the body. They permit us to elevate, depress, abduct, adduct, and medially and laterally rotate the eyeballs.
3. Muscles that move the mandible (lower jaw) are also known as the muscles of mastication because they are involved in chewing.
4. The extrinsic muscles that move the tongue are important in chewing, swallowing, and speech.
5. Muscles of the floor of the anterior neck, called suprahyoid muscles, are located above the hyoid bone. They elevate the hyoid bone, oral cavity, and tongue during swallowing.
6. Muscles that move the head alter its position and help balance the head on the vertebral column.
7. Most muscles of the larynx (voice box) depress the hyoid bone and larynx during swallowing and speech and move the internal portion of the larynx during speech.

8. Muscles of the pharynx are arranged into a circular layer, which functions in swallowing, and a longitudinal layer, which functions in swallowing and speaking.
9. Muscles that act on the abdominal wall help contain and protect the abdominal viscera, move the vertebral column, compress the abdomen, and produce the force required for defecation, urination, vomiting, and childbirth.
10. Muscles used in breathing alter the size of the thoracic cavity so that ventilation can occur and assist in venous return of blood to the heart.
11. Muscles of the pelvic floor support the pelvic viscera, resist the thrust that accompanies increases in intra-abdominal pressure, and function as sphincters at the anorectal junction, urethra, and vagina.
12. Muscles of the perineum assist in urination, erection of the penis and clitoris, ejaculation, and defecation.
13. Muscles that move the pectoral (shoulder) girdle stabilize the scapula so it can function as a stable point of origin for most of the muscles that move the humerus.
14. Muscles that move the humerus (arm bone) originate for the most part on the scapula (scapular muscles); the remaining muscles originate on the axial skeleton (axial muscles).
15. Muscles that move the radius and ulna (forearm bones) are involved in flexion and extension at the elbow joint and are organized into flexor and extensor compartments.
16. Muscles that move the wrist, hand, thumb, and fingers are many and varied; those muscles that act on the digits are called extrinsic muscles.
17. The intrinsic muscles of the hand are important in skilled activities and provide humans with the ability to grasp and manipulate objects precisely.
18. Muscles that move the vertebral column are quite complex because they have multiple origins and insertions and because there is considerable overlap among them.
19. Muscles that move the femur (thigh bone) originate for the most part on the pelvic girdle and insert on the femur; these muscles are larger and more powerful than comparable muscles in the upper limb.
20. Muscles that move the femur (thigh bone) and tibia and fibula (leg bones) are separated into medial (adductor), anterior (extensor), and posterior (flexor) compartments.
21. Muscles that move the foot and toes are divided into anterior, lateral, and posterior compartments.
22. Intrinsic muscles of the foot, unlike those of the hand, are limited to the functions of support and locomotion.

Choose the one best answer to the following questions.

1. Which of the following muscles that act on the thumb is *not* located in the thenar eminence?
   a. abductor pollicis brevis
   b. flexor pollicis brevis
   c. opponens pollicis
   d. flexor pollicis longus
   e. *Two* of the above

2. The thenar muscles act on:
   a. the wrist.
   b. the thumb.
   c. the little finger.
   d. all fingers except the thumb.
   e. the great toe.

3. During exhalation, the diaphragm:
   a. contracts and flattens.
   b. relaxes and flattens.
   c. contracts and forms a dome.
   d. relaxes and forms a dome.
   e. does not contract or relax.

4. Choose the muscle(s) that depress(es) the mandible:
   a. lateral pterygoid
   b. digastric
   c. medial pterygoid
   d. Both a and b are correct.
   e. All three muscles depress the mandible.

5. The action of the pectoralis major muscle is to:
   a. abduct the arm and rotate the arm laterally.
   b. flex, adduct, and rotate the arm medially.
   c. adduct the arm and rotate the arm laterally.
   d. abduct the arm and rotate the arm medially.
   e. abduct and raise the arm.

6. Which of the following is part of the medial compartment of muscles that act on the femur?
   a. semitendinosus   b. biceps femoris
   c. rectus femoris   d. gluteus medius
   e. gracilis

7. Muscles of which group share a common origin on the ischium and act to extend the thigh and flex the leg?
   a. gluteal muscles   b. quadriceps muscles
   c. adductor muscles   d. hamstring muscles
   e. peroneal muscles

8. In compartment syndrome, external or internal pressure constricts the structures within a compartment, resulting in damaged blood vessels and subsequent reduction of blood supply. If the syndrome is localized to the posterior crural compartment, what function is most significantly affected?
   a. flexion of the wrist   b. abduction of the pollex
   c. extension of the knee joint   d. dorsiflexion of the ankle joint
   e. plantar flexion of the ankle joint

Complete the following.

9. A muscle that contracts to cause the desired action is called the _____. Muscles that assist or cooperate with the muscle that causes the desired action are known as _____.

10. The attachment of a muscle's tendon to the stationary bone is called the _____, and the attachment of the muscle's other tendon to the movable bone is called the _____.

11. In an anatomical lever system, _____ act as levers, _____ act as fulcrums, and the effort is provided by _____.

12. The muscles of the posterior compartment of muscles also known as the hamstrings are the _____, the _____, and the _____.

13. The sternocleidomastoid and longissimus capitis muscles insert on the _____ of the temporal bone.

14. The diaphragm inserts onto the _____ tendon.

15. Three posterior thoracic muscles that adduct the scapula are the _____, the _____, and the _____.

16. The diamond-shaped area that extends from the pubic symphysis anteriorly, to the coccyx posteriorly, and to the ischial tuberosities laterally is the _____.

17. A circular muscle that decreases the size of an opening, such as the anus, is known as a _____.

18. The pectineus, psoas major, adductor longus, and iliacus muscles have one action in common: _____ of the thigh.

19. The quadriceps femoris is the main muscle mass on the _____ surface of the thigh. The name of this muscle mass indicates it has _____ heads of origin. All converge to insert on the _____ (bone) by means of the patellar ligament.

20. The three muscles that utilize the calcaneal tendon to insert on the calcaneus are the _____, the _____, and the _____.

Are the following statements true or false? If the statement is False, *find and fix the mistake to make the statement* True.

21. The flexor retinaculum is located on the anterior aspect of the wrist and serves to hold the tendons, nerves, and blood vessels close to the bones of the wrist.

22. The action of the biceps brachii on the forearm at the elbow joint illustrates the structure and action of a first-class lever system.

23. The muscles that plantar flex the foot at the ankle joint are located in the anterior compartment of the leg.

24. The inferior free border of the internal oblique aponeurosis forms the inguinal ligament, which runs from the anterior superior iliac spine to the pubic tubercle.

Matching

25. Match each of these muscles with the clue(s) given by the parts of its name.

    _____ (a) transversus abdominis   (1) action
    _____ (b) gluteus minimus   (2) direction of fibers
    _____ (c) biceps brachii   (3) location
    _____ (d) sternocleidomastoid   (4) number of tendons of origin
    _____ (e) adductor magnus   (5) sites of origin and/or insertion
    _____ (f) tibialis anterior
    _____ (g) rhomboid major   (6) size or shape

26. Match the muscles with the actions:

    _____ (a) presses cheeks against teeth and lips   (1) latissimus dorsi
    _____ (b) moves eyeball inferiorly and laterally, rotates eyeball medially   (2) quadriceps femoris
    (3) superior oblique
    (4) buccinator
    (5) trapezius
    _____ (c) elevates and adducts scapula   (6) external intercostals
    _____ (d) extends leg at knee joint   (7) triceps brachii
    _____ (e) extends forearm at shoulder joint   (8) deltoid
    (9) external oblique
    (10) gluteus maximus
    _____ (f) extends, adducts, and medially rotates arm at shoulder joint
    _____ (g) flexes and rotates vertebral column
    _____ (h) extends and laterally rotates thigh at hip joint
    _____ (i) abducts arm at shoulder joint
    _____ (j) elevates ribs during inspiration

27. Match the following muscles with the appropriate nerves:

    _____ (a) most of the muscles of facial expression   (1) hypoglossal (XII) nerve
    _____ (b) muscles that move the mandible   (2) thoracic nerves 7–12
    (3) facial (VII) nerve
    _____ (c) most muscles that move the tongue   (4) tibial nerve
    _____ (d) most muscles of the pharynx   (5) femoral nerve
    (6) trigeminal (V) nerve, mandibular division
    _____ (e) muscles that move the head   (7) deep peroneal nerve
    _____ (f) muscles of the anterolateral abdominal wall   (8) accessory (XI) and cervical spinal nerves
    (9) superficial peroneal nerve
    (10) phrenic nerve
    (11) vagus (X) nerve
    _____ (g) diaphragm
    _____ (h) quadriceps femoris muscles
    _____ (i) muscles of the anterior compartment of the leg
    _____ (j) muscles of the lateral compartment of the leg
    _____ (k) muscles of the posterior compartment of the leg

CRITICAL THINKING QUESTIONS

1. Three-year-old Ming likes to form her tongue into a cylinder shape and use it as a straw when she drinks her milk. Name the muscles that Ming uses to protrude her lips and tongue and to suck up the milk.

2. Baby Eddie weighed 13 lb 4 oz at birth and he's been gaining ever since. His mom has noticed a sharp pain between her shoulder blades when she picks up Eddie to put him in his high chair (again). Which muscle may Eddie's mother have strained?

3. Anthropology graduate students are learning about "bite force" in Neanderthals. Since this is a significant feature of chewing, it is critical to understand muscle attachments onto the mandible, and specifically what actions these muscles produce. Distinguish the attachment sites and actions of the four muscles of mastication.

4. Perry was leading his fantasy baseball league until his best pitcher went on the disabled list with a shoulder injury. What is the most likely injury to this player? What specific muscles are likely to be involved?

5. Wyman has been doing a lot of heavy lifting lately. He has suddenly noticed a bulge on the anterior aspect of his torso down near his groin. What has probably happened to him? Do you think he needs to see a doctor? Why or why not?

? ANSWERS TO FIGURE QUESTIONS

11.1 The belly of the muscle that extends the forearm, the triceps brachii, is located posterior to the humerus.

11.2 Second-class levers produce the most force.

11.3 For muscles named after their various characteristics, here are possible correct responses (for others, see Table 11.2): direction of fibers: external oblique; shape: deltoid; action: extensor digitorum; size: gluteus maximus; origin and insertion: sternocleidomastoid; location: tibialis anterior; number of tendons of origin: biceps brachii.

11.4 The frontal belly of the occipitofrontalis muscle is involved in frowning; the zygomaticus major muscle contracts when you smile; the platysma muscle contributes to pouting; the orbicularis oculi muscle contributes to squinting.

11.5 The inferior oblique muscle moves the eyeball superiorly and laterally because it originates at the anteromedial aspect of the floor of the orbit and inserts on the posterolateral aspect of the eyeball.

11.6 The masseter is the strongest muscle of mastication.

11.7 Functions of the tongue include chewing, perception of taste, swallowing, and speech.

11.8 The suprahyoid and infrahyoid muscles stabilize the hyoid bone to assist in tongue movements.

11.9 The triangles in the neck formed by the sternocleidomastoid muscles are important anatomically and surgically because of the structures that lie within their boundaries.

11.10 The extrinsic muscles move the larynx as a whole, while the intrinsic muscles move parts of the larynx.

11.11 The longitudinal muscles elevate the larynx, and the infrahyoid muscles depress the larynx.

11.12 The rectus abdominis muscle aids in urination.

11.13 The diaphragm is innervated by the phrenic nerve.

11.14 The borders of the pelvic diaphragm are the pubic symphysis anteriorly, the coccyx posteriorly, and the walls of the pelvis laterally.

11.15 The borders of the perineum are the pubic symphysis anteriorly, the coccyx posteriorly, and the ischial tuberosities laterally.

11.16 The main action of the muscles that move the pectoral girdle is to stabilize the scapula to assist in movements of the humerus.

11.17 The rotator cuff consists of the flat tendons of the subscapularis, supraspinatus, infraspinatus, and teres minor muscles that form a nearly complete circle around the shoulder joint.

11.18 The brachialis is the most powerful forearm flexor; the triceps brachii is the most powerful forearm extensor.

11.19 Flexor tendons of the digits and wrist and the median nerve pass through the flexor retinaculum.

11.20 Muscles of the thenar eminence act on the thumb.

11.21 The splenius muscles arise from the midline and extend laterally and superiorly to their insertion.

11.22 Free upper limb muscles exhibit diversity of movement; free lower limb muscles function in stability, locomotion, and maintenance of posture. In addition, free lower limb muscles usually cross two joints and act equally on both.

11.23 The quadriceps femoris consists of the rectus femoris, vastus lateralis, vastus medialis, and vastus intermedius; the hamstrings consist of the biceps femoris, semitendinosus, and semimembranosus.

11.24 The superior and inferior extensor retinacula firmly hold the tendons of the anterior compartment muscles to the ankle.

11.25 The plantar aponeurosis supports the longitudinal arch and encloses the flexor tendons of the foot.

12 | SURFACE ANATOMY

INTRODUCTION In Chapter 1 we introduced several branches of anatomy and noted their relationship to our understanding of the body's structure. In this chapter we will take a closer look at **surface anatomy**. Recall that surface anatomy is the study of the anatomical landmarks on the exterior of the body. A knowledge of surface anatomy will not only help you to identify structures on the body's exterior, but it will also assist you in locating the positions of various internal structures.

The study of surface anatomy involves two related, although distinct, activities: visualization and palpation. **Visualization** involves looking in a very selective and purposeful manner at a specific part of the body. **Palpation** (pal-PĀ-shun) means using the sense of touch to determine the location of an internal part of the body through the skin. Like visualization, palpation is performed selectively and purposefully, and it supplements information already gained through other methods, including visualization. Because of the varying depth of different structures from the surface of the body and due to gender differences in the thickness of the dermis and subcutaneous layer over different parts of the body, palpations may range from light, to moderate, to deep. •

OVERVIEW OF SURFACE ANATOMY

◉ **OBJECTIVE**

- Explain how surface anatomy can be used in a physical examination.

Knowledge of surface anatomy has many applications, both anatomically and clinically. From an anatomical viewpoint, a knowledge of surface anatomy can provide valuable information regarding the location of structures such as bones, muscles, blood vessels, nerves, lymph nodes, and internal organs. Clinically, a knowledge of surface anatomy is the basis for conducting a physical examination and performing certain diagnostic tests. Health care professionals use their understanding of surface anatomy to learn where to take the pulse, measure blood pressure, draw blood, stop bleeding, insert needles and tubes, make surgical incisions, *reduce* (straighten) fractured bones, and listen to sounds made by the heart, lungs, and intestines. Clinicians also draw on a knowledge of surface anatomy to assess the status of lymph nodes and to identify the presence of tumors or other unusual masses in the body.

EXHIBIT 12.1 Surface Anatomy of the Head

◉ **OBJECTIVE**

- Describe the surface features of the head.

The **head** (*cephalic region*, or *caput*) contains the brain and sense organs (eyes, ears, nose, and tongue) and is divided into the cranium and face. The **cranium** (*skull*, or *brain case*) is that portion of the head that surrounds and protects the brain; the **face** is the anterior portion of the head.

Regions of the Cranium and Face

The cranium and face are divided into several regions, which are described here and illustrated in Figure 12.1.

- **Frontal region:** forms the front of the skull and includes the frontal bone.
- **Parietal region:** forms the crown of the skull and includes the parietal bones.
- **Temporal region:** forms the side of the skull and includes the temporal bones.
- **Occipital region:** forms the base of the skull and includes the occipital bone.
- **Orbital (ocular) region:** includes the eyeball, eyebrow, and eyelid.
- **Infraorbital region:** the region inferior to the orbit.
- **Zygomatic region:** the region inferolateral to the orbit that includes the zygomatic (cheek) bone.
- **Nasal region:** region of the nose.
- **Oral region:** region of the mouth.
- **Mental region:** anterior portion of the mandible.
- **Buccal region** (BUK-al): region of the cheek.
- **Auricular region** (aw-RIK-ū-lar): region of the external ear.

In addition to locating the various regions of the head, it is also possible to visualize and/or palpate certain structures within each region. We will examine various bone and muscle structures of the head and then look at specific areas such as the eyes, ears, and nose.

Many of the bony landmarks of the head and muscles of facial expression are labeled in Figure 12.2 on page 416. As you read about each feature, refer to the figure, and try to visualize and/or palpate each feature on your own head or on the head of a study partner.

Bony Landmarks of the Head

Several bony structures of the skull can be detected through palpation, including the following:

- **Sagittal suture.** Move the fingers from side to side over the superior aspect of the scalp in order to palpate this suture (see Figure 7.2a on page 166).
- **Coronal** and **lambdoid sutures.** Use the same side-to-side technique to palpate these structures located on the frontal and occipital regions of the skull (see Figure 7.3a on page 168).
- **External occipital protuberance.** This is the most prominent bony landmark on the occipital region of the skull (see Figure 7.3a on page 168).
- **Orbit.** The entire circumference of the orbit can be palpated. Deep to the eyebrow, the superior aspect of the orbit, you can palpate the **supraorbital margin** of the frontal bone.
- **Nasal bones.** These can be palpated in the nasal region between the orbits on either side of the midline. If you wear glasses, the bridge of the glasses sit on these bones.
- **Mandible.** The **ramus** (vertical portion), **body** (horizontal portion), and **angle** (area where the ramus meets the body) of the mandible can be easily palpated at the mental and buccal regions of the head.
- **Zygomatic arch and zygomatic bones.** These can be palpated in the zygomatic region (see Figure 7.3a on page 168).
- **Mastoid process.** This is the prominent bony landmark situated posterior to the ear that is easily palpated in the auricular region (see Figure 7.3a on page 168).

Muscles of Facial Expression

Several muscles of facial expression can be palpated while they are contracting:

- **Occipitofrontalis.** By raising and lowering the eyebrows, it is possible to palpate the frontal belly and the occipital belly in the frontal and occipital regions, respectively, as they alternately contract and the scalp moves forward and backward.
- **Orbicularis oculi.** By closing the eyes and placing the fingers on the eyelids, the muscle can be palpated in the orbital region by tightly squeezing the eyes shut.
- **Corrugator supercilii.** By frowning and drawing the eyebrows toward each other, this muscle can be felt above the nose near the medial end of the eyebrow.
- **Zygomaticus major.** By smiling, this muscle can be palpated between the corner of the mouth and zygomatic bone.
- **Depressor labii inferioris.** This muscle can be palpated in the mental region between the lower lip and chin when moving the lower lip inferiorly to expose the lower teeth.
- **Orbicularis oris.** By closing the lips tightly, this muscle can be palpated in the oral region around the margin of the lips.

Recall that there are five principal regions of the body: (1) **head**, (2) **neck**, (3) **trunk**, (4) **upper limbs**, and (5) **lower limbs**. These are discussed in this chapter in Exhibits 12.1 through 12.5. Using these regions as our guide, we will begin our study of the surface anatomy with the head and then move to the other regions, concluding with the lower limbs. Because living bodies are best suited for studying surface anatomy, you will find it helpful to visualize and/or palpate the various structures described on your own body, or on a study partner, as you study each region. Following a brief introduction to a region of the body, you will be presented with a list of the prominent structures to locate. This directed approach will help to organize your learning efforts. Labeled photographs illustrate most of the structures listed in each region.

◉ CHECKPOINT
1. What is surface anatomy? Why are visualization and palpation important in learning surface anatomy?
2. What are some of the applications of a knowledge of surface anatomy?

Figure 12.1 **Principal regions of the cranium and face.**

🔑 ⚿ The head consists of an anterior face and a cranium that surrounds the brain.

(a) Anterior view

FRONTAL REGION
TEMPORAL REGION
NASAL REGION
AURICULAR REGION
BUCCAL REGION
MENTAL REGION
ORBITAL REGION
ZYGOMATIC REGION
INFRAORBITAL REGION
ORAL REGION

(b) Right lateral view

PARIETAL REGION
TEMPORAL REGION
OCCIPITAL REGION
AURICULAR REGION
BUCCAL REGION
FRONTAL REGION
ORBITAL REGION
NASAL REGION
ZYGOMATIC REGION
INFRAORBITAL REGION
ORAL REGION
MENTAL REGION

❓ **Which regions are named for bones that are deep to them?**

◻ **EXHIBIT 12.1 CONTINUES** ▶

EXHIBIT 12.1 **415**

EXHIBIT 12.1 Surface Anatomy of the Head CONTINUED

- **Temporalis.** By alternately clenching the teeth and then opening the mouth, it is possible to palpate this muscle in the temporal region just superior to the zygomatic arch.
- **Masseter.** Again, by alternately clenching the teeth and then opening the mouth, this muscle can be palpated over the ramus of the mandible.

Surface Features of the Eyes

Now we will examine the surface features of the eyes, illustrated in Figure 12.3.

- **Iris** (=rainbow). Circular pigmented muscular structure behind the cornea (transparent covering over the iris).
- **Pupil.** Opening in the center of the iris through which light travels.

- **Sclera** (*skleros*=hard). "White" of the eye; a coat of fibrous tissue that covers the entire eyeball except for the cornea.
- **Conjunctiva** (kon-junk-TĪ-va). Membrane that covers the exposed surface of the eyeball and lines the eyelids.
- **Eyelids (palpebrae).** Folds of skin and muscle lined on their inner aspect by conjunctiva.
- **Palpebral fissure.** Space between the eyelids when they are open; when the eye is closed, it lies just inferior to the level of the pupil.
- **Medial commissure** (KOM-i-syūr). Medial (near the nose) site of the union of the upper and lower eyelids.
- **Lateral commissure.** Lateral (away from the nose) site of the union of the upper and lower eyelids.
- **Lacrimal** (*lacrim*=tears) **caruncle.** Fleshy, yellowish projection of the medial commissure that contains modified sudoriferous (sweat) and sebaceous (oil) glands.

Figure 12.2 **Surface anatomy of the head.**

Several muscles of facial expression can be palpated while they are contracting.

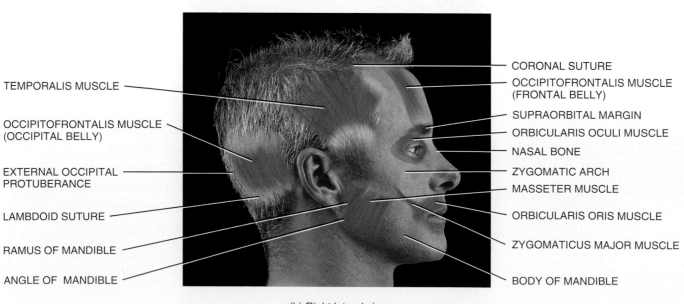

(a) Anterior view

(b) Right lateral view

? When palpating the sagittal suture, what sheetlike tendon is also palpated deep to the skin?

- **Eyelashes.** Hairs on the margin of the eyelids, usually arranged in two or three rows.
- **Eyebrows (supercilia).** Several rows of hairs superior to the upper eyelids.

Figure 12.3 Surface anatomy of the right eye.

The eyeball is protected by the eyebrow, eyelashes, and eyelids.

Anterior view

? **Through which part of the eye does light enter?**

Figure 12.4 Surface anatomy of the right ear.

The most conspicuous surface feature of the ear is the auricle.

Right lateral view

? **Which part of the temporal bone can be palpated just posterior to the auricle?**

Surface Features of the Ears

Next, locate the surface features of the ears in Figure 12.4.

- **Auricle** (AW-ri-kul). Shell-shaped portion of the external ear; also called the *pinna*. It funnels sound waves into the external auditory canal and plays an important role in helping to *localize* sound (determine its location). Just posterior to the auricle, the mastoid process of the temporal bone can be palpated.

EXHIBIT **12.1** CONTINUES

EXHIBIT 12.1 **417**

EXHIBIT 12.1   Surface Anatomy of the Head CONTINUED

- **Tragus** (TRĀ-gus; *tragos*=goat; when hair grows on the tragus it is thought to resemble the beard on a goat's chin). Cartilaginous projection anterior to the external auditory canal. Just anterior to the tragus and posterior to the neck of the mandible you can palpate the **superficial temporal artery** and feel the pulse in the vessel. Inferior to this point, it is possible to feel the **temporomandibular joint (TMJ)**. As you open and close your jaw, you can feel the movement of the mandibular condyle.
- **Antitragus.** Cartilaginous projection opposite the tragus.
- **Concha** (KON-ka). Hollow of the auricle.
- **Helix.** Superior and posterior free margin of the auricle.
- **Antihelix.** Semicircular ridge superior and posterior to the tragus.
- **Triangular fossa.** Depression in the superior portion of the antihelix.
- **Lobule.** Inferior portion of the auricle; it does not contain cartilage. Commonly referred to as the earlobe.
- **External auditory canal (meatus).** Canal about 3 cm (1 in.) long extending from the external ear to the eardrum (tympanic membrane). It contains ceruminous glands that secrete cerumen (earwax). The **condylar process** of the mandible can be palpated by placing your little finger in the canal and opening and closing your mouth.

Surface Features of the Nose and Mouth

To complete the discussion of the surface anatomy of the head, locate the following features of the nose and mouth (Figure 12.5):

- **Root.** Superior attachment of the nose at the forehead, between the eyes.
- **Apex.** Tip of the nose.
- **Dorsum nasi.** Rounded anterior border connecting the root and apex; in profile, it may be straight, convex, concave, or wavy.
- **External naris.** External opening into the nose.
- **Ala.** Convex flared portion of the inferior lateral surfaces of the nose.
- **Bridge.** Superior portion of the dorsum nasi, superficial to and formed by the nasal bones.
- **Philtrum** (FIL-trum=love charm or potion). The vertical groove on the upper lip that extends along the midline to the inferior portion of the nose.
- **Lips (labia).** Superior and inferior fleshy borders of the oral cavity.

◉ **CHECKPOINT**
- List the regions of the head and name one structure in each.
- List and define five surface features of the eyes, ears, and nose.
- How would you locate the superficial temporal artery, temporomandibular joint, zygomatic arch, mastoid process, and condylar process of the mandible?

Figure 12.5 Surface anatomy of the nose and lips.

The external nares permit air to move into and out of the nose during breathing.

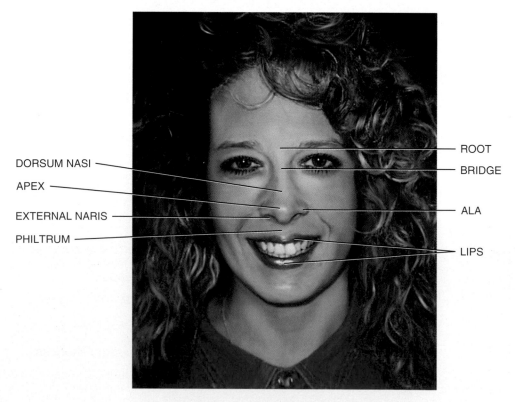

DORSUM NASI
APEX
EXTERNAL NARIS
PHILTRUM
ROOT
BRIDGE
ALA
LIPS

Anterior view

❓ **What is the name of the anterior border of the nose that connects the root and apex?**

EXHIBIT 12.2   Surface Anatomy of the Neck

◉ **OBJECTIVE**

• Describe the surface features of the neck.

The **neck** (*cervical*) is the region of the body that connects the head to the trunk and is divided into an *anterior cervical region*, two *lateral cervical regions*, and a *posterior cervical region (nucha)*.

The following are the major surface features of the neck, most of which are illustrated in Figure 12.6:

• **Thyroid cartilage (Adam's apple).** The largest of the cartilages that compose the larynx (voice box). It is the most prominent structure in the midline of the anterior cervical region. The common carotid artery *bifurcates* (divides) at the level of the superior border of the thyroid cartilage to form the internal and external carotid arteries.

• **Hyoid bone.** Located just superior to the thyroid cartilage. It is the first structure palpated in the midline inferior to the chin. It is easily palpated laterally as you move posterior from its midline body onto the greater cornu (see Figure 7.14 on page 192).

• **Cricoid cartilage.** A laryngeal cartilage located just inferior to the thyroid cartilage, it attaches the larynx to the trachea (windpipe). After you pass your finger over the cricoid cartilage moving inferiorly, your fingertip sinks in. The cricoid cartilage is used as a landmark for performing a tracheotomy.

• **Thyroid gland.** Two-lobed gland just inferior to the larynx with one lobe on either side.

• **Sternocleidomastoid muscle.** Forms the major portion of the lateral aspect of the neck. If you rotate your head to either side, you can palpate the muscle from its origin on the sternum and clavicle to its insertion on the mastoid process of the temporal bone.

• **Common carotid artery.** Lies just deep to the sternocleidomastoid muscle along its anterior border.

• **Internal jugular vein.** Located lateral to the common carotid artery.

• **Subclavian artery.** Located just lateral to the inferior portion of the sternocleidomastoid muscle. Pressure on this artery can stop bleeding in the upper limb because this vessel supplies blood to the entire limb.

• **External carotid artery.** Located superior to the larynx, just anterior to the sternocleidomastoid muscle, this artery is the site of the carotid (neck) pulse.

• **External jugular vein.** Located superficial to the sternocleidomastoid muscle, this vessel is readily seen if you are angry or if your collar is too tight.

• **Trapezius muscle.** Extends inferiorly and laterally from the base of the skull and occupies a portion of the lateral cervical region. A "stiff neck" is frequently associated with inflammation of this muscle.

• **Vertebral spines.** The spinous processes of the cervical vertebrae may be felt along the midline of the posterior region of the neck. Especially prominent at the base of the neck is the spinous process of the seventh cervical vertebra, called the vertebra prominens, which is followed by the prominent spinous process of the first thoracic vertebra (see Figure 12.7). The sternocleidomastoid muscle is not only a landmark for several arteries and veins, it is also the muscle that divides the neck into two major triangles: anterior and posterior (see Figure 11.9 on page 332). The triangles are important because of the structures that lie within their boundaries (see Exhibit 11.6 on pages 332–334).

◉ **CHECKPOINT**

• What is the neck?
• How would you palpate the hyoid bone, cricoid cartilage, thyroid gland, common carotid artery, external carotid artery, and external jugular vein?

Figure 12.6 Surface anatomy of the neck.

🔑 The anatomical subdivisions of the neck are the anterior cervical region, lateral cervical regions, and posterior cervical region.

THYROID GLAND —
TRAPEZIUS MUSCLE —
JUGULAR NOTCH —
CLAVICLE —

— HYOID BONE
— THYROID CARTILAGE
— EXTERNAL JUGULAR VEIN
— CRICOID CARTILAGE
— STERNOCLEIDOMASTOID MUSCLE
— SUBCLAVIAN ARTERY

❓ **Which muscle divides the neck into anterior and posterior triangles?**

EXHIBIT 12.2 **419**

EXHIBIT 12.3   Surface Anatomy of the Trunk

OBJECTIVE

- Describe the surface anatomy of the various features of the trunk—back, chest, abdomen, and pelvis.

The **trunk** is divided into the back, chest, abdomen, and pelvis. We will consider each region separately.

Surface Features of the Back

Among the prominent features of the **back**, or *dorsum*, are several superficial bones and muscles (Figure 12.7).

- **Vertebral spines.** The spinous processes of vertebrae, especially the thoracic and lumbar vertebrae, are quite prominent when the vertebral column is flexed.
- **Scapulae.** These easily identifiable surface landmarks on the back lie between ribs 2–7. In fact, it is also possible to palpate some ribs on the back. Depending on how lean a person is, it might be possible to palpate various parts of the scapula, such as the **vertebral border**, **axillary border**, **inferior angle**, **spine**, and **acromion**. The **spinous process of T3** is at about the same level as the spine of the scapula, and the **spinous process of T7** is approximately opposite the inferior angle of the scapula.
- **Latissimus dorsi muscle.** A broad, flat, triangular muscle of the lumbar region that extends superiorly to the axilla and, when well developed, gives a V-shape to the trunk of the body.
- **Erector spinae (sacrospinalis) muscle.** Located on either side of the vertebral column between the skull and iliac crests.
- **Infraspinatus muscle.** Located inferior to the spine of the scapula.
- **Trapezius muscle.** Extends from the cervical and thoracic vertebrae to the spine and acromion of the scapula and the lateral end of the clavicle. It also occupies a portion of the lateral region of the neck and forms the posterior border of the posterior triangle of the neck.
- **Teres major muscle.** Located inferior to the infraspinatus muscle; together with the latissimus dorsi muscle it forms the inferior border of the posterior axillary fold (posterior wall of the axilla or armpit region).
- **Posterior axillary fold.** Formed by the latissimus dorsi and teres major muscles, the posterior axillary fold can be palpated between the fingers and thumb at the posterior aspect of the axilla (armpit region); forms the posterior wall of the axilla.
- **Triangle of auscultation** (aw-skul-TĀ-shun; *ausculto-*=listening). A triangular region of the back just medial to the inferior part of the scapula, where the rib cage is not covered by superficial muscles. It is bounded by the latissimus dorsi and trapezius muscles and vertebral border of the scapula. The triangle of auscultation is a landmark of clinical significance because in this area respiratory sounds can be heard clearly through a stethoscope pressed against the skin. If a patient folds the arms across the chest and bends forward, the lung sounds can be heard clearly in the intercostal space between ribs 6 and 7.

Figure 12.7 Surface anatomy of the back.

The posterior boundary of the axilla, the posterior axillary fold, is formed mainly by the latissimus dorsi and teres major muscles.

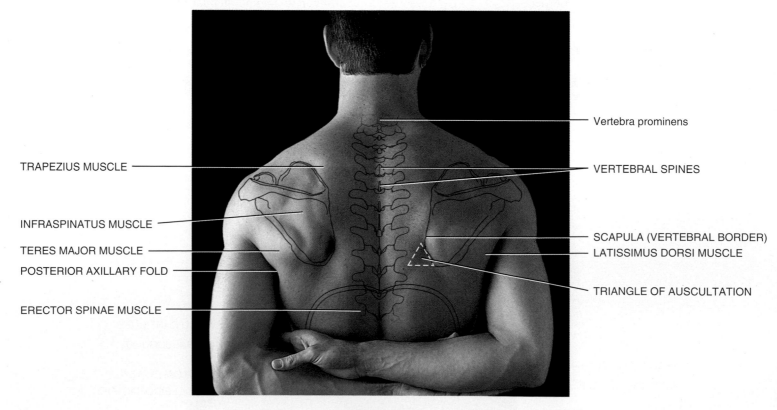

Posterior view

? What is the clinical significance of the triangle of auscultation?

Surface Features of the Chest

The surface anatomy of the **chest**, or *thorax*, includes several distinguishing surface landmarks (Figure 12.8), as well as some surface markings used to identify the location of the heart and lungs within the chest.

- **Clavicle.** Visible at the junction of the neck and thorax. Inferior to the clavicle, especially where it articulates with the manubrium (superior portion) of the sternum, the **first rib** can be palpated. In a depression superior to the medial end of the clavicle, just lateral to the sternocleidomastoid muscle, the **trunks of the brachial plexus** and axillary artery can be palpated (see Figure 18.9a on page 604).
- **Sternoclavicular joint.** Formed where the medial end of the clavicle articulates with the manubrium of the sternum. Move your upper limb around and feel the subtle movements that occur at this joint. Just posterior to these joints the internal jugular and subclavian veins unite to form the brachiocephalic veins.
- **Suprasternal (jugular) notch of sternum.** A depression on the superior border of the manubrium of the sternum between the medial ends of the clavicles; the **trachea** (windpipe) can be palpated posterior to the notch. A horizontal line through this landmark is at the level of the second thoracic vertebra.
- **Manubrium of sternum.** Superior portion of the sternum at the same levels as the bodies of the third and fourth thoracic vertebrae and anterior to the arch of the aorta.
- **Sternal angle of sternum.** Formed at the junction between the manubrium and body of the sternum, located about 4 cm (1.5 in.) inferior to the suprasternal notch. It is palpable deep to the skin and locates the costal cartilage of the second rib. It is the most reliable landmark of the chest and the starting

point from which ribs are counted. At or inferior to the sternal angle and slightly to the right, the trachea divides into right and left primary bronchi. It also represents the top of the heart, the branching of the pulmonary trunk, and the level where the azygous vein joins the superior vena cava. A horizontal line through this landmark is at the level of the fourth thoracic vertebra.

- **Body of sternum.** Midportion of the sternum anterior to the heart and the vertebral bodies of T5–T8.
- **Xiphoid process of sternum.** Inferior portion of the sternum, medial to the seventh costal cartilages. The joint between the xiphoid process and body of the sternum is called the **xiphisternal joint.** The heart lies on the diaphragm deep to this joint. A horizontal line through this joint is at the level of the ninth thoracic vertebra.
- **Costal margin.** Inferior edges of the costal cartilages of ribs 7 through 10. The first costal cartilage lies inferior to the medial end of the clavicle; the seventh costal cartilage is the most inferior costal cartilage to articulate directly with the sternum; the tenth costal cartilage forms the most inferior part of the costal margin when viewed anteriorly. At the superior end of the costal margin is the xiphisternal joint.
- **Serratus anterior muscle.** Inferior and lateral to the pectoralis major muscle.
- **Ribs.** Twelve pairs of ribs help to form the bony cage of the thoracic cavity. Depending on body leanness, they may or may not be visible. One site for listening to the heartbeat in adults is the left fifth intercostal space, just medial to the left midclavicular line. This point marks the apex of the heart.
- **Mammary glands.** Accessory organs of the female reproductive system located inside the breasts. These glands overlie the pectoralis major muscle (two-thirds) and serratus anterior muscle (one-third). After puberty, the mammary glands en-

Figure 12.8 Surface anatomy of the chest.

🔒🗝 Thoracic viscera are protected by the sternum, ribs, and thoracic vertebrae, which form the skeleton of the thorax.

STERNOCLAVICULAR JOINT

MANUBRIUM OF STERNUM

CORACOID PROCESS

STERNAL ANGLE

PECTORALIS MAJOR MUSCLE

ANTERIOR AXILLARY FOLD

NIPPLE

SERRATUS ANTERIOR MUSCLE

SUPRASTERNAL NOTCH

CLAVICLE

BODY OF STERNUM

XIPHISTERNAL JOINT

XIPHOID PROCESS OF STERNUM

RIB

COSTAL MARGIN

Anterior view of surface features of the chest

❓ What is the most reliable landmark of the chest?

EXHIBIT 12.3 Surface Anatomy of the Trunk CONTINUED

large to their hemispherical shape, and in young adult females they extend from the second through the sixth ribs and from the lateral margin of the sternum to the midaxillary line. The **midaxillary line** is a vertical line that extends downward on the lateral thoracic wall from the center of the axilla.

- **Nipples.** Superficial to the fourth intercostal space or fifth rib, about 10 cm (4 in.) from the midline in males and most females. The position of the nipples in females is variable depending on the size and pendulousness of the breasts. The right dome of the diaphragm is just inferior to the right nipple, the left dome is about 2–3 cm (1 in.) inferior to the left nipple, and the central tendon is at the level of the xiphisternal joint.
- **Anterior axillary fold.** Formed by the lateral border of the pectoralis major muscle; can be palpated between the fingers and thumb; forms the anterior wall of the axilla (armpit region).
- **Pectoralis major muscle.** Principal upper chest muscle; in the male, the inferior border of the muscle forms a curved line leading to the anterior wall of the axilla and serves as a guide to the fifth rib. In the female, the inferior border is mostly covered by the breast.

Although structures within the chest are almost totally hidden by the sternum, ribs, and thoracic vertebrae, it is important to indicate the surface markings of the heart and lungs. The projection (shape of an organ on the surface of the body) of the heart on the anterior surface of the chest is indicated by four points as follows (see Figure 14.1c on page 454). The *inferior left point* is the apex (inferior, pointed end) of the heart, which projects downward and to the left and can be palpated in the fifth intercostal space, about 9 cm (3.5 in) to the left of the midline, or in the fifth intercostal space at the midclavicular line (a vertical line extended down from the middle of the clavicle). The *inferior right point* is at the lower border of the costal cartilage of the right sixth rib, about 3 cm to the right of the midline (one finger breadth to the right of the lateral edge of the sternum). The *superior right point* is located at the superior border of the costal cartilage of the right

third rib, about 3 cm to the right of the midline (one finger breadth to the right of the lateral edge of the sternum). The *superior left point* is located at the inferior border of the costal cartilage of the left second rib, about 3cm to the left of the midline (one finger breadth to the left of the lateral edge of the sternum). If you connect the four points, you can determine the heart's location and size (about the size of a closed fist).

The lungs almost totally fill the thorax. The *apex* (superior, pointed end) of the lungs lies just superior to the medial third of the clavicles and is the only area that can be palpated (see Figure 24.9a on page 768). The anterior, lateral, and posterior surfaces of the lungs lie against the ribs. The *base* (inferior, broad end) of the lungs is concave and fits over the convex area of the diaphragm. It extends from the sixth costal cartilage anteriorly to the spinous process of the tenth thoracic vertebra posteriorly. (The base of the right lung is usually slightly higher than that of the left due to the position of the liver below it.) The covering around the lungs is called the *pleura*. At the base of each lung, the pleura extends about 5 cm below the base from the sixth costal cartilage anteriorly to the twelfth rib posteriorly. Thus, the lungs do not completely fill the pleural cavity in this area. A needle is typically inserted into this space in the pleural cavity to withdraw fluid from the chest.

Surface Features of the Abdomen and Pelvis

Following are some of the prominent surface anatomy features of the **abdomen** and **pelvis** (Figure 12.9):

- **Umbilicus.** Also called the *navel*; it marks the site of attachment of the umbilical cord to the fetus. It is level with the intervertebral disc between the bodies of vertebrae L3 and L4. The **abdominal aorta**, which branches into the right and left common iliac arteries anterior to the body of vertebra L4, can be palpated with deep pressure through the upper part of the anterior abdominal wall just to the left of the midline. The **in-**

Figure 12.9 Surface anatomy of the abdomen and pelvis.

 The linea alba is a frequent site for an abdominal incision because cutting through it severs no muscles and few blood vessels and nerves.

RECTUS ABDOMINIS MUSCLE

EXTERNAL OBLIQUE MUSCLE

Iliac crest

ANTERIOR SUPERIOR ILIAC SPINE

Pectoralis major muscle

Serratus anterior muscle

LINEA ALBA

TENDINOUS INTERSECTION

LINEA SEMILUNARIS

UMBILICUS

McBURNEY'S POINT

(a) Anterior view of abdomen

ferior vena cava lies to the right of the abdominal aorta and is wider; it arises anterior to the body of vertebra L5.

- **External oblique muscle.** Located inferior to the serratus anterior muscle. The aponeurosis of the muscle on its inferior border is the **inguinal ligament**, a structure along which hernias frequently occur.
- **Rectus abdominis muscles.** Located just lateral to the midline of the abdomen. They can be seen by raising the shoulders off the ground while in the supine position without using the arms (doing an abdominal crunch exercise).
- **Linea alba.** Flat, tendinous *raphe* (intersection of muscle tendons) forming a furrow along the midline between the rectus abdominis muscles. The furrow extends from the xiphoid process to the pubic symphysis. It is broad superior to the umbilicus and narrow inferior to it. The linea alba is a frequently selected site for abdominal surgery because an incision through it severs no muscles and only a few blood vessels and nerves.
- **Tendinous intersections of rectus abdominis muscle.** Fibrous bands that run transversely or obliquely across the rectus abdominis muscles. One intersection is at the level of the umbilicus, one is at the level of the xiphoid process, and one is midway between the two. The rectus abdominis muscles thus contain four prominent bulges and three tendinous intersections.
- **Linea semilunaris.** The lateral edge of the rectus abdominis muscle can be seen as a curved line that extends from the costal margin at the top of the ninth costal cartilage to the pubic tubercle.
- **McBurney's point.** A clinically important site (see next section) located two-thirds of the way down an imaginary line drawn between the umbilicus and anterior superior iliac spine. McBurney's point is an important landmark related to the appendix. Pressure of the finger on McBurney's point produces pain in acute **appendicitis**, inflammation of the appendix, aiding in diagnosis.
- **Iliac crest.** Superior margin of the ilium of the hip bone. It forms the outline of the superior border of the buttock. When you rest your hands on your hips, they rest on the iliac crests.

(b) Anterior view of pelvis

(c) Posterior view of pelvis

? **What is the clinical significance of McBurney's point?**

EXHIBIT **12.3** CONTINUES ▶

EXHIBIT 12.3 **423**

EXHIBIT 12.3   Surface Anatomy of the Trunk CONTINUED

A horizontal line drawn across the highest point of each iliac crest is called the **supracristal line**, which intersects the spinous process of the fourth lumbar vertebra. This vertebra is a landmark for performing a spinal tap (see page 589).

- **Anterior superior iliac spine.** The anterior end of the iliac crest that lies at the upper lateral end of the fold of the groin.
- **Posterior superior iliac spine.** The posterior end of the iliac crest, indicated by a dimple in the skin that coincides with the middle of the sacroiliac joint, where the hip bone attaches to the sacrum.
- **Pubic tubercle.** Projection on the superior border of the pubis of the hip bone. Attached to it is the medial end of the **inguinal ligament**, the inferior free edge of the aponeurosis of the external oblique muscle that forms the **inguinal canal**. The lateral end of the ligament is attached to the anterior superior iliac spine. The spermatic cord in males and the round ligament of the uterus in females pass through the inguinal canal.
- **Pubic symphysis.** Anterior joint of the hip bones; palpated as a firm resistance in the midline at the inferior portion of the anterior abdominal wall.
- **Mons pubis.** An elevation of adipose tissue covered by skin and pubic hair that is anterior to the pubic symphysis.
- **Sacrum.** The fused spinous processes of the sacrum, called the **median sacral crest**, can be palpated beneath the skin superior to the **gluteal cleft**, a depression along the midline that separates the buttocks (and is part of the discussion on the buttocks on page 430).
- **Coccyx.** The inferior surface of the tip of the coccyx can be palpated in the gluteal cleft, about 2.5 cm (1 in.) posterior to the anus.

Chapter 1 discussed the division of the abdomen and pelvis into nine regions (see Figure 1.8 on page 18). The nine-region designation subdivided the abdomen and pelvis by drawing two vertical lines (left and right clavicular), an upper horizontal line (subcostal), and a lower horizontal line (transtubercular). Take time to examine Figure 1.8 carefully, so that you can see which organs or parts of organs lie within each region.

In Chapter 11, as part of your study of skeletal muscles, you were introduced to the **perineum**, the diamond-shaped region medial to the thigh and buttocks. It is bounded by the pubic symphysis anteriorly, the ischial tuberosities laterally, and the coccyx posteriorly. A transverse line drawn between the ischial tuberosities divides the perineum into an anterior *urogenital triangle* that contains the external genitals, and a posterior *anal triangle* that contains the anus. Details concerning the functions of the perineum are provided in Chapter 11. At this point keep in mind that the musculature of the perineum constitutes the floor of the pelvic cavity.

◉ **CHECKPOINT**

- What are the divisions of the trunk?
- What is the triangle of auscultation?
- What is the significance of the xiphisternal joint as a landmark?
- Why is the midaxillary line an important anatomical landmark?
- How would you indicate the surface markings of the heart?
- Describe the position of the lungs in the thorax.
- Why do the lungs not completely fill the pleural cavity? Why is this important clinically?
- Why are the linea alba, linea semilunaris, McBurney's point, and supracristal line important landmarks?
- What is the inguinal ligament, and why is it important?

EXHIBIT 12.4 Surface Anatomy of the Upper Limb (Extremity)

OBJECTIVE

- Describe the surface anatomy of the various regions of the upper limb.

The **upper limb** consists of the pectoral (shoulder) girdle and free upper limb.

Surface Features of the Shoulder

The **shoulder**, or *acromial region*, is located at the lateral aspect of the clavicle, where the clavicle joins the scapula and the scapula joins the humerus. The region presents several conspicuous surface features (Figure 12.10).

- **Acromioclavicular joint.** A slight elevation at the lateral end of the clavicle. It is the joint between the acromion of the scapula and the clavicle.
- **Acromion.** The expanded lateral end of the spine of the scapula that forms the top of the shoulder. It can be palpated about 2.5 cm (1 in.) distal to the acromioclavicular joint.
- **Humerus.** The **greater tubercle** of the humerus may be palpated on the superior aspect of the shoulder. It is the most laterally palpable bony structure.
- **Deltoid muscle.** Triangular muscle that forms the rounded prominence of the shoulder. The *deltoid muscle* is a frequent site for an intramuscular injection. To avoid injury to major blood vessels and nerves, the injection is given in the midportion of the muscle about 2–3 finger-widths inferior to the acromion of the scapula and lateral to the axilla.

Figure 12.10 Surface anatomy of the shoulder.

The deltoid muscle gives the shoulder its rounded prominence.

- **Coracoid process.** Anterior projection of the scapula that can be palpated at the medial border of the deltoid muscle just inferior to the clavicle.

Surface Features of the Armpit

The armpit region, or **axilla**, is a pyramid-shaped area at the junction of the arm and the chest that enables blood vessels and nerves to pass between the neck and the free upper limbs (Figure 12.11a, b).

- **Apex.** The apex of the axilla is surrounded by the clavicle, scapula, and first rib.
- **Base.** The base of the axilla is formed by the concave skin and fascia that extends from the arm to the chest wall. It contains hair, and deep to the base the axillary lymph nodes can be palpated.
- **Anterior wall.** The anterior wall of the axilla is composed mainly of the pectoralis major muscle (anterior axillary fold; see also Figure 12.8).
- **Posterior wall.** The posterior wall of the axilla is formed mainly by the teres major and latissimus dorsi muscles (posterior axillary fold; see also Figure 12.7).
- **Medial wall.** The medial wall of the axilla is formed by ribs 1–4 and their corresponding intercostal muscles, plus the overlying serratus anterior muscle.
- **Lateral wall.** Finally, the lateral wall of the axilla is formed by the coracobrachialis and biceps brachii muscles and the superior portion of the shaft of the humerus. Passing through the axilla are the axillary artery and vein, branches of the brachial plexus, and axillary lymph nodes. All these structures are surrounded by a considerable amount of axillary fat.

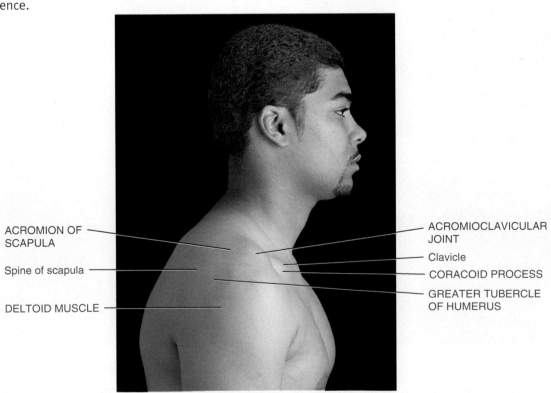

ACROMION OF SCAPULA

Spine of scapula

DELTOID MUSCLE

ACROMIOCLAVICULAR JOINT

Clavicle

CORACOID PROCESS

GREATER TUBERCLE OF HUMERUS

Right Lateral View

? **Which structure forms the top of the shoulder?**

EXHIBIT 12.4 CONTINUES ▶

EXHIBIT 12.4 **425**

EXHIBIT 12.4   Surface Anatomy of the Upper Limb (Extremity) CONTINUED

Surface Features of the Arm and Elbow

The **arm**, or *brachium*, is the region between the shoulder and elbow. The **elbow**, or *cubitus*, is the region where the arm and forearm join. The arm and elbow present several surface anatomy features (Figure 12.11b–d).

- **Humerus.** This arm bone may be palpated along its entire length, especially near the elbow (see descriptions of the medial and lateral epicondyles that follow).

- **Biceps brachii muscle.** Forms the bulk of the anterior surface of the arm. On the medial side of the muscle is a groove that contains the **brachial artery**.
- **Triceps brachii muscle.** Forms the bulk of the posterior surface of the arm.
- **Medial epicondyle.** Medial projection of the humerus near the elbow.
- **Lateral epicondyle.** Lateral projection of the humerus near the elbow.

Figure 12.11 Surface anatomy of the axilla, arm, and elbow. The location of the muscles that form the walls of the axilla are shown in Figure 11.17a on page 358.

🔑 The biceps brachii and triceps brachii muscles form the bulk of the musculature of the arm.

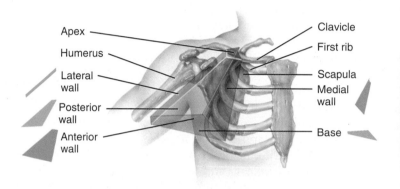

(a) Location and parts of the axilla

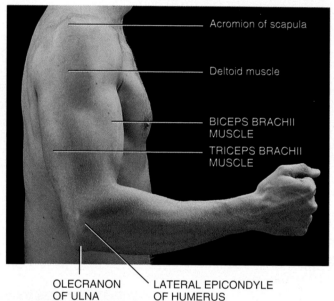

(c) Right lateral view of arm

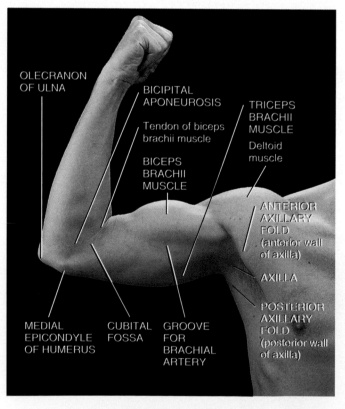

(b) Medial view of arm

(d) Anterior view of cubital fossa

❓ Which blood vessel in the cubital fossa is frequently used to withdraw blood?

- **Olecranon.** Projection of the proximal end of the ulna between and slightly superior to the epicondyles when the forearm is extended; it forms the elbow.
- **Ulnar nerve.** Can be palpated in a groove posterior to the medial epicondyle. The "funny bone" is the region where the ulnar nerve rests against the medial epicondyle. Hitting the nerve at this point produces a sharp pain along the medial side of the forearm that almost everyone would agree isn't very funny.
- **Cubital fossa.** Triangular space in the anterior region of the elbow bounded proximally by an imaginary line between the humeral epicondyles, laterally by the medial border of the brachioradialis muscle, and medially by the lateral border of the pronator teres muscle; contains the tendon of the biceps brachii muscle, the median cubital vein, brachial artery and its terminal branches (radial and ulnar arteries), and parts of the median and radial nerves.
- **Median cubital vein.** Crosses the cubital fossa obliquely and connects the lateral cephalic with the medial basilic veins of the arm. The median cubital vein is frequently used to withdraw blood from a vein for diagnostic purposes or to introduce substances into blood, such as medications, contrast media for radiographic procedures, nutrients, and blood cells and/or plasma for transfusion.
- **Brachial artery.** Continuation of the axillary artery that passes posterior to the coracobrachialis muscle and then medial to the biceps brachii muscle. It enters the middle of the cubital fossa and passes deep to the bicipital aponeurosis, which separates it from the median cubital vein. **Blood pressure** is usually measured in the brachial artery, when the cuff of a *sphygmomanometer* (blood pressure instrument) is wrapped around the arm and a stethoscope is placed over the brachial artery in the cubital fossa. Pulse can also be detected in the artery in the cubital fossa. However, blood pressure can be measured at any artery where you can obstruct blood flow. This becomes important in case the brachial artery cannot be utilized. In such situations, the radial or popliteal arteries might be used to obtain a blood pressure reading.
- **Bicipital aponeurosis.** An aponeurosis that inserts the biceps brachii muscle into the deep fascia in the medial aspect of the forearm. It can be felt when the muscle contracts.

Surface Features of the Forearm and Wrist

The **forearm**, or *antebrachium*, is the region between the elbow and wrist. The **wrist**, or carpus, is between the forearm and palm. Following are some prominent surface anatomy features of the forearm and wrist (Figure 12.12).

- **Ulna.** The medial bone of the forearm. It can be palpated along its entire length from the olecranon (see Figure 12.11a, b) to the **styloid process**, a projection on the distal end of the bone at the medial (little finger) side of the wrist. The **head of the ulna** is a conspicuous enlargement just proximal to the styloid process.
- **Radius.** The distal half of the radius can be palpated proximal to the thumb side of the hand. The proximal half is covered by muscles. The **styloid process** of the radius is a projection on the distal end of the bone at the lateral (thumb) side of the wrist.

- **Muscles.** Because of their close proximity it is difficult to identify individual muscles of the forearm. Instead, it is much easier to identify tendons as they approach the wrist and then trace them proximally to the following muscles:
  - **Brachioradialis muscle.** Located at the superior and lateral aspect of the forearm.
  - **Flexor carpi radialis muscle.** The tendon of this muscle is on the lateral side of the forearm about 1 cm medial to the styloid process of the radius.
  - **Palmaris longus muscle.** The tendon of this muscle is medial to the flexor carpi radialis tendon and can be seen quite prominently if the wrist is slightly flexed and the base of the thumb and little finger are drawn together. About 15 to 20% of individuals do not have this muscle in at least one arm.
  - **Flexor digitorum superficialis muscle.** The tendon of this muscle is medial to the palmaris longus tendon and can be palpated by flexing the fingers at the metacarpophalangeal and proximal interphalangeal joints.
  - **Flexor carpi ulnaris muscle.** The tendon of this muscle is on the medial aspect of the forearm.
- **Radial artery.** Located on the lateral aspect of the wrist between the flexor carpi radialis tendon and styloid process of the radius. It is frequently used to take a pulse.
- **Pisiform bone.** Medial bone of the proximal row of carpals that can be palpated as a projection distal and anterior to the styloid process of the ulna.
- **"Anatomical snuffbox."** A triangular depression between tendons of extensor pollicis brevis and extensor pollicis longus muscles. It derives its name from a habit of previous centuries in which a person would take a pinch of snuff (powdered tobacco or scented powder) and place it in the depression before sniffing it into the nose. The styloid process of the radius, the base of the first metacarpal, trapezium, scaphoid, and deep branch of the radial artery can all be palpated in the depression.
- **Wrist creases.** Three more or less constant lines on the anterior aspect of the wrist (named *proximal, middle,* and *distal*) where the skin is firmly attached to underlying deep fascia.

Surface Features of the Hand

The **hand**, or *manus*, is the region from the wrist to the termination of the upper limb; it has several conspicuous surface features (Figure 12.13 on page 429).

- **Knuckles.** Commonly refers to the dorsal aspect of the heads of metacarpals II–V (or 2–5), but also includes the dorsal aspects of the metacarpophalangeal and interphalangeal joints.
- **Dorsal venous network of the hand (dorsal venous arch).** Superficial veins on the dorsum of the hand that drain blood into the cephalic vein. It can be displayed by compressing the blood vessels at the wrist for a few moments as the hand is opened and closed.
- **Tendon of extensor digiti minimi muscle.** This can be seen on the dorsum of the hand in line with the phalanx of the little finger.
- **Tendons of extensor digitorum muscle.** These can be seen on the dorsum of the hand in line with the phalanges of the ring, middle, and index fingers.

EXHIBIT **12.4** CONTINUES ▶

EXHIBIT 12.4 **427**

EXHIBIT 12.4   Surface Anatomy of the Upper Limb (Extremity) CONTINUED

Figure 12.12 Surface anatomy of the forearm and wrist.

🔑 Muscles of the forearm are most easily identified by locating their tendons near the wrist and tracing them proximally.

HEAD OF ULNA

STYLOID PROCESS
OF ULNA

Tributary to cephalic vein

STYLOID PROCESS OF
RADIUS

"ANATOMICAL SNUFFBOX"

TENDON OF EXTENSOR
POLLICIS BREVIS MUSCLE

TENDON OF EXTENSOR
POLLICIS LONGUS MUSCLE

(a) Dorsum of wrist

BRACHIORADIALIS
MUSCLE

TENDON OF FLEXOR CARPI
RADIALIS MUSCLE

TENDON OF PALMARIS
LONGUS MUSCLE

TENDON OF FLEXOR CARPI
ULNARIS MUSCLE

TENDON OF FLEXOR
DIGITORUM SUPERFICIALIS
MUSCLE

(b) Anterior aspect of forearm and wrist

TENDON OF PALMARIS
LONGUS MUSCLE

TENDON OF FLEXOR
CARPI RADIALIS MUSCLE

RADIAL ARTERY

WRIST CREASE

Thenar eminence

TENDON OF FLEXOR
DIGITORUM SUPERFICIALIS
MUSCLE

TENDON OF FLEXOR CARPI
ULNARIS MUSCLE

PISIFORM BONE

Hypothenar eminence

(c) Anterior aspect of wrist

❓ Which blood vessel is frequently used to take a pulse?

Figure 12.13 Surface anatomy of the hand.

🔑 Several tendons on the dorsum of the hand can be identified by their alignment with the phalanges of the digits.

KNUCKLES (proximal interphalangeal joints)

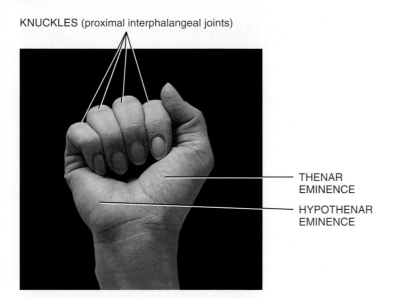

THENAR EMINENCE

HYPOTHENAR EMINENCE

(a) Palmar and dorsal view

KNUCKLES (metacarpal-phalangeal joints)

(b) Dorsal view

DIGITAL FLEXION CREASES

PALMAR FLEXION CREASES

THENAR EMINENCE

HYPOTHENAR EMINENCE

(c) Palmar view

DORSAL VENOUS NETWORK OF THE HAND (DORSAL VENOUS ARCH)

TENDON OF EXTENSOR DIGITI MINIMI MUSCLE

TENDONS OF EXTENSOR DIGITORUM MUSCLE

(d) Dorsum

❓ **Muscles that form the thenar eminence move which digit?**

- **Tendon of the extensor pollicis brevis muscle.** This tendon (described previously) is in line with the phalanx of the thumb (see Figure 12.12a).
- **Thenar eminence.** Larger, rounded contour on the lateral aspect of the palm formed by muscles that move the thumb. Also called the ball of the thumb.
- **Hypothenar eminence.** Smaller, rounded contour on the medial aspect of the palm formed by muscles that move the little finger. Also called the ball of the little finger.
- **Palmar flexion creases.** Skin creases on the palm.
- **Digital flexion creases.** Skin creases on the anterior surface of the fingers.

◉ CHECKPOINT
- What are the major divisions of the free upper limb?
- What is the clinical significance of the deltoid muscle?
- In which arteries of the upper limb can a pulse be detected?
- What is the "funny bone"?
- What is the clinical importance of the median cubital vein?
- What artery is normally used to measure blood pressure?
- Explain the easiest way to identify the muscles of the forearm.
- What are the knuckles?
- Why are the thenar and hypothenar eminences important?

EXHIBIT 12.4 **429**

EXHIBIT 12.5 Surface Anatomy of the Lower Limb (Extremity)

■ OBJECTIVE

- Describe the surface features of the various regions of the lower limb.

The **lower limb** consists of the pelvic (hip) girdle and free lower limb.

Surface Features of the Buttock

The **buttock**, or *gluteal region*, is formed mainly by the gluteus maximus muscle. The outline of the superior border of the buttock is formed by the iliac crests (see Figure 12.9c). Following are some surface features of the buttock (Figure 12.14).

- **Gluteus maximus muscle.** Forms the major portion of the prominence of the buttocks. The sciatic nerve is deep to this muscle.
- **Gluteus medius muscle.** Superior and lateral to the gluteus maximus muscle. Another common site for an intramuscular injection is the gluteus medius muscle. In order to give this injection, the buttock is divided into quadrants and the upper outer quadrant is used as an injection site. The iliac crest serves as the landmark for this quadrant. This site is chosen because the gluteus medius muscle in this area is quite thick, and there is less chance of injury to the sciatic nerve or major blood vessels.
- **Gluteal (natal) cleft.** Depression along the midline that separates the left and right buttocks.
- **Gluteal fold.** Inferior limit of the buttock that roughly corresponds to the inferior margin of the gluteus maximus muscle.
- **Ischial tuberosity.** Just superior to the medial side of the gluteal fold, the tuberosity bears the weight of the body when seated.

- **Greater trochanter.** A projection of the proximal end of the femur on the lateral side of the thigh. It is about 20 cm (8 in.) inferior to the highest point of the iliac crest.

Surface Features of the Thigh and Knee

The **thigh**, or *femoral region*, is the region from the hip to the knee. The **knee**, or *genu*, is the region where the thigh and leg join. Several muscles are clearly visible in the thigh. Following are several surface features of the thigh and knee (Figure 12.15).

- **Sartorius muscle.** Superficial anterior muscle that can be traced from the lateral aspect of the thigh to the medial aspect of the knee.
- **Quadriceps femoris muscle.** Three of the four components of the muscle can be seen: **rectus femoris** at the midpoint of the anterior aspect of the thigh; **vastus medialis** at the anteromedial aspect of the thigh; and **vastus lateralis** at the anterolateral aspect of the thigh. The fourth component, the vastus intermedius, is deep to the rectus femoris (see Figure 11.22a on page 388). The vastus lateralis muscle of the quadriceps femoris group is another site that may be used for an intramuscular injection. This site is located at a point midway between the greater trochanter (see Figure 12.14) of the femur and the patella (knee cap). Injection in this area reduces the chance of injury to major blood vessels and nerves.
- **Adductor longus muscle.** Located at the superior aspect of the medial thigh. It is the most anterior of the three adductor muscles (adductor magnus, adductor brevis, and adductor longus; see Figure 11.22b on page 389).
- **Femoral triangle.** A space at the proximal end of the thigh formed by the inguinal ligament superiorly, the sartorius

Figure 12.14 Surface anatomy of the buttock.

⚷ The gluteus medius muscle is a frequent site for an intramuscular injection.

Iliac crest

GLUTEUS MEDIUS MUSCLE

GLUTEUS MAXIMUS MUSCLE

GLUTEAL (NATAL) CLEFT

GREATER TROCHANTER

ISCHIAL TUBEROSITY

GLUTEAL FOLD

Posterior view

? Which muscle forms the bulk of the buttock?

Figure 12.15 Surface anatomy of the thigh and knee.

The quadriceps femoris and hamstrings form the bulk of the musculature of the thigh.

Femoral triangle

SARTORIUS MUSCLE

ADDUCTOR LONGUS MUSCLE

RECTUS FEMORIS MUSCLE

Gracilis muscle

VASTUS LATERALIS MUSCLE

VASTUS MEDIALIS MUSCLE

(a) Anterior view of thigh

SEMITENDINOSUS AND SEMIMEMBRANOSUS MUSCLES

VASTUS LATERALIS MUSCLE

BICEPS FEMORIS MUSCLE

POPLITEAL FOSSA

Tendon of semitendinosus muscle

Gastrocnemius muscle (medial and lateral heads)

(b) Posterior view of popliteal fossa

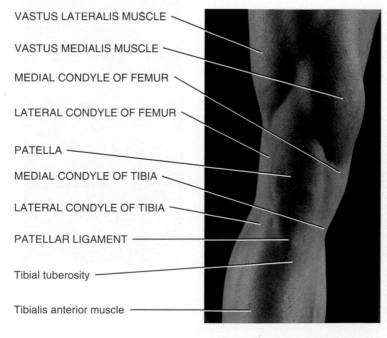

VASTUS LATERALIS MUSCLE

VASTUS MEDIALIS MUSCLE

MEDIAL CONDYLE OF FEMUR

LATERAL CONDYLE OF FEMUR

PATELLA

MEDIAL CONDYLE OF TIBIA

LATERAL CONDYLE OF TIBIA

PATELLAR LIGAMENT

Tibial tuberosity

Tibialis anterior muscle

(c) Anterior view of knee

? Which muscles form the borders of the popliteal fossa?

■ EXHIBIT **12.5** CONTINUES ▶

EXHIBIT 12.5 **431**

EXHIBIT 12.5 Surface Anatomy of the Lower Limb (Extremity) CONTINUED

Figure 12.16 Surface anatomy of the leg, ankle, and foot.

The calcaneal tendon is a common tendon for the gastrocnemius and soleus (calf muscles) and inserts into the calcaneus (heel bone).

Vastus lateralis muscle

Rectus femoris muscle

Vastus medialis muscle

Patella

Patellar ligament

FIBULARIS LONGUS MUSCLE

TIBIAL TUBEROSITY

GASTROCNEMIUS MUSCLE

ANTERIOR BORDER OF TIBIA (SHIN)

SOLEUS MUSCLE

TIBIALIS ANTERIOR MUSCLE

GREAT SAPHENOUS VEIN

Medial malleolus of tibia

Lateral malleolus of fibula

(a) Anterior view of leg, ankle, and foot

Popliteal fossa

GASTROCNEMIUS MUSCLE (MEDIAL AND LATERAL HEADS)

SOLEUS MUSCLE

CALCANEAL (ACHILLES) TENDON

MEDIAL MALLEOLUS OF TIBIA

LATERAL MALLEOLUS OF FIBULA

CALCANEUS

(b) Posterior view of leg and ankle

TENDONS OF EXTENSOR DIGITORUM LONGUS MUSCLE

LATERAL MALLEOLUS OF FIBULA

MEDIAL MALLEOLUS OF TIBIA

Tendon of tibialis anterior muscle

CALCANEUS

TENDON OF EXTENSOR HALLUCIS LONGUS MUSCLE

(c) Dorsum of foot

MEDIAL MALLEOLUS OF TIBIA

LATERAL MALLEOLUS OF FIBULA

DORSAL VENOUS ARCH

TENDONS OF EXTENSOR DIGITORUM LONGUS MUSCLE

TENDON OF EXTENSOR HALLUCIS LONGUS MUSCLE

(d) Dorsum of foot

? Which artery is just lateral to the tendon of the extensor digitorum muscle and is a site where a pulse may be detected?

muscle laterally, and the adductor longus muscle medially. The triangle contains, from lateral to medial, the femoral nerve, artery, vein, and deep inguinal lymph nodes. The triangle is an important arterial pressure point in cases of severe hemorrhage of the lower limb. Hernias frequently occur in this area as they follow the vessels through the abdominal wall into the thigh.

- **Hamstring muscles.** Superficial, posterior thigh muscles located below the gluteal folds. They are the **biceps femoris**, which lies more laterally as it passes inferiorly to the knee, and the **semitendinosus** and **semimembranosus**, which lie medially as they pass inferiorly to the knee. The tendons of the hamstring muscles can be palpated laterally and medially on the posterior aspect of the knee.

The knee presents several distinguishing surface features.

- **Patella.** Also called the *kneecap*, this large sesamoid bone is located within the tendon of the quadriceps femoris muscle on the anterior surface of the knee along the midline.
- **Patellar ligament.** Continuation of the quadriceps femoris tendon inferior to the patella. Infrapatellar fat pads cushion it on both sides.
- **Medial condyle of femur.** Medial projection on the distal end of the femur.
- **Medial condyle of tibia.** Medial projection on the proximal end of the tibia.
- **Lateral condyle of femur.** Lateral projection on the distal end of the femur.
- **Lateral condyle of tibia.** Lateral projection on the proximal end of the tibia. All four condyles can be palpated just inferior to the patella on either side of the patellar ligament.
- **Popliteal fossa** (pop-LIT-ē-al). A diamond-shaped area on the posterior aspect of the knee that is clearly visible when the knee is flexed. The fossa is bordered superolaterally by the biceps femoris muscle, superomedially by the semimembranosus and semitendinosus muscles, and inferolaterally and inferomedially by the lateral and medial heads of the gastrocnemius muscle, respectively. The **head of the fibula** can be easily palpated on the lateral side of the popliteal fossa. The fossa also contains the popliteal artery and vein. It is sometimes possible to detect a pulse in the popliteal artery.

Surface Features of the Leg, Ankle, and Foot

The **leg**, or *crus*, is the region between the knee and ankle. The **ankle**, or *tarsus*, is between the leg and foot. The **foot** is the region from the ankle to the termination of the free lower limb. Following are several surface anatomy features of the leg, ankle, and foot (Figure 12.16 on page 432).

- **Tibial tuberosity.** Bony prominence on the superior, anterior surface of the tibia into which the patellar ligament inserts.

- **Tibialis anterior muscle.** Lies against the lateral surface of the tibia, where it is easy to palpate, particularly when the foot is dorsiflexed. The distal tendon of the muscle can be traced to its insertion on the medial cuneiform and base of the first metatarsal.
- **Tibia.** The medial surface and anterior border (shin) of the tibia are subcutaneous and can be palpated throughout the length of the bone.
- **Fibularis (peroneus) longus muscle.** A superficial lateral muscle that overlies the fibula.
- **Gastrocnemius muscle.** Forms the bulk of the midportion and superior portion of the posterior aspect of the leg. The medial and lateral heads can be seen clearly in a person standing on tiptoe.
- **Soleus muscle.** Located deep to the gastrocnemius muscle; it and the gastrocnemius together are referred to as the calf muscles.
- **Calcaneal (Achilles) tendon.** Prominent tendon of the gastrocnemius and soleus muscles on the posterior aspect of the ankle; inserts into the **calcaneus** (heel) bone of the foot.
- **Lateral malleolus of fibula.** Projection of the distal end of the fibula that forms the lateral prominence of the ankle. The head of the fibula, at the proximal end of the bone, lies at the same level as the tibial tuberosity.
- **Medial malleolus of tibia.** Projection of the distal end of the tibia that forms the medial prominence of the ankle.
- **Dorsal venous arch.** Superficial veins on the dorsum of the foot that unite to form the small and great saphenous veins. The great saphenous vein is the longest vein of the body, extending from the foot to the groin, where it joins the femoral vein in the femoral triangle.
- **Tendons of extensor digitorum longus muscle.** Visible in line with phalanges II–V (2–5).
- **Tendon of extensor hallucis longus muscle.** Visible in line with phalanx I (great toe). Pulsations in the dorsalis pedis artery may be felt in most people just lateral to this tendon where the blood vessel passes over the navicular and cuneiform bones of the tarsus.

■ **CHECKPOINT**
- What are the divisions of the free lower limb?
- Why is the greater trochanter of the femur important?
- What are the major anterior and posterior muscles of the thigh?
- Define the femoral triangle. Why is it important?
- What are the borders of the popliteal fossa? What structures pass through it?
- Define the buttock. What is the gluteal cleft? Gluteal fold?

Now that you have studied some of the surface features of the principal regions of the body, you can appreciate their importance in locating the positions of many internal structures, especially bones, joints, and muscles. In later chapters you will learn to relate surface features to various blood vessels, lymph nodes, nerves, and organs in the chest, abdomen, and pelvis. Once you have visualized and palpated as many of these structures as possible on your own body, you will gain a better appreciation of the application of surface anatomy to your anatomical and clinical studies.

EXHIBIT 12.5 **433**

Overview of Surface Anatomy (p. 414)

1. Surface anatomy is the study of anatomical landmarks on the exterior of the body.
2. Surface features may be noted by visualization and/or palpation.
3. Knowledge of surface anatomy has many anatomical and clinical applications.
4. The regions of the body include the head, neck, trunk, upper limbs, and lower limbs.

Surface Anatomy of the Head (p. 414)

1. The head is divided into a cranium and face, each of which is divided into distinct regions; surface anatomy of the head is discussed in Exhibit 12.1.
2. Many muscles of facial expression and skull bones are easily palpated.
3. The eyes, ears, and nose present numerous readily identifiable surface features.

Surface Anatomy of the Neck (p. 419)

1. The neck connects the head to the trunk; surface anatomy of the neck is discussed in Exhibit 12.2.
2. The neck contains several important arteries and veins.
3. The neck is divisible into several triangles by specific muscles and bones.

Surface Anatomy of the Trunk (p. 420)

1. The trunk is divided into the back, chest, abdomen, and pelvis; surface anatomy of the trunk is discussed in Exhibit 12.3.
2. The back presents several muscles and superficial bones that serve as landmarks.
3. The triangle of auscultation is a region of the back not covered by superficial muscles where respiratory sounds can be heard clearly.
4. The skeleton of the chest protects the organs within and provides several surface landmarks for identifying the position of the heart and lungs.
5. The abdomen and pelvis contain several important landmarks, such as the linea alba, linea semilunaris, McBurney's point, and supracristal line.
6. The perineum forms the floor of the pelvis.

Surface Anatomy of the Upper Limb (Extremity) (p. 425)

1. The upper limb consists of the pectoral (shoulder) girdle and free upper limb; the surface anatomy of the upper limb is discussed in Exhibit 12.4.
2. The prominence of the shoulder is formed by the deltoid muscle, a frequent site of an intramuscular injection.
3. The axilla is a pyramid-shaped area at the junction of the arm and chest that contains blood vessels and nerves that pass between the neck and upper limbs.
4. The arm contains several blood vessels used for withdrawing blood, introducing fluids, taking the pulse, and measuring blood pressure.
5. Several muscles of the forearm are more easily identifiable by their tendons. The radial artery is a principal blood vessel for detecting a pulse.
6. The hand contains the origin of the cephalic vein and extensor tendons on the dorsum, and the thenar and hypothenar eminences on the palm.

Surface Anatomy of the Lower Limb (Extremity) (p. 430)

1. The lower limb consists of the pelvic (hip) girdle and free lower limb; the surface anatomy of the lower limb is discussed in Exhibit 12.5.
2. The buttock is formed mainly by the gluteus maximus muscle.
3. The major muscles of the thigh are the anterior quadriceps femoris and the posterior hamstrings. An important landmark in the thigh is the femoral triangle.
4. The popliteal fossa is a diamond-shaped area on the posterior aspect of the knee.
5. The gastrocnemius muscle forms the bulk of the midportion of the posterior aspect of the leg and together with the soleus muscle forms the calf of the leg.
6. The foot contains the origin of the great saphenous vein, the largest vein in the body.

SELF-QUIZ QUESTIONS

Choose the one best answer to the following questions.

1. Which of the following is true of the sternal angle?
   a. It is palpated to locate the costal cartilage of the second rib.
   b. It is the junction between the manubrium and body of the sternum.
   c. It is external to the point at which the trachea branches into the right and left primary bronchi.
   d. Only a and b are correct.
   e. a, b, and c are all correct.

2. Which of the following terms refers to the region of the external ear?
   a. temporal   b. oral
   c. mental   d. auricular
   e. buccal

3. The apex of what structure can be palpated just superior to the medial third of the clavicle?
   a. heart   b. lung
   c. larynx   d. diaphragm
   e. hyoid bone

4. The muscle that can be palpated just superior and anterior to the angle of the mandible by alternately clenching the teeth and then opening the mouth is the:
   a. masseter.   b. sternocleidomastoid.
   c. temporalis.   d. zygomaticus major.
   e. orbicularis oris.

5. The mastoid process can be palpated just posterior to the:
   a. lateral commissure.   b. tragus.
   c. auricle.   d. dorsum nasi.
   e. cricoid cartilage.

6. The pulse is commonly taken in the neck just anterior to the sternocleidomastoid muscle from a blood vessel called the:
   a. external jugular vein.   b. subclavian artery.
   c. internal jugular vein.   d. common carotid artery.
   e. external carotid artery.

7. The femoral triangle is bounded by the adductor longus, the sartorius, and the:
   a. psoas major.
   b. iliac crest.
   c. ischial tuberosity.
   d. inguinal ligament.
   e. semitendinosus.

Complete the following.

8. Arrange the following limb landmarks in order from proximal to distal: popliteal fossa, pisiform, antecubital fossa, styloid process, calcaneus, axillary artery pulse, malleolus, femoral artery pulse.
   Free upper limb: _____, _____, _____, _____
   Free lower limb: _____, _____, _____, _____

9. Use the directional terms (anterior, posterior, superior, inferior, proximal, distal, medial, lateral) to complete the following:
   a. The mental region is _____ to the oral region.
   b. The orbital region is _____ to the frontal region.
   c. The antebrachium is _____ to the brachium.
   d. The sternal angle is _____ to the xiphoid process.
   e. The tragus of the ear is _____ to the external auditory canal.
   f. The iris of the eye is _____ to the medial commissure.
   g. The lobule is the _____ portion of the auricle.

10. The triangle of auscultation is bounded by the vertebral border of the scapula, the _____ muscle, and the _____ muscle.

11. The heart lies on the diaphragm deep to the _____ joint.

Are the following statements true or false? If the statement is False, *find and fix the mistake to make the statement* True.

12. The ulnar nerve lies in a groove posterior to the medial epicondyle.

13. All of the following surface features are located on the free lower limbs: tibial tuberosity, calcaneal (Achilles) tendon, lateral malleolus.

14. McBurney's point is located at the halfway point along a line between the umbilicus and the anterior superior iliac spine.

15. A landmark for performing a tracheostomy is the cricoid cartilage.

16. A vein of the free upper limb often used for intravenous therapy is located in the popliteal fossa.

Matching.

17. Match the following:
    ___ (a) frequent site for intramuscular injections in the upper limb
    ___ (b) a "stiff neck" results from an inflammation of this muscle
    ___ (c) divides neck into anterior and posterior (lateral) triangles
    ___ (d) contributes to the anterior axillary fold
    ___ (e) occupies most of the posterior surface of the arm
    ___ (f) tendon is visible on the anterior surface of the wrist when making a fist
    ___ (g) tendon forms one border of the "anatomical snuffbox"
    ___ (h) frequently used as a site for insulin injection

    (1) palmaris longus
    (2) triceps brachii
    (3) trapezius
    (4) deltoid
    (5) vastus lateralis
    (6) extensor pollicis longus
    (7) sternocleidomastoid
    (8) pectoralis major

18. Match the following terms with their descriptions:
    ___ (a) prominent posterior projection in the midline at the base of the neck
    ___ (b) most prominent structure in the midline of the anterior cervical region
    ___ (c) superior border of the buttock
    ___ (d) forms the tip of the shoulder
    ___ (e) located deep to the dimple that coincides with the middle of the sacroiliac joint
    ___ (f) bears most of the weight of a person when sitting
    ___ (g) attachment point for calcaneal (Achilles) tendon
    ___ (h) protuberance at the distal end of the ulna

    (1) thyroid cartilage
    (2) iliac crest
    (3) ischial tuberosity
    (4) acromion
    (5) vertebra prominens
    (6) calcaneus
    (7) styloid process
    (8) posterior superior iliac spine

CRITICAL THINKING QUESTIONS

1. Fred, the football fanatic, ran into the goal post and landed on his back. He told the coach that he felt all right but when he reached behind his head, he felt a bump on the back of his skull and another at the base of his neck. The team physician examined Fred and told him that the bumps were nothing to worry about; that they had always been there. What were these bumps?

2. Great Aunt Edith heard you're bringing your new boyfriend, the one with an anatomically correct heart tattooed on his bicep (he can even make it beat) and "I brake for ambulances" bumper sticker, to Thanksgiving dinner. She told your mother that she's having "palpations." Your new boyfriend, a first-year medical student, said she probably meant "palpitations." Is there a difference?

3. After Great Aunt Edith meets your new boyfriend, she decides he's not so bad after all, and says to him, "Why did you tattoo that heart on your arm? Why didn't you put it on your chest where it belongs?" Your boyfriend thinks this is a great idea. Describe the projection points on the surface of the chest that the artist should use to be anatomically correct.

4. Randy has been told that he needs to receive a large dose of an antibiotic to treat his infection. When he hears that this will be administered by a gluteal intramuscular injection, he says he isn't worried because, "They're injecting me there because it's so padded with fat, I won't feel it." If the nurse injected him in the fatty part of his buttocks, would the injection have been administered correctly? What muscle should the nurse be aiming for, and what is the major landmark for locating the muscle? What are the risks of not injecting the correct muscle?

5. An ad for a new telephoto camera appears in a medical journal. It states "you'll see the pores on your neighbor's ala, the sweat on his philtrum, the hairs in his external nares, the studs in his helix, the ring in his supercilia, and the color of his sclera." Is this a great camera? Explain.

12.1 The frontal, parietal, temporal, occipital, zygomatic, and nasal regions are all named for bones that are deep to them.

12.2 When palpating the sagittal suture, the sheetlike epicranial aponeurosis (galea aponeurotica) is also palpated deep to the skin.

12.3 Light enters the eye through the pupil.

12.4 The mastoid process of the temporal bone can be palpated just posterior to the auricle.

12.5 The anterior border of the nose that connects the root and apex is the dorsum nasi.

12.6 The sternocleidomastoid muscle divides the neck into anterior and posterior triangles.

12.7 With the aid of a stethoscope, respiratory sounds can be heard clearly in the triangle of auscultation.

12.8 The sternal angle is the most reliable landmark of the chest.

12.9 Pain produced by pressure of the finger on McBurney's point indicates acute appendicitis.

12.10 The acromion of the scapula forms the top of the shoulder.

12.11 Blood is usually withdrawn from the median cubital vein; this vein is also used to administer medication.

12.12 A pulse is frequently taken by placing pressure on the radial artery in the wrist.

12.13 Muscles that form the thenar eminence move the thumb.

12.14 The gluteus maximus muscle forms the bulk of the buttock.

12.15 The popliteal fossa is bordered superolaterally by the biceps femoris, superomedially by the semimembranosus and semitendinosus, and inferolaterally and inferomedially by the lateral and medial heads of the gastrocnemius.

12.16 A pulse may be felt in the dorsalis pedis artery, located just lateral to the tendon of the extensor digitorum muscle.

13 | THE CARDIOVASCULAR SYSTEM: BLOOD

INTRODUCTION The **cardiovascular system** (*cardio-*=heart; *vascular*=blood vessels) consists of three interrelated components: blood, the heart, and blood vessels. The focus of this chapter is blood; the next two chapters will examine the heart and blood vessels, respectively. Blood transports various substances, helps regulate several life processes, and affords protection against disease. For all of its similarities in origin, composition, and functions, blood is as unique from one person to another as are skin, bone, and hair. Health-care professionals routinely examine and analyze its differences through various blood tests when trying to determine the cause of different diseases. Despite these differences, blood is the most easily and widely shared of human tissues, saving many thousands of lives every year through blood transfusions. The branch of science concerned with the study of blood, blood-forming tissues, and the disorders associated with them is **hematology** (hēm-a-TOL-ō-jē; *hema-* or *hemato-*=blood;-*logy*=study of). Because of their common origins, the development of blood in the embryo and fetus is considered in Chapter 15 with the development of the blood vessels. Most cells of a multicellular organism cannot move around to obtain oxygen and nutrients and get rid of carbon dioxide and other wastes. Instead, these needs are met by two fluids: blood and interstitial fluid. **Blood** is a connective tissue composed of a liquid portion called plasma (the extracellular matrix) and a cellular portion consisting of various cells and cell fragments. **Interstitial fluid** is the watery fluid that bathes body cells and is constantly renewed by the blood. Oxygen brought into the lungs and water and nutrients brought into the gastrointestinal tract are transported by the blood, diffuse from the blood into the interstitial fluid, and then diffuse into body cells. Carbon dioxide and other wastes move in the reverse direction, from the body cells into the interstitial fluid and then into the blood. Blood then transports the wastes to various organs—the lungs, kidneys, skin, and digestive system—for elimination from the body.

CONTENTS AT A GLANCE

In order for blood to reach all the cells, it must be moved throughout the body. The *heart* is the pump that circulates the blood. *Blood vessels* convey blood from the heart to body cells and from body cells back to the heart. The largest blood vessels, which take blood away from the heart, are called *arteries*. Arteries branch into smaller vessels called *arterioles*. As an arteriole enters a tissue, it divides into numerous microscopic vessels called *capillaries*. Substances exchanged between the blood and interstitial fluid pass through the thin walls of capillaries. Before leaving a tissue, capillaries unite to form small vessels called *venules*, which in turn merge to form progressively larger vessels called *veins*. Veins convey blood back to the heart. •

FUNCTIONS OF BLOOD

◉ **OBJECTIVE**
• List and describe the functions of blood.

Blood, a liquid connective tissue, is closely related to other fluids within the body. In fact most of the other body fluids, such as interstitial fluid, lymph, cerebrospinal fluid, and aqueous humor, arise during development from the blood and are continually replenished by the blood. These extracellular fluids nourish, protect, and exchange materials with every cell of the body. This watery environment is derived from the blood, renewed by the blood, and returned to the blood. Based on these relationships, the blood has three general functions:

1. **Transportation.** Blood transports oxygen from the lungs to the cells of the body and carbon dioxide from the body cells to the lungs for exhalation. It carries nutrients from the gastrointestinal tract to body cells and hormones from endocrine glands to other body cells. Blood also transports heat and waste products to the lungs, kidneys, and skin for elimination from the body.
2. **Regulation.** Circulating blood helps maintain homeostasis in all body fluids. Blood helps regulate pH through buffers. It also helps adjust body temperature through the heat-absorbing and coolant properties of the water in blood plasma and its variable rate of flow through the skin, where excess heat can be lost from the blood to the environment. Blood osmotic pressure also influences the water content of cells, mainly through interactions of dissolved ions and proteins.
3. **Protection.** Blood can clot, which protects against its excessive loss from the cardiovascular system after an injury. In addition, white blood cells protect against disease by carrying on phagocytosis. Several types of blood proteins, including antibodies, interferons, and complement, help protect against disease in a variety of ways.

◉ **CHECKPOINT**
1. What substances does blood transport?

PHYSICAL CHARACTERISTICS OF BLOOD

◉ **OBJECTIVE**
• List the principal physical characteristics of blood.

Blood is denser and more viscous (thicker) than water, which is part of the reason it flows more slowly than water. The temperature of blood is about 38°C (100.4°F), which is slightly higher than normal body temperature, and it has a slightly alkaline pH ranging from 7.35 to 7.45. The color of blood varies. When saturated with oxygen it is bright red; when unoxygenated, the blood is dark red to purple. Blood constitutes about 8% of the total body weight. The blood volume is 5–6 liters (1.5 gal) in an average-sized adult male and 4–5 liters (1.2 gal) in an average-sized adult female. This difference in volume is due to the difference in body size.

◉ **CHECKPOINT**
2. How much does the blood of a 150-pound person weigh?

COMPONENTS OF BLOOD

◉ **OBJECTIVE**
• Describe the principal components of blood.

Whole blood is composed of (1) blood plasma, a watery liquid extracellular matrix that contains dissolved substances, and (2) formed elements, which are cells and cell fragments. If a sample of blood is centrifuged (spun) in a small glass tube, the cells sink to the bottom of the tube and the lighter-weight plasma forms a layer on top (Figure 13.1a). Blood is about 45% formed elements and about 55% blood plasma. Normally, more than 99% of the formed elements are red-colored red blood cells (RBCs). Pale or colorless white blood cells (WBCs) and platelets occupy less than 1% of the formed elements. They form a very thin layer, called the *buffy coat*, between the packed RBCs and blood plasma in centrifuged blood. Figure 13.1b shows the composition of blood plasma and the numbers of the various types of formed elements in blood.

Blood Plasma

When the formed elements are removed from blood, the straw-colored liquid called **blood plasma** (or simply **plasma**) is left. Plasma is about 91.5% water and 8.5% solutes, most of which (7% by weight) are proteins. Some of the proteins in plasma are also found elsewhere in the body, but those confined to blood are called **plasma proteins**. Hepatocytes (liver cells) synthesize most of the plasma proteins, which include the **albumins** (54% of plasma proteins), **globulins** (38%), and **fibrinogen** (7%). Certain blood cells develop into cells that produce gamma globulins, an important type of globulin. These plasma proteins are also called **antibodies** or **immunoglobulins** because they are produced during certain immune responses. Foreign substances (antigens) such as bacteria and viruses stimulate production of antibodies. An antibody binds specifically to the antigen that stimulated its production and this disables the invading antigen.

Figure 13.1 Components of blood in a normal adult.

Blood is a connective tissue that consists of blood plasma (liquid) plus formed elements (red blood cells, white blood cells, and platelets).

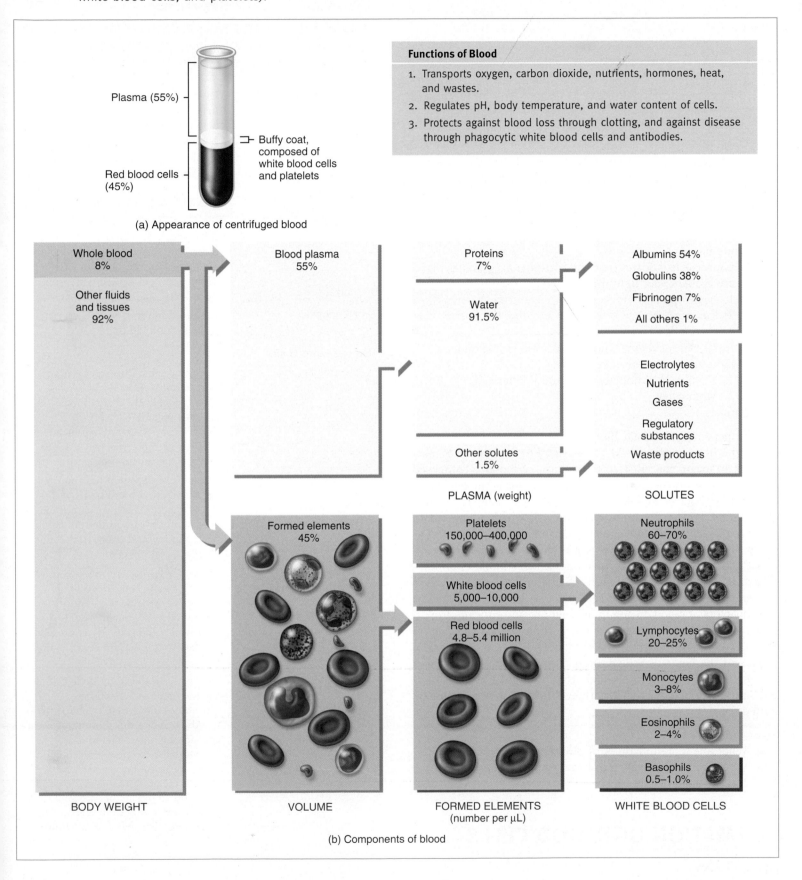

(a) Appearance of centrifuged blood

Plasma (55%)

Buffy coat, composed of white blood cells and platelets

Red blood cells (45%)

Functions of Blood

1. Transports oxygen, carbon dioxide, nutrients, hormones, heat, and wastes.
2. Regulates pH, body temperature, and water content of cells.
3. Protects against blood loss through clotting, and against disease through phagocytic white blood cells and antibodies.

Whole blood 8%

Other fluids and tissues 92%

Blood plasma 55%

Proteins 7%

Albumins 54%

Globulins 38%

Fibrinogen 7%

All others 1%

Water 91.5%

Electrolytes

Nutrients

Gases

Regulatory substances

Other solutes 1.5%

Waste products

PLASMA (weight)

SOLUTES

Formed elements 45%

Platelets 150,000–400,000

Neutrophils 60–70%

White blood cells 5,000–10,000

Red blood cells 4.8–5.4 million

Lymphocytes 20–25%

Monocytes 3–8%

Eosinophils 2–4%

Basophils 0.5–1.0%

BODY WEIGHT

VOLUME

FORMED ELEMENTS (number per μL)

WHITE BLOOD CELLS

(b) Components of blood

What is the approximate volume of blood in your body?

Besides proteins, other solutes in plasma include electrolytes, nutrients, regulatory substances such as enzymes and hormones, gases, and waste products such as urea, uric acid, creatinine, ammonia, and bilirubin.

Table 13.1 describes the chemical composition of blood plasma.

Formed Elements

The **formed elements** of the blood include **red blood cells (RBCs)**, **white blood cells (WBCs)**, and **platelets** (Figure 13.2). Although RBCs and WBCs are living cells, platelets are cell fragments. Unlike RBCs and platelets, which perform limited roles, WBCs have a number of more general functions. There are several distinct types of WBCs—neutrophils, lymphocytes, monocytes, eosinophils, and basophils—each having a unique microscopic appearance. The roles of each type of WBC are discussed later in this chapter.

I. Red blood cells
II. White blood cells
   A. Granular leukocytes (contain conspicuous granules that are visible under light microscope after staining)
      1. Neutrophils
      2. Eosinophils
      3. Basophils
   B. Agranular leukocytes (no granules are visible under a light microscope after staining)
      1. T and B lymphocytes and natural killer cells
      2. Monocytes
III. Platelets

The percentage of total blood volume occupied by RBCs is called the **hematocrit** (he-MAT-ō-krit). For example, a hematocrit of 40 means that 40% of the volume of blood is composed of RBCs. The normal range of hematocrit for adult females is about 38–46% (average=42); for adult males it is about 40–54% (average=47). The hormone testosterone, which is present in much higher concentration in males than in females, stimulates synthesis of a hormone called **erythropoietin** (e-rith′-rō-POY-e-tin) or **EPO** by the kidneys. This hormone, which stimulates production of RBCs, contributes to higher hematocrits in males. Lower values in women during their reproductive years may be due to excessive loss of blood during menstruation. A significant drop in hematocrit indicates anemia, a lower-than-normal number of RBCs. In *polycythemia* the percentage of RBCs is abnormally high, and the hematocrit may be 65% or higher. Polycythemia may be caused by conditions such as an unregulated increase in RBC production, tissue hypoxia, dehydration, or blood doping by athletes.

◉ **CHECKPOINT**
  3. What are some functions of blood plasma proteins?
  4. What is the significance of lower-than-normal or higher-than-normal hematocrit?

FORMATION OF BLOOD CELLS

◉ OBJECTIVE
• Explain the origin of blood cells.

Although some lymphocytes have a lifetime measured in years, most formed elements of the blood are continually dying and being replaced within hours, days, or weeks. Negative feedback systems

TABLE 13.1	
Substances in Blood Plasma	
CONSTITUENT	**DESCRIPTION**
Water (91.5%)	Liquid portion of blood. Acts as solvent and suspending medium for components of blood; absorbs, transports, and releases heat.
Plasma proteins (7.0%)	Exert colloid osmotic pressure, which helps maintain water balance between blood and tissues and regulates blood volume.
Albumins	Smallest and most numerous plasma proteins; produced by liver. Function as transport proteins for several steroid hormones and for fatty acids.
Globulins	Produced by liver and by plasma cells, which develop from B lymphocytes. Antibodies (immunoglobulins) help attack viruses and bacteria. Alpha and beta globulins transport iron, lipids, and fat-soluble vitamins.
Fibrinogen	Produced by liver. Plays essential role in blood clotting.
Other solutes (1.5%)	
Electrolytes	Inorganic salts. Positively charged ions (cations) include Na^+, K^+, Ca^{2+}, Mg^{2+}; negatively charged ions (anions) include Cl^-, HPO_4^{2-}, SO_4^{2-} and HCO_3^-. Help maintain osmotic pressure and play essential roles in the function of cells.
Nutrients	Products of digestion pass into blood for distribution to all body cells. Include amino acids (from proteins), glucose (from carbohydrates), fatty acids and glycerol (from triglycerides), vitamins, and minerals.
Gases	Oxygen (O_2), carbon dioxide (CO_2), and nitrogen (N_2). More O_2 is associated with hemoglobin inside red blood cells, but more CO_2 is dissolved in plasma. N_2 is present but has no known function in the body.
Regulatory substances	Enzymes, produced by body cells, catalyze chemical reactions. Hormones, produced by endocrine glands, regulate metabolism, growth, and development. Vitamins are cofactors for enzymatic reactions.
Waste products	Most are breakdown products of protein metabolism and are carried by blood to organs of excretion. Include urea, uric acid, creatine, creatinine, bilirubin, and ammonia.

regulate the total number of RBCs and platelets in circulation, and their numbers normally remain steady. The abundance of the different types of WBCs, however, varies in response to challenges by invading pathogens and other foreign antigens.

Figure 13.2 Scanning electron micrograph and photomicrograph of the formed elements of blood.

The formed elements of blood are red blood cells (RBCs), white blood cells (WBCs), and platelets.

White blood cell

Platelet

Red blood cell

SEM 3500x

(a)

White blood cell
(leukocyte–neutrophil)

Blood plasma

Red blood cell
(erythrocyte)

Platelet

White blood cell
(leukocyte–monocyte)

LM 400x

(b)

? **Which formed elements of the blood are cell fragments?**

• CLINICAL CONNECTION | **Bone Marrow Examination**

Sometimes a sample of red bone marrow must be obtained in order to diagnose certain blood disorders, such as leukemia and severe anemias. **Bone marrow examination** may involve *bone marrow aspiration* (withdrawal of a small amount of red bone marrow with a fine needle and syringe) or a *bone marrow biopsy* (removal of a core of red bone marrow with a larger needle). Both types of samples are usually taken from the iliac crest of the hip bone, although samples are sometimes aspirated from the sternum. In young children, bone marrow samples are taken from a vertebra or tibia (shin bone). The tissue or cell sample is then sent to a pathology lab for analysis. Specifically, the lab technicians look for signs of neoplastic (cancer) cells or other diseased cells to assist in diagnosis. •

Illiac crest

Withdrawal of red bone marrow
from posterior iliac crest

Biopsy needle

Bone marrow

Detailed view

The process by which the formed elements of blood develop is called **hemopoiesis** (hē-mō-poy-Ē-sis; *-poiesis*=making) or *hematopoiesis*. As you will learn in Chapter 15, before birth, hemopoiesis first occurs in the yolk sac of an embryo and later in the liver, spleen, thymus, and lymph nodes of a fetus. In the last three months before birth, red bone marrow becomes the primary site of hemopoiesis and continues to be the source of blood cells after birth and throughout life.

Red bone marrow is a highly vascularized connective tissue located in the microscopic spaces between trabeculae of spongy bone tissue. It is present chiefly in bones of the axial skeleton, pectoral and pelvic girdles, and the proximal epiphyses of the humerus and femur. About 0.05–0.1% of red bone marrow cells are derived from mesenchyme and are called **pluripotent stem cells** (ploo-RIP-ō-tent; *pluri-*=several). Pluripotent stem cells are cells that have the capacity to develop into many different types of cells (Figure 13.3).

In newborns all bone marrow is red and thus active in blood cell production. As an individual grows and in adulthood, the rate of blood cell formation decreases; the red bone marrow in the medullary (marrow) cavity of long bones becomes inactive and is replaced by yellow bone marrow, which is largely fat cells. Under certain conditions, such as severe bleeding, yellow bone marrow can revert to red bone marrow by extension of red bone marrow into yellow bone marrow and repopulation of yellow bone marrow by pluripotent stem cells.

Stem cells in red bone marrow reproduce themselves, proliferate, and differentiate into cells that give rise to blood cells, macrophages, reticular cells, mast cells, and adipocytes. Some of the stem cells can also form osteoblasts, chondroblasts, and muscle cells, and may someday be used as a source of bone, cartilage, and muscular tissue for tissue and organ replacement. The reticular cells produce reticular fibers, which form the stroma (framework) that supports red bone marrow cells. Blood from nutrient

Figure 13.3 Origin, development, and structure of blood cells. Some of the generations of some cell lines have been omitted.

🔑 Blood cell production is called hemopoiesis and occurs mainly in red bone marrow after birth.

❓ **From which connective tissue do pluripotent stem cells develop?**

and metaphyseal arteries (see Figure 6.4 on pages 146–147) enters a bone and passes into the enlarged and leaky capillaries, called *sinuses*, that surround red bone marrow cells and fibers. After blood cells form, they enter the sinuses and other blood vessels and leave the bone through nutrient and periosteal veins (see Figure 6.4 on pages 146–147). With the exception of lymphocytes, formed elements do not divide once they leave red bone marrow.

In order to form blood cells, pluripotent stem cells produce two further types of stem cells, which have the capacity to develop into several kinds of cells. These stem cells are called *myeloid stem cells* and *lymphoid stem cells*. Myeloid stem cells begin and complete their development in red bone marrow and give rise to red blood cells, platelets, monocytes, neutrophils, eosinophils, and basophils. Lymphoid stem cells begin their development in red bone marrow but complete it in lymphatic tissues; they give rise to lymphocytes. Although stem cells have distinctive cell identity markers in their plasma membranes, they cannot be distinguished histologically and resemble lymphocytes.

During hemopoiesis, the myeloid stem cells differentiate into **progenitor cells** (prō-JEN-i-tor). Progenitor cells are no longer capable of reproducing themselves and are committed to giving rise to more specific elements of blood. Some progenitor cells are known as *colony-forming units (CFUs)*. After the CFU designation is an abbreviation that designates the mature elements in blood that they will produce: CFU-E ultimately produces erythrocytes (red blood cells), CFU-Meg produces megakaryocytes, the source of platelets, and CFU-GM ultimately produces granulocytes (specifically, neutrophils) and monocytes. Progenitor cells, like stem cells, resemble lymphocytes and cannot be distinguished by their microscopic appearance alone. Other myeloid stem cells develop directly into cells called precursor cells (described next). Lymphoid stem cells differentiate into T lymphoblasts and B lymphoblasts, which ultimately develop into T lymphocytes (T cells) and B lymphocytes (B cells), respectively.

In the next generation, the cells are called **precursor cells**, also known as **blasts**. Over several cell divisions they develop into the actual formed elements of blood. For example, monoblasts develop into monocytes, eosinophilic myeloblasts develop into eosinophils, and so on. Precursor cells have recognizable microscopic appearances.

Several hormones called **hemopoietic growth factors** regulate the differentiation and proliferation of particular progenitor cells. **Erythropoietin** or **EPO** increases the number of red blood cell precursors. The main producers of EPO are cells in the kidneys that lie between the kidney tubules (peritubular interstitial cells). With renal failure, EPO release slows and RBC production is inadequate. **Thrombopoietin** (throm′-bō-POY-ē-tin) or **TPO** is a hormone produced by the liver that stimulates the formation of platelets (thrombocytes) from megakaryocytes. Several different **cytokines**, small glycoproteins produced by red bone marrow cells, leukocytes, macrophages, fibroblasts, and endothelial cells, regulate development of different blood cell types. They typically act as local hormones (autocrines or paracrines). Cytokines stimulate proliferation of progenitor cells in red bone marrow and regulate the activities of cells involved in nonspecific defenses (such as phagocytes) and immune responses (such as B cells and T cells). Two important families of cytokines that stimulate the formation of white blood cell formation are **colony-stimulating factors (CSFs)** and **interleukins**.

⊙ **CHECKPOINT**

5. Which hemopoietic growth factors regulate differentiation and proliferation of CFU-E and formation of platelets from megakaryocytes?

RED BLOOD CELLS

⊙ **OBJECTIVE**

- Describe the structure, functions, life cycle, and production of red blood cells.

Red blood cells (RBCs) or **erythrocytes** (e-RITH-rō-sīts; *erythro-*=red; *-cyte*=cell) contain the oxygen-carrying protein **hemoglobin**, which is a pigment that gives whole blood its red color. A healthy adult male has about 5.4 million red blood cells per microliter (μL) of blood,* and a healthy adult female has about 4.8 million. (One drop of blood is about 50μL.) To maintain normal numbers of RBCs, new mature cells must enter the circulation at the astonishing rate of at least 2 million per second, a pace that balances the equally high rate of RBC destruction.

RBC Anatomy

RBCs are biconcave discs with a diameter of 7–8μm (Figure 13.4a). Mature red blood cells have a simple structure. Their plasma membrane is both strong and flexible, which allows them to deform without rupturing as they squeeze through narrow capillaries. As you will see later, certain glycolipids in the plasma membrane of RBCs are antigens that account for the various blood groups such as the ABO and Rh groups. RBCs lack a nucleus and other organelles and can neither reproduce nor carry on extensive metabolic activities. The cytosol of RBCs contains hemoglobin molecules, which were synthesized before loss of the nucleus during RBC production and which constitute about 33% of the cell's weight.

---

*$1\mu L = 1mm^3 = 10^{-6}$ liter.

Figure 13.4 **The shapes of a red blood cell (RBC) and a hemoglobin molecule.** In (b), note that each of the four polypeptide chains of a hemoglobin molecule (blue) has one heme group (gold), which contains an iron ion (Fe^{2+}), shown in red.

🔑 The iron portion of a heme group binds oxygen for transport by hemoglobin.

8μm

Surface view

Sectioned view

(a) RBC shape

(b) Hemoglobin molecule

Polypeptide

Iron

Heme

❓ How many molecules of O_2 can one hemoglobin molecule transport?

RBC Functions

Red blood cells are highly specialized for their oxygen transport function. Because mature RBCs have no nucleus, all their internal space is available for oxygen transport. Because RBCs lack mitochondria and generate ATP anaerobically (without oxygen), they do not use up any of the oxygen they transport. Even the shape of an RBC facilitates its function. A biconcave disc has a much greater surface area for its volume than, say, a sphere or a cube. Its shape thus provides a large surface area for the diffusion of gas molecules into and out of the RBC; in addition, the thinness of the cell enhances the rapid diffusion of O_2 between the exterior and innermost regions of the cell.

Each RBC contains about 280 million hemoglobin molecules; each hemoglobin molecule can carry up to four oxygen molecules. A hemoglobin molecule consists of a protein called **globin**, composed of four polypeptide chains (two alpha and two beta chains), plus four nonprotein pigments called **hemes** (Figure 13.4b). One ringlike heme binds to each polypeptide chain. At the center of each heme ring is an iron ion (Fe^{2+}) that can combine reversibly with one oxygen molecule. The oxygen picked up in the lungs is transported bound to the iron of the heme group. As blood flows through tissue capillaries, the iron–oxygen reaction reverses. Hemoglobin releases oxygen, which diffuses first into the interstitial fluid and then into cells.

Hemoglobin also transports about 13% of the total carbon dioxide, a waste product of metabolism. Blood flowing through tissue capillaries picks up carbon dioxide, some of which combines with amino acids in the globin part of hemoglobin. As blood flows through the lungs, the carbon dioxide is released from hemoglobin and then exhaled.

In addition to its key role in transporting oxygen and carbon dioxide, hemoglobin also plays a role in regulation of blood flow and blood pressure. The gaseous hormone **nitric oxide (NO)**, produced by the endothelial cells that line blood vessels, binds to hemoglobin. Under some circumstances, hemoglobin releases NO. The released NO causes *vasodilation*, an increase in blood vessel diameter that occurs when the smooth muscle in the vessel wall relaxes. Vasodilation improves blood flow and enhances oxygen delivery to cells near the site of NO release.

RBC Life Cycle

Red blood cells live only about 120 days because of the wear and tear their plasma membranes undergo as they squeeze through blood capillaries. Without a nucleus and other organelles, RBCs cannot synthesize new components to replace damaged ones. The plasma membrane becomes more fragile with age, and the cells are more likely to burst, especially as they squeeze through narrow channels in the spleen. Ruptured red blood cells are removed from circulation and destroyed by fixed phagocytic macrophages in the spleen and liver, and the breakdown products are recycled.

Erythropoiesis: Production of RBCs

Erythropoiesis (e-rith-rō-poy-Ē-sis), the production of RBCs, starts in the red bone marrow with a precursor cell called a **proerythroblast** (see Figure 13.3). The proerythroblast divides several times, producing cells that begin to synthesize hemoglobin. Ultimately, a cell near the end of the developmental sequence ejects its nucleus and becomes a **reticulocyte** (re-TIK-ū-lō-sīt). Loss of the nucleus causes the center of the cell to in-

dent, producing the distinctive biconcave shape. Reticulocytes, which are composed of about 34% hemoglobin and retain some mitochondria, ribosomes, and endoplasmic reticulum, pass from red bone marrow into the bloodstream by squeezing through holes in the plasma membrane of the endothelial cells of blood capillaries. Reticulocytes usually develop into erythrocytes, or mature red blood cells, within 1–2 days after their release from red bone marrow.

Normally, erythropoiesis and red blood cell destruction proceed at roughly the same pace. If the oxygen-carrying capacity of the blood falls because erythropoiesis is not keeping up with RBC destruction, RBC production is increased. Cellular oxygen deficiency, called **hypoxia** (hī-POKS-ē-a), may occur if not enough oxygen enters the blood. For example, the reduced oxygen content of air at high altitudes results in a reduced level of oxygen in the blood. Oxygen delivery may also fall due to anemia, which has many causes. Lack of iron, certain amino acids, and vitamin B_{12} are but a few. Circulatory problems that reduce blood flow to tissues may also reduce oxygen delivery. Whatever the cause, hypoxia stimulates the kidneys to step up the release of erythropoietin. This hormone circulates through the blood to the red bone marrow, where it speeds the development of proerythroblasts into reticulocytes. When the number of circulating RBCs increases, more oxygen can be delivered to body tissues.

Premature newborns often exhibit anemia, due in part to inadequate production of erythropoietin. During the first weeks after birth, the liver, not the kidneys, produces most EPO. Because the liver is less sensitive than the kidneys to hypoxia, newborns have a smaller EPO response to anemia than do adults. Because *fetal hemoglobin* (hemoglobin present at birth) carries up to 30% more oxygen, the loss of fetal hemoglobin makes the anemia worse.

Blood Group Systems

More than 100 kinds of genetically determined antigens have been detected on the surface of red blood cells. Many of these antigens appear in characteristic patterns, a fact that enables scientists or health care professionals to identify a person's blood as belonging to one or more blood grouping systems; at least 14 are currently recognized. Each system is characterized by the presence or absence of specific antigens on the surface of a red blood cell's plasma membrane. The two major blood groups distinguished on the basis of these antigens are called the ABO and Rh blood groups.

The *ABO blood grouping system* is based on two antigens, symbolized as *A* and *B*. Individuals whose erythrocytes manufacture only antigen *A* are said to have blood type A. Those who manufacture only antigen *B* are type B. Individuals who manufacture both *A* and *B* are type AB. Those who manufacture neither are type O.

The *Rh blood grouping system* is so named because it was first worked out using the blood of the *Rh*esus monkey. Individuals whose erythrocytes have the Rh antigens (D antigens) are designated Rh^+. Those who lack Rh antigens are designated Rh^-.

As just noted, the presence or absence of certain antigens on red blood cells is the basis for classifying blood into several different groups. Such information is very important when a transfusion is given. A **transfusion** (trans-FŪ-zhun) is the transfer of whole blood or blood components (red blood cells or blood plasma, for example) into the bloodstream. A transfusion may be given to treat low blood volume, anemia, or a low platelet count. However, in an incompatible blood transfusion, one person's red blood cell antigens may be recognized as foreign by antibodies in

• CLINICAL CONNECTION Sickle-Cell Disease

The RBCs of a person with **sickle-cell disease** (SCD) contain Hb-S, an abnormal kind of hemoglobin. When Hb-S gives up oxygen to the interstitial fluid, it forms long, stiff, rodlike structures that bend the erythrocyte into a sickle shape. The sickled cells rupture easily. Even though erythropoiesis is stimulated by the loss of the cells, it cannot keep pace with hemolysis. People with sickle-cell disease always have some degree of anemia and mild jaundice and may experience joint or bone pain, breathlessness, rapid heart rate, abdominal pain, fever, and fatigue as a result of tissue damage caused by prolonged recovery oxygen uptake (oxygen debt). Any activity that reduces the amount of oxygen in the blood, such as vigorous exercise, may produce a **sickle-cell crisis** (worsening of the anemia, pain in the abdomen and long bones of the limbs, fever, and shortness of breath).

Sickle-cell disease is inherited. People with two sickle-cell genes have severe anemia; those with only one defective gene have minor problems. Sickle-cell genes are found primarily among populations, or descendants of populations, that live in the malaria belt around the world, including parts of Mediterranean Europe, sub-Saharan Africa, and tropical Asia. The genes responsible for the tendency of the RBCs to sickle also alter the permeability of the plasma membranes of sickled cells, causing potassium ions to leak out. Low levels of potassium kill the malaria parasites that may infect sickled cells. Because of this effect, a person with one normal gene and one sickle-cell gene has higher-than-average resistance to malaria. The possesssion of a single sickle-cell gene thus confers a survival benefit. Treatment of SCD consists of administration of analgesics to relieve pain, fluid therapy to

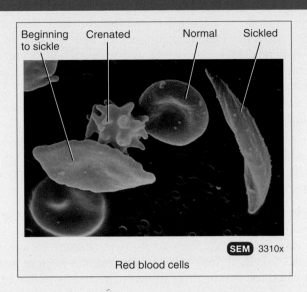

Beginning to sickle   Crenated   Normal   Sickled

SEM 3310x

Red blood cells

maintain hydration, oxygen to reduce the possibility of oxygen debt, antibiotics to counter infections, and blood transfusions. People who suffer from SCD have normal fetal hemoglobin (Hb-F), a slightly different form of hemoglobin that predominates at birth and is present in small amounts thereafeter. In some patients with sickle-cell disease, a drug called hydroxyurea promotes transcription of the normal Hb-F gene, elevates that level of Hb-F, and reduces the chance that the RBCs will sickle. Unfortunately, this drug also has toxic effects on the bone marrow; thus, its safety for long-term use is questionable. •

---

their blood plasma when transfused into someone with a different blood type. In this case, the transfused cells are attacked by antibodies and undergo hemolysis (burst), releasing hemoglobin into the plasma; the liberated hemoglobin may cause kidney damage.

◉ CHECKPOINT

6. Describe the size, microscopic appearance, and functions of RBCs.
7. Define erythropoiesis. Relate erythropoiesis to hematocrit. What factors accelerate and slow erythropoiesis?
8. What is the basis for the ABO and Rh blood grouping systems?

WHITE BLOOD CELLS

◉ OBJECTIVE

• Describe the structure, functions, and production of white blood cells.

WBC Anatomy and Types

Unlike red blood cells, white blood cells, or **leukocytes** (LOO-kō-sīts; *leuko-*=white), have a nucleus and do not contain hemoglobin (Figure 13.5). WBCs are classified as either granular or agranular, depending on whether they contain conspicuous cyto-

---

Figure 13.5 Structure of white blood cells.

 White blood cells are distinguished from one another by the shape of their nuclei and the staining properties of their cytoplasmic granules.

            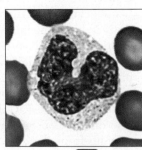

LM all 1600x

(a) Eosinophil   (b) Basophil   (c) Neutrophil   (d) Lymphocyte   (e) Monocyte

 Which WBCs are called granular leukocytes? Why?

plasmic vesicles (originally called *granules*) that are made visible by staining when viewed through a light microscope. *Granular leukocytes* include neutrophils, eosinophils, and basophils; *agranular leukocytes* include lymphocytes and monocytes. As shown in Figure 13.3, monocytes and granular leukocytes develop from myeloid stem cells. In contrast, lymphocytes develop from lymphoid stem cells.

Granular Leukocytes

After staining, each of the three types of granular leukocytes display conspicuous granules with distinctive coloration that can be recognized under a light microscope. Granular leukocytes can be distinguished as follows:

- **Eosinophil** (ē-ō-SIN-ō-fil). The large, uniform-sized granules within an eosinophil are *eosinophilic* (=eosin-loving)—they stain red-orange with acidic dyes (Figure 13.5a). The granules usually do not cover or obscure the nucleus, which most often has two or three lobes connected by either a thin strand or a thick strand of nuclear material.
- **Basophil** (BĀ-sō-fil). The round, variable-sized granules of a basophil are *basophilic* (=basic loving)—they stain blue-purple with basic dyes (Figure 13.5b). The granules commonly obscure the nucleus, which has two lobes.
- **Neutrophil** (NOO-trō-fil). The granules of a neutrophil are smaller, evenly distributed, and pale lilac in color (Figure 13.5c). Because the granules do not strongly attract either the acidic (red) or basic (blue) stain, these WBCs are neutrophilic (=neutral loving). The nucleus has two to five lobes, connected by very thin strands of nuclear material. As the cells age, the number of nuclear lobes increases. Because older neutrophils thus have several differently shaped nuclear lobes, they are often called *polymorphonuclear leukocytes (PMNs)*, or "polys."

Agranular Leukocytes

Even though so-called agranular leukocytes possess cytoplasmic granules, the granules are not visible under a light microscope because of their small size and poor staining qualities. Characteristics of agranular leukocytes are as follows:

- **Lymphocyte** (LIM-fō-sīt). The nucleus of a lymphocyte stains dark and is round or slightly indented (Figure 13.5d). The cytoplasm stains sky blue and forms a rim around the nucleus. The larger the cell, the more cytoplasm is visible. Lymphocytes are classified by cell diameter as large lymphocytes (10–14 μm) or small lymphocytes (6–9 μm). Although the functional significance of the size difference between small and large lymphocytes is unclear, the distinction is still clinically useful because an increase in the number of large lymphocytes has diagnostic significance in acute viral infections and in some immunodeficiency diseases.
- **Monocyte** (MON-ō-sīt). The nucleus of a monocyte is usually kidney-shaped or horseshoe-shaped, and the cytoplasm is blue-gray and has a foamy appearance (Figure 13.5e). The cytoplasm's color and appearance are due to very fine *azurophilic granules* (az'-ū-rō-FIL-ik; *azur*=blue; *philos*=loving), which are lysosomes. Blood is merely a conduit for monocytes, which migrate from the blood into the tissues, where they enlarge and differentiate into **macrophages** (=large eaters). Some become **fixed macrophages**, which means they reside in a particular tissue; examples are alveolar macrophages in the lungs or macrophages in the spleen. Others become **wandering macrophages**, which roam the tissues and gather at sites of infection or inflammation.

White blood cells and other nucleated body cells have proteins, called *major histocompatibility (MHC) antigens*, protruding from their plasma membranes into the extracellular fluid. These cell identity markers are unique for each person (except identical twins). Although RBCs possess blood group antigens, they lack the MHC antigens.

WBC Functions

The skin and mucous membranes of the body are continuously exposed to microscopic organisms, such as bacteria, some of which are capable of invading deeper tissues and causing disease. Once pathogens enter the body, the general function of white blood cells is to combat them by phagocytosis or immune responses. To accomplish these tasks, many WBCs leave the bloodstream and collect at points of pathogen invasion or inflammation. Once granular leukocytes and monocytes leave the bloodstream to fight injury or infection, they never return to it. Lymphocytes, on the other hand, continually recirculate—from blood to interstitial spaces of tissues to lymphatic fluid and back to blood. Only 2% of the total lymphocyte population is circulating in the blood at any given time; the rest are in lymphatic fluid and organs such as skin, lungs, lymph nodes, and spleen.

RBCs are contained within the bloodstream, but WBCs are able to cross capillary walls by a process termed **emigration** (em'-i-GRĀ-shun; *e-*=out; *migra-*=wander), formerly called *diapedesis*. During emigration, WBCs roll along the endothelium that forms capillary walls, stick to it, and then squeeze between endothelial cells. The precise signals that stimulate emigration through a particular blood vessel vary for the different types of WBCs.

Neutrophils and macrophages are active in **phagocytosis**; they can ingest bacteria and dispose of dead matter. Several different chemicals released by microbes and inflamed tissues attract phagocytes, a phenomenon called **chemotaxis**.

Among WBCs, neutrophils respond most quickly to tissue destruction by bacteria. After engulfing a pathogen during phagocytosis, a neutrophil unleashes several destructive chemicals to destroy the ingested pathogen. These chemicals include the enzyme **lysozyme**, which destroys certain bacteria, and **strong oxidants**, such as hydrogen peroxide (H_2O_2) and the hypochlorite anion (OCl^-), which is similar to household bleach. Neutrophils also contain **defensins**, proteins that exhibit a broad range of antibiotic activity against bacteria and fungi. Within a neutrophil, granules containing defensins form peptide "spears" that poke holes in microbe membranes; the resulting loss of cellular contents kills the invader.

Monocytes take longer to reach a site of infection than neutrophils, but they arrive in large numbers. Upon their arrival monocytes enlarge and differentiate into wandering macrophages, which phagocytize many more microbes than neutrophils can. They also clean up cellular debris following an infection.

At sites of inflammation, basophils leave capillaries, enter tissues, and release heparin, histamine, and serotonin. These substances intensify the inflammatory reaction and are involved in hypersensitivity (allergic) reactions.

Eosinophils leave the capillaries and enter tissue fluid. They are believed to release enzymes, such as histaminase, that combat the effects of histamine and other mediators of inflammation in allergic reactions. Eosinophils also phagocytize antigen–antibody complexes and are effective against certain parasitic worms. A

TABLE 13.2

Significance of *High* and *Low* White Blood Cell Counts

WBC TYPE	*HIGH* COUNT MAY INDICATE	*LOW* COUNT MAY INDICATE
Neutrophils	Bacterial infection, burns, stress, inflammation.	Radiation exposure, drug toxicity, vitamin B_{12} deficiency, systemic lupus erythematosus (SLE).
Lymphocytes	Viral infections, some leukemias.	Prolonged illness, immunosuppression, treatment with cortisol.
Monocytes	Viral or fungal infections, tuberculosis, some leukemias, other chronic diseases.	Bone marrow suppression, treatment with cortisol.
Eosinophils	Allergic reactions, parasitic infections, autoimmune diseases.	Drug toxicity, stress.
Basophils	Allergic reactions, leukemias, cancers, hypothyroidism.	Pregnancy, ovulation, stress, hyperthyroidism.

high eosinophil count often indicates an allergic condition or a parasitic infection.

Lymphocytes are the major combatants in immune responses (described in detail in Chapter 16). During a life span of approximately 100 to 300 days, the majority of lymphocytes continually move among the lymphoid tissues, lymph, and blood, spending only a few hours at a time in the blood. Therefore, only a small proportion of the total lymphocytes are present in the blood at any given moment. Three main types of lymphocytes are B cells, T cells, and natural killer (NK) cells. B cells develop into plasma cells, which produce antibodies that help destroy bacteria and inactivate their toxins. T cells attack viruses, fungi, transplanted cells, cancer cells, and some bacteria. Immune responses carried out by B cells and T cells help combat infection and provide protection against some diseases. T cells also are responsible for transfusion reactions, allergies, and the rejection of transplanted organs. Natural killer cells attack a wide variety of infectious microbes and certain spontaneously arising tumor cells.

An increase in the number of circulating WBCs usually indicates inflammation or infection. A physician may order a **differential white blood cell count** to detect infection or inflammation, determine the effects of possible poisoning by chemicals or drugs, monitor blood disorders (such as leukemia) and the effects of chemotherapy, or detect allergic reactions and parasitic infections. Because each type of white blood cell plays a different role, determining the *percentage* of each type in the blood assists in diagnosing the condition.

Table 13.2 lists the significance of both high and low WBC counts.

CHECKPOINT

9. Explain the importance of emigration, chemotaxis, and phagocytosis in fighting bacterial invaders.
10. What is a differential white blood cell count?
11. What functions are performed by B cells, T cells, and natural killer cells?

PLATELETS

OBJECTIVE
• Describe the structure, function, and origin of platelets.

Besides the immature cell types that develop into erythrocytes and leukocytes, hemopoietic stem cells also differentiate into cells that produce platelets. Under the influence of the hormone **thrombopoietin**, myeloid stem cells develop into megakaryocyte-colony-forming cells that in turn develop into precursor cells

called megakaryoblasts (see Figure 13.3). Megakaryoblasts transform into *megakaryocytes*, huge cells that splinter into 2000–3000 fragments. Each fragment, enclosed by a piece of the cell membrane, is a **platelet (thrombocyte)**. Platelets break off from the megakaryocytes in red bone marrow and then enter the blood circulation. Between 150,000 and 400,000 platelets are present in each μL of blood. They are disc-shaped, 2–4μm in diameter, and exhibit many vesicles but no nucleus. Platelets help stop blood loss from damaged blood vessels by coming together to form a platelet plug that fills the gap in the blood vessel wall. Their vesicles also contain chemicals that, once released, promote blood clotting. Platelets can initiate a series of chemical reactions that culminates in the formation of a network of insoluble protein threads called *fibrin*. A **blood clot** is a gel-like mass that consists of fibrin threads, platelets, and any blood cells trapped in the fibrin (Figure 13.6). The blood clot not only provides a seal in the damaged area of a blood vessel to prevent blood loss, but also

Figure 13.6 Scanning electron micrograph (SEM) of a portion of a blood clot, showing a platelet and red blood cells trapped by fibrin threads.

A blood clot is a gel that contains formed elements of the blood entangled in fibrin threads.

Platelet

Red blood cell

Fibrin thread

SEM 15,000x

? What are the two functions of a blood clot?

pulls the edges of the damaged vessel together to help heal the damage. Platelets have a short life span, normally just 5–9 days. Aged and dead platelets are removed from the circulation by fixed macrophages in the spleen and liver.

Table 13.3 summarizes the formed elements in blood.

⊙ **CHECKPOINT**

12. Compare RBCs, WBCs, and platelets with respect to size, number per μL, and life span.

TABLE 13.3

Summary of Formed Elements in Blood

NAME AND APPEARANCE	NUMBER	CHARACTERISTICS*	FUNCTIONS
Red Blood Cells (RBCs) or Erythrocytes	4.8 million/μL in females; 5.4 million/μL in males	7–8 mm diameter, biconcave discs, without nuclei; live for about 120 days.	Hemoglobin within RBCs transports most of the oxygen and part of the carbon dioxide in the blood.
White Blood Cells (WBCs) or Leukocytes	5000–10,000/μL	Most live for a few hours to a few days.†	Combat pathogens and other foreign substances that enter the body.
Granular leukocytes			
Neutrophils	60–70% of all WBCs	10–12 μm diameter; nucleus has 2–5 lobes connected by thin strands of chromatin; cytoplasm has very fine, pale lilac granules.	Phagocytosis. Destruction of bacteria with lysozyme, defensins, and strong oxidants, such as superoxide anion, hydrogen peroxide, and hypochlorite anion.
Eosinophils	2–4% of all WBCs	10–12 μm diameter; nucleus usually has 2 lobes connected by thick strand of chromatin; large, red-orange granules fill the cytoplasm.	Combat the effects of histamine in allergic reactions, phagocytize antigen–antibody complexes, and destroy certain parasitic worms.
Basophils	0.5–1% of all WBCs	8–10 μm diameter; nucleus has 2 lobes; large cytoplasmic granules appear deep blue-purple.	Liberate heparin, histamine, and serotonin in allergic reactions that intensify the overall inflammatory response.
Agranular leukocytes			
Lymphocytes (T cells, B cells, and natural killer cells)	20–25% of all WBCs	Small lymphocytes are 6–9 μm in diameter; large lymphocytes are 10–14 μm in diameter; nucleus is round or slightly indented; cytoplasm forms a rim around the nucleus that looks sky blue; the larger the cell, the more cytoplasm is visible.	Mediate immune responses, including antigen–antibody reactions. B cells develop into plasma cells, which secrete antibodies. T cells attack invading viruses, cancer cells, and transplanted tissue cells. Natural killer cells attack a wide variety of infectious microbes and certain spontaneously arising tumor cells.
Monocytes	3–8% of all WBCs	12–20 μm diameter; nucleus is kidney shaped or horseshoe shaped; cytoplasm is blue-gray and has foamy appearance.	Phagocytosis (after transforming into fixed or wandering macrophages).
Platelets (Thrombocytes)	150,000–400,000/μL	2–4 μm diameter cell fragments that live for 5–9 days; contain many vesicles but no nucleus.	Form platelet plug in homeostasis; release chemicals that promote vascular spasm and blood clotting.

*Colors are those seen when using Wright's stain.

†Some lymphocytes, called T and B memory cells, can live for many years once they are established.

STEM CELL TRANSPLANTS FROM BONE MARROW AND CORD-BLOOD

◉ OBJECTIVE

- Explain the importance of bone marrow transplants and stem cell transplants.

A **bone marrow transplant** is the replacement of cancerous or abnormal red bone marrow with healthy red bone marrow in order to establish normal blood cell counts. In patients with cancer or certain genetic diseases, the defective red bone marrow is destroyed by high doses of chemotherapy and whole body radiation just before the transplant takes place. These treatments kill the cancer cells and destroy the patient's immune system in order to decrease the chance of transplant rejection.

Healthy red bone marrow for transplanting may be supplied by a donor or by the patient when the underlying disease is inactive. The red bone marrow from a donor is usually removed from the iliac crest of the hip bone under general anesthesia with a syringe and is then injected into the recipient's vein, much like a blood transfusion. The injected marrow migrates to the recipient's red bone marrow cavities where the donor's stem cells multiply. If all goes well, the recipient's red bone marrow is replaced entirely by healthy, noncancerous cells.

Bone marrow transplants have been used to treat aplastic anemia, certain types of leukemia, severe combined immunodeficiency disease (SCID), Hodgkin's disease, non-Hodgkin's lymphoma, multiple myeloma, thalassemia, sickle-cell disease, breast cancer, ovarian cancer, testicular cancer, and hemolytic anemia. However, there are some drawbacks. Since the recipient's white blood cells have been completely destroyed by chemotherapy and radiation, the patient is extremely vulnerable to infection. (It takes about 2–3 weeks for transplanted bone marrow to produce enough white blood cells to protect against infection.) Another problem is that transplanted red bone marrow may produce T cells that attack the recipient's tissues, a reaction called *graft-versus-host disease*. Moreover, any of the recipient's T cells that survived the chemotherapy and radiation can attack donor transplant cells. Another drawback is that patients must take immunosuppressive drugs for life. Because these drugs reduce the level of immune system activity, they increase the risk of infection. Immunosuppressive drugs also have side effects such as fever, muscle aches, headache, nausea, fatigue, depression, high blood pressure, and kidney and liver damage.

A more recent advance in obtaining stem cells involves a **cord-blood transplant**. Recall from Chapter 4 that the connection between the placenta and embryo (and later the fetus) is the umbilical cord. The placenta contains stem cells, which may be obtained from the umbilical cord shortly after birth. The stem cells are removed from the cord with a syringe and then frozen. Stem cells from the cord have several advantages over those obtained from red bone marrow:

1. They are easily collected following permission of the newborn's parents.
2. They are more abundant than stem cells in red bone marrow.
3. They are less likely to cause graft-versus-host disease, so the match between donor and recipient does not have to be as close as in a bone marrow transplant. This provides a larger number of potential donors.
4. They are less likely to transmit infections.
5. They can be stored indefinitely in cord-blood banks.

◉ CHECKPOINT

13. Compare and contrast cord-blood transplants and bone marrow transplants.

KEY MEDICAL TERMS ASSOCIATED WITH BLOOD

Acute normovolemic hemodilution (nor-mō-vō-LĒ-mik hē-mō-dī-LOO-shun) Removal of blood immediately before surgery and its replacement with a cell-free solution to maintain sufficient blood volume for adequate circulation. At the end of surgery, once bleeding has been controlled, the collected blood is returned to the body.

Autologous preoperative transfusion (aw-TOL-o-gus trans-FŪ-zhun; *auto-*=self) Donating one's own blood; can be done up to 6 weeks before elective surgery. Also called *predonation*. This procedure eliminates the risk of incompatibility and blood-borne disease.

Blood bank A facility that collects and stores a supply of blood for future use by the donor or others. Because blood banks have additional and diverse functions (immunohematology reference work, continuing medical education, bone and tissue storage, and clinical consultation), they are more appropriately referred to as **centers of transfusion medicine.**

Cyanosis (sī-a-NŌ-sis; *cyano-*=blue) Slightly bluish/dark-purple skin discoloration, most easily seen in the nail beds and mucous membranes, due to an increased quantity of reduced hemoglobin (hemoglobin not combined with oxygen) in systemic blood.

Embolus (EM-bō-lus=plug) A blood clot, bubble of air or particle of fat from broken bones, mass of bacteria, or other foreign material transported by the blood.

Gamma globulin (GLOB-ū-lin) Solution of immunoglobulins from blood consisting of antibodies that react with specific pathogens, such as viruses. It is prepared by injecting the specific virus into animals, removing blood from the animals after antibodies have accumulated, isolating the antibodies, and injecting them into a human to provide short-term immunity.

Hemochromatosis (hē-mō-krō-ma-TŌ-sis; *chroma*=color) Disorder of iron metabolism characterized by excessive absorption of ingested iron and excess deposits of iron in tissues (especially the liver, heart, pituitary gland, gonads, and pancreas) that result in bronze discoloration of the skin, cirrhosis, diabetes mellitus, and bone and joint abnormalities.

Hemorrhage (HEM-or-ij; *rhegnynai*=bursting forth) Loss of a large amount of blood; can be either internal (from blood vessels into tissues) or external (from blood vessels directly to the surface of the body).

Jaundice (*jaund-*=yellow) An abnormal yellowish discoloration of the sclerae (white of the eyes), skin, and mucous membranes due to excess bilirubin (yellow-orange pigment) in the blood. The three main categories of jaundice are *prehepatic jaundice*, due to excess production of bilirubin; *hepatic jaundice*, due to abnormal bilirubin processing by the liver caused by congenital liver disease, cirrhosis (scar tissue formation) of the liver, or hepatitis (liver inflammation); and *extrahepatic jaundice*, due to blockage of bile drainage by gallstones or cancer of the bowel or pancreas.

Multiple myeloma (mī-e-LŌ-ma) Malignant disorder of plasma cells in red bone marrow; symptoms (pain, osteoporosis, hypercalcemia, thrombocytopenia, kidney damage) are caused by the growing tumor cell mass or antibodies produced by malignant cells.

Phlebotomist (fle-BOT-ō-mist; *phlebo-*=vein; *-tom*=cut) A technician who specializes in withdrawing blood.

Polycythemia (pol′-ē-sī-THĒ-mē-a) An abnormal increase in the number of red blood cells in which hematocrit is above 55%, the upper limit of normal.

Septicemia (sep′-ti-SĒ-mē-a; *septic-*=decay; *-emia*=condition of blood) Toxins or disease-causing bacteria in the blood. Also called "blood poisoning."

Thrombocytopenia (throm′-bō-sī′-tō-PĒ-nē-a; *-penia*=poverty) Very low platelet count that results in a tendency to bleed from capillaries.

Thrombus (THROM-bus=clot) A clot in the cardiovascular system formed in an unbroken blood vessel (usually a vein). The clot consists of a network of insoluble fibrin threads in which the formed elements of blood are trapped.

Transfusion (trans-FŪ-zhun) Transfer of whole blood, blood components (red blood cells only or plasma only), or red bone marrow directly into the bloodstream. Transfusions are given to increase the oxygen-carrying capacity of the blood, improve immunity, assist blood clotting, and restore blood volume.

Venesection (vē-ne-SEK-shun; *ven*=vein) Opening of a vein for withdrawal of blood. Although **phlebotomy** (fle-BOT-ō-me) is a synonym for venesection, in clinical practice phlebotomy refers to therapeutic bloodletting, such as the removal of some blood to lower its viscosity in a patient with polycythemia.

Whole blood Blood containing all formed elements, plasma, and plasma solutes in natural concentrations.

STUDY OUTLINE

Introduction (p. 437)
1. The cardiovascular system consists of the blood, heart, and blood vessels.
2. Blood is a connective tissue composed of blood plasma (liquid portion) and formed elements (cells and cell fragments).

Functions of Blood (p. 438)
1. Blood transports oxygen, carbon dioxide, nutrients, wastes, and hormones.
2. It helps regulate pH, body temperature, and water content of cells.
3. It provides protection through clotting and by combating toxins and microbes, a function of certain phagocytic white blood cells or specialized plasma proteins.

Physical Characteristics of Blood (p. 438)
1. Physical characteristics of blood include a viscosity greater than that of water; a temperature of 38°C (100.4°F); and a pH of 7.35–7.45.
2. Blood constitutes about 8% of body weight, and its volume is 4–6 liters in adults.

Components of Blood (p. 438)
1. Blood consists of about 55% blood plasma and about 45% formed elements.
2. The hematocrit is the percentage of total blood volume occupied by red blood cells.
3. Blood plasma consists of 91.5% water and 8.5% solutes.
4. Principal solutes include proteins (albumins, globulins, fibrinogen), nutrients, vitamins, hormones, respiratory gases, electrolytes, and waste products.
5. The formed elements in blood include red blood cells (erythrocytes), white blood cells (leukocytes), and platelets.

Formation of Blood Cells (p. 440)
1. Hemopoiesis is the formation of blood cells from hemopoietic stem cells in red bone marrow.
2. Myeloid stem cells form RBCs, platelets, granulocytes, and monocytes. Lymphoid stem cells give rise to lymphocytes.
3. Several hemopoietic growth factors stimulate differentiation and proliferation of the various blood cells.

Red Blood Cells (p. 443)
1. Mature RBCs are biconcave discs that lack nuclei and contain hemoglobin.
2. The function of the hemoglobin in red blood cells is to transport oxygen and some carbon dioxide.

3. RBCs live about 120 days. A healthy male has about 5.4 million RBCs/μL of blood; a healthy female, about 4.8 million/μL.
4. After phagocytosis of aged RBCs by macrophages, hemoglobin is recycled.
5. RBC formation, called erythropoiesis, occurs in adult red bone marrow of certain bones. It is stimulated by hypoxia, which stimulates the release of erythropoietin by the kidneys.
6. A reticulocyte count is a diagnostic test that indicates the rate of erythropoiesis.

White Blood Cells (p. 445)
1. WBCs are nucleated cells. The two principal types are granulocytes (neutrophils, eosinophils, and basophils) and agranulocytes (lymphocytes and monocytes).
2. The general function of WBCs is to combat inflammation and infection. Neutrophils and macrophages (which develop from monocytes) act by phagocytosis.
3. Eosinophils combat the effects of histamine in allergic reactions, phagocytize antigen–antibody complexes, and combat parasitic worms; basophils liberate heparin, histamine, and serotonin in allergic reactions that intensify the inflammatory response.
4. B lymphocytes, in response to the presence of foreign substances called antigens, differentiate into plasma cells that produce antibodies. Antibodies attach to the antigens and render them harmless. This antigen–antibody response combats infection and provides immunity. T lymphocytes destroy foreign invaders directly. Natural killer cells attack infectious microbes and tumor cells.
5. Except for lymphocytes, which may live for years, WBCs usually live for only a few hours or a few days. Normal blood contains 5000–10,000 WBCs/μL.

Platelets (p. 447)
1. Platelets (thrombocytes) are disc-shaped structures without nuclei.
2. They are fragments derived from megakaryocytes and are involved in clotting.
3. Normal blood contains 150,000–400,000 platelets/μL.

Stem Cell Transplants from Bone Marrow and Cord-Blood (p. 449)
1. Bone marrow transplants involve removal of marrow as a source of stem cells from the iliac crest.
2. In a cord-blood transplant, stem cells from the placenta are removed from the umbilical cord.
3. Cord-blood transplants have several advantages over bone marrow transplants.

SELF-QUIZ QUESTIONS

Choose the one best answer to the following questions.

1. Which of the following is *not* a site of hemopoiesis in the adult body?
   a. sternum
   b. liver
   c. vertebrae
   d. head of humerus
   e. parietal bones

2. Which of the following would *not* lead to a type of anemia?
   a. insufficient dietary intake of iron
   b. hemorrhage
   c. inability of the stomach to produce intrinsic factor
   d. leukemia
   e. destruction of red bone marrow by radiation

3. The normal red blood cell count in healthy adult males is:
   a. 5.4 million RBCs/μL.
   b. 2 million RBCs/μL.
   c. 0.5 million RBCs/μL.
   d. 250,000 RBCs/μL.
   e. 8,000 RBCs/μL.

4. The *buffy coat* of a centrifuged sample of blood contains mostly:
   a. erythrocytes.
   b. platelets.
   c. blood plasma.
   d. white blood cells.
   e. gamma globulins.

5. The predominant type of blood plasma protein is:
   a. albumins.
   b. fibrinogen.
   c. immunoglobulins.
   d. erythropoietin.
   e. colony-stimulating factors.

6. Thrombopoietin is a hormone that stimulates the formation of:
   a. albumins.
   b. B lymphocytes.
   c. platelets.
   d. fibrinogen.
   e. erythrocytes.

7. Healthy red marrow for transplanting may be supplied by a donor or by the patient when the underlying disease is inactive. The marrow from a donor is usually removed from the
   a. iliac crest of the hip bone.
   b. liver.
   c. bodies of vertebrae.
   d. clavicle.
   e. lateral malleolus of the tibia.

8. A gaseous hormone that can bind to hemoglobin and be released to cause localized vasodilation is:
   a. erythropoietin.
   b. nitric oxide.
   c. thrombopoietin.
   d. fibrinogen.
   e. lysozyme.

9. Place the following cells in the correct order for the production of red blood cells: (1) proerythroblasts, (2) CFU-E cells, (3) myeloid stem cells, (4) erythroblasts, (5) pluripotent stem cells, (6) reticulocytes.
   a. 5, 3, 2, 1, 4, 6
   b. 3, 2, 5, 6, 1, 4
   c. 4, 5, 3, 6, 1, 2
   d. 5, 2, 1, 3, 4, 6
   e. 2, 3, 5, 1, 4, 6

10. Which of the following best describes the structure of erythrocytes?
    a. irregular fragments of larger cells
    b. biconcave discs that lack nuclei
    c. spherical cells with lobed nuclei and large granules
    d. spherical, slightly indented cells with large, round nuclei
    e. large, irregularly shaped cells with kidney-bean-shaped nuclei

11. Platelets function in the:
    a. transport of carbon dioxide.
    b. production of vitamin K.
    c. utilization of calcium and phosphorus.
    d. destruction of bacteria.
    e. clotting of blood.

12. Place the following cells in the correct order for the process of neutrophil production: (1) myeloblasts, (2) pluripotent stem cells, (3) CFU-GM cells, (4) myeloid stem cells.
    a. 2, 3, 4, 1
    b. 2, 4, 3, 1
    c. 4, 2, 1, 3
    d. 3, 2, 4, 1
    e. 3, 1, 2, 4

Complete the following.

13. A test to determine the percentage of each type of white blood cell is known as a _____ count.

14. Blood is a connective tissue that consists of about _____ % extracellular material and about _____ % formed elements.

15. The process by which white blood cells squeeze through capillaries to reach an injured or infected body tissue is called _____.

16. _____, commonly known as "blood poisoning," is an extremely dangerous condition in which toxins or disease-causing bacteria enter the blood.

17. The rate of red blood cell production is measured by doing a _____ count.

18. Two families of cytokines that stimulate white blood cell formation are the _____ and the _____.

19. The center of a heme ring is a(n) _____ ion.

Are the following statements true or false? If the statement is False, *find and fix the mistake to make the statement* True.

20. Nucleated cells (including white blood cells) have surface MHC antigens.

21. Red blood cells have lobed nuclei and abundant mitochondria.

22. Embolus and thrombus are types of clots.

23. Gamma globulin is a specific type of virus.

Matching.

24. Match the following terms with their descriptions (answers are used more than once):
    ____ **(a)** constitute the largest percentage (about 60–70%) of leukocytes
    ____ **(b)** make up 20–25% of leukocytes
    ____ **(c)** are involved in immunity; form plasma cells for antibody production
    ____ **(d)** are involved in allergic response; release serotonin, heparin, and histamine
    ____ **(e)** are involved in allergic responses; release antihistamines
    ____ **(f)** develop into wandering macrophages that clean up sites of infection
    ____ **(g)** are important in phagocytosis (two answers)
    ____ **(h)** contain large reddish granules in the cytoplasm
    ____ **(i)** contain large deep blue or purple granules in cytoplasm
    ____ **(j)** cytoplasm is blue-gray and foamy in appearance

    (1) basophils
    (2) eosinophils
    (3) lymphocytes
    (4) monocytes
    (5) neutrophils

CRITICAL THINKING QUESTIONS

1. A professional bicyclist finished second in a grueling cross-country bike race. Although he appeared to be in great physical condition, he suffered a heart attack a few hours after the race. The press speculated that his condition was caused by blood doping. Explain.

2. Maddy has been sniffling and sneezing her way through Human Anatomy class all semester. While checking a smear of her own blood, she noticed that her blood had a lot more of the bluish-black granular cells than her lab partner's blood. Maddy's eyes are watery and itchy so she asked her instructor to confirm her finding. What are these cells and why does Maddy have a higher number of them than usual?

3. Gus suffers from chronic renal (kidney) failure and undergoes dialysis regularly. One of his associated problems is chronic anemia, which the doctor says is directly related to the renal failure. What is likely to be the connection between the two, and how might it be treated?

4. Millie was at her grandmother's house and found a very old bottle of a tonic that promised to relieve the symptoms of "iron-poor blood." What do you think the symptoms of "iron-poor blood" would be, and how would enriching the iron content of blood relieve these symptoms?

5. Raoul plans on having some elective surgery performed about a month after the semester is over. His doctor suggested that he donate some of his own blood now. Raoul's not sure about this; he figures he's going to need all the blood he has for the surgery. Why should he give up some of his blood now?

? ANSWERS TO FIGURE QUESTIONS

13.1 Blood volume is about 6 liters in males and 4–5 liters in females, representing about 8% of body weight.

13.2 Platelets are cell fragments.

13.3 Pluripotent stem cells develop from mesenchyme.

13.4 One hemoglobin molecule can transport four O_2 molecules—one bound to each heme group.

13.5 Neutrophils, eosinophils, and basophils are called granular leukocytes because all have cytoplasmic granules that are visible through a light microscope when stained.

13.6 The two functions of a blood clot are to seal a damaged blood vessel and pull the edges of the vessel together to prevent blood loss.

14

THE CARDIOVASCULAR SYSTEM: THE HEART

INTRODUCTION In Chapter 13 we examined the composition and functions of blood. For blood to reach body cells and exchange materials with them, it must be constantly pumped by the heart through the body's blood vessels. The heart beats about 100,000 times every day, which adds up to about 35 million beats in a year, and approximately 2.5 billion times in an average lifetime. The left side of the heart pumps blood through an estimated 100,000 km (60,000 mi) of blood vessels. The right side of the heart pumps blood through the lungs, enabling blood to pick up oxygen and unload carbon dioxide. Even while you are sleeping, your heart pumps 30 times its own weight each minute, which amounts to about 5 liters (5.3 qt) to the lungs and the same volume to the rest of the body. At this rate, the heart pumps more than 14,000 liters (3,600 gal) of blood in a day, or 10 million liters (2.6 million gal) in a year. You don't spend all your time sleeping, however, and your heart pumps more vigorously when you are active. Thus, the actual blood volume the heart pumps in a single day is much larger.

The scientific study of the normal heart and the diseases associated with it is **cardiology** (kar-dē-OL-ō-jē; *cardio-*=heart; *-logy*=study of). This chapter explores the structure of the heart and the unique properties that permit it to pump for a lifetime without rest. •

LOCATION AND SURFACE PROJECTION OF THE HEART

⦿ **OBJECTIVE**

- Describe the location of the heart, and trace its outline on the surface of the chest.

For all its might, the cone-shaped heart is relatively small, roughly the same size as a closed fist—about 12 cm (5 in.) long, 9 cm (3.5 in.) wide at its broadest point, and 6 cm (2.5 in.) thick. Its mass averages 250 g (8 oz) in adult females and 300 g (10 oz) in adult males. The heart rests on the diaphragm, near the midline of the thoracic cavity in the **mediastinum** (mē-dē-a-STĪ-num), an anatomical region that extends from the sternum to the vertebral column, the first rib to the diaphragm, and between the coverings (pleurae) of the lungs (Figure 14.1a, b).

Figure 14.1 **Position of the heart and associated structures in the mediastinum (dashed outline) in part (a), and the points of the heart that correspond to its surface projection.** (See Tortora, *A Photographic Atlas of the Human Body*, 2e, Figures 6.5 and 6.6.)

 The heart is located in the mediastinum; two-thirds of its mass is to the left of the midline.

ANTERIOR

Heart

Right lung

Aorta

Sternum

Muscle

Left lung

Esophagus

Sixth thoracic vertebra

POSTERIOR

(a) Inferior view of transverse section of thoracic cavity showing the heart in the mediastinum

Transverse plane

View

Superior vena cava

SUPERIOR BORDER

RIGHT BORDER

Right lung

Pleura (cut to reveal lung inside)

Diaphragm

INFERIOR SURFACE

Arch of aorta

Pulmonary trunk

Left lung

LEFT BORDER

APEX OF HEART

(b) Anterior view of the heart in the thoracic cavity

Superior vena cava

Rib (cut)

Right lung

Arch of aorta

Pulmonary trunk

Left lung

Heart

Apex of heart

Diaphragm

(c) Anterior view

About two-thirds of the mass of the heart lies to the left of the body's midline. The position of the heart in the mediastinum is more readily appreciated by examining its ends, surfaces, and borders (Figure 14.1b). Visualize the heart as a cone lying on its side. The pointed end of the heart is the **apex**, which is formed by the tip of the left ventricle (a lower chamber of the heart), and rests on the diaphragm. It is directed anteriorly, inferiorly, and to the left. The **base** of the heart is its posterior surface. It is formed by the atria (upper chambers of the heart), mostly the left atrium, into which the four pulmonary veins open, and a portion of the right atrium that receives the superior and inferior vena cavae (see Figure 14.3c, d). In addition to the apex and base, the heart has several surfaces and borders (margins) that are useful in determining its surface projection (described shortly). The **anterior surface** is deep to the sternum and ribs. The **inferior surface** is the portion of the heart that rests mostly on the diaphragm and is found between the apex and right border. The **right border** faces the right lung and extends from the inferior surface to the base; the **left border**, also called the *pulmonary border*, faces the **left lung** and extends from base to the apex.

Determining an organ's **surface projection** means outlining its dimensions with respect to landmarks on the surface of the body. This practice is useful when conducting diagnostic procedures (such as a lumbar puncture), auscultation (listening to heart and lung sounds), and anatomical studies. We can project the heart on the anterior surface of the chest by locating the fol-lowing landmarks (Figure 14.1d): The **superior right point** is located at the superior border of the third right costal cartilage, about 3 cm (1 in.) to the right of the midline. The **superior left point** is located at the inferior border of the second left costal cartilage, about 3 cm to the left of the midline. A line connecting these two points corresponds to the superior margin of the base of the heart. The **inferior left point** is located at the apex of the heart in the fifth left intercostal space, about 9 cm (3.5 in.) to the left of the midline. A line connecting the superior and inferior left points corresponds to the left border of the heart. The **inferior right point** is located at the superior border of the sixth right costal cartilage, about 3 cm to the right of the midline. A line connecting the inferior left and right points corresponds to the inferior surface of the heart, and a line connecting the inferior and superior right points corresponds to the right border of the heart. When all four points are connected, they form an outline that roughly reveals the size and shape of the heart.

● **CHECKPOINT**

1. Describe the position of the heart in the mediastinum by defining its apex, base, anterior and posterior surfaces, and right and left borders.
2. Explain the location of the superior right point, superior left point, inferior left point, and inferior right point. Why are these points significant?

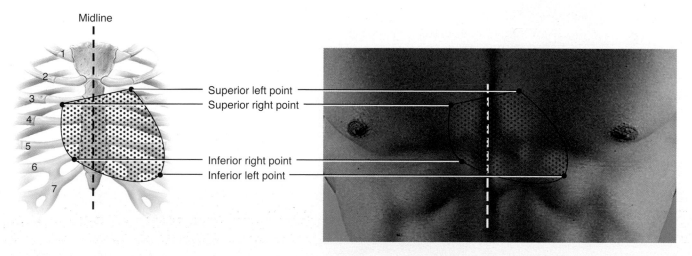

(d) Surface projection of the heart

❓ **What is the mediastinum?**

STRUCTURE AND FUNCTION OF THE HEART

• Describe the structure of the pericardium and the heart wall.
• Discuss the external and internal anatomy of the chambers of the heart.
• List the structure and function of each valve of the heart.

Pericardium

The membrane that surrounds and protects the heart is the **pericardium** (*peri-*=around). It confines the heart to its position in the mediastinum while allowing sufficient freedom of movement for vigorous and rapid contraction. The pericardium consists of two principal portions: (1) the fibrous pericardium and (2) the serous pericardium (Figure 14.2a, c). (See also Figure 1.7e on page 17.) The superficial **fibrous pericardium** is a tough, inelastic, dense irregular connective tissue. It resembles a bag that rests on and attaches to the diaphragm; the open end of the bag is fused to the connective tissues of the blood vessels entering and leaving the heart. The fibrous pericardium prevents overstretching of the heart, provides protection, and anchors the heart in the mediastinum.

The deeper **serous pericardium** is a thinner, more delicate membrane that forms a double layer around the heart (Figure 14.2a, c). The outer **parietal layer** of the serous pericardium is fused to the fibrous pericardium. The inner **visceral layer** of the serous pericardium, also called the **epicardium** (*epi-*=on top of) when combined with the underlying areolar or adipose tissue, adheres tightly to the surface of the heart. Between the parietal and visceral layers of the serous pericardium is a thin film of lubricating fluid. This fluid, known as **pericardial fluid**, is a slippery secretion of the pericardial cells that reduces friction between the membranes as the heart moves. The space that contains the few milliliters of pericardial fluid is called the **pericardial cavity**.

Inflammation of the pericardium is called **pericarditis** (per-i-kar-DĪ-tis). The most common type, *acute pericarditis*, begins suddenly and has no known cause in most cases but is sometimes linked to a viral infection. As a result of irritation to the pericardium, there is chest pain that may extend to the left shoulder and down the left arm (often mistaken for a heart attack) and pericardial friction rub (a scratchy or creaking sound heard through a stethoscope as the visceral layer of the serous pericardium rubs against the parietal layer of the serous pericardium). Acute pericarditis usually lasts for about one week and is treated with drugs that reduce inflammation and pain, such as ibuprofen or aspirin.

Figure 14.2 Pericardium and heart wall.

🔑 ⚍ The pericardium is a sac that surrounds and protects the heart.

Pericardium
Epicardium
Myocardium
Endocardium

PERICARDIUM
Heart wall

FIBROUS PERICARDIUM

ENDOCARDIUM

Trabeculae carneae

PARIETAL LAYER OF
SEROUS PERICARDIUM

Coronary blood vessels

Pericardial cavity

MYOCARDIUM (CARDIAC MUSCLE)

VISCERAL LAYER OF
SEROUS PERICARDIUM
(EPICARDIUM)

(a) Portion of pericardium and right ventricular heart wall showing the divisions of the pericardium and layers of the heart wall

Layers of the Heart Wall

The wall of the heart consists of three layers (Figure 14.2a, b): the epicardium (external layer), the myocardium (middle layer), and the endocardium (inner layer). The **epicardium** is composed of two tissue layers. The outermost, as you just learned, is called the *visceral layer of the serous pericardium.* This is the thin, transparent mesothelial outer layer of the heart wall. Beneath the mesosthelium is a variable layer of delicate connective tissue or adipose tissue. The adipose tissue predominates and becomes thickest over the ventricular surfaces, where it houses the major coronary and cardiac vessels of the heart. The amount of fat varies from person to person and corresponds to the general extent of body fat in an individual, typically increasing with age. The epicardium imparts a smooth, slippery texture to the outermost surface of the heart.

The middle **myocardium** (*myo-* =muscle) is responsible for the pumping action of the heart and is composed of cardiac muscle tissue. It makes up approximately 95% of the heart wall. The muscle fibers (cells), like those of striated skeletal muscle tissue, are wrapped and bundled with endomysial and perimysial connective tissue sheaths. The muscle is organized in a pattern of struts and weaves (criss-crossing bundles of muscular beams) that generate the strong pumping actions of the heart (Figure 14.2d). Although it is striated like skeletal muscle, cardiac muscle is involuntary like smooth muscle. Recall what you learned in chapter 10 about cardiac muscle.

1. Cardiac muscle fibers are shorter in length and less circular in transverse section than skeletal muscle fibers (see Figure 14.2c).
2. They exhibit branching, so that individual cardiac muscle fibers have a "stair-step" appearance (see Table 3.5B on page 86).
3. The ends of cardiac muscle fibers connect to neighboring fibers by **intercalated discs** containing **desmosomes,** which hold fibers together, and **gap junctions,** which allow muscle action potentials to conduct from one muscle fiber to its neighbors.

Myocarditis (mī-ō-kar-DĪ-tis) is an inflammation of the myocardium that usually occurs as a complication of a viral infection, rheumatic fever, or exposure to radiation or certain chemicals or medications.

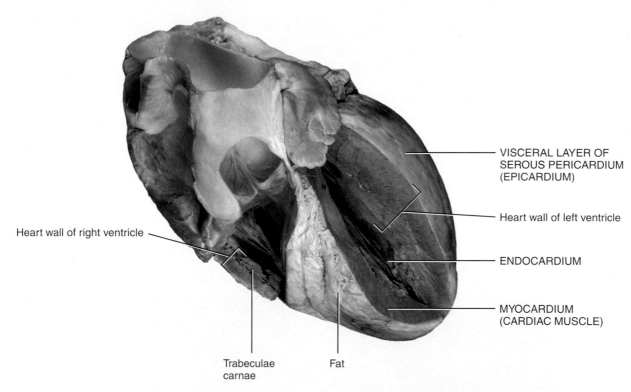

Heart wall of right ventricle

VISCERAL LAYER OF SEROUS PERICARDIUM (EPICARDIUM)

Heart wall of left ventricle

ENDOCARDIUM

MYOCARDIUM (CARDIAC MUSCLE)

Trabeculae carnae

Fat

(b) Anterior view of layers of the heart wall

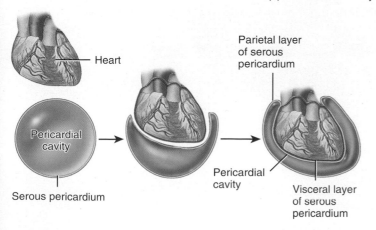

Heart

Serous pericardium

Pericardial cavity

Parietal layer of serous pericardium

Pericardial cavity

Visceral layer of serous pericardium

(c) Simplified relationship of the serous pericardium to the heart

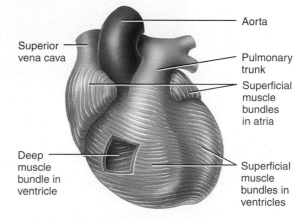

Superior vena cava

Aorta

Pulmonary trunk

Superficial muscle bundles in atria

Deep muscle bundle in ventricle

Superficial muscle bundles in ventricles

(d) Cardiac muscle bundles of the myocardium

CONTINUES ▶

Mitochondria are larger and more numerous in cardiac muscle fibers than in skeletal muscle fibers (Figure 14.2f). In a cardiac muscle fiber, mitochondria take up 25% of the fiber. Recall that actin, myosin, and bands, zones, and Z discs are present. There is one transverse tubule per sarcomere, located at the Z disc. Because the sarcoplasmic reticulum of cardiac muscle fibers is smaller than the SR of skeletal muscle fibers, there is a smaller intracellular reserve of Ca^{2+} in cardiac muscle.

The innermost **endocardium** (*endo-*=within) is a thin layer of endothelium overlying a thin layer of connective tissue. It provides a smooth lining for the chambers of the heart and covers the valves of the heart. The smooth endothelial lining minimizes the surface friction as blood passes through the heart. The endocardium is continuous with the endothelial lining of the large blood vessels attached to the heart. **Endocarditis** (en′-dō-kar-DĪ-tis) refers to an inflammation of the endocardium and typically involves the heart valves. Most cases are caused by bacteria (bacterial endocarditis).

FIGURE 14.2 CONTINUED

(e) Cardiac muscle fibers

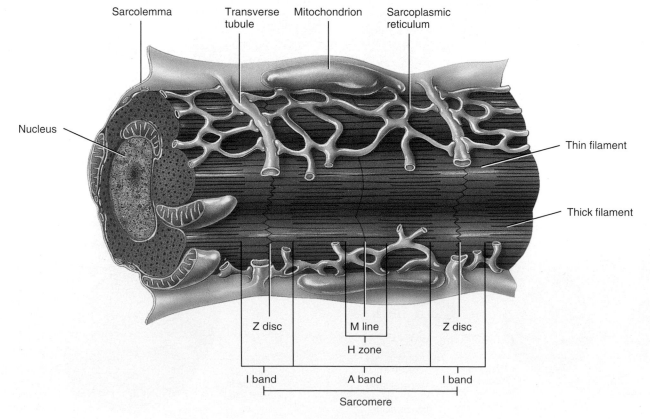

(f) Arrangement of components in a cardiac muscle fiber

? Which layer is both a part of the pericardium and a part of the heart wall?

Chambers of the Heart

The heart is a dual pump that contains four chambers, two upper or receiving chambers called the **atria** (=entry halls or chambers) and the two lower or pumping chambers called the **ventricles** (=little bellies). The paired atria receive blood from blood vessels returning blood to the heart, called veins, while the ventricles eject the blood from the heart into blood vessels called arteries. The right pump, consisting of the right atrium and right ventricle, is the weaker **pulmonary pump**. The pulmonary pump moves deoxygenated blood through the blood vessels of the lungs. The left pump, comprised of the left atrium and left ventricle, is the stronger **systemic pump**. The systemic pump circulates oxygenated blood to all the systems of the body. On the anterior surface of each atrium is a wrinkled pouchlike structure called an **auricle** (OR-i-kul; *auri-*=ear), so named because of its resemblance to a dog's ear (Figure 14.3a, b). Each auricle increases the capacity of an atrium slightly so that it can

Figure 14.3 Structure of the heart: surface features.

 Sulci are grooves that contain blood vessels and fat and mark the external boundaries between the various chambers.

(a) Anterior external view showing surface features

(b) Anterior external view

CONTINUES ▶

hold a greater volume of blood. Also on the surface of the heart are a series of grooves, called **sulci** (SUL-sē), which contain coronary blood vessels and a variable amount of fat (Figure 14.3a–c). Each sulcus (*SUL-kus;* singular) marks the external boundary between two chambers of the heart. The deep **coronary sulcus** (*coron-=*resembling a crown) encircles most of the heart and marks the external boundary between the superior atria and inferior ventricles. The **anterior interventricular sulcus** is a shallow groove on the anterior surface of the heart that marks the external external boundary between the right and left ventricles. This sulcus continues around to the posterior surface

of the heart as the **posterior interventricular sulcus**, which marks the external boundary between the ventricles on the posterior aspect of the heart.

Right Atrium

The **right atrium** forms the right border of the heart (see Figure 14.1b) and is about 2 to 3 mm (0.08 to 0.12 in.) in average thickness. It receives blood from three veins: *superior vena cava, inferior vena cava,* and *coronary sinus* (Figure 14.4a-d). The anterior and posterior walls within the right atrium differ considerably. During development of the heart the atrial

FIGURE 14.3 CONTINUED ▶

(c) Posterior external view showing surface features

(d) Posterior external view

? **The coronary sulcus forms an external boundary between which chambers of the heart?**

chambers enlarge by absorbing a significant portion of the associated entry veins into their walls. This posterior venous portion of the adult atria is characterized by a smooth internal wall. This contrasts with the parallel ridges, the **pecti-** **nate muscles** (*pectin*=comb), that line the remainder of the atrial wall. The ridged regions of the atria signify the original embryonic atrial chambers, the majority of which correspond to the muscular auricles (Figure 14.4b). Between the

Figure 14.4 **Structure of the heart: internal anatomy.**

The thickness of the four chambers varies according to their functions.

Left common carotid artery
Left subclavian artery
Brachiocephalic trunk

Frontal plane

Arch of aorta
Ligamentum arteriosum

Superior vena cava
Right pulmonary artery
PULMONARY VALVE

Left pulmonary artery
Pulmonary trunk

Right pulmonary veins

Left pulmonary veins
LEFT ATRIUM
AORTIC VALVE
BICUSPID (MITRAL) VALVE

Opening of superior vena cava

CHORDAE TENDINEAE

Fossa ovalis
RIGHT ATRIUM
Opening of coronary sinus
Opening of inferior vena cava

LEFT VENTRICLE
INTERVENTRICULAR SEPTUM

PAPILLARY MUSCLE
TRABECULAE CARNEAE

TRICUSPID VALVE

RIGHT VENTRICLE

Inferior vena cava

Descending aorta

(a) Anterior view of frontal section showing internal anatomy

Brachiocephalic trunk
Left subclavian artery

Left common carotid artery

Superior vena cava

Arch of aorta
Ligamentum anteriosum

Right pulmonary vein

RIGHT AURICLE
(cut open)

Pulmonary trunk

Pectinate muscles

Left pulmonary vein
LEFT AURICLE

RIGHT ATRIUM

Cusp of tricuspid valve

Chordae tendineae

LEFT VENTRICLE

Papillary muscle

INTERVENTRICULAR SEPTUM

RIGHT VENTRICLE

TRABECULAE CARNEAE

(b) Anterior view of partially sectioned heart

CONTINUES ▶

■ **FIGURE 14.4 CONTINUED** ▶

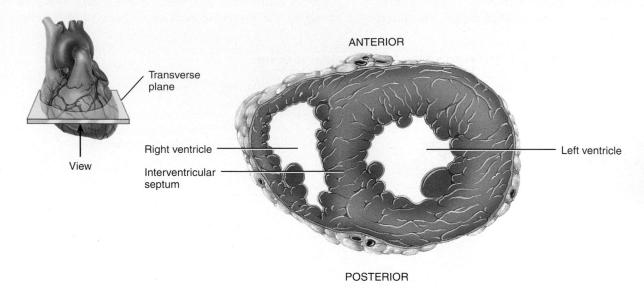

(c) Inferior view of transverse section showing differences in thickness of ventricular walls

? **Which chamber has the thickest wall?**

right atrium and left atrium is a thin partition called the **interatrial septum** (*inter-*=between; *septum*=a dividing wall or partition). A prominent feature of this septum is an oval depression called the **fossa ovalis**, which is the remnant of the *foramen ovale*, an opening in the interatrial septum of the fetal heart that directs blood from the right atrium to the left atrium in order to bypass the nonfunctioning fetal lungs. The foramen ovale normally closes soon after birth (see Figure 15.17 on page 533). Blood passes from the right atrium into the right ventricle through a valve called the **tricuspid valve** (trī-KUS-pid; *tri -* = three; *cuspid*=point) because it consists of three leaflets or cusps (Figure 14.4a and 14.5). It is also called the **right atrioventricular valve**. The valves of the heart are composed of dense connective tissue covered by endocardium.

Right Ventricle

The **right ventricle**, about 4 to 5 mm (0.16 to 0.2 inches) in average thickness, forms most of the anterior surface of the heart (see Figure 14.3a). The inside of the right ventricle contains a series of ridges formed by raised bundles of cardiac muscle fibers called **trabeculae carneae** (tra-BEK-ū-lē KAR-nē-ē; *trabeculae*= little beams; *carneae*=fleshy) (Figure 14.4a-b). Some of the trabeculae carneae contain part of the conduction system of the heart (described on page 470). The cusps of the tricuspid valve are connected to tendonlike cords, the **chordae tendineae** (KOR-dē ten-DIN-ē-ē; *chord-*=cord; *tend-*=tendon), which in turn are connected to cone-shaped trabeculae carneae called **papillary muscles** (*papill-*=nipple) (see Figure 14.6c). Internally, the right ventricle is separated from

the left ventricle by a partition called the **interventricular septum**. Blood passes from the right ventricle through the **pulmonary valve** into a large vessel called the *pulmonary trunk*, which divides into right and left *pulmonary arteries*.

Left Atrium

The **left atrium** forms most of the base of the heart (see Figure 14.3c, d). It receives blood from the lungs through four *pulmonary veins*. Like the right atrium, the inside of the left atrium has a smooth posterior wall where the pulmonary veins were absorbed into the heart during development. Because the ridged pectinate muscles are confined to the auricle of the left atrium, the anterior wall of the left atrium is smooth. Blood passes from the left atrium into the left ventricle through the **bicuspid (mitral) valve** (*bi-*=two), which has two cusps. The term mitral refers to its resemblance to a bishop's miter (hat), which is two-sided. It is also called the **left atrioventricular valve**.

Left Ventricle

The **left ventricle**, the thickest part of the heart (averaging 10 to 15 mm; 0.4 to 0.6 in.), forms the apex of the heart (see Figure 14.1b). Like the right ventricle, the left ventricle contains trabeculae carneae and has chordae tendineae that anchor the cusps of the bicuspid valve to papillary muscles. Blood passes from the left ventricle through the **aortic valve** into the largest artery of the body, the *ascending aorta* (*aorte*=to suspend, because the aorta once was believed to lift up the heart). Some of the blood in the aorta flows into the *coronary arteries*, which branch from the ascending aorta and carry blood to the heart wall; the remainder of the blood passes into the *arch of the aorta* and *descending aorta*. Branches of the arch of the aorta and descending aorta carry blood throughout the body.

During fetal life, a temporary blood vessel, called the *ductus arteriosus*, shunts blood from the pulmonary trunk into the aorta. Hence, only a small amount of blood enters the nonfunctioning fetal lungs (see Figure 15.7 on page 494). The ductus arteriosus normally closes shortly after birth, leaving a remnant known as the **ligamentum arteriosum**, which connects the arch of the aorta and pulmonary trunk (Figure 14.4a, b).

Myocardial Thickness and Function

The thickness of the myocardium of the four chambers varies according to function. The atria are thin-walled because they deliver blood under less pressure into the adjacent ventricles; since the ventricles pump blood under higher pressure over greater distances, their walls are thicker (Figure 14.4a-b). Although the right and left ventricles act as two separate pumps that simultaneously eject equal volumes of blood, the right side has a much smaller workload. It pumps blood under lower pressure only a short distance to the lungs. The left ventricle pumps blood under higher pressure over great distances to all other parts of the body. Thus, the left ventricle works harder than the right ventricle to maintain the same rate of blood flow. The anatomy of the two ventricles confirms this

functional difference: The muscular wall of the left ventricle is considerably thicker than that of the right ventricle (Figure 14.4c; see also 14.2b). Note also that the perimeter of the lumen (space) of the left ventricle is circular, in contrast to that of the right ventricle, which is crescent-shaped.

Fibrous Skeleton of the Heart

In addition to cardiac muscle tissue, the heart wall also contains dense connective tissue that forms the **fibrous skeleton of the heart** (Figure 14.5). Essentially, the fibrous skeleton consists of four dense connective tissue rings that surround the valves of the heart, fuse with one another, and merge with the interventricular septum. In addition to forming a structural foundation for the heart valves, the fibrous skeleton prevents overstretching of the valves as blood passes through them. It also serves as a point of insertion for bundles of cardiac muscle fibers and acts as an electrical insulator between the atria and ventricles.

Heart Valves

As each chamber of the heart contracts, it pushes a volume of blood into a ventricle or out of the heart into an artery. Valves open and close in response to pressure changes as the heart contracts and relaxes. Each of the four valves helps to ensure one-way flow of blood by opening to let blood through and then closing to prevent its backflow.

Figure 14.5 **Fibrous skeleton of the heart.**

The fibrous skeleton provides a base for the attachment of heart valves, prevents overstretching of the valves, serves as a point of insertion for cardiac muscle bundles, and prevents the direct spread of action potentials from the atria to the ventricles.

Superior view (the atria have been removed)

? **What is the composition of the fibrous skeleton of the heart?**

Atrioventricular Valves

Because they are located between an atrium and a ventricle, the tricuspid and bicuspid valves are termed **atrioventricular (AV) valves**. When an AV valve is open, the rounded ends of the cusps (leaflets) project into the ventricle (Figure 14.6a, d). Blood moves from the atria into the ventricles through open AV valves when atrial pressure is higher than ventricular pressure. At this time, the papillary muscles are relaxed, and the chordae tendineae are slack. When the ventricles contract, the pressure of the blood drives the cusps upward until their edges meet and close the opening (Figure 14.6b, e). At the same time, the papillary muscles are also contracting, which pulls on and tightens the chordae tendineae, preventing the valve cusps from everting (opening in the opposite direction into the atria due to the high ventricular pressure). If the AV valves or chordae tendineae are damaged, blood may regurgitate (flow back) into the atria when the ventricles contract.

Semilunar Valves

The aortic and pulmonary valves are known as the **semilunar (SL) valves** (*semi*=half; *lunar*=moon-shaped) because

Figure 14.6 **Valves of the heart.**

Heart valves prevent the backflow of blood.

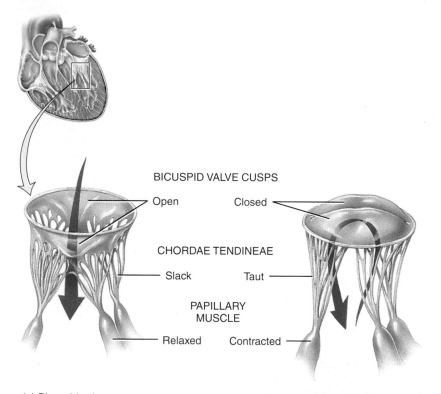

BICUSPID VALVE CUSPS

Open — Closed

CHORDAE TENDINEAE

Slack — Taut

PAPILLARY MUSCLE

Relaxed — Contracted

(a) Bicuspid valve open

(b) Bicuspid valve closed

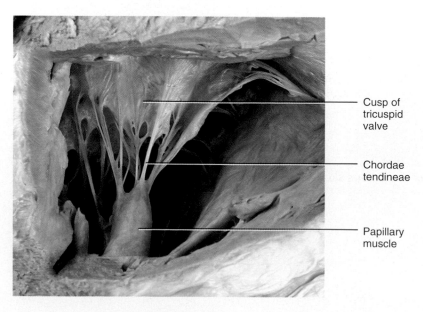

Cusp of tricuspid valve

Chordae tendineae

Papillary muscle

(c) Tricuspid valve open

they are made up of three crescent moon-shaped cusps (Figure 14.6d–g). Each cusp attaches to the arterial wall by its convex outer margin. The SL valves allow ejection of blood from the heart into arteries, but prevent backflow of blood into the ventricles. The free borders of the cusps project into the lumen of the artery. When the ventricles contract, pressure builds up within the chambers. The semilunar valves open when pressure in the ventricles exceeds the pressure in the arteries, permitting ejection of blood from the ventricles into the pulmonary trunk and aorta (Figure 14.6e). As the ventricles relax, blood starts to flow back toward the heart. This backflowing blood fills the valve cusps, which causes the semilunar valves to close tightly (Figure 14.6d).

Surprisingly perhaps, there are no valves guarding the junctions between the venae cavae and the right atrium or the pulmonary veins and the left atrium. As the atria contract, a small amount of blood does flow backward from the atria into these vessels. However, backflow is minimized by a different mechanism; as the atrial muscle contracts, it compresses and nearly collapses the venous entry points.

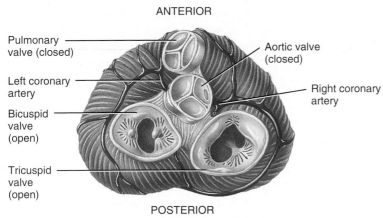

(d) Superior view with atria removed: pulmonary and aortic valves closed, bicuspid and tricuspid valves open.

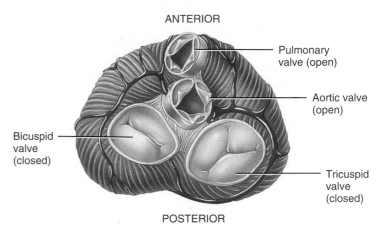

(e) Superior view with atria removed: pulmonary and aortic valves open, bicuspid and tricuspid valves closed.

(f) Superior view of atrioventricular and semilunar valves

(g) Superior view of aortic valve

? **How do papillary muscles prevent atrioventricular valve cusps from everting or swinging upward into the atria?**

• CLINICAL **CONNECTION** | Heart Valve Disorders

Mitral valve

When the valves of the heart operate normally, they open fully and close completely at the proper times. A narrowing of a heart valve opening that restricts blood flow is known as **stenosis** (ste-NO-sis=a narrowing); failure of a valve to close completely is termed **insufficiency** or **incompetence**. In **mitral stenosis**, scar formation or a congenital defect causes narrowing of the mitral valve. One cause of **mitral insufficiency**, in which there is backflow of blood from the left ventricle into the left atrium, is **mitral valve prolapse (MVP)**. In MVP one or both cusps of the mitral valve protrude into the left atrium during ventricular contraction. Mitral valve prolapse is one of the most common valvular disorders, affecting as much as 30% of the population. It is more prevalent in women than in men, and does not always pose a serious threat. In **aortic stenosis** the aortic valve is narrowed, and in **aortic insufficiency** there is backflow of blood from the aorta into the left ventricle.

Certain infectious diseases can damage or destroy the heart valves. One example is **rheumatic fever**, an acute systemic inflammatory disease that usually occurs after a streptococcal infection of the throat. The bacteria trigger an immune response in which antibodies produced to destroy the bacteria instead attack and inflame the connective tissues in joints, heart valves, and other organs. Even though rheumatic fever may weaken the entire heart wall, most often it damages the mitral and aortic valves.

If daily activities are affected by symptoms and if a heart valve cannot be repaired surgically, then the valve must be replaced. Tissue valves may be provided by human donors or pigs; sometimes, mechanical replacements are used. In any case, valve replacement involves open heart surgery. The aortic valve is the most commonly replaced heart valve. •

◉ **CHECKPOINT**

3. Define the following external features of the heart: auricle, coronary sulcus, anterior interventricular sulcus, and posterior interventricular sulcus.

4. Describe the characteristic internal features of each chamber of the heart.

5. List the blood vessels that deliver blood to or receive ejected blood from each chamber of the heart, and name the valve that blood passes through on its way to the next heart chamber or blood vessel.

6. Describe the relationship between wall thickness and function for each heart chamber.

7. How does the fibrous skeleton of the heart assist the operation of heart valves?

8. What causes the heart valves to open and to close?

CIRCULATION OF BLOOD

◉ **OBJECTIVES**

• Describe the flow of blood through the chambers of the heart and through systemic and pulmonary circulations.

• Discuss coronary circulation.

Systemic and Pulmonary Circulations

In postnatal (after birth) circulation, the heart pumps blood into two circuits with each beat—**systemic circulation** and **pulmonary circulation** (*pulmon-*=lung). As you will see later, the two circuits are arranged in series so that the output of one becomes the input of the other, as would happen if you attached two garden hoses together. The left side of the heart, which receives bright red freshly oxygenated (oxygen-rich) blood from the lungs, is the pump for systemic circulation (Figure 14.7a). The left ventricle ejects blood into the *aorta*, which branches into progressively smaller *systemic arteries* that carry the blood to all organs throughout the body—except for the air sacs (alveoli) of the lungs, which are supplied by the pulmonary circulation. In systemic tissues, arteries give rise to smaller-diameter *arterioles*, which finally lead into extensive beds of *systemic capillaries*. Exchange of nutrients and gases occurs across the thin capillary walls: In the tissues, blood unloads O_2 (oxygen) and picks up CO_2 (carbon dioxide). In most cases, blood flows through only one capillary and then enters a *systemic venule*. Venules carry deoxygenated (oxygen-poor) blood away from tissues and merge to form larger *systemic veins*, and ultimately the blood flows back to the right atrium.

The right side of the heart is the pump for pulmonary circulation (Figure 14.7a); it receives all the dark red deoxygenated blood returning from the systemic circulation. Blood ejected from the right ventricle flows into the *pulmonary trunk*, which branches into *pulmonary arteries* that carry blood to the right and left lungs. In pulmonary capillaries, blood unloads CO_2 which is exhaled, and picks up O_2. The freshly oxygenated blood then flows into pulmonary veins and returns to the left atrium. The flowchart in Figure 14.7b shows the route of blood flow through the chambers and valves of the heart and the pulmonary and systemic circulations.

Coronary Circulation

Nutrients could not possibly diffuse from blood in the chambers of the heart through all the layers of cells that make up the heart wall. For this reason, the wall of the heart has its own blood supply. The flow of blood through the many vessels that pierce the myocardium is called the **coronary (cardiac) circulation** because the arteries of the heart encircle it like a crown encircles the head (*corona*=crown). While it is contracting, the heart receives little oxygenated blood by way of the **coronary arteries**, which branch from the ascending aorta (Figure 14.8a, c, d on pages 468–469). When the heart relaxes, however, the high pressure of blood in the aorta propels blood through the coronary arteries, into capillaries, and then into **coronary veins** (Figure 14.8b, c, d).

Figure 14.7 **Systemic and pulmonary circulations.** Throughout this book, blood vessels that carry oxygenated blood (which looks bright red in the body) are colored red, and those that carry deoxygenated blood (which looks dark red in the body) are colored blue.

The left side of the heart pumps freshly oxygenated blood into systemic circulation—to all tissues of the body except the air sacs (alveoli) of the lungs; the right side of the heart pumps deoxygenated blood into pulmonary circulation—to the air sacs (alveoli) of the lungs.

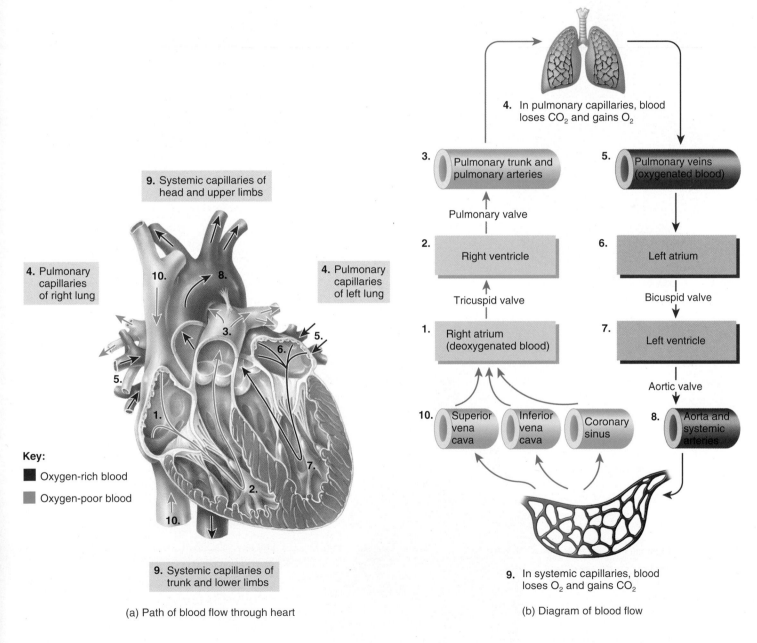

9. Systemic capillaries of head and upper limbs

4. Pulmonary capillaries of right lung

4. Pulmonary capillaries of left lung

Key:
Oxygen-rich blood
Oxygen-poor blood

9. Systemic capillaries of trunk and lower limbs

(a) Path of blood flow through heart

4. In pulmonary capillaries, blood loses CO_2 and gains O_2

3. Pulmonary trunk and pulmonary arteries

5. Pulmonary veins (oxygenated blood)

Pulmonary valve

2. Right ventricle

6. Left atrium

Tricuspid valve

Bicuspid valve

1. Right atrium (deoxygenated blood)

7. Left ventricle

Aortic valve

10. Superior vena cava | Inferior vena cava | Coronary sinus

8. Aorta and systemic arteries

9. In systemic capillaries, blood loses O_2 and gains CO_2

(b) Diagram of blood flow

? In part (b), which numbers represent pulmonary circulation? Which represent systemic circulation?

Figure 14.8 Coronary (cardiac) circulation. The views of the heart from the anterior aspect in (a) and (b) are drawn as if the heart were transparent to reveal blood vessels on the posterior aspect. (See Tortora, *A Photographic Atlas of the Human Body*, 2e, Figure 6.9.)

The right and left coronary arteries deliver blood to the heart; the coronary veins drain blood from the heart into the coronary sinus.

(a) Anterior view of coronary arteries

(b) Anterior view of coronary veins

(c) Anterior view

Coronary Arteries

Two coronary arteries, the right and left coronary arteries, branch from the ascending aorta and supply oxygenated blood to the myocardium (Figure 14.8a, c). The **left coronary artery** passes inferior to the left auricle and divides into the anterior interventricular and circumflex branches. The **anterior interventricular branch** or *left anterior descending (LAD) artery* is in the anterior interventricular sulcus and supplies oxygenated blood to the walls of both ventricles. The **circumflex branch** lies in the coronary sulcus and distributes oxygenated blood to the walls of the left ventricle and left atrium.

The **right coronary artery** supplies small branches (*atrial branches*) to the right atrium. It continues inferior to the right auricle and ultimately divides into the posterior interventricular and marginal branches. The **posterior interventricular branch** follows the posterior interventricular sulcus and supplies the walls of the two ventricles with oxygenated blood. The **marginal branch** in the coronary sulcus transports oxygenated blood to the myocardium of the right ventricle.

Most parts of the body receive blood from branches of more than one artery, and where two or more arteries supply the same region, they usually connect. These connections, called **anastomoses** (a-nas-tō-MŌ-sēs), provide alternate routes called **collateral circuits**, for blood to reach a particular organ or tissue. The myocardium contains many anastomoses that connect branches of a given coronary artery or extend between branches of different coronary arteries. They provide detours for arterial blood if a main route becomes obstructed. Thus, the heart muscle may receive sufficient oxygen even if one of its coronary arteries is partially blocked.

Coronary Veins

After blood passes through the arteries of the coronary circulation, it passes into capillaries where it delivers oxygen and nutrients to the heart muscle and collects carbon dioxide and wastes. From the capillaries the blood then enters into veins. Most of the deoxygenated blood from the myocardium drains into a large vascular sinus in the coronary sulcus on the posterior surface of the heart, called the **coronary sinus** (Figure 14.8b, d), which empties into the right atrium. (A *vascular sinus* is a thin-walled vein that has no smooth muscle to alter its diameter.) The principal tributaries carrying blood into the coronary sinus include the following:

- **Great cardiac vein** in the anterior interventricular sulcus, which drains the areas of the heart supplied by the left coronary artery (left and right ventricles and left atrium)
- **Middle cardiac vein** in the posterior interventricular sulcus, which drains the areas supplied by the posterior interventricular branch of the right coronary artery (left and right ventricles)
- **Small cardiac vein** in the coronary sulcus, which drains the right atrium and right ventricle
- **Anterior cardiac veins,** which drain the right ventricle and open directly into the right atrium

◉ **CHECKPOINT**

9. In correct sequence, list the heart chambers, heart valves, and blood vessels encountered by a drop of blood as it flows out of the right atrium until it reaches the aorta.
10. Which arteries deliver oxygenated blood to the myocardium of the left and right ventricles?

Descending aorta

Superior vena cava

Left pulmonary artery

Right pulmonary artery

Right pulmonary veins

LEFT ATRIUM

RIGHT ATRIUM

Left pulmonary veins

OBLIQUE VEIN

Inferior vena cava

CIRCUMFLEX BRANCH OF LEFT CORONARY ARTERY

CORONARY SINUS

MIDDLE CARDIAC VEIN

Right ventricle

Left ventricle

POSTERIOR INTERVENTRICULAR BRANCH

POSTERIOR VEINS OF THE LEFT VENTRICLE

(d) Posterior view

❓ **Which coronary blood vessel delivers oxygenated blood to the walls of the left atrium and left ventricle?**

CARDIAC CONDUCTION SYSTEM

● OBJECTIVES

- Explain the structural and functional features of the conduction system of the heart.
- Define an electrocardiogram and indicate its diagnostic importance.

During embryonic development, approximately 1% of the cardiac muscle fibers become *autorhythmic cells* (*auto-*=self), that is, cells that repeatedly and rhythmically generate action potentials. Autorhythmic cells act as a **pacemaker**, setting the rhythm for the contraction of the entire heart, and they form the **conduction system**, the route that delivers action potentials throughout the heart muscle. The conduction system ensures that cardiac chambers are stimulated to contract in a coordinated manner, which makes the heart an effective pump. Cardiac action potentials propagate through the following components of the conduction system (Figure 14.9a):

Figure 14.9 **The conduction system of the heart and a normal electrocardiogram.** The route of action potentials through the numbered components of the conduction system in part (a) is described in the text.

🔑 The conduction system ensures that cardiac chambers contract in a coordinated manner.

Frontal plane

Right atrium

1 SINOATRIAL (SA) NODE

2 ATRIOVENTRICULAR (AV) NODE

3 ATRIOVENTRICULAR (AV) BUNDLE (BUNDLE OF HIS)

4 RIGHT AND LEFT BUNDLE BRANCHES

Right ventricle

5 PURKINJE FIBERS

Left atrium

Left ventricle

(a) Anterior view of frontal section

Key:

Atrial contraction
Ventricular contraction

(b) Waves associated with a normal electrocardiogram of a single heartbeat

❓ Which component of the conduction system provides the only electrical connection between the atria and the ventricles?

❶ Normally, cardiac excitation begins in the **sinoatrial (SA) node**, located in the right atrial wall just inferior and lateral to the opening of the superior vena cava. Each action potential from the SA node conducts throughout both atria via gap junctions in the intercalated discs of atrial fibers. With the arrival of the action potential, the atria contract.

❷ By conducting along atrial muscle fibers, the action potential reaches the **atrioventricular (AV) node,** located in the interatrial septum, just anterior to the opening of the coronary sinus. At the AV node, the action potential slows considerably, providing time for the atria to empty their blood into the ventricles.

❸ From the AV node, the action potential enters the **atrioventricular (AV) bundle** (also known as the **bundle of His**), the only site where action potentials can conduct from the atria to the ventricles. (Elsewhere, the fibrous skeleton of the heart electrically insulates the atria from the ventricles.)

❹ After conducting along the AV bundle, the action potential then enters both the **right** and **left bundle branches** that course through the interventricular septum toward the apex of the heart.

❺ Finally, the large-diameter **Purkinje fibers** rapidly conduct the action potential, beginning at the apex of the ventricular myocardium, pushing the blood upward to the semilunar valves. About 0.20 sec (200 milliseconds) after the atria contract, the ventricles contract.

The SA node initiates action potentials 90 to 100 times per minute, faster than any other region of the conducting system. Thus, the SA node sets the rhythm for contraction of the heart—it is the *natural pacemaker* of the heart. Various hormones and neurotransmitters can speed or slow pacing of the heart by SA node fibers. In a person at rest, for example, acetylcholine released by the parasympathetic division of the autonomic nervous system typically slows SA node pacing to about 75 action potentials per minute, causing 75 heartbeats per minute.

Transmission of action potentials through the conduction system generates an electric current that can be detected on the body's surface. A recording of the electrical changes that accompany the heartbeat is called an **electrocardiogram** (e-lek-trō-KAR-dē-ō-gram), which is abbreviated as either *ECG* or *EKG*. The action potentials are graphed as a series of up-and-down waves during an ECG. Three clearly recognizable waves normally accompany each cardiac cycle (Figure 14.9b). The first, called the **P wave**, is the spread of an action potential from the SA node through the two atria. A fraction of a second after the P wave begins, the atria contract. The second wave, called the **QRS wave**, is the spread of the action potential through the ventricles. Shortly after the QRS wave begins, the ventricles contract. The third wave, the **T wave**, indicates ventricular relaxation. There is no wave to show atrial relaxation because the stronger QRS wave masks this event.

• CLINICAL **CONNECTION** Artificial Pacemakers

Artificial pacemaker

Leads

Heart

If the SA node becomes damaged or diseased, the slower AV node can pick up the pacemaking task. Its rate of spontaneous depolarization is 40 to 60 times per minute. If the activity of both nodes is suppressed, the heartbeat may still be maintained by autorhythmic fibers in the ventricles—the AV bundle, a bundle branch, or Purkinje fibers. However, the pacing rate is so slow (20–35 beats per minute) that blood flow to the brain is inadequate. When this condition occurs, normal heart rhythm can be restored and maintained by surgically implanting an **artificial pacemaker**, a device that sends out small electrical currents to stimulate the heart to contract. A pacemaker consists of a battery and impulse generator and is usually implanted beneath the skin just inferior to the clavicle. The pacemaker is connected to one or two flexible leads (wires) that are threaded through the superior vena cava and then passed into various chambers of the heart. Many of the newer pacemakers, referred to as *activity-adjusted pacemakers,* automatically speed up the heartbeat during exercise. •

Variations in the size and duration of deflection waves of an ECG are useful in diagnosing abnormal cardiac rhythms, detecting an enlarged heart, determining if certain regions of the heart are damaged, and identifying the cause of chest pain.

◉ CHECKPOINT

11. What are autorhythmic cells?
12. Trace an action potential through the conduction system of the heart.
13. What is an electrocardiogram? What is its diagnostic significance?

CARDIAC CYCLE (HEARTBEAT)

◉ **OBJECTIVE**

- Describe the phases associated with a cardiac cycle.

The **cardiac cycle** comprises all the events associated with one heartbeat. In a normal cardiac cycle, the two atria contract while the two ventricles relax. Then while the two ventricles contract, the two atria relax. **Systole** (SIS-tō-lē= contraction) refers to the phase of contraction of a chamber of the heart; **diastole** (di-AS-tō-lē=dilation or expansion) is the phase of relaxation. For the purposes of our discussion, we will divide the cardiac cycle into the following phases illustrated in Figure 14.10:

❶ Relaxation period. At the end of a cardiac cycle when the ventricles start to relax, all four chambers are in diastole. This is the beginning of the relaxation period. As the ventricles relax, pressure within the chambers drops, and blood starts to flow from the pulmonary trunk and aorta back toward the ventricles. As this blood becomes trapped in the semilunar cusps, the semilunar valves close. As the ventricles continue to relax, the space inside expands, and the pressure falls.

When ventricular pressure drops below atrial pressure, the AV valves open and ventricular filling begins. The major part of ventricular filling (75%) occurs just after the AV valves open, *without atrial systole.*

❷ Atrial systole (contraction). Atrial systole marks the end of the relaxation period and accounts for the remaining 25% of the blood that fills the ventricles. Throughout the period of ventricular filling, the AV valves are still open and the semilunar valves are still closed.

❸ Ventricular systole (contraction). Ventricular contraction pushes blood up against the AV valves, forcing them shut. For a very brief period, all four valves are closed again. As ventricular contraction continues, pressure inside the chambers rises sharply. When left ventricular pressure rises above the pressure in the arteries, both semilunar valves open, and ejection of blood from the heart begins. This lasts until the ventricles start to relax. Then, the semilunar valves close and another relaxation period begins.

◉ **CHECKPOINT**

14. What is a cardiac cycle?
15. Describe the major events of each phase of the cardiac cycle.

Figure 14.10 The cardiac cycle (heartbeat).

🔑 A cardiac cycle comprises all the events associated with a single heartbeat.

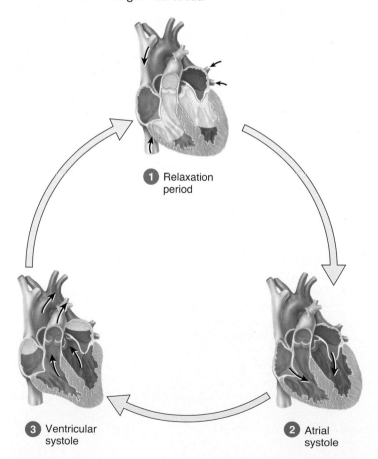

❶ Relaxation period

❸ Ventricular systole

❷ Atrial systole

❓ **What is the status of the four heart valves during ventricular filling?**

HEART SOUNDS

◉ **OBJECTIVE**

- Describe how heart sounds are produced.

Auscultation (aws-kul-TĀ-shun; *ausculta-*=listening), as you learned in Chapter 1, is the act of listening to sounds within the body, and it is usually done with a stethoscope. The sound of the heartbeat comes primarily from blood turbulence caused by the closing of the heart valves. Smoothly flowing blood is silent. Compare the sounds made by whitewater rapids or a waterfall to the silence of a smoothly flowing river. During each cardiac cycle, there are four **heart sounds**, but in a normal heart only the first and second heart sounds (S1 and S2) are loud enough to be heard through a stethoscope.

The first sound (S1), which can be described as a **lubb** sound, is louder and a bit longer than the second sound. S1 is caused by blood turbulence associated with closure of the AV valves soon after ventricular systole begins. The second sound (S2), which is shorter and not as loud as the first, can be described as a **dupp** sound. S2 is caused by blood turbulence associated with closure of the SL valves at the beginning of ventricular diastole. Although S1 and S2 are due to blood turbulence associated with the closure of valves, they are best heard at the surface of the chest in locations that are slightly different from the locations of the valves (Figure 14.11). This is because the sound is carried by the blood flow away from the valves. Normally not loud enough to be heard, S3 is due to blood turbulence during rapid ventricular filling, and S4 is due to blood turbulence during atrial systole.

◉ **CHECKPOINT**

16. What is the basis for heart sounds?
17. Where are the two heart sounds best heard on the surface of the chest?

Figure 14.11 Location of valves (purple) and auscultation sites (red) for heart sounds.

 Listening to sounds within the body is called auscultation; it is usually done with a stethoscope.

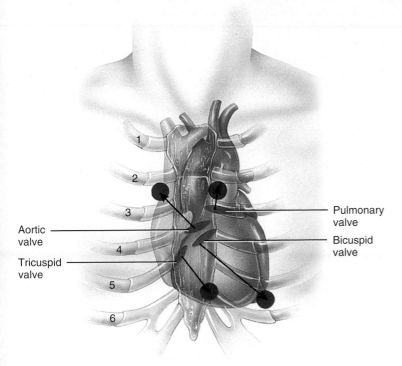

Aortic valve

Tricuspid valve

Pulmonary valve

Bicuspid valve

Anterior view of heart valve locations and auscultation sites

? **Which heart sound is related to blood turbulence associated with closure of the atrioventricular valves?**

EXERCISE AND THE HEART

◉ OBJECTIVE

• Explain the relationship between exercise and the heart.

Regardless of current fitness level, a person's cardiovascular fitness can be improved at any age with regular exercise. Some types of exercise are more effective than others for improving the health of the cardiovascular system. **Aerobics**, any activity that works large body muscles for at least 20 minutes, elevates cardiac output and accelerates metabolic rate. Three to five such sessions a week are usually recommended for improving the health of the cardiovascular system. Brisk walking, running, bicycling, cross-country skiing, and swimming are examples of aerobic activities.

Sustained exercise increases the oxygen demand of the muscles. Whether the demand is met depends mainly on the adequacy of cardiac output and proper functioning of the respiratory system. After several weeks of training, a healthy person increases maximal cardiac output (the amount of blood ejected from the ventricles into their respective arteries per minute), thereby increasing the maximal rate of oxygen delivery to the tissues. Oxygen delivery also rises because skeletal muscles develop more capillary networks in response to long-term training.

During strenuous activity, a well-trained athlete can achieve a cardiac output double that of a sedentary person, in part because training causes hypertrophy (enlargement) of the heart. Even though the heart of a well-trained athlete is larger, *resting* cardiac output is about the same as in a healthy untrained person, because *stroke volume* (volume of blood pumped by each beat of a ventricle) is increased while heart rate is decreased. The resting heart rate of a trained athlete often is only 40–60 beats per minute *(resting bradycardia)*. Regular exercise also helps reduce blood pressure, anxiety, and depression; control weight; and increase the body's ability to dissolve blood clots.

◉ CHECKPOINT

18. What are some of the cardiovascular benefits of regular exercise?

• CLINICAL CONNECTION Coronary Artery Disease

Coronary artery disease (CAD) is a serious medical problem that annually affects about 7 million people. Responsible for nearly three quarters of a million deaths in the United States each year, it is the leading cause of death for both men and women. CAD is defined as the effects of the accumulation of atherosclerotic plaques (described shortly) in coronary arteries that lead to a reduction in blood flow to the myocardium. Some individuals have no signs or symptoms; others experience angina pectoris (chest pain), and still others suffer heart attacks.

Risk Factors for CAD

People who possess combinations of certain risk factors are more likely to develop CAD. *Risk factors* (characteristics, symptoms, or signs present in a disease-free person that are statistically associated with a greater chance of developing a disease) include smoking, high blood pressure, diabetes, high cholesterol levels, obesity, "type A"

personality, sedentary lifestyle, and family history of CAD. Most of these are *modifiable*; that is, they can be altered by changing diet and other habits or can be controlled by taking medications. However, other risk factors are *unmodifiable* (beyond our control), including genetic predisposition (family history of CAD at an early age), age, and gender. (For example, adult males are more likely than adult females to develop CAD; after age 70 the risks are roughly equal.) Smoking is undoubtedly the number-one risk factor in all CAD-associated disease (ischemic heart disease, strokes, etc.). Smoking roughly doubles the risk of morbidity and mortality.

Development of Atherosclerotic Plaques

Although the following discussion of atherosclerosis applies to coronary arteries, the process can also occur in arteries outside the heart. Thickening of the walls of arteries and loss of elasticity are the main characteristics of a group of diseases referred to as **arteriosclerosis**

LM 20x   LM 20x

Partially obstructed lumen (space through which blood flows)

Atherosclerotic plaque

Normal artery   Obstructed artery

(A) Normal and obstructed arteries

(ar-tē-rē-ō-skle-RŌ-sis; *sclero-*=hardening). One form of arteriosclerosis is **atherosclerosis** (ath-er-ō-skle-RŌ-sis), a progressive disease characterized by the formation in the walls of large and medium-sized arteries of lesions called **atherosclerotic plaques** (Figure A, right).

To understand how atherosclerotic plaques develop, you will need to know about molecules produced by the liver and small intestine called **lipoproteins**. These spherical particles consist of an inner core of triglycerides and other lipids and an outer shell of proteins, phospholipids, and cholesterol. Most lipids, including cholesterol, do not dissolve in water and must be made water-soluble in order to be transported in the blood. This is accomplished by combining them with lipoproteins. Two major lipoproteins are called **low-density lipoproteins (LDLs)** and **high-density lipoproteins (HDLs)**. LDLs transport cholesterol from the liver to body cells for use in cell membrane repair and the production of steroid hormones and bile salts. However, excessive amounts of LDLs promote atherosclerosis, so the cholesterol in these particles is known as "bad cholesterol." HDLs, on the other hand, remove excess cholesterol from body cells and transport it to the liver for elimination. Because HDLs decrease blood cholesterol level, the cholesterol in HDLs is known as "good cholesterol." Basically, when you get your blood test results, you want your LDL to be low and your HDL to be high.

It has recently been learned that inflammation, a defensive response of the body to tissue damage, plays a key role in the development of atherosclerotic plaques. As a result of tissue damage, blood vessels dilate and increase their permeability, and phagocytes, including macrophages, appear in larger numbers. The formation of atherosclerotic plaques begins when excess LDLs from the blood accumulate in the inner layer of an artery wall (layer closest to the bloodstream), the lipids and proteins in the LDLs undergo oxidation, and the proteins bind to sugars (*glycation*). In response, endothelial and smooth muscle cells of the artery secrete substances that attract monocytes from the blood and convert them into macrophages. The macrophages then ingest and become so filled with the oxidized LDL particles that they have a foamy appearance when viewed microscopically (foam cells). T cells (lymphocytes) follow monocytes into the inner lining of an artery and there release chemicals that intensify the inflammatory response. Together, the foam cells, macrophages, and T cells form a **fatty streak,** the beginning of an atherosclerotic plaque. If the plaque is large enough, it can significantly decrease or stop the flow of blood and result in a heart attack.

In most inflammatory responses, macrophages release chemicals that promote healing following fatty streak formation. However, macrophages and endothelial cells also secrete chemicals that cause smooth muscle cells of the middle layer of an artery to migrate to the top of the atherosclerotic plaque, forming a cap over it and thus walling it off from the blood.

Because most atherosclerotic plaques expand away from the bloodstream rather than into it, blood can still flow through the affected artery with relative ease, often for decades. Only about 15% of heart attacks occur when plaque in a coronary artery expands into the bloodstream and restricts blood flow. Most heart attacks occur when the cap over the plaque breaks open in response to chemicals produced by foam cells. In addition, T cells induce foam cells to produce tissue factor (TF), a chemical that begins the cascade of reactions that results in blood clot formation.

In recent years, a number of new risk factors (all modifiable) have been identified as significant predictors of CAD. **C-reactive proteins (CRPs)** are proteins produced by the liver or present in blood in an inactive form that are converted to an active form during inflammation. CRPs may play a direct role in the development of atherosclerosis by promoting the uptake of LDL by macrophages. **Lipoprotein (a)** is an LDL-like particle that attaches to endothelial cells, macrophages, and blood platelets, may promote the proliferation of smooth muscle fibers, and inhibits the breakdown of blood clots. **Fibrinogen** is a glycoprotein involved in blood clotting that may help regulate cellular proliferation, vasoconstriction, and platelet aggregation. **Homocysteine** is an amino acid that may induce blood vessel damage by promoting platelet aggregation and smooth muscle fiber proliferation.

Diagnosis of CAD

Many procedures may be employed to diagnose CAD; the specific procedure used depends on the signs and symptoms of the affected individual.

A resting electrocardiogram is the standard test employed to diagnose CAD. **Stress testing** can also be performed. In an *exercise stress test*, the functioning of the heart is monitored when placed under physical stress by exercising using a treadmill, an exercise bicycle, or arm exercises. During the procedure, ECG recordings are monitored continuously and blood pressure is monitored at intervals. A *nonexercise (pharmacologic) stress test* is used for individuals who cannot exercise due to conditions such as arthritis. A medication is injected that stresses the heart to mimic the effects of exercise. During both exercise and nonexercise stress testing, radionuclide imaging may be performed to evaluate blood flow through heart muscle.

Diagnosis of CAD may also involve **echocardiography** (ek'-ō-kar-dē-OG-ra-fē), a technique that uses ultrasound waves to image the interior of the heart. Echocardiography allows the heart to be seen in motion and can be used to determine the size, shape, and functions of the chambers of the heart; the volume and velocity of blood pumped from the heart and the status of heart valves; the presence of birth defects; and abnormalities of the pericardium. A fairly recent technique for evaluating CAD is **electron beam computerized tomography (EBCT)**, which detects calcium deposits in coronary arteries. These calcium deposits are indicators of atherosclerosis.

Cardiac catheterization (kath'-e-ter-i-ZĀ-shun) is an invasive procedure used to visualize the heart's chambers, valves, and great vessels in order to diagnose and treat disease not related to abnormalities of the coronary arteries. It may also be used to measure pressure in the heart and great vessels; to assess cardiac output; to measure the flow of blood through the heart and great vessels; to identify the location of septal and valvular defects; and to take tissue and blood samples. The basic procedure involves inserting a long, flexible, radiopaque catheter (plastic tube) into a peripheral vein (for *right heart catheterization*) or a peripheral artery (for *left heart catheterization*) and guiding it under *fluoroscopy* (x-ray observation).

Coronary angiography (an'-jē-OG-ra-fē; *angio-*=blood vessel; *-grapho*=to write) is an invasive procedure used to obtain information about the coronary arteries. A catheter is inserted into an artery in the groin or wrist and threaded under fluoroscopy toward the heart and then into the coronary arteries. After the tip of the catheter is in place, a radiopaque contrast medium is injected into the coronary arteries. The radiographs of the arteries, called *angiograms*, appear in motion on a monitor, and the information is recorded on a videotape or computer disc. Coronary angiography may be used to visualize coronary arteries and to inject clot-dissolving drugs, such as streptokinase or tissue plasminogen activator (t-PA), into a coronary artery to dissolve an obstructing thrombus.

Treatment of CAD

Treatment options for CAD include drugs (antihypertensives, nitroglycerine, beta blockers, cholesterol-lowering drugs, and clot-dissolving agents) and various surgical and nonsurgical procedures designed to increase the blood supply to the heart.

Coronary artery bypass grafting (CABG) is a surgical procedure in which a blood vessel from another part of the body is attached (*grafted*) to a coronary artery to bypass an area of blockage (Figure B). A piece of the grafted blood vessel is sutured between the aorta and the unblocked portion of the coronary artery (Figure B).

A nonsurgical procedure used to treat CAD is **percutaneous transluminal coronary angioplasty (PTCA)** (*percutaneous*=through the skin; *trans-*=across; *lumen*=an opening or channel in a tube; *angio-*=blood vessel; *-plasty*=to mold or to shape). In this procedure, a balloon catheter is inserted into an artery of an arm or a leg and gently guided into a coronary artery (Figure C). While dye is released, angiograms (x-rays of blood vessels) are taken to locate the plaques. Next, the catheter is advanced to the point of obstruction, and a balloonlike device is inflated with air to squash the plaque against the blood vessel wall. Because 30 to 50% of PTCA-opened arteries fail due to restenosis (renarrowing) within six months after the procedure, a stent may be inserted via a catheter. A *stent* is a metallic, fine wire tube that is permanently placed in an artery to keep the artery patent

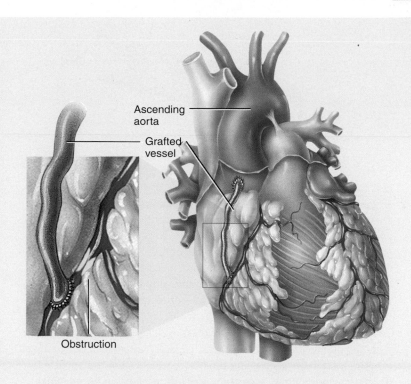

Ascending aorta

Grafted vessel

Obstruction

(B) Coronary artery bypass grafting (CABG)

Balloon | Atherosclerotic plaque | Narrowed lumen of artery | Coronary artery

Balloon catheter with uninflated balloon is threaded to obstructed area in artery

When balloon is inflated, it stretches arterial wall and squashes atherosclerotic plaque

After lumen is widened, balloon is deflated and catheter is withdrawn

(C) Percutaneous transluminal coronary angioplasty (PTCA)

(D) Stent in an artery

(E) Angiogram showing a stent in the circumflex artery

(open), permitting blood to circulate (Figures D, E). Restenosis may be due to damage from the procedure itself, for PTCA may damage the arterial wall, leading to platelet activation, proliferation of smooth muscle fibers, and plaque formation. Recently, *drug-coated (drug-eluting) coronary stents* have been used to prevent restenosis. The stents are coated with one of several *antiproliferative drugs* (drugs that inhibit the proliferation of smooth muscle fibers of the middle layer of an artery) and anti-inflammatory drugs. It has been shown that drug-coated stents reduce the rate of restenosis when compared to bare-metal (noncoated) stents.

One area of current research involves cooling the body's core temperature during procedures such as coronary artery bypass grafting (CABG). There have been some promising results from the application of cold therapy during a cerebral vascular accident (CVA or stroke). This research stemmed from observations of people who had suffered a hypothermic incident (such as cold water drowning) and recovered with relatively minimal neurologic deficits. •

DEVELOPMENT OF THE HEART

◉ OBJECTIVE

• Describe the development of the heart.

The cardiovascular system is the first system to form and function in an embryo, and the heart is the first functional organ. This is necessary because the rapidly growing embryo needs an efficient way to obtain oxygen and nutrients and get rid of wastes. Recall that oxygen and nutrients in the mother's intervillous spaces diffuse into embryonic chorionic villi and wastes diffuse in the opposite direction as early as 21 days after fertilization (see Figure 4.10 on page 104. Blood vessels in the chorionic villi connect to the embryonic heart by way of the umbilical arteries and umbilical vein (see Figure 4.10c on page 104).

As we trace the development of the heart, keep in mind that many congenital (present at birth) disorders of the heart develop during embryonic life. Such disorders are responsible for almost half of all deaths from birth defects.

The *heart* begins its development from **mesoderm** on day 18 or 19 following fertilization. In the head end of the embryo, the heart develops from a group of mesodermal cells called the **cardiogenic area** (kar-dē-ō-JEN-ik; *cardio-*=heart; *-genic*= producing) (Figure 14.12a). In response to induction signals from the underlying endoderm, the mesoderm in the cardiogenic area forms a pair of elongated strands called **cardiogenic cords**. Shortly after, these cords develop a hollow center and then become known as **endocardial tubes** (Figure 14.12b). With lateral folding of the embryo, the paired endocardial tubes approach each other and fuse into a single tube

called the **primitive heart tube** on day 21 following fertilization (Figure 14.12c).

On the twenty-second day, the primitive heart tube develops into five distinct regions and begins to pump blood. From tail end to head end (and the direction of blood flow) they are the (1) **sinus venosus**, (2) **atrium**, (3) **ventricle**, (4) **bulbus cordis**, and (5) **truncus arteriosus**. The sinus venosus initially receives blood from all the veins in the embryo; contractions of the heart begin in this region and follow sequentially in the other regions. Thus, at this stage, the heart consists of a series of unpaired regions. The fates of the regions are as follows:

1. The sinus venosus develops into part of the *right atrium, coronary sinus*, and *sinoatrial (SA) node*.
2. The atrium develops into part of the *right atrium* and the *left atrium*.
3. The ventricle gives rise to the *left ventricle*.
4. The bulbus cordis develops into the *right ventricle*.
5. The truncus arteriosus gives rise to the *ascending aorta* and *pulmonary trunk*.

On day 23, the primitive heart tube elongates. Because the bulbus cordis and ventricle grow more rapidly than other parts of the tube and because the atrial and venous ends of the tube are confined by the pericardium, the tube begins to loop and fold. At first, the primitive heart tube assumes a U-shape; later it becomes S-shaped (Figure 14.12e). As a result of these movements, which are completed by day 28, the atria and ventricles of the future heart are reoriented to assume their final adult positions (Figure 14.12f). The remainder of heart development consists of reconstruction of the chambers and the formation of septa and valves to form a four-chambered heart.

Figure 14.12 Development of the heart. Arrows within the structures indicate the direction of blood flow.

🔑 The heart begins its development from mesoderm on day 18 or 19 following fertilization.

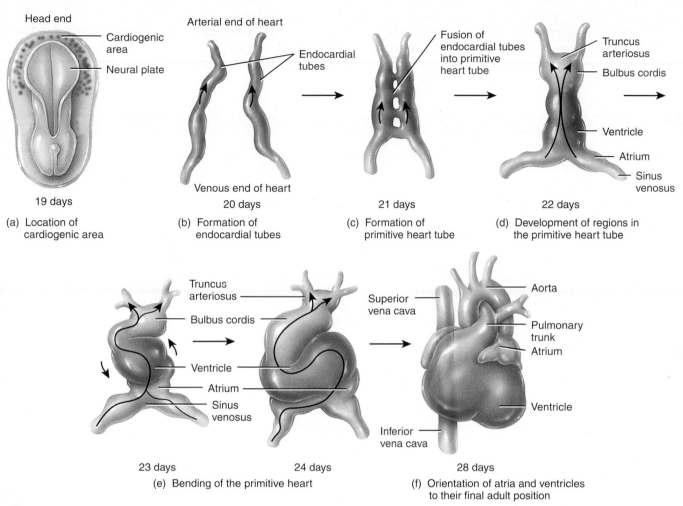

Head end
- Cardiogenic area
- Neural plate

19 days
(a) Location of cardiogenic area

Arterial end of heart
- Endocardial tubes

Venous end of heart

20 days
(b) Formation of endocardial tubes

Fusion of endocardial tubes into primitive heart tube

21 days
(c) Formation of primitive heart tube

- Truncus arteriosus
- Bulbus cordis
- Ventricle
- Atrium
- Sinus venosus

22 days
(d) Development of regions in the primitive heart tube

- Truncus arteriosus
- Bulbus cordis
- Ventricle
- Atrium
- Sinus venosus

- Superior vena cava
- Aorta
- Pulmonary trunk
- Atrium
- Ventricle
- Inferior vena cava

23 days   **24 days**   **28 days**
(e) Bending of the primitive heart   (f) Orientation of atria and ventricles to their final adult position

❓ **When during embryonic development does the primitive heart begin to contract?**

On about day 28, thickenings of mesoderm of the inner lining of the heart wall, called **endocardial cushions**, appear (Figure 14.13). They grow toward each other, fuse, and divide the single **atrioventricular canal** (region between atria and ventricles) into smaller, separate left and right atrioventricular canals. Also, the *interatrial septum* begins its growth toward the fused endocardial cushions. Ultimately, the interatrial septum and endocardial cushions unite and an opening in the septum, the **foramen ovale**, develops. The interatrial septum divides the atrial region into a *right atrium* and a *left*

Figure 14.13 Partitioning of the heart into four chambers.

🔑 Partitioning of the heart begins on about the 28th day after fertilization.

Superior vena cava
- Pulmonary veins
- Atrium
Future interatrial septum
- Atrioventricular canals
Ventricle
- Endocardial cushion
Future interventricular septum
- Inferior vena cava

About 28 days

Right atrium
- Foramen ovale
- Left atrium
Tricuspid valve
- Bicuspid valve
Right ventricle
- Left ventricle

About 8 weeks

❓ **When is partitioning of the heart complete?**

atrium. Before birth, the foramen ovale allows most blood entering the right atrium to pass into the left atrium. After birth, it normally closes so that the interatrial septum is a complete partition. The remnant of the foramen ovale is the fossa ovalis (see Figure 14.4a). Formation of the *interventricular septum* partitions the ventricular region into a *right ventricle* and a *left ventricle*. Partitioning of the atrioventricular canal, atrial region, and ventricular region is basically com-

plete by the end of the fifth week. The *atrioventricular valves* form between the fifth and eighth weeks. The *semilunar valves* form between the fifth and ninth weeks.

◉ CHECKPOINT

19. Why is the cardiovascular system the first of all the systems to develop?
20. From which primary germ layer does the heart develop?

KEY MEDICAL TERMS ASSOCIATED WITH THE HEART

Asystole (ā-SIS-tō-lē; *a*=without) Failure of the myocardium to contract.

Cardiac arrest (KAR-dē-ak a-REST) A clinical term meaning cessation of an effective heartbeat. The heart may be completely stopped or in ventricular fibrillation.

Cardiac rehabilitation (rē-ha-bil-i-TĀ-shun) A supervised program of progressive exercise, psychological support, education, and training to enable a patient to resume normal activities following a myocardial infarction.

Cardiomegaly (kar-dē-ō-MEG-a-lē; *mega*=large) Heart enlargement.

Cardiomyopathy (kar-dē-ō-mī-OP-a-thē; *myo*=muscle; *-pathos*=disease) A progressive disorder in which ventricular structure or function is impaired. In **dilated cardiomyopathy**, the ventricles enlarge (stretch) and become weaker and reduce the heart's pumping action. In **hypertrophic cardiomyopathy**, the ventricular walls thicken and the pumping efficiency of the ventricle is reduced.

Cor pulmonale (CP) (kor pul-mōn-ALE; *cor-*=heart; *pulmon-*=lung) A term referring to right ventricular hypertrophy from disorders that bring about hypertension (high blood pressure) in the pulmonary circulation.

Echocardiography (ek-ō-kar-dē-OG-ra-fē; *cardio*=heart; *-grapho*=to record) The use of ultrasound to study the structure of the heart and its motions and the presence of pericardial fluid.

Electrocardiography (e-LEK-trō-kar-dē-og-ra-fē) (ECG or EKG). A method of recording electrical currents generated by the heart muscle by means of an electrocardiograph.

Electrophysiological testing (ē-lek-trō-fiz-ē-ol-Ō-ji-kal) A procedure in which a catheter with an electrode is passed through blood vessels and introduced into the heart. It is used to detect the exact locations of abnormal electrical conduction pathways. Once an abnormal pathway is located, it can be destroyed by sending a current through the electrode, a procedure called *radio frequency ablation*.

Palpitation (pal-pi-TĀ-shun) A fluttering of the heart or an abnormal rate or rhythm of the heart.

Paroxysmal tachycardia (par-ok-SIZ-mal tak-e-KAR-dē-a; *tachy-*=swift or fast) A period of rapid heartbeats that begins and ends suddenly.

Radionuclide imaging (rā-dē-ō-NOO-klīd; *-radio*=x-ray or gamma ray) A procedure in which a radioactively labeled substance (radionuclide) is introduced into a vein and distributed throughout the body. Gamma cameras provide an image that provides information about organ structure and function. One type of radionuclide imaging, called **thallium scanning**, is used to look at the activity of the heart during exercise. It reveals areas of heart muscle with a poor blood supply.

Sick sinus syndrome An abnormally functioning SA node that initiates heartbeats too slowly or rapidly, pauses too long between heartbeats, or stops producing heartbeats. Symptoms include lightheadedness, shortness of breath, loss of consciousness, and palpitations.

Sudden cardiac death The unexpected cessation of circulation and breathing due to an underlying heart disease such as ischemia, myocardial infarction, or a disturbance in cardiac rhythm.

STUDY OUTLINE

Location and Surface Projection of the Heart (p. 454)

1. The heart is located in the mediastinum; about two-thirds of its mass is to the left of the midline.
2. The heart is basically a cone lying on its side; it consists of an apex, a base, anterior and inferior surfaces, and right and left borders.
3. Four points are used to project the heart's location to the surface of the chest.

Structure and Function of the Heart (p. 456)

1. The pericardium is the membrane that surrounds and protects the heart; it consists of an outer fibrous layer and an inner serous pericardium, which is composed of a parietal layer and a visceral layer.
2. Between the parietal and visceral layers of the serous pericardium is the pericardial cavity, a potential space filled with a few milliliters of lubricating pericardial fluid that reduces friction between the two membranes.
3. The wall of the heart has three layers: epicardium (visceral layer of the serous pericardium plus the underlying adipose tissue), myocardium, and endocardium.
4. The epicardium consists of mesothelium and connective tissue, the myocardium is composed of cardiac muscle tissue, and the endocardium consists of endothelium and connective tissue.

5. Cardiac muscle fibers usually contain a single centrally located nucleus. Compared to skeletal muscle fibers, cardiac muscle fibers have more and larger mitochondria, slightly smaller sarcoplasmic reticulum, and wider transverse tubules, which are located at Z discs.
6. Cardiac muscle fibers are connected via end-to-end intercalated discs. Desmosomes in the discs provide strength, and gap junctions allow muscle action potentials to conduct from one muscle fiber to its neighbors.
7. The heart chambers include two superior chambers, the right and left atria, and two inferior chambers, the right and left ventricles.
8. External features of the heart include the auricles (flaps of each atrium that increase their volume), the coronary sulcus between the atria and ventricles, and the anterior and posterior sulci between the ventricles on the anterior and posterior surfaces of the heart, respectively.
9. The right atrium receives blood from the superior vena cava, inferior vena cava, and coronary sinus. The right atrium is separated internally from the left atrium by the interatrial septum, which contains the fossa ovalis. Blood exits the right atrium through the tricuspid valve.
10. The right ventricle receives blood from the right atrium. It is separated internally from the left ventricle by the interventricular

septum and pumps blood to the lungs through the pulmonary valve and pulmonary trunk.

11. Oxygenated blood enters the left atrium from the pulmonary veins and exits through the bicuspid (mitral) valve.

12. The left ventricle pumps oxygenated blood into the systemic circulation through the aortic valve and aorta.

13. The thickness of the myocardium of the four chambers varies according to the chamber's function. The left ventricle has the thickest wall because of its high workload.

14. The fibrous skeleton of the heart is dense connective tissue that surrounds and supports the valves of the heart.

15. Heart valves prevent the backflow of blood within the heart.

16. The atrioventricular (AV) valves, which lie between atria and ventricles, are the tricuspid valve on the right side of the heart and the bicuspid (mitral) valve on the left. The chordae tendineae and papillary muscles stabilize the flaps of the AV valves and stop blood from backing into the atria.

17. Each of the two arteries that leaves the heart has a semilunar valve (aortic and pulmonary).

Circulation of Blood (p. 466)

1. The left side of the heart is the pump for systemic circulation, the circulation of blood throughout the body except for the air sacs of the lungs. The left ventricle ejects blood into the aorta, and blood then flows into systemic arteries, arterioles, capillaries, venules, and veins, which carry it back to the right atrium.

2. The right side of the heart is the pump for pulmonary circulation, the circulation of blood through the lungs. The right ventricle ejects blood into the pulmonary trunk, and blood then flows into pulmonary arteries, pulmonary capillaries, and pulmonary veins, which carry it back to the left atrium.

3. The flow of blood through the vessels that supply the heart is called coronary (cardiac) circulation.

4. The principal arteries of coronary circulation are the left and right coronary arteries; the principal veins are the cardiac veins and the coronary sinus.

Cardiac Conduction System (p. 470)

1. Autorhythmic cells form the conduction system; these are modified cardiac muscle fibers that spontaneously generate action potentials.

2. Components of the conduction system are the sinoatrial (SA) node (pacemaker), atrioventricular (AV) node, atrioventricular (AV) bundle (bundle of His), bundle branches, and Purkinje fibers.

3. The record of electrical changes during the course of cardiac cycles is called an electrocardiogram (ECG).

4. A normal ECG consists of a P wave (atrial depolarization), a QRS wave (onset of ventricular depolarization), and a T wave (ventricular repolarization).

Cardiac Cycle (Heartbeat) (p. 472)

1. A cardiac cycle consists of the systole (contraction) and diastole (relaxation) of both atria, plus the systole and diastole of both ventricles.

2. The phases of the cardiac cycle are (a) relaxation period, (b) ventricular filling, and (c) ventricular systole.

Heart Sounds (p. 472)

1. S1, the first heart sound (lubb), is caused by blood turbulence associated with the closing of the atrioventricular valves.

2. S2, the second sound (dupp), is caused by blood turbulence associated with the closing of semilunar valves.

Exercise and the Heart (p. 473)

1. Sustained exercise increases oxygen demand on muscles.

2. Among the benefits of aerobic exercise are increased cardiac output, decreased blood pressure, weight control, and increased fibrinolytic activity.

Development of the Heart (p. 476)

1. The heart develops from mesoderm.

2. The endothelial tubes develop into the four-chambered heart and great vessels of the heart.

SELF-QUIZ QUESTIONS

Choose the one best answer to the following questions.

1. On a surface projection, the apex of the heart is found at the:
   a. superior right point.
   b. superior left point.
   c. inferior right point.
   d. inferior left point.
   e. a line connecting the inferior right and left points.

2. The layer of the heart wall that is composed of cardiac muscle tissue is the:
   a. endocardium.
   b. myocardium.
   c. epicardium.
   d. pericardium.
   e. a, b, and c are all correct.

3. Which of the following is the correct route of blood through the heart from systemic circulation to pulmonary circulation and back to systemic circulation?
   a. right atrium, tricuspid valve, right ventricle, pulmonary valve, left atrium, mitral valve, left ventricle, aortic valve
   b. left atrium, triscuspid valve, left ventricle, pulmonary valve, right atrium, mitral valve, right ventricle, aortic valve
   c. left atrium, pulmonary valve, right atrium, tricuspid valve, left ventricle, aortic valve, right ventricle, mitral valve
   d. left ventricle, mitral valve, left atrium, pulmonary valve, right ventricle, tricuspid valve, right atrium, aortic valve
   e. right atrium, mitral valve, right ventricle, pulmonary valve, left atrium, tricuspid valve, left ventricle, aortic valve

4. Which of the following vessels carries the most highly oxygenated blood?
   a. superior vena cava
   b. coronary sinus
   c. inferior vena cava
   d. pulmonary veins
   e. pulmonary arteries

5. Which chamber(s) of the heart can occupy a greater volume of blood?
   a. right atrium   b. right ventricle   c. left atrium
   d. left ventricle   e. two of the above

6. Which of the following sequences correctly represents the conduction of an impulse through the heart?
   a. SA node, AV node, AV bundle, bundle branches
   b. SA node, AV bundle, AV node, bundle branches
   c. AV node, SA node, AV bundle, bundle branches
   d. SA node, bundle branches, AV node, AV bundle
   e. AV node, AV bundle, SA node, bundle branches

Complete the following.

7. The outermost layer of the heart, the _____, is also the innermost layer of the serous pericardium, the _____.

8. The first two vessels to branch from the ascending aorta are the _____ and the _____.

9. The first heart sound is created by turbulence of blood at the closing of the _____ valves.

10. The branch of the left coronary artery that lies in the coronary sulcus and distributes oxygenated blood to the walls of the left atrium and ventricle is the _____.

11. During the period of ventricular filling, the AV valves are in the _____ position, and the semilunar valves are in the _____ position.

12. The term _____ refers to the contraction of a heart chamber, and the word _____ refers to the relaxation of a heart chamber.

13. The heart chamber with the thickest wall is the _____.

14. The _____ point of the heart is located in the fifth left intercostal space, about 9 cm to the left of the midline.

Are the following statements true or false? If the statement is False, *find and fix the mistake to make the statement* True.

15. Trabeculae carneae are ridges located on the inner surface of the right ventricle.

16. The superior vena cava passes through the diaphragm.

17. Blood passes into the aorta from the right ventricle.

18. Blood in the coronary sinus flows into the right atrium.

19. Most of ventricular filling occurs during atrial systole.

20. The Purkinje fibers distribute action potentials first to the apex of the ventricles, then superiorly to the remainder of the ventricles.

21. Chordae tendineae attach AV valve cusps to pectinate muscles.

22. The visceral layer of the serous pericardium is also called the endocardium.

Matching

23. Match the following (answers may be used more than once):

____ **(a)** also called the mitral valve

____ **(b)** prevents backflow of blood into right atrium

____ **(c)** prevents backflow of blood into right ventricle

____ **(d)** prevents backflow of blood into left ventricle

____ **(e)** located between the right atrium and right ventricle

____ **(f)** located between the left atrium and left ventricle

(1) aortic valve
(2) pulmonary valve
(3) bicuspid valve
(4) tricuspid valve

CRITICAL THINKING QUESTIONS

1. Explain why the AV valves don't flip open backward when the ventricles contract.

2. Mr. Williams was diagnosed with blockages in his anterior interventricular branch or left anterior descending (LAD) and circumflex branch of the left coronary artery. What regions of the heart may be affected by these blockages?

3. The heart is constantly beating, which means it's constantly moving. Why doesn't the heart beat its way out of position? Why doesn't the muscle contraction pull the muscle fibers apart?

4. Aleesha's three-year-old daughter has a sore throat. Her pediatrician wants Aleesha to bring her daughter into the office so she can be tested for a streptococcal infection. Aleesha wonders why this is necessary, since children get so many sore throats, and they usually recover quickly. What would you tell her?

5. Franz has been diagnosed with aortic valve stenosis. What does this mean, and how do you think it will affect his heart function?

? ANSWERS TO FIGURE QUESTIONS

14.1 The mediastinum is the anatomical region that extends from the sternum to the vertebral column from the first rib to the diaphragm and between the pleurae of the lungs.

14.2 The visceral layer of the serous pericardium (epicardium) is both a part of the pericardium and a part of the heart wall.

14.3 The coronary sulcus forms an external boundary between the superior atria and inferior ventricles.

14.4 The left ventricle is the chamber of the heart that has the thickest wall.

14.5 The fibrous skeleton is composed of dense connective tissue.

14.6 The papillary muscles contract, which pulls on the chordae tendineae and prevents valve cusps of the atrioventricular valves from everting and letting blood flow back into the atria.

14.7 In part (b), numbers 2 (right ventricle) through 6 depict pulmonary circulation, and numbers 7 (left ventricle) through 10 and 1 (right atrium) depict systemic circulation.

14.8 The circumflex branch delivers oxygenated blood to the walls of the left atrium and left ventricle.

14.9 The only electrical connection between the atria and the ventricles is the atrioventricular bundle.

14.10 The atrioventricular valves are open and the semilunar valves are closed during ventricular filling.

14.11 The first heart sound (S1), or lubb, is associated with the closure of atrioventricular valves.

14.12 The heart begins to contract by the twenty-second day of gestation.

14.13 Partitioning of the heart is complete by the end of the fifth week.

15 | THE CARDIOVASCULAR SYSTEM: BLOOD VESSELS

INTRODUCTION If you have ever planted a garden of significant size, you probably know all about the importance of an irrigation system. At its simplest, an irrigation system is a network of channels or furrows that deliver needed water from one main source to the roots of all the garden's plants. Like a garden irrigation system, the body's blood vessels form an extensive network of "irrigation channels" to deliver needed fluid—in this case the homeostatically maintained blood—to all the body's cells. In fact, this delivery system is probably the most phenomenal irrigation network imaginable. Emanating from a muscular pump, the heart, these vessels form an extensive system of tubular roadways that carry nourishing blood away from the heart toward the tissues, then make a U-turn through small permeable vessels at the tissue level. Here, life-supporting exchanges, such as O_2 and nutrient delivery and waste pickup, are made between the blood and surrounding cells. This irrigation fluid then flows back to the heart through a set of return vessels that parallel the delivery vessels. This circular pattern of flow to and from the heart constitutes the vascular (blood vessel) component of the cardiovascular (circulatory) system. This system of tubular roadways is so extensive that, if all of the individual vessels were placed end-to-end, they would conservatively extend about 75,000 miles, roughly three times around the earth. Furthermore, the small permeable vessels that supply the tissues are so intimately distributed among the trillions of body cells that even the most minor tissue damage leads to the rupture of small vessels. This chapter focuses on the structure and functions of the various types of blood vessels and on the vessels that constitute the major circulatory routes. •

ANATOMY OF BLOOD VESSELS

⦿ **OBJECTIVE**

- Contrast the structure and function of arteries, arterioles, capillaries, venules, and veins.

The five main types of blood vessels are arteries, arterioles, capillaries, venules, and veins. **Arteries** (AR-ter-ēz) carry blood *away from the heart* to other organs. A learning trick to help you remember this is artery=**a**way. Large, elastic arteries leave the heart and divide into medium-sized, muscular arteries that branch out into the various regions of the body. Medium-sized arteries then divide into small arteries, which in turn divide into still smaller arteries called **arterioles** (ar-TER-ē-ōls). As the arterioles enter a tissue, they branch into myriad tiny vessels called **capillaries** (KAP-i-lar′-ēz=hairlike). The thin walls of capillaries allow exchange of substances between the blood and body tissues. Groups of capillaries within a tissue reunite to form small veins called **venules** (VEN-ūls). These in turn merge to form progressively larger blood vessels called veins. **Veins** (VĀNZ) are the blood vessels that convey blood from the tissues *back to the heart.*

Angiogenesis (an′-jē-ō-JEN-e-sis; *angio-*=blood vessel; *-genesis*= production) refers to the growth of new blood vessels. It is an important process in embryonic and fetal development, and in postnatal life serves important functions such as wound healing, formation of a new uterine lining after menstruation, formation of the corpus luteum after ovulation, and development of blood vessels around obstructed arteries in the coronary circulation. Several proteins (peptides) are known to promote and inhibit angiogenesis.

Clinically, angiogenesis is important because cells of a malignant tumor secrete proteins called tumor angiogenesis factors (TAFs) that stimulate blood vessel growth to provide nourishment for the tumor cells. Scientists are seeking chemicals that would inhibit angiogenesis and thus stop the growth of tumors. In diabetic retinopathy, angiogenesis may be important in the development of blood vessels that acturally cause blindness, so finding inhibitors of angiogenesis may also prevent the blindness associated with diabetes.

Basic Structure of a Blood Vessel

The wall of a blood vessel consists of three layers, or tunics, of different tissues: an epithelial inner lining, a middle layer consisting of a smooth muscle and elastic connective tissue, and a connective tissue outer covering. The three structural layers of a generalized blood vessel from innermost to outermost are the *tunica interna (intima)*, *tunica media*, and *tunica externa* (Figure 15.1). Subtle modifications of this basic design account for the five types of blood vessels and the structural and functional differences among the various vessel types. Always remember that structural variations correlate to the differences in function that occur throughout the cardiovascular system.

Tunica Interna (Intima)

The **tunica interna (intima)** (*tunic*=garment; *interna* or *intima*= innermost) forms the inner lining of a blood vessel and is in direct contact with the blood as it flows through the *lumen*, or interior opening, of the vessel (Figure 15.1a, b). Although this layer has multiple parts, these tissue components contribute minimally to the thickness of the vessel wall. Its innermost layer is a simple squamous epithelium, called **endothelium**, which is continuous with the endocardial lining of the heart. The endothelium is a thin layer of flattened cells that lines the inner surface of the entire cardiovascular system (heart and blood vessels). Until recently, endothelial cells were regarded as little more than a passive barrier between the blood and the remainder of the vessel wall. It is now known that endothelial cells are active participants in a variety of vessel-related activities, including physically influencing blood flow, secreting locally acting chemical mediators that influence the contractile state of the vessel's overlying smooth muscle, and assisting with capillary permeability.

The second component of the tunica interna is a *basement membrane* deep to the endothelium. It provides a physical support base for the epithelial layer. Its framework of collagen fibers affords the basal lamina significant tensile strength, yet its properties also provide resilience for stretching and recoil. The basal lamina anchors the endothelium to the underlying connective tissue while also regulating molecular movement. It appears to play an important role in guiding cell movements during tissue repair of blood vessel walls. The outermost part of the tunica interna, which forms the boundary between the tunica interna and tunica media, is the **internal elastic lamina** (*lamina*=thin plate). The internal elastic lamina is a thin sheet of elastic fibers with a variable number of window-like openings that give it the look of Swiss cheese. These openings facilitate diffusion of materials through the tunica interna to the thicker tunica media.

Tunica Media

The **tunica media** (*media*=middle) is a muscular and connective tissue layer that displays the greatest variation among the different vessel types (Figure 15.1a, b). In most vessels, it is a relatively thick layer comprised mainly of smooth muscle cells and substantial amounts of elastic fibers. The primary role of the smooth muscle cells, which extend circularly around the lumen like a ring encircles your finger, is to regulate the diameter of the lumen wall. As you will learn in more detail shortly, the rate of blood flow through different parts of the vascular tree is regulated by the extent of smooth muscle contraction in the walls of particular vessels. Furthermore, the extent of smooth muscle contraction in particular vessel types is crucial in the regulation of blood pressure.

In addition to regulating blood flow and blood pressure, smooth muscle contracts when vessels are damaged to help limit loss of blood through the injured vessel, and help produce the elastic fibers within the tunica media that allow the vessels to stretch and recoil under the applied pressure of the blood.

The tunica media is the most variable of the tunics. As you study the different types of blood vessels in the remainder of this chapter, you will see that the structural differences in this layer account for the many variations in function among the different vessel types.

Tunica Externa

The outer covering of a blood vessel, the **tunica externa** (*externa*=outermost) consists of elastic and collagenous fibers (Figure 15.1a, b). It ranges from a thin connective tissue wrapping to the thickest layer of the blood vessel. Separating the tunica externa from the tunica media is a less prominent network of elastic fibers, the **external elastic lamina,** which is part of the tunica media. The tunica externa contains numerous nerves and, especially in larger vessels, tiny blood vessels that supply the tissue of the vessel wall. These small vessels that supply blood to the tissues of the vessel are called **vasa vasorum** (*vas*=vessel), or vessels to the vessels. They are easily seen on large vessels such as the aorta. In addition to the important role of supplying the vessel wall with nerves and self-vessels, the tunica externa helps anchor the vessels to surrounding tissues.

Figure 15.1 Comparative structure of blood vessels. The relative size of the capillary in (c) is enlarged.

Arteries carry blood from the heart to tissues; veins carry blood from tissues to the heart.

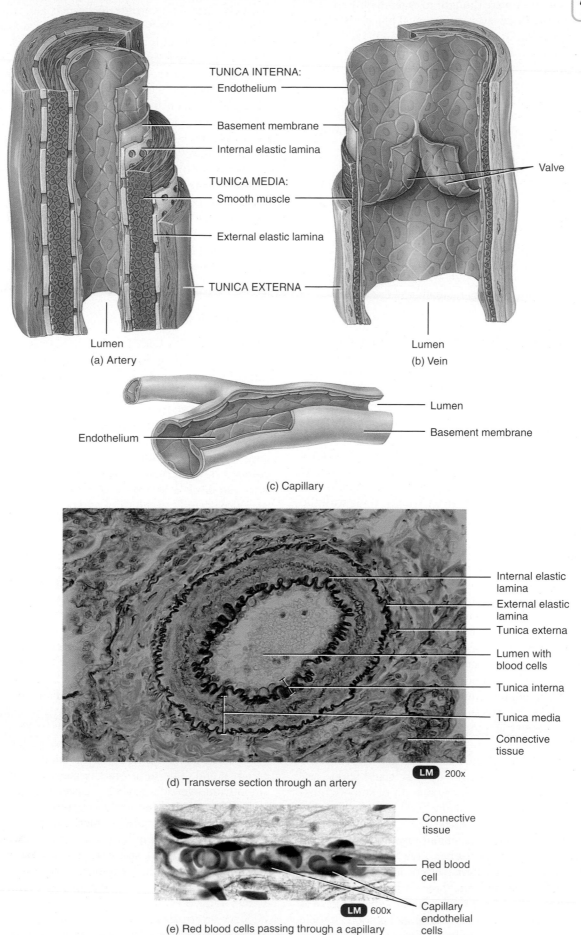

TUNICA INTERNA:
Endothelium

Basement membrane

Internal elastic lamina

TUNICA MEDIA:
Smooth muscle

External elastic lamina

TUNICA EXTERNA

Valve

Lumen
(a) Artery

Lumen
(b) Vein

Lumen

Basement membrane

Endothelium

(c) Capillary

Internal elastic lamina

External elastic lamina

Tunica externa

Lumen with blood cells

Tunica interna

Tunica media

Connective tissue

LM 200x

(d) Transverse section through an artery

Connective tissue

Red blood cell

Capillary endothelial cells

LM 600x

(e) Red blood cells passing through a capillary

Which vessel—the femoral artery or the femoral vein—has a thicker wall? Which has a wider lumen?

Arteries

Because **arteries** (*ar-*=air; *ter-*=to carry) were found empty at death, in ancient times they were thought to contain only air. The wall of an artery has the three layers of a typical blood vessel, but has a thick muscular-to-elastic tunica media (Figure 15.1a). Due to their plentiful elastic fibers, arteries normally have high *compliance*, which means that their walls stretch easily or expand without tearing in response to a small increase in pressure.

Sympathetic fibers of the autonomic nervous system innervate the smooth muscle of blood vessels. An increase in sympathetic stimulation typically stimulates the smooth muscle to contract, squeezing the vessel wall and narrowing the lumen. Such a decrease in the diameter of the lumen of a blood vessel is called **vasoconstriction**. In contrast, when sympathetic stimulation decreases, or in the presence of certain chemicals (such as nitric oxide, H^+, and lactic acid), smooth muscle fibers relax. The resulting increase in lumen diameter is called **vasodilation**. Additionally, when an artery or arteriole is damaged, its smooth muscle contracts, producing vascular spasm of the vessel. Such a vasospasm limits blood flow through the damaged vessel and helps reduce blood loss if the vessel is small.

Elastic Arteries

Elastic arteries are the largest arteries in the body, ranging from the garden hose-sized aorta and pulmonary trunk to the finger-sized branches of the aorta. They have the largest diameter among arteries, but their vessel walls (approximately one-tenth of the vessel's total diameter) are relatively thin compared to the overall size of the vessel. These vessels are characterized by well-defined internal and external elastic laminae, along with a thick tunica media that is dominated by elastic fibers, the **elastic lamellae** (*lamellae*=little plate). The elastic lamellae give the wall a yellowish tint. Elastic arteries include the two major trunks that exit the heart (the aorta and the pulmonary trunk), along with the aorta's major initial branches, such as the brachiocephalic, subclavian, common carotid, and common iliac arteries (see Figure 15.6). Elastic arteries perform an important function: They help propel blood onward while the ventricles are relaxing. As blood is ejected from the heart into elastic arteries, their walls stretch, easily accommodating the surge of blood. As they stretch, the elastic fibers momentarily store mechanical energy, functioning as a **pressure reservoir.** Then, the elastic fibers recoil and convert stored (potential) energy in the vessel into kinetic energy of the blood. Thus, blood continues to move through the arteries even while the ventricles are relaxed. Because they conduct blood from the heart to medium-sized, more muscular arteries, elastic arteries also are called *conducting arteries*.

Muscular Arteries

Medium-sized arteries are called **muscular arteries** because their tunica media contains more smooth muscle and fewer elastic fibers than elastic arteries. The large amount of smooth muscle, approximately three-quarters of the total mass, makes the walls of muscular arteries relatively thick. Thus, muscular arteries are capable of greater vasoconstriction and vasodilation to adjust the rate of blood flow. Muscular arteries have a well-defined internal elastic lamina, but a thin external elastic lamina. These two elastic laminae form the inner and outer boundaries of the muscular tunica media. In large arteries, the thick tunica media can have as many as 40 layers of circumferentially arranged smooth muscle cells; in smaller arteries there are as few as three layers.

Muscular arteries span a range of sizes from the pencil-sized femoral and axillary arteries to string-sized arteries, which enter organs, measuring as little as 0.5 mm in diameter. Compared to elastic arteries, the vessel wall of muscular arteries comprises a larger percentage (25%) of the total vessel diameter. Because the muscular arteries continue to branch and ultimately distribute blood to each of the various organs, they are called **distributing arteries**. Examples include the brachial artery in the arm and radial artery in the forearm (see Figure 15.6).

The tunica externa is often thicker than the tunica media in muscular arteries. This outer layer contains fibroblasts, collagen fibers, and elastic fibers all oriented longitudinally. The loose structure of this layer permits changes in the diameter of the vessel to take place but also prevents shortening or retraction of the vessel when it is cut.

Because of the reduced amount of elastic tissue in the walls of muscular arteries, these vessels do not have the ability to recoil and help propel the blood like the elastic arteries. Instead, the thick, muscular tunica media is primarily responsible for the functions of the muscular arteries. The ability of the muscle to contract and maintain a state of partial contraction is referred to as *vascular tone*. Vascular tone stiffens the vessel wall and is important in maintaining vessel pressure and efficient blood flow.

Anastomoses

Most tissues of the body receive blood from more than one artery. The union of the branches of two or more arteries supplying the same body region is called an **anastomosis** (a-nas-tō-MŌ-sis=connecting; plural is *anastomoses*). Anastomoses between arteries provide alternate routes for blood to reach a tissue or organ. If blood flow stops momentarily when normal movements compress a vessel, or if a vessel is blocked by disease, injury, or surgery, then circulation to a part of the body is not necessarily stopped. The alternate route of blood flow to a body part through an anastomosis is known as **collateral circulation**. Anastomoses may also occur between veins and between arterioles and venules. Arteries that do not anastomose are known as **end arteries**. Obstruction of an end artery interrupts the blood supply to a whole segment of an organ, producing necrosis (death) of that segment. Alternate blood routes may also be provided by nonanastomosing vessels that supply the same region of the body.

Arterioles

Literally meaning small arteries, **arterioles** are abundant microscopic vessels that regulate the flow of blood into the capillary networks of the body's tissues (see Figure 15.2). The approximately 400 million arterioles have diameters that range in size from 30 μm to 15 μm. The wall thickness of arterioles is one-half of the total vessel diameter.

Arterioles have a thin tunica interna with a thin, fenestrated (with small pores) internal elastic lamina that disappears at the terminal end. The tunica media consists of one to two layers of smooth muscle cells having a circular orientation in the vessel wall. The terminal end of the arteriole, the region called the **metarteriole** (*meta*=after), tapers toward the capillary junction. At the metarteriole–capillary junction, the distal-most muscle cell forms the **precapillary sphincter** (=to bind tight), which monitors the blood flow into the capillary; the other muscle cells in the arteriole regulate the resistance (opposition) to blood flow (see Figure 15.2).

Since arterioles play a key role in regulating blood flow from arteries into capillaries by regulating *resistance*, the opposition to blood flow, they are known as **resistance vessels**. In a blood vessel, resistance is due mainly to friction between blood and the inner walls of blood vessels. When the blood vessel diameter is smaller, the friction is greater, so there is more resistance. Contraction of arteriolar smooth muscle causes vasoconstriction, which increases resistance even more and decreases blood flow into capillaries supplied by that arteriole. By contrast, relaxation of arteriolar smooth muscle causes vasodilation, which decreases resistance and increases blood flow into capillaries. A change in arteriole diameter can also affect blood pressure. Vasoconstriction of arterioles increases blood pressure, and vasodilation of arterioles decreases blood pressure.

The tunica externa of an arteriole consists loose collagenous connective tissue containing abundant unmyelinated sympathetic nerves. This sympathetic nerve supply, along with the actions of local chemical mediators, can alter the diameter of arterioles and thus vary the rate of blood flow and resistance through these vessels.

Capillaries

Capillaries (*capillus*=little hair), the smallest of blood vessels, have diameters of 5–10 μm, and form the "U-turns" that connect the arterial outflow to the venous return (Figure 15.2). Since red blood cells have a diameter of 8 μm, they must often fold upon themselves in order to pass single file through the lumens of these vessels. Capillaries form an extensive network, approximately 20 billion in number, of short (hundreds of μm in length), branched, interconnecting vessels that course among the individual cells of the body. This network forms an enormous surface area to make contact with the body's cells. The flow of blood from a metarteri-

ole through capillaries and into a **postcapillary venule** (a venule that receives blood from a capillary) is called the **microcirculation** (*micro*=small) of the body. The primary function of capillaries is the exchange of substances between the blood and interstitial fluid. Because of this, these thin-walled vessels are referred to as **exchange vessels**.

Capillaries are found near almost every cell in the body, but their number varies with the metabolic activity of the tissue they serve. Body tissues with high metabolic requirements, such as muscles, the brain, the liver, the kidneys, and the nervous system, use more O_2 and nutrients and thus have extensive capillary networks. Tissues with lower metabolic requirements, such as tendons and ligaments, contain fewer capillaries. Capillaries are absent in a few tissues, such as all covering and lining epithelia, the cornea and lens of the eye, and cartilage.

The structure of capillaries is well suited to their function as exchange vessels because they lack both a tunica media and a tunica externa. Because capillary walls are composed of only a single layer of endothelial cells (see Figure 15.1e) and a basement membrane, a substance in the blood must pass through just one cell layer to reach the interstitial fluid and tissue cells. Exchange of materials occurs only through the walls of capillaries and the beginning of venules; the walls of arteries, arterioles, most venules, and veins present too thick a barrier. Capillaries form extensive branching networks that increase the surface area available for rapid exchange of materials. In most tissues, blood flows through only a small part of the capillary network when metabolic needs are low. However, when a tissue is active, such as contracting muscle, the entire capillary network fills with blood.

Figure 15.2 Arteriole, capillaries, and venules.

🔑 Arterioles regulate blood flow into capillaries, where nutrients, gases, and wastes are exchanged between the blood and interstitial fluid.

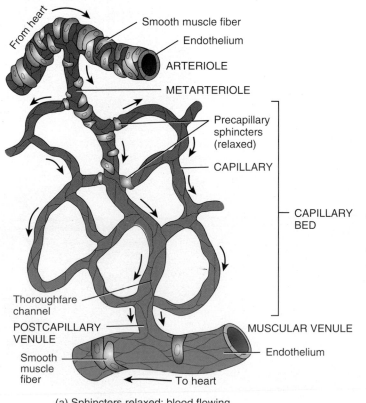

(a) Sphincters relaxed: blood flowing through capillaries

(b) Sphincters contracted: blood flowing through thoroughfare channel

❓ **Why do metabolically active tissues have extensive capillary networks?**

Throughout the body, capillaries function as part of a **capillary bed** (Figure 15.2a), a network of 10-100 capillaries that arises from a single metarteriole. In most parts of the body, blood can flow through a capillary network from an arteriole into a venule as follows:

1. **Capillaries**. In this route, blood flows from an arteriole into **capillaries** and then into venules as noted earlier, at the junctions between the metarteriole and the capillaries are rings of smooth muscle fibers called precapillary sphincters that control the flow of blood through the capillaries. When the precapillary sphincters are relaxed (open), blood flows into the capillaries (Figure 15.2a); when precapillary sphincters contract (close or partially close), blood flow through the capillaries ceases or decreases (Figure 15.2b). Typically, blood flows intermittently through capillaries due to alternating contraction and relaxation of the smooth muscle of metarterioles and the precapillary sphincters. This intermittent contraction and relaxation, which may occur 5 to 10 times per minute, is called **vasomotion**. In part, vasomotion is due to chemicals released by the endothelial cells; nitric oxide is one example. At any given time, blood flows through only about 25% of the capillaries.

2. **Thoroughfare channel**. The proximal end of a metarteriole is surrounded by scattered smooth muscle fibers whose contraction and relaxation help regulate blood flow. The distal end of the vessel has no smooth muscle and resembles a capillary and is called a **thoroughfare channel**. Such a channel provides a direct route for blood from an arteriole to a venule, thus bypassing capillaries.

The body contains three different types of capillaries: continuous capillaries, fenestrated capillaries, and sinusoids (Figure 15.3). Most capillaries are **continuous capillaries**, in which the plasma membranes of endothelial cells form a continuous tube that is interrupted only by **intercellular clefts**, gaps between neighboring endothelial cells (Figure 15.3a). Continuous capillaries are found in the brain, skeletal and smooth muscle, connective tissues, and the lungs.

Other capillaries of the body are **fenestrated capillaries** (*fenestr-*=window). The plasma membranes of the endothelial cells in these capillaries have many **fenestrations**, small pores (holes) ranging from 70 to 100 nm in diameter (Figure 15.3b). Fenestrated capillaries are found in the kidneys, villi of the small intestine, choroid plexuses of the ventricles in the brain, ciliary processes of the eyes, and endocrine glands.

Sinusoids (*sinus*=curve) are wider and more winding than other capillaries. Their endothelial cells may have unusually large fenestrations. In addition to having an incomplete or absent basement membrane (Figure 15.3c), sinusoids have very large intercellular clefts that allow proteins and in some cases even blood cells to pass from a tissue into the bloodstream. For example, newly formed blood cells enter the bloodstream through the sinusoids of red bone marrow. In addition, sinusoids contain specialized lining cells that are adapted to the function of the tissue. Sinusoids in the liver, for example, contain phagocytic cells that remove bacteria and other debris from the blood. The spleen, anterior pituitary, and parathyroid glands also have sinusoids.

Usually, blood passes from the heart and then in sequence through arteries, arterioles, capillaries, venules, and veins and then back to the heart. In some parts of the body, however, blood passes from one capillary network into another through a vein called a *portal vein*. Such a circulation of blood is called a **portal system**. The name of the portal system gives the name of the second capillary location. As you will see later, there are portal

systems associated with the liver (hepatic portal circulation) and the pituitary gland (hypophyseal portal system).

Venules

Unlike their thick-walled arterial counterparts, **venules** (=little vein) and veins have thin walls that do not readily maintain their shape. Venules drain the capillary blood and begin the return flow of blood back toward the heart (see Figure 15.2). Because they

Figure 15.3 Types of capillaries (shown in transverse sections).

Capillaries are microscopic blood vessels that connect arterioles and venules.

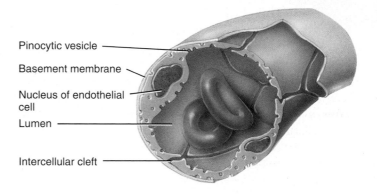

(a) Continuous capillary formed by endothelial cells

Pinocytic vesicle
Basement membrane
Nucleus of endothelial cell
Lumen
Intercellular cleft

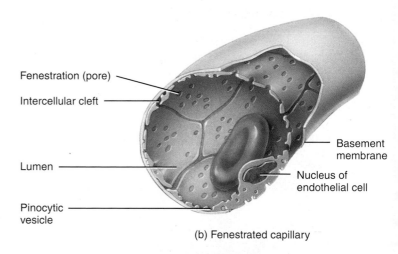

Fenestration (pore)
Intercellular cleft
Lumen
Pinocytic vesicle
Basement membrane
Nucleus of endothelial cell

(b) Fenestrated capillary

Incomplete basement membrane
Lumen
Nucleus of endothelial cell
Intercellular cleft

(c) Sinusoid

? How do materials cross capillary walls?

carry blood toward the heart, veins are referred to as *afferent* (*af-*=toward; *-ferent*=carried) *vessels*.

As noted earlier, venules that initially receive blood from capillaries are called **postcapillary venules**. They are the smallest venules, measuring 10 μm to 50 μm in diameter and have loosely organized intercellular junctions (the weakest endothelial contacts encountered along the entire vascular tree) and thus are very porous. They function as significant sites of exchange of nutrients and wastes and white blood cell emigration, and for this reason form part of the microcirculatory exchange unit along with the capillaries.

As the postcapillary venules continue to enlarge, they acquire one or two layers of circularly arranged smooth muscle cells. These **muscular venules** (50 μm to 200 μm) have thicker walls across which exchanges with the interstitial fluid can no longer occur. The thin walls of the postcapillary and muscular venules are the most distensible elements of the vascular system; this allows them to expand and serve as excellent reservoirs for accumulating large volumes of blood. Blood volume increases of 360% have been measured in the postcapillary and muscular venules.

Veins

While **veins** do show structural changes as they increase in size from small to medium to large, the structural changes are not as distinct as they are in arteries. Veins, in general, have very thin walls relative to their total diameter (average thickness is less than one-tenth of the vessel diameter). They range in size from 0.5 mm in diameter for small veins to 3 cm in the large caval veins entering the heart.

Although veins are composed of essentially the same three layers as arteries, the relative thicknesses of the layers are different. The tunica interna of veins is thinner than that of arteries; the tunica media of veins is much thinner than in arteries, with relatively little smooth muscle and elastic fibers. The tunica externa of a vein is its thickest layer and consists of collagen and elastic fibers. Veins lack the external or internal elastic laminae found in arteries (see Figure 15.1b). They are distensible enough to adapt to variations in the volume and pressure of blood passing through them, but are not designed to withstand high pressure. The lumen of a vein is larger than that of a comparable artery, and veins often appear collapsed (flattened) when sectioned.

The pumping action of the heart is a major factor in moving venous blood back to the heart. The contraction of skeletal muscles in the free lower limbs also helps boost venous return to the heart (Figure 15.4). The average blood pressure in veins is considerably lower than in arteries. The difference in pressure can be noticed when blood flows from a cut vessel. Blood leaves a cut vein in an even, slow flow but spurts rapidly from a cut artery. Most of the structural differences between arteries and veins reflect this pressure difference. For example, the walls of veins are not as strong as those of arteries.

Figure 15.4 Role of skeletal muscle contractions and venous valves in returning blood to the heart. (a) When skeletal muscles contract, the proximal valve opens, and blood is forced toward the heart. (b) Sections of a venous valve.

Venous return depends on the pumping action of the heart, skeletal muscle contractions, and valves in veins.

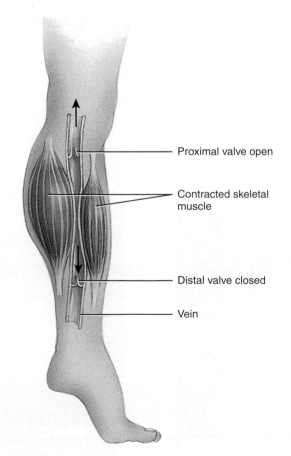

(a) Diagram of contracted skeletal muscles

- Proximal valve open
- Contracted skeletal muscle
- Distal valve closed
- Vein

- Transverse plane
- Cusps of valve
- Transverse section
- Vein
- Frontal plane
- Cusps of valve
- Longitudinally cut

(b) Photographs of a valve in a vein

❓ **Why are valves more important in arm veins and leg veins than in neck veins?**

Many veins, especially those in the free limbs, also contain **valves**, thin folds of tunica interna that form flaplike cusps. The valve cusps project into the lumen, pointing toward the heart (see Figure 15.1b and 15.4a and b). The low blood pressure in veins allows blood returning to the heart to slow and even back up; the valves aid in venous return by preventing the backflow of blood.

A **vascular (venous) sinus** is a vein with a thin endothelial wall that has no smooth muscle to alter its diameter. In a vascular sinus, the surrounding dense connective tissue replaces the tunica media and tunica externa in providing support. For example, dural venous sinuses, which are supported by the dura mater, convey deoxygenated blood from the brain to the heart. Another example of a vascular sinus is the coronary sinus of the heart (see Figure 14.3c on page 460).

While veins follow paths similar to those of their arterial counterparts, they differ from arteries in a number of ways, aside from the structures of their walls. First, veins are more numerous than arteries for several reasons. Some veins are paired and accompany medium- to small-sized muscular arteries. These double sets of veins escort the arteries and connect with one another via venous channels called **anastomotic veins**. The anastomotic veins cross the accompanying artery to form ladder-like rungs between the paired veins (see Figure 15.12c). The greatest number of paired veins occurs within the free limbs. The subcutaneous layer deep to the skin is another source of veins. These veins, called **superficial veins**, course through the subcutaneous layer unaccompanied by parallel arteries. Along their course, the superficial veins form small connections (anastomoses) with the

deep veins that travel below the fascia between the skeletal muscles. These connections allow communication between the deep and superficial blood flows. The amount of blood flow through superficial veins varies from location to location within the body. In the free upper limb, the superficial veins are much larger than the deep veins and serve as the major pathways from the capillaries of the free upper limb back to the heart. In the free lower limb, the opposite is true; the deep veins serve as the principal return pathways. In fact, one-way valves in small anastomosing vessels allow blood to pass from the superficial to the deep veins, but prevent the blood from passing in the reverse direction. This design has important implications in the development of varicose veins.

Blood Distribution

The largest portion of your blood volume at rest—about 64%—is in systemic veins and venules. Systemic arteries hold about 13% of the blood volume, systemic capillaries hold about 7%, pulmonary blood vessels hold about 9%, and the heart holds about 7%. Because systemic veins and venules contain a large percentage of the blood volume, they function as **blood reservoirs** from which blood can be diverted quickly if the need arises. For example, when there is increased muscular activity, the cardiovascular center in the brain stem sends more sympathetic impulses to veins. The result is *venoconstriction*, constriction of veins, which reduces the volume of blood in reservoirs and allows a greater blood volume to flow to skeletal muscles, where it is needed most. A similar mechanism operates in cases of hemorrhage, when blood volume and pressure decrease; in this case, venoconstriction helps coun-

---

• CLINICAL **CONNECTION** Varicose Veins

Leaky venous valves can cause veins to become dilated and twisted in appearance, a condition called **varicose veins** or **varices** (VAR-i-sēz; *varic-*=a swollen vein). The singular is *varix* (VAR-iks). The condition may occur in the veins of almost any body part, but it is most common in the esophagus and in superficial veins of the lower limbs. Those in the lower limbs can range from cosmetic problems to serious medical conditions. The valvular defect may be congenital or may result from mechanical stress (prolonged standing or pregnancy) or aging. The leaking venous valves allow the backflow of blood from the deep veins to the less efficient superficial veins, where the blood pools. This creates pressure that distends the vein and allows fluid to leak into surrounding tissue. As a result, the affected vein and the tissue around it may become inflamed and painfully tender. Veins close to the surface of the legs, especially the saphenous vein, are highly susceptible to varicosities; deeper veins are not as vulnerable because surrounding skeletal muscles prevent their walls from stretching excessively. Varicose veins in the anal canal are referred to as *hemorrhoids*. Esophageal varices result from dilated veins in the walls of the lower part of the esophagus and sometimes the upper part of the stomach. Bleeding esophageal varices are life-threatening and are usually a result of chronic liver disease.

Several treatment options are available for varicose veins in the lower limbs. *Elastic stockings* (support hose) may be used for individuals with mild symptoms or for whom other options are not recommended. *Sclerotherapy* involves injection of a solution into varicosed veins that damages the tunica interna by producing a harmless superficial thrombophlebitis (inflammation involving a blood clot). Healing of the damaged part leads to scar formation that occludes the vein.

Radiofrequency endovenous occlusion involves the application of radiofrequency energy to heat up and close off varicosed veins. *Laser occlusion* uses laser therapy to shut down veins. In a surgical procedure called *stripping*, veins may be removed. In this procedure, a flexible wire is threaded through the vein and then pulled out to strip (remove) it from the body. •

Dilated and twisted appearance of varicose veins in the leg

teract the drop in blood pressure. Among the principal blood reservoirs are the veins of the abdominal organs (especially the liver and spleen) and the veins of the skin.

A summary of the distinguishing features of blood vessels is presented in Table 15.1.

⊙ CHECKPOINT

1. Discuss the importance of elastic fibers and smooth muscle in the tunica media of arteries.

2. Distinguish between elastic and muscular arteries in terms of location, histology, and function.
3. Describe the structural features of capillaries that allow exchange of materials between blood and body cells.
4. What are the main structural and functional differences between arteries and veins?
5. Describe the relationship between anastomoses and collateral circulation.

TABLE 15.1

Distinguishing Features of Blood Vessels

BLOOD VESSEL	SIZE	TUNICA INTERNA	TUNICA MEDIA	TUNICA EXTERNA	FUNCTION
Elastic arteries	Largest arteries in the body.	Well-defined internal elastic lamina.	Thick and dominated by elastic fibers; well-defined external elastic lamina.	Thinner than tunica media.	Conduct blood from the heart to the muscular arteries.
Muscular arteries	Medium-sized arteries.	Well-defined internal elastic lamina.	Thick and dominated by smooth muscle; thin external elastic lamina.	Thicker than tunica media.	Distribute blood to arterioles.
Arterioles	Microscopic (15–30 μm in diameter).	Thin with a fenestrated internal elastic lamina that disappears distally.	One or two layers of circularly oriented smooth muscle; distal-most smooth muscle cell forms a precapillary sphincter.	Loose collagenous connective tissue and sympathetic nerves.	Deliver blood to capillaries and help regulate blood flow to capillaries.
Capillaries	Microscopic; smallest blood vessels (5–10 μm in diameter).	Endothelium and basement membrane.	None.	None.	Permit exchange of nutrients and wastes between blood and interstitial fluid; distribute blood to postcapillary venules.
Postcapillary venules	Microscopic (10–50 μm in diameter).	Endothelium and basement membrane.	None.	Sparse.	Pass blood into muscular venules; permit exchange of nutrients and wastes between blood and interstitial fluid and function in white blood cell emigration.
Muscular venules	Microscopic (50–200 μm in diameter).	Endothelium and basement membrane.	One or two layers of circularly oriented smooth muscle.	Sparse.	Pass blood into veins; reservoirs for accumulating large volumes of blood (along with postcapillary venules).
Veins	Range from 0.5 mm to 3 cm in diameter.	Endothelium and basement membrane; no internal elastic lamina; contain valves; lumen is much larger than in accompanying artery.	Much thinner than in arteries; no external elastic lamina.	Thickest of three layers.	Return blood to the heart, facilitated by valves in veins in free limbs.

CIRCULATORY ROUTES

◉ OBJECTIVE

- Describe and compare the major routes that blood takes through various regions of the body.

Arteries, arterioles, capillaries, venules, and veins are organized into routes that deliver blood throughout the body. We can now look at the basic routes the blood takes as it is transported through its vessels.

Figure 15.5 shows the **circulatory routes** for blood flow. The routes are parallel; that is, in most cases a portion of the cardiac output flows separately to each tissue of the body. Thus each organ receives its own supply of freshly oxygenated blood. The two basic postnatal (after birth) routes for blood flow are the systemic and pulmonary circulations. The **systemic circulation** includes all the arteries and arterioles that carry oxygenated blood from the left ventricle to systemic capillaries, plus the veins and venules that carry deoxygenated blood returning to the right atrium after flowing through body organs. Blood leaving the aorta and flowing through the systemic arteries is a bright red color. As it moves through capillaries, it loses some of its oxygen and picks up carbon dioxide, so that blood in systemic veins is a dark red color.

Some subdivisions of the systemic circulation are the **coronary (cardiac) circulation** (see Figure 14.8 on page 468), which supplies the myocardium of the heart; **cerebral circulation**, which supplies the brain (see Figure 15.7c); and the **hepatic portal circulation**, which extends from the gastrointestinal tract to the liver (see Figure 15.15). The nutrient arteries to the lungs, such as the bronchial arteries, also are part of the systemic circulation.

When blood returns to the heart from the systemic route, it is pumped out of the right ventricle through the **pulmonary circulation** to the lungs (see Figure 15.16). In capillaries of the air sacs (alveoli) of the lungs, the blood loses some of its carbon dioxide and takes on oxygen. Bright red again, it returns to the left atrium of the heart and reenters the systemic circulation as it is pumped out by the left ventricle.

Another major route—the **fetal circulation**—exists only in the fetus and contains special structures that allow the developing fetus to exchange materials with its mother (see Figure 15.17).

Systemic Circulation

The systemic circulation carries oxygen and nutrients to body tissues and removes carbon dioxide and other wastes and heat from the tissues. All systemic arteries branch from the aorta. Deoxygenated blood returns to the heart through the systemic veins. All the veins of the systemic circulation drain into the **superior vena cava**, **inferior vena cava**, or **coronary sinus**, which in turn empty into the right atrium.

The principal arteries and veins of the systemic circulation are described and illustrated in Exhibits 15.1–15.12 and Figures 15.6–15.14 to assist you in learning their names. The blood vessels are organized in the exhibits according to regions of the body. Figure 15.6a shows an overview of the major arteries, and Figure 15.10 shows an overview of the major veins. As you study the various blood vessels in the exhibits, refer to these two figures to see the relationships of the blood vessels under consideration to other regions of the body.

Each of the exhibits contains the following information:

- **An overview.** This information provides a general orientation to the blood vessels under consideration, with emphasis on how the blood vessels are organized into various regions and on distinguishing and/or interesting features about the blood vessels.
- **Blood vessel names.** Students often have difficulty with the pronunciations and meanings of blood vessels' names. To learn them more easily, study the phonetic pronunciations and word derivations that indicate how blood vessels get their names.
- **Region supplied or drained.** For each artery listed, there is a description of the parts of the body that receive blood from the vessel. For each vein listed, there is a description of the parts of the body that are drained by the vessel.
- **Illustrations and photographs.** The figures that accompany the exhibits contain several elements. There is an illustration of the blood vessels under consideration. In many cases, flow diagrams are provided to indicate the patterns of blood distribution or drainage. Cadaver photographs are also included in selected exhibits to provide more realistic views of the blood vessels.

Figure 15.5 **Circulatory routes.** Long black arrows indicate the systemic circulation (detailed in Exhibits 15.1–15.12), short black arrows the pulmonary circulation (detailed in Figure 15.16), and red arrows the hepatic portal circulation (detailed in Figure 15.15). Refer to Figure 14.8 on page 468 for details of the coronary circulation, and to Figure 15.17 for details of the fetal circulation.

Blood vessels are organized into various routes that deliver blood to tissues of the body.

= Oxygenated blood
= Deoxygenated blood

Systemic capillaries of head, neck, and upper limbs
Left pulmonary artery
Aorta
Pulmonary trunk
Pulmonary capillaries of left lung
Left pulmonary veins
Superior vena cava
Left atrium
Right atrium
Right ventricle
Left ventricle
Inferior vena cava
Celiac trunk
Common hepatic artery
Hepatic vein
Splenic artery
Left gastric artery
Sinusoids of liver
Capillaries of spleen
Capillaries of stomach
Hepatic portal vein
Common iliac vein
Superior mesenteric artery
Systemic capillaries of gastrointestinal tract
Inferior mesenteric artery
Internal iliac vein
External iliac vein
Common iliac artery
Internal iliac artery
Systemic capillaries of pelvis
External iliac artery
Arterioles
Venules
Systemic capillaries of lower limbs

? What are the two main circulatory routes?

EXHIBIT 15.1   The Aorta and Its Branches

OBJECTIVES

- Identify the four principal divisions of the aorta.
- Locate the major arterial branches arising from each division.

The **aorta** (=to lift up) is the largest artery of the body, with a diameter of 2–3 cm (about 1 in.). Its four principal divisions are the ascending aorta, arch of the aorta, thoracic aorta, and abdominal aorta. The portion of the aorta that emerges from the left ventricle posterior to the pulmonary trunk is the **ascending aorta**. The beginning of the aorta contains the aortic valve (see Figure 14.4a on page 461). The ascending aorta gives off two coronary artery branches that supply the myocardium of the heart. Then the ascending aorta arches to the left, forming the **arch of the aorta**, which descends and ends at the level of the intervertebral disc between the fourth and fifth thoracic vertebrae.

As the aorta continues to descend, it lies close to the vertebral bodies, passes through the aortic hiatus of the diaphragm, and divides at the level of the fourth lumbar vertebra into two **common iliac arteries**, which carry blood to the free lower limbs. The section of the aorta between the arch of the aorta and the diaphragm is called the **thoracic aorta**; the section between the diaphragm and the common iliac arteries is the **abdominal aorta**. Each division of the aorta gives off arteries that branch into distributing arteries that lead to various organs. Within the organs, the arteries divide into arterioles and then into capillaries that service the systemic tissues (all tissues except the alveoli of the lungs).

CHECKPOINT

- What general regions do each of the four principal divisions of the aorta supply?

DIVISION AND BRANCHES	REGION SUPPLIED
ASCENDING AORTA	
Right and left coronary arteries	Heart.
ARCH OF THE AORTA	
Brachiocephalic trunk (brā′-kē-ō-se-FAL-ik)	
Right common carotid artery (ka-ROT-id)	Right side of head and neck.
Right subclavian artery (sub-KLā-vē-an)	Right free upper limb.
Left common carotid artery	Left side of head and neck.
Left subclavian artery	Left free upper limb.
THORACIC AORTA (*thorac-*=chest)	
Pericardial arteries (per-i-KAR-dē-al)	Pericardium.
Bronchial arteries (BRONG-kē-al)	Bronchi of lungs.
Esophageal arteries (e-sof′-a-JĒ-al)	Esophagus.
Mediastinal arteries (mē′-dē-as-TĪ-nal)	Structures in mediastinum.
Posterior intercostal arteries (in′-ter-KOS-tal)	Intercostal and chest muscles.
Subcostal arteries (sub-KOS-tal)	Same as posterior intercostals.
Superior phrenic arteries (FREN-ik)	Superior and posterior surfaces of diaphragm.
ABDOMINAL AORTA	
Inferior phrenic arteries (FREN-ik)	Inferior surface of diaphragm.
Celiac trunk (SĒ-lē-ak)	
Common hepatic artery (he-PAT-ik)	Liver, stomach, duodenum, pancreas.
Left gastric artery (GAS-trik)	Stomach and esophagus.
Splenic artery (SPLĒN-ik)	Spleen, pancreas, and stomach.
Superior mesenteric artery (MES-en-ter′-ik)	Small intestine, cecum, ascending and transverse colons, and pancreas.
Suprarenal arteries (soo-pra-RĒ-nal)	Adrenal (suprarenal) glands.
Renal arteries (RĒ-nal)	Kidneys.
Gonadal arteries (gō-NAD-al)	
Testicular arteries (tes-TIK-ū-lar)	Testes (male).
Ovarian arteries (ō-VAR-ē-an)	Ovaries (female).
Inferior mesenteric artery	Transverse, descending, and sigmoid colons; rectum.
Common iliac arteries (IL-ē-ak)	
External iliac arteries	Lower limbs.
Internal iliac arteries	Uterus (female), prostate (male), muscles of buttocks, and urinary bladder.

Figure 15.6 Aorta and its principal branches.

🔓 All systemic arteries branch from the aorta.

Right internal carotid — Right external carotid

Right vertebral — Left common carotid

Right common carotid —

Right subclavian — Left subclavian

Brachiocephalic trunk — ARCH OF AORTA

ASCENDING AORTA — Left axillary

Right brachial — THORACIC AORTA

— Diaphragm

ABDOMINAL AORTA —

Celiac trunk — Left gastric

— Splenic

Common hepatic — Left renal

Right radial — Superior mesenteric

Left gonadal (testicular or ovarian)

Right renal — Inferior mesenteric

Right ulnar — Left common iliac

Right deep palmar arch — Left external iliac

Right superficial palmar arch — Left internal iliac

Left common palmar digital

Left proper palmar digital

Left femoral

Right deep femoral — Left deep femoral

Left popliteal

Left anterior tibial

Left posterior tibial

Left fibular (peroneal)

Left dorsal artery of foot (dorsalis pedis)

Left arcuate

Left dorsal metatarsal

Left dorsal digital

(a) Overall anterior view of the principal branches of the aorta

⬛ **EXHIBIT 15.1 CONTINUES** ▶

EXHIBIT 15.1 **493**

EXHIBIT 15.1 The Aorta and Its Branches CONTINUED

FIGURE 15.6 CONTINUED

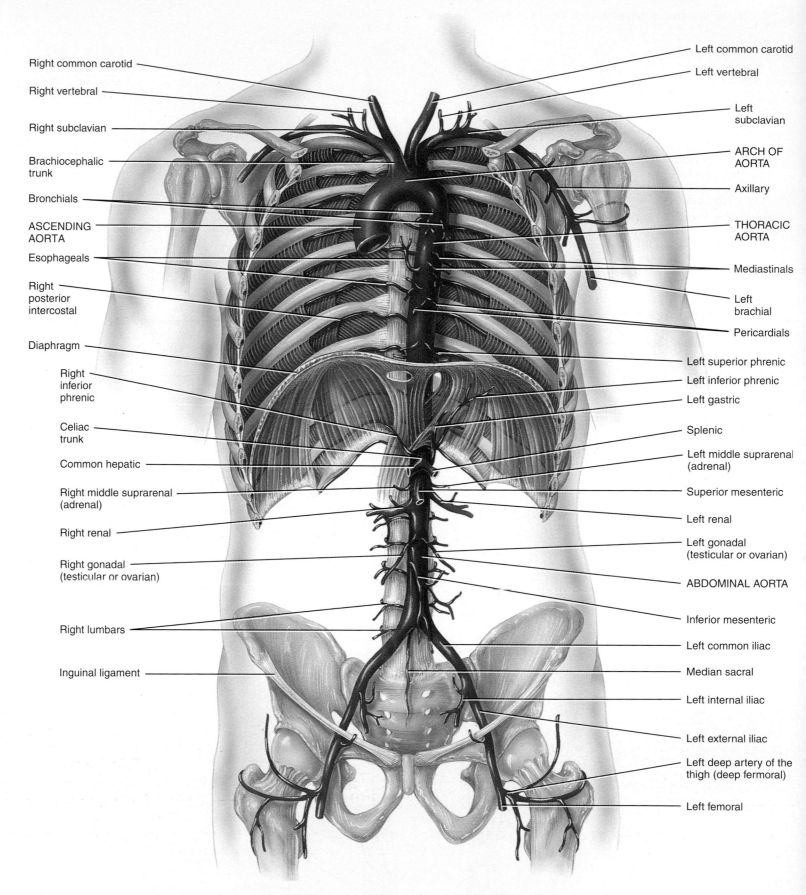

Right common carotid

Right vertebral

Right subclavian

Brachiocephalic trunk

Bronchials

ASCENDING AORTA

Esophageals

Right posterior intercostal

Diaphragm

Right inferior phrenic

Celiac trunk

Common hepatic

Right middle suprarenal (adrenal)

Right renal

Right gonadal (testicular or ovarian)

Right lumbars

Inguinal ligament

Left common carotid

Left vertebral

Left subclavian

ARCH OF AORTA

Axillary

THORACIC AORTA

Mediastinals

Left brachial

Pericardials

Left superior phrenic

Left inferior phrenic

Left gastric

Splenic

Left middle suprarenal (adrenal)

Superior mesenteric

Left renal

Left gonadal (testicular or ovarian)

ABDOMINAL AORTA

Inferior mesenteric

Left common iliac

Median sacral

Left internal iliac

Left external iliac

Left deep artery of the thigh (deep fermoral)

Left femoral

(b) Detailed anterior view of the principal branches of the aorta

SUPERIOR

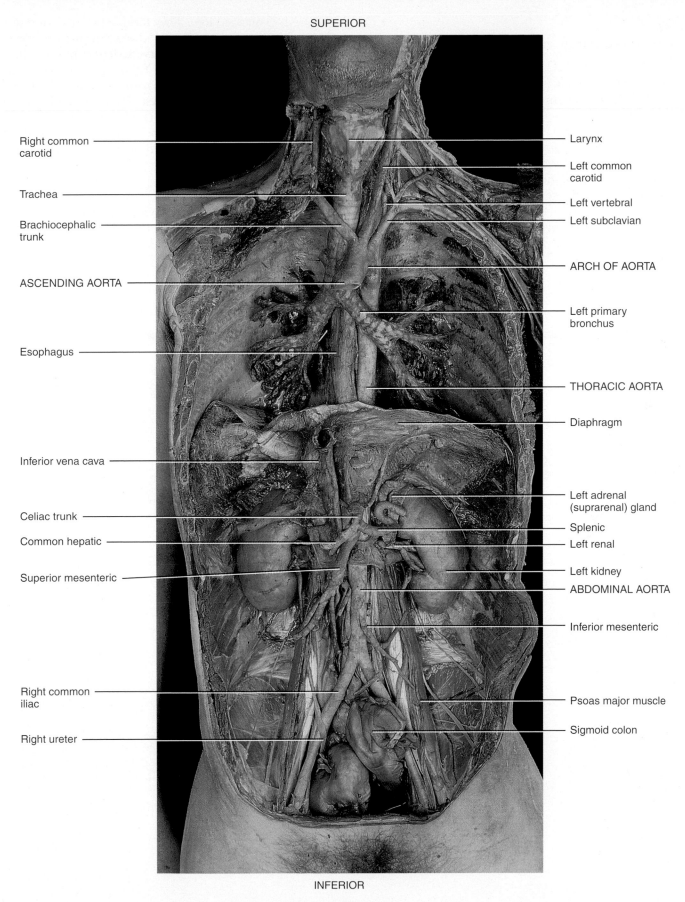

Right common carotid

Trachea

Brachiocephalic trunk

ASCENDING AORTA

Esophagus

Inferior vena cava

Celiac trunk

Common hepatic

Superior mesenteric

Right common iliac

Right ureter

Larynx

Left common carotid

Left vertebral

Left subclavian

ARCH OF AORTA

Left primary bronchus

THORACIC AORTA

Diaphragm

Left adrenal (suprarenal) gland

Splenic

Left renal

Left kidney

ABDOMINAL AORTA

Inferior mesenteric

Psoas major muscle

Sigmoid colon

INFERIOR

(c) Anterior view of the principal branches of the aorta

? What are the four principal subdivisions of the aorta?

EXHIBIT 15.1 **495**

EXHIBIT 15.2   Ascending Aorta

◉ OBJECTIVE

- Identify the two primary arterial branches of the ascending aorta.

The **ascending aorta** is about 5 cm (2 in.) in length and begins at the aortic valve. It is directed superiorly, slightly anteriorly, and to the right. It ends at the level of the sternal angle, where it becomes the arch of the aorta. The beginning of the ascending aorta is posterior to the pulmonary trunk and right auricle; the right pulmonary artery is posterior to it. At its origin, the ascending aorta contains three dilations called *aortic sinuses*. Two of these, the right and left sinuses, give rise to the right and left coronary arteries, respectively.

The right and left **coronary arteries** (*coron-*=crown) arise from the ascending aorta just superior to the aortic valve (see Figure 14.8 on page 468). They form a crownlike ring around the heart, giving off branches to the atrial and ventricular myocardium. The **posterior interventricular branch** (in-ter-ven-TRIK-ū-lar; *inter-*=between) of the right coronary artery supplies both ventricles, and the **marginal branch** supplies the right ventricle. The **anterior interventricular branch,** also known as the **left anterior descending (LAD) branch,** of the left coronary artery supplies both ventricles, and the **circumflex branch** (SER-kum-flex; *circum-*=around; *-flex*=to bend) supplies the left atrium and left ventricle.

◉ CHECKPOINT

- Which branches of the coronary arteries supply the left ventricle? Why does the left ventricle have such an extensive arterial blood supply?

SCHEME OF DISTRIBUTION

EXHIBIT 15.3   The Arch of the Aorta

◉ OBJECTIVE

- Identify the three principal arteries that branch from the arch of the aorta.

The **arch of the aorta** is 4–5 cm (almost 2 in.) in length and is the continuation of the ascending aorta. It emerges from the pericardium posterior to the sternum at the level of the sternal angle. The arch of the aorta is directed superiorly and posteriorly to the left and then inferiorly; it ends at the intervertebral disc between the fourth and fifth thoracic vertebrae, where it becomes the thoracic aorta. Three major arteries branch from the superior aspect of the arch of the aorta: the brachiocephalic trunk, the left common carotid, and the left subclavian. The first and largest branch from the arch of the aorta is the **brachiocephalic trunk** (brā′-kē-ō-se-FAL-ik; *brachio-*=arm; *-cephalic*=head). It extends superiorly, bending slightly to the right, and divides at the right

sternoclavicular joint to form the right subclavian artery and right common carotid artery. The second branch from the arch of the aorta is the **left common carotid artery** (ka-ROT-id), which divides into the same branches with the same names as the right common carotid artery. The third branch from the arch of the aorta is the **left subclavian artery** (sub-KLĀ-vē-an), which distributes blood to the left vertebral artery and vessels of the left upper limb. Arteries branching from the left subclavian artery are similar in distribution and name to those branching from the right subclavian artery. The following description focuses on the principal arteries originating from the brachiocephalic trunk.

◉ CHECKPOINT

- What general regions do the arteries that arise from the arch of the aorta supply?

SCHEME OF DISTRIBUTION

EXHIBIT 15.3   The Arch of the Aorta CONTINUED

BRANCH	DESCRIPTION AND REGION SUPPLIED
BRACHIOCEPHALIC TRUNK	The brachiocephalic trunk divides to form the right subclavian artery and right common carotid artery (**Figure 15.7a**).
Right subclavian artery (sub-KLĀ-vē-an)	The **right subclavian artery** extends from the brachiocephalic trunk to the first rib and then passes into the armpit (axilla). The general distribution of the artery is to the brain and spinal cord, neck, shoulder, thoracic viscera and wall, and scapular muscles.
Internal thoracic or **mammary artery** (thor-AS-ik; *thorac-*=chest)	The **internal thoracic artery** arises from the first part of the subclavian artery and descends posterior to the costal cartilages of the superior six ribs. It terminates at the sixth intercostal space. It supplies the anterior thoracic wall and structures in the mediastinum. In coronary artery bypass grafting, if only a single vessel is obstructed, the internal thoracic (usually the left) is used to create the bypass. The upper end of the artery is left attached to the subclavian artery and the cut end is connected to the coronary artery at a point distal to the blockage. The lower end of the internal thoracic artery is tied off. Artery grafts are preferred over vein grafts because arteries can withstand the greater pressure of blood flowing through coronary arteries and are less likely to become obstructed over time.
Vertebral artery (VER-te-bral)	Before passing into the axilla, the right subclavian artery gives off a major branch to the brain called the **right vertebral artery** (Figure 15.7b). The right vertebral artery passes through the foramina of the transverse processes of the sixth through first cervical vertebrae and enters the skull through the foramen magnum to reach the inferior surface of the brain. Here it unites with the left vertebral artery to form the **basilar** (BAS-i-lar) **artery.** The vertebral artery supplies the posterior portion of the brain with blood. The basilar artery passes along the midline of the anterior aspect of the brain stem. It gives off several branches (**posterior cerebral** and **cerebellar arteries**) that supply the cerebellum and pons of the brain and the inner ear.
Axillary artery (AK-sil-ār-ē=armpit)	The continuation of the right subclavian artery into the axilla is called the **axillary artery**. (Note that the right subclavian artery, which passes deep to the clavicle, is a good example of the practice of giving the same vessel different names as it passes through different regions.) Its general distribution is the shoulder, thoracic and scapular muscles, and humerus.
Brachial artery (BRĀ-kē-al=arm)	The **brachial artery** is the continuation of the axillary artery into the arm. The brachial artery provides the main blood supply to the arm and is superficial and palpable along its course. It begins at the tendon of the teres major muscle and ends just distal to the bend of the elbow. At first, the brachial artery is medial to the humerus, but as it descends it gradually curves laterally and passes through the cubital fossa, a triangular depression anterior to the elbow where you can easily detect the pulse of the brachial artery and listen to the various sounds when taking a person's blood pressure. Just distal to the bend in the elbow, the brachial artery divides into the radial artery and ulnar artery. Blood pressure is usually measured in the brachial artery. In order to control hemorrhage, the best place to compress the brachial artery is near the middle of the arm.
Radial artery (RĀ-dē-al=radius)	The **radial artery** is the smaller branch and is a direct continuation of the brachial artery. It passes along the lateral (radial) aspect of the forearm and then through the wrist and hand, supplying these structures with blood. At the wrist, the radial artery makes contact with the distal end of the radius, where it is covered only by fascia and skin. Because of its superficial location at this point, it is a common site for measuring the radial pulse.

Ulnar artery (UL-nar=ulna)	The **ulnar artery,** the larger branch of the brachial artery, passes along the medial (ulnar) aspect of the forearm and then into the wrist and hand, supplying these structures with blood. In the palm, branches of the radial and ulnar arteries anastomose to form the superficial palmar arch and the deep palmar arch.
Superficial palmar arch (*palma*=palm)	The **superficial palmar arch** is formed mainly by the ulnar artery, with a contribution from a branch of the radial artery. The arch is superficial to the long flexor tendons of the fingers and extends across the palm at the bases of the metacarpals. It gives rise to **common palmar digital arteries,** which supply the palm. Each divides into a pair of **proper palmar digital arteries,** which supply the fingers.
Deep palmar arch	Mainly the radial artery forms the **deep palmar arch,** with a contribution from a branch of the ulnar artery. The arch is deep to the long flexor tendons of the fingers and extends across the palm, just distal to the bases of the metacarpals. Arising from the deep palmar arch are **palmar metacarpal arteries,** which supply the palm and anastomose with the common palmar digital arteries of the superficial palmar arch.
Right common carotid artery	The **right common carotid artery** begins at the bifurcation (division into two branches) of the brachiocephalic trunk, posterior to the right sternoclavicular joint, and passes superiorly in the neck to supply structures in the head (Figure 15.7b). At the superior border of the larynx (voice box), it divides into the right external and right internal carotid arteries. Pulse may be detected in the common carotid artery, just lateral to the larynx. It is convenient to detect a carotid pulse when exercising or when administering cardiopulmonary resuscitation.
External carotid artery	The **external carotid artery** begins at the superior border of the larynx and terminates near the temporomandibular joint of the parotid gland, where it divides into two branches: the superficial temporal and maxillary arteries. The carotid pulse can be detected in the external carotid artery just anterior to the sternocleidomastoid muscle at the superior border of the larynx. The general distribution of the external carotid artery is to structures external to the skull.
Internal carotid artery	The **internal carotid artery** has no branches in the neck and supplies structures internal to the skull. It enters the cranial cavity through the carotid foramen in the temporal bone. The internal carotid artery supplies blood to the eyeball and other orbital structures, ear, most of the cerebrum of the brain, pituitary gland, and external nose. The terminal branches of the internal carotid artery are the **anterior cerebral artery,** which supplies most of the medial surface of the cerebrum and deep masses of gray matter within the cerebrum, and the **middle cerebral artery,** which supplies most of the lateral surface of the cerebrum (Figure 15.7c). Inside the cranium, anastomoses of the left and right internal carotid arteries, along with the basilar artery, form an arrangement of blood vessels at the base of the brain near the hypophyseal fossa called the **cerebral arterial circle (circle of Willis).** From this circle (Figure 15.7c) arise arteries supplying most of the brain. Essentially, the cerebral arterial circle is formed by the union of the **anterior cerebral arteries** (branches of internal carotids) and **posterior cerebral arteries** (branches of basilar artery). The posterior cerebral arteries supply the inferolateral surface of the temporal lobe and lateral and medial surfaces of the occipital lobe of the cerebrum, deep masses of gray matter within the cerebrum, and midbrain. The posterior cerebral arteries are connected with the internal carotid arteries by the **posterior communicating arteries.** The **anterior communicating arteries** connect the anterior cerebral arteries. The **internal carotid arteries** are also considered part of the cerebral arterial circle. The functions of the cerebral arterial circle are to equalize blood pressure to the brain and provide alternate routes for blood flow to the brain, should the arteries become damaged.
LEFT COMMON CAROTID ARTERY	**See description in the introduction to this exhibit.**
LEFT SUBCLAVIAN ARTERY	**See description in the introduction to this exhibit.**

◼ EXHIBIT **15.3** CONTINUES ▶

EXHIBIT 15.3   **499**

EXHIBIT 15.3 The Arch of the Aorta CONTINUED

Figure 15.7 Arch of the aorta and its branches. Note in (c) the arteries that constitute the cerebral arterial circle (circle of Willis). (See Tortora, *A Photographic Atlas of the Human Body*, 2e, Figure 6.12.)

The arch of the aorta ends at the level of the intervertebral disc between the fourth and fifth thoracic vertebrae.

(a) Anterior view of branches of brachiocephalic trunk in upper limb

(b) Right lateral view of branches of brachiocephalic trunk in neck and head

(c) Inferior view of base of brain showing cerebral arterial circle

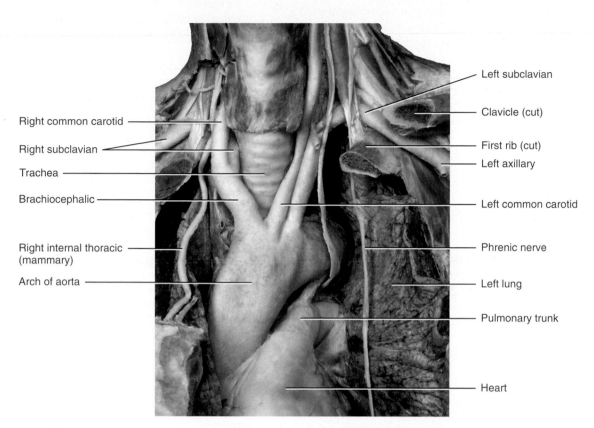

Right common carotid

Right subclavian

Trachea

Brachiocephalic

Right internal thoracic
(mammary)

Arch of aorta

Left subclavian

Clavicle (cut)

First rib (cut)

Left axillary

Left common carotid

Phrenic nerve

Left lung

Pulmonary trunk

Heart

(d) Anterior view, branches of arch of aorta

Right common carotid

Right subclavian

Pectoralis major
(cut and reflected)

Right axillary

Right subscapular

Right deep brachial

Right brachial

Left subclavian

Left common carotid

Brachiocephalic

Arch of aorta

Pulmonary trunk

Heart

(e) Anterior view of arteries of shoulder and brachium

? **What are the three major branches of the arch of the aorta, in order of their origination?**

EXHIBIT 15.3   **501**

EXHIBIT 15.4 Thoracic Aorta

OBJECTIVE

- Identify the visceral and parietal branches of the thoracic aorta.

The **thoracic aorta** is about 20 cm (8 in.) long and is a continuation of the arch of the aorta (see Figure 15.6b). It begins at the level of the intervertebral disc between the fourth and fifth thoracic vertebrae, where it lies to the left of the vertebral column. As it descends, it moves closer to the midline and extends through an opening in the diaphragm (aortic hiatus), which is located anterior to the vertebral column at the level of the intervertebral disc between the twelfth thoracic and first lumbar vertebrae.

Along its course, the thoracic aorta sends off numerous small arteries, **visceral branches** to viscera, and **parietal branches** to body wall structures.

CHECKPOINT

- What general regions do the visceral and parietal branches of the thoracic aorta supply?

BRANCH	DESCRIPTION AND REGION SUPPLIED
VISCERAL	
Pericardial arteries (per'-i-KAR-dē-al; *peri-*=around; *cardia-*=heart)	Two or three tiny **pericardial arteries** supply blood to the pericardium.
Bronchial arteries (BRONG-kē-al=windpipe)	One right and two left **bronchial arteries** supply the bronchial tubes, pleurae, bronchial lymph nodes, and esophagus. (The right bronchial artery arises from the third posterior intercostal artery; the two left bronchial arteries arise from the thoracic aorta.)
Esophageal arteries (e-sof'-a-JĒ-al; *eso-*=to carry; *phage-*=food)	Four or five **esophageal arteries** supply the esophagus.
Mediastinal arteries (mē'-dē-as-TĪ-nal)	Numerous small **mediastinal arteries** supply blood to structures in the mediastinum.
PARIETAL	
Posterior intercostal arteries (in'-ter-KOS-tal; *inter-*=between; *costa*=rib)	Nine pairs of **posterior intercostal arteries** supply the intercostal, pectoralis major and minor, and serratus anterior muscles; overlying subcutaneous tissue and skin; mammary glands; and vertebrae, meninges, and spinal cord.
Subcostal arteries (sub-KOS-tal; *sub-*=under)	The left and right **subcostal arteries** have a distribution similar to that of the posterior intercostals.
Superior phrenic arteries (FREN-ik=pertaining to the diaphragm)	Small **superior phrenic arteries** supply the superior and posterior surfaces of the diaphragm.

SCHEME OF DISTRIBUTION

EXHIBIT 15.5   Abdominal Aorta

- Identify the visceral and parietal branches of the abdominal aorta.

The **abdominal aorta** is the continuation of the thoracic aorta. It begins at the aortic hiatus in the diaphragm and ends at about the level of the fourth lumbar vertebra, where it divides into the right and left common iliac arteries. The abdominal aorta lies anterior to the vertebral column.

As with the thoracic aorta, the abdominal aorta gives off visceral and **parietal branches.** The unpaired visceral branches arise from the anterior surface of the aorta and include the **celiac trunk** and the **superior mesenteric** and **inferior mesenteric arteries** (see Figure 15.6b). The paired visceral branches arise from the lateral surfaces of the aorta and include the **suprarenal, renal,** and **gonadal arteries.** The unpaired parietal branch is the **median sacral artery.** The paired parietal branches arise from the posterolateral surfaces of the aorta and include the **inferior phrenic** and **lumbar arteries.**

● **CHECKPOINT**

- Name the paired visceral and parietal branches and the unpaired visceral and parietal branches of the abdominal aorta, and indicate the general regions they supply.

BRANCH	DESCRIPTION AND REGION SUPPLIED
UNPAIRED VISCERAL BRANCHES	
Celiac trunk (SĒ-lē-ak)	The **celiac trunk (artery)** is the first visceral branch from the aorta inferior to the diaphragm, at about the level of the twelfth thoracic vertebra (Figure 15.8a). Almost immediately, the celiac trunk divides into three branches: the left gastric, splenic, and common hepatic arteries (Figure 15.8a). **1.** The **left gastric artery** (GAS-trik=stomach) is the smallest of the three branches. It passes superiorly to the left toward the esophagus and then turns to follow the lesser curvature of the stomach. It supplies the stomach and esophagus. **2.** The **splenic artery** (SPLĒN-ik=spleen) is the largest branch of the celiac trunk. It arises from the left side of the celiac trunk distal to the left gastric artery, and passes horizontally to the left along the pancreas. Before reaching the spleen, it gives rise to three arteries: • **Pancreatic artery** (pan-krē-AT-ik), which supplies the pancreas. • **Left gastroepiploic artery** (gas'-trō-ep'-i-PLŌ-ik; *epiplo-*=omentum), which supplies the stomach and greater omentum. • **Short gastric artery,** which supplies the stomach. **3.** The **common hepatic artery** (he-PAT-ik=liver) is intermediate in size between the left gastric and splenic arteries. Unlike the other two branches of the celiac trunk, the common hepatic artery arises from the right side. It gives rise to three arteries: • **Proper hepatic artery,** which supplies the liver, gallbladder, and stomach. • **Right gastric artery,** which supplies the stomach. • **Gastroduodenal artery** (gas'-trō-doo'-ō-DĒ-nal), which supplies the stomach, duodenum of the small intestine, pancreas, and greater omentum.
Superior mesenteric artery (MES-en-ter'-ik; *meso-*=middle; *-enteric*=pertaining to the intestines)	The **superior mesenteric artery** (Figure 15.8b) arises from the anterior surface of the abdominal aorta about 1 cm inferior to the celiac trunk at the level of the first lumbar vertebra. It extends inferiorly and anteriorly and between the layers of mesentery, which is a portion of the peritoneum that attaches the small intestine to the posterior abdominal wall. It anastomoses extensively and has five branches: **1.** The **inferior pancreaticoduodenal artery** (pan'-krē-at'-i-kō-doo'-ō-DĒ-nal) supplies the pancreas and duodenum. **2.** The **jejunal** (je-JOO-nal) and **ileal arteries** (IL-ē-al) supply the jejunum and ileum of the small intestine, respectively. **3.** The **ileocolic artery** (il'-ē-ō-KŌL-ik) supplies the ileum and ascending colon of the large intestine. **4.** The **right colic artery** (KŌL-ik) supplies the ascending colon. **5.** The **middle colic artery** supplies the transverse colon of the large intestine.
Inferior mesenteric artery	The **inferior mesenteric artery** (Figure 15.8c) arises from the anterior aspect of the abdominal aorta at the level of the third lumbar vertebra and then passes inferiorly to the left of the aorta. It anastomoses extensively and has three branches: **1.** The **left colic artery** supplies the transverse colon and descending colon of the large intestine. **2.** The **sigmoid arteries** (SIG-moyd) supply the descending colon and sigmoid colon of the large intestine. **3.** The **superior rectal artery** (REK-tal) supplies the rectum of the large intestine.

● **EXHIBIT 15.5 CONTINUES** ▶

EXHIBIT 15.5   **503**

EXHIBIT 15.5 Abdominal Aorta CONTINUED

PAIRED VISCERAL BRANCHES

Suprarenal arteries
(soo′-pra-RĒ-nal;
supra-=above;
ren-=kidney)

Although there are three pairs of **suprarenal (adrenal) arteries** that supply the adrenal (suprarenal) glands (superior, middle, and inferior), only the middle pair originates directly from the abdominal aorta (see Figure 15.6b). The middle suprarenal arteries arise at the level of the first lumbar vertebra at or superior to the renal arteries. The superior suprarenal arteries arise from the inferior phrenic artery, and the inferior suprarenal arteries originate from the renal arteries.

Renal arteries
(RĒ-nal; *ren-*=kidney)

The right and left **renal arteries** usually arise from the lateral aspects of the abdominal aorta at the superior border of the second lumbar vertebra, about 1 cm inferior to the superior mesenteric artery (see Figure 15.6b). The right renal artery, which is longer than the left, arises slightly lower than the left and passes posterior to the right renal vein and inferior vena cava. The left renal artery is posterior to the left renal vein and is crossed by the inferior mesenteric vein. The renal arteries carry blood to the kidneys, adrenal (suprarenal) glands, and ureters. Their distribution within the kidneys is discussed in Chapter 26.

Gonadal (gō-NAD-al;
gon-=seed) [testicular
(test-TIK-ū-lar) or **ovarian
arteries** (ō-VAR-ē-an)]

The **gonadal arteries** arise from the abdominal aorta at the level of the second lumbar vertebra just inferior to the renal arteries (see Figure 15.6b). In males, the gonadal arteries are specifically referred to as the **testicular arteries.** They pass through the inguinal canal and supply the testes, epididymis, and ureters. In females, the gonadal arteries are called the **ovarian arteries.** They are much shorter than the testicular arteries and supply the ovaries, uterine (fallopian) tubes, and ureters.

UNPAIRED PARIETAL BRANCH

Median sacral artery
(SĀ-kral=pertaining to
the sacrum)

The **median sacral artery** arises from the posterior surface of the abdominal aorta about 1 cm superior to the bifurcation (division into two branches) of the aorta into the right and left common iliac arteries (see Figure 15.6b). The median sacral artery supplies the sacrum and coccyx.

PAIRED PARIETAL BRANCHES

Inferior phrenic arteries
(FREN-ik=pertaining to
the diaphragm)

The **inferior phrenic arteries** are the first paired branches of the abdominal aorta, immediately superior to the origin of the celiac trunk (see Figure 15.6b). (They may also arise from the renal arteries.) The inferior phrenic arteries are distributed to the inferior surface of the diaphragm and adrenal (suprarenal) glands.

Lumbar arteries
(LUM-bar=pertaining to
the loin)

The four pairs of **lumbar arteries** arise from the posterolateral surface of the abdominal aorta (see Figure 15.6b). They supply the lumbar vertebrae, spinal cord and its meninges, and the muscles and skin of the lumbar region of the back.

SCHEME OF DISTRIBUTION

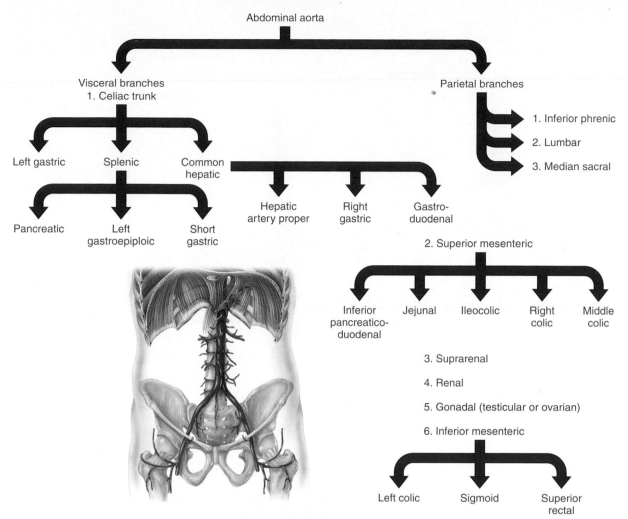

Abdominal aorta

Visceral branches
1. Celiac trunk

- Left gastric
- Splenic
- Common hepatic

- Pancreatic
- Left gastroepiploic
- Short gastric

- Hepatic artery proper
- Right gastric
- Gastro-duodenal

Parietal branches

1. Inferior phrenic
2. Lumbar
3. Median sacral

2. Superior mesenteric

- Inferior pancreatico-duodenal
- Jejunal
- Ileocolic
- Right colic
- Middle colic

3. Suprarenal

4. Renal

5. Gonadal (testicular or ovarian)

6. Inferior mesenteric

- Left colic
- Sigmoid
- Superior rectal

Figure 15.8 Abdominal aorta and principal branches.

🔑 The abdominal aorta is the continuation of the thoracic aorta.

- Stomach (turned upward)
- Left gastroepiploic
- Right gastroepiploic
- Short gastrics
- Right gastric
- Diaphragm
- Celiac trunk
- Left gastric
- Proper hepatic
- Splenic
- Spleen
- Common hepatic
- Pancreatic
- Gastroduodenal
- Pancreas
- Superior pancreaticoduodenal
- Superior mesenteric
- Jejunum of small intestine
- Duodenum of small intestine
- Abdominal aorta

(a) Anterior view of celiac trunk and its branches

⬤ **EXHIBIT 15.5 CONTINUES** ▶

EXHIBIT 15.5 **505**

EXHIBIT 15.5 Abdominal Aorta CONTINUED

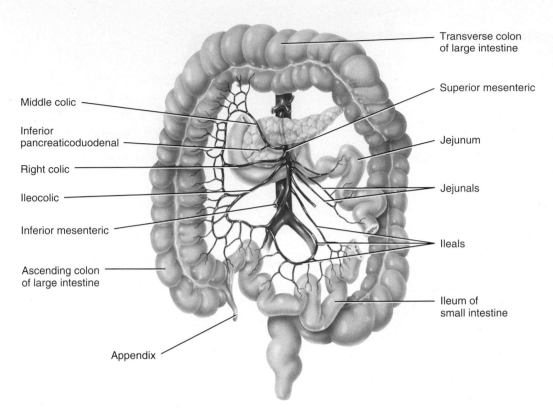

Middle colic

Inferior pancreaticoduodenal

Right colic

Ileocolic

Inferior mesenteric

Ascending colon of large intestine

Appendix

Transverse colon of large intestine

Superior mesenteric

Jejunum

Jejunals

Ileals

Ileum of small intestine

(b) Anterior view of superior mesenteric artery and its branches

Superior mesenteric

Inferior mesenteric

Abdominal aorta

Common iliac

Sigmoids

Superior rectal

Transverse colon of large intestine

Left colic

Descending colon of large intestine

Sigmoid colon

Rectum of large intestine

(c) Anterior view of inferior mesenteric artery and its branches

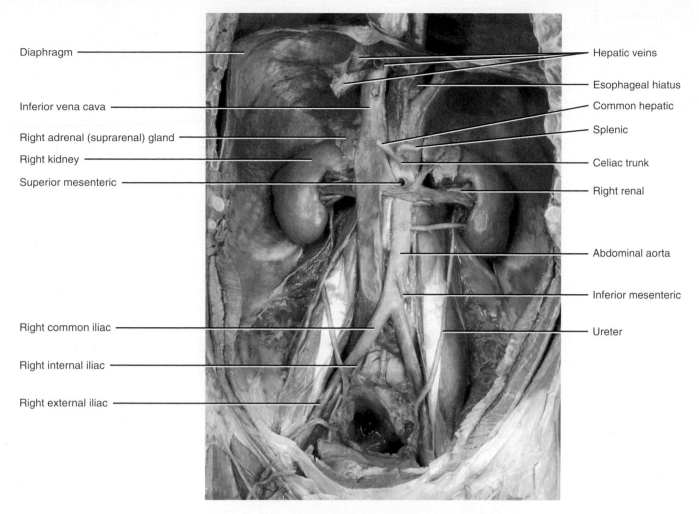

Diaphragm

Inferior vena cava

Right adrenal (suprarenal) gland

Right kidney

Superior mesenteric

Right common iliac

Right internal iliac

Right external iliac

Hepatic veins

Esophageal hiatus

Common hepatic

Splenic

Celiac trunk

Right renal

Abdominal aorta

Inferior mesenteric

Ureter

(d) Anterior view of arteries of abdomen and pelvis

Liver

Right gastroepiploic

Right gastric

Common hepatic

Proper hepatic

Gastroduodenal

Superior pancreaticoduodenal

Diaphragm

Stomach (turned upward)

Left gastroepiploic

Spleen

Left gastric

Splenic

Celiac trunk

Pancreas

(e) Anterior view of celiac trunk and its branches

EXHIBIT 15.5 CONTINUES ▶

EXHIBIT 15.5   **507**

EXHIBIT 15.5 Abdominal Aorta CONTINUED

Diaphragm
Transverse colon

Descending colon

Middle colic

Ascending colon

Superior mesenteric

Right colic

Cecum

Ileocolic

Jejunals

Ileals

Ileum

Sigmoid colon

(f) Anterior view of superior mesenteric artery and its branches

Diaphragm

Transverse colon

Ascending colon

Descending colon

Left colic

Inferior mesenteric

Cecum

Superior rectal

Sigmoid

Sigmoid colon

Ileum

(g) Anterior view of inferior mesenteric artery and its branches

❓ **Where does the abdominal aorta begin?**

EXHIBIT 15.6 Arteries of the Pelvis and Lower Limbs

- Identify the two major branches of the common iliac arteries.

The abdominal aorta ends by dividing into the right and left **common iliac arteries.** These, in turn, divide into the **internal** and **external iliac arteries.** In sequence, the external iliacs become the **femoral arteries** in the thighs, the **popliteal arteries** posterior to the knee, and the **anterior** and **posterior tibial arteries** in the legs.

- What general regions do the internal and external iliac arteries supply?

BRANCH	DESCRIPTION AND REGION SUPPLIED
COMMON ILIAC ARTERIES (IL-ē-ak=pertaining to the ilium)	
	At about the level of the fourth lumbar vertebra, the abdominal aorta divides into the right and left **common iliac arteries,** the terminal branches of the abdominal aorta. Each passes inferiorly about 5 cm (2 in.) and gives rise to two branches: internal iliac and external iliac arteries. The general distribution of the common iliac arteries is to the pelvis, external genitals, and lower limbs.
Internal iliac arteries	The **internal iliac (hypogastric) arteries** are the primary arteries of the pelvis. They begin at the bifurcation (division into two branches) of the common iliac arteries anterior to the sacroiliac joint at the level of the lumbosacral intervertebral disc. They pass posteromedially as they descend in the pelvis and divide into anterior and posterior divisions. The general distribution of the internal iliac arteries is to the pelvis, buttocks, external genitals, and thigh.
External iliac arteries	The **external iliac arteries** are larger than the internal iliac arteries. Like the internal iliac arteries, they begin at the bifurcation of the common iliac arteries. They descend along the medial border of the psoas major muscles following the pelvic brim, pass posterior to the midportion of the inguinal ligaments, and become the femoral arteries. The general distribution of the external iliac arteries is to the lower limbs. Specifically, branches of the external iliac arteries supply the muscles of the anterior abdominal wall, the cremaster muscle in males and the round ligament of the uterus in females, and the lower limbs.
Femoral arteries (FEM-o-ral= pertaining to the thigh)	The **femoral arteries** descend along the anteromedial aspects of the thighs to the junction of the middle and lower third of the thighs. Here they pass through an opening in the tendon of the adductor magnus muscle, then emerge posterior to the femurs as the popliteal arteries. A pulse may be felt in the femoral artery just inferior to the inguinal ligament. Recall from Chapter 11 that the femoral artery, along with the femoral vein and nerve and deep inguinal lymph nodes, are located in the *femoral triangle* (see Figure 11.22a on page 388). The general distribution of the femoral arteries is to the lower abdominal wall, groin, external genitals, and muscles of the thigh. A major branch of the femoral artery, the **deep artery of the thigh (deep femoral),** supplies most of the muscles of the thigh: quadriceps femoris, adductors, and hamstrings.
	Recall that in cardiac catheterization a catheter is inserted through a blood vessel and advanced into the major vessels and heart chamber. A catheter often contains a measuring instrument or other device at its tip. To reach the left side of the heart, the catheter is inserted into the femoral artery and passed into the aorta to the coronary arteries or heart chamber.
Popliteal arteries (pop′-li-TĒ-al= posterior surface of the knee)	The **popliteal arteries** are the continuation of the femoral arteries through the popliteal fossa (space behind the knee). They descend to the inferior border of the popliteus muscles, where they divide into the anterior and posterior tibial arteries. A pulse may be detected in the popliteal arteries. In addition to supplying the adductor magnus and hamstring muscles and the skin on the posterior aspect of the legs, branches of the popliteal arteries also supply the gastrocnemius, soleus, and plantaris muscles of the calf, knee joint, femur, patella, and fibula.
Anterior tibial arteries (TIB-ē-al) =pertaining to the shin bone	The **anterior tibial arteries** descend from the bifurcation of the popliteal arteries. They are smaller than the posterior tibial arteries. The anterior tibial arteries descend through the anterior muscular compartment of the leg. They pass through the interosseous membrane that connects the tibia and fibula, lateral to the tibia. The anterior tibial arteries supply the knee joints, anterior compartment muscles of the legs, skin over the anterior aspects of the legs, and ankle joints. At the ankles, the anterior tibial arteries become the **dorsal arteries of the foot (dorsalis pedis arteries).** A pulse in this artery may be taken to evaluate the peripheral vascular system. The dorsal arteries of the foot supply the muscles, skin, and joints on the dorsal aspects of the feet. On the dorsum of the feet, the dorsal arteries of the foot give off a transverse branch at the first (medial) cuneiform bone called the **arcuate arteries** (*arcuat-*=bowed) that run laterally over the bases of the metatarsals. From the arcuate arteries branch the **dorsal metatarsal arteries,** which supply the feet. The dorsal metatarsal arteries terminate by dividing into the **dorsal digital arteries,** which supply the toes.
Posterior tibial arteries	The **posterior tibial arteries,** the direct continuations of the popliteal arteries, descend from the bifurcation of the popliteal arteries. They pass down the posterior muscular compartment of the legs posterior to the medial malleolus of the tibia. They terminate by dividing into the medial and lateral plantar arteries. Their general distribution is to the muscles, bones, and joints of the leg and foot. Major branches of the posterior tibial arteries are the **fibular (peroneal) arteries,** which supply the fibularis, soleus, tibialis posterior, and flexor hallucis muscles. They also supply the fibula, tarsus, and lateral aspect of the heel. The bifurcation of the posterior tibial arteries into the medial and lateral plantar arteries occurs deep to the flexor retinaculum on the medial side of the feet. The **medial plantar arteries** (PLAN-tar=sole) supply the abductor hallucis and flexor digitorum brevis muscles and the toes. The **lateral plantar arteries** unite with a branch of the dorsal arteries of the foot to form the **plantar arch.** The arch begins at the base of the fifth metatarsal and extends medially across the metacarpals. As the arch crosses the foot, it gives off **plantar metatarsal arteries,** which supply the feet. These terminate by dividing into **plantar digital arteries,** which supply the toes.

EXHIBIT 15.6 CONTINUES

EXHIBIT 15.6 **509**

EXHIBIT 15.6   Arteries of the Pelvis and Lower Limbs CONTINUED

SCHEME OF DISTRIBUTION

Abdominal aorta

Right common iliac

Left common iliac
(gives rise to same branches
as right common iliac, except
that arteries are labeled left
instead of right)

Right external iliac

Right internal iliac

Right femoral

Right deep artery of the thigh (deep peroneal)

Right popliteal

Right anterior tibial

Right posterior tibial

Right dorsal artery
of foot (dorsalis pedis)

Right fibular (peroneal)

Right arcuate

Right dorsal
metatarsal

Right lateral
plantar

Right medial
plantar

Right dorsal
digital

Right plantar
arch

Right plantar
metatarsal

Right plantar
digital

Figure 15.9 Arteries of the pelvis and right free lower limb.

The internal iliac arteries carry most of the blood supply to the pelvic viscera and wall.

Right common iliac

Right internal iliac

Right external iliac

Abdominal aorta

Left common iliac

Right deep artery of the thigh (deep femoral)

Right femoral

Right popliteal

Right anterior tibial

Right posterior tibial

Right fibular (peroneal)

Right arcuate

Right dorsal artery of foot (dorsalis pedis)

Right dorsal metatarsal

Right dorsal digital

Right lateral plantar

Right medial plantar

Right plantar arch

Right plantar metatarsal

Right plantar digital

(a) Anterior view

(b) Posterior view

EXHIBIT 15.6 CONTINUES

EXHIBIT 15.6   **511**

EXHIBIT 15.6   Arteries of the Pelvis and Lower Limbs CONTINUED

Inguinal ligament

Femoral vein

Femoral artery

Abdominal aorta

Common iliac artery

Internal iliac artery

External iliac artery

Sartorius muscle

Deep artery of the thigh
(deep femoral)

Adductor longus muscle

(c) Anterior view of arteries of pelvis and thigh

Rectus abdominis

Gluteus medius

Gluteus minimus

Iliopsoas

Spermatic cord (cut)

Muscular branches
of femoral artery

Deep femoral artery

Femoral artery

Adductor longus

Gracilis

Vastus lateralis

Vastus intermedius

Vastus medialis

Inguinal ligament

Femoral nerve

Sartorius (cut)

Tensor fasciae
latae

Penis (cut)

Femoral vein

Femoral artery

Rectus femoris

(d) Anterior view of arteries of thigh with right side deep and
left side more superficial

● At what point does the abdominal aorta divide into the common iliac arteries?

EXHIBIT 15.7   Veins of the Systemic Circulation

EXHIBIT 15.7

⊙ **OBJECTIVE**

• Identify the three systemic veins that return deoxygenated blood to the heart.

Figure 15.10   Principal veins.

🔑 Deoxygenated blood returns to the heart via the superior and inferior venae cavae and the coronary sinus.

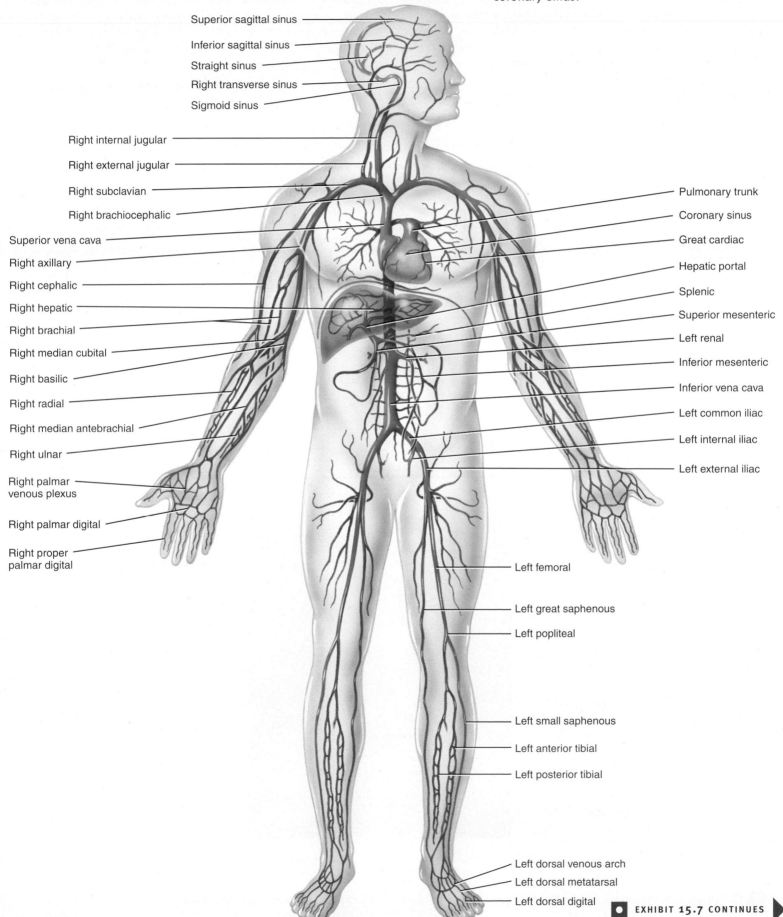

Superior sagittal sinus

Inferior sagittal sinus

Straight sinus

Right transverse sinus

Sigmoid sinus

Right internal jugular

Right external jugular

Right subclavian

Right brachiocephalic

Superior vena cava

Right axillary

Right cephalic

Right hepatic

Right brachial

Right median cubital

Right basilic

Right radial

Right median antebrachial

Right ulnar

Right palmar venous plexus

Right palmar digital

Right proper palmar digital

Pulmonary trunk

Coronary sinus

Great cardiac

Hepatic portal

Splenic

Superior mesenteric

Left renal

Inferior mesenteric

Inferior vena cava

Left common iliac

Left internal iliac

Left external iliac

Left femoral

Left great saphenous

Left popliteal

Left small saphenous

Left anterior tibial

Left posterior tibial

Left dorsal venous arch

Left dorsal metatarsal

Left dorsal digital

(a) Overall anterior view of the principal veins

⬛ EXHIBIT **15.7** CONTINUES ▶

EXHIBIT 15.7   **513**

EXHIBIT 15.7   Veins of the Systemic Circulation CONTINUED

As you have already learned, arteries distribute blood to various parts of the body and veins drain blood away from them. For the most part, arteries are deep; veins may be superficial or deep. Superficial veins are located just beneath the skin and can be seen easily. Because there are no large superficial arteries, the names of superficial veins do not correspond to those of arteries. Superficial veins are clinically important as sites for withdrawing blood or giving injections. Deep veins generally travel alongside arteries and usually bear the same name. Arteries usually follow definite pathways; veins are more difficult to follow because they connect in irregular networks in which many tributaries merge to form a large vein. Although only one systemic artery, the aorta, takes oxygenated blood away from the heart (left ventricle), three systemic veins, the **coronary sinus, superior vena cava,** and **inferior vena cava,** return deoxygenated blood to the heart (right atrium). The coronary sinus receives blood from the cardiac veins; the superior vena cava receives blood from other veins superior to the diaphragm, except the air sacs (alveoli) of the lungs; the inferior vena cava receives blood from veins inferior to the diaphragm.

◉ CHECKPOINT

• What are the three tributaries of the coronary sinus?

VEIN	DESCRIPTION AND REGION DRAINED
Coronary sinus (KOR-ō-nar-ē; *corona*=crown)	The **coronary sinus** is the main vein of the heart; it receives almost all venous blood from the myocardium. It is located in the coronary sulcus (see Figure 14.3c on page 460) and opens into the right atrium between the orifice of the inferior vena cava and the tricuspid valve. It is a wide venous channel into which three veins drain. It receives the **great cardiac vein** (in the anterior interventricular sulcus) into its left end, and the **middle cardiac vein** (in the posterior interventricular sulcus) and the **small cardiac vein** into its right end. Several **anterior cardiac veins** drain directly into the right atrium.
Superior vena cava (SVC) (VĒ-na CĀ-va; *vena*=vein; *cava*=cavelike)	The **superior vena cava** is about 7.5 cm (3 in.) long and 2 cm (1 in.) in diameter and empties its blood into the superior part of the right atrium. It begins posterior to the right first costal cartilage by the union of the right and left brachiocephalic veins and ends at the level of the right third costal cartilage, where it enters the right atrium. The SVC drains the head, neck, chest, and free upper limbs.
Inferior vena cava (IVC)	The **inferior vena cava** is the largest vein in the body, about 3.5 cm (1.4 in.) in diameter. It begins anterior to the fifth lumbar vertebra by the union of the common iliac veins, ascends behind the peritoneum to the right of the midline, pierces the caval opening of the diaphragm at the level of the eighth thoracic vertebra, and enters the inferior part of the right atrium. The IVC drains the abdomen, pelvis, and free lower limbs. The inferior vena cava is commonly compressed during the later stages of pregnancy by the enlarging uterus, producing edema of the ankles and feet and temporary varicose veins.

◉ **FIGURE 15.10 CONTINUED** ▶

Right internal jugular
External jugular
Trachea
Right subclavian
Right brachiocephalic
Left brachiocephalic
First rib (cut)
Superior vena cava
Aorta
Right atrium (cut)
Diaphragm
Inferior vena cava

(b) Anterior view of superior vena cava and its tributaries

❓ **Which general regions of the body are drained by the superior vena cava and the inferior vena cava?**

EXHIBIT 15.8 Veins of the Head and Neck

• Identify the three major veins that drain blood from the head.

Most blood draining from the head passes into three pairs of veins: the **internal jugular**, **external jugular**, and **vertebral veins**. Within the brain, all veins drain into dural venous sinuses and then into the internal jugular veins. **Dural venous sinuses** are endothelial-lined venous channels between layers of the cranial dura mater.

• Which general areas are drained by the internal jugular, external jugular, and vertebral veins?

VEIN	DESCRIPTION AND REGION DRAINED
Internal jugular veins (JUG-ū-lar; *jugular*=throat)	The flow of blood from the dural venous sinuses into the internal jugular veins is as follows (Figure 15.11): The **superior sagittal sinus** (SAJ-i-tal=straight) begins at the frontal bone, where it receives a vein from the nasal cavity, and passes posteriorly to the occipital bone. Along its course, it receives blood from the superior, medial, and lateral aspects of the cerebral hemispheres, meninges, and cranial bones. The superior sagittal sinus usually turns to the right and drains into the right transverse sinus. The **inferior sagittal sinus** is much smaller than the superior sagittal sinus; it begins posterior to the attachment of the falx cerebri and receives the great cerebral vein to become the straight sinus. The great cerebral vein drains the deeper parts of the brain. Along its course the inferior sagittal sinus also receives tributaries from the superior and medial aspects of the cerebral hemispheres.
	The **straight sinus** runs in the tentorium cerebelli and is formed by the union of the inferior sagittal sinus and the great cerebral vein. The straight sinus also receives blood from the cerebellum and usually drains into the left transverse sinus.
	The **transverse sinuses** begin near the occipital bone, pass laterally and anteriorly, and become the sigmoid sinuses near the temporal bone. The transverse sinuses receive blood from the cerebrum, cerebellum, and cranial bones.
	The **sigmoid sinuses** (SIG-moyd=S-shaped) are located along the temporal bone. They pass through the jugular foramina, where they terminate in the internal jugular veins. The sigmoid sinuses drain the transverse sinuses.
	The **cavernous sinuses** (KAV-er-nus=cavelike) are located on either side of the sphenoid bone. They receive blood from the ophthalmic veins from the orbits, and from the cerebral veins from the cerebral hemispheres. They ultimately empty into the transverse sinuses and internal jugular veins. The cavernous sinuses are unique because they have nerves and a major blood vessel passing through them on their way to the orbit and face. The oculomotor (III) nerve, trochlear (IV) nerve, and ophthalmic and maxillary branches of the trigeminal (V) nerve as well as the internal carotid arteries pass through the cavernous sinuses.
	The right and left **internal jugular veins** pass inferiorly on either side of the neck lateral to the internal carotid and common carotid arteries. They then unite with the subclavian veins posterior to the clavicles at the sternoclavicular joints to form the right and left brachiocephalic veins (brā'-kē-ō-se-FAL-ik; *brachio-*=arm; *cephal-*=head). From here blood flows into the superior vena cava. The general structures drained by the internal jugular veins are the brain (through the dural venous sinuses), face, and neck.
External jugular veins	The right and left **external jugular veins** begin in the parotid glands near the angle of the mandible. They are superficial veins that descend through the neck across the sternocleidomastoid muscles. They terminate at a point opposite the middle of the clavicles, where they empty into the subclavian veins. The general structures drained by the external jugular veins are external to the cranium, such as the scalp and superficial and deep regions of the face. When venous pressure rises, for example, during heavy coughing or straining or in cases of heart failure, the external jugular veins become very prominent along the side of the neck.
Vertebral veins (VER-te-bral; *vertebra*=vertebrae)	The right and left **vertebral veins** originate inferior to the occipital condyles. They descend through successive transverse foramina of the first six cervical vertebrae and emerge from the foramina of the sixth cervical vertebra to enter the brachiocephalic veins in the root of the neck. The vertebral veins drain deep structures in the neck such as the cervical vertebrae, cervical spinal cord, and some neck muscles.

■ EXHIBIT 15.8 CONTINUES ▶

EXHIBIT 15.8 **515**

EXHIBIT 15.8 Veins of the Head and Neck CONTINUED

SCHEME OF DRAINAGE

Figure 15.11 Principal veins of the head and neck.

Blood draining from the head passes into the internal jugular, external jugular, and vertebral veins.

(a) Right lateral view

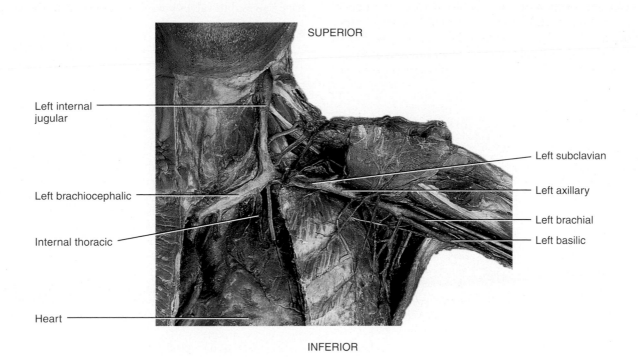

SUPERIOR

Left internal jugular

Left subclavian

Left axillary

Left brachiocephalic

Left brachial

Internal thoracic

Left basilic

Heart

INFERIOR

(b) Anterior view

Cerebrum

Superior sagittal venous sinus

Dura mater

Cerebral veins

Junction of the sinuses

Transverse venous sinus

Cerebellar veins

Sigmoid venous sinus

Cerebellum

Dura mater covering spinal cord

(c) Posterior view of brain and spinal cord with meningeal coverings dissected to show dural sinuses and veins on brain

? **Into which veins in the neck does all venous blood in the brain drain?**

EXHIBIT 15.8 **517**

EXHIBIT 15.9   Veins of the Upper Limbs

◉ **OBJECTIVE**
- Identify the principal veins that drain the upper limbs.

Both superficial and deep veins return blood from the upper limbs to the heart. **Superficial veins** are located just deep to the skin and are often visible. They anastomose extensively with one another and with deep veins, and they do not accompany arteries. Superficial veins are larger than deep veins and return most of the blood from the upper limbs. **Deep veins** are located deep in the body. They usually accompany arteries and have the same names as the corresponding arteries. Both superficial and deep veins have valves, but valves are more numerous in the deep veins.

◉ **CHECKPOINT**
- Where do the cephalic, basilic, median antebrachial, radial, and ulnar veins originate?

VEIN	DESCRIPTION AND REGION DRAINED
SUPERFICIAL	
Cephalic veins (se-FAL-ik=pertaining to the head)	The principal superficial veins that drain the upper limbs are the cephalic and basilic veins. They originate in the hand and convey blood from the smaller superficial veins into the axillary veins. The **cephalic veins** begin on the lateral aspect of the **dorsal venous networks of the hands (dorsal venous arches),** networks of veins on the dorsum of the hands formed by the **dorsal metacarpal veins** (Figure 15.12a). These veins, in turn, drain the **dorsal digital veins,** which pass along the sides of the fingers. Following their formation from the dorsal venous networks of the hands, the cephalic veins arch around the radial side of the forearms to the anterior surface and ascend through the entire limbs along the anterolateral surface. The cephalic veins end where they join the axillary veins, just inferior to the clavicles. **Accessory cephalic veins** originate either from a venous plexus on the dorsum of the forearms or from the medial aspects of the dorsal venous networks of the hands, and unite with the cephalic veins just inferior to the elbow. The cephalic veins drain blood from the lateral aspect of the upper limbs.
Basilic veins (ba-SIL-ik=royal, of prime importance)	The **basilic veins** begin on the medial aspects of the dorsal venous networks of the hands and ascend along the posteromedial surface of the forearm and anteromedial surface of the arm (Figure 15.12b). They drain blood from the medial aspects of the upper limbs. Anterior to the elbow, the basilic veins are connected to the cephalic veins by the **median cubital veins** (*cubital*=pertaining to the elbow), which drain the forearm. If veins must be punctured for an injection, transfusion, or removal of a blood sample, the medial cubital veins are preferred. After receiving the median cubital veins, the basilic veins continue ascending until they reach the middle of the arm. There they penetrate the tissues deeply and run alongside the brachial arteries until they join the brachial veins. As the basilic and brachial veins merge in the axillary area, they form the axillary veins.
Median antebrachial veins (an'-tē-BRĀ-kē-al; *ante-*=before, in front of; *brachi-*=arm)	The **median antebrachial veins (median veins of the forearm)** begin in the **palmar venous plexuses,** networks of veins on the palms. The plexuses drain the **palmar digital veins** in the fingers. The median antebrachial veins ascend anteriorly in the forearms to join the basilic or median cubital veins, sometimes both. They drain the palms and forearms.
DEEP	
Radial veins (RĀ-dē-al=pertaining to the radius)	The paired **radial veins** begin at the **deep palmar venous arches** (Figure 15.12c). These arches drain the **palmar metacarpal veins** in the palms. The radial veins drain the lateral aspects of the forearms and pass alongside the radial arteries. Just inferior to the elbow joint, the radial veins unite with the ulnar veins to form the brachial veins.
Ulnar veins (UL-nar=pertaining to the ulna)	The paired **ulnar veins,** which are larger than the radial veins, begin at the **superficial palmar venous arches.** These arches drain the **common palmar digital veins** and the **proper palmar digital veins** in the fingers. The ulnar veins drain the medial aspect of the forearms, pass alongside the ulnar arteries, and join with the radial veins to form the brachial veins.
Brachial veins (BRĀ-kē-al; *brachi-*=arm)	The paired **brachial veins** accompany the brachial arteries. They drain the forearms, elbow joints, arms, and humerus. They pass superiorly and join with the basilic veins to form the axillary veins.
Axillary veins (AK-sil-ār-ē; *axilla*=armpit)	The **axillary veins** ascend to the outer borders of the first ribs, where they become the subclavian veins. The axillary veins receive tributaries that correspond to the branches of the axillary arteries. The axillary veins drain the arms, axillas, and superolateral chest wall.
Subclavian veins (sub-KLĀ-vē-an; *sub-*=under; *clavian*=pertaining to the clavicle)	The **subclavian veins** are continuations of the axillary veins that terminate at the sternal end of the clavicles, where they unite with the internal jugular veins to form the brachiocephalic veins. The subclavian veins drain the arms, neck, and thoracic wall. The thoracic duct of the lymphatic system delivers lymph into the junction between the left subclavian and the left internal jugular veins. The right lymphatic duct delivers lymph into the junction between the right subclavian and right internal jugular veins (see Figure 16.3 on page 545). In a procedure called *central line placement*, the right subclavian vein is frequently used to administer nutrients and medication and measure venous pressure.

SCHEME OF DRAINAGE

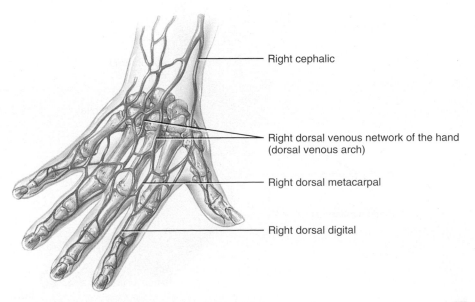

Figure 15.12 **Principal veins of the right upper limb.** (See Tortora, *A Photographic Atlas of the Human Body,* 2e, Figure 6.20.)

⑥ 🔑 Deep veins usually accompany arteries that have similar names.

Right cephalic

Right dorsal venous network of the hand (dorsal venous arch)

Right dorsal metacarpal

Right dorsal digital

(a) Posterior view of superficial veins of the hand

EXHIBIT **15.9** CONTINUES ▶

EXHIBIT 15.9 **519**

EXHIBIT 15.9   Veins of the Upper Limbs CONTINUED

● **FIGURE 15.12 CONTINUED**

Right external jugular

Right subclavian

Right internal jugular

Right brachiocephalic

Right axillary

Right basilic

Right cephalic

Superior vena cava

Sternum

Right accessory cephalic

Right median cubital

Right cephalic

Right basilic

Right median antebrachial

Right palmar venous plexus

Right palmar digital

(b) Anterior view of superficial veins

Right external jugular

Right subclavian

Right internal jugular

Right brachiocephalic

Right axillary

Right brachial

Superior vena cava

Right radials

Right ulnars

Right deep palmar venous arch

Right superficial palmar venous arch

Right common palmar digital

Right proper palmar digital

Right palmar metacarpal

(c) Anterior view of deep veins

Biceps brachii muscle

Brachial

Median cubital

Basilic

Cephalic

Median antebrachial

(d) Anteromedial view of superficial veins of arm and forearm

? From which vein in the upper limb is a blood sample often taken?

EXHIBIT 15.10   Veins of the Thorax

OBJECTIVE

- Identify the components of the azygos system of veins.

Although the brachiocephalic veins drain some portions of the thorax, most thoracic structures are drained by a network of veins, called the **azygos system,** that runs on either side of the vertebral column. The system consists of three veins—the **azygos, hemiazygos,** and **accessory hemiazygos veins**—that show considerable variation in origin, course, tributaries, anastomoses, and termination. Ultimately they empty into the superior vena cava.

CHECKPOINT

- What is the importance of the azygos system relative to the inferior vena cava?

VEIN	DESCRIPTION AND REGION DRAINED
Brachiocephalic vein (brā′-kē-ō-se-FAL-ik; *brachio-*=arm; *cephalic*= pertaining to the head)	The right and left **brachiocephalic veins,** formed by the union of the subclavian and internal jugular veins, drain blood from the head, neck, upper limbs, mammary glands, and superior thorax. The brachiocephalic unite to form the superior vena cava. Because the superior vena cava is to the right of the body's veins midline, the left brachiocephalic vein is longer than the right. The right brachiocephalic vein is anterior and to the right of the brachiocephalic trunk. The left brachiocephalic vein is anterior to the brachiocephalic trunk, the left common carotid and left subclavian arteries, the trachea, the left vagus (X) nerve, and phrenic nerve.
Azygos system (az-Ī-gos=unpaired)	The **azygos system,** besides collecting blood from the thorax and abdominal wall, may serve as a bypass for the inferior vena cava that drains blood from the lower body. Several small veins directly link the azygos system with the inferior vena cava. Large veins that drain the lower limbs and abdomen conduct blood into the azygos system. If the inferior vena cava or hepatic portal vein becomes obstructed, the azygos system can return blood from the lower body to the superior vena cava.
Azygos vein	The **azygos vein** is anterior to the vertebral column, slightly to the right of the midline. It usually begins at the junction of the right ascending lumbar and right subcostal veins near the diaphragm. At the level of the fourth thoracic vertebra, it arches over the root of the right lung to end in the superior vena cava. Generally, the azygos vein drains the right side of the thoracic wall, thoracic viscera, and abdominal wall. Specifically, the azygos vein receives blood from most of the **right posterior intercostal, hemiazygos, accessory hemiazygos, esophageal, mediastinal, pericardial,** and **bronchial veins.**
Hemiazygos vein (HEM-ē-az-ī-gos; *hemi-*=half)	The **hemiazygos vein** is anterior to the vertebral column and slightly to the left of the midline. It often begins at the junction of the left ascending lumbar and left subcostal veins. It terminates by joining the azygos vein at about the level of the ninth thoracic vertebra. Generally, the hemiazygos vein drains the left side of the thoracic wall, thoracic viscera, and abdominal wall. Specifically, the hemiazygos vein receives blood from the ninth through eleventh **left posterior intercostal, esophageal, mediastinal,** and sometimes the **accessory hemiazygos veins.**
Accessory hemiazygos vein	The **accessory hemiazygos vein** is also anterior to the vertebral column and to the left of the midline. It begins at the fourth or fifth intercostal space and descends from the fifth to the eighth thoracic vertebra or ends in the hemiazygos vein. It terminates by joining the azygos vein at about the level of the eighth thoracic vertebra. The accessory hemiazygos vein drains the left side of the thoracic wall. It receives blood from the fourth through eighth **left posterior intercostal veins** (the first through third left posterior intercostal veins open into the left brachiocephalic vein), **left bronchial,** and **mediastinal veins.**

SCHEME OF DRAINAGE

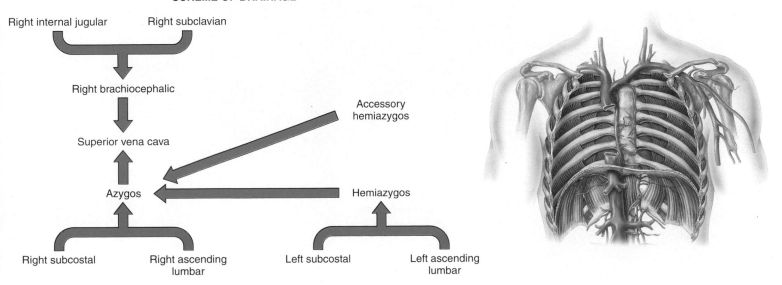

EXHIBIT 15.10 CONTINUES ▶

EXHIBIT 15.10   **521**

EXHIBIT 15.10 Veins of the Thorax CONTINUED

Figure 15.13 Principal veins of the thorax, abdomen, and pelvis.

Most thoracic structures are drained by the azygos system of veins.

Right internal jugular
Right external jugular
Right brachiocephalic
Right superior intercostal
Superior vena cava
Right posterior intercostal
Azygos
Mediastinals
Bronchial
Pericardial
Diaphragm
Hepatics
Right suprarenal
Right subcostal
Right renal
Right ascending lumbar
Right gonadal (testicular or ovarian)
Right lumbar
Right common iliac
Right internal iliac
Right external iliac

Left internal jugular
Left external jugular
Left subclavian
Left brachiocephalic
Left superior intercostal
Left axillary
Left cephalic
Left posterior intercostal
Left brachial
Accessory hemiazygos
Left basilic
Esophageals
Left inferior phrenics
Hemiazygos
Left suprarenal
Left renal
Left ascending lumbar
Left gonadal (testicular or ovarian)
Inferior vena cava
Left common iliac
Inguinal ligament
Middle sacral
Left internal iliac
Left external iliac
Left femoral

(a) Anterior view

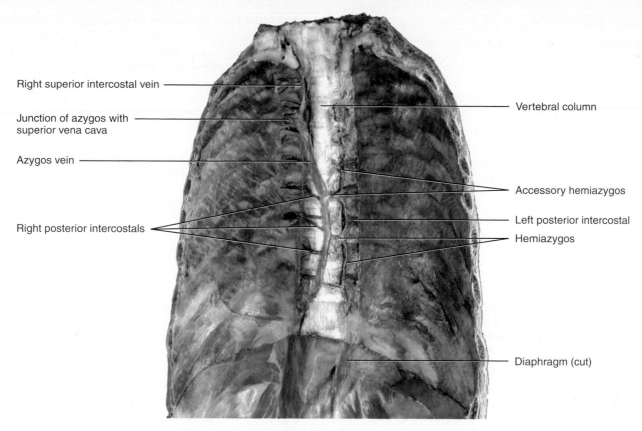

Right superior intercostal vein

Junction of azygos with superior vena cava

Azygos vein

Right posterior intercostals

Vertebral column

Accessory hemiazygos

Left posterior intercostal

Hemiazygos

Diaphragm (cut)

(b) Anterior view of posterior thoracic wall

SUPERIOR

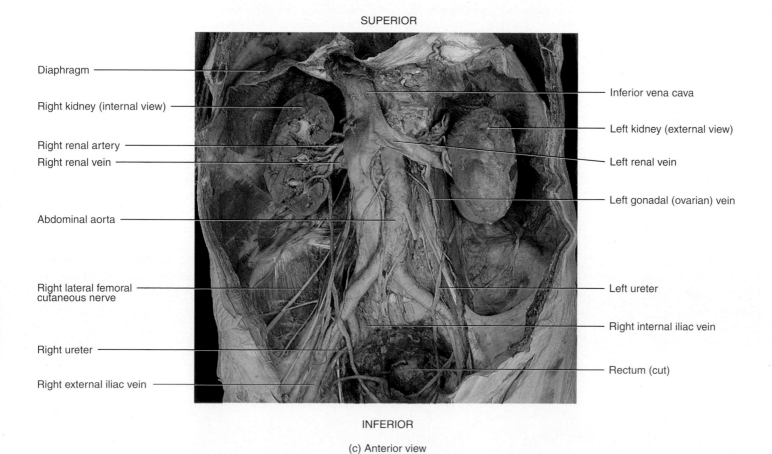

Diaphragm

Right kidney (internal view)

Right renal artery

Right renal vein

Abdominal aorta

Right lateral femoral cutaneous nerve

Right ureter

Right external iliac vein

Inferior vena cava

Left kidney (external view)

Left renal vein

Left gonadal (ovarian) vein

Left ureter

Right internal iliac vein

Rectum (cut)

INFERIOR

(c) Anterior view

❓ **Which vein returns blood from the abdominopelvic viscera to the heart?**

EXHIBIT 15.10   **523**

EXHIBIT 15-11 Veins of the Abdomen and Pelvis

OBJECTIVE

- Identify the principal veins that drain the abdomen and pelvis.

Blood from the abdominal and pelvic viscera and abdominal wall returns to the heart via the inferior vena cava. Many small veins enter the inferior vena cava. Most carry return flow from parietal branches of the abdominal aorta, and their names correspond to the names of the arteries.

The inferior vena cava does not receive veins directly from the gastrointestinal tract, spleen, pancreas, and gallbladder. These organs pass their blood into a common vein, the **hepatic portal vein,** which delivers the blood to the liver. The superior mesenteric and splenic veins unite to form the hepatic portal vein (see Figure 15.16). This special flow of venous blood, called the **hepatic portal circulation,** is described shortly. After passing through the liver for processing, blood drains into the hepatic veins, which empty into the inferior vena cava.

CHECKPOINT

- What structures do the lumbar, gonadal, renal, suprarenal, inferior phrenic, and hepatic veins drain?

VEIN	DESCRIPTION AND REGION DRAINED
Inferior vena cava (VĒ-na CĀ-va; *vena*=vein; *cava*=cavelike)	The two common iliac veins that drain the free lower limbs, pelvis, and abdomen unite to form the **inferior vena cava.** The inferior vena cava extends superiorly through the abdomen and thorax to the right atrium.
Common iliac veins (IL-ē-ak=pertaining to the ilium)	The **common iliac veins** are formed by the union of the internal and external iliac veins anterior to the sacroiliac joint and represent the distal continuation of the inferior vena cava at their bifurcation. The right common iliac vein is much shorter than the left and is also more vertical. Generally, the common iliac veins drain the pelvis, external genitals, and lower limbs.
Internal iliac veins	The **internal iliac veins** begin near the superior portion of the greater sciatic notch and run medial to their corresponding arteries. Generally, the veins drain the thigh, buttocks, external genitals, and pelvis.
External iliac veins	The **external iliac veins** are companions of the internal iliac arteries and begin at the inguinal ligaments as continuations of the femoral veins. They end anterior to the sacroiliac joint where they join with the internal iliac veins to form the common iliac veins. The external iliac veins drain the lower limbs, cremaster muscle in males, and the abdominal wall.
Lumbar veins (LUM-bar=pertaining to the loin)	A series of parallel **lumbar veins,** usually four on each side, drain blood from both sides of the posterior abdominal wall, vertebral canal, spinal cord, and meninges. The lumbar veins run horizontally with the lumbar arteries. The lumbar veins connect at right angles with the right and left **ascending lumbar veins,** which form the origin of the corresponding azygos or hemiazygos vein. The lumbar veins drain blood into the ascending lumbars and then run to the inferior vena cava, where they release the remainder of the flow.
Gonadal veins (gō-NAD-al; *gono*=seed) [**testicular** (tes-TIK-ū-lar) or **ovarian** (ō-VAR-ē-an)]	The **gonadal veins** ascend with the gonadal arteries along the posterior abdominal wall. In the male, the gonadal veins are called the testicular veins. The **testicular veins** drain the testes (the left testicular vein empties into the left renal vein, and the right testicular vein drains into the inferior vena cava). In the female, the gonadal veins are called the ovarian veins. The **ovarian veins** drain the ovaries. The left ovarian vein empties into the left renal vein, and the right ovarian vein drains into the inferior vena cava.
Renal veins (RĒ-nal; *ren-*=kidney)	The large **renal veins** pass anterior to the renal arteries. The left renal vein is longer than the right renal vein and passes anterior to the abdominal aorta. It receives the left testicular (or ovarian), left inferior phrenic, and usually left suprarenal veins. The right renal vein empties into the inferior vena cava posterior to the duodenum. The renal veins drain the kidneys.
Suprarenal veins (soo'-pra-RĒ-nal; *supra-*=above)	The **suprarenal veins** drain the adrenal (suprarenal) glands (the left suprarenal vein empties into the left renal vein, and the right suprarenal vein empties into the inferior vena cava).
Inferior phrenic veins (FREN-ik=pertaining to the diaphragm)	The **inferior phrenic veins** drain the diaphragm (the left inferior phrenic vein usually sends one tributary to the left suprarenal vein, which empties into the left renal vein, and another tributary that empties into the inferior vena cava; the right inferior phrenic vein empties into the inferior vena cava).
Hepatic veins (he-PAT-ik=pertaining to the liver)	The **hepatic veins** drain the liver.

SCHEME OF DRAINAGE

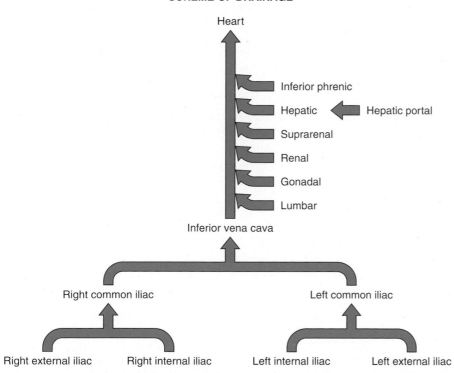

Heart

Inferior phrenic

Hepatic ← Hepatic portal

Suprarenal

Renal

Gonadal

Lumbar

Inferior vena cava

Right common iliac

Left common iliac

Right external iliac   Right internal iliac

Left internal iliac   Left external iliac

EXHIBIT 15.11   **525**

EXHIBIT 15.12 Veins of the Lower Limbs

OBJECTIVE

- Identify the principal superficial and deep veins that drain the lower limbs.

As with the upper limbs, blood from the lower limbs is drained by both **superficial** and **deep veins.** The superficial veins often anastomose with one another and with deep veins along their length. Deep veins, for the most part, have the same names as corresponding arteries. All veins of the lower limbs have valves, which are more numerous than in veins of the upper limbs.

CHECKPOINT

- What is the clinical importance of the great saphenous veins?

VEIN	DESCRIPTION AND REGION DRAINED
SUPERFICIAL VEINS	
Great saphenous veins (sa-FĒ-nus; *saphen-*= clearly visible)	The **great (long) saphenous veins,** the longest veins in the body, ascend from the foot to the groin in the subcutaneous layer. They begin at the medial end of the dorsal venous arches of the foot. The **dorsal venous arches** (VĒ-nus) are networks of veins on the dorsum of the foot formed by the **dorsal digital veins,** which collect blood from the toes, and then unite in pairs to form the **dorsal metatarsal veins,** which parallel the metatarsals. As the dorsal metatarsal veins approach the foot, they combine to form the dorsal venous arches. The great saphenous veins pass anterior to the medial malleolus of the tibia and then superiorly along the medial aspect of the leg and thigh just deep to the skin. They receive tributaries from superficial tissues and connect with the deep veins as well. They empty into the femoral veins at the groin. Generally, the great saphenous veins drain mainly the medial side of the leg and thigh, the groin, external genitals, and abdominal wall. Along their length, the great saphenous veins have from 10 to 20 valves, with more located in the leg than the thigh. These veins are more likely to be subject to varicosities than other veins in the lower limbs because they must support a long column of blood and are not well supported by skeletal muscles. The great saphenous veins are often used for prolonged administration of intravenous fluids. This is particularly important in very young children and in patients of any age who are in shock and whose veins are collapsed. In coronary artery bypass grafting, if multiple blood vessels need to be grafted, sections of the great saphenous vein are used along with at least one artery as a graft. After the great saphenous vein is removed and divided into sections, the sections are used to bypass the blockages. The vein grafts are reversed so that the valves do not obstruct the flow of blood.
Small saphenous veins	The **small (short) saphenous veins** begin at the lateral aspect of the dorsal venous arches of the foot. They pass posterior to the lateral malleolus of the fibula and ascend deep to the skin along the posterior aspect of the leg. They empty into the popliteal veins in the popliteal fossa, posterior to the knee. Along their length, the small saphenous veins have from 9 to 12 valves. The small saphenous veins drain the foot and posterior aspect of the leg. They may communicate with the great saphenous veins in the proximal thigh.
DEEP VEINS	
Posterior tibial veins (TIB-ē-al)	The **plantar digital veins** on the plantar surfaces of the toes unite to form the **plantar metatarsal veins,** which parallel the metatarsals. They in turn unite to form the **deep plantar venous arches.** From each arch emerges the **medial** and **lateral plantar veins.** The medial and lateral plantar veins, posterior to the medial malleolus of the tibia, form the paired **posterior tibial veins,** which sometimes merge into a single vessel. They accompany the posterior tibial artery through the leg. They ascend deep to the muscles in the posterior aspect of the leg and drain the foot and posterior compartment muscles. About two-thirds of the way up the leg, the posterior tibial veins drain blood from the **fibular (peroneal) veins,** which drain the lateral and posterior leg muscles. The posterior tibial veins unite with the anterior tibial veins just inferior to the popliteal fossa to form the popliteal veins.
Anterior tibial veins	The paired **anterior tibial veins** arise in the dorsal venous arch and accompany the anterior tibial artery. They ascend in the interosseous membrane between the tibia and fibula and unite with the posterior tibial veins to form the popliteal vein. The anterior tibial veins drain the ankle joint, knee joint, tibiofibular joint, and anterior portion of the leg.
Popliteal veins (pop'-li-TĒ-al=pertaining to the hollow behind knee)	The **popliteal veins,** formed by the union of the anterior and posterior tibial veins, also receive blood from the small saphenous veins and tributaries that correspond to branches of the popliteal artery. The popliteal veins drain the knee joint and the skin, muscles, and bones of portions of the calf and thigh around the knee joint.
Femoral veins (FEM-o-ral)	The **femoral veins** accompany the femoral arteries and are the continuations of the popliteal veins just superior to the knee. The femoral veins extend up the posterior surface of the thighs and drain the muscles of the thighs, femurs, external genitals, and superficial lymph nodes. The largest tributaries of the femoral veins are the **deep veins of the thigh (deep femoral veins).** Just before penetrating the abdominal wall, the femoral veins receive the deep femoral veins and the great saphenous veins. The veins formed from this union penetrate the body wall and enter the pelvic cavity. Here they are known as the **external iliac veins.** In order to take blood samples or pressure recordings from the right side of the heart, a catheter is inserted into the femoral vein as it passes through the femoral triangle. The catheter passes through the external and common iliac veins and inferior vena cava into the right atrium.

SCHEME OF DRAINAGE

Inferior vena cava

Right common iliac

Right external iliac   Right internal iliac

Right femoral   Right deep vein of the thigh (deep femoral)

Right popliteal

Right posterior tibial   Right anterior tibial

Right great saphenous   Right small saphenous   Right fibular (peroneal)

Right medial and lateral plantar   Right dorsal venous arch

Right dorsal venous arch

Right deep plantar venous arch

Right dorsal metatarsal

Right plantar metatarsal

Right dorsal digital

Right plantar digital

EXHIBIT **15.12** CONTINUES

EXHIBIT 15.12   **527**

EXHIBIT 15.12 Veins of the Lower Limbs CONTINUED

Figure 15.14 **Principal veins of the pelvis and free lower limbs.** (See Tortora, *A Photographic Atlas of the Human Body*, 2e, Figure 6.22.)

Deep veins usually bear the names of their companion arteries.

Inferior vena cava

Right common iliac

Right internal iliac

Right external iliac

Left common iliac

Right deep vein of the thigh (deep femoral)

Right femoral

Right accessory saphenous

Right great saphenous

Right popliteal

Right small saphenous

Right fibular (peroneal)

Right anterior tibial

Right great saphenous

Right small saphenous

Right posterior tibial

Right dorsal venous arch

Right dorsal metatarsal

Right dorsal digital

Right medial plantar

Right lateral plantar

Right deep plantar venous arch

Right plantar metatarsal

Right plantar digital

(a) Anterior view

(b) Posterior view

Femoral artery

Inguinal ligament

Femoral vein

Femoral nerves

Accessory saphenous vein

Fascia lata

Great saphenous vein

Tributary to great
saphenous vein

(c) Anteromedial view of thigh

? **Which veins of the lower limb are superficial?**

EXHIBIT 15.12   **529**

The Hepatic Portal Circulation

The **hepatic portal circulation** carries venous blood from the gastrointestinal organs and spleen to the liver. A vein that carries blood from one capillary network to another is called a **portal vein**. The **hepatic portal vein** (*hepat-*=liver) receives blood from capillaries of gastrointestinal organs and the spleen and delivers it to the sinusoids of the liver (Figure 15.15). After a meal, hepatic portal blood is rich in nutrients absorbed from the gastrointestinal tract. The liver stores some of them and modifies others before they pass into the general circulation. For example, the liver converts glucose into glycogen for storage, reducing blood glucose level shortly after a meal. The liver also detox-ifies harmful substances, such as alcohol, that have been absorbed from the gastrointestinal tract and destroys bacteria by phagocytosis.

The superior mesenteric and splenic veins unite to form the hepatic portal vein. The **superior mesenteric vein** drains blood from the small intestine and portions of the large intestine, stomach, and pancreas through the *jejunal, ileal, ileocolic, right colic, middle colic, pancreaticoduodenal,* and *right gastroepiploic veins.* The **splenic vein** drains blood from the stomach, pancreas, and portions of the large intestine through the *short gastric, left gastroepiploic, pancreatic,* and *inferior mesenteric veins.* The inferior mesenteric vein, which passes into the splenic vein, drains portions of the large intestine through the superior *rectal, sigmoidal,*

Figure 15.15 Hepatic portal circulation. A schematic diagram of blood flow through the liver, including arterial circulation, is shown in (b); deoxygenated blood is indicated in blue, oxygenated blood in red.

 The hepatic portal circulation delivers venous blood from the organs of the gastrointestinal tract and spleen to the liver.

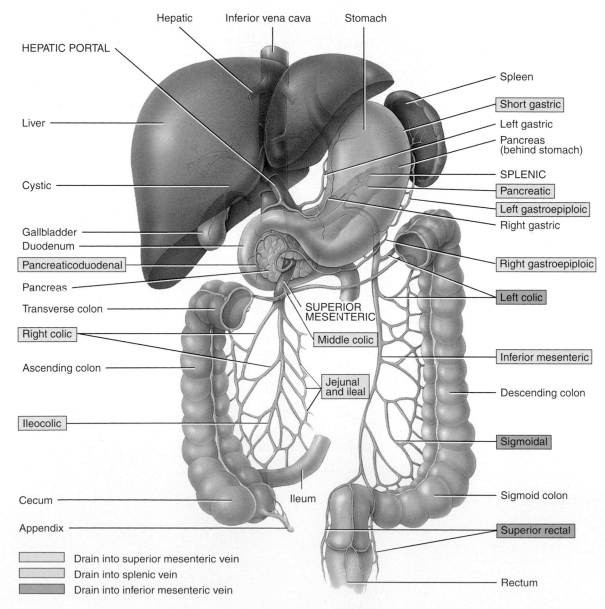

(a) Anterior view of veins draining into the hepatic portal vein and *left colic veins.* The *right* and *left gastric veins,* which open directly into the hepatic portal vein, drain the stomach. The *cystic vein,* which also opens into the hepatic portal vein, drains the gallbladder.

At the same time the liver is receiving nutrient-rich but deoxygenated blood via the hepatic portal vein, it also is receiving oxygenated blood via the hepatic artery, a branch of the celiac trunk. The oxygenated blood mixes with the deoxygenated blood in sinusoids. Eventually, blood leaves the sinusoids of the liver through the **hepatic veins,** which drain into the inferior vena cava.

The Pulmonary Circulation

The **pulmonary circulation** (*pulmo*=lung) carries deoxygenated blood from the right ventricle to the air sacs (alveoli) within the lungs and returns oxygenated blood from the air sacs to the left atrium (Figure 15.16). The **pulmonary trunk** emerges from the right ventricle and passes superiorly, posteriorly, and to the left. It then divides into two branches: the **right pulmonary artery** to the right lung and the **left pulmonary artery** to the left lung. After birth, the pulmonary arteries are the only arteries that carry deoxygenated blood. On entering the lungs, the branches divide and subdivide until finally they form capillaries around the air sacs (alveoli) within the lungs. CO_2 passes from the blood into the air sacs and is exhaled. Inhaled O_2 passes from the air within the lungs into the blood. The pulmonary capillaries unite to form venules and eventually **pulmonary veins,** which exit the lungs and carry the oxygenated blood to the left atrium. Two left and two right pulmonary veins enter the left atrium. After birth, the pulmonary veins are the only veins that carry oxygenated blood. Contractions of the left ventricle then eject the oxygenated blood into the systemic circulation.

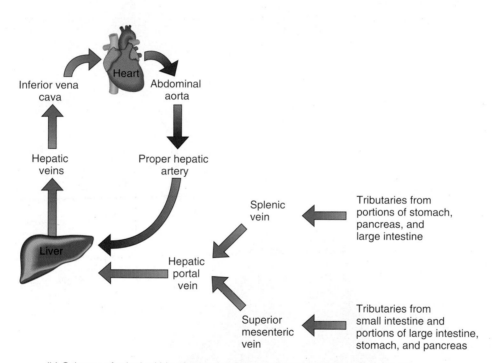

(b) Scheme of principal blood vessels of hepatic portal circulation and arterial supply and venous drainage of liver

? Which veins carry blood away from the liver?

Figure 15.16 Pulmonary circulation.

The pulmonary circulation brings deoxygenated blood from the right ventricle to the lungs and returns oxygenated blood from the lungs to the left atrium.

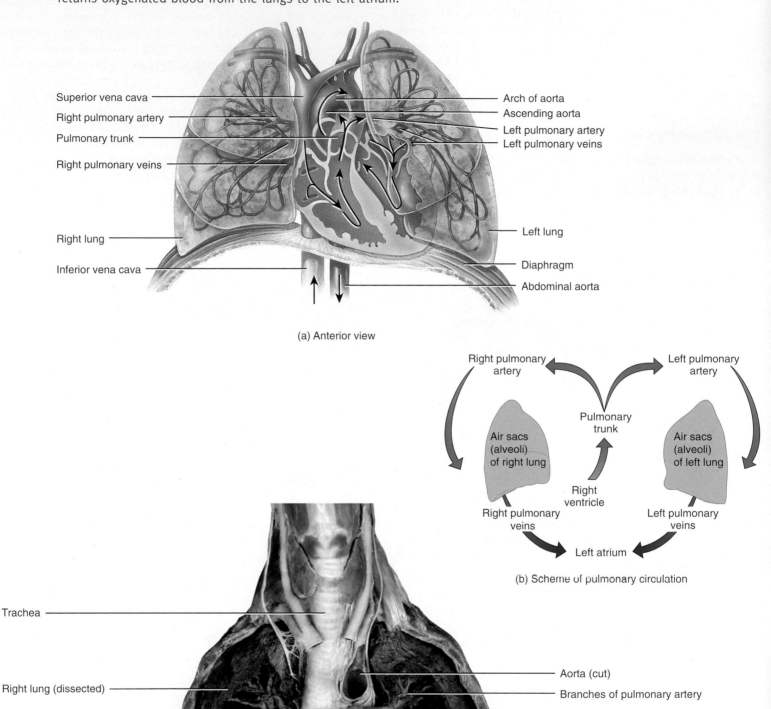

(a) Anterior view

(b) Scheme of pulmonary circulation

(c) Pulmonary circulation into the lungs

? **After birth, which are the only arteries that carry deoxygenated blood?**

The Fetal Circulation

The circulatory system of a fetus, called the **fetal circulation**, exists only in the fetus and contains special structures that allow the developing fetus to exchange materials with its mother (Figure 15.17). It differs from the postnatal (after birth) circulation be- cause the lungs, kidneys, and gastrointestinal organs do not begin to function until birth. The fetus obtains O_2 and nutrients from and eliminates CO_2 and other wastes into the maternal blood.

The exchange of materials between fetal and maternal circulations occurs through the **placenta** (pla-SEN-ta), which forms inside the mother's uterus and attaches to the umbilicus (navel) of

Figure 15.17 Fetal circulation and changes at birth. The gold boxes between parts (a) and (b) describe the fate of certain fetal structures once postnatal circulation is established.

The lungs and gastrointestinal organs do not begin to function until birth.

Right atrium

Right ventricle

Arch of aorta

Superior vena cava

DUCTUS ARTERIOSUS becomes Ligamentum arteriosum

Lung

Pulmonary artery

Pulmonary veins

Heart

FORAMEN OVALE becomes Fossa ovalis

Inferior vena cava

Liver

DUCTUS VENOSUS becomes Ligamentum venosum

Hepatic portal vein

UMBILICAL VEIN becomes Ligamentum teres

Umbilicus

Inferior vena cava

Abdominal aorta

Common iliac artery

UMBILICAL ARTERIES become Medial umbilical ligaments

Urinary bladder

Urethra

UMBILICAL CORD

Placenta

Left atrium

Left ventricle

(a) Fetal circulation

(b) Circulation at birth

Oxygenated blood

Mixed oxygenated and deoxygenated blood

Deoxygenated blood

FIGURE **15.17** CONTINUES

FIGURE **15.17** CONTINUED ▶

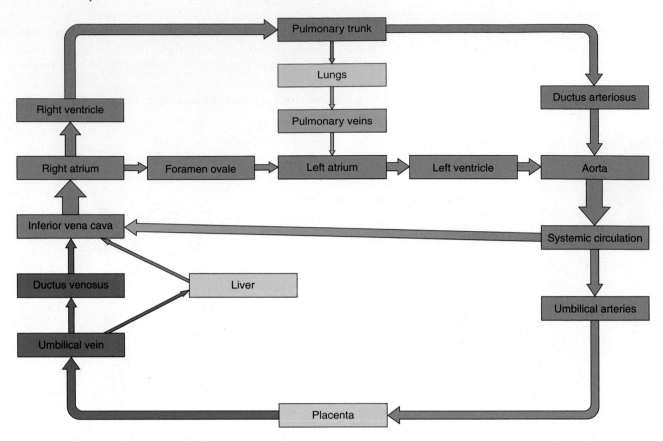

(c) Scheme of fetal circulation

❓ **Through which structure does the exchange of materials between mother and fetus occur?**

the fetus by the **umbilical cord** (um-BIL-i-kal). The placenta communicates with the mother's cardiovascular system through many small blood vessels that emerge from the uterine wall. The umbilical cord contains blood vessels that branch into capillaries in the placenta. Wastes from the fetal blood diffuse out of the capillaries, into spaces containing maternal blood (intervillous spaces) in the placenta, and finally into the mother's uterine veins. Nutrients travel the opposite route—from the maternal blood vessels to the intervillous spaces to the fetal capillaries. Normally, there is no direct mixing of maternal and fetal blood because all exchanges occur by diffusion through capillary walls.

Blood passes from the fetus to the placenta via two **umbilical arteries** (Figure 15.17). These branches of the internal iliac (hypogastric) arteries are within the umbilical cord. At the placenta, fetal blood picks up O_2 and nutrients and eliminates CO_2 and wastes. The oxygenated blood returns from the placenta via a single **umbilical vein**. This vein ascends to the liver of the fetus, where it divides into two branches. Some blood flows through the branch that joins the hepatic portal vein and enters the liver, but most of the blood flows into the second branch, the **ductus venosus** (DUK-tus ve-NŌ-sus), which drains into the inferior vena cava.

Deoxygenated blood returning from lower body regions of the fetus mingles with oxygenated blood from the ductus venosus in the inferior vena cava. This mixed blood then enters the right atrium. Deoxygenated blood returning from upper body regions of the fetus enters the superior vena cava and also passes into the right atrium.

Most of the fetal blood does not pass from the right ventricle to the lungs, as it does in postnatal circulation, because an opening called the **foramen ovale** (fō-RĀ-men ō-VAL-ē) exists in the septum between the right and left atria. Most of the blood that enters the right atrium passes through the foramen ovale into the left atrium and joins the systemic circulation. The blood that does pass into the right ventricle is pumped into the pulmonary trunk, but little of this blood reaches the nonfunctioning fetal lungs. Instead, most is sent through the **ductus arteriosus** (ar-tē-rē-ō-sus), a vessel that connects the pulmonary trunk with the aorta. The blood in the aorta is carried to all fetal tissues through the systemic circulation. When the common iliac arteries branch into the external and internal iliacs, part of the blood flows into the internal iliacs, into the umbilical arteries, and back to the placenta for another exchange of materials.

After birth, when pulmonary (lung), renal (kidney), and digestive functions begin, the following vascular changes occur (Figure 15.17b):

1. When the umbilical cord is tied off, blood no longer flows through the umbilical arteries, they fill with connective tissue, and the distal portions of the umbilical arteries become fibrous cords called the **medial umbilical ligaments**. Although the arteries are closed functionally only a few minutes after birth, complete obliteration of the lumens may take 2 to 3 months.
2. The umbilical vein collapses but remains as the **ligamentum teres (round ligament)**, a structure that attaches the umbilicus to the liver.

3. The ductus venosus collapses but remains as the **ligamentum venosum**, a fibrous cord on the inferior surface of the liver.

4. The placenta is expelled as the **afterbirth**.

5. The foramen ovale normally closes shortly after birth to become the **fossa ovalis**, a depression in the interatrial septum. When an infant takes its first breath, the lungs expand and blood flow to the lungs increases. Blood returning from the lungs to the heart increases pressure in the left atrium. This closes the foramen ovale by pushing the valve that guards it against the interatrial septum. Permanent closure occurs in about a year.

6. The ductus arteriosus closes by vasoconstriction almost immediately after birth and becomes the **ligamentum arteriosum**. Complete anatomical obliteration of the lumen takes 1 to 3 months.

⊙ CHECKPOINT

6. What is the purpose of the systemic circulation?
7. Diagram the hepatic portal circulation. Why is this route important?
8. Diagram the route of the pulmonary circulation.
9. Discuss the anatomy and physiology of the fetal circulation. Indicate the function of the umbilical arteries, umbilical vein, ductus venosus, foramen ovale, and ductus arteriosus.

DEVELOPMENT OF BLOOD VESSELS AND BLOOD

⊙ **OBJECTIVE**

- Describe the development of blood vessels and blood.

The development of blood cells and the formation of blood vessels begins outside the embryo as early as 15 to 16 days in the **mesoderm** of the wall of the yolk sac, chorion, and connecting stalk. About 2 days later, blood vessels form within the embryo. The early formation of the cardiovascular system is linked to the small amount of yolk in the ovum and yolk sac. As the embryo develops rapidly during the third week, there is a greater need to develop a cardiovascular system to supply sufficient nutrients to the embryo and remove wastes from it.

Blood vessels and blood cells develop from the same precursor cell, called a **hemangioblast** (hē-MAN-jē-ō-blast; *hema-*=blood; *-blast*=immature stage). Once mesenchyme develops into hemangioblasts, they can give rise to cells that produce blood vessels or cells that produce blood cells.

Blood vessels develop from **angioblasts** derived from hemangioblasts (Figure 15.18). Angioblasts aggregate to form isolated masses and cords throughout the embryonic disc called **blood islands** (Figure 15.18). Spaces soon appear in the islands and become the lumens of the blood vessels. Some of the angioblasts immediately around the spaces give rise to the *endothelial lining of the blood vessels*. Angioblasts around the endothelium form the *tunics* (interna, media, and externa) of the larger blood vessels. Growth and fusion of blood islands form an extensive network of blood vessels throughout the embryo. By continuous branching, blood vessels outside the embryo connect with those inside the embryo, linking the embryo with the placenta.

Blood cells develop from **pluripotent stem cells** derived from hemangioblasts. This development occurs in the walls of blood vessels in the yolk sac, chorion, and allantois at about three weeks

Figure 15.18 Development of blood vessels and blood cells from blood islands.

⑥ ⎍ Blood vessel development begins in the embryo on about day 17 or 18.

Blood islands

Angioblasts

Wall of yolk sac:
— Mesoderm
— Endoderm

Endothelial lining

Blood cells

Space (future lumen)

Blood cells

Lumen of blood vessel

❓ **From which germ cell layer are blood vessels and blood derived?**

after fertilization. Blood formation in the embryo itself begins at about the fifth week in the liver and the twelfth week in the spleen, red bone marrow, and thymus.

◉ **CHECKPOINT**

10. What are the sites of blood cell production outside the embryo and within the embryo?

AGING AND THE CARDIOVASCULAR SYSTEM

◉ **OBJECTIVE**

• Explain the effects of aging on the cardiovascular system.

General changes in the cardiovascular system associated with aging include decreased compliance of the aorta, reduction in cardiac muscle fiber size, progressive loss of cardiac muscular strength, reduced cardiac output, a decline in maximum heart rate, and an increase in systolic blood pressure. Total blood cholesterol tends to increase with age, as does low-density lipoprotein (LDL); high-density lipoprotein (HDL) tends to decrease. There is an increase in the incidence of coronary artery disease (CAD), the major cause of heart disease and death in older Americans. Congestive heart failure, a set of symptoms associated with impaired pumping of the heart, is also prevalent in older individuals. Changes in blood vessels that serve brain tissue—for example, atherosclerosis—reduce nourishment to the brain and result in the malfunction or death of brain cells. By age 80, cerebral blood flow is 20% less and renal blood flow is 50% less than in the same person at age 30.

◉ **CHECKPOINT**

11. How does aging affect the heart?

---

wall of the left ventricle. An increased afterload, however, leads to myocardial hypertrophy that is accompanied by muscle damage and fibrosis (a buildup of collagen fibers between the muscle fibers). As a result, the left ventricle enlarges, weakens, and dilates. Because arteries in the brain are usually less protected by surrounding tissues than are the major arteries in other parts of the body, prolonged hypertension can eventually cause them to rupture, resulting in a stroke. Hypertension also damages kidney arterioles, causing them to thicken, which narrows the lumen; because the blood supply to the kidneys is thereby reduced, the kidneys secrete more renin, which elevates the blood pressure even more.

Lifestyle Changes to Reduce Hypertension

Although several categories of drugs (described next) can reduce elevated blood pressure, the following lifestyle changes are also effective in managing hypertension:

- *Lose weight.* This is the best treatment for high blood pressure short of using drugs. Loss of even a few pounds helps reduce blood pressure in overweight hypertensive individuals.
- *Limit alcohol intake.* Drinking in moderation may lower the risk of coronary heart disease, mainly among males over 45 and females over 55. Moderation is defined as no more than one 12-oz beer per day for females and no more than two 12-oz beers per day for males.
- *Exercise.* Becoming more physically fit by engaging in moderate activity (such as brisk walking) several times a week for 30 to 45 minutes can lower systolic blood pressure by about 10 mmHg.
- *Reduce intake of sodium (salt).* Roughly half the people with hypertension are "salt sensitive." For them, a high-salt diet appears to promote hypertension, and a low-salt diet can lower their blood pressure.
- *Maintain recommended dietary intake of potassium, calcium, and magnesium.* Higher levels of potassium, calcium, and magnesium in the diet are associated with a lower risk of hypertension.
- *Don't smoke.* Smoking has devastating effects on the heart and can augment the damaging effects of high blood pressure by promoting vasoconstriction.
- *Manage stress.* Various meditation and biofeedback techniques help some people reduce high blood pressure. These methods may work by decreasing the daily release of epinephrine and norepinephrine by the adrenal medulla.

Drug Treatment of Hypertension

Drugs having several different mechanisms of action are effective in lowering blood pressure. Many people are successfully treated with *diuretics,* agents that decrease blood pressure by decreasing blood volume because they increase elimination of water and salt in the urine. *ACE (angiotensin converting enzyme) inhibitors* block formation of angiotensin II and thereby promote vasodilation and decrease the secretion of aldosterone. *Beta blockers* reduce blood pressure by inhibiting the secretion of renin and by decreasing heart rate and contractility. *Vasodilators* relax the smooth muscle in arterial walls, causing vasodilation and lowering blood pressure by lowering systemic vascular resistance. An important category of vasodilators are the *calcium channel blockers,* which slow the inflow of Ca^{2+} into vascular smooth muscle cells. They reduce the heart's workload by slowing Ca^{2+} entry into pacemaker cells and regular myocardial fibers, thereby decreasing heart rate and the force of myocardial contraction.

KEY MEDICAL TERMS ASSOCIATED WITH BLOOD VESSELS

Aneurysm (AN-ū-rizm) A thin, weakened section of the wall of an artery or a vein that bulges outward, forming a balloonlike sac. Common causes are atherosclerosis, syphilis, congenital blood vessel defects, and trauma.

Angiography (an′-jē-OG-ra-fē; *angio-*=vessel; *-grapho*=to write) A diagnostic procedure in which a radiopaque dye is injected through a catheter that has been introduced into a blood vessel and guided to the blood vessel to be examined. The dye flows into the appropriate blood vessel, making abnormalities such as blockages visible on x-rays. Angiography is used to study the blood vessels of the heart (*coronary angiography*), aorta and its branches (*aortography*), and blood vessels in the legs (*femoral angiography*).

Arteritis (ar′-te-RĪ-tis; *-itis*=inflammation of) Inflammation of an artery, probably due to an autoimmune response.

Carotid endarterectomy (ka-ROT-id end′-ar-ter-EK-tō-mē) The removal of atherosclerotic plaque from the carotid artery to restore greater blood flow to the brain.

Claudication (klaw′-di-KĀ-shun) Pain and lameness or limping caused by defective circulation of the blood in the vessels of the limbs.

Deep vein thrombosis The presence of a thrombus (blood clot) in a deep vein of the lower limbs. It may lead to (1) pulmonary embolism, if the thrombus dislodges and then lodges within the pulmonary arterial blood flow, and (2) postphlebitic syndrome, which consists of edema, pain, and skin changes due to destruction of venous valves.

Doppler ultrasound scanning Imaging technique commonly used to measure blood flow. A transducer is placed on the skin and an image is displayed on a monitor that provides the exact position and severity of a blockage.

Hypotension (hī-pō-TEN-shun) Low blood pressure; most commonly used to describe an acute drop in blood pressure, as occurs during excessive blood loss.

Normotensive (nor′-mō-TEN-siv) Characterized by normal blood pressure.

Occlusion (ō-KLOO-shun) The closure or obstruction of the lumen of a structure such as a blood vessel. An example is an atherosclerotic plaque in an artery.

Orthostatic hypotension (or′-thō-STAT-ik; *ortho-*=straight; *-static*=causing to stand) An excessive lowering of systemic blood pressure when a person assumes an erect or semierect posture. Also called **postural hypotension**.

Phlebitis (fle-BĪ-tis; *phleb-*=vein) Inflammation of a vein, often in a leg.

Syncope (SIN-kō-pē=cutting short) Fainting. A sudden, temporary loss of consciousness followed by spontaneous recovery. It is usually due to decreased blood flow to the brain.

Thrombectomy (throm-BEK-tō-mē; *thrombo-*=clot) An operation to remove a blood clot from a blood vessel.

Thrombophlebitis (throm′-bō-fle-BĪ-tis) Inflammation of a vein involving clot formation.

Venipuncture (VEN-i-punk-chur; *vena-*=vein) The puncture of a vein, usually to withdraw blood for analysis or introduce a solution, for example, an antibiotic. The median cubital vein is frequently used.

White coat (office) hypertension A stress-induced syndrome found in patients who have elevated blood pressure when being examined by health-care personnel, but otherwise have normal blood pressure.

Anatomy of Blood Vessels (p. 482)

1. Arteries carry blood away from the heart. The wall of an artery consists of a tunica interna, a tunica media (which maintains elasticity and contractility), and a tunica externa.
2. Large arteries are termed elastic (conducting) arteries, and medium-sized arteries are called muscular (distributing) arteries.
3. Many arteries anastomose, which means the distal ends of two or more vessels unite. An alternate blood route from an anastomosis is called collateral circulation. Arteries that do not anastomose are called end arteries.
4. Arterioles are small arteries that deliver blood to capillaries.
5. Through constriction and dilation, arterioles assume key roles in regulating blood flow from arteries into capillaries and in altering arterial blood pressure.
6. Capillaries are microscopic blood vessels through which materials are exchanged between blood and tissue cells. Some capillaries are continuous; others are fenestrated.
7. Capillaries branch to form an extensive network throughout a tissue. This network increases the surface area available for the exchange of materials between the blood and the body tissue; it also allows rapid exchange of large quantities of materials.
8. Precapillary sphincters regulate blood flow through capillaries.
9. Microscopic blood vessels in the liver are called sinusoids.
10. Venules are small vessels that form from the merging capillaries; venules merge to form veins.
11. Veins consist of the same three tunics as arteries but have a thinner tunica interna and tunica media. The lumen of a vein is also larger than that of a comparable artery.
12. Veins contain valves to prevent backflow of blood.
13. Weak valves can lead to varicose veins.
14. Vascular (venous) sinuses are veins with very thin walls.
15. Systemic veins are collectively called blood reservoirs because they hold a large volume of blood. If the need arises, this blood can be shifted into other blood vessels through vasoconstriction.
16. The principal blood reservoirs are the veins of the abdominal organs (liver and spleen) and skin.

Circulatory Routes (p. 490)

1. The two basic postnatal circulatory routes are the systemic and pulmonary circulations.
2. Among the subdivisions of the systemic circulation are the coronary (cardiac) and the hepatic portal circulations.

3. Fetal circulation exists only in the fetus.
4. The systemic circulation carries oxygenated blood from the left ventricle through the aorta to all parts of the body (including some lung tissue, but *not* the air sacs of the lungs) and returns the deoxygenated blood to the right atrium.
5. The aorta is divided into the ascending aorta, the arch of the aorta, and the descending aorta. Each section gives off arteries that branch to supply the whole body.
6. Blood returns to the heart through the systemic veins. All veins of the systemic circulation drain into the superior or inferior venae cavae or the coronary sinus; these in turn empty into the right atrium.
7. The principal blood vessels of the systemic circulation may be reviewed in Exhibits 15.1–15.12.
8. The hepatic portal circulation detours venous blood from the gastrointestinal organs and spleen and directs it into the hepatic portal vein of the liver before it is returned to the heart. It enables the liver to utilize nutrients and detoxify harmful substances in the blood.
9. The pulmonary circulation takes deoxygenated blood from the right ventricle to the alveoli within the lungs and returns oxygenated blood from the alveoli to the left atrium. It allows blood to be oxygenated for systemic circulation.
10. The fetal circulation involves the exchange of materials between fetus and mother.
11. The fetus derives O_2 and nutrients and eliminates CO_2 and wastes through the maternal blood supply via the placenta.
12. At birth, when pulmonary (lung), digestive, and liver functions begin, the special structures of fetal circulation are no longer needed.

Development of Blood Vessels and Blood (p. 535)

1. Blood vessels develop from mesenchyme (hemangioblasts → angioblasts → blood islands) in mesoderm called blood islands.
2. Blood cells also develop from mesenchyme (hemangioblasts → pluripotent stem cells).

Aging and the Cardiovascular System (p. 536)

1. General changes associated with aging include reduced elasticity of blood vessels, reduction in cardiac muscle size, reduced cardiac output, and increased systolic blood pressure.
2. The incidence of coronary artery disease (CAD), congestive heart failure (CHF), and atherosclerosis increases with age.

Choose the one best answer to the following questions.

1. Which statement best describes arteries?
   a. All carry oxygenated blood to the heart.
   b. All contain valves to prevent the backflow of blood.
   c. All carry blood away from the heart.
   d. Only large arteries are lined with endothelium.
   e. All branch from the descending aorta.

2. Which statement is true of veins?
   a. Their tunica interna is thicker than in arteries.
   b. Their tunica externa is thicker than in arteries.
   c. Most veins in the limbs have valves.
   d. They always carry deoxygenated blood.
   e. All ultimately empty into the inferior vena cava.

3. Put the following vessels in the correct order to trace the route of a drop of blood moving from the right side of the heart to the left side of the heart. (1) pulmonary vein, (2) pulmonary artery, (3) arterioles, (4) venules, (5) capillaries.
   a. 1,4,5,3,2   b. 1,3,5,4,2   c. 2,4,5,3,1
   d. 2,5,4,1,3   e. 2,3,5,4,1

4. Vessels that function as a pressure reservoir to ensure blood flow while the ventricles are relaxed are the:
   a. venules.   b. elastic arteries.   c. capillaries.
   d. sinusoids.   e. aorta.

5. In fetal circulation, blood bypasses pulmonary circulation by passing through the:
   a. ductus venosus.   b. ductus arteriosus.   c. fossa ovalis.
   d. All of the above.   e. b and c.

6. *All* of the following are elastic arteries *except* the:
   a. aorta.   b. brachiocephalic.   c. common carotid.
   d. vertebral.   e. femoral.

7. Metarterioles lead into:
   a. the aorta.   b. vasa recta.   c. capillary beds.
   d. elastic arteries.   e. Both b and c are correct.

8. The hepatic portal vein is formed by the union of the:
   a. left and right common iliac veins.
   b. superior and inferior mesenteric veins.
   c. superior mesenteric vein and splenic vein.
   d. inferior mesenteric vein and splenic vein.
   e. right and left hepatic veins.

9. The cerebral arterial circle (circle of Willis) forms anastomoses between the:
   a. middle cerebral arteries.
   b. internal and external jugular veins.
   c. internal and external carotid arteries.
   d. internal carotid arteries and the basilar artery.
   e. middle cerebral arteries and the internal carotid arteries.

10. *All* of the following vessels branch directly from the aorta *except* the:
    a. right subclavian artery.   b. left subclavian artery.
    c. left common carotid artery.   d. brachiocephalic artery.
    e. right coronary artery.

11. Blood in the internal jugular veins has just come from the:
    a. brachiocephalic veins.   b. external jugular veins.
    c. sinuses of the brain.   d. superior vena cava.
    e. vertebral veins.

Complete the following.

12. The three branches of the celiac trunk are the _____, the _____, and the _____.

13. The walls of large blood vessels receive their oxygen and nutrients from blood in vessels known as _____.

14. The layer of an artery wall that contains endothelium is the _____.

15. The two common iliac veins unite to form the _____.

16. Blood from gastrointestinal organs and the spleen is transported to the liver via the _____ vein.

17. The union of the distal branches of two or more arteries supplying the same body region is called a(n) _____.

18. A decrease in the size of the blood vessel lumen due to contraction of smooth muscle present in the tunica _____ is called _____.

19. Between 90% and 95% of all cases of hypertension are _____ hypertension, which cannot be attributed to any identifiable cause. The remaining 5–10% of cases are _____ hypertension, which has an identifiable underlying cause.

20. Three types of capillaries include _____ capillaries found in skeletal and smooth muscle; _____ capillaries found in the kidneys and choroid plexuses of the brain ventricles; and _____, which are wide capillaries seen in red bone marrow and the liver.

21. The _____ veins are the only veins that carry oxygenated blood after birth.

Are the following statements true or false? If the statement is False, *find and fix the mistake to make the statement* True.

22. At rest, the largest portion of blood is located in the capillaries.

23. The external jugular veins join with the internal jugular veins to form the superior vena cava.

24. Arterioles play a key role in regulating blood pressure because of their role in regulating resistance to flow.

25. The median cubital veins connect the basilic and cephalic veins anterior to the elbow.

26. An increase in stimulation by sympathetic nerves typically stimulates vasodilation of blood vessels.

27. A thrombus lodged in the ulnar artery would prevent any blood from reaching the hand.

Matching

28. Match the following:
    ____ **(a)** passes through cervical transverse foramina
    ____ **(b)** continues as left axillary artery
    ____ **(c)** gives rise to the left common carotid artery
    ____ **(d)** is a branch of the thoracic aorta
    ____ **(e)** abdominal vessel that branches to supply blood to the pancreas, small intestine, and part of the large intestine
    ____ **(f)** becomes the dorsalis pedis artery at the ankle
    ____ **(g)** receives venous blood from the myocardium
    ____ **(h)** is located in the lateral arm
    ____ **(i)** is the longest vein in the body
    ____ **(j)** is located near the gastrocnemius muscle
    ____ **(k)** becomes the internal jugular vein outside of the skull

    (1) anterior tibial artery
    (2) esophageal artery
    (3) cephalic vein
    (4) sigmoid sinus
    (5) coronary sinus
    (6) posterior tibial vein
    (7) left subclavian artery
    (8) arch of aorta
    (9) great saphenous vein
    (10) vertebral artery
    (11) superior mesenteric artery

CRITICAL THINKING QUESTIONS

1. Which structures present in the fetal circulation are absent in the adult circulatory system? Why do these changes occur?

2. The word "sinus" has more than one anatomical meaning. In the skeletal system, what are "sinuses"? Name them. In the circulatory system, what does this same term mean? Identify the path of blood taken from the superior sagittal sinus to the superior vena cava.

3. You've read about varicose veins. Why aren't there varicose arteries?

4. Enrique is scheduled for a medical procedure called a *cardiac catheterization*. During this procedure, a tube will be inserted into

Enrique's femoral artery and snaked through the blood vessels to reach the heart. If the ultimate target is the pulmonary arteries, what vessels will the tube need to pass through?

5. Gina is 45 years old and works as an executive secretary in the college's biology department. She is 5′3″ tall, weighs 163 pounds, and has smoked since she was 15 years old. Her doctor measured her blood pressure at "132 over 86." What do you think the doctor will be advising Gina to do, and why?

? ANSWERS TO FIGURE QUESTIONS

15.1 The femoral artery has the thicker wall; the femoral vein has the wider lumen.

15.2 Metabolically active tissues have extensive capillary networks because they use O_2 and produce wastes more rapidly than inactive tissues.

15.3 Materials cross capillary walls through intercellular clefts and fenestrations, via transcytosis in pinocytic vesicles, and through the plasma membranes of endothelial cells.

15.4 Valves are more important in arm veins and leg veins than in neck veins because gravity tends to cause pooling of blood in the veins of the free limbs when you are standing. The valves prevent backflow into the free limbs so the blood proceeds toward the right atrium after each heartbeat. When you are erect, gravity aids the flow of blood in neck veins back toward the heart.

15.5 The two main circulatory routes are the systemic and pulmonary circulations.

15.6 The four principal subdivisions of the aorta are the ascending aorta, arch of the aorta, thoracic aorta, and abdominal aorta.

15.7 The three major branches of the arch of the aorta, in order of their origination, are the brachiocephalic trunk, left common carotid artery, and left subclavian artery.

15.8 The abdominal aorta begins at the aortic hiatus in the diaphragm.

15.9 The abdominal aorta divides into the common iliac arteries at about the level of L4.

15.10 The superior vena cava drains regions above the diaphragm, and the inferior vena cava drains regions below the diaphragm.

15.11 All venous blood in the brain drains into the internal jugular veins.

15.12 The median cubital vein of the free upper limb is often used for withdrawing blood.

15.13 The inferior vena cava returns blood from abdominopelvic viscera to the heart.

15.14 The common iliac veins unite to form the inferior vena cava.

15.15 The superficial veins of the free lower limbs are the dorsal venous arch and the great saphenous and small saphenous veins.

15.16 The hepatic veins carry blood away from the liver.

15.17 After birth, the only arteries that carry deoxygenated blood are the pulmonary arteries.

15.18 Exchange of materials between mother and fetus occurs across the placenta.

15.19 Blood vessels and blood cells are derived from mesoderm.

16 | THE LYMPHATIC SYSTEM AND IMMUNITY

INTRODUCTION *War!* What comes to mind when you think of this word—great battles, devastation, destruction, and death? Anyone who has studied human history can tell you that war has been a recurring theme since the beginning. Think for a moment about the tremendous number of wars that have stormed through the pages of history. However, all these wars pale in comparison to the constant battles being waged within our bodies.

In an epic medieval battle scene, wave after wave of foot soldiers advance toward the enemy in hand-to-hand combat. Stationed behind this advancing wave, rows of archers shoot arrow after arrow, some targeting individuals, others simply aiming for the onrushing horde. This is not unlike the day-to-day battle waged by the body's lymphatic system against the foreign invaders that attempt to inhabit and damage our tissues. Some of the body's immune cells engage in hand-to-hand combat, and like foot soldiers, tend to have brief life spans. Like the arrows from the archers' bows, the antibodies produced by other body cells mark individuals for death and destruction. These "archer" cells, not being on the front lines, have a longer life span. Sometimes these arrows even miss their mark and destroy self rather than the enemy, the lymphatic system's version of "friendly fire." As you study the invading "armies" and the defense mechanisms deployed in response by your body's immune system, you will learn that there are many strategies to defend your "soil."

Maintaining physical health requires continuous combat against harmful agents in our internal and external environments. Despite constant exposure to a variety of **pathogens** (PATH-ō-jens), disease-producing microbes such as bacteria and viruses, most people remain healthy. The body's surface also responds to cuts and bumps, exposure to ultraviolet rays in sunlight, chemical toxins, and minor burns with an array of defensive ploys. **Immunity** or **resistance** is the ability to ward off damage or disease through our defenses. Lack of resistance is termed **susceptibility**.

The two general types of immunity are (1) innate (non-specific) immunity and (2) adaptive (specific) immunity. **Innate (nonspecific) immunity** refers to defenses that are present from birth and are always available to protect us against disease. Innate immunity does not involve specific recognition of a microbe and acts against all microbes in the same way. In addition, innate immunity does not have a memory component, that is, it cannot recall a previous contact with a foreign molecule. Among the components of innate immunity are the first line of defense (skin and mucous membranes) and the second line of defense (natural killer cells and phagocytes, inflammation, fever, and antimicrobial substances). Innate immune responses represent immunity's early-warning system; they are designed to prevent microbes from gaining access into the body and to help eliminate those that do gain access.

Adaptive (specific) immunity refers to defenses that involve specific recognition of a microbe once it has breached the defenses of innate immunity. Adaptive immunity is based on a specific response to a specific microbe; that is, it adapts or adjusts to handle a single type of invader. Unlike innate immunity, adaptive immunity is slower to respond but it does have a memory component. Adaptive immunity involves lymphocytes called T lymphocytes (T cells) and B lymphocytes (B cells). •

LYMPHATIC SYSTEM STRUCTURE AND FUNCTIONS

◉ **OBJECTIVES**
- Describe the components and major functions of the lymphatic system.
- Describe the organization of lymphatic vessels.
- Describe the formation and flow of lymph.
- Compare the structure and functions of the primary and secondary lymphatic organs and tissues.

The **lymphatic system** (lim-FAT-ik) consists of a fluid called lymph, vessels called lymphatic vessels that transport the lymph, a number of structures and organs containing lymphatic tissue, and red bone marrow, where stem cells develop into the various types of blood cells, including lymphocytes (Figure 16.1). The lymphatic system assists in circulating body fluids and helps defend the body against disease-causing agents. Most components of blood plasma filter through blood capillary walls to form interstitial fluid. After interstitial fluid passes into lymphatic vessels, it is referred to as **lymph** (LIMF=clear fluid). So, interstitial fluid and lymph are very similar; the major difference between the two is location. Interstitial fluid is found between cells, but lymph is located within lymphatic vessels and lymphatic tissue.

Lymphatic tissue is a specialized form of reticular connective tissue (see Table 3.4C on page 78) that contains large numbers of lymphocytes. Recall from Chapter 13 that lymphocytes are agranular white blood cells. Two types of lymphocytes participate in adaptive immune responses: B cells and T cells.

Functions of the Lymphatic System

The lymphatic system has three primary functions:

1. **Drains excess interstitial fluid.** Lymphatic vessels drain excess interstitial fluid from tissue spaces and return it to the blood.
2. **Transports dietary lipids.** Lymphatic vessels transport to the blood the lipids and lipid-soluble vitamins (A, D, E, and K) absorbed by the gastrointestinal tract.
3. **Carries out immune responses.** Lymphatic tissue initiates highly specific responses directed against particular microbes or abnormal cells. T and B cells, aided by macrophages, recognize foreign cells, microbes, toxins, and cancer cells and respond to them in two basic ways: (1) T cells destroy the intruders by causing them to rupture or by releasing cytotoxic (cell-killing) substances. (2) B cells differentiate into plasma cells that protect us against disease by producing antibodies, proteins that combine with and cause destruction of specific foreign substances.

Lymphatic Vessels and Lymph Circulation

Lymphatic vessels begin as **lymphatic capillaries**. These tiny vessels are closed at one end and located in the spaces between cells (Figure 16.2 on page 544). Just as blood capillaries converge to form venules and then veins, lymphatic capillaries unite to form larger **lymphatic vessels** (see Figure 16.1), which resemble veins in structure but have thinner walls and more valves. At intervals along the lymphatic vessels, lymph flows through **lymph nodes**, encapsulated masses of B cells and T cells. In the skin, lymphatic vessels lie in the subcutaneous tissue and generally follow the same route as veins; lymphatic vessels of the viscera generally follow arteries, forming plexuses (networks) around them. Tissues that lack lymphatic capillaries include avascular tissues (such as cartilage, the epidermis, and the cornea of the eye), the central nervous system, portions of the spleen, and red bone marrow.

Lymphatic Capillaries

Lymphatic capillaries are slightly larger in diameter than blood capillaries and have a unique structure that permits interstitial fluid to flow into them but not out. The ends of endothelial cells that make up the wall of a lymphatic capillary overlap (Figure 16.2b on page 544). When pressure is greater in the interstitial fluid than in lymph, the cells separate slightly, like the opening of a one-way swinging door, and interstitial fluid enters the lymphatic capillary. When pressure is greater inside the lymphatic capillary, the cells adhere more closely, and lymph cannot escape back into interstitial fluid. Attached to the lymphatic capillaries are *anchoring filaments*, which contain elastic fibers. They extend out from the lymphatic capillary, attaching lymphatic endothelial cells to surrounding tissues. When excess interstitial fluid accumulates and causes tissue swelling, the anchoring filaments are

Figure 16.1 **Components of the lymphatic system.**

The lymphatic system consists of lymph, lymphatic vessels, lymphatic tissues, and red bone marrow.

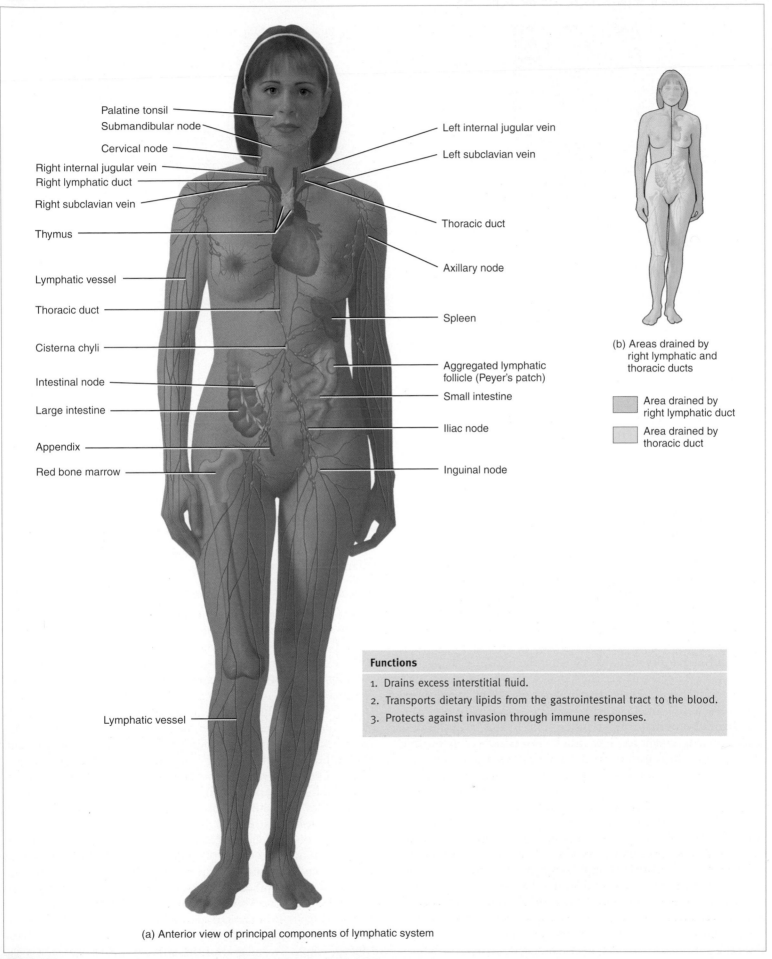

(a) Anterior view of principal components of lymphatic system

(b) Areas drained by right lymphatic and thoracic ducts

- Area drained by right lymphatic duct
- Area drained by thoracic duct

Functions

1. Drains excess interstitial fluid.
2. Transports dietary lipids from the gastrointestinal tract to the blood.
3. Protects against invasion through immune responses.

? **What tissue contains stem cells that develop into lymphocytes?**

Figure 16.2 Lymphatic capillaries.

Lymphatic capillaries are found throughout the body except in avascular tissues, the central nervous system, portions of the spleen, and red bone marrow.

(a) Relationship of lymphatic capillaries to tissue cells and blood capillaries

(b) Details of a lymphatic capillary

? Is lymph more similar to blood plasma or to interstitial fluid? Why?

pulled, making the openings between cells even larger so that more fluid can flow into the lymphatic capillary.

In the small intestine, specialized lymphatic capillaries called **lacteals** (LAK-tē-als; *lact-*=milky) carry dietary lipids into lymphatic vessels and ultimately into the blood. The presence of these lipids causes the lymph draining from the small intestine to appear creamy white; such lymph is referred to as **chyle** (KĪL=juice). Elsewhere, lymph is a clear, pale-yellow fluid.

Lymph Trunks and Ducts

Lymph passes from lymphatic capillaries into lymphatic vessels and then through lymph nodes. As lymphatic vessels exit lymph nodes in a particular region of the body, they unite to form **lymph trunks**. The principal trunks are the lumbar, intestinal, bronchomediastinal, subclavian, and jugular trunks (Figure 16.3). The **lumbar trunks** drain lymph from the free lower limbs, the wall and viscera of the pelvis, the kidneys, the adrenal glands, and the abdominal wall. The **intestinal trunk** drains lymph from the stomach, intestines, pancreas, spleen, and part of the liver. The **bronchomediastinal trunks** drain lymph from the thoracic wall, lung, and heart. The **subclavian trunks** drain the free upper limbs. The **jugular trunks** drain the head and neck.

Lymph passes from lymph trunks into two main channels, the thoracic duct and the right lymphatic duct, and then drains into venous blood. The **thoracic (left lymphatic) duct** is about 38–45 cm (15–18 in.) long and begins as a dilation called the **cisterna chyli** (sis-TER-na KĪ-lē; *cisterna*=cavity or reservoir) anterior to the second lumbar vertebra. The thoracic duct is the main duct for the return of lymph to blood. The cisterna chyli receives lymph from the right and left lumbar trunks and from the intestinal trunk. In the neck, the thoracic duct also receives lymph from the left jugular, left subclavian, and left bronchomediastinal trunks. Therefore, the thoracic duct receives lymph from the left side of the head, neck, and chest, the left free upper

limb, and the entire body inferior to the ribs (see Figure 16.1b). The thoracic duct in turn drains lymph into venous blood at the junction of the left internal jugular and left subclavian veins.

The **right lymphatic duct** (Figure 16.3) is about 1.2 cm (0.5 in.) long and receives lymph from the right jugular, right subclavian, and right bronchomediastinal trunks. Thus, the right lymphatic duct receives lymph from the upper-right side of the body (see Figure 16.1b). From the right lymphatic duct, lymph drains into venous blood at the junction of the right internal jugular and right subclavian veins.

Formation and Flow of Lymph

Most components of blood plasma filter freely through the capillary walls to form interstitial fluid. More fluid filters out of blood capillaries, however, than returns to them by reabsorption. The excess filtered fluid—about 3 liters per day—drains into lymphatic vessels and becomes lymph. Because most blood plasma proteins are too large to leave blood vessels, interstitial fluid contains only a small amount of protein. Proteins that do leave blood plasma cannot return to the blood directly by diffusion because the concentration gradient (high level of proteins inside blood capillaries, low level outside) opposes such movement. The proteins can, however, move readily through the more permeable lymphatic capillaries into lymph. Thus, an important function of lymphatic vessels is to return lost blood plasma proteins to the bloodstream.

Like veins, lymphatic vessels contain valves, which ensure the one-way movement of lymph. As noted previously, lymph drains into venous blood through the right lymphatic duct and the thoracic duct at the junction of the internal jugular and subclavian veins (Figure 16.3). Thus, the sequence of fluid flow is blood capillaries (blood) → interstitial spaces (interstitial fluid) → lymphatic capillaries (lymph) → lymphatic vessels (lymph) → lymphatic ducts (lymph) → junction of the internal jugular and subclavian

Figure 16.3 **Routes for drainage of lymph from lymph trunks into the thoracic and right lymphatic ducts.** The arrows indicate the direction of lymph flow.

All lymph returns to the bloodstream through the thoracic (left) lymphatic duct and right lymphatic duct.

(a) Overall anterior view

(b) Detailed anterior view

? **Which lymphatic vessels empty into the cisterna chyli, and which duct receives lymph from the cisterna chyli?**

Figure 16.4 Schematic diagram showing the relationship of the lymphatic system to the cardiovascular system. Arrows show direction of flow of lymph and blood.

 The sequence of fluid flow is blood capillaries (blood) → interstitial spaces (interstitial fluid) → lymphatic capillaries (lymph) → lymphatic vessels (lymph) → lymphatic ducts (lymph) → junction of internal jugular and subclavian veins (blood).

Does inhalation promote or hinder the flow of lymph?

veins (blood). Figure 16.4 illustrates this sequence, along with the relationship of the lymphatic and cardiovascular systems.

The same two "pumps" that aid the return of venous blood to the heart maintain the flow of lymph.

1. **Skeletal muscle pump.** The "milking action" of skeletal muscle contractions compresses lymphatic vessels (as well as veins) and forces lymph toward the junction of the internal jugular and subclavian veins.

2. **Respiratory pump.** Lymph flow is also maintained by pressure changes that occur during inhalation (breathing in). Lymph flows from the abdominal region, where the pressure is higher, toward the thoracic region, where it is lower. When the pressures reverse during exhalation (breathing out), the valves prevent backflow of lymph. In addition, when a lymphatic vessel distends, the smooth muscle in its wall contracts, which helps move lymph from one segment of the vessel to the next.

Lymphatic Organs and Tissues

Lymphatic organs and tissues, which are widely distributed throughout the body, are classified into two groups based on their functions. **Primary lymphatic organs** are the sites where stem cells divide and become *immunocompetent*, that is, capable of mounting an immune response. The primary lymphatic organs are the red bone marrow (in flat bones and the epiphyses of some long bones of adults) and the thymus. Pluripotent stem cells in red bone marrow give rise to mature, immunocompetent B cells and to pre-T cells (immature T cells), which migrate to and

become immunocompetent T cells in the thymus. The **secondary lymphatic organs** and **tissues** are the sites where most immune responses occur. They include lymph nodes, the spleen, and lymphatic nodules (follicles). The thymus, lymph nodes, and spleen are considered organs because each is surrounded by a connective tissue capsule; lymphatic nodules, in contrast, are not organs because they lack such a capsule.

Thymus

The **thymus** is a bilobed organ located in the mediastinum between the sternum and the aorta. It extends from the top of the sternum, sometimes even projecting into the inferior cervical region, to the level of the fourth costal cartilages, anterior to the top of the heart and its great vessels (Figure 16.5a). An enveloping layer of connective tissue holds the two lobes closely together, but a connective tissue **capsule** encloses each lobe separately. Extensions of the capsule, called **trabeculae** (tra-BEK-ū-lē=little beams), penetrate inward and divide each lobe into **lobules** (Figure 16.5b).

Each thymic lobule consists of a deeply staining outer cortex and a lighter-staining central medulla (Figure 16.5b). The **cortex** is composed of large numbers of T cells and scattered dendritic cells, epithelial cells, and macrophages. Immature T cells migrate from red bone marrow to the cortex of the thymus, where they proliferate and begin to mature. **Dendritic cells** (*dendr-*=a tree), so named because they have long, branched projections that resemble the dendrites of a neuron, assist the maturation process. As you will see shortly, dendritic cells in other parts of the body, such as lymph nodes, play another key role in immune responses. Each of the specialized **epithelial cells** in the cortex has several

Figure 16.5 **Thymus.** (See Tortora, *A Photographic Atlas of the Human Body*, 2e, Figure 7.2a.)

The bilobed thymus is largest at puberty and then atrophies with age.

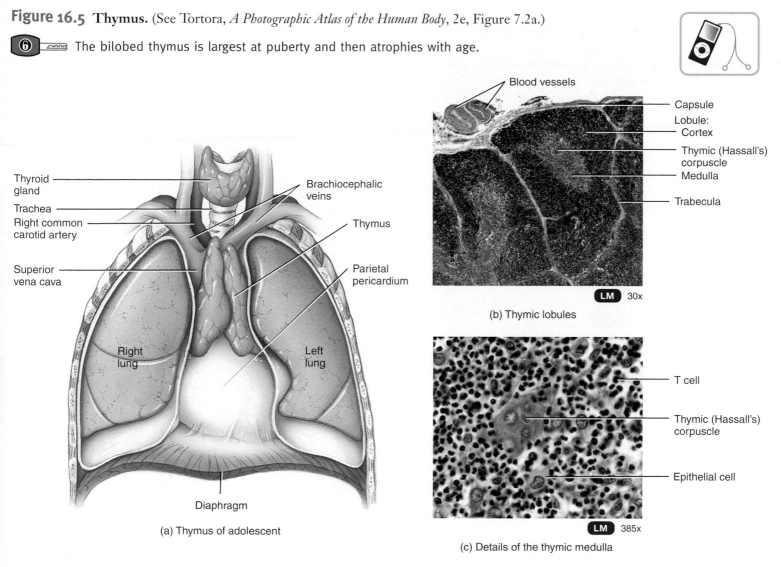

(a) Thymus of adolescent

(b) Thymic lobules

LM 30x

(c) Details of the thymic medulla

LM 385x

? **Which types of lymphocytes mature in the thymus?**

long processes that surround and serve as a framework for as many as 50 T cells. These epithelial cells help pre-T cells mature into T cells. Additionally, the epithelial cells produce thymic hormones that are thought to aid in the maturation of T cells. Only about 2% of developing T cells survive in the cortex. The remaining cells die via **apoptosis** (programmed cell death). Thymic macrophages help clear out the debris of dead and dying cells. The surviving T cells enter the medulla.

The **medulla** consists of widely scattered, more mature T cells, epithelial cells, dendritic cells, and macrophages (Figure 16.5c). Some of the epithelial cells become arranged into concentric layers of flat cells that degenerate and become filled with keratohyalin granules and keratin. These clusters are called **thymic (Hassall's) corpuscles**. Although their role is uncertain, they may serve as sites of T cell death in the medulla. T cells that leave the thymus via the blood are carried to lymph nodes, the spleen, and other lymphatic tissues where they colonize parts of these organs and tissues.

Because of its high content of lymphoid tissue and a rich blood supply, the thymus has a reddish appearance in a living body. With age, however, fatty infiltrations replace the lymphoid tissue and it takes on more of the yellowish color of the invading fat, giving the false impression of reduced size. In infants, the thymus has a mass of about 70 g (2.3 oz). It is after puberty that adipose and areolar connective tissue begin to replace the thymic tissue. By the time a person reaches maturity, the functional portion of the gland is reduced considerably, and in old age the functional portion may weigh only 3 g (0.1 oz). Before the thymus atrophies, it populates the secondary lymphatic organs and tissues with T cells. However, some T cells continue to proliferate in the thymus throughout an individual's lifetime.

Lymph Nodes

Located along lymphatic vessels are about 600 bean-shaped **lymph nodes**. They are scattered throughout the body, both superficially and deep, and usually occur in groups (see Figure 16.1). Large groups of lymph nodes are present near the mammary glands and in the axillae and groin. Later in the chapter, the principal groups of lymph nodes in various regions of the body will be presented in a series of exhibits, starting on page 552.

Lymph nodes are 1–25 mm (0.04–1 in.) long and, like the thymus, are covered by a **capsule** of dense connective tissue that extends into the node (Figure 16.6). The capsular extensions, called **trabeculae**, divide the node into compartments, give support, and provide a route for blood vessels into the interior of a node. Internal to the capsule is a supporting network of reticular fibers and fibroblasts. The capsule, trabeculae, reticular fibers, and fibroblasts constitute the *stroma*, or framework, of a lymph node.

The *parenchyma* of a lymph node is divided into a superficial cortex and a deep medulla. Within the **outer cortex** are egg-shaped aggregates of B cells called **lymphatic nodules** (follicles). A lymphatic nodule consisting chiefly of B cells is called a *primary lymphatic nodule.* Most lymphatic nodules in the outer cortex are *secondary lymphatic nodules* (Figure 16.6), which form in response to a challenge by an antigen and are sites of plasma cell and memory B cell formation. An **antigen** (meaning **anti**body-**gen**erator) is a substance that is recognized as foreign and provokes an immune response. After B cells in a primary lymphatic nodule recognize an antigen, the primary lymphatic nodule develops

into a secondary lymphatic nodule. The center of a secondary lymphatic nodule contains a region of light-staining cells called a *germinal center.* In the germinal center are B cells, follicular dendritic cells (a special type of dendritic cell), and macrophages. When follicular dendritic cells "present" an antigen, B cells proliferate and develop into antibody-producing plasma cells or memory B cells. Memory B cells persist after an immune

Figure 16.6 Structure of a lymph node. Arrows indicate the direction of lymph flow through a lymph node.

Lymph nodes are present throughout the body, usually clustered in groups.

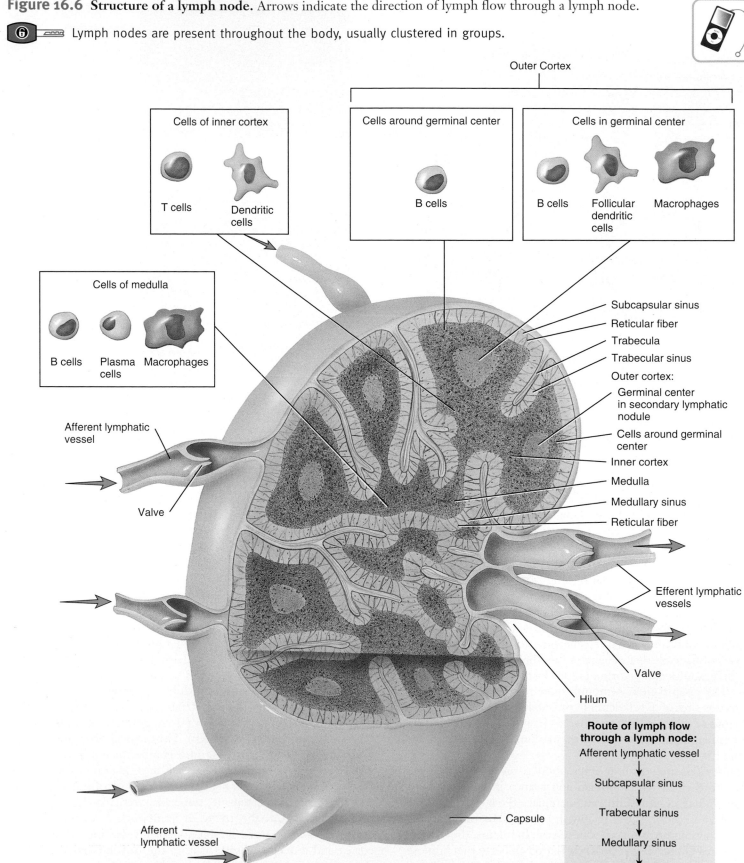

(a) Partially sectioned lymph node

response and remember having encountered a specific antigen. B cells that do not develop properly undergo apoptosis and are destroyed by macrophages. The region of a secondary lymphatic nodule surrounding the germinal center is composed of dense accumulations of B cells that have migrated away from their site of origin within the nodule.

The **inner cortex**, also called the *paracortex*, does not contain lymphatic nodules. It consists mainly of T cells and dendritic cells that enter a lymph node from other tissues. The dendritic cells present antigens to T cells, causing their proliferation. The newly formed T cells then migrate from the lymph node to areas of the body where there is antigenic activity.

The **medulla** of a lymph node contains B cells, antibody-producing plasma cells that have migrated out of the cortex into the medulla, and macrophages. The various cells are embedded in a network of reticular fibers and reticular cells.

As you have already learned, lymph flows through a lymph node in one direction only (Figure 16.6a). It enters through **afferent lymphatic vessels** (*afferent*=to carry toward), which penetrate the convex surface of the node at several points. The afferent vessels contain valves that open toward the center of the node so that the lymph is directed *inward*. Within the node,

lymph enters **sinuses**, a series of irregular channels that contain branching reticular fibers, lymphocytes, and macrophages. From the afferent lymphatic vessels, lymph flows into the **subcapsular sinus**, immediately beneath the capsule. From there the lymph flows through **trabecular sinuses**, which extend through the cortex parallel to the trabeculae, and into **medullary sinuses**, which extend through the medulla. The medullary sinuses drain into one or two **efferent lymphatic vessels** (*efferent*=to carry away), which are wider than afferent vessels and fewer in number. They contain valves that open away from the center of the lymph node to convey lymph, antibodies secreted by plasma cells, and activated T cells *out* of the lymph node. Efferent lymphatic vessels emerge from one side of the lymph node at a slight depression called a **hilum** (HI-lum). Blood vessels also enter and leave the node at the hilum.

Lymph nodes function as a type of filter. As lymph enters one end of a lymph node, foreign substances are trapped by the reticular fibers within the sinuses of the node. Then macrophages destroy some foreign substances by phagocytosis while lymphocytes destroy others by immune responses. The filtered lymph then leaves the other end of the lymph node.

Capsule
Subcapsular sinus

Outer cortex

Trabecular sinus

Germinal center in secondary lymphatic nodule

Trabecula

Inner cortex

Medullary sinus

Medulla

LM 40x

(b) Portion of a lymph node

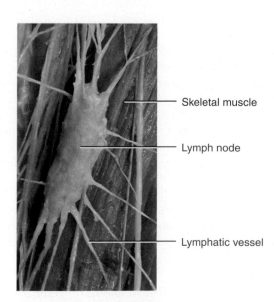

Macrophage

Lymphocyte

Medullary sinus

Reticular fiber

SEM 100x

(c) Details of a lymph node

Skeletal muscle

Lymph node

Lymphatic vessel

(d) Anterior view of an inguinal lymph node

? **What happens to foreign substances in the lymph when they enter a lymph node?**

• CLINICAL **CONNECTION** Metastasis Through Lymphatic Vessels

Metastasis (me-TAS-ta-sis; *meta-*=beyond; *stasi-*=to stand), the spread of a disease from one part of the body to another, can occur via lymphatic vessels. All malignant tumors eventually metastasize. Cancer cells may travel in the blood or lymph and establish new tumors where they lodge. When metastasis occurs via lymphatic vessels, secondary tumor sites can be predicted according to the direction of lymph flow from the primary tumor site. Cancerous lymph nodes feel enlarged, firm, nontender, and fixed to underlying structures. By contrast, most lymph nodes that are enlarged due to an infection are softer, tender, and movable. •

Cancerous lymph node

Cancerous lymph node in neck due to metastasis from a carrier in the pharynx (throat)

Spleen

The oval **spleen** is the largest single mass of lymphatic tissue in the body. It is a soft, encapsulated organ of variable size, but on average, it fits in a person's open hand and measures about 12 cm (5 in.) in length (Figure 16.7a). It is located in the left hypochondriac region between the stomach and diaphragm. The superior surface of the spleen is smooth and convex and conforms to the concave surface of the diaphragm. Neighboring organs make indentations in the visceral surface of the spleen—the *gastric impression* (stomach), the *renal impression* (left kidney), and the *colic impression* (left colic flexure of the large intestine). Like lymph nodes, the spleen has a hilum. Through it pass the large, tortuous splenic artery and splenic vein, along with efferent lymphatic vessels and sympathetic nerves that regulate the blood flow in the vessels.

A capsule of dense connective tissue surrounds the spleen and is covered in turn by a serous membrane, the visceral peritoneum. Trabeculae extend inward from the capsule. The capsule plus trabeculae, reticular fibers, and fibroblasts constitute the stroma of the spleen; the parenchyma of the spleen consists of two different kinds of tissue called white pulp and red pulp (Figure 16.7c, d). **White pulp** is lymphatic tissue, consisting mostly of lymphocytes and macrophages arranged around branches of the splenic artery called **central arteries**. The **red pulp** consists of blood-filled **venous sinuses** and cords of splenic tissue called **splenic (Billroth's) cords**. Splenic cords consist of red blood cells, macrophages, lymphocytes, plasma cells, and granulocytes. Veins are closely associated with the red pulp.

Blood flowing into the spleen through the splenic artery enters the central arteries of the white pulp. Within the white pulp, B cells and T cells carry out immune functions, similar to lymph nodes, while spleen macrophages destroy blood-borne pathogens by phagocytosis. Within the red pulp, the spleen performs three functions related to blood cells: (1) removal by macrophages of ruptured, worn out, or defective blood cells and platelets; (2) storage of platelets, up to one-third of the body's supply; and (3) production of blood cells (hemopoiesis) during fetal life.

Lymphatic Nodules

Lymphatic nodules are egg-shaped masses of lymphatic tissue that are not surrounded by a capsule. Because they are scattered throughout the lamina propria (connective tissue) of mucous membranes lining the gastrointestinal, urinary, reproductive tracts, and the respiratory airways, lymphatic nodules in these areas are also referred to as **mucosa-associated lymphatic tissue (MALT)**.

Although many lymphatic nodules are small and solitary, some occur in multiple large aggregations in specific parts of the body. Among these are the tonsils in the pharyngeal region and the aggregated lymphatic follicles (Peyer's patches) in the ileum of the small intestine. Aggregations of lymphatic nodules also occur in the appendix. Usually there are five **tonsils**, which form a ring at the junction of the oral cavity and oropharynx and at the junction of the nasal cavity and nasopharynx (see Figure 24.2b on page 755). The single **pharyngeal tonsil** (fa-RIN-jē-al) or **adenoid** is embedded in the posterior wall of the nasopharynx. This pyramid-shaped mass of lymphoid tissue is covered with a mucous membrane. The two **palatine tonsils** (PAL-a-tīn) lie in the lateral wall of the oropharynx in the **tonsilar fossa** just inferior to the soft palate; these are the tonsils commonly removed in a tonsillectomy. The almond-shaped palatine tonsils have numerous branched crypts that form a surface area of approximately 300 cm² (118 in.²). The paired **lingual tonsils** (LIN-gwal), located at the base of the tongue, may also require removal during a tonsillectomy. The tonsils are strategically positioned to participate in immune responses against inhaled or ingested foreign substances. Tonsils are masses of lymphoid tissue covered with mucosal epithelium. The epithelium forms narrow invaginations, called **crypts**, into the lymphoid tissue below. The crypts greatly increase the mucosal surface associated with the lymphoid tissue. In the crypts the mucosa becomes very thin, forming patches of **reticulated epithelium**. This specialized epithelium is well-designed for the transfer of antigen from the environment of the oral cavity and pharynx to the lymphoid cells of the tonsils.

◉ CHECKPOINT

1. How are interstitial fluid and lymph similar, and how do they differ?
2. How do lymphatic vessels differ in structure from veins?
3. Construct a diagram that shows the route of lymph circulation.
4. What is the role of the thymus in immunity?
5. What functions do lymph nodes serve?
6. Describe the locations and functions of the spleen and tonsils.

Figure 16.8 Principal lymph nodes of the head and neck.

🔓 ▭ The facial lymph nodes include the infraorbital, buccal, and mandibular lymph nodes.

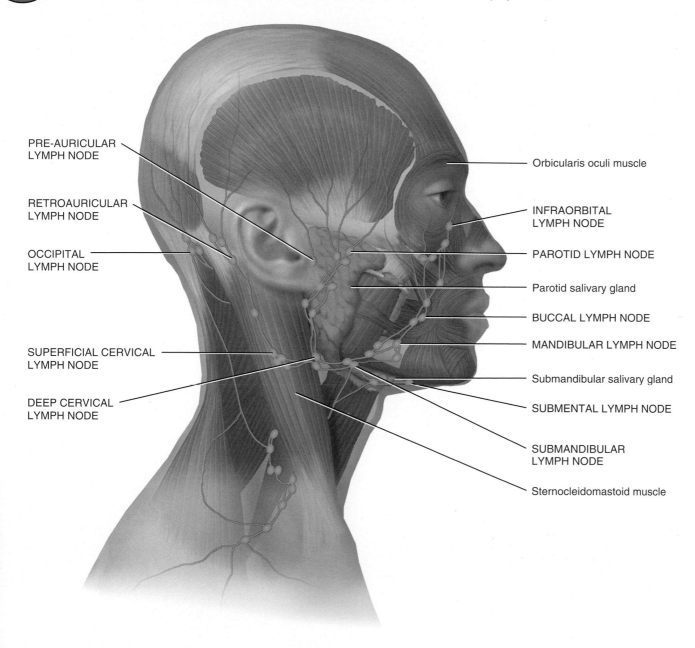

PRE-AURICULAR LYMPH NODE

RETROAURICULAR LYMPH NODE

OCCIPITAL LYMPH NODE

SUPERFICIAL CERVICAL LYMPH NODE

DEEP CERVICAL LYMPH NODE

Orbicularis oculi muscle

INFRAORBITAL LYMPH NODE

PAROTID LYMPH NODE

Parotid salivary gland

BUCCAL LYMPH NODE

MANDIBULAR LYMPH NODE

Submandibular salivary gland

SUBMENTAL LYMPH NODE

SUBMANDIBULAR LYMPH NODE

Sternocleidomastoid muscle

Lateral view of lymph nodes of the head and neck

❓ Which is the largest group of lymph nodes in the neck?

EXHIBIT 16.1   **553**

EXHIBIT 16.2   Principal Lymph Nodes of the Upper Limbs

LYMPH NODES	LOCATION	DRAINAGE
Supratrochlear nodes (soo-pra-TROK-lē-ar; *supra-*=above; *-trochlea*=pulley)	Superior to medial epicondyle of humerus.	Medial fingers, palm, and forearm.
Deltopectoral nodes (del′-tō-PEK-tō-ral; *-pectus*=breast)	Inferior to clavicle.	Lymphatic vessels on radial side of upper limb.
Axillary nodes (AK-sil-ār-ē; *axil-*=armpit)*	Most deep lymph nodes of the upper limbs are in the axilla and are called the axillary nodes. They are large.	
Lateral nodes	Medial and posterior aspects of axillary artery.	Most of entire upper limb.
Pectoral (anterior) nodes	Along inferior border of the pectoralis minor muscle.	Skin and muscles of anterior and lateral thoracic walls and central and lateral portions of mammary gland.
Subscapular (posterior) nodes	Along subscapular artery.	Skin and muscles of posterior part of neck and thoracic wall.
Central (intermediate) nodes	Base of axilla embedded in adipose tissue.	Lateral, pectoral (anterior), and subscapular (posterior) nodes.
Subclavicular (apical) nodes	Posterior and superior to pectoralis minor muscle.	Deltopectoral nodes.

*Because infection or malignancy of the upper limbs may cause tenderness and swelling in the axilla, the axillary nodes, especially the lateral group, are clinically important in that they filter lymph from much of the upper limb. More than 75% of the lymph drainage of the breast is to the pectoral (anterior) group of axillary lymph nodes. In breast cancer, it is possible that cancer cells leave the breast and lodge in the pectoral nodes. From here, metastasis may develop in other axillary nodes. Most of the remaining lymphatic drainage is to the sternal (parasternal) lymph nodes (see Figure 16.12 also).

Figure 16.9 Principal lymph nodes of the upper limbs. In (b) the direction of drainage is indicated by arrows.

Most of the lymph drainage of the breast is to the pectoral group of axillary lymph nodes.

DELTOPECTORAL LYMPH NODE

Deltoid muscle

LATERAL LYMPH NODE

Pectoralis major muscle

Cephalic vein

Basilic vein

SUPRATROCHLEAR LYMPH NODE

SUBCLAVICULAR LYMPH NODE

CENTRAL (INTERMEDIATE) LYMPH NODE

STERNAL LYMPH NODE

Pectoralis major muscle

LATERAL LYMPH NODE

SUBSCAPULAR LYMPH NODE

PECTORAL LYMPH NODE

Mammary gland

(a) Anterior view of lymph nodes of the upper limb

(b) Anterior view of mostly axillary lymph nodes

? Which lymph nodes drain most of the upper limb?

EXHIBIT 16.2   **555**

EXHIBIT 16.3   Principal Lymph Nodes of the Lower Limbs

LYMPH NODES	LOCATION	DRAINAGE
Popliteal nodes (pop-LIT-ē-al; *poples-*=ham of the knee)	In adipose tissue in popliteal fossa. (Not shown)	Knee and portions of leg and foot, especially heel.
Superficial inguinal nodes (ING-gwi-nal; *inguen-*=groin)	Parallel to saphenous vein.	Anterior and lateral abdominal wall to level of umbilicus, gluteal region, external genitals, perineal region, and entire superficial lymphatics of lower limb.
Deep inguinal nodes	Medial to femoral vein.	Deep lymphatics of lower limb, penis, and clitoris.

Figure 16.10 Principal lymph nodes of the lower limbs.

The inguinal lymph nodes drain the lymphatics of the lower limbs.

SUPERFICIAL INGUINAL LYMPH NODES

Inguinal ligament

Femoral vein

DEEP INGUINAL LYMPH NODE

Great saphenous vein

Great saphenous vein

(a) Anterior view of inguinal lymph nodes

Inguinal ligament

Femoral artery

Femoral vein

SUPERFICIAL INGUINAL LYMPH NODES

Accessory saphenous vein

LYMPHATIC VESSELS

Great saphenous vein

(b) Anterior view of inguinal lymph nodes and lymphatic vessels of the thigh

? **Which lymph nodes are parallel to the saphenous vein?**

EXHIBIT 16.3 **557**

EXHIBIT 16.4 Principal Lymph Nodes of the Abdomen and Pelvis

PARIETAL LYMPH NODES	LOCATION	DRAINAGE
Parietal nodes are located retroperitoneally (behind the parietal peritoneum) and in close association with larger blood vessels.		
External iliac nodes (IL-ē-ak=ilium)	Along external iliac vessels.	Deep lymphatics of abdominal wall inferior to umbilicus, adductor region of thigh, urinary bladder, prostate, ductus (vas) deferens, seminal vesicles, prostatic and membranous urethra, uterine (fallopian) tubes, uterus, and vagina.
Common iliac nodes	Along course of common iliac vessels.	Pelvic viscera.
Internal iliac nodes	Near internal iliac artery.	Pelvic viscera, perineum, gluteal region, and posterior surface of thigh.
Sacral nodes (SĀ-krum=holy bone)	In hollow of sacrum.	Rectum, prostate, and posterior pelvic wall.
Lumbar nodes (LUM-bar; *lumb-*=loin)	From aortic bifurcation to diaphragm; arranged around aorta and designated as *right lateral aortic nodes*, *left lateral aortic nodes*, *preaortic nodes* and *retroaortic nodes*.	Efferents from testes, ovaries, uterine (fallopian) tubes, uterus, kidneys, adrenal (suprarenal) glands, abdominal surface of diaphragm, and lateral abdominal wall.

Figure 16.11 Principal lymph nodes of the abdomen and pelvis.

The parietal lymph nodes are retroperitoneal and in close association with larger blood vessels.

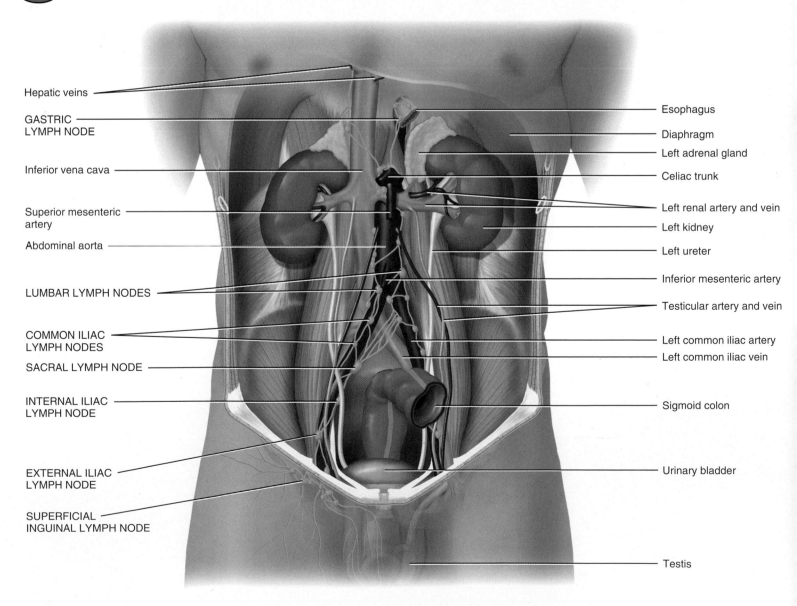

(a) Anterior view of abdominal and pelvic lymph nodes

PARIETAL LYMPH NODES	LOCATION	DRAINAGE
Visceral nodes are found in association with visceral arteries.		
Celiac nodes (SE-lē-ak; *koilia-*=belly)	Consist of three groups: gastric, hepatic, and pancreaticosplenic.	
Gastric nodes	Along lesser curvature of stomach.	Lesser curvature of stomach; inferior, anterior, and posterior aspects of stomach; esophagus.
Hepatic nodes	Along hepatic artery. (Not shown)	Stomach, duodenum, liver, gallbladder, and pancreas.
Pancreaticosplenic nodes (pan-krē-at′-i-kō-SPLEN-ik)	Along splenic artery. (Not shown)	Stomach, spleen, and pancreas.
Superior mesenteric nodes (MES-en-ter′-ik; *meso-*=middle; *-enteric*=intestines)	Consist of three groups: mesenteric, ileocolic, and transverse mesocolic.	
Mesenteric nodes	Along superior mesenteric artery.	Jejunum and all parts of ileum, except for terminal portion.
Ileocolic nodes (il′-ē-ō-KŌL-ik)	Along ileocolic artery.	Terminal portion of ileum, appendix, cecum, and ascending colon.
Transverse mesocolic nodes	Between layers of transverse mesocolon.	Descending iliac and sigmoid parts of colon.
Inferior mesenteric nodes	Near left colic, sigmoid, and superior rectal arteries.	Descending, iliac, and sigmoid parts of colon; superior part of rectum; superior anal canal.

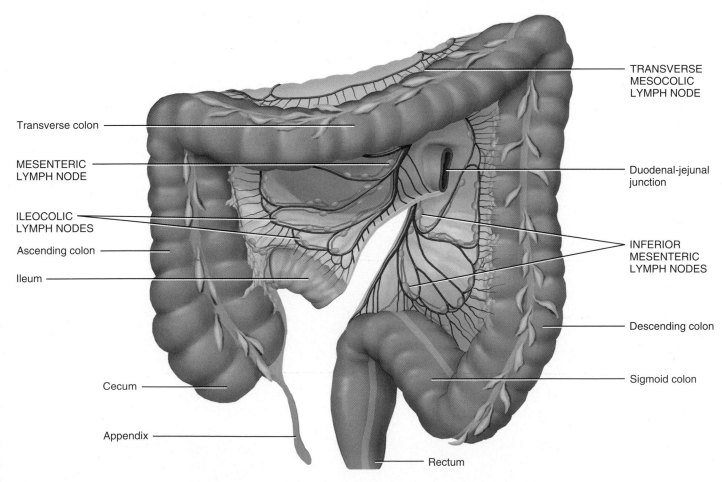

(b) Anterior view of superior and inferior mesenteric lymph nodes

? **What are the three groups of celiac lymph nodes?**

EXHIBIT 16.4 **559**

EXHIBIT 16.5   Principal Lymph Nodes of the Thorax

PARIETAL LYMPH NODES	LOCATION	DRAINAGE
Parietal nodes drain the wall of the thorax.		
Sternal (parasternal) nodes	Alongside internal thoracic artery.	Central and lateral parts of mammary gland, deeper structures of anterior abdominal wall superior to umbilicus, diaphragmatic surface of liver, and deeper parts of anterior portion of thoracic wall.
Intercostal nodes (in'-ter-KOS-tal; inter-=between; costa=rib)	Near heads of ribs at posterior parts of intercostal spaces.	Posterolateral aspect of thoracic wall.
Phrenic (diaphragmatic) nodes (FREN-ik=diaphragm)	On thoracic aspect of diaphragm and divisible into three sets called anterior phrenic, middle phrenic, and posterior phrenic.	
Anterior phrenic nodes	Posterior to base of xiphoid process.	Convex surface of liver, diaphragm, and anterior abdominal wall.
Middle phrenic nodes	Close to phrenic nerves where they pierce diaphragm.	Medial part of diaphragm and convex surface of liver.
Posterior phrenic nodes	Posterior surface of diaphragm near aorta.	Posterior part of diaphragm.
VISCERAL LYMPH NODES	**LOCATION**	**DRAINAGE**
Visceral nodes drain the viscera in the thorax.		
Anterior mediastinal nodes (mē'-dē-as-TĪ-nal; media-=middle; -stinum=partition)	Anterior part of superior mediastinum, anterior to arch of aorta.	Thymus and pericardium.
Posterior mediastinal nodes	Posterior to pericardium.	Esophagus, posterior aspect of the pericardium, diaphragm, and convex surface of liver.
Tracheobronchial nodes	Are divided into five groups: tracheal, superior and inferior tracheobronchial, bronchopulmonary, and pulmonary nodes.	
Tracheal nodes	Either side of trachea.	Trachea and upper esophagus.
Superior tracheobronchial nodes (trā'-kē-ō-BRONG-kē-al)	Between trachea and bronchi.	Trachea and bronchi.
Inferior tracheobronchial nodes	Between bronchi.	Trachea and bronchi.
Bronchopulmonary nodes (brong-kō-PUL-mō-nar-ē)	In hilus of each lung.	Lungs and bronchi.
Pulmonary nodes	Within lungs on larger bronchial tube branches.	Lungs and bronchi.

Figure 16.12 Principal lymph nodes of the thorax.

The parietal lymph nodes drain the thoracic wall, while the visceral lymph nodes drain the viscera of the thorax.

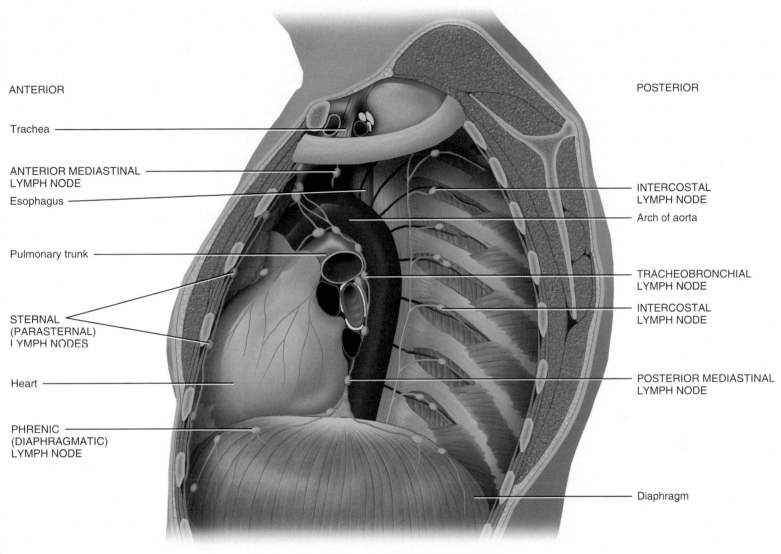

ANTERIOR

Trachea

ANTERIOR MEDIASTINAL
LYMPH NODE

Esophagus

Pulmonary trunk

STERNAL
(PARASTERNAL)
LYMPH NODES

Heart

PHRENIC
(DIAPHRAGMATIC)
LYMPH NODE

POSTERIOR

INTERCOSTAL
LYMPH NODE

Arch of aorta

TRACHEOBRONCHIAL
LYMPH NODE

INTERCOSTAL
LYMPH NODE

POSTERIOR MEDIASTINAL
LYMPH NODE

Diaphragm

View from left side showing thoracic lymph nodes

Which group of lymph nodes drains most of the diaphragm?

EXHIBIT 16.5 **561**

DEVELOPMENT OF LYMPHATIC TISSUES

◉ **OBJECTIVE**

• Describe the development of lymphatic tissues.

Lymphatic tissues begin to develop by the end of the fifth week of embryonic life. *Lymphatic vessels* develop from **lymph sacs** that arise from developing veins, which are derived from **mesoderm**.

The first lymph sacs to appear are the paired **jugular lymph sacs** at the junction of the internal jugular and subclavian veins (Figure 16.13). From the jugular lymph sacs, lymphatic capillary plexuses spread to the thorax, upper limbs, neck, and head. Some of the plexuses enlarge and form lymphatic vessels in their respective regions. Each jugular lymph sac retains at least one connection with its jugular vein, the left one developing into the superior portion of the thoracic duct (left lymphatic duct).

The next lymph sac to appear is the unpaired **retroperitoneal lymph sac** at the root of the mesentery of the intestine. It develops from the primitive vena cava and mesonephric (primitive kidney) veins. Capillary plexuses and lymphatic vessels spread from the retroperitoneal lymph sac to the abdominal

viscera and diaphragm. The sac establishes connections with the cisterna chyli but loses its connections with neighboring veins.

At about the time the retroperitoneal lymph sac is developing, another lymph sac, the **cisterna chyli**, develops inferior to the diaphragm on the posterior abdominal wall. It gives rise to the inferior portion of the *thoracic duct* and the *cisterna chyli* of the thoracic duct. Like the retroperitoneal lymph sac, the cisterna chyli also loses its connections with surrounding veins.

The last of the lymph sacs, the paired **posterior lymph sacs**, develop from the iliac veins. The posterior lymph sacs produce capillary plexuses and lymphatic vessels of the abdominal wall, pelvic region, and lower limbs. The posterior lymph sacs join the cisterna chyli and lose their connections with adjacent veins.

With the exception of the anterior part of the sac from which the cisterna chyli develops, all lymph sacs become invaded by **mesenchymal cells** and are converted into groups of *lymph nodes*.

The *spleen* develops from mesenchymal cells between layers of the dorsal mesentery of the stomach. The *thymus* arises as an outgrowth of the **third pharyngeal pouch** (see Figure 23.8 on page 747).

◉ **CHECKPOINT**

7. Name the four lymph sacs from which lymphatic vessels develop.

Figure 16.13 Development of lymphatic tissue.

The lymphatic system is derived from mesoderm.

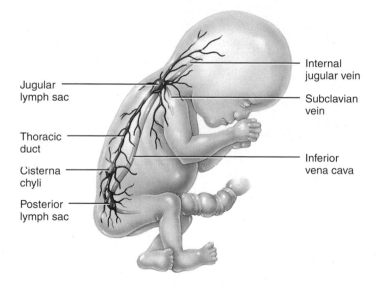

Jugular lymph sac

Thoracic duct

Cisterna chyli

Posterior lymph sac

Internal jugular vein

Subclavian vein

Inferior vena cava

❓ **When do lymphatic tissues begin to develop?**

AGING AND THE LYMPHATIC SYSTEM

◉ **OBJECTIVE**

• Describe the effects of aging on the lymphatic system and the immune response.

With advancing age, elderly individuals become more susceptible to all types of infections and malignancies. Their response to vaccines is decreased, and they tend to produce more autoantibodies (antibodies against their body's own molecules). In addition, the immune system exhibits lowered levels of function. For example, T cells become less responsive to antigens (foreign substances), and fewer T cells respond to infections. This may result from age-related atrophy of the thymus or decreased production of thymic hormones. Because the T cell population decreases with age, B cells are also less responsive. Consequently, antibody levels do not increase as rapidly in response to a challenge by an antigen, resulting in increased susceptibility to various infections. It is for this key reason that elderly individuals are encouraged to get influenza (flu) vaccinations each year.

◉ **CHECKPOINT**

8. What changes occur in the lymphatic system with advancing age?

• CLINICAL CONNECTION | AIDS: Acquired Immunodeficiency Syndrome

Acquired immunodeficiency syndrome (AIDS) is a condition in which a person experiences a telltale assortment of infections due to the progressive destruction of immune system cells by the **human immunodeficiency virus (HIV)**. AIDS represents the end stage of infection by HIV. A person who is infected with HIV may be symptom-free for many years, even while the virus is actively attacking the immune system. In the two decades after the first five cases were reported in 1981, 22 million

people died of AIDS. Worldwide, 35 to 40 million people are currently infected with HIV.

HIV Transmission

Because HIV is present in the blood and some body fluids, it is most effectively transmitted (spread from one person to another) by actions or practices that involve the exchange of blood or body fluids between people. HIV is transmitted in semen or vaginal fluid during unprotected (without a condom) anal, vaginal, or oral sex. HIV also is transmitted by direct blood-to-blood contact, such as occurs among intravenous drug users who share hypodermic needles or health-care professionals who may be accidentally stuck by HIV-contaminated hypodermic needles. In addition, HIV can be transmitted from an HIV-infected mother to her baby at birth or during breast-feeding.

The chance of transmitting or of being infected by HIV during vaginal or anal intercourse can be greatly reduced—although not entirely eliminated—by the use of latex condoms. Public health programs aimed at encouraging drug users not to share needles have proven effective at checking the increase in new HIV infections in this population. Also, giving certain drugs to pregnant HIV-infected women greatly reduces the risk of transmission of the virus to their babies.

HIV is a very fragile virus; it cannot survive for long outside the human body. The virus is not transmitted by insect bites. One cannot become infected by casual physical contact with an HIV-infected person, such as by hugging or sharing household items. The virus can be eliminated from personal care items and medical equipment by exposing them to heat (135°F for 10 minutes) or by cleaning them with common disinfectants such as hydrogen peroxide, rubbing alcohol, household bleach, or germicidal cleansers such as Betadine® or Hibiclens®. Standard dishwashing and clothes washing also kills HIV.

HIV: Structure and Infection

HIV consists of an inner core of ribonucleic acid (RNA) covered by a protein coat (capsid) surrounded by an outer layer, the envelope, composed of a lipid bilayer penetrated by glycoproteins (Figure A). HIV is classified as a *retrovirus* since its genetic information is carried in RNA instead of DNA. Glycoproteins assist HIV's binding to and entering host cells. Outside a living host cell, a virus is unable to replicate. However, when the virus infects and enters a host cell, its RNA uses the host cell's resources to make thousands of copies of the virus. New viruses eventually leave and then infect other cells.

HIV mainly damages T cells. Over 10 billion viral copies may be made each day. The viruses bud so rapidly from an infected cell's plasma membrane that the cell ruptures and dies. In most HIV-infected people, T cells are initially replaced as fast as they are destroyed. After several years, however, the body's ability to replace T cells is slowly exhausted, and the number of T cells in circulation gradually declines.

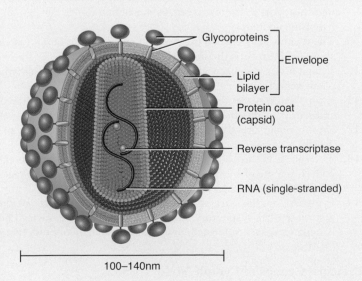

(A) Human immunodeficiency virus (HIV)

Glycoproteins
Envelope
Lipid bilayer
Protein coat (capsid)
Reverse transcriptase
RNA (single-stranded)

100–140nm

Signs, Symptoms, and Diagnosis of HIV Infection

Soon after being infected with HIV, most people experience a brief flu-like illness. Common signs and symptoms are fever, fatigue, rash, headache, joint pain, sore throat, and swollen lymph nodes. About 50% of infected people also experience night sweats. As early as three to four weeks after HIV infection, plasma cells begin secreting antibodies against HIV. These antibodies are detectable in blood plasma and form the basis for some of the screening tests for HIV. When people test "HIV-positive," it usually means they have antibodies to HIV antigens in their bloodstream.

Progression to AIDS

After a period of 2 to 10 years, the virus destroys enough helper T cells that most infected people begin to experience symptoms of immunodeficiency. HIV-infected people commonly have enlarged lymph nodes and experience persistent fatigue, involuntary weight loss, night sweats, skin rashes, diarrhea, and various lesions of the mouth and gums. In addition, the virus may begin to infect neurons in the brain, affecting the person's memory and producing visual disturbances.

As the immune system slowly collapses, an HIV-infected person becomes susceptible to a host of *opportunistic infections*. These are diseases caused by microorganisms that are normally held in check but now proliferate because of the defective immune system. AIDS is diagnosed when the helper T cell count drops below 200 cells per microliter (=cubic millimeter) of blood or when opportunistic infections arise, whichever occurs first. In time, opportunistic infections usually are the cause of death.

Treatment of HIV Infection

At present, infection with HIV cannot be cured. Vaccines designed to block new HIV infections and to reduce the viral load (the number of copies of HIV RNA in a microliter of blood plasma) in those who are already infected are in clinical trials. Meanwhile, two categories of drugs have proved successful in extending the life of many HIV-infected people:

1. **Reverse transcriptase inhibitors** interfere with the action of reverse transcriptase, the enzyme that the virus uses to convert its RNA into a DNA copy. Among the drugs in this category are zidovudine (ZDV, previously called AZT), didanosine (ddI), and stavudine (d4T). Trizivir®, approved in 2000 for treatment of HIV infection, combines three reverse transcriptase inhibitors in one pill.
2. **Protease inhibitors** interfere with the action of protease, a viral enzyme that cuts proteins into pieces to assemble the coat of newly produced HIV particles. Drugs in this category include nelfinavir, saquinavir, ritonavir, and indinavir.

In 1996, physicians treating HIV-infected patients widely adopted *highly active antiretroviral therapy (HAART)*—a combination of two differently acting reverse transcriptase inhibitors and one protease inhibitor. Most HIV-infected individuals receiving HAART experience a drastic reduction in viral load and an increase in the number of T cells in their blood. Not only does HAART delay the progression of HIV infection to AIDS, but many people with AIDS have seen the remission or disappearance of opportunistic infections and an apparent return to health. Unfortunately, HAART is very costly (exceeding $10,000 per year), the dosing schedule is grueling, and not all people can tolerate the toxic side effects of these drugs. Although HIV may virtually disappear from the blood with drug treatment (and thus a blood test may be "negative" for HIV), the virus typicallly still lurks in various lymphatic tissues. In such cases, the infected person can still transmit the virus to another person.

KEY MEDICAL TERMS ASSOCIATED WITH THE LYMPHATIC SYSTEM AND IMMUNITY

Adenitis (ad-e-NĪ-tis; *aden-*=gland; *-itis*=inflammation of) Enlarged, tender, and inflamed lymph nodes resulting from an infection.

Allograft (AL-ō-graft; *allo-*=other) A transplant between genetically distinct individuals of the same species. Skin transplants from other people and blood transfusions are allografts.

Autograft (AW-tō-graft; *auto-*=self) A transplant in which one's own tissue is grafted to another part of the body (such as skin grafts for burn treatment or plastic surgery).

Autoimmune disease (aw-tō-i-MŪN) A disease in which the immune system fails to recognize self-antigens and attacks the person's own cells. Examples are rheumatoid arthritis (RA), systemic lupus erythematosus (SLE), rheumatic fever, hemolytic and pernicious anemias, Addison's disease, Graves' disease, insulin-dependent diabetes mellitus, myasthenia gravis, multiple sclerosis (MS), and ulcerative colitis. Also called *autoimmunity*.

Chronic fatigue syndrome (CFS) A disorder, usually occurring in young adults and primarily in females, characterized by (1) extreme fatigue that impairs normal activities for at least 6 months and (2) the absence of other known diseases (cancer, infections, drug abuse, toxicity, or psychiatric disorders) that might produce similar symptoms.

Gamma globulin (GLOB-ū-lin) Suspension of immunoglobulins from blood consisting of antibodies that react with a specific pathogen. It is prepared by injecting the pathogen into animals, removing blood from the animals after antibodies have been produced, isolating the antibodies, and injecting them into a human to provide short-term immunity.

Graft Any tissue or organ used for transplantation or a transplant of such structures.

Hypersplenism (hī′-per-SPLĒN-izm; *hyper-*=over) Abnormal splenic activity due to splenic enlargement; associated with an increased rate of destruction of normal blood cells.

Lymphadenopathy (lim-fad′-e-NOP-a-thē; *lymph-*=clear fluid; *-pathy*=disease) Enlarged, sometimes tender lymph nodes as a response to infection; also called **swollen glands**.

Lymphedema (lim′-fe-DĒ-ma; *edema*=swelling) Accumulation of lymph in lymphatic vessels, causing painless swelling of a limb.

Severe combined immunodeficiency disease (SCID) A rare inherited disorder in which both B cells and T cells are missing or inactive. Scientists have now identified mutations in several genes that are responsible for some types of SCID. In some cases, an infusion of red bone marrow cells from a sibling having very similar MHC (HLA) antigens can provide normal stem cells that give rise to normal B and T cells.

Splenomegaly (splē′-nō-MEG-a-lē; *mega-*=large) Enlarged spleen.

Systemic lupus erythematosus (er-e′-thēm-a-TŌ-sus) [*SLE*, or *lupus* (*lupus*=wolf)] An autoimmune, noncontagious, inflammatory disease of connective tissue, occurring mostly in young women. In SLE, damage to blood vessel walls results in the release of chemicals that mediate inflammation. Symptoms of SLE include joint pain, slight fever, fatigue, oral ulcers, weight loss, enlarged lymph nodes and spleen, photosensitivity, rapid loss of large amounts of scalp hair, and sometimes an eruption across the bridge of the nose and cheeks called a "butterfly rash."

Tonsillectomy (ton′-si-LEK-tō-mē; *-ectomy*=excision) Removal of a tonsil.

Transplantation (tranz-plan-TĀ-shun; *trans-*=across; *-planto*=to plant) The transfer of living cells, tissues, or organs from a donor to a recipient or from one part of the body to another part of the same body in order to restore a lost function. The success of a transplant depends on *major histocompatibility (MHC) antigens* on the surfaces of white blood cells and other body cells (except for red blood cells). MHC antigens are unique for each person; the better the match between donor and recipient MHC antigens, the greater the probability that graft rejection will be avoided.

Xenograft (ZEN-ō-graft; *xeno-*=strange or foreign) A transplant between animals of different species. Xenografts from porcine (pig) or bovine (cow) tissue may be used in a human as a physiological dressing for severe burns. Other xenografts include pig heart valves and baboon hearts.

STUDY OUTLINE

Introduction (p. 541)

1. The ability to ward off disease is called immunity or resistance. Lack of resistance is called susceptibility.
2. Innate immunity refers to defenses that are present at birth; they are always present and provide immediate or general protection against a wide variety of pathogens. Adaptive immunity refers to defenses that respond to a particular invader; it involves activation of specific lymphocytes that can combat a specific invader.

Lymphatic System Structure and Functions (p. 542)

1. The lymphatic system carries out immune responses and consists of lymph, lymphatic vessels, and structures and organs that contain lymphatic tissue (specialized reticular tissue containing many lymphocytes).
2. The lymphatic system drains interstitial fluid, transports dietary lipids, and protects against pathogens through immune responses.
3. Lymphatic vessels begin as lymph capillaries with one closed end in tissue spaces between cells.
4. Interstitial fluid drains into lymphatic capillaries, thus forming lymph.
5. Lymph capillaries merge to form larger vessels, called lymphatic vessels, which convey lymph into and out of structures called lymph nodes.
6. The route of lymph flow is from lymph capillaries to lymphatic vessels to lymph trunks to the thoracic duct or right lymphatic duct to the subclavian veins.

7. Lymph flows as a result of skeletal muscle contractions and respiratory movements. It is also aided by valves in lymphatic vessels.
8. The primary lymphatic organs are red bone marrow and the thymus. Secondary lymphatic organs are lymph nodes, spleen, and lymphatic nodules.
9. The thymus lies between the sternum and the large blood vessels above the heart. It is the site of T cell maturation.
10. Lymph nodes are encapsulated, oval structures located along lymphatic vessels.
11. Lymph enters nodes through afferent lymphatic vessels, is filtered, and exits through efferent lymphatic vessels.
12. Lymph nodes are the site of proliferation of plasma cells and T cells.
13. The spleen is the largest single mass of lymphatic tissue in the body. It is a site of B cell proliferation into plasma cells and phagocytosis of bacteria and worn-out red blood cells.
14. Lymphatic nodules are scattered throughout the mucosa of the gastrointestinal, respiratory, urinary, and reproductive tracts. This lymphatic tissue is termed mucosa-associated lymphoid tissue (MALT).

Principal Groups of Lymph Nodes (p. 552)

1. Lymph nodes are scattered throughout the body in superficial and deep groups.
2. The principal groups of lymph nodes are found in the head and neck, upper limbs, lower limbs, abdomen and pelvis, and thorax.

Development of Lymphatic Tissues (p. 562)

1. Lymphatic vessels develop from lymph sacs, which arise from developing veins. Thus, they are derived from mesoderm.
2. Lymph nodes develop from lymph sacs that become invaded by mesenchymal cells.

Aging and the Lymphatic System (p. 562)

1. With advancing age, individuals become more susceptible to infections and malignancies, respond less well to vaccines, and produce more autoantibodies.
2. Immune responses also diminish with age.

SELF-QUIZ QUESTIONS

Choose the one best answer to the following questions.

1. Put the following items in the correct order for the pathway of lymph from the stomach to the blood: (1) thoracic duct, (2) lymphatic vessels and lymph nodes, (3) lymph capillaries, (4) interstitial fluid, (5) junction of left internal jugular and left subclavian veins.
   - **a.** 2,3,4,5,1
   - **b.** 4,3,2,1,5
   - **c.** 1,5,3,2,4
   - **d.** 4,2,1,5,3
   - **e.** 3,2,4,1,5

2. This secondary lymphatic nodule is strategically positioned in the nasopharynx so it can participate in immune responses against inhaled foreign substances.
   - **a.** palatine tonsils
   - **b.** MALT
   - **c.** pharyngeal tonsil
   - **d.** adenoid
   - **e.** Two of the above

3. Which of the following is an important function of the lymphatic system?
   - **a.** controlling body temperature by evaporation of sweat
   - **b.** manufacturing all white blood cells
   - **c.** transporting fluids out to, and back from, the body tissues
   - **d.** returning fluid and proteins to the cardiovascular system
   - **e.** All of the above are correct.

4. The spleen:
   - **a.** serves as storage site for blood platelets.
   - **b.** is where phagocytosis of aged erythrocytes takes place.
   - **c.** is where phagocytosis of aged or damaged platelets occurs.
   - **d.** is a site of blood formation in the fetus.
   - **e.** All of the above are correct.

5. Plasma cells are a differentiated form of:
   - **a.** T lymphocyte.
   - **b.** B lymphocyte.
   - **c.** natural killer cell.
   - **d.** dendritic cell.
   - **e.** endothelial cell.

6. Which of the following statements about HIV transmission is true?
   - **a.** Standard dishwashing and clothes washing kills HIV.
   - **b.** The transmission of HIV requires the transfer of, or direct contact with, infected body fluids.
   - **c.** Organ transplants and artificial insemination are routes for HIV transmission.
   - **d.** HIV may be present in semen, blood, and breast milk.
   - **e.** All of the above are correct.

7. The thoracic duct drains lymph into the venous blood at the junction of the:
   - **a.** left brachiocephalic vein and the superior vena cava.
   - **b.** left internal jugular vein and the left subclavian vein.
   - **c.** right internal jugular vein and the right subclavian vein.
   - **d.** left internal and external jugular veins.
   - **e.** superior vena cava and right atrium.

8. Which of the following statements is true about the flow of lymph? (1) It is aided by a pressure difference—toward the thoracic region. (2) It is maintained primarily by the milking action of muscles. (3) It is possible because of valves in lymphatic vessels.
   - **a.** 1 only
   - **b.** 2 only
   - **c.** 3 only
   - **d.** Both 2 and 3 are correct.
   - **e.** 1, 2, and 3 are all correct.

Complete the following.

9. Specialized lymphatic capillaries in the small intestine are called _____.

10. Lymphatic vessels are similar in structure to _____ but have thinner walls and more valves.

11. The primary lymphatic organs, so called because they produce the lymphocytes for the immune system, are the _____ and the _____.

12. The tonsils include the _____ tonsil in the posterior wall of the nasopharynx; the _____ in the posterior region of the oral cavity; and the _____ tonsils at the base of the tongue.

13. Disease-producing microbes are collectively known as _____.

14. Lymphatic nodules in lamina propria of mucous membranes are collectively referred to as _____.

15. In the treatment of HIV infection, reverse transcriptase inhibitors have proven effective because they interfere with the action of an enzyme that the virus uses to convert its _____ into _____.

Are the following statements true or false? If the statement is False, *find and fix the mistake to make the statement* True.

16. B cells are located in the medulla of a lymph node.

17. Lymph is conveyed into a node at several points by afferent lymphatic vessels.

18. The main difference between lymph and interstitial fluid is location.

19. There are no lymphatic capillaries in cartilage.

20. Because the T cell population increases with age, elderly persons should not be encouraged to get flu (influenza) shots every year.

Matching

21. Match the following lymph nodes with their descriptions:
    - _____ **(a)** facial nodes
    - _____ **(b)** axillary nodes
    - _____ **(c)** popliteal nodes
    - _____ **(d)** common iliac nodes
    - _____ **(e)** celiac nodes
    - _____ **(f)** phrenic nodes
    - _____ **(g)** sacral nodes
    - _____ **(h)** deep cervical nodes

    - **(1)** drain the knee, leg, and heel
    - **(2)** three groups of nodes that drain the stomach, liver, pancreas, and spleen
    - **(3)** largest group of nodes in the neck
    - **(4)** three groups of nodes that drain the mucous membranes of the cheeks, eyelids, and nose
    - **(5)** anterior to sacrum; drain the rectum, prostate gland, and pelvic wall
    - **(6)** three groups of nodes that drain parts of the liver, diaphragm, and abdominal wall
    - **(7)** deep nodes; drain the upper limbs, skin and muscles of the thorax, and part of the neck
    - **(8)** arranged along blood vessels of the same name; drain the pelvic viscera

CRITICAL THINKING QUESTIONS

1. Jamal got a splinter in his right heel while he was playing beach volleyball. He neglected to clean it properly and it became infected. The first aid station warned him to have the wound taken care of before the infection spreads to his blood. How can an infection travel from his foot to his cardiovascular system?

2. Four-year old Kelsey had a history of repeated throat infections, averaging 10 per year. She breathed loudly through her mouth and even snored. Following the pediatrician's recommendation, she had an operation to remove the troublesome organs. When she returned to preschool, she told the other kids about all the ice cream she got to eat following her "tonsil-X-me." Where are the tonsils located, and what is their function?

3. After several years of struggling with a chronic autoimmune disease, Kelly's doctor recommended a splenectomy. Since then, every time she goes to the dentist, she is advised to take antibiotics for a period of time prior to her appointment. Why would the doctor

have recommended removal of such an important organ, and why would the dentist recommend the antibiotics?

4. Nan was ordering her dinner at an Italian restaurant. She informed the waiter that she was "highly allergic to eggplant" and that it was extremely important that her meal be eggplant-free. Dinner arrived, and shortly after she started eating her grilled vegetables, she started having trouble breathing and talking. There were no visible pieces of eggplant on her plate, so she announced to her dinner companion that the vegetables had been cooked on the same grill as the eggplant for other diners. How would her difficulty breathing lead her to this conclusion? How does being allergic to eggplant relate to difficulty breathing? If the difficulty breathing continues to worsen, what treatment might be required? Explain why such treatment would help.

5. An infection with a tropical parasite may block a lymphatic vessel. What would be the effect of blockage of the left subclavian trunk?

? ANSWERS TO FIGURE QUESTIONS

16.1 Red bone marrow contains stem cells that develop into lymphocytes.

16.2 Lymph is more similar to interstitial fluid than to blood plasma because the protein content of lymph is low.

16.3 The left and right lumbar trunks and the intestinal trunk empty into the cisterna chyli, which then drains into the thoracic duct.

16.4 Inhalation promotes the movement of lymph from abdominal lymphatic vessels toward the thoracic region.

16.5 T cells mature in the thymus.

16.6 When foreign substances enter a lymph node, they may be phagocytized by macrophages or attacked by lymphocytes that mount immune responses.

16.7 After birth, white pulp of the spleen functions in immunity, and red pulp of the spleen removes worn-out blood cells and stores platelets.

16.8 The deep cervical nodes are the largest group of lymph nodes in the neck.

16.9 The lateral lymph nodes drain most of the upper limb.

16.10 The superficial inguinal lymph nodes are parallel to the saphenous vein.

16.11 The three groups of celiac lymph nodes are the gastric, hepatic, and pancreaticosplenic nodes.

16.12 The phrenic lymph nodes drain most of the diaphragm.

16.13 Lymphatic tissues begin to develop by the end of the fifth week of gestation.

17 | NERVOUS TISSUE

INTRODUCTION As it did throughout the 1980s and 1990s, the computer continues to revolutionize our world today. Those of you old enough to remember have witnessed tremendous changes in computing power during the last 25 years. The first desktop computers operated with a total RAM of 16 KB. Today it is not uncommon to have a desktop or even a notebook with 1 gig of RAM; some quick math will tell you that RAM capacity has increased *one million times* over the last 25 years. In fact, the advances in computing technology seem to change so quickly that the machine you buy today is outdated tomorrow. In the mid-1970s a computer game called "Pong" was the zenith of computer game technology. Many people spent countless hours mesmerized in front of their computers, moving a small vertical line up and down one side of the screen in an attempt to hit a moving white dot to an opponent. Compare that with today's computer game enthusiast who can interact with and manipulate fantastic three-dimensional visual landscapes. A combination of factors makes these phenomenal advances in computing power possible. One of paramount significance is ever-increasing sophistication in the microcircuitry, which creates increased speed, greater processing power, better integration, and a large number of programming pathways.

However, even the greatest supercomputer pales in comparison with the sophisticated circuitry and functionality of the human nervous system, which allows humans to create things, like the amazing machines we call computers. In this chapter we will introduce the basic organization of this human computer and study the fundamental components that function as its wires and circuitry.

Because the nervous system is quite complex, we will consider different aspects of its structure and function in several related chapters. This chapter focuses on the organization of the nervous system and the properties of the cells that make up nervous tissue—neurons (nerve cells) and neuroglia (cells that support the activities of neurons). In chapters that follow, we will examine the structure and functions of the spinal cord and spinal nerves (Chapter 18), and of the brain and cranial nerves (Chapter 19). Then we will discuss the autonomic nervous system, the part of the nervous system that operates without voluntary control (Chapter 20). Next, we examine the somatic senses—touch, pressure, warmth, cold, pain, and others—and the sensory and motor pathways to understand how nerve impulses pass into the spinal cord and brain or from the spinal cord and brain to muscles and glands (Chapter 21). Our exploration of this complex yet fascinating system concludes with a discussion of the special senses: smell, taste, vision, hearing, and equilibrium (Chapter 22).

The branch of medical science that deals with the normal functioning and disorders of the nervous system is **neurology** (noo-ROL-ō-jē; *neuro-*=nerve or nervous system; *-logy*=study of). A **neurologist** is a physician who specializes in the diagnosis and treatment of disorders of the nervous system. •

OVERVIEW OF THE NERVOUS SYSTEM

◉ OBJECTIVES

- List the structures and basic functions of the nervous system.
- Describe the organization of the nervous system.

Structures of the Nervous System

With a mass of only 2 kg (4.5 lb), about 3% of total body weight, the **nervous system** is one of the smallest and yet the most complex of the 11 body systems. The nervous system is a highly organized network of billions of neurons and even more neuroglia, the two cell types in nervous tissue. The structures that make up the nervous system include the brain, cranial nerves and their branches, the spinal cord, spinal nerves and their branches, ganglia, enteric plexuses, and sensory receptors (Figure 17.1). These will be discussed shortly.

Organization of the Nervous System

Although we have only one nervous system, it is convenient to organize it into various components for ease of study. Broadly speaking, the nervous system can be organized both anatomically and functionally.

Anatomical Organization

The nervous system differs from other body systems because many of its cells are extremely long. Thus it is difficult to divide the nervous system into discrete organs with their own unique cell populations like we do for the other body systems. Anatomically the nervous system consists of two intimately interconnected divisions: (1) the central nervous system and (2) the peripheral nervous system (Figure 17.1).

CENTRAL NERVOUS SYSTEM   The **central nervous system (CNS)** is composed of the brain and spinal cord. The **brain** is enclosed and protected by the skull in the cranial cavity and contains about 100 billion (10^{11}) neurons. The **spinal cord** is enclosed and protected by the bones of the vertebral column in the vertebral canal and contains about 100 million neurons. The brain and spinal cord are continuous with one another through the foramen magnum of the occipital bone. The CNS processes many different kinds of incoming sensory information. It is also the source of thoughts, emotions, and memories. Most nerve impulses that stimulate muscles to contract and glands to secrete originate in the CNS.

PERIPHERAL NERVOUS SYSTEM   The **peripheral** (pe-RIF-e-ral) **nervous system (PNS)** is composed of all the nervous structures outside the CNS, such as cranial nerves and their branches, spinal nerves and their branches, ganglia, enteric plexuses, and sensory receptors. These structures link all parts of the body to the CNS. Twelve pairs (right and left) of **cranial nerves**, numbered I through XII, emerge from the base of the brain. A **nerve** is a bundle of hundreds to thousands of axons (nerve fibers) plus associated connective tissue and blood vessels that lies outside the brain and spinal cord. Each nerve follows a defined path and serves a specific region of the body. For example, cranial nerve I carries signals for the sense of smell from the nose to the brain. Thirty-one pairs of **spinal nerves** emerge from the spinal cord, each serving a specific region on the right or left side of the body. **Ganglia** (GANG-lē-a=swelling or knot; singular is *ganglion*) are small masses of nervous tissue, consisting primarily of neuron cell bodies, that are located outside the brain and spinal cord. Ganglia are closely associated with cranial and spinal nerves. In the walls of organs of the gastrointestinal tract, extensive networks of neurons, called **enteric plexuses**, help regulate the digestive system. **Sensory receptors** are structures that monitor changes in the internal and external environment, such as sensory receptors in the skin that detect touch sensations.

Functional Organization

The nervous system carries out a complex array of tasks. It allows us to sense various smells, produce speech, and remember past events; it also provides signals that control body movements, and regulates the operation of internal organs. These diverse activities can be grouped into three basic functions: sensory (input), integrative (control), and motor (output).

Figure 17.1 Components of the nervous system and anatomical organization of the nervous system.

The major structures of the nervous system are the brain and cranial nerves, spinal cord, and spinal nerves, ganglia, enteric plexuses, and sensory receptors.

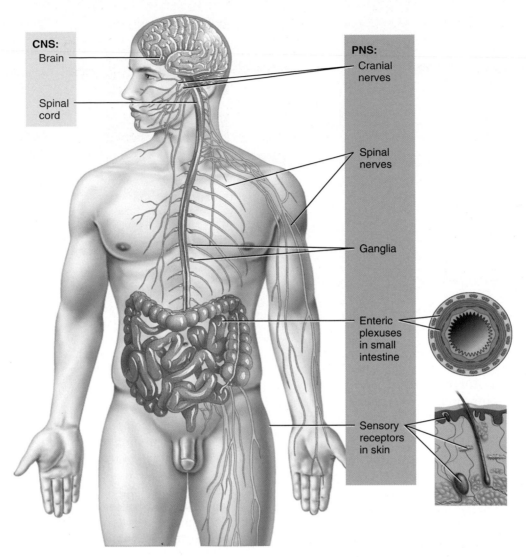

CNS:
Brain

Spinal cord

PNS:
Cranial nerves

Spinal nerves

Ganglia

Enteric plexuses in small intestine

Sensory receptors in skin

? How is the nervous system organized anatomically?

- **Sensory function.** Sensory receptors *detect* internal stimuli, such as an increase in blood pressure, and external stimuli, such as a raindrop landing on your arm. Neurons called **sensory** or **afferent neurons** (AF-er-ent NOO-ronz; *af-*=toward; *-ferrent*=carried) carry this sensory information into the brain and spinal cord through cranial and spinal nerves.

- **Integrative function.** The nervous system *integrates* (processes) sensory information by analyzing and storing some of it and by making decisions for appropriate responses. An important integrative function is **perception**, the conscious awareness of sensory stimuli. Perception occurs in the brain. Many of the neurons that participate in integration are **interneurons** (neurons that interconnect with other neurons), with axons that extend for only a short distance and contact nearby neurons in the brain or spinal cord. The vast majority of neurons in the body are interneurons.

- **Motor function.** Once sensory information is integrated, the nervous system may elicit an appropriate motor response, such as muscular contraction or glandular secretion. The neurons

that serve this function are called **motor** or **efferent neurons** (EF-er-ent; *ef-*=away from). Motor neurons carry information from the brain toward the spinal cord or out of the brain and spinal cord to **effectors** (muscles and glands) through cranial and spinal nerves. Stimulation of the effectors by motor neurons causes muscles to contract and glands to secrete.

Let's now take a closer look at the sensory and motor functions of the nervous system. In order to discuss the sensory and motor functions, we will further divide the peripheral nervous system into three components: (1) somatic nervous system, (2) autonomic nervous system, and (3) enteric nervous system. (Figure 17.2).

SOMATIC NERVOUS SYSTEM The **somatic** (*somat-*=body) **nervous system (SNS)** of the PNS consists of sensory neurons, called *somatic sensory neurons*, that convey information from sensory receptors in the skin, skeletal muscles, joints, and for the special senses (vision, hearing, taste, and smell) to the CNS. These *somatic sensory pathways* are involved in the *input* of information to the CNS for integration (processing). The SNS also consists of

Figure 17.2 Functional organization of the nervous system.

Sensory pathways are involved in the input of information to the CNS; motor pathways are involved in the output of information from the CNS.

? Which motor component of the peripheral nervous system is voluntary?

motor neurons, called *somatic motor neurons*, that convey information from the CNS to *skeletal muscles only*. These *somatic motor pathways* are involved in the *output* of information from the CNS that results in a muscular contraction. Because these motor responses can be consciously controlled, the actions of these parts of the SNS are *voluntary*.

AUTONOMIC NERVOUS SYSTEM The **autonomic** (*auto-*=self; *-nomic*=law) **nervous system (ANS)** of the PNS also has sensory and motor components. Sensory neurons, called *autonomic (visceral) sensory neurons*, convey information from autonomic sensory receptors, located primarily in the visceral organs (smooth muscle organs in the thorax, abdomen, and pelvis) to the CNS. *Autonomic motor neurons* convey information from the CNS to *smooth muscle*, *cardiac muscle*, and *glands* and cause the muscles to contract and the glands to secrete. Because its motor responses are not normally under conscious control, the action of the ANS is involuntary. The motor part of the ANS consists of two branches, the **sympathetic division** and the **parasympathetic division**. With a few exceptions, effectors receive nerves from both divisions, and usually the two divisions have opposing actions. For example, sympathetic neurons increase heart rate, and parasympathetic neurons slow it down. In general, the sympathetic division helps support exercise or emergency actions, so-called "fight-or-flight" responses, and the parasympathetic division takes care of "rest-and-digest" activities. Also, the sympathetic division is the major regulator of smooth muscle of the cardiovascular system and the parasympathetic division is the major regulator of the smooth muscle of the digestive and respiratory systems.

ENTERIC NERVOUS SYSTEM The **enteric** (*enteron*=intestines) **nervous system (ENS)** of the PNS is called the "brain of the gut" and consists of over 100 million neurons that occur throughout most of the length of the gastrointestinal (GI) tract. The ENS also has both sensory and motor components and can operate independently from the CNS. *Sensory neurons* of the ENS monitor chemical changes within the GI tract as well as the stretching of its walls. Motor neurons of the ENS govern contraction of GI tract smooth muscle to propel food through the GI tract and

secretions of the GI tract organs such as acid from the stomach, and endocrine cells, which secrete hormones. The ENS, like the ANS, is involuntary.

◉ CHECKPOINT

1. What are the components of the CNS and PNS?
2. What types of problems would result from damage to each of the following: sensory neurons, interneurons, and motor neurons?
3. Describe the components and functions of the SNS, ANS, and ENS, and indicate which of these subdivisions of the PNS control voluntary actions and which regulate involuntary actions.

HISTOLOGY OF NERVOUS TISSUE

◉ OBJECTIVES

• Compare the histological characteristics and functions of neurons and neuroglia.
• Distinguish between gray matter and white matter.

Nervous tissue is a vascular tissue comprised of two types of cells—*neurons* and *neuroglia*. These cells combine in a variety of ways in different regions of the nervous system. In addition to forming the complex processing networks within the brain and spinal cord, neurons comprise the organic circuitry that connects all regions of the body to this central processing unit. As highly specialized cells capable of reaching great lengths and making extremely intricate connections with other cells, neurons provide most of the unique functions of the nervous system, such as sensing, thinking, remembering, controlling muscle activity, and regulating glandular secretions. As a result of their specialization, they have lost the ability to undergo mitotic divisions. Neuroglia are smaller cells but they greatly outnumber neurons (according to some estimates by as much as 25 times). Neuroglia support, nourish, and protect neurons, and maintain the interstitial fluid that bathes them. Unlike neurons, neuroglia continue to divide throughout an individual's lifetime. Both neurons and neuroglia differ structurally depending on whether they are located in the central nervous system or the peripheral nervous system. These differences in structure correlate with the differences in function of the central nervous system and the peripheral nervous system.

Neurons

Neurons or **nerve cells** possess **electrical excitability**, the ability to respond to a stimulus and convert it into a nerve impulse (a *stimulus* is any change in the environment that is strong enough to initiate a nerve impulse). Very simply, a **nerve impulse (action potential)** is an electrical signal that propagates (travels) along the surface of the membrane of a neuron. It begins and travels due to the movement of ions (such as sodium and potassium) between interstitial fluid and the inside of a neuron through specific ion channels in the plasma membrane of a neuron. Once begun, a nerve impulse travels rapidly and at a constant strength.

Some neurons are tiny and propagate impulses over a short distance (less than 1 mm) within the CNS. Others are the longest cells in your body. Motor neurons that cause muscles to wiggle your toes, for example, extend from the lumbar region of your spinal cord (just above waist level) to your foot. Some sensory neurons are even longer. Those that allow you to feel sensations in your toes stretch all the way from your foot to the lower portion of your brain. Nerve impulses travel these great distances at speeds ranging from 0.5 to 130 meters per second (1 to 280 mi/hr).

Parts of a Neuron

While neurons come in a wide variety of shapes and sizes, there exists a general pattern of design shared by all of them. A neuron typically consists of two basic parts: (1) the cell body and (2) a variable number of processes called nerve fibers. The nerve fibers

---

• CLINICAL **CONNECTION** | Neurotoxins and Local Anesthetics

Certain shellfish and other organisms contain **neurotoxins** (noo-rō-TOK-sins), substances that produce their poisonous effects by acting on the nervous system. One particularly lethal neurotoxin is tetrodotoxin (TTX), present in the viscera of Japanese pufferfish. TTX effectively blocks nerve impulses.

Local anesthetics are drugs that block pain and other somatic sensations. Examples include procaine (Novocaine®) and lidocaine, which may be used to produce anesthesia in the skin during suturing of a gash, in the mouth during dental work, or in the lower body during childbirth. Like TTX, these drugs act by blocking nerve impulses, so pain signals do not reach the CNS.

Localized cooling of a nerve can also produce an anesthetic effect because axons propagate impulses at lower speeds when cooled. The application of ice to injured tissue can reduce pain because propagation of the pain sensations along axons is partially blocked. •

---

exhibit great variation in length and size and are classified, based on distinct structural and functional differences, as dendrites or an axon (Figure 17.3). The **cell body (perikaryon)** contains a nucleus surrounded by cytoplasm that includes typical cellular organelles such as lysosomes, mitochondria, and a Golgi complex. Neuronal cell bodies also contain prominent clusters of rough endoplasmic reticulum, termed *Nissl bodies.* The Nissl bodies are responsible for high levels of protein synthesis, an activity that is important in the proper function of the neuron, as proteins play important roles in maintenance and repair, transmission of nerve impulses, and reception of stimuli. The cytoskeleton includes both *neurofibrils*, composed of bundles of intermediate filaments that provide the cell shape and support, and *microtubules*, which assist in moving materials between the cell body and axon. Many neurons also contain *lipofuscin*, a pigment that occurs as clumps of yellowish brown granules in the cytoplasm. Lipofuscin is a product of neuronal lysosomes that accumulates in a neuron, but does not seem to cause harm to the neuron as it ages. The plasma membrane of the neuronal cell body ranges from smooth to very bumpy; these bumps are caused by many small projections called **gemmules**, which increase the surface area available for interactions with other nerve cells.

A **nerve fiber** is a general term for any neuronal process (extension) that emerges from the cell body of a neuron. Most neurons have two kinds of processes: multiple dendrites and a single axon. **Dendrites** (=little trees) are the receiving or input portions of a neuron. They usually are short, tapering, and highly branched: the greater the branching, the greater the surface area of the neuron for receiving synaptic communication from other neurons. In many neurons the dendrites form a tree-shaped array of processes extending from the cell body. While there are many different dendritic branching patterns, each type of neuron has a similar branching pattern. Their cytoplasm contains Nissl bodies, mitochondria, and other organelles. The plasma membranes of the dendrites and the cell body contain numerous receptor sites for binding chemical messengers from other cells. These receptor sites include gemmules and small projections on the plasma membrane of the small branches of dendrites called **dendritic spines**.

The nerve fiber called the **axon** (=axis) forms the other type of neuronal process and it varies significantly from the dendrites. Axons vary in length from less than a millimeter in neurons that communicate only with neighboring cells, to longer than a meter in neurons that communicate with distant parts of the nervous system or with peripheral organs. The single axon of a neuron carries nerve impulses toward another neuron, a muscle fiber, or a gland cell. An axon is a long, thin, cylindrical projection that often joins the cell body at a cone-shaped elevation called the **axon hillock** (=small hill). The part of the axon closest to the axon hillock is the **initial segment**. In most neurons, impulses arise at the junction of the axon hillock and the initial segment, an area called the **trigger zone**, and then travel along the axon. The trigger zone is free of Nissl bodies and has numerous voltage-sensitive channels in the plasma membrane. An axon contains mitochondria, microtubules, and neurofibrils. Because rough endoplasmic reticulum is not present in the axon, protein synthesis does not occur. The cytoplasm of an axon, called **axoplasm**, is surrounded by a plasma membrane known as the **axolemma** (*lemma*=sheath or husk). Along the length of an axon, side branches called **axon collaterals** may branch off, typically at a right angle to the axon. The axon and its collaterals end by dividing into many fine processes called **axon terminals (telodendria)**.

The site of communication between two neurons or between a neuron and an effector cell is called a **synapse** (SIN-aps). The term **presynaptic neuron** refers to a nerve cell that carries a

Figure 17.3 Structure of a typical neuron. Arrows indicate the direction of information flow: dendrites → cell body → axon → axon terminals. The axon is actually much longer than shown.

🔑 The basic parts of a neuron are several dendrites, a cell body, and an axon.

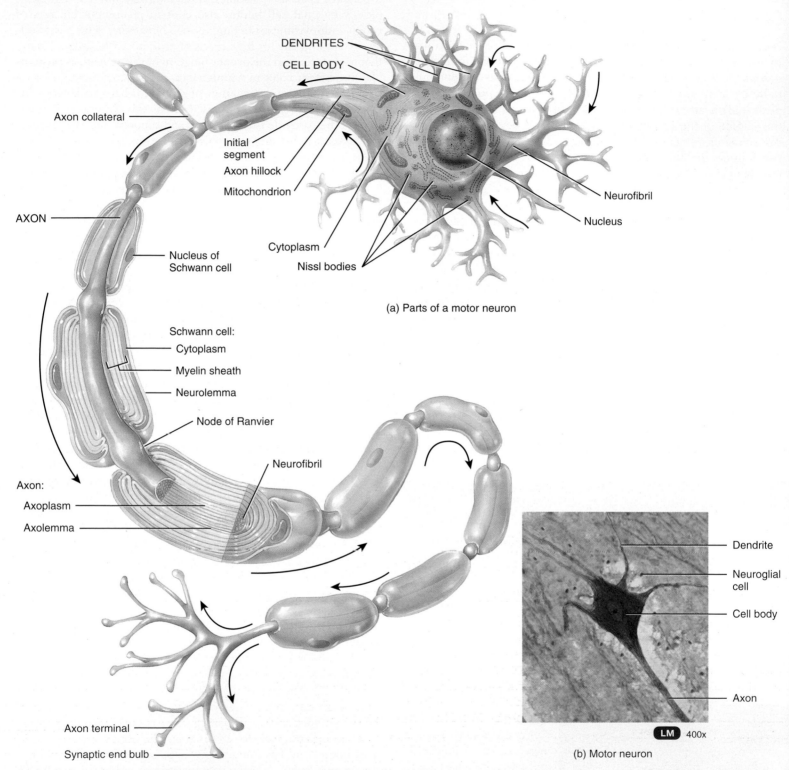

(a) Parts of a motor neuron

(b) Motor neuron

❓ **What roles do the dendrites, cell body, and axon play in communication of nerve impulses?**

nerve impulse toward a synapse. It is the cell that sends a signal. A **postsynaptic cell** is a nerve cell or effector (muscle or gland) that carries a nerve impulse away from a synapse. It is the cell that receives the signal. The tips of some axon terminals swell into bulb-shaped structures called **synaptic end bulbs (synaptic boutons)**; others exhibit a string of swollen bumps called **varicosities**. Both synaptic end bulbs and varicosities contain many tiny membrane-enclosed sacs called **synaptic vesicles** that

store a chemical called a neurotransmitter. A **neurotransmitter** is a molecule released from a synaptic vesicle that excites or inhibits postsynaptic neurons, muscle fibers, or gland cells. Neurons were long thought to liberate just one type of neurotransmitter at all synaptic end bulbs. We now know that many neurons contain two or even three neurotransmitters and that the release of the second or third neurotransmitter from a presynaptic neuron may be due to frequency of activation of the neuron.

The synapse between a motor neuron and a muscle fiber is called a **neuromuscular junction** (see Figure 10.6 on pages 295–296); the synapse between a neuron and a glandular cell is called a **neuroglandular junction**. Recall from Chapter 10 that most synapses contain a small gap between cells called a **synaptic cleft** (see Figure 10.6c) and that a nerve impulse cannot jump the gap to excite the next cell. Instead, the first cell communicates with the second cell indirectly by releasing a neurotransmitter. For example, at a neuromuscular junction, synaptic vesicles release the neurotransmitter acetylcholine (ACh) (see Figure 10.6c). The ACh is picked up by ACh receptors (integral transmembrane proteins) on muscle fibers, thus triggering an action potential in the muscle fibers. This causes the muscle fibers to contract. Once ACh has accomplished its task, it is broken down rapidly by an enzyme in the synaptic cleft called *acetylcholinesterase (AChE).*

Neurotransmitters

About 100 substances are either known or suspected neurotransmitters. **Acetylcholine (ACh)** is released by many PNS neurons and by some CNS neurons. ACh is an excitatory neurotransmitter at the neuromuscular junction. It is also known to be an inhibitory neurotransmitter at other synapses. For example, parasympathetic neurons slow heart rate by releasing ACh at inhibitory synapses.

Several amino acids are neurotransmitters in the CNS. **Glutamate** and **aspartate** have powerful excitatory effects. Two other amino acids, **gamma aminobutyric acid** (GAM-ma amē′-nō-bū-TER-ik) **(GABA)** and **glycine**, are important inhibitory neurotransmitters. Antianxiety drugs such as diazepam (Valium) enhance the action of GABA.

Some neurotransmitters are modified amino acids. These include norepinephrine, dopamine, and serotonin. **Norepinephrine (NE)** plays roles in arousal (awakening from deep sleep), dreaming, and regulating mood. Brain neurons containing the neurotransmitter **dopamine (DA)** are active during emotional responses, addictive behaviors, and pleasurable experiences. In addition, dopamine-releasing neurons help regulate skeletal muscle tone and some aspects of movement due to contraction of skeletal muscles. One form of schizophrenia is due to accumulation of excess dopamine. **Serotonin** is thought to be involved in sensory perception, temperature regulation, control of mood, appetite, and the onset of sleep.

Neurotransmitters consisting of amino acids linked by peptide bonds are called **neuropeptides** (noor-ō-PEP-tīds). The **endorphins** (en-DOR-fins) are neuropeptides that serve as the body's natural painkillers. Acupuncture may produce analgesia (loss of pain sensation) by increasing the release of endorphins. Endorphins have also been linked to improved memory and learning and to feelings of pleasure or euphoria.

An important newcomer to the ranks of recognized neurotransmitters is the simple gas **nitric oxide (NO)**, which is different from all previously known neurotransmitters because it is not synthesized in advance and packaged into synaptic vesicles. Rather, it is formed on demand, diffuses out of cells that produce it and into neighboring cells, and acts immediately. Some research suggests that NO plays a role in learning and memory.

Structural Diversity in Neurons

Neurons display great diversity in size and shape. For example, their cell bodies range in diameter from 5 micrometers (μm) (slightly smaller than a red blood cell) up to 135 μm (barely large enough to see with the unaided eye). The pattern of dendritic branching is varied and distinctive for neurons in different parts of the nervous system. A few small neurons lack an axon, and many others have very short axons. As noted earlier in the chapter, your longest axons are almost as long as you are tall, extending from your toes to the lowest part of your brain.

Both structural and functional features are used to classify the various neurons in the body. Structurally, neurons are classified according to the number of processes extending from the cell body (Figure 17.4).

Figure 17.4 Structural classification of neurons. Breaks indicate that axons are longer than shown.

A multipolar neuron has many processes extending from the cell body; a bipolar neuron has two and a unipolar neuron has one.

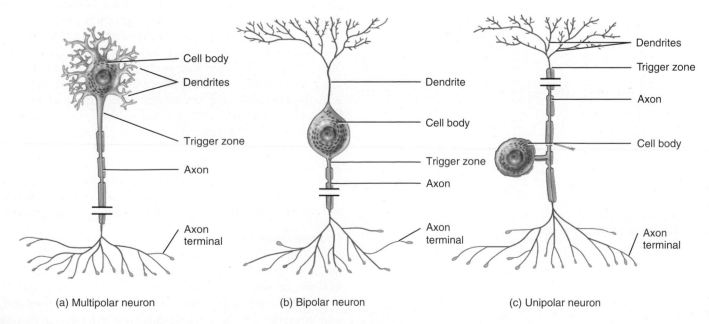

(a) Multipolar neuron

(b) Bipolar neuron

(c) Unipolar neuron

? **What occurs at a trigger zone?**

- **Multipolar neurons** usually have several dendrites and one axon (see also Figure 17.3). Most neurons in the brain and spinal cord are of this type, as well as all motor (efferent) neurons.
- **Bipolar neurons** have one main dendrite and one axon. They are found in the retina of the eye, in the inner ear, and in the olfactory area of the brain.
- **Unipolar neurons** are sensory neurons that begin in the embryo as bipolar neurons. During development, the axon and dendrite fuse into a single process that divides into two branches a short distance from the cell body. Both branches have the characteristic structure and function of an axon. They are long, cylindrical processes that propagate action potentials. However, the axon branch that extends into the periphery has dendrites at its distal tip, whereas the axon branch that extends into the CNS ends in synaptic end bulbs. The dendrites monitor a sensory stimulus such as touch or stretching. The trigger zone for nerve impulses in a unipolar neuron is at the junction of the dendrites and axon (Figure 17.4c). The impulses then propagate toward the synaptic end bulbs.

Some neurons are named for the histologist who first described them or for an aspect of their shape or appearance; examples include **Purkinje cells** (pur-KIN-jē) in the cerebellum (Figure 17.5a) and **pyramidal cells** (pi-RAM-i-dal), found in the cerebral cortex of the brain, which have pyramid-shaped cell bodies (Figure 17.5b). Often, a distinctive pattern of dendritic branching allows identification of a particular type of neuron in the CNS.

Figure 17.5 Two examples of CNS neurons. Arrows indicate the direction of information flow.

The dendritic branching pattern often is distinctive for a particular type of neuron.

(a) Purkinje cell   (b) Pyramidal cell

? How did pyramidal cells get their name?

Neuroglia

Neuroglia (noo-RŌG-lē-a; -*glia*=glue) or **glia** (GLĪ-a) constitute about half the volume of the CNS. Their name derives from the idea of early histologists that they were the "glue" that held nervous tissue together. We now know that neuroglia are not merely passive bystanders but rather active participants in the function of nervous tissue. Generally, neuroglia are smaller than neurons, and as stated previously they are much more numerous. In contrast to neurons, glia do not generate or propagate nerve impulses, and they have the ability to multiply and divide in the mature nervous system. In cases of injury or disease, neuroglia multiply to fill in the spaces formerly occupied by neurons. Brain tumors derived from glia, called **gliomas**, tend to be highly malignant and rapidly growing. Of the six types of neuroglia, four—astrocytes, oligodendrocytes, microglia, and ependymal cells—are found only in the CNS. The remaining two types—Schwann cells (neurolemmocytes) and satellite cells— are present in the PNS.

Neuroglia of the CNS

Neuroglia of the CNS can be distinguished on the basis of size, cytoplasmic processes, and intracellular organization into four types: astrocytes, oligodendrocytes, microglia, and ependymal cells (Figure 17.6).

ASTROCYTES (AS-trō-sīts; *astro-*=star; *-cyte*=cell). These are star-shaped cells with many processes and are the largest and most numerous of the neuroglia. There are two types of astrocytes. *Protoplasmic astrocytes* have many short branching processes and are found in gray matter (described shortly). *Fibrous astrocytes* have many long unbranched processes and are located mainly in white matter (also described shortly). The processes of astrocytes make contact with blood capillaries, neurons, and the pia mater (a thin membrane around the brain and spinal cord).

Following are the functions of astrocytes: (1) Astrocytes contain microfilaments that provide them with considerable strength, which enables them to support neurons. (2) Since neurons of the CNS must be isolated from various potentially harmful substances in blood, the endothelial cells of CNS blood capillaries have very selective permeability characteristics. In effect, the endothelial cells create a *blood–brain barrier*, which restricts the movement of substances between the blood and interstitial fluid of the CNS. Processes of astrocytes wrapped around blood capillaries secrete chemicals that maintain the unique permeability characteristics of the endothelial cells. Details of the blood–brain barrier are discussed in Chapter 19. (3) In the embryo, astrocytes secrete chemicals that appear to regulate the growth, migration, and interconnections among neurons in the brain. (4) Astrocytes help to maintain the appropriate chemical environment for the generation of nerve impulses. For example, they regulate the concentration of important ions such as K^+; take up excess neurotransmitters; and serve as a conduit for the passage of nutrients and other substances between blood capillaries and neurons. (5) Astrocytes may also play a role in learning and memory by influencing the formation of neural synapses.

OLIGODENDROCYTES (OL-i-gō-den′-drō-sīts; *oligo-*=few; *dendro-*=tree). These resemble astrocytes, but are smaller and contain fewer processes. Oligodendrocyte processes are responsible for forming and maintaining the myelin sheath around CNS axons. As you will see shortly, the myelin sheath is a lipid and protein covering around some axons that insulates the axon and

Figure 17.6 **Neuroglia of the central nervous system (CNS).**

 Neuroglia of the CNS are distinguished on the basis of size, cytoplasmic processes, and intracellular organization.

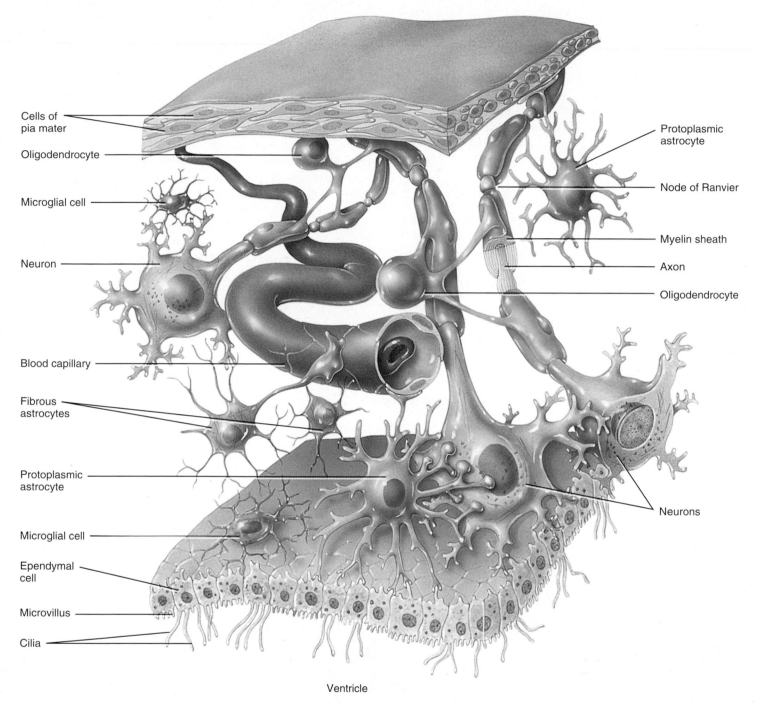

Cells of pia mater

Oligodendrocyte

Microglial cell

Neuron

Blood capillary

Fibrous astrocytes

Protoplasmic astrocyte

Microglial cell

Ependymal cell

Microvillus

Cilia

Protoplasmic astrocyte

Node of Ranvier

Myelin sheath

Axon

Oligodendrocyte

Neurons

Ventricle

? **Which CNS neuroglia function as phagocytes?**

increases the speed of nerve impulse conduction. Such axons are said to be myelinated. Oligodendrocyte processes that make contact with cell bodies of neurons are thought to influence the chemistry of the interstitial fluid of neurons.

MICROGLIA (mī-KROG-lē-a; *micro-*=small). These neuroglia are small cells with slender processes that give off numerous spinelike projections. Unlike other neuroglial cells, which develop from the neural tube, microglia originate in red bone marrow and migrate into the CNS as it develops. Microglia function as phagocytes. Like tissue macrophages, they remove cellular

debris formed during normal development of the nervous system and phagocytize microbes and damaged nervous tissue.

EPENDYMAL CELLS (ep-EN-de-mal; *epen-*=above; *dym-*=garment). Ependymal cells are cuboidal to columnar cells arranged in a single layer that possess microvilli and cilia. These cells line the ventricles of the brain and central canal of the spinal cord (spaces filled with cerebrospinal fluid). Functionally, ependymal cells produce, possibly monitor, and assist in the circulation of cerebrospinal fluid. They also form the blood–cerebrospinal fluid barrier, which is discussed in Chapter 19.

Neuroglia of the PNS

Neuroglia of the PNS completely surround axons and cell bodies. The two types of glial cells in the PNS are Schwann cells (neurolemmocytes) and satellite cells (Figure 17.7).

SCHWANN CELLS (SCHVON or SCHWON)   These flat cells, also called **neurolemmocytes,** encircle PNS axons. Like the oligodendrocytes of the CNS, they form the myelin sheath around axons. However, a single oligodendrocyte myelinates several axons, but each Schwann cell myelinates a single axon (Figure 17.7a; see also Figure 17.8a, c). A single Schwann cell can also enclose as many as 20 or more unmyelinated axons (axons that lack a myelin sheath) (Figure 17.7b). Schwann cells participate in axon regeneration, which is more easily accomplished in the PNS than in the CNS.

SATELLITE CELLS (SAT-i-līt).   These flat cells surround the cell bodies of neurons of PNS ganglia (Figure 17.7c). (Recall that ganglia are collections of neuronal cell bodies outside the CNS.) Besides providing structural support, satellite cells regulate the exchange of materials between neuronal cell bodies and interstitial fluid.

Myelination

Axons that are surrounded by a multilayered lipid and protein covering, called the **myelin sheath,** are said to be **myelinated**

(Figure 17.8a). The sheath electrically insulates the axon of a neuron and increases the speed of nerve impulse conduction. Axons without such a covering are said to be **unmyelinated** (Figure 17.8b).

Two types of neuroglia produce myelin sheaths: Schwann cells (in the PNS) and oligodendrocytes (in the CNS). In the PNS, Schwann cells begin to form myelin sheaths around axons during fetal development. Each Schwann cell wraps about 1 millimeter (1 mm=0.04 in.) of a single axon's length by spiraling many times around the axon (Figure 17.8a). Eventually, multiple layers of glial plasma membrane surround the axon, with the Schwann cell's cytoplasm and nucleus forming the outermost layer. The inner portion, consisting of up to 100 layers of Schwann cell membrane, is the myelin sheath. The outer nucleated cytoplasmic layer of the Schwann cell, which encloses myelinated or unmyelinated axons, is the **neurolemma (sheath of Schwann).** A neurolemma is found only around axons in the PNS. When an axon is injured, the neurolemma aids regeneration by forming a regeneration tube that guides and stimulates regrowth of the axon. Gaps in the myelin sheath, called **nodes of Ranvier** (RON-vē-ā) **(nerve fiber nodes),** appear at intervals along the axon (see Figures 17.3 and 17.7a). Each Schwann cell wraps one axon segment between two nodes.

In the CNS, an oligodendrocyte myelinates parts of several axons. Each oligodendrocyte puts forth about 15 broad, flat processes that spiral around CNS axons, forming a myelin sheath

Figure 17.7 Neuroglia of the peripheral nervous system (PNS).

Neuroglia of the PNS completely surround axons and cell bodies of neurons.

(a)   (b)   (c)

❓ **How do Schwann cells and oligodendrocytes differ with respect to number of axons myelinated?**

(see Figure 17.6). A neurolemma is not present, however, because the oligodendrocyte cell body and nucleus do not envelop the axon. Nodes of Ranvier are present, but they are fewer in number. Axons in the CNS display little regrowth after injury. This is thought to be due, in part, to the absence of a neurolemma, and in part to an inhibitory influence on axon regrowth exerted by the oligodendrocytes.

The amount of myelin increases from birth to maturity, and its presence greatly increases the speed of nerve impulse conduction.

An infant's responses to stimuli are neither as rapid nor as coordinated as those of an older child or an adult, in part because myelination is still in progress during infancy. **Demyelination** (dē-mī'-e-li-NĀ-shun) refers to the loss or destruction of myelin sheaths around axons. It may result from disorders such as multiple sclerosis or Tay-Sachs disease, or from medical treatments such as radiation therapy and chemotherapy. Any single episode of demyelination may cause deterioration of affected nerves.

Figure 17.8 Myelinated and unmyelinated axons.

Axons surrounded by a myelin sheath produced either by Schwann cells or by oligodendrocytes are said to be myelinated.

(a) Transverse sections of stages in the formation of a myelin sheath

(b) Transverse section of unmyelinated axons

(c) Transverse section of myelinated axon

(d) Transverse section of unmyelinated axons

? What is the functional advantage of myelination?

• CLINICAL CONNECTION | Multiple Sclerosis

Multiple sclerosis (MS) is a disease that causes a progressive destruction of myelin sheaths surrounding neurons in the CNS. It afflicts about 350,000 people in the United States and 2 million people worldwide. It usually appears between the ages of 20 and 40, affecting females twice as often as males. MS is most common in whites, less common in blacks, and rare in Asians. MS is an autoimmune disease—the body's own immune system spearheads the attack. The condition's name describes the anatomical pathology: In *multiple* regions the myelin sheaths deteriorate to *scleroses*, which are hardened scars or plaques. Magnetic resonance imaging (MRI) studies reveal numerous plaques in the white matter of the brain and spinal cord. The destruction of myelin sheaths slows and then short-circuits propagation of nerve impulses.

The most common form of the condition is *relapsing-remitting MS*, which usually appears in early adulthood. The first symptoms may include a feeling of heaviness or weakness in the muscles, abnormal sensations, or double vision. An attack is followed by a period of remission during which the symptoms temporarily disappear. One attack follows another over the years, usually every year or two. The result is a progressive loss of function interspersed with remission periods, during which symptoms abate.

Although the cause of MS is unclear, both genetic susceptibility and exposure to some environmental factor (perhaps a herpes virus) appear to contribute. Many patients with relapsing-remitting MS are treated with injections of beta interferon. This treatment lengthens the time between relapses, decreases the severity of relapses, and slows formation of new lesions in some cases. Unfortunately, not all MS patients can tolerate beta interferon, and therapy becomes less effective as the disease progresses. •

Nerve cell bodies

Oligodendrocyte

Myelin disruption

MS affected axon

Normal axon with myelin sheath

Brain

Spinal cord

Gray and White Matter

In a freshly dissected section of the brain or spinal cord, some regions look white and glistening, and others appear gray (Figure 17.9). The **white matter** is aggregations of myelinated and unmyelinated axons of many neurons. The whitish color of myelin gives white matter its name. The **gray matter** of the nervous system contains neuronal cell bodies, dendrites, unmyelinated axons, axon terminals, and neuroglia. It appears grayish, rather than white, because the Nissl bodies impart a gray color and there is little or no myelin in these areas. Blood vessels are present in both white and gray matter.

In the spinal cord, the white matter surrounds an inner core of gray matter shaped like a butterfly or the letter H; in the brain, a thin shell of gray matter covers the surface of the largest portions of the brain, the cerebrum and cerebellum (Figure 17.9). When used to describe nervous tissue, a **nucleus** is a cluster of neuronal cell bodies within the CNS. (Recall that the term *ganglion* refers to a similar arrangement within the PNS.) Many nuclei of gray matter also lie deep within the brain. The arrangements of gray and white matter in the spinal cord and brain are discussed more extensively in Chapters 18 and 19, respectively.

◉ CHECKPOINT

4. Describe the parts of a neuron and the functions of each.
5. Give several examples of the structural diversity of neurons.
6. What is a neurolemma, and why is it important?
7. With reference to the central nervous system, what is a nucleus?

NEURAL CIRCUITS

◉ OBJECTIVE

• Identify the structure and function of the various types of neural circuits in the nervous system.

The CNS contains billions of neurons organized into complicated networks called **neural circuits**, each of which is a functional group of neurons that processes a specific kind of information. In a **simple series circuit**, a presynaptic neuron stimulates a single postsynaptic neuron. The second neuron then stimulates another, and so on. However, most neural circuits are more complex.

Figure 17.9 Distribution of gray matter and white matter in the spinal cord and brain.

White matter consists of myelinated and unmyelinated axons of many neurons. Gray matter consists of neuron cell bodies, dendrites, axon terminals, bundles of unmyelinated axons, and neuroglia.

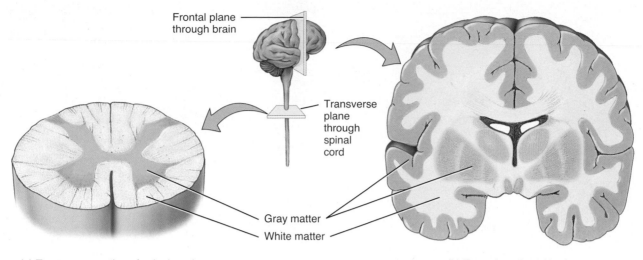

(a) Transverse section of spinal cord

(b) Frontal section of brain

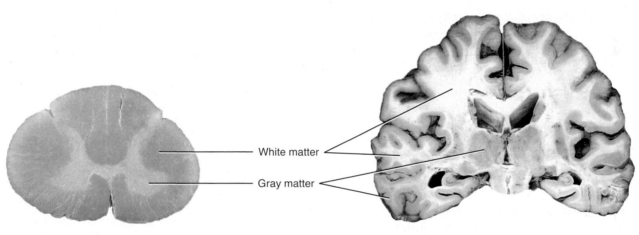

(c) Transverse section of spinal cord

(d) Frontal section of brain

? What is responsible for the white appearance of white matter?

A single presynaptic neuron may synapse with several postsynaptic neurons. Such an arrangement, called **divergence**, permits one presynaptic neuron to influence several postsynaptic neurons (or several muscle fibers or gland cells) at the same time. In a **diverging circuit**, the nerve impulse from a single presynaptic neuron causes the stimulation of increasing numbers of cells along the circuit (Figure 17.10a). For example, a small number of neurons in the brain that govern a particular body movement stimulate a much larger number of neurons in the spinal cord. Sensory signals also spread into diverging circuits and are often relayed to several regions of the brain. This arrangement has the effect of amplifying the signal.

In another arrangement, called **convergence**, several presynaptic neurons synapse with a single postsynaptic neuron. This arrangement permits more effective stimulation or inhibition of the postsynaptic neuron. In a **converging circuit** (Figure 17.10b), the postsynaptic neuron receives nerve impulses from several different sources. For example, a single motor neuron that synapses with skeletal muscle fibers at neuromuscular junctions receives input from several pathways that originate in different brain regions.

Some circuits are constructed so that stimulation of the presynaptic cell causes the postsynaptic cell to transmit a series of nerve impulses. One such circuit is called a **reverberating circuit** (Figure 17.10c). In this pattern, the incoming impulse stimulates the first neuron, which stimulates the second, which stimulates the third, and so on. Branches from later neurons synapse with earlier ones. This arrangement sends impulses back through the circuit again and again. The output signal may last from a few seconds to many hours, depending on the number of synapses and the arrangement of neurons in the circuit. Inhibitory neurons may turn off a reverberating circuit after a period of time. Among the body responses thought to be the result of output signals from reverberating circuits are breathing, coordinated muscular activities, waking up, sleeping (when reverberation stops), and short-term memory.

Figure 17.10 Examples of neural circuits.

🔑 A neural circuit is a functional group of neurons that processes a specific kind of information.

(a) Diverging circuit   (b) Converging circuit   (c) Reverberating circuit   (d) Parallel after-discharge circuit

❓ **A motor neuron in the spinal cord typically receives input from neurons that originate in several different regions of the brain. Is this an example of convergence or divergence?**

A fourth type of circuit is the **parallel after-discharge circuit** (Figure 17.10d). In this circuit, a single presynaptic cell stimulates a group of neurons, each of which synapses with a common postsynaptic cell. Parallel after-discharge circuits may be involved in precise activities such as mathematical calculations.

⊙ CHECKPOINT

8. What is a neural circuit?
9. What are the functions of diverging, converging, reverberating, and parallel after-discharge circuits?

REGENERATION AND NEUROGENESIS

⊙ OBJECTIVE

• Define regeneration and neurogenesis.

Throughout your life, your nervous system exhibits **plasticity**, the capability to change based on experience. At the level of individual neurons, the changes that can occur include the sprouting of new dendrites, synthesis of new proteins, and changes in synaptic contacts with other neurons. Undoubtedly, both chemical and electrical signals drive the changes that occur. Despite this plasticity, mammalian neurons have very limited powers of **regeneration**, the capability to replicate or repair themselves. In the PNS, damage to dendrites and myelinated axons may be repaired if the cell body remains intact and if the Schwann cells that produce myelination remain active. In the CNS, little or no repair of damage to neurons occurs. Even when the cell body remains intact, a severed axon cannot be repaired or regrown.

Neurogenesis—the birth of new neurons from undifferentiated stem cells—occurs regularly in some animals. For example, new neurons appear and disappear every year in some songbirds. Until recently, the dogma was "no new neurons" in the adult brains of humans and other primates. Then, in 1992, Canadian researchers published their unexpected finding that **epidermal growth factor (EGF)** stimulated cells taken from the brains of adult mice to proliferate into both neurons and astrocytes. Previously, EGF was known to trigger mitosis in a variety of nonneuronal cells and to promote wound healing and tissue regeneration. In 1998 scientists discovered that significant numbers of new neurons do arise in the adult human hippocampus, an area of the brain that is crucial for learning.

The nearly complete lack of neurogenesis in other regions of the brain and spinal cord seems to result from two factors: (1) inhibitory influences from neuroglia, particularly oligodendrocytes, and (2) absence of growth-stimulating cues that were present during fetal development. Axons in the CNS are myelinated by oligodendrocytes rather than Schwann cells, and this CNS myelin is one of the factors inhibiting regeneration of neurons. Perhaps this same mechanism stops axonal growth once a target region has been reached during development. Also, after axonal damage, nearby astrocytes proliferate rapidly, forming a type of scar tissue that acts as a physical barrier to regeneration. Thus, injury of the brain or spinal cord usually is permanent. Ongoing research seeks ways to improve the environment for existing spinal cord axons to bridge the injury gap. Scientists also are trying to find ways to stimulate dormant stem cells to replace neurons lost through damage or disease and to develop tissue-cultured neurons that can be used for transplantation purposes.

⊙ CHECKPOINT

10. What factors contribute to the absence of neurogenesis in most parts of the brain?

KEY MEDICAL TERMS ASSOCIATED WITH NERVOUS TISSUE

Guillain-Barré Syndrome (GBS) (GĒ-an ba-RĀ) An acute demyelinating disorder in which macrophages strip myelin from axons in the PNS. It is the most common cause of acute paralysis in North America and Europe and may result from the immune system's response to a bacterial infection.

Neuroblastoma (noor-ō-blas-TŌ-ma) A malignant tumor that consists of immature nerve cells (neuroblasts); occurs most commonly in the abdomen and most frequently in the adrenal glands.

Neuropathy (noo-ROP-a-thē; *neuro-*=a nerve; *-pathy*=disease) Any disorder that affects the nervous system but particularly a disorder of a cranial or spinal nerve. An example is *facial neuropathy* (Bell's palsy), a disorder of the facial (VII) nerve.

Rabies (RĀ-bēz; *rabi-*=mad, raving) A fatal disease caused by a virus that reaches the CNS via an axon, usually transmitted by the bite of a dog or other meat-eating animal. The symptoms are excitement, aggressiveness, and madness, followed by paralysis and death.

STUDY OUTLINE

Overview of the Nervous System (p. 567)

1. Structures that make up the nervous system are the brain, 12 pairs of cranial nerves and their branches, the spinal cord, 31 pairs of spinal nerves and their branches, ganglia, enteric plexuses, and sensory receptors.
2. Anatomically, the nervous system consists of two divisions. (1) The central nervous system (CNS) consists of the brain and spinal cord. (2) The peripheral nervous system (PNS) consists of all nervous tissue outside the CNS.
3. Functionally, the nervous system integrates all body activities by sensing changes (sensory function), interpreting them (integrative function), and reacting to them (motor function).
4. Sensory (afferent) neurons carry sensory information from cranial and spinal nerves into the brain and spinal cord or from a lower to a higher level in the spinal cord and brain. Interneurons have short axons that contact nearby neurons in the brain, spinal cord, or a ganglion. Motor (efferent) neurons carry information from the brain toward the spinal cord or out of the brain and spinal cord into cranial or spinal nerves.
5. Components of the PNS include the somatic nervous system (SNS), autonomic nervous system (ANS), and enteric nervous system (ENS).
6. The SNS consists of somatic sensory neurons that conduct impulses from somatic and special sense receptors to the CNS (input) and somatic motor neurons from the CNS to skeletal muscles (output).
7. The ANS contains autonomic sensory neurons from visceral organs (input) and autonomic motor neurons that convey impulses from the CNS to smooth muscle tissue, cardiac muscle tissue, and glands (output).
8. The ENS consists of neurons in enteric plexuses in the gastrointestinal (GI) tract that function independently of the ANS and CNS to some extent. Sensory neurons of the ENS monitor chemical changes and stretching of the GI tract (input) and motor neurons of the ENS generate contractions and secretions of the GI tract (output).

Histology of Nervous Tissue (p. 570)

1. Nervous tissue consists of neurons (nerve cells) and neuroglia. Neurons have the property of electrical excitability and are responsible for most unique functions of the nervous system: sensing, thinking, remembering, controlling muscle activity, and regulating glandular secretions.
2. Most neurons have three parts. The dendrites are the main receiving or input region. Integration occurs in the cell body, which includes typical cellular organelles. The output part typically is a single axon, which propagates nerve impulses toward another neuron, a muscle fiber, or a gland cell.
3. Synapses are the sites of functional contact between excitable cells. Axon terminals contain synaptic vesicles filled with neurotransmitter molecules.
4. On the basis of their structure, neurons are classified as multipolar, bipolar, or unipolar.
5. Neuroglia support, nurture, and protect neurons and maintain the interstitial fluid that bathes them. Neuroglia in the CNS include astrocytes, oligodendrocytes, microglia, and ependymal cells. Neuroglia in the PNS include Schwann cells and satellite cells.
6. Two types of neuroglia produce myelin sheaths: oligodendrocytes myelinate axons in the CNS, and Schwann cells myelinate axons in the PNS.
7. White matter consists of aggregates of myelinated processes; gray matter contains cell bodies, dendrites, and axon terminals of neurons, unmyelinated axons, and neuroglia.
8. In the spinal cord, gray matter forms an H-shaped inner core that is surrounded by white matter. In the brain, a thin, superficial shell of gray matter covers the cerebral and cerebellar hemispheres.

Neural Circuits (p. 578)

1. Neurons in the central nervous system are organized into networks called neural circuits.
2. Neural circuits include simple series, diverging, converging, reverberating, and parallel after-discharge circuits.

Regeneration and Neurogenesis (p. 580)

1. The nervous system exhibits plasticity (the capability to change based on experience), but it has very limited powers of regeneration (the capability to replicate or repair damaged neurons).
2. Neurogenesis, the birth of new neurons from undifferentiated stem cells, is normally very limited in adult humans. Repair of damaged axons does not occur in most regions of the CNS.

SELF-QUIZ QUESTIONS

Choose the one best answer to the following questions.

1. Which of the following statements is *true*?
   a. A neuron usually has many axons that are connected to other neurons.
   b. Most unipolar neurons are found in the central nervous system.
   c. Multipolar neurons possess numerous dendrites and only one axon.
   d. Sensory neurons are usually bipolar.
   e. All of the above are true.

2. The myelin sheath of peripheral nervous system neurons is produced by:
   a. Schwann cells. b. astrocytes.
   c. oligodendrocytes. d. microglia.
   e. ependymal cells.

3. Endorphins are neuropeptides that:
   a. stimulate skeletal muscle contraction.
   b. increase heart rate.
   c. stimulate vasoconstriction.
   d. reduce pain sensations.
   e. regulate body temperature.

4. The point of contact between a nerve fiber and a muscle is called the:
   a. axon hillock.
   b. node of Ranvier.
   c. trigger zone.
   d. neuromuscular junction.
   e. ganglion.

5. Which of the following is an effector of somatic efferent neurons?
   a. thyroid gland
   b. blood vessel smooth muscle
   c. biceps brachii muscle
   d. pain receptor
   e. sudoriferous (sweat) gland

6. Neuroglia that are important components of the blood–brain barrier are the:
   a. astrocytes.
   b. oligodendrocytes.
   c. microglia.
   d. ependymal cells.
   e. Schwann cells.

7. Neuroglia that are phagocytic are the:
   a. astrocytes.
   b. oligodendrocytes.
   c. microglia.
   d. ependymal cells.
   e. Schwann cells.

8. Which of the following best describes a diverging circuit?
   a. Several presynaptic neurons stimulate one postsynaptic neuron.
   b. One presynaptic neuron stimulates several postsynaptic neurons.
   c. Neuron 1 stimulates Neuron 2, which stimulates Neuron 3. Neurons 2 and 3 stimulate Neuron 1.
   d. One presynaptic neuron stimulates one postsynaptic neuron.
   e. None of these fits the definition of a diverging circuit.

Complete the following.

9. Cerebrospinal fluid is produced by neuroglia known as _____.

10. The peripheral nervous system is subdivided into the _____ nervous system, the _____ nervous system, and the _____ nervous system.

11. A bundle of nerve fibers located in the central nervous system is known as a(n) _____.

12. Cells and organs contacted by motor neurons in cranial and spinal nerves are termed _____.

13. Two amino acids that are important inhibitory neurotransmitters are _____ and _____.

14. Sacs in the axon terminals that store neurotransmitters are called _____.

15. The myelin sheath of axons in the central nervous system is produced by cells called _____.

16. Multiple sclerosis (MS) is a demyelinating disease which strips the myelin from axons in the _____ nervous system, in contrast to Guillain-Barré Syndrome (GBS), which strips the myelin from axons in the _____ nervous system.

Are the following statements true or false? If the statement is False, *find and fix the mistake to make the statement* True.

17. A unipolar neuron originates as a bipolar neuron, the axon and dendrite of which fuse close to the cell body.

18. Enteric motor neurons control secretions from organs in the gastrointestinal tract.

19. Astrocytes are responsible for regulating concentrations of potassium ions needed for generation of nerve impulses.

20. White matter refers to aggregations of myelinated processes from many axons.

21. The nucleus of a neuron is located in the axon terminal.

22. Skeletal muscles are effectors in the autonomic nervous system.

Matching

23. Match the following:

    _____ (a) connects an axon to the cell body
    _____ (b) yellowish pigment that increases with age; appears to be a byproduct of lysosomes
    _____ (c) plasma membrane of the axon
    _____ (d) long, thin fibrils composed of bundles of intermediate filaments
    _____ (e) orderly arrangement of rough endoplasmic reticulum; site of protein synthesis
    _____ (f) conducts nerve impulses toward cell body
    _____ (g) conducts nerve impulses away from cell body; has synaptic end bulbs that contain neurotransmitters
    _____ (h) fine filaments that are branching ends of axons and axon collaterals

    (1) axon
    (2) axon terminal
    (3) axon hillock
    (4) Nissl bodies
    (5) dendrite
    (6) lipofuscin
    (7) axolemma
    (8) neurofibrils

24. Match the following (answers are used more than once):

    _____ (a) association neurons
    _____ (b) unipolar neurons
    _____ (c) afferent neurons
    _____ (d) efferent neurons
    _____ (e) most central nervous system neurons are of this type
    _____ (f) these neurons conduct impulses away from the central nervous system
    _____ (g) these neurons conduct impulses toward the central nervous system

    (1) sensory neurons
    (2) motor neurons
    (3) interneurons

CRITICAL THINKING QUESTIONS

1. Varudhini was reviewing her anatomy notes when she heard a catchy advertising jingle on the radio. Hours later, on her way to class, the jingle was still running through her mind. Varudhini was suddenly reminded of a neuronal circuit from her anatomy notes. Name the type of circuit and explain its uses in the body.

2. Following a brain injury in her early thirties, Jaqueline, an artist, could no longer draw pictures with her hands, but within a few years was able to produce beautiful artwork using her feet "as her hands." What characteristic of the nervous system made this possible?

3. Long-distance runners often describe a "runner's high." What neurotransmitter is responsible for producing this "high," and how does it work?

4. The buzzing of the alarm clock woke Mohamed. He yawned, stretched, and started to salivate as he smelled the coffee brewing. He could feel his stomach rumble. List the divisions of the nervous system involved in each of these actions.

5. Althea is in research and development for a large pharmaceutical company. She has just been pleasantly surprised by the results of her recent experiments on a new drug to treat a viral infection of the brain. Her secretary exclaims, "Great! With my shares of stock in the company, I'll be rich!" Althea told her secretary, "Don't buy that new Porsche just yet. What works in the Petri dish in the lab may not work in a real person. First of all, the drug has to be able to reach the infection." The secretary was perplexed, and said, "Well, why not? Just inject it into a vein, and it will get there." What would your response be?

? ANSWERS TO FIGURE QUESTIONS

17.1 The anatomical subdivisions of the nervous system are the CNS (brain and spinal cord) and PNS (all nervous structures outside the CNS).

17.2 The somatic motor component of the peripheral nervous system is voluntary.

17.3 Dendrites receive (motor neurons or interneurons) or generate (sensory neurons) inputs; the cell body also receives input signals; the axon conducts nerve impulses (action potentials) and transmits the message to another neuron or effector cell by releasing a neurotransmitter at its synaptic end bulbs.

17.4 Nerve impulses arise at the trigger zone.

17.5 The cell body of a pyramidal cell is shaped like a pyramid.

17.6 Microglia function as phagocytes in the CNS.

17.7 One Schwann cell myelinates a single axon; one oligodendrocyte myelinates several axons.

17.8 Myelination increases the speed of nerve impulse conduction.

17.9 Myelin makes white matter look shiny and white.

17.10 A motor neuron receiving input from several other neurons is an example of convergence.

18

THE SPINAL CORD AND THE SPINAL NERVES

INTRODUCTION The spinal cord and spinal nerves contain neural circuits that control some of your quickest reactions to environmental changes. If you pick up something hot, for example, you may find that the grasping muscles in your hand relax and you drop it even before the sensation of extreme heat or pain reaches your conscious perception. This is an example of a **spinal cord reflex**—a quick, automatic response to certain kinds of stimuli that involves neurons only in the spinal nerves and spinal cord. In addition to processing reflexes, the spinal cord is the site for integrating neuronal stimulation that arises locally or is triggered by nerve impulses from the peripheral nervous system (the periphery) and the brain. The spinal cord is also the highway traveled by sensory nerve impulses headed for the brain, as well as motor nerve impulses from the brain to skeletal muscles and other effectors. Keep in mind that the spinal cord is continuous with the brain and that together they constitute the central nervous system (CNS). •

SPINAL CORD ANATOMY

⊙ **OBJECTIVE**

- Describe the protective structures and the gross anatomical features of the spinal cord.

Protective Structures

Recall from the previous chapter that central nervous system tissue is very delicate and does not respond well to injury or damage. These characteristics make protection imperative. The first line of defense is the hard bony skull and vertebral column. The skull encases the brain and the vertebral column surrounds the spinal cord, providing strong protective cases against damaging blows or bumps. The second line of defense is the meninges, three protective membranes that lie between the bony encasement and the nervous tissue. Finally, central nervous tissue "floats" in cerebrospinal fluid, which cushions both the brain and spinal cord.

Vertebral Column

The spinal cord is located within the vertebral canal of the vertebral column. The vertebral foramina of all the vertebrae, stacked one on top of the other, form the vertebral canal. The surrounding vertebrae provide a sturdy shelter for the enclosed spinal cord

(see Figure 18.1b). The vertebral ligaments provide additional protection.

Meninges

The **meninges** [(me-NIN-jēz; singular is **menix** (ME-ninks)] are three connective tissue coverings that encircle the spinal cord and brain. From superficial to deep they are the: (1) dura mater; (2) arachnoid mater; and (3) pia mater. These connective tissue coverings form protective, vascular coats. Between the inner two layers, cerebrospinal fluid serves as a shock absorber and suspension system for the brain and spinal cord. The **spinal meninges** surround the spinal cord (Figure 18.1a) and are continuous with the **cranial meninges**, which encircle the brain (shown in Figure 19.4a on page 617). The most superficial of the three spinal meninges is composed of dense, irregular connective tissue and is called the **dura mater** (DOO-ra MĀ-ter=tough mother; early anatomists erroneously believed that all the tissues of the body arose from these tissues). The dura mater forms a sac from the level of the foramen magnum in the occipital bone, where it is continuous with the dura mater of the brain, to the second sacral vertebra. The spinal cord is also protected by a cushion of fat and connective tissue located in the **epidural space**, a space between the dura mater and the wall of the vertebral canal (Figure 18.1b).

Figure 18.1 **Gross anatomy of the spinal cord.** The spinal meninges are evident in both views.

🔑 ▭ Meninges are connective tissue coverings that surround the spinal cord and brain.

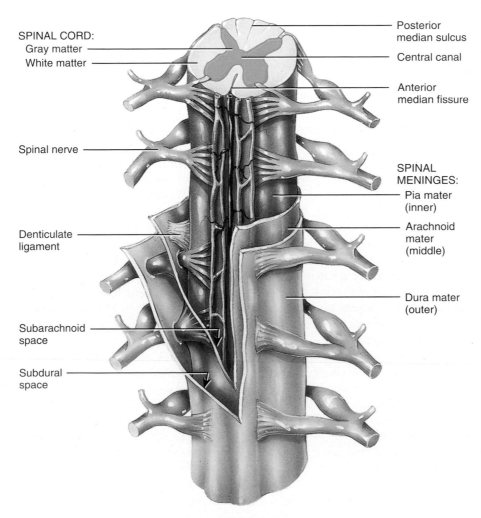

SPINAL CORD:
Gray matter
White matter

Posterior
median sulcus

Central canal

Anterior
median fissure

Spinal nerve

SPINAL
MENINGES:
Pia mater
(inner)

Arachnoid
mater
(middle)

Denticulate
ligament

Dura mater
(outer)

Subarachnoid
space

Subdural
space

(a) Anterior view and transverse section through spinal cord

🔘 **FIGURE 18.1 CONTINUES** ▶

■ **FIGURE 18.1 CONTINUED** ▶

(b) Transverse section of the spinal cord within a cervical vertebra

? What are the superior and inferior boundaries of the spinal dura mater?

The other two meningeal membranes, often referred to collectively as the **leptomeninges**, are thin, somewhat transparent layers comprised of arachnoid cells and thin, loose arrays of collagen. The middle meninx is an avascular covering called the **arachnoid mater** (a-RAK-noyd MĀ-ter; *arachn-*=spider; *-oid*=similar to) because of its spider's web arrangement of delicate collagen fibers and some elastic fibers. It is deep to the dura mater and is continuous with the arachnoid mater of the brain. Between the dura mater and the arachnoid mater is a thin **subdural space**, which contains interstitial fluid.

The innermost meninx is the **pia mater** (PĒ-a MĀ-ter; *pia*=delicate), a thin transparent connective tissue layer that adheres to the surface of the spinal cord and brain. It consists of thin squamous to cuboidal cells within interlacing bundles of collagen fibers and some fine elastic fibers. Within the pia mater are many blood vessels that supply oxygen and nutrients to the spinal cord. Between the arachnoid mater and the pia mater is the **subarachnoid space**, which contains cerebrospinal fluid that serves as a shock absorber and suspension system for the spinal cord and brain.

All three spinal meninges cover the spinal nerves up to the point of exit from the spinal column through the intervertebral foramina. Triangular-shaped membranous extensions of the pia mater suspend the spinal cord in the middle of its dural sheath. These extensions, called **denticulate ligaments** (den-TIK-ū-lāt =small tooth), are thickenings of the pia mater. They project laterally and fuse with the arachnoid mater and inner surface of the dura mater between the anterior and posterior nerve roots of spinal nerves on either side (Figure 18.1a, b). Extending all along the length of the spinal cord, the denticulate ligaments protect the spinal cord against sudden displacement that could result in shock.

External Anatomy of the Spinal Cord

The **spinal cord**, although roughly cylindrical, is flattened slightly anteriorly and posteriorly. In adults, it extends from the medulla oblongata, the inferior part of the brain, to the superior border of the second lumbar vertebra (Figure 18.2a). In newborn infants, it extends to the third or fourth lumbar vertebra. During early childhood, both the spinal cord and the vertebral column grow longer as part of overall body growth. Elongation of the spinal cord stops around age 4 or 5, but growth of the vertebral column continues. Thus, the spinal cord does not extend the entire length of the vertebral column. The length of the adult spinal cord ranges from 42 to 45 cm (16–18 in.). Its diameter is about 2 cm (0.75 in.) in the midthoracic region, somewhat larger in the lower cervical and midlumbar regions, and smallest at the inferior tip.

When the spinal cord is viewed externally, two conspicuous enlargements can be seen (Figure 18.2a). The superior enlargement, the **cervical enlargement**, extends from the fourth cervical vertebra to the first thoracic vertebra. Nerves to and from the upper limbs arise from the cervical enlargement. The inferior enlargement, called the **lumbar enlargement**, extends from the ninth to the twelfth thoracic vertebra. Nerves to and from the lower limbs arise from the lumbar enlargement.

Inferior to the lumbar enlargement, the spinal cord terminates as a tapering, conical structure called the **conus medullaris**

■ TEXT CONTINUES ON P. 58●

Figure 18.2 **External anatomy of the spinal cord and the spinal nerves.**

 The spinal cord extends from the medulla oblongata of the brain to the superior border of the second lumbar vertebra.

CERVICAL PLEXUS (C1–C5):
  Lesser occipital nerve
  Great auricular
  Ansa cervicalis
  Transverse cervical nerve
  Supraclavicular nerve
  Phrenic nerve

BRACHIAL PLEXUS (C5–T1):
  Musculocutaneous nerve
  Axillary nerve
  Median nerve
  Radial nerve
  Ulnar nerve

Intercostal
(thoracic) nerves

Subcostal nerve
(intercostal nerve 12)

LUMBAR PLEXUS (L1–L4):
  Iliohypogastric nerve
  Ilioinguinal nerve
  Genitofemoral nerve
  Lateral femoral
  cutaneous nerve

  Femoral nerve
  Obturator nerve

SACRAL PLEXUS (L4–S4):
  Superior gluteal nerve
  Inferior gluteal nerve

  Sciatic nerve:
    Common fibular
    nerve
    Tibial nerve

  Posterior cutaneous
  nerve of thigh
  Pudendal nerve

Medulla oblongata
Atlas (first cervical vertebra)
CERVICAL NERVES (8 pairs)
Cervical enlargement

First thoracic vertebra

THORACIC NERVES (12 pairs)

Lumbar enlargement

First lumbar vertebra
Conus medullaris

LUMBAR NERVES (5 pairs)
Cauda equina

Ilium of hip bone

Sacrum
SACRAL NERVES (5 pairs)

COCCYGEAL NERVES (1 pair)

C1 C2 C3 C4 C5 C6 C7 C8 T1 T2 T3 T4 T5 T6 T7 T8 T9 T10 T11 T12 L1 L2 L3 L4 L5 S1 S2 S3 S4 S5

(a) Posterior view of entire spinal cord and portions of spinal nerves

● FIGURE **18.2** CONTINUES ▶

● FIGURE **18.2** CONTINUED ▶

SUPERIOR

Fourth ventricle

Cerebellum of brain (cut)

Glossopharyngeal (IX) and vagus (X) nerves

Accessory (XI) nerve

Occipital bone (cut)

Gracile fasciculus

Posterior median sulcus

Cuneate fasciculus

Vertebral artery

Denticulate ligament

Dura mater and arachnoid mater (cut)

Posterior (dorsal) rootlets of spinal nerve

INFERIOR

(b) Posterior view of cervical region of spinal cord

SUPERIOR

Conus medullaris

Posterior (dorsal) rami of spinal nerves

Cauda equina

Sacrum

Gluteus maximus

Filum terminale

Right coccygeal nerve

INFERIOR

(c) Posterior view of inferior portion of spinal cord

? **What portion of the spinal cord connects with sensory and motor nerves of the upper limbs?**

In a **spinal tap (lumbar puncture)**, a local anesthetic is given, and a long needle is inserted into the subarachnoid space. During this procedure, the patient lays on the side with the vertebral column flexed. Flexion of the vertebral column increases the distance between the spinous processes of the vertebrae, which allows easy access to the subarachnoid space. The spinal cord ends around the second lumbar vertebra (L2); however, the spinal meninges extend to the second sacral vertebra (S2). Between vertebrae L2 and S2 the spinal meninges are present, but the spinal cord is absent. Consequently, a spinal tap is normally performed in adults between the L3 and L4 or L4 and L5 lumbar vertebrae because this region provides safe access to the subarachnoid space without the risk of damaging the spinal cord. (A line drawn across the highest points of the iliac crests, called the *supracristal line*, passes through the spinous process of the fourth lumbar vertebra.) A spinal tap is used to withdraw cerebrospinal fluid (CSF) for diagnostic purposes; to introduce antibiotics, contrast media for myelography, or anesthetics; to administer chemotherapy; to measure CSF pressure; and/or to evaluate the effects of treatment for diseases such as meningitis. •

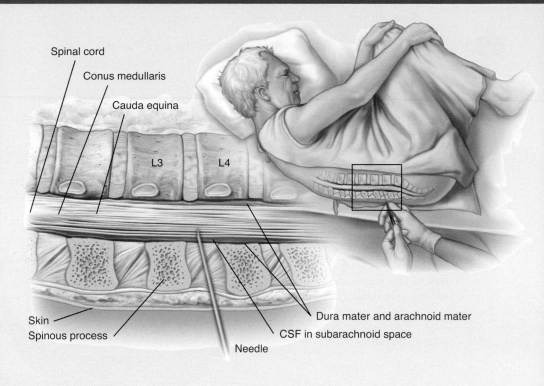

Spinal cord
Conus medullaris
Cauda equina
L3   L4
Skin
Spinous process
Needle
Dura mater and arachnoid mater
CSF in subarachnoid space

(KŌ-nus med-ū-LAR-is; *conus*=cone). The conus medullaris ends at the level of the intervertebral disc between the first and second lumbar vertebrae in adults (Figure 18.2a, c). Arising from the conus medullaris is the **filum terminale** (FĪ-lum ter-mi-NAL-ē=terminal filament), an extension of the pia mater that extends inferiorly and blends with the arachnoid mater and dura mater to anchor the spinal cord to the coccyx.

Because the spinal cord is shorter than the vertebral column, nerves that arise from the lumbar, sacral, and coccygeal regions of the spinal cord do not leave the vertebral column at the same level they exit the cord. The roots of these spinal nerves angle inferiorly in the vertebral canal from the end of the spinal cord like wisps of hair. Appropriately, the roots of these nerves are collectively named the **cauda equina** (KAW-da ē-KWĪ-na), meaning "horse's tail" (Figure 18.2a, c).

Internal Anatomy of the Spinal Cord

Two grooves penetrate the white matter of the spinal cord and divide it into right and left sides (Figure 18.3). The **anterior median fissure** is a wide groove on the anterior (ventral) side. The **posterior median sulcus** is a narrow groove on the posterior (dorsal) side. The gray matter of the spinal cord is shaped like the letter H or a butterfly and is surrounded by white matter. The gray matter consists primarily of the cell bodies of neurons, neuroglia, unmyelinated axons, and the dendrites of interneurons and motor neurons. The white matter consists of bundles of myelinated axons of sensory neurons, interneurons, and motor

neurons. The **gray commissure** (KOM-mi-shur) forms the crossbar of the H. In the center of the gray commissure is a small space called the **central canal**; it extends the entire length of the spinal cord and contains cerebrospinal fluid. At its superior end, the central canal is continuous with the fourth ventricle (a space that also contains cerebrospinal fluid) in the medulla oblongata of the brain. Anterior to the gray commissure is the **anterior (ventral) white commissure**, which connects the white matter of the right and left sides of the spinal cord.

In the gray matter of the spinal cord and brain, clusters of neuronal cell bodies form functional groups called **nuclei**. *Sensory nuclei* receive input from receptors via sensory neurons, and *motor nuclei* provide output to effector tissues via motor neurons. The gray matter on each side of the spinal cord is subdivided into regions called **horns**. The **anterior (ventral) gray horns** contain *somatic motor nuclei*, which are clusters of cell bodies of somatic motor neurons that provide nerve impulses for contraction of skeletal muscles. The **posterior (dorsal) gray horns** contain cell bodies and axons of interneurons as well as axons of incoming sensory neurons. Between the anterior and posterior gray horns are the **lateral gray horns**, which are present only in the thoracic, upper lumbar, and sacral segments of the spinal cord. The lateral horns contain the cell bodies of *autonomic motor nuclei* that regulate activity of smooth muscle, cardiac muscle, and glands.

The white matter, like the gray matter, is organized into regions. The anterior and posterior gray horns divide the white matter on each side into three broad areas called **columns**: (1) **anterior (ventral) white columns**, (2) **posterior (dorsal) white**

Figure 18.3 Internal anatomy of the spinal cord: the organization of gray matter and white matter. For simplicity, dendrites are not shown in this and several other illustrations of transverse sections of the spinal cord. Blue and red arrows in (a) indicate the direction of nerve impulse propagation.

In the spinal cord, white matter surrounds the gray matter.

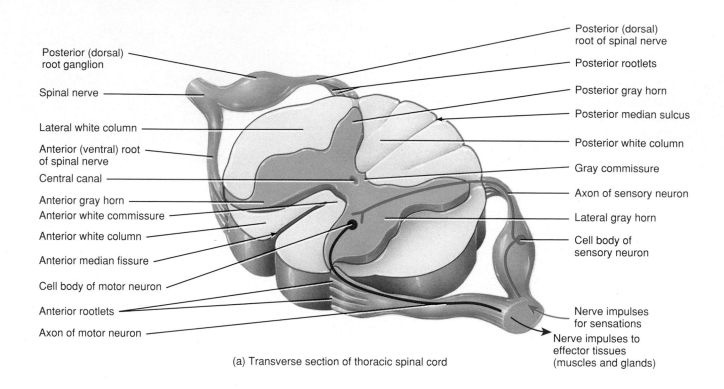

(a) Transverse section of thoracic spinal cord

Labels (left, top to bottom):
Posterior (dorsal) root ganglion
Spinal nerve
Lateral white column
Anterior (ventral) root of spinal nerve
Central canal
Anterior gray horn
Anterior white commissure
Anterior white column
Anterior median fissure
Cell body of motor neuron
Anterior rootlets
Axon of motor neuron

Labels (right, top to bottom):
Posterior (dorsal) root of spinal nerve
Posterior rootlets
Posterior gray horn
Posterior median sulcus
Posterior white column
Gray commissure
Axon of sensory neuron
Lateral gray horn
Cell body of sensory neuron
Nerve impulses for sensations
Nerve impulses to effector tissues (muscles and glands)

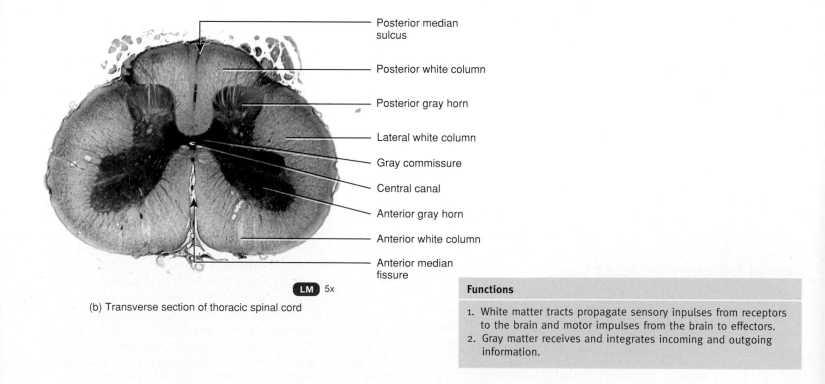

(b) Transverse section of thoracic spinal cord

LM 5x

Labels:
Posterior median sulcus
Posterior white column
Posterior gray horn
Lateral white column
Gray commissure
Central canal
Anterior gray horn
Anterior white column
Anterior median fissure

Functions

1. White matter tracts propagate sensory impulses from receptors to the brain and motor impulses from the brain to effectors.
2. Gray matter receives and integrates incoming and outgoing information.

? What is the difference between a horn and a column in the spinal cord?

columns, and (3) **lateral white columns**. Each column in turn contains distinct bundles of axons having a common origin or destination and carrying similar information. These bundles, which may extend long distances up or down the spinal cord, are called **tracts**. **Sensory (ascending) tracts** consist of axons that conduct nerve impulses from the spinal cord toward the brain. Tracts consisting of axons that carry nerve impulses away from the brain down the spinal cord are called **motor (descending) tracts**. Sensory and motor tracts of the spinal cord are continuous with sensory and motor tracts in the brain.

The white matter and the gray matter exhibit differences depending on the region of the spinal cord (Table 18.1). There is more white matter at the cranial end of the spinal cord than at the caudal end. Because all ascending tracts are headed toward the brain, the ascending tracts become thicker as they progress from caudal to cranial. When you think about it, it makes sense that the descending tracts are thickest in the cranial region as well. The gray matter, especially the ventral horn, is the largest at lower cervical levels and at lower lumbar upper sacral levels. These levels correspond to upper and lower limb anatomy respectively, where large amounts of muscle tissue require motor innervation.

◉ **CHECKPOINT**

1. Where are the spinal meninges located? Where are the epidural, subdural, and subarachnoid spaces located?
2. What are the cervical and lumbar enlargements?
3. Define conus medullaris, filum terminale, and cauda equina.
4. What is a spinal segment? How is the spinal cord partially divided into right and left sides?
5. Define each of the following terms: gray commissure, central canal, anterior gray horn, lateral gray horn, posterior gray horn, anterior white column, lateral white column, posterior white column, ascending tract, descending tract.

TABLE 18.1

Comparison of Various Spinal Cord Segments

SEGMENT		DISTINGUISHING CHARACTERISTICS
Cervical	 (Segment C6)	Relatively large diameter, relatively large amounts of white matter, oval in shape; in upper cervical segments (C1–C4), posterior gray horn is large, but anterior gray horn is relatively small; in lower cervical segments (C5 and below), posterior gray horns are enlarged and anterior gray horns are well-developed.
Thoracic	 (Segment T5)	Small diameter is due to relatively small amounts of gray matter; except for first thoracic segment, anterior and posterior gray horns are relatively small; a small lateral gray horn is present.
Lumbar	 (Segment L4)	Nearly circular; very large anterior and posterior gray horns; relatively less white matter than cervical segments.
Sacral	 (Segment S3)	Relatively small, but with relatively large amounts of gray matter; relatively small amounts of white matter; anterior and posterior gray horns are large and thick.
Coccygeal		Resemble lower sacral spinal segments, but much smaller.

•CLINICAL **CONNECTION** | Spinal Cord Injury

Most **spinal cord injuries** are due to trauma as a result of factors such as automobile accidents, falls, contact sports, diving, and acts of violence. The effects of the injury depend on the extent of direct trauma to the spinal cord or compression of the cord by fractured or displaced vertebrae or blood clots. Although any segment of the spinal cord may be involved, most common sites of injury are in the cervical, lower thoracic, and upper lumbar regions. Depending on the location and extent of spinal cord damage, paralysis may occur. **Monoplegia** (*mono-*=one; *-plegia*=blow or strike) is paralysis of one limb only. **Diplegia** (*di-*=two) is paralysis of both upper limbs or both lower limbs. **Paraplegia** (*para-*= beyond) is paralysis of both lower limbs. **Hemiplegia** (*hemi-*=half) is paralysis of the upper limb, trunk, and lower limb on one side of the body, and **quadriplegia** (*quad-*=four) is paralysis of all four limbs.

Complete transection (tran-SEK-shun; *trans-*=across; *-section*=a cut) of the spinal cord means that the cord is severed from one side to the other, thus cutting all sensory and motor tracts. It results in a loss of all sensations and voluntary movement below the level of the transection. A person will have permanent loss of all sensations in dermatomes below the injury because ascending nerve impulses cannot propagate past the transection to reach the brain. At the same time, all voluntary muscle contractions will be lost below the transection because nerve impulses descending from the brain also cannot pass. The extent of paralysis of skeletal muscles depends on the level of injury. The following list outlines which muscle functions may be *retained* at progressively lower levels of spinal cord transection.

- C1–C3: no function maintained from the neck down; ventilator needed for breathing
- C4–C5: diaphragm, which allows breathing
- C6–C7: some arm and chest muscles, which allows feeding, some dressing, and propelling wheelchair
- T1–T3: intact arm function
- T4–T9: control of trunk above the umbilicus
- T10–L1: most thigh muscles, which allows walking with long leg braces
- L1–L2: most leg muscles, which allows walking with short leg braces

Hemisection is a partial transection of the cord on either the right or the left side. After hemisection, three main symptoms, known together as *Brown-Séquard syndrome* (sē-KAR), occur below the level of the injury: (1) Damage of the posterior column causes loss of proprioception and fine touch sensations on the *ipsilateral* (same) side as the injury. (2) Damage of the lateral corticospinal tract causes ipsilateral paralysis. (3) Damage of the spinothalamic tracts causes loss of pain and temperature sensations on the *contralateral* (opposite) side.

Following complete transection, and to varying degrees after hemisection, spinal shock occurs. **Spinal shock** is an immediate response to spinal cord injury characterized by temporary **areflexia** (ā'-rē-FLEK-sē-a), loss of reflex function. The areflexia occurs in parts of the body served by spinal nerves below the level of the injury. Signs of acute spinal shock include slow heart rate, low blood pressure, flaccid paralysis of skeletal muscles, loss of somatic sensations, and urinary bladder dysfunction. Spinal shock may begin within 1 hour after injury and may last from several minutes to several months, after which reflex activity gradually returns.

In many cases of traumatic injury of the spinal cord, the patient may have an improved outcome if an anti-inflammatory corticosteroid drug called methylprednisolone is given within 8 hours of the injury. This is because the degree of neurologic deficit is greatest immediately following traumatic injury as a result of edema (collection of fluid within tissues) as the immune system responds to the injury. •

SPINAL NERVES

⦿ OBJECTIVES

- Describe the components, connective tissue coverings, and branching of a spinal nerve.
- Define a plexus, and identify the distribution of nerves of the cervical, brachial, lumbar, and sacral plexuses.
- Describe the clinical significance of dermatomes.

Spinal nerves and the nerves that branch from them are part of the peripheral nervous system (PNS). They connect the CNS to sensory receptors, muscles, and glands in all parts of the body. There are 31 pairs of spinal nerves. The spinal cord appears to be segmented because the 31 pairs of spinal nerves emerge at regular intervals from intervertebral foramina (see Figure 18.2a). Indeed, each pair of spinal nerves is said to arise from a *spinal segment*. Within the spinal cord there is no obvious segmentation but, for convenience, the naming of spinal nerves is based on the segment in which they are located. There are 8 pairs of *cervical nerves* (represented in Figure 18.2a as C1–C8), 12 pairs of *thoracic nerves* (T1–T12), 5 pairs of *lumbar nerves* (L1–L5), 5 pairs of *sacral nerves* (S1–S5), and 1 pair of *coccygeal nerves* (Co1). The first cervical pair emerges between the atlas (first cervical vertebra) and the occipital bone. All other spinal nerves emerge from the vertebral column through the intervertebral foramina between adjoining vertebrae.

Not all spinal cord segments are aligned with their corresponding vertebrae. Recall that the spinal cord ends near the level of the superior border of the second lumbar vertebra, and that the roots of the lumbar, sacral, and coccygeal nerves descend at an angle to reach their respective foramina before emerging from the vertebral column. This arrangement constitutes the cauda equina (see Figure 18.2a, c).

Distribution of Spinal Nerves

The design of a spinal nerve is like that of a tree. Beneath the ground, a tree consists of tiny rootlets that converge to form large roots. These large roots come together at the surface to form the trunk, which then divides into numerous large branches that fork into smaller and smaller branches. Spinal nerves follow the same pattern and use much of the same terminology (see Figures 18.3 and 18.4).

The spinal nerves arise from the spinal cord as a series of small **rootlets**, of which there are two types: *anterior (ventral) rootlets* and *posterior (dorsal) rootlets*. From the anterolateral aspect of the cord, anterior rootlets emerge in two or three irregular rows. The anterior rootlets contain the axons of multipolar motor neurons arising from cell bodies in the anterior regions of the spinal cord gray matter. These axons transmit action potentials to the muscles and glands of the body. Projecting from the *posterolateral sulcus* of the spinal cord is another series of rootlets, the

Figure 18.4 Branches of a typical spinal nerve. (See also Figure 18.1b.)

🔑 ⊏⊐ The principal branches of a spinal nerve are the posterior (dorsal) ramus, the anterior (ventral) ramus, the meningeal branch, and the rami communicantes.

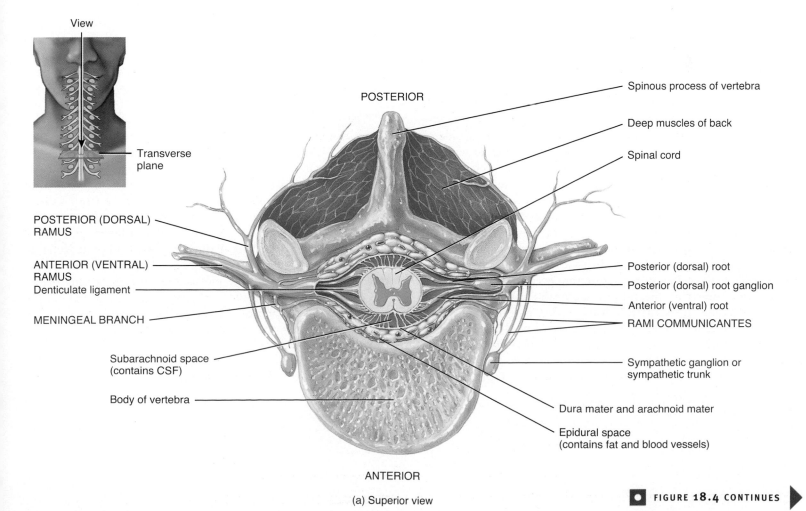

View

Transverse plane

POSTERIOR

Spinous process of vertebra

Deep muscles of back

Spinal cord

POSTERIOR (DORSAL) RAMUS

ANTERIOR (VENTRAL) RAMUS

Denticulate ligament

MENINGEAL BRANCH

Subarachnoid space (contains CSF)

Body of vertebra

Posterior (dorsal) root

Posterior (dorsal) root ganglion

Anterior (ventral) root

RAMI COMMUNICANTES

Sympathetic ganglion or sympathetic trunk

Dura mater and arachnoid mater

Epidural space (contains fat and blood vessels)

ANTERIOR

(a) Superior view

⬛ **FIGURE 18.4 CONTINUES** ▶

FIGURE 18.4 CONTINUED ▶

(b) Anterior view and oblique section of spinal cord

? **Which spinal nerve branches serve the upper and lower limbs?**

posterior rootlets, which contain the central processes of the sensory unipolar neurons. These neurons transmit action potentials from peripheral receptor organs to the central nervous system.

Branches of Spinal Nerves

Each series of anterior rootlets converge to form larger **anterior (ventral) roots**. Likewise each series of posterior rootlets converge to form larger **posterior (dorsal) roots**. Each posterior root has a swelling, the **posterior (dorsal) root ganglion**, which contain the cell bodies of sensory neurons. The anterior and posterior roots form an anterior and posterior pair on each side of the spinal cord that corresponds to one developmental segment or level of the body. As the sensory posterior root and motor anterior root project laterally from the spinal cord they converge to form a **mixed nerve**, called the **spinal nerve trunk**. (A mixed nerve contains both motor and sensory axons.) The spinal nerve trunk runs for a short distance before branching into two large branches and a variable series of smaller branches (Figure 18.4). Each large branch, named a **ramus** [(=branch; plural is **rami** (RĀ-mī)], follows a specific course to different peripheral regions. The two largest branches, the anterior (ventral) ramus and posterior (dorsal) ramus, are somatic branches that run in the musculoskeletal wall of the body. The **posterior (dorsal) ramus** serves the deep muscles and skin of the posterior surface of the trunk. The **anterior (ventral) ramus** serves the muscles and structures of the upper and lower limbs and the muscles and skin of the

lateral and anterior regions of the trunk. Smaller visceral branches, such as the meningeal branch and the communicating rami, form the autonomic pathways to smooth muscle and glandular tissue. The **meningeal branch** reenters the vertebral cavity through the intervertebral foramen and supplies the vertebrae, vertebral ligaments, blood vessels of the spinal cord, and meninges. The communicating rami or **rami communicantes** (kō-mū-ni-KAN-tēz) are components of the autonomic nervous system and will be discussed in Chapter 20.

Plexuses

Axons from the anterior rami of spinal nerves, except for thoracic nerves T2–T12, do not go directly to the body structures they supply. Instead, they form networks on both the left and right sides of the body by joining with various numbers of axons from anterior rami of adjacent nerves. Such a network of axons is called a **plexus** (=braid or network). The principal plexuses are the **cervical plexus, brachial plexus, lumbar plexus**, and **sacral plexus**. A smaller **coccygeal plexus** is also present. Refer to Figure 18.2 to see the relationships of the plexuses to one another. Emerging from the plexuses are nerves bearing names that are often descriptive of the general regions they serve or the course they take. Each of the nerves in turn may have several branches named for the specific structures they innervate.

Exhibits 18.1–18.4 (Figures 18.5–18.8) summarize the principal plexuses. The anterior rami of spinal nerves T2–T12 are called intercostal nerves and will be discussed next.

EXHIBIT 18.1 Cervical Plexus

OBJECTIVE

- Describe the origin and distribution of the cervical plexus.

The **cervical plexus** (SER-vi-kul) is formed by the roots (anterior rami) of the first four cervical nerves (C1–C4), with contributions from C5 (Figure 18.5). There is one on each side of the neck alongside the first four cervical vertebrae.

The cervical plexus supplies the skin and muscles of the head, neck, and superior part of the shoulders and chest. The phrenic nerves arise from the cervical plexuses and supply motor fibers to the diaphragm. Branches of the cervical plexus also run parallel to two cranial nerves, the accessory (XI) nerve and hypoglossal (XII) nerve.

NERVE	ORIGIN	DISTRIBUTION
SUPERFICIAL (SENSORY) BRANCHES		
Lesser occipital	C2	Skin of scalp posterior and superior to ear.
Great auricular (aw-RIK-ū-lar)	C2–C3	Skin anterior, inferior, and over ear, and over parotid glands.
Transverse cervical	C2–C3	Skin over anterior aspect of neck.
Supraclavicular	C3–C4	Skin over superior portion of chest and shoulder.
DEEP (LARGELY MOTOR) BRANCHES		
Ansa cervicalis (AN-sa ser-vi-KAL-is)		This nerve divides into superior and inferior roots.
Superior root	C1	Infrahyoid and geniohyoid muscles of neck.
Inferior root	C2–C3	Infrahyoid muscles of neck.
Phrenic (FREN-ik)	C3–C5	Diaphragm.
Segmental branches	C1–C5	Prevertebral (deep) muscles of neck, levator scapulae, and middle scalene muscles.

Figure 18.5 **Cervical plexus in anterior view.** (See Tortora, *A Photographic Atlas of the Human Body*, 2e, Figure 8.7.)

The cervical plexus supplies the skin and muscles of the head, neck, superior portion of the shoulders and chest, and diaphragm.

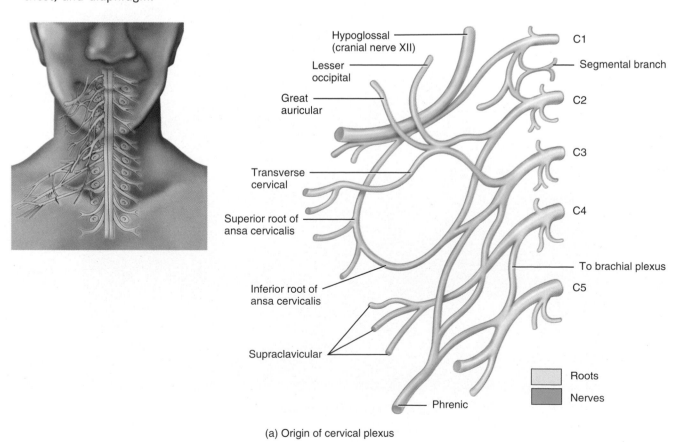

(a) Origin of cervical plexus

EXHIBIT 18.1 CONTINUES ▶

EXHIBIT 18.1   **595**

EXHIBIT 18.1   Cervical Plexus CONTINUED

● CLINICAL **CONNECTION** | Injuries to the Phrenic Nerves

The phrenic nerves originate from C3, C4, and C5 and supply the diaphragm. Complete severing of the spinal cord above the origin of the phrenic nerves causes respiratory arrest. Breathing stops because the phrenic nerves no longer send impulses to the diaphragm. The phrenic nerves may also be paralyzed due to pressure from malignant tumors in the mediastinum ●

◉ CHECKPOINT
- Which nerve from the cervical plexus causes contraction of the diaphragm?

◉ FIGURE **18.5** CONTINUED ▶

Lesser occipital

Great auricular

Transverse cervical

Nerve to sterno-cleidomastoid

Nerves to trapezius

Supraclavicular

Nerve to inferior belly of omohyoid

Hypoglossal (XII)

Nerve to geniohyoid

Nerve to thyrohyoid

Nerve to sternohyoid

Nerve to superior belly of omohyoid

Superior root of ansa cervicalis

Inferior root of ansa cervicalis

Nerves to sternothyroid

Vagus (X)

Phrenic

(b) Anterior view of cervical plexus

❓ Why does complete severing of the spinal cord at level C2 cause respiratory arrest?

EXHIBIT 18.2   Brachial Plexus

OBJECTIVE

• Describe the origin, distribution, and effects of damage to the brachial plexus.

The roots (anterior rami) of spinal nerves C5–C8 and T1 form the **brachial plexus** (BRĀ-kē-al), which extends inferiorly and laterally on either side of the last four cervical and first thoracic vertebrae (Figure 18.6a). It passes above the first rib posterior to the clavicle and then enters the axilla.

Since the brachial plexus is so complex, an explanation of its various parts is helpful. As with the cervical and other plexuses, the **roots** are the anterior rami of the spinal nerves. The roots of several spinal nerves unite to form **trunks** in the inferior part of the neck. These are the *superior*, *middle*, and *inferior trunks*. Posterior to the clavicles, the trunks divide into **divisions** called the *anterior* and *posterior divisions*. In the axillae, the divisions unite to form **cords** called the *lateral*, *medial*, and *posterior cords*. The cords are named for their relationship to the axillary artery, a large artery that supplies blood to the upper limb. The principal **nerves** of the brachial plexus branch from the cords.

The brachial plexus provides almost the entire nerve supply of the shoulders and upper limbs (Figure 18.6b). Five important nerves arise from the brachial plexus: (1) The **axillary nerve** supplies the deltoid and teres minor muscles. (2) The **musculocutaneous nerve** supplies the anterior muscles of the arm. (3) The **radial nerve** supplies the muscles on the posterior aspect of the arm and forearm. (4) The **median nerve** supplies most of the muscles of the anterior forearm (6½ out of 8) and some of the muscles of the hand. (5) The **ulnar nerve** supplies the anteromedial muscles of the forearm (the other 1½) and most of the muscles of the hand.

Figure 18.6   **Brachial plexus in anterior view.** (See Tortora, *A Photographic Atlas of the Human Body*, 2e, Figure 8.9.)

The brachial plexus supplies the shoulders and upper limbs.

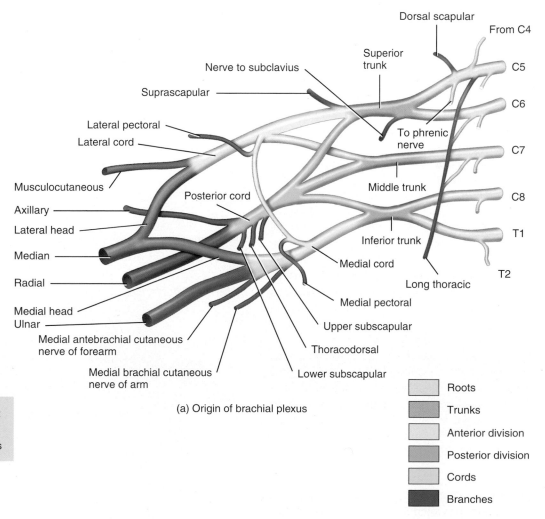

(a) Origin of brachial plexus

Roots
Trunks
Anterior division
Posterior division
Cords
Branches

MNEMONIC for subunits of the brachial plexus:
Risk Takers Don't Cautiously Behave.
Roots, Trunks, Divisions, Cords, Branches

EXHIBIT 18.2 CONTINUES ▶

EXHIBIT 18.2   **597**

EXHIBIT 18.2   Brachial Plexus CONTINUED

● CLINICAL CONNECTION
Injuries to Nerves Emerging from the Brachial Plexus

Injury to the superior roots of the brachial plexus (C5–C6) may result from forceful pulling away of the head from the shoulder, as might occur from a heavy fall on the shoulder or excessive stretching of an infant's neck during childbirth. The presentation of this injury is characterized by an upper limb in which the shoulder is adducted, the arm is medially rotated, the elbow is extended, the forearm is pronated, and the wrist is flexed. This condition is called **Erb-Duchenne palsy** or **waiter's tip position**. There is loss of sensation along the lateral side of the arm.

Radial (and axillary) **nerve injury** can be caused by improperly administered intramuscular injections into the deltoid muscle. The radial nerve may also be injured when a cast is applied too tightly around the mid-humerus. Radial nerve injury is indicated by **wrist drop**, the inability to extend the wrist and fingers. Sensory loss is minimal due to the overlap of sensory innervation by adjacent nerves.

Median nerve injury may result in **median nerve palsy**, which is indicated by numbness, tingling, and pain in the palm and fingers. There is also inability to pronate the forearm and flex the proximal interphalangeal joints of all digits and the distal interphalangeal joints of the second and third digits. In addition, wrist flexion is weak and is accompanied by adduction, and thumb movements are weak.

Ulnar nerve injury may result in **ulnar nerve palsy**, which is indicated by an inability to abduct or adduct the fingers, atrophy of the interosseus muscles of the hand, hyperextension of the metacarpophalangeal joints, and flexion of the interphalangeal joints, a condition called **clawhand**. There is also loss of sensation over the little finger.

Long thoracic nerve injury results in paralysis of the serratus anterior muscle. The medial border of the scapula protrudes, giving it the appearance of a wing. When the arm is raised, the vertebral border and inferior angle of the scapula pull away from the thoracic wall and protrude outward, causing the medial border of the scapula to protrude; because the scapula looks like a wing, this condition is called **winged scapula**. The arm cannot be abducted beyond the horizontal position. •

◉ CHECKPOINT
- Injury of which nerve could cause paralysis of the serratus anterior muscle?

● FIGURE 18.6 CONTINUED

(b) Distribution of nerves from the brachial plexus

NERVE	ORIGIN	DISTRIBUTION
Dorsal scapular (SKAP-ū-lar)	C5	Levator scapulae, rhomboid major, and rhomboid minor muscles.
Long thoracic (thor-RAS-ik)	C5–C7	Serratus anterior muscle.
Nerve to subclavius (sub-KLĀ-vē-us)	C5–C6	Subclavius muscle.
Suprascapular	C5–C6	Supraspinatus and infraspinatus muscles.
Musculocutaneous (mus'-kū-lō-kū-TĀN-ē-us)	C5–C7	Coracobrachialis, biceps brachii, and brachialis muscles.
Lateral pectoral (PEK-to-ral)	C5–C7	Pectoralis major muscle.
Upper subscapular	C5–C6	Subscapularis muscle.
Thoracodorsal (tho-RĀ-kō-dor-sal)	C6–C8	Latissimus dorsi muscle.
Lower subscapular	C5–C6	Subscapularis and teres major muscles.
Axillary (AK-si-lar-ē)	C5–C6	Deltoid and teres minor muscles; skin over deltoid and superior posterior aspect of arm.
Median	C5–T1	Flexors of forearm (6½), except flexor carpi ulnaris and the ulnar half of the flexor digitorum profundus, and some muscles of the hand (5 muscles of the lateral palm; thin ½ LOAF, which stands for the first and second lumbricals, opponens pollicis, abductor pollicis brevis, and flexor pollicis brevis); skin of lateral two-thirds of palm of hand and fingers.
Radial		Triceps brachii, anconeus, and extensor muscles of forearm; skin of posterior arm and forearm, lateral two-thirds of dorsum of hand, and fingers over proximal and middle phalanges.
Medial pectoral	C8–T1	Pectoralis major and pectoralis minor muscles.
Medial cutaneous nerve of arm (kū'-TĀ-nē-us)	C8–T1	Skin of medial and posterior aspects of distal third of arm.
Medial cutaneous nerve of forearm	C8–T1	Skin of medial and posterior aspects of forearm.
Ulnar	C8–T1	Flexor carpi ulnaris, ulnar half of the flexor digitorum profundus (1½), and most muscles of the hand; skin of medial side of hand, little finger, and medial half of ring finger.

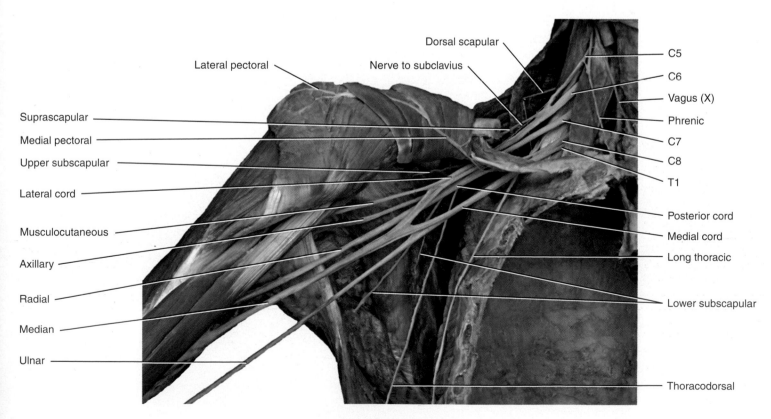

(c) Anterior view of brachial plexus

? What five important nerves arise from the brachial plexus?

EXHIBIT 18.2 **599**

EXHIBIT 18.3   Lumbar Plexus

● **OBJECTIVE**

• Describe the origin and distribution of the lumbar plexus.

The roots (anterior rami) of spinal nerves L1–L4 form the **lumbar plexus** (LUM-bar) (Figure 18.7). Unlike the brachial plexus, there is minimal intricate intermingling of fibers in the lumbar plexus. On either side of the first four lumbar vertebrae, the lumbar plexus passes obliquely outward, between the superficial and deep heads of the psoas major muscle and anterior to the quadratus lumborum muscle. It then gives rise to its peripheral nerves.

NERVE	ORIGIN	DISTRIBUTION
Iliohypogastric (Il'-ē-ō-hī-pō-GAS-trik)	L1	Muscles of anterolateral abdominal wall; skin of inferior abdomen and buttock.
Ilioinguinal (Il'-ē-ō-IN-gwi-nal)	L1	Muscles of anterolateral abdominal wall; skin of superior medial aspect of thigh, root of penis and scrotum in male, and labia majora and mons pubis in female.
Genitofemoral (jen'-i-tō-FEM-or-al)	L1–L2	Cremaster muscle; skin over middle anterior surface of thigh, scrotum in male, and labia majora in female.
Lateral cutaneous nerve of thigh	L2–L3	Skin over lateral, anterior, and posterior aspects of thigh.
Femoral	L2–L4	This is the largest nerve arising from the lumbar plexus. It is distributed to the flexor muscles of hip joint and extensor muscles of knee joint; skin over anterior and medial aspect of thigh and medial side of leg and foot.
Obturator (OB-too-rā-tor)	L2–L4	Adductor muscles of hip joint; skin over medial aspect of thigh.

Figure 18.7   **Lumbar plexus in anterior view.**

🔑 ▭▭ The lumbar plexus supplies the anterolateral abdominal wall, external genitals, and part of the lower limbs.

From T12

L1

Iliohypogastric

L2

Ilioinguinal

Genitofemoral

L3

Lateral cutaneous nerve of thigh

L4

L5

Femoral

Accessory obturator (present in only about 10% of population)

Lumbosacral trunk

Obturator

	Roots
	Anterior division
	Posterior division

(a) Origin of lumbar plexus

The lumbar plexus supplies the anterolateral abdominal wall, external genitals, and part of the lower limbs.

◉ CHECKPOINT
- What structures are supplied by the lumbar plexus?

● CLINICAL **CONNECTION** | Lumbar Plexus Injuries

The largest nerve arising from the lumbar plexus is the femoral nerve. **Femoral nerve injury,** which can occur in stab or gunshot wounds, is indicated by an inability to extend the leg and by loss of sensation in the skin over the anteromedial aspect of the thigh.

Obturator nerve injury results in paralysis of the adductor muscles of the thigh and loss of sensation over the medial aspect of the thigh. It may result from pressure on the nerve by the fetal head during pregnancy.   ●

L2
L3
L4

Hip bone

Sacrum

Obturator nerve

Pudendal nerve

Femoral nerve

Sciatic nerve

Femur

Tibial nerve

Common fibular nerve

Fibula

Tibia

Deep fibular nerve

Superficial fibular nerve

Tibial nerve

Medial plantar nerve

Lateral plantar nerve

Anterior view

Posterior view

(b) Distribution of nerves from the lumbar and sacral plexuses

Iliohypogastric

Ilioinguinal

Subcostal

Lateral cutaneous nerve of thigh

Nerve to iliacus

Femoral

Lateral cutaneous nerve of thigh

Sympathetic trunk

Genitofemoral

Iliohypogastric

Ilioinguinal

(c) Anterior view of lumbar plexus

❓ What is the origin of the lumbar plexus?

EXHIBIT 18.3   **601**

EXHIBIT 18.4   Sacral and Coccygeal Plexuses

◉ **OBJECTIVE**

- Describe the origin and distribution of the sacral and coccygeal plexuses.

The roots (anterior rami) of spinal nerves L4–L5 and S1–S4 form the **sacral plexus** (SĀ-kral) (Figure 18.8). This plexus is situated largely anterior to the sacrum. The sacral plexus supplies the buttocks, perineum, and lower limbs. The largest nerve in the body—the sciatic nerve—arises from the sacral plexus.

The roots (anterior rami) of spinal nerves S4–S5 and the coccygeal nerves form a small **coccygeal plexus**, which supplies a small area of skin in the coccygeal region.

● **CLINICAL CONNECTION** | **Sciatic Nerve Injury**

The most common form of back pain is caused by compression or irritation of the sciatic nerve, the longest nerve in the human body. Injury to the sciatic nerve and its branches results in **sciatica**, pain that may extend from the buttock down the posterior and lateral aspect of the leg and the lateral aspect of the foot. The sciatic nerve may be injured because of a herniated (slipped) disc, dislocated hip, osteoarthritis of the lumbosacral spine, pressure from the uterus during pregnancy, inflammation, irritation, or an improperly administered gluteal intramuscular injection.

In the majority of sciatic nerve injuries, the common fibular portion is the most affected, frequently from fractures of the fibula or by pressure from casts or splints. Damage to the common fibular nerve causes the foot to be plantar flexed, a condition called **footdrop**, and inverted, a condition called **equinovarus**. There is also loss of function along the anterolateral aspects of the leg and dorsum of the foot and toes. Injury to the tibial portion of the sciatic nerve results in dorsiflexion of the foot plus eversion, a condition called **calcaneovalgus**. Loss of sensation on the sole also occurs. Treatments for sciatica are similar to those outlined earlier for a herniated (slipped) disc—rest, pain medications, exercises, ice or heat, and massage. ●

◉ **CHECKPOINT**

- Injury of which nerve causes foot drop?

Figure 18.8 Sacral plexus in anterior view. The distribution of the nerves of the sacral plexus is shown in Figure 18.11b. (See Tortora, *A Photographic Atlas of the Human Body*, 2e, Figure 8.11.)

🔑 The sacral plexus supplies the buttocks, perineum, and lower limbs.

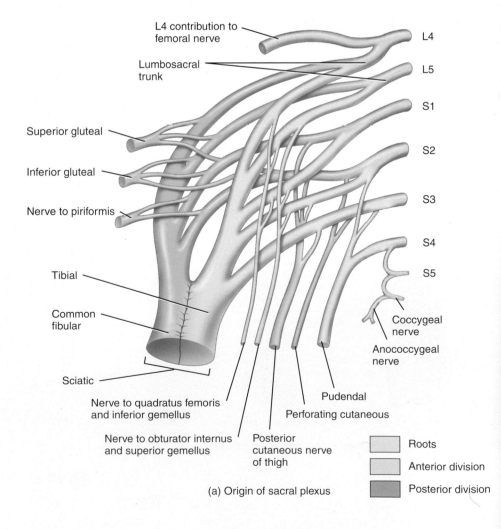

(a) Origin of sacral plexus

Roots
Anterior division
Posterior division

NERVE	ORIGIN	DISTRIBUTION
Superior gluteal (GLOO-tē-al)	L4–L5 and S1	Gluteus minimus and gluteus medius muscles and tensor fasciae latae.
Inferior gluteal	L5–S2	Gluteus maximus muscle.
Nerve to piriformis (pir-i-FORM-is)	S1–S2	Piriformis muscle.
Nerve to quadratus femoris (quad-RĀ-tus FEM-or-is) and inferior gemellus (jem-EL-us)	L4–L5 and S1	Quadratus femoris and inferior gemellus muscles.
Nerve to obturator internus (OB-too-rā′-tor in-TER-nus) and superior gemellus	L5–S2	Obturator internus and superior gemellus muscles.
Perforating cutaneous (kū′-TĀ-ne-us)	S2–S3	Skin over inferior medial aspect of buttock.
Posterior cutaneous nerve of thigh	S1–S3	Skin over anal region, inferior lateral aspect of buttock, superior posterior aspect of thigh, superior part of calf, scrotum in male, and labia majora in female.
Sciatic (sī-AT-ik)	L4–S3	Actually two nerves—tibial and common fibular—bound together by common sheath of connective tissue. It splits into its two divisions, usually at the knee. (See below for distributions.) As the sciatic nerve descends through the thigh, it sends branches to hamstring muscles and the adductor magnus.
Tibial (TIB-ē-al)	L4–S3	Gastrocnemius, plantaris, soleus, popliteus, tibialis posterior, flexor digitorum longus, and flexor hallucis longus muscles. Branches of tibial nerve in foot are medial plantar nerve and lateral plantar nerve.
Medial plantar (PLAN-tar) (see Figure 18.7b)		Abductor hallucis, flexor digitorum brevis, and flexor hallucis brevis muscles; skin over medial two-thirds of plantar surface of foot.
Lateral plantar (see Figure 18.7b)		Remaining muscles of foot not supplied by medial plantar nerve; skin over lateral third of plantar surface of foot.
Common fibular (FIB-ū-lar)	L4–S2	Divides into a superficial fibular and a deep fibular branch.
Superficial fibular		Fibularis longus and fibularis brevis muscles; skin over distal third of anterior aspect of leg and dorsum of foot.
Deep fibular		Tibialis anterior, extensor hallucis longus, fibularis tertius, and extensor digitorum longus and extensor digitorum brevis muscles; skin on adjacent sides of great and second toes.
Pudendal (pū-DEN-dal)	S2–S4	Muscles of perineum; skin of penis and scrotum in male and clitoris, labia majora, labia minora, and vagina in female.

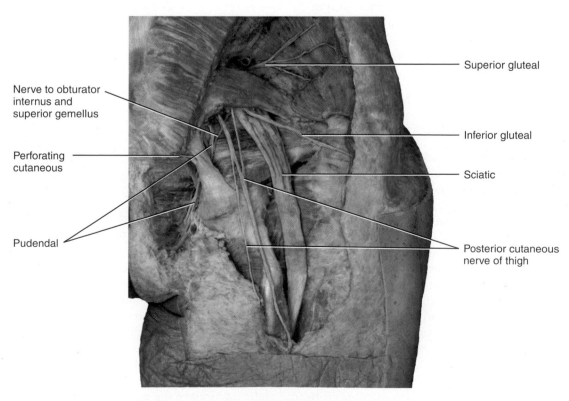

(b) Posterior view of sacral plexus

❓ What is the origin of the sacral plexus?

EXHIBIT 18.4 **603**

Intercostal Nerves

The anterior rami of spinal nerves T2–T12 do not enter into the formation of plexuses and are known as **intercostal** or **thoracic nerves** (see Figure 18.2a). These nerves connect directly to the structures they supply in the intercostal spaces. After leaving its intervertebral foramen, the anterior ramus of nerve T2 innervates the intercostal muscles of the second intercostal space and supplies the skin of the axilla and posteromedial aspect of the arm. Nerves T3–T6 extend along the costal grooves of the ribs and then to the intercostal muscles and skin of the anterior and lateral chest wall. Nerves T7–T12 supply the intercostal muscles and abdominal muscles, and the overlying skin. The posterior rami of the intercostal nerves supply the deep back muscles and skin of the posterior aspect of the thorax.

Dermatomes

The skin over the entire body is supplied by somatic sensory neurons that carry nerve impulses from the skin into the spinal cord and brain stem. Likewise, somatic motor neurons that carry impulses out of the spinal cord innervate the underlying skeletal muscles. Each spinal nerve contains sensory neurons that serve a specific, predictable segment of the body. The trigeminal (V) nerve serves most of the skin of the face and scalp. The area of the skin that provides sensory input to the CNS via one pair of spinal nerves or the trigeminal (V) nerve is called a **dermatome** (*derma-* =skin; *-tome*=thin segment) (Figure 18.9). The nerve supply in adjacent dermatomes overlaps somewhat. Knowing which spinal cord segments supply each dermatome makes it possible to locate damaged regions of the spinal cord. If the skin in a particular region is stimulated but the sensation is not perceived, the nerves supplying that dermatome are probably damaged. In regions where the overlap is considerable, little loss of sensation may result if only one of the nerves supplying the dermatome is damaged. Information about the innervation patterns of spinal nerves can also be used therapeutically. Cutting posterior roots or infusing local anesthetics can block pain either permanently or transiently. Because dermatomes overlap, deliberate production of a region of complete anesthesia may require that at least three adjacent spinal nerves be cut or blocked by an anesthetic drug.

Connective Tissue Coverings of Spinal Nerves

Nerves, like skeletal muscles, consist of long cells. While the function of nerves and muscles is quite different, the structure is similar in many ways. Remember the design of a skeletal muscle. The long muscle cells have a loose connective tissue surrounding, the endomysium, that distributes capillaries throughout the muscle tissue. Thicker sheets of connective tissue, the perimysium, bundle groups of muscle fibers together into units within the muscle called fasciculi. Many fasciculi are held together into a larger bundle of fibers, the individual muscle, by the thicker connective tissue epimysium.

Nerves have an identical design except the connective tissue wraps have the root word *neurium*, meaning *nerve*, instead of *mysium*, which means *muscle*. Figure 18.10 illustrates the structure of a nerve. Within the nerve are many axons with their surrounding neurilemma and myelin sheaths. The axon and its associated glial cells form the **nerve fiber**. Each nerve fiber sits in a loose connective tissue called the endoneurium. The **endoneurium** (en′-dō-NOO-rē-um; *endo-*=within or inner) consists of a mesh of collagen fibers, fibroblasts, and macrophages surrounded by endoneurial fluid (extracellular fluid) derived from the capillaries.

Figure 18.9 Distribution of dermatomes.

A dermatome is an area of skin that provides sensory input via the posterior roots of one pair of spinal nerves or via the trigeminal (V) nerve.

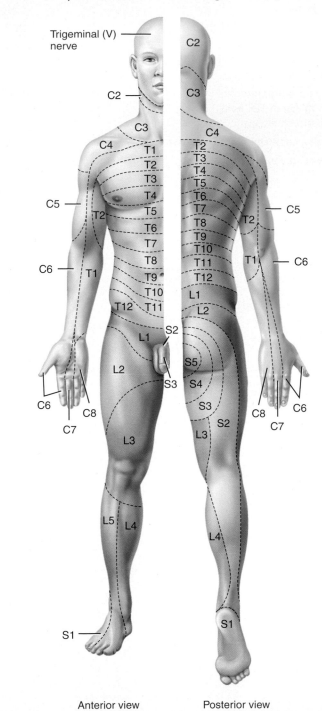

Anterior view   Posterior view

? Which is the only spinal nerve that does not have a corresponding dermatome?

This fluid nourishes the neuron and provides the necessary environment for its function of propagating action potentials.

A thicker sheath of connective tissue, the **perineurium** (per′-i-NOO-rē-um; *peri-*=around), packages groups of nerve fibers together into bundles called **fasciculi** (singular=**fasciculus**). The perineurium consists of collagenous sheaths with up to 15 layers of fibroblasts distributed in a meshwork of collagen. This important sheath functions as a diffusion barrier that, along with the tight junctions in the capillaries (blood–nerve barrier), maintains the osmotic environment and fluid pressure within the

Figure 18.10 **Organization and connective tissue coverings of a spinal nerve.** (Part b: © Dr. Richard Kessel and Dr. Randy Kardon/Visuals Unlimited, *Tissues and Organs: A Text-Altas of Scanning Electron Microscopy.* © 1979 by W. H. Freeman and Company. Reprinted by permission.)

 Three layers of connective tissue wrappings protect axons: endoneurium surrounds individual axons, perineurium surrounds bundles of axons (fascicles), and epineurium surrounds an entire nerve.

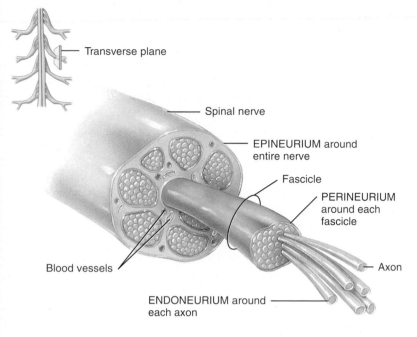

(a) Transverse section showing the coverings of a spinal nerve

(b) Transverse section of 12 nerve fascicles

? **Why are all spinal nerves classified as mixed nerves?**

endoneurium. The perineurium terminates by blending with the connective tissue capsules of various types of nerve endings and muscle junctions.

Completing the structure of the nerve is an outer connective tissue sheath, the **epineurium** (ep´-i-NOO-rē-um; *epi-*=over), which bundles all the fasciculi together to form a single nerve. The epineurium consists of fibroblasts and thick collagen strands that primarily parallel the long axis of the nerve. The epineurium makes up, on average, 50% of the cross-sectional area of a nerve. This important sheath gives the nerve the necessary tensile strength to resist the tensile forces that can so easily damage the delicate nervous tissue. This strength factor of a peripheral nerve cannot be overemphasized. Within any nerve in the body there is a greater amount of collagenous connective tissue than nervous tissue, which makes it possible for the nervous tissue to leave the protection of the bony skull and vertebral column and course through the peripheral tissues of the body. The neurons are not torn apart by the strong tensile activity generated by muscular activities and movements of the body, because their strong collagenous sheaths protect them. The epineurium also contains the small blood vessels and lymphatics for the nerve.

◉ CHECKPOINT

6. How are the spinal nerves named and numbered? Why are spinal nerves classified as mixed nerves?
7. Describe how a spinal nerve is connected to the spinal cord.
8. Describe the branches and innervations of a typical spinal nerve.
9. Describe the principal plexuses and the regions they supply.

SPINAL CORD FUNCTIONS

◉ OBJECTIVES

- Describe the functions of the major sensory and motor tracts of the spinal cord.
- Describe the functional components of a reflex arc and the ways reflexes maintain homeostasis.

The spinal cord has two principal functions in maintaining homeostasis: nerve impulse propagation and integration of information. The *white matter tracts* in the spinal cord are highways for nerve impulse propagation. Along these tracts, sensory impulses from receptors flow toward the brain, and motor impulses flow from the brain toward skeletal muscles and other effector tissues. The *gray matter* of the spinal cord receives and integrates incoming and outgoing information.

Sensory and Motor Tracts

The first way the spinal cord promotes homeostasis is by conducting nerve impulses along tracts. Often, the name of a tract indicates its position in the white matter and where it begins and ends. For example, the anterior spinothalamic tract is located in the *anterior* white column; it begins in the *spinal cord* and ends in the *thalamus* (a region of the brain). Notice that the location of the axon terminals comes last in the name. This regularity in naming allows you to determine the direction of information flow along any tract named according to this convention. Thus, because the anterior spinothalamic tract conveys nerve impulses from the spinal cord toward the brain, it is a sensory (ascending) tract. Figure 18.11 highlights the major sensory and motor tracts in the spinal cord. These tracts are described in detail in Chapter 21 and summarized in Tables 21.3 and 21.4 on pages 688 and 691.

Nerve impulses from sensory receptors propagate up the spinal cord to the brain along two main routes on each side: the spinothalamic tracts and the posterior columns. The **lateral** and **anterior spinothalamic tracts** convey nerve impulses for sensing pain, warmth, coolness, itching, tickling, deep pressure, and a crude, poorly localized sense of touch. The right and left **posterior columns** carry nerve impulses for several kinds of sensations. These include (1) *proprioception*, awareness of the positions and movements of muscles, tendons, and joints; (2) *discriminative touch*, the ability to feel exactly what part of the body is touched; (3) *two-point discrimination*, the ability to distinguish the touching of two different points on the skin, even though they are close together; (4) *light pressure sensations*; and (5) *vibration sensations*.

The sensory systems keep the CNS informed of changes in the external and internal environments. Responses to this information are brought about by motor systems, which enable us to move about and change our physical relationship to the world around us. As sensory information is conveyed to the CNS, it becomes part of a large pool of sensory input. Each piece of incoming information is integrated with all the other information arriving from activated sensory neurons.

Through the activity of interneurons, integration occurs in several regions of the spinal cord and brain. As a result, motor impulses to make a muscle contract or a gland secrete can be initiated at several levels. Most regulation of involuntary activities of smooth muscle, cardiac muscle, and glands by the autonomic nervous system (ANS) originates in the brain stem (the lower part of the brain that is continuous with the spinal cord) and in a nearby brain region called the hypothalamus.

The cerebral cortex (superficial gray matter of the cerebrum) plays a major role in controlling precise, voluntary muscular movements. Other brain regions integrate automatic movements, such as arm swinging during walking. Motor output to skeletal muscles travels down the spinal cord in two types of descending pathways: direct and indirect. The **direct pathways** include the *lateral corticospinal, anterior corticospinal,* and *corticobulbar tracts.* They convey nerve impulses that originate in the cerebral cortex and are destined to cause precise, *voluntary* movements of skeletal muscles. **Indirect pathways** include the *rubrospinal, tectospinal,* and *vestibulospinal tracts.* They convey nerve impulses from the brain stem and other parts of the brain that govern *automatic movements* and help coordinate body movements with visual stimuli. Indirect pathways also maintain skeletal muscle tone, sustain contraction of postural muscles, and play a major role in equilibrium by regulating muscle tone in response to movements of the head.

Reflexes and Reflex Arcs

The second way the spinal cord promotes homeostasis is by serving as an integrating center for some reflexes. A **reflex** is a fast, involuntary, unplanned sequence of actions that occurs in response to a particular stimulus. Some reflexes are inborn, such as pulling your hand away from a hot surface before you even feel that it is hot. Other reflexes are learned or acquired. For instance, you learn many reflexes while acquiring driving expertise. Slamming on the brakes in an emergency is one example. When integration takes place in the spinal cord gray matter, the reflex is a **spinal reflex.** An example is the familiar patellar reflex (knee jerk). By contrast, if integration occurs in the brain stem rather than the spinal cord, the reflex is a **cranial reflex.** An example is the tracking movements of your eyes as you read this sentence. You are probably most aware of **somatic reflexes,** which involve contraction of skeletal muscles. Equally important, however, are the **autonomic**

Figure 18.11 The locations of selected sensory and motor tracts, shown in a transverse section of the spinal cord. Sensory tracts are indicated on one half and motor tracts on the other half of the cord, but in fact all tracts are present on both sides.

The name of a tract often indicates its location in the white matter and where it begins and ends.

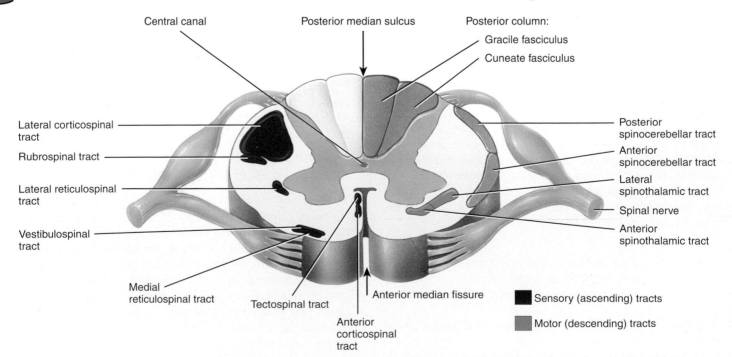

? **Based on its name, what are the position in the spinal cord, origin, and destination of the anterior corticospinal tract? Is this a sensory or a motor tract?**

(visceral) reflexes, which generally are not consciously perceived. They involve responses of smooth muscle, cardiac muscle, and glands. As you will see in Chapter 20, body functions such as heart rate, digestion, urination, and defecation are controlled by the autonomic nervous system through autonomic reflexes.

Nerve impulses propagating into, through, and out of the CNS follow specific pathways, depending on the kind of information, its origin, and its destination. The pathway followed by nerve impulses that produce a reflex is a **reflex arc (reflex circuit)**. Using the **patellar reflex** (knee jerk) as an example, the basic components of a reflex arc are as follows (Figure 18.12):

❶ **Sensory receptor.** The distal end of a sensory neuron (dendrite) or an associated sensory structure serves as a sensory receptor. Sensory receptors respond to a specific type of *stimulus* (a change in the internal or external environment) by generating one or more nerve impulses. In the patellar reflex, sensory receptors known as *muscle spindles* detect slight stretching of the quadriceps femoris muscle (anterior thigh) when the patellar (knee cap) ligament is tapped with a reflex hammer.

❷ **Sensory neuron.** The nerve impulses conduct from the sensory receptor along the axon of a sensory neuron to its axon terminals, which are located in the CNS gray matter. Axon branches of the sensory neuron also relay nerve impulses to the brain, allowing conscious awareness that the reflex has occurred.

❸ **Integrating center.** One or more regions of gray matter in the CNS act as an integrating center. In the simplest type of reflex, such as the patellar reflex, the integrating center is a single synapse between a sensory neuron and a motor neuron. In other types of reflexes, the integrating center includes one or more interneurons.

❹ **Motor neuron.** Impulses triggered by the integrating center pass out of the spinal cord (or brain stem, in the case of a cranial reflex) along a motor neuron to the part of the body that will respond. In the patellar reflex, the axon of the motor neuron extends to the quadriceps femoris muscle.

❺ **Effector.** The part of the body that responds to the motor nerve impulse, such as a muscle or gland, is the effector. The patellar reflex is a *somatic reflex* because its effector is a skeletal muscle, the quadriceps femoris muscle, which contracts and thereby relieves the stretching that initiated the reflex. In sum, the patellar reflex causes extension of the knee by contraction of the quadriceps femoris muscle in response to tapping the patellar ligament. If the effector is smooth muscle, cardiac muscle, or a gland, the reflex is an *autonomic (visceral) reflex.*

Figure 18.12 Patellar reflex showing the general components of a reflex arc. The arrows show the direction of nerve impulse propagation.

Reflexes are fast, involuntary responses to particular stimuli.

? Why is this reflex a somatic reflex?

Reflexes are often used for diagnosing disorders of the nervous system and locating injured tissue. If a reflex ceases to function or functions abnormally, the physician may suspect that the damage lies somewhere along a particular conduction pathway. Simply tapping or stroking the body can test many somatic reflexes. Among the somatic reflexes of clinical significance are the patellar reflex (knee reflex), Achilles reflex (ankle jerk), Babinski sign, and the abdominal reflex.

■ CHECKPOINT

10. What are the functions of the anterior spinothalamic tract and the posterior columns?
11. Describe the components of the patellar reflex arc.

KEY MEDICAL TERMS ASSOCIATED WITH THE SPINAL CORD AND SPINAL NERVES

Epidural block Injection of an anesthetic drug into the epidural space, the space between the dura mater and the vertebral column, in order to cause a temporary loss of sensation. Such injections in the lower lumbar region are used to control pain during childbirth.

Meningitis (men-in-JĪ-tis; *-itis*=inflammation) Inflammation of the meninges due to an infection, usually caused by a bacterium or virus. Symptoms include fever, headache, stiff neck, vomiting, confusion, lethargy, and drowsiness.

Myelitis (mī-e-LĪ-tis; *myel-*=spinal cord) Inflammation of the spinal cord.

Myelography (mī-e-LOG-ra-fē, *myel-*=marrow, *-graph*=to write) A procedure in which a CT scan or x-ray image of the spinal cord is taken after injection of a radiopaque dye (contrast medium) to diagnose abnormalities such as tumors and herniated intervertebral discs.

Nerve block Loss of sensation in a region due to injection of a local anesthetic; an example is local dental anesthesia.

Neuralgia (noo-RAL-jē-a; *neur-*=nerve; *-algia*=pain) Attacks of pain along the entire course or a branch of a sensory nerve.

Neuritis (*-itis*=inflammation) Inflammation of one or several nerves that may result from irritation to the nerve produced by direct blows, bone fractures, contusions, or penetrating injuries. Additional causes include infections, vitamin deficiency (usually thiamine), and poisons such as carbon monoxide, carbon tetrachloride, heavy metals, and some drugs.

Paresthesia (par-es-THĒ-zē-a; *par-*=departure from normal; *-esthesia*=sensation) An abnormal sensation such as burning, pricking, tickling, or tingling resulting from a disorder of a sensory nerve.

STUDY OUTLINE

Spinal Cord Anatomy (p. 585)

1. The spinal cord is protected by the vertebral column, the meninges, cerebrospinal fluid, and denticulate ligaments.
2. The three meninges are coverings that run continuously around the spinal cord and brain. They are the dura mater, arachnoid mater, and pia mater.
3. The spinal cord begins as a continuation of the medulla oblongata and ends at about the second lumbar vertebra in an adult.
4. The spinal cord contains cervical and lumbar enlargements that serve as points of origin for nerves to the limbs.
5. The tapered inferior portion of the spinal cord is the conus medullaris, from which arise the filum terminale and cauda equina.
6. Spinal nerves connect to each segment of the spinal cord by two roots. The posterior or dorsal root contains sensory axons, and the anterior or ventral root contains motor neuron axons.
7. The anterior median fissure and the posterior median sulcus partially divide the spinal cord into right and left sides.
8. The gray matter in the spinal cord is divided into horns, and the white matter into columns. In the center of the spinal cord is the central canal, which runs the length of the spinal cord and is filled with cerebrospinal fluid.
9. Parts of the spinal cord observed in transverse section are the gray commissure; central canal; anterior, posterior, and lateral gray horns; and anterior, posterior, and lateral white columns, which contain ascending and descending tracts. Each part has specific functions.
10. The spinal cord conveys sensory and motor information by way of ascending and descending tracts, respectively.

Spinal Nerves (p. 593)

1. The 31 pairs of spinal nerves are named and numbered according to the region and level of the spinal cord from which they emerge.
2. There are 8 pairs of cervical, 12 pairs of thoracic, 5 pairs of lumbar, 5 pairs of sacral, and 1 pair of coccygeal nerves.
3. Spinal nerves typically are connected with the spinal cord by a pos-

terior root and an anterior root. All spinal nerves contain both sensory and motor axons (are mixed nerves).
4. Three connective tissue coverings associated with spinal nerves are the endoneurium, perineurium, and epineurium.
5. Branches of a spinal nerve include the posterior ramus, anterior ramus, meningeal branch, and rami communicantes.
6. The anterior rami of spinal nerves, except for T2–T12, form networks of nerves called plexuses.
7. Emerging from the plexuses are nerves bearing names that typically describe the general regions they supply or the route they follow.
8. Nerves of the cervical plexus supply the skin and muscles of the head, neck, and upper part of the shoulders; they connect with some cranial nerves and innervate the diaphragm.
9. Nerves of the brachial plexus supply the upper limbs and several neck and shoulder muscles.
10. Nerves of the lumbar plexus supply the anterolateral abdominal wall, external genitals, and part of the lower limbs.
11. Nerves of the sacral plexus supply the buttocks, perineum, and part of the lower limbs.
12. Nerves of the coccygeal plexus supply the skin of the coccygeal region.
13. Anterior rami of nerves T2–T12 do not form plexuses and are called intercostal (thoracic) nerves. They are distributed directly to the structures they supply in intercostal spaces.
14. Sensory neurons within spinal nerves and the trigeminal (V) nerve serve specific, constant segments of the skin called dermatomes.
15. Knowledge of dermatomes helps a physician determine which segment of the spinal cord or which spinal nerve is damaged.

Spinal Cord Functions (p. 605)

1. The white matter tracts in the spinal cord are highways for nerve impulse propagation. Along these tracts, sensory input travels toward the brain, and motor output travels from the brain toward skeletal muscles and other effector tissues.
2. Sensory input travels along two main routes in the white matter of the spinal cord: the posterior columns and the spinothalamic tracts.

3. Motor output travels along two main routes in the white matter of the spinal cord: direct pathways and indirect pathways.
4. A second major function of the spinal cord is to serve as an integrating center for spinal reflexes. This integration occurs in the gray matter.
5. A reflex is a fast, predictable sequence of involuntary actions, such as muscle contractions or glandular secretions, which occurs in response to certain changes in the environment.
6. Reflexes may be spinal or cranial and somatic or autonomic (visceral).
7. The components of a reflex arc are sensory receptor, sensory neuron, integrating center, motor neuron, and effector.

SELF-QUIZ QUESTIONS

Choose the one best answer to the following questions.

1. Choose the *true* statement:
   a. Tracts are located only in the spinal cord and not in the brain.
   b. Tracts appear white because they consist of bundles of unmyelinated nerve fibers.
   c. Ascending tracts are all motor tracts.
   d. Ascending tracts are in the white matter; descending tracts are in the gray matter.
   e. None of the above is true.

2. Which of the following lists the meninges and spaces in the correct order from most superficial to deepest?
   a. pia mater, subarachnoid space, arachnoid mater, subdural space, dura mater, epidural space
   b. dura mater, epidural space, arachnoid mater, subdural space, pia mater, subarachnoid space
   c. arachnoid mater, subarachnoid space, epidural space, dura mater, subdural space, pia mater
   d. subdural space, dura mater, epidural space, arachnoid mater, subarachnoid space, pia mater
   e. epidural space, dura mater, subdural space, arachnoid mater, subarachnoid space, pia mater

3. The posterior (dorsal) root ganglion contains:
   a. cell bodies of motor neurons.
   b. cell bodies of sensory neurons.
   c. cranial nerve axons.
   d. synapses.
   e. All of the above are correct.

4. Which sequence best represents the course of a nerve impulse over a reflex arc?
   a. receptor, integrating center, sensory neuron, motor neuron, effector
   b. effector, sensory neuron, integrating center, motor neuron, receptor
   c. receptor, sensory neuron, integrating center, motor neuron, effector
   d. receptor, motor neuron, integrating center, sensory neuron, effector
   e. effector, integrating center, sensory neuron, motor neuron, receptor

5. The spinothalamic tracts convey sensory information to the brain regarding:
   a. pain.
   b. two-point discrimination.
   c. temperature.
   d. proprioception.
   e. Both a and c are correct.

6. The extension of the pia mater that anchors the spinal cord to the coccyx is the:
   a. cauda equina.
   b. denticulate ligament.
   c. conus medullaris.
   d. filum terminale.
   e. dorsal root.

7. Which of the following is *true* regarding the number of spinal nerves?
   a. There are 7 pairs of cervical nerves, 12 pairs of thoracic nerves, 4 pairs of lumbar nerves, 1 pair of sacral nerves, and 1 pair of coccygeal nerves.
   b. There are 8 pairs of cervical nerves, 12 pairs of thoracic nerves, 5 pairs of lumbar nerves, 5 pairs of sacral nerves, and 5 pairs of coccygeal nerves.
   c. There are 7 pairs of cervical nerves, 12 pairs of thoracic nerves, 4 pairs of lumbar nerves, 5 pairs of sacral nerves, and 1 pair of coccygeal nerves.
   d. There are 7 pairs of cervical nerves, 14 pairs of thoracic nerves, 4 pairs of lumbar nerves, 5 pairs of sacral nerves, and 1 pair of coccygeal nerves.
   e. There are 8 pairs of cervical nerves, 12 pairs of thoracic nerves, 5 pairs of lumbar nerves, 5 pairs of sacral nerves, and 1 pair of coccygeal nerves.

8. What fills the central canal of the spinal cord?
   a. blood
   b. bone
   c. cerebrospinal fluid
   d. axon bundles
   e. lymph

Complete the following.

9. The spinal cord extends from the _____ of the brain to the level of the _____ of the vertebral column.

10. Lateral extensions of the pia mater that help to protect the spinal cord against displacement are called _____.

11. The effectors in somatic reflexes are _____, while the effectors for autonomic reflexes are _____, _____, and _____.

12. The brachial plexus is constructed from the _____ rami of spinal nerves_____.

13. Cell bodies of motor neurons leading to skeletal muscles are located in the _____ gray horns of the spinal cord, while cell bodies of neurons supplying smooth muscle, cardiac muscle, or glands lie in the _____ gray horns of the spinal cord.

14. The area of skin that provides sensory input to the central nervous system via one pair of spinal nerves *or* the trigeminal nerve is called a(n) _____.

Are the following statements true or false? If the statement is False, *find and fix the mistake to make the statement* True.

15. Cell bodies of motor neurons are located in the posterior (dorsal) root ganglia.

16. The lateral spinothalamic tract is a motor tract.

17. The tapering inferior end of the spinal cord at the level of the first two lumbar vertebrae is called the conus medullaris.

18. The main functions of the spinal cord are to propagate nerve impulses and to integrate sensory input and motor output.

19. The rubrospinal and vestibulospinal tracts are examples of indirect motor pathways.

20. Individual axons are protected by a connective tissue wrapping known as the epineurium.

21. Inability to feel touch around the umbilicus could indicate an injury to the T5 spinal nerve.

22. Dermatomes of the anterior lower limb correspond to nerves arising from the lumbar plexus.

Matching

23. Match the following:

_____ **(a)** provides the entire nerve supply for the shoulder and upper limbs

_____ **(b)** gives rise to the phrenic nerve that supplies the diaphragm

_____ **(c)** gives rise to the median, radial, and axillary nerves

_____ **(d)** not a plexus at all, but rather segmentally arranged nerves

_____ **(e)** supplies fibers to the scalp, neck, and part of the shoulder and chest

_____ **(f)** supplies fibers to the femoral nerve that innervates the flexor muscles of the thigh and the extensor muscles of the leg

_____ **(g)** gives rise to the largest nerve in the body, the sciatic nerve, that supplies the posterior aspect of the thigh and the leg

(1) brachial plexus

(2) cervical plexus

(3) sacral plexus

(4) lumbar plexus

(5) intercostal nerves

24. Match the following:

_____ **(a)** the joining together of the anterior rami of adjacent nerves

_____ **(b)** spinal nerve branches that serve the deep muscles and skin of the posterior surface of the trunk

_____ **(c)** spinal nerve branches that serve the muscles and structures of the upper and lower limbs and the lateral and ventral trunk

_____ **(d)** area of the spinal cord from which nerves to the upper limbs arise

_____ **(e)** area of the spinal cord from which nerves to the lower limbs arise

_____ **(f)** contains motor neuron axons and conducts impulses from the spinal cord to the periphery

_____ **(g)** contains sensory nerve fibers and conducts impulses from the periphery into the spinal cord

(1) cervical enlargement

(2) lumbar enlargement

(3) posterior root

(4) anterior root

(5) posterior ramus

(6) anterior ramus

(7) plexus

25. Match the following functions (motor/sensory) to the correct (peripheral) nerve (answers can be used more than once):

_____ **(a)** innervates adductor muscles of the thigh

_____ **(b)** provides sensory input from the posterior arm and forearm

_____ **(c)** innervates upper limb muscles used to move hand into a fist

_____ **(d)** provides sensory input from the skin surrounding the upper thorax and shoulder

_____ **(e)** injury to this nerve results in "wrist drop"

_____ **(f)** innervates most intrinsic muscles of the hand

_____ **(g)** innervates the main respiratory muscle (diaphragm)

_____ **(h)** provides sensory input from the skin of the abdomen

_____ **(i)** provides sensation from the perineum (e.g., childbirth)

_____ **(j)** provides efferent response in patellar tendon reflex (e.g., extension of knee)

_____ **(k)** innervates the hamstring muscles

_____ **(l)** provides sensory input from the skin of the thorax

_____ **(m)** provides sensory input from the skin of the palm

_____ **(n)** provides sensory input from the anterior thigh

_____ **(o)** innervates anterior abdominal wall musculature

(1) median nerve

(2) phrenic nerve

(3) femoral nerve

(4) radial nerve

(5) sciatic nerve

(6) intercostal nerve

(7) pudendal nerve

(8) ulnar nerve

(9) supraclavicular nerve

(10) obturator nerve

CRITICAL THINKING QUESTIONS

1. A high school senior dove headfirst into a murky pond. Unfortunately, his head hit a submerged log and he is now paralyzed from the neck down (quadriplegia). Can you deduce the location of the injury? What is the likelihood of recovery from his injury?

2. The spinal cord is covered in protective layers and enclosed by the vertebral column. How is it able to send and receive messages from the periphery of the body?

3. Why doesn't the spinal cord "creep" up toward the head every time you bend over? Why doesn't it get all twisted out of position when you exercise?

4. Nischal slipped on the ice and broke his "tailbone." However, he did no damage to his spinal cord. How is it possible that he broke part of his vertebral column without damaging the spinal cord?

5. Not long after Nischal fell on the ice, he began experiencing pain in his lower back and down his leg, as well as some numbness in his foot. When he walks, he has trouble controlling the flexion of his foot. He told his friend that "the doctor said something about compression of the lumbar vertebrae." What has probably happened to Nischal?

? ANSWERS TO FIGURE QUESTIONS

18.1 The superior boundary of the spinal dura mater is the foramen magnum of the occipital bone. The inferior boundary is the second sacral vertebra.

18.2 The cervical enlargement of the spinal cord connects with sensory and motor nerves of the upper limbs.

18.3 A horn is an area of gray matter, and a column is a region of white matter in the spinal cord.

18.4 The anterior rami serve the upper and lower limbs.

18.5 Severing the spinal cord at level C2 causes respiratory arrest because it prevents descending nerve impulses from reaching the phrenic nerve, which stimulates contraction of the diaphragm, the main muscle needed for breathing.

18.6 The axillary, musculocutaneous, radial, median, and ulnar nerves are five important nerves that arise from the brachial plexus.

18.7 The lumbar plexus originates from the roots of spinal nerves L1–L4.

18.8 The origin of the sacral plexus is the anterior rami of spinal nerves L4–L5 and S1–S4.

18.9 The only spinal nerve without a corresponding dermatome is C1.

18.10 All spinal nerves are mixed because they contain a posterior root with sensory axons and an anterior root with motor axons.

18.11 The anterior corticospinal tract is located on the anterior side of the spinal cord, originates in the cortex of the cerebrum, and ends in the spinal cord. It contains descending axons and thus is a motor tract.

18.12 It is a somatic reflex because the effector is a skeletal muscle.

19 | THE BRAIN AND THE CRANIAL NERVES

INTRODUCTION Solving an equation, feeling hungry, laughing—the neural processes needed for each of these activities occur in different regions of the **brain**, that portion of the central nervous system contained within the cranium. About 100 billion neurons and 10–50 trillion neuroglia make up the brain, which has a mass of about 1300 g (almost 3 lb) in adults. On average, each neuron forms 1000 synapses with other neurons. Thus, the total number of synapses, about a thousand trillion (10^{15}), is larger than the number of stars in the galaxy.

The brain is the center for registering sensations, correlating them with one another and with stored information, making decisions, and taking actions. It also is the center for the intellect, emotions, behavior, and memory. But the brain encompasses yet a larger domain: It directs our behavior toward others. With ideas that excite, artistry that dazzles, or rhetoric that mesmerizes, one person's thoughts and actions may influence and shape the lives of many others. As you will see shortly, different regions of the brain are specialized for different functions. Different parts of the brain also work together to accomplish certain shared functions. This chapter explores how the brain is protected and nourished, what functions occur in the major regions of the brain, and how the spinal cord and the 12 pairs of cranial nerves connect with the brain to form the control center of the human body. •

BRAIN ORGANIZATION, PROTECTION, AND BLOOD SUPPLY

⊙ **OBJECTIVES**
- Identify the major parts of the brain.
- Explain how the brain is protected.
- Describe the blood supply of the brain.

Knowledge of the embryological development of the brain is necessary to understand the terminology used for the principal parts of the adult brain. The development of the brain is introduced briefly here, and discussed in more detail at the end of the chapter.

Recall from Chapter 4 that the brain and spinal cord develop from ectoderm arranged in a tubular structure called the neural tube (see also Figure 19.26). The anterior part of the neural tube expands, and constrictions appear that create three regions called *primary brain vesicles*: prosencephalon, mesencephalon, and rhombencephalon (see Figure 19.27). The walls of these brain regions develop into the nervous tissue of the brain, while the hollow of the tube remains as the ventricular spaces of the brain. The mesencephalon gives rise to the midbrain and aqueduct of the midbrain (cerebral aqueduct). Both the prosencephalon and rhombencephalon subdivide further, forming secondary brain vesicles. The prosencephalon gives rise to the telencephalon and diencephalon, and the rhombencephalon develops into the metencephalon and myelencephalon. The telencephalon develops into the cerebrum and lateral ventricles. The diencephalon forms the thalamus, hypothalamus, epithalamus, and the third ventricle. The metencephalon becomes the pons, cerebellum, and upper part of the fourth ventricle. Finally, the myelencephalon forms the medulla oblongata and lower part of the fourth ventricle. The various parts of the brain will be described shortly.

These relationships are summarized in Table 19.1.

Major Parts of the Brain

The adult brain consists of four major parts: brain stem, cerebellum, diencephalon, and cerebrum (Figure 19.1). The **brain stem** is continuous with the spinal cord and consists of the medulla oblongata, pons, and midbrain. Posterior to the brain stem is the **cerebellum** (ser′-e-BEL-um=little brain). Superior to the brain stem is the **diencephalon** (dī-en-SEF-a-lon; *di-*=through; *-encephalon*=brain), which as noted previously consists of the thalamus, hypothalamus, and epithalamus. Supported on the diencephalon and brain stem is the **cerebrum** (se-RĒ-brum=brain), the largest part of the brain.

Protective Coverings of the Brain

The cranium and the cranial meninges surround and protect the brain. The **cranial meninges** (me-NIN-jēz) are continuous with the spinal meninges, have the same basic structure, and bear the same names: the outer **dura mater**, the middle **arachnoid mater**, and the inner **pia mater** (Figure 19.2 on page 615). Note that the cranial dura mater has two layers; the spinal dura mater has only one. The two dural layers around the brain are fused together except where they separate to enclose the dural venous sinuses (endothelial-lined venous channels) that drain venous blood from the brain and deliver it into the internal jugular veins. Also, there is no epidural space around the brain. Blood vessels that enter brain tissue pass along the surface of the brain, and as they penetrate inward, they are sheathed by a loose-fitting sleeve of pia mater. Three extensions of the dura mater separate parts of the brain. (1) The **falx cerebri** (FALKS CER-e-brē; *falx*=sickle-shaped) separates the two hemispheres (sides) of the cerebrum. (2) The **falx cerebelli** (cer-e-BEL-ī) separates the two hemispheres of the cerebellum. (3) The **tentorium cerebelli** (ten-TŌ-rē-um=tent) separates the cerebrum from the cerebellum.

TABLE 19.1

Development of the Brain

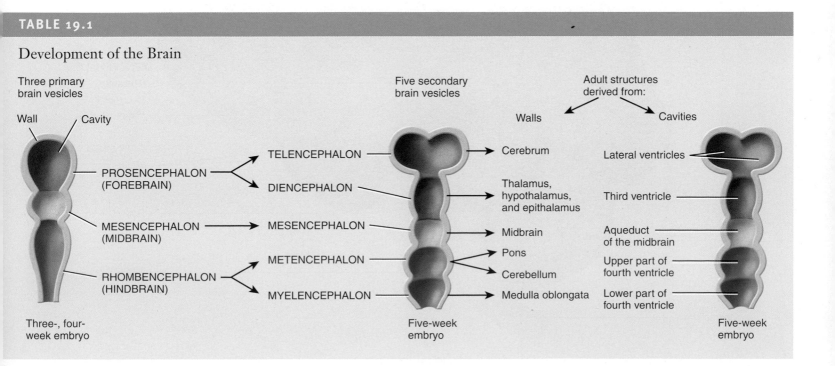

Three primary brain vesicles	Five secondary brain vesicles	Adult structures derived from: Walls	Cavities
PROSENCEPHALON (FOREBRAIN)	TELENCEPHALON	Cerebrum	Lateral ventricles
	DIENCEPHALON	Thalamus, hypothalamus, and epithalamus	Third ventricle
MESENCEPHALON (MIDBRAIN)	MESENCEPHALON	Midbrain	Aqueduct of the midbrain
RHOMBENCEPHALON (HINDBRAIN)	METENCEPHALON	Pons / Cerebellum	Upper part of fourth ventricle
	MYELENCEPHALON	Medulla oblongata	Lower part of fourth ventricle

Three-, four-week embryo   Five-week embryo   Five-week embryo

Figure 19.1 The brain. The pituitary gland is discussed with the endocrine system in Chapter 23. (See Tortora, *A Photographic Atlas of the Human Body*, 2e, Figures 8.12 and 8.13.)

🔑 The four principal parts of the brain are the brain stem, cerebellum, diencephalon, and cerebrum.

(a) Sagittal section, medial view

(b) Sagittal section, medial view

❓ **Which part of the brain is the largest?**

Figure 19.2 The protective coverings of the brain.

Cranial bones and the cranial meninges protect the brain.

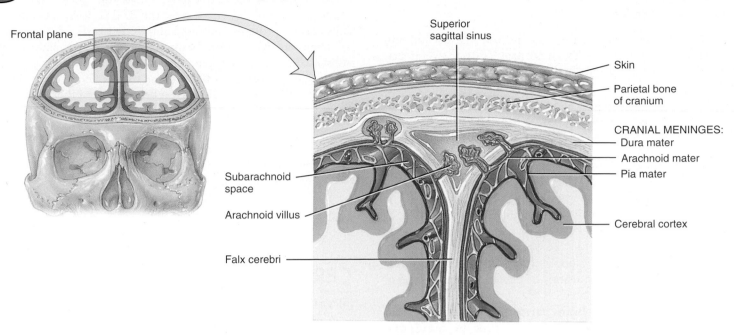

Frontal plane

Superior sagittal sinus

Skin

Parietal bone of cranium

CRANIAL MENINGES:
Dura mater
Arachnoid mater
Pia mater

Subarachnoid space

Arachnoid villus

Cerebral cortex

Falx cerebri

(a) Frontal section through skull showing the cranial meninges

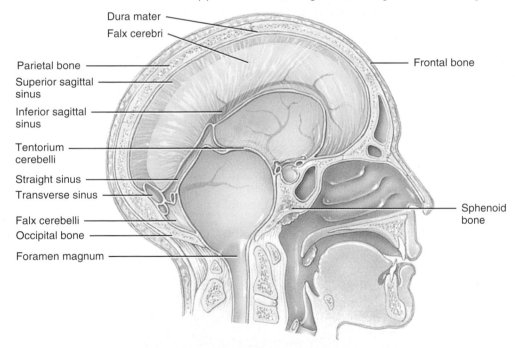

Dura mater
Falx cerebri

Parietal bone
Superior sagittal sinus
Inferior sagittal sinus

Tentorium cerebelli

Straight sinus
Transverse sinus

Falx cerebelli
Occipital bone

Foramen magnum

Frontal bone

Sphenoid bone

(b) Extensions of the dura mater

Superior sagittal sinus

Skin

Parietal bone of cranium

CRANIAL MENINGES:
Dura mater

Arachnoid mater

Subarachnoid space

Pia mater

Cerebral cortex

Falx cerebri

(c) Frontal section

? What are the three layers of the cranial meninges, from superficial to deep?

Brain Blood Flow and the Blood–Brain Barrier

Blood flows to the brain mainly via the internal carotid and vertebral arteries (see Figure 15.7b on page 500); the veins that return blood from the head to the heart are shown in Figure 15.11a (page 516).

In an adult, the brain represents only 2% of total body weight, but consumes about 20% of the oxygen and glucose used even at rest. Neurons synthesize ATP almost exclusively from glucose via reactions that use oxygen. When activity of neurons and neuroglia increases in a region of the brain, blood flow to that area also increases. Even a brief slowing of brain blood flow may cause unconsciousness. Typically, an interruption in blood flow for 1 or 2 minutes impairs neuronal function, and total deprivation of oxygen for about 4 minutes causes permanent injury. Because virtually no glucose is stored in the brain, the supply of glucose also must be continuous. If blood entering the brain has a low level of glucose, mental confusion, dizziness, convulsions, and loss of consciousness may occur.

The existence of a **blood–brain barrier (BBB)** protects brain cells from harmful substances and pathogens by preventing passage of many substances from blood into brain tissue. The blood–brain barrier consists mainly of tight junctions that seal together the endothelial cells of brain capillaries and partly of a thick basement membrane around the capillaries. The processes of many astrocytes, which as you learned in Chapter 17 are one type of neuroglia, press up against the capillaries and secrete chemicals that maintain the permeability characteristics of the tight junctions. A few water-soluble substances, such as glucose, cross the BBB by active transport. Other substances, such as creatinine, urea, and most ions, cross the BBB very slowly. Still other substances—proteins and most antibiotic drugs—do not pass at all from the blood into brain tissue. However, lipid-soluble substances, such as oxygen, carbon dioxide, alcohol, and most anesthetic agents, access brain tissue freely. Trauma, certain toxins, and inflammation can cause a breakdown of the blood–brain barrier.

☑ CHECKPOINT

1. Compare the sizes and locations of the cerebrum and cerebellum.
2. Describe the locations of the cranial meninges.
3. Explain the blood supply to the brain and the importance of the blood–brain barrier.

CEREBROSPINAL FLUID

☑ OBJECTIVE

• Explain the formation and circulation of cerebrospinal fluid.

Cerebrospinal fluid (CSF) is a clear, colorless liquid comprised primarily of water that protects the brain and spinal cord against chemical and physical injuries. It also carries oxygen, glucose, and other needed chemicals from the blood to neurons and neuroglia. CSF continuously circulates through cavities in the brain and spinal cord and around the brain and spinal cord in the subarachnoid space (between the arachnoid mater and pia mater).

Figure 19.3 Locations of ventricles within a "transparent" brain. One interventricular foramen on each side connects a lateral ventricle to the third ventricle, and the aqueduct of the midbrain connects the third ventricle to the fourth ventricle.

🔑 Ventricles are cavities within the brain that are filled with cerebrospinal fluid.

Right lateral view of brain

❓ **Which brain region is anterior to the fourth ventricle? Which is posterior to it?**

Formation of CSF in the Ventricles

Figure 19.3 on page 616 shows the four CSF-filled cavities within the brain, which are called **ventricles** (VEN-tri-kuls=little cavities). A **lateral ventricle** is located in each hemisphere of the cerebrum. Anteriorly, the lateral ventricles are separated by a thin membrane, the **septum pellucidum** (SEP-tum pe-LOO-si-dum; *pellucid*=transparent). The **third ventricle** is a narrow cavity along the midline superior to the hypothalamus and between the right and left halves of the thalamus. The **fourth ventricle** lies between the brain stem and the cerebellum.

The total volume of CSF is 80 to 150 mL (3 to 5 oz) in an adult. CSF contains glucose, proteins, lactic acid, urea, cations (Na^+, K^+, Ca^{2+}, Mg^{2+}), and anions (Cl^- and HCO_3^-); it also contains some white blood cells. The CSF contributes to homeostasis in three main ways:

1. **Mechanical protection.** CSF serves as a shock-absorbing medium that protects the delicate tissues of the brain and spinal cord from jolts that would otherwise cause them to hit the bony walls of the cranial cavity and vertebral canal. This is the primary function of this important fluid. The fluid also buoys the brain so that it "floats" in the cranial cavity.
2. **Chemical protection.** CSF provides an optimal chemical environment for accurate neuronal signaling in neurons that border the fluid. Even slight changes in the ionic composition of CSF within the brain can seriously disrupt production of nerve impulses.
3. **Circulation.** CSF is a medium for minor exchange of nutrients and waste products between the blood and nervous tissue.

The sites of CSF production are the **choroid plexuses** (KŌ-royd=membrane-like), networks of blood capillaries in the walls of the ventricles (Figure 19.4a). The capillaries are covered by ependymal cells joined tightly by tight junctions. The ependymal cells form cerebrospinal fluid from blood plasma by filtration and secretion. Because of the tight junctions between ependymal cells, materials entering CSF from choroid capillaries cannot leak between these cells; instead, they must pass through the ependymal cells. This **blood–cerebrospinal fluid barrier** permits certain substances to enter the CSF but excludes others, protecting the brain and spinal cord from potentially harmful blood-borne substances. In contrast to the blood–brain barrier, which is formed mainly by tight junctions of brain capillary endothelial cells, the blood–cerebrospinal fluid barrier is formed by tight junctions of ependymal cells.

Circulation of CSF

The CSF formed in the choroid plexuses of each lateral ventricle flows into the third ventricle through two narrow, oval openings, the **interventricular foramina** (Figure 19.4b). More CSF is added by the choroid plexus in the roof of the third ventricle. The fluid then flows through the **aqueduct of the midbrain (cerebral aqueduct)**, which passes through the midbrain, into the fourth ventricle. The choroid plexus of the fourth ventricle contributes more fluid. CSF enters the subarachnoid space through three openings in the roof of the fourth ventricle: a **median aperture** and the paired **lateral apertures**, one on each side. CSF then circulates in the central canal of the spinal cord

Figure 19.4 Pathways of circulating cerebrospinal fluid.

CSF is formed from blood plasma by ependymal cells that cover the choroid plexuses of the ventricles.

ANTERIOR

View

Transverse plane

Falx cerebri

Cerebrum

Lateral ventricle

Septum pellucidum

Choroid plexus

Falx cerebri

Superior sagittal sinus

POSTERIOR

(a) Superior view of transverse section of brain showing choroid plexuses

Ependymal cell

Blood capillary

Tight junction

CSF   Ventricle

Details of a section through a choroid plexus

FIGURE 19.4 CONTINUES ▶

FIGURE 19.4 CONTINUED

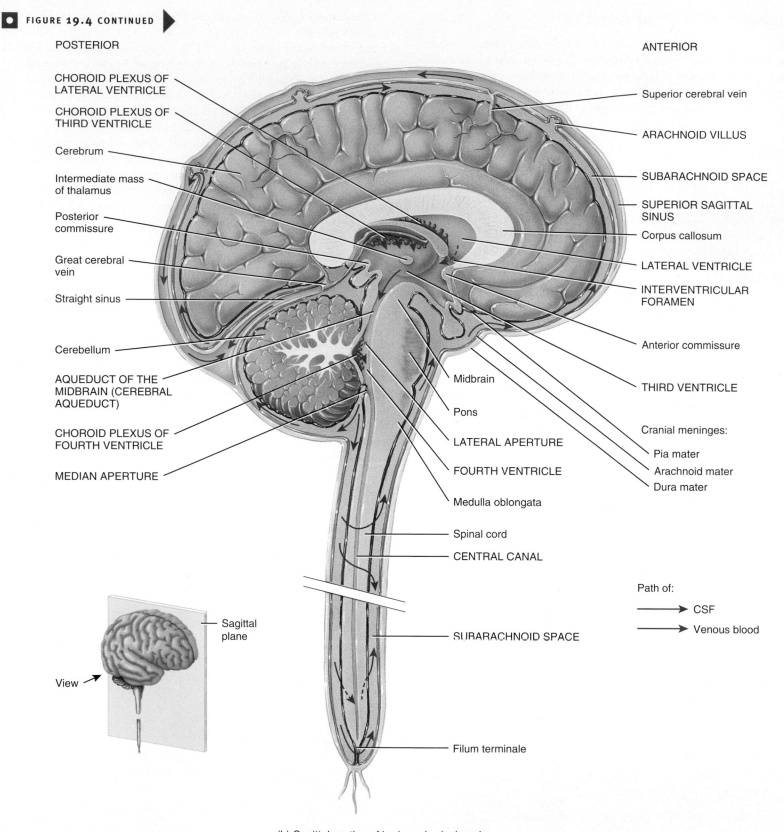

POSTERIOR

CHOROID PLEXUS OF
LATERAL VENTRICLE

CHOROID PLEXUS OF
THIRD VENTRICLE

Cerebrum

Intermediate mass
of thalamus

Posterior
commissure

Great cerebral
vein

Straight sinus

Cerebellum

AQUEDUCT OF THE
MIDBRAIN (CEREBRAL
AQUEDUCT)

CHOROID PLEXUS OF
FOURTH VENTRICLE

MEDIAN APERTURE

ANTERIOR

Superior cerebral vein

ARACHNOID VILLUS

SUBARACHNOID SPACE

SUPERIOR SAGITTAL
SINUS

Corpus callosum

LATERAL VENTRICLE

INTERVENTRICULAR
FORAMEN

Anterior commissure

THIRD VENTRICLE

Cranial meninges:

Pia mater
Arachnoid mater
Dura mater

Midbrain

Pons

LATERAL APERTURE

FOURTH VENTRICLE

Medulla oblongata

Spinal cord

CENTRAL CANAL

SUBARACHNOID SPACE

Filum terminale

Sagittal
plane

View

Path of:

CSF

Venous blood

(b) Sagittal section of brain and spinal cord

and in the subarachnoid space around the surface of the brain and spinal cord.

CSF is gradually reabsorbed into the blood through **arachnoid villi**, fingerlike extensions of the arachnoid that project into the dural venous sinuses, especially the **superior sagittal sinus** (see also Figure 19.2). (A *sinus* is an endothelial lined space in the thick-walled dura mater.) Unlike most veins these are noncollapsible spaces. Normally, CSF is reabsorbed as rapidly as it is formed by the choroid plexuses, at a rate of about

20 mL/hr (480 mL/day). Because the rates of formation and reabsorption are the same, the pressure of CSF is normally constant.

CHECKPOINT

4. What structures are the sites of CSF production, and where are they located?

5. What is the difference between the blood–brain barrier and the blood–cerebrospinal fluid barrier?

Superior sagittal sinus

ARACHNOID VILLUS

Falx cerebri

LATERAL VENTRICLE

CHOROID PLEXUS

Cerebrum

AQUEDUCT OF
THE MIDBRAIN
(cerebral aqueduct)

Cerebellum

FOURTH VENTRICLE

Corpus callosum

Septum pellucidum

THIRD VENTRICLE

SUBARACHNOID SPACE
(surrounding brain)

Tentorium cerebelli

LATERAL APERTURE

MEDIAN APERTURE

Frontal
plane

View

SPINAL CORD

SUBARACHNOID SPACE
(surrounding spinal cord)

(c) Frontal section of brain and spinal cord

? **Where is CSF reabsorbed?**

● CLINICAL **CONNECTION** | Hydrocephalus

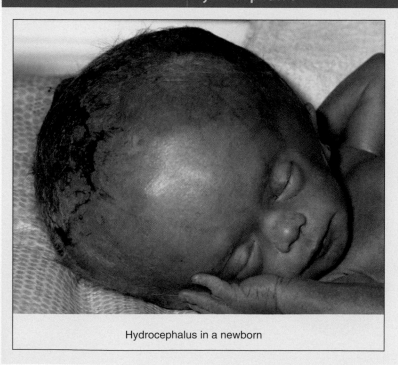

Hydrocephalus in a newborn

Abnormalities in the brain—tumors, inflammation, or developmental malformations—can interfere with the drainage of CSF from the ventricles into the subarachnoid space. When excess CSF accumulates in the ventricles, the CSF pressure rises. Elevated CSF pressure causes a condition called **hydrocephalus** (hī-drō-SEF-a-lus; *hydro-*=water; *cephal-*=head). In a baby whose fontanels have not yet closed, the head bulges due to the increased pressure. If the condition persists, the fluid buildup compresses and damages the delicate nervous tissue. Hydrocephalus is relieved by draining the excess CSF. In one procedure, called *endoscopic third ventriculostomy (ETV)*, a neurosurgeon makes a hole in the floor of the third ventricle and the CSF drains directly into the subarachnoid space. In adults, hydrocephalus may occur after head injury, meningitis, or subarachnoid hemorrhage. This condition can quickly become life-threatening and requires immediate intervention since the skull bones have already fused. ●

THE BRAIN STEM

- Describe the structures and functions of the brain stem.

The brain stem is the part of the brain between the spinal cord and the diencephalon. It consists of three structures: (1) medulla oblongata, (2) pons, and (3) midbrain. Extending through the brain stem is the reticular formation, a netlike region of interspersed gray and white matter.

Medulla Oblongata

The **medulla oblongata** (me-DOOL-la ob′-long-GA-ta), or more simply the **medulla**, is a continuation of the superior part of the spinal cord; it forms the inferior part of the brain stem (Figure 19.5; see also Figure 19.1). The medulla begins at the foramen magnum and extends to the inferior border of the pons, a distance of about 3 cm (1.2 in.). This short region of the central nervous system resembles the spinal cord in many ways. It gives rise to many nerve roots like the spinal cord; however, these are the roots of cranial nerves rather than spinal nerves. Six of the 12 pairs of cranial nerves arise from this region. While its surface features are similar to those of the spinal cord, its internal anatomy shows significant differences in the arrangement of the gray and white matter. Key landmarks that distinguish the medulla oblongata from the spinal cord are the pyramid and olive at its slightly expanded dorsal end, which will be discussed shortly.

Within the medulla's white matter are all sensory (ascending) and motor (descending) tracts extending between the spinal cord and other parts of the brain. Some of the white matter forms bulges on the anterior aspect of the medulla. These protrusions are the **pyramids** (Figure 19.6 on page 622; see also Figure 19.5a), formed by the largest motor tracts that pass from the cerebrum to the spinal cord. Just superior to the junction of the medulla with the spinal cord, 90% of the axons in the left pyramid cross to the right side, and 90% of the axons in the right pyramid cross to the left side. This crossing is called the **decussation of pyramids** (d ē′-ka-SĀ-shun; *decuss-*=crossing) and explains why each side of the brain controls movements on the opposite side of the body.

Figure 19.5 **Medulla oblongata in relation to the rest of the brain stem.**

🔑 The brain stem consists of the medulla oblongata, pons, and midbrain.

View

ANTERIOR

Cerebrum

Olfactory bulb

Olfactory tract

Pituitary gland

Optic tract

CEREBRAL PEDUNCLE OF MIDBRAIN

PONS

Cerebellar peduncle

MEDULLA OBLONGATA

Spinal nerve C1

Spinal cord

Cerebellum

CRANIAL NERVES:

Olfactory (I) nerve fibers

Optic (II) nerve

Oculomotor (III) nerve

Trochlear (IV) nerve

Trigeminal (V) nerve

Abducens (VI) nerve

Facial (VII) nerve

Vestibulocochlear (VIII) nerve

Glossopharyngeal (IX) nerve

Vagus (X) nerve

Accessory (XI) nerve

Hypoglossal (XII) nerve

POSTERIOR

(a) Inferior aspect of brain

Cerebrum

CRANIAL NERVES:

Olfactory bulb

Olfactory (I) nerve

Olfactory tract

Optic (II) nerve

Optic chiasm

Oculomotor (III) nerve

Optic tract

Trochlear (IV) nerve

Pituitary stalk

Trigeminal (V) nerve

Mammillary body

Abducens (VI) nerve

Ophthalmic nerve

Facial (VII) nerve

Maxillary nerve

Vestibulocochlear (VIII) nerve

Cerebral peduncle of midbrain

Glossopharyngeal (IX) nerve

Mandibular nerve

Middle cerebellar peduncle

Vagus (X) nerve

PONS

Accessory (XI) nerve

Medulla oblongata

Hypoglossal (XII) nerve

Olive

Cerebellum

Pyramids

Spinal cord

Decussation of pyramids

(b) Inferior aspect of brain

Cerebrum

Olfactory (I) nerve

Optic (II) nerve

Oculomotor (III) nerve

Abducens (VI) nerve

Trochlear (IV) nerve

Trigeminal (V) nerve

Cerebellum

Facial (VII) nerve

Vestibulocochlear (VIII) nerve

Glossopharyngeal (IX) nerve

Vagus (X) nerve

Accessory (XI) nerve

Hypoglossal (XII) nerve

(c) Posterolateral view of dissected brain

? **What part of the brain stem contains the pyramids? The cerebral peduncles? Literally means "bridge"?**

The medulla also contains several **nuclei**, masses of gray matter where neurons form synapses with one another. Several of these nuclei control vital body functions. The **cardiovascular center** regulates the rate and force of the heartbeat and the diameter of blood vessels. The **medullary rhythmicity area** of the respiratory center adjusts the basic rhythm of breathing. Other nuclei in the medulla control reflexes for vomiting, coughing, and sneezing.

Just lateral to each pyramid is an oval-shaped swelling called an **olive** (see Figures 19.5a and 19.6). Within the olive is the **inferior olivary nucleus**. Neurons here relay impulses from proprioceptors (monitoring joint and muscle positions) to the cerebellum.

Nuclei associated with sensations of touch, conscious proprioception, pressure, and vibration are located in the posterior part of the medulla. These nuclei are the right and left **gracile nucleus** (GRAS-il=slender) and **cuneate nucleus** (KŪ-nē-āt=wedge). Many ascending sensory axons form synapses in these nuclei, and postsynaptic neurons then relay the sensory information to the thalamus on the opposite side of the brain (see Figure 21.4a on page 686). The axons ascend to the thalamus in a band of white matter called the **medial lemniscus** (lem-NIS-kus=ribbon), which extends through the medulla, pons, and midbrain (see Figure 19.7b on page 624).

Finally, the medulla contains nuclei associated with the following five pairs of cranial nerves (see Figure 19.5):

1. **Vestibulocochlear (VIII) nerves.** Several nuclei in the medulla receive sensory input from and provide motor output to the cochlea of the internal ear via the cochlear branches of the vestibulocochlear nerves. These nerves convey impulses related to hearing.
2. **Glossopharyngeal (IX) nerves.** Nuclei in the medulla relay sensory and motor impulses related to taste, swallowing, and salivation via the glossopharyngeal nerves.
3. **Vagus (X) nerves.** Nuclei in the medulla receive sensory impulses from and provide motor impulses to the pharynx and larynx and many thoracic and abdominal viscera via the vagus nerves.
4. **Accessory (XI) nerves (cranial portion,** these fibers are actually vagal neurons). Nuclei in the medulla are the origin for nerve impulses that control swallowing via the cranial portion of the accessory nerves.
5. **Hypoglossal (XII) nerves.** Nuclei in the medulla are the origin for nerve impulses that control tongue movements during speech and swallowing via the hypoglossal nerves.

Pons

The **pons** (=bridge) lies directly superior to the medulla and anterior to the cerebellum and is about 2.5 cm (1 in.) long (see Figures 19.1 and 19.5). Like the medulla, the pons consists of both nuclei and tracts. As its name implies, the pons is a bridge that connects parts of the brain with one another. These connections are provided by bundles of axons. Some axons of the pons connect the right and left sides of the cerebellum. Others are part of ascending sensory tracts and descending motor tracts.

The pons has two major structural components: a ventral region and a dorsal region. The ventral region of the pons forms a large synaptic relay station consisting of scattered gray centers called the **pontine nuclei** (PON-tīn). Entering and exiting these

Figure 19.6 Internal anatomy of the medulla oblongata.

🔒 The pyramids of the medulla contain the largest motor tracts that run from the cerebrum to the spinal cord.

View

Transverse plane

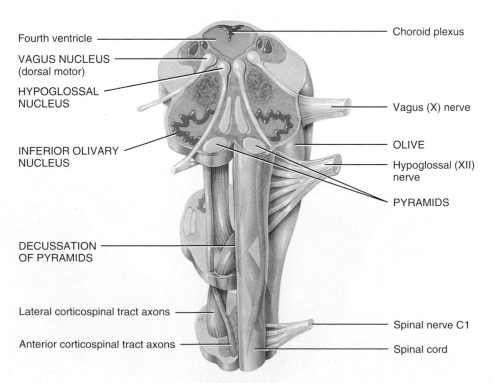

Fourth ventricle

VAGUS NUCLEUS (dorsal motor)

HYPOGLOSSAL NUCLEUS

INFERIOR OLIVARY NUCLEUS

DECUSSATION OF PYRAMIDS

Lateral corticospinal tract axons

Anterior corticospinal tract axons

Choroid plexus

Vagus (X) nerve

OLIVE

Hypoglossal (XII) nerve

PYRAMIDS

Spinal nerve C1

Spinal cord

Transverse section and anterior surface of medulla oblongata

❓ **What does decussation mean? What is the functional consequence of decussation of the pyramids?**

nuclei are numerous white matter tracts, each of which provides a connection between the cortex of a cerebral hemisphere and that of the opposite hemisphere of the cerebellum. This complex circuitry plays an essential role in coordinating and maximizing the effiency of voluntary motor output throughout the body. The dorsal region of the pons is more like the other regions of the brainstem, the medulla and midbrain. It contains ascending and descending tracts along with the nuclei of cranial nerves.

Other nuclei located in the pons are the **pneumotaxic area** (noo-mō-TAK-sik) and the **apneustic area** (ap-NOO-stik), shown in Figure 24.14 on page 775. Together with the medullary rhythmicity area, the pneumotaxic and apneustic areas help control breathing.

The pons also contains nuclei associated with the following four pairs of cranial nerves (see Figure 19.5):

1. **Trigeminal (V) nerves.** Nuclei in the pons receive sensory impulses for somatic sensations from the head and face and provide motor impulses that govern chewing via the trigeminal nerves.
2. **Abducens (VI) nerves.** Nuclei in the pons provide motor impulses that control eyeball movement via the abducens nerves.
3. **Facial (VII) nerves.** Nuclei in the pons receive sensory impulses for taste and provide motor impulses to regulate secretion of saliva and tears and contraction of muscles of facial expression via the facial nerves.
4. **Vestibulocochlear (VIII) nerves.** Nuclei in the pons receive sensory impulses from and provide motor impulses to the vestibular apparatus via the vestibular branches of the vestibulocochlear nerves. These nerves convey impulses related to balance and equilibrium.

Midbrain

This short segment of the brainstem sits just superior to the pons where it is obscured by the large, overlapping cerebral hemispheres. The **midbrain (mesencephalon)** extends from the pons to the diencephalon (see Figures 19.1 and 19.5) and is about 2.5 cm (1 in.) long. The aqueduct of the midbrain (cerebral aqueduct) passes through the midbrain, connecting the third ventricle above with the fourth ventricle below. Like the medulla and the pons, the midbrain contains both tracts and nuclei.

The anterior part of the midbrain contains a pair of tracts called **cerebral peduncles** (pe-DUNK-kuls or PĒ-dung-kuls=little feet; see Figures 19.5 and 19.7a, c). They contain axons of corticospinal, corticopontine, and corticobulbar motor neurons, which conduct nerve impulses from the cerebrum to the spinal cord, medulla, and pons. The cerebral peduncles also contain axons of sensory neurons that extend from the medulla to the thalamus.

The posterior part of the midbrain, called the **tectum** (TEK-tum=roof), contains four rounded elevations (Figure 19.7a). The two superior elevations are known as the **superior colliculi** (kō-LIK-ū-lī=little hills, singular is *colliculus*). These nuclei serve as reflex centers for certain visual activities. Through neural circuits from the retina of the eye to the superior colliculi to the extrinsic eye muscles, visual stimuli elicit eye movements for tracking moving images (such as a moving car) and scanning stationary images (as you are doing to read this sentence). Other superior colliculi reflexes are the accommodation reflex (adjusts shape of lens for close versus far vision) and reflexes that govern movements of the eyes, head, and neck in response to visual stimuli. The two inferior elevations, the **inferior colliculi**, are part of the

Figure 19.7 Midbrain.

The midbrain connects the pons to the diencephalon.

(a) Posterior view of midbrain in relation to brain stem

FIGURE **19.7** CONTINUES

FIGURE 19.7 CONTINUED ▶

View

Transverse plane

POSTERIOR

Tectum

Reticular formation

Medial lemniscus

Cerebral peduncle

ANTERIOR

SUPERIOR COLLICULUS
Periaqueductal gray matter
Aqueduct of the midbrain (cerebral aqueduct)
Medial geniculate nucleus
Oculomotor nucleus
RED NUCLEUS
SUBSTANTIA NIGRA
Corticospinal, corticopontine, and corticobulbar axons
Oculomotor (III) nerve

(b) Transverse section of midbrain

View

Midbrain
Pons
Medulla

SUPERIOR

Internal capsule

CEREBRAL PEDUNCLE
SUPERIOR COLLICULI
Trochlear (IV) nerve
Superior cerebellar peduncle
Middle cerebellar peduncle
Inferior cerebellar peduncle

Thalamus
Pineal gland
Medial geniculate nucleus
INFERIOR COLLICULI
Median eminence

Posterior median sulcus

INFERIOR

(c) Posterior view of midbrain in relation to brain stem

? **What is the importance of the cerebral peduncles?**

auditory pathway, relaying impulses from the receptors for hearing in the ear to the thalamus. These two nuclei also are reflex centers for the *startle reflex*, sudden movements of the head and body that occur when you are surprised by a loud noise such as a gunshot.

The midbrain contains several nuclei, including the left and right **substantia nigra** (sub-STAN-shē-a=substance; NĪ-gra= black), which are large, darkly pigmented nuclei. Neurons that release dopamine extend from the substantia nigra to the basal

ganglia and help control subconscious muscle activities (Figure 19.7b). Loss of these neurons is associated with Parkinson disease (see page 691). Also present are the left and right **red nuclei** (Figure 19.7b), which look reddish due to their rich blood supply and an iron-containing pigment in their neuronal cell bodies. Axons from the cerebellum and cerebral cortex form synapses in the red nuclei, which function with the cerebellum to coordinate muscular movements.

Finally, nuclei in the midbrain are associated with two pairs of cranial nerves (see Figure 19.5):

1. **Oculomotor (III) nerves.** Nuclei in the midbrain provide motor impulses that control movements of the eyeball, constriction of the pupil, and changes in shape of the lens via the oculomotor nerves.
2. **Trochlear (IV) nerves.** Nuclei in the midbrain provide motor impulses that control movements of the eyeball via the trochlear nerves.

Reticular Formation

In addition to the well-defined nuclei already described, much of the brain stem consists of small clusters of neuronal cell bodies (gray matter) interspersed among small bundles of myelinated axons (white matter). The broad region where white matter and gray matter exhibit a netlike arrangement is known as the **reticular formation** (*ret-*=net; Figure 19.7b). It extends from the superior part of the spinal cord, throughout the brain stem, and into the inferior part of the diencephalon. Neurons within the reticular formation have both ascending (sensory) and descending (motor) functions. Part of the reticular formation, called the **reticular activating system (RAS)**, consists of sensory axons that project to the cerebral cortex. The RAS helps maintain consciousness and is active during awakening from sleep. For example, we awaken to the sound of an alarm clock, to a flash of lightning, or to a painful pinch because of RAS activity that arouses the cerebral cortex. The reticular formation's main descending function is to help regulate *muscle tone*, which is the slight degree of contraction in normal resting muscles.

The functions of the brain stem are summarized in Table 19.2 on page 635.

⊙ **CHECKPOINT**

6. Where are the medulla, pons, and midbrain located relative to one another?
7. Define decussation of pyramids. Why is it important?
8. What body functions are governed by nuclei in the brain stem?
9. What are two important functions of the reticular formation?

THE CEREBELLUM

⊙ **OBJECTIVE**

• Describe the structure and functions of the cerebellum.

The **cerebellum**, the second-largest part of the brain, occupies the inferior and posterior aspects of the cranial cavity. Like the cerebrum, it has a highly folded surface that greatly increases the surface area of its outer gray matter cortex, allowing for a greater number of neurons. The cerebellum accounts for about a tenth of the brain mass yet contains nearly half of the neurons in the brain. The cerebellum is posterior to the medulla and pons and inferior to the posterior portion of the cerebrum (see Figure 19.1). A deep groove known as the **transverse fissure**, and the **tentorium cerebelli**, which supports the posterior part of the cerebrum, separate the cerebellum from the cerebrum (see Figure 19.4b). The tentorium cerebelli is a tentlike fold of the dura mater that is attached to the temporal and occipital bones.

In superior or inferior views, the shape of the cerebellum resembles a butterfly. The central constricted area is the **vermis** (=worm), and the lateral "wings" or lobes are the **cerebellar hemispheres** (Figure 19.8a, b). Each hemisphere consists of lobes separated by deep and distinct fissures. The **anterior lobe**

Figure 19.8 Cerebellum.

🔑 ⚏ The cerebellum coordinates skilled movements and regulates posture and balance.

(a) Superior view

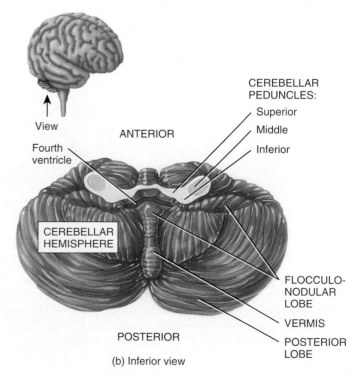

(b) Inferior view

FIGURE 19.8 CONTINUES ▶

and **posterior lobe** govern subconscious aspects of skeletal muscle movements. The **flocculonodular lobe** (flok-ū-lō-NOD-ū-lar; *flocculo-*=wool-like tuft) on the inferior surface contributes to equilibrium and balance.

The superficial layer of the cerebellum, called the **cerebellar cortex**, consists of gray matter in a series of slender, parallel ridges called **folia** (=leaves). Deep to the gray matter are tracts of white matter called **arbor vitae** (=tree of life) that resemble branches of a tree. Even deeper, within the white matter, are the **cerebellar nuclei**, regions of gray matter that give rise to axons carrying impulses from the cerebellum to other brain centers and to the spinal cord.

Three paired **cerebellar peduncles** (pe-DUNG-kls) attach the cerebellum to the brain stem (see also Figure 19.7c). These bundles of white matter consist of axons that conduct impulses between the cerebellum and other parts of the brain. The **inferior cerebellar peduncles** carry sensory information from the vestibular apparatus of the inner ear and from proprioceptors throughout the body into the cerebellum; their axons extend from the inferior olivary nucleus of the medulla and from the spinocerebellar tracts of the spinal cord into the cerebellum. The **middle cerebellar peduncles** are the largest peduncles; their axons carry commands for voluntary movements (those that originate in motor areas of the cerebral cortex) from the nuclei in the pons into the cerebellum. The **superior cerebellar peduncles** contain axons that extend from the cerebellum to the red nuclei of the midbrain and to several nuclei of the thalamus.

The primary function of the cerebellum is to evaluate how well movements initiated by motor areas in the cerebrum are actually being carried out. When movements initiated by the cerebral motor areas are not being carried out correctly, the cerebellum detects the discrepancies. It then sends feedback signals to motor areas of the cerebral cortex, via its connections to the red nucleus and thalamus. The feedback signals help correct the errors, smooth the movements, and coordinate complex sequences of skeletal muscle contractions. The cerebellum is also

■ **FIGURE 19.8 CONTINUED** ▶

Midsagittal plane

View

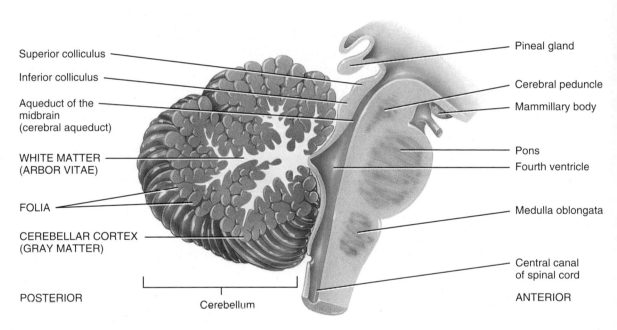

(c) Midsagittal section of cerebellum and brain stem

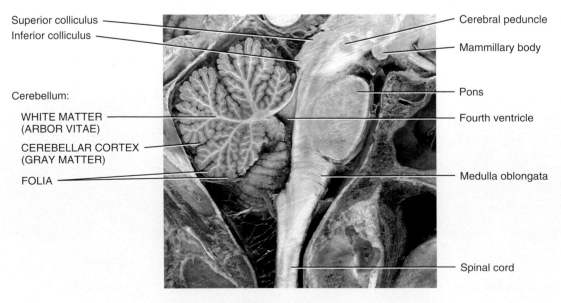

(d) Midsagittal section

❓ **Which fiber tracts carry information into and out of the cerebellum?**

the main brain region that regulates posture and balance. These aspects of cerebellar function make possible all skilled muscular activities, from catching a baseball to dancing to speaking. The presence of reciprocal connections between the cerebellum and association areas of the cerebral cortex suggest that the cerebellum may also have nonmotor functions such as cognition (acquisition of knowledge) and language processing. This view is supported by imaging studies like MRI and PET. Studies also suggest that the cerebellum may play a role in processing sensory information.

The functions of the cerebellum are summarized in Table 19.2 on page 635.

◉ CHECKPOINT

10. Describe the location and principal parts of the cerebellum.
11. Where do the axons of each of the three pairs of cerebellar peduncles begin and end? What are their functions?

Figure 19.9 Thalamus. Note the position of the thalamus in (a), the lateral view, and (b), the medial view. Various thalamic nuclei in (c) and (d) are correlated by color to the cortical regions to which they project in (a) and (b).

 The thalamus is the principal relay station for sensory impulses that reach the cerebral cortex from other parts of the brain and the spinal cord.

THE DIENCEPHALON

◉ OBJECTIVE

• Describe the components and functions of the diencephalon.

The **diencephalon** forms a central core of brain tissue just superior to the midbrain. It is almost completely surrounded by the cerebral hemispheres and contains numerous nuclei involved in a wide variety of both sensory and motor processing between higher and lower brain centers. The diencephalon extends from the brain stem to the cerebrum and surrounds the third ventricle; it includes the thalamus, hypothalamus, and epithalamus. Projecting from the hypothalamus is the hypophysis, or pituitary gland. The optic tracts carrying neurons from the retina enter this region of the brain.

Thalamus

The **thalamus** (THAL-a-mus=inner chamber), which measures about 3 cm (1.2 in.) in length, makes up 80% of the diencephalon. Thin layers of white matter partially outline the thalamus, which consists of paired oval masses of gray matter organized into nuclei with interspersed tracts of white matter (Figure 19.9). A bridge of gray matter called the **intermediate mass (interthalamic adhesion)** joins the right and left halves of the thalamus in about 70% of human brains.

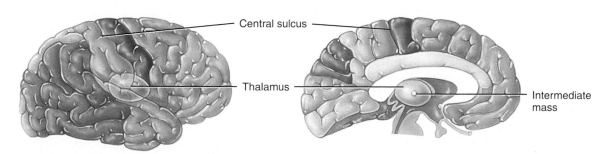

(a) Lateral view of right cerebral hemisphere   (b) Medial view of left cerebral hemisphere

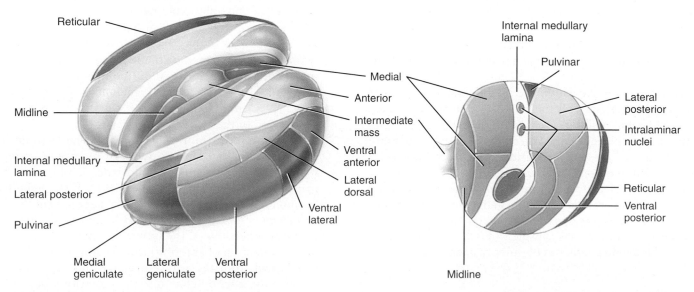

(c) Superolateral view of thalamus showing locations of thalamic nuclei (reticular nucleus is shown on the left side only; all other nuclei are shown on the right side)

(d) Transverse section of right side of thalamus showing locations of thalamic nuclei

❓ **What structure connects the right and left halves of the thalamus?**

The thalamus is the major relay station for most sensory impulses that reach the primary sensory areas of the cerebral cortex from the spinal cord, brain stem, and midbrain. Although crude perception of painful, thermal, and pressure sensations arises at the level of the thalamus, precise localization of these sensations depends on nerve impulses arriving at the cerebral cortex.

The thalamus contributes to motor functions by transmitting information from the cerebellum and basal ganglia to the primary motor area of the cerebral cortex. It also relays nerve impulses between different areas of the cerebrum. The thalamus also plays a role in the regulation of autonomic activities and the maintenance of consciousness. Axons that connect the thalamus and cerebral cortex pass through the **internal capsule**, a thick band of white matter lateral to the thalamus (see Figure 19.13b).

A vertical Y-shaped sheet of white matter called the **internal medullary lamina** divides the gray matter of the right and left sides of the thalamus (Figure 19.9c). It consists of myelinated axons that enter and leave the various thalamic nuclei.

Based on their positions and functions, there are seven major groups of nuclei on each side of the thalamus (Figure 19.9c, d).

1. The **anterior nucleus** connects to the hypothalamus and limbic system (described on page 634). It functions in emotions, regulation of alertness, and memory.
2. The **medial nuclei** connect to the cerebral cortex, limbic system, and basal ganglia. They function in emotions, learning, memory, awareness, and cognition (thinking and knowing).
3. Nuclei in the **lateral group** connect to the superior colliculi, limbic system, and cortex in all lobes of the cerebrum. The **lateral dorsal nucleus** functions in the expression of emotions. The **lateral posterior nucleus** and **pulvinar nucleus** help integrate sensory information.
4. Five nuclei are part of the **ventral group**. The **ventral anterior nucleus** contributes to motor functions, possibly movement planning. The **ventral lateral nucleus** connects to the cerebellum and motor parts of the cerebral cortex. Its neurons are active during movements on the opposite side of the body. The **ventral posterior nucleus** relays impulses for somatic sensations such as touch, pressure, proprioception, vibration, heat, cold, and pain from the face and body to the cerebral cortex. The **lateral geniculate nucleus** (je-NIK-ū-lat=bent like a knee) relays visual impulses for sight from the retina to the primary visual area of the cerebral cortex. The **medial geniculate nucleus** relays auditory impulses for hearing from the ear to the primary auditory area of the cerebral cortex.
5. **Intralaminar nuclei** lie within the internal medullary lamina and make connections with the reticular formation, cerebellum, basal ganglia, and wide areas of the cerebral cortex. They function in pain perception, integration of sensory and motor information, and arousal (activation of the cerebral cortex from the brain stem reticular formation).
6. The **midline nucleus** forms a thin band adjacent to the third ventricle and has a presumed function in memory and olfaction.
7. The **reticular nucleus** surrounds the lateral aspect of the thalamus, next to the internal capsule. This nucleus monitors, filters, and integrates activities of other thalamic nuclei.

Hypothalamus

The **hypothalamus** (*hypo-*=under) is a small part of the diencephalon located inferior to the thalamus. Weighing approximately 4 grams (0.14 ounces) this small area is much more important than

its small size suggests. It is composed of a dozen or so nuclei in four major regions:

1. The **mammillary region** (*mammill-*=nipple-shaped), adjacent to the midbrain, is the most posterior part of the hypothalamus. It includes the mammillary bodies and posterior hypothalamic nuclei (Figure 19.10). The **mammillary bodies** are two small, rounded projections that serve as relay stations for reflexes related to the sense of smell (see also Figure 19.5).
2. The **tuberal region**, the widest part of the hypothalamus, includes the dorsomedial, ventromedial, and arcuate nuclei, plus the stalklike **infundibulum** (in-fun-DIB-ū-lum=funnel), which connects the pituitary gland to the hypothalamus (Figure 19.10). The **median eminence** is a slightly raised region that encircles the infundibulum.
3. The **supraoptic region** (*supra-*=above; *-optic*=eye) lies superior to the optic chiasm (point of crossing of optic nerves) and contains the paraventricular nucleus, supraoptic nucleus, anterior hypothalamic nucleus, and suprachiasmatic nucleus (Figure 19.10). Axons from the paraventricular and supraoptic nuclei form the hypothalamohypophyseal tract, which extends through the infundibulum to the posterior lobe of the pituitary.
4. The **preoptic region** anterior to the supraoptic region is usually considered part of the hypothalamus because it participates with the hypothalamus in regulating certain autonomic activities. The preoptic region contains the medial and lateral preoptic nuclei (Figure 19.10).

The hypothalamus controls many body activities and is one of the major regulators of homeostasis. Sensory impulses related to both somatic and visceral senses arrive at the hypothalamus, as do impulses from receptors for vision, taste, and smell. Other receptors within the hypothalamus itself continually monitor osmotic pressure, glucose level, certain hormone concentrations, and the temperature of blood. The hypothalamus has several very important connections with the pituitary gland and produces a variety of hormones, which are described in more detail in Chapter 23. Some functions can be attributed to specific hypothalamic nuclei, but others are not so precisely localized. Important functions of the hypothalamus include the following:

1. **Control of the ANS.** The hypothalamus controls and integrates activities of the autonomic nervous system, which regulates contraction of smooth and cardiac muscle and the secretions of many glands. Axons extend from the hypothalamus to sympathetic and parasympathetic nuclei in the brain stem and spinal cord. Through the ANS, the hypothalamus is a major regulator of visceral activities, including regulation of heart rate, movement of food through the gastrointestinal tract, and contraction of the urinary bladder.
2. **Production of hormones.** The hypothalamus produces several hormones and has two types of important connections with the pituitary gland, an endocrine gland located inferior to the hypothalamus (see Figure 19.1). First, hypothalamic hormones are released into capillary networks in the median eminence. The bloodstream carries these hormones directly to the anterior lobe of the pituitary, where they stimulate or inhibit secretion of anterior pituitary hormones. Second, axons extend from the paraventricular and supraoptic nuclei through the infundibulum into the posterior lobe of the pituitary. The cell bodies of these neurons make one of two hormones (oxytocin or antidiuretic hormone). Their axons transport the hormones to the posterior pituitary, where they are released.

Figure 19.10 Hypothalamus. Selected portions of the hypothalamus and a three-dimensional representation of hypothalamic nuclei are shown (after Netter).

 The hypothalamus controls many body activities and is an important regulator of homeostasis.

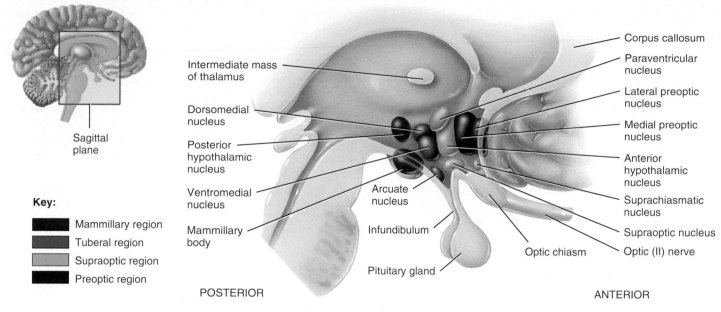

Sagittal section of brain showing hypothalamic nuclei

? What are the four major regions of the hypothalamus, from posterior to anterior?

3. **Regulation of emotional and behavioral patterns.** Together with the limbic system (described shortly), the hypothalamus participates in expressions of rage, aggression, pain, and pleasure, and the behavioral patterns related to sexual arousal.
4. **Regulation of eating and drinking.** The hypothalamus regulates food intake through the arcuate and paraventricular nuclei. It also contains a thirst center. When certain cells in the hypothalamus are stimulated by rising osmotic pressure of the extracellular fluid, they cause the sensation of thirst. The intake of water by drinking restores the osmotic pressure to normal, removing the stimulation and relieving the thirst.
5. **Control of body temperature.** If the temperature of blood flowing through the hypothalamus is above normal, the hypothalamus directs the autonomic nervous system to stimulate activities that promote heat loss. When blood temperature is below normal, the hypothalamus generates impulses that promote heat production and retention.
6. **Regulation of circadian rhythms and states of consciousness.** The suprachiasmatic nucleus establishes patterns of awakening and sleep that occur on a circadian (daily) schedule. This nucleus receives input from the eyes (retina) and sends output to other hypothalamic nuclei, the reticular formation, and the pineal gland.

Epithalamus

The **epithalamus** (*epi-*=above), a small region superior and posterior to the thalamus, consists of the pineal gland and habenular nuclei. The **pineal gland** (PĪN-ē-al=pinecone-like) is about the size of a small pea and protrudes from the posterior midline of the third ventricle (see Figure 19.1). The pineal gland is part of

the endocrine system because it secretes the hormone **melatonin**. As more melatonin is liberated during darkness than in light, this hormone is thought to promote sleepiness. Melatonin also appears to contribute to the setting of the body's biological clock. The **habenular nuclei** (ha-BEN-ū-lar), shown in Figure 19.7a, are involved in olfaction, especially emotional responses to odors such as a loved one's cologne or Mom's chocolate chip cookies baking in the oven.

The functions of the three parts of the diencephalon are summarized in Table 19.2 on page 635.

Circumventricular Organs

Parts of the diencephalon, called **circumventricular** (ser'-kum-ven-TRIK-ū-lar) **organs (CVOs)** because they lie in the wall of the third **ventricle**, can monitor chemical changes in the blood because they lack a blood–brain barrier. CVOs include part of the hypothalamus, the pineal gland, the pituitary gland, and a few other nearby structures. Functionally, these regions coordinate homeostatic activities of the endocrine and nervous systems, such as the regulation of blood pressure, fluid balance, hunger, and thirst. CVOs are also thought to be the sites of entry into the brain of HIV, the virus that causes AIDS. Once in the brain, HIV may cause dementia (irreversible deterioration of mental state) and other neurological disorders.

◉ CHECKPOINT
12. Why is the thalamus considered a "relay station" in the brain?
13. Why is the hypothalamus considered part of both the nervous system and the endocrine system?

THE CEREBRUM

⊙ OBJECTIVES
- Describe the cortex, convolutions, fissures, and sulci of the cerebrum.
- List and locate the lobes of the cerebrum.
- Describe the nuclei that comprise the basal ganglia.
- List the structures and describe the functions of the limbic system.

The cerebrum is the "seat of intelligence." It provides us with the ability to read, write, and speak; to make calculations and compose music; and to remember the past, plan for the future, and imagine things that have never existed before.

By far the largest portion of the human brain, the cerebrum consists of the cerebral hemispheres and the basal ganglia. The right and left halves of the cerebrum, called **cerebral hemispheres**, are separated by the falx cerebri. The hemispheres consist of an outer rim of gray matter, an internal region of cerebral white matter, and gray matter nuclei deep within the white matter. The outer rim of gray matter is the **cerebral cortex** (*cortex*=rind or bark) (Figure 19.11a). Although only 2–4 mm (0.08–0.16 in.) thick, the cerebral cortex contains billions of neurons. Deep to the cerebral cortex lies the cerebral white matter.

During embryonic development, when brain size increases rapidly, the gray matter of the cortex enlarges much faster than the deeper white matter. As a result, the cortical region rolls and folds upon itself. The folds are called **gyri** (JĪ-rī=circles) or **convolutions** (Figure 19.11a, b). The deepest grooves between folds are known as **fissures**; the shallower grooves between folds are termed **sulci** (SUL-sī=grooves). (The singular terms are *gyrus* and *sulcus*.) The most prominent fissure, the **longitudinal fissure**, separates the cerebrum into right and left halves called **cerebral hemispheres**. The hemispheres are connected internally by the **corpus callosum** (kal-LŌ-sum; *corpus*=body; *callosum*= hard), a broad band of white matter containing axons that extend between the hemispheres (see Figure 19.12b, c).

Figure 19.11 Cerebrum. Because the insula cannot be seen externally, it has been projected to the surface in (b).

⊙ The cerebrum is the "seat of intelligence"; it provides us with the ability to read, write, and speak; to make calculations and compose music; to remember the past and plan for the future; and to create.

Gyrus
Sulcus
Cerebral cortex
Cerebral white matter
Fissure

Details of a gyrus, sulcus, and fissure

ANTERIOR

Frontal lobe
Longitudinal fissure
Precentral gyrus
Central sulcus
Postcentral gyrus
Parietal lobe
Occipital lobe

Left hemisphere
Right hemisphere

POSTERIOR

(a) Superior view

Postcentral gyrus
Parietal lobe
Parieto-occipital sulcus
Occipital lobe
Transverse fissure
Cerebellum

Central sulcus
Precentral gyrus
Frontal lobe
Insula (projected to surface)
Lateral cerebral sulcus
Temporal lobe

(b) Right lateral view

POSTERIOR ANTERIOR

Central sulcus — — Precentral gyrus

Postcentral gyrus — — Frontal lobe

Parietal lobe —

— Insula

— Temporal lobe (cut)

Occipital lobe —

— Medulla oblongata

Cerebellum — — Spinal cord

(c) Right lateral view with temporal lobe cut away

? During development, does the gray matter or the white matter enlarge more rapidly? What are the brain folds, shallow grooves, and deep grooves called?

Lobes of the Cerebrum

Each cerebral hemisphere can be further subdivided into four lobes. The lobes are named after the bones that cover them: frontal, parietal, temporal, and occipital lobes (see Figure 19.11a–c). The **central sulcus** (SUL-kus) separates the **frontal lobe** from the **parietal lobe**. A major gyrus, the **precentral gyrus**— located immediately anterior to the central sulcus—contains the primary motor area of the cerebral cortex. Another major gyrus, the **postcentral gyrus**, which is located immediately posterior to the central sulcus, contains the primary somatosensory area of the cerebral cortex. The **lateral cerebral sulcus (fissure)** separates the **frontal lobe** from the **temporal lobe**. The **parieto-occipital sulcus** separates the **parietal lobe** from the **occipital lobe**. A fifth part of the cerebrum, the **insula**, cannot be seen at the surface of the brain because it lies within the lateral cerebral sulcus, deep to the parietal, frontal, and temporal lobes (Figure 19.11b, c).

Cerebral White Matter

The **cerebral white matter** consists primarily of myelinated axons in three types of tracts (Figure 19.12 and see also Figure 19.4a):

Figure 19.12 Organization of fibers into white matter tracts of the left cerebral hemisphere.

🔓 Association tracts, commissural tracts, and projection tracts form white matter tracts in the cerebral hemispheres.

Midsagittal plane — Cerebral cortex

— COMMISSURAL AND PROJECTION TRACTS

ASSOCIATION TRACTS — COMMISSURAL TRACTS:

View — CORPUS CALLOSUM

Septum pellucidum —

— ANTERIOR COMMISSURE

Mammillary body —

POSTERIOR ANTERIOR

Medial view of tracts revealed by removing gray matter from a midsagittal section

? Which fibers carry impulses between gyri of the same hemisphere? Between gyri in opposite hemispheres? Between the cerebrum and thalamus, brain stem, and spinal cord?

1. **Association tracts** contain axons that conduct nerve impulses between gyri in the same hemisphere.
2. **Commissural tracts** contain axons that conduct nerve impulses from gyri in one cerebral hemisphere to corresponding gyri in the other cerebral hemisphere. Three important groups of commissural tracts are the **corpus callosum** (the largest fiber bundle in the brain, containing about 300 million fibers), **anterior commissure**, and **posterior commissure**.
3. **Projection tracts** contain axons that conduct nerve impulses from the cerebrum to lower parts of the CNS (thalamus, brainstem, or spinal cord) or from lower parts of the CNS to the cerebrum. An example is the **internal capsule**, a thick band of white matter that contains both ascending and descending axons (see Figure 19.13b).

Basal Ganglia

Deep within each cerebral hemisphere are three nuclei (masses of gray matter) that are collectively termed the **basal ganglia** (Figure 19.13). (Recall that "ganglion" usually means a collection of neuronal cells bodies *outside* the CNS. The name here is the one exception to that general rule. An alternative term used in some textbooks—*basal nuclei*—is not used by most neuroscientists. It can be confused with the names of other brain regions, such as the nucleus basalis, which deteriorates in people who suffer from Alzheimer disease.)

Two of the basal ganglia are side-by-side, just lateral to the thalamus. They are the **globus pallidus** (GLŌ-bus PAL-i-dus; *globus*=ball; *pallidus*=pale), which is closer to the thalamus, and

Figure 19.13 Basal ganglia. In (a) and (b), the basal ganglia have been projected to the surface and are shown in aqua. (See Tortora, *A Photographic Atlas of the Human Body*, 2e, Figure 8.22.)

🔑 The basal ganglia control automatic movements of skeletal muscles and muscle tone.

(a) Lateral view of right side of brain

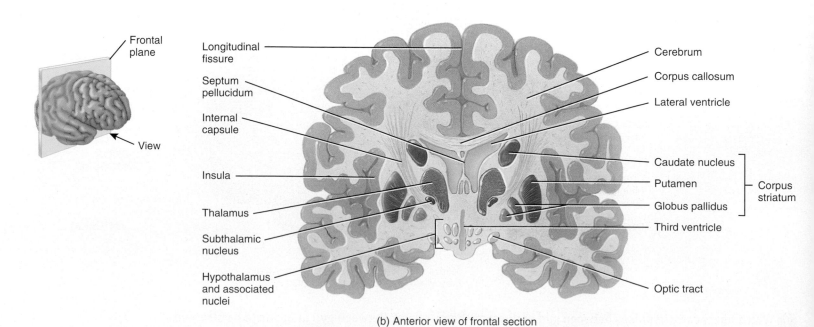

(b) Anterior view of frontal section

the **putamen** (pū-TĀ-men=shell), which is closer to the cerebral cortex. Together, the globus pallidus and putamen are referred to as the **lentiform nucleus** (LEN-ti-form=shaped like a lens.) The third of the basal ganglia is the **caudate nucleus** (KAW-dāt; *caud-*=tail), which has a large "head" connected to a smaller "tail" by a long comma-shaped "body." Together, the lentiform and caudate nuclei are known as the **corpus striatum** (strī-Ā-tum; *corpus*=body; *striatum*=striated). The term corpus striatum refers to the striated (striped) appearance of the internal capsule as it passes among the basal ganglia. Nearby structures that are functionally linked to the basal ganglia are the *substantia nigra* of the midbrain (see Figure 19.7b) and the *subthalamic nuclei* of the diencephalon (see Figure 19.13b). Axons from the substantia nigra terminate in the caudate nucleus and putamen. The subthalamic nuclei interconnect with the globus pallidus.

The basal ganglia receive input from the cerebral cortex and provide output to motor parts of the cortex via the medial and ventral group nuclei of the thalamus. In addition, the nuclei of the basal ganglia have extensive connections with one another. A major function of the basal ganglia is to help regulate initiation and termination of movements. Activity of neurons in the putamen precedes or anticipates body movements; activity of neurons in the caudate nucleus occurs prior to eye movements. The globus pallidus helps regulate the muscle tone required for specific body movements. The basal ganglia also control subconscious contractions of skeletal muscles. Examples include automatic arm swings while walking and true laughter in response to a joke (not the kind you consciously initiate to humor your anatomy instructor).

In addition to influencing motor function, the basal ganglia have other roles. They help initiate and terminate some cognitive processes, such as attention, memory, and planning, and may act with the limbic system to regulate emotional behaviors. Some psychiatric disorders, such as obsessive-compulsive disorder, schizophrenia, and chronic anxiety, are thought to involve dysfunction of circuits between the basal ganglia and the limbic system.

The Limbic System

Encircling the upper part of the brain stem and the corpus callosum is a ring of structures on the inner border of the cerebrum and floor of the diencephalon that constitutes the **limbic system** (*limbic*=border). The main components of the limbic system are as follows (Figure 19.14):

1. The so-called **limbic lobe** is a rim of cerebral cortex on the medial surface of each hemisphere. It includes the **cingulate gyrus** (*cingul-*=belt), which lies above the corpus callosum, and the **parahippocampal gyrus** (par′-a-hip-ō-KAM-pal), which is in the temporal lobe below. The **hippocampus** (=seahorse) is a portion of the parahippocampal gyrus that extends into the floor of the lateral ventricle.
2. The **dentate gyrus** (*dentate*=toothed) lies between the hippocampus and parahippocampal gyrus.
3. The **amygdala** (*amygda-*=almond-shaped) is composed of several groups of neurons located close to the tail of the caudate nucleus.
4. The **septal nuclei** are located within the septal area formed by the regions under the corpus callosum and the paraterminal gyrus (a cerebral gyrus).
5. The **mammillary bodies of the hypothalamus** are two round masses close to the midline near the cerebral peduncles.

Longitudinal fissure

Corpus callosum

Lateral ventricle

Internal capsule

Fornix

Insula

Thalamus

Hypothalamus

Optic chiasm

Cerebral cortex

Cerebral white matter

Septum pellucidum

Caudate nucleus

Putamen

Globus pallidus

Corpus striatum

Anterior commissure

Third ventricle

(c) Anterior view of frontal section

? **Where are the basal ganglia located relative to the thalamus?**

Figure 19.14 Components of the limbic system and surrounding structures.

The limbic system governs emotional aspects of behavior.

Sagittal plane

View

Fornix

Stria medullaris

Stria terminalis

Hippocampus (in temporal lobe)

Dentate gyrus

Anterior nucleus of thalamus

Mammillothalamic tract

Corpus callosum

Cingulate gyrus (in frontal lobe)

Anterior commissure

Septal nuclei

Mammillary body in hypothalamus

Olfactory bulb

Amygdala

Parahippocampal gyrus (in temporal lobe)

POSTERIOR

Sagittal section

ANTERIOR

? **Which part of the limbic system functions with the cerebrum in memory?**

6. Two nuclei of the thalamus, the **anterior nucleus** and the **medial nucleus**, participate in limbic circuits.
7. The **olfactory bulbs** are flattened bodies of the olfactory pathway that rest on the cribriform plate.
8. The **fornix, stria terminalis, stria medullaris, medial forebrain bundle**, and **mammillothalamic tract** (mam-i-lō-tha-LAM-ik) are linked by bundles of interconnecting myelinated axons.

The limbic system is sometimes called the "emotional brain" because it plays a primary role in a range of emotions, including pain, pleasure, docility, affection, and anger. It also is involved in olfaction (smell) and memory. Experiments have shown that when different areas of an animals' limbic system are stimulated, the animals' reactions indicate that they are experiencing intense pain or extreme pleasure. Stimulation of other limbic system areas in animals produces tameness and signs of affection. Stimulation

---

•CLINICAL CONNECTION | Brain Injuries

Brain injuries are commonly associated with head trauma and result in part from displacement and distortion of neuronal tissue at the moment of impact. Additional tissue damage occurs when normal blood flow is restored after a period of ischemia (reduced blood flow). The sudden increase in oxygen level produces large numbers of oxygen free radicals (charged oxygen molecules with an unpaired electron). Brain cells recovering from the effects of a stroke or cardiac arrest also release free radicals. Free radicals cause damage by disrupting cellular DNA and enzymes and by altering plasma membrane permeability. Brain injuries can also result from hypoxia.

Various degrees of brain injury are described by specific terms. A **concussion** is an injury characterized by an abrupt, but temporary, loss of consciousness (from seconds to hours), disturbances of vision, and problems with equilibrium. It is caused by a blow to the head or the sudden stopping of a moving head and is the most common brain injury. A concussion produces no obvious bruising of the brain. Signs of a concussion are headache, drowsiness, nausea and/or vomiting, lack of concentration, confusion, or post-traumatic amnesia (memory loss).

A **contusion** is bruising of the brain due to trauma, and includes the leakage of blood from microscopic vessels. It is usually associated with a concussion. In a contusion, the pia mater may be torn, allowing blood to enter the subarachnoid space. The area most commonly affected is the frontal lobe. A contusion usually results in an immediate loss of consciousness (generally lasting no longer than 5 minutes), loss of reflexes, transient cessation of respiration, and decreased blood pressure. Vital signs typically stabilize in a few seconds.

A **laceration** is a tear of the brain, usually from a skull fracture or a gunshot wound. A laceration results in rupture of large blood vessels, with bleeding into the brain and subarachnoid space. Consequences include cerebral hematoma (localized pool of blood, usually clotted, that swells against the brain tissue), edema, and increased intracranial pressure. If the blood clot is small enough, it may pose no major threat and may be absorbed. If the blood clot is large it may require surgical evacuation. Swelling infringes on the limited space that the brain occupies in the cranial cavity. Swelling causes excruciating headaches. Brain tissue can become necrotic due to the swelling; if the swelling is severe enough, the brain can herniate through the foramen magnum, resulting in death. •

of a cat's amygdala or certain nuclei of the hypothalamus produces a behavioral pattern called rage—the cat extends its claws, raises its tail, opens its eyes wide, hisses, and spits. By contrast, removal of the amygdala produces an animal that lacks fear and aggression. Likewise, a person whose amygdala is damaged fails to recognize fearful expressions in others or to express fear in situations where this emotion would normally be appropriate.

The hippocampus, together with other parts of the cerebrum, functions in memory. People with damage to certain limbic system structures forget recent events and cannot commit anything to memory.

The functions of the cerebrum are summarized in Table 19.2.

◉ **CHECKPOINT**

14. How are the lobes of the cerebrum separated from one another? What is the insula?

15. How is cerebral white matter organized?
16. What are the functions of the basal ganglia?
17. What system is the amygdala part of and what is its function?
18. Describe the cortex, convolutions, fissures, and sulci of the cerebrum.
19. List and locate the lobes of the cerebrum. How are they separated from one another? What is the insula?
20. Describe the organization of cerebral white matter and indicate the function of each major group of fibers.
21. Give the name and function of each of the nuclei that form basal ganglia, and describe the effects of basal ganglia damage.
22. Define the limbic system and list several of its functions.

TABLE 19.2

Summary of Functions of Principal Parts of the Brain

PART	FUNCTION	PART	FUNCTION
Brain stem Medulla oblongata Pons Midbrain	*Medulla oblongata:* Relays sensory input and motor output between other parts of the brain and the spinal cord. Reticular formation (also in pons, midbrain, and diencephalon) functions in consciousness and arousal. Vital centers regulate heartbeat, blood vessel diameter, and breathing (together with pons). Other centers coordinate swallowing, vomiting, coughing, sneezing, and hiccupping. Contains nuclei of origin for cranial nerves VIII, IX, X, XI, and XII. *Pons:* Relays impulses from one side of the cerebellum to the other and between the medulla and midbrain. Pneumotaxic area and apneustic area, together with the medulla, help control breathing. Contains nuclei of origin for cranial nerves V, VI, VII, and VIII. *Midbrain:* Relays motor output from the cerebral cortex to the pons and sensory input from the spinal cord to the thalamus. Superior colliculi coordinate movements of the eyeballs in response to visual and other stimuli, and the inferior colliculi coordinate movements of the head and trunk in response to auditory stimuli. Most of substantia nigra and red nucleus contribute to control of movement. Contains nuclei of origin for cranial nerves III and IV.	**Diencephalon** Epithalamus Thalamus Hypothalamus	*Thalamus:* Relays almost all sensory input to the cerebral cortex. Provides crude perception of touch, pressure, pain, and temperature. Includes nuclei involved in movement planning and control. *Hypothalamus:* Controls and integrates activities of the autonomic nervous system and pituitary gland. Regulates emotional and behavioral patterns and circadian rhythms. Controls body temperature and regulates eating and drinking behavior. Helps maintain the waking state and establishes patterns of sleep. Produces the hormones oxytocin and antidiuretic hormone (ADH). *Epithalamus:* Consists of pineal gland, which secretes melatonin, and the habenular nuclei.
Cerebellum Cerebellum	Compares intended movements with what is actually happening to smooth and coordinate complex, skilled movements. Regulates posture and balance. May have a role in cognition and language processing.	**Cerebrum** Cerebrum	Sensory areas are involved in the perception of sensory information; motor areas control muscular movement; and association areas deal with more complex integrative functions such as memory, personality traits, and intelligence. Basal ganglia coordinate gross, automatic muscle movements and regulate muscle tone. Limbic system functions in emotional aspects of behavior related to survival.

FUNCTIONAL ORGANIZATION OF THE CEREBRAL CORTEX

◉ OBJECTIVE

• Describe the locations and functions of the sensory, association, and motor areas of the cerebral cortex.

Specific types of sensory, motor, and integrative signals are processed in certain cerebral regions (Figure 19.15). Generally, **sensory areas** receive and interpret sensory impulses, **motor areas** initiate movements, and **association areas** deal with more complex integrative functions such as memory, emotions, reasoning, will, judgment, personality traits, and intelligence.

Sensory Areas

Sensory impulses arrive mainly in the posterior half of both cerebral hemispheres, in regions behind the central sulci. In the cerebral cortex, primary sensory areas have the most direct connections with peripheral sensory receptors.

Secondary sensory areas and sensory association areas often are adjacent to the primary areas. They usually receive input both from the primary areas and from other brain regions. Secondary sensory areas and sensory association areas integrate sensory experiences to generate meaningful patterns of recognition and awareness. A person with damage in the *primary* visual area would be blind in at least part of his visual field, but a person with damage to a visual *association* area might see normally yet be unable to recognize a friend.

The following are some important sensory areas (Figure 19.15):

• The **primary somatosensory area** (areas 1, 2, and 3) is located directly posterior to the central sulcus of each cerebral hemisphere in the postcentral gyrus of each parietal lobe. It extends from the lateral cerebral sulcus, along the lateral surface of the parietal lobe to the longitudinal fissure, and then along the medial surface of the parietal lobe within the longitudinal fissure.

The primary somatosensory area receives nerve impulses for touch, proprioception (joint and muscle position), pain, itching, tickle, and thermal sensations. A "map" of the entire body is present in the primary somatosensory area: Each point within the area receives impulses from a specific part of the body (see Figure 21.5a on page 687). The size of the cortical area receiving impulses from a particular part of the body depends on the number of receptors present there rather than on the size of the body part. For example, a larger region of the somatosensory area receives impulses from the lips and fingertips than from the thorax or hip. The major function of the primary somatosensory area is to pinpoint the areas where sensations originate, so that you know exactly where on your body to swat at that mosquito.

Figure 19.15 Functional areas of the cerebrum. Broca's speech area and Wernicke's area are in the left cerebral hemisphere of most people; they are shown here to indicate their relative locations. The numbers, still used today, are from K. Brodmann's map of the cerebral cortex, first published in 1909.

🔑 Particular areas of the cerebral cortex process sensory, motor, and integrative signals.

Lateral view of right cerebral hemisphere

❓ **What area(s) of the cerebrum integrate(s) interpretation of visual, auditory, and somatic sensations? Translates thoughts into speech? Controls skilled muscular movements? Interprets sensations related to taste? Interprets pitch and rhythm? Interprets shape, color, and movement of objects? Controls voluntary scanning movements of the eyes?**

- The **primary visual area** (area 17), located at the posterior tip of the occipital lobe mainly on the medial surface (next to the longitudinal fissure), receives visual information and is involved in visual perception.
- The **primary auditory area** (areas 41 and 42), located in the superior part of the temporal lobe near the lateral cerebral sulcus, receives information for sound and is involved in auditory perception.
- The **primary gustatory area** (area 43), located at the base of the postcentral gyrus superior to the lateral cerebral sulcus in the parietal cortex, receives impulses for taste and is involved in gustatory perception.
- The **primary olfactory area** (area 28), located in the temporal lobe on the medial aspect (and thus not visible in Figure 19.15), receives impulses for smell and is involved in olfactory perception.

Motor Areas

Motor output from the cerebral cortex flows mainly from the anterior part of each hemisphere. Among the most important motor areas are the following (Figure 19.15):

- The **primary motor area** (area 4) is located in the precentral gyrus of the frontal lobe. Each region in the primary motor area controls voluntary contractions of specific muscles or groups of muscles (see Figure 21.5b on page 687). Electrical stimulation of any point in the primary motor area causes contraction of specific skeletal muscle fibers on the opposite side of the body. As is true for the primary somatosensory area, body parts do not "map" to the primary motor area in proportion to their size. More cortical area is devoted to those muscles involved in skilled, complex, or delicate movements. For instance, the cortical region devoted to muscles that move the fingers is much larger than the region for muscles that move the toes.
- **Broca's** (BRŌ-kaz) **speech area** (areas 44 and 45) is located in the frontal lobe close to the lateral cerebral sulcus. Speaking and understanding language are complex activities that involve several sensory, association, and motor areas of the cortex. In about 97% of the population, these language areas are localized in the *left* hemisphere. The planning and production of speech occurs in the *left* frontal lobe in most people. From Broca's speech area, nerve impulses pass to the premotor regions that control the muscles of the larynx, pharynx, and mouth. The impulses from the premotor area result in specific, coordinated muscle contractions. Simultaneously, impulses propagate from Broca's speech area to the primary motor area. From here, impulses also control the breathing muscles to regulate the proper flow of air past the vocal cords. The coordinated contractions of your speech and breathing muscles enable you to speak your thoughts. People who suffer a cerebrovascular accident (CVA) or stroke in this area can still have clear thoughts, but are unable to form words, a phenomenon referred to as *nonfluent aphasia*.

Association Areas

The association areas of the cerebrum consist of some motor and sensory areas, plus large areas on the lateral surfaces of the occipital, parietal, and temporal lobes and on the frontal lobes anterior to the motor areas. Association areas are connected with one another by association tracts and include the following (Figure 19.15):

- The **somatosensory association area** (areas 5 and 7) is just posterior to and receives input from the primary somatosensory area, as well as from the thalamus and other parts of the brain. This area permits you to determine the exact shape and texture of an object without looking at it, to determine the orientation of one object with respect to another as they are felt, and to sense the relationship of one body part to another. Another role of the somatosensory association area is the storage of memories of past sensory experiences, enabling you to compare current sensations with previous experiences. For example, the somatosensory association area allows you to recognize objects such as a pencil and a paperclip simply by touching them.
- The **prefrontal cortex (frontal association area)** is an extensive area in the anterior portion of the frontal lobe that is well-developed in primates, especially humans (areas 9, 10, 11, and 12; area 12 is not illustrated since it can be seen only in a medial view). This area has numerous connections with other areas of the cerebral cortex, thalamus, hypothalamus, limbic system, and cerebellum. The prefrontal cortex is concerned with the makeup of a person's personality, intellect, complex learning abilities, recall of information, initiative, judgment, foresight, reasoning, conscience, intuition, mood, planning for the future, and development of abstract ideas. A person with bilateral damage to the prefrontal cortices typically becomes rude, inconsiderate, incapable of accepting advice, moody, inattentive, less creative, unable to plan for the future, and incapable of anticipating the consequences of rash or reckless words or behavior.
- The **visual association area** (areas 18 and 19), located in the occipital lobe, receives sensory impulses from the primary visual area and the thalamus. It relates present and past visual experiences and is essential for recognizing and evaluating what is seen. For example, the visual association area allows you to recognize an object such as a spoon simply by looking at it.
- The **auditory association area** (area 22), located inferior and posterior to the primary auditory area in the temporal cortex, allows you to recognize a particular sound as speech, music, or noise.
- **Wernicke's (posterior language) area** (VER-ni-kēz) (area 22, and possibly areas 39 and 40), a broad region in the *left* temporal and parietal lobes, interprets the meaning of speech by recognizing spoken words. It is active as you translate words into thoughts. The regions in the *right* hemisphere that correspond to Broca's and Wernicke's areas in the left hemisphere also contribute to verbal communication by adding emotional content, such as anger or joy, to spoken words. Unlike those who have CVAs in Broca's area, people who suffer strokes in Wernicke's area can still speak, but cannot arrange words in a coherent fashion, a disorder called *fluent aphasia*, or "word salad."
- The **common integrative area** (areas 5, 7, 39, and 40) is bordered by somatosensory, visual, and auditory association areas. It receives nerve impulses from these areas and from the primary gustatory area, primary olfactory area, the thalamus, and parts of the brain stem. This area integrates sensory interpretations from the association areas and impulses from other areas, allowing the formation of thoughts based on a variety of sensory inputs. It then transmits signals to other parts of the brain for the appropriate response to the sensory signals it has interpreted.
- The **premotor area** (area 6) is a motor association area that is immediately anterior to the primary motor area. Neurons in

this area communicate with the primary motor cortex, the sensory association areas in the parietal lobe, the basal ganglia, and the thalamus. The premotor area deals with learned motor activities of a complex and sequential nature. It generates nerve impulses that cause specific groups of muscles to contract in a specific sequence, as when you write your name. The premotor area also serves as a memory bank for such movements.

• The **frontal eye field area** (area 8) in the frontal cortex is sometimes included in the premotor area. It controls voluntary scanning movements of the eyes—like those you just used in reading this sentence.

Hemispheric Lateralization

Although the brain is almost symmetrical on its right and left sides, subtle anatomical differences between the two hemispheres exist. For example, in about two-thirds of the population, the planum temporale, a region of the temporal lobe that includes Wernicke's area, is 50% larger on the left side than on the right side. This asymmetry appears in the human fetus at about 30 weeks of gestation. Physiological differences also exist; although the two hemispheres share performance of many functions, each hemisphere also specializes in performing certain unique functions. This functional asymmetry is termed **hemispheric lateralization**.

In the most obvious example of hemispheric lateralization, the left hemisphere receives somatic sensory signals from and controls muscles on the right side of the body, and the right hemisphere receives sensory signals from and controls the left side of the body. In most people the left hemisphere is more important for reasoning, numerical and scientific skills, spoken and written language, and the ability to use and understand sign language. Patients with damage in the left hemisphere often exhibit aphasia. Conversely, the right hemisphere is more specialized for musical and artistic awareness; spatial and pattern perception; recognition of faces and emotional content of language; discrimination of different smells; and generating mental images of sight, sound, touch, taste, and smell to compare relationships among them.

Patients with damage in right hemisphere regions that correspond to Broca's and Wernicke's areas in the left hemisphere speak in a monotonous voice, having lost the ability to impart emotional inflection to what they say.

Despite some dramatic differences in functions of the two hemispheres, there is considerable variation from one person to another. Also, lateralization seems less pronounced in females than in males, both for language (left hemisphere) and for visual and spatial skills (right hemisphere). For instance, females are less likely than males to suffer aphasia after damage to the left hemisphere. A possibly related observation is that the anterior commissure is 12% larger and the corpus callosum has a broader posterior portion in females. Recall that both the anterior commissure and the corpus callosum are commissural tracts that provide communication between the two hemispheres.

Table 19.3 summarizes some of the functional differences between the two cerebral hemispheres.

Brain Waves

At any instant, brain neurons are generating millions of nerve impulses (action potentials). Taken together, these electrical signals are called **brain waves**. Brain waves generated by neurons close to the brain surface, mainly neurons in the cerebral cortex, can be detected by sensors called electrodes placed on the forehead and scalp. A record of such waves is called an **electroencephalogram** (e-lek′-trō-en-SEF-a-lō-gram; *electro-*=electricity; *-gram*=recording) or **EEG**. Electroencephalograms are useful both in studying normal brain functions, such as changes that occur during sleep, and in diagnosing a variety of brain disorders, such as epilepsy, tumors, metabolic abnormalities, sites of trauma, and degenerative diseases. The EEG is also utilized to determine if "life" is present, that is, to establish or confirm that death or brain death has occurred.

◉ **CHECKPOINT**

23. Compare the functions of the sensory, motor, and association areas of the cerebral cortex.
24. What is hemispheric lateralization?
25. What is the diagnostic value of an EEG?

TABLE 19.3
Functional Differences Between the Two Cerebral Hemispheres

LEFT HEMISPHERE FUNCTIONS	RIGHT HEMISPHERE FUNCTIONS
Receives somatic sensory signals from and controls muscles on right side of body.	Receives somatic sensory signals from and controls muscles on left side of body.
Reasoning.	Musical and artistic awareness.
Numerical and scientific skills.	Space and pattern perception.
Ability to use and understand sign language.	Recognition of faces and emotional content of facial expressions.
Spoken and written language.	Generating emotional content of language.
	Generating mental images to compare spatial relationships.
	Identifying and discriminating among odors.

• CLINICAL CONNECTION | Cerebrovascular Accident

The most common brain disorder is a **cerebrovascular accident (CVA)**, also called a **stroke** or **brain attack**. CVAs affect 500,000 people a year in the United States and represent the third leading cause of death, behind heart attacks and cancer. A CVA is characterized by abrupt onset of persistent neurological symptoms, such as paralysis or loss of sensation, that arise from destruction of brain tissue. Common causes of CVAs are intracerebral hemorrhage (from a blood vessel in the pia mater or brain), emboli (blood clots), and atherosclerosis (formation of cholesterol-containing plaques that block blood flow) of the cerebral arteries.

Among the risk factors implicated in CVAs are high blood pressure, high blood cholesterol, heart disease, narrowed carotid arteries, transient ischemic attacks (TIAs; discussed next), diabetes, smoking, obesity, and excessive alcohol intake.

A clot-dissolving drug called *tissue plasminogen activator* (*t-PA*) is now being used to open up blocked blood vessels in the brain. The drug is most effective when administered within three hours of the onset of the CVA, however, and is helpful only for CVAs due to a blood clot. Use of t-PA can decrease the permanent disability associated with these types of CVAs by 50%. New studies show that "cold therapy" might be successful in limiting the amount of residual damage from a CVA. These "cooling" therapies developed from knowledge obtained following examination of cold water drowning victims. States of hypothermia seem to trigger a survival response in which the body requires less oxygen; this principle has been showing promise in the treatment of stroke victims. Some commercial companies now provide "CVA survival kits," which include cooling blankets that can be kept in the home. •

Ischemic stroke

Hemorrhagic stroke

Atherosclerotic blood vessel

Ruptured blood vessel

CRANIAL NERVES

◉ OBJECTIVE

- Identify the cranial nerves by name, number, and type, and give the function of each.

The 12 pairs of **cranial nerves** are so named because they pass through various foramina in the bones of the cranium and arise from the brain inside the cranial cavity. Like the 31 pairs of spinal nerves, they are part of the peripheral nervous system (PNS). Each cranial nerve has both a number, designated by a roman numeral, and a name (see Figure 19.5). The numbers indicate the order, from anterior to posterior, in which the nerves arise from the brain. The names designate a nerve's distribution or function.

Cranial nerves emerge from the nose (cranial nerve I), the eyes (cranial nerve II), the inner ear (cranial nerve VIII), the brain stem (cranial nerves III–XII), and the spinal cord (cranial nerve XI). Three cranial nerves (I, II, and VIII) carry axons of sensory neurons and thus are called **sensory nerves**. Five cranial nerves (III, IV, VI, XI, and XII) contain only axons of motor neurons as they leave the brain stem and are called **motor nerves.** The other four cranial nerves (V, VII, IX, and X) are **mixed nerves** because they contain axons of both sensory and motor neurons. The cell bodies of sensory neurons are located in ganglia outside the brain. The cell bodies of motor neurons lie in nuclei within the brain.

Note: Table 19.4 on pages 649–652 contains a summary of the cranial nerves, including clinical connections related to their dysfunction.

Cranial nerves III, VII, IX, and X include both somatic and autonomic motor axons. The somatic axons innervate skeletal muscles; the autonomic axons, which are part of the parasympathetic division, innervate glands, smooth muscle, and cardiac muscle. Although the cranial nerves are mentioned singly in the following descriptions of their type, location, and function, remember that they are paired structures. The cell bodies of sensory neurons are located in ganglia outside the brain (with one exception, all proprioceptive sensory neurons in the head region have their cell bodies in the mesencephalic (ganglion) nucleus); the cell bodies of motor neurons lie in nuclei within the brain.

Olfactory (I) Nerve

The **olfactory (I) nerve** (ol-FAK-tō-rē; *olfact-*=to smell) is entirely sensory; it contains axons that conduct nerve impulses for olfaction, the sense of smell (Figure 19.16). The olfactory epithelium occupies the superior part of the nasal cavity, covering the inferior surface of the cribriform plate and extending down along the superior nasal concha. The olfactory receptors within the olfactory epithelium are bipolar neurons. Each has a single odor-sensitive dendrite projecting from one side of the cell body and an unmyelinated axon extending from the other side. Bundles of axons of olfactory receptors extend through about 20 olfactory foramina in the cribriform plate of the ethmoid bone on each side of the nose. These 40 or so bundles of axons collectively form the right and left olfactory nerves.

Olfactory nerves end in the brain in paired masses of gray matter called the **olfactory bulbs**, two extensions of the brain that rest on the cribriform plate. Within the olfactory bulbs, the axon terminals of olfactory receptors form synapses with the dendrites and cell bodies of the next neurons in the olfactory pathway. The axons of these neurons make up the **olfactory tracts**, which extend posteriorly from the olfactory bulbs (see Figure 19.5). Axons in the olfactory tracts end in the primary olfactory area in the temporal lobe of the cerebral cortex.

Optic (II) Nerve

The **optic (II) nerve** (OP-tik; *opti-*=the eye, vision) is entirely sensory; it contains axons that conduct nerve impulses for vision (Figure 19.17). In the retina, rods and cones initiate visual signals and relay them to bipolar cells, which pass the signals on to ganglion cells. Axons of all the ganglion cells in the retina of each eye join to form an optic nerve, which passes through the optic foramen. About 10 mm (0.4 in.) posterior to the eyeball, the two optic nerves merge to form the **optic chiasm** (KĪ-azm=a crossover, as in the letter X). Within the chiasm, axons from the medial half of each eye cross to the opposite side; axons from the lateral half remain on the same side. Posterior to the chiasm, the regrouped axons, some from each eye, form the **optic tracts**. Most axons in the optic tracts end in the lateral geniculate nucleus of the thalamus. There they synapse with neurons whose axons extend to the primary visual area in the occipital lobe of the cerebral cortex (area 17 in Figure 19.15). A few axons pass through the optic chiasm and then extend to the superior colliculi of the midbrain. They synapse with motor neurons that control the extrinsic and intrinsic eye muscles.

Oculomotor (III) Nerve

The **oculomotor (III) nerve** (ok′-ū-lō-MŌ-tor; *oculo-*=eye; *-motor*=a mover) is a motor cranial nerve. Its motor nucleus is in the anterior part of the midbrain (Figure 19.18 on page 642). The oculomotor nerve extends anteriorly and divides into superior and inferior branches, both of which pass through the superior orbital fissure into the orbit. Axons in the superior branch innervate the superior rectus (an extrinsic eyeball muscle) and the levator palpebrae superioris (the muscle of the upper eyelid). Axons in the inferior branch supply the medial rectus, inferior rectus, and inferior oblique muscles—all extrinsic eyeball muscles. These somatic motor neurons control movements of the eyeball and upper eyelid.

The inferior branch of the oculomotor nerve also provides parasympathetic innervation to intrinsic eyeball muscles, which consist of smooth muscle. They include the ciliary muscle of the eyeball and the circular muscles (sphincter pupillae) of the iris. Parasympathetic impulses propagate from the accessory oculomotor nucleus in the midbrain to the **ciliary ganglion**, a relay center of the autonomic nervous system. From the ciliary ganglion,

Figure 19.16 Olfactory (I) nerve.

The olfactory epithelium is located on the inferior surface of the cribriform plate and superior nasal conchae.

Olfactory tract
Olfactory bulb
Bundle of axons of olfactory receptors
Cribriform plate
Olfactory epithelium
Olfactory receptor
Olfactory bulb
Cribriform plate
Olfactory epithelium
Anterior
Olfactory (I) nerve
Olfactory bulb
Olfactory tract
Posterior

? Where do axons in the olfactory tracts terminate?

parasympathetic axons extend to the ciliary muscle, which adjusts the lens for near vision. Other parasympathetic axons stimulate the circular muscles of the iris to contract when bright light stimulates the eye, causing a decrease in the size of the pupil (constriction).

Proprioceptive sensory axons from the extrinsic eyeball muscles begin their course toward the brain in the oculomotor nerve but eventually leave the nerve to join the ophthalmic branch of the trigeminal nerve. They *do not* return to the brain in the oculomotor nerve. The cell bodies of the sensory axons reside in the trigeminal ganglion and they enter the midbrain via the trigeminal nerve. These axons convey nerve impulses for **proprioception**, the nonvisual perception of the movements and position of the body, from the extrinsic eyeball muscles.

Trochlear (IV) Nerve

The **trochlear (IV) nerve** (TRŌ K-lē-ar; *trochle-*=a pulley) is a motor cranial nerve. It is the smallest of the 12 cranial nerves and is the only one that arises from the posterior aspect of the brain stem.

The motor neurons originate in the trochlear nucleus in the midbrain, and axons from the nucleus exit the brain on its posterior aspect and pass through the superior orbital fissure of the orbit (Figure 19.18b). These somatic motor axons innervate the superior oblique muscle of the eyeball, another extrinsic eyeball muscle that controls movement of the eyeball.

Proprioceptive sensory axons from the superior oblique muscle begin their course toward the brain in the trochlear nerve but

Figure 19.17 Optic (II) nerve.

In sequence, visual signals are relayed from rods and cones to bipolar cells to ganglion cells.

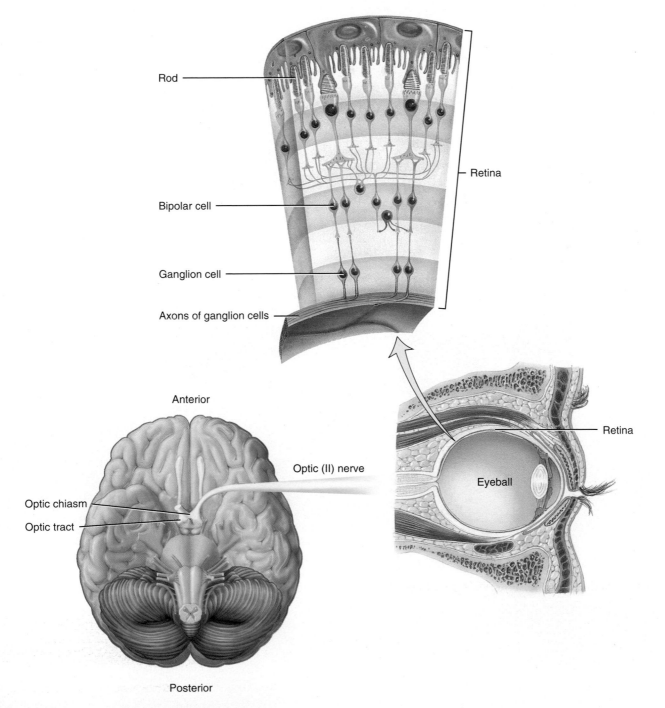

Rod

Retina

Bipolar cell

Ganglion cell

Axons of ganglion cells

Anterior

Optic (II) nerve

Optic chiasm

Optic tract

Retina

Eyeball

Posterior

? **Where do most axons in the optic tracts terminate?**

eventually leave the nerve to join the ophthalmic branch of the trigeminal nerve. They *do not* return to the brain in the trochlear nerve. The cell bodies of the sensory neurons reside in the trigeminal ganglion and they enter the midbrain via the trigeminal nerve. Like those of the oculomotor nerve, these axons convey nerve impulses for proprioception, the nonvisual perception of the movements and position of the body, from the extrinsic eyeball muscles.

Trigeminal (V) Nerve

The **trigeminal (V) nerve** (trī-JEM-i-nal=triple, for its three branches) is a mixed cranial nerve and the largest of the cranial

nerves. The trigeminal nerve emerges from two roots on the ventrolateral surface of the pons. The large sensory root has a swelling called the **trigeminal ganglion,** which is located in a fossa on the inner surface of the petrous portion of the temporal bone. The ganglion contains cell bodies of most of the primary sensory neurons. The smaller motor root originates in a nucleus in the pons.

As indicated by its name, the trigeminal nerve has three branches: ophthalmic, maxillary, and mandibular (Figure 19.19). The **ophthalmic nerve** (of-THAL-mik; *ophthalm-*=the eye), the smallest branch, passes into the orbit via the superior orbital fissure. The **maxillary nerve** (*maxilla*=upper jaw bone) is inter-

Figure 19.18 Oculomotor (III), trochlear (IV), and abducens (VI) nerves.

The oculomotor nerve has the widest distribution among extrinsic eye muscles.

Which branch of the oculomotor nerve is distributed to the superior rectus muscle? Which is the smallest cranial nerve?

mediate in size between the ophthalmic and mandibular nerves and passes through the foramen rotundum. The **mandibular nerve** (*mandibula*=lower jaw bone), the largest branch, passes through the foramen ovale.

Sensory axons in the trigeminal nerve carry nerve impulses for touch, pain, and thermal sensations. The ophthalmic nerve contains sensory axons from the skin over the upper eyelid, eyeball, lacrimal glands, upper part of the nasal cavity, side of the nose, forehead, and anterior half of the scalp. The maxillary nerve includes sensory axons from the mucosa of the nose, palate, part of the pharynx, upper teeth, upper lip, and lower eyelid. The mandibular nerve contains sensory axons from the anterior two-thirds of the tongue (not taste), cheek and mucosa deep to it, lower teeth, skin over the mandible and side of the head anterior to the ear, and mucosa of the floor of the mouth. The sensory axons from the three branches enter the semilunar ganglion and terminate in nuclei in the pons. The trigeminal nerve also contains sensory fibers from proprioceptors located in the muscles of mastication.

Branchial (referring to the embryonic branchial or pharyngeal arches from which the muscles that these nerves innervate arise) **motor axons** of the trigeminal nerve are part of the mandibular nerve and supply muscles of mastication (masseter, temporalis, medial pterygoid, lateral pterygoid, anterior belly of digastric, and mylohyoid muscles, as well as the tensor veli palatini muscle in the soft palate and tensor tympani muscle in the middle ear). These motor neurons mainly control chewing movements.

Abducens (VI) Nerve

The **abducens (VI) nerve** (ab-DOO-senz; *ab-*=away; *-ducens*=to lead) is a motor cranial nerve that originates from the abducens nucleus in the pons (see Figure 19.18c). Somatic motor axons extend from the nucleus to the lateral rectus muscle of the eyeball, an extrinsic eyeball muscle, through the superior orbital fissure of the orbit. The abducens nerve is so named because nerve impulses

Figure 19.19 Trigeminal (V) nerve.

The three branches of the trigeminal nerve leave the cranium through the superior orbital fissure, foramen rotundum, and foramen ovale.

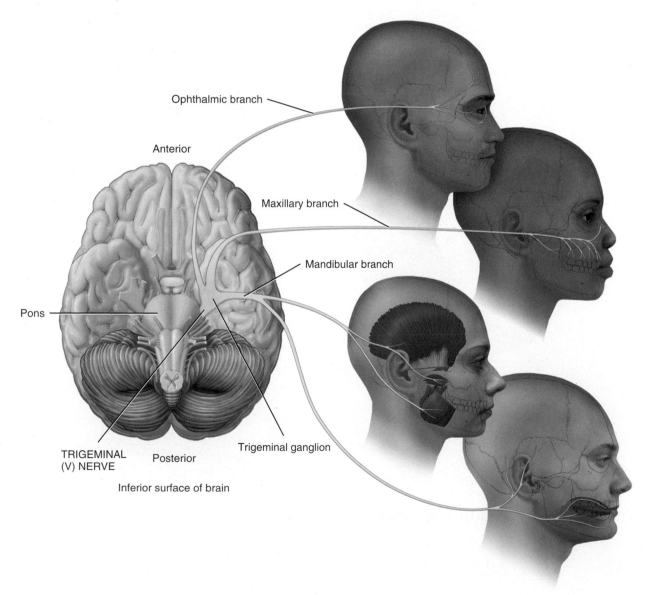

Ophthalmic branch

Anterior

Maxillary branch

Mandibular branch

Pons

TRIGEMINAL (V) NERVE   Posterior   Trigeminal ganglion

Inferior surface of brain

? How does the trigeminal nerve compare in size to the other cranial nerves?

cause abduction (lateral rotation) of the eyeball. Sensory axons from proprioceptors in the lateral rectus muscle begin their course toward the brain in the abducens nerve but eventually leave it to join the ophthalmic branch of the trigeminal nerve. They *do not* return to the brain in the abducens nerve, the cell bodies of the sensory neurons reside in the trigeminal ganglion and they enter the pons via the trigeminal nerve. These axons convey nerve impulses for proprioception, the nonvisual perception of the movements and position of the body, from the extrinsic eyeball muscles.

Facial (VII) Nerve

The **facial (VII) nerve** (FĀ-shal=face) is a mixed cranial nerve. Its sensory axons extend from the taste buds of the anterior two-thirds of the tongue through the **geniculate ganglion** (je-NIK-ū-lāt), a cluster of cell bodies of sensory neurons that lie beside the facial nerve, and end in the pons (Figure 19.20). The sensory portion of the facial nerve also contains axons from proprioceptors in muscles of the face and scalp and from skin in the ear canal.

Axons of branchial motor neurons arise from a nucleus in the pons, pass through the petrous portion of the temporal bone, and innervate facial, scalp, and neck muscles. Nerve impulses propagating along these axons cause contraction of the muscles of facial expression plus the stylohyoid muscle, the posterior belly of the digastric muscle, and the stapedius muscle.

Axons of parasympathetic neurons that are part of the facial nerve end in two parasympathetic ganglia: the **pterygopalatine** (ter'-i-gō-PAL-a-tīn) **ganglion** and the **submandibular ganglion**. From the two ganglia, other parasympathetic axons extend

Figure 19.20 Facial (VII) nerve.

The facial nerve causes contraction of the muscles of facial expression.

Anterior

Pons

Posterior

Inferior surface or brain

Geniculate ganglion

FACIAL (VII) NERVE

Tongue

Salivary glands

? Where do the motor fibers for the facial nerve originate?

to lacrimal glands (which secrete tears), nasal glands, palatine glands, and saliva-producing sublingual and submandibular glands.

Vestibulocochlear (VIII) Nerve

The **vestibulocochlear (VIII) nerve** (vest-tib-ū-lō-KOK-lē-ar; *ves-tibulo-*=small cavity; *-cochlear*=a spiral, snail-like) was formerly known as the **acoustic** or **auditory nerve**. It is a sensory cranial nerve and has two branches, the vestibular branch and the cochlear branch (Figure 19.21). The **vestibular branch** carries impulses for equilibrium and the **cochlear branch** carries impulses for hearing.

Sensory axons in the vestibular branch extend from the semicircular canals, the saccule, and the utricle of the inner ear to the **vestibular ganglion**, where their cell bodies are located (see Figure 22.13b on page 715), and end in vestibular nuclei in the pons. Some sensory axons also enter the cerebellum via the inferior cerebellar peduncle.

Sensory axons in the cochlear branch arise in the spiral organ (organ of Corti) in the cochlea of the internal ear. The cell bodies of cochlear branch sensory neurons are located in the **spiral ganglion** of the cochlea (see Figure 22.13b). From there, axons extend to nuclei in the medulla oblongata.

Figure 19.21 Vestibulocochlear (VIII) nerve.

The vestibular branch of the vestibulocochlear nerve carries impulses for equilibrium, while the cochlear branch carries impulses for hearing.

Anterior

Vestibular ganglion

Semicircular canal

Vestibular branch

Cochlear branch

Vestibule (contains saccule and utricle)

VESTIBULOCOCHLEAR (VIII) NERVE

Pons

Cochlea (contains spiral organ)

Posterior

? **Where do sensory axons of the cochlear branch originate?**

Glossopharyngeal (IX) Nerve

The **glossopharyngeal (IX) nerve** (glos′-Ō-fa-RIN-jē-al; *glosso-* =tongue; *-pharyngeal*=throat) is a mixed cranial nerve. Sensory axons of the glossopharyngeal nerve arise from taste buds and somatic sensory receptors on the posterior one-third of the tongue, from proprioceptors in swallowing muscles supplied by the motor portion, from baroreceptors (stretch receptors) in the carotid sinus, and from chemoreceptors in the carotid body near the carotid arteries (Figure 19.22; see also Figure 19.23). The cell bodies of these sensory neurons are located in the **superior** and **inferior ganglia**. From the ganglia, sensory axons pass through the jugular foramen and end in the medulla.

Axons of motor neurons in the glossopharyngeal nerve arise in nuclei of the medulla and exit the skull through the jugular foramen. Branchial motor neurons innervate the stylopharyngeus muscle, and autonomic motor neurons (parasympathetic) stimulate the parotid gland to secrete saliva. Some of the cell bodies of parasympathetic motor neurons are located in the **otic ganglion**.

Vagus (X) Nerve

The **vagus (X) nerve** (VĀ-gus=vagrant or wandering) is a mixed cranial nerve that is distributed from the head and neck into the thorax and abdomen (Figure 19.23). The nerve derives its name from its wide distribution. In the neck, it lies medial and posterior to the internal jugular vein and common carotid artery.

Sensory axons in the vagus nerve arise from the skin of the external ear, a few taste buds in the epiglottis and pharynx, and proprioceptors in muscles of the neck and throat. Also, sensory axons come from baroreceptors (stretch receptors) in the carotid sinus and chemoreceptors in the carotid body near the carotid arteries and in the aortic bodies near the arch of the aorta, and visceral sensory receptors in most organs of the thoracic and abdominal cavities. These axons pass through the jugular foramen and end in the medulla and pons.

The somatic motor neurons, which run briefly with the accessory nerve (historically these motor neurons have been called the cranial accessory nerve, but these fibers actually belong to the

---

Figure 19.22 Glossopharyngeal (IX) nerve.

Sensory fibers of the glossopharyngeal nerve supply the taste buds.

Anterior

Medulla oblongata

Posterior

Inferior surface of brain

Parotid gland

Stylopharyngeus muscle (behind parotid gland)

Otic ganglion

Inferior ganglion

Superior ganglion

GLOSSOPHARYNGEAL (IX) NERVE

Soft palate

Palatine tonsil

Tongue

Carotid body

Carotid sinus

Through which foramen does the glossopharyngeal nerve exit the skull?

vagus nerve), arise from nuclei in the medulla oblongata and supply muscles of the pharynx, larynx, and soft palate that are used in swallowing and vocalization.

Axons of autonomic motor neurons (parasympathetic) in the vagus nerve originate in nuclei of the medulla and end in the lungs and heart. Vagal parasympathetic axons also supply glands of the gastrointestinal (GI) tract and smooth muscle of the respiratory passageways, esophagus, stomach, gallbladder, small intestine, and most of the large intestine (see Figure 20.3 on page 664).

Accessory (XI) Nerve

The **accessory (XI) nerve** (ak-SES-ō-rē=assisting) is a motor cranial nerve (Figure 19.24). Historically it has been divided into two parts, a cranial accessory nerve and a spinal accessory nerve. The cranial portion actually is part of the vagus nerve and should be classified with it. The accessory nerve is the "old" spinal part of the nerve. Its motor axons arise in the anterior gray horn of the first five segments of the cervical portion of the spinal cord. The axons from the segments exit the spinal cord laterally and come together, pass through the foramen magnum, and then exit through the jugular foramen along with axons in the vagus. The accessory nerve conveys motor impulses to the sternocleidomastoid and trapezius muscles to coordinate head movements. Sensory axons in the accessory nerve that originate from proprioceptors in the muscles supplied by its motor neurons begin their course toward the brain in the accessory nerve but eventually leave the nerve to join nerves of the cervical plexus. From here they enter the spinal cord via the posterior roots of cervical spinal nerves to pass to and end in the medulla oblongata. The sensory axons *do not* return to the brain in the accessory nerve and like all sensory axons have cell bodies in posterior root sensory ganglia.

Figure 19.23 Vagus (X) nerve.

The vagus nerve is widely distributed in the head, neck, thorax, and abdomen.

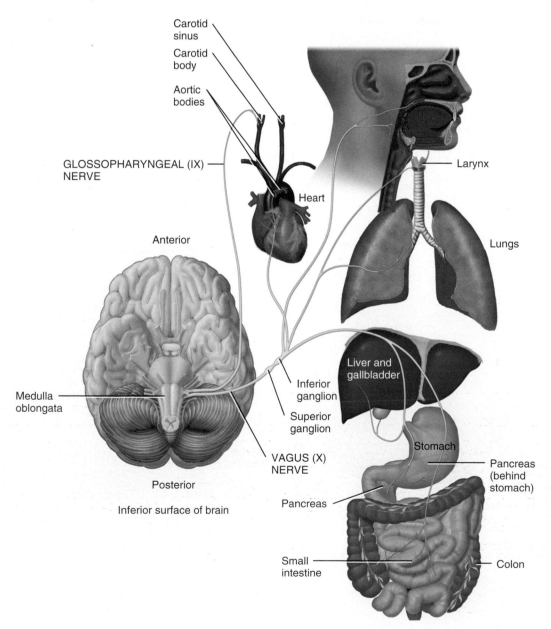

Carotid sinus

Carotid body

Aortic bodies

GLOSSOPHARYNGEAL (IX) NERVE

Larynx

Heart

Anterior

Lungs

Medulla oblongata

Inferior ganglion

Liver and gallbladder

Superior ganglion

VAGUS (X) NERVE

Stomach

Posterior

Pancreas (behind stomach)

Pancreas

Inferior surface of brain

Small intestine

Colon

? Where is the vagus nerve located in the neck?

Figure 19.24 Accessory (XI) nerve.

🔓 �== The accessory nerve exits the cranium through the jugular foramen.

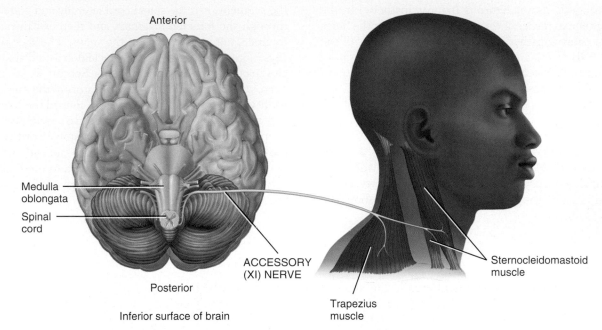

Inferior surface of brain

❓ **How does the accessory nerve differ from the other cranial nerves?**

Figure 19.25 Hypoglossal (XII) nerve.

🔓 �== The hypoglossal nerve exits the cranium through the hypoglossal canal.

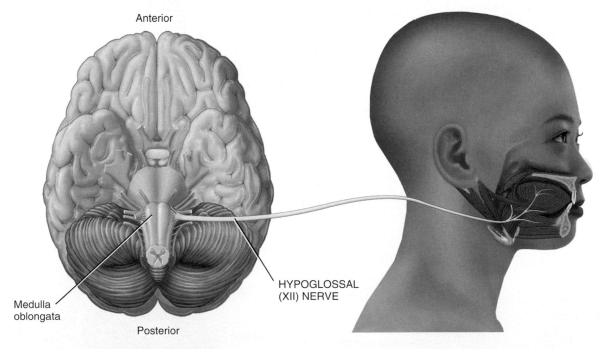

Inferior surface of brain

❓ **What important motor functions are related to the hypoglossal nerve?**

Hypoglossal (XII) Nerve

The **hypoglossal (XII) nerve** (hī-pō-GLOS-al; *hypo-*=below; *-glossal*=tongue) is a motor cranial nerve. The somatic motor axons originate in the hypoglossal nucleus in the medulla oblongata, pass through the hypoglossal canal, and supply the muscles of the tongue. These axons conduct nerve impulses for speech and swallowing. Sensory axons that originate from proprioceptors in the tongue muscles begin their course toward the brain in the hypoglossal nerve but leave the nerve to join cervical spinal nerves and end in the medulla oblongata, again entering the central nervous system via posterior roots of cervical spinal nerves

the sensory axon *do not* return to the brain in the hypoglossal nerve (Figure 19.25 on page 648).

Table 19.4 presents a summary of the cranial nerves including clinical connections related to their dysfunction.

◉ **CHECKPOINT**

26. How are cranial nerves named and numbered?
27. What is the difference between a mixed cranial nerve and a sensory cranial nerve?
28. What sort of test could reveal damage to each of the 12 cranial nerves?

TABLE 19.4*

Summary of Cranial Nerves

NUMBER AND NAME	TYPE AND LOCATION	FUNCTION AND CLINICAL CONNECTION
Olfactory (I) nerve Olfactory bulb / Olfactory nerve / Olfactory tract	**Sensory** Arises in olfactory mucosa, passes through foramina in the cribriform plate of the ethmoid bone, and ends in the olfactory bulb. The olfactory tract extends via two pathways to olfactory areas of cerebral cortex.	*Function:* Smell. *Clinical connection:* Loss of the sense of smell, called *anosmia* (an-OZ-mē-a), may result from head injuries in which the cribriform plate of the ethmoid bone is fractured or from lesions along the olfactory pathway.
Optic (II) nerve Optic nerve / Optic tract	**Sensory** Arises in the retina of the eye, passes through the optic foramen, forms the optic chiasm and then the optic tracts, and terminates in the geniculate nuclei of thalamus. From the thalamus, axons extend to the primary visual area (area 17) of the cerebral cortex.	*Function:* Vision *Clinical connection:* Fractures in the orbit, damage along the visual pathway, and diseases of the nervous system may result in visual field defects and loss of visual acuity. Blindness due to a defect in or loss of one or both eyes is called *anopia* (an-Ō-pē-a).
Oculomotor (III) nerve Oculomotor nerve	**Motor** Originates in the midbrain and passes through the superior orbital fissure. Axons of somatic motor neurons innervate the levator palpebrae superioris muscle of the upper eyelid and four extrinsic eyeball muscles (superior rectus, medial rectus, inferior rectus, and inferior oblique). Parasympathetic axons innervate the ciliary muscle of the eyeball and the circular muscles (sphincter pupillae) of the iris.	*Somatic motor function:* Movement of upper eyelid and eyeball. *Autonomic motor function (parasympathetic):* Accommodation of lens for near vision and constriction of pupil. *Clinical connection:* Nerve damage causes *strabismus* (a deviation of the eye in which both eyes do not fix on the same object), *ptosis* (drooping) of the upper eyelid, dilation of the pupil, movement of the eyeball downward and outward on the damaged side, loss of accommodation for near vision, and *diplopia* (di-PLŌ-pē-a) (double vision).

*See Figure 19.5c on page 621 for a photo of all of the cranial nerves.

CONTINUES

TABLE 19.4 CONTINUED

Summary of Cranial Nerves

NUMBER AND NAME	TYPE AND LOCATION	FUNCTION AND CLINICAL CONNECTION
Trochlear (IV) nerve Trochlear nerve	**Motor** Originates in the midbrain and passes through the superior orbital fissure. Innervates the superior oblique muscle, an extrinsic eyeball muscle.	*Function:* Movement of the eyeball. *Clinical connection:* In trochlear nerve paralysis, diplopia and strabismus occur.
Trigeminal (V) nerve Trigeminal nerve	**Mixed** *Sensory portion:* Consists of three branches, all of which end in the pons. (1) The **ophthalmic nerve** (*ophthalm-*=the eye) contains axons from the skin over the upper eyelid, eyeball, lacrimal glands, nasal cavity, side of nose, forehead, and anterior half of scalp that pass through superior orbital fissure. (2) The **maxillary nerve** (*maxilla*=upper jaw bone) contains axons from the mucosa of the nose, palate, parts of the pharynx, upper teeth, upper lip, and lower eyelid that pass through the foramen rotundum. (3) The **mandibular nerve** (*mandibula*= lower jaw bone) contains axons from the anterior two-thirds of the tongue (somatic sensory axons but not axons for the special sense of taste), the lower teeth, skin over mandible, cheek and mucosa deep to it, and side of head in front of ear that pass through the foramen ovale. *Motor portion:* Is part of the mandibular branch, which originates in the pons, passes through the foramen ovale, and innervates muscles of mastication (masseter, temporalis, medial pterygoid, lateral pterygoid, anterior belly of digastric, and mylohyoid muscles, as well as the tensor veli palatini and tensor tympani muscles).	*Sensory function:* Conveys impulses for touch, pain, and temperature sensations and proprioception. *Somatic motor function:* Chewing. *Clinical connection: Neuralgia* (pain) of one or more branches of the trigeminal nerve is called *trigeminal neuralgia (tic douloureux)*. Injury of the mandibular nerve may cause paralysis of the chewing muscles and a loss of the sensations of touch, temperature, and proprioception in the lower part of the face. Dentists apply anesthetic drugs to branches of the maxillary nerve for anesthesia of upper teeth and to branches of the mandibular nerve for anesthesia of lower teeth.
Abducens (VI) nerve Abducens nerve	**Motor** Originates in the pons, passes through the superior orbital fissure, and innervates the lateral rectus muscle, an extrinsic eyeball muscle.	*Function:* Movement of the eyeball. *Clinical connection:* With damage to this nerve, the affected eyeball cannot move laterally beyond the midpoint, and the eye usually is directed medially.

NUMBER AND NAME	TYPE AND LOCATION	FUNCTION AND CLINICAL CONNECTION
Facial (VII) nerve	**Mixed** *Sensory portion:* Arises from taste buds on the anterior two-thirds of the tongue, passes through the stylomastoid foramen and geniculate ganglion (located beside the facial nerve), and ends in the pons. From there, axons extend to the thalamus, and then to the gustatory areas of the cerebral cortex. Also contains axons from proprioceptors in muscles of the face and scalp. *Motor portion:* Originates in the pons and passes through the stylomastoid foramen. Axons of somatic motor neurons innervate facial, scalp, and neck muscles. Parasympathetic axons innervate lacrimal, sublingual, submandibular, nasal, and palatine glands.	*Sensory function:* Touch, pain, and temperature sensations, and proprioception and taste. *Somatic motor function:* Facial expression. *Autonomic motor function (parasympathetic):* Secretion of saliva and tears. *Clinical connection:* Damage due to viral infection (shingles) or a bacterial infection (Lyme disease) produces *Bell's palsy* (paralysis of the facial muscles), loss of taste, decreased salivation, and loss of ability to close the eyes, even during sleep.
Vestibulocochlear (VIII) nerve	**Sensory** *Vestibular branch:* Arises in the semicircular canals, saccule, and utricle and forms the vestibular ganglion. Axons end in the pons and cerebellum. *Cochlear branch:* Arises in the spiral organ (organ of Corti), forms the spiral ganglion, passes through nuclei in the medulla, and ends in the thalamus. Axons synapse with thalamic neurons that relay impulses to the primary auditory area (areas 41 and 42) of the cerebral cortex.	*Vestibular branch function:* Conveys impulses related to equilibrium. *Cochlear branch function:* Conveys impulses for hearing. *Clinical connection:* Injury to the vestibular branch may cause *vertigo*, a subjective feeling that one's own body or the environment is rotating, *ataxia* (muscular incoordination), and *nystagmus* (involuntary rapid movement of the eyeball). Injury to the cochlear branch may cause *tinnitus* (ringing in the ears) or deafness.
Glossopharyngeal (IX) nerve	**Mixed** *Sensory portion:* Consists of axons from taste buds and somatic sensory receptors on posterior one-third of the tongue, from proprioceptors in swallowing muscles supplied by the motor portion, and from baroreceptors in carotid sinus and chemoreceptors in carotid body near the carotid arteries. Axons pass through the jugular foramen and end in the medulla. *Motor portion:* Originates in the medulla and passes through the jugular foramen. Axons of somatic motor neurons innervate the stylopharyngeus muscle, a muscle of the pharynx that elevates the larynx during swallowing. Parasympathetic axons innervate the parotid salivary gland.	*Sensory function:* Taste and somatic sensations (touch, pain, temperature) from posterior third of tongue; proprioception in swallowing muscles; monitoring of blood pressure; monitoring of O_2 and CO_2 in blood for regulation of breathing rate and depth. *Somatic motor function:* Elevates the pharynx during swallowing and speech. *Autonomic motor function (parasympathetic):* Stimulates secretion of saliva. *Clinical connection:* Injury causes difficulty in swallowing, reduced secretion of saliva, loss of sensation in the throat, and loss of taste sensation.

CONTINUES

TABLE 19.4 CONTINUED

Summary of Cranial Nerves

NUMBER AND NAME	TYPE AND LOCATION	FUNCTION AND CLINICAL CONNECTION
Vagus (X) nerve — Vagus nerve	**Mixed** *Sensory portion:* Consists of axons from small number of taste buds in the epiglottis and pharynx, proprioceptors in muscles of the neck and throat, baroreceptors in the arch of the aorta, chemoreceptors in the aortic bodies near the arch of the aorta, and visceral sensory receptors in most organs of the thoracic and abdominal cavities. Axons pass through the jugular foramen and end in the medulla and pons. *Motor portion:* Originates in medulla and passes through the jugular foramen. Axons of somatic motor neurons innervate skeletal muscles in the throat and neck. Parasympathetic axons innervate smooth muscle in the airways, esophagus, stomach, small intestine, most of large intestine, and gallbladder; cardiac muscle in the heart; and glands of the gastrointestinal (GI) tract.	*Sensory function:* Taste and somatic sensations (touch, pain, temperature, and proprioception) from epiglottis and pharynx; monitoring of blood pressure; monitoring of O_2 and CO_2 in blood for regulation of breathing rate and depth; sensations from visceral organs in thorax and abdomen. *Somatic motor function:* Swallowing, coughing, and voice production. *Autonomic motor function (parasympathetic):* Smooth muscle contraction and relaxation in organs of the GI tract; slowing of the heart rate; secretion of digestive fluids. *Clinical connection:* Injury interrupts sensations from many organs in the thoracic and abdominal cavities, interferes with swallowing, paralyzes vocal cords, and causes heart rate to increase.
Accessory (XI) nerve — Accessory nerve	**Motor** Originates in the anterior gray horn of the first five cervical segments of the spinal cord and emerges laterally from the cord and then ascends through the foramen magnum into the cranial cavity. It then arches inferiorly to leave the jugular foramen and supply sternocleidomastoid and trapezius muscles to coordinate head movements.	*Function:* Mediates movement of head and pectoral girdle. *Clinical connection:* If nerve is damaged, the sternocleidomastoid and trapezius muscles become paralyzed, with resulting inability to raise the shoulders and difficulty in turning the head.
Hypoglossal (XII) nerve — Hypoglossal nerve	**Motor** Originates in the medulla, passes through the hypoglossal canal, and supplies muscles of the tongue.	*Function:* Movement of tongue during speech and swallowing. *Clinical connection:* Injury results in difficulty in chewing, speaking, and swallowing. The tongue, when protruded, curls toward the affected side, and the affected side atrophies.

*MNEMONIC for cranial nerves:

Oh	**Oh**	**Oh**	**To**	**Touch**	**And**	**Feel**	**Very**	**Green**	**Vegetables**	**AH!**
Olfactory	**O**ptic	**O**culomotor	**T**rochlear	**T**rigeminal	**A**bducens	**F**acial	**V**estibulocochlear	**G**lossopharyngeal	**V**agus	**A**ccessory **H**ypoglossal

DEVELOPMENT OF THE NERVOUS SYSTEM

◉ OBJECTIVE

- Describe how the parts of the brain develop.

As you learned in Chapter 4, development of the nervous system begins in the third week of gestation with a thickening of the **ectoderm** called the **neural plate** (Figure 19.26). The plate folds inward and forms a longitudinal groove, the **neural groove**. The raised edges of the neural plate are called **neural folds**. As development continues, the neural folds increase in height and meet to form a tube called the **neural tube**.

Three layers of cells differentiate from the wall that encloses the neural tube. The outer or **marginal layer** cells develop into the *white matter* of the nervous system. The middle or **mantle layer** cells develop into the *gray matter*. The inner or **ependymal layer** cells eventually form the *lining of the central canal of the spinal cord* and *ventricles of the brain*.

The **neural crest** is a mass of tissue between the neural tube and the skin ectoderm (Figure 19.26b). It differentiates and eventually forms the *posterior (dorsal) root ganglia of spinal nerves, spinal nerves, ganglia of cranial nerves, cranial nerves, ganglia of the autonomic nervous system, adrenal medulla*, and *meninges*.

As you learned earlier in this chapter on page 613, during the third to fourth week of embryonic development, the anterior part of the neural tube develops into three enlarged areas called **primary brain vesicles** that are named for their relative positions. These are the **prosencephalon** (prōs′-en-SEF-a-lon; *pros*-=before) or forebrain, **mesencephalon** (mes′-en-SEF-a-lon; *mes*-=middle) or midbrain, and **rhombencephalon** (rom′-ben-SEF-a-lon; *rhomb*-=behind) or hindbrain (Figure 19.27a on page 654; see also Table 19.1). During the fifth week of development, **secondary brain vesicles** begin to develop. The prosencephalon develops into two secondary brain vesicles called the **telencephalon** (tel′-en-SEF-a-lon; *tel*-=distant) and the **diencephalon** (dī-en-SEF-a-lon; *di*-=through) (Figure 19.27b). The rhombencephalon also develops into two secondary brain

---

Figure 19.26 Origin of the nervous system. (a) Dorsal view of an embryo in which the neural folds have partially united, forming the early neural tube. (b) Transverse sections through the embryo showing the formation of the neural tube.

🔑 The nervous system begins developing in the third week from a thickening of ectoderm called the neural plate.

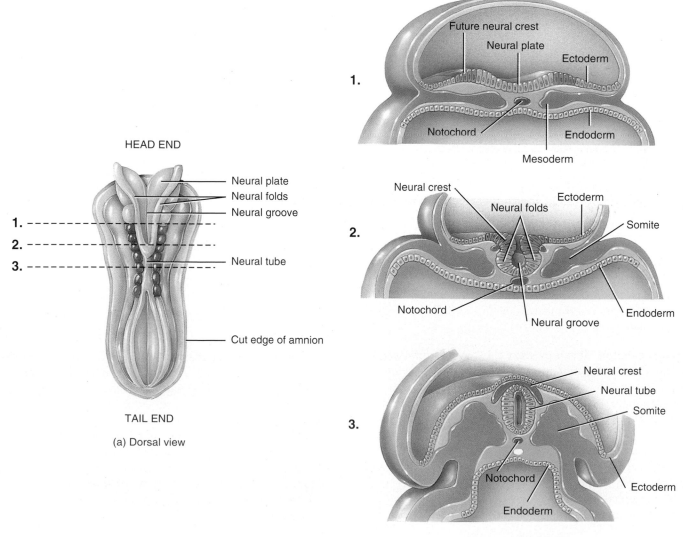

(a) Dorsal view

(b) Transverse sections

❓ **What is the origin of the gray matter of the nervous system?**

Figure 19.27 Development of the brain and spinal cord.

(6) 🔑 The various parts of the brain develop from the primary brain vesicles.

MESENCEPHALON (MIDBRAIN)

RHOMBENCEPHALON (HINDBRAIN)

PROSENCEPHALON (FOREBRAIN)

Developing ear

Developing eye

Pharyngeal arches

Spinal cord

Developing heart

Lateral view of right side

(a) Three–four week embryo showing primary brain vesicles

METENCEPHALON

MYELENCEPHALON

MESENCEPHALON (MIDBRAIN)

DIENCEPHALON

TELENCEPHALON

Spinal cord

Developing upper limb

(b) Seven-week embryo showing secondary brain vesicles

Midbrain

Cerebellum

Pons

Medulla oblongata

Outline of diencephalon

Cerebral hemisphere

Spinal cord

(c) Eleven-week fetus showing expanding cerebral hemispheres overgrowing the diencephalon

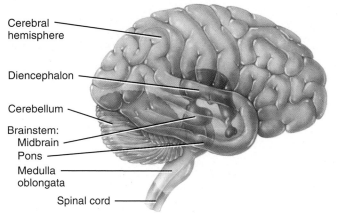

Cerebral hemisphere

Diencephalon

Cerebellum

Brainstem:
  Midbrain
  Pons
  Medulla oblongata

Spinal cord

(d) Brain at birth (the diencephalon and superior portion of the brain stem have been projected to the surface)

❓ **Which primary brain vesicle does not develop into a secondary brain vesicle?**

---

vesicles called the **metencephalon** (met′-en-SEF-a-lon; *met-* = after) and the **myelencephalon** (mī-el-en-SEF-a-lon; *myel-* = marrow). The area of the neural tube inferior to the myelencephalon gives rise to the *spinal cord*.

The brain vesicles continue to develop as follows (Figure 19.27c, d; see also Table 19.1):

- The telencephalon develops into the *cerebral hemispheres*, including the *basal ganglia*, and houses the paired *lateral ventricles*.
- The diencephalon develops into the *epithalamus, thalamus, hypothalamus*, and *pineal gland* and houses the third ventricle.
- The mesencephalon develops into the *midbrain*, which surrounds the *aqueduct of the midbrain (cerebral aqueduct)*.
- The metencephalon becomes the *pons* and *cerebellum* and houses the upper part of the *fourth ventricle*.
- The myelencephalon develops into the *medulla oblongata* and houses the remainder of the *fourth ventricle*.

◉ **CHECKPOINT**

29. What parts of the brain develop from each primary brain vesicle?

AGING AND THE NERVOUS SYSTEM

◉ **OBJECTIVE**

• Describe the effects of aging on the nervous system.

The brain grows rapidly during the first few years of life. Growth is due mainly to an increase in the size of neurons already present, the proliferation and growth of neuroglia, the development of dendritic branches and synaptic contacts, and continuing myelination of axons. From early adulthood onward, brain mass declines. By the time a person reaches age 80, the brain weighs about 7% less than it did in young adulthood. Although the number of neurons present does not decrease very much, the number of synaptic contacts declines. Associated with the decrease in brain mass is a decreased capacity for sending nerve impulses to and from the brain. As a result, processing of information diminishes, conduction velocity decreases, voluntary motor movements slow down, and reflex times increase.

◉ **CHECKPOINT**

30. How is brain mass related to age?

654

KEY MEDICAL TERMS ASSOCIATED WITH THE BRAIN AND THE CRANIAL NERVES

Agnosia (ag-NŌ-zē-a; *a-*=without; *-gnosia*=knowledge) Inability to recognize the significance of sensory stimuli such as sounds, sights, smells, tastes, and touch.

Apraxia (a-PRAK-sē-a; *-praxia*=coordinated) Inability to carry out purposeful movements in the absence of paralysis.

Arousal (a-ROW-zal) Awakening from sleep, a response due to stimulation of the reticular activating system (RAS).

Coma A state of deep unconsciousness from which a person cannot be aroused due to damage to the RAS or other parts of the brain. A comatose patient lies in a sleeplike state with eyes closed. In the lightest stages of coma, brain stem and spinal cord reflexes persist, but in the deepest stages, even these reflexes are lost. If respiratory and cardiovascular controls are lost, the patient dies.

Consciousness (KON-shus-nes) A state of wakefulness in which an individual is fully alert, aware, and oriented, partly as a result of feedback between the cerebral cortex and reticular activating system.

Delirium (de-LIR-ē-um=off the track) A transient disorder of abnormal cognition and disordered attention accompanied by disturbances of the sleep–wake cycle and psychomotor behavior (hyperactivity or hypoactivity of movements and speech). Also called **acute confusional state (ACS)**.

Dementia (de-MEN-shē-a; *de-*=away from; *-mentia*=mind) Permanent or progressive general loss of intellectual abilities, including impairment of memory, judgment and abstract thinking, and changes in personality.

Encephalitis (en′-sef-a-LĪ-tis) An acute inflammation of the brain caused by either a direct attack by any of several viruses or an allergic reaction to any of the many viruses that are normally harmless to the central nervous system. If the virus affects the spinal cord as well, the condition is called **encephalomyelitis**.

Encephalopathy (en-sef′-a-LOP-a-thē; *encephalo*=brain, *-pathos*=disease) Any disorder of the brain.

Learning The ability to acquire new knowledge or skills through instruction or experience.

Lethargy (LETH-ar-jē) A condition of functional sluggishness.

Memory The process by which knowledge acquired through learning is retained over time.

Microcephaly (mī-krō-SEF-a-lē; *micro-*=small; *-cephal*=head) A congenital condition that involves the development of a small brain and skull and frequently results in mental retardation.

Reye's (RĪZ) syndrome Occurs after a viral infection, particularly chickenpox or influenza, most often in children or teens who have taken aspirin; characterized by vomiting and brain dysfunction (disorientation, lethargy, and personality changes) that may progress to coma and death.

Sleep A state of altered consciousness or partial unconsciousness from which an individual can be aroused by many different stimuli.

Stupor (STOO-por) Unresponsiveness from which a patient can be aroused only briefly and only by vigorous and repeated stimulation.

STUDY OUTLINE

Brain Organization, Protection, and Blood Supply (p. 613)

1. The major parts of the brain are the brain stem, cerebellum, diencephalon, and cerebrum.
2. The brain is protected by cranial bones and the cranial meninges.
3. The cranial meninges are continuous with the spinal meninges. From superficial to deep the cranial meninges are the dura mater, arachnoid mater, and pia mater.
4. Blood flow to the brain is mainly via the internal carotid and vertebral arteries.
5. Any interruption of the oxygen or glucose supply to the brain can result in weakening of, permanent damage to, or death of brain cells.
6. The blood–brain barrier (BBB) causes different substances to move between the blood and the brain tissue at different rates and prevents the movement of some substances from blood into the brain.

Cerebrospinal Fluid (p. 616)

1. Cerebrospinal fluid (CSF) is formed in the choroid plexuses and circulates through the lateral ventricles, third ventricle, fourth ventricle, subarachnoid space, and central canal. Most of the fluid is absorbed into the blood across the arachnoid villi of the superior sagittal blood sinus.
2. Cerebrospinal fluid provides mechanical protection, chemical protection, and circulation of nutrients.

The Brain Stem (p. 620)

1. The medulla oblongata is continuous with the superior part of the spinal cord and contains both motor and sensory tracts. It contains nuclei that are reflex centers for regulation of heart rate, respiratory rate, vasoconstriction, swallowing, coughing, vomiting, and sneezing. It also contains nuclei associated with cranial nerves VIII–XII.
2. The pons is superior to the medulla. It connects the spinal cord with the brain and links parts of the brain with one another by way of tracts. Pontine nuclei relay nerve impulses related to voluntary skeletal movements from the cerebral cortex to the cerebellum. The pons contains the pneumotaxic and apneustic centers, which

help control breathing. It contains nuclei associated with cranial nerves V–VII and the vestibular branch of cranial nerve VIII.
3. The midbrain connects the pons and diencephalon and surrounds the cerebral aqueduct. It conveys motor impulses from the cerebrum to the cerebellum and spinal cord, sends sensory impulses from the spinal cord to the thalamus, and regulates auditory and visual reflexes. It also contains nuclei associated with cranial nerves III and IV.
4. A large part of the brain stem consists of small areas of gray matter and white matter called the reticular formation, which helps maintain consciousness, causes awakening from sleep, and contributes to regulating muscle tone.

The Cerebellum (p. 625)

1. The cerebellum occupies the inferior and posterior aspects of the cranial cavity. It consists of two lateral hemispheres and a medial, constricted vermis.
2. The cerebellum connects to the brain stem by three pairs of cerebellar peduncles.
3. The cerebellum coordinates contractions of skeletal muscles and maintains normal muscle tone, posture, and balance.

The Diencephalon (p. 627)

1. The diencephalon surrounds the third ventricle and consists of the thalamus, hypothalamus, and epithalamus.
2. The thalamus is superior to the midbrain and contains nuclei that serve as relay stations for all sensory impulses to the cerebral cortex. It also allows crude appreciation of pain, temperature, and pressure and mediates some motor activities.
3. The hypothalamus is inferior to the thalamus. It controls and integrates the autonomic nervous system, connects the nervous and endocrine systems, functions in rage and aggression, controls body temperature, regulates food and fluid intake, and establishes circadian rhythms.
4. The epithalamus consists of the pineal gland and the habenular nuclei. The pineal gland secretes melatonin, which is thought to promote sleep and to help set the body's biological clock.

5. The subthalamus connects to motor areas of the cerebrum.
6. Circumventricular organs (CVOs) can monitor chemical changes in the blood because they lack the blood–brain barrier.

The Cerebrum (p. 630)

1. The cerebrum is the largest part of the brain. Its cortex contains gyri (convolutions), fissures, and sulci.
2. The cerebral hemispheres are divided into four lobes: frontal, parietal, temporal, and occipital.
3. The white matter of the cerebrum is deep to the cortex and consists of myelinated and unmyelinated axons extending to other regions as association, commissural, and projection fibers.
4. The basal ganglia are several groups of nuclei in each cerebral hemisphere. They help control large, automatic movements of skeletal muscles and help regulate muscle tone.
5. The limbic system encircles the upper part of the brain stem and the corpus callosum. It functions in emotional aspects of behavior and memory.
6. Table 19.2 on page 635 summarizes the functions of various parts of the brain.

Functional Organization of the Cerebral Cortex (p. 636)

1. The sensory areas of the cerebral cortex allow perception of sensory impulses. The motor areas are the regions that govern muscular movement. The association areas are concerned with more complex integrative functions.
2. The primary somatosensory area (areas 1, 2, and 3) receives nerve impulses from somatic sensory receptors for touch, proprioception, pain, and temperature. Each point within the area receives impulses from a specific part of the face or body.
3. The primary visual area (area 17) receives impulses that convey visual information. The primary auditory area (areas 41 and 42) interprets the basic characteristics of sound such as pitch and rhythm. The primary gustatory area (area 43) receives impulses for taste. The primary olfactory area (area 28) receives impulses for smell.
4. Motor areas include the primary motor area (area 4), which controls voluntary contractions of specific muscles or groups of muscles, and Broca's speech area (areas 44 and 45), which controls production of speech.
5. The somatosensory association area (areas 5 and 7) permits you to determine the exact shape and texture of an object without looking at it and to sense the relationship of one body part to another. The visual association area (areas 18 and 19) relates present to past visual experiences and is essential for recognizing and evaluating what is seen. The auditory association area (area 22) deals with the meanings of sounds.

6. Wernicke's area (area 22 and possibly 39 and 40) interprets the meaning of speech by translating words into thoughts. The common integrative area (areas 5, 7, 39, and 40) integrates sensory interpretations from the association areas and impulses from other areas, allowing thoughts based on sensory inputs.
7. The premotor area (area 6) generates nerve impulses that cause specific groups of muscles to contract in specific sequences. The frontal eye field area (area 8) controls voluntary scanning movements of the eyes.
8. Subtle anatomical differences exist between the two hemispheres, and each has unique functions. Each hemisphere receives sensory signals from and controls the opposite side of the body. The left hemisphere is more important for language, numerical and scientific skills, and reasoning. The right hemisphere is more important for musical and artistic awareness, spatial and pattern perception, recognition of faces, emotional content of language, identifying odors, and generating mental images of sight, sound, touch, taste, and smell.
9. Brain waves generated by the cerebral cortex are recorded from the surface of the head in an electroencephalogram (EEG). The EEG may be used to diagnose epilepsy, infections, and tumors.

Cranial Nerves (p. 639)

1. Twelve pairs of cranial nerves originate from the nose, eyes, inner ear, brain stem, and spinal cord.
2. They are named primarily based on their distribution and are numbered I–XII in order of attachment to the brain. Table 19.4 on pages 649–652 summarizes the types, locations, functions, and disorders of the cranial nerves.

Development of the Nervous System (p. 653)

1. The development of the nervous system begins with a thickening of a region of the ectoderm called the neural plate.
2. During embryological development, primary brain vesicles form from the neural tube and serve as forerunners of various parts of the brain.
3. The telencephalon forms the cerebrum, the diencephalon develops into the thalamus and hypothalamus, the mesencephalon develops into the midbrain, the metencephalon develops into the pons and cerebellum, and the myelencephalon forms the medulla.

Aging and the Nervous System (p. 654)

1. The brain grows rapidly during the first few years of life.
2. Age-related effects involve loss of brain mass and decreased capacity for sending nerve impulses.

SELF-QUIZ QUESTIONS

Choose the one best answer to the following questions.

1. The diencephalon consists of the:
   a. insula and hypothalamus.
   b. pons and hypothalamus.
   c. hypothalamus and midbrain.
   d. thalamus and hypothalamus.
   e. thalamus and midbrain.

2. Blood flows into the brain mainly via the:
   a. superior sagittal sinus.
   b. internal and external jugular veins.
   c. internal carotid and vertebral arteries.
   d. anterior and posterior communicating arteries.
   e. external carotid and subclavian arteries.

3. The blood–cerebrospinal fluid barrier is formed primarily by the tight junctions between:
   a. microglia.   b. oligodendrocytes.
   c. neurons.   d. endothelial cells.
   e. ependymal cells.

4. Which of the following is not located in the cerebrum?
   a. globus pallidus   b. corpora quadrigemina
   c. corpus striatum   d. caudate nucleus
   e. lentiform nucleus

5. Which of the following does not contribute to the formation of a blood–brain barrier?
   a. tight junctions between the endothelial cells
   b. the continuous, thick basement membrane of the capillary endothelium
   c. a layer of oligodendrocytes
   d. processes of astrocytes that contact the capillary walls
   e. All of the above are correct.

6. Damage to the occipital lobe of the cerebrum would most likely cause:
   a. loss of hearing.   b. loss of vision.
   c. loss of ability to smell.   d. paralysis.
   e. loss of muscle sense (proprioception).

7. The medullary rhythmicity center in the medulla oblongata controls:
   a. the basic rhythm of breathing.
   b. cardiac rhythm.
   c. blood pressure.
   d. the pattern of hormone secretion throughout the lifespan.
   e. swinging the arms while walking.

8. Which of the following is *not* a function of the hypothalamus?
   a. control of conscious skeletal movements
   b. regulation of hormone secretion from the pituitary gland
   c. regulation of body temperature
   d. regulation of water intake
   e. establishment of sleep/wake cycles

9. Which cranial nerve is responsible for opening the eyes?
   a. cranial nerve VI   **b.** trigeminal
   c. cranial nerve VIII   **d.** oculomotor
   e. cranial nerve II

Complete the following.

10. The structures responsible for the production of cerebrospinal fluid are the _____.

11. The ventricle located between the right and left halves of the thalamus is the _____.

12. Two areas of the pons that help control breathing are the _____ center and the _____ center.

13. A group of sensory axons that projects from the brain stem to the cerebral cortex and helps maintain consciousness is known as the _____.

14. White matter fibers that transmit nerve impulses between gyri in the same hemisphere are called _____ fibers.

15. The outer layer of the cerebrum is called the _____. It is composed of _____ matter, which means that it contains mainly neuron cell bodies. The surface of the cerebrum is a series of tightly packed ridges called _____, with shallow grooves between them called _____.

16. Write "w" for white matter or "g" for gray matter:
    (a) corpus callosum: _____
    (b) olive: _____
    (c) cerebral cortex: _____
    (d) corpora quadrigemina: _____
    (e) arbor vitae: _____
    (f) caudate and lenticular nuclei: _____

17. Number the following in the correct order for the circulation of CSF, from its production to its reabsorption:
    (a) arachnoid villi: _____
    (b) median and lateral apertures: _____
    (c) aqueduct of the midbrain: _____
    (d) lateral ventricle: _____
    (e) third ventricle: _____
    (f) fourth ventricle: _____
    (g) interventricular foramen: _____
    (h) subarachnoid space: _____
    (i) choroid plexuses: _____
    (j) superior sagittal sinus: _____

18. The most posterior part of the hypothalamus, which serves as a relay station for reflexes related to the sense of smell, is a pair of small, rounded projections called the _____.

19. The pineal gland secretes the hormone _____, which is thought to promote sleepiness and to help set the body's biological clock.

20. The _____ cranial nerve is responsible for closing the eyes.

Are the following statements true or false? If the statement is False, *find and fix the mistake to make the statement* True.

21. Circumventricular organs coordinate homeostatic activities of the endocrine and nervous systems.

22. The limbic system functions in the control of behavior.

23. The three parts of the brain stem are the medulla oblongata, pons, and cerebellum.

24. The basal ganglia contain the supraoptic nucleus and the geniculate nuclei.

25. The cranial nerve responsible for abducting your right eye as you read this sentence is cranial nerve III.

Matching

26. Match the following (answers may be used more than once):
    ____ **(a)** has two main parts, separated by a longitudinal fissure, but joined internally by the corpus callosum
    ____ **(b)** responsible for coordination of skilled movements and regulation of posture and balance
    ____ **(c)** cranial nerves III–IV attach to this brain part
    ____ **(d)** cranial nerves V–VIII attach to this brain part
    ____ **(e)** cranial nerves VIII–XII attach to this brain part
    ____ **(f)** regulates food and fluid intake and body temperature
    ____ **(g)** all sensations are relayed through here
    ____ **(h)** centers for control of heart rate and respiration are located here
    ____ **(i)** constitutes four-fifths of the diencephalon
    ____ **(j)** connects to the pituitary gland via the infundibulum
    ____ **(k)** part of the brain stem; contains tracts that connect the cerebellum, midbrain, and medulla oblongata
    ____ **(l)** surrounds the aqueduct of the midbrain, and contains nuclei that serve as reflex centers for head and eye movements

    (1) hyopothalamus
    (2) medulla
    (3) midbrain
    (4) pons
    (5) thalamus
    (6) cerebellum
    (7) cerebrum

27. Match the cranial nerve and the description:
    ____ **(a)** accommodation of lens for near vision
    ____ **(b)** control of secretion of digestive fluids; slow heart rate
    ____ **(c)** chewing
    ____ **(d)** facial expression, secretion of saliva, and tears
    ____ **(e)** movement of tongue during speech and swallowing
    ____ **(f)** taste; sensations from tongue; monitoring of blood pressure and blood oxygen levels
    ____ **(g)** smell
    ____ **(h)** movements of head and shoulders
    ____ **(i)** movement of eyeball via lateral rectus muscle
    ____ **(j)** vision
    ____ **(k)** equilibrium
    ____ **(l)** movement of eyeball via superior oblique muscle

    (1) olfactory
    (2) optic
    (3) oculomotor
    (4) trochlear
    (5) trigeminal
    (6) abducens
    (7) facial
    (8) vestibulocochlear
    (9) glossopharyngeal
    (10) vagus
    (11) accessory
    (12) hypoglossal

CRITICAL THINKING QUESTIONS

1. An elderly relative suffered a stroke and now has difficulty moving her right arm and also has speech problems. What areas of the brain were damaged by the stroke?

2. Wolfgang partied a little too hard one night and passed out drunk in his bathroom at home. He awoke with a lump on his head from hitting the sink, but he thought he was okay. However, when Tony came over a day later, he said, "Man, are you still drunk? You're not making any sense, and you look like you're going to pass out again!" Before Wolfgang could answer, he vomited. What do you think—is Wolfgang still drunk? Why, or why not?

3. Dr. M. D. Hatter has developed a drug that inhibits the activity of the amygdala of the limbic system. Do you think this is good or a bad thing? Explain your answer.

4. Alzheimer disease (AD) "is a disabling senile dementia, the loss of reasoning and ability to care for oneself, that afflicts about 11% of the population over age 65." One of the postmortem findings in the brains of Alzheimer patients is exaggerated widening of cerebral sulci as well as narrowing of cerebral gyri. Can you explain the relationship between this anatomical change and the symptoms?

5. Dwayne's first trip to the dentist after a 10-year absence resulted in extensive dental work. He received numbing injections of anesthetic in several locations during the session. While having lunch right after the appointment, soup dribbled out of Dwayne's mouth because he had no feeling in his left upper lip, right lower lip, and tip of his tongue. What happened to Dwayne?

? ANSWERS TO FIGURE QUESTIONS

19.1 The largest part of the brain is the cerebrum.

19.2 From superficial to deep, the three cranial meninges are the dura mater, arachnoid mater, and pia mater.

19.3 The brain stem is anterior to the fourth ventricle, and the cerebellum is posterior to it.

19.4 Cerebrospinal fluid is reabsorbed by the arachnoid villi that project into the dural venous sinuses.

19.5 The medulla oblongata contains the pyramids; the midbrain contains the cerebral peduncles; pons means "bridge."

19.6 Decussation means crossing to the opposite side. The functional consequence of decussation of the pyramids is that each side of the cerebrum controls muscles on the opposite side of the body.

19.7 The cerebral peduncles are the main sites where tracts extend and nerve impulses are conducted between the superior parts of the brain and the inferior parts of the brain and spinal cord.

19.8 The cerebellar peduncles carry information into and out of the cerebellum.

19.9 In about 70% of human brains, the intermediate mass connects the right and left halves of the thalamus.

19.10 From posterior to anterior, the four major regions of the hypothalamus are the mammillary, tuberal, supraoptic, and preoptic regions.

19.11 The gray matter enlarges more rapidly during development, in the process producing convolutions or gyri (folds), sulci (shallow grooves), and fissures (deep grooves).

19.12 Association tracts connect gyri of the same hemisphere; commissural tracts connect gyri in opposite hemispheres; projection tracts connect the cerebrum with the thalamus, brain stem, and spinal cord.

19.13 The basal ganglia are lateral, superior, and inferior to the thalamus.

19.14 The hippocampus is the component of the limbic system that functions with the cerebrum in memory.

19.15 The common integrative area integrates interpretation of visual, auditory, and somatic sensations; the motor speech area translates thoughts into speech; the premotor area controls skilled muscular movements; the gustatory areas interpret sensations related to taste; the auditory areas interpret pitch and rhythm; the visual areas interpret shape, color, and movement of objects; the frontal eye field area controls voluntary scanning movements of the eyes.

19.16 Axons in the olfactory tracts terminate in the primary olfactory area in the temporal lobe of the cerebral cortex.

19.17 Most axons in the optic tracts terminate in the primary visual area in the occipital lobe of the cerebral cortex.

19.18 The superior branch of the oculomotor nerve is distributed to the superior rectus muscle; the trochlear nerve is the smallest cranial nerve.

19.19 The trigeminal nerve is the largest cranial nerve.

19.20 Motor axons of the facial nerve originate in the pons.

19.21 Sensory axons of the cochlear branch originate in the spiral organ, the organ of hearing.

19.22 The glossopharyngeal nerve exits the skull through the jugular foramen.

19.23 The vagus nerve is located medial and posterior to the internal jugular vein and common carotid artery in the neck.

19.24 The accessory nerve is the only cranial nerve that originates from both the brain and spinal cord.

19.25 Two important motor functions of the hypoglossal nerve are speech and swallowing.

19.26 The gray matter of the nervous system derives from the mantle layer cells of the neural tube.

19.27 The mesencephalon does not develop into a secondary brain vesicle.

20 THE AUTONOMIC NERVOUS SYSTEM

INTRODUCTION It is the end of the semester, you have studied diligently for your anatomy final, and now it is time to take the examination. As you enter the crowded lecture hall to find a seat, you sense tension in the room as the other students nervously chatter about last-minute details they think will be important to know for the test. Suddenly you feel your heart race with excitement—or is that apprehension? You notice that your mouth becomes somewhat dry as a cold sweat surfaces on your skin. You also seem to breathe a little bit faster and deeper. As you wait for the professor to pass out the test, these symptoms become more and more pronounced. Finally the test arrives at your desk. You stretch and take a deep breath, as you slowly flip through the exam to get a feel for the questions being asked. As you scan the questions, you recognize that you can answer them all with confidence. What a relief! The symptoms you were experiencing begin to disappear as you focus on transferring your knowledge from your brain to the paper.

Most of the effects just described fall under the control of the autonomic nervous system. As you learned in Chapter 17, the **autonomic nervous system (ANS)** is a division of the peripheral nervous system that consists of (1) *autonomic sensory neurons* in visceral organs and in blood vessels that convey information to (2) *integrating centers* in the central nervous system (CNS), (3) *autonomic motor neurons* that propagate from the CNS to various effector tissues to regulate the activity of smooth muscle, cardiac muscle, and many glands, and (4) the *enteric division*, a specialized network of nerves and ganglia forming an independent nerve network within the wall of the gastrointestinal (GI) tract. Functionally, the ANS usually operates without conscious control. The system was originally named *autonomic* because it was thought to operate autonomously (in a self-governing manner), without control by the CNS. However, centers in the hypothalamus and brain stem do regulate ANS reflexes.

In this chapter, we compare structural and functional features of the autonomic nervous system with those of the somatic nervous system, which was introduced in Chapter 17. Then we discuss the anatomy of the ANS and compare the organization and actions of its two major parts, the sympathetic and parasympathetic divisions. •

COMPARISON OF SOMATIC AND AUTONOMIC NERVOUS SYSTEMS

◉ OBJECTIVE

- Outline the structural and functional differences between the somatic and autonomic nervous systems.

As you learned in Chapter 17, the somatic nervous system includes both sensory and motor neurons. Sensory neurons convey input from receptors for somatic senses (pain, thermal, tactile, and proprioceptive sensations; see Chapter 21) and from receptors for the special senses (vision, hearing, taste, smell, and equilibrium; see Chapter 22). All these sensations normally are consciously perceived. In turn, somatic motor neurons innervate skeletal muscles—the effectors of the somatic nervous system—and produce voluntary movements. When a somatic motor neuron stimulates a skeletal muscle, the muscle contracts; the effect always is excitation. If somatic motor neurons cease to stimulate a muscle, the result is a paralyzed, limp muscle that has no muscle tone. In addition, even though we are generally not conscious of breathing, the muscles that generate respiratory movements are skeletal muscles controlled by somatic motor neurons. If the respiratory motor neurons become inactive, breathing stops. A few skeletal muscles, such as those in the middle ear, are controlled by reflexes and cannot be contracted voluntarily.

The main input to the ANS comes from **autonomic (visceral) sensory neurons**. Mostly, these neurons are associated with **interoceptors** (receptors inside the body), such as chemoreceptors that monitor blood CO_2 level, and mechanoreceptors that detect the degree of stretch in the walls of organs or blood vessels. These sensory signals are not consciously perceived most of the time, although intense activation may produce conscious sensations. Two examples of perceived visceral sensations are sensations of pain from damaged viscera and angina pectoris (chest pain) from inadequate blood flow to the heart. Some sensations monitored by somatic sensory (Chapter 21) and special sensory neurons (Chapter 22) also influence the ANS. For example, pain can produce dramatic changes in some autonomic activities.

Autonomic motor neurons regulate visceral activities by either increasing (exciting) or decreasing (inhibiting) activities in their effector tissues, which are cardiac muscle, smooth muscle, and glands. Changes in the diameter of the pupils, dilation and constriction of blood vessels, and adjustment of the rate and force of the heartbeat are examples of autonomic motor responses. Unlike skeletal muscle, tissues innervated by the ANS often function to some extent even if their nerve supply is damaged. For example, the heart continues to beat when it is removed for transplantation. Single-unit smooth muscle, like that found in the lining of the gastrointestinal tract, contracts rhythmically on its own, and glands produce some secretions in the absence of ANS control.

Most autonomic responses cannot be consciously altered or suppressed to any great degree. You probably cannot voluntarily slow your heartbeat to half its normal rate. For this reason, some autonomic responses are the basis for polygraph ("lie detector") tests. Nevertheless, practitioners of yoga or other techniques of meditation may learn how to regulate at least some of their autonomic activities through long practice. (Biofeedback, in which monitoring devices display information about a body function such as heart rate or blood pressure, enhances the ability to learn such conscious control.) Signals from the general somatic and special senses, acting via the limbic system, also influence responses of autonomic motor neurons. For example, seeing a bike about to hit you, hearing the squealing brakes of a nearby car as you cross the street, or being grabbed from behind by an attacker would increase the rate and force of your heartbeat.

Recall from Chapter 10 that the axon of a single, myelinated somatic motor neuron extends from the CNS all the way to the skeletal muscle fibers in its motor unit (Figure 20.1a). By contrast, most autonomic motor pathways consist of two motor neurons *in series*, one following the other (Figure 20.1b). The first neuron has its cell body in the CNS; its myelinated axon extends from the CNS to an **autonomic ganglion**. (Recall that a ganglion is a collection of neuronal cell bodies outside the CNS.) The cell body of the second neuron is also in that autonomic ganglion; its unmyelinated axon extends directly from the ganglion to the effector (smooth muscle, cardiac muscle, or a gland). In some autonomic pathways, the first motor neuron extends to the adrenal medullae (inner portions of the adrenal glands) rather than an autonomic ganglion. All somatic motor neurons release only acetylcholine (ACh) as their neurotransmitter; autonomic motor neurons release either ACh or norepinephrine (NE).

The output (motor) part of the ANS has two divisions: the **sympathetic division** and the **parasympathetic division**. Most organs have **dual innervation**, that is, they receive impulses from both sympathetic and parasympathetic neurons. In general, nerve impulses from one division of the ANS stimulate the organ to increase its activity (excitation), and impulses from the other division decrease the organ's activity (inhibition). For example, an increased rate of nerve impulses from the *sympathetic* division increases heart rate; an increased rate of nerve impulses from the *parasympathetic* division decreases heart rate. The sympathetic division is often called the *flight-or-fight division* and the majority of its output is directed at the smooth muscle of blood vessels. Sympathetic activities result in increased alertness and metabolic activities in order to prepare the body for an emergency situation. Responses to such situations, which may occur during physical activity or emotional stress, include a rapid heart rate, faster breathing rate, dilation of the pupils, dry mouth, sweaty but cool skin, dilation of blood vessels to organs involved in combating stress (such as the heart and skeletal muscles), constriction of blood vessels to organs not involved in combating stress (for example, the gastrointestinal tract and kidneys), and the release of glucose from the liver.

The parasympathetic division is often referred to as the *rest-and-digest division* because its activities conserve and restore body energy during times of rest or digesting a meal and the majority of its output is directed to the smooth muscle and glandular tissue

Figure 20.1 Motor neuron pathways in the (a) somatic nervous system and (b) autonomic nervous system (ANS). Note that autonomic motor neurons release either acetylcholine (ACh) or norepinephrine (NE); somatic motor neurons release ACh.

 Somatic nervous system stimulation always excites its effectors (skeletal muscle fibers); stimulation by the autonomic nervous system either excites or inhibits visceral effectors.

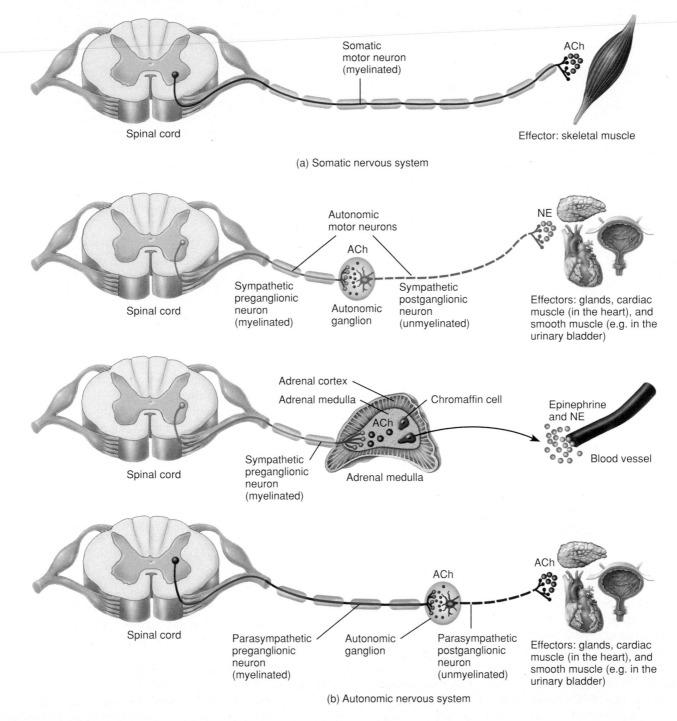

(a) Somatic nervous system

(b) Autonomic nervous system

? **What does dual innervation mean?**

of the gastrointestinal and respiratory tracts. The parasympathetic division conserves energy and replenishes nutrient stores. Although both the sympathetic and parasympathetic divisions are concerned with maintaining health, they do so in dramatically different ways.

Table 20.1 on page 662 compares the somatic and autonomic nervous systems.

◉ CHECKPOINT

1. Why is the autonomic nervous system so named?
2. What are the main input and output components of the autonomic nervous system?

TABLE 20.1

Comparison of the Somatic and Autonomic Nervous Systems

	SOMATIC NERVOUS SYSTEM	AUTONOMIC NERVOUS SYSTEM
Sensory input	Special senses and somatic senses.	Mainly from interoceptors; some from special senses and somatic senses.
Control of motor output	Voluntary control from cerebral cortex, with contributions from basal ganglia, cerebellum, spinal cord.	Involuntary control from limbic system, hypothalamus, brain stem, and spinal cord; limited control from brain stem, and cerebral cortex.
Motor neuron pathway	One-neuron pathway: Somatic motor neurons extending from CNS synapse directly with effector.	Usually two-neuron pathway: Preganglionic neurons extending from CNS synapse with postganglionic neurons in an autonomic ganglion, and postganglionic neurons extending from ganglion synapse with a visceral effector. Alternatively, preganglionic neurons may extend from CNS to synapse with cells of adrenal medullae.
Neurotransmitters and hormones	All somatic motor neurons release ACh.	All preganglionic axons release acetylcholine (ACh); most sympathetic postganglionic neurons release norepinephrine (NE); those to most sweat glands release ACh; all parasympathetic postganglionic neurons release ACh; adrenal medullae release epinephrine and norepinephrine.
Effectors	Skeletal muscle.	Smooth muscle, cardiac muscle, and glands.
Responses	Contraction of skeletal muscle.	Contraction or relaxation of smooth muscle; increased or decreased rate and force of contraction of cardiac muscle; increased or decreased secretions of glands.

ANATOMY OF AUTONOMIC MOTOR PATHWAYS

◉ **OBJECTIVES**

• Describe preganglionic and postganglionic neurons of the autonomic nervous system.
• Compare the anatomical components of the sympathetic and parasympathetic divisions of the autonomic nervous system.

Anatomical Components of an Autonomic Motor Pathway

The first of the two motor neurons in any autonomic motor pathway is called a **preganglionic neuron** (Figure 20.1b). Its cell body is in the brain or spinal cord, and its axon exits the CNS as part of a cranial or spinal nerve. The axon of a preganglionic neuron is a small-diameter, myelinated fiber that usually extends to an **autonomic ganglion**. There it synapses with a **postganglionic neuron**, the second neuron in the autonomic motor pathway (Figure 20.1b). Notice that the postganglionic neuron lies entirely outside the CNS. Its cell body and dendrites are located in an autonomic ganglion, where it forms one or more synapses with preganglionic neurons. The axon of a postganglionic neuron is a small-diameter, unmyelinated type C fiber that terminates in a visceral effector. Thus, preganglionic neurons convey nerve impulses from the CNS to autonomic ganglia, and postganglionic neurons relay the impulses from autonomic ganglia to visceral effectors.

Preganglionic Neurons

In the sympathetic division, the preganglionic neurons have their cell bodies in the lateral horns of the gray matter in the 12 thoracic segments and the first two lumbar segments of the spinal cord (Figure 20.2). For this reason, the sympathetic division is

also called the **thoracolumbar division** (thōr'-a-kō-LUM-bar), and the axons of the sympathetic preganglionic neurons are known as the **thoracolumbar outflow**.

Cell bodies of preganglionic neurons of the parasympathetic division are located in the nuclei of four cranial nerves in the brain stem (III, VII, IX, and X) and in the lateral gray horns of the second through fourth sacral segments of the spinal cord (see Figure 20.3). Hence, the parasympathetic division is also known as the **craniosacral division** (krā'-nē-ō-SĀ-kral), and the axons of the parasympathetic preganglionic neurons are referred to as the **craniosacral outflow**. As you will see later, sympathetic preganglionic axons are relatively short while parasympathetic preganglionic axons are relatively long.

Autonomic Ganglia

The autonomic ganglia may be divided into three general groups: Two of the groups are components of the sympathetic division, and one group is a component of the parasympathetic division.

SYMPATHETIC GANGLIA The sympathetic ganglia are the sites of synapses between sympathetic preganglionic and postganglionic neurons. The two groups of sympathetic ganglia are the sympathetic trunk ganglia and prevertebral ganglia. **Sympathetic trunk ganglia** (also called *vertebral chain ganglia* or *paravertebral ganglia*) lie in a vertical row on either side of the vertebral column. These ganglia extend from the base of the skull to the coccyx (Figure 20.2). Because the sympathetic trunk ganglia are near the spinal cord, most sympathetic preganglionic axons are short. Postganglionic axons from sympathetic trunk ganglia mostly innervate organs above the diaphragm. Sympathetic trunk ganglia in the neck have specific names—the **superior**, **middle**, and **inferior cervical ganglia** (Figure 20.2). The remaining sympathetic trunk ganglia do not have individual names.

The second group of sympathetic ganglia, the **prevertebral** (*collateral*) **ganglia**, lies anterior to the vertebral column and close

Figure 20.2 Structure of the sympathetic division of the autonomic nervous system. Solid lines represent preganglionic axons; dashed lines represent postganglionic axons. Although the innervated structures are shown only for one side of the body for diagrammatic purposes, the sympathetic division actually innervates tissues and organs on both sides.

Cell bodies of sympathetic preganglionic neurons are located in the lateral horns of gray matter in the 12 thoracic and first two lumbar segments of the spinal cord.

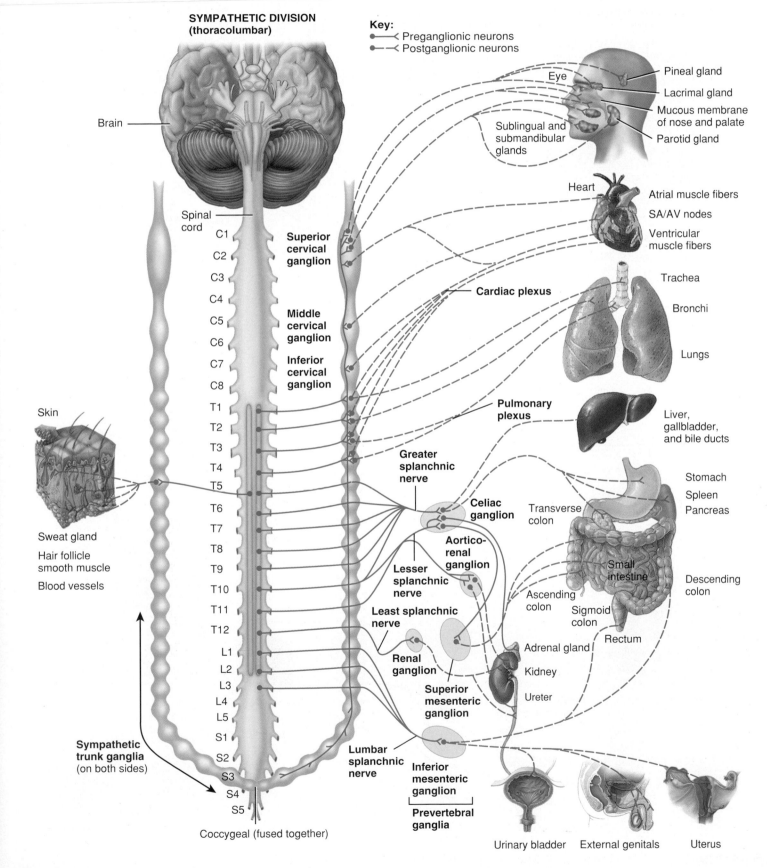

? **Which division, sympathetic or parasympathetic, has longer preganglionic axons? Why?**

Figure 20.3 Structure of the parasympathetic division of the autonomic nervous system. Solid lines represent preganglionic axons; dashed lines represent postganglionic axons. Although the innervated structures are shown only for one side of the body for diagrammatic purposes, the parasympathetic division actually innervates tissues and organs on both sides.

Cell bodies of parasympathetic preganglionic neurons are located in brain stem nuclei and in the lateral horns of gray matter in the second through fourth sacral segments of the spinal cord.

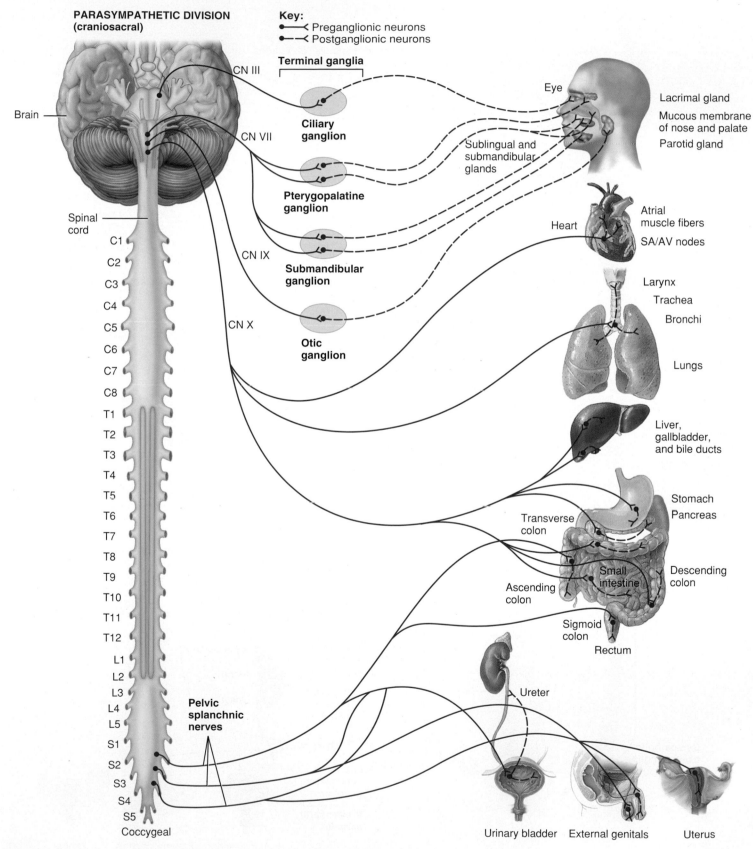

Which ganglia are associated with the parasympathetic division? Sympathetic division?

to the large abdominal arteries. In general, postganglionic axons from prevertebral ganglia innervate organs below the diaphragm. There are four major prevertebral ganglia (Figure 20.2; see also Figure 20.4): (1) The **celiac ganglion** (SĒ-lē-ak) is on either side of the celiac artery just inferior to the diaphragm. (2) The **superior mesenteric ganglion** is near the beginning of the superior mesenteric artery in the upper abdomen. (3) The **inferior mesenteric ganglion** is near the beginning of the inferior mesenteric artery in the middle of the abdomen. (4) The **aorticorenal ganglion** is near the renal artery as it branches from the aorta.

PARASYMPATHETIC GANGLIA Preganglionic axons of the parasympathetic division synapse with postganglionic neurons in **terminal** (*intramural*) **ganglia.** Most of these ganglia are located close to or actually within the wall of a visceral organ. Because the axons of parasympathetic preganglionic neurons extend from the CNS to a terminal ganglion in an innervated organ, they are

longer than most of the axons of sympathetic preganglionic neurons. Terminal ganglia in the head have specific names. They are the **ciliary ganglion, pterygopalatine ganglion, submandibular ganglion,** and **otic ganglion** (Figure 20.3). The remaining terminal ganglia do not have individual names. Typically, parasympathetic ganglia are near or within the walls of visceral effectors.

Autonomic Plexuses

In the thorax, abdomen, and pelvis, axons of both sympathetic and parasympathetic preganglionic neurons form tangled networks called **autonomic plexuses,** many of which lie along major arteries. The autonomic plexuses also may contain sympathetic ganglia and axons of autonomic sensory neurons. The major plexuses in the thorax are the **cardiac plexus,** which supplies the heart, and the **pulmonary plexus,** which supplies the bronchial tree (Figure 20.4; see also Figure 20.2).

Figure 20.4 Autonomic plexuses in the thorax, abdomen, and pelvis.

 An autonomic plexus is a network of sympathetic and parasympathetic axons that sometimes also includes autonomic sensory axons and sympathetic ganglia.

(a) Anterior view

FIGURE **20.4** CONTINUES ▶

◼ FIGURE 20.4 CONTINUED ▶

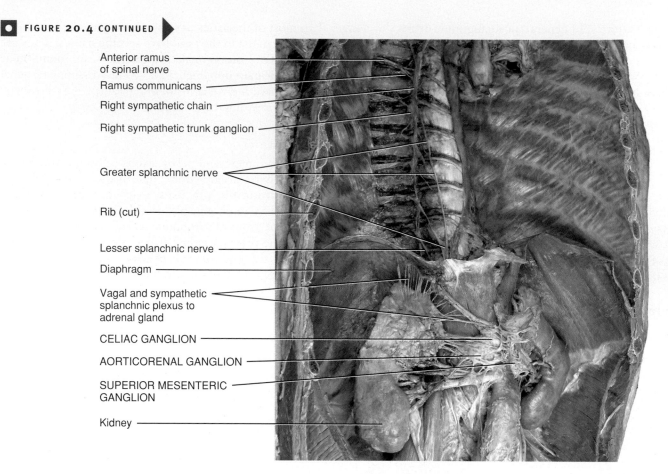

Anterior ramus of spinal nerve
Ramus communicans
Right sympathetic chain
Right sympathetic trunk ganglion
Greater splanchnic nerve
Rib (cut)
Lesser splanchnic nerve
Diaphragm
Vagal and sympathetic splanchnic plexus to adrenal gland
CELIAC GANGLION
AORTICORENAL GANGLION
SUPERIOR MESENTERIC GANGLION
Kidney

(b) Anterior view

? Which is the largest autonomic plexus?

The abdomen and pelvis also contain major autonomic plexuses and often are named after the artery along which they are distributed (Figure 20.4). The **celiac** (*solar*) **plexus** is the largest autonomic plexus and surrounds the celiac and superior mesenteric arteries. It contains two large celiac ganglia, two aorticorenal ganglia, and a dense network of autonomic axons and is distributed to the liver, gallbladder, stomach, pancreas, spleen, kidneys, medulla (inner region) of the adrenal gland, testes, and ovaries. The **superior mesenteric plexus** contains the superior mesenteric ganglion and supplies the small and large intestine. The **inferior mesenteric plexus** contains the inferior mesenteric ganglion, which innervates the large intestine. The **hypogastric plexus** supplies pelvic viscera. The **renal plexuses** contain the renal ganglion and supply the renal arteries within the kidneys and the ureters.

Postganglionic Neurons

Unlike other neurons, postganglionic autonomic fibers do not end in a single terminal swelling like a synaptic knob or end plate. The terminal branches of autonomic fibers contain numerous swellings, called **varicosities**, that simultaneously release neurotransmitter over a large area of the innervated organ rather than on single cells. This extensive release of neurotransmitter and the greater number of postganglionic neurons means that entire organs, rather than discrete cells, are typically influenced by autonomic activity.

SYMPATHETIC POSTGANGLIONIC NEURONS Once axons of sympathetic preganglionic neurons pass to sympathetic trunk ganglia, they may connect with postganglionic neurons in one of the following ways (Figure 20.5):

❶ An axon may synapse with postganglionic neurons in the first ganglion it reaches.

❷ An axon may ascend or descend to a higher or lower ganglion before synapsing with postganglionic neurons.

❸ An axon may continue, without synapsing, through the sympathetic trunk ganglion to end at a prevertebral ganglion and synapse with postganglionic neurons there.

In addition, some preganglionic sympathetic axons extend to and terminate in the adrenal medullae.

A single sympathetic preganglionic fiber has many axon collaterals (branches) and may synapse with 20 or more postganglionic neurons. This pattern of projection is an example of divergence (see Chapter 17) and helps explain why many sympathetic responses affect almost the entire body simultaneously. After exiting their ganglia, the postganglionic axons typically terminate in several visceral effectors (see Figure 20.2). Sympathetic postganglionic axons are relatively long.

PARASYMPATHETIC POSTGANGLIONIC NEURONS Axons of preganglionic neurons of the parasympathetic division pass to terminal ganglia near or within a visceral effector (see Figure 20.3). In the ganglion, the presynaptic neuron usually synapses with only four or five postsynaptic neurons, all of which supply a single visceral effector. Thus, parasympathetic responses can be localized to a single effector. Parasympathetic postganglionic axons are rel-

Figure 20.5 Types of connections between ganglia and postganglionic neurons in the sympathetic division of the ANS. Also illustrated are the gray and white rami communicantes.

 Sympathetic ganglia lie in two chains on either side of the vertebral column (sympathetic trunk ganglia) and near large abdominal arteries anterior to the vertebral column (prevertebral ganglia).

Preganglionic neuron

Postganglionic neurons

Anterior view

❓ **What substance gives the white rami their white appearance?**

---

atively short. With this background in mind, we can now examine some specific structural features of the sympathetic and parasympathetic divisions of the ANS.

Structure of the Sympathetic Division

Cell bodies of sympathetic preganglionic neurons are part of the lateral horns of all thoracic segments and of the first two lumbar segments of the spinal cord (see Figure 20.2). The preganglionic axons leave the spinal cord along with the somatic motor neurons at the same segmental level. After exiting through the intervertebral foramina, the myelinated preganglionic sympathetic axons pass into the anterior root of a spinal nerve and enter a short pathway called a **white ramus** before passing to the nearest sympathetic trunk ganglion on the same side (Figure 20.5). Collectively, the white rami are called the **white rami communicantes**

(kō-mū-ni-KAN-tēz; singular is *ramus communicans*). The "white" in their name indicates that they contain myelinated axons. Only the thoracic and first two or three lumbar nerves have white rami communicantes. The white rami communicantes connect the anterior ramus of the spinal nerve with the ganglia of the sympathetic trunk. The paired sympathetic trunk ganglia are arranged anterior and lateral to the vertebral column, one on either side. Typically, there are 3 cervical, 11 or 12 thoracic, 4 or 5 lumbar, and 4 or 5 sacral sympathetic trunk ganglia, and 1 coccygeal ganglion. The right and left coccygeal ganglia are fused together and usually lie at the midline. Although the sympathetic trunk ganglia extend inferiorly from the neck, chest, and abdomen to the coccyx, they receive preganglionic axons only from the thoracic and lumbar segments of the spinal cord (see Figure 20.2).

The cervical portion of each sympathetic trunk ganglion is located in the neck and is subdivided into superior, middle, and inferior ganglia (see Figure 20.2). Postganglionic neurons leaving the **superior cervical ganglion** serve the head and heart. They are distributed to sweat glands, smooth muscle of the eye, blood vessels of the face, lacrimal glands, nasal mucosa, salivary glands (submandibular, sublingual, and parotid), and the heart. Gray rami communicantes (described shortly) from the superior cervical ganglion also pass to the upper two to four cervical spinal nerves. Postganglionic neurons leaving the **middle cervical ganglion** and **inferior cervical ganglion** innervate the heart.

The thoracic portion of each sympathetic trunk ganglion lies anterior to the necks of the corresponding ribs. This region of the sympathetic trunk receives most of the sympathetic preganglionic axons, and its postganglionic neurons innervate the heart, lungs, bronchi, and other thoracic viscera. In the skin, these neurons also innervate sweat glands, blood vessels, and arrector pili muscles of hair follicles.

The lumbar portion of each sympathetic trunk ganglion lies lateral to the corresponding lumbar vertebrae. The sacral region of the sympathetic trunk ganglion lies in the pelvic cavity on the medial side of the anterior sacral foramina. Unmyelinated postganglionic axons from the lumbar and sacral sympathetic trunk ganglia enter a short pathway called a **gray ramus** and then merge with a spinal nerve or join the hypogastric plexus via direct visceral branches. The **gray rami communicantes** are structures containing the postganglionic axons that connect the ganglia of the various portions of the sympathetic trunk ganglion to spinal nerves (Figure 20.5). The axons are unmyelinated. Gray rami communicantes outnumber the white rami because there is a gray ramus leading to each of the 31 pairs of spinal nerves that carries sympathetic output to the smooth muscle and glands of the body wall and limbs.

As preganglionic axons extend from a white ramus communicans into the sympathetic trunk ganglion, they give off several axon collaterals (branches). These collaterals terminate and synapse in several ways. Some synapse in the first ganglion at the level of entry. Others pass up or down the sympathetic trunk for a variable distance to form the **sympathetic chains**, the fibers on which the ganglia are strung (Figure 20.5). Many postganglionic axons rejoin the spinal nerves through gray rami and supply peripheral visceral effectors such as sweat glands, smooth muscle in blood vessels, and arrector pili muscles of hair follicles.

Some preganglionic axons pass through the sympathetic trunk without terminating in it. Beyond the trunk, they form nerves known as **splanchnic nerves** (SPLANK-nik; see Figure 20.2), which extend to and terminate in the outlying prevertebral ganglia. Splanchnic nerves from the thoracic area terminate in the **celiac ganglion**, where the preganglionic neurons synapse with postganglionic cell bodies. Preganglionic axons from the fifth through ninth or tenth thoracic ganglia (T5–T9 or T10) form the **greater splanchnic nerve**, which pierces the diaphragm and enters the celiac ganglion of the celiac plexus. From there, postganglionic neurons follow and innervate blood vessels to the stomach, spleen, liver, kidneys, and small intestine. Preganglionic axons from the tenth and eleventh thoracic ganglia (T10–T11) form the **lesser splanchnic nerve**, which pierces the diaphragm and passes through the celiac plexus to enter the aorticorenal ganglion and superior mesenteric ganglion of the superior mesenteric plexus. Postganglionic neurons from the superior mesenteric ganglion follow and innervate blood vessels of the small intestine and proximal colon. The **least** or **lowest splanchnic nerve,** which is not always present, is formed by preganglionic axons from the twelfth thoracic ganglia (T12) or a branch of the lesser splanchnic nerve. It passes through the diaphragm, and enters the

renal plexus near the kidney. Postganglionic neurons from the renal plexus supply kidney arterioles and the ureter.

Preganglionic axons that form the **lumbar splanchnic nerve** from the first through fourth lumbar ganglia (L1–L4) enter the inferior mesenteric plexus and terminate in the inferior mesenteric ganglion, where they synapse with postganglionic neurons. Axons of postganglionic neurons extend through the hypogastric plexus and principally supply blood vessels of the distal colon and rectum, urinary bladder, and genital organs. Postganglionic axons leaving the prevertebral ganglia follow the course of various arteries to abdominal and pelvic visceral effectors.

Sympathetic preganglionic neurons also extend to the adrenal medullae. Developmentally, the adrenal medullae and sympathetic ganglia are derived from the same tissue, the neural crest (see Figure 19.26 on page 653). The adrenal medullae are modified sympathetic ganglia, and their cells are similar to sympathetic postganglionic neurons. Rather than extending to another organ, however, these cells release hormones into the blood. Upon stimulation by sympathetic preganglionic neurons, the adrenal medullae release a mixture of hormones—about 80% **epinephrine**, 20% **norepinephrine**, and a trace amount of **dopamine**.

Structure of the Parasympathetic Division

Cell bodies of parasympathetic preganglionic neurons are found in nuclei in the brain stem and in the lateral horns of the second through fourth sacral segments of the spinal cord (see Figure 20.3). Their axons emerge as part of a cranial nerve or as part of the anterior root of a spinal nerve. The **cranial parasympathetic outflow** consists of preganglionic axons that extend from the brain stem in four cranial nerves. The **sacral parasympathetic outflow** consists of preganglionic axons in anterior roots of the second through fourth sacral nerves. The preganglionic axons of both the cranial and sacral outflows end in terminal ganglia, where they synapse with postganglionic neurons.

The cranial outflow has five components: four pairs of ganglia and the plexuses associated with the vagus (X) nerve. The four pairs of cranial parasympathetic ganglia innervate structures in the head and are located close to the organs they innervate (see Figure 20.3).

1. The **ciliary ganglia** lie lateral to each optic (II) nerve near the posterior aspect of the orbit. Preganglionic axons pass with the oculomotor (III) nerves to the ciliary ganglia. Postganglionic axons from the ganglia innervate smooth muscle fibers in the eyeball.
2. The **pterygopalatine ganglia** (ter′-i-gō-PAL-a-tīn) are located lateral to the sphenopalatine foramen in the pterygopalatine fossa, between the sphenoid and palatine bones. They receive preganglionic axons from the facial (VII) nerve and send postganglionic axons to the nasal mucosa, palate, pharynx, and lacrimal glands.
3. The **submandibular ganglia** are found near the ducts of the submandibular salivary glands. They receive preganglionic axons from the facial nerves and send postganglionic axons to the submandibular and sublingual salivary glands.
4. The **otic ganglia** are situated just inferior to each foramen ovale. They receive preganglionic axons from the glossopharyngeal (IX) nerves and send postganglionic axons to the parotid salivary glands.

Preganglionic axons that leave the brain as part of the vagus (X) nerves carry nearly 80% of the total craniosacral outflow.

Vagal axons extend to many terminal ganglia in the thorax and abdomen. Because the terminal ganglia are close to or in the walls of their visceral effectors, postganglionic parasympathetic axons are very short. As the vagus nerve passes through the thorax, it sends axons to the heart and the airways of the lungs. In the abdomen, it supplies the liver, gallbladder, stomach, pancreas, small intestine, and part of the large intestine.

The sacral parasympathetic outflow consists of preganglionic axons from the anterior roots of the second through fourth sacral nerves (S2–S4), which form the **pelvic splanchnic nerves** (see Figure 20.3). These nerves synapse with parasympathetic postganglionic neurons located in terminal ganglia in the walls of the innervated viscera. From the ganglia, parasympathetic postganglionic axons innervate smooth muscle and glands in the walls of the colon, ureters, urinary bladder, and reproductive organs.

Structure of the Enteric Division

To understand and appreciate the **enteric division** of the autonomic nervous system it is important to realize that the gastrointestinal tract, like the surface of the body, forms an extensive area of contact with the environment. Although this gastrointestinal environment is inside the body, it is still considered part of the external environment. Just as the surface of the body must respond to important environmental stimuli in order to function properly, the surface of the gastrointestinal tract must respond to surrounding stimuli to generate proper homeostatic controls. In fact, these responses and controls are so important that the gastrointestinal tract has its own nervous system with intrinsic input, processing, and output. The enteric division can and does function independent of central nervous system activity, but can also receive controlling input from the central nervous system.

The enteric division is the specialized network of nerves and ganglia forming a complex, integrated neuronal network within the wall of the gastrointestinal tract, pancreas, and gallbladder. This incredible nerve network contains in the neighborhood of 100 million neurons, approximately the same number as the spinal cord, and is capable of continued function without input from the central nervous system. The enteric network of nerves and ganglia contains sensory neurons capable of monitoring tension in the intestinal wall and accessing the composition of the intestinal contents. These sensory neurons relay their input signals to interneurons within the enteric ganglia. The interneurons establish an integrative network that processes the incoming signals and generates regulatory output signals to motor neurons throughout plexuses within the wall of the digestive organs. The motor neurons carry the output signals to the smooth muscle and glands of the gastrointestinal tract to exert control over its motility and secretory activities.

Most of the nerve fibers that innervate the digestive organs arise from two plexuses within the enteric system. The largest, the **myenteric plexus**, is positioned between the outer longitudinal and circular muscle layers from the upper esophagus to the anus. The myenteric plexus communicates extensively with a somewhat smaller plexus, the **submucous plexus**, which occupies the gut wall between the circular muscle layer and the muscularis mucosae (see page 784) and runs from the stomach to the anus. Neurons emerge from the ganglia of these two plexuses to form smaller plexuses around blood vessels and within the muscle layers and mucosa of the gut wall. It is this system of nerves that makes possible the normal motility and secretory functions of the gastrointestinal tract.

◉ CHECKPOINT
 3. Why is the sympathetic division called the thoracolumbar division even though its ganglia extend from the cervical to the sacral region?
 4. Prepare a list of the organs served by each sympathetic and parasympathetic ganglion.
 5. Describe the locations of sympathetic trunk ganglia, prevertebral ganglia, and terminal ganglia. Which types of autonomic neurons synapse in each type of ganglion?
 6. How can the sympathetic division produce simultaneous effects throughout the body, while parasympathetic effects typically are localized to specific organs?

ANS NEUROTRANSMITTERS AND RECEPTORS

◉ OBJECTIVE

• Describe the neurotransmitters and receptors involved in autonomic responses.

Autonomic neurons are classified based on the neurotransmitter they produce and release. The receptors for the neurotransmitters are integral membrane proteins that are located in the plasma membrane of the postsynaptic neuron or effector cell.

Cholinergic Neurons and Receptors

Cholinergic neurons (kō′-lin-ER-jik) release the neurotransmitter **acetylcholine** (as′-ē-til-KŌ-lēn) or **ACh**. In the ANS, the cholinergic neurons include (1) all sympathetic and parasympathetic preganglionic neurons, (2) sympathetic postganglionic neurons that innervate most sweat glands, and (3) all parasympathetic postganglionic neurons (Figure 20.6).

ACh is stored in synaptic vesicles and released by exocytosis. It then diffuses across the synaptic cleft and binds with specific **cholinergic receptors**, integral membrane proteins in the *postsynaptic* plasma membrane. The two types of cholinergic receptors, both of which bind ACh, are nicotinic receptors and muscarinic receptors. **Nicotinic receptors** are present in the plasma membranes of dendrites and cell bodies of both sympathetic and parasympathetic postganglionic neurons (Figure 20.6a, b) and in the motor end plate at the neuromuscular junction. They are so named because nicotine mimics the action of ACh by binding to these receptors. (Nicotine, a natural substance in tobacco leaves, is not normally present in the bodies of nonsmokers.) **Muscarinic receptors** are present in the plasma membranes of all effectors (smooth muscle, cardiac muscle, and glands) innervated by parasympathetic postganglionic axons. In addition, most sweat glands, which receive their innervation from cholinergic sympathetic postganglionic neurons, possess muscarinic receptors (see Figure 20.6). These receptors are so named because a mushroom poison called muscarine mimics the actions of ACh by binding to muscarinic receptors.

Activation of nicotinic receptors by ACh causes depolarization and thus excitation of the postsynaptic cell—which can be a postganglionic neuron, an autonomic effector, or a skeletal muscle fiber. Activation of muscarinic receptors by ACh sometimes causes depolarization (excitation) and sometimes causes hyperpolarization (inhibition), depending on which particular cell bears the muscarinic receptors. For example, binding of ACh to mus-

Figure 20.6 Cholinergic neurons and adrenergic neurons in the sympathetic and parasympathetic divisions. Cholinergic neurons release acetylcholine; adrenergic neurons release norepinephrine. Cholinergic and adrenergic receptors all are integral membrane proteins located in the plasma membrane of a postsynaptic neuron or an effector cell.

 Most sympathetic postganglionic neurons are adrenergic; other autonomic neurons are cholinergic.

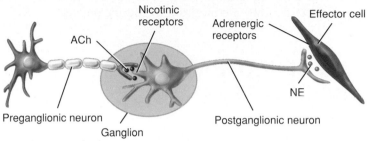

(a) Sympathetic division—innervation to most effector tissues

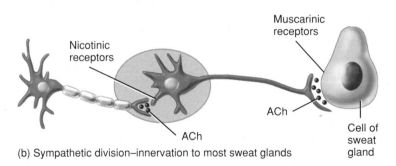

(b) Sympathetic division—innervation to most sweat glands

(c) Parasympathetic division

? Which neurons are cholinergic and possess nicotinic ACh receptors? What type of receptors for ACh do the effector tissues innervated by these neurons possess?

carinic receptors inhibits (relaxes) smooth muscle sphincters in the gastrointestinal tract. By contrast, ACh excites smooth muscle fibers in the circular muscles of the iris of the eye, causing them to contract. Because acetylcholine is quickly inactivated by the enzyme **acetylcholinesterase (AChE)**, effects triggered by cholinergic neurons are brief.

Adrenergic Neurons and Receptors

In the ANS, **adrenergic neurons** (ad'-ren-ER-jik) release **norepinephrine** (nor'-ep-i-NEF-rin) or **NE**, also known as **noradrenalin** (Figure 20.6a). Most sympathetic postganglionic neurons are adrenergic. Like ACh, NE is synthesized and stored in synaptic vesicles and released by exocytosis. Molecules of NE diffuse across the synaptic cleft and bind to specific adrenergic receptors on the postsynaptic membrane, causing either excitation or inhibition of the effector cell.

Adrenergic receptors bind both NE and epinephrine, a hormone with actions similar to NE. As noted previously, NE is released as a neurotransmitter by sympathetic postganglionic neurons. Both epinephrine and NE are released as hormones into the blood by the adrenal medullae. The two main types of adrenergic receptors are **alpha (α) receptors** and **beta (β) receptors**, which are found on visceral effectors innervated by most sympathetic postganglionic axons. These receptors are further classified into subtypes—α_1, α_2, β_1, β_2, and β_3—based on the specific responses they elicit and by their selective binding of the drugs that activate or block them. Although there are some exceptions, activation of α_1 and β_1 receptors generally produces excitation, in contrast to activation of α_2 and β_2 receptors, which causes inhibition of effector tissues. β_3 receptors are present only on cells of brown adipose tissue, where their activation causes thermogenesis (heat production). Cells of most effectors contain either α or β receptors; some visceral effector cells contain both. NE stimulates alpha receptors more strongly than beta receptors; epinephrine is a potent stimulator of both alpha and beta receptors.

The activity of NE at a synapse is terminated either when the NE is taken up by the axon that released it or when the NE is enzymatically inactivated by either **catechol-O-methyltransferase (COMT)** or **monoamine oxidase (MAO)**. Compared to ACh, NE lingers in the synaptic cleft for a longer time. Thus, effects triggered by adrenergic neurons typically are longer lasting than those triggered by cholinergic neurons.

◉ CHECKPOINT

7. Why are cholinergic and adrenergic neurons so named?
8. What neurotransmitters and hormones bind to adrenergic receptors?

FUNCTIONS OF THE ANS

◉ OBJECTIVE

• Describe the major responses of the body to stimulation by the sympathetic and parasympathetic divisions of the ANS.

As noted earlier, most body organs receive innervation from both divisions of the ANS, which typically work in opposition to one another. The balance between sympathetic and parasympathetic activity is regulated by the hypothalamus. Typically, the hypothalamus turns up sympathetic activity at the same time it turns down parasympathetic activity, and vice versa. As you learned in the last section of this chapter, the two divisions affect body organs differently because of the different neurotransmitters released by

A large variety of drugs and natural products can selectively activate or block specific cholinergic or adrenergic receptors. An **agonist** is a substance that binds to and activates a receptor, in the process mimicking the effect of a natural neurotransmitter or hormone. Phenylephrine, an adrenergic agonist at α_1 receptors, is a common ingredient in cold and sinus medications. Because it constricts blood vessels in the nasal mucosa, phenylephrine reduces production of mucus, thus relieving nasal congestion. An **antagonist** is a substance that binds to and blocks a receptor, thereby preventing a natural neurotransmitter or hormone from exerting its effect. For example, atropine, which blocks muscarinic ACh receptors, dilates the pupils, reduces glandular secretions, and relaxes smooth muscle in the gastrointestinal tract. It is used to dilate the pupils during eye examinations, in the treatment of smooth muscle disorders such as iritis and intestinal hypermotility, and as an antidote for chemical warfare agents that inactivate AChE.

Propranolol (Inderal®) often is prescribed for patients with hypertension (high blood pressure). It is a nonselective beta blocker, meaning it binds to all types of beta receptors and prevents their activation by epinephrine and norepinephrine. The desired effects of propranolol are due to its *blockade* of β_1 receptors—namely, decreased heart rate and force of contraction and a consequent decrease in blood pressure. Undesired

Dilated pupil

Atropine, an antagonist, blocks muscarinic ACh receptors and dilates the pupils

effects due to blockade of β_2 receptors may include hypoglycemia (low blood glucose), resulting from decreased glycogen breakdown and decreased gluconeogenesis (the conversion of a noncarbohydrate into glucose in the liver), and mild bronchoconstriction (narrowing of the airways). If these side effects pose a threat to the patient, a selective β_1 blocker such as metoprolol (Lopressor®) can be prescribed.

their postganglionic neurons and the different adrenergic and cholinergic receptors on the cells of their effector organs. A few structures receive only sympathetic innervation—sweat glands, arrector pili muscles attached to hair follicles in the skin, the kidneys, the spleen, most blood vessels, and the adrenal medullae (see Figure 20.2). In these structures there is no opposition from the parasympathetic division; increases and decreases in sympathetic activity are responsible for the changes.

Sympathetic Responses

During physical or emotional stress, the sympathetic division dominates the parasympathetic division. High sympathetic activity favors body functions that can support vigorous physical activity and rapid production of ATP. At the same time, the sympathetic division decreases body functions that favor the storage of energy. Physical exertion and a variety of emotions—such as fear, embarrassment, or rage—stimulate the sympathetic division. Visualizing body changes that occur during "E situations" such as exercise, emergency, excitement, and embarrassment will help you remember most of the sympathetic responses. Activation of the sympathetic division and release of hormones by the adrenal medullae set in motion a series of physiological responses collectively called the **fight-or-flight response**, which includes the following effects (many of which were experienced by the student in the introduction to this chapter):

1. The pupils of the eyes dilate.
2. Heart rate, force of heart contraction, and blood pressure increase.
3. The airways dilate, allowing faster movement of air into and out of the lungs.
4. Blood vessels that supply organs involved in exercise or fighting off danger—skeletal muscles, cardiac muscle, liver, and adipose tissue—dilate, allowing greater blood flow through these tissues.

5. Liver cells perform glycogenolysis (breakdown of glycogen to glucose), and adipose tissue cells perform lipolysis (breakdown of triglycerides to fatty acids and glycerol).
6. Release of glucose by the liver increases blood glucose level.
7. Processes that are not essential for meeting the stressful situation are inhibited. For example, the blood vessels that supply the kidneys and gastrointestinal tract constrict, which decreases blood flow through these tissues. The result is a slowing of urine formation and digestive activities, which are not essential during exercise.

The effects of sympathetic stimulation are longer lasting and more widespread than the effects of parasympathetic stimulation for three reasons: (1) Sympathetic postganglionic axons diverge more extensively; as a result, many tissues are activated simultaneously. (2) AChE quickly inactivates ACh, but NE lingers in the synaptic cleft for a longer period. (3) Epinephrine and NE secreted as hormones into the blood from the adrenal medulla intensify and prolong the responses caused by NE released as a neurotransmitter from sympathetic postganglionic axons. These blood-borne hormones circulate throughout the body, affecting all tissues that have α and β receptors. In time, blood-borne NE and epinephrine are inactivated by enzymatic destruction in the liver.

Parasympathetic Responses

In contrast to the "fight-or-flight" activities of the sympathetic division, the parasympathetic division enhances "rest-and-digest" activities. Parasympathetic responses support body functions that conserve and restore body energy during times of rest and recovery. In the quiet intervals between periods of exercise, parasympathetic impulses to the digestive glands and the smooth muscle of the gastrointestinal tract predominate over sympathetic impulses. This allows energy-supplying food to be digested and absorbed. At the same time, parasympathetic responses decrease body functions that support physical activity.

The acronym "SLUDD" can be helpful in remembering five parasympathetic responses. It stands for salivation (S), lacrima-

tion (L), urination (U), digestion (D), and defecation (D). All of these activities are stimulated mainly by the parasympathetic division. Besides the increasing SLUDD responses, other important parasympathetic responses are "three decreases": decreased

---

TABLE 20.2

Structure and Function of Sympathetic and Parasympathetic Divisions of the ANS

	SYMPATHETIC (THORACOLUMBAR)	PARASYMPATHETIC (CRANIOSACRAL)
Distribution	Wide regions of the body: skin, sweat glands, arrector pili muscles of hair follicles, adipose tissue, smooth muscle of blood vessels.	Limited mainly to head and to viscera of thorax, abdomen, and pelvis; some blood vessels.
Location of preganglionic neuron cell bodies and site of outflow	Cell bodies of preganglionic neurons are located in lateral gray horns of spinal cord segments T1–L2. Axons of preganglionic neurons constitute thoracolumbar outflow.	Cell bodies of preganglionic neurons are located in the nuclei of cranial nerves III, VII, IX, and X and the lateral gray matter of spinal cord segments S2–S4. Axons of preganglionic neurons constitute craniosacral outflow.
Associated ganglia	Two types: sympathetic trunk ganglia and prevertebral ganglia.	One type: terminal ganglia.
Ganglia locations	Close to CNS and distant from visceral effectors.	Typically near or within wall of visceral effectors.

TABLE 20.3

Effects of Sympathetic and Parasympathetic Divisions of the ANS

VISCERAL EFFECTOR	EFFECT OF SYMPATHETIC STIMULATION (α OR β ADRENERGIC RECEPTORS, EXCEPT AS NOTED)*	EFFECT OF PARASYMPATHETIC STIMULATION (MUSCARINIC ACH RECEPTORS)
Glands		
Adrenal medullae	Secretion of epinephrine and norepinephrine (nicotinic ACh receptors).	No known effect.
Lacrimal (tear)	Slight secretion of tears (α).	Secretion of tears.
Pancreas	Inhibits secretion of digestive enzymes and the hormone insulin (α_2); promotes secretion of the hormone glucagon (β_2).	Secretion of digestive enzymes and the hormone insulin.
Posterior pituitary	Secretion of antidiuretic hormone (ADH) (β_1).	No known effect.
Pineal	Increases synthesis and release of melatonin (β).	No known effect.
Sweat	Increases sweating in most body regions (muscarinic ACh receptors); sweating on palms and soles (α_1).	No known effect.
Adipose tissue[†]	Lipolysis (breakdown of triglycerides into fatty acids and glycerol) (β_1); release of fatty acids into blood (β_1 and β_3).	No known effect.
Liver[†]	Glycogenolysis (conversion of glycogen into glucose); gluconeogenesis (conversion of noncarbohydrates into glucose); decreased bile secretion (α and β_2).	Glycogen synthesis; increased bile secretion.
Kidney, juxtaglomerular cells[†]	Secretion of renin (β_1).	No known effect.
Cardiac (Heart) Muscle	Increased heart rate and force of atrial and ventricular contractions (β_1).	Decreased heart rate; decreased force of atrial contraction.
Smooth Muscle		
Iris, radial muscle	Contraction → dilation of pupil (α_1).	No known effect.
Iris, circular muscle	No known effect.	Contraction → constriction of pupil.
Ciliary muscle of eye	Relaxation for distant vision (β_2).	Contraction for close vision.
Lungs, bronchial muscle	Relaxation → airway dilation (β_2).	Contraction → airway constriction.
Gallbladder and ducts	Relaxation (β_2).	Contraction → increased release of bile into small intestine.
Stomach and intestines	Decreased motility and tone (α_1, α_2, β_2); contraction of sphincters (α_1).	Increased motility and tone; relaxation of sphincters.
Spleen	Contraction and discharge of stored blood into general circulation (α_1).	No known effect.

heart rate, decreased diameter of airways (bronchoconstriction), and decreased diameter (constriction) of the pupils.

Table 20.2 compares the structural and functional features of the sympathetic and parasympathetic divisions of the ANS.

Table 20.3 lists the responses of glands, cardiac muscle, and smooth muscle to stimulation by the sympathetic and parasympathetic divisions of the ANS.

TABLE 20.2 CONTINUED

	SYMPATHETIC (THORACOLUMBAR)	PARASYMPATHETIC (CRANIOSACRAL)
Axon length and divergence	Preganglionic neurons with short axons synapse with many postganglionic neurons with long axons that pass to many visceral effectors.	Preganglionic neurons with long axons usually synapse with four to five postganglionic neurons with short axons that pass to a single visceral effector.
Rami communicantes	Both present; white rami communicantes contain myelinated preganglionic axons, and gray rami communicantes contain unmyelinated postganglionic axons.	Neither present.
Neurotransmitters	Preganglionic neurons release acetylcholine (ACh), which is excitatory and stimulates postganglionic neurons; most postganglionic neurons release norepinephrine (NE); postganglionic neurons that innervate most sweat glands and some blood vessels in skeletal muscle release ACh.	Preganglionic neurons release acetylcholine (ACh), which is excitatory and stimulates postganglionic neurons; postganglionic neurons release ACh.
Physiological effects	Fight-or-flight responses.	Rest-and-digest activities.

TABLE 20.3 CONTINUED

VISCERAL EFFECTOR	EFFECT OF SYMPATHETIC STIMULATION (α OR β ADRENERGIC RECEPTORS, EXCEPT AS NOTED)*	EFFECT OF PARASYMPATHETIC STIMULATION (MUSCARINIC ACH RECEPTORS)
Ureter	Increases motility (α_1).	Increases motility (?).
Urinary bladder	Relaxation of muscular wall (β_2); contraction of sphincter (α_1).	Contraction of muscular wall; relaxation of sphincter.
Uterus	Inhibits contraction in nonpregnant women (β_2); promotes contraction in pregnant women (α_1).	Minimal effect.
Sex organs	In males: contraction of smooth muscle of ductus (vas) deferens, seminal vesicle, prostate → ejaculation of semen (α_1).	Vasodilation; erection of clitoris (females) and penis (males).
Hair follicles, arrector pili muscle	Contraction → erection of hairs (α_1).	No known effect.
Vascular Smooth Muscle		
Salivary gland arterioles	Vasoconstriction, which decreases secretion (α_1).	Vasodilation, which increases K^+ and water secretion.
Gastric gland arterioles	Vasoconstriction, which inhibits secretion (α_1).	Secretion of gastric juice.
Intestinal gland arterioles	Vasoconstriction, which inhibits secretion (α_1).	Secretion of intestinal juice.
Coronary (heart) arterioles	Relaxation → vasodilation (β_2); contraction → vasoconstriction (α_1, α_2); contraction → vasoconstriction (muscarinic ACh receptors).	Contraction → vasoconstriction.
Skin and mucosal arterioles	Contraction → vasoconstriction (α_1).	Vasodilation, which may not be physiologically significant.
Skeletal muscle arterioles	Contraction → vasoconstriction (α_1); relaxation → vasodilation (β_2); relaxation → vasodilation (muscarinic ACh receptors).	No known effect.
Abdominal viscera arterioles	Contraction → vasoconstriction (α_1, β_2).	No known effect.
Brain arterioles	Slight contraction → vasoconstriction (α_1).	No known effect.
Kidney arterioles	Constriction of blood vessels → decreased urine volume (α_1).	No known effect.
Systemic veins	Contraction → constriction (α_1); relaxation → dilation (β_2).	No known effect.

*Subcategories of α and β receptors are listed if known.   †Grouped with glands because they release substances into the blood.

⊙ **CHECKPOINT**

9. Define cholinergic and adrenergic neurons and receptors.
10. What are some examples of the antagonistic effects of the sympathetic and parasympathetic divisions of the autonomic nervous system?
11. What happens during the fight-or-flight response?
12. Why is the parasympathetic division of the ANS called an energy conservation/restoration system?
13. Using Table 20.3 as a reference, describe the sympathetic response in a frightening situation for each of the following body parts: hair follicles, iris of eye, lungs, spleen, adrenal medullae, urinary bladder, stomach, intestines, gallbladder, liver, heart, arterioles of the abdominal viscera, and arterioles of skeletal muscles.

INTEGRATION AND CONTROL OF AUTONOMIC FUNCTIONS

⊙ **OBJECTIVES**

• Describe the components of an autonomic reflex.
• Explain the relationship of the hypothalamus to the ANS.

Autonomic Reflexes

Autonomic reflexes are responses that occur when nerve impulses pass over an autonomic reflex arc. These reflexes play a key role in regulating controlled conditions in the body, such as *blood pressure*, by adjusting heart rate, force of ventricular contraction, and blood vessel diameter; *digestion*, by adjusting the motility (movement) and muscle tone of the gastrointestinal tract; and *defecation* and *urination*, by regulating the opening and closing of sphincters.

The components of an autonomic reflex arc are as follows:

1. **Receptor.** Like the receptor in a somatic reflex arc (see Figure 18.12 on page 607), the receptor in an autonomic reflex arc is the distal end of a sensory neuron, which responds to a stimulus and produces a change that will ultimately trigger nerve impulses. Autonomic sensory receptors are mostly associated with interoceptors.
2. **Sensory neuron.** Conducts nerve impulses from receptors to the CNS.
3. **Integrating center.** Interneurons within the CNS relay signals from sensory neurons to motor neurons. The main integrating centers for most autonomic reflexes are located in the hypothalamus and brain stem. Some autonomic reflexes, such as those for urination and defecation, have integrating centers in the spinal cord.
4. **Motor neurons.** Nerve impulses triggered by the integrating center propagate out of the CNS along motor neurons to an effector. In an autonomic reflex arc, two motor neurons connect the CNS to an effector: The preganglionic neuron conducts motor impulses from the CNS to an autonomic ganglion, and the postganglionic neuron conducts motor impulses from an autonomic ganglion to an effector (see Figure 20.1).
5. **Effector.** In an autonomic reflex arc, the effectors are smooth muscle, cardiac muscle, and glands, and the reflex is called an autonomic reflex.

Autonomic Control by Higher Centers

Normally, we are not aware of muscular contractions of the digestive organs, heartbeat, changes in the diameter of blood vessels, and pupil dilation and constriction because the integrating centers for these autonomic responses are in the spinal cord or the lower regions of the brain. Somatic or autonomic sensory neurons deliver input to these centers, and autonomic motor neurons provide output that adjusts activity in the visceral effector, usually without our conscious perception.

The hypothalamus is the major control and integration center of the ANS. The hypothalamus receives sensory input related to visceral functions, olfaction (smell), and gustation (taste), as well as input related to changes in temperature, and levels of various substances in blood. In addition, it receives input relating to emotions from the limbic system. Output from the hypothalamus influences autonomic centers in the brain stem (such as the cardiovascular, salivation, swallowing, and vomiting centers) and in the spinal cord (such as the defecation and urination reflex centers in the sacral spinal cord).

Anatomically, the hypothalamus is connected to both the sympathetic and parasympathetic divisions of the ANS by axons of neurons whose dendrites and cell bodies are in various hypothalamic nuclei. The axons form tracts from the hypothalamus to sympathetic and parasympathetic nuclei in the brain stem and spinal cord through relays in the reticular formation. The posterior and lateral parts of the hypothalamus control the sympathetic division. As you might expect, stimulation of the posterior or lateral hypothalamus produces an increase in heart rate and force of contraction, a rise in blood pressure due to constriction of blood vessels, an increase in body temperature, dilation of the pupils, and inhibition of the gastrointestinal tract. In contrast, stimulation of the anterior and medial parts of the hypothalamus, which control the parasympathetic division, results in a decrease in heart rate, lowering of blood pressure, constriction of the pupils, and increased secretion and motility of the gastrointestinal tract.

⊙ **CHECKPOINT**

14. Give three examples of activities in the body that are controlled by autonomic reflexes.
15. How does an autonomic reflex arc differ from a somatic reflex arc?

KEY MEDICAL TERMS ASSOCIATED WITH THE AUTONOMIC NERVOUS SYSTEM

Autonomic nerve neuropathy (noo-ROP-a-thē) A neuropathy (a disorder of a cranial or spinal nerve) that affects one or more autonomic nerves, with multiple effects on the autonomic nervous system including fainting and low blood pressure when standing (orthostatic hypotension) due to decreased sympathetic control of the cardiovascular system, constipation, urinary incontinence, and impotence. This type of neuropathy is often caused by long-term diabetes mellitus and is known as *diabetic neuropathy*.

Biofeedback A technique in which an individual is provided with information regarding an autonomic response such as heart rate, blood pressure, or skin temperature via various electronic monitoring devices. By concentrating on positive thoughts, individuals learn to alter autonomic responses. For example, biofeedback has been used to decrease heart rate and blood pressure and increase skin temperature in order to decrease the severity of migraine headaches.

Dysautonomia (dis-aw-tō-NŌ-mē-a; *dys-*=difficult; *autonomia*=self-governing) An inherited disorder in which the autonomic nervous system functions abnormally, resulting in reduced tear gland secretions, poor vasomotor control, motor incoordination, skin blotching, absence of pain sensation, difficulty in swallowing, hyporeflexia, excessive vomiting, and emotional instability.

Hyperhydrosis (hī'-per-hī-DRŌ-sis; *hyper-*=above or too much; *hidrosis*=sweat; *-osis*=condition) Excessive or profuse sweating due to intense stimulation of sweat glands.

Mass reflex In cases of severe spinal cord injury above the level of the sixth thoracic vertebra, stimulation of the skin or overfilling of a visceral organ (such as the urinary bladder or colon) below the level of the injury results in intense activation of autonomic and somatic output from the spinal cord as reflex activity returns. The exaggerated response occurs because there is no inhibitory input from the brain. The mass reflex consists of flexor spasms of the lower limbs, evacuation of the urinary bladder and colon, and profuse sweating below the level of the lesion.

Megacolon (*mega-*=big) An abnormally large colon. In congenital megacolon, parasympathetic nerves to the distal segment of the colon do not develop properly. Loss of motor function in the segment causes massive dilation of the normal proximal colon. The condition results in extreme constipation, abdominal distention, and, occasionally, vomiting.

Reflex sympathetic dystrophy (RSD) A syndrome that includes spontaneous pain, painful hypersensitivity to stimuli such as light touch, and excessive coldness and sweating in the involved body part. The disorder frequently involves the forearms, hands, knees, and feet. It appears that activation of the sympathetic division of the autonomic nervous system due to traumatized nociceptors as a result of trauma or surgery on bones or joints is involved. Also called **complex regional pain syndrome type 1**.

Vagotomy (vā-GOT-ō-mē; *-tome*=incision) Cutting the vagus (X) nerve. It is frequently done to decrease the production of hydrochloric acid in persons with ulcers.

STUDY OUTLINE

Comparison of Somatic and Autonomic Nervous Systems (p. 660)

1. The somatic nervous system operates under conscious control; the ANS usually operates without conscious control.
2. Sensory input to the somatic nervous system is mainly from the special senses and somatic senses; sensory input to the ANS is primarily from interoceptors, with some contributions from the special senses and somatic senses.
3. The axons of somatic motor neurons extend from the CNS and synapse directly with an effector. Autonomic motor pathways consist of two motor neurons in series. The axon of the first motor neuron extends from the CNS and synapses in a ganglion with the second motor neuron; the second neuron synapses with an effector.
4. The output (motor) portion of the ANS has two divisions: sympathetic and parasympathetic. Most body organs receive dual innervation; usually one ANS division causes excitation and the other causes inhibition.
5. Somatic nervous system effectors are skeletal muscles; ANS effectors include cardiac muscle, smooth muscle, and glands.
6. Table 20.1 on page 662 compares the somatic and autonomic nervous systems.

Anatomy of Autonomic Motor Pathways (p. 662)

1. Preganglionic neurons are myelinated; postganglionic neurons are unmyelinated.
2. The cell bodies of sympathetic preganglionic neurons are in the lateral gray horns of the 12 thoracic and the first two or three lumbar segments of the spinal cord; the cell bodies of parasympathetic preganglionic neurons are in four cranial nerve nuclei (III, VII, IX, and X) in the brain stem and lateral gray horns of the second through fourth sacral segments of the spinal cord.
3. Autonomic ganglia are classified as sympathetic trunk ganglia (on both sides of the vertebral column), prevertebral ganglia (anterior to the vertebral column), and terminal ganglia (near or inside visceral effectors).
4. Sympathetic preganglionic neurons synapse with postganglionic neurons in ganglia of the sympathetic trunk ganglia or in prevertebral ganglia; parasympathetic preganglionic neurons synapse with postganglionic neurons in terminal ganglia.
5. The enteric nervous system consists of nerves and ganglia in the wall of the gastrointestinal tract, pancreas, and gallbladder.

ANS Neurotransmitters and Receptors (p. 669)

1. Cholinergic neurons release acetylcholine, which binds to nicotinic or muscarinic cholinergic receptors.
2. In the ANS, the cholinergic neurons include all sympathetic and parasympathetic preganglionic neurons, all parasympathetic postganglionic neurons, and sympathetic postganglionic neurons that innervate most sweat glands.
3. In the ANS, adrenergic neurons release norepinephrine. Both epinephrine and norepinephrine bind to alpha and beta adrenergic receptors.
4. Most sympathetic postganglionic neurons are adrenergic.
5. An agonist is a substance that binds to and activates a receptor, mimicking the effect of a natural neurotransmitter or hormone. An antagonist is a substance that binds to and blocks a receptor, thereby preventing a natural neurotransmitter or hormone from exerting its effect.

Functions of the ANS (p. 670)

1. The sympathetic division favors body functions that can support vigorous physical activity and rapid production of ATP in a series of physiological responses called the fight-or-flight response; the parasympathetic division regulates activities that conserve and restore body energy.
2. The effects of sympathetic stimulation are longer-lasting and more widespread than the effects of parasympathetic stimulation.
3. Table 20.2 on pages 672–673 compares structural and functional features of the sympathetic and parasympathetic divisions.
4. Table 20.3 on pages 672–673 lists the effects of sympathetic and parasympathetic stimulation on effectors throughout the body.

Integration and Control of Autonomic Functions (p. 674)

1. An autonomic reflex adjusts the activities of smooth muscle, cardiac muscle, and glands.
2. An autonomic reflex arc consists of a receptor, a sensory neuron, an integrating center, two autonomic motor neurons, and a visceral effector.
3. The hypothalamus is the major control and integration center of the ANS. It is connected to both the sympathetic and the parasympathetic divisions.

SELF-QUIZ QUESTIONS

Choose the one best answer to the following questions.

1. Which of the following statements is *true* for the parasympathetic division of the autonomic nervous system?
   a. Ganglia are close to the central nervous system and distant from the effector.
   b. The PNS forms the craniosacral outflow.
   c. The PNS supplies nerves to blood vessels, sweat glands, and adrenal (suprarenal) glands.
   d. Some nerve fibers of the PNS pass through paravertebral ganglia.
   e. All of the above statements are true.

2. Which of the following cranial nerve nuclei is *not* a site of cell bodies of preganglionic neurons of the parasympathetic division?
   a. III   b. V   c. VII   d. IX   e. X

3. Which of the following statements about gray rami is *false*?
   a. They carry preganglionic neurons from the anterior rami of spinal nerves to trunk ganglia.
   b. They contain only sympathetic nerve fibers.
   c. They carry impulses from sympathetic trunk ganglia to spinal nerves.
   d. They are located at all levels of the vertebral column.
   e. None of the above is false.

4. Which of the following is *not* a sympathetic ganglion?
   a. ciliary
   b. celiac
   c. inferior cervical
   d. superior mesenteric
   e. All of the above are sympathetic ganglia.

5. The ANS provides the chief nervous control in which of these activities?
   a. following a moving object with the eyes
   b. moving a hand reflexively from a hot object
   c. typing
   d. digesting food
   e. surfing the Internet

6. White rami communicantes connect:
   a. preganglionic and postganglionic parasympathetic fibers.
   b. preganglionic and postganglionic sympathetic fibers.
   c. sympathetic and parasympathetic ganglia.
   d. sympathetic trunk and prevertebral ganglia.
   e. the anterior ramus of a spinal nerve with sympathetic trunk ganglia.

7. Cholinergic fibers are thought to include:
   a. all preganglionic axons.
   b. all postganglionic parasympathetic axons.
   c. a few postganglionic sympathetic axons.
   d. all axons of somatic motor neurons.
   e. All of the above are correct.

8. Which of the following statements would *not* be included in a description of the sympathetic trunks?
   a. The trunks lie anterior and lateral to the spinal cord.
   b. Each trunk contains 24 ganglia.

   c. Each trunk receives preganglionic fibers from the thoracic and sacral regions of the spinal cord.
   d. The ganglia of the trunks are called paravertebral ganglia.
   e. White rami direct nerve fibers into the trunk; gray rami direct nerve fibers out of the trunk.

9. Which of the following could *not* be the result of the binding of acetylcholine to nicotinic receptors?
   a. contraction of a skeletal muscle
   b. stimulation of a postganglionic sympathetic neuron
   c. stimulation of a postganglionic parasympathetic neuron
   d. increase in heart rate
   e. All of the above are possible.

Complete the following.

10. _____ nerves are extensions of preganglionic sympathetic axons that pass through sympathetic trunk ganglia and extend to prevertebral ganglia.

11. Organs such as the heart or stomach that receive impulses from both divisions of the ANS are said to have _____ innervation.

12. Norepinephrine is enzymatically inactivated by COMT or _____.

13. The two types of adrenergic receptors are _____ receptors and _____ receptors.

14. The two types of cholinergic receptors are _____ receptors and _____ receptors.

15. The endocrine gland stimulated directly by sympathetic preganglionic neurons is the _____.

16. Nearly 80% of the craniosacral outflow of the parasympathetic division is conducted via preganglionic axons leaving the brain with the _____ nerve.

Are the following statements true or false? If the statement is False, *find and fix the mistake to make the statement* True.

17. In the fight-or-flight response, blood glucose levels increase.

18. A major difference between autonomic ganglia and posterior root ganglia is that only autonomic ganglia contain synapses.

19. The ganglia that make up the sympathetic trunks are also called prevertebral or collateral ganglia.

20. Postganglionic neurons begin in the spinal cord and extend to effectors.

21. An event occurring during the fight-or-flight response is increased urine production due to blood being shunted to the kidneys.

22. All sympathetic and parasympathetic preganglionic neurons release the neurotransmitter norepinephrine.

Matching

23. Match the following plexuses with their descriptions (answers may be used more than once):

____ **(a)** plexus located mostly posterior to each lung

____ **(b)** the largest autonomic plexus; consists of 2 large ganglia and a network of fibers that surround the celiac and superior mesenteric arteries

____ **(c)** plexus located anterior to the fifth lumbar vertebra; supplies pelvic viscera

____ **(d)** plexus containing the inferior mesenteric ganglion; supplies the large intestine

____ **(e)** plexus at the base of the heart; surrounds the large blood vessels emerging from the heart

____ **(f)** plexus containing the superior mesenteric ganglion; supplies the small and large intestine

(1) applies to the sympathetic division of the ANS

(2) applies to the parasympathetic division of the ANS

(3) applies to both divisions of the ANS

24. Match the following (answers may be used more than once):

____ **(a)** thoracolumbar outflow

____ **(b)** has long preganglionic fibers and short postganglionic fibers

____ **(c)** has short preganglionic fibers and long postganglionic fibers

____ **(d)** sends some preganglionic fibers through cranial nerves

____ **(e)** has some preganglionic fibers synapsing in sympathetic trunk ganglia

____ **(f)** has more widespread and longer-lasting effect in the body

____ **(g)** has some fibers running in gray rami to supply sweat glands, hair follicle muscles, and blood vessels

____ **(h)** has fibers in white rami (connecting spinal nerves with sympathetic trunk ganglia)

____ **(i)** contains fibers that supply viscera with motor impulses

____ **(j)** celiac and superior mesenteric ganglia are sites of postganglionic neuron cell bodies

(1) cardiac plexus

(2) pulmonary plexus

(3) superior mesenteric plexus

(4) inferior mesenteric plexus

(5) celiac (solar) plexus

(6) hypogastric plexus

CRITICAL THINKING QUESTIONS

1. Skydiving, hang gliding, and bungee jumping can all give you a great rush (or get you killed). How do these activities cause this rush?

2. "The quickest way to a man's heart is through his stomach," Sophia's grandmother is fond of saying. Trace the pathway that an impulse would follow from a full stomach to a happy heart.

3. Mando has been living in a refugee camp for two years. The guards are hostile to the refugees, and they regularly fire off rounds from their machine guns at night and frequently make threatening re-

marks and gestures to the women in the camp. Every time a guard comes near, Mando feels her heart race and her mouth go dry. Physiologically, what is happening to Mando, and what do you think some of the long-term physiological effects might be?

4. The autonomic and the enteric divisions of the nervous system both control the digestive system. How do they compare?

5. While reviewing the chapter on the brain, your study partner says, "The hypothalamus can't be very important. It's so small." Would you agree with your friend? Why or why not?

? ANSWERS TO FIGURE QUESTIONS

20.1 Dual innervation means that a body organ receives nerve impulses from both sympathetic and parasympathetic neurons of the ANS.

20.2 Most parasympathetic preganglionic axons are longer than most sympathetic preganglionic axons because most parasympathetic ganglia are in the walls of visceral organs, and most sympathetic ganglia are close to the spinal cord in the sympathetic trunk.

20.3 Terminal ganglia are associated with the parasympathetic division; sympathetic trunk and prevertebral ganglia are associated with the sympathetic division.

20.4 The celiac (solar) plexus is the largest autonomic plexus.

20.5 White rami look white due to the presence of myelin.

20.6 Cholinergic neurons with nicotinic ACh receptors include sympathetic postganglionic neurons innervating sweat glands and all parasympathetic postganglionic neurons. The effectors innervated by these cholinergic neurons possess muscarinic receptors.

21 | SOMATIC SENSES AND MOTOR CONTROL

INTRODUCTION Imagine a camping trip to a beautiful rocky coastline surrounding a sandy patch of beach. As you rouse from your night of slumber on the packed sand, you slowly stretch your stiffened joints and cautiously climb out of your sleeping bag to greet the crisp morning air. You rub the sleep from your eyes and see the fog rolling in off the white crests of the slapping waves. As you walk toward the ocean you take a deep breath, you smell the salty scent of the tide and feel individual grains of sand between your wriggling toes. Suddenly you stop to rub your exposed arms vigorously, as the cool, crisp air sends a chill through your still-sleepy body. You see and hear noisy gulls gliding through the air overhead, and listen to a distant boat sounding its horn. As you turn to walk toward the coastline where the sounds of the water are playing their tune against the rocks, you glance down into the tide pools left behind by the receding waves, and notice a colorful array of intertidal life—sea stars, mussels, anemones, and scurrying crabs. Bending down to take a closer look, your face is splashed by an incoming wave, giving you a taste of the salty sea. You think for a minute about the beauty you have *sensed* in the past few minutes. Your mind is flooded with what you have *seen*, what you have *felt*, what you have *smelled*, what you have *heard*, and what you have *tasted*.

The previous four chapters described the organization of the nervous system. Now we will see how certain parts cooperate to carry out its three basic functions: (1) receiving sensory input; (2) integrating the sensory input, that is, processing and interpreting the input, deciding a course of action, and storing information; and (3) transmitting motor impulses that result in a response (muscular contraction or glandular secretion). In this chapter we will explore the nature and types of sensations, the pathways that convey the sensory input from the body to the brain, and the pathways that carry motor commands from the brain to effectors. Chapter 22 deals with the special senses of smell, taste, vision, hearing, and equilibrium. •

OVERVIEW OF SENSATIONS

◉ **OBJECTIVES**
- Define a sensation and describe the conditions necessary for a sensation to occur.
- Describe the different ways to classify sensory receptors.

Definition of Sensations

A **sensation** is the conscious or subconscious awareness of changes in the external or internal conditions of the body. For a sensation to occur, four conditions must be satisfied:

1. A *stimulus*, or change in the environment, capable of activating certain sensory neurons, must occur.
2. A *sensory receptor* must convert the stimulus to nerve impulses.
3. The nerve impulses must be *conducted* along a neural pathway from the sensory receptor to the brain.
4. A region of the brain must receive and *integrate* the nerve impulses, producing a sensation.

A stimulus that activates a sensory receptor may be in the form of light, heat, pressure, mechanical energy, or chemical energy. A sensory receptor responds to a stimulus by altering its membrane's permeability to small ions. In most types of sensory receptors, the resulting flow of ions across the membrane produces a change that triggers one or more nerve impulses. The impulses are then conducted along the sensory neuron toward the CNS.

Characteristics of Sensations

A **perception** is the conscious awareness and interpretation of a sensation. Perceptions are integrated in the cerebral cortex. You seem to see with your eyes, hear with your ears, and feel pain in an injured part of your body. This is because sensory impulses from each part of the body arrive in a specific region of the cerebral cortex, which interprets the sensation as coming from the stimulated sensory receptors.

Each unique type of sensation, such as touch, pain, vision, or hearing, is called a **sensory modality** (mō-DAL-i-tē). Based on the receptor stimulated, a sensory neuron carries information for one sensory modality only. Neurons relaying impulses for touch, for example, do not also transmit impulses for pain. The specialization of sensory neurons enables nerve impulses from the eyes to be perceived as sight, and those from the ears to be perceived as sounds.

A characteristic of most sensory receptors is **adaptation,** a decrease in sensation during a prolonged stimulus. Adaptation is caused in part by a decrease in the responsiveness of sensory receptors. As a result of adaptation, the perception of a sensation decreases or disappears even though the stimulus persists. For example, when you first step into a hot shower, the water may feel very hot, but soon the sensation decreases to one of comfortable warmth even though the stimulus (the high temperature of the water) does not change. Receptors vary in how quickly they adapt.

Classification of Sensations

The senses can be grouped into two classes: general senses and special senses.

1. The **general senses** refer to both somatic senses and visceral senses. **Somatic senses** (*somat-*=of the body) are tactile sensations (touch, pressure, and vibration); thermal sensations (warm and cold); pain sensations; and proprioceptive sensations, which allow perception of both the static (nonmoving) positions of limbs and body parts (joint and muscle position sense) and movements of the limbs and head. **Visceral senses** provide information about conditions within internal organs.
2. The **special senses** include smell, taste, vision, hearing, and equilibrium (balance).

In addition to introducing the mechanism of sensation, this chapter deals with the somatic senses; the special senses will be covered in Chapter 22.

Types of Sensory Receptors

Several structural and functional characteristics of sensory receptors can be used to group them into different classes. On a microscopic level, sensory receptors may be (1) free nerve endings of sensory neurons, (2) encapsulated nerve endings of sensory neurons, or (3) separate cells that synapse with sensory neurons (see Figure 21.1 on page 681).

Free nerve endings are bare dendrites; when viewed under a light microscope they lack structural specialization. Receptors for pain, thermal, tickle, itch, and some touch sensations are free nerve endings. Receptors for other somatic sensations and visceral sensations, such as pressure, vibration, and some touch sensations are **encapsulated nerve endings**. Their dendrites are enclosed in a connective tissue capsule that has a distinctive microscopic structure—for example, lamellated (pacinian) corpuscles (see Figure 21.1). The different types of capsules enhance the sensitivity or specificity of the receptor. Sensory receptors for some special senses are specialized, **separate cells** that synapse with sensory neurons. These include *hair cells* for hearing and equilibrium in the inner ear and *gustatory receptor cells* in taste buds (see Figure 22.2c on page 700), and *photoreceptors*, sensory receptors for vision located in the retina.

Sensory receptors can be grouped by the type of energy they respond to and are categorized as follows:

- **Photoreceptors** are responsive to light.
- **Mechanoreceptors** are sensitive to mechanical stimuli such as deformation, stretching, or bending of cells. Examples include skeletal muscle receptors sensitive to stretch, the receptors in the ear containing fine hair cells that are bent as a result of sound waves, and blood pressure–monitoring baroreceptors.
- **Thermoreceptors** are sensitive to heat and cold.
- **Osmoreceptors** detect changes in the concentration of solutes in the body fluids and the resultant changes in osmotic activity.

- **Chemoreceptors** are sensitive to specific chemicals. Chemoreceptors include the receptors for smell and taste, as well as those located deeper within the body that detect O_2 and CO_2 concentrations in the blood or the chemical content of the gastrointestinal tract.
- **Nociceptors** (pain receptors) are sensitive to tissue damage such as pinching or burning, or to distortion of tissue. Intense stimulation of any receptor is also perceived as painful.

Another way to group sensory receptors is based on the location of the receptors and the origin of the stimuli that activate them:

- **Exteroceptors** (EKS-ter-ō-sep′-tors) are located at or near the external surface of the body; they are sensitive to stimuli originating outside the body and provide information about the *external* environment. The sensations of hearing, vision, smell, taste, touch, pressure, vibration, temperature, and pain are conveyed by exteroceptors.
- **Interoceptors** (IN-ter-ō-sep′-tors) or **visceroceptors** are located in blood vessels, visceral organs, muscles, and the nervous system and monitor conditions in the *internal* environment. The nerve impulses produced by interoceptors usually are not consciously perceived; occasionally, however, activation of interoceptors by strong stimuli may be felt as pain or pressure.
- **Proprioceptors** (PRŌ-prē-ō-sep′-tors; *proprio-*=one's own) are located in muscles, tendons, joints, and the inner ear. They provide information about body position, muscle length and tension, and the position and movement of our joints.

Table 21.1 summarizes the classification of sensory receptors and lists different types of sensory receptors according to the type of stimulus they detect.

◉ CHECKPOINT

1. Distinguish between sensation and perception.
2. Define sensory modality and adaptation.
3. What events are needed for a sensation to occur?
4. Describe the classification of sensory receptors.

SOMATIC SENSATIONS

◉ OBJECTIVES
- Describe the location and function of the receptors for tactile, thermal, and pain sensations.
- Identify the receptors for proprioception and describe their functions.

Somatic sensations arise from stimulation of sensory receptors embedded in the skin or subcutaneous layer; in mucous membranes of the mouth, vagina, and anus; in muscles, tendons, and joints; and in the inner ear. The sensory receptors for somatic sensations are distributed unevenly—some parts of the body surface are densely populated with receptors, but other parts contain only a few. The areas with the highest density of somatic sensory receptors are the tip of the tongue, the lips, and the fingertips. Somatic sensations that arise from stimulating the skin surface are **cutaneous sensations** (kū-TĀ-nē-us; *cutane-*=skin). Somatic sensations are of four sensory modalities: tactile, thermal, pain, and proprioceptive.

Tactile Sensations

The **tactile sensations** (TAK-tīl; *tact-*=touch) include touch, pressure, vibration, itch, and tickle. Several types of encapsulated

TABLE 21.1	
Classification of Sensory Receptors	
BASIS OF CLASSIFICATION	**DESCRIPTION**
MICROSCOPIC FEATURES	
Free nerve endings	Bare dendrites associated with pain, thermal, tickle, itch, and some touch sensations.
Encapsulated nerve endings	Dendrites enclosed in a connective tissue capsule for pressure, vibration, and some touch sensations.
Separate cells	Receptor cells synapse with first-order sensory neurons; located in the retina of the eye (photoreceptors), inner ear (hair cells), and taste buds of the tongue (gustatory receptor cells).
RECEPTOR LOCATION AND ACTIVATING STIMULI	
Exteroceptors	Located at or near body surface; sensitive to stimuli originating outside body; provide information about external environment; convey visual, smell, taste, touch, pressure, vibration, thermal, and pain sensations.
Interoceptors	Located in blood vessels, visceral organs, and nervous system; provide information about internal environment; impulses produced usually are not consciously perceived but occasionally may be felt as pain or pressure.
Proprioceptors	Located in muscles, tendons, joints, and inner ear; provide information about body position, muscle length and tension, position and motion of joints, and equilibrium (balance).
TYPE OF STIMULUS DETECTED	
Mechanoreceptors	Detect mechanical stimuli, provide sensations of touch, pressure, vibration, proprioception, and hearing and equilibrium; also monitor stretching of blood vessels and internal organs.
Thermoreceptors	Detect changes in temperature.
Nociceptors	Respond to painful stimuli resulting from physical or chemical damage to tissue.
Photoreceptors	Detect light that strikes the retina of the eye.
Chemoreceptors	Detect chemicals in mouth (taste), nose (smell), and body fluids.
Osmoreceptors	Sense the osmotic pressure of body fluids.

mechanoreceptors mediate sensations of touch, pressure, and vibration. Other touch sensations, as well as itch and tickle sensations, are detected by free nerve endings. Tactile receptors in the skin or subcutaneous layer include corpuscles of touch, hair root plexuses, type I and II cutaneous mechanoreceptors, lamellated corpuscles, and free nerve endings (Figure 21.1).

Touch

Sensations of **touch** generally result from stimulation of tactile receptors in the skin or subcutaneous layer. **Crude touch** is the ability to perceive that something has contacted the skin, even though its exact location, shape, size, or texture cannot be determined. **Fine touch** provides specific information about a touch sensation, such as exactly what point on the body is touched plus the shape, size, and texture of the source of stimulation.

There are two types of rapidly adapting touch receptors. **Corpuscles of touch**, or **Meissner corpuscles** (MĪS-ner), are receptors for fine touch that are located in the dermal papillae of hairless skin. Each corpuscle is a mass of dendrites enclosed by an egg-shaped capsule of connective tissue. Because corpuscles of touch are rapidly adapting receptors, they generate nerve impulses mainly at the onset of a touch. They are abundant in the hands, eyelids, tip of the tongue, lips, nipples, soles, clitoris, and tip of the penis. **Hair root plexuses** are rapidly adapting touch receptors found in hairy skin; they consist of free nerve endings wrapped around hair follicles. Hair root plexuses detect movements on the surface of the skin that disturb hairs. For example, an insect landing on a hair causes movement of the hair shaft that stimulates the free nerve endings.

There also are two types of slowly adapting touch receptors. **Type I cutaneous mechanoreceptors**, also known as **Merkel discs**, function in fine touch. Merkel discs are saucer-shaped, flattened free nerve endings that make contact with Merkel cells of the stratum basale (see Figure 5.2d on page 120). These mechanoreceptors are plentiful in the fingertips, hands, lips, and external genitalia. **Type II cutaneous mechanoreceptors**, or

Ruffini corpuscles, are elongated, encapsulated receptors located deep in the dermis, and in ligaments and tendons. Present in the hands and abundant on the soles, they are most sensitive to stretching that occurs as digits or free limbs are moved.

Pressure and Vibration

Pressure is a sustained sensation that is felt over a larger area than touch. It occurs with deformation of deeper tissues. Receptors that contribute to sensations of pressure include corpuscles of touch, type I mechanoreceptors, and lamellated corpuscles. **Lamellated**, or **pacinian**, **corpuscles** (pa-SIN-ē-an) are large oval structures composed of a multilayered connective tissue capsule that encloses a dendrite. Like corpuscles of touch, lamellated corpuscles adapt rapidly. They are widely distributed in the body: in the dermis and subcutaneous layer; in submucosal tissues that underlie mucous and serous membranes; around joints, tendons, and muscles; in the periosteum; and in the mammary glands, external genitalia, and certain viscera, such as the pancreas and urinary bladder.

Sensations of **vibration** result from rapidly repetitive sensory signals from tactile receptors. The receptors for vibration sensations are corpuscles of touch and lamellated corpuscles. Corpuscles of touch can detect lower-frequency vibrations; lamellated corpuscles detect higher-frequency vibrations.

Itch and Tickle

The **itch** sensation results from stimulation of free nerve endings by certain chemicals, such as bradykinin, often because of a local inflammatory response. Free nerve endings and lamellated corpuscles are thought to mediate the **tickle** sensation. This intriguing

---

Figure 21.1 Structure and location of sensory receptors in the skin and subcutaneous layer.

 The somatic sensations of touch, pressure, vibration, warmth, cold, and pain arise from sensory receptors in the skin, subcutaneous layer, and mucous membranes.

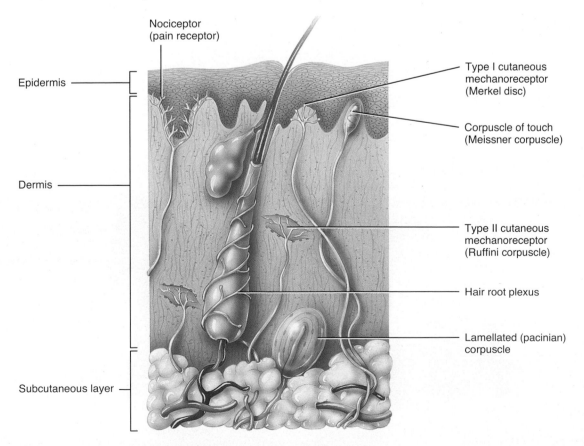

? **Which sensations can arise when free nerve endings are stimulated?**

sensation typically arises only when someone else touches you, not when you touch yourself. The explanation of this puzzle seems to lie in the impulses that conduct to and from the cerebellum when you are moving your fingers and touching yourself that don't occur when someone else is tickling you.

Thermal Sensations

Thermoreceptors are free nerve endings. Two distinct **thermal sensations**—coldness and warmth—are detected by different receptors. **Cold receptors** are located in the stratum basale of the epidermis. Temperatures between 10° and 40°C (50–105°F) activate cold receptors. **Warm receptors** are located in the dermis and are activated by temperatures between 32° and 48°C (90–118°F). Cold and warm receptors both adapt rapidly at the onset of a stimulus but continue to generate impulses at a lower frequency throughout a prolonged stimulus. Temperatures below 10°C and above 48°C stimulate mainly pain receptors rather than thermoreceptors, producing painful sensations.

Pain Sensations

Pain is indispensable for survival. It serves a protective function by signaling the presence of noxious, tissue-damaging conditions. From a medical standpoint, the subjective description of pain along with an indication of where it occurs may help pinpoint the underlying cause of disease.

Nociceptors (nō-sī-SEP-tors; *noci*=harmful), the receptors for pain, are free nerve endings found in every tissue of the body except the brain (Figure 21.1). Intense thermal, mechanical, or chemical stimuli can activate nociceptors. Tissue irritation or injury releases chemicals such as prostaglandins, kinins, and potassium ions (K^+) that stimulate nociceptors. Pain may persist even after a pain-producing stimulus is removed because pain-mediating chemicals linger, and because nociceptors exhibit very little adaptation. Conditions that elicit pain include excessive distention (stretching) of a structure, prolonged muscular contractions, muscle spasms, or ischemia (inadequate blood flow to an organ).

Types of Pain

There are two types of pain: fast and slow. The perception of **fast pain** occurs very rapidly, usually within 0.1 second after a stimulus is applied. This type of pain is also known as acute, sharp, or pricking pain. The pain felt from a needle puncture or knife cut to the skin are examples of fast pain. Fast pain is not felt in deeper tissues of the body. The perception of **slow pain**, by contrast, begins a second or more after a stimulus is applied. It then gradually increases in intensity over a period of several seconds or minutes. This type of pain, which may be excruciating, is also referred to as *chronic, burning, aching,* or *throbbing* pain. Slow pain can occur both in the skin and in deeper tissues or internal organs. An example is the pain associated with a toothache.

Pain that arises from stimulation of receptors in the skin is called **superficial somatic pain**. Stimulation of receptors in skeletal muscles, joints, tendons, and fascia causes **deep somatic pain**. **Visceral pain** results from stimulation of nociceptors in visceral organs.

Localization of Pain

Fast pain is very precisely localized to the stimulated area. For example, if someone pricks you with a pin, you know exactly which part of your body was stimulated. Slow somatic pain also is well localized but more diffuse (involves large areas); it usually appears to come from a larger area of the skin. In some instances of visceral slow pain, the pain is felt in the affected area. If the pleural membranes around the lungs are inflamed, for example, you experience chest pain.

In many instances of visceral pain, the pain is felt in or just deep to the skin that overlies the stimulated organ, or in a surface area far from the stimulated organ. This phenomenon is called **referred pain**. Figure 21.2 shows skin regions to which visceral pain may be referred. In general, the visceral organ involved and the area to which the pain is referred are served by the same segment of the spinal cord. For example, sensory fibers from the heart, the skin over the heart, and the skin along the medial aspect of the left arm enter spinal cord segments T1 to T5. Thus, the pain of a heart attack typically is felt in the skin over the heart and along the left arm.

Figure 21.2 Distribution of referred pain. The colored portions of the diagrams indicate skin areas to which visceral pain is referred.

Nociceptors are present in almost every tissue of the body.

Liver and gallbladder
Lung and diaphragm
Heart
Gallbladder
Pancreas
Stomach
Small intestine
Ovary
Kidney
Ovary
Appendix
Colon
Ureter
Urinary bladder

Liver and gallbladder
Stomach
Liver and gallbladder
Kidney

(a) Anterior view

(b) Posterior view

? Which visceral organs have the broadest area for referred pain?

Proprioceptive Sensations

Proprioceptive sensations allow us to know where our head and free limbs are located and how they are moving even if we are not looking at them, so that we can walk, type, or put a shirt on without using our eyes. **Kinesthesia** (kin′-es-THĒ-zē-a; *kin-*=motion; *-esthesia*=perception) is the perception of body movements. Proprioceptive sensations arise in receptors termed **proprioceptors**. Those proprioceptors embedded in muscles (especially postural muscles) and tendons inform us of the degree to which muscles are contracted, the amount of tension on tendons, and the positions of joints. Hair cells of the vestibular apparatus in the inner ear are proprioceptors that monitor the orientation of the head relative to the ground and head position during movements. They provide information for maintaining balance and equilibrium (described in Chapter 22). Because proprioceptors adapt slowly and only slightly, the brain continually receives nerve impulses related to the position of different body parts and makes adjustments to ensure coordination.

Proprioceptive sensations also allow us to estimate the weight of objects and determine the muscular effort necessary to perform a task. For example, as you pick up a bag you quickly realize whether it contains popcorn or books, and you then exert the correct amount of force needed to lift it. Here we discuss three types of proprioceptors: muscle spindles within skeletal muscles, tendon organs within tendons, and joint kinesthetic receptors within synovial joint capsules.

Muscle Spindles

A muscle spindle consists of slowly adapting sensory nerve endings that wrap around 3 to 10 specialized muscle fibers, called **intrafusal muscle fibers** (*intrafusal*=within a spindle) (Figure 21.3). A connective tissue capsule encloses the sensory nerve endings and intrafusal fibers and anchors the spindle to endomysium and

Figure 21.3 Two types of proprioceptors: a muscle spindle and a tendon organ. In muscle spindles, which monitor changes in skeletal muscle length, sensory nerve endings wrap around the central portion of intrafusal muscle fibers. In tendon organs, which monitor the force of muscle contraction, sensory nerve endings are activated by increasing tension on a tendon.

Proprioceptors provide information about body position and movement.

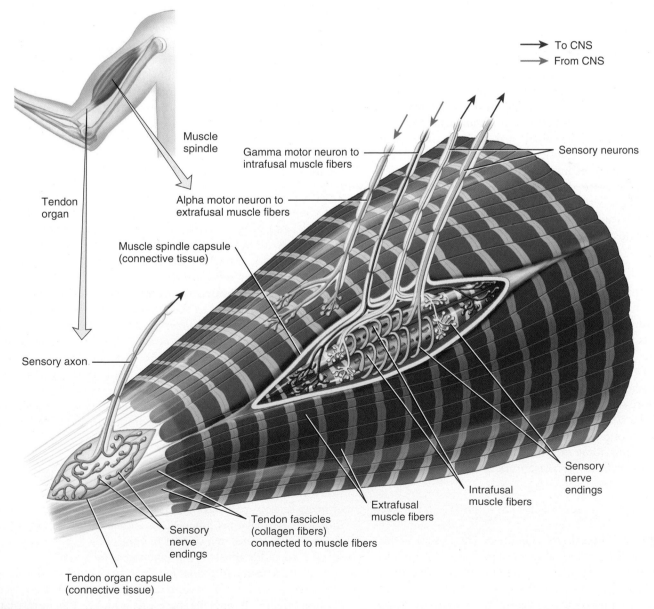

? How is a muscle spindle activated?

perimysium. Muscle spindles are interspersed among ordinary skeletal muscle fibers and aligned parallel to them. In muscles that produce finely controlled movements, such as movements of the fingers or eyes, muscle spindles are plentiful. Muscles involved in coarser but more forceful movements, such as the quadriceps femoris and hamstrings muscles of the thigh, have fewer muscle spindles. The only skeletal muscles that lack spindles are the tiny muscles of the middle ear.

The main function of muscle spindles is to measure *muscle length*—how much a muscle is being stretched. Either sudden or prolonged stretching of the central areas of the intrafusal muscle fibers stimulates the sensory nerve endings. The resulting nerve impulses propagate into the CNS. Information from muscle spindles arrives quickly at the somatic sensory areas of the cerebral cortex, which allows conscious perception of free limb positions and movements. At the same time, impulses from muscle spindles also pass to the cerebellum, where the input is used to coordinate muscle contractions.

In addition to their sensory nerve endings near the middle of intrafusal fibers, muscle spindles contain motor neurons called **gamma motor neurons**. The motor neurons terminate near both ends of the intrafusal fibers and adjust the tension in a muscle spindle to variations in the length of the muscle. For example, when a muscle shortens, gamma motor neurons stimulate the ends of the intrafusal fibers to contract slightly. This keeps the intrafusal fibers taut and maintains the sensitivity of the muscle spindle to stretching of the muscle. As the frequency of impulses in its gamma motor neuron increases, a muscle spindle becomes more sensitive to stretching of its midregion.

Surrounding muscle spindles are ordinary skeletal muscle fibers, called **extrafusal muscle fibers** (*extrafusal*=outside a spindle), which are innervated by **alpha motor neurons**. The cell bodies of both gamma and alpha motor neurons are located in the anterior gray horn of the spinal cord (or in the brain stem for muscles in the head). During stretching of a muscle, impulses in muscle spindle sensory axons propagate into the spinal cord and brain stem and activate alpha motor neurons that innervate extrafusal muscle fibers in the same muscle. In this way, activation of its muscle spindles causes contraction of an entire skeletal muscle, which relieves the stretching.

Tendon Organs

Tendon organs are located at the junction of a tendon and a muscle and protect tendons and their associated muscles from damage due to excessive tension. (When a muscle contracts, it exerts a force that tends to pull the points of attachment at either end toward each other. This force is the *muscle tension*.) Each **tendon organ** consists of a thin capsule of connective tissue that encloses a few tendon fascicles (bundles of collagen fibers) (see Figure 21.3). One or more sensory nerve endings penetrate the capsule, winding among the collagen fibers. When tension is applied to a muscle, the tendon organs generate nerve impulses that propagate into the CNS, providing information about changes in muscle tension. Tendon reflexes decrease muscle tension by causing relaxation.

Joint Kinesthetic Receptors

Several types of **joint kinesthetic receptors** are present within and around the articular capsules of synovial joints (see Chapter 9). Free nerve endings and type II cutaneous mechanoreceptors (Ruffini corpuscles) in the capsules of joints respond to pressure. Small lamellated (pacinian) corpuscles in the connective tissue outside articular capsules respond to acceleration and deceleration in the movement of joints. Articular ligaments contain receptors similar to tendon organs that adjust the adjacent muscles when excessive strain is placed on the joint.

Table 21.2 summarizes the somatic sensory receptors and the sensations they convey.

◉ **CHECKPOINT**

5. Which somatic sensory receptors are encapsulated?
6. Which somatic sensory receptors adapt slowly, and which adapt rapidly?
7. Which somatic sensory receptors mediate fine touch sensations?
8. How does fast pain differ from slow pain?
9. What is referred pain, and how is it useful in diagnosing internal disorders?
10. What aspects of muscle function are monitored by muscle spindles and tendon organs?

SOMATIC SENSORY PATHWAYS

◉ **OBJECTIVE**

• Describe the neuronal components and functions of the posterior column–medial lemniscus pathway, the anterolateral pathway, and the spinocerebellar pathway.

On reaching the spinal cord, sensory information has two possible destinations (1) it may become part of a reflex arc, bringing about an appropriate effector response, or (2) it may be relayed upward to the brain via ascending pathways for further processing and possible conscious awareness. **Somatic sensory pathways** relay information from the somatic sensory receptors just described to the primary somatosensory area in the cerebral cortex and to the cerebellum. The pathways to the cerebral cortex consist of thousands of sets of three neurons: first-order neurons, second-order neurons, and third-order neurons.

1. **First-order neurons** conduct impulses from somatic receptors into the brain stem or spinal cord. From the face, mouth, teeth, and eyes, somatic sensory impulses propagate along *cranial nerves* into the brain stem. From the neck, trunk, limbs, and posterior aspect of the head, somatic sensory impulses propagate along *spinal nerves* into the spinal cord.
2. **Second-order neurons** conduct impulses from the brain stem and spinal cord to the thalamus or cerebellum. Axons of second-order neurons *decussate* (cross over to the opposite side) in the brain stem or spinal cord before ascending to the ventral posterior nucleus of the thalamus. Thus, all somatic sensory information from one side of the body reaches the thalamus on the opposite side.
3. **Third-order neurons** conduct impulses from the thalamus to the primary somatosensory area of the cortex on the same side.

Regions within the CNS where neurons synapse with other neurons that are a part of a particular sensory or motor pathway are known as **relay stations** because neural signals are being relayed from one region of the CNS to another. For example, the neurons of many sensory pathways synapse with neurons in the thalamus; therefore the thalamus functions as a major relay station. In addition to the thalamus, many other regions of the CNS including the spinal cord and brain stem can function as relay stations.

Somatic sensory impulses entering the spinal cord ascend to the cerebral cortex via two general pathways: (1) the posterior column–medial lemniscus pathway and (2) the anterolateral

TABLE 21.2

Summary of Receptors for Somatic Sensations

RECEPTOR TYPE	RECEPTOR STRUCTURE AND LOCATION	SENSATIONS	ADAPTATION RATE
TACTILE RECEPTORS			
Corpuscles of touch (Meissner corpuscles)	Capsule surrounds mass of dendrites in dermal papillae of hairless skin.	Fine touch, pressure, and slow vibrations.	Rapid.
Hair root plexuses	Free nerve endings wrapped around hair follicles in skin.	Crude touch.	Rapid.
Type I cutaneous mechanoreceptors (tactile or Merkel discs)	Saucer-shaped free nerve endings make contact with Merkel cells in epidermis.	Fine touch and pressure.	Slow.
Type II cutaneous mechanoreceptors (Ruffini corpuscles)	Elongated capsule surrounds dendrites deep in dermis and in ligaments and tendons.	Stretching of skin.	Slow.
Lamellated (pacinian) corpuscles	Oval, layered capsule surrounds dendrites; present in dermis and subcutaneous layer, submucosal tissues, joints, periosteum, and some viscera.	Pressure, fast vibrations, and tickling.	Rapid.
Itch and tickle receptors	Free nerve endings and lamellated (pacinian) corpuscles in skin and mucous membranes.	Itching and tickling.	Both slow and rapid.
THERMORECEPTORS			
Warm receptors and cold receptors	Free nerve endings in skin and mucous membranes of mouth, vagina, and anus.	Warmth or cold.	Initially rapid, then slow.
PAIN RECEPTORS			
Nociceptors	Free nerve endings in every tissue of the body except the brain.	Pain.	Slow.
PROPRIOCEPTORS			
Muscle spindles	Sensory nerve endings wrap around central area of encapsulated intrafusal muscle fibers within most skeletal muscles.	Muscle length.	Slow.
Tendon organs	Capsule encloses collagen fibers and sensory nerve endings at junction of tendon and muscle.	Muscle tension.	Slow.
Joint kinesthetic receptors	Lamellated (pacinian) corpuscles, type II cutaneous mechanoreceptors (Ruffini corpuscles), tendon organs, and free nerve endings.	Joint position and movement.	Rapid.

(spinothalamic) pathways. Somatic sensory impulses entering the spinal cord reach the cerebellum via the spinocerebellar tracts.

Posterior Column–Medial Lemniscus Pathway to the Cortex

Nerve impulses for conscious proprioception and most tactile sensations ascend to the cerebral cortex along the **posterior column–medial lemniscus pathway** (Figure 21.4a). The name of the pathway comes from the names of two white-matter tracts that convey the impulses: the posterior column of the spinal cord and the medial lemniscus of the brain stem.

First-order neurons extend from sensory receptors in the trunk and limbs into the spinal cord and ascend to the medulla oblongata on the same side of the body. These are the longest neurons in the body. The cell bodies of these first-order neurons are in the posterior (dorsal) root ganglia of spinal nerves. In the spinal cord, their axons form the **posterior (dorsal) columns**, which consist of two parts: the **gracile fasciculus** (GRAS-īl fa-SIK-ū-lus) and the **cuneate fasciculus** (KŪ-nē-āt). (See Table 21.3 on page 688.) The axon terminals synapse with second-order neurons whose cell bodies are located in the gracile nucleus or the cuneate nucleus of the medulla. Impulses from the posterior head, neck, free upper limbs, and upper thorax (above T6) propagate along axons in the cuneate fasciculus and arrive at the cuneate nucleus. Impulses from the lower thorax (below T6), abdomen,

pelvis, and lower limbs propagate along axons in the gracile fasciculus and arrive at the gracile nucleus. (The exception to this rule is the proprioceptive neurons from the lower limbs. They relay in the nucleus dorsalis or thoracic nucleus of Clarke in the thoracic spinal cord.) Impulses for somatic sensations from the head arrive in the brain stem via sensory axons that are part of cranial nerves V, VII, IX, and X. Most somatic sensory axons carrying impulses from the face are part of the trigeminal (V) nerve.

The axons of the second-order neurons cross to the opposite side of the medulla and enter the **medial lemniscus** (lem-NIS-kus=ribbon), a thin ribbonlike projection tract that extends from the medulla to the ventral posterior nucleus of the thalamus. In the thalamus, the axon terminals of second-order neurons synapse with third-order neurons, which project their axons to the primary somatosensory area of the cerebral cortex.

Impulses conducted along the posterior column–medial lemniscus pathway give rise to several highly evolved and refined sensations:

- **Fine touch** is the ability to recognize specific information about a touch sensation, such as what point on the body is touched plus the shape, size, and texture of the source of stimulation.
- **Stereognosis** is the ability to recognize the size, shape, and texture of an object by feeling it. Examples are reading Braille or identifying a paperclip by feeling it.

- **Proprioception** is the awareness of the precise position of body parts, and **kinesthesia** is the awareness of directions of movement. Proprioceptors also allow **weight discrimination**, the ability to assess the weight of an object.
- **Vibratory sensations** arise when rapidly fluctuating touch stimuli are present.

Anterolateral Pathways to the Cortex

Like the posterior column–medial lemniscus pathway just described, the **anterolateral** or **spinothalamic pathways** (spī-nō-tha-LAM-ik) are composed of three-neuron sets (Figure 21.4b). The first-order neurons connect a receptor of the neck, trunk, or free limbs with the spinal cord. The cell bodies of the first-order neurons are in the posterior root ganglion. The axon terminals of the first-order neurons synapse with the second-order neurons, whose cell bodies are located in the posterior gray horn of the spinal cord.

The axons of the second-order neurons cross to the opposite side of the spinal cord. Then, they pass upward to the brain stem in either the **lateral spinothalamic tract** or the **anterior spinothalamic tract**. The lateral spinothalamic tract conveys sensory impulses for pain and temperature; the anterior spinothalamic tract conveys impulses for tickle, itch, crude touch, pressure, and vibrations. The axons of the second-order neurons end in the ventral posterior nucleus of the thalamus, where they synapse with the third-order neurons. The axons of the third-order neurons project to the primary somatosensory area on the same side of the cerebral cortex as the thalamus.

Figure 21.4 Somatic sensory pathways.

Nerve impulses are conducted along sets of first-order, second-order, and third-order neurons to the primary somatosensory area (postcentral gyrus) of the cerebral cortex.

(a) Posterior column–medial lemniscus pathway

(b) Anterolateral (spinothalamic) pathways

? What types of sensory deficits could be produced by damage to the right lateral spinothalamic tract?

Mapping the Primary Somatosensory Area

Specific areas of the cerebral cortex receive somatic sensory input from particular parts of the body and other areas of the cerebral cortex provide output in the form of instructions for movement of particular parts of the body. The *somatic sensory map* and the *somatic motor map* relate body parts to these cortical areas.

Precise localization of somatic sensations occurs when nerve impulses arrive at the **primary somatosensory area** (areas 1, 2, and 3 in Figure 19.15 on page 636), which occupies the postcentral gyri of the parietal lobes of the cerebral cortex. Each region in this area receives sensory input from a different part of the body. Figure 21.5a maps the destination of somatic sensory signals from different parts of the left side of the body in the somatosensory area of the right cerebral hemisphere. The left cere-

bral hemisphere has a similar primary somatosensory area that receives sensory input from the right side of the body.

Note that some parts of the body—chiefly the lips, face, tongue, and hand—provide input to large regions in the somatosensory area. Other parts of the body, such as the trunk and lower limbs, project to much smaller cortical regions. The relative sizes of these regions in the somatosensory area are proportional to the number of specialized sensory receptors within the corresponding part of the body. For example, there are many sensory receptors in the skin of the lips but few in the skin of the trunk. The size of the cortical region that represents a body part may expand or shrink somewhat, depending on the quantity of sensory impulses received from that body part. For example, people who learn to read Braille eventually have a larger cortical region in the somatosensory area to represent the fingertips.

Figure 21.5 **Somatic sensory and somatic motor maps in the cerebral cortex.** (a) Primary somatosensory area (postcentral gyrus) and (b) primary motor area (precentral gyrus) of the right cerebral hemisphere. The left hemisphere has similar representation. (After Penfield and Rasmussen.)

🔑 Each point on the body surface maps to a specific region in both the primary somatosensory area and the primary motor area.

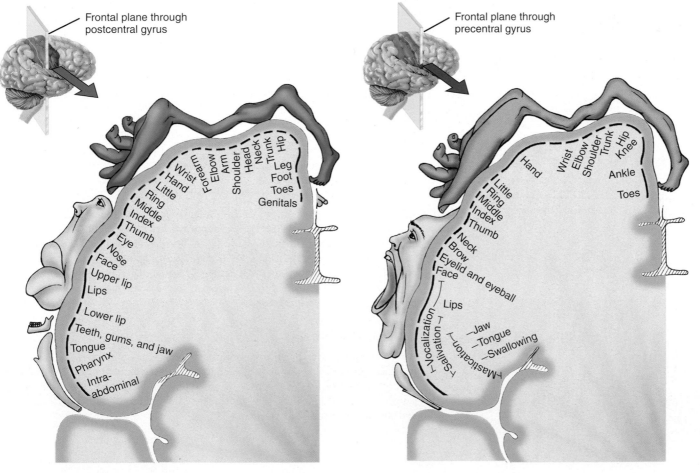

(a) Frontal section of primary somatosensory area in right cerebral hemisphere

(b) Frontal section of primary motor area in right cerebral hemisphere

❓ How do the somatic sensory and somatic motor representations compare for the hand, and what does this difference imply?

Somatic Sensory Pathways to the Cerebellum

Two tracts in the spinal cord—the **posterior spinocerebellar tract** (spī-nō-ser-e-BEL-ar) and the **anterior spinocerebellar tract**—are the major routes whereby proprioceptive impulses reach the cerebellum. Although not consciously perceived, sensory impulses conveyed to the cerebellum along these two pathways are critical for posture, balance, and coordination of skilled movements. Table 21.3 summarizes the major somatic sensory tracts in the spinal cord and pathways in the brain.

⦿ CHECKPOINT

11. What are three differences between the posterior column–medial lemniscus pathway and the anterolateral pathways?
12. Which body parts have the largest representation in the primary somatosensory area?
13. What type of sensory information is carried in the spinocerebellar tracts, and what is its function?

TABLE 21.3

Major Somatic Sensory Tract Pathways

TRACTS AND LOCATIONS

Posterior column:
Gracile fasciculus
Cuneate fasciculus

Lateral spinothalamic tract

Anterior spinothalamic tract

Posterior spinocerebellar tract

Anterior spinocerebellar tract

PATHWAY FUNCTIONS

Posterior Column Pathway

Posterior column tract: Conveys nerve impulses for the sensations of discriminative touch, proprioception, weight discrimination, and vibration. Axons of first-order neurons from one side of the body form the posterior column on the same side and end in the medulla, where they synapse with dendrites and cell bodies of second-order neurons. Axons of second-order neurons decussate (cross to the opposite side) and extend to the thalamus. Third-order neurons transmit nerve impulses from the thalamus to the primary somatosensory cortex on the side opposite the site of stimulation.

Anterolateral Pathway

Lateral spinothalamic tract: Conveys nerve impulses for pain and temperature sensations. Axons of first-order neurons from one side of the body synapse with second-order neurons in the posterior gray horn on the same side. Axons of second-order neurons decussate, enter the lateral spinothalamic tract on the opposite side, and extend to the thalamus. Third-order neurons transmit nerve impulses from the thalamus to the primary somatosensory cortex on the side opposite the site of stimulation.

Anterior spinothalamic tract: Conveys nerve impulses for itch, tickle, pressure, vibrations, and crude, poorly localized touch sensations. Axons of first-order neurons from one side of the body synapse with dendrites and cell bodies of second-order neurons in the posterior gray horn on the same side. Axons of second-order neurons decussate, enter the anterior spinothalamic tract on the opposite side, and extend to the thalamus. Third-order neurons transmit nerve impulses from the thalamus to the primary somatosensory cortex on the side opposite the site of stimulation.

Spinocerebellar Pathway

Anterior and posterior spinocerebellar tracts: Convey nerve impulses from proprioceptors in the trunk and lower limb of one side of the body to the same side of the cerebellum. The proprioceptive input informs the cerebellum of actual movements, allowing it to coordinate, smooth, and refine skilled movements and maintain posture and balance.

SOMATIC MOTOR PATHWAYS

◉ **OBJECTIVES**

- Identify the locations and functions of the different types of neurons in the somatic motor pathways.
- Compare the locations and functions of the direct and indirect motor pathways.
- Explain how the basal ganglia and cerebellum contribute to movements.

Neural circuits in the brain and spinal cord orchestrate all voluntary and involuntary movements. Ultimately, all excitatory and inhibitory signals that control movement converge on the alpha motor neurons that extend out of the brain stem and spinal cord to innervate skeletal muscles in the head and body. These neurons, also known as **lower motor neurons (LMNs)**, have their cell bodies in the brain stem and spinal cord. Their axons extend from the motor nuclei of cranial nerves to skeletal muscles of the face and head and from the anterior gray horns at all levels of the spinal cord to skeletal muscles of the limbs and trunk. Only lower motor neurons provide output from the CNS to skeletal muscle fibers. For this reason, they are also called the *final common pathway*.

Several groups of neurons participate in control of movement by providing input to lower motor neurons.

1. **Local circuit neurons.** Input arrives at lower motor neurons from nearby interneurons called *local circuit neurons*. These neurons are located close to the lower motor neuron cell bodies in the brain stem and spinal cord. Local circuit neurons receive input from somatic sensory receptors, such as nociceptors and muscle spindles, as well as from higher centers in the brain. They help coordinate rhythmic activity in specific muscle groups, such as alternating flexion and extension of the free lower limbs during walking.

2. **Upper motor neurons.** Both local circuit neurons and lower motor neurons receive input from *upper motor neurons (UMNs)*. Most upper motor neurons synapse with local circuit neurons, which in turn synapse with lower motor neurons. (A few upper motor neurons synapse directly with lower motor neurons.) UMNs from the cerebral cortex are essential for planning, initiating, and directing sequences of voluntary movements. Other UMNs originate in motor centers of the brain stem: the red nucleus, the vestibular nucleus, the superior colliculus, and the reticular formation. UMNs from the brain stem regulate muscle tone, control postural muscles, and help maintain balance and orientation of the head and body. Both the basal ganglia and cerebellum exert an influence on upper motor neurons.

3. **Basal ganglia neurons.** *Basal ganglia neurons* assist movement by providing input through the thalamus to upper motor neurons. Neural circuits interconnect the basal ganglia with motor areas of the cerebral cortex, thalamus, subthalamic nucleus, and substantia nigra. These circuits help initiate and terminate movements, suppress unwanted movements, and establish a normal level of muscle tone.

4. **Cerebellar neurons.** *Cerebellar neurons* also aid movement by controlling the activity of upper motor neurons through the thalamus. Neural circuits interconnect the cerebellum with motor areas of the cerebral cortex and the brain stem. A prime function of the cerebellum is to monitor differences between intended movements and movements actually performed. Then, it issues commands to upper motor neurons to reduce errors in movement. The cerebellum thus coordinates body movements and helps maintain normal posture and balance.

The axons of upper motor neurons extend from the brain to lower motor neurons via two types of somatic motor pathways—direct and indirect. **Direct motor pathways** provide input to lower motor neurons via axons that extend directly from the cerebral cortex. **Indirect motor pathways** provide input to lower motor neurons from motor centers in the brain stem. These brain stem centers in turn receive signals from neurons in the basal ganglia, cerebellum, and cerebral cortex. Direct and indirect pathways both govern generation of nerve impulses in the lower motor neurons, the neurons that stimulate contraction of skeletal muscles.

Before we examine these pathways we consider the role of the motor cortex in voluntary movement.

● **CLINICAL CONNECTION** | Paralysis

Damage or disease of *lower* motor neurons produces **flaccid paralysis** of muscles on the same side of the body. There is neither voluntary nor reflex action of the innervated muscle fibers, muscle tone is decreased or lost, and the muscle remains limp or flaccid. Injury or disease of *upper* motor neurons in the cerebral cortex causes **spastic paralysis** of muscles on the opposite side of the body. In this condition muscle tone is increased, reflexes are exaggerated, and pathological reflexes such as the Babinski sign appear. ●

Mapping the Motor Areas

Control of body movements occurs via neural circuits in several regions of the brain. The **primary motor area** (area 4 in Figure 19.15 on page 636), located in the precentral gyrus of the frontal lobe (Figure 21.5b) of the cerebral cortex, is a major control region for planning and initiating voluntary movements. The adjacent **premotor area** (area 6) and even the primary somatosensory area in the postcentral gyrus also contribute axons to the descending motor pathways. As is true for somatic sensory representation in the somatosensory area, different muscles are represented unequally in the primary motor area. The cortical area devoted to a muscle is proportional to the number of motor units in that muscle. Muscles in the thumb, fingers, lips, tongue, and vocal cords have large representations, in contrast to the much smaller representation of the trunk. By comparing Figures 21.5a and b, you can see that somatosensory and somatic motor representations are similar but not identical for most parts of the body.

Direct Motor Pathways

Nerve impulses for voluntary movements propagate from the cerebral cortex to lower motor neurons via the direct motor pathways (Figure 21.6 on page 690), also known as the *pyramidal pathways*. Areas of the cerebral cortex that contain large, pyramid-shaped cell bodies of upper motor neurons include not only the primary motor area in the precentral gyrus (area 4 in Figure 19.15 on page 636) but also the premotor area (area 6) and even the primary somatosensory area in the postcentral gyrus (areas 1, 2, and 3). Axons of these cortical UMNs descend through the internal capsule of the cerebrum. In the medulla oblongata, the axon bundles form the ventral bulges known as the pyramids.

About 90% of the axons of upper motor neurons *decussate* (cross over) to the *contralateral* (opposite) side in the medulla oblongata. The 10% that remain on the *ipsilateral* (same) side eventually decussate at the spinal cord levels where they synapse with an interneuron or lower motor neuron. Thus, the right cerebral

Figure 21.6 **Direct motor pathways in which signals initiated by the primary motor area in the right hemisphere control skeletal muscles on the left side of the body.** Spinal cord tracts carrying impulses of direct motor pathways are the lateral corticospinal tract and anterior corticospinal tract.

🔑 Direct pathways convey impulses that result in precise, voluntary movements.

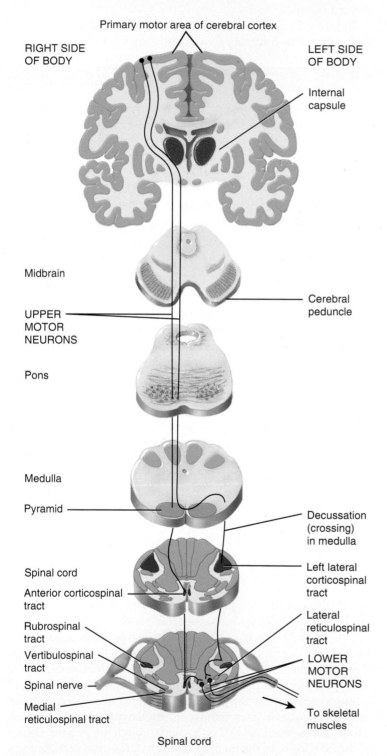

What other tracts (not shown in the figure) convey impulses that result in precise, voluntary movements?

cortex controls muscles on the left side of the body, and the left cerebral cortex controls muscles on the right side of the body.

Three tracts contain axons of upper motor neurons belonging to the direct motor pathways:

1. **Lateral corticospinal tracts.** Axons of UMNs that decussate in the medulla form the *lateral corticospinal tracts* (kor'-ti-kō-SPĪ-nal) in the right and left lateral white columns of the spinal cord (Figure 21.6 and Table 21.4). These motor neurons control muscles located in distal parts of the free limbs. The distal muscles are responsible for precise, agile, and highly skilled movements of the hands, and feet. Examples include the movements needed to button a shirt or play the piano.

2. **Anterior corticospinal tracts.** Axons of cortical UMNs that do not decussate in the medulla form the *anterior corticospinal tracts* in the right and left anterior white columns (Figure 21.6 and Table 21.4). At each spinal cord level, some of these axons decussate via the anterior white commissure. Then they synapse with interneurons or lower motor neurons in the anterior gray horn. Axons of these lower motor neurons exit the cervical and upper thoracic segments of the cord in the anterior roots of spinal nerves. They terminate in skeletal muscles that control movements of the trunk and proximal parts of the free limbs, thus coordinating movements of the axial skeleton.

3. **Corticobulbar tracts.** Some axons of upper motor neurons that conduct impulses for the control of skeletal muscles in the head extend through the internal capsule to the midbrain, where they join the *corticobulbar tracts* (kor'-ti-kō-BUL-bar) in the right and left cerebral peduncles (see Table 21.4). Some of the axons in the corticobulbar tracts have decussated, whereas others have not. The axons terminate in the motor nuclei of nine pairs of cranial nerves in the pons and medulla oblongata: the oculomotor (III), trochlear (IV), trigeminal (V), abducens (VI), facial (VII), glossopharyngeal (IX), vagus (X), accessory (XI), and hypoglossal (XII). The lower motor neurons of cranial nerves convey impulses that control precise, voluntary movements of the eyes, tongue, and neck, plus chewing, facial expression, and speech.

Table 21.4 summarizes the functions and pathways of the tracts in the direct motor pathways.

Indirect Motor Pathways

The **indirect motor pathways** or **extrapyramidal pathways** include all somatic motor tracts other than the corticospinal and corticobulbar tracts. Nerve impulses conducted along the indirect pathways follow complex, polysynaptic circuits that involve the motor cortex, basal ganglia, thalamus, cerebellum, reticular formation, and nuclei in the brain stem. Axons of upper motor neurons that carry nerve impulses from the indirect pathways descend from various nuclei of the brain stem into five major tracts of the spinal cord and terminate on local circuit neurons or lower motor neurons. These tracts are the **rubrospinal** (ROO-brō-spī-nal), **tectospinal** (TEK-tō-spī-nal), **vestibulospinal** (ves-TIB-ū-lō-spī-nal), **lateral reticulospinal** (re-TIK-ū-lō-spī-nal), and **medial reticulospinal tracts**.

Table 21.4 summarizes the functions and pathways of the tracts in the indirect motor pathways.

TABLE 21.4

Major Somatic Motor Tract Pathways

TRACTS AND LOCATIONS	PATHWAY FUNCTIONS
 Lateral corticospinal tract Anterior corticospinal tract Spinal cord	**Direct (Pyramidal) Pathways** **Lateral corticospinal:** Conveys nerve impulses from motor cortex of cerebrum to skeletal muscles on opposite side of body for precise, voluntary movements of distal parts of free limbs. Most axons of upper motor neurons (UMNs) decussate (cross over) in pyramids of medulla oblongata, then enter opposite side of spinal cord to form this tract. At their level of termination, these UMNs provide input to lower motor neurons in anterior gray horn on same side. **Anterior corticospinal:** Conveys nerve impulses from motor cortex to skeletal muscles on opposite side of body for movements of trunk and proximal parts of free limbs. Axons of UMNs descend from cortex into medulla oblongata, then enter spinal cord on same side to form this tract. At their level of termination, these UMNs decussate across spinal cord to anterior gray horn on the opposite side where lower motor neurons receive their input.
 Cerebral peduncle Corticobulbar tract Midbrain of brain stem	**Corticobulbar:** Conveys nerve impulses from motor cortex to skeletal muscles of head and neck to coordinate precise, voluntary movements. Axons of UMNs descend from cortex into brain stem, where some decussate and others do not. They provide input to lower motor neurons in cranial nerves controlling voluntary movements of eyes, tongue, and neck; chewing; facial expression; and speech.
 Rubrospinal tract Tectospinal tract Medial reticulospinal tract Lateral reticulospinal tract Vestibulospinal tract Spinal cord	**Indirect (Extrapyramidal) Pathways** **Rubrospinal:** Conveys nerve impulses from the red nucleus (which receives input from the cerebral cortex and cerebellum) to contralateral skeletal muscles that govern precise, voluntary movements of the distal parts of the free upper limbs. **Tectospinal:** Conveys nerve impulses from the superior colliculus to contralateral skeletal muscles that reflexively move the head, eyes, and trunk in response to visual or auditory stimuli. **Vestibulospinal:** Conveys nerve impulses from the vestibular nucleus (which receives input about head movements from the inner ear) to ipsilateral skeletal muscles of the trunk and proximal parts of the free limbs for maintaining posture and balance in response to head movements. **Medial and lateral reticulospinal:** Conveys nerve impulses from reticular formation to ipsilateral skeletal muscles of the trunk and proximal parts of the free limbs for maintaining posture and regulating muscle tone in response to ongoing body movements.

• CLINICAL CONNECTION | Parkinson Disease

Parkinson disease (PD) is a progressive disorder of the CNS that typically affects its victims around age 60. Neurons that extend from the substantia nigra to the putamen and caudate nucleus, where they release the neurotransmitter dopamine (DA), degenerate in PD. In the caudate nucleus of the basal ganglia are neurons that liberate the neurotransmitter acetylcholine (ACh). Although the level of ACh does not change as the level of DA declines, the imbalance of neurotransmitter activity—too little DA and too much ACh—is thought to cause most of the symptoms. The cause of PD is unknown, but toxic environmental chemicals, such as pesticides, herbicides, and carbon monoxide, are suspected contributing agents. Only 5% of PD patients have a family history of the disease.

In PD patients, involuntary skeletal muscle contractions often interfere with voluntary movement. For instance, the muscles of the upper limb may alternately contract and relax, causing the hand to shake. This shaking, called **tremor**, is the most common symptom of PD. Also, muscle tone may increase greatly, causing rigidity of the involved body part. Rigidity of the facial muscles gives the face a masklike appearance. The expression is characterized by a wide-eyed, unblinking stare and a slightly open mouth with uncontrolled drooling.

Motor performance is also impaired by **bradykinesia** (*brady-*=slow), slowness of movements. Activities such as shaving, cutting food, and buttoning a blouse take longer and become increasingly more difficult as the disease progresses. Muscular movements also exhibit **hypokinesia** (*hypo-*=under), decreasing range of motion. For example, words are written smaller, letters are poorly formed, and eventually handwriting becomes illegible. Often, walking is impaired; steps become shorter and shuffling, and arm swing diminishes. Even speech may be affected.

CONTINUED ▶

• CLINICAL CONNECTION Parkinson Disease CONTINUED

Treatment of PD is directed toward increasing levels of DA and decreasing levels of ACh. Although people with PD do not manufacture enough dopamine, taking it orally is useless because DA cannot cross the blood–brain barrier. Even though symptoms are partially relieved by a drug developed in the 1960s called levodopa (L-dopa), a precursor of DA, the drug does not slow the progression of the disease. As more and more affected brain cells die, the drug becomes useless. Another drug called selegiline (Deprenyl®) is used to inhibit monoamine oxidase, an enzyme that degrades catecholamine neurotransmitters such as dopamine. This drug slows progression of PD and may be used together with levodopa. Anticholinergic drugs such as benzotropine and trihexyphenidyl can also be used to block the effects of ACh at some of the synapses between basal ganglia neurons, which helps to restore the balance between ACh and DA. Anticholinergic drugs effectively reduce symptomatic tremor, rigidity, and drooling.

For more than a decade, surgeons have sought to reverse the effects of Parkinson disease by transplanting dopamine-rich fetal nervous tissue into the basal ganglia (usually the putamen) of patients with severe PD. Only a few postsurgical patients have shown any degree of improvement, such as less rigidity and improved quickness of motion. Another surgical technique that has produced improvement for some patients is *pallidotomy*, in which a part of the globus pallidus that generates tremors and produces muscle rigidity is destroyed. In addition, some patients are being treated with a surgical procedure called *deep-brain stimulation (DBS)*, which involves the implantation of electrodes into the subthalamic nucleus. The electrical currents released by the implanted electrodes reduce many of the symptoms of PD. •

Basal ganglia; decreased dopaminergic neurons

Thalamus

Substantia nigra; depigmentation of nucleus

Cerebral peduncles; midbrain

Roles of the Basal Ganglia

As previously noted, the basal ganglia and cerebellum influence movement through their effects on upper motor neurons. The functions of the basal ganglia include the following:

1. The basal ganglia play a major role in the initiation and termination of movements. Two parts of the basal ganglia, the caudate nucleus and the putamen, receive input from sensory, association, and motor areas of the cerebral cortex and from the substantia nigra. Output from the basal ganglia comes from the globus pallidus and substantia nigra, which send feedback signals to the upper motor neurons in the motor cortex by way of the thalamus. (Figure 19.13b on page 632 shows these parts of the basal ganglia.) This circuit—from cortex to basal ganglia to thalamus to cortex—appears to function in the initiation and termination of movements. Neurons in the putamen generate impulses just before body movements occur; and neurons in the caudate nucleus generate impulses just before eye movements occur.

2. The basal ganglia suppress unwanted movements by their inhibitory effects on the thalamus and superior colliculus.

3. The basal ganglia influence muscle tone. The globus pallidus sends impulses into the reticular formation that reduce muscle tone. Damage or destruction of some basal ganglia connections causes a generalized increase in muscle tone.

4. The basal ganglia influence many aspects of cortical function, including sensory, limbic, cognitive, and linguistic functions. For example, the basal ganglia help initiate and terminate some cognitive processes, such as attention, memory, and planning. In addition, the basal ganglia may act with the limbic system to regulate emotional behaviors.

Roles of the Cerebellum

In addition to maintaining proper posture and balance, the cerebellum is active in learning and performing rapid, coordinated, highly skilled movements such as hitting a golf ball, speaking, and swimming. Cerebellar function involves four activities.

1. The cerebellum *monitors intentions for movement* by receiving impulses from the motor cortex and basal ganglia via the pontine nuclei in the pons regarding what movements are planned.

2. The cerebellum *monitors actual movement* by receiving input from proprioceptors in joints and muscles that reveals what actually is happening. These nerve impulses travel in the anterior and posterior spinocerebellar tracts. Nerve impulses from the vestibular (equilibrium-sensing) apparatus in the inner ear and from the eyes also enter the cerebellum.

3. The cerebellum *compares the command signals* (intentions for movement) *with sensory information* (actual movement performed).

4. If there is a discrepancy between intended and actual movement, the cerebellum *sends out corrective feedback* to upper motor neurons. This information travels via the thalamus to UMNs in the cerebral cortex but goes directly to UMNs in brain stem motor centers. As movements occur, the cerebellum continuously provides error corrections to upper motor neurons, which decreases errors and smoothes the motion. It also contributes over longer periods to the learning of new motor skills.

Skilled activities such as tennis or volleyball provide good examples of the contribution of the cerebellum to movement. To make a good serve or to block a spike, you must bring your racket or arms forward just far enough to make solid contact. How do you stop at exactly the right point? For example, before you even hit a tennis ball, the cerebellum has sent nerve impulses to the cerebral cortex and basal ganglia informing them where your arm or racket swing must stop. In response to impulses from the cerebellum, the cortex and basal ganglia transmit motor impulses to opposing body muscles to stop the swing.

◉ **CHECKPOINT**

14. Which parts of the body have the largest representation in the motor cortex? Which have the smallest?
15. Explain why the two main somatic motor pathways are named "direct" and "indirect."
16. Compare the effects of disease of the basal ganglia with damage to the cerebellum.

INTEGRATION OF SENSORY INPUT AND MOTOR OUTPUT

◉ **OBJECTIVE**

• Explain how sensory input and motor output are integrated.

Sensory pathways provide the input that keeps the central nervous system informed of changes in the external and internal environment. Output from the CNS is then conveyed through motor pathways, which enable movement and glandular secretions to occur. As sensory information reaches the CNS, it becomes part of a large pool of sensory information. However, the CNS does not necessarily respond to every impulse. Rather, the incoming sensory information is **integrated**; that is, it is processed and interpreted and a course of action is taken.

The integration process occurs not just once but at many places along the pathways of the CNS and at both conscious and subconscious levels. It occurs within the spinal cord, brain stem, cerebellum, basal ganglia, and cerebral cortex. As a result, the output descending along a motor pathway that makes a muscle contract or a gland secrete can be modified and responded to at any of these levels. Motor portions of the cerebral cortex play the major role for initiating and controlling precise movements of muscles. The basal ganglia largely integrate semivoluntary, automatic movements such as walking, swimming, and laughing. The cerebellum assists the motor cortex and basal ganglia by making body movements smooth and coordinated and by contributing significantly to the maintenance of normal posture and balance.

◉ **CHECKPOINT**

17. What does integration of sensory input mean?

KEY MEDICAL TERMS ASSOCIATED WITH SOMATIC SENSES

Acupuncture (ak-ū-PUNK-chur) The use of fine needles (lasers, ultrasound, or electricity) inserted into specific exterior body locations (acupoints) and manipulated to relieve pain and provide therapy for various conditions. The placement of needles may cause the release of neurotransmitters such as endorphins, painkillers that may inhibit pain pathways.

Cerebral palsy (CP) A motor disorder that results in the loss of muscle control and coordination; caused by damage of the motor areas of the brain during fetal life, birth, or infancy. Radiation during fetal life, temporary lack of oxygen during birth, and hydrocephalus during infancy may also cause cerebral palsy.

Coma (KŌ-ma) A state of unconsciousness in which a person's responses to stimuli are reduced or absent. In a *light coma*, an individual may respond to certain stimuli, such as sound, touch, or light, and move his or her eyes, cough, and even murmur. In a *deep coma*, a person does not respond to any stimuli and does not make any movements. Causes of coma include head injuries, cardiac arrest, stroke, brain tumors, infections (encephalitis and meningitis), seizures, alcoholic intoxication, drug overdose, severe lung disorders (chronic obstructive pulmonary disease, pulmonary edema, pulmonary embolism), inhalation of large amounts of carbon monoxide, liver or kidney failure, low or high blood sugar or sodium levels, and low or high body temperature.

Dysarthria (dis-AR-thrē-a; *dys-*=difficult; *-arthro*=to articulate) Difficult or imperfect speech due to loss of muscular control as a result of disorders of motor pathways.

Pain threshold The smallest intensity of a painful stimulus at which a person perceives pain. All individuals have the same pain threshold.

Pain tolerance The greatest intensity of painful stimulation that a person is able to tolerate. Individuals vary in their tolerance to pain.

Synesthesia (sin-es-THĒ-zē-a; *syn-*=together; *-aisthesis*=sensation) A condition in which sensations of two or more modalities accompany one another. In some cases, a stimulus for one sensation is perceived as a stimulus for another; for example, a sound produces a sensation of color. In other cases, a stimulus from one part of the body is experienced as coming from a different part.

STUDY OUTLINE

Overview of Sensations (p. 679)

1. Sensation is the awareness of changes in the external and internal conditions of the body.
2. For a sensation to occur, a stimulus must reach a sensory receptor, the stimulus must be converted to a nerve impulse, and the impulse conducted to the brain; finally, the impulse must be integrated by a region of the brain.
3. When stimulated, most sensory receptors ultimately produce one or more nerve impulses.
4. Sensory impulses from each part of the body arrive in specific regions of the cerebral cortex.
5. Modality is the distinct quality that makes one sensation different from others.
6. Adaptation is a decrease in sensation during a prolonged stimulus.
7. Two classes of senses are general senses, which include somatic senses and visceral senses, and special senses, which include the modalities of smell, taste, vision, hearing, and equilibrium (balance).
8. Types of sensory receptors are based on microscopic structure, location and origin of the stimuli that activate them, and type of stimulus they detect.
9. Table 21.1 on page 680 summarizes the classification of sensory receptors.

Somatic Sensations (p. 680)

1. Somatic sensations include tactile sensations (touch, pressure, vibration, itch, and tickle), thermal sensations (warmth and cold), pain, and proprioception.
2. Receptors for tactile, thermal, and pain sensations are located in the skin, subcutaneous layer, and mucous membranes of the mouth, vagina, and anus.
3. Receptors for proprioceptive sensations (position and movement of body parts) are located in muscles, tendons, joints, and the inner ear.
4. Receptors for touch are (a) hair root plexuses and corpuscles of touch (Meissner corpuscles), which are rapidly adapting, and (b) slowly adapting type I cutaneous mechanoreceptors (tactile or Merkel discs). Type II cutaneous mechanoreceptors (Ruffini corpuscles), which are slowly adapting, are sensitive to stretching. Receptors for pressure include corpuscles of touch, type I mechanoreceptors, and lamellated (pacinian) corpuscles. Receptors for vibration are corpuscles of touch and lamellated corpuscles. Itch receptors are free nerve endings. Both free nerve endings and lamellated corpuscles mediate the tickle sensation.
5. Thermoreceptors are free nerve endings. Cold receptors are located in the stratum basale of the epidermis; warm receptors are located in the dermis.
6. Pain receptors (nociceptors) are free nerve endings that are located in nearly every body tissue.
7. Proprioceptors include muscle spindles, tendon organs, joint kinesthetic receptors, and hair cells of the inner ear.
8. Table 21.2 on page 685 summarizes the somatic sensory receptors and the sensations they convey.

Somatic Sensory Pathways (p. 684)

1. Somatic sensory pathways from receptors to the cerebral cortex involve three-neuron sets: first-order, second-order, and third-order.
2. Axon collaterals (branches) of somatic sensory neurons simultaneously carry signals into the cerebellum and the reticular formation of the brain stem.
3. Impulses propagating along the posterior column–medial lemniscus pathway relay fine touch, stereognosis, proprioception, and vibratory sensations.
4. The neural pathway for pain and thermal sensations is the lateral spinothalamic tract.
5. The neural pathway for tickle, itch, crude touch, and pressure sensations is the anterior spinothalamic pathway.

6. The pathways to the cerebellum are the anterior and posterior spinocerebellar tracts, which transmit impulses for subconscious muscle and joint position sense from the trunk and free lower limbs.
7. Table 21.3 on page 688 summarizes the major somatic sensory pathways.
8. Specific regions of the primary somstosensory area (postcentral gyrus) of the cerebral cortex receive somatic sensory input from different parts of the body.

Somatic Motor Pathways (p. 689)

1. All excitatory and inhibitory signals that control movement converge on the alpha motor neurons, also known as lower motor neurons (LMNs) or the final common pathway.
2. Several groups of neurons participate in control of movement by providing input to lower motor neurons: local circuit neurons, upper motor neurons, basal ganglia neurons, and cerebellar neurons.
3. The primary motor area (precentral gyrus) of the cortex is the major control region for planning and initiating voluntary movements.
4. The axons of upper motor neurons extend from the brain to lower motor neurons via direct and indirect motor pathways. The direct (pyramidal) pathways include the lateral and anterior corticospinal tracts and corticobulbar tracts. Indirect (extrapyramidal) pathways extend from several motor centers of the brain stem into the spinal cord.
5. Neurons of the basal ganglia assist movement by providing input to the upper motor neurons. They help initiate and terminate movements, suppress unwanted movements, and establish a normal level of muscle tone.
6. The cerebellum is active in learning and performing rapid, coordinated, highly skilled movements. It also contributes to the maintenance of balance and posture.
7. Table 21.4 on page 691 summarizes the major somatic motor pathways.

Integration of Sensory Input and Motor Output (p. 693)

1. Sensory input keeps the CNS informed of changes in the environment.
2. Incoming sensory information is integrated at many stations along the CNS at both conscious and subconscious levels.
3. A motor response makes a muscle contract or a gland secrete.

SELF-QUIZ QUESTIONS

Choose the one best answer to the following questions.

1. Which of the following is the receptor for pain?
   a. chemoreceptor
   b. photoreceptor
   c. nociceptor
   d. thermoreceptor
   e. mechanoreceptor

2. Which of the following is *not* classified as a somatic sense?
   a. pain
   b. joint position
   c. touch
   d. temperature (thermal)
   e. equilibrium

3. Proprioceptive input reaches the cerebellum by which two routes?
   a. anterior and posterior spinocerebellar tracts
   b. anterior and lateral spinothalamic tracts
   c. anterior and lateral corticospinal tracts
   d. direct and indirect pathways
   e. gracile fasciculus and cuneate to fasciculus tracts

4. Which of the following statements is *true* for muscle spindles?
   a. They are located within and around articular capsules of synovial joints.

   b. They are located at the junction of a tendon and a muscle.
   c. They respond to acceleration and deceleration of joint movement, and they monitor pressure at joints.
   d. They monitor changes in the length of a skeletal muscle.
   e. All of these statements are true.

5. Gamma motor neurons stimulate:
   a. lamellated (pacinian) corpuscles
   b. corpuscles of touch (Meissner corpuscles)
   c. cutaneous mechanoreceptors
   d. intrafusal muscle fibers
   e. extrafusal muscle fibers

6. The region of head structures (throat, lips, face) is represented on the postcentral gyrus (primary somatosensory cortex) lying closest to the lateral sulcus. Which spinal cord nerve/tract is most responsible for delivering afferent information to this region?
   a. posterior spinocerebellar
   b. anterior spinocerebellar
   c. trigeminal (V) nerve
   d. cuneate fasciculus
   e. gracile fasciculus

7. The region of free lower limb structures is represented on the precentral gyrus (primary motor cortex) lying closest to the longitudinal fissure. Which spinal cord tract is most responsible for getting efferent information delivered from this region?
   a. cuneate fasciculus   **b.** gracile fasciculus
   c. anterior spinocerebellar   **d.** lateral corticospinal
   e. corticobulbar

Complete the following:

8. A change in the environment capable of activating certain sensory neurons is called a(n) _____.

9. Sensory receptors may be classified by location into three groups: _____, _____, and _____.

10. The receptors for touch, pressure, and vibration are all _____-receptors, based on the type of stimulus to which they respond.

11. Upper motor neurons of the indirect pathways begin in the _____ region of the brain and synapse with either lower motor neurons or _____ neurons.

12. The pain felt in the skin of the thorax and left arm during a heart attack is known as _____ pain.

13. The axons of sensory receptors in the trunk and free limbs form the _____ of the spinal cord.

14. The ability to recognize size, shape, and texture of an object by feeling it is known as _____.

15. Second- and third-order neurons of the posterior column–medial lemniscus pathway synapse in the _____.

16. All direct and indirect motor pathways descending within the spinal cord converge on _____.

Are the following statements true or false? If the statement is False, *find and fix the mistake to make the statement* True.

17. Stimulation of pain receptors in skeletal muscles, joints, and tendons results in visceral pain.

18. A sensory pathway, from receptor to the cerebral cortex, involves three neurons.

19. The precentral gyrus of the cerebral cortex is the premotor area.

20. Cold receptors are located in the epidermis.

21. Third-order neurons conduct impulses from somatic receptors to the central nervous system.

22. Indirect (extrapyramidal) motor pathways arise from cerebral white matter.

Matching

23. Match the following (answers are used more than once):
    ____ **(a)** located in skeletal muscles
    ____ **(b)** located within and around articular capsules of synovial joints
    ____ **(c)** located at the junction of a tendon and a muscle
    ____ **(d)** help protect tendons and their associated muscles from damage due to excessive tension
    ____ **(e)** respond to acceleration and deceleration of joint movement; monitor pressure at joints
    ____ **(f)** monitor changes in the length of a skeletal muscle

    (1) tendon organ
    (2) muscle spindle
    (3) joint kinesthetic receptor

24. Match the following:
    ____ **(a)** sensitive to movement of hair shaft
    ____ **(b)** egg-shaped masses located in dermal papillae; plentiful in fingertips, palms, and soles; receptors for discriminative touch
    ____ **(c)** onion-shaped structures sensitive to pressure in skin, membranes, joints, and some viscera
    ____ **(d)** dendrites that contact epidermal cells in the stratum basale of skin; function in discriminative touch

    (1) corpuscles of touch (Meissner corpuscles)
    (2) tactile (Merkel) discs
    (3) lamellated (pacinian) corpuscles
    (4) hair root plexuses

25. Match the following:
    ____ **(a)** conveys nerve impulses from the superior colliculus to contralateral skeletal muscles that move the head and eyes in response to visual stimuli
    ____ **(b)** conveys nerve impulses from the motor cortex to skeletal muscles on opposite side of body for movements of the axial skeleton; upper motor neurons decussate at their level of termination in the spinal cord
    ____ **(c)** conveys nerve impulses from the reticular formation to facilitate flexor reflexes, inhibit extensor reflexes, and decrease muscle tone in muscles of the axial skeleton and proximal part of the free limbs
    ____ **(d)** conveys nerve impulses from the red nucleus (which receives input from the cerebral cortex and cerebellum) to contralateral skeletal muscles that govern precise movements of the distal parts of the free limbs
    ____ **(e)** conveys nerve impulses from the motor cortex to skeletal muscles of the head and neck to coordinate precise, voluntary movements; upper motor neurons provide input to cranial nerve nuclei
    ____ **(f)** conveys nerve impulses from the motor cortex to skeletal muscle on the opposite side of the body for precise, voluntary movements of the free limbs, hands, and feet; upper motor neurons decussate in the medulla
    ____ **(g)** conveys nerve impulses from the reticular formation to facilitate extensor reflexes, inhibit flexor reflexes, and increase muscle tone in muscles of the axial skeleton and proximal parts of the free limbs
    ____ **(h)** conveys nerve impulses from the vestibular nucleus (which receives input about head movements from the vestibular apparatus in the inner ear) to regulate ipsilateral muscle tone for maintaining balance in response to head movements

    (1) lateral corticospinal tract
    (2) vestibulospinal tract
    (3) tectospinal tract
    (4) corticobulbar tract
    (5) medial reticulospinal tract
    (6) rubrospinal tract
    (7) anterior corticospinal tract
    (8) lateral reticulospinal tract

CRITICAL THINKING QUESTIONS

1. When Sau Lan held the cup of hot cocoa in her hands, at first it felt comfortably warm and within minutes, she didn't notice the temperature at all. Absently she took a big gulp and almost choked when she felt the hot cocoa burn her mouth and throat. What happened—did the hot cocoa get hotter?

2. Jenny was scheduled for surgery. As a part of her preparation, she was to take a course of antibiotic treatment for 10 days prior to the surgery. Unfortunately, in a rare reaction to the drugs, the antibiotics destroyed her muscle spindles. How will this affect Jenny's life?

3. Stella was visiting a long-term care facility as a part of her observational experiences in her allied health program. There she saw a patient who was having great difficulty controlling his body movements. She was not allowed to look at the patient's charts, so she was trying to figure out all the possible nervous system problems he could have. Damage to what specific parts of the brain might result in such loss of control?

4. Very young children are usually given only spoons (no forks or knives) and cups with lids at the dinner table. Why?

5. Jon used to predict the weather by how much the bunion (abnormally swollen joint on big toe) on his left foot was bothering him. Last year, Jon's left foot was amputated due to complications from diabetes. But sometimes, he still thinks he feels that bunion. Explain Jon's weather toe.

? ANSWERS TO FIGURE QUESTIONS

21.1 Pain, thermal sensations, tickle, and itch involve activation of different free nerve endings.

21.2 The kidneys have the broadest area for referred pain.

21.3 Muscle spindles are activated when the central areas of their intrafusal fibers are stretched.

21.4 Damage to the right lateral spinothalamic tract could result in loss of pain and thermal sensations on the left side of the body.

21.5 The hand has a larger representation in the motor area than in the somatosensory area, which implies greater precision in the hand's movement control than discriminative ability in its sensation.

21.6 The corticobulbar and rubrospinal tracts (see Table 21.4 on page 691) convey impulses that result in precise, voluntary movements.

22 | SPECIAL SENSES

INTRODUCTION Recall from Chapter 21 that the *general senses* include somatic senses (tactile, thermal, pain, and proprioceptive) and visceral sensations. Receptors for the general senses are scattered throughout the body and are relatively simple in structure. They are mostly modified dendrites of sensory neurons. Receptors for the *special senses*—smell, taste, vision, hearing, and equilibrium—are anatomically distinct from one another and are concentrated in specific locations in the head. They are usually embedded in the epithelial tissue within complex sensory organs such as the eyes and ears. Neural pathways for the special senses are more complex than those for the general senses.

In this chapter we examine the structure and function of the special sense organs, and the pathways involved in conveying their information to the central nervous system. **Ophthalmology** (of-thal-MOL-ō-jē; *opthalmo-*=eye; *-logy*=study of) is the science that deals with the eyes and their disorders. The other special senses are, in large part, the concern of **otorhinolaryngology** (ō′-tō-rī′-nō-lar-in-GOL-ō-jē; *oto-*=ear; *rhino-*=nose; *laryngo-*=larynx), the science that deals with the ears, nose, pharynx (throat), and larynx and their disorders. •

CONTENTS AT A GLANCE

OLFACTION: SENSE OF SMELL

⊙ **OBJECTIVE**

- Describe the olfactory receptors and the neural pathway for olfaction.

Both smell and taste are chemical senses; the sensations arise from the interaction of molecules with smell or taste receptors. To be detected by either sense, the stimulating molecules must be dissolved. Because impulses for smell and taste propagate to the limbic system (and to higher cortical areas as well), certain odors and tastes can evoke strong emotional responses or a flood of memories.

Anatomy of Olfactory Receptors

The nose contains 10–100 million receptors for the sense of smell or **olfaction** (ol-FAK-shun; *olfact-*=smell), contained within an area called the *olfactory epithelium*. With a total area of 5 cm² (a little less than 1 in.²), the olfactory epithelium occupies the superior part of the nasal cavity, covering the inferior surface of the cribriform plate and extending along the superior nasal concha (Figure 22.1a). The olfactory epithelium consists of three kinds of cells: olfactory receptors, supporting cells, and basal cells (Figure 22.1b, c).

Olfactory receptors are the first-order neurons of the olfactory pathway. Each olfactory receptor is a bipolar neuron with an exposed knob-shaped dendrite and an axon projecting through the cribriform plate and ending in the olfactory bulb. The parts of the olfactory receptors that respond to inhaled chemicals are the **olfactory hairs**, cilia that project from the dendrite. Chemicals that have an odor and can therefore stimulate the olfactory hairs are called **odorants**. Olfactory receptors respond to the chemical stimulation of an odorant molecule by initiating the olfactory response.

Supporting cells are columnar epithelial cells of the mucous membrane lining the nose. They provide physical support, nourishment, and electrical insulation for the olfactory receptors, and they help detoxify chemicals that come in contact with the olfactory epithelium. **Basal cells** are stem cells located between the bases of the supporting cells. They continually undergo cell division to produce new olfactory receptors, which live for only a month or so before being replaced. This process is remarkable considering that olfactory receptors are neurons, and as you have already learned, mature neurons are generally not replaced.

Within the connective tissue that supports the olfactory epithelium are **olfactory (Bowman's) glands**, which produce mucus that is carried to the surface of the epithelium by ducts. The secretion moistens the surface of the olfactory epithelium and

Figure 22.1 Olfactory receptors and olfactory pathway. (a) Location of olfactory epithelium in nasal cavity. (b) Anatomy of olfactory receptors, which are neurons whose axons extend through the cribriform plate and terminate in the olfactory bulb. (c) Histology of the olfactory epithelium. (d) Olfactory pathway.

🔑 The olfactory epithelium consists of olfactory receptors, supporting cells, and basal cells.

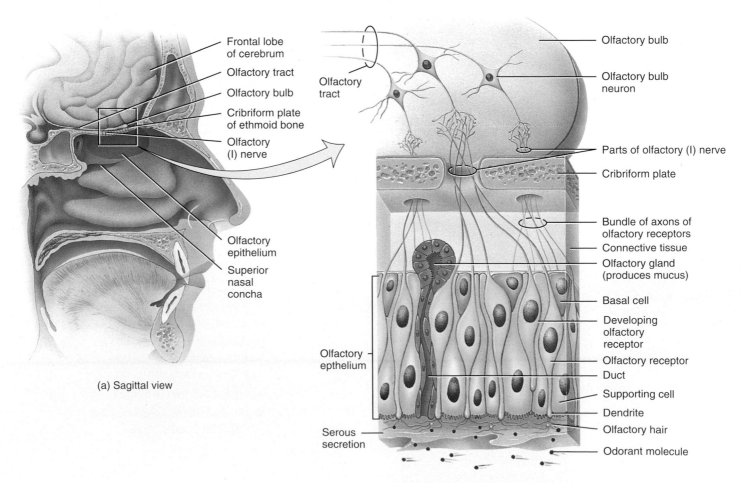

(a) Sagittal view

(b) Enlarged aspect of olfactory receptors

dissolves odorants so that transduction can occur. Both supporting cells of the nasal epithelium and olfactory glands are innervated by branches of the facial (VII) nerve, which can be stimulated by certain chemicals. Impulses in these nerves in turn stimulate the lacrimal glands in the eyes and nasal mucous glands. The result is tears and a runny nose after inhaling substances such as pepper or the vapors of household ammonia.

The Olfactory Pathway

On each side of the nose about 40 bundles of the slender, un-myelinated axons of olfactory receptors extend through about 20 olfactory foramina in the cribriform plate of the ethmoid bone (Figure 22.1b). These bundles of axons collectively form the right and left **olfactory (I) nerves**. The olfactory nerves terminate in the brain in paired masses of gray matter called the **olfactory bulbs**, which are located below the frontal lobes of the cerebrum

and lateral to the crista galli of the ethmoid bone. Within the olfactory bulbs, the axon terminals of olfactory receptors form synapses with the dendrites and cell bodies of neurons in the olfactory pathway.

Axons of olfactory bulb neurons extend posteriorly and form the **olfactory tract** (Figure 22.1a,b). Olfactory sensations are the only sensations that reach the cerebral cortex without first synapsing in the thalamus. Some of the axons of the olfactory tract project to the primary olfactory area; located at the inferior and medial surface of the temporal lobe, the primary olfactory area (area 28) is where conscious awareness of smell begins (Figure 22.1d). Other axons of the olfactory tract project to the limbic system and hypothalamus; these connections account for our emotional and memory-evoked responses to odors. Examples include sexual excitement upon smelling a certain perfume, nausea upon smelling a food that once made you violently ill, or an odor-evoked memory of a childhood experience.

(c) Histology of olfactory epithelium

(d) Olfactory pathway

? **Which part of an olfactory receptor detects an odorant molecule?**

From the primary olfactory area, pathways also extend to the frontal lobe. An important region for odor identification and discrimination is the orbitofrontal area (area 11 in Figure 19.15 on page 636). People who suffer damage in this area have difficulty identifying different odors. Positron emission tomography (PET) studies suggest some degree of hemispheric lateralization: The orbitofrontal area of the *right* hemisphere exhibits greater activity during olfactory processing.

⊙ CHECKPOINT

1. How do basal cells contribute to olfaction?
2. What is the sequence of events from the binding of an odorant molecule to an olfactory hair to the arrival of a nerve impulse in an olfactory bulb?

GUSTATION: SENSE OF TASTE

⊙ OBJECTIVE

- Describe the gustatory receptors and the neural pathway for gustation.

Taste or **gustation** (gus-TĀ-shun; *gust*=taste), like olfaction, is a chemical sense. However, it is much simpler than olfaction in that only five primary tastes can be distinguished: *sour, sweet, bitter, salty,* and *umami* (ū-MAM-ē). The umami taste, recently reported

by Japanese scientists, is described as "meaty" or "savory." Umami is believed to arise from taste receptors that are stimulated by monosodium glutamate (MSG), a substance naturally present in many foods and added to others as a flavor enhancer. All other flavors, such as chocolate, pepper, and coffee, are combinations of the five primary tastes, plus accompanying olfactory and tactile (touch) sensations. Odors from food can pass upward from the mouth into the nasal cavity, where they stimulate olfactory receptors. Because olfaction is much more sensitive than taste, a given concentration of a food substance may stimulate the olfactory system thousands of times more strongly than it stimulates the gustatory system. When you have a cold or are suffering from allergies and cannot taste your food, it is actually olfaction that is blocked, not taste.

Anatomy of Gustatory Receptors

The receptors for taste are located in the taste buds (Figure 22.2). The nearly 10,000 taste buds of a young adult are mainly on the tongue, but they are also found on the soft palate (posterior portion of roof of mouth), pharynx (throat), and epiglottis (cartilage lid over voice box). The number of taste buds declines with age. Each **taste bud** is an oval body consisting of three kinds of epithelial cells: supporting cells, gustatory receptor cells, and basal cells (Figure 22.2c, d). The **supporting cells** contain microvilli

Figure 22.2 Gustatory receptor cells and gustatory pathway.

⊙ Gustatory receptor cells are located in taste buds.

(a) Dorsum of tongue showing location of papillae

(b) Details of papillae

(c) Structure of a taste bud

and surround about 50 **gustatory receptor cells**. A single, long microvillus, called a **gustatory hair**, projects from each gustatory receptor cell to the external surface through the **taste pore**, an opening in the taste bud. **Basal cells**, stem cells found at the periphery of the taste bud near the connective tissue layer, produce supporting cells. The supporting cells then develop into gustatory receptor cells, which have a life span of about 10 days. At their base, the gustatory receptor cells synapse with dendrites of the first-order sensory neurons that form the first part of the gustatory pathway. Each first-order neuron has many dendrites, which receive input from many gustatory receptor cells in several taste buds.

Taste buds are found in elevations on the tongue called **papillae** (pa-PIL-ē), which provide a rough texture to the upper surface of the tongue (Figure 22.2a, b). Three types of papillae contain taste buds:

1. About 12 very large, circular **vallate (circumvallate) papillae** (VAL-āt=wall-like) form an inverted V-shaped row at the back of the tongue. Each of these papillae houses 100–300 taste buds.
2. **Fungiform papillae** (FUN-ji-form=mushroomlike) are mushroom-shaped elevations scattered over the entire surface of the tongue that contain about five taste buds each.
3. **Foliate papillae** (FŌ-lē-āt=leaflike) are located in small trenches on the lateral margins of the tongue, but most of their taste buds degenerate in early childhood.

In addition, the entire surface of the tongue has **filiform papillae** (FIL-i-form=threadlike). These pointed, threadlike structures contain tactile receptors but no taste buds. They increase friction between the tongue and food, making it easier for the tongue to move food in the oral cavity.

Chemicals that stimulate gustatory receptor cells are known as **tastants**. Once a tastant is dissolved in saliva, it can make contact with the gustatory hairs, ultimately triggering nerve impulses in the first-order sensory neurons that synapse with gustatory receptor cells.

The Gustatory Pathway

Three cranial nerves contain axons of sensory neurons from taste buds (Figure 22.2e). The facial (VII) nerve serves the anterior two-thirds of the tongue, the glossopharyngeal (IX) nerve serves the posterior one-third of the tongue, and the vagus (X) nerve serves the throat and epiglottis (see Figures 19.20, 19.22, and 19.23 on pages 644–647). From taste buds, impulses propagate along these cranial nerves to the gustatory nucleus in the medulla

(e) Gustatory pathway

LM 200x

LM 700x

(d) Histology of a taste bud from a circumvallate papilla

Taste pore
Gustatory hair
Stratified squamous epithelium

Gustatory receptor cell

Supporting cell
Basal cell

Connective tissue

? Beginning at the gustatory receptor cells, what structures form the gustatory pathway?

oblongata. From the medulla, some axons carrying taste signals project to the limbic system and the hypothalamus; others project to the thalamus. Taste signals that project from the thalamus to the primary gustatory area in the parietal lobe of the cerebral cortex (see area 43 in Figure 19.15 on page 636) give rise to the conscious perception of taste.

◉ **CHECKPOINT**
3. How do olfactory receptors and gustatory receptor cells differ in structure and function?
4. Compare the olfactory and gustatory pathways.

VISION

◉ **OBJECTIVES**
- List and describe the accessory structures of the eye and the structural components of the eyeball.
- Describe the neural pathway for vision.

Vision is extremely important to human survival. More than half the sensory receptors in the human body are located in the eyes, and a large part of the cerebral cortex is devoted to processing visual information. In this section of the chapter, we examine the accessory structures of the eye, the eyeball itself, the formation of visual images, and the visual pathway from the eye to the brain.

Accessory Structures of the Eye

The **accessory structures of the eye** include the eyelids, eyelashes, eyebrows, lacrimal (tearing) apparatus, and extrinsic eye muscles.

Eyelids

The upper and lower **eyelids**, or **palpebrae** (PAL-pe-brē; singular is *palpebra*), shade the eyes during sleep, protect the eyes from excessive light and foreign objects, and spread lubricating secretions over the eyeballs (Figure 22.3; see also Figure 12.3 on page 417). The upper eyelid is more movable than the lower and contains in its superior region the **levator palpebrae superioris muscle**. Sometimes a person may experience an annoying *twitch* in an eyelid, an involuntary quivering similar to muscle twitches in the hand, forearm, leg, or foot. Twitches are almost always harmless and usually last for only a few seconds. They are often associated with stress and fatigue. The space between the upper and lower eyelids that exposes the eyeball is the **palpebral fissure** (PAL-pe-bral). Its angles are known as the **lateral commissure** (KOM-i-shur), which is narrower and closer to the temporal bone, and the **medial commissure**, which is broader and nearer the nasal bone. In the medial commissure is a small, reddish elevation, the **lacrimal caruncle** (KAR-ung-kul), which contains sebaceous (oil) glands and sudoriferous (sweat) glands. The whitish material that sometimes collects in the medial commissure comes from these glands.

From superficial to deep, each eyelid consists of epidermis, dermis, subcutaneous tissue, fibers of the orbicularis oculi muscle, a tarsal plate, tarsal glands, and conjunctiva (Figure 22.3). The **tarsal plate** is a thick fold of connective tissue that gives form and support to the eyelids. Embedded in each tarsal plate is a row of elongated modified sebaceous glands, known as **tarsal** or **Meibomian glands** (mī-BŌ-mē-an), that secrete a fluid that helps keep the eyelids from adhering to each other. Infection of the tarsal glands produces a tumor or cyst on the eyelid called a **chalazion** (ka-LĀ-zē-on=small bump). The **conjunctiva** (kon′-

junk-TĪ-va) is a thin, protective mucous membrane composed of nonkeratinized stratified columnar epithelium with numerous goblet cells that is supported by areolar connective tissue. The **palpebral conjunctiva** lines the inner aspect of the eyelids, and the **bulbar conjunctiva** passes from the eyelids onto the surface of the eyeball. Over the sclera (the "white" of the eye) the conjunctiva is vascular, but over the cornea (a transparent region that forms the outer anterior surface of the eyeball) it loses its connective tissue component and consists of clear avascular epithelium. Both the sclera and the cornea will be discussed in more detail shortly. Dilation and congestion of the blood vessels of the bulbar conjunctiva due to local irritation or infection are the cause of **bloodshot eyes**.

Eyelashes and Eyebrows

The **eyelashes**, which project from the border of each eyelid, and the **eyebrows**, which arch transversely above the upper eyelids, help protect the eyeballs from foreign objects, perspiration, and the direct rays of the sun. Sebaceous glands at the base of the hair follicles of the eyelashes, called **sebaceous ciliary glands**, release a lubricating fluid into the follicles. Infection of these glands, usually by bacteria, cause a painful, pus-filled swelling called a **sty**.

The Lacrimal Apparatus

The **lacrimal apparatus** (*lacrim-*=tears) is a group of structures that produces and drains **lacrimal fluid** or **tears**. The **lacrimal glands** are each about the size and shape of an almond. They are supplied by parasympathetic fibers of the facial (VII) nerves and secrete lacrimal fluid, which drains into 6–12 **lacrimal ducts** that empty tears onto the surface of the conjunctiva of the upper lid (Figure 22.3b). From here the tears pass medially over the anterior surface of the eyeball to enter two small openings called **lacrimal puncta**. Tears then pass into two ducts, the **lacrimal canals**, which lead into the **lacrimal sac** and then into the **nasolacrimal duct**. This duct carries the lacrimal fluid into the nasal cavity just inferior to the inferior nasal concha. An infection of the lacrimal sacs is called **dacryocystitis** (dak′-rē-ō-sis-TĪ-tis; *dacryo-*=lacrimal sac;*-itis*=inflammation of). It is usually caused by a bacterial infection and results in blockage in the nasolacrimal ducts.

Lacrimal fluid is a watery solution containing salts, some mucus, and **lysozyme**, a protective bactericidal enzyme. The fluid protects, cleans, lubricates, and moistens the eyeball. After being secreted, lacrimal fluid is spread medially over the surface of the eyeball by the blinking of the eyelids. Each gland produces about 1 mL of lacrimal fluid per day.

Normally, tears are cleared away as fast as they are produced, either by evaporation or by passing into the lacrimal canals and then into the nasal cavity. If an irritating substance makes contact with the conjunctiva, however, the lacrimal glands are stimulated to oversecrete, and tears accumulate (*watery eyes*). This is a protective mechanism, as the tears dilute and wash away the irritating substance. Watery eyes also occur when an inflammation of the nasal mucosa, such as occurs with a cold, obstructs the nasolacrimal ducts and blocks drainage of tears. Humans are unique in expressing emotions, both happiness and sadness, by **crying**. In response to parasympathetic stimulation, the lacrimal glands produce excessive lacrimal fluid that may spill over the edges of the eyelids and even fill the nasal cavity with fluid. Thus, crying often produces a runny nose.

Extrinsic Eye Muscles

Positioned at the front of the face, the eyes sit in the bony depressions of the skull called the *orbits*. The orbits help protect the eyes, stabilize them in three-dimensional space, and anchor them to the muscles that produce their essential movements. The ex-

Figure 22.3 Accessory structures of the eye.

Accessory structures of the eye include the eyelids, eyelashes, eyebrows, lacrimal apparatus, and extrinsic eye muscles.

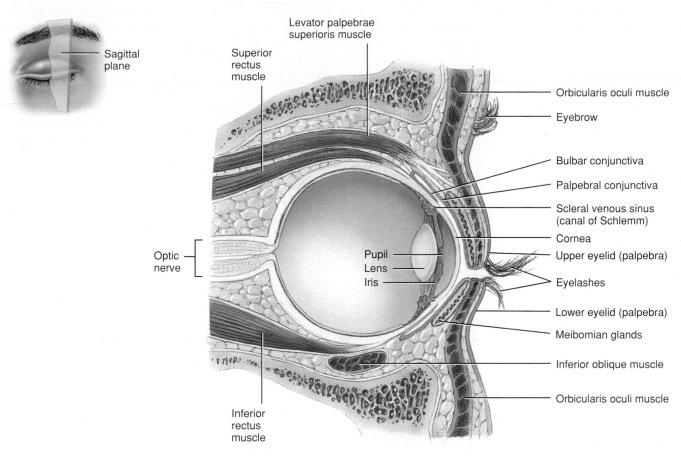

(a) Sagittal section of eye and its accessory structures

(b) Anterior view of the lacrimal apparatus

? What is lacrimal fluid, and what are its functions?

trinsic eye muscles extend from the walls of the orbit to the sclera (white) of the eye and are surrounded in the orbit by a significant quantity of **periorbital fat**. These muscles are capable of moving the eye in almost any direction. Six extrinsic eye muscles move each eye: the **superior rectus**, **inferior rectus**, **lateral rectus**, **medial rectus**, **superior oblique**, and **inferior oblique** (see Figures 22.3a and 22.4). They receive innervation from cranial nerves III, IV, and VI. In general, the motor units in these mus-

cles are small. Some motor neurons serve only two or three muscle fibers—fewer than in any other part of the body except the larynx (voice box). Such small motor units permit smooth, precise, and rapid movement of the eyes. As indicated in Exhibit 11.2 on page 324, the extrinsic eye muscles move the eyeball laterally, medially, superiorly, and inferiorly. For example, looking to the right requires simultaneous contraction of the lateral rectus of the right eyeball and medial rectus of the left eyeball and relaxation of the lateral rectus of the left eyeball and medial rectus of the right eyeball. The oblique muscles preserve rotational stability of the eyeball. Circuits in the brain stem and cerebellum coordinate and synchronize the movements of the eyes.

The surface anatomy of the accessory structures of the eye and the eyeball may be reviewed in Figure 12.3 on page 417.

Anatomy of the Eyeball

The adult **eyeball** measures about 2.5 cm (1 in.) in diameter. Of its total surface area, only the anterior one-sixth is exposed; the remainder is recessed and protected by the orbit, into which it fits. Anatomically, the wall of the eyeball consists of three layers: (1) fibrous tunic, (2) vascular tunic, and (3) retina.

Fibrous Tunic

The **fibrous tunic**, the outer layer of the eyeball, is a strong, dense collagenous connective tissue layer (Figure 22.4). The greater part of this tunic forms a tough outer covering, the **sclera** (SKLE-ra; *sclera*=hard), which forms the visible white part of the

eye. The sclera provides protection for the more delicate internal structures of the eye, while helping maintain the shape of the eyeball against the pressure in the eye. It also serves as the important site of muscle attachment for the extrinsic muscles of the eye, allowing the eye to undergo the critical movements necessary for proper vision. At its anterior surface the outer tunic becomes transparent and has a more exaggerated curvature. This dome-like region, called the **cornea** (KOR-nē-a), is superficial to the colored iris and allows light to enter the interior of the eye. Its curved surface is the major refractory structure for incoming light. The outer surface of the cornea is covered by and continuous with the same nonkeratinized stratified squamous epithelium of the bulbar conjunctiva. While both the sclera and cornea consist of dense collagenous connective tissue, the sclera is opaque and the cornea is transparent because the corneal tissue is avascular and has regularly spaced collagen fibers that are smaller than the wavelength of light. At the corneoscleral junction a hollow channel forms within the tissue. This channel, the **scleral venous sinus (canal of Schlemm)**, connects with the aqueous fluid in the anterior aspect of the eye and veins within the sclera. It plays

Figure 22.4 Anatomy of the eyeball.

The wall of the eyeball consists of three layers: the fibrous tunic, the vascular tunic, and the retina.

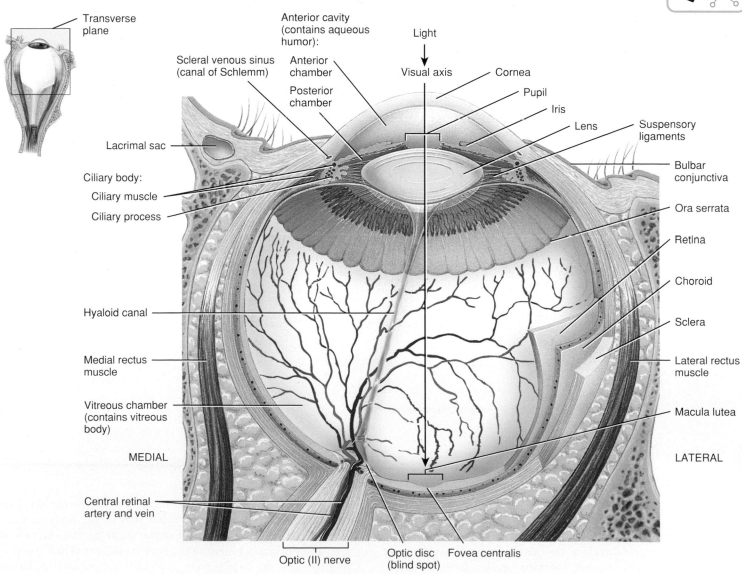

Superior view of transverse section of right eyeball

? What are the components of the fibrous tunic and vascular tunic?

an important role in the drainage of the aqueous fluid (see Figure 22.8).

Vascular Tunic

The **vascular tunic** or **uvea** (Ū-vē-a) is the middle layer of the eyeball. It is composed of three parts: choroid, ciliary body, and iris (Figure 22.4). The highly vascularized **choroid** (KŌ-royd), which is the posterior portion of the vascular tunic, lines most of the internal surface of the sclera. Its numerous blood vessels provide nutrients to the posterior surface of the retina. The choroid also contains melanocytes that produce the pigment melanin, which causes this layer to appear dark brown in color. Melanin in the choroid absorbs stray light rays, which prevents reflection and scattering of light within the eyeball. As a result, the image cast on the retina by the cornea and lens remains sharp and clear. Albinos lack melanin in all parts of the body, including the eye. They often need to wear sunglasses, even indoors, because even moderately bright light is perceived as bright glare due to light scattering.

In the anterior portion of the vascular tunic, the choroid becomes the **ciliary body** (SIL-ē-ar'-ē). It extends from the **ora serrata** (Ō-ra ser-RĀ-ta), the jagged anterior margin of the retina, to a point just posterior to the junction of the sclera and cornea. Like the choroid, the ciliary body appears dark brown in color because it contains melanin-producing melanocytes. In addition, the ciliary body consists of ciliary processes and ciliary muscle. The **ciliary processes** are protrusions or folds on the internal surface of the ciliary body. They contain blood capillaries that secrete aqueous humor. Extending from the ciliary process are **zonular fibers (suspensory ligaments)** that attach to the lens. The smooth muscle of the ciliary body is called the **ciliary muscle**. The ciliary muscle with its complex arrangement of longitudinal, oblique, and circularly oriented fibers produces a narrowing or sphincter action of the ringlike ciliary body. Contraction or relaxation of the ciliary muscle changes the tightness of the zonular fibers, which alters the shape of the lens, adapting it for near or far vision.

The **iris** (=rainbow), the colored portion of the eyeball, is shaped like a flattened donut and its opening (the donut hole) is called the **pupil**. It is suspended between the cornea and the lens as an anterior projection of the ciliary body. It consists of melanocytes and circular and radial smooth muscle fibers. The amount of melanin in the iris determines the eye color. The eyes appear brown to black when the iris contains a large amount of melanin, blue when melanin concentration is very low, and green when melanin concentration is moderate.

The two distinct fiber arrangements of the smooth muscle form the **sphincter pupillae muscle**, which is a flat, thin band of circularly oriented muscle fibers at the pupillary boundary of the iris and the **dilator pupillae muscle**, which attaches to the outer circumference of the sphincter pupillae and projects like the spokes of a wheel toward the base of the iris. A principal function of the iris is to regulate the amount of light entering the eyeball through the pupil. The pupil appears black because, as you look through the lens, you see the heavily pigmented back of the eye (choroid and retina). However, if bright light is directed into the pupil, the reflected light is red because of the blood vessels on the surface of the retina. It is for this reason that a person's eyes appear red in a photograph ("red eye") when the flash is directed into the pupil. Autonomic reflexes regulate pupil diameter in response to light levels (Figure 22.5). When bright light stimulates the eye, parasympathetic fibers of the oculomotor (III) nerve stimulate the sphincter pupillae muscle of the iris to contract, causing a decrease in the size of the pupil (constriction). In dim

Figure 22.5 Responses of the pupil to light of varying brightness.

 Contraction of the circular muscles causes constriction of the pupil; contraction of the radial muscles causes dilation of the pupil.

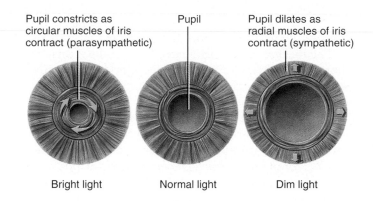

Pupil constricts as circular muscles of iris contract (parasympathetic)

Pupil

Pupil dilates as radial muscles of iris contract (sympathetic)

Bright light   Normal light   Dim light

? Which division of the autonomic nervous system causes pupillary constriction? Which causes pupillary dilation?

light, sympathetic neurons stimulate the dilator pupillae muscle of the iris to contract, causing an increase in the pupil's size (dilation).

Retina

The third and inner coat of the eyeball, the **retina**, lines the posterior three-quarters of the eyeball and is the beginning of the visual pathway (see Figure 22.4). The anatomy of this layer can be viewed with an *ophthalmoscope* (of-THAL-mō-skōp; *ophthalmos-* =eye; *-skopeo*=to examine), an instrument that shines light into the eye and allows an observer to peer through the pupil, providing a magnified image of the retina and its blood vessels as well as the optic (II) nerve (Figure 22.6). The surface of the retina is the only place in the body where blood vessels can be viewed directly and examined for pathological changes, such as those that occur with hypertension, diabetes mellitus, cataracts, and age-related macular disease. Several landmarks are visible through an ophthalmoscope. The **optic disc** is the site where the optic (II) nerve

Figure 22.6 A normal retina, as seen through an ophthalmoscope.

 Blood vessels in the retina can be viewed directly and examined for pathological changes.

NASAL SIDE   TEMPORAL SIDE

Optic disc   Macula lutea

Fovea centralis

Retinal blood vessels

Left eye

? Evidence of what diseases may be seen through an ophthalmoscope?

exits the eyeball. Bundled together with the optic nerve are the **central retinal artery**, a branch of the ophthalmic artery, and the **central retinal vein** (see Figure 22.4). Branches of the central retinal artery fan out to nourish the anterior surface of the retina; the central retinal vein drains blood from the retina through the optic disc. Also visible are the macula lutea and fovea centralis, which are described shortly.

The retina consists of a pigmented layer and a neural layer. The **pigmented layer** is a sheet of melanin-containing epithelial cells located between the choroid and the neural part of the retina. The melanin in the pigmented layer of the retina, like in the choroid, also helps to absorb stray light rays. The **neural (sensory) layer** of the retina is a multilayered outgrowth of the brain that processes visual data extensively before sending nerve impulses into axons that form the optic nerve. Three distinct layers of retinal neurons—the **photoreceptor layer**, the **bipolar cell layer**, and the **ganglion cell layer**—are separated by two zones, the *outer* and *inner synaptic layers*, where synaptic contacts are made (Figure 22.7). Note that light passes through the ganglion and bipolar cell layers and both synaptic layers before it reaches the photoreceptor layer. Two other types of cells present in the bipolar cell layer of the retina are called **horizontal cells** and **amacrine cells.** These cells form laterally directed neural circuits that modify the signals being transmitted along the pathway from photoreceptors to bipolar cells to ganglion cells.

Photoreceptors are specialized to begin the process by which light rays are ultimately converted to nerve impulses. There are two types of photoreceptors: rods and cones. Each retina has about 6 million cones and 120 million rods. **Rods** allow us to see in dim light, such as moonlight. Because they do not provide color vision, in dim light we see only black, white, and all shades of gray in between. Bright lights stimulate the **cones**, which produce color vision. Most of our visual experiences are mediated by the cone system, the loss of which produces legal blindness. A person who loses rod vision mainly has difficulty seeing in dim light and thus should not drive at night. The photoreceptive rods and cones are situated next to the outer pigment layer and their photoreceptive ends face away from the incoming light. They consist of three parts (Figure 22.7): (1) an outer segment, which detects the light stimulus; (2) an inner segment forming the

---

● CLINICAL **CONNECTION** Major Causes of Blindness

Cataracts

A common cause of blindness is a loss of transparency of the lens known as a **cataract** (CAT-a-rakt=waterfall) (Figure A). The lens becomes cloudy (less transparent) due to changes in the structure of the lens proteins. Cataracts often occur with aging but may also be caused by injury, excessive exposure to ultraviolet rays, certain medications (such as long-term use of steroids), or complications of other diseases (for example, diabetes). People who smoke also have increased risk of developing cataracts. Fortunately, sight can usually be restored by surgical removal of the old lens and implantation of a new artificial one.

Glaucoma

Glaucoma (glaw-KŌ-ma) is the most common cause of blindness in the United States, afflicting about 2% of the population over age 40. Glaucoma is an abnormally high intraocular pressure due to a buildup of aqueous humor within the anterior cavity. The fluid compresses the lens into the vitreous body and puts pressure on the neurons of the retina. Persistent pressure results in a progression from mild visual impairment to irreversible destruction of neurons of the retina, damage to the optic nerve, and blindness (Figure B). Glaucoma is painless, and the other eye compensates largely, so a person may experience considerable retinal damage and loss of vision before the condition is diagnosed. Because glaucoma occurs more often with advancing age, regular measurement of intraocular pressure is an increasingly important part of an eye exam as people grow older. Risk factors include race (blacks are more susceptible), increasing age, family history, and past eye injuries and disorders.

Age-Related Macular Disease

Age-related macular disease (AMD), also known as *macular degeneration,* is a degenerative disorder of the retina and pigmented layer in persons 50 years of age and older. In AMD, abnormalities occur in the region of the macula lutea, which is ordinarily the area of most acute vision. Victims of advanced AMD retain their peripheral vision but lose the ability to see straight ahead. For instance, they cannot see facial features to identify a person in front of them (Figure C). AMD is the leading cause of blindness in those over age 75, afflicting 13 million Americans, and is 2.5 times more common in pack-a-day smokers than in nonsmokers. Initially, a person may experience blurring and distortion at the center of the visual field. In "dry" AMD, central vision gradually diminishes because the pigmented layer atrophies and degenerates. There is no effective treatment. In about 10% of cases, dry AMD progresses to "wet" AMD, in which new blood vessels form in the choroid and leak plasma or blood under the retina. Vision loss can be slowed by using laser surgery to destroy leaking blood vessels. •

(A) A scene as it might be viewed by a person with a cataract.

(B) A scene as it might be viewed by a person with glaucoma.

(C) A scene as it might be viewed by a person with age-related macular degeneration.

Figure 22.7 Microscopic structure of the retina. Eventually, nerve impulses arise in ganglion cells and propagate along their axons, which make up the optic (II) nerve.

In the retina, visual signals pass from photoreceptors to bipolar cells to ganglion cells.

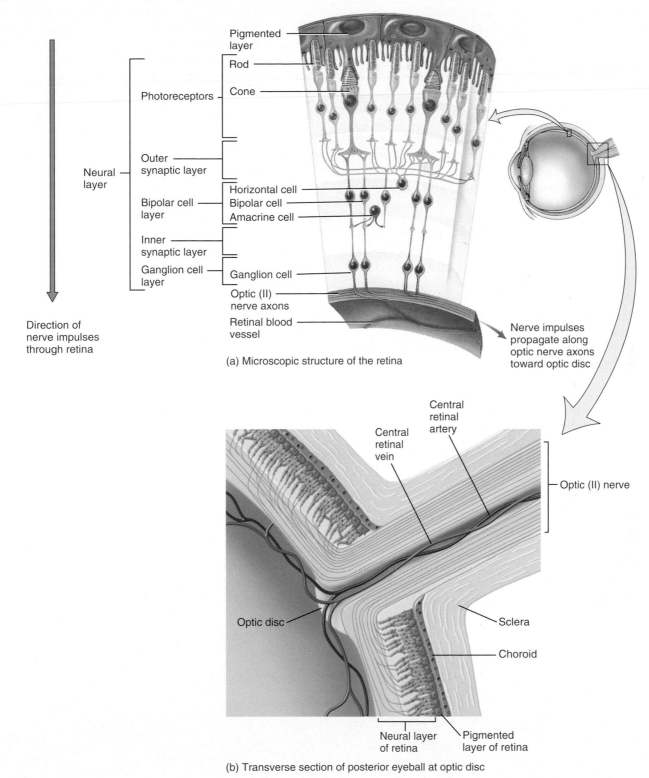

(a) Microscopic structure of the retina

(b) Transverse section of posterior eyeball at optic disc

? What are the two types of photoreceptors, and how do their functions differ?

midregion of the cell, which contains the metabolic machinery; and (3) a synaptic terminal, which lies closest to the eye's interior, facing the bipolar neurons. The outer segment, which is rod-shaped in rods and cone-shaped in cones, is composed of stacked, flattened, membranous discs containing an abundance of **photopigment** molecules. Over a billion of these molecules may be packed into the outer segment of each photoreceptor.

The **macula lutea** (MAK-ū-la LOO-tē-a; *macula*=a small, flat spot; *lute-*=yellowish) is in the exact center of the posterior portion of the retina, at the visual axis of the eye (see Figure 22.6). The **fovea centralis** (see Figure 22.4), a small depression in the center of the macula lutea, contains only cones. In addition, the layers of bipolar and ganglion cells, which scatter light to some extent, do not cover the cones here; these layers are displaced to the periphery of the fovea centralis. As a result, the fovea centralis is the area of highest **visual acuity** or **resolution** (sharpness of vision). A main reason that you move your head and eyes while looking at something is to place images of interest on your fovea centralis—as you do to read the words in this sentence! Rods are absent in the fovea centralis and are more plentiful toward the periphery of the retina. Because rod vision is more sensitive than cone vision, you can see a faint object (such as a dim star) better if you gaze slightly to one side rather than looking directly at it.

From photoreceptors, information flows through the outer synaptic layer to bipolar cells and then from bipolar cells through the inner synaptic layer to ganglion cells. The axons of ganglion cells extend posteriorly to the optic disc and exit the eyeball as the optic (II) nerve. The optic disc is also called the **blind spot**. Because it contains no rods or cones, we cannot see an image that strikes the blind spot. Normally, you are not aware of having a blind spot, but you can easily demonstrate its presence. Hold this page about 20 in. from your face with the cross shown below directly in front of your right eye. You should be able to see the cross and the square when you close your left eye. Now, keeping your left eye closed, slowly bring the page closer to your face while keeping your right eye focused on the cross. At a certain distance the square will disappear from your field of vision because its image falls on the blind spot.

Lens

Posterior to the pupil and iris, within the cavity of the eyeball, is the lens (see Figure 22.4). Within the cells of the lens, proteins called *crystallins*, arranged like the layers of an onion, make up the refractive media of the lens, which normally is perfectly transparent and lacks blood vessels. It is enclosed by a clear connective tissue capsule and held in position by encircling zonular fibers, which attach to the ciliary processes. The lens helps focus images on the retina to facilitate clear vision.

Interior of the Eyeball

In addition to the cornea and the lens, other features of the eye transmit and refract the light to the retina. The interior of the eye consists of two fluid-filled cavities, separated by the lens, each of

Figure 22.8 **The anterior and posterior chambers of the eye, seen in a section through the anterior portion of the eyeball at the junction of the cornea and sclera.** Arrows indicate the direction of flow of aqueous humor.

The lens separates the posterior chamber of the anterior cavity from the vitreous chamber.

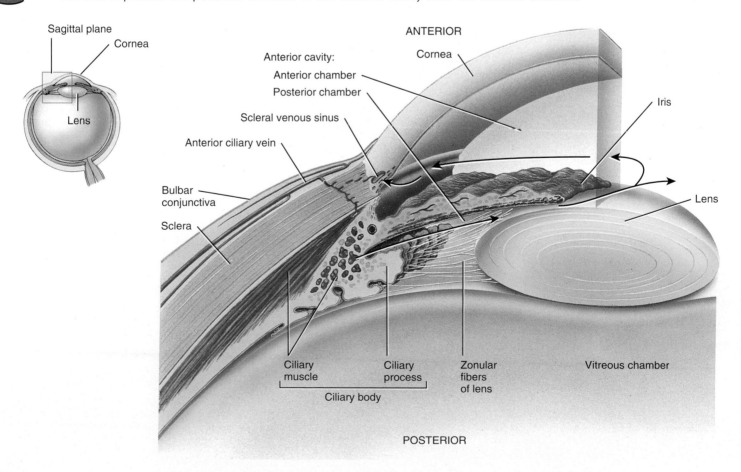

Where is aqueous humor produced, what is its circulation path, and where does it drain from the eyeball?

which is transparent to permit light to pass through the eye from the cornea to the retina. The **anterior cavity**, between the cornea and lens, contains a clear watery fluid, the **aqueous humor** (*aqua*=water). The anterior cavity consists of two chambers. The **anterior chamber** lies between the cornea and the iris, and the **posterior chamber** lies behind the iris and in front of the zonular fibers and lens (Figure 22.8 on page 708). Aqueous humor continually filters out of blood capillaries in the ciliary processes of the ciliary body and enters the posterior chamber. It then flows forward between the iris and the lens, through the pupil, and into the anterior chamber. From the anterior chamber, aqueous humor drains into the scleral venous sinus (canal of Schlemm) and then into the blood. Normally, aqueous humor is completely replaced about every 90 minutes.

The pressure in the eye, called **intraocular pressure**, is produced mainly by the aqueous humor and partly by the vitreous body (described shortly); normally it is about 16 mmHg (millimeters of mercury). The intraocular pressure maintains the shape of the eyeball and prevents the eyeball from collapsing. Puncture wounds to the eyeball may cause the loss of aqueous humor and the vitreous body. This can cause a decrease in intraocular pressure, a detached retina, and even blindness.

The larger posterior cavity of the eyeball is the **vitreous chamber**, which lies between the lens and the retina. Within the vitreous chamber is the **vitreous body**, a jellylike substance that contributes to intraocular pressure. The vitreous body, which occupies approximately four-fifths of the eyeball and plays an important role in maintaining the spherical shape of the eye, is an oval structure with an anterior depression that houses the lens and is firmly affixed to all neighboring structures. Formed during embryonic life, this colorless body consists of approximately 99% water interspersed with collagen fibers and hyaluronic acid. Structurally the vitreous body has a gel-like periphery surrounding a fluid center. Around the perimeter of the lens, the outer layer of the vitreous body forms the fibrous suspension network of the lens, the zonular fibers (see Figure 22.4). The pressure of the vitreous body holds the retina flush against the choroid, so that the retina provides an even surface for the reception of clear images. The vitreous body also contains phagocytic cells that remove debris, keeping this part of the eye clear for unobstructed vision. Occasionally, collections of debris may cast a shadow on the retina and create the appearance of specks, hairs, and fine strings that dart in and out of the field of vision. These *vitreal floaters* result from changes in the vitreous body related to aging and from certain conditions such as nearsightedness and inflammation. Vitreal floaters are usually harmless and do not require treatment. The **hyaloid canal** (see Figure 22.4) is a narrow channel that is inconspicuous in adults and runs through the vitreous body from the optic disc to the posterior aspect of the lens. In the fetus, this channel is occupied by the hyaloid artery (see Figure 22.19).

Table 22.1 summarizes the structures associated with the eyeball.

TABLE 22.1

Summary of the Structures of the Eyeball

STRUCTURE	FUNCTION	STRUCTURE	FUNCTION
Fibrous tunic	*Cornea:* Admits and refracts (bends) light. *Sclera:* Provides shape and protects inner parts.	**Lens**	Refracts light.
Vascular tunic	*Iris:* Regulates amount of light that enters eyeball. *Ciliary body:* Secretes aqueous humor and alters shape of lens for near or far vision. *Choroid:* Provides blood supply and absorbs scattered light.	**Anterior cavity**	Contains aqueous humor that helps maintain shape of eyeball and supplies oxygen and nutrients to lens and cornea.
Retina	Receives light and ultimately converts it into nerve impulses. Output to brain is via axons of ganglion cells, which form the optic (II) nerve.	**Vitreous chamber**	Contains vitreous body that helps maintain shape of eyeball and keeps the retina attached to the choroid.

The Visual Pathway

After considerable processing of visual signals at synapses among the various types of neurons in the neural layer of the retina, the axons of retinal ganglion cells provide output from the retina to the brain. They exit the eyeball as the optic (II) nerve (see Figure 22.7).

Processing of Visual Input in the Retina

Within the retina, certain features of visual input are enhanced while other features may be discarded. Input from several cells may either converge upon a smaller number of postsynaptic neurons or diverge to a larger number. On the whole, convergence predominates: 1 million ganglion cells receive input from about 126 million photoreceptor cells.

Chemicals (neurotransmitters) released by rods and cones induce changes in both bipolar cells and horizontal cells that lead to the generation of nerve impulses (see Figure 22.7). Amacrine cells synapse with ganglion cells and also transmit information to them. When bipolar, horizontal, or amacrine cells transmit signals to ganglion cells, the ganglion cells initiate nerve impulses.

Pathway in the Brain

The axons of the optic (II) nerve pass through the **optic chiasm** (kī-AZM=a crossover, as in the letter X), a crossing point of the optic nerves (Figure 22.9). Some fibers cross to the opposite side; others remain uncrossed. After passing through the optic chiasm, the fibers, now part of the **optic tract**, enter the brain and terminate in the lateral geniculate nucleus of the thalamus. Here they synapse with neurons whose axons form the **optic radiations**, which project to the primary visual areas in the occipital lobes of the cerebral cortex (area 17 in Figure 19.15 on page 636).

Figure 22.9 The visual pathway. Partial dissection of the brain in part (a) reveals the optic radiations (axons extending from the thalamus to the occipital lobe), along with the rest of the visual pathway, which is diagrammed in part (b).

🔑 The optic chiasm is the crossing point of the optic nerves.

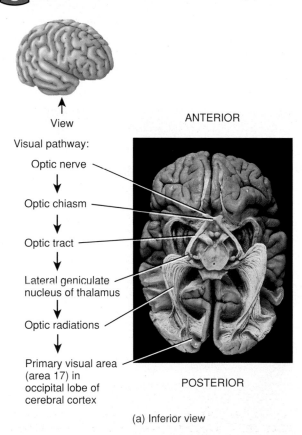

View

Visual pathway:

Optic nerve → Optic chiasm → Optic tract → Lateral geniculate nucleus of thalamus → Optic radiations → Primary visual area (area 17) in occipital lobe of cerebral cortex

ANTERIOR

POSTERIOR

(a) Inferior view

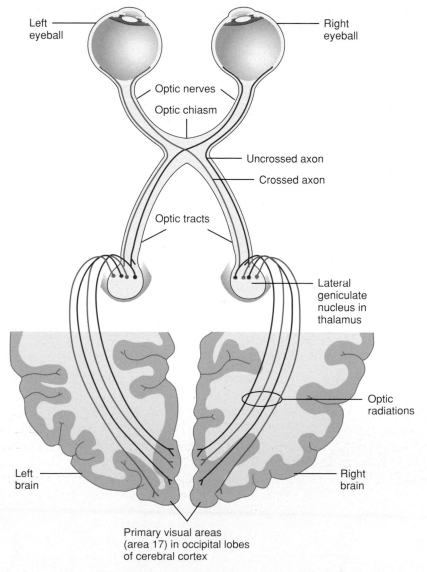

Left eyeball

Right eyeball

Optic nerves

Optic chiasm

Uncrossed axon

Crossed axon

Optic tracts

Lateral geniculate nucleus in thalamus

Optic radiations

Left brain

Right brain

Primary visual areas (area 17) in occipital lobes of cerebral cortex

(b) Superior view of transverse section through eyeballs and brain

❓ Where does the optic tract terminate?

● CHECKPOINT

5. Describe the structure and importance of the eyelids, eyelashes, and eyebrows.
6. What is the function of the lacrimal apparatus?
7. What are the components of the fibrous tunic, vascular tunic, and retina, and what are their functions?
8. Describe the histology of the neural portion of the retina.
9. Describe the pathway for vision, beginning with light entering the eye and ending in the brain.

HEARING AND EQUILIBRIUM

● OBJECTIVES

• Describe the anatomy of the structures in the three principal regions of the ear.
• List the principal events involved in hearing.
• Identify the receptor organs for equilibrium, and describe how they function.
• Describe the auditory and equilibrium pathways.

The ear is an engineering marvel because its sensory receptors can transduce sound vibrations with amplitudes as small as the diameter of an atom of gold (0.3 nm) into electrical signals 1000 times faster than photoreceptors can respond to light. Besides receptors for sound waves, the ear also contains receptors for equilibrium.

Anatomy of the Ear

The **ear** is divided into three main regions: (1) the external ear, which collects sound waves and channels them inward; (2) the middle ear, which conveys sound vibrations to the oval window; and (3) the internal ear, which houses the receptors for hearing and equilibrium.

External (Outer) Ear

The **external (outer) ear** consists of the auricle, external auditory canal, and tympanic membrane (eardrum) (Figure 22.10). The **auricle (pinna)** is a flap of elastic cartilage shaped like the flared end of a trumpet and covered by skin. The rim of the auricle is the **helix**; the inferior portion is the **lobule**. Ligaments and muscles attach the auricle to the head. The **external auditory canal** (*audit-*=hearing) is a curved tube about 2.5 cm (1 in.) long that lies in the cartilage of the auricle and the temporal bone and leads to the tympanic membrane. The **tympanic membrane** (tim-PAN-ik; *tympan-*=a drum) or **eardrum** is a thin, semitransparent partition between the external auditory canal and middle ear. It attaches by a fibrocartilaginous ring to the temporal bone at the base of the external acoustic meatus and stretches across the opening as a three-layered window. The three layers are, from outside to inside, a covering of epidermis, a middle layer of dense connective tissue containing collagen, elastic fibers, and fibroblasts, and a lining of simple cuboidal epithelium. Between

Figure 22.10 Anatomy of the ear.

The ear has three principal regions: the external (outer) ear, the middle ear, and the internal (inner) ear.

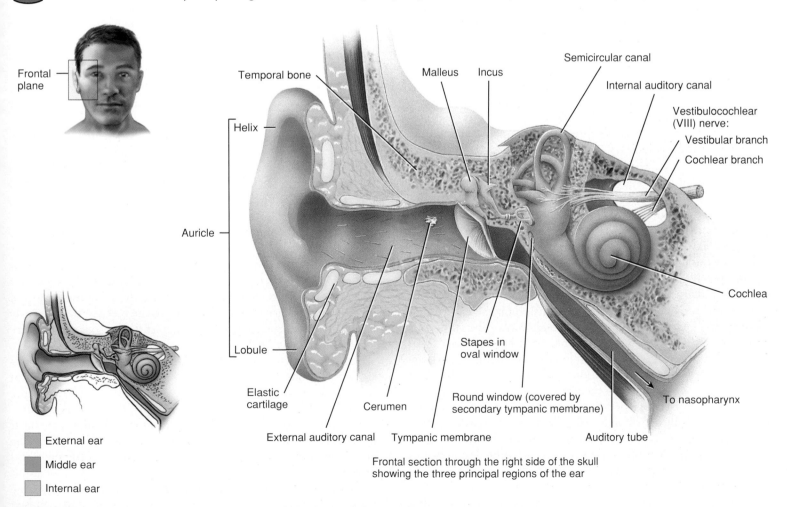

Frontal section through the right side of the skull showing the three principal regions of the ear

? To which structure of the external ear does the malleus of the middle ear attach?

the epithelial layers is connective tissue composed of collagen, elastic fibers, and fibroblasts. The tympanic membrane is somewhat convex toward the middle ear cavity with the apex forming the **umbo**. Attached along the superior inner surface of the tympanic membrane, to the point of the umbo, is the first middle ear ossicle, the malleus. The handle of this bone can be easily viewed through an **otoscope** (oto-=ear; -skopeo=to view), a viewing instrument that illuminates and magnifies the external ear and tympanic membrane. Tearing of the tympanic membrane is called a **perforated eardrum**. It may be due to pressure from a cotton swab, trauma, or a middle ear infection, and usually heals within a month.

Near the exterior, the external auditory canal contains a few hairs and specialized sebaceous (oil) glands called **ceruminous glands** (se-ROO-mi-nus) that secrete earwax or **cerumen** (se-ROO-men). The combination of hairs and cerumen helps prevent dust and foreign objects from entering the ear. Cerumen also prevents damage by water and adventurous insects to the delicate skin of the external auditory canal. Cerumen usually dries up and falls out of the ear canal. Some people, however, produce a large amount of cerumen, which can become impacted and can muffle incoming sounds. The treatment for **impacted cerumen** is usually periodic ear irrigation or removal of wax with a blunt instrument; the latter should be attempted only by trained medical personnel.

Middle Ear

The **middle ear** is a small, air-filled cavity in the petrous portion of the temporal bone that is lined by epithelium (Figure 22.11). It is separated from the external ear by the tympanic membrane and from the internal ear by a thin bony partition that contains two small membrane-covered openings: the oval window and the round window. Extending across the middle ear and attached to it by ligaments are the three smallest bones in the body, the **auditory ossicles** (OS-si-kuls), which are connected to one another by synovial joints. The bones, named for their shapes, are the malleus, incus, and stapes—commonly called the hammer, anvil, and stirrup, respectively. The "handle" of the **malleus** (MAL-ē-us) attaches to the internal surface of the tympanic membrane. The head of the malleus articulates with the body of the incus. The **incus** (ING-kus), the middle bone in the series, articulates with the head of the stapes. The base or footplate of the **stapes** (STĀ-pēz) fits into the **oval window (fenestra vestibuli)**. Directly below the oval window is another opening, the **round window (fenestra cochlea)**, which is enclosed by a membrane called the **secondary tympanic membrane**.

Besides the ligaments, two tiny skeletal muscles also attach to the ossicles (Figure 22.11). The **tensor tympani muscle**, which is innervated by the mandibular branch of the trigeminal (V) nerve, limits movement and increases tension on the eardrum to prevent damage to the inner ear from loud noises. The **stapedius muscle** (sta-PĒ-dē-us), which is innervated by the facial (VII) nerve, is the smallest of all skeletal muscles. By dampening large vibrations of the stapes due to loud noises, it protects the oval window, but it also decreases the sensitivity of hearing. For this reason, paralysis of the stapedius muscle is associated with **hyperacusia** (abnormally sensitive hearing). Because it takes a fraction of a second for the tensor tympani and stapedius muscles to contract, they can protect the inner ear from prolonged loud noises such as thunder, but not from brief ones such as a gunshot.

The anterior wall of the middle ear contains an opening that leads directly into the **auditory (pharyngotympanic) tube**, commonly known as the **eustachian tube**. The auditory tube, which consists of both bone and elastic cartilage, connects the middle ear with the nasopharynx (superior portion of the throat). It is normally closed at its medial (pharyngeal) end. During swallowing and yawning, it opens, allowing air to enter or leave the middle ear until the pressure in the middle ear equals the atmospheric pressure. When the pressures are balanced, the tympanic membrane vibrates freely as sound waves strike it. If the pressure is not equalized, intense pain, hearing impairment, ringing in the ears, and vertigo could develop. The auditory tube also is a route whereby pathogens may travel from the nose and throat to the middle ear, causing the most common type of ear infection.

Internal (Inner) Ear

The **internal (inner) ear** is also called the **labyrinth** (LAB-i-rinth) because of its complicated series of canals (Figure 22.12 on page 714). Structurally, it consists of two main divisions: an outer bony labyrinth that encloses an inner membranous labyrinth. The **bony labyrinth** is a series of cavities in the petrous part of the temporal bone divided into three areas: (1) the semicircular canals and (2) the vestibule, both of which contain receptors for equilibrium, and (3) the cochlea, which contains receptors for hearing. The bony labyrinth is lined with periosteum and contains **perilymph**. This fluid, which is chemically similar to cerebrospinal fluid, surrounds the **membranous labyrinth**, a series of epithelial sacs and tubes inside the bony labyrinth that have the same general form as the bony labyrinth. The epithelial membranous labyrinth contains **endolymph**. The level of potassium ions (K^+) in endolymph is unusually high for an interstitial fluid, and potassium ions play a role in the generation of auditory signals (described shortly).

The **vestibule** (VES-ti-būl) is the oval central portion of the bony labyrinth. The membranous labyrinth in the vestibule consists of two sacs called the **utricle** (Ū-tri-kl=little bag) and the **saccule** (SAK-ūl=little sac), which are connected by a small duct. Projecting superiorly and posteriorly from the vestibule are the three bony **semicircular canals**, each of which lies at approximately right angles to the other two. Based on their positions, they are named the anterior, posterior, and lateral semicircular canals. The anterior and posterior semicircular canals are vertically oriented; the lateral one is horizontally oriented. At one end of each canal is a swollen enlargement called the **ampulla** (am-PUL-la=saclike duct). The portions of the membranous labyrinth that lie inside the bony semicircular canals are called the **semicircular ducts**. These structures communicate with the utricle of the vestibule.

The vestibular branch of the vestibulocochlear (VIII) nerve consists of *ampullary, utricular,* and *saccular nerves*. These nerves contain both first-order sensory neurons and motor neurons that synapse with receptors for equilibrium. The first-order sensory neurons carry sensory information from the receptors, and the motor neurons carry feedback signals to the receptors, apparently to modify their sensitivity. Cell bodies of the sensory neurons are located in the **vestibular ganglia** (see Figure 22.13b).

Anterior to the vestibule is the **cochlea** (KOK-lē-a=snail-shaped), a bony spiral canal (Figure 22.13a on page 715) that resembles a snail's shell and makes almost three turns around a central bony core called the **modiolus** (mō-DĪ-ō-lus; Figure 22.13b). The cochlea is divided into three channels: cochlear duct, scala vestibuli, and scala tympani (Figure 22.13a–c). The **cochlear duct** or **scala media** is a continuation of the membranous labyrinth and is filled with endolymph. The channel above the cochlear duct is the **scala vestibuli**, which ends at the oval window. The channel below is the **scala tympani**, which ends at the round window. Because both the scala vestibuli and scala tympani are part of the bony labyrinth of the cochlea, these chambers

Figure 22.11 The right middle ear and the auditory ossicles.

Common names for the malleus, incus, and stapes are the hammer, anvil, and stirrup, respectively.

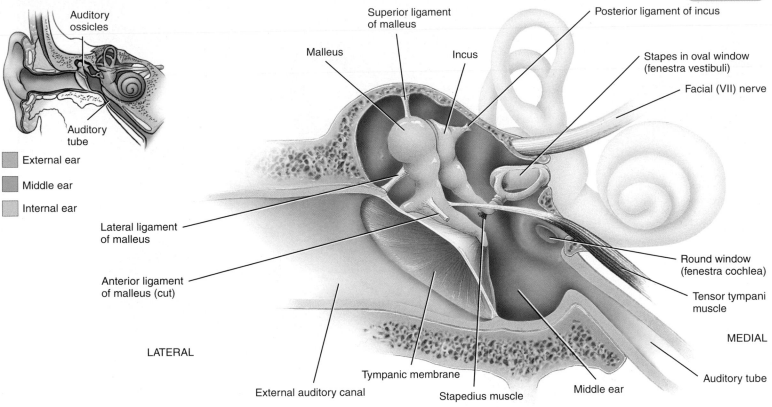

External ear

Middle ear

Internal ear

Auditory ossicles

Auditory tube

Superior ligament of malleus

Malleus

Incus

Posterior ligament of incus

Stapes in oval window (fenestra vestibuli)

Facial (VII) nerve

Lateral ligament of malleus

Anterior ligament of malleus (cut)

LATERAL

External auditory canal

Tympanic membrane

Stapedius muscle

Middle ear

Round window (fenestra cochlea)

Tensor tympani muscle

MEDIAL

Auditory tube

(a) Frontal section showing location of auditory ossicles

Cochlea

Foramen spinosum

Foramen ovale

Internal carotid artery

Superior ligament of malleus

Malleus

Incus

Tensor tympani muscle

Tympanic membrane

Stapes

Round window

Oval window

(b) Medial view of the right middle and inner ear

(c) Posterosuperior view of right middle ear with incus and stapes removed

What structures separate the middle ear from the internal ear?

are filled with perilymph. The scala vestibuli and scala tympani are completely separated by the cochlear duct, except for an opening at the apex of the cochlea, the **helicotrema** (hel-i-kō-TRĒ-ma; see Figure 22.14). The cochlea adjoins the wall of the vestibule, into which the scala vestibuli opens. The perilymph in the vestibule is continuous with that of the scala vestibuli.

The **vestibular membrane** separates the cochlear duct from the scala vestibuli, and the **basilar membrane** separates the

cochlear duct from the scala tympani. Resting on the basilar membrane is the **spiral organ** or **organ of Corti** (Figure 22.13c, d on page 716). The spiral organ is a coiled sheet of epithelial cells, including supporting cells and about 16,000 hair cells, which are the receptors for hearing. There are two groups of hair cells: The *inner hair cells* are arranged in a single row and the *outer hair cells* are arranged in three rows. At the apical tip of each hair cell is a hair bundle, consisting of 40–80 **stereocilia** (microvilli) and a

TEXT CONTINUES ON PAGE **717**

Figure 22.12 **The right internal ear.** The outer, cream-colored area is part of the bony labyrinth; the inner, pink-colored area is the membranous labyrinth.

🔑 The bony labyrinth contains perilymph, and the membranous labyrinth contains endolymph.

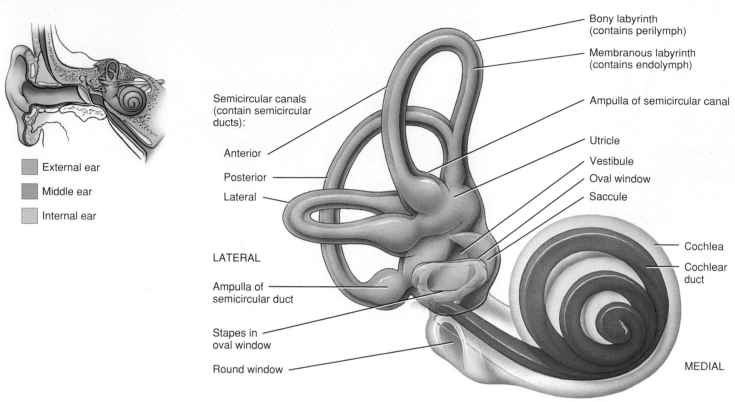

External ear

Middle ear

Internal ear

Semicircular canals (contain semicircular ducts):

Anterior

Posterior

Lateral

LATERAL

Ampulla of semicircular duct

Stapes in oval window

Round window

Bony labyrinth (contains perilymph)

Membranous labyrinth (contains endolymph)

Ampulla of semicircular canal

Utricle

Vestibule

Oval window

Saccule

Cochlea

Cochlear duct

MEDIAL

(a) Components of the right internal ear

Incus

Malleus

Stapes

Cochlea

Semicircular canals

Utricle

(b) Auditory ossicles (left) and cast of internal ear (right) compared to a dime for size

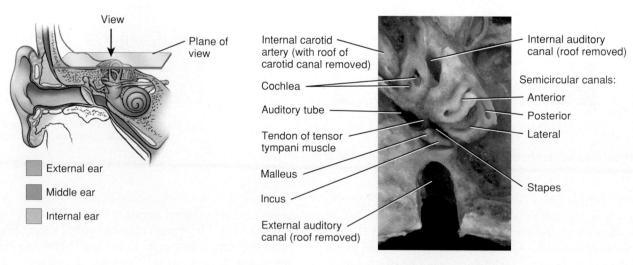

View

Plane of view

External ear

Middle ear

Internal ear

Internal carotid artery (with roof of carotid canal removed)

Cochlea

Auditory tube

Tendon of tensor tympani muscle

Malleus

Incus

External auditory canal (roof removed)

Internal auditory canal (roof removed)

Semicircular canals:

Anterior

Posterior

Lateral

Stapes

(c) Superior view of petrous temporal bone

❓ **What are the names of the two sacs that lie in the vestibule?**

Figure 22.13 **Semicircular canals, vestibule, and cochlea of the right ear.** Note that the cochlea makes almost three complete turns.

The three channels in the cochlea are the scala vestibuli, the scala tympani, and the cochlear duct.

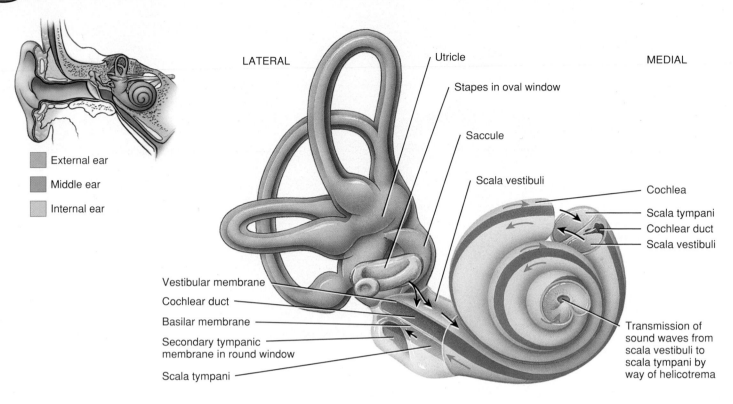

(a) Sections through the cochlea

(b) Components of the vestibulocochlear nerve (cranial nerve VIII)

CONTINUES ▶

⬤ FIGURE 22.13 CONTINUED ▶

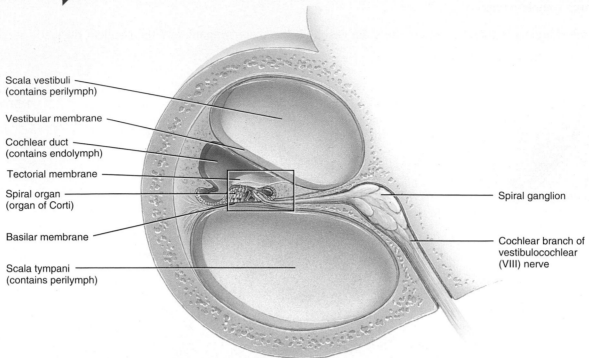

Scala vestibuli
(contains perilymph)

Vestibular membrane

Cochlear duct
(contains endolymph)

Tectorial membrane

Spiral organ
(organ of Corti)

Basilar membrane

Scala tympani
(contains perilymph)

Spiral ganglion

Cochlear branch of
vestibulocochlear
(VIII) nerve

(c) Section through one turn of the cochlea

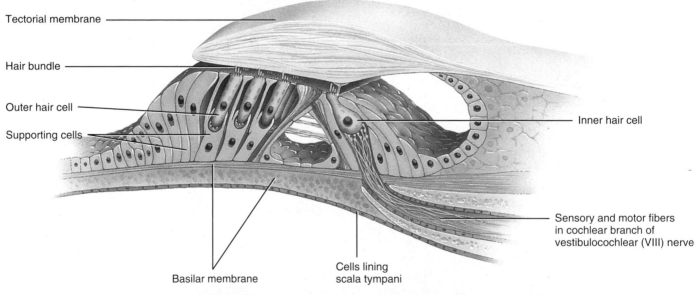

Tectorial membrane

Hair bundle

Outer hair cell

Supporting cells

Inner hair cell

Sensory and motor fibers
in cochlear branch of
vestibulocochlear (VIII) nerve

Basilar membrane

Cells lining
scala tympani

(d) Enlargement of spiral organ (organ of Corti)

Cochlear duct

Tectorial membrane

Outer hair cells

Inner hair cell

Supporting cells

Basilar membrane

Scala tympani

LM 140x

(e) Histology of the spiral organ (organ of Corti)

❓ **What are the three subdivisions of the bony labyrinth?**

kinocilium (cilium) that extend into the endolymph of the cochlear duct. Despite their name, stereocilia are actually long, hairlike microvilli arranged in several rows of graded height.

At their basal ends, inner and outer hair cells synapse both with first-order sensory neurons and with motor neurons from the cochlear branch of the vestibulocochlear (VIII) nerve. Cell bodies of the sensory neurons are located in the **spiral ganglion** (Figure 22.13b, c). Although outer hair cells outnumber them by 3 to 1, the inner hair cells synapse with 90–95% of the first-order sensory neurons in the cochlear nerve that relay auditory information to the brain. By contrast, 90% of the motor neurons in the cochlear nerve synapse with outer hair cells. The **tectorial membrane** (*tector-* =covering), a flexible gelatinous membrane, overlies the hair cells of the spiral organ (Figure 22.13d). The ends of the stereocilia of the hair cells are embedded in the tectorial membrane while the bodies of the hair cells rest on the basilar membrane.

Mechanism of Hearing

Sound waves are a series of alternating high- and low-pressure regions traveling in the same direction through some medium (such as air). They originate from a vibrating object in much the same way that ripples arise and travel over the surface of a pond when you toss a stone into it.

The following events are involved in hearing (Figure 22.14):

❶ The auricle directs sound waves into the external auditory canal.

❷ When sound waves strike the tympanic membrane, the alternating waves of high and low pressure in the air cause the tympanic membrane to vibrate back and forth. The tympanic membrane vibrates slowly in response to low-frequency (low-pitched) sounds and rapidly in response to high-frequency (high-pitched) sounds.

❸ The central area of the tympanic membrane connects to the malleus, which vibrates along with the tympanic membrane. This vibration is transmitted from the malleus to the incus and then to the stapes.

❹ As the stapes moves back and forth, it pushes the membrane of the oval window in and out. The oval window vibrates about 20 times more vigorously than the tympanic membrane because the auditory ossicles efficiently transmit small vibrations spread over a large surface area (the tympanic membrane) into larger vibrations on a smaller surface (the oval window).

❺ The movement of the oval window sets up fluid pressure waves in the perilymph of the cochlea. As the oval window bulges inward, it pushes on the perilymph of the scala vestibuli.

❻ Pressure waves are transmitted from the scala vestibuli to the scala tympani and eventually to the round window, causing it to bulge outward into the middle ear. (See ❾ in the figure.)

❼ As the pressure waves deform the walls of the scala vestibuli and scala tympani, they move into the endolymph inside the cochlear duct.

Figure 22.14 Events in the stimulation of auditory receptors in the right ear. The cochlea has been uncoiled to more easily visualize the transmission of sound waves and their distortion of the vestibular and basilar membranes of the cochlear duct.

The function of hair cells of the spiral organ (organ of Corti) is to ultimately convert a mechanical vibration (stimulus) into an electrical signal (nerve impulse).

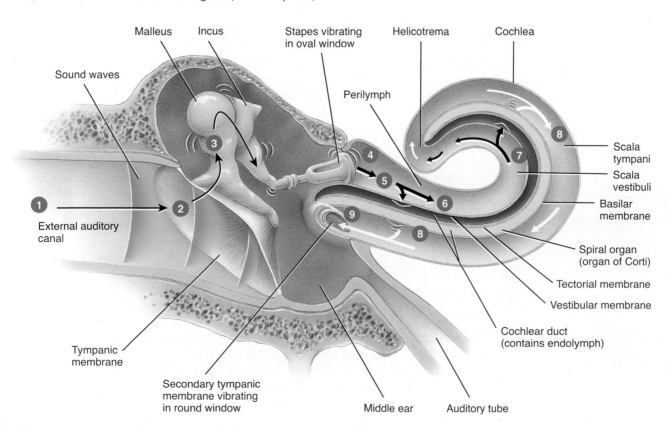

? What structure vibrates and sets up pressure waves in the perilymph?

❽ The pressure waves in the endolymph cause the basilar membrane to vibrate, which moves the hair cells of the spiral organ against the tectorial membrane. This leads to bending of the stereocilia that ultimately leads to the generation of nerve impulses in first-order neurons in cochlear nerve fibers.

❾ Sound waves of various frequencies cause certain regions of the basilar membrane to vibrate more intensely than other regions. Each segment of the basilar membrane is "tuned" for a particular pitch. Because the membrane is narrower and stiffer at the base of the cochlea (closer to the oval window), high-frequency (high-pitched) sounds induce maximal vibrations in this region. Toward the apex of the cochlea, the basilar membrane is wider and more flexible; low-frequency (low-pitched) sounds cause maximal vibration of the basilar membrane there. Loudness is determined by the intensity of sound waves. High-intensity sound waves cause larger vibrations of the basilar membrane, which leads to a higher frequency of nerve impulses reaching the brain. Louder sounds also may stimulate a larger number of hair cells.

Besides its role in detecting sounds, the cochlea has the surprising ability to produce sounds. These usually inaudible sounds, called *otoacoustic emissions*, can be picked up by placing a sensitive microphone next to the eardrum. They are caused by vibrations of the outer hair cells that occur in response to sound waves and to signals from motor neurons. This vibratory behavior appears to change the stiffness of the tectorial membrane and is thought to enhance the movement of the basilar membrane, which amplifies the responses of the inner hair cells. At the same time, the outer hair cell vibrations set up a traveling wave that goes back toward the stapes and leaves the ear as an otoacoustic emission. Detection of these inner ear–produced sounds is a fast, inexpensive, and noninvasive way to screen newborns for hearing defects. In deaf babies, otoacoustic emissions are not produced or are greatly reduced in size.

The Auditory Pathway

Bending of the stereocilia of the hair cells of the spiral organ causes the release of a neurotransmitter (probably glutamine), which generates nerve impulses in the sensory neurons attached to the hair cells. The cell bodies of the sensory neurons are located in the spiral ganglia. Nerve impulses pass along the axons of the neurons, which form the cochlear branch of the vestibulocochlear (VIII) nerve (Figure 22.15). These axons synapse with neurons in the cochlear nuclei in the medulla oblongata on the same side. Some of the axons from the cochlear nuclei decussate (cross over) in the medulla, ascend in a tract called the lateral meniscus on the opposite side, and terminate in the inferior colliculus in the midbrain. Other axons from the cochlear nuclei end in the superior olivary nucleus in the pons on each side. Slight differences in the timing of nerve impulses arriving from the two ears at the superior olivary nuclei allow us to locate the source of a sound. Axons from the superior olivary nuclei also ascend in the lateral meniscus tracts on both sides and end in the inferior collicule. From each inferior colliculus, nerve impulses are conveyed to the medial geniculate nucleus in the thalamus and finally to the primary auditory area of the cerebral cortex in the temporal lobe of the cerebrum (see areas 41 and 42 in Figure 19.15 on page 636). Because many auditory axons decussate in the medulla while others remain on the same side, the right and left primary auditory areas receive nerve impulses from both ears.

Mechanism of Equilibrium

There are two types of **equilibrium** (balance). **Static equilibrium** refers to the maintenance of the position of the body (mainly the head) relative to the force of gravity. **Dynamic equilibrium** is the maintenance of body position (mainly the head) in response to sudden movements such as rotation, acceleration,

---

Figure 22.15 The auditory pathway.

 Since many auditory axons decussate in the medulla while others stay on the same side, the right and left primary auditory areas receive nerve impulses from both ears.

Medial geniculate nucleus in thalamus

Cochlear branch of vestibulocochlear (VIII) nerve

Cochlear nuclei in medulla oblongata

Primary auditory area in cerebral cortex

Inferior colliculus in midbrain

Lateral menisci

Superior olivary nucleus in pons

Cerebellum

? Where are the cell bodies of the first sensory neurons in the auditory pathway located?

and deceleration. Collectively, the receptor organs for equilibrium are called the **vestibular apparatus** (ves-TIB-ū-lar), which includes the saccule, utricle, and semicircular ducts.

Otolithic Organs: Saccule and Utricle

The walls of both the utricle and the saccule contain a small, thickened region called a **macula** (MAK-ū-la; Figure 22.16 on page 720). The two *maculae* (MAK-ū-lē; plural), which are perpendicular to one another, are the receptors for static equilibrium, and contribute to some aspects of dynamic equilibrium as well. Their role in static equilibrium is to provide sensory information on the position of the head in space, essential to maintaining appropriate posture and balance. The maculae detect linear acceleration and deceleration—for example, the sensations you feel while in an elevator or a car that is speeding up or slowing down.

The maculae consist of two kinds of cells: **hair cells**, which are the sensory receptors, and **supporting cells**. Hair cells have on their surface 40–80 *stereocilia* of graded height, which are actually microvilli, plus one *kinocilium*, a conventional cilium anchored firmly to its basal body and extending beyond the longest stereocilium. Collectively, the stereocilia and kinocilium are called a **hair bundle**. Scattered among the hair cells are columnar supporting cells that probably secrete the thick, gelatinous, glycoprotein layer, called the **otolithic membrane**, that rests on the hair cells. A layer of dense calcium carbonate crystals, called **otoliths** (*oto*=ear; *-liths*=stones), extends over the entire surface of the otolithic membrane.

Because the otolithic membrane sits on top of the macula, when you tilt your head forward, the otolithic membrane and otoliths are pulled by gravity and slide downhill over the hair cells in the direction of the tilt, bending the hair bundles. The movement of the hair bundles initiates responses that ultimately lead to the generation of nerve impulses. The hair cells synapse with first-order sensory neurons in the vestibular branch of the vestibulocochlear (VIII) nerve.

Semicircular Ducts

The three semicircular ducts function in dynamic equilibrium. The ducts lie at right angles to one another in three planes (Figure 22.17 on page 721). The two vertical ducts are the anterior and posterior semicircular ducts, and the horizontal one is the lateral semicircular duct (see also Figure 22.12). This positioning permits detection of rotational acceleration or deceleration. In the **ampulla**, the dilated portion of each duct, is a small elevation called the **crista** (KRIS-ta=crest; plural is cristae; KRIS-tē). Each crista contains a group of **hair cells** and **supporting cells** covered by a mass of gelatinous material called the **cupula** (KŪ-pū-la). When the head moves, the attached semicircular ducts and hair cells move with it. The endolymph within the ampulla, however, is not attached and lags behind due to its inertia. As the moving hair cells drag along the stationary fluid, the hair bundles bend. Bending of the hair bundles produces responses that lead to nerve impulses that pass along the ampullary nerve, a branch of the vestibular division of the vestibulocochlear (VIII) nerve.

TEXT CONTINUES ON PAGE 722 ▶

Figure 22.16 Location and structure of receptors in the maculae of the right ear. Both sensory neurons (blue) and motor neurons (red) synapse with the hair cells.

The movement of stereocilia initiates responses that ultimately lead to the generation of nerve impulses.

Key:
— Sensory fiber
— Motor fiber

(a) Overall structure of a section of the macula

(c) Position of macula with head upright (left) and tilted forward (right)

(b) Details of two hair cells

? With which type of equilibrium are the maculae mainly concerned?

Figure 22.17 **Location and structure of the membranous semicircular ducts of the right ear.** Both sensory neurons (blue) and motor neurons (red) synapse with the hair cells. The ampullary nerves are branches of the vestibular division of the vestibulocochlear (VIII) nerve.

The positions of the membranous semicircular ducts permit detection of rotational movements.

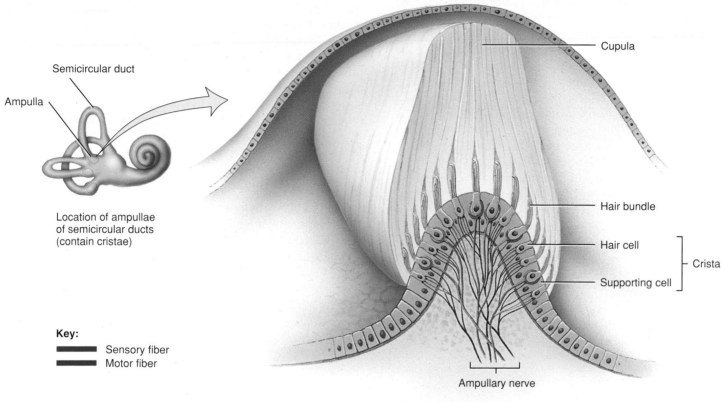

(a) Details of a crista

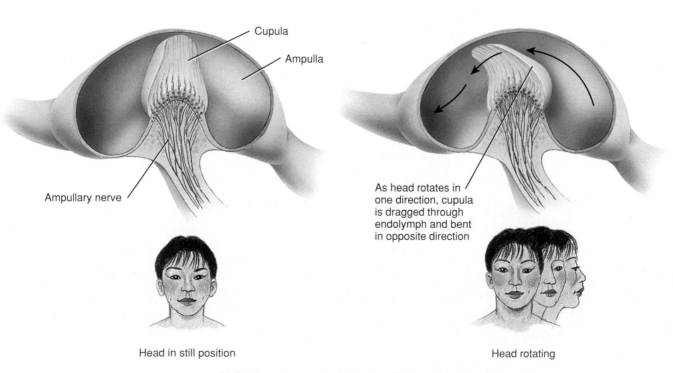

(b) Position of a cupula with the head in the still position (left) and when the head rotates (right)

? **With which type of equilibrium are the membranous semicircular ducts associated?**

Equilibrium Pathways

Bending of hair bundles of the hair cells in the semicircular canals, utricle, or saccule causes the release of a neurotransmitter (probably glutamine), which generates nerve impulses in the sensory neurons attached to hair cells. The cell bodies of the hair cells are located in the vestibular ganglia. Nerve impulses pass along the axons of the neurons, which form the vestibular branch of the vestibulocochlear (VIII) nerve (Figure 22.18). Most of these axons synapse with sensory neurons in vestibular nuclei, the major integrating centers for equilibrium, in the medulla oblongata and pons. The vestibular nuclei also receive input from the eyes and somatic receptors, especially proprioceptors in the neck muscles that indicate the positions of the head. The remaining axons enter the cerebellum through the inferior cerebellar peduncles (see Figure 19.7 on page 623). Bidirectional pathways connect the cerebellum and vestibular nuclei.

The vestibular nuclei integrate information from vestibular, visual, and somatic receptors and then send commands to: (1) the nuclei of cranial nerves—oculomotor (III), trochlear (IV), and abducens (VI)—that control coupled movements of the eyes with those of the head to help maintain focus on the visual field; (2) nuclei of the accessory (XI) nerves to help control head and neck movements to assist in maintaining equilibrium; (3) the vestibulospinal tract, which conveys impulses down the spinal cord to maintain muscle tone in skeletal muscles to help maintain equilibrium; and (4) the ventral posterior nucleus in the thalamus and then to the vestibular area in the parietal lobe of the cerebral cortex (which is part of the primary somatosensory area; see areas 1, 2, and 3 in Figure 19.15 on page 636) to provide us with the conscious awareness of the position and movements of the head.

Various pathways among the vestibular nuclei, cerebellum, and cerebrum enable the cerebellum to play a key role in maintaining static and dynamic equilibrium. The cerebellum continuously receives updated sensory information from the utricle and saccule. It monitors this information and makes corrective adjustments. Essentially, in response to input from the utricle, saccule, and semicircular ducts, the cerebellum continuously sends nerve impulses to the motor areas of the cerebrum. This feedback allows correction of signals from the motor cortex to specific skeletal muscles to smooth movements and coordinate complex sequences of muscle contractions to help maintain equilibrium.

Table 22.2 summarizes the structures of the ear related to hearing and equilibrium.

◉ CHECKPOINT

10. List the components of the external, middle, and internal ear and their functions.
11. Explain the mechanism of hearing.
12. Describe the auditory pathway.
13. Compare the function of the maculae in maintaining static equilibrium with the role of the cristae in maintaining dynamic equilibrium.
14. Outline the equilibrium pathways.
15. Describe the role of vestibular input to the cerebellum.

Figure 22.18 Equilibrium pathway.

🔑 The vestibular nuclei are the major integrating centers for equilibrium.

Ventral posterior nucleus in thalamus

Vestibular branch of vestibulocochlear (VIII) nerve

Vestibular ganglion

Vestibular area in cerebral cortex

Oculomotor nerve (III) nucleus

Trochlear nerve (IV) motor nucleus

Abducens nerve (V) motor nucleus

Cerebellum

Accessory nerve (XI) nucleus

Spinal cord

Vestibulospinal tract

❓ What is the function of the vestibulospinal tract?

TABLE 22.2

Summary of Structures of the Ear

REGIONS OF THE EAR AND KEY STRUCTURES	FUNCTION
External (outer) ear	*Auricle (pinna):* Collects sound waves. *External auditory canal (meatus):* Directs sound waves to the tympanic membrane. *Tympanic membrane (eardrum):* Sound waves cause it to vibrate, which, in turn, causes the malleus to vibrate.
Middle ear	*Auditory ossicles:* Transmit and amplify vibrations from tympanic membrane to oval window. *Auditory (eustachian) tube:* Equalizes air pressure on both sides of the tympanic membrane.
Internal (inner) ear 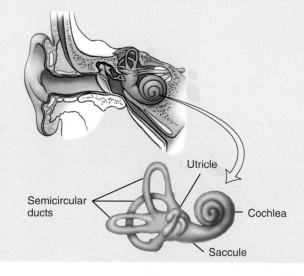	*Cochlea:* Contains a series of fluids, channels, and membranes that transmit vibrations to the spiral organ (organ of Corti), the organ of hearing; hair cells in the spiral organ ultimately produce nerve impulses in the cochlear branch of the vestibulocochlear (VIII) nerve. *Vestibular apparatus:* Includes semicircular ducts, utricle, and saccule, which generate nerve impulses that propagate along the vestibular branch of the vestibulocochlear (VIII) nerve. *Semicircular ducts:* Contain cristae, sites of hair cells for dynamic equilibrium maintenance of the position of the body (mainly the head) relative to the force of gravity. *Utricle:* Contains macula, site of hair cells for static equilibrium, maintenance of the position of the body (mainly the head) in response to movements such as rotation, acceleration, and deceleration. *Saccule:* Contains macula, site of hair cells for static equilibrium, maintenance of the position of the body (mainly the head) in response to movements such as rotation, acceleration and deceleration.

Labels in External (outer) ear image: External auditory canal, Auricle, Tympanic membrane

Labels in Middle ear image: Auditory ossicles, Auditory tube

Labels in Internal (inner) ear image: Utricle, Semicircular ducts, Cochlea, Saccule

DEVELOPMENT OF THE EYES AND EARS

◉ **OBJECTIVE**
• Describe the development of the eyes and the ears.

Development of the Eyes

The *eyes* begin to develop about 22 days after fertilization when the **ectoderm** of the lateral walls of the prosencephalon (forebrain) bulges out to form a pair of shallow grooves called the **optic grooves** (Figure 22.19a). Within a few days, as the neural tube is closing, the optic grooves enlarge and grow toward the surface ectoderm and become known as the **optic vesicles** (Figure 22.19b). When the optic vesicles reach the surface ectoderm, the surface ectoderm thickens to form the **lens placodes**. In addition, the distal portions of the optic vesicles invaginate (Figure 22.19c), forming the **optic cups**; they remain attached to the prosencephalon by narrow, hollow proximal structures called **optic stalks** (Figure 22.19d).

The lens placodes also invaginate and develop into lens vesicles that sit in the optic cups. The lens vesicles eventually develop into the *lenses*. Blood is supplied to the developing lenses (and retina) by the hyaloid arteries. These arteries gain access to the developing eyes through a groove on the inferior surface of the optic cup and optic stalk called the **choroid fissure**. As the lenses mature, part of the hyaloid arteries that pass through the vitreous chamber degenerate; the remaining portions of the hyaloid arteries become the *central retinal arteries*.

The inner wall of the optic cup forms the *neural layer* of the retina, while the outer layer forms the *pigmented layer* of the retina. Axons from the neural layer grow through the optic stalk to the brain, converting the optic stalk to the *optic (II) nerve*. Although myelination of the optic nerves begins late in fetal life, it is not completed until the tenth week after birth.

The anterior portion of the optic cup forms the epithelium of the *ciliary body, iris,* and *circular and radial muscles* of the iris. The connective tissue of the ciliary body, *ciliary muscle,* and *zonular fibers* of the lens develop from **mesenchyme** around the anterior portion of the optic cup.

Mesenchyme surrounding the optic cup and optic stalk differentiates into an inner layer that gives rise to the *choroid* and an outer layer that develops into the *sclera* and part of the *cornea*. The remainder of the cornea is derived from surface **ectoderm**.

The *anterior chamber* develops from a cavity that forms in the **mesenchyme** between the iris and cornea; the *posterior chamber* develops from a cavity that forms in the **mesenchyme** between the iris and lens.

Some mesenchyme around the developing eye enters the optic cup through the choroid fissure. This mesenchyme occupies the space between the lens and retina and differentiates into a delicate network of fibers. Later the spaces between the fibers fill with a jellylike substance, thus forming the *vitreous body* in the vitreous chamber.

The *eyelids* form from surface **ectoderm** and **mesenchyme**. The upper and lower eyelids meet and fuse at about eight weeks of development and remain closed until about 26 weeks of development.

Figure 22.19 Development of the eyes.

The eyes begin to develop about 22 days after fertilization from ectoderm of the prosencephalon.

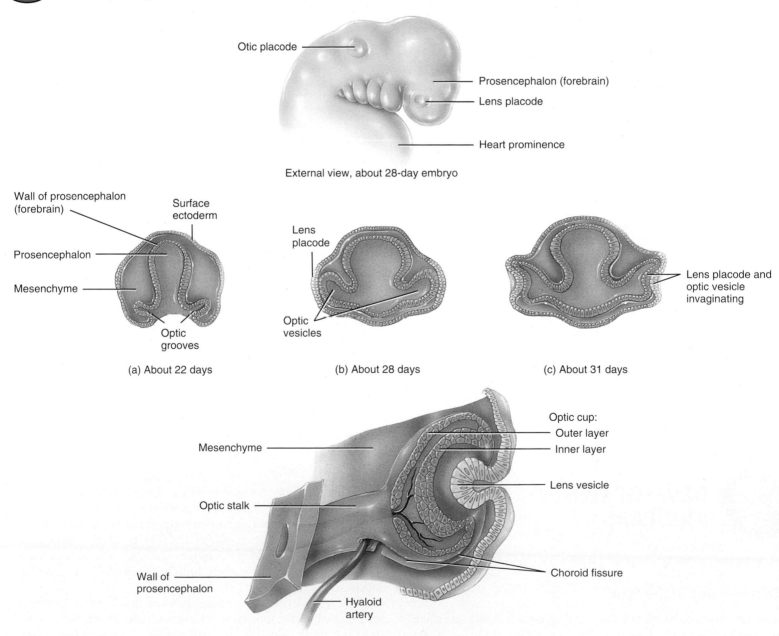

External view, about 28-day embryo

(a) About 22 days

(b) About 28 days

(c) About 31 days

(d) About 32 days

? **Which structure gives rise to the neural and pigmented layers of the retina?**

Development of the Ears

The first portion of the ear to develop is the *internal ear*. It begins to form about 22 days after fertilization as a thickening of the surface ectoderm, called **otic placodes** (Figure 22.20a), which appear on either side of the rhombencephalon (hindbrain). The otic placodes invaginate quickly (Figure 22.20b) to form the **otic pits** (Figure 22.20c). Next, the otic pits pinch off from the surface ectoderm to form the **otic vesicles** within the mesenchyme of the head (Figure 22.20d). During later development, the otic vesicles will form the structures associated with the *membranous labyrinth* of the internal ear. **Mesenchyme** around the otic vesicles produces cartilage that later ossifies to form the bone associated with the *bony labyrinth* of the internal ear.

The *middle ear* develops from a structure called the first **pharyngeal (branchial) pouch,** an **endoderm**-lined outgrowth of the primitive pharynx (see the inset in Figure 22.20). The pharyngeal pouches were discussed in detail in Chapter 4 on page 108. The *auditory ossicles* develop from the first and second pharyngeal arches.

The *external ear* develops from the first **pharyngeal cleft,** an **endoderm**-lined groove between the first and second pharyngeal arches (see the inset in Figure 22.20). The pharyngeal clefts were discussed in detail in Chapter 4 on page 108.

◉ CHECKPOINT

16. How do the origins of the eyes and ears differ?

Figure 22.20 Development of the ears.

🔑 ⚿ The first part of the ears to develop are the internal ears, which begin to form about 22 days after fertilization as thickenings of surface ectoderm.

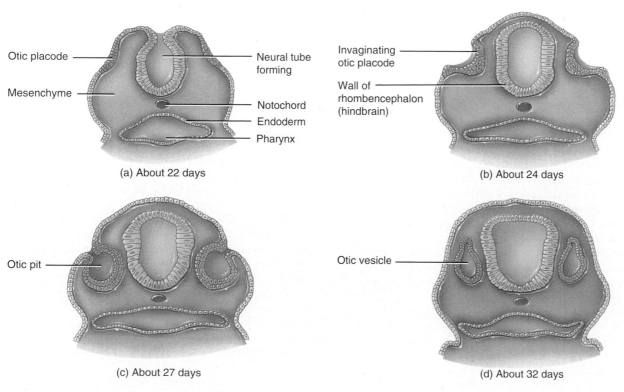

❓ **How do the three parts of the ear differ in origin?**

AGING AND THE SPECIAL SENSES

◉ **OBJECTIVE**

• Describe the age-related changes that occur in the eyes and ears.

Most people do not experience any problems with the senses of smell and taste until about age 50. This is due to a gradual loss of olfactory receptors and gustatory receptor cells coupled with their slower rate of replacement as we age.

Several age-related changes occur in the eyes. As noted earlier, the lens loses some of its elasticity and thus cannot change shape as easily, resulting in presbyopia. Cataracts (loss of transparency of the lenses) also occur with aging. In old age, the sclera ("white" of the eye) becomes thick and rigid and develops a yellowish or brownish coloration due to many years of exposure to ultraviolet light, wind, and dust. The sclera may also develop random splotches of pigment, especially in people with dark complexions. The iris fades or develops irregular pigment. The muscles that regulate the size of the pupil weaken with age and the pupils become smaller, react more slowly to light, and dilate more slowly in the dark. For these reasons, elderly people find that objects are not as bright, their eyes may adjust more slowly when going out-doors, and they have problems going from brightly lit to darkly lit places. Some diseases of the retina are more likely to occur in old age, including age-related macular disease and detached retina. A disorder called glaucoma develops in the eyes of aging people as a result of the buildup of aqueous humor. Tear production and the number of mucous cells in the conjunctiva may decrease with age, resulting in dry eyes. The eyelids lose their elasticity, becoming baggy and wrinkled. The amount of fat around the orbits may decrease, causing the eyeballs to sink into the orbits. Finally, as we age the sharpness of vision decreases, color and depth perception are reduced, and "vitreal floaters" increase.

By about age 60, around 25 percent of individuals experience a noticeable hearing loss, especially for higher-pitched sounds. The age-associated progressive loss of hearing in both ears is called **presbycusis** (pres′-bī-KŪ-sis; *presby-*=old; *-acou*=hearing; *-sis*=condition). It may be related to damaged and lost hair cells in the spiral organ or degeneration of the nerve pathway for hearing. Tinnitus (ringing in the ears) and vestibular imbalance also occur more frequently in the elderly.

◉ **CHECKPOINT**

17. What changes in the eyes are related to the aging process, and how do they take place?

KEY MEDICAL TERMS ASSOCIATED WITH SPECIAL SENSES

Ageusia (a-GŪ-sē-a; *a-*=without; *-geusis*=taste) Loss of the sense of taste.

Amblyopia (am′-blē-Ō-pē-a; *ambly-*=dull or dim) Term used to describe the loss of vision in an otherwise normal eye that, because of muscle imbalance, cannot focus in synchrony with the other eye. Sometimes called "wandering eyeball" or a "lazy eye."

Anosmia (an-OZ-mē-a; *a-*=without; *osmi*=smell, odor) Total lack of the sense of smell.

Barotrauma (bar′-ō-TRAW-ma; *baros-*=weight) Damage or pain, mainly affecting the middle ear, as a result of pressure changes. It occurs when pressure on the outer side of the tympanic membrane is higher than on the inner side, for example, when flying in an airplane or driving. Swallowing or holding your nose and exhaling with your mouth closed usually opens the auditory tubes, allowing air into the middle ear to equalize the pressure.

Blepharitis (blef-a-RĪ-tis; *blephar-*=eyelid; *-itis*=inflammation of) An inflammation of the eyelid.

Conjunctivitis (pinkeye) An inflammation of the conjunctiva; when caused by bacteria such as pneumococci, staphylococci, or *Hemophilus influenzae*, it is very contagious and more common in children. Conjunctivitis may also be caused by irritants, such as dust, smoke, or pollutants in the air, in which case it is not contagious.

Corneal abrasion (KOR-nē-al a-BRĀ-zhun) A scratch on the surface of the cornea, for example, from a speck of dirt or damaged contact lenses. Symptoms include pain, redness, watering, blurry vision, sensitivity to bright light, and frequent blinking.

Corneal transplant A procedure in which a defective cornea is removed and a donor cornea of similar diameter is sewn in. It is the most common and most successful transplant operation. Since the cornea is avascular, antibodies in the blood that might cause rejection do not enter the transplanted tissue, and rejection rarely occurs.

Diabetic retinopathy (ret-i-NOP-a-thē; *retino-*=retina; *-pathos*=suffering) Degenerative disease of the retina due to diabetes mellitus, in which blood vessels in the retina are damaged or new ones grow and interfere with vision.

Exotropia (ek′-sō-TRŌ-pē-a; *ex-*=out; *-tropia*=turning) Turning outward of the eyes.

Keratitis (ker′-a-TĪ-tis; *kerat-*=cornea) An inflammation or infection of the cornea.

Miosis (mī-Ō-sis) Constriction of the pupil.

Motion sickness Paleness, restlessness, nausea, weakness, dizziness, and malaise that may progress to vomiting caused by increased activity of the semicircular canals. It occurs during motion, for example, in a car, on a boat, on a train, or in an airplane. Usually, when motion stops, symptoms improve.

Mydriasis (mi-DRĪ-a-sis) Dilation of the pupil.

Nystagmus (nis-TAG-mus; *nystagm-*=nodding or drowsy) A rapid involuntary movement of the eyeballs, possibly caused by a disease of the central nervous system. It is associated with conditions that cause vertigo.

Otalgia (ō-TAL-jē-a; *oto-*=ear; *-algia*=pain) Earache.

Photophobia (fō′-tō-FŌ-bē-a; *photo-*=light; *-phobia*=fear) Abnormal visual intolerance to light.

Ptosis (TŌ-sis=fall) Falling or drooping of the eyelid (or slippage of any organ below its normal position).

Retinoblastoma (ret-i-nō-blas-TŌ-ma; *-oma*=tumor) A highly malignant tumor arising from immature retinal cells; it accounts for 2% of childhood cancers.

Scotoma (skō-TŌ-ma=darkness) An area of reduced or lost vision in the visual field.

Strabismus (stra-BIZ-mus; *strabismos*=squinting) Misalignment of the eyeballs so that the eyes do not move in unison when viewing an object; the affected eye turns either medially or laterally with respect to the normal eye and the result is double vision (diplopia). It may be caused by physical trauma, vascular injuries, or tumors of the extrinsic eye muscle or the oculomotor (III), trochlear (IV), or abducens (VI) nerves.

Tinnitus (ti-NĪ-tus) A ringing, roaring, or clicking in the ears.

Tonometer (tō-NOM-ē-ter; *tono-*=tension or pressure; *-metron*=measure) An instrument for measuring pressure, especially intraocular pressure.

Trachoma (tra-KŌ-ma) A serious form of conjunctivitis and the greatest single cause of blindness in the world. It is caused by the bacterium *Chlamydia trachomatis*. The disease produces an excessive growth of subconjunctival tissue and invasion of blood vessels into the cornea, which progresses until the entire cornea is opaque.

Vertigo (VER-ti-gō=dizziness) A sensation of spinning or movement in which the world seems to revolve or the person seems to revolve in space, often associated with nausea and, in some cases, vomiting. It may be caused by arthritis of the neck or an infection of the vestibular apparatus.

STUDY OUTLINE

Olfaction: Sense of Smell (p. 698)

1. The olfactory receptors, which are bipolar neurons, are in the nasal epithelium.
2. Axons of olfactory receptors form the olfactory (I) nerves, which convey nerve impulses to the olfactory bulbs, olfactory tracts, limbic system, and cerebral cortex (temporal and frontal lobes).

Gustation: Sense of Taste (p. 700)

1. The receptors for gustation, the gustatory receptor cells, are located in taste buds.
2. Dissolved chemicals, called tastants, stimulate gustatory receptor cells.
3. Gustatory receptor cells trigger nerve impulses in cranial nerves VII, IX, and X. Taste signals then pass to the medulla oblongata, thalamus, and cerebral cortex (parietal lobe).

Vision (p. 702)

1. Accessory structures of the eyes include the eyebrows, eyelids, eyelashes, lacrimal apparatus, and extrinsic eye muscles.
2. The lacrimal apparatus consists of structures that produce and drain tears.
3. The eye is constructed of three layers: (a) fibrous tunic (sclera and cornea), (b) vascular tunic (choroid, ciliary body, and iris), and (c) retina.
4. The retina consists of a pigmented layer and a neural layer that includes a photoreceptor layer, bipolar cell layer, ganglion cell layer, horizontal cells, and amacrine cells.
5. The anterior cavity contains aqueous humor; the vitreous chamber contains the vitreous body.
6. Chemicals (neurotransmitters) released by rods and cones induce changes in bipolar cells and horizontal cells that ultimately lead to the generation of nerve impulses.
7. Impulses from ganglion cells are conveyed into the optic (II) nerve, through the optic chiasm and optic tract, to the thalamus. From the thalamus, impulses for vision propagate to the cerebral cortex (occipital lobe). Axon collaterals of retinal ganglion cells extend to the midbrain and hypothalamus.

Hearing and Equilibrium (p. 711)

1. The external (outer) ear consists of the auricle, external auditory canal, and tympanic membrane (eardrum).
2. The middle ear consists of the auditory (eustachian) tube, ossicles, oval window, and round window.
3. The internal (inner) ear consists of the bony labyrinth and membranous labyrinth. The internal ear contains the spiral organ (organ of Corti), the organ of hearing.
4. Sound waves enter the external auditory canal and strike the tympanic membrane, causing it to vibrate. The vibrations pass through

the ossicles, strike the oval window, set up waves in the perilymph, strike the vestibular membrane and scala tympani, increase pressure in the endolymph, vibrate the basilar membrane, and stimulate hair bundles on the spiral organ (organ of Corti).
5. Bending of stereocilia ultimately leads to generation of nerve impulses in sensory neurons.
6. Impulses pass into the cochlear branch of the vestibulocochlear (VIII) nerve and then the cochlear nuclei in the medulla. Most fibers decussate in the medulla, ascend in the lateral lemniscus, and terminate in the inferor colliculus in the midbrain. Some axons from the cochlear nuclei end in the superior olivary nuclei in the pons on both sides. Axons from the superior olivary nuclei ascend and end in the inferior colliculi. Nerve impulses from the inferior colliculi are passed to the medial geniculate nuclei in the thalamus and auditory areas in the temporal lobes of the cerebrum.
7. Static equilibrium is the orientation of the body relative to the pull of gravity. The maculae of the utricle and saccule are the sense organs of static equilibrium.
8. Dynamic equilibrium is the maintenance of body position in response to rotational acceleration or declaration. The cristae in the membranous semicircular ducts are the principal sense organs of dynamic equilibrium.
9. Impulses pass into the vestibular branch of the vestibulocochlear (VIII) nerve. Most of the axons synapse with neurons in vestibular nuclei in the medulla and pons. The remaining axons enter the cerebellum. The vestibular nuclei send impulses to cranial nerve nuclei that control eye, head, and neck movements, muscle tone in skeletal muscles, and the posterior nuclei in the thalamus and vestibular areas in the cerebral cortex.

Development of the Eyes and Ears (p. 723)

1. The eyes begin their development about 22 days after fertilization from ectoderm of the lateral walls of the prosencephalon (forebrain).
2. The ears begin their development about 22 days after fertilization from a thickening of ectoderm on either side of the rhombencephalon (hindbrain). The sequence of development of the ear is internal ear, middle ear, and external ear.

Aging and the Special Senses (p. 726)

1. Most people do not experience problems with the senses of smell and taste until about age 50.
2. Among the age-related changes to the eyes are presbyopia, cataracts, difficulty adjusting to light, macular disease, glaucoma, dry eyes, and decreased sharpness of vision.
3. With age there is a progressive loss of hearing and tinnitus occurs more frequently.

SELF-QUIZ QUESTIONS

Choose the one best answer to the following questions:

1. The lateral (primary) olfactory area is located in the:
   a. occipital lobe of the cerebrum.
   b. cerebellum.
   c. frontal lobe of the cerebrum.
   d. temporal lobe of the cerebrum.
   e. midbrain.

2. Which of the following sensory cells are also first-order sensory neurons?
   a. olfactory receptors
   b. gustatory receptor cells
   c. hair cells of the spiral organ (organ of Corti)
   d. hair cells of the otolithic organs
   e. All of the above are correct.

3. The shape of the lens is altered by the contraction and relaxation of the:
   a. superior rectus muscle.
   b. ciliary muscle.
   c. circular muscles.
   d. radial muscles.
   e. Both c and d are correct.

4. In the eye, only cones are present in the:
   a. lens.
   b. bipolar cell layer.
   c. central fovea.
   d. ganglion cell layer.
   e. vitreous body.

5. The bony labyrinth of the inner ear:
   a. is lined with ceruminous glands.
   b. consists of semicircular canals, vestibule, and tympanic antrum.
   c. contains a fluid called perilymph.
   d. houses hearing and equilibrium receptors.
   e. Both c and d are correct.

6. The iris separates the:
   a. anterior and posterior chambers.
   b. anterior cavity and the vitreous chamber.
   c. pupil and the lens.
   d. posterior chamber and the vitreous chamber.
   e. sclera and the choroid.

7. Increased sympathetic nervous stimulation of the radial muscles of the eye causes:
   a. increased production of tears.
   b. the eyeballs to turn toward the midline.
   c. the lens to become flatter in shape.
   d. improved color vision.
   e. dilation of the pupil.

8. The optic tract extends from the:
   a. retina to the optic chiasm.
   b. optic chiasm to the occipital lobe.
   c. thalamus to the occipital lobe.
   d. macula lutea to the optic disc.
   e. pupil to the retina.

9. Which of the following nerve–ganglion pairings is correct?
   a. cochlear nerve–spiral ganglion
   b. vestibular nerve–modiolus ganglion
   c. olfactory nerve–optic chiasm
   d. optic nerve–vestibular ganglion
   e. glossopharyngeal nerve–geniculate ganglion

Complete the following.

10. The two skeletal muscles that attach to the ossicles of the ear are the _____ muscle and the _____ muscle.

11. Taste buds are located on elevated projections of the tongue called _____. The _____ type is scattered over the entire surface of the tongue.

12. Aqueous humor is produced by the _____ body, and is reabsorbed into the _____ sinus.

13. The cochlear duct is separated from the scala tympani by the _____, on which rests the _____ that contains the receptors for hearing.

14. The mucous membrane that lines the eyelids and covers the anterior surface of the eyeball is called the _____.

15. _____ equilibrium is the maintenance of body position in response to sudden movement, and _____ equilibrium is the maintenance of body position in response to gravity.

16. Number the following in the correct sequence for the conduction pathway for the sense of smell. (a) olfactory bulbs: _____; (b) olfactory receptors: _____; (c) olfactory (I) nerves: _____; (d) olfactory tract: _____; (e) cortical region of temporal lobe: _____

17. Number the following in the correct sequence for the pathway of sound waves. (a) external auditory canal: _____; (b) stapes: _____; (c) malleus: _____; (d) incus: _____; (e) oval window: _____; (f) tympanic membrane (eardrum): _____

18. Otoliths are calcium carbonate crystals that are essential components of the receptors for the sense of _____ found in the _____ and the _____.

Are the following statements true or false? If the statement is False, find and fix the mistake to make the statement True.

19. Dendrites of olfactory receptors collectively make up the olfactory (I) nerves.

20. Cones are most important for color vision, and rods are most important for seeing shades of gray in dim light.

21. The membranous labyrinth of the internal (inner) ear contains perilymph.

22. The primary function of the auditory (eustachian) tube is to conduct sound waves from the exterior to the tympanic membrane.

23. The optic nerve is comprised of axons from retinal bipolar cells.

Matching

24. Match the following terms to their descriptions:
    ____ (a) white of the eye
    ____ (b) a clear structure, composed of protein layers arranged like an onion
    ____ (c) blind spot; area in which there are no cones or rods
    ____ (d) exact center of the posterior portion of the retina
    ____ (e) nonvascular, transparent, fibrous coat; most anterior eye structure
    ____ (f) layer containing neurons; if detached, causes blindness
    ____ (g) dark brown layer; prevents reflection of light rays; also nourishes eyeball because it is vascular
    ____ (h) a hole; appears black, like a circular doorway leading into a dark room
    ____ (i) regulates the amount of light entering the eye; colored part of the eye
    ____ (j) attaches to the lens by means of radially arranged fibers called the suspensory ligaments
    ____ (k) serrated margin of the retina
    ____ (l) located at the junction of sclera and cornea; drains aqueous humor

    (1) macula lutea
    (2) choroid
    (3) cornea
    (4) ciliary muscle
    (5) iris
    (6) lens
    (7) optic disc
    (8) ora serrata
    (9) pupil
    (10) retina
    (11) sclera
    (12) scleral venous sinus (canal of Schlemm)

25. Match the following terms to their descriptions:
    ____ (a) organs of dynamic equilibrium; located in the ampullae of the semicircular canals
    ____ (b) tympanic membrane
    ____ (c) air-filled cavity of temporal bone
    ____ (d) opening between the middle and inner ear; receives base of stapes
    ____ (e) oval central portion of the bony labyrinth
    ____ (f) connects the middle ear to the nasopharynx, allowing for equalization of pressure on both sides of the eardrum
    ____ (g) organ for hearing
    ____ (h) ear bones: malleus, incus, and stapes
    ____ (i) receptor for static equilibrium; also contributes to some aspects of dynamic equilibrium; consists of hair cells and supporting cells
    ____ (j) opening into the middle ear; is enclosed by a membrane called the secondary tympanic membrane
    ____ (k) contains the spiral organ (organ of Corti)
    ____ (l) receptor organs for equilibrium; the saccule, utricle, and semicircular canals

    (1) macula
    (2) round window
    (3) oval window
    (4) eardrum
    (5) cristae
    (6) auditory (eustachian) tube
    (7) spiral organ (organ of Corti)
    (8) auditory ossicles
    (9) middle ear
    (10) vestibular apparatus
    (11) cochlea
    (12) vestibule

26. Match the following (answers may be used more than once):
    ___**(a)** occipital lobe
    ___**(b)** saccule, utricle, and semicircular ducts
    ___**(c)** primary gustatory area
    ___**(d)** geniculate ganglion
    ___**(e)** injury can cause hyposmia
    ___**(f)** axons of retinal ganglion cells
    ___**(g)** spiral ganglion
    ___**(h)** conveys taste sensation from the posterior one-third of the tongue
    ___**(i)** organ of Corti
    ___**(j)** provides somatic efferent innervation to the stapedius muscle
    ___**(k)** vestibular ganglion
    ___**(l)** conveys taste sensation from the anterior two-thirds of the tongue
    ___**(m)** consists of axons from a small number of taste buds in the epiglottis and pharynx
    ___**(n)** anterior cranial fossa

    (1) olfactory (I) nerve
    (2) optic (II) nerve
    (3) facial (VII) nerve
    (4) vestibular branch of the vestibulocochlear (VIII) nerve
    (5) cochlear branch of the vestibulocochlear (VIII) nerve
    (6) glossopharyngeal (IX) nerve
    (7) vagus (X) nerve

CRITICAL THINKING QUESTIONS

1. Suestia was diagnosed with a brain tumor. Early symptoms had included false sensations of smell. How/why might this have occurred?

2. Three-year old Atticus developed sinusitis from his original rhinitis. This eventually developed into a severe pharyngitis and laryngitis. Can you explain how an otitis would finally arise? What would eventually happen if this otitis continued untreated?

3. Lucas noticed that the colored ring in one of his mother's eyes was a different shape than the other eye. His mother explained that she had gotten poked in the eye with a stick when she was a little girl about his age and that had caused the damage to her eye. What structure of the eye was injured by the stick? What effect would the damage have on her vision?

4. Reuben was on his first cruise and he felt miserable! The ship's doctor mentioned something about messages from his eyes confusing his ears (or the other way around) but Reuben was too sick to listen. He took his medication and went to bed. Explain the cause of Reuben's seasickness.

5. Why do the eyes in some flash photographs appear to be red?

? ANSWERS TO FIGURE QUESTIONS

22.1 The olfactory hairs detect odorant molecules.

22.2 The gustatory pathway is as follows: gustatory receptor cells → cranial nerves VII, IX, or X → medulla oblongata → either limbic system and hypothalamus, or thalamus → primary gustatory area in the parietal lobe of the cerebral cortex.

22.3 Lacrimal fluid, or tears, is a watery solution containing salts, some mucus, and lysozyme that protects, cleans, lubricates, and moistens the eyeball.

22.4 The fibrous tunic consists of the cornea and sclera; the vascular tunic consists of the choroid, ciliary body, and iris.

22.5 The parasympathetic division of the ANS causes pupillary constriction, and the sympathetic division causes pupillary dilation.

22.6 An ophthalmoscopic examination of the blood vessels of the eye can reveal evidence of hypertension, diabetes mellitus, cataracts, and age-related macular disease.

22.7 The two types of photoreceptors are rods and cones. Rods provide black-and-white vision in dim light; cones provide high visual acuity and color vision in bright light.

22.8 After its secretion by the ciliary process, aqueous humor flows into the posterior chamber, around the iris, into the anterior chamber, and out of the eyeball through the scleral venous sinus.

22.9 The optic tract terminates in the lateral geniculate nucleus of the thalamus.

22.10 The malleus of the middle ear is attached to the tympanic membrane, which is considered part of the external ear.

22.11 The oval and round windows separate the middle ear from the internal ear.

22.12 The two sacs that lie in the vestibule are the utricle and saccule.

22.13 The three subdivisions of the bony labyrinth are the semicircular canals, vestibule, and cochlea.

22.14 The oval window vibrates and sets up pressure waves in the perilymph.

22.15 The cell bodies of the first sensory neurons in the auditory pathway are located in the spiral ganglia.

22.16 The maculae are associated primarily with static equilibrium. They provide sensory information on the position of the head in space.

22.17 The semicircular ducts are associated with dynamic equilibrium.

22.18 The vestibulospinal tract conveys impulses down the spinal cord to maintain muscle tone in skeletal muscles and to help maintain equilibrium.

22.19 The optic cup forms the neural and pigmented layers of the optic part of the retina.

22.20 The internal ear develops from surface ectoderm, the middle ear develops from pharyngeal pouches, and the external ear develops from a pharyngeal cleft.

INTRODUCTION Together, the nervous and endocrine systems coordinate functions of all body systems. As you learned in the last several chapters, the nervous system exerts its control through nerve impulses conducted along axons of neurons. At synapses, nerve impulses trigger the release of mediator (messenger) molecules called *neurotransmitters*. In contrast, the endocrine system releases regulating molecules called **hormones** (*hormon*=to excite or get moving) into interstitial fluid and then the bloodstream. The circulating blood delivers the hormones to virtually all the cells of the body; cells that recognize a particular hormone then respond. The science that deals with the structure and function of the endocrine glands and the diagnosis and treatment of disorders of the endocrine system is **endocrinology** (en'-dō-kri-NOL-ō-jē; *endo-*=within; *-crino*=to secrete; *-logy*=study of).

The nervous and endocrine systems are both coordinated as an interlocking supersystem termed the *neuroendocrine system*. Certain parts of the nervous system stimulate or inhibit the release of hormones, which in turn may promote or inhibit the generation of nerve impulses. The nervous system causes muscles to contract and glands to secrete either more or less of their product. The endocrine system not only helps regulate the activity of smooth muscle, cardiac muscle, and some glands; it affects virtually all other tissues as well. Hormones alter metabolism, regulate growth and development, and influence reproductive processes.

The nervous and endocrine systems respond to stimuli at different rates. Nerve impulses most often produce an effect within a few milliseconds; some hormones can act within seconds, but others can take several hours or more to cause a response. The effects of nervous system activation are generally briefer than the effects produced by the endocrine system. Finally, the nervous system produces more localized responses, whereas the endocrine system has more widespread effects. Table 23.1 compares the characteristics of the nervous and endocrine systems.

In this chapter we examine the principal endocrine glands and hormone-producing tissues, along with their roles in coordinating body activities. •

ENDOCRINE GLANDS DEFINED

◉ **OBJECTIVE**

- Distinguish between exocrine glands and endocrine glands.

Recall from Chapter 3 that the body contains two kinds of glands: exocrine glands and endocrine glands. **Exocrine glands** (*exo-*=outside) secrete their products into ducts that carry the secretions into body cavities, into the lumen of an organ, or to the outer surface of the body. Exocrine glands include sudoriferous (sweat), sebaceous (oil), mucous, and digestive glands. **Endocrine glands**, by contrast, secrete their products (hormones) into the interstitial fluid surrounding the secretory cells, rather than into ducts. From the interstitial fluid, hormones diffuse into blood capillaries and blood carries them throughout the body. Very small amounts of most hormones are required in most cases, so circulating levels typically are low. The endocrine glands include the pituitary, thyroid, parathyroid, adrenal, and pineal glands (Figure 23.1 on page 732). In addition, several other organs and tissues do not function exclusively as endocrine glands, but contain cells that secrete hormones. These include the hypothalamus, thymus, pancreas, ovaries, testes, kidneys, stomach, liver, small intestine, skin, heart, adipose tissue, and placenta. Taken together, all endocrine glands and hormone-secreting cells constitute the **endocrine system**.

◉ **CHECKPOINT**

1. Which of the following are exclusively exocrine glands or endocrine glands: thymus, pancreas, placenta, pituitary, kidneys, sudoriferous glands, mucous glands, testes, ovaries?

HORMONES

◉ **OBJECTIVE**

- Describe how hormones interact with target-cell receptors.

Hormones have powerful effects, even when present in very low concentrations. As a rule, most of the body's 50 or so hormones affect only a few types of cells. The reason that some cells respond to a particular hormone and others do not depends on whether the cells have hormone receptors.

Although a given hormone travels throughout the body in the blood, it affects only certain **target cells**. Hormones, like neurotransmitters, influence their target cells by chemically binding to specific protein **receptors**. Only the target cells for a given hormone have receptors that bind and recognize that hormone. For

---

● **CLINICAL CONNECTION** | **Blocking Hormone Receptors**

Synthetic hormones that **block the receptors** for certain naturally occurring hormones are available as drugs. For example, RU486 (mifepristone), which is used to induce abortion, binds to the receptors for progesterone (a female sex hormone) and prevents progesterone from exerting its normal effects. When RU486 is given to a pregnant woman, the uterine conditions needed for nurturing an embryo are not maintained; embryonic development stops, and the embryo is sloughed off along with the uterine lining. RU486 is often used in conjunction with one or two other drugs, methotrexate and misoprostol. This example illustrates an important endocrine principle: If a hormone is prevented from interacting with its receptors, the hormone cannot perform its normal functions. •

TABLE 23.1

Comparison of Control by the Nervous and Endocrine Systems

CHARACTERISTIC	NERVOUS SYSTEM	ENDOCRINE SYSTEM
Mediator molecules	Neurotransmitters released locally in response to nerve impulses.	Hormones delivered to tissues throughout the body by the blood.
Site of mediator action	Close to site of release, at a synapse; binds to receptors in postsynaptic membrane.	Far from site of release (usually); binds to receptors on or in target cells.
Types of target cells	Muscle (smooth, cardiac, and skeletal), cells, gland cells, other neurons.	Cells throughout the body.
Time to onset of action	Typically within milliseconds (thousandths of a second).	Seconds to hours or days.
Duration of action	Generally briefer (milliseconds.)	Generally longer (seconds to days).

Figure 23.1 Location of many endocrine glands. Also shown are other organs that contain endocrine tissue, and associated structures.

🔑 ▭ Endocrine glands secrete hormones, which circulating blood delivers to target tissues.

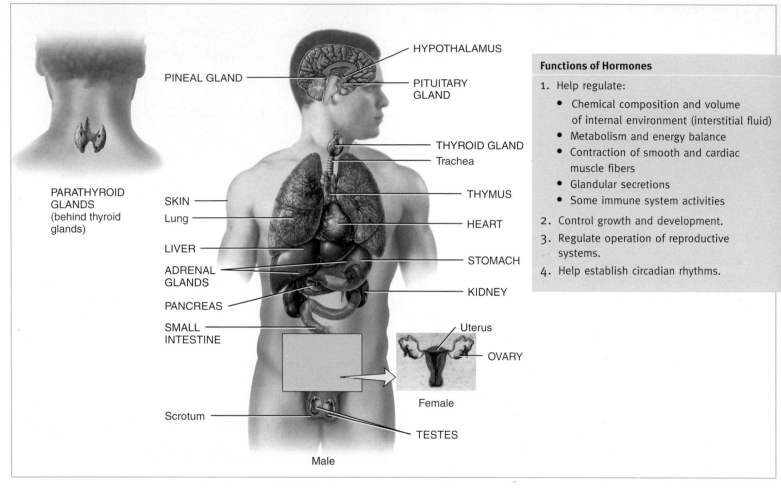

Functions of Hormones

1. Help regulate:
   • Chemical composition and volume of internal environment (interstitial fluid)
   • Metabolism and energy balance
   • Contraction of smooth and cardiac muscle fibers
   • Glandular secretions
   • Some immune system activities
2. Control growth and development.
3. Regulate operation of reproductive systems.
4. Help establish circadian rhythms.

❓ **What is the basic difference between endocrine glands and exocrine glands?**

example, thyroid-stimulating hormone (TSH) binds to receptors on cells of the thyroid gland, but it does not bind to cells of the ovaries because ovarian cells do not have TSH receptors.

Receptors, like other cellular proteins, are constantly being synthesized and broken down. Generally, a target cell has 2000–100,000 receptors for a particular hormone. When a hormone is present in excess, the number of target-cell receptors may decrease. This decreases the responsiveness of target cells to the hormone. In contrast, when a hormone (or neurotransmitter) is deficient, the number of receptors may increase to make the target tissue more sensitive.

◉ **CHECKPOINT**

2. Explain the relationship between target-cell receptors and hormones.

HYPOTHALAMUS AND PITUITARY GLAND

◉ **OBJECTIVES**

• Explain why the hypothalamus is classified as an endocrine gland.
• Describe the location, histology, hormones, and functions of the anterior and posterior pituitary.

For many years the **pituitary** (pi-TOO-i-tar-ē) **gland** or **hypophysis** (hī-POF-i-sis) was called the "master" endocrine gland because it secretes several hormones that control other endocrine glands. We now know that the pituitary gland itself has a master—the **hypothalamus**. This small region of the brain, inferior to the thalamus, is the major integrating link between the nervous and endocrine systems. It receives input from several other regions of the brain, including the limbic system, cerebral cortex, thalamus, and reticular activating system. It also receives sensory signals from internal organs and from the retina.

Painful, stressful, and emotional experiences all cause changes in hypothalamic activity. In turn, the hypothalamus controls the autonomic nervous system, regulating body temperature, thirst, hunger, sexual behavior, and defensive reactions such as fear and rage. Thus, not only is the hypothalamus an important regulatory center in the nervous system, it is also a crucial endocrine gland. Hormones secreted by the hypothalamus (described shortly) and the pituitary gland play important roles in the regulation of virtually all aspects of growth, development, metabolism, and homeostasis.

The pituitary gland is a pea-shaped structure measuring about 1–1.5 cm (0.5 in.) in diameter that lies in the hypophyseal fossa of the sphenoid bone and attaches to the hypothalamus by a stalk, the **infundibulum** (=a funnel; Figure 23.2a). The pituitary gland has two anatomically and functionally separate portions, the an-

Figure 23.2 **Hypothalamus and pituitary gland, and their blood supply.** Figure 23.2b indicates that releasing and inhibiting hormones synthesized by hypothalamic neurons are transported within axons and released from the axon terminals. The hormones diffuse into capillaries of the primary plexus of the hypophyseal portal system and are carried by the hypophyseal portal veins to the secondary plexus of the hypophyseal portal system for distribution to target cells in the anterior pituitary.

Hypothalamic hormones are an important link between the nervous and endocrine systems.

Sagittal section of pituitary gland

(b) Path of releasing and inhibiting hormones

(a) Relationship of the hypothalamus to the pituitary gland

LM all about 65x

(c) Histology of anterior pituitary

What is the functional importance of the hypophyseal portal veins?

terior pituitary and posterior pituitary. The **anterior pituitary (anterior lobe)**, which accounts for about 75% of the total weight of the gland, develops from an outgrowth of ectoderm called the *hypophyseal pouch* in the roof of the mouth (see Figure 23.8b on page 747). The anterior pituitary consists of two parts in an adult: The **pars distalis** is the larger bulbar portion, and the **pars tuberalis** forms a sheath around the infundibulum. The **posterior pituitary (posterior lobe)** also develops from an ectodermal outgrowth, this one called the *neurohypophyseal bud* (see Figure 23.8b). The posterior pituitary also consists of two parts: the **pars nervosa**, the larger bulbar portion, and the infundibulum. The posterior pituitary contains axons and axon terminals of more than 10,000 neurons whose cell bodies are located in the supraoptic and paraventricular nuclei of the hypothalamus (see Figure 19.10 on page 629). The axon terminals in the posterior pituitary gland are associated with specialized neuroglia called **pituicytes** (pi-TOO-i-sītz). These cells have a supporting role similar to that of astrocytes (see Chapter 17).

A third region, called the **pars intermedia**, atrophies during fetal development and ceases to exist as a separate lobe in adults (see Figure 23.8b). However, some of its cells migrate into adjacent parts of the anterior pituitary, where they persist.

Anterior Pituitary

The **anterior pituitary**, or **adenohypophysis** (ad'-e-nō-hī-POF-i-sis; *adeno-*=gland; *-hypophysis*=undergrowth), secretes hormones that regulate a wide range of bodily activities, from growth to reproduction. Release of anterior pituitary hormones is stimulated by **releasing hormones** and suppressed by **inhibiting hormones** from the hypothalamus (Figure 23.2b).

Hypothalamic hormones reach the anterior pituitary through a portal system. As you learned in Chapter 13, blood usually passes from the heart through an artery to a capillary to a vein and back to the heart. In a portal system, blood flows from one capillary network into a portal vein, and then into a second capillary network without passing through the heart. The name of the portal system gives the location of the *second* capillary network. In the **hypophyseal portal system** (hī'-pō-FIZ-ē-al), blood flows from capillaries in the hypothalamus into portal veins that carry blood to capillaries of the anterior pituitary.

The **superior hypophyseal arteries**, branches of the internal carotid arteries, bring blood into the hypothalamus (Figure 23.2a). At the juncture of the median eminence of the hypothalamus and the infundibulum, these arteries divide into a capillary network called the **primary plexus of the hypophyseal portal system**. From the primary plexus, blood drains into **hypophyseal portal veins** that pass down the surface of the infundibulum. In the anterior pituitary, the hypophyseal portal veins divide again and form another capillary network called the **secondary plexus of the hypophyseal portal system**.

Near the median eminence and above the optic chiasm are clusters of specialized neurons, called **neurosecretory cells**. They synthesize the hypothalamic releasing and inhibiting hormones in their cell bodies and package the hormones inside vesicles, which reach the axon terminals by axonal transport. When nerve impulses reach the axon terminals, they stimulate the vesicles to undergo exocytosis. The hormones then diffuse into the primary plexus of the hypophyseal portal system.

Quickly, the hypothalamic hormones flow with the blood through the portal veins and into the secondary plexus. This direct route permits hypothalamic hormones to act immediately on anterior pituitary cells, before they are diluted or destroyed in the general circulation. Hormones secreted by anterior

pituitary cells pass into the secondary plexus capillaries, which drain into the anterior hypophyseal veins. Now in the general circulation, anterior pituitary hormones travel to target tissues throughout the body.

The following list describes the seven major hormones secreted by five types of anterior pituitary cells (Figure 23.2c):

1. **Human growth hormone (hGH)**, or **somatotropin** (sō'-ma-tō-TRŌ-pin; *somato-*=body; *-tropin*=change), is secreted by cells called **somatotrophs**. Human growth hormone in turn stimulates several tissues to secrete **insulinlike growth factors**, hormones that stimulate general body growth and regulate various aspects of metabolism.

2. **Thyroid-stimulating hormone (TSH)**, or **thyrotropin** (thī-rō-TRŌ-pin; *thyro-*=shield), which controls the secretions and other activities of the thyroid gland, is secreted by **thyrotrophs**.

3. **Follicle-stimulating hormone (FSH)** and **luteinizing hormone (LH)** (LOO-tē-in'-īz-ing) are secreted by **gonadotrophs** (*gonado-*=seed). FSH and LH both act on the gonads: They stimulate secretion of estrogens and progesterone and the maturation of oocytes in the ovaries, and they stimulate secretion of testosterone and sperm production in the testes.

4. **Prolactin (PRL)**, which initiates milk production in the mammary glands, is released by **lactotrophs** (*lacto-*=milk).

5. **Adrenocorticotropic hormone (ACTH)**, or **corticotropin** (*cortico-*=rind or bark), which stimulates the adrenal cortex to secrete glucocorticoids, is synthesized by **corticotrophs**. Some corticotrophs also secrete **melanocyte-stimulating hormone (MSH)**.

The five different types of cells of the anterior pituitary may be classified according to their staining reaction. **Basophils** (thyrotrophs, gonadotrophs, and corticotrophs) make up about 10% of the cells, stain blue with basic dyes, and contain secretory granules. **Acidophils** (somatotrophs and lactotrophs) make up about 40% of the cells, stain red with acidic dyes, and also contain secretory granules. **Chromophobes** (KRŌ-mō-fōbs) make up about 50% of the cells and have little affinity for basic or acidic dyes. They have few or no secretory granules and do not secrete hormones. They are believed to be basophils or acidophils that have already released their granules.

Four of the anterior pituitary hormones influence the activity of another endocrine gland and are called **tropic hormones**, or **tropins**. These include TSH, ACTH, FSH, and LH. The two **gonadotropins**, FSH and LH, are hormones that specifically regulate the functions of the gonads (ovaries and testes).

The principal actions of the hormones of the anterior pituitary gland are summarized in Table 23.2.

Posterior Pituitary

Although the **posterior pituitary**, or **neurohypophysis**, does not *synthesize* hormones, it does *store* and *release* two hormones. As noted earlier, it consists of pituicytes and axon terminals of hypothalamic neurosecretory cells. The cell bodies of the neurosecretory cells are in the paraventricular and supraoptic nuclei of the hypothalamus; their axons form the **hypothalamohypophyseal tract** (hī'-pō-thal'-a-mō-hī-pō-FIZ-ē-al), which begins in the hypothalamus and ends near blood capillaries in the posterior pituitary gland (Figure 23.3). Different neurosecretory cells produce two hormones: **oxytocin (OT**; ok'-sē-TŌ-sin; *oxytoc-*=quick birth) and **antidiuretic hormone (ADH)**, also called **vasopressin** (vā-sō-PRES-in; *vaso-*=blood vessel; *-pressus*=to press).

arranged in clusters called **acini** (singular is *acinus*); these cells produce digestive enzymes, which flow into the gastrointestinal tract through a network of ducts. Scattered among the exocrine acini are 1–2 million tiny clusters of endocrine cells called **pancreatic islets** or **islets of Langerhans** (LAHNG-er-hanz; Figure 23.7b,c). Abundant capillaries serve both the exocrine and endocrine portions of the pancreas.

Each pancreatic islet contains four types of hormone-secreting cells:

1. **Alpha** or **A cells** constitute about 15% of pancreatic islet cells and secrete **glucagon** (GLOO-ka-gon) (see also Table 23.7).
2. **Beta** or **B cells** constitute about 80% of pancreatic islet cells and secrete **insulin** (IN-soo-lin) (see also Table 23.7).
3. **Delta** or **D cells** constitute about 5% of pancreatic islet cells and secrete **somatostatin** (identical to growth hormone–inhibiting hormone secreted by the hypothalamus) (see also Table 23.7).
4. **F cells** constitute the remainder of pancreatic islet cells and secrete **pancreatic polypeptide** (see also Table 23.7).

The superior and inferior pancreaticoduodenal arteries and the splenic and superior mesenteric arteries supply blood to the pancreas (Figure 23.7a). The veins, in general, correspond to the arteries. Venous blood reaches the hepatic portal vein by means of the splenic and superior mesenteric veins.

The nerves to the pancreas are autonomic nerves derived from the celiac and superior mesenteric plexuses. Included are preganglionic vagal, postganglionic sympathetic, and sensory fibers. Parasympathetic vagal fibers are said to terminate at both acinar (exocrine) and islet (endocrine) cells. Although nervous innervation is presumed to influence enzyme formation, pancreatic secretion is controlled largely by the hormones secretin and cholecystokinin (CCK) released by the small intestine. The sympathetic fibers that enter the islets are vasomotor (innervate blood vessels) and are accompanied by sensory fibers that transmit impulses, especially for pain.

⊙ CHECKPOINT

9. Why is the pancreas classified as both an exocrine gland and an endocrine gland?

TABLE 23.7

Summary of Pancreatic Islet Hormones

HORMONE AND SOURCE	PRINCIPAL ACTIONS
Glucagon from alpha cells of pancreatic islets Alpha cell	Raises blood glucose level by accelerating breakdown of glycogen into glucose in liver (glycogenolysis), converting other nutrients into glucose in liver (gluconeogenesis), and releasing glucose into the blood.
Insulin from beta cells of pancreatic islets Beta cell	Lowers blood glucose level by accelerating transport of glucose into cells, converting glucose into glycogen (glycogenesis), and decreasing glycogenolysis and gluconeogenesis; also increases lipogenesis and stimulates protein synthesis.
Somatostatin from delta cells of pancreatic islets Delta cell	Inhibits secretion of insulin and glucagon and slows absorption of nutrients from the gastrointestinal tract.
Pancreatic polypeptide from F cells of pancreatic islets F cell	Inhibits somatostatin secretion, gallbladder contraction, and secretion of pancreatic digestive enzymes.

● CLINICAL **CONNECTION** Diabetes Mellitus

The most common endocrine disorder is **diabetes mellitus** (MEL-i-tus; *melli-*=honey sweetened), caused by an inability to produce or use insulin (Figure A). Diabetes mellitus is the fourth leading cause of death by disease in the United States, primarily because of its damage to the cardiovascular system. Because insulin is unavailable to aid transport of glucose into body cells, blood glucose level is high and glucose "spills" into the urine (glucosuria). Hallmarks of diabetes mellitus are the three "polys": *polyuria,* excessive urine production due to an inability of the kidneys to reabsorb water; *polydipsia,* excessive thirst; and *polyphagia,* excessive eating.

Both genetic and environmental factors contribute to onset of the two types of diabetes mellitus—type 1 and type 2—but the exact mechanisms are still unknown. In **type 1 diabetes** insulin level is low because the person's immune system destroys the pancreatic beta cells. It is also called **insulin-dependent diabetes mellitus (IDDM)** because insulin injections are required to prevent death. Most commonly, IDDM develops in people younger than age 20, though it persists throughout life. By the time symptoms of IDDM arise, 80–90% of the islet beta cells have been destroyed. IDDM is most common in northern Europe, especially in Finland where nearly 1% of the population develops IDDM by 15 years of age. In the United States, IDDM is 1.5–2.0 times more common in whites than in African American or Asian populations.

The cellular metabolism of an untreated type 1 diabetic is similar to that of a starving person. Because insulin is not present to aid the entry of glucose into body cells, most cells use fatty acids to produce ATP. Stores of triglycerides in adipose tissue are catabolized to yield fatty acids and glycerol. The byproducts of fatty acid breakdown—organic acids called *ketones* or *ketone bodies*—accumulate. Buildup of ketones causes blood pH to fall, a condition known as **ketoacidosis**. Unless treated quickly, ketoacidosis can cause death.

The breakdown of stored triglycerides also causes weight loss. As lipids are transported by the blood from storage depots to cells, lipid particles are deposited on the walls of blood vessels, leading to atherosclerosis and a multitude of cardiovascular problems, including cerebrovascular insufficiency, ischemic heart disease, peripheral vascular

(A) Beta cells are the source of insulin

LM 240x

Beta cell (produces insulin)

Pancreatic islet

Alpha cell

disease, and gangrene. A major complication of diabetes is loss of vision due either to cataracts (excessive glucose attaches to lens proteins, causing cloudiness) or damage to blood vessels of the retina. Severe kidney problems also may result from damage to renal blood vessels.

Type 1 diabetes is treated through self-monitoring of blood glucose level (up to 7 times daily) (Figure B), regular meals containing 45–50% carbohydrates and less than 30% fats, exercise, and periodic insulin injections (up to 3 times a day) (Figure C). Several implantable pumps are available to provide insulin without the need for repeated injections. Because they lack a reliable glucose sensor, however, the person must self-monitor blood glucose level to determine insulin doses. It is also possible to successfully transplant a pancreas, but immunosuppressive drugs must then be taken for life. Most pancreatic transplants are done in people who also need a kidney transplant because of renal failure. Another promising approach under investigation is transplantation of isolated islets in semipermeable hollow tubes. The tubes allow glucose and insulin to enter and leave but prevent entry of immune system cells that might attack the islet cells.

OVARIES AND TESTES

◉ OBJECTIVE

- Describe the location, hormones, and functions of the male and female gonads.

The female gonads, called the **ovaries**, are paired oval bodies located in the pelvic cavity. The ovaries produce female sex hormones called **estrogens** and **progesterone**. Along with the gonadotropic hormones of the pituitary gland, the sex hormones regulate the female reproductive cycle, maintain pregnancy, and prepare the mammary glands for lactation. These hormones are also responsible for the development and maintenance of female secondary sexual characteristics. The ovaries also produce **inhibin**, a hormone that inhibits secretion of follicle-stimulating hormone (FSH) from the anterior pituitary. During pregnancy, the ovaries and placenta produce a hormone called **relaxin**, which increases the flexibility of the pubic symphysis during pregnancy

and helps dilate the uterine cervix during labor and delivery. These actions help ease the baby's passage by enlarging the birth canal.

The male has two oval gonads, called **testes**, that produce **testosterone**, the primary androgen. Testosterone stimulates descent of the testes before birth, regulates production of sperm, and stimulates the development and maintenance of male secondary sexual characteristics such as beard growth. The testes also produce inhibin, which inhibits secretion of FSH. The specific roles of gonadotropic hormones and sex hormones are discussed in Chapter 27.

Table 23.8 summarizes the hormones produced by the ovaries and testes and their principal actions.

◉ CHECKPOINT

10. Explain why the ovaries and testes are considered endocrine glands.

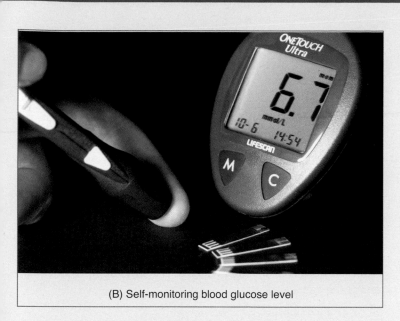

(B) Self-monitoring blood glucose level

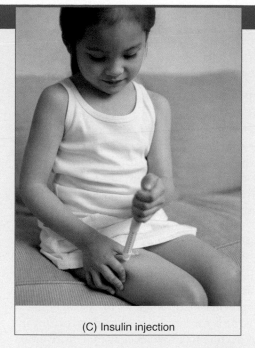

(C) Insulin injection

Type 2 diabetes, also called **non-insulin-dependent diabetes mellitus (NIDDM)**, is much more common than type 1, representing more than 90% of all cases. Type 2 diabetes most often occurs in obese people who are over age 35. Clinical symptoms are mild, and the high glucose levels in the blood often can be controlled by diet, exercise, and weight loss. Sometimes, a drug such as *glyburide* (DiaBeta) is used to stimulate secretion of insulin by pancreatic beta cells. Although some type 2 diabetics need insulin, many have a sufficient amount (or even a surplus) of insulin in the blood. For these people, diabetes arises not from a shortage of insulin but because target cells become less sensitive to it due to down-regulation of insulin receptors.

Hyperinsulinism most often results when a diabetic injects too much insulin. The main symptom is **hypoglycemia**, decreased blood glucose level, which occurs because the excess insulin stimulates too much uptake of glucose by body cells. The resulting hypoglycemia stimulates the secretion of epinephrine, glucagon, and human growth hormone. As a consequence, anxiety, sweating, tremor, increased heart rate, hunger, and weakness occur. When blood glucose falls, brain cells are deprived of the steady supply of glucose they need to function effectively. Severe hypoglycemia leads to mental disorientation, convulsions, unconsciousness, and shock. Shock due to an insulin overdose is termed **insulin shock**. Death can occur quickly unless blood glucose level is raised. From a clinical standpoint, a diabetic suffering from either a hyperglycemia or a hypoglycemia crisis can have very similar symptoms—mental changes, coma, seizures, and so on. It is important to quickly and correctly identify the cause of the underlying symptoms and treat them appropriately.

TABLE 23.8

Summary of Hormones of the Ovaries and Testes

HORMONE	PRINCIPAL ACTIONS
OVARIAN HORMONES	
Estrogens and progesterone	Together with gonadotropic hormones of the anterior pituitary, regulate the female reproductive cycle, regulate oogenesis, maintain pregnancy, prepare the mammary glands for lactation, and promote development and maintenance of female secondary sex characteristics.
Ovaries	
Relaxin	Increases flexibility of pubic symphysis during pregnancy and helps dilate uterine cervix during labor and delivery.
Inhibin	Inhibits secretion of FSH from the anterior pituitary.
TESTICULAR HORMONES	
Testosterone	Stimulates descent of the testes before birth, regulates production of sperm, and promotes development and maintenance of male secondary sex characteristics.
Testes	
Inhibin	Inhibits secretion of FSH from the anterior pituitary.

PINEAL GLAND

◉ OBJECTIVE

• Describe the location, histology, hormones, and functions of the pineal gland.

The **pineal gland** (PĪN-ē-al;=pinecone shape) is a small endocrine gland attached to the roof of the third ventricle of the brain at the midline (see Figure 23.1). It is part of the epithalamus, positioned between the two superior colliculi, and it weighs 0.1–0.2 g. The gland, which is covered by a capsule formed by the pia mater, consists of masses of neuroglia and secretory cells called **pinealocytes** (pin-ē-AL-ō-sīts). Sympathetic postganglionic fibers from the superior cervical ganglion terminate in the pineal gland.

Although many anatomical features of the pineal gland have been known for years, its physiological role is still unclear. One hormone secreted by the pineal gland is **melatonin**. Melatonin contributes to the setting of the body's biological clock, which is controlled from the suprachiasmatic nucleus of the hypothalamus. During sleep, levels of melatonin in the bloodstream increase tenfold and then decline to a low level again before awakening. Small doses of melatonin given orally can induce sleep and reset daily rhythms, which might benefit workers whose shifts alternate between daylight and nighttime hours. Melatonin also is a potent antioxidant that may provide some protection against damaging oxygen free radicals. In animals with specific breeding seasons, melatonin inhibits reproductive functions outside the breeding season. What effect, if any, melatonin exerts on human reproductive function is still unclear.

The posterior cerebral artery supplies the pineal gland with blood, and the great cerebral vein drains it.

◉ CHECKPOINT

11. What is the relationship between melatonin and sleep?

THYMUS

◉ OBJECTIVE

• Describe the role of the thymus in immunity.

Because of its role in immunity, the details of the structure and functions of the **thymus** are discussed in Chapter 16, which examines the lymphatic system and immunity. In this chapter, only its hormonal role in immunity will be discussed. As you learned in Chapter 13, lymphocytes are one type of white blood cell. There are two types of lymphocytes, called T cells and B cells, based on their specific roles in immunity. Hormones produced by the thymus, called **thymosin**, **thymic humoral factor (THF)**, **thymic factor (TF)**, and **thymopoietin**, promote the proliferation and maturation of T cells, which destroy foreign substances and microbes. There is also some evidence that thymic hormones may retard the aging process.

◉ CHECKPOINT

12. Which thymic hormones assume a role in immunity?

OTHER ENDOCRINE TISSUES

◉ OBJECTIVE

• List the hormones secreted by cells in other tissues and organs, and describe their functions.

As you learned at the beginning of this chapter, cells in organs other than those usually classified as endocrine glands have an en-

---

TABLE 23.9

Summary of Hormones Produced by Other Organs and Tissues that Contain Endocrine Cells

HORMONE	PRINCIPAL ACTIONS
GASTROINTESTINAL TRACT	
Gastrin	Promotes secretion of gastric juice and increases movements of the stomach.
Glucose-dependent insulinotropic peptide (GIP)	Stimulates release of insulin by pancreatic beta cells.
Secretin	Stimulates secretion of pancreatic juice and bile.
Cholecystokinin (CCK)	Stimulates secretion of pancreatic juice, regulates release of bile from the gallbladder, and brings about a feeling of fullness after eating.
PLACENTA	
Human chorionic gonadotropin (hCG)	Stimulates the corpus luteum in the ovary to continue the production of estrogens and progesterone to maintain pregnancy.
Estrogens and **progesterone**	Maintain pregnancy and help prepare mammary glands to secrete milk.
Human chorionic somatomammotropin (hCS)	Stimulates the development of the mammary glands for lactation.
KIDNEYS	
Renin	Part of a sequence of reactions that raises blood pressure by bringing about vasoconstriction and secretion of aldosterone.
Erythropoietin (EPO)	Increases rate of red blood cell formation.
Calcitriol* (active form of vitamin D)	Aids in the absorption of dietary calcium and phosphorus.
HEART	
Atrial natriuretic peptide (ANP)	Decreases blood pressure.
ADIPOSE TISSUE	
Leptin	Suppresses appetite and may increase the activity of FSH and LH.

*Synthesis begins in the skin, continues in the liver, and ends in the kidneys.

docrine function and secrete hormones. You learned about several of these in this chapter: the hypothalamus, thymus, pancreas, ovaries, and testes. Other tissues with endocrine functions are summarized in Table 23.9.

◉ CHECKPOINT

13. List the hormones secreted by the gastrointestinal tract, placenta, kidneys, skin, adipose tissue, and heart, and indicate their functions.

DEVELOPMENT OF THE ENDOCRINE SYSTEM

◉ OBJECTIVE

• Describe the development of endocrine glands.

The development of the endocrine system is not as localized as the development of other systems because endocrine organs develop in widely separated parts of the embryo.

About three weeks after fertilization, the *pituitary gland (hypophysis)* begins to develop from two different regions of the **ectoderm**. The *posterior pituitary (neurohypophysis)* is derived from an outgrowth of ectoderm called the **neurohypophyseal bud**, located on the floor of the hypothalamus (Figure 23.8). The *infundibulum*, also an outgrowth of the neurohypophyseal bud, connects the posterior pituitary to the hypothalamus. The *anterior pituitary (adenohypophysis)* is derived from an outgrowth of ectoderm from the roof of the mouth called the **hypophyseal (Rathke's) pouch**. The pouch grows toward the neurohypophyseal bud and eventually loses its connection with the roof of the mouth.

The *thyroid gland* develops during the fourth week as a midventral outgrowth of **endoderm**, called the **thyroid diverticulum**, from the floor of the pharynx at the level of the second pair of pharyngeal pouches. The outgrowth projects inferiorly and differentiates into the right and left lateral lobes and the isthmus of the gland.

The *parathyroid glands* develop during the fourth week from **endoderm** as outgrowths from the third and fourth **pharyngeal pouches**, which help to form structures of the head and neck.

The adrenal cortex and adrenal medulla develop during the fifth week and have completely different embryological origins. The *adrenal cortex* is derived from intermediate **mesoderm** from the same region that produces the gonads. Endocrine tissues that secrete steroid hormones all are derived from mesoderm. The *adrenal medulla* is derived from **ectoderm** from **neural crest** cells

---

Figure 23.8 Development of the endocrine system.

◉ ⚿ Glands of the endocrine system develop from all three primary germ layers.

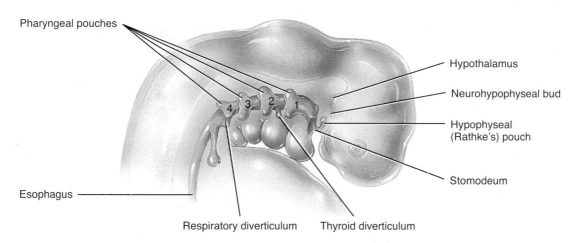

(a) Location of the neurohypophyseal bud, hypophyseal (Rathke's) pouch, thyroid diverticulum and pharyngeal pouches in a 28-day embryo

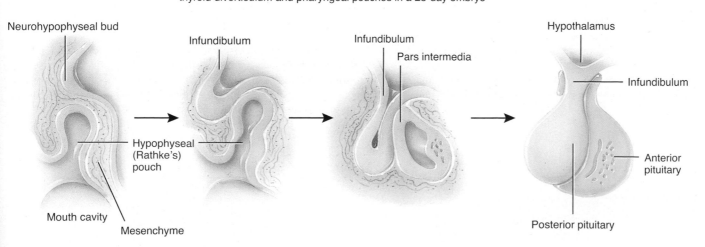

(b) Development of the pituitary gland between five and sixteen weeks

❓ **Which two endocrine glands develop from tissues with two different embryological origins?**

that migrate to the superior portion of the kidney. Recall that neural crest cells also give rise to sympathetic ganglia and other structures of the nervous system (see Figure 19.26 on page 653).

The *pancreas* develops during the fifth through seventh weeks from two outgrowths of **endoderm** from the part of the **foregut** that later becomes the duodenum (see Figure 4.12c on page 107). The two outgrowths eventually fuse to form the pancreas. The origin of the ovaries and testes is discussed in the section on the reproductive system.

The *pineal gland* arises during the seventh week as an outgrowth between the thalamus and colliculi of the midbrain from **ectoderm** associated with the **diencephalon** (see Figure 19.27 on page 654).

The *thymus* arises during the fifth week from **endoderm** of the third pharyngeal pouches.

◉ **CHECKPOINT**

14. Compare the origins of the adrenal cortex and adrenal medulla.

AGING AND THE ENDOCRINE SYSTEM

◉ **OBJECTIVE**

• Describe the effects of aging on the endocrine system.

Although some endocrine glands shrink as we get older, their performance may or may not be compromised. Production of human growth hormone by the anterior pituitary decreases, which is one cause of muscle atrophy as aging proceeds. The thyroid gland often decreases its output of thyroid hormones with age, causing a decrease in metabolic rate, an increase in body fat, and hypothyroidism, which is seen more often in older people.

Due to less negative feedback (lower levels of thyroid hormones), thyroid-stimulating hormone level increases with age.

With aging, the blood level of PTH rises, perhaps due to inadequate dietary intake of calcium. In a study of older women who took 2,400 mg/day of supplemental calcium, blood levels of PTH were as low as those of younger women. Both calcitriol and calcitonin levels are lower in older persons. Together, the rise in PTH and the fall in calcitonin level heighten the age-related decrease in bone mass that leads to osteoporosis and increased risk of fractures.

The adrenal glands contain increasingly more fibrous tissue and thus produce less cortisol and aldosterone with advancing age. However, production of epinephrine and norepinephrine remains normal. The pancreas releases insulin more slowly with age, and receptor sensitivity to glucose declines. As a result, blood glucose levels in older people increase faster and return to normal more slowly than in younger individuals.

The thymus is largest in infancy. After puberty, its size begins to decrease, and thymic tissue is replaced by adipose and areolar connective tissue. In older adults, the thymus has atrophied significantly. However, it still produces new T cells for immune responses.

The ovaries decrease in size with age, and they no longer respond to gonadotropins. The resultant decreased output of estrogens leads to conditions such as osteoporosis, high blood cholesterol, and atherosclerosis. FSH and LH levels are high due to less negative feedback inhibition of estrogens. Although testosterone production by the testes decreases with age, the effects are not usually apparent until very old age, and many elderly males can still produce active sperm in normal numbers.

◉ **CHECKPOINT**

15. Which hormone is related to muscle atrophy associated with aging?

KEY MEDICAL TERMS ASSOCIATED WITH THE ENDOCRINE SYSTEM

Gynecomastia (gī-ne′-kō-MAS-tē-a; *gyneco-*=woman; *mast-*=breast) Excessive development of mammary glands in a male, sometimes caused by a tumor of the adrenal gland.

Hirsutism (HER-soo-tizm; *hirsut-*=shaggy) Presence of excessive bodily and facial hair in a male pattern, especially in women; may be caused by excess androgen production due to tumors or drugs.

Thyroid crisis (storm) A severe, potentially life-threatening, state of hyperthyroidism characterized by high body temperature, rapid heart rate, high blood pressure, gastrointestinal symptoms (abdominal pain, vomiting, diarrhea), agitation, tremors, confusion, seizures, and possibly coma.

Virilism (VIR-il-izm; *virilis*=masculine) The presence of mature masculine characteristics in females or prepubescent males. In a female, virile characteristics include growth of a beard, development of a much deeper voice, occasionally the development of baldness, development of a masculine distribution of hair on the body and on the pubis, growth of the clitoris such that it may resemble a penis, atrophy of the breasts, infrequent or absent menstruation, and increased muscularity that produces a male-like physique. In prepubescent males the syndrome causes the same characteristics as in females, plus rapid development of the male sexual organs and emergence of sexual desires.

Virilizing adenoma (*aden*=gland; *oma*=tumor) Tumor of the adrenal gland that liberates excessive androgens, causing virilism (masculinization) in females. Occasionally, adrenal tumor cells liberate estrogens to the extent that a male patient develops gynecomastia. Such a tumor is called a **feminizing adenoma**.

Introduction (p. 730)

1. The nervous system controls homeostasis through nerve impulses; the endocrine system uses hormones.
2. The nervous system causes muscles to contract and glands to secrete; the endocrine system affects virtually all body tissues.
3. Table 23.1 on page 731 compares the characteristics of the nervous and endocrine systems.

Endocrine Glands Defined (p. 731)

1. Exocrine (sudoriferous, sebaceous, and digestive) glands secrete their products through ducts into body cavities or onto body surfaces.
2. Endocrine glands secrete hormones into the blood.
3. The endocrine system consists of endocrine glands and several organs that contain endocrine tissue.
4. Hormones regulate the internal environment, metabolism, and energy balance.
5. Hormones also help regulate muscular contraction, glandular secretion, and certain immune responses.
6. Hormones affect growth, development, and reproduction.

Hormones (p. 731)

1. The amount of hormone released is determined by the body's need for the hormone.
2. Cells that respond to the effects of hormones are called target cells.
3. The combination of hormone and receptor activates a chain of events in a target cell that produce the physiological effects of the hormone.

Hypothalamus and Pituitary Gland (p. 732)

1. The hypothalamus is the major integrating link between the nervous and endocrine systems.
2. The hypothalamus and pituitary gland regulate virtually all aspects of growth, development, and metabolism, and they also affect other body activities.
3. The pituitary gland is located in the hypophyseal fossa and is divided into the anterior pituitary (glandular portion), the posterior pituitary (nervous portion), and pars intermedia (avascular zone in between).
4. Hormones of the anterior pituitary are controlled by releasing or inhibiting hormones produced by the hypothalamus.
5. The blood supply to the anterior pituitary is from the superior hypophyseal arteries. It carries releasing and inhibiting hormones from the hypothalamus.
6. Histologically, the anterior pituitary consists of somatotrophs that produce human growth hormone (hGH); lactotrophs that produce prolactin (PRL); corticotrophs that secrete adrenocorticotropic hormone (ACTH) and melanocyte-stimulating hormone (MSH); thyrotrophs that secrete thyroid-stimulating hormone (TSH); and gonadotrophs that synthesize follicle-stimulating hormone (FSH) and luteinizing hormone (LH).
7. hGH stimulates body growth. TSH regulates thyroid gland activities. FSH and LH both regulate the activities of the ovaries and testes. PRL helps initiate milk secretion. MSH increases skin pigmentation. ACTH regulates the activities of the adrenal cortex.
8. The neural connection between the hypothalamus and posterior pituitary is via the supraopticohypophyseal tract.
9. Hormones made by the hypothalamus and stored in the posterior pituitary are oxytocin (OT), which stimulates contraction of the uterus and ejection of milk, and antidiuretic hormone (ADH), which stimulates water reabsorption by the kidneys and arteriole constriction.
10. The hormones of the anterior pituitary are summarized in Table 23.2 on page 735, and the posterior pituitary hormones are summarized in Table 23.3 on page 736.

Thyroid Gland (p. 736)

1. The thyroid gland is located inferior to the larynx.
2. Histologically, the thyroid gland consists of thyroid follicles composed of follicular cells, which secrete the thyroid hormones thyroxine (T4) and triiodothyronine (T3), and parafollicular cells, which secrete calcitonin (CT).
3. Thyroid hormones regulate the rate of metabolism, growth and development, and the reactivity of the nervous system.
4. Calcitonin (CT) lowers the blood level of calcium.
5. A summary of thyroid gland hormones and their actions is presented in Table 23.4 on page 738.

Parathyroid Glands (p. 738)

1. The parathyroid glands are embedded on the posterior surfaces of the lateral lobes of the thyroid gland.
2. The parathyroid glands consist of chief cells and oxyphil cells.
3. Parathyroid hormone (PTH) increases blood calcium level and decreases blood phosphate level.
4. A summary of parathyroid hormone and its actions is presented in Table 23.5 on page 738.

Adrenal Glands (p. 740)

1. The adrenal glands are located superior to the kidneys. They consist of an outer adrenal cortex and an inner adrenal medulla.
2. Histologically, the adrenal cortex is divided into a zona glomerulosa, zona fasciculata, and zona reticularis; the adrenal medulla consists of chromaffin cells and large blood vessels.
3. Cortical secretions include mineralocorticoids, glucocorticoids, and gonadocorticoids.
4. The medullary secretions epinephrine and norepinephrine (NE) produce effects similar to sympathetic responses and are released during stress.
5. A summary of adrenal gland hormones and their actions is presented in Table 23.6 on page 741.

Pancreas (p. 742)

1. The pancreas is posterior and slightly inferior to the stomach.
2. Histologically, the pancreas consists of pancreatic islets, or islets of Langerhans (endocrine cells), and clusters of enzyme-producing cells (acini) (exocrine cells). The four types of cells in the endocrine portion are alpha, beta, delta, and F cells.
3. Alpha cells secrete glucagon, beta cells secrete insulin, delta cells secrete somatostatin, and F cells secrete pancreatic polypeptide.
4. Glucagon increases blood sugar level.
5. Insulin decreases blood sugar level.
6. A summary of pancreatic hormones and their actions is presented in Table 23.7 on page 743.

Ovaries and Testes (p. 744)

1. The ovaries are located in the pelvic cavity and produce sex hormones that function in the development and maintenance of female secondary sexual characteristics, the reproductive cycle, pregnancy, lactation, and normal reproductive functions.
2. The testes lie inside the scrotum and produce sex hormones that function in the development and maintenance of male secondary sexual characteristics and normal reproductive functions.
3. Table 23.8 on page 745 summarizes the hormones produced by the ovaries and testes and their principal actions.

Pineal Gland (p. 746)

1. The pineal gland is attached to the roof of the third ventricle.
2. Histologically, the pineal gland consists of secretory cells called pinealocytes, neuroglial cells, and scattered postganglionic sympathetic fibers.

3. The pineal gland secretes melatonin, which contributes to setting the body's biological clock, a process controlled by the suprachiasmatic nucleus. During sleep, levels of melatonin in the bloodstream increase tenfold and then decline to a low level again before awakening.

Thymus (p. 746)

1. The thymus secretes several hormones related to immunity.
2. Thymosin, thymic humoral factor (THF), thymic factor (TF), and thymopoietin promote the maturation of T cells.

Other Endocrine Tissues (p. 746)

1. The gastrointestinal tract synthesizes several hormones, including gastrin, gastric inhibitory peptide (GIP), secretin, and cholecystokinin (CCK).
2. The placenta produces human chorionic gonadotropin (hCG), estrogens, progesterone, and human chorionic somatomammotropin (hCS).
3. The kidneys release erythropoietin.
4. The skin begins the synthesis of vitamin D.
5. The atria of the heart produce atrial natriuretic peptide (ANP).
6. Adipose tissue produces leptin.
7. A summary of hormones secreted by other endocrine tissues is included in Table 23.9 on page 746.

Development of the Endocrine System (p. 747)

1. The development of the endocrine system is not as localized as in other systems because endocrine organs develop in widely separated parts of the embryo.
2. The pituitary gland, adrenal medulla, and pineal gland develop from ectoderm; the adrenal cortex develops from mesoderm; and the thyroid gland, parathyroid glands, pancreas, and thymus develop from endoderm.

Aging and the Endocrine System (p. 748)

1. Although some endocrine glands shrink as we get older, their performance may or may not be compromised.
2. Production of human growth hormone, thyroid hormones, cortisol, aldosterone, and estrogens decrease with advancing age.
3. With aging, the blood levels of TSH, LH, FSH, and PTH rise.
4. The pancreas releases insulin more slowly with age, and receptor sensitivity to glucose declines.
5. After puberty, thymus size begins to decrease, and thymic tissue is replaced by adipose and areolar connective tissue.

SELF-QUIZ QUESTIONS

Choose the one best answer to the following questions.

1. A generalized anti-inflammatory effect is most closely associated with:
   a. glucocorticoids.
   b. mineralocorticoids.
   c. parathyroid hormone (PTH).
   d. insulin.
   e. melatonin.

2. A chemical produced by the hypothalamus that causes the anterior pituitary to secrete a hormone is called a:
   a. gonadotropic hormone.
   b. tropic hormone.
   c. releasing hormone.
   d. target hormone.
   e. neurotransmitter.

3. Which of the following is *not* released by the adenohypophysis?
   a. ADH
   b. hGH
   c. TSH
   d. FSH
   e. ACTH

4. The thyroid gland is located:
   a. anterior to the sternum.
   b. in the abdominal cavity, inferior to the liver.
   c. in the hypophyseal fossa of the sphenoid bone.
   d. inferior to the larynx with lobes on either side of the trachea.
   e. in the roof of the third ventricle of the brain.

5. The primary mineralocorticoid is:
   a. ACTH.
   b. antidiuretic hormone.
   c. cortisol.
   d. aldosterone.
   e. epinephrine.

6. Which one of these glands is called the "emergency gland" and helps the body meet sudden stress?
   a. pituitary gland
   b. pancreas
   c. thyroid gland
   d. thymus
   e. adrenal (suprarenal) glands

Complete the following.

7. Place numbers in the blanks to indicate the correct order of the vessels that supply blood to the anterior pituitary. (a) hypophyseal portal veins: ____, (b) primary plexus: ____, (c) superior hypophyseal arteries:____, (d) secondary plexus:____

8. Cell bodies of neurons whose axons make up the posterior pituitary are located in the ____.

9. Thyroid stimulating hormone (TSH) controls the secretion of hormones by follicular cells of the thyroid gland. Therefore TSH is a ____ hormone and the follicular cells are the ____ cells.

10. Glucagon, produced by ____ cells of the pancreatic islets (islets of Langerhans) causes blood sugar level to ____.

11. Insulin-like growth factors are secreted by tissues in response to ____, which is produced by the anterior pituitary.

12. The superior thyroid artery is a branch of the ____ artery.

13. Increased production of calcitriol, the active form of vitamin D, by the kidney is an effect of ____.

14. A fracture of the ____ bone has the potential of damaging the pituitary gland resulting in *polyuria* (increased urine output) due to a decrease in ____ hormone production.

Are the following statements true or false? If the statement is False, *find and fix the mistake to make the statement* True.

15. Increasing the number of receptors on a target cell decreases the cell's sensitivity to a hormone.

16. The secretory portion of the posterior pituitary is the axons of neurosecretory cells; the secretory portion of the anterior pituitary is glandular epithelium.

17. Pituicytes are the hormone-secreting cells of the posterior pituitary.

18. Oxytocin stimulates milk production.

19. Chromaffin cells are the principal secreting cells of the pineal gland.

20. The outer region of the adrenal (suprarenal) gland is called the adrenal cortex.

21. The tail of the pancreas is situated closer to the duodenum of the small intestine than to the spleen.

Matching

22. Match the following glands with the hormones they produce:

____ **(a)** adrenal cortex	**(1)** ACTH
____ **(b)** adrenal medulla	**(2)** inhibin
____ **(c)** anterior pituitary	**(3)** epinephrine
____ **(d)** pineal gland	**(4)** glucagon
____ **(e)** posterior pituitary	**(5)** melatonin
____ **(f)** testes	**(6)** cortisol
____ **(g)** pancreas	**(7)** oxytocin

23. Match the following hormone-secreting cells to the hormones they secrete:

____ **(a)** insulin	**(1)** corticotrophs
____ **(b)** glucagon	**(2)** somatotrophs
____ **(c)** calcitonin	**(3)** thyrotrophs
____ **(d)** TSH	**(4)** gonadotrophs
____ **(e)** hGH	**(5)** beta cells of pancreatic islets (islets of Langerhans)
____ **(f)** testosterone	**(6)** alpha cells of pancreatic islets (islets of Langerhans)
____ **(g)** ACTH	
____ **(h)** progesterone	**(7)** follicular cells of the thyroid gland
____ **(i)** thyroxine and triiodothyronine	**(8)** parafollicular cells of the thyroid gland
____ **(j)** FSH and LH	**(9)** testis
	(10) ovary

24. Match the endocrine gland to the description of its location.

____ **(a)** posterior and slightly inferior to the stomach	**(1)** pineal gland
	(2) adrenal gland
	(3) pituitary gland
____ **(b)** hypophyseal fossa (sella turcica)	**(4)** thyroid gland
	(5) pancreas
____ **(c)** superior to each kidney and inferior to the diaphragm	
____ **(d)** roof of the third ventricle	
____ **(e)** directly anterior and slightly inferior to the larynx (voicebox)	

CRITICAL THINKING QUESTIONS

1. You've won a trip to beautiful Tropicanaland, a 12-hour time difference from where you live. Your coworkers gave you a bottle of melatonin, a bottle of melanocyte-stimulating hormone, and a very bright flashlight as a bon voyage present. You'll be arriving at 8 P.M. Tropicanaland time, which is 8 A.M. your time. How can you adjust to Tropicanaland time most quickly?

2. Amadu, who has just arrived in the United States from Africa, has what appears to be a tumor in his neck. His doctor says it is not a tumor but a goiter. After some blood work, the doctor determines that a diet rich in seafoods and iodized salt should be sufficient to solve the problem. What is a goiter, and how is a change of diet going to help Amadu?

3. Lester's son isn't growing as tall as he had hoped, although he is not abnormally short. "He's not going to be any taller than me," said Lester. "He'll never make it in the NBA and be able to support me in my old age!" Lester thinks that maybe he should find a doctor who will treat his son with human growth hormone. Besides stimulating growth of tissues, human growth hormone raises blood glucose. Do you think this treatment is a good idea (physiologically speaking) in the long-term? Explain your answer.

4. For several years, military pilots were given radiation treatments in their nasal cavity to reduce sinus problems that interfered with flying. Years later, some of these former pilots began to exhibit problems with their pituitary gland hormones. Can you propose an explanation for this relationship?

5. A patient was found to have markedly elevated blood sugar. Tests showed that his insulin level was actually a bit elevated. How can someone with elevated insulin have elevated blood sugar?

? ANSWERS TO FIGURE QUESTIONS

23.1 Secretions of endocrine glands diffuse into interstitial fluid and then into the blood; exocrine secretions flow into ducts that lead into body cavities or to the body surface.

23.2 The hypophyseal portal veins carry blood from the median eminence of the hypothalamus, where hypothalamic releasing and inhibiting hormones are secreted, to the anterior pituitary, where these hormones act.

23.3 Functionally, both the hypothalamohypophyseal tract and the hypophyseal portal veins carry hypothalamic hormones to the pituitary gland. Structurally, the tract is composed of axons of neurons that extend from the hypothalamus to the posterior pituitary; the portal veins are blood vessels that extend to the anterior pituitary.

23.4 Follicular cells secrete T_3 and T_4, also known as thyroid hormones. Parafollicular cells secrete calcitonin.

23.5 Parafollicular cells of the thyroid gland secrete calcitonin; chief cells of the parathyroid gland secrete PTH.

23.6 The adrenal glands are superior to the kidneys in the retroperitoneal space.

23.7 The pancreas is both an endocrine and an exocrine gland.

23.8 The pituitary gland and the adrenal glands both include tissues having two different embryological origins.

24 THE RESPIRATORY SYSTEM

INTRODUCTION Have you ever swallowed something and had it go down the wrong tube, making you cough or choke uncontrollably until it cleared? This uncomfortable result occurs because the respiratory system and digestive system both arise from the embryonic gut tube and share the nose, mouth, and throat as a common initial pathway. The respiratory system is an anterior outgrowth of the embryonic gut tube in the pharyngeal region. While the majority of the gut tube gives rise to the digestive system (Chapter 25), the tube that will become the respiratory system forms a highly branched network of respiratory airways. The respiratory tubes, like other tubular structures in the body, have the basic design features shared by all tubular anatomy: an epithelial inner lining, a muscular and connective tissue middle layer, and an outer covering layer of connective tissue. Adaptations of this basic structural plan account for the principal functions associated with the respiratory system—gas transport and gas exchange.

Like the digestive and urinary systems that will be covered in the next two chapters, the respiratory system forms an extensive area of contact between the external environment and capillary blood vessels. This area of contact allows the body to constantly renew and replenish the metabolic gases in the internal fluid environment that surrounds and nourishes every body cell.

Cells continually use oxygen (O_2) for the metabolic reactions that release energy from nutrient molecules and produce ATP. At the same time, these reactions release carbon dioxide (CO_2). Because an excessive amount of CO_2 produces acidity that can be toxic to cells, excess CO_2 must be eliminated quickly and efficiently. The two systems that cooperate to supply O_2 to and eliminate CO_2 from body cells are the cardiovascular and respiratory systems. The respiratory system is responsible for gas exchange—intake of O_2 and elimination of CO_2—and the cardiovascular system transports blood containing the gases between the lungs and body cells. Failure of either system disrupts homeostasis by causing rapid death of cells from oxygen starvation and buildup of waste products. In addition to functioning in gas exchange, the respiratory system also participates in regulating blood pH, contains receptors for the sense of smell, filters inhaled air, produces sounds, and rids the body of small amounts of water and heat in exhaled air.

The branch of medicine that deals with the diagnosis and treatment of diseases of the ears, nose, and throat (ENT) is called **otorhinolaryngology** (ō′-tō-rī′-nō-lar′-in-GOL-ō-jē; *oto-*=ear; *rhino-*=nose; *laryngo-*=voice box; *-logy*=study of). A **pulmonologist** (pul-mō-NOL-ō-gist; *pulmo-*=lung) is a specialist in the diagnosis and treatment of diseases of the lungs. •

RESPIRATORY SYSTEM ANATOMY

◉ OBJECTIVES

• Describe the anatomy and histology of the nose, pharynx, larynx, trachea, bronchi, and lungs.
• Identify the functions of each respiratory system structure.

The **respiratory system** consists of the nose, pharynx (throat), larynx (voice box), trachea (windpipe), bronchi, and lungs (Figure 24.1). *Structurally*, the respiratory system consists of two parts: (1) The **upper respiratory system** includes the nose, pharynx, and associated structures. (2) The **lower respiratory system** includes the larynx, trachea, bronchi, and lungs.

Functionally, the respiratory system also consists of two parts: (1) The **conducting zone** consists of a series of interconnecting cavities and tubes both outside and within the lungs. These include the nose, pharynx, larynx, trachea, bronchi, bronchioles, and terminal bronchioles and their function is to filter, warm, and moisten air and conduct it into the lungs. (2) The **respiratory zone** consists of tissues within the lungs where gas exchange occurs. These include the respiratory bronchioles, alveolar ducts, alveolar sacs, and alveoli, and they are the main sites of gas exchange between air and blood.

Nose

The head has two openings through which substances such as air and food can enter the body—the nose and the mouth. While air can enter through either of these passageways, it is the nose that forms the primary entryway for inhaled air. The nose consists of much more than what you see on someone's face. In fact, the visible part of the nose makes up only about one-fourth of the entire nasal region. The **nose** is divided into external and internal portions. The **external nose**, the skin and muscle-covered portion of the nose visible on the face, is an extension of bone and cartilage with an internal dividing wall and two entryways. An anterior view of the bony skull reveals a pear-shaped opening, the *anterior nasal aperture*, bordered laterally and inferiorly by the maxillary bones and superiorly by the nasal bones (see Figure 7.2a on page 166). The nasal bones project anteriorly to form the upper bony framework or "bridge" of the external nose on which a pair of glasses rest. Attached to the edges of the anterior nasal aperture are the soft, but sturdy, cartilages that project from the skull to form the more extensive cartilaginous framework of the external nose. Three larger cartilages—the unpaired **septal nasal car-** tilage and the paired **major alar cartilages**—and a series of smaller cartilages—the **minor alar cartilages**—define the cartilage skeleton of the nose. The major and minor alar cartilages form the inferior part of the lateral sides of the nose. The septal nasal cartilage forms the superior portion of the lateral sides of the nose, and the cartilaginous septum that divides the external nose into right and left chambers (Figure 24.2a on page 755). Because it consists of pliable hyaline cartilage, the cartilaginous framework of the external nose is somewhat flexible. The surface anatomy of the nose is shown in Figure 12.5 on page 418.

The entryways into the external nose are the **external nares** (NĀ-rez; singular is *naris*) or **nostrils**, which lead into the cavities about the size of a finger tip, called the **nasal vestibules**. The lower half of each nasal vestibule is lined with skin continuous with the skin of the face. This skin has numerous hairs, with sebaceous and sweat glands that secrete onto its surface. The upper lining of each nasal vestibule transitions into a mucous membrane that continues deeper into the nasal cavity.

Deeper into the skull, beyond the region of the nasal vestibules, is the internal nose. The **internal nose** is a large cavity in the anterior aspect of the skull that lies inferior to the nasal bone and superior to the mouth. Anteriorly, the internal nose merges with the external nose, and posteriorly it communicates with the pharynx through two openings called the **internal nares** or **choanae** (kō-Ā-nē) (Figure 24.2b). Ducts from the *paranasal sinuses* (frontal, sphenoidal, maxillary, and ethmoidal paranasal sinuses) and the *nasolacrimal ducts* which drain tears from the lacrimal glands, also open into the internal nose (see Figure 7.11 on pages 188 and 189). The lateral walls of the internal nose are formed by the ethmoid, maxillae, lacrimal, palatine, and inferior nasal conchae bones (see Figure 7.7 on page 178); the ethmoid also forms the roof. The palatine bones and palatine processes of the maxillae, which together constitute the hard palate, form the floor of the internal nose. This large cavity within the skull, called the **nasal cavity**, forms the majority of the nose and is divided into two regions—the large inferior *respiratory region* and the small superior *olfactory region*. The nasal cavity, like the nasal vestibules, is divided by an intermediate **nasal septum** into right and left halves. The anterior part of the nasal septum is formed by the septal nasal cartilage, which attaches to the vomer and perpendicular plate of the ethmoid bone. These bones then form the posterior bony part of the nasal septum (see Figure 7.9 on page 185). A strong impact to the nasal region can break these delicate nasal septal bones or separate the cartilage portion of the nasal septum from the bony portion.

TEXT CONTINUES ON PAGE 757 ▶

Figure 24.1 Structures of the respiratory system.

The upper respiratory system includes the nose, pharynx, and associated structures; the lower respiratory system includes the larynx, trachea, bronchi, and lungs.

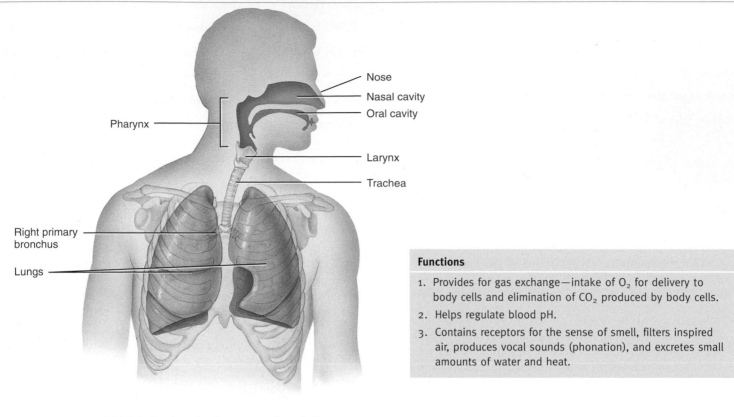

Nose

Nasal cavity

Oral cavity

Pharynx

Larynx

Trachea

Right primary bronchus

Lungs

Functions

1. Provides for gas exchange—intake of O_2 for delivery to body cells and elimination of CO_2 produced by body cells.
2. Helps regulate blood pH.
3. Contains receptors for the sense of smell, filters inspired air, produces vocal sounds (phonation), and excretes small amounts of water and heat.

(a) Anterior view showing organs of respiration

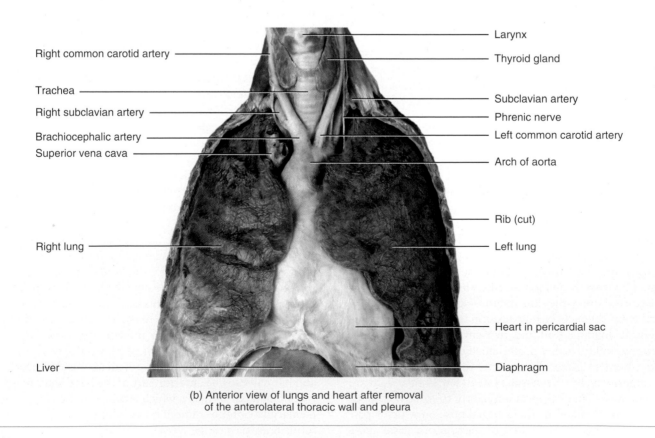

Right common carotid artery

Trachea

Right subclavian artery

Brachiocephalic artery

Superior vena cava

Right lung

Liver

Larynx

Thyroid gland

Subclavian artery

Phrenic nerve

Left common carotid artery

Arch of aorta

Rib (cut)

Left lung

Heart in pericardial sac

Diaphragm

(b) Anterior view of lungs and heart after removal
of the anterolateral thoracic wall and pleura

Which structures are part of the conducting portion of the respiratory system?

Figure 24.2 **Respiratory structures in the head and neck.** (See Tortora, *A Photographic Atlas of the Human Body*, 2e, Figure 11.2.)

As air passes through the nose, it is warmed, filtered, and moistened and olfaction occurs.

Bony framework:
Frontal bone
Nasal bones
Maxilla

Cartilaginous framework:
Lateral nasal cartilages
Septal cartilage
Alar cartilage

Dense fibrous connective and adipose tissue

(a) Anterolateral view of external portion of nose showing cartilaginous and bony framework

Sagittal plane

Nasal meatuses
Superior
Middle
Inferior

Frontal sinus
Frontal bone
Olfactory epithelium

Sphenoid bone
Sphenoidal sinus
Internal naris
Pharyngeal tonsil
NASOPHARYNX
Opening of auditory tube
Uvula
Palatine tonsil
Fauces
OROPHARYNX
Epiglottis
LARYNGOPHARYNX

Superior
Middle
Inferior
Nasal conchae
Nasal vestibule
External naris
Maxilla
Oral cavity
Palatine bone
Soft palate
Lingual tonsil
Mandible
Hyoid bone

Tongue

Esophagus
Trachea

Ventricular fold (false vocal cord)
Vocal fold (true vocal cord)
Larynx
Thyroid cartilage
Cricoid cartilage
Thyroid gland

Nasopharynx
Oropharynx
Laryngopharynx

Regions of the pharynx

(b) Sagittal section of the left side of the head and neck showing the location of respiratory structures

CONTINUES ▶

FIGURE 24.2 CONTINUED ▶

Brain

Ethmoidal cells

Sphenoidal sinus

Nasal meatuses:
 Superior
 Middle
 Inferior

Internal nares

Opening of auditory tube

Nasopharynx

Soft palate

Hard palate:
 Palatine bone
 Maxilla

Epiglottis

Laryngopharynx

Esophagus

Trachea

Frontal bone

Frontal sinus

Nasal bone

Olfactory bulb

Nasal conchae:
 Superior
 Middle
 Inferior

Nasal vestibule

Upper lip

Oral cavity

Lower lip

Tongue

Oropharynx

Mandible

Larynx

(c) Sagittal section

Frontal plane

View

Eyeball

Nasal septum:
 Perpendicular plate of ethmoid

Vomer

Hard palate

Periorbital fat

Ethmoidal cell

Superior nasal concha

Middle nasal concha

Inferior nasal concha

Maxillary sinus

(d) Frontal section showing nasal conchae

? **What is the path taken by air molecules into and through the nose?**

During the healing process the bones and cartilage can become displaced to one side, resulting in a *deviated septum*. This displaced septum can lead to a narrowing of one side of the nasal cavity. This makes it more difficult to breathe through that side of the nose.

Three shelves formed by projections of the superior, middle, and inferior nasal conchae extend out of each lateral wall of the nasal cavity. The conchae, almost reaching the bony nasal septum, subdivide each side of the nasal cavity into a series of groovelike passageways—the **superior**, **middle**, and **inferior meatuses** (mē-Ā-tus-ēz=openings or passages). Mucous membrane lines the cavity and its shelves. The arrangement of conchae and meatuses increases surface area in the internal nose and prevents dehydration by acting as a baffle that traps water droplets during exhalation.

The olfactory receptors lie in the membrane lining the superior nasal conchae and adjacent nasal septum. This region is called the **olfactory epithelium**. Inferior to the olfactory epithelium, the mucous membrane contains capillaries and pseudostratified ciliated columnar epithelium with many goblet cells. As inhaled air whirls around the conchae and meatuses, it is warmed by blood circulating in the abundant capillaries. Mucus secreted by the goblet cells moistens the air and traps dust particles. Drainage from the nasolacrimal ducts and perhaps secretions from the paranasal sinuses also help moisten the air. The cilia move the mucus and trapped dust particles toward the pharynx, at which point they can be removed (i.e., swallowed or spit out) from the respiratory tract.

In summary, the interior structures of the nose have three functions: (1) warming, moistening, and filtering incoming air; (2) detecting olfactory (smell) stimuli; and (3) modifying speech vibrations as they pass through the large, hollow resonating chambers. *Resonance* refers to prolonging, amplifying, or modifying a sound by vibration.

● CLINICAL **CONNECTION** | Coryza and Influenza

(A) Sneezing is one symptom of the common cold

Hundreds of viruses can cause **coryza** (ko-RĪ-za) or the **common cold**, but a group of viruses called *rhinoviruses* is responsible for about 40% of all colds in adults. Typical symptoms include sneezing, excessive nasal secretion, dry cough, and congestion (Figure A). The uncomplicated common cold is not usually accompanied by a fever. Complications include sinusitis, asthma, bronchitis, ear infections, and laryngitis. Recent investigations suggest an association between emotional stress and the common cold. The higher the stress level, the greater the frequency and duration of colds.

Influenza (flu) is also caused by a virus. Its symptoms include chills, fever (usually higher than 101°F=39°C), headache, and muscular aches. Influenza can become life-threatening and may develop into pneumonia. It is important to recognize that influenza is a respiratory disease, not a gastrointestinal (GI) disease. Many people mistakenly report having "the flu" when they are suffering from a GI illness. •

● CHECKPOINT

1. What functions do the respiratory and cardiovascular systems have in common?
2. What structural and functional features are different in the upper and lower respiratory systems?
3. How do the structure and functions of the external nose and the internal nose compare?

Pharynx

The **pharynx** (FAIR-inks), or throat, is a funnel-shaped tube about 13 cm (5 in.) long that starts at the internal nares and extends to the level of the cricoid cartilage, the most inferior cartilage of the larynx (voice box) (Figure 24.2). The pharynx lies just posterior to the nasal and oral cavities, superior to the larynx and esophagus, and just anterior to the cervical vertebrae. Its wall is composed of skeletal muscles and is lined with a mucous membrane. Contraction of the skeletal muscles assists in deglutition (swallowing). The pharynx functions as a passageway for air and food, provides a resonating chamber for speech sounds, and houses the tonsils, which participate in immunological reactions against foreign invaders. The pharynx can be divided into three anatomical regions: (1) nasopharynx, (2) oropharynx, and (3) laryngopharynx. (See the lower orientation diagram in Figure 24.2b.)

The superior portion of the pharynx, called the **nasopharynx**, lies posterior to the nasal cavity and extends to the plane of the soft palate. The *soft palate*, which forms the posterior portion of the roof of the mouth, is an arch-shaped muscular partition between the nasopharynx and oropharynx that is covered by mucous membrane. There are five openings in the wall of the nasopharynx: two internal nares, two openings that lead into the *auditory (pharyngotympanic) tubes* (commonly known as the *eustachian tubes*), and the opening into the oropharynx. The posterior wall also contains the **pharyngeal tonsil (adenoid)**. Through the internal nares, the nasopharynx receives air from the nasal cavity and receives packages of dust-laden mucus. The nasopharynx is lined with pseudostratified ciliated columnar epithelium, and the cilia move the mucus down toward the most inferior part of the pharynx. The nasopharynx also exchanges small amounts of air with the auditory tubes to equalize air pressure between the pharynx and the middle ear.

The intermediate portion of the pharynx, the **oropharynx**, lies posterior to the oral cavity and extends from the soft palate inferiorly to the level of the hyoid bone. In addition to communicating upward with the nasopharynx and downward with the laryngopharynx, it has an anterior opening, the **fauces** (FAW-sēz= throat), the opening from the mouth. This portion of the pharynx has both respiratory and digestive functions because it is a common passageway for air, food, and drink. Because the oropharynx is subject to abrasion by food particles, it is lined with nonkeratinized stratified squamous epithelium. Two pairs of tonsils, the **palatine** and **lingual tonsils**, are found in the oropharynx.

The inferior portion of the pharynx, the **laryngopharynx** (la-rin'-gō-FAIR-inks), or **hypopharynx**, begins at the level of the hyoid bone. At its inferior end, it opens into the esophagus (food tube) posteriorly and the larynx (voice box) anteriorly. Like the oropharynx, the laryngopharynx is both a respiratory and a digestive pathway and is lined by nonkeratinized stratified squamous epithelium.

The arterial supply of the pharynx includes the ascending pharyngeal artery, the ascending palatine branch of the facial artery, the descending palatine and pharyngeal branches of the maxillary artery, and the muscular branches of the superior thyroid artery.

The veins of the pharynx drain into the pterygoid plexus and the internal jugular veins.

Most of the muscles of the pharynx are innervated by nerve branches from the pharyngeal plexus supplied by the glossopharyngeal (IX) and vagus (X) nerves.

Larynx

The **larynx** (LAIR-inks), or voice box, is a short passageway that connects the laryngopharynx with the trachea. It lies in the midline of the neck anterior to the fourth through sixth cervical vertebrae (C4–C6).

The wall of the larynx is composed of nine pieces of cartilage (Figure 24.3). Three occur singly (thyroid cartilage, epiglottis, and cricoid cartilage), and three occur in pairs (arytenoid, cuneiform, and corniculate cartilages). Of the paired cartilages, the arytenoid cartilages are the most important because they influence the positions and tensions of the vocal folds (true vocal cords). The extrinsic muscles of the larynx connect the cartilages to other structures in the throat; the intrinsic muscles connect the cartilages to each other (see Figure 11.10 on page 336).

The **thyroid cartilage (Adam's apple)** consists of two fused plates of hyaline cartilage that form the anterior wall of the larynx and give it a triangular shape. It is usually larger in males than in

Figure 24.3 Larynx. (See Tortora, *A Photographic Atlas of the Human Body*, 2e, Figures 11.5 and 11.6.)

The larynx is composed of nine pieces of cartilage.

Larynx   Thyroid gland

(a) Anterior view

(b) Posterior view

Sagittal plane

(c) Sagittal section

females due to the influence of male sex hormones on its growth during puberty. The ligament that connects the thyroid cartilage to the hyoid bone just superior to it is called the **thyrohyoid membrane**.

The **epiglottis** (*epi-*=over; *glottis*=tongue) is a large, leaf-shaped piece of elastic cartilage that is covered with epithelium (Figure 24.3b,c,g). The "stem" of the epiglottis is the tapered inferior portion that is attached to the anterior rim of the thyroid cartilage and hyoid bone. The broad superior "leaf" portion of the epiglottis is unattached and is free to move up and down like a trap door. During swallowing, the pharynx and larynx rise. Elevation of the pharynx widens it to receive food or drink; elevation of the larynx causes the epiglottis to move down and form a lid over the opening into the larynx, closing it off. The narrowed passageway through the larynx is called the **glottis**. The glottis consists of a pair of folds of mucous membrane, the vocal folds in the larynx, and the space between them called the **rima glottidis** (RĪ-ma GLOT-ti-dis; Figure 24.3g). The closing of the larynx in this way during swallowing routes liquids and foods into the esophagus and keeps them out of the larynx and airways. When small particles of dust, smoke, food, or liquids pass into the larynx, a cough reflex occurs, usually expelling the material.

The **cricoid cartilage** (KRĪ-koyd=ringlike) is a ring of hyaline cartilage that forms the inferior wall of the larynx. It is attached to the first ring of cartilage of the trachea by the **cricotracheal ligament**. The thyroid cartilage is connected to the cricoid cartilage by the **cricothyroid ligament**. The cricoid cartilage is

the landmark for making an emergency airway called a tracheotomy.

The paired **arytenoid cartilages** (ar'-i-TĒ-noyd=ladle-like) are triangular pieces of mostly hyaline cartilage located at the posterior, superior border of the cricoid cartilage. They form synovial joints with the cricoid cartilage and have a wide range of mobility.

The paired **corniculate cartilages** (kor-NIK-ū-lāt=shaped like a small horn), horn-shaped pieces of elastic cartilage, are located at the apex of each arytenoid cartilage. The paired **cuneiform cartilages** (KŪ-nē-i-form=wedge-shaped), club-shaped elastic cartilages anterior to the corniculate cartilages, support the vocal folds and lateral aspects of the epiglottis.

The lining of the larynx superior to the vocal folds is nonkeratinized stratified squamous epithelium. The lining of the larynx inferior to the vocal folds is pseudostratified ciliated columnar epithelium consisting of ciliated columnar cells, goblet cells, and basal cells. The mucus secreted by these cells helps trap dust not removed in the upper passages. In contrast to the action of the cilia in the upper respiratory tract, which move mucus and trapped particles *down* toward the pharynx, the cilia in the lower respiratory tract move the mucus *up* toward the pharynx.

The Structures of Voice Production

The mucous membrane of the larynx forms two pairs of folds (Figure 24.3c): a superior pair called the **ventricular folds (false**

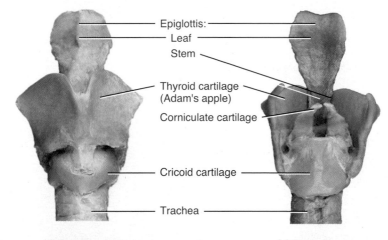

(d) Anterior view

(e) Posterior vew

(f) Lateral view

(g) Superior view

❓ **How does the epiglottis prevent aspiration of foods and liquids?**

vocal cords) and an inferior pair called simply the **vocal folds (true vocal cords)**. The space between the ventricular folds is known as the **rima vestibuli**. The **laryngeal sinus (ventricle)** is a lateral expansion of the middle portion of the laryngeal cavity; it is bordered superiorly by the ventricular folds and inferiorly by the vocal folds (see Figure 24.2b).

When the ventricular folds are brought together, they function in holding the breath against pressure in the thoracic cavity, such as might occur when a person strains to lift a heavy object. Deep to the mucous membrane of the vocal folds, which is lined by nonkeratinized stratified squamous epithelium, are bands of elastic ligaments stretched between pieces of rigid cartilage like

the strings on a guitar. Intrinsic laryngeal muscles attach to both the rigid cartilage and the vocal folds. When the muscles contract, they pull the elastic ligaments tight and stretch the vocal folds out into the airways so that the rima glottidis is narrowed. This varies the tension in the vocal folds, much like loosening or tightening a guitar string. Air passing through the larynx vibrates the folds and produces sound (phonation) by setting up sound waves in the column of air in the pharynx, nose, and mouth. The variation in the pitch of the sound is related to the tension in the vocal folds. The greater the pressure of air, the louder the sound.

When the intrinsic muscles of the larynx contract, they pull on the arytenoid cartilages, which causes them to pivot and slide.

Figure 24.4 Movement of the vocal folds.

The glottis consists of a pair of folds of mucous membrane, the vocal folds in the larynx, and the space between them (the rima glottidis).

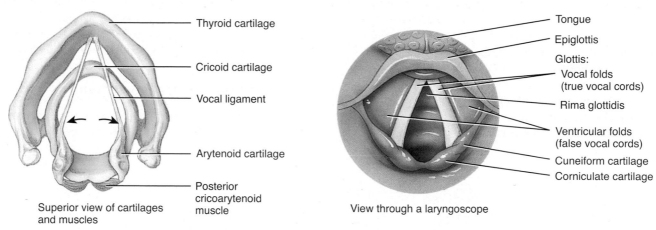

Thyroid cartilage
Cricoid cartilage
Vocal ligament
Arytenoid cartilage
Posterior cricoarytenoid muscle

Superior view of cartilages and muscles

Tongue
Epiglottis
Glottis:
Vocal folds (true vocal cords)
Rima glottidis
Ventricular folds (false vocal cords)
Cuneiform cartilage
Corniculate cartilage

View through a laryngoscope

(a) Movement of vocal folds apart (abduction)

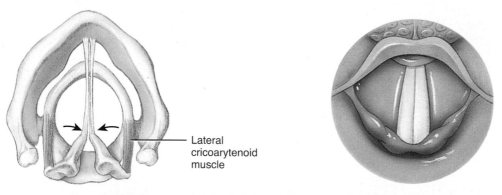

Lateral cricoarytenoid muscle

(b) Movement of vocal folds together (adduction)

View
Larynx

Ventricular folds (false vocal folds)

Epiglottis
Vocal folds (true vocal cords)
Rima glottidis
Cuneiform cartilage
Corniculate cartilage

(c) Superior view

? What is the main function of the vocal folds?

Contraction of the posterior cricoarytenoid muscles, for example, moves the vocal folds apart (abduction), thereby opening the rima glottidis (Figure 24.4). By contrast, contraction of the lateral cricoarytenoid muscles moves the vocal folds together (adduction), thereby closing the rima glottidis (Figure 24.4b). Other intrinsic muscles can elongate (and place tension on) or shorten (and relax) the vocal folds.

Pitch is controlled by the tension on the vocal folds. If they are pulled taut by the muscles, they vibrate more rapidly, and a higher pitch results. Decreasing the muscular tension on the vocal folds produces lower-pitch sounds. Due to the influence of androgens (male sex hormones), the vocal folds are usually thicker and longer in males than in females, and therefore they vibrate more slowly. Thus, men's voices generally have a lower range of pitch than women's.

Sound originates from the vibration of the vocal folds, but other structures are necessary for converting the sound into recognizable speech. The pharynx, mouth, nasal cavity, and paranasal sinuses all act as resonating chambers that give the voice its human and individual quality. We produce the vowel sounds by constricting and relaxing the muscles in the wall of the pharynx. Muscles of the face, tongue, and lips help us enunciate words.

Whispering is accomplished by closing all but the posterior portion of the rima glottidis. Because the vocal folds do not vibrate during whispering, there is no pitch to this form of speech. However, we can still produce intelligible speech while whispering by changing the shape of the oral cavity as we enunciate. As the size of the oral cavity changes, its resonance qualities change, which imparts a vowel-like pitch to the air as it rushes toward the lips.

The arteries of the larynx are the superior and inferior laryngeal arteries. The superior and inferior laryngeal veins accompany the arteries. The superior laryngeal vein empties into the superior thyroid vein, and the inferior laryngeal vein empties into the inferior thyroid vein.

The nerves of the larynx are the superior and recurrent (inferior) laryngeal branches of the vagus (X) nerves.

Trachea

The **trachea** (TRĀ-kē-a=sturdy), or windpipe, is a tubular passageway for air that is about 12 cm (5 in.) long and 2.5 cm (1 in.) in diameter. It is located anterior to the esophagus (Figure 24.5) and extends from the larynx to the superior border of the fifth thoracic vertebra (T5), where it divides into the right and left primary bronchi (see Figure 24.6).

The layers of the tracheal wall, from deep to superficial, are (1) the mucosa, (2) the submucosa, (3) the media, or middle tunic, and (4) the adventitia. The mucosa of the trachea consists of an epithelial layer of pseudostratified ciliated columnar epithelium and an underlying layer of lamina propria that contains elastic and reticular fibers. The epithelium consists of ciliated columnar cells and goblet cells that reach the luminal surface, plus basal cells that do not (see Table 3.1E on page 66). The epithelium provides the same protection against dust as the membrane lining the nasal cavity and larynx. The submucosa consists of areolar connective tissue that contains seromucous glands and their ducts.

In the middle tunic the 16–20 incomplete horizontal rings of hyaline cartilage resemble the letter C. The rings are stacked one above another and are connected together by dense connective tissue. They may be felt through the skin inferior to the larynx. The open part of each C-shaped cartilage ring faces posteriorly toward the esophagus (Figure 24.5) and is spanned by a fibromuscular membrane. Within this membrane are transverse smooth muscle fibers, called the **trachealis muscle**, and elastic connective tissue that allow the diameter of the trachea to change subtly during inhalation and exhalation, which is important in maintaining efficient air flow. The solid C-shaped cartilage rings provide a semirigid support so that the tracheal wall does not collapse inward (especially during inhalation) and

Figure 24.5 Location of the trachea in relation to the esophagus.

🔑 The trachea is anterior to the esophagus and extends from the larynx to the superior border of the fifth thoracic vertebra.

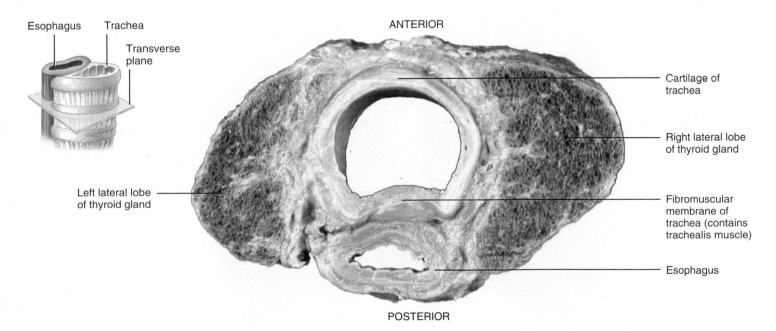

Superior view of transverse section of thyroid gland, trachea, and esophagus

❓ **What is the benefit of not having complete rings of tracheal cartilage between the trachea and the esophagus?**

obstruct the air passageway. The adventitia of the trachea consists of areolar connective tissue that joins the trachea to surrounding tissues.

The arteries of the trachea are branches of the inferior thyroid, internal thoracic, and bronchial arteries. The veins of the trachea terminate in the inferior thyroid veins.

The smooth muscle and glands of the trachea are innervated parasympathetically via the vagus (X) nerves directly and by their recurrent laryngeal branches. Sympathetic innervation is through branches from the sympathetic trunk and its ganglia.

Bronchi

At the superior border of the fifth thoracic vertebra, the trachea divides into a **right primary bronchus** (BRON-kus=windpipe), which goes into the right lung, and a **left primary bronchus**, which goes into the left lung (Figure 24.6). The right primary bronchus is more vertical, shorter, and wider than the left. As a result, an aspirated object is more likely to enter and lodge in the right primary bronchus than the left. Like the trachea, the primary bronchi (BRON-kī) contain incomplete rings of cartilage and are lined by pseudostratified ciliated columnar epithelium.

At the point where the trachea divides into right and left primary bronchi is an internal ridge called the **carina** (ka-RĪ-na=keel of a boat). It is formed by a posterior and somewhat inferior projection of the last tracheal cartilage. The mucous membrane of the carina is one of the most sensitive areas of the entire larynx and trachea for triggering a cough reflex. Widening and distortion of the carina is a serious sign because it usually indicates a carcinoma of the lymph nodes around the region where the trachea divides.

On entering the lungs, the primary bronchi divide to form smaller bronchi—the **secondary (lobar) bronchi**, one for each lobe of the lung. (The right lung has three lobes; the left lung has two.) The secondary bronchi continue to branch, forming still smaller bronchi, called **tertiary (segmental) bronchi**, that divide into **bronchioles**. Bronchioles in turn branch repeatedly, and the smallest ones branch into even smaller tubes called **terminal bronchioles**. This extensive branching from the trachea resembles an inverted tree and is commonly referred to as the **bronchial tree**.

As the branching becomes more extensive in the bronchial tree, several structural changes may be noted.

1. The mucous membrane in the bronchial tree changes from pseudostratified ciliated columnar epithelium in the primary bronchi, secondary bronchi, and tertiary bronchi to ciliated simple columnar epithelium with some goblet cells

Figure 24.6 **Branching of airways from the trachea: the bronchial tree.**

The bronchial tree begins at the trachea and ends at the terminal bronchioles.

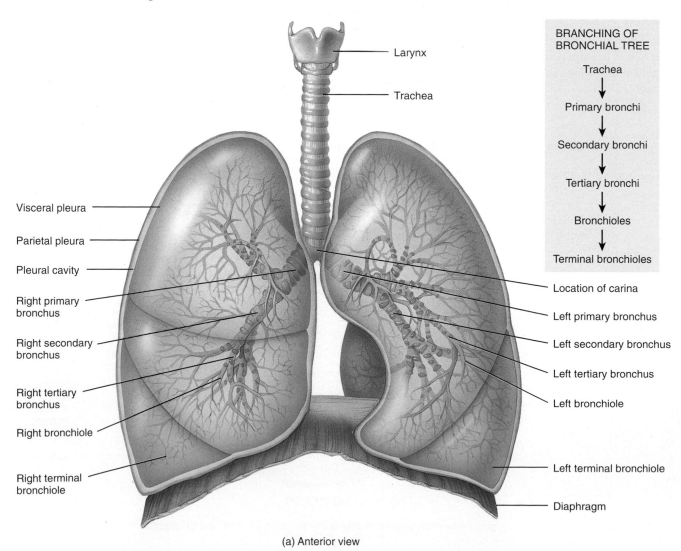

BRANCHING OF
BRONCHIAL TREE

Trachea
↓
Primary bronchi
↓
Secondary bronchi
↓
Tertiary bronchi
↓
Bronchioles
↓
Terminal bronchioles

Larynx

Trachea

Visceral pleura

Parietal pleura

Pleural cavity

Right primary bronchus

Right secondary bronchus

Right tertiary bronchus

Right bronchiole

Right terminal bronchiole

Location of carina

Left primary bronchus

Left secondary bronchus

Left tertiary bronchus

Left bronchiole

Left terminal bronchiole

Diaphragm

(a) Anterior view

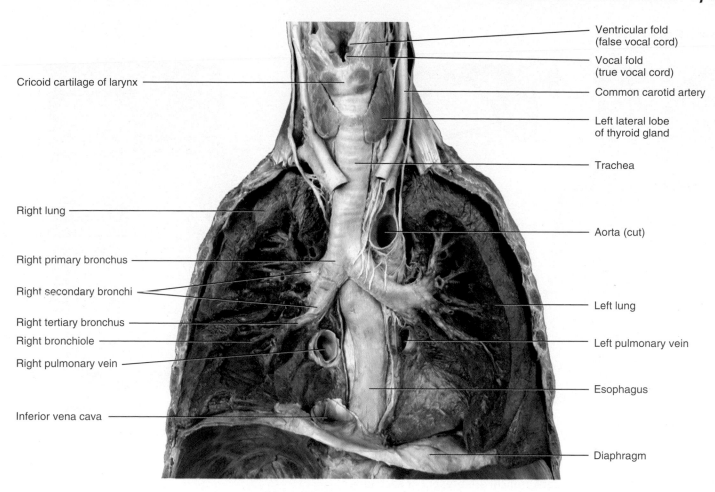

Ventricular fold (false vocal cord)

Vocal fold (true vocal cord)

Cricoid cartilage of larynx

Common carotid artery

Left lateral lobe of thyroid gland

Trachea

Right lung

Aorta (cut)

Right primary bronchus

Right secondary bronchi

Right tertiary bronchus

Left lung

Right bronchiole

Right pulmonary vein

Left pulmonary vein

Esophagus

Inferior vena cava

Diaphragm

(b) Anterior view of bronchial tree in lungs

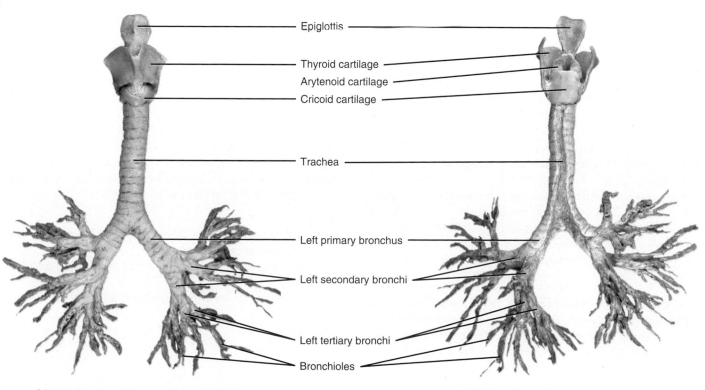

Epiglottis

Thyroid cartilage
Arytenoid cartilage
Cricoid cartilage

Trachea

Left primary bronchus

Left secondary bronchi

Left tertiary bronchi

Bronchioles

(c) Anterior view of larynx and bronchial tree

(d) Posterior view of larynx and bronchial tree

CONTINUES ▶

◉ FIGURE 24.6 CONTINUED ▶

Bronchiole

Lumen of
bronchiole

Smooth muscle

Epithelium
(ciliated simple
columnar)

LM 175x

Transverse section of a bronchiole

Cilia

Cartilage

Smooth
muscle

Epithelium
(ciliated
simple
columnar)

Goblet cell

LM 350x

Details of epithelium

(e) Histology of a bronchiole

❓ How many lobes and secondary bronchi are present in each lung?

in larger bronchioles, to mostly ciliated simple cuboidal epithelium with no goblet cells in smaller bronchioles, to mostly nonciliated simple cuboidal epithelium in terminal bronchioles. (In regions where simple nonciliated cuboidal epithelium is present, inhaled particles are removed by macrophages.)

2. Plates of cartilage gradually replace the incomplete rings of cartilage in primary bronchi and finally disappear in the distal bronchioles.

3. As the amount of cartilage decreases, the amount of smooth muscle increases. Smooth muscle encircles the lumen in spiral bands. Because there is no supporting cartilage, however, muscle spasms, such as those that occur during an asthma attack, can close off the airways, a potentially life-threatening situation.

During exercise, activity in the sympathetic division of the autonomic nervous system (ANS) increases and causes the adrenal medullae to release the hormones epinephrine and norepinephrine, both of which cause relaxation of smooth muscle in the bronchioles, which dilates the airways. The result is im-

proved lung ventilation because air reaches the alveoli more quickly. The parasympathetic division of the ANS and mediators of allergic reactions such as histamine cause contraction of bronchiolar smooth muscle and result in constriction of distal bronchioles.

The blood supply to the bronchi is via the left bronchial and right bronchial arteries. The veins that drain the bronchi are the right bronchial vein, which enters the azygos vein, and the left bronchial vein, which empties into the accessory hemiazygos vein or the left superior intercostal vein.

◉ CHECKPOINT

4. List the roles of the three anatomical regions of the pharynx in respiration.

5. How does the larynx function in respiration and voice production?

6. Describe the location, structure, and function of the trachea.

7. What is the structure and function of the bronchial tree?

Lungs

The **lungs** (=lightweights, because they float) are paired cone-shaped organs in the thoracic cavity. They are separated from each other by the heart and other structures in the mediastinum, which separates the thoracic cavity into two anatomically distinct chambers. Because of this separation, if trauma causes one lung to collapse, the other may remain expanded. Each lung is surrounded by a protective, double-layered serous membrane called the **pleural membrane** (PLOOR-al; *pleur-*=side). The superficial layer lines the wall of the thoracic cavity and is called the **parietal pleura**; the deep layer, the **visceral pleura**, is attached to the lungs themselves (Figure 24.7). The two layers are continuous with one another where the bronchi enter the lung. Between the visceral and parietal pleurae is a small space, the **pleural cavity,** which contains a small amount of lubricating fluid secreted by the two layers. This fluid reduces friction between the membranes, allowing them to slide easily over one another during breathing. Pleural fluid also causes the pleurae to adhere to one another just as a film of water causes two glass microscope slides to stick together, a phenomenon called *surface tension*. Separate pleural cavities surround the left and right lungs. Inflammation of the pleural membrane, called **pleurisy** or **pleuritis**, may in its early stages cause pain due to friction between the parietal and visceral layers of the pleura. If the inflammation persists, excess fluid accumulates in the pleural space, a condition known as **pleural effusion**.

The lungs extend from the diaphragm to just slightly superior to the clavicles and lie against the ribs anteriorly and posteriorly (Figure 24.8a). The broad inferior portion of the lung, the **base**, is concave and fits over the convex area of the diaphragm. The narrow superior portion of the lung is the **apex**. The surface of the lung lying against the ribs, the **costal surface**, matches the rounded curvature of the ribs. The **mediastinal (medial) surface** of each lung contains a region, the **hilum**, through which bronchi, pulmonary blood vessels, lymphatic vessels, and nerves enter and exit (Figure 24.8e,g). These structures are held together by the pleura and connective tissue and constitute the **root** of the lung. Medially, the left lung also contains a concavity, the **cardiac notch**, in which the heart lies. Due to the space occupied by the heart, the left lung is about 10% smaller than the right lung. The right lung is thicker and broader, but it is also somewhat shorter than the left lung because the diaphragm is higher on the right side to accommodate the liver, which lies inferior to it.

The lungs almost fill the thorax (Figure 24.8a). The apex of the lungs lies superior to the medial third of the clavicles and is the only area that can be palpated. The anterior, lateral, and posterior surfaces of the lungs lie against the ribs. The base of the lungs extends from the sixth costal cartilage anteriorly to the spinous process of the tenth thoracic vertebra posteriorly. The pleura extends about 5 cm (2 in.) below the base from the sixth costal cartilage anteriorly to the twelfth rib posteriorly. Thus, the lungs do not completely fill the pleural cavity in this area. Removal of excessive fluid in the pleural cavity can be accomplished without injuring lung tissue by inserting a needle anteriorly through the seventh intercostal space, a procedure termed **thoracentesis** (thor′-a-sen-TĒ-sis;-*centesis*=puncture). The needle is passed along the superior border of the eighth rib to avoid damage to the intercostal nerves and blood vessels. Inferior to the seventh intercostal space there is danger of penetrating the diaphragm.

Figure 24.7 Relationship of the pleural membranes to the lungs.

The parietal pleura lines the thoracic cavity; the visceral pleura covers the lungs.

Transverse plane

View

Visceral pleura

Superior vena cava

Parietal pleura

Right lung

Pleural cavity

Sternum

Left lung

Ascending aorta

Pulmonary arteries

Pulmonary vein

Esophagus

Thoracic aorta

Body of T4

Spinal cord

LATERAL

MEDIAL

POSTERIOR

Inferior view of a transverse section through the thoracic cavity showing the pleural cavity and pleural membranes

? What type of membrane is the pleural membrane?

Lobes, Fissures, and Lobules

One or two fissures divide each lung into lobes (Figure 24.8b–g). Both lungs have an **oblique fissure**, which extends inferiorly and anteriorly; the right lung also has a **horizontal fissure**. The oblique fissure in the left lung separates the **superior lobe** from the **inferior lobe**. In the right lung, the superior part of the oblique fissure separates the superior lobe from the inferior lobe; the inferior part of the oblique fissure separates the inferior lobe from the **middle lobe**, which is bordered superiorly by the horizontal fissure.

Each lobe receives its own secondary (lobar) bronchus. Thus, the right primary bronchus gives rise to three secondary (lobar) bronchi called the **superior, middle,** and **inferior secondary (lobar) bronchi,** and the left primary bronchus gives rise to **superior** and **inferior secondary (lobar) bronchi.** Within the lung, the secondary bronchi give rise to the **tertiary (segmental)**

Figure 24.8 Surface anatomy of the lungs. (See Tortora, *A Photographic Atlas of the Human Body*, 2e, Figure 11.14.)

The oblique fissure divides the left lung into two lobes. The oblique and horizontal fissures divide the right lung into three lobes.

(a) Anterior view of lungs and pleurae in thorax

(b) Lateral view of right lung

(c) Lateral view of left lung

(d) Medial view of right lung

(e) Medial view of left lung

bronchi; there are 10 tertiary bronchi in each lung. The segment of lung tissue that each tertiary bronchus supplies is called a **bronchopulmonary segment** (Figure 24.9). Bronchial and pulmonary disorders (such as tumors or abscesses) that are localized in a particular bronchopulmonary segment may be surgically removed without seriously disrupting the surrounding lung tissue.

Each bronchopulmonary segment of the lungs has many small compartments called **lobules**; each lobule is wrapped in elastic connective tissue and contains a lymphatic vessel, an arteriole, a venule, and a branch from a terminal bronchiole (Figure 24.10a). Terminal bronchioles subdivide into microscopic branches called **respiratory bronchioles** (Figure 24.10b). As the respiratory bronchioles penetrate more deeply into the lungs, the epithelial lining changes from simple cuboidal to simple squamous. Respiratory bronchioles in turn subdivide into several (2–11) **alveolar ducts**. The respiratory passages from the trachea to the alveolar ducts contain about 25 orders of branching; that is, branching occurs about 25 times—from the trachea into primary bronchi (first-order branching) into secondary bronchi (second-order branching) and so on down to the alveolar ducts.

Alveoli

Around the circumference of the alveolar ducts are numerous alveoli and alveolar sacs. An **alveolus** (al-VĒ-ō-lus) is a cup-shaped outpouching lined by simple squamous epithelium and supported by a thin elastic basement membrane; an **alveolar sac** consists of two or more alveoli that share a common opening (see Figure 24.10a, b). The walls of alveoli consist of two types of alveolar epithelial cells (Figure 24.11 on page 770). **Type I alveolar cells**, the predominant cells, are simple squamous epithelial cells that form a nearly continuous lining of the alveolar wall. **Type II alveolar cells**, also called **septal cells**, are fewer in number and are found between type I alveolar cells. The thin type I alveolar cells are the main sites of gas exchange. Type II alveolar cells, which are rounded or cuboidal epithelial cells with free surfaces containing

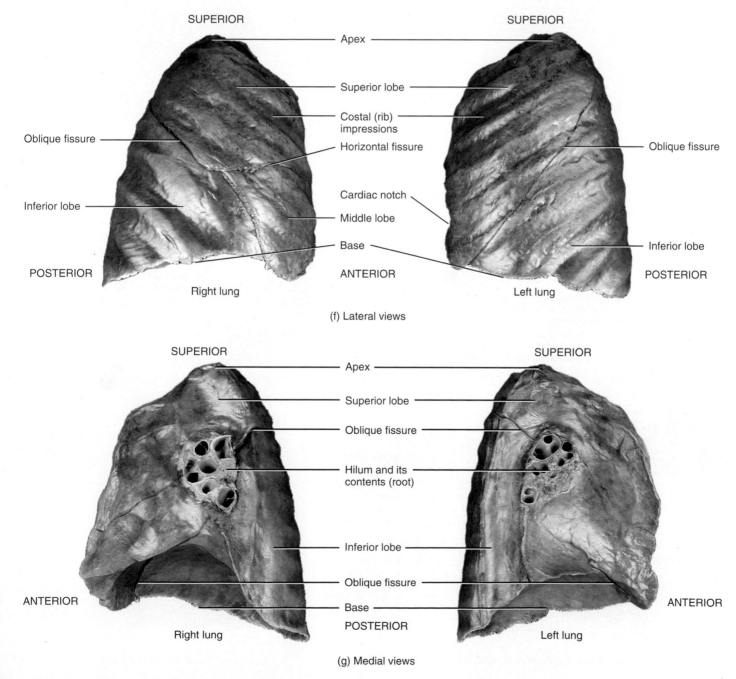

(f) Lateral views

(g) Medial views

? **Why are the right and left lungs slightly different in size and shape?**

microvilli, secrete alveolar fluid. This fluid keeps the surface between the cells and the air moist. Included in the alveolar fluid is **surfactant** (sur-FAK-tant), a complex mixture of phospholipids and lipoproteins. Surfactant lowers the surface tension of alveolar fluid, which reduces the tendency of alveoli to collapse (described later).

Associated with the alveolar wall are **alveolar macrophages (dust cells)**, wandering phagocytes that remove fine dust parti-

cles and other debris in the alveolar spaces. Also present are fibroblasts that produce reticular and elastic fibers. Underlying the layer of type I alveolar cells is an elastic basement membrane. On the outer surface of the alveoli, the pulmonary arterioles and venules disperse into a network of blood capillaries (see Figure 24.10a) that consist of a single layer of endothelial cells and basement membrane.

Figure 24.9 Bronchopulmonary segments of the lungs. The bronchial branches are shown in the center of the figure. The bronchopulmonary segments within the lungs are numbered and named for convenience.

There are 10 tertiary (segmental) bronchi in each lung; each is composed of smaller compartments called lobules.

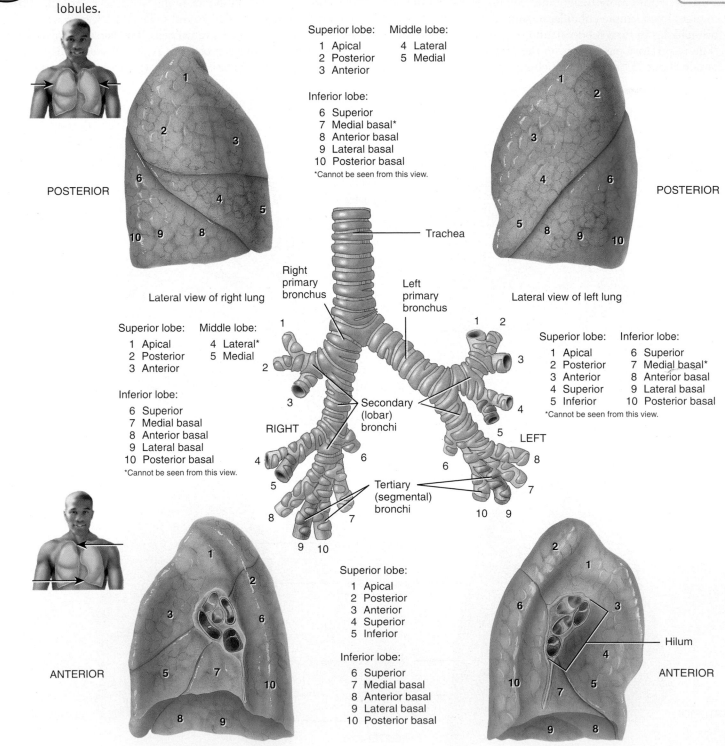

Superior lobe: Middle lobe:
1 Apical   4 Lateral
2 Posterior   5 Medial
3 Anterior

Inferior lobe:
6 Superior
7 Medial basal*
8 Anterior basal
9 Lateral basal
10 Posterior basal
*Cannot be seen from this view.

POSTERIOR

Lateral view of right lung

POSTERIOR

Lateral view of left lung

Superior lobe: Middle lobe:
1 Apical   4 Lateral*
2 Posterior   5 Medial
3 Anterior

Inferior lobe:
6 Superior
7 Medial basal
8 Anterior basal
9 Lateral basal
10 Posterior basal
*Cannot be seen from this view.

Trachea
Right primary bronchus
Left primary bronchus
Secondary (lobar) bronchi
Tertiary (segmental) bronchi
RIGHT
LEFT

Superior lobe: Inferior lobe:
1 Apical   6 Superior
2 Posterior   7 Medial basal*
3 Anterior   8 Anterior basal
4 Superior   9 Lateral basal
5 Inferior   10 Posterior basal
*Cannot be seen from this view.

Superior lobe:
1 Apical
2 Posterior
3 Anterior
4 Superior
5 Inferior

Inferior lobe:
6 Superior
7 Medial basal
8 Anterior basal
9 Lateral basal
10 Posterior basal

ANTERIOR

Medial and basal views of right lung

Hilum

ANTERIOR

Medial and basal views of left lung

? Which bronchi supply a bronchopulmonary segment?

Figure 24.10 **Microscopic anatomy of a lobule of the lungs.**

Alveolar sacs consist of two or more alveoli that share a common opening.

(a) Diagram of a portion of a lobule of the lung

(b) Lung lobule

LM about 30x

LM 1000x

(c) Details of several alveoli

? **What types of cells make up the wall of an alveolus?**

Figure 24.11 Structural components of an alveolus. The respiratory membrane consists of a layer of type I and type II alveolar cells, an epithelial basement membrane, a capillary basement membrane, and the capillary endothelium.

The exchange of respiratory gases occurs by diffusion across the respiratory membrane.

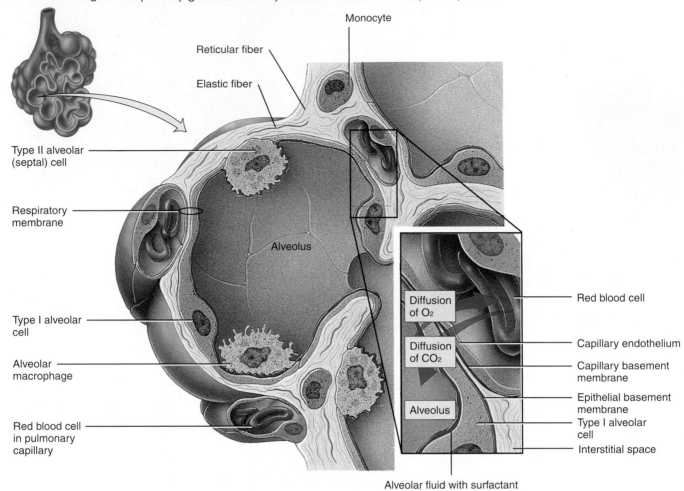

(a) Section through an alveolus showing its cellular components

(b) Details of respiratory membrane

? How thick is the respiratory membrane?

The exchange of O_2 and CO_2 between the air spaces in the lungs and the blood takes place by diffusion across the alveolar and capillary walls, which together form the **respiratory membrane**. Extending from the alveolar air space to blood plasma, the respiratory membrane consists of four layers (see Figure 24.11b):

1. A layer of type I and type II alveolar cells and associated alveolar macrophages that constitutes the **alveolar wall**.
2. An **epithelial basement membrane** underlying the alveolar wall.
3. A **capillary basement membrane** that is often fused to the epithelial basement membrane.
4. The **endothelial cells** of the capillary.

Despite having several layers, the respiratory membrane is very thin—only 0.5 μm thick, about one-sixteenth the diameter of a red blood cell. This thinness allows rapid diffusion of gases. It has been estimated that the lungs contain 300 million alveoli, providing an immense surface area of 70 m^2 (750 ft^2)—about the size of a handball court—for the exchange of gases.

Blood Supply to the Lungs

The lungs receive blood via two sets of arteries: pulmonary arteries and bronchial arteries. Deoxygenated blood passes through the pulmonary trunk, which divides into a left pulmonary artery that enters the left lung and a right pulmonary artery that enters the right lung. Return of the oxygenated blood to the heart occurs by way of the four pulmonary veins, which drain into the left atrium (see Figure 15.16 on page 532). A unique feature of pulmonary blood vessels is their constriction in response to localized hypoxia (low O_2 level). In all other body tissues, hypoxia causes dilation of blood vessels, which serves to increase blood flow. In the lungs, however, blood vessels constrict in response to hypoxia, diverting oxygenated blood from poorly ventilated areas to well-ventilated regions of the lungs.

Bronchial arteries, which branch from the aorta, deliver oxygenated blood to the lungs. This blood mainly passes to the muscular walls of the bronchi and bronchioles. Connections do exist between branches of the bronchial arteries and branches of the pulmonary arteries; most blood returns to the heart via pulmonary veins. Some blood, however, drains into bronchial veins, branches of the azygos system, and returns to the heart via the superior vena cava.

The nerve supply of the lungs is derived from the pulmonary plexus, located anterior and posterior to the roots of the lungs. The pulmonary plexus is formed by branches of the vagus (X) nerves and sympathetic trunks. Motor parasympathetic fibers

(A) Cigarette smoke is the most common cause of lung cancer

In the United States, **lung cancer** is the leading cause of cancer death in both males and females, accounting for 160,000 deaths annually. At the time of diagnosis, lung cancer is usually well advanced, with distant metastases present in about 55% of patients, and regional lymph node involvement in an additional 25%. Most people with lung cancer die within a year of the initial diagnosis; the overall survival rate is only 10–15%. Cigarette smoke is the most common cause of lung cancer (Figure A). Roughly 85% of lung cancer cases are related to smoking, and the disease is 10 to 30 times more common in smokers than nonsmokers. Exposure to secondhand smoke is also associated with lung cancer and heart disease. In the United States, secondhand smoke causes an estimated 4000 deaths a year from lung cancer, and nearly 40,000 deaths a year from heart dis-

ease. Other causes of lung cancer are ionizing radiation and inhaled irritants, such as asbestos and radon gas. Emphysema is a common precursor to the development of lung cancer.

The most common type of lung cancer, **bronchogenic carcinoma**, starts in the epithelium of the bronchial tubes. Bronchogenic tumors are named based on where they arise. For example, *adenocarcinomas* develop in peripheral areas of the lungs from bronchial glands and alveolar cells, *squamous cell carcinomas* develop from the epithelium of larger bronchial tubes, and *small (oat) cell carcinomas* develop from epithelial cells in primary bronchi near the hilus of the lungs and tend to involve the mediastinum early on. Depending on the type of bronchogenic tumors, they may be aggressive, locally invasive, and undergo widespread metastasis. The tumors begin as epithelial lesions that grow to form masses that obstruct the bronchial tubes or invade adjacent lung tissue. Bronchogenic carcinomas metastasize to lymph nodes, the brain, bones, liver, and other organs.

Symptoms of lung cancer are related to the location of the tumor. These may include a chronic cough, spitting blood from the respiratory tract, wheezing, shortness of breath, chest pain, hoarseness, difficulty swallowing, weight loss, anorexia, fatigue, bone pain, confusion, problems with balance, headache, anemia, thrombocytopenia, and jaundice.

Treatment consists of partial or complete surgical removal of a diseased lung (pulmonectomy), radiation therapy, and chemotherapy. •

arise from the dorsal nucleus of the vagus (X) nerve, and motor sympathetic fibers are postganglionic fibers of the second to fifth thoracic paravertebral ganglia of the sympathetic trunk.

◉ CHECKPOINT

8. Where are the lungs located in relation to the clavicles, ribs, and heart? Where can the lungs be palpated?
9. Distinguish the parietal pleura from the visceral pleura.
10. Define each of the following parts of a lung: base, apex, costal surface, medial surface, hilus, root, cardiac notch, lobe, and lobule.
11. What is a bronchopulmonary segment?
12. Describe the histology and function of the respiratory membrane.

MECHANICS OF PULMONARY VENTILATION (BREATHING)

◉ OBJECTIVES

• Distinguish among pulmonary ventilation, external respiration, and internal respiration.
• Describe how inspiration and expiration occur.

Respiration is the exchange of gases between the atmosphere, blood, and body cells. It takes place in three basic steps:

1. **Pulmonary ventilation.** The first process, pulmonary (*pulmo*=lung) ventilation, or breathing, consists of inhalation (inflow) and exhalation (outflow) of air and is the exchange of air between the atmosphere and the air spaces of the lungs.
2. **External (pulmonary) respiration.** This is the exchange of gases between the air spaces of the lungs and blood in pulmonary capillaries across the respiratory membrane. The blood gains O_2 and loses CO_2.

3. **Internal (tissue) respiration.** This is the exchange of gases between blood in systemic capillaries and tissue cells. The blood loses O_2 and gains CO_2.

The flow of air between the atmosphere and lungs occurs for the same reason that blood flows through the body: A pressure gradient (difference) exists. Air moves into the lungs when the pressure inside the lungs is less than the air pressure in the atmosphere. Air moves out of the lungs when the pressure inside the lungs is greater than the pressure in the atmosphere.

Inhalation

Breathing in is called **inhalation (inspiration).** Just before each inhalation, the air pressure inside the lungs is equal to the pressure of the atmosphere, which at sea level is about 760 millimeters of mercury (mmHg), or 1 atmosphere (atm). For air to flow into the lungs, the pressure inside the alveoli must become lower than the atmospheric pressure. This condition is achieved by increasing the size of the lungs.

For inhalation to occur, the lungs must expand. This increases lung volume and thus decreases the pressure in the lungs below atmospheric pressure. The first step in expanding the alveoli of the lungs during normal quiet breathing involves contraction of the principal muscles of inhalation—the diaphragm and/or external intercostals (Figure 24.12).

The diaphragm, the most important muscle of inhalation, is a dome-shaped skeletal muscle that forms the floor of the thoracic cavity. It is innervated by fibers of the phrenic nerves, which emerge from both sides of the spinal cord at cervical levels 3, 4, and 5. Contraction of the diaphragm causes it to flatten, lowering its dome. This increases the vertical diameter of the thoracic cavity and accounts for the movement of about 75% of the air that enters the lungs during normal quiet inhalation. The distance the diaphragm moves during inspiration ranges from 1 cm (0.4 in.)

during normal quiet breathing up to about 10 cm (4 in.) during strenuous exercise. Advanced pregnancy, excessive obesity, or confining abdominal clothing can prevent a complete descent of the diaphragm. At the same time the diaphragm contracts, the external intercostals contract. These skeletal muscles run obliquely downward and forward between adjacent ribs, and when these muscles contract, the ribs are pulled superiorly and the sternum is pushed anteriorly. This increases the anteroposterior and lateral diameters of the thoracic cavity.

As the diaphragm and external intercostals contract and the overall size of the thoracic cavity increases, the walls of the lungs are pulled outward. The parietal and visceral pleurae normally adhere strongly to each other because of the below-atmospheric pressure between them and because of the surface tension created by their moist adjoining surfaces. As the thoracic cavity expands, the parietal pleura lining the cavity is pulled outward in all directions, and the visceral pleura and lungs are pulled along with it, increasing the volume of the lungs.

When the volume of the lungs increases, alveolar pressure decreases from 760 to 758 mmHg. A pressure gradient is thus established between the atmosphere and the alveoli. Air rushes from the atmosphere into the lungs due to a gas pressure difference, and inhalation takes place. Air continues to move into the lungs as long as the pressure difference exists.

During deep, forceful inhalations, accessory muscles of inspiration also participate in increasing the size of the thoracic cavity (Figure 24.12a). The muscles are so named because they make little, if any, contribution during normal quiet inhalation, but during exercise or forced ventilation they may contract vigorously. The accessory muscles of inhalation include the sternocleidomas-

Figure 24.12 Muscles of inhalation and exhalation and their actions. The pectoralis minor muscle (not shown here) is illustrated in Figure 11.13a on page 346.

During deep, labored breathing, accessory muscles of inhalation (sternocleidomastoids, scalenes, and pectoralis minors) participate.

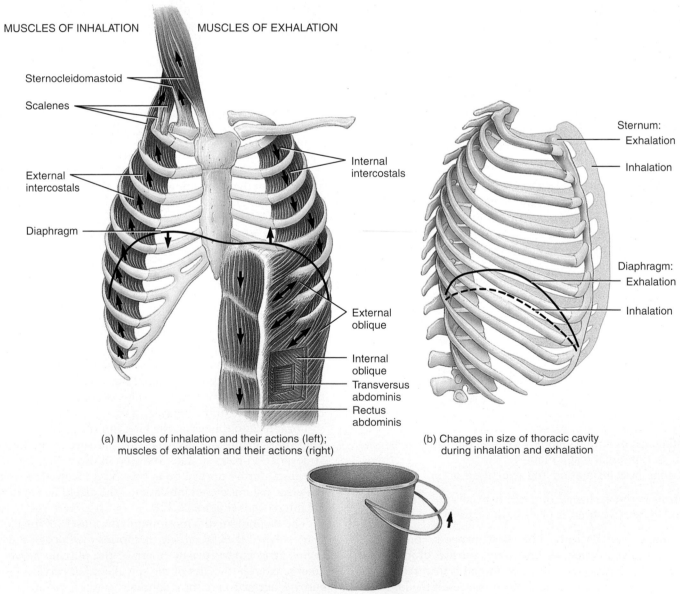

MUSCLES OF INHALATION   MUSCLES OF EXHALATION

Sternocleidomastoid

Scalenes

External intercostals

Diaphragm

Internal intercostals

External oblique

Internal oblique

Transversus abdominis

Rectus abdominis

(a) Muscles of inhalation and their actions (left); muscles of exhalation and their actions (right)

Sternum:
Exhalation
Inhalation

Diaphragm:
Exhalation
Inhalation

(b) Changes in size of thoracic cavity during inhalation and exhalation

(c) During inhalation, the ribs move upward and outward like the handle on a bucket

? **Right now, what is the main muscle that powers your breathing?**

toid muscles, which elevate the sternum; the scalene muscles, which elevate the first two ribs; and the pectoralis minor muscles, which elevate the third through fifth ribs.

Exhalation

Breathing out, called **exhalation (expiration)**, is also achieved by a pressure gradient, but in this case the gradient is reversed: The pressure in the lungs is greater than the pressure of the atmosphere. Normal exhalation during quiet breathing depends on two factors: (1) the recoil of elastic fibers that were stretched during inspiration and (2) the inward pull of surface tension due to the film of alveolar fluid.

Exhalation starts when the inspiratory muscles relax. As the external intercostals relax, the ribs move inferiorly; as the diaphragm relaxes, its dome moves superiorly owing to its elasticity. These movements decrease the vertical, anteroposterior, and lateral diameters of the thoracic cavity. Also, surface tension exerts an inward pull between the parietal and visceral pleurae and the elastic basement membranes of the alveoli and elastic fibers in bronchioles and alveolar ducts recoil. As a result, lung volume decreases and the alveolar pressure increases to 762 mmHg. Air then flows from the area of higher pressure in the alveoli to the area of lower pressure in the atmosphere.

During labored breathing and when air movement out of the lungs is impeded, muscles of exhalation—abdominal and internal intercostals—contract. Contraction of the abdominal muscles moves the inferior ribs inferiorly and compresses the abdominal viscera, thus forcing the diaphragm superiorly. Contraction of the internal intercostals, which extend inferiorly and posteriorly between adjacent ribs, also pulls the ribs inferiorly.

Respirations also provide humans with methods for expressing emotions such as laughing, sighing, and sobbing. Moreover, respiratory air can be used to expel foreign matter from the lower air passages through actions such as sneezing and coughing. Respiratory movements can also be modified and controlled when you talk or sing. Some of the modified respiratory movements that express emotion or clear the airways are listed in Table 24.1. All of these movements are reflexes, but some of them can be initiated voluntarily.

◉ **CHECKPOINT**

13. What are the basic differences among pulmonary ventilation, external respiration, and internal respiration?
14. Compare what happens during quiet ventilation and forceful ventilation.
15. List and describe the various types of modified respiratory movements.

REGULATION OF RESPIRATION

◉ **OBJECTIVE**

• Describe the various factors that regulate the rate and depth of respiration.

Although breathing can be controlled voluntarily for short periods, the nervous system usually controls respirations automatically to meet the body's demand without conscious effort.

Role of the Respiratory Center

The size of the thorax is altered by the action of the respiratory muscles, which contract and relax as a result of nerve impulses transmitted to them from centers in the brain. The area from

TABLE 24.1	
Modified Respiratory Movements	
MOVEMENT	**DESCRIPTION**
Coughing	A long-drawn and deep inhalation followed by a complete closure of the rima glottidis, which results in a strong exhalation that suddenly pushes the rima glottidis open and sends a blast of air through the upper respiratory passages. Stimulus for this reflex act may be a foreign body lodged in the larynx, trachea, or epiglottis.
Sneezing	Spasmodic contraction of muscles of exhalation that forcefully expels air through the nose and mouth. Stimulus may be an irritation of the nasal mucosa.
Sighing	A long-drawn and deep inhalation immediately followed by a shorter but forceful exhalation.
Yawning	A deep inhalation through the widely opened mouth producing an exaggerated depression of the mandible. It may be stimulated by drowsiness, or someone else's yawning, but the precise cause is unknown.
Sobbing	A series of convulsive inhalations followed by a single prolonged exhalation. The rima glottidis closes earlier than normal after each inhalation so only a little air enters the lungs with each inhalation.
Crying	An inhalation followed by many short convulsive exhalations, during which the rima glottidis remains open and the vocal folds vibrate; accompanied by characteristic facial expressions and tears.
Laughing	The same basic movements as crying, but the rhythm of the movements and the facial expressions usually differ from those of crying. Laughing and crying are sometimes indistinguishable.
Hiccupping	Spasmodic contraction of the diaphragm followed by a spasmodic closure of the rima glottidis, which produces a sharp sound on inhalation. Stimulus is usually irritation of the sensory nerve endings of the gastrointestinal tract.
Valsalva (val-SAL-va) maneuver	Forced exhalation against a closed rima glottidis as may occur during periods of straining while defecating.

which nerve impulses are sent to respiratory muscles consists of clusters of neurons located bilaterally in the medulla oblongata and pons of the brain stem. This area, called the **respiratory center**, consists of a widely dispersed group of neurons functionally divided into three areas: (1) the medullary rhythmicity area in the medulla oblongata; (2) the pneumotaxic area in the pons; and (3) the apneustic area, also in the pons (Figure 24.13).

Medullary Rhythmicity Area

The function of the **medullary rhythmicity area** (rith-MIS-i-tē) is to control the basic rhythm of respiration, which in the normal resting state is about 2 seconds of inhalation and 3 seconds of exhalation. Within the medullary rhythmicity area are inspiratory and expiratory neurons that constitute inspiratory and expiratory areas, respectively. We first consider the role of the inspiratory neurons in respiration.

Figure 24.13 Location of the respiratory center in the brain.

 The respiratory center is composed of neurons in the medullary rhythmicity area in the medulla oblongata plus the pneumotaxic and apneustic areas in the pons.

Sagittal plane

RESPIRATORY CENTER:

Pneumotaxic area

Apneustic area

Medullary rhythmicity area:
Inspiratory area

Expiratory area

Midbrain

Pons

Medulla oblongata

Spinal cord

Sagittal section of brain stem

? Which area contains neurons that are active and then inactive in a repeating cycle?

Nerve impulses generated in the **inspiratory area** establish the basic rhythm of breathing. While the inspiratory area is active, it generates nerve impulses for about 2 seconds. The impulses propagate to the external intercostal muscles via intercostal nerves and to the diaphragm via the phrenic nerves. When the nerve impulses reach the diaphragm and external intercostal muscles, the muscles contract and inhalation occurs. Even when all incoming nerve connections to the inspiratory area are cut or blocked, neurons in this area still rhythmically discharge impulses that cause inhalation. At the end of 2 seconds, the inspiratory area becomes inactive and nerve impulses cease. With no impulses arriving, the diaphragm and external intercostal muscles relax for about 3 seconds, allowing passive elastic recoil of the lungs and thoracic wall. Then, the cycle repeats.

The neurons of the **expiratory area** remain inactive during quiet breathing. However, during forceful breathing nerve impulses from the inspiratory area activate the expiratory area. Impulses from the expiratory area cause contraction of the internal intercostal and abdominal muscles, which decreases the size of the thoracic cavity and causes forceful exhalation.

Pneumotaxic Area

Although the medullary rhythmicity area controls the basic rhythm of respiration, other sites in the brain stem help coordinate the transition between inhalation and exhalation. One of these sites is the **pneumotaxic area** (noo-mō-TAK-sik; *pneumo-* =air or breath; *-taxic*=arrangement) in the upper pons (see Figure 24.13), which transmits inhibitory impulses to the inspiratory area. The major effect of these nerve impulses is to help turn off

the inspiratory area before the lungs become too full of air. In other words, the impulses shorten the duration of inhalation. When the pneumotaxic area is more active, breathing rate is more rapid.

Apneustic Area

Another part of the brain stem that coordinates the transition between inhalation and exhalation is the **apneustic area** (ap-NOO-stik) in the lower pons (see Figure 24.13). This area sends stimulatory impulses to the inspiratory area that activate it and prolong inhalation. The result is a long, deep inhalation. When the pneumotaxic area is active, it overrides the signals from the apneustic area.

Regulation of the Respiratory Center

Although the basic rhythm of respiration is set and coordinated by the inspiratory area, the rhythm can be modified in response to inputs from other brain regions, receptors in the peripheral nervous system, and other factors.

Cortical Influences on Respiration

Because the cerebral cortex has connections with the respiratory center, we can voluntarily alter our pattern of breathing. We can even refuse to breathe at all for a short time. Voluntary control is protective because it enables us to prevent water or irritating gases from entering the lungs. The ability to not breathe, however, is limited by the buildup of CO_2 and H^+ in the body. When CO_2 and H^+ concentrations increase to a certain level, the inspiratory area is strongly stimulated, nerve impulses are sent along the phrenic and intercostal nerves to inspiratory muscles, and breathing resumes, whether you want it to or not. It is impossible for people to kill themselves by voluntarily holding their breath. Even if you hold your breath long enough that you faint, breathing resumes when consciousness is lost. Nerve impulses from the hypothalamus and limbic system also stimulate the respiratory center, allowing emotional stimuli to alter respirations as, for example, when you laugh or cry.

Chemoreceptor Regulation of Respiration

Certain chemical stimuli determine how quickly and how deeply we breathe. The respiratory system functions to maintain proper levels of CO_2 and O_2 and is very responsive to changes in the levels of either in body fluids. Sensory neurons that are responsive to chemicals are termed **chemoreceptors**. Chemoreceptors in two locations monitor levels of CO_2, H^+, and O_2 and provide input to the respiratory center (Figure 24.14). **Central chemoreceptors** are located in the medulla oblongata in the *central* nervous system. They respond to changes in H^+ or CO_2 concentration, or both, in cerebrospinal fluid. **Peripheral chemoreceptors** are located in the **aortic bodies**, clusters of chemoreceptors located in the wall of the arch of the aorta, and in the **carotid bodies**, which are oval nodules in the wall of the left and right common carotid arteries where they divide into the internal and external carotid arteries. These chemoreceptors are part of the *peripheral* nervous system and are sensitive to changes in O_2, H^+, and CO_2 in the blood. Axons of sensory neurons from the aortic bodies are part of the vagus (X) nerves, and those from the carotid bodies are part of the right and left glossopharyngeal (IX) nerves.

If there is even a slight increase in CO_2, central and peripheral chemoreceptors are stimulated. The chemoreceptors send nerve impulses to the brain that cause the inspiratory area to become highly active, and the rate of respiration increases. This allows the body to expel more CO_2 until the CO_2 is lowered to normal. If arterial CO_2 is lower than normal, the chemoreceptors are not stimulated, and stimulatory impulses are not sent to the inspiratory area. Consequently, the rate of respiration decreases until CO_2 accumulates and the CO_2 level rises to normal.

The Inflation Reflex

Located in the walls of bronchi and bronchioles are stretch-sensitive receptors called **baroreceptors** or **stretch receptors**. When these receptors become stretched during overinflation of

the lungs, nerve impulses are sent along the vagus (X) nerves to the inspiratory and apneustic areas. In response, the inspiratory area is inhibited, and the apneustic area is inhibited from activating the inspiratory area. As a result, expiration begins. As air leaves the lungs during expiration, the lungs deflate and the stretch receptors are no longer stimulated. Thus, the inspiratory and apneustic areas are no longer inhibited, and a new inspiration begins. Some evidence suggests that this reflex, referred to as the **inflation (Hering-Breuer) reflex**, is mainly a protective mechanism for preventing excessive inflation of the lungs rather than a key component in the normal regulation of respiration.

◉ CHECKPOINT

16. How does the medullary rhythmicity area function in regulating respiration? How are the apneustic and pneumotaxic areas related to the control of respiration?
17. Explain how each of the following modifies respiration: cerebral cortex, inflation reflex, CO_2 levels, and O_2 levels.

Figure 24.14 **Locations of peripheral chemoreceptors.**

 Chemoreceptors are sensory neurons that respond to changes in the levels of certain chemicals in the body.

- Medulla oblongata
- Sensory axons in glossopharyngeal nerve (cranial nerve IX)
- Internal carotid artery
- Carotid body
- External carotid artery
- Carotid sinus
- Common carotid artery
- Sensory axons in vagus nerve (cranial nerve X)
- Arch of aorta
- Aortic bodies
- Heart

? **Which chemicals stimulate peripheral chemoreceptors?**

Pneumonia is an acute infection or inflammation of the alveoli. It is the most common infectious cause of death in the United States, where an estimated 4 million cases occur annually. When certain microbes enter the lungs of susceptible individuals, they release damaging toxins, stimulating inflammation and immune responses that have damaging side effects. The toxins and immune response damage alveoli and bronchial mucous membranes; inflammation and edema cause the alveoli to fill with fluid, interfering with ventilation and gas exchange.

The most common cause of pneumonia is the pneumococcal bacterium *Streptococcus pneumoniae* (Figure A), but other microbes may also cause pneumonia. Those who are most susceptible to pneumonia are the elderly, infants, immunocompromised individuals (AIDS or cancer patients, or those taking immunosuppressive drugs), cigarette smokers, and individuals with an obstructive lung disease. Most cases of pneumonia are preceded by an upper respiratory infection that often is viral. Individuals then develop fever, chills, productive or dry cough, malaise, chest pain, and sometimes dyspnea (difficult breathing) and hemoptysis (spitting blood).

Treatment may involve antibiotics, bronchodilators, oxygen therapy, increased fluid intake, and chest physiotherapy (percussion, vibration, and postural drainage). ●

SEM

(A) *Streptococcus pneumoniae*, the most common cause of pneumonia

EXERCISE AND THE RESPIRATORY SYSTEM

◉ **OBJECTIVE**

• Describe the effects of exercise on the respiratory system.

The respiratory and cardiovascular systems make adjustments in response to both intensity and duration of exercise. The effects of exercise on the heart are discussed in Chapter 14. Here we focus on how exercise affects the respiratory system.

The heart pumps the same amount of blood to the lungs as to all the rest of the body. Thus, as cardiac output rises, the rate of blood flow through the lungs also increases. As blood flows more rapidly through the lungs, it picks up more O_2. In addition, the rate at which O_2 diffuses from alveolar air into the blood increases during maximal exercise because blood flows through a larger percentage of the pulmonary capillaries, providing a greater surface area for the diffusion of O_2 into blood.

When muscles contract during exercise, they consume large amounts of O_2 and produce large amounts of CO_2. During vigorous exercise, O_2 consumption and pulmonary ventilation both increase dramatically. At the onset of exercise, an abrupt increase in pulmonary ventilation is due to neural changes that send excitatory impulses to the inspiratory area in the medulla oblongata. The more gradual increase in ventilation during moderate exercise is due to chemical and physical changes in the bloodstream. During moderate exercise, the depth of ventilation (bigger breaths) changes more than the breathing rate. When exercise is more strenuous, the frequency of breathing also increases.

At the end of an exercise session, an abrupt decrease in pulmonary ventilation is followed by a more gradual decline to the resting level. The initial decrease is due mainly to changes in neural factors when movement stops or slows; the more gradual phase reflects the slower return of blood chemistry levels and temperature to the resting state.

◉ **CHECKPOINT**

18. How does exercise affect the inspiratory area?
19. Describe the changes in pulmonary ventilation caused by a brisk walk in the park.

DEVELOPMENT OF THE RESPIRATORY SYSTEM

◉ **OBJECTIVE**

• Describe the development of the respiratory system.

The development of the mouth and pharynx are discussed in Chapter 25. Here we consider the development of the remainder of the respiratory system.

At about four weeks of development, the respiratory system begins as an outgrowth of the foregut just anterior to the pharynx. This outgrowth is called the **respiratory diverticulum** (lung bud) (Figure 24.15; see also Figure 23.8a on page 747). The **endoderm** lining the respiratory diverticulum gives rise to the epithelium and glands of the trachea, bronchi, and alveoli. **Splanchnic mesoderm** (see Figure 4.9d on page 103) surrounding the respiratory diverticulum gives rise to the connective tissue, cartilage, and smooth muscle of these structures.

The epithelial lining of the *larynx* develops from the **endoderm** of the respiratory diverticulum; the cartilages and muscles originate from the **fourth** and **sixth pharyngeal arches** (see Figure 4.13 on page 108).

Figure 24.15 Development of the bronchial tubes and lungs.

 The respiratory system develops from endoderm and mesoderm.

? When does the respiratory system begin to develop in an embryo?

As the respiratory diverticulum elongates, its distal end enlarges to form a globular **tracheal bud**, which gives rise to the *trachea*. Soon after, the tracheal bud divides into **bronchial buds**, which branch repeatedly and develop with the *bronchi*. By 24 weeks, 17 orders of branches have formed and *respiratory bronchioles* have developed.

During weeks 6 to 16, all major elements of the *lungs* have formed, except for those involved in gaseous exchange (respiratory bronchioles, alveolar ducts, and alveoli). Since respiration is not possible at this stage, fetuses born during this time cannot survive.

During weeks 16 to 26, lung tissue becomes highly vascular and respiratory bronchioles, alveolar ducts, and some primitive alveoli develop. Although it is possible for a fetus born near the end of this period to survive if given intensive care, death frequently occurs due to the immaturity of the respiratory and other systems.

From 26 weeks to birth, many more primitive alveoli develop; they consist of type I alveolar cells (main sites of gaseous exchange) and type II surfactant-producing cells. Blood capillaries also establish close contact with the primitive alveoli. Recall that surfactant is necessary to lower surface tension of alveolar fluid and thus reduce the tendency of alveoli to collapse on exhalation. Although surfactant production begins by 20 weeks, it is present in only small quantities. Amounts sufficient to permit survival of a premature (preterm) infant are not produced until 26 to 28 weeks' gestation. Infants born before 26–28 weeks are severely at risk of **respiratory distress syndrome (RDS)**, in which the alveoli collapse during exhalation and must be reinflated during inhalation. The condition is treated by employing respirators that force air into the lungs and by administering surfactant.

At about 30 weeks, mature alveoli develop. However, it is estimated that only about one-sixth of the full complement of alveoli develop before birth; the remainder develop after birth during the first eight years.

As the lungs develop, they acquire their *pleural sacs*. The *visceral pleura* develops from **splanchnic mesoderm** and the *parietal pleura* develops from **somatic mesoderm** (see Figure 4.9d on page 103). The space between the pleural layers is the *pleural cavity*.

During development, breathing movements of the fetus cause the aspiration of fluid into the lungs. The fluid is a mixture of amniotic fluid, mucus from the bronchial glands, and surfactant. At birth, the lungs are about half-filled with fluid. When breathing begins at birth, most of the fluid is rapidly reabsorbed by blood and lymph capillaries and a small amount is expelled through the nose and mouth during delivery.

◉ CHECKPOINT

20. What structures develop from the respiratory diverticulum?
21. Which respiratory structures develop from endoderm? From mesoderm?
22. How many weeks old must a fetus be for it to survive as a preterm infant? Why?

AGING AND THE RESPIRATORY SYSTEM

◉ OBJECTIVE

- Describe the effects of aging on the respiratory system.

With advancing age, the airways and tissues of the respiratory tract, including the alveoli, become less elastic and more rigid; the chest wall becomes more rigid as well. The result is a decrease in lung capacity. In fact, vital capacity (the maximum amount of air that can be expired after maximal inhalation) can decrease as much as 35% by age 70. A decrease in blood level of O_2, decreased activity of alveolar macrophages, and diminished ciliary action of the epithelium lining the respiratory tract also occur. Because of all these age-related factors, elderly people are more susceptible to pneumonia, bronchitis, emphysema, and other pulmonary disorders. Age-related changes in the structure and functions of the lung can also contribute to an older person's reduced ability to perform vigorous exercises, such as running.

◉ CHECKPOINT

23. What accounts for the decrease in lung capacity with aging?

KEY MEDICAL TERMS ASSOCIATED WITH THE RESPIRATORY SYSTEM

Abdominal thrust (Heimlich) maneuver (HĪM-lik ma-NOO-ver) First-aid procedure designed to clear the airways of obstructing objects. It is performed by applying a quick upward thrust between the navel and costal margin that causes sudden elevation of the diaphragm and forceful, rapid expulsion of air from the lungs; this action forces air out the trachea to eject the obstructing object. The abdominal thrust maneuver is also used to expel water from the lungs of near-drowning victims before resuscitation is begun.

Apnea (AP-nē-a; *a*=without; *pnoia*=air or breath) Absence of ventilatory movements.

Asphyxia (as-FIK-sē-a; *sphyxia*=pulse) Oxygen starvation due to low atmospheric oxygen or interference with ventilation, external respiration, or internal respiration.

Aspiration (as′-pi-RĀ-shun) Inhalation of a foreign substance such as water, food, or a foreign body into the bronchial tree; also, the drawing of a substance in or out by suction.

Bronchiectasis (bron′-kē-EK-ta-sis; *-ektasis*=stretching) A chronic dilation of the bronchi or bronchioles resulting from damage to the bronchial wall, for example, from respiratory infections.

Bronchography (bron-KOG-ra-fē) An imaging technique used to visualize the bronchial tree using x-rays. After an opaque contrast medium is inhaled through an intratracheal catheter, radiographs of the chest in various positions are taken, and the developed film, a **bronchogram** (BRON-kō-gram), provides a picture of the bronchial tree.

Bronchoscopy (bron-KOS-kō-pē) Visual examination of the bronchi through a **bronchoscope**, an illuminated, flexible tubular instrument that is passed through the mouth (or nose), larynx, and trachea into the bronchi. The examiner can view the interior of the trachea and bronchi to biopsy a tumor, clear an obstructing object or secretions from an airway, take cultures or smears for microscopic examination, stop bleeding, or deliver drugs.

Cheyne-Stokes respiration (CHĀN STŌKS res′-pi-RĀ-shun) A repeated cycle of irregular breathing that begins with shallow breaths that increase in depth and rapidity and then decrease and cease altogether for 15–20 seconds. Cheyne-Stokes is normal in infants; it is also often seen just before death from pulmonary, cerebral, cardiac, or kidney disease.

Dyspnea (DISP-nē-a; *dys-*=painful, difficult) Painful or labored breathing.

Epistaxis (ep′-i-STAK-sis) Loss of blood from the nose due to trauma, infection, allergy, malignant growths, or bleeding disorders. It can be arrested by cautery with silver nitrate, electrocautery, or firm packing. Also called **nosebleed**.

Hemoptysis (hē-MOP-ti-sis; *hemo-*=blood; *-ptysis*=spit) Spitting of blood from the respiratory tract.

Hyperventilation (*hyper-*=above) Rapid and deep breathing.

Hypoventilation (*hypo-*=below) Slow and shallow breathing.

Mechanical ventilation The use of an automatically cycling device (ventilator or respirator) to assist breathing. A plastic tube is inserted into the nose or mouth and the tube is attached to a device that forces air into the lungs. Exhalation occurs passively due to the elastic recoil of the lungs.

Pulmonary embolism (EM-bō-lizm) The blockage of a pulmonary artery or its branches by a blood clot that travels to the lungs, usually from a vein in a leg or the pelvis.

Rales (RĀLS) Sounds sometimes heard in the lungs that resemble bubbling or rattling due to the presence of an abnormal type or amount of fluid or mucus within the bronchi or alveoli, or to bronchoconstriction that causes turbulent airflow.

Respirator (RES-pi-rā′-tor) An apparatus fitted to a mask over the nose and mouth, or hooked directly to an endotracheal or tracheotomy tube, that is used to assist or support ventilation or to provide nebulized medication to the air passages.

Respiratory failure A condition in which the respiratory system either cannot supply sufficient O_2 to maintain metabolism or cannot eliminate enough CO_2 to prevent respiratory *acidosis* (a lower-than-normal pH in interstitial fluid).

Rhinitis (rī-NĪ-tis; *rhin-*=nose) Chronic or acute inflammation of the mucous membrane of the nose due to viruses, bacteria, or irritants. Excessive mucus production produces a runny nose, nasal congestion, and postnasal drip.

Sleep apnea (AP-nē-a; *a-*=without; *-pnea*=breath) A disorder in which a person repeatedly stops breathing for 10 or more seconds while sleeping, most often because loss of muscle tone in pharyngeal muscles allows the airway to collapse.

Sputum (SPŪ-tum=to spit) Mucus and other fluids from the air passages that is *expectorated* (expelled by coughing).

Strep throat Inflammation of the pharynx caused by the bacterium *Streptococcus pyogenes*. It may also involve the tonsils and middle ear.

Sudden infant death syndrome (SIDS) Death of infants between the ages of 1 week and 12 months thought to be due to hypoxia while sleeping in a prone position (on the stomach) and the rebreathing of exhaled air trapped in a depression of the mattress. It is now recommended that normal newborns be placed on their backs for sleeping: "back to sleep."

Tachypnea (tak′-ip-NĒ-a; *tachy-*=rapid; *-pnea*=breath) Rapid breathing rate.

Wheeze (HWĒZ) A whistling, squeaking, or musical high-pitched sound during breathing resulting from a partially obstructed airway.

STUDY OUTLINE

Respiratory System Anatomy (p. 753)

1. The respiratory system consists of the nose, pharynx, larynx, trachea, bronchi, and lungs. They act with the cardiovascular system to supply oxygen (O_2) to and remove carbon dioxide (CO_2) from the blood.
2. The external nose is made of cartilage and skin and is lined with a mucous membrane. Openings to the exterior are the external nares.
3. The internal nose communicates with the paranasal sinuses and nasopharynx through the internal nares.
4. The nasal cavity is divided by a septum. The anterior portion of the cavity is called the nasal vestibule. The nose warms, moistens, and filters air and functions in olfaction and speech.
5. The pharynx (throat) is a muscular tube lined by a mucous membrane. The anatomical regions of the pharynx are the nasopharynx, oropharynx, and laryngopharynx.
6. The nasopharynx functions in respiration. The oropharynx and laryngopharynx function both in digestion and in respiration.
7. The larynx (voice box) is a passageway that connects the pharynx with the trachea. It contains the thyroid cartilage (Adam's apple); the epiglottis, which prevents food from entering the larynx; the cricoid cartilage, which connects the larynx and trachea; and the paired arytenoid, corniculate, and cuneiform cartilages.
8. The larynx contains vocal folds, which produce sound as they vibrate. Taut folds produce high pitches, and relaxed ones produce low pitches.
9. The trachea (windpipe) extends from the larynx to the primary bronchi. It is composed of C-shaped rings of cartilage and smooth muscle and is lined with pseudostratified ciliated columnar epithelium.
10. The bronchial tree consists of the trachea, primary bronchi, secondary bronchi, tertiary bronchi, bronchioles, and terminal bronchioles. Walls of bronchi contain rings of cartilage; walls of bronchioles contain increasingly smaller plates of cartilage and increasing amounts of smooth muscle.
11. Lungs are paired organs in the thoracic cavity enclosed by the pleural membrane. The parietal pleura is the superficial layer that lines the thoracic cavity; the visceral pleura is the deep layer that covers the lungs.
12. The right lung has three lobes separated by two fissures; the left lung has two lobes separated by one fissure, along with a depression, the cardiac notch.
13. Secondary bronchi give rise to branches called segmental bronchi, which supply segments of lung tissue called bronchopulmonary segments.
14. Each bronchopulmonary segment consists of lobules, which contain lymphatics, arterioles, venules, terminal bronchioles, respiratory bronchioles, alveolar ducts, alveolar sacs, and alveoli.
15. Alveolar walls consist of type I alveolar cells, type II alveolar cells, and associated alveolar macrophages.
16. Gas exchange occurs across the respiratory membranes.

Mechanics of Pulmonary Ventilation (Breathing) (p. 771)

1. Pulmonary ventilation, or breathing, consists of inhalation and exhalation.
2. Inhalation occurs when alveolar pressure falls below atmospheric pressure. Contraction of the diaphragm and external intercostals increases the size of the thorax, thereby decreasing the intrapleural pressure so that the lungs expand. Expansion of the lungs decreases alveolar pressure so that air moves down a pressure gradient from the atmosphere into the lungs.
3. During forceful inhalation, accessory muscles of inhalation (sternocleidomastoids, scalenes, and pectoralis minors) are also used.
4. Exhalation occurs when alveolar pressure is higher than atmospheric pressure. Relaxation of the diaphragm and external intercostals results in elastic recoil of the chest wall and lungs, which increases intrapleural pressure; lung volume decreases and alveolar pressure increases, so air moves from the lungs to the atmosphere.
5. Forceful exhalation involves contraction of the internal intercostal and abdominal muscles.

Regulation of Respiration (p. 773)

1. The respiratory center consists of a medullary rhythmicity area, a pneumotaxic area, and an apneustic area.
2. The inspiratory area sets the basic rhythm of respiration.
3. The pneumotaxic and apneustic areas coordinate the transition between inspiration and expiration.
4. Respiration may be modified by a number of factors, including cortical influences; the inflation reflex; and chemical stimuli, such as O_2, CO_2, and H^+ levels.

Exercise and the Respiratory System (p. 776)

1. The rate and depth of ventilation change in response to both the intensity and duration of exercise.
2. An increase in pulmonary blood flow and O_2-diffusing capacity occurs during exercise.
3. The abrupt increase in ventilation at the start of exercise is due to neural changes that send excitatory impulses to the inspiratory area in the medulla oblongata. The more gradual increase in ventilation during moderate exercise is due to chemical and physical changes in the bloodstream.

Development of the Respiratory System (p. 776)

1. The respiratory system begins as an outgrowth of endoderm called the respiratory diverticulum.
2. Smooth muscle, cartilage, and connective tissue of the bronchial tubes and pleural sacs develop from mesoderm.

Aging and the Respiratory System (p. 777)

1. Aging results in decreased vital capacity, decreased blood level of O_2, and diminished alveolar macrophage activity.
2. Elderly people are more susceptible to pneumonia, emphysema, bronchitis, and other pulmonary disorders.

SELF-QUIZ QUESTIONS

Choose the one best answer to the following questions.

1. Which of the following is a part of the cartilaginous framework of the nose?
   a. arytenoid cartilage
   b. cricoid cartilage
   c. alar cartilage
   d. cartilage of the carina
   e. cuneiform cartilage

2. Which is the correct order, from superficial to deep, of the following five structures? (1) parietal pleura, (2) visceral pleura, (3) pleural cavity, (4) lung, (5) wall of thoracic cavity.
   a. 5,1,3,2,4   b. 5,3,1,2,4   c. 4,2,3,1,5
   d. 5,1,2,4,3   e. 5,2,3,1,4

3. The nasopharynx extends from the
   a. exterior of the body to the nasal conchae.
   b. oropharynx to the laryngopharynx.
   c. nasal cavity to the soft palate.
   d. soft palate to the hyoid bone.
   e. mouth to the esophagus.

4. The trachea extends from the
   a. laryngopharynx to the larynx.
   b. epiglottis to the bronchi.
   c. hyoid bone to the fourth cervical vertebra.
   d. fourth to sixth cervical vertebrae.
   e. larynx to the fifth thoracic vertebra.

5. Pseudostratified ciliated columnar epithelium is found in the
   a. visceral pleura.   b. alveoli.   c. oropharynx.
   d. trachea.   e. Both c and d are correct.

6. Which of the following is *not* a component of the conducting portion of the respiratory system?
   a. terminal bronchiole   b. larynx   c. bronchus
   d. trachea   e. respiratory bronchiole

7. Which of the following structures is the smallest in diameter?
   a. left primary bronchus
   b. respiratory bronchioles
   c. secondary bronchi
   d. alveolar ducts
   e. right primary bronchus

8. Inhalation (inspiration) occurs when:
   a. the phrenic nerve stimulates contraction of the diaphragm, which then flattens, thus increasing the volume of the thoracic cavity.
   b. the diaphragm relaxes and flattens, thus increasing the volume of the thoracic cavity.
   c. the phrenic nerve stimulates contraction of the diaphragm, which then forms a dome, thus decreasing the volume of the thoracic cavity.
   d. the vagus (X) nerve stimulates contraction of the diaphragm, which then flattens, thus increasing the volume of the thoracic cavity.
   e. the vagus (X) nerve stimulates contraction of the diaphragm, which then forms a dome, thus decreasing the volume of the thoracic cavity.

9. The spasmodic contraction of the diaphragm followed by a spasmodic closure of the rima glottidis produces a sharp sound on inhalation. The stimulus for this respiratory movement is usually irritation of the sensory nerve endings of the gastrointestinal tract. What particular respiratory movement has just been described?
   a. sneezing   b. yawning   c. laughing
   d. hiccupping   e. coughing

Complete the following.

10. C-shaped cartilage rings support the _____.

11. The flap-like structure that prevents aspiration of solid and liquid substances into the trachea is called the _____.

12. The space between the vocal folds is called the rima _____, whereas the space between the ventricular folds is called the rima _____.

13. A section of lung tissue supplied by a tertiary bronchus is called a _____.

14. Place numbers in the blanks to indicate the route through which air passes as it enters a lobule en route to alveoli. (a) alveolar ducts: _____; (b) respiratory bronchiole: _____; (c) terminal bronchiole: _____.

15. Collapse of all or part of a lung is called _____.

16. The component of alveolar fluid that reduces the tendency for alveoli to collapse is _____.

17. Yawning is a deep _____ through the widely opened mouth producing an exaggerated depression of the mandible. It may be stimulated by drowsiness or fatigue but the precise cause is unknown.

Are the following statements true or false? If the statement is False, *find and fix the mistake to make the statement* True.

18. Bronchioles have more smooth muscle and less cartilage than bronchi.

19. The pharynx, larynx, and trachea are parts of the upper respiratory system.

20. The trachea is posterior to the esophagus.

21. The branches of the aorta that carry blood to the lungs are the pulmonary arteries.

22. Forced exhalation against a closed rima glottidis as may occur during periods of straining while defecating is referred to as the *Valsalva maneuver.*

Matching

23. Match the following terms to their descriptions:
    ___ (a) thicker, broader, and shorter than the other lung
    ___ (b) has a cardiac notch
    ___ (c) has only two secondary bronchi
    ___ (d) has a horizontal fissure
    ___ (e) has a primary bronchus that is shorter, wider, and more vertical

    (1) right lung
    (2) left lung

24. Match the following terms to their descriptions:
    ___ **(a)** nasal conchae and meatuses are located here
    ___ **(b)** contains pharyngeal tonsil
    ___ **(c)** palatine tonsils are located here
    ___ **(d)** this structure leads directly into the esophagus
    ___ **(e)** cricoid, epiglottis, and thyroid cartilages are located here
    ___ **(f)** vocal cords located here enable voice production
    ___ **(g)** both a respiratory and a digestive pathway (two answers)
    ___ **(h)** fauces opens into this structure

    (1) nasal cavity
    (2) oropharynx
    (3) laryngopharynx
    (4) larynx
    (5) nasopharynx

25. Match the following:
    ___ **(a)** cuboidal epithelium with microvilli that secrete alveolar fluid
    ___ **(b)** found in the medulla oblongata; sensitive to H^+ and CO_2 levels in the cerebrospinal fluid
    ___ **(c)** simple squamous epithelial cells forming the lining of the alveolar wall
    ___ **(d)** located in the walls of bronchi and bronchioles; sensitive to stretch
    ___ **(e)** make up the capillary wall of the respiratory membrane
    ___ **(f)** found in the wall of the aortic arch; sensitive to levels of O_2, CO_2, and H^+ in blood
    ___ **(g)** phagocytic cells

    (1) type I alveolar cells
    (2) type II alveolar cells
    (3) alveolar macrophages
    (4) endothelial cells
    (5) central chemoreceptors
    (6) peripheral chemoreceptors
    (7) baroreceptors

CRITICAL THINKING QUESTIONS

1. Your friend Hedge wants to pierce his nose to go along with the 6 earrings in his ear. He thinks a ring through the center would be awesome but wonders if there's a difference between piercing the center versus the side of the nose. Is there?

2. Suzanne is traveling high into the Andes Mountains of South America to visit some ancient Incan ruins. Though she is in excellent physical condition, she finds herself breathing rapidly. What is happening to Suzanne and why/how?

3. Upon placement of an endotracheal tube (ET) into an anesthetized patient, the anesthesiology resident realized that air sounds were coming from the epigastric region rather than from the lungs. What went wrong?

4. Gretchen, who is nine years old, is upset because after finally persuading her daddy to play soccer with her, he had to sit down and rest after only 15 minutes. She indignantly berated him, saying, "You know, Daddy, if you'd stop smoking, you could play longer." Why is it so hard for Gretchen's father to catch his breath? Be specific.

5. LaTasha is losing her patience with her little sister, LaTonya. "I'm going to hold my breath 'til I turn blue and die and then you're gonna get it!" screams LaTonya. LaTasha is not too worried. Why not?

? ANSWERS TO FIGURE QUESTIONS

24.1 The conducting portion of the respiratory system includes the nose, pharynx, larynx, trachea, bronchi, and bronchioles (except the respiratory bronchioles).

24.2 The path of air is external nares → nasal vestibule → nasal cavity → internal nares.

24.3 During swallowing, the epiglottis closes over the rima glottidis, the entrance to the trachea, to prevent aspiration of food and liquids into the lungs.

24.4 The main function of the vocal folds is voice production.

24.5 Because the tissues between the esophagus and trachea are soft, the esophagus can bulge and press against the trachea during swallowing.

24.6 The left lung has two lobes and two secondary bronchi; the right lung has three of each.

24.7 The pleural membrane is a serous membrane.

24.8 Because two-thirds of the heart lies to the left of the midline, the left lung contains a cardiac notch to accommodate the position of the heart. The right lung is shorter than the left because the diaphragm is higher on the right side to accommodate the liver.

24.9 Tertiary bronchi supply a bronchopulmonary segment.

24.10 The wall of an alveolus is made up of type I alveolar cells, type II alveolar cells, and associated alveolar macrophages.

24.11 The respiratory membrane averages 0.5 μm in thickness.

24.12 If you are at rest while reading, your diaphragm is responsible for about 75% of each inhalation.

24.13 The medullary inspiratory area contains autorhythmic neurons that have cycles of activity/inactivity.

24.14 Peripheral chemoreceptors are responsive to changes in the partial pressures of oxygen and carbon dioxide, and concentrations of H^+ in the blood.

24.15 The respiratory system begins to develop about 4 weeks after fertilization.

25 | THE DIGESTIVE SYSTEM

INTRODUCTION Food contains a variety of nutrients—molecules needed for building new body tissues, repairing damaged tissues, and sustaining needed chemical reactions. Food is also vital to life because it is the source of the energy that drives the chemical reactions occurring in every cell of the body. As consumed, however, most food cannot be used as a source of cellular energy. It must first be broken down into molecules small enough to cross the plasma membranes of cells. This breakdown process is known as **digestion**. The passage of these smaller molecules through cells into the blood and lymph is termed **absorption**. The organs that perform these functions—collectively called the **digestive system**—are the focus of this chapter. Like the respiratory system, the digestive system forms from the embryonic gut tube. This tubular system, which extends from the mouth to the anus, forms an extensive surface area in contact with the external environment and is closely associated with the cardiovascular system. The combination of extensive environmental exposure and close association with blood vessels is essential for processing the food that we eat.

The medical specialty that deals with the structure, function, diagnosis, and treatment of diseases of the stomach and intestines is called **gastroenterology** (gas'-trō-en'-ter-OL-ō-jē; *gastro-*=stomach; *enter-*=intestines; *-ology*= study of). The medical specialty that deals with the diagnosis and treatment of disorders of the rectum and anus is called **proctology** (prok-TOL-ō-jē; *proct-* = rectum). •

OVERVIEW OF THE DIGESTIVE SYSTEM

◉ **OBJECTIVES**

- Identify the organs of the digestive system.
- Describe the basic processes performed by the digestive system.

The digestive system (Figure 25.1) is composed of two groups of organs: the gastrointestinal (GI) tract and the accessory digestive organs. The **gastrointestinal (GI) tract**, or **alimentary canal** (*alimentary*=nourishment), is a continuous tube that extends from the mouth to the anus through the thoracic and abdominopelvic cavities. Organs of the gastrointestinal tract include the mouth, pharynx, esophagus, stomach, small intestine, and large intestine. The length of the GI tract is variable. It is about 5-7 meters (16.5-23 ft.) in a living person. It is longer in a cadaver (about 7-9 meters or 23-29.5 ft.) because the muscles along the walls of GI tract organs are in a state of *tonus* (sustained contraction). The **accessory digestive organs** include the teeth, tongue, salivary glands, liver, gallbladder, and pancreas. Teeth aid in the physical breakdown of food, and the tongue assists in chewing and swallowing. The other accessory digestive organs never come into direct contact with food. They produce or store secretions that flow into the GI tract through ducts and aid in the chemical breakdown of food.

The GI tract contains food and its byproducts from the time it is eaten until it is digested and absorbed or eliminated. Muscular contractions in the wall of the GI tract physically break down the food by churning it, and propel the food along the tract, from the esophagus to the anus. The contractions also help to dissolve foods by mixing them with fluids secreted into the tract. Enzymes secreted by accessory digestive organs and cells that line the tract break down the food chemically.

Overall, the digestive system performs six basic functions:

1. **Ingestion.** This process involves taking foods and liquids into the mouth (eating).
2. **Secretion.** Each day, cells within the walls of the GI tract and accessory digestive organs secrete a total of about 7 liters of water, acid, buffers, and enzymes into the lumen (interior space) of the tract.
3. **Mixing and propulsion.** Alternating contraction and relaxation of smooth muscle in the walls of the GI tract mix food and secretions and propel them toward the anus. This capability of the GI tract to mix and move material along its length is termed **motility**.
4. **Digestion.** Mechanical and chemical processes break down ingested food into small molecules. In **mechanical digestion** the teeth cut and grind food before it is swallowed, and then smooth muscles of the stomach and small intestine churn the food. As a result, food molecules become dissolved and thoroughly mixed with digestive enzymes. **Chemical digestion** is the breakdown of the large carbohydrate, lipid, protein, and nucleic acid molecules in food into smaller molecules. Digestive enzymes produced by the salivary glands, tongue, stomach, pancreas, and small intestine speed up these breakdown reactions. A few substances in food can be absorbed without chemical digestion, including amino acids, cholesterol, glucose, vitamins, minerals, and water.
5. **Absorption.** The entrance of ingested and secreted fluids, ions, and the products of digestion into the epithelial cells lining the lumen of the GI tract is called **absorption**. The absorbed substances pass into the blood or lymph and circulate to cells throughout the body.

6. **Defecation.** Wastes, indigestible substances, bacteria, cells sloughed from the lining of the GI tract, and digested materials that were not absorbed leave the body through the anus in a process called **defecation**. The eliminated material is called **feces**.

◉ **CHECKPOINT**

1. Which components of the digestive system are GI tract organs and which are accessory digestive organs?
2. Which organs of the digestive system come in contact with food, and how do they help break it down?

LAYERS OF THE GI TRACT

◉ **OBJECTIVE**

- Describe the layers that form the wall of the gastrointestinal tract.

Like other tubular systems in the body, the wall of the GI tract, from the lower esophagus to the anal canal, has a basic arrangement of four layers. The four layers of the tract, from deep to superficial, are (1) the mucosa, (2) the submucosa, (3) the muscularis, and (4) the serosa (Figure 25.2 on page 784).

Mucosa

The **mucosa**, or inner lining of the GI tract, is a mucous membrane. It is composed of (1) a layer of epithelium in direct contact with the contents of the GI tract, (2) areolar connective tissue, and (3) a thin layer of smooth muscle (muscularis mucosae).

1. The **epithelium** in the mouth, pharynx, esophagus, and anal canal is mainly nonkeratinized stratified squamous epithelium that serves a protective function. Simple columnar epithelium, which functions in secretion and absorption, lines the stomach and intestines. Neighboring simple columnar epithelial cells are firmly sealed to one another by tight junctions that restrict leakage between the cells. The rate of renewal of GI tract epithelial cells is rapid: Every 5–7 days they slough off and are replaced by new cells. Located among the absorptive epithelial cells are exocrine cells that secrete mucus and fluid into the lumen of the tract, and several types of endocrine cells, collectively called **enteroendocrine cells**, that secrete hormones.
2. The **lamina propria** (*lamina*=thin, flat plate; *propria*=one's own) is an areolar connective tissue layer containing many blood and lymphatic vessels that carry the nutrients absorbed by the GI tract to the other tissues of the body. This layer supports the epithelium and binds it to the muscularis mucosae (discussed next). The lamina propria also contains most of the cells of the **mucosa-associated lymphoid tissue (MALT)**. These prominent lymphatic nodules contain immune system cells that protect against disease. MALT is present all along the GI tract, especially in the tonsils, small intestine, appendix, and large intestine, and it contains about as many immune cells as are present in all the rest of the body. The lymphocytes and macrophages in MALT mount immune responses against microbes, such as bacteria, that may penetrate the epithelium.
3. A thin layer of smooth muscle fibers called the **muscularis mucosae** causes the mucous membrane of the stomach and small intestine to form many small folds, increasing the surface area for digestion and absorption. Movements of the muscularis mucosae ensure that all absorptive cells are fully exposed to the contents of the GI tract.

Figure 25.1 Organs of the digestive system.

The organs of the gastrointestinal (GI) tract are the mouth, pharynx, esophagus, stomach, small intestine, and large intestine. Accessory digestive organs include the teeth, tongue, salivary glands, liver, gallbladder, and pancreas.

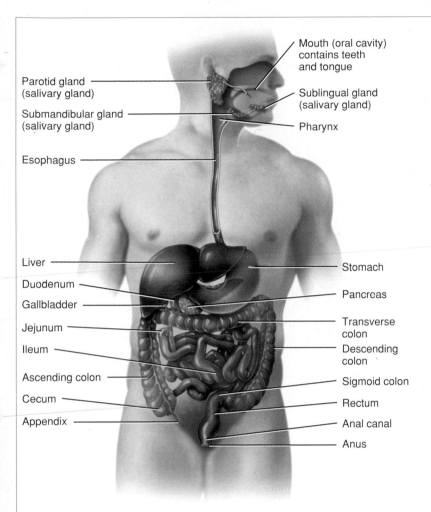

Functions

1. *Ingestion:* taking food into the mouth.
2. *Secretion:* release of water, acid, buffers, and enzymes into the lumen of the GI tract.
3. *Mixing and propulsion:* churning and propulsion of food through the GI tract.
4. *Digestion:* mechanical and chemical breakdown of food.
5. *Absorption:* passage of digested products from the GI tract into the blood and lymph.
6. *Defecation:* the elimination of feces from the GI tract.

(a) Right lateral view of head and neck and anterior view of trunk

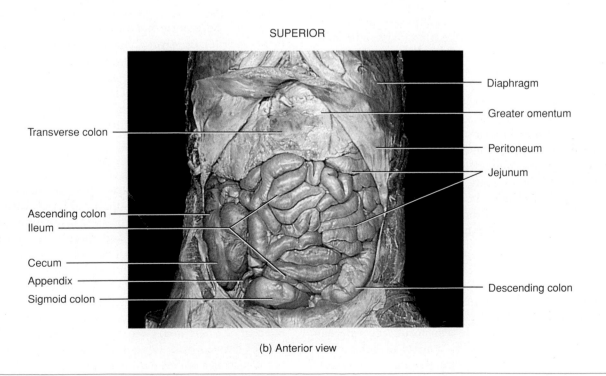

(b) Anterior view

Which structures of the digestive system secrete digestive enzymes?

Figure 25.2 Layers of the gastrointestinal tract.

 The four layers of the gastrointestinal tract, from deep to superficial, are the mucosa, submucosa, muscularis, and serosa.

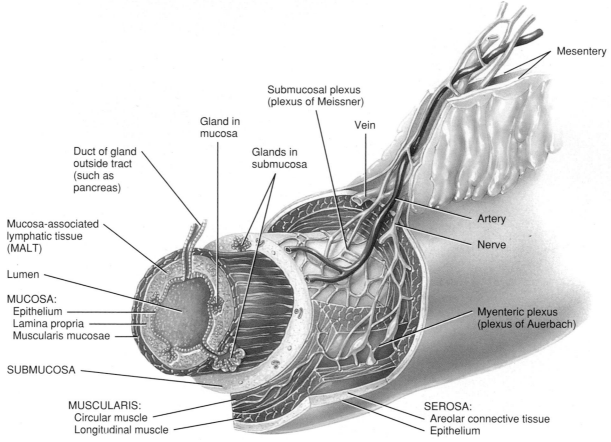

❓ **What is the function of the enteric plexuses in the wall of the gastrointestinal tract?**

Submucosa

The **submucosa** is a thin meshwork of collagenous fibers, nerves, and blood vessels. It consists of areolar connective tissue that binds the mucosa to the middle layer, the muscularis. The submucosa is highly vascular and contains the **submucosal plexus**, or *plexus of Meissner*, a portion of the **enteric nervous system (ENS)**. The ENS is the "brain of the gut" and consists of approximately 100 million neurons in two enteric plexuses that extend the entire length of the GI tract. The submucosal plexus contains sensory and motor enteric neurons, plus parasympathetic and sympathetic postganglionic fibers that innervate the mucosa and submucosa. It regulates movements of the mucosa and vasoconstriction of blood vessels. Because it also innervates secretory cells of mucosal glands, the submucosal plexus is important in controlling secretions of the GI tract. The submucosa may also contain glands and lymphatic tissue.

Muscularis

The **muscularis** of the mouth, pharynx, and superior and middle parts of the esophagus contains *skeletal muscle* that produces voluntary swallowing. Skeletal muscle also forms the external anal sphincter, which permits voluntary control of defecation. Throughout the rest of the tract, the muscularis consists of *smooth muscle* that is generally found in two sheets: an inner sheet of circular fibers and an outer sheet of longitudinal fibers. Involuntary

contractions of the smooth muscles assist in the mechanical breakdown of food, mix it with digestive secretions, and propel it along the tract. The muscularis also contains the second major plexus of the enteric nervous system—the **myenteric plexus** (*my-*=muscle), or *plexus of Auerbach*, which contains enteric neurons, parasympathetic ganglia and postganglionic fibers, and sympathetic postganglionic fibers that are vasomotor to the blood vessels of this layer. This plexus mostly controls GI tract motility (movement), in particular the frequency and strength of the contractions of the muscularis.

Serosa

Those portions of the GI tract that are suspended in the abdominopelvic cavity have a superficial layer called the **serosa**. As its name implies, the serosa is a serous membrane composed of areolar connective tissue and simple squamous epithelium (mesothelium). The epithelial portion of the serosa is also called the *visceral peritoneum*, as it forms the portion of the peritoneum that surrounds the organs that are suspended in the peritoneal cavity, which we will examine in detail shortly. The name *serous membrane* arises from the fact that the epithelium has a lubricating coat of *serous fluid*, a watery solution of electrolytes and other solutes derived from interstitial fluid of the adjacent tissues and blood plasma from local capillaries. The serous fluid also contains various white blood cells. When the highly coiled digestive tube rubs against neighboring areas of the tube or against the inside of the body wall, the serous membrane protects the outer wall of the gut from abrasion. The esophagus lacks a serosa; instead, only a single layer of areolar connective tissue called the *adventitia* surrounds it.

● CHECKPOINT

3. Where along the GI tract is the muscularis composed of skeletal muscle? Is control of this skeletal muscle voluntary or involuntary?

4. What two plexuses form the enteric nervous system, and where are they located?

PERITONEUM

● OBJECTIVE

• Describe the peritoneum and its folds.

The **peritoneum** (per′-i-tō-NĒ-um; *peri-*=around) is the largest serous membrane of the body; it consists of a layer of simple squamous epithelium (mesothelium) with an underlying supporting layer of connective tissue. The peritoneum is divided into the **parietal peritoneum**, which lines the wall of the abdominopelvic cavity, and the **visceral peritoneum** or serosa, which as you just learned covers some of the organs in the cavity (Figure 25.3a). The slim space between the parietal and visceral portions of the peritoneum, called the **peritoneal cavity**, contains lubricating serous fluid. In certain diseases, the peritoneal cavity may become

distended by the accumulation of several liters of fluid, a condition called **ascites** (a-SĪ-tēz).

As we will see, some organs lie against the posterior abdominal wall and do not project into the peritoneal cavity. These organs, called **retroperitoneal organs** (*retro*=behind), are covered by peritoneum only on their anterior surfaces (ascending and descending colon, duodenum, and pancreas) or are separated from the peritoneum by fat and have no peritoneum on them at all (kidney and adrenal glands).

Unlike the pericardium and pleurae, which smoothly cover the heart and lungs, the peritoneum contains large fat-filled folds where the mesothelium reflects (folds back) from the parietal peritoneum to the visceral peritoneum or from visceral peritoneum of one organ to that of another organ. The folds bind the organs to one another and to the walls of the abdominal cavity. They also contain blood vessels, lymphatic vessels, and nerves that supply the abdominal organs. There are five major peritoneal folds: the greater omentum, falciform ligament, lesser omentum, mesentery, and mesocolon.

The **greater omentum** (ō-MEN-tum=fat skin), the largest peritoneal fold, drapes over the transverse colon and coils of the small intestine like a "fatty apron" (Figure 25.3a, d). The greater omentum is a double sheet that folds back upon itself, giving it a total of four layers. From attachments along the stomach and duodenum, the greater omentum extends downward anterior to the small intestine, then turns and extends upward and attaches to the transverse colon. The greater omentum normally contains

---

Figure 25.3 Relationship of the peritoneal folds to each other and to organs of the digestive system. The size of the peritoneal cavity in (a) is exaggerated for emphasis.

🔑 The peritoneum is the largest serous membrane in the body.

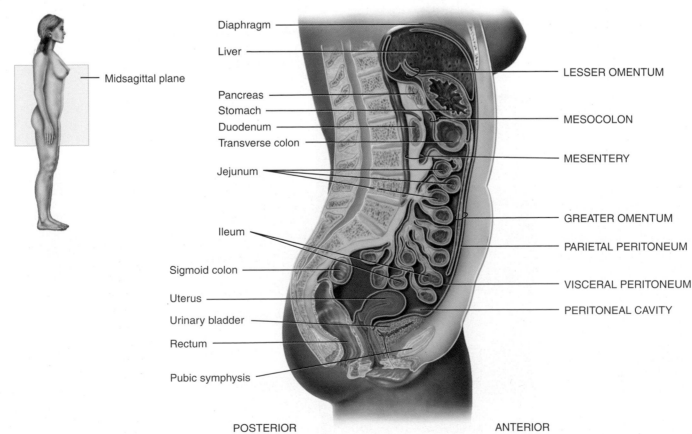

(a) Midsagittal section showing the peritoneal folds

CONTINUES ▶

◾ FIGURE **25.3** CONTINUED ▶

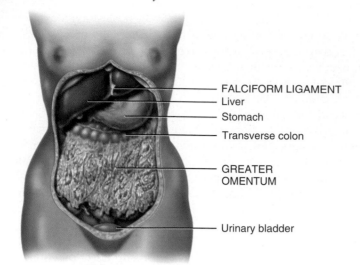

FALCIFORM LIGAMENT
Liver
Stomach
Transverse colon

GREATER OMENTUM

Urinary bladder

(b) Anterior view

Gallbladder

Liver (reflected upward)

Transverse colon

Ascending colon

LESSER OMENTUM

Stomach

Duodenum

Descending colon

Sigmoid colon

(c) Lesser omentum, anterior view (liver and gallbladder lifted)

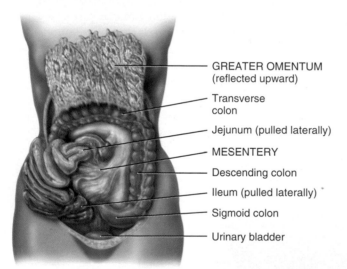

GREATER OMENTUM (reflected upward)

Transverse colon

Jejunum (pulled laterally)

MESENTERY

Descending colon

Ileum (pulled laterally)

Sigmoid colon

Urinary bladder

(d) Anterior view (greater omentum lifted and small intestine reflected to right side)

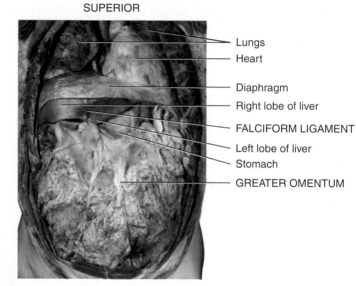

SUPERIOR

Lungs

Heart

Diaphragm

Right lobe of liver

FALCIFORM LIGAMENT

Left lobe of liver

Stomach

GREATER OMENTUM

(e) Anterior view

GREATER OMENTUM (reflected upward)

Transverse colon

Descending colon

Small intestine

Ascending colon

Sigmoid colon

(f) Anterior view

❓ Which peritoneal fold binds the small intestine to the posterior abdominal wall?

a considerable amount of adipose tissue. Its adipose tissue content can greatly expand with weight gain, giving rise to the characteristic "beer belly" seen in some overweight individuals. The many lymph nodes of the greater omentum contribute macrophages and antibody-producing plasma cells that help combat and contain infections of the GI tract.

The **falciform ligament** (FAL-si-form; *falc-*=sickle-shaped) attaches the liver to the anterior abdominal wall and diaphragm (Figure 25.3b, e). The liver is the only digestive organ that is attached to the anterior abdominal wall.

The **lesser omentum** arises as an anterior fold of the serosa of the stomach and duodenum. It suspends the stomach and duodenum from the liver (Figure 25.3a, c). It is the pathway of blood vessels entering the liver and contains the hepatic portal vein, the common hepatic artery, and the common bile duct, along with some lymph nodes.

Another fold of the peritoneum, called the **mesentery** (MEZ-en-ter'-ē; *mes-*=middle), is fan-shaped and binds the jejunum and ileum of the small intestine to the posterior abdominal wall (Figure 25.3a, d). It extends from the posterior abdominal wall to wrap around the small intestine and then returns to its origin, forming a double-layered structure. Between the two layers are blood vessels (branches and tributaries of the superior mesenteric artery and vein), lymphatic vessels, and lymph nodes associated with the jejunum and ileum.

Two separate folds of peritoneum called the **mesocolon** (mez'-ō-KO-lon) bind the transverse colon (transverse mesocolon) and sigmoid colons (sigmoid mesocolon) of the large intestine to the posterior abdominal wall (Figure 25.3a); these folds carry blood (superior and inferior mesenteric vessels) and lymphatic vessels to the intestines. The mesentery and mesocolon

hold the intestines loosely in place, allowing for a great amount of movement as muscular contractions mix and move the intestinal contents along the GI tract.

◉ CHECKPOINT

5. Describe the locations of the visceral peritoneum and parietal peritoneum.
6. Describe the attachment sites and functions of the mesentery, mesocolon, falciform ligament, lesser omentum, and greater omentum.

MOUTH

◉ OBJECTIVES

- Identify the locations of the salivary glands, and describe the functions of their secretions.
- Describe the structure and functions of the tongue.
- Identify the parts of a typical tooth, and compare deciduous and permanent dentitions.

The **mouth**, also referred to as the **oral** or **buccal cavity** (BUK-al; *bucca*=cheeks), is formed by the cheeks, hard and soft palates, and tongue (Figure 25.4). The **cheeks** form the lateral walls of the oral cavity. They are covered externally by skin and internally by a mucous membrane, which consists of nonkeratinized stratified squamous epithelium. Buccinator muscles and connective tissue lie between the skin and mucous membranes of the cheeks. The anterior portions of the cheeks end at the lips.

The **lips** or **labia** (=fleshy borders) are fleshy folds surrounding the opening of the mouth. They contain the orbicularis oris muscle and are covered externally by skin and internally by a

Figure 25.4 Structures of the mouth (oral cavity).

◉ ▱ The mouth is formed by the cheeks, hard and soft palates, and tongue.

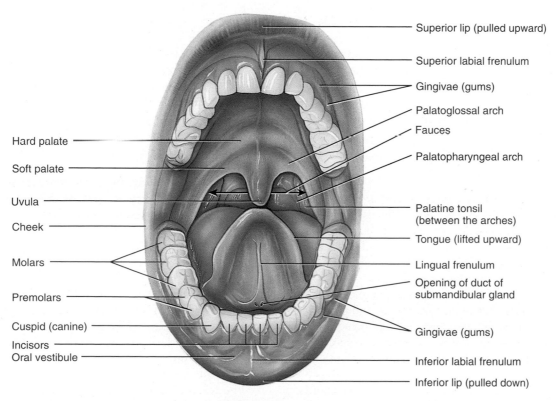

- Superior lip (pulled upward)
- Superior labial frenulum
- Gingivae (gums)
- Palatoglossal arch
- Fauces
- Palatopharyngeal arch
- Palatine tonsil (between the arches)
- Tongue (lifted upward)
- Lingual frenulum
- Opening of duct of submandibular gland
- Gingivae (gums)
- Inferior labial frenulum
- Inferior lip (pulled down)

- Hard palate
- Soft palate
- Uvula
- Cheek
- Molars
- Premolars
- Cuspid (canine)
- Incisors
- Oral vestibule

Anterior view

❓ **What is the function of the uvula?**

mucous membrane. The inner surface of each lip is attached to its corresponding gum by a midline fold of mucous membrane called the **labial frenulum** (LĀ-bē-al FREN-ū-lum; *frenulum*=small bridle). During chewing, contraction of the buccinator muscles in the cheeks and orbicularis oris muscle in the lips helps keep food between the upper and lower teeth. These muscles also assist in speech.

The cheeks and lips play important roles in procuring food and keeping it in the mouth. It would be impossible to suckle milk without these structures. Look at a newborn puppy. It has a short snout, with a reduced mouth opening and definite cheeks, much different from an adult dog. This arrangement allows the puppy to suckle. As the dog grows, the snout and mouth opening increase in size, while the cheeks are greatly reduced. Without the cheeks and lips it is impossible to create a tight seal to generate suction. The cheeks and lips also serve important nondigestive functions. For example, the creation of many sounds we make during speech depends on certain lip formations. Blowing air from your mouth is also a function of your cheeks and lips.

The **oral vestibule** (=entrance to a canal) of the cavity is a space bounded externally by the cheeks and lips and internally by the gums and teeth. The **oral cavity proper** is a space that extends from the gums and teeth to the **fauces** (FAW-sēs=passages), the opening between the oral cavity and the oropharynx.

The **palate** is a wall or septum that separates the oral cavity from the nasal cavity, forming the roof of the mouth. This important structure makes it possible to chew and breathe at the same time. The **hard palate**—the anterior two-thirds of the palate—is formed by the maxillae and palatine bones and is covered by a mucous membrane. The **soft palate**, which forms the posterior portion of the roof of the mouth, is an arch-shaped muscular partition between the oropharynx and nasopharynx that is lined with mucous membrane.

Hanging from the free border of the soft palate is a conical muscular process called the **uvula** (Ū-vū-la=little grape). During swallowing, the soft palate and uvula are drawn superiorly, closing off the nasopharynx and preventing swallowed foods and liquids from entering the nasal cavity. Lateral to the base of the uvula are two muscular folds that run down the lateral sides of the soft palate: Anteriorly, the **palatoglossal arch** extends to the side of the base of the tongue; posteriorly, the **palatopharyngeal arch** (PAL-a-tō-fa-rin'-jē-al) extends to the side of the pharynx. The palatine tonsils are situated between the arches, and the lingual tonsils are situated at the base of the tongue. At the posterior border of the soft palate, the mouth opens into the oropharynx through the fauces (see Figure 25.4).

Salivary Glands

A **salivary gland** is a gland that releases a secretion called saliva into the oral cavity. Ordinarily, just enough saliva is secreted to keep the mucous membranes of the mouth and pharynx moist and to cleanse the mouth and teeth. When food enters the mouth, however, secretion of saliva increases to lubricate, dissolve, and begin the chemical breakdown of the food.

The mucous membrane of the mouth and tongue contains many small salivary glands that open directly, or indirectly via short ducts, to the oral cavity. These glands, which include *labial*, *buccal*, and *palatal glands* in the lips, cheeks, and palate, respectively, and *lingual glands* in the tongue, all make a small contribution to saliva.

However, most saliva is secreted by the **major salivary glands**, which lie beyond the oral mucosa; their secretions empty into ducts that lead to the oral cavity. The three pairs of major salivary glands are the parotid, submandibular, and sublingual glands (Figure 25.5a). The **parotid glands** (*par-*=near; *ot-*=ear) are located inferior and anterior to the ears, between the skin and the masseter muscle. Each secretes saliva into the oral cavity via a **parotid (Stensen's) duct** that pierces the buccinator muscle to open into the vestibule opposite the second maxillary (upper) molar tooth. The **submandibular glands** are found in the floor of the oral cavity beneath the base of the tongue; they are medial and partly inferior to the mandible. Their ducts, the **submandibular (Wharton's) ducts**, run under the mucosa on either side of the midline of the floor of the mouth and enter the oral cavity proper lateral to the lingual frenulum. The **sublingual glands** are superior to the submandibular glands. Their ducts, the **lesser sublingual (Rivinus') ducts**, open into the floor of the mouth in the oral cavity proper.

The parotid gland receives its blood supply from branches of the external carotid artery and is drained by tributaries of the external jugular vein. The submandibular gland is supplied by branches of the facial artery and drained by tributaries of the facial vein. The sublingual gland is supplied by the sublingual branch of the lingual artery and the submental branch of the facial artery and is drained by tributaries of the sublingual and submental veins.

The salivary glands receive both sympathetic and parasympathetic innervation. The sympathetic fibers form plexuses on the blood vessels that supply the glands and initiate vasoconstriction, which decreases the production of saliva. The parotid gland receives sympathetic fibers from the plexus on the external carotid artery; the submandibular and sublingual glands receive sympathetic fibers that contribute to the sympathetic plexus and accompany the facial artery to the glands. The parasympathetic fibers of the glands produce vasodilation and stimulate the glands, thus increasing the production of saliva.

The fluids secreted by the buccal glands, minor salivary glands, and the three pairs of major salivary glands constitute **saliva**. Amounts of saliva secreted daily vary considerably but range from 1000 to 1500 mL (1 to 1.6 qt). Chemically, saliva is 99.5% water and 0.5% solutes and has a slightly acidic pH (6.35 to 6.85). The solute portion includes mucus, an enzyme that destroys bacteria (*lysozyme*), the digestive enzymes salivary amylase and lingual lipase, and traces of salts, proteins, and other organic compounds. **Salivary amylase** plays a minor role in the breakdown of starch in the mouth. **Lingual lipase** and mucus are secreted by **lingual glands** on the dorsum (upper surface) of the tongue. This enzyme, which is active in the stomach, can digest as much as 30% of dietary triglycerides (fats) into simpler fatty acids and monoglycerides.

The secretion of saliva, called **salivation** (sal-i-VĀ-shun), is controlled by the nervous system. Normally, parasympathetic stimulation promotes continuous secretion of a moderate amount of saliva, which keeps the mucous membranes moist and lubricates the movements of the tongue and lips during speech. The saliva is then swallowed and helps moisten the esophagus. Eventually, most components of saliva are reabsorbed, which prevents fluid loss. Sympathetic stimulation dominates during stress, resulting in dryness of the mouth. During dehydration, the salivary glands stop secreting saliva to conserve water; the resulting dryness in the mouth contributes to the sensation of thirst. Drinking will then not only restore the homeostasis of body water but also moisten the mouth.

The feel and taste of food also are potent stimulators of salivary gland secretions. Chemicals in the food stimulate receptors

Figure 25.5 **The three major salivary glands—parotid, sublingual, and submandibular.** The submandibular glands, shown in the light micrograph in (b), consist mostly of serous acini (serous-fluid-secreting portions of gland) and a few mucous acini (mucus-secreting portions of gland); the parotid glands consist of serous acini only; and the sublingual glands consist of mostly mucous acini and a few serous acini.

Saliva lubricates and dissolves foods and begins the chemical breakdown of carbohydrates and lipids.

(a) Location of salivary glands

(b) Portion of submandibular gland

LM 240x

(c) Right lateral view

The ducts of which salivary glands empty on either side of the lingual frenulum?

in taste buds on the tongue, and impulses are conveyed from the taste buds to two salivary nuclei in the brain stem (**superior** and **inferior salivatory nuclei**). Returning parasympathetic impulses in fibers of the facial (VII) and glossopharyngeal (IX) nerves stimulate the secretion of saliva. Saliva continues to be heavily se-

creted for some time after food is swallowed; this flow of saliva washes out the mouth and dilutes and buffers the remnants of irritating chemicals. The smell, sight, sound, or thought of food may also stimulate secretion of saliva.

Tongue

The **tongue** is an accessory digestive organ composed of skeletal muscle covered with mucous membrane. Together with its associated muscles, it forms the floor of the oral cavity. The tongue is divided into symmetrical lateral halves by a median septum that extends its entire length, and it is attached to the hyoid bone, styloid process of the temporal bone, and mandible. Each half of the tongue consists of an identical complement of extrinsic and intrinsic muscles.

The **extrinsic muscles** of the tongue, which originate outside the tongue (attach to bones in the area) and insert into connective tissues in the tongue, include the *hyoglossus, genioglossus,* and *styloglossus* muscles (see Figure 11.7 on page 329). The extrinsic muscles move the tongue from side to side and in and out to maneuver food for chewing, shape the food into a rounded mass, and force the food to the back of the mouth for swallowing. They also form the floor of the mouth, hold the tongue in position, and assist in speech. The **intrinsic muscles** originate in and insert into connective tissue within the tongue and alter the shape and size of the tongue for speech and swallowing. The intrinsic muscles include the *longitudinalis superior, longitudinalis inferior, transversus linguae,* and *verticalis linguae* muscles. The **lingual frenulum** (*lingua*=the tongue), a fold of mucous membrane in the midline of the undersurface of the tongue, is attached to the floor of the mouth and aids in limiting the movement of the tongue posteriorly (see Figures 25.4 and 25.5). If a lingual frenulum is abnormally short or rigid—a condition called **ankyloglossia** (ang′-kē-lō-GLOSS-ē-a)—the person is said to be "tongue-tied" because of the resulting impairment to speech.

The dorsum and lateral surfaces of the tongue are covered with **papillae** (pa-PIL-ē=nipple-shaped), projections of the lamina propria covered with stratified squamous epithelium (see Figure 22.2 on pages 700–701). Many papillae contain taste buds, the receptors for gustation (taste). As their name implies, **fungiform papillae** (FUN-ji-form=mushroomlike) are mushroomlike elevations distributed among the more numerous filiform papillae. They are scattered over the dorsum of the tongue, but are concentrated mainly around the margins of the tongue. They appear as red dots on the surface of the tongue resulting from dense capillaries beneath a thin, nonkeratinized epithelium, and most of them contain taste buds. **Vallate (circumvallate) papillae** (VAL-āt=wall-like) are arranged in an inverted V shape on the posterior surface of the tongue; all of them contain taste buds. **Foliate papillae** (FŌ-lē-āt=leaflike) are located in small trenches on the lateral margins of the tongue, but most of their taste buds degenerate in early childhood. **Filiform papillae** (FIL-i-form=threadlike) are pointed, threadlike projections distributed in parallel rows over the anterior two-thirds of the tongue. Although filiform papillae lack taste buds, they contain receptors for touch and increase friction between the tongue and food, making it easier for the tongue to move food in the oral cavity. The mottled white color of the tongue results from the dead keratinized cells on the apex of these plentiful filiform papillae.

Teeth

The **teeth**, or **dentes** (Figure 25.6), are accessory digestive organs located in sockets of the alveolar processes of the mandible and maxillae. The alveolar processes are covered by the **gingivae** (jin-JI-vē), or gums, which extend slightly into each socket to form the **gingival sulcus**. The sockets are lined by the **periodontal ligament** or **membrane** (*odont-*=tooth), which consists of dense fibrous connective tissue and is attached to the socket walls and cemental surface covering the roots of the teeth. The periodontal ligament anchors the teeth in position and acts as a shock absorber during chewing.

Figure 25.6 A typical tooth and surrounding structures.

Teeth are anchored in sockets of the alveolar processes of the mandible and maxillae.

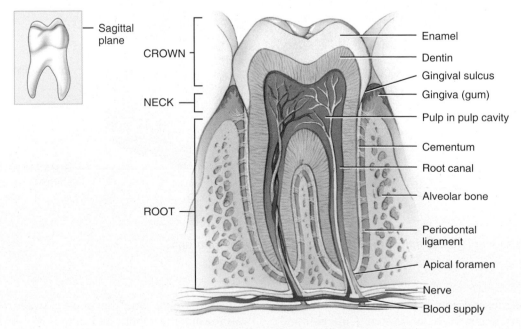

Sagittal section of a mandibular (lower) molar

? **What type of tissue is the main component of teeth?**

A typical tooth consists of three principal regions. The **crown** is the visible portion above the level of the gums. One to three **roots** are embedded in each socket. The **neck** is the constricted junction of the crown and root near the gum line.

Internally, **dentin** forms the majority of the tooth. Dentin consists of a calcified connective tissue that gives the tooth its basic shape and rigidity. It is harder than bone because of its higher content of calcium salts (70% of dry weight). It also differs from bone in that it lacks blood vessels, gets deposited incrementally (in successive waves a little at a time), and does not remodel. Dentin contains *dentinal tubules*, parallel microscopic tubules radiating through the dentin from the pulp cavity. Cells that produce the dentin, odontoblasts, line the pulp cavity and send cytoplasmic processes into the dentinal tubules.

The dentin of the crown is covered by **enamel**, which consists primarily of calcium phosphate and calcium carbonate. Enamel is also harder than bone because of its even higher content of calcium salts (about 95% of dry weight). In fact, enamel is the hardest substance in the body. It is thickest over the cusps and tapers to a thin edge at the neck of the tooth where it terminates. Enamel arises as a cellular secretion during development, after which the body cannot replace or repair it. Enamel protects the tooth from the wear and tear of chewing and from acids that could easily dissolve the softer dentin. The dentin of the root is covered by **cementum**, another bonelike substance, which attaches the root to the periodontal ligament.

The dentin of a tooth encloses a space. The enlarged part of the space, the **pulp cavity**, lies within the crown and is filled with **pulp**, a connective tissue containing blood vessels, nerves, and lymphatic vessels. Narrow extensions of the pulp cavity, called **root canals**, run through the root of the tooth. Each root canal has an opening at its base, the **apical foramen**, through which blood vessels, lymphatic vessels, and nerves extend. The vessels bring nourishment, lymphatic vessels offer protection, and nerves provide sensation.

The arteries that supply blood to the teeth are distributed to the pulp cavity and surrounding periodontal ligament. These include the superior alveolar branches of the maxillary artery (anterior and posterior) and the incisive and dental branches of the inferior alveolar artery.

The teeth receive sensory fibers from branches of the maxillary and mandibular divisions of the trigeminal (V) nerve—the maxillary teeth from branches of the maxillary division and the mandibular teeth from branches of the mandibular division.

The branch of dentistry that is concerned with the prevention, diagnosis, and treatment of diseases that affect the pulp, root, periodontal ligament, and alveolar bone is known as **endodontics** (en′-dō-DON-tiks; *endo-*=within). **Orthodontics** (or′-thō-

Figure 25.7 Dentitions and times of eruptions (indicated in parentheses). A designated letter (deciduous teeth) or number (permanent teeth) uniquely identifies each tooth. Deciduous teeth begin to erupt at 6 months of age, and one pair of teeth appears about each month thereafter, until all 20 are present.

🔑 There are 20 teeth in a complete deciduous set and 32 teeth in a complete permanent set.

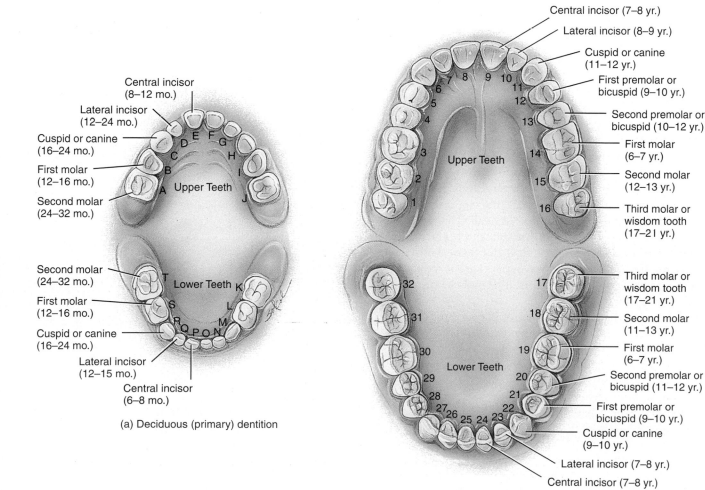

Central incisor (8–12 mo.)
Lateral incisor (12–24 mo.)
Cuspid or canine (16–24 mo.)
First molar (12–16 mo.)
Second molar (24–32 mo.)

Second molar (24–32 mo.)
First molar (12–16 mo.)
Cuspid or canine (16–24 mo.)
Lateral incisor (12–15 mo.)
Central incisor (6–8 mo.)

Upper Teeth
Lower Teeth

(a) Deciduous (primary) dentition

Central incisor (7–8 yr.)
Lateral incisor (8–9 yr.)
Cuspid or canine (11–12 yr.)
First premolar or bicuspid (9–10 yr.)
Second premolar or bicuspid (10–12 yr.)
First molar (6–7 yr.)
Second molar (12–13 yr.)
Third molar or wisdom tooth (17–21 yr.)

Third molar or wisdom tooth (17–21 yr.)
Second molar (11–13 yr.)
First molar (6–7 yr.)
Second premolar or bicuspid (11–12 yr.)
First premolar or bicuspid (9–10 yr.)
Cuspid or canine (9–10 yr.)
Lateral incisor (7–8 yr.)
Central incisor (7–8 yr.)

Upper Teeth
Lower Teeth

(b) Permanent (secondary) dentition

CONTINUES ▶

◉ FIGURE **25.7** CONTINUED ▶

Right lateral view

Deciduous teeth:
P - Central incisor
Q - Lateral incisor
R - Cuspid (canine)
S - First molar (bicuspid)
T - Second molar

Permanent teeth:
25 - Central incisor
26 - Lateral incisor
27 - Cuspid (canine)
28 - First premolar (bicuspid)
29 - Second premolar
30 - First molar
31 - Second molar

(c) Mandible of a six year old child showing erupted
deciduous teeth and unerupted permanent teeth

Right lateral view

8, 25 - Central incisor
7, 26 - Lateral incisor
6, 27 - Cuspid (canine)
5, 28 - First premolar (bicuspid)

4, 29 - Second premolar
3, 30 - First molar
2, 31 - Second molar
1, 32 - Third molar (wisdom tooth)

(d) Mandible (and maxilla) showing permanent teeth and blood
and nerve supply to them

❓ Which permanent teeth do not replace any deciduous teeth?

DON-tiks; *ortho-*=straight) is a branch of dentistry that is concerned with the prevention and correction of abnormally aligned teeth, and **periodontics** (per′-ē-ō-DON-tiks) is a branch of dentistry concerned with the treatment of abnormal conditions of the tissues immediately surrounding the teeth.

Humans have two **dentitions**, or sets of teeth: deciduous and permanent. The first of these—the **deciduous teeth** (*decidu-*= falling out), also called **primary teeth**, **milk teeth**, or **baby teeth**—begin to erupt (emerge) at about 6 months of age, and one pair of teeth appears at about each month thereafter, until all 20 are present (Figure 25.7a, c). The incisors, which are closest to the midline, are chisel-shaped and adapted for cutting into food. They are referred to as either **central** or **lateral incisors** on the basis of their position. Next to the incisors, moving posteriorly, are the **canines**, which have a pointed surface called a cusp. Canines are used to tear and shred food. Incisors and canines have only one root. Posterior to them lie the **first** and **second molars**, which have four cusps. Maxillary (upper) molars have three roots; mandibular (lower) molars have two roots. The molars crush and grind food to prepare for swallowing.

All the deciduous teeth are lost—generally between the ages of 6 and 12 years—and are replaced by the **permanent (secondary) teeth** (Figure 25.7b, d). The permanent dentition contains 32 teeth that erupt between age 6 and adulthood. The pattern resembles the deciduous dentition, with the following exceptions. The deciduous molars are replaced by the **first** and **second premolars (bicuspids)**, which have two cusps and one root (upper first premolars have two roots) and are used for crushing and grinding. The permanent molars, which erupt into the mouth posterior to the premolars, do not replace any deciduous teeth and erupt as the jaw grows to accommodate them—the **first molars** at age 6, the **second molars** at age 12, and the **third molars (wisdom teeth)** after age 17 or not at all.

Often the human jaw does not have enough room posterior to the second molars to accommodate the eruption of the third molars. In this case, the third molars remain embedded in the alveolar bone and are said to be *impacted*. They often cause pressure and pain and must be removed surgically. In some people, third molars may be dwarfed in size or may not develop at all.

Table 25.1 lists terms used to describe the surface orientation of teeth.

TABLE 25.1

Surface Orientation of Teeth

TOOTH SURFACE	DESCRIPTION
Labial	Contacting the lips.
Buccal	Contacting or facing the cheeks.
Lingual	Facing the tongue (teeth of the mandible only).
Palatal	Facing the palate (teeth of the maxillae only).
Mesial	Anterior or medial side relative to dental arch.
Distal	Posterior or lateral side relative to dental arch.
Occlusal	The biting surface.

◉ CHECKPOINT

7. What structures form the mouth (oral cavity)?
8. How are the major salivary glands distinguished on the basis of location and structure?
9. How do the extrinsic and intrinsic muscles of the tongue differ in function?
10. Contrast the functions of incisors, cuspids, premolars, and molars.

PHARYNX

◉ **OBJECTIVE**

- Describe the structure and function of the pharynx.

Through chewing, or **mastication** (mas'-ti-KĀ-shun; *masticare*= to chew), the tongue manipulates food, the teeth grind it, and the food is mixed with saliva. As a result, the food is reduced to a soft, flexible mass called a **bolus** (*bolos*=lump) that is easily swallowed. When food is first swallowed, it passes from the mouth into the pharynx.

The **pharynx** (*pharynx*=throat) is a funnel-shaped tube that extends from the internal nares to the esophagus posteriorly and the larynx anteriorly (see Figure 24.3 on pages 758–759). The pharynx is composed of skeletal muscle and lined by mucous membrane. The nasopharynx functions only in respiration, but the oropharynx and laryngopharynx have both digestive and respiratory functions. Swallowing, or **deglutition** (dē-gloo-TISH-un), is a mechanism that moves food from the mouth to the stomach. It is helped by saliva and mucus and involves the mouth, pharynx, and esophagus. Food that is swallowed passes from the mouth into the oropharynx and laryngopharynx before passing into the esophagus. Muscular contractions of the oropharynx and laryngopharynx help propel food into the esophagus and then into the stomach.

◉ **CHECKPOINT**

11. What is a bolus? How is it formed?
12. Where does the pharynx begin and end?

ESOPHAGUS

◉ **OBJECTIVE**

- Describe the location, anatomy, histology, and function of the esophagus.

The **esophagus** (e-SOF-a-gus=eating gullet) is a collapsible muscular tube that lies posterior to the trachea. It is about 25 cm (10 in.) long. It begins at the inferior end of the laryngopharynx, passes through the mediastinum anterior to the vertebral column, pierces the diaphragm through an opening called the **esophageal hiatus**, and ends in the superior portion of the stomach (see Figure 25.1). Sometimes, a portion of the stomach protrudes above the diaphragm through the esophageal hiatus. This condition, called a *hiatal hernia*, will be described later in the chapter under the listing for hernia in the Key Medical Terms list.

The arteries of the esophagus, the esophageal arteries, arise from the following arteries along its length: inferior thyroid, thoracic aorta, intercostal arteries, phrenic, and left gastric arteries. It is drained by the adjacent esophageal veins, which primarily drain into the various azygos veins. The esophagus is innervated by recurrent laryngeal branches of the vagus nerves, the cervical sympathetic chain, and other branches of the vagus (X) nerves.

Histology

The **mucosa** of the esophagus consists of nonkeratinized stratified squamous epithelium, lamina propria (areolar connective tissue), and a muscularis muscosae (smooth muscle) (Figure 25.8a). Near the stomach, the mucosa of the esophagus also contains mucous glands. The stratified squamous epithelium

Figure 25.8 **Histology of the esophagus.** A higher magnification of nonkeratinized, stratified squamous epithelium is shown in Table 3.1f on page 67. (See Tortora, *A Photographic Atlas of the Human Body*, 2e, Figure 12.8a.)

 The esophagus secretes mucus and transports food to the stomach.

Lumen of esophagus

Mucosa:
- Nonkeratinized stratified squamous epithelium
- Lamina propria
- Muscularis mucosae

Submucosa

Muscularis (circular layer)

Muscularis (circular layer)

Adventitia

Transverse plane

LM 20x

(a) Wall of the esophagus

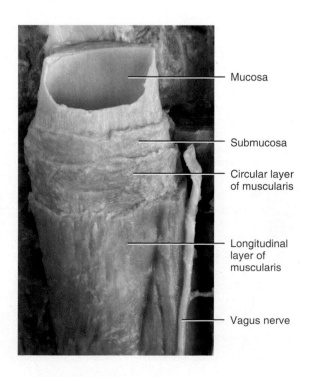

Mucosa

Submucosa

Circular layer of muscularis

Longitudinal layer of muscularis

Vagus nerve

(b) Anterosuperior view of step dissection of esophagus

❓ **In which layers of the esophagus are the glands that secrete lubricating mucus located?**

associated with the lips, mouth, tongue, oropharynx, laryn-gopharynx, and esophagus affords considerable protection against abrasion and wear-and-tear from food particles that are chewed, mixed with secretions, and swallowed. The **submucosa** contains areolar connective tissue with blood vessels and mucous glands and numerous elastic fibers that assist in closing the distended tube. The **muscularis** of the superior third of the esophagus is skeletal muscle, the intermediate third is skeletal and smooth muscle, and the inferior third is smooth muscle. The superficial layer of the esophagus is known as the **adventitia** (ad-ven-TISH-a), rather than serosa as in the stomach and intestines, because the areolar connective tissue of this layer is not covered by mesothelium (the esophagus does not push into a serous lined cavity) and because the connective tissue merges with the connective tissue of surrounding structures of the mediastinum, through which it passes. The adventitia attaches the esophagus to surrounding structures.

Functions

The esophagus secretes mucus and transports food into the stomach. It does not produce digestive enzymes, and it does not carry on absorption. The passage of food from the laryngopharynx into the esophagus is regulated at the entrance to the esophagus by a sphincter (a circular band or ring of muscle that is normally contracted) called the **upper esophageal sphincter** (e-sof′-a-JĒ-al) **(UES)**. It consists of skeletal muscle (cricopharyngeus muscle) attached to the cricoid cartilage. The elevation of the larynx causes the sphincter to relax, allowing the bolus to enter the esophagus. This sphincter also relaxes during exhalation.

Food is pushed through the esophagus by a progression of involuntary coordinated contractions and relaxations of the circular and longitudinal layers of the muscularis called **peristalsis** (per′-i-STAL-sis; *stalsis*=constriction) (Figure 25.9). Peristalsis occurs in other tubular structures, including other portions of the GI

Figure 25.9 Peristalsis during deglutition (swallowing).

🔑 Peristalsis consists of progressive, wavelike contractions of the muscularis.

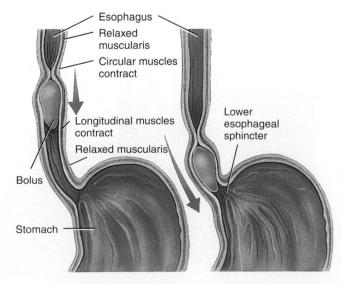

Anterior view of frontal sections of peristalsis in esophagus

- Esophagus
- Relaxed muscularis
- Circular muscles contract
- Longitudinal muscles contract
- Relaxed muscularis
- Bolus
- Stomach
- Lower esophageal sphincter

❓ **Does peristalsis "push" or "pull" food along the gastrointestinal tract?**

tract and the ureters, bile ducts, and uterine tubes; in the esophagus it is controlled by the medulla oblongata. In the section of the esophagus lying just superior to the bolus, the circular muscle fibers contract, constricting the esophageal wall and squeezing the bolus toward the stomach. Meanwhile, longitudinal fibers inferior to the bolus also contract, which shortens this inferior section and pushes its walls outward so it can receive the bolus. The contractions are repeated in a wave that pushes the food toward the stomach. Mucus secreted by esophageal glands lubricates the bolus and reduces friction.

Just superior to the level of the diaphragm, the esophagus narrows slightly. This narrowing is a physiological sphincter in the inferior part of the esophagus composed of smooth muscle known as the **lower esophageal (gastroesophageal** or **cardiac) sphincter (LES)**. (A *physiological sphincter* is a section of a tubular structure, in this case the esophagus, that functions like a sphincter even though no sphincter muscle is actually present.) The lower esophageal sphincter relaxes during swallowing and thus allows the bolus to pass from the esophagus into the stomach.

◉ CHECKPOINT

13. Describe the location and histology of the esophagus.
14. What is the role of the esophagus in digestion?
15. Compare the operation of the upper and lower esophageal sphincters. Which one is a physiological sphincter?

STOMACH

◉ OBJECTIVE

- Describe the location, anatomy, histology, and functions of the stomach.

The **stomach** is typically a J-shaped enlargement of the GI tract directly inferior to the diaphragm in the epigastric, umbilical, and left hypochondriac regions of the abdomen (see Figure 1.8 on page 18). The stomach connects the esophagus to the duodenum, the first part of the small intestine (Figure 25.10). Because a meal can be eaten much more quickly than the intestines can digest and absorb it, the stomach functions as a mixing area and holding reservoir. At appropriate intervals after food is ingested, the stomach forces a small quantity of material into the first portion of the small intestine. The position and size of the stomach vary continually; the diaphragm pushes it inferiorly with each inhalation and pulls it superiorly with each exhalation. Empty, it is about the size of a large sausage, but it is the most distensible portion of the GI tract and can accommodate a large quantity of food. In the stomach, the digestion of starch that began in the mouth continues, digestion of proteins and triglycerides begins, the semisolid bolus is converted to a liquid, and certain substances are absorbed.

Anatomy

The stomach has four main regions: the cardia, fundus, body, and pylorus (see Figure 25.10). The **cardia** (CAR-dē-a) surrounds the superior opening of the stomach. The rounded portion superior and to the left of the cardia is the **fundus** (FUN-dus). Inferior to the fundus is the large central portion of the stomach, called the **body**. The region of the stomach that connects to the duodenum is the **pylorus** (pī-LOR-us; *pyl-*=gate; *-orus*=guard); it has two parts, the **pyloric antrum** (AN-trum=cave), which connects to the body of the stomach, and the **pyloric canal**, which leads

Figure 25.10 **External and internal anatomy of the stomach.**

The four regions of the stomach are the cardia, fundus, body, and pylorus.

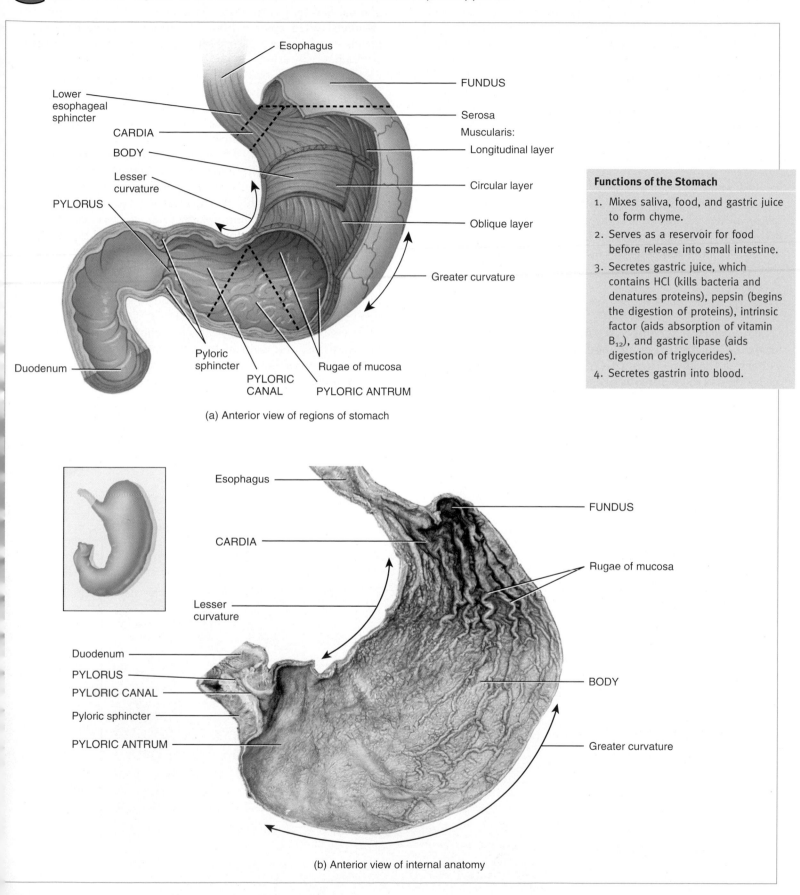

Functions of the Stomach

1. Mixes saliva, food, and gastric juice to form chyme.
2. Serves as a reservoir for food before release into small intestine.
3. Secretes gastric juice, which contains HCl (kills bacteria and denatures proteins), pepsin (begins the digestion of proteins), intrinsic factor (aids absorption of vitamin B_{12}), and gastric lipase (aids digestion of triglycerides).
4. Secretes gastrin into blood.

(a) Anterior view of regions of stomach

(b) Anterior view of internal anatomy

After a very large meal, does your stomach have rugae?

into the duodenum. When the stomach is empty, the mucosa lies in large folds, called **rugae** (ROO-gē=wrinkles), that can be seen with the unaided eye. The pylorus communicates with the duodenum of the small intestine via a smooth muscle sphincter called the **pyloric sphincter**. The concave medial border of the stomach is called the **lesser curvature**, and the convex lateral border is called the **greater curvature**.

The arterial supply of the stomach arises from the celiac trunk. The right and left gastric arteries form an anastomosing arch along the lesser curvature, and the right and left gastroepiploic arteries form a similar arch on the greater curvature. Short gastric arteries supply the fundus. The veins of the same name accompany the arteries and drain, directly or indirectly, into the hepatic portal vein.

The vagus (X) nerves convey parasympathetic fibers to the stomach. These fibers form synapses within the submucosal plexus in the submucosa and the myenteric plexus in the muscularis. The sympathetic nerves arise from the celiac ganglia, and the nerves reach the stomach along the branches of the celiac artery.

Histology

The stomach wall is composed of the same basic layers as the rest of the GI tract, with certain modifications (Figure 25.11a). The surface of the **mucosa** is a layer of simple columnar epithelial cells called **surface mucous cells**. The mucosa contains a **lamina**

propria (areolar connective tissue) and a **muscularis mucosae** (smooth muscle). Epithelial cells extend down into the lamina propria, where they form columns of secretory cells called **gastric glands**. Several gastric glands open into the bottom of narrow channels called **gastric pits**. Secretions from several gastric glands flow into each gastric pit and then into the lumen of the stomach.

The gastric glands contain three types of *exocrine gland cells* that secrete their products into the stomach lumen: mucous neck cells, chief cells, and parietal cells. Both mucous surface cells and **mucous neck cells** secrete mucus (Figure 25.11b). **Parietal cells** produce intrinsic factor (needed for absorption of vitamin B_{12}) and hydrochloric acid. The **chief (zymogenic) cells** secrete pepsinogen and gastric lipase. The secretions of the mucous, parietal, and chief cells form **gastric juice**, which totals 2000–3000 mL (roughly 2–3 qt) per day. In addition, gastric glands include a type of enteroendocrine cell, the **G cell**, which is located mainly in the pyloric antrum and secretes the hormone gastrin into the bloodstream. Gastrin stimulates growth of the gastric glands and secretion of large amounts of gastric juice. It also strengthens contraction of the lower esophageal sphincter, increases motility of the stomach, and relaxes the pyloric and ileocecal sphincters (described later).

The **submucosa** of the stomach is composed of areolar connective tissue, as it is throughout the entire GI tract. The **muscularis** has three layers of smooth muscle (rather than the two found in the lower esophagus and small and large intestines): an

Figure 25.11 Histology of the stomach.

The muscularis of the stomach has three layers of smooth muscle tissue.

(a) Three-dimensional view of layers of the stomach

outer longitudinal layer, a middle circular layer, and an inner oblique layer. The oblique layer is limited primarily to the body of the stomach. This arrangement of muscle allows the stomach to more effectively churn and mix the food. The **serosa** is composed of simple squamous epithelium (mesothelium), the visceral peritoneum, and areolar connective tissue. At the lesser curvature, the stomach's mesothelium extends superiorly to the liver as the lesser omentum. At the greater curvature, the stomach's mesothelium continues inferiorly as the greater omentum and drapes over the intestines.

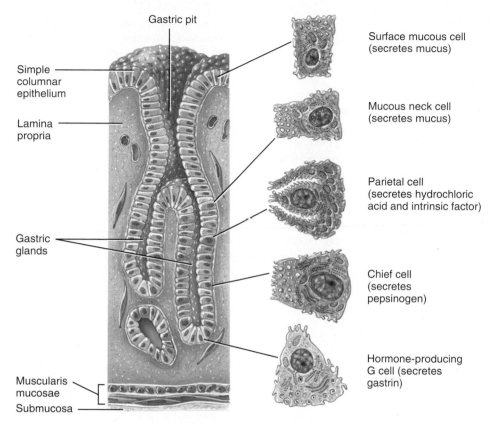

(b) Sectional view of the stomach mucosa showing gastric glands and cell types

(c) Fundic mucosa

? What types of cells are found in gastric glands, and what does each secrete?

16. Describe the location and anatomical features of the stomach.
17. Compare the epithelium of the esophagus with that of the stomach. How is each adapted to the function of the organ?
18. Describe the importance of rugae, surface mucous cells, mucous neck cells, chief cells, parietal cells, and G cells in the stomach.

Functions

Several minutes after food enters the stomach, gentle, rippling, peristaltic movements called **mixing waves** pass over the stomach every 15–25 seconds. These waves macerate food, mix it with secretions of the gastric glands, and reduce it to a soupy liquid called **chyme** (KĪM=juice). Few mixing waves are observed in the fundus, which primarily has a storage function. As digestion proceeds, more vigorous mixing waves begin at the body of the stomach and intensify as they reach the pylorus. The pyloric sphincter normally remains almost, but not completely, closed; as food reaches the pylorus, each mixing wave forces several milliliters of chyme into the duodenum through the pyloric sphincter. Most of the chyme is forced back into the body of the stomach, where mixing continues. The next wave pushes the chyme forward again and forces a little more into the duodenum.

The enzymatic digestion of proteins begins in the stomach. In the adult, this is achieved mainly through the enzyme **pepsin**, secreted by chief cells in an inactive form called *pepsinogen*. Pepsin breaks certain peptide bonds between the amino acids making up proteins. Thus, a protein chain of many amino acids is broken down into smaller fragments called **peptides**. Pepsin also brings about the clumping and digestion of milk proteins. Another enzyme of the stomach is **gastric lipase**. Gastric lipase splits the short-chain triglycerides (fats) in butterfat molecules found in milk. The enzyme has a limited role in the adult stomach. To digest fats, adults rely almost exclusively on the lingual lipase secreted by salivary glands in the mouth, and **pancreatic lipase**, an enzyme secreted by the pancreas into the small intestine.

Within 2–4 hours after eating a meal, the stomach has emptied its contents into the duodenum. Foods rich in carbohydrate spend the least time in the stomach; high-protein foods remain somewhat longer; and emptying is slowest after a fat-laden meal containing large amounts of triglycerides.

● CLINICAL **CONNECTION** | Vomiting

Vomiting, or *emesis*, is the forcible expulsion of the contents of the upper GI tract (stomach and sometimes duodenum) through the mouth. The strongest stimuli for vomiting are irritation and distension of the stomach; other stimuli include unpleasant sights, general anesthesia, dizziness, and certain drugs, such as morphine and derivatives of digitalis. Nerve impulses are transmitted to the vomiting center in the medulla oblongata, and returning impulses travel to the upper GI tract organs, diaphragm, and abdominal muscles. Vomiting basically involves squeezing the stomach between the diaphragm and abdominal muscles and expelling the contents through open esophageal sphincters. Prolonged vomiting, especially in infants and elderly people, can be serious because the loss of acidic gastric juice can lead to alkalosis (higher than normal blood pH), and dehydration. •

The stomach wall is impermeable to the passage of most materials into the blood; most substances are not absorbed until they reach the small intestine. However, the stomach does participate in the absorption of some water, electrolytes, certain drugs (especially aspirin), and alcohol.

◉ CHECKPOINT
19. How does mechanical digestion occur in the stomach?
20. What are the functions of gastric lipase and lingual lipase in the stomach?
21. Describe the role of the stomach in absorption.

PANCREAS

◉ OBJECTIVE
• Describe the location, anatomy, histology, and function of the pancreas.

From the stomach, chyme passes into the small intestine. Because chemical digestion in the small intestine depends on activities of the pancreas, liver, and gallbladder, we first consider the activities of these accessory digestive organs and their contributions to digestion in the small intestine.

Anatomy

The **pancreas** (*pan-*=all; *-creas*=flesh), a retroperitoneal gland that is about 12–15 cm (5–6 in.) long and 2.5 cm (1 in.) thick, lies posterior to the greater curvature of the stomach (see Figure 25.1). It, along with the liver and gallbladder, develops as an embryonic epithelial outgrowth of the duodenum. The pancreas consists of a head, a body, and a tail and is usually connected to the duodenum by two ducts (Figure 25.12a, c, d). The **head** is the expanded portion of the organ near the curve of the duodenum. Projecting from the lower portion of the head is the hooklike **uncinate process** that arches behind the superior mesenteric artery and vein, encircling them with pancreatic tissue. Superior to and to the left of the head are the central **body** and the tapering **tail**.

Pancreatic secretions pass from the secreting cells into small ducts that ultimately unite to form two larger ducts that convey the secretions into the small intestine. The larger of the two ducts is called the **pancreatic duct (duct of Wirsung)**. In most people, the pancreatic duct joins the common bile duct from the liver and gallbladder and enters the duodenum as a dilated common duct called the **hepatopancreatic ampulla (ampulla of Vater)**. The ampulla opens onto an elevation of the duodenal mucosa, the **major duodenal papilla**, that lies about 10 cm (4 in.) inferior to the pyloric sphincter of the stomach. The smaller of the two ducts, the **accessory duct (duct of Santorini)**, leads from the pancreas and empties into the duodenum about 2.5 cm (1 in.) superior to the hepatopancreatic ampulla.

The arterial supply of the pancreas is from the superior and inferior pancreaticoduodenal arteries and from the splenic and superior mesenteric arteries. The veins, in general, correspond to the arteries. Venous blood reaches the hepatic portal vein by means of the splenic and superior mesenteric veins.

The nerves to the pancreas are autonomic nerves that branch from the celiac and superior mesenteric plexuses. Included are preganglionic vagal, postganglionic sympathetic, and sensory fibers. Parasympathetic vagal fibers are said to terminate at both acinar (exocrine) and islet (endocrine) cells. Although the innervation is presumed to influence enzyme formation, pancreatic secretion is controlled largely by the hormones secretin and

Figure 25.12 **Relation of the pancreas to the liver, gallbladder, and duodenum.** The inset shows details of the common bile duct and pancreatic duct forming the hepatopancreatic ampulla (ampulla of Vater) and emptying into the duodenum.

6 Pancreatic enzymes digest starches (polysaccharides), proteins, triglycerides, and nucleic acids.

(a) Anterior view

(b) Details of hepatopancreatic ampulla

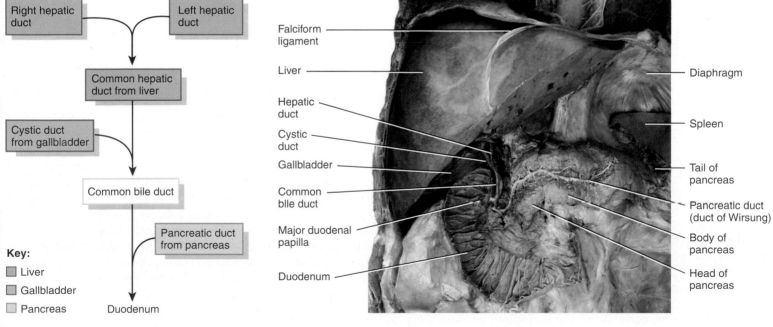

(c) Ducts carrying bile from liver and gallbladder and pancreatic juice from pancreas to the duodenum

(d) Anterior view

CONTINUES ▶

FIGURE **25.12** CONTINUED

SUPERIOR

Duodenum (cut open)

Common bile duct

Major duodenal papilla

Head of pancreas

Uncinate process

Tail of pancreas

Body of pancreas

Pancreatic duct (duct of Wirsung)

MEDIAL

LATERAL

(e) Anterior view dissected to reveal pancreatic duct

? What type of fluid is found in the pancreatic duct? The common bile duct? The hepatopancreatic ampulla?

cholecystokinin (CCK) released by the small intestine. The sympathetic fibers enter the islets and also end on blood vessels; these fibers are vasomotor and accompanied by sensory fibers, especially for pain.

Histology

The pancreas is made up of small clusters of glandular epithelial cells, about 99% of which are arranged in clusters called **acini** (AS-i-nē) and constitute the *exocrine* portion of the organ (see Figure 23.7b, c on page 742). The cells within acini secrete a mixture of fluid and digestive enzymes called **pancreatic juice**. The remaining 1% of the cells are organized into clusters called **pancreatic islets (islets of Langerhans)**, the *endocrine* portion of the pancreas. These cells secrete the hormones glucagon, insulin, somatostatin, and pancreatic polypeptide. The functions of these hormones are discussed in Table 23.7 on page 743.

Functions

Each day the pancreas produces 1200–1500 mL (about 1.2–1.5 qt) of **pancreatic juice**, a clear, colorless liquid consisting mostly of water, some salts, sodium bicarbonate, and several enzymes. The sodium bicarbonate gives pancreatic juice a slightly alkaline pH (7.1–8.2) that buffers acidic gastric juice in chyme, stops the action of pepsin from the stomach, and creates the proper pH for the action of digestive enzymes in the small intestine. The enzymes in pancreatic juice include a carbohydrate-digesting enzyme called **pancreatic amylase**; several protein-digesting enzymes called **trypsin** (TRIP-sin), **chymotrypsin** (kī'-mō-TRIP-sin), **carboxypeptidase** (kar-bok'-sē-PEP-ti-dās), and **elastase** (ē-LAS-tās); the principal triglyceride-digesting enzyme in adults, called **pancreatic lipase**; and nucleic acid–digesting enzymes called **ribonuclease** and **deoxyribonuclease**.

☉ CHECKPOINT

22. Describe the duct system connecting the pancreas to the duodenum.
23. What are pancreatic acini? Contrast their functions with those of the pancreatic islets (islets of Langerhans).
24. Describe the composition and functions of pancreatic juice.

LIVER AND GALLBLADDER

☉ OBJECTIVE

• Describe the location, anatomy, histology, and functions of the liver and gallbladder.

The **liver** is the largest internal organ and heaviest gland of the body, weighing about 1.4 kg (about 3 lb) in an average adult. Of the organs of the body, it is second in size only to the skin. The liver is inferior to the diaphragm and occupies most of the right hypochondriac and part of the epigastric regions of the abdominopelvic cavity (see Figure 1.8 on page 18).

The **gallbladder** (*gall-*=bile) is a pear-shaped sac that is located on the inferior surface of the liver. It is 7–10 cm (3–4 in.) long and part of it typically hangs below the anterior inferior margin of the liver (Figure 25.12a).

Anatomy

The liver is almost completely covered by visceral peritoneum and *is* completely covered by a capsule composed of dense irregular connective tissue that lies deep to the peritoneum. The liver is divided into two principal lobes—a large **right lobe** and a smaller **left lobe**—by the **falciform ligament**, a mesenteric fold from the parietal peritoneum of the diaphragm and anterior abdominal wall to the visceral peritoneum of the liver (Figures 25.12 and 25.13). The right lobe is considered by many anatomists to include an inferior **quadrate lobe** and a posterior **caudate lobe**; however, on the basis of internal morphology (primarily the distribution of blood vessels), the quadrate and caudate lobes more appropriately belong to the left lobe. The falciform ligament extends from the undersurface of the diaphragm between the two principal lobes of the liver to the superior surface of the liver, helping to suspend the liver in the abdominal cavity. The free border of the falciform ligament is the **ligamentum teres (round ligament)**, a fibrous cord that is a remnant of the umbilical vein of the fetus; it extends from the liver to the umbilicus. The right and left **coronary ligaments** are narrow extensions of the parietal peritoneum that suspend the liver from the diaphragm.

The parts of the gallbladder are the broad **fundus**, which projects downward beyond the inferior border of the liver; the central portion, called the **body**; and a tapered portion called the **neck**. The body and neck project superiorly.

Figure 25.13 **External anatomy of the liver.** The anterior view is illustrated in Figure 25.12a.

The two principal lobes of the liver, the right and left lobes, are separated by the falciform ligament.

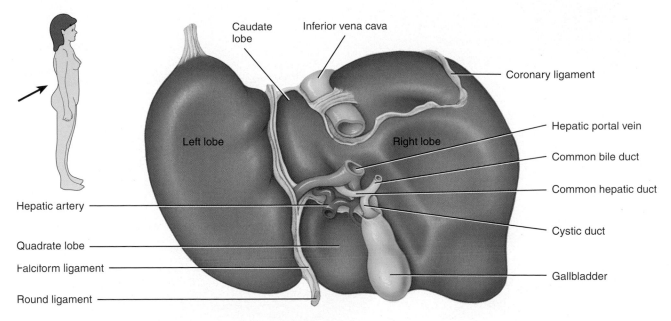

(a) Posteroinferior surface of liver

PATH OF BILE FLOW FROM THE LIVER INTO THE DUODENUM
Hepatocytes → Bile capillaries → Small bile ducts → Right and left hepatic ducts →
Common hepatic duct → Common bile duct (or cystic duct for storage in gallbladder) →
Hepatopancreatic ampulla (ampulla of Vater)

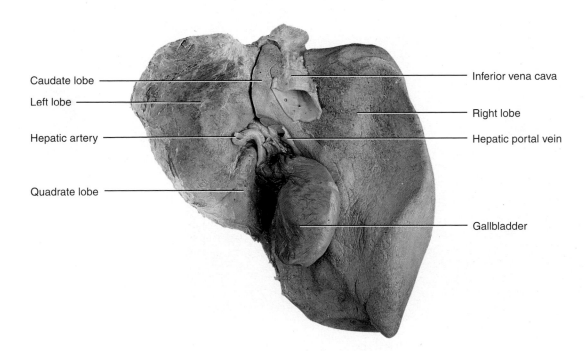

(b) Posteroinferior surface of liver

? **Within which abdominopelvic region (see Figure 1.8a on page 18) could you palpate (feel) most of the liver to decide if it is enlarged?**

Histology

Histologically, the liver is composed of several components (Figure 25.14a–c):

1. Hepatocytes (*hepat-*=liver; *-cytes*=cells). Hepatocytes are the major functional cells of the liver and perform a wide array of metabolic, secretory, and endocrine functions. These are specialized epithelial cells with 5 to 12 sides that make up about 80% of the volume of the liver. Hepatocytes form complex three-dimensional arrangements called **hepatic laminae.** The hepatic laminae are plates of hepatocytes one cell thick bordered on either side by the endothelial-lined vascular spaces

Figure 25.14 Histology of the liver.

Histologically, the lobule is composed of hepatocytes, bile canaliculi, and hepatic sinusoids.

Right lobe

Left lobe

Inferior vena cava
Hepatic artery
Hepatic portal vein

Connective tissue

Hepatocyte

Hepatic laminae

Portal triad:
Bile duct
Branch of hepatic artery
Branch of hepatic portal vein
Central vein

Hepatic sinusoids

(a) Overview of histological components of liver

Central vein

To hepatic vein

Hepatic sinusoid

Bile canaliculi

Portal triad:
Bile duct
Branch of hepatic portal vein
Branch of hepatic artery
Hepatic laminae
Hepatocyte
Stellate reticuloendothelial (Kupffer) cell
Connective tissue

(b) Details of histological components of liver

Hepatocyte

Central vein

Sinusoid

LM 100x

LM 50x

Portal triad:
Branch of hepatic artery

Bile duct

Branch of hepatic portal vein

LM 150x

(c) Photomicrograph

CONTINUES

called hepatic sinusoids. The hepatic laminae are highly branched, irregular structures. Grooves in the cell membranes between neighboring hepatocytes provide spaces for canaliculi (described next) into which the hepatocytes secrete bile. Bile, a yellow, brownish, or olive-green liquid secreted by hepatocytes, serves as both an excretory product and a digestive secretion.

2. **Bile canaliculi** (kan-a-LIK-ū-li=small canals). These are small ducts between hepatocytes that collect bile produced by the hepatocytes. From bile canaliculi, bile passes into **bile ductules** and then **bile ducts.** The bile ducts merge and eventually form the larger **right** and **left hepatic ducts,** which unite and exit the liver as the **common hepatic duct** (see Figure 25.12a-b). The common hepatic duct joins the **cystic duct** (cystic=bladder) from the gallbladder to form the **common bile duct.** From here, bile enters the small intestine to participate in digestion.

3. **Hepatic sinusoids.** These are highly permeable blood capillaries between rows of hepatocytes that receive oxygenated blood from branches of the hepatic artery and nutrient-rich deoxygenated blood from branches of the hepatic portal vein. Recall that the hepatic portal vein brings venous blood from the gastrointestinal organs and spleen into the liver. Hepatic sinusoids converge and deliver blood into a **central vein.** From central veins the blood flows into the **hepatic veins,** which drain into the inferior vena cava (see Figure 15.5 on page 491). Whereas blood flows toward a central vein, bile flows in the opposite direction. Also present in the hepatic sinusoids are fixed phagocytes called **stellate reticuloendothelial (Kupffer) cells,** which destroy worn-out white and red blood cells, bacteria, and other foreign matter in the venous blood draining from the gastrointestinal tract.

Together, a bile duct, branch of the hepatic artery, and branch of the hepatic vein are referred to as a **portal triad** (tri=three).

The hepatocytes, bile duct system, and hepatic sinusoids can be organized into anatomical and functional units in three different ways:

1. **Hepatic lobule.** For years, anatomists described the hepatic lobule as the functional unit of the liver. According to this model, each hepatic lobule is shaped like a hexagon (six-sided structure) (Figure 25.14d, left). At its center is the central vein and radiating out from it are rows of hepatocytes and hepatic sinusoids. Located at three corners of the hexagon is a portal triad. This model is based on a description of the liver of adult pigs. In the human liver it is difficult to find such well-defined hepatic lobules surrounded by thick layers of connective tissue.

2. **Portal lobule.** This model emphasizes the exocrine function of the liver, that is, bile secretion. Accordingly, the bile duct of a portal triad is taken as the center of the portal lobule. The triangular shape of the portal lobule is defined by three imaginary straight lines that connect three central veins closest to the portal triad (Figure 25.14d, center). This model has not gained widespread acceptance.

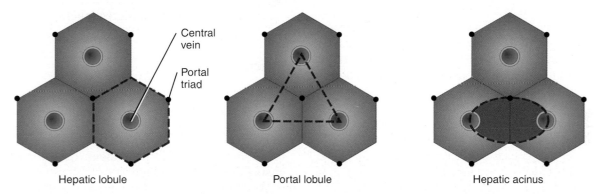

(d) Comparison of three units of liver structure and function

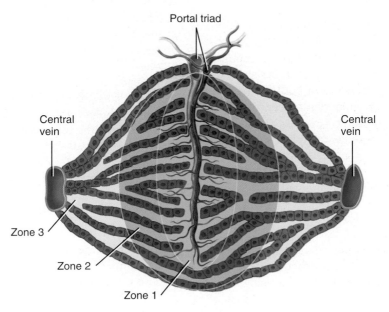

(e) Details of hepatic acinus

? Which type of liver cell is phagocytic?

3. Hepatic acinus. In recent years, the preferred structural and functional unit of the liver is the hepatic acinus. Each hepatic acinus is an approximately oval mass that includes portions of two neighboring hepatic lobules. The short axis of the hepatic acinus is defined by branches of the portal triad—branches of the hepatic artery, vein, and bile ducts—that run along the border of the hepatic lobules. The long axis of the acinus is defined by two imaginary curved lines, which connect the two central veins closest to the short axis (Figure 25.14d, right). Hepatocytes in the hepatic acinus are arranged in three zones around the short axis, with no sharp boundaries between them (Figure 25.14e). Cells in zone 1 are closest to the branches of the portal triad and the first to receive incoming oxygen, nutrients, and toxins from incoming blood. These cells are the first ones to take up glucose and store it as glycogen after a meal and break down glycogen to glucose during fasting. They are also the first to show morphological changes following bile duct obstruction or exposure to toxic substances. Zone 1 cells are the last ones to die if circulation is impaired and the first ones to regenerate. Cells in zone 3 are farthest from branches of the portal triad and are the last to show the effects of bile obstruction or exposure to toxins, the first ones to show the effects of impaired circulation, and the last ones to regenerate. Zone 3 cells also are the first to show evidence of fat accumulation. Cells in zone 2 have structural and functional characteristics intermediate between the cells in zones 1 and 3.

The hepatic acinus is the smallest structural and functional unit of the liver. Its popularity and appeal are based on the fact that it provides a logical description and interpretation of (1) patterns of glycogen storage and release as well as, (2) toxic effects, degeneration, and regeneration in the three zones of an hepatic acinus relative to the proximity of the zones to branches of the portal triad.

The mucosa of the gallbladder consists of simple columnar epithelium arranged in rugae resembling those of the stomach. The wall of the gallbladder lacks a submucosa. The middle, muscular coat consists of smooth muscle fibers; the contraction of these fibers ejects the contents of the gallbladder into the cystic duct. The gallbladder's outer coat is the visceral peritoneum. The functions of the gallbladder are to store bile and concentrate it (up to tenfold) until the bile is needed in the small intestine.

Blood and Nerve Supply

The liver receives blood from two sources (Figure 25.15). From the hepatic artery it obtains oxygenated blood, and from the hepatic portal vein it receives deoxygenated blood containing newly absorbed nutrients, drugs, and possibly microbes and toxins from the gastrointestinal tract. Branches of both the hepatic artery and the hepatic portal vein carry blood into liver sinusoids, where oxygen, most of the nutrients, and certain toxic substances are taken up by the hepatocytes. Products manufactured by the hepatocytes and nutrients needed by other cells are secreted back into the blood, which then drains into the central vein and eventually passes into a hepatic vein. Because blood from the gastrointestinal tract passes through the liver as part of the hepatic portal circulation, the liver is often a site for metastasis of cancer that originates in the GI tract.

The nerve supply to the liver consists of parasympathetic innervation from the vagus (X) nerves and sympathetic innervation from the greater splanchnic nerves through the celiac ganglia.

The gallbladder is supplied by the cystic artery, which usually arises from the right hepatic artery. The cystic veins drain the gallbladder. The nerves to the gallbladder include branches from the celiac plexus and the vagus (X) nerve.

Figure 25.15 Hepatic blood flow: sources, path through the liver, and return to the heart.

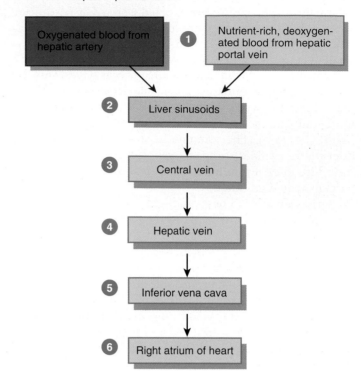

The liver receives oxygenated blood via the hepatic artery and nutrient-rich deoxygenated blood via the hepatic portal vein.

? Why is the liver often a site for metastasis of cancer that originates in the gastrointestinal tract?

Functions

Hepatocytes continuously secrete 800–1000 mL (about 1 qt) of bile per day. Bile salts, which are sodium salts and potassium salts of bile acids (mostly cholic acid and chenodeoxycholic acid), play roles in (1) **emulsification**, the breakdown of large lipid globules into a suspension of droplets about 1 μm in diameter, and (2) the **absorption** of digested lipids.

Between meals, bile flows into the gallbladder for storage because the **sphincter of the hepatopancreatic ampulla (sphincter of Oddi;** see Figure 25.12b) closes off the entrance to the duodenum. After a meal, several neural and hormonal stimuli promote the production and release of bile. Parasympathetic impulses along the vagus (X) nerve fibers can stimulate the liver to increase bile production to more than twice the baseline rate. Fatty acids and amino acids in chyme entering the duodenum stimulate some duodenal enteroendocrine cells to secrete the hormone cholecystokinin (CCK) into the blood. CCK causes contraction of the walls of the gallbladder, which squeezes stored bile out of the gallbladder into the cystic duct and through the common bile duct. CCK also causes relaxation of the sphincter of the hepatopancreatic ampulla, which allows bile to flow into the duodenum.

In addition to secreting bile, the liver performs many other vital functions:

• **Carbohydrate metabolism.** The liver is especially important in maintaining a normal blood glucose level. When blood glucose is low, the liver can break down glycogen to glucose and release glucose into the bloodstream. The liver can also convert certain amino acids and lactic acid to glucose, and it can convert other sugars, such as fructose and galactose, into glucose. When blood glucose is high, as occurs just after eating a meal, the liver converts glucose to glycogen and triglycerides for storage.

- **Lipid metabolism.** Hepatocytes store some triglycerides; break down fatty acids to generate ATP; synthesize lipoproteins, which transport fatty acids, triglycerides, and cholesterol to and from body cells; synthesize cholesterol; and use cholesterol to make bile salts.

- **Protein metabolism.** Hepatocytes *deaminate* [remove the amino group ($-NH_2$) from] amino acids so that the amino acids can be used for ATP production or converted to carbohydrates or fats. The resulting toxic ammonia (NH_3) is then converted into the much less toxic urea, which is excreted in urine. Hepatocytes also synthesize most plasma proteins, such as alpha and beta globulins, albumin, prothrombin, and fibrinogen.

- **Processing of drugs and hormones.** The liver can detoxify substances such as alcohol or secrete drugs such as penicillin, erythromycin, and sulfonamides into bile. It can also inactivate hormones such as thyroid hormones, estrogens, and aldosterone.

- **Excretion of bilirubin.** As previously noted, bilirubin, derived from the heme of aged red blood cells, is absorbed by the liver from the blood and secreted into bile. Most of the bilirubin in bile is metabolized in the small intestine by bacteria and eliminated in feces.

- **Synthesis of bile salts.** Bile salts are used in the small intestine for the emulsification and absorption of lipids, cholesterol, phospholipids, and lipoproteins.

- **Storage.** In addition to glycogen, the liver is a prime storage site for certain vitamins (A, B_{12}, D, E, and K) and minerals (iron and copper), which are released from the liver when needed elsewhere in the body.

- **Phagocytosis.** The stellate reticuloendothelial (Kupffer) cells of the liver phagocytize aged red blood cells and white blood cells and some bacteria.

- **Activation of vitamin D.** The skin, liver, and kidneys participate in synthesizing the active form of vitamin D.

CHECKPOINT

25. Draw and label a diagram of a liver lobule.
26. Describe the pathways of blood flow into, through, and out of the liver.
27. How are the liver and gallbladder connected to the duodenum?
28. Describe the functions of the liver and gallbladder.

SMALL INTESTINE

OBJECTIVE

- Describe the location, anatomy, histology, and function of the small intestine.

Most digestion and absorption of nutrients occur in a long tube called the **small intestine**. Because of this, its structure is specially adapted for this function. Its length alone provides a large surface area for digestion and absorption, and that area is further increased by circular folds, villi, and microvilli. The small intestine begins at the pyloric sphincter of the stomach, coils through the central and inferior part of the abdominal cavity, and eventually opens into the large intestine. It averages 2.5 cm (1 in.) in diameter; its length is about 3 m (10 ft) in a living person and about 6.5 m (21 ft) in a cadaver due to the loss of smooth muscle tone after death.

Anatomy

The small intestine is divided into three regions (Figure 25.16a). The **duodenum** (doo-ō′-DĒ-num), the shortest region, is retroperitoneal. *Duodenum* means "12"; it is so named because it is about as long as the width of 12 fingers. It starts at the pyloric sphincter of the stomach and extends about 25 cm (10 in.) until it merges with the next section, called the jejunum. The **jejunum** (jē-JOO-num) is about 1 m (3 ft) long and extends to the ileum. *Jejunum* means "empty," which is how it is found at death. The

Figure 25.16 Regions of the small intestine. See also Figure 25.1b.

Most digestion and absorption occur in the small intestine.

Functions of the Small Intestine

1. Segmentations mix chyme with digestive juices and bring food into contact with the mucosa for absorption; peristalsis propels chyme through the small intestine.

2. Completes the digestion of carbohydrates (starches), proteins, and lipids; begins and completes the digestion of nucleic acids.

3. Absorbs about 90% of nutrients and water.

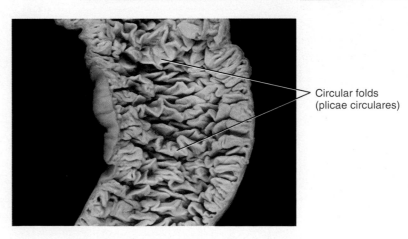

(a) Anterior view of external anatomy

(b) Internal anatomy of jejunum

? Which portion of the small intestine is the longest?

final and longest region of the small intestine, the **ileum** (IL-ē-um=twisted), measures about 2 m (6 ft) and joins the large intestine at a smooth muscle sphincter called the **ileocecal** (il′-ē-ō-SĒ-kal) **sphincter**.

The arterial blood supply of the small intestine is from the superior mesenteric artery and the gastroduodenal artery, which arises from the hepatic artery of the celiac trunk. Blood is returned by way of the superior mesenteric vein, which anastomoses with the splenic vein to form the hepatic portal vein.

The nerves to the small intestine are supplied by the superior mesenteric plexus. The branches of the plexus contain postganglionic sympathetic fibers, preganglionic parasympathetic fibers, and sensory fibers. The sensory fibers are components of the vagus (X) nerves and spinal nerves via the sympathetic pathways. In the wall of the small intestine are two autonomic plexuses: the myenteric plexus between the muscular layers and the submucosal plexus in the submucosa. The nerve fibers to the smooth muscle of the blood vessels arise chiefly from the sympathetic division of the autonomic nervous system, while the nerve fibers to the smooth muscle of the intestinal wall originate from the vagus (X) nerves.

Histology

The wall of the small intestine is composed of the same four layers that make up most of the GI tract: mucosa, submucosa, muscularis, and serosa (Figure 25.17a). The **mucosa** is composed of a layer of epithelium, lamina propria, and muscularis mucosae. The epithelial layer of the small intestinal mucosa consists of simple columnar epithelium that contains many types of cells (Figure 25.17b). **Absorptive cells** of the epithelium digest and absorb

nutrients in small intestinal chyme. Also present in the epithelium are **goblet cells**, which secrete mucus. The small intestinal mucosa contains many deep crevices lined with glandular epithelium. Cells lining the crevices form the **intestinal glands (crypts of Lieberkühn)** and secrete intestinal juice (to be discussed shortly). Besides absorptive cells and goblet cells, the intestinal glands also contain paneth cells and enteroendocrine cells. **Paneth cells** secrete lysozyme, a bactericidal enzyme, and are capable of phagocytosis. Paneth cells may have a role in regulating the microbial population in the small intestine. Three types of **enteroendocrine cells** are found in the intestinal glands of the small intestine: **S cells**, **CCK cells**, and **K cells**, which secrete the hormones **secretin** (se-KRĒ-tin), **cholecystokinin** (kō-lē-sis′-tō-KĬN-in) **(CCK)**, and **glucose-dependent insulinotropic peptide (GIP)**, respectively.

The lamina propria of the small intestinal mucosa contains areolar connective tissue and has an abundance of mucosa-associated lymphoid tissue (MALT). **Solitary lymphatic nodules** are most numerous in the distal part of the ileum (Figure 25.18c on page 809). Groups of lymphatic nodules referred to as **aggregated lymphatic follicles (Peyer's patches)** are also present in the ileum. As it does throughout the digestive tract, the muscularis mucosae consists of smooth muscle.

The **submucosa** of the duodenum contains **duodenal (Brunner's) glands** (Figure 25.18a on page 808), which secrete an alkaline mucus that helps neutralize gastric acid in the chyme. Sometimes the lymphatic tissue of the lamina propria extends through the muscularis mucosae into the submucosa.

The **muscularis** of the small intestine consists of two layers of smooth muscle. The outer, thinner layer contains longitudinal fibers; the inner, thicker layer contains circular fibers. Except for

Figure 25.17 Histology of the small intestine.

Circular folds, villi, and microvilli increase the surface area of the small intestine for digestion and absorption.

Lumen of small intestine

Villus Blood Lacteal
capillary

Simple columnar epithelium
Lamina propria
Opening of intestinal gland
Lymphatic nodule
Muscularis mucosae
Arteriole
Venule
Lymphatic vessel
Circular layer of muscle
Myenteric plexus
Longitudinal layer of muscle

MUCOSA
SUBMUCOSA
MUSCULARIS
SEROSA

(a) Three-dimensional view of layers of the small intestine showing villi a major portion of the duodenum, the **serosa** (or visceral peritoneum) completely surrounds the small intestine.

Even though the wall of the small intestine is composed of the same four basic layers as the rest of the GI tract, special structural features of the small intestine facilitate the processes of digestion and absorption. These structural features include circular folds, villi, and microvilli. **Circular folds** or *plicae circulares* (PLĪ-sē ser-kyū-LA-rēs) are folds of the mucosa and submucosa (see Figure 25.16b). These permanent ridges, which are about 10 mm (0.4 in.) long, begin near the proximal portion of the duodenum and end at about the midportion of the ileum. Some extend all the way around the circumference of the intestine; others extend only part of the way around. Circular folds enhance absorption by increasing surface area and causing the chyme to spiral, rather than move in a straight line, as it passes through the small intestine.

Also present in the small intestine are **villi** (=tufts of hair), which are fingerlike projections of the mucosa that are 0.5–1 mm long (Figure 25.18). The large number of villi (20–40 per square millimeter) vastly increases the surface area of the epithelium available for absorption and digestion and gives the intestinal mucosa a velvety appearance. Each villus (singular form) is covered by epithelium and has a core of lamina propria; embedded in the connective tissue of the lamina propria are an arteriole, a venule, a blood capillary network, and a **lacteal** (LAK-tē-al=milky), which is a lymphatic capillary. Nutrients absorbed by the epithelial cells covering the villus pass through the wall of a capillary or a lacteal to enter blood or lymph, respectively.

Besides circular folds and villi, the small intestine also has **microvilli** (mī-krō-VIL-ī; *micro-*=small), which are projections of the apical (free) membrane of the absorptive cells. Each microvillus is a 1 μm-long cylindrical, membrane-covered projection that contains a bundle of 20–30 actin filaments. When viewed through a light microscope, the microvilli are too small to be seen individually; instead they form a fuzzy line, called the **brush border**, extending into the lumen of the small intestine (Figure 25.18d). There are an estimated 200 million microvilli per square millimeter of small intestine. Because the microvilli greatly increase the surface area of the plasma membrane, larger amounts of digested nutrients can diffuse into absorptive cells in a given period. The brush border also contains several brush-border enzymes that have digestive functions (discussed shortly).

Functions

Chyme entering the small intestine contains partially digested carbohydrates, proteins, and lipids (mostly triglycerides). The completion of the digestion of carbohydrates, proteins, and lipids is the result of the collective action of pancreatic juice, bile, and intestinal juice in the small intestine.

Intestinal juice is a clear yellow fluid secreted in amounts of 1 to 2 liters (about 1 to 2 quarts) each day. It has a pH of 7.6, which is slightly alkaline, and contains water and mucus. Together, pancreatic and intestinal juice provide a vehicle for the absorption of substances from chyme as they come in contact with the villi.

The absorptive epithelial cells synthesize several digestive enzymes, called **brush-border enzymes**, and insert them in the plasma membrane of the microvilli. Thus, some enzymatic digestion occurs at the surface of the epithelial cells that line the villi; in other parts of the GI tract, enzymatic digestion occurs in the lumen exclusively. Among the brush-border enzymes are four carbohydrate-digesting enzymes called **α-dextrinase**, **maltase**,

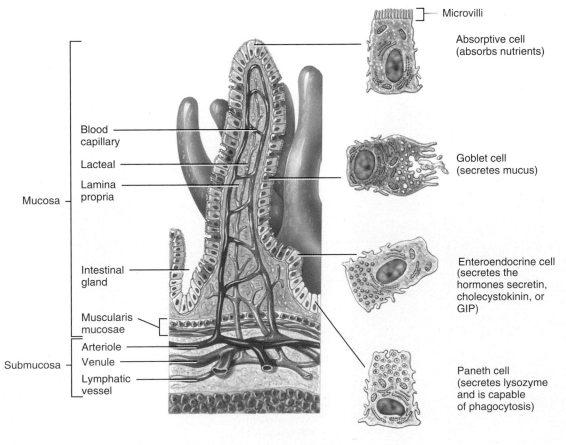

(b) Enlarged villus showing lacteal, capillaries, intestinal glands, and cell types

? What is the functional significance of the blood capillary network and lacteal in the center of each villus?

sucrase, and **lactase**; protein-digesting enzymes called **peptidases** (**aminopeptidase** and **dipeptidase**); and two types of nucleotide-digesting enzymes, **nucleosidases** and **phosphatases**. Also, as cells slough off into the lumen of the small intestine, they break apart and release enzymes that help digest nutrients in the chyme.

The two types of movements of the small intestine—segmentations and a type of peristalsis called migrating motility complexes—are governed mainly by the myenteric plexus. **Segmentations** are localized, mixing contractions that occur in portions of the intestine distended by a large volume of chyme. Segmentations mix chyme with the digestive juices and bring the particles of food into contact with the mucosa for absorption; they do not push the intestinal contents along the tract. A segmentation starts with the contractions of circular muscle fibers in

Figure 25.18 Histology of the duodenum and ileum.

Microvilli greatly increase the surface area of the small intestine for digestion and absorption.

(a) Wall of the duodenum

(b) Three villi from the duodenum

• CLINICAL **CONNECTION** Peptic Ulcer Disease

In the United States, 5–10% of the population develops **peptic ulcer disease (PUD)**. An **ulcer** is a craterlike lesion in a membrane; ulcers that develop in areas of the GI tract exposed to acidic gastric juice are called **peptic ulcers**. The most common complication of peptic ulcers is bleeding, which can lead to anemia if enough blood is lost. In acute cases, peptic ulcers can lead to shock and death. Three distinct causes of PUD are recognized: (1) the bacterium *Helicobacter pylori*; (2) nonsteroidal anti-inflammatory drugs (NSAIDs) such as aspirin; and (3) hypersecretion of HCl, as occurs in Zollinger-Ellison syndrome, a gastrin-producing tumor, usually of the pancreas.

Helicobacter pylori (previously named *Campylobacter pylori*) is the most frequent cause of PUD. The bacterium produces an enzyme called *urease*, which splits urea into ammonia and carbon dioxide. While shielding the bacterium from the acidity of the stomach, the ammonia also damages the protective mucous layer of the stomach and the underlying gastric cells. *H. pylori* also produces catalase, an enzyme that may protect the microbe from phagocytosis by neutrophils, plus several adhesion proteins that allow the bacterium to attach itself to gastric cells.

Several therapeutic approaches are helpful in the treatment of PUD. Because cigarette smoke, alcohol, caffeine, and NSAIDs can impair mucosal defensive mechanisms, in the process increasing mucosal susceptibility to the damaging effects of HCl, these substances should be

avoided. In cases associated with *H. pylori*, treatment with an antibiotic drug often resolves the problem. Oral antacids such as Tums® or Maalox® can help temporarily by buffering gastric acid. When hypersecretion of HCl is the cause of PUD, H₂ blockers (such as Tagamet®) or proton pump inhibitors such as omeprazole (Prilosec®), which block secretion of H⁺ from parietal cells, may be used. •

---

a portion of the small intestine, an action that constricts the intestine into segments. Next, muscle fibers that encircle the middle of each segment also contract, dividing each segment again. Finally, the fibers that first contracted relax, and each small segment unites with an adjoining small segment so that large segments are formed again. As this sequence of events repeats, the chyme sloshes back and forth. Segmentations occur most rapidly in the duodenum, about 12 times per minute, and progressively decrease to about 8 times per minute in the ileum. This movement is similar to alternately squeezing the middle and then the ends of a capped tube of toothpaste.

After most of a meal has been absorbed, which lessens distention of the wall of the small intestine, segmentation stops and

peristalsis begins. The type of peristalsis that occurs in the small intestine, termed a **migrating motility complex (MMC)**, begins in the lower portion of the stomach and pushes chyme forward along a short stretch of small intestine before dying out. The MMC slowly migrates down the small intestine, reaching the end of the ileum in 90–120 minutes. Then another MMC begins in the stomach. Altogether, chyme remains in the small intestine for 3–5 hours.

All the chemical and mechanical phases of digestion from the mouth through the small intestine are directed toward changing food into forms that can pass through the epithelial cells lining the mucosa into the underlying blood and lymphatic vessels. These forms are monosaccharides (glucose, fructose, and galac-

(c) Lymphatic nodules in the ileum

(d) Several microvilli from the duodenum

? What is the function of the fluid secreted by duodenal (Brunner's) glands?

tose) from carbohydrates; single amino acids, dipeptides, and tripeptides from proteins; fatty acids, glycerol, and monoglycerides from lipids; and pentoses and nitrogenous bases from nucleic acids. Passage of these digested nutrients from the gastrointestinal tract into the blood or lymph is called **absorption**. Absorption occurs by diffusion, facilitated diffusion, osmosis, and active transport.

About 90% of all absorption of nutrients takes place in the small intestine. The other 10% occurs in the stomach and large intestine. Any undigested or unabsorbed material left in the small intestine passes on to the large intestine.

◉ CHECKPOINT

29. Name and describe the regions of the small intestine.
30. In what ways are the mucosa and submucosa of the small intestine adapted for digestion and absorption?
31. Describe the types of movement in the small intestine.
32. What is absorption? In what form are the products of carbohydrate, protein, and lipid digestion absorbed?

LARGE INTESTINE

◉ OBJECTIVE

- Describe the anatomy, histology, and functions of the large intestine.

The large intestine is the terminal portion of the GI tract and is divided into four principal regions. As chyme moves through the

large intestine, bacteria act on it and water, ions, and vitamins are absorbed. As a result, feces are formed and then eliminated from the body.

Anatomy

The **large intestine**, which is about 1.5 m (5 ft) long and 6.5 cm (2.5 in.) in diameter, extends from the ileum to the anus. The ascending colon and descending colon are retroperitoneal, while the remaining parts of the colon and cecum are attached to the posterior abdominal wall by their **mesocolon**, a double layer of peritoneum connecting the parietal peritoneum to the visceral peritoneum (see Figure 25.3a). Structurally, the four principal regions of the large intestine are the cecum, colon, rectum, and anal canal (Figure 25.19a).

The opening from the ileum into the large intestine is guarded by a fold of mucous membrane called the **ileocecal sphincter** or **valve**, which allows materials from the small intestine to pass into the large intestine. Hanging inferior to the ileocecal valve is the **cecum**, a small pouch about 6 cm (2.4 in.) long. Attached to the cecum is a twisted, coiled tube, measuring about 8 cm (3 in.) in length, called the **appendix** or **vermiform appendix** (*vermiform*=worm-shaped; *appendix*=appendage). The mesentery of the appendix, called the **mesoappendix**, attaches the appendix to the inferior part of the mesentery of the ileum.

The open end of the cecum merges with a long tube called the **colon** (=food passage), which is divided into ascending, transverse, descending, and sigmoid portions. Both the ascending and descending colon are retroperitoneal; the transverse and sigmoid

Figure 25.19 Anatomy of the large intestine.

◉ ⌐▭ The regions of the large intestine are the cecum, colon, rectum, and anal canal.

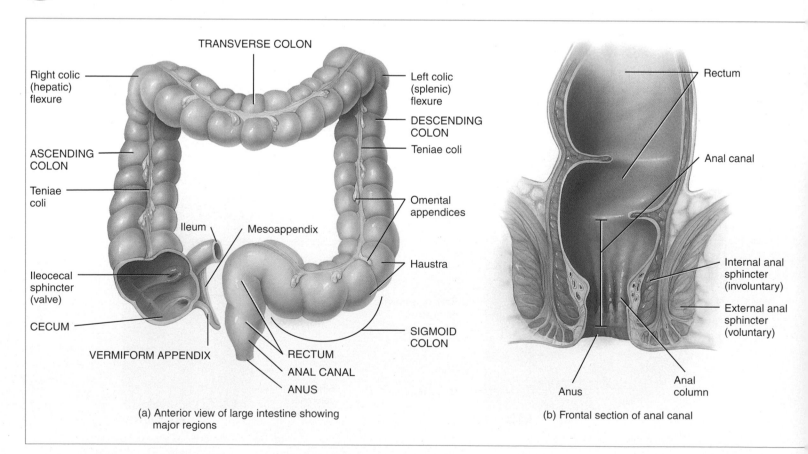

(a) Anterior view of large intestine showing major regions

(b) Frontal section of anal canal

colon are not. The **ascending colon** ascends on the right side of the abdomen, reaches the inferior surface of the liver, and turns abruptly to the left to form the **right colic (hepatic) flexure**. The colon continues across the abdomen to the left side as the **transverse colon**. It curves beneath the inferior end of the spleen on the left side as the **left colic (splenic) flexure** and passes inferiorly to the level of the left iliac crest as the **descending colon**. The **sigmoid colon** (*sigm-*=S-shaped) begins near the left iliac crest, projects medially to the midline, and terminates as the rectum at about the level of the third sacral vertebra.

The **rectum**, the last 20 cm (8 in.) of the GI tract, lies anterior to the sacrum and coccyx. The terminal 2–3 cm (1in.) of the rectum is called the **anal canal** (Figure 25.19b). The mucous membrane of the anal canal is arranged in longitudinal folds called **anal columns** that contain a network of arteries and veins. The opening of the anal canal to the exterior, called the **anus**, is guarded by an **internal anal sphincter** of smooth muscle (involuntary) and an **external anal sphincter** of skeletal muscle (voluntary). Normally the anus is closed except during the elimination of feces.

The arterial supply of the cecum and colon is derived from branches of the superior mesenteric and inferior mesenteric arteries. The venous return is by way of the superior and inferior mesenteric veins ultimately to the hepatic portal vein and into the liver. The arterial supply of the rectum and anal canal branches from the superior, middle, and inferior rectal arteries. The rectal veins correspond to the rectal arteries.

The nerves to the large intestine consist of sympathetic, parasympathetic, and sensory components. The sympathetic innervation arises from the celiac, superior, and inferior mesenteric ganglia and superior and inferior mesenteric plexuses. The fibers reach the viscera by way of the thoracic and lumbar splanchnic nerves. The parasympathetic innervation is derived from the vagus (X) and pelvic splanchnic nerves.

Histology

The wall of the large intestine contains the typical four layers found in the rest of the GI tract: mucosa, submucosa, muscularis, and serosa. The **mucosa** consists of simple columnar epithelium, lamina propria (areolar connective tissue), and muscularis mucosae (smooth muscle) (Figure 25.20a). The epithelium contains mostly absorptive and goblet cells (Figure 25.20b and c). The absorptive cells function primarily in water absorption; the goblet cells secrete mucus that lubricates the passage of the colonic contents. Both absorptive and goblet cells are located in long, straight, tubular **intestinal glands (crypts of Lieberkühn)** that extend the full thickness of the mucosa. Solitary lymphatic nodules are also found in the lamina propria of the mucosa and may extend through the muscularis mucosae into the submucosa. Compared to the small intestine, the mucosa of the large intestine does not have as many structural adaptations that increase surface area. There are no circular folds or villi; however, microvilli are present on the absorptive cells. Consequently, much more absorption occurs in the small intestine than in the large intestine.

The **submucosa** of the large intestine consists of areolar connective tissue. The **muscularis** consists of an external layer of longitudinal smooth muscle and an internal layer of circular smooth muscle. Unlike other parts of the GI tract, portions of the longitudinal muscles are thickened, forming three conspicuous bands called the **teniae coli** (TĒ-nē-ē KŌ-lī; *teniae*=flat bands) that run most of the length of the large intestine (see Figure

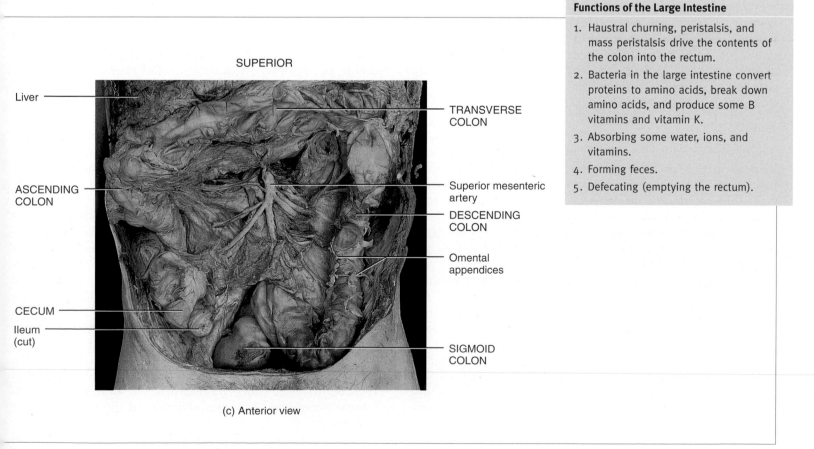

SUPERIOR

Liver

TRANSVERSE COLON

ASCENDING COLON

Superior mesenteric artery

DESCENDING COLON

Omental appendices

CECUM

Ileum (cut)

SIGMOID COLON

(c) Anterior view

Functions of the Large Intestine

1. Haustral churning, peristalsis, and mass peristalsis drive the contents of the colon into the rectum.
2. Bacteria in the large intestine convert proteins to amino acids, break down amino acids, and produce some B vitamins and vitamin K.
3. Absorbing some water, ions, and vitamins.
4. Forming feces.
5. Defecating (emptying the rectum).

? **Which portions of the colon are retroperitoneal?**

Figure 25.20 Histology of the large intestine.

Intestinal glands formed by simple columnar epithelial cells and goblet cells extend the full thickness of the mucosa.

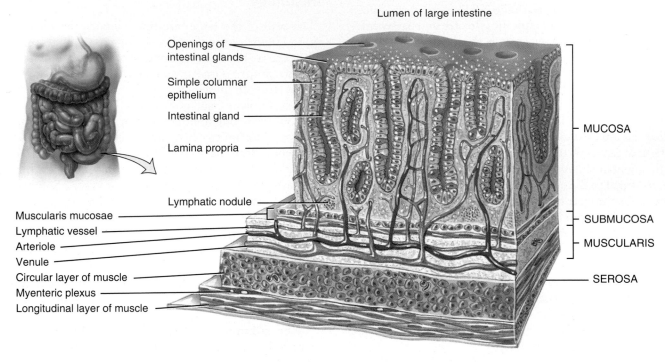

Lumen of large intestine

Openings of
intestinal glands

Simple columnar
epithelium

Intestinal gland

Lamina propria

MUCOSA

Lymphatic nodule

Muscularis mucosae

Lymphatic vessel

Arteriole

Venule

Circular layer of muscle

Myenteric plexus

Longitudinal layer of muscle

SUBMUCOSA

MUSCULARIS

SEROSA

(a) Three-dimensional view of layers of the large intestine

25.19a). The portions of the wall between the teniae coli have little or no longitudinal muscle. Tonic contractions of the bands gather the colon into a series of pouches called **haustra** (HAWS-tra=shaped like pouches; singular is **haustrum**), which give the colon a puckered appearance. A single layer of circular smooth muscle lies deep to the teniae coli. The **serosa** of the large intestine is part of the visceral peritoneum. Small pouches of visceral peritoneum filled with fat are attached to teniae coli and are called **omental (fatty) appendices** (see Figure 25.19a).

Functions

The passage of chyme from the ileum into the cecum is regulated by the action of the ileocecal sphincter. Normally, the valve remains partially closed, and the passage of chyme into the cecum is a slow process. Immediately after a meal, ileal peristalsis intensifies, the sphincter relaxes, and chyme is forced from the ileum into the cecum. As food passes through the ileocecal sphincter, it fills the cecum and accumulates in the ascending colon, and movements of the colon begin.

One movement characteristic of the large intestine is **haustral churning**. In this process, the haustra remain relaxed and distended while they fill up. When the distension reaches a certain point, the wall contracts and squeezes the contents into the next haustrum. **Peristalsis** also occurs, although at a slower rate (3 to 12 contractions per minute) than in other portions of the tract. A final type of movement is **mass peristalsis**, a strong peristaltic wave that begins at about the middle of the transverse colon and quickly drives the colonic contents into the rectum. Mass peristalsis usually takes place three or four times a day, during or immediately after a meal.

The final stage of digestion occurs in the colon through the activity of bacteria that inhabit the lumen. Mucus is secreted by

the glands of the large intestine, but no enzymes are secreted. Chyme is prepared for elimination by the action of bacteria, which ferment any remaining carbohydrates and release hydrogen, carbon dioxide, and methane gases. These gases contribute to flatus (gas) in the colon, termed *flatulence* when it is excessive. Bacteria also convert any remaining proteins to amino acids and break down the amino acids into simpler substances: indole, skatole, hydrogen sulfide, and fatty acids. Some of the indole and skatole is eliminated in the feces and contributes to their odor; the rest is absorbed and transported to the liver, where these compounds are converted to less toxic compounds and excreted in the urine. Bacteria also decompose bilirubin to simpler pigments, including stercobilin, which give feces their brown color. Several vitamins needed for normal metabolism, including some B vitamins and vitamin K, are bacterial products that are absorbed in the colon.

By the time chyme has remained in the large intestine 3–10 hours, it has become solid or semisolid as a result of water absorption and is now called **feces**. Chemically, feces consist of water, inorganic salts, sloughed-off epithelial cells from the mucosa of the gastrointestinal tract, bacteria, products of bacterial decomposition, unabsorbed digested materials, and indigestible parts of food.

Although 90% of all water absorption occurs in the small intestine, the large intestine absorbs enough to make it an important organ in maintaining the body's water balance. Of the 0.5–1.0 liter of water that enters the large intestine, all but about 100–200 mL is absorbed via osmosis. The large intestine also absorbs electrolytes, including sodium and chloride, and some vitamins.

Mass peristaltic movements push fecal material from the sigmoid colon into the rectum. The resulting distension of the rectal wall stimulates stretch receptors, which initiates a **defeca-**

Openings of intestinal glands

Lamina propria

Microvilli

Intestinal gland

Absorptive cell (absorbs water)

Goblet cell (secretes mucus)

Lymphatic nodule

Muscularis mucosae

Submucosa

(b) Sectional view of intestinal glands and cell types

Mucosa

Submucosa

Muscularis

Serosa

Lumen of large intestine

Lamina propria

Intestinal gland

Lymphatic nodule

Muscularis mucosae

LM 315x

(c) Portion of the wall of the large intestine

Opening of intestinal gland

Lumen of large intestine

Absorptive cell

Goblet cell

Lamina propria

Intestinal gland

LM 300x

(d) Details of mucosa of large intestine

? **What is the function of the goblet cells of the large intestine?**

tion reflex that empties the rectum. The defecation reflex occurs as follows: In response to distension of the rectal wall, the receptors send sensory nerve impulses to the sacral spinal cord. Motor impulses from the cord travel along parasympathetic nerves back to the descending colon, sigmoid colon, rectum, and anus. The resulting contraction of the longitudinal rectal muscles shortens the rectum, thereby increasing the pressure within it. This pressure, along with voluntary contractions of the diaphragm and abdominal muscles, plus parasympathetic stimulation, opens the internal sphincter.

The external sphincter is voluntarily controlled. If it is voluntarily relaxed, defecation occurs and the feces are expelled through the anus; if it is voluntarily constricted, defecation can be postponed. Voluntary contractions of the diaphragm and abdominal muscles aid defecation by increasing the pressure within the abdomen, which pushes the walls of the sigmoid colon and rectum inward. If defecation does not occur, the feces back up into the sigmoid colon until the next wave of mass peristalsis again stimulates the stretch receptors, further creating the urge to defecate. In infants, the defecation reflex causes automatic emptying of the rectum because voluntary control of the external anal sphincter has not yet developed.

Diarrhea (dī-a-RĒ-a; *dia-*=through; *rrhea*=flow) is an increase in the frequency, volume, and fluid content of the feces caused by increased motility of and decreased absorption by the intestines. When chyme passes too quickly through the small intestine and feces pass too quickly through the large intestine, there is not enough time for absorption. Frequent diarrhea can result in dehydration and electrolyte imbalances. Excessive motility may be caused by lactose intolerance, stress, or microbes that irritate the gastrointestinal mucosa.

Constipation (kon-sti-PĀ-shun; *con-*=together; *stip-*=to press) refers to infrequent or difficult defecation caused by decreased motility of the intestines. Because the feces remain in the colon for prolonged periods of time, excessive water absorption occurs, and the feces become dry and hard. Constipation may be caused by poor habits (delaying defecation), spasms of the colon, insufficient fiber in the diet, inadequate fluid intake, lack of exercise, emotional stress, and certain drugs. A common treatment is a mild laxative, such as milk of magnesia, which induces defecation. However, many physicians maintain that laxatives are habit-forming, and that adding fiber to the diet, increasing the amount of exercise, and increasing fluid intake are safer ways of controlling this common problem.

Dietary fiber consists of indigestible plant carbohydrates—such as cellulose, lignin, and pectin—found in fruits, vegetables, grains, and beans. **Insoluble fiber**, which does not dissolve in water, includes the woody or structural parts of plants such as the skins of fruits and vegetables and the bran coating around wheat and corn kernels. Insoluble fiber passes through the GI tract largely unchanged and speeds up the passage of material through the tract. **Soluble fiber** dissolves in water and forms a gel, which slows the passage of material through the tract; it is found in abundance in beans, oats, barley, broccoli, prunes, apples, and citrus fruits.

People who choose a fiber-rich diet may reduce their risk of developing obesity, diabetes, atherosclerosis, gallstones, hemorrhoids, diverticulitis, appendicitis, and colorectal cancer. Soluble fiber also may help lower blood cholesterol. The liver normally converts cholesterol to bile salts, which are released into the small intestine to help fat digestion. Having accomplished their task, the bile salts are reabsorbed by the small intestine and recycled back to the liver. Since soluble fiber binds to bile salts to prevent their reabsorption, the liver makes more bile salts to replace those lost in feces. Thus, the liver uses more cholesterol to make more bile salts and blood cholesterol level is lowered.

A summary of the digestive organs and their functions is presented in Table 25.2.

◉ CHECKPOINT

33. What are the principal regions of the large intestine?
34. How does the muscularis of the large intestine differ from that of the rest of the gastrointestinal tract? What are haustra?
35. Describe the mechanical movements that occur in the large intestine.
36. Define defecation. How does it occur?
37. Explain the activities of the large intestine that change its contents into feces.

DEVELOPMENT OF THE DIGESTIVE SYSTEM

◉ OBJECTIVE

• Describe the development of the digestive system.

During the fourth week of development, the cells of the **endoderm** form a cavity called the **primitive gut**, the forerunner of the gastrointestinal tract (see Figure 4.12b on page 107). Soon afterward the mesoderm forms and splits into two layers (somatic and splanchnic), as shown in Figure 4.9d on page 103. The splanchnic mesoderm associates with the endoderm of the primitive gut; as a result, the primitive gut has a double-layered wall. The **endodermal layer** gives rise to the *epithelial lining* and *glands* of most of the gastrointestinal tract; the **mesodermal layer** produces the *smooth muscle* and *connective tissue* of the tract.

The primitive gut elongates and differentiates into an anterior **foregut**, an intermediate **midgut**, and a posterior **hindgut** (see Figure 4.12c on page 107). Until the fifth week of development, the midgut opens into the yolk sac; after that time, the yolk sac constricts and detaches from the midgut, and the midgut seals. In the region of the foregut, a depression consisting of ectoderm, the **stomodeum** (stō-mō-DĒ-um), appears (see Figure 4.12d on page 107). This develops into the *oral cavity*. The **oropharyngeal membrane** is a depression of fused ectoderm and endoderm on the surface of the embryo that separates the foregut from the stomodeum. The membrane ruptures during the fourth week of development, so that the foregut is continuous with the outside of

TABLE 25.2

Summary of Organs of the Digestive System and their Functions

ORGANS	FUNCTIONS
Tongue	Maneuvers food for mastication, shapes food into a bolus, maneuvers food for deglutition, detects sensations for taste, and initiates digestion of triglycerides.
Salivary glands	Saliva produced by these glands softens, moistens, and dissolves foods; cleanses mouth and teeth; initiates the digestion of starch.
Teeth	Cut, tear, and pulverize food to reduce solids to smaller particles for swallowing.
Pancreas	Pancreatic juice buffers acidic gastric juice in chyme, stops the action of pepsin from the stomach, creates the proper pH for digestion in the small intestine, and participates in the digestion of carbohydrates, proteins, triglycerides, and nucleic acids.
Liver	Produces bile, which is required for the emulsification and absorption of lipids in the small intestine.
Gallbladder	Stores and concentrates bile and releases it into the small intestine.
Mouth	See the functions of the tongue, salivary glands, and teeth, all of which are in the mouth. Additionally, the lips and cheeks keep food between the teeth during mastication, and buccal glands lining the mouth produce saliva.
Pharynx	Receives a bolus from the oral cavity and passes it into the esophagus.
Esophagus	Receives a bolus from the pharynx and moves it into the stomach. This requires relaxation of the upper esophageal sphincter and secretion of mucus.
Stomach	Mixing waves combine saliva, food, and gastric juice, which activates pepsin, initiates protein digestion, kills microbes in food, helps absorb vitamin B_{12}, contracts the lower esophageal sphincter, increases stomach motility, relaxes the pyloric sphincter, and moves chyme into the small intestine.
Small intestine	Segmentation mixes chyme with digestive juices; peristalsis propels chyme toward the ileocecal sphincter; digestive secretions from the small intestine, pancreas, and liver complete the digestion of carbohydrates, proteins, lipids, and nucleic acids; circular folds, villi, and microvilli help absorb about 90% of digested nutrients.
Large intestine	Haustral churning, peristalsis, and mass peristalsis drive the colonic contents into the rectum; bacteria produce some B vitamins and vitamin K; absorption of some water, ions, and vitamins occurs; defecation.

the embryo through the oral cavity. Another depression consisting of ectoderm, the **proctodeum** (prok-tō-DĒ-um), forms in the hindgut and goes on to develop into the *anus*. (See Figure 4.12d on page 107.) The **cloacal membrane** (klō-Ā-kul) is a fused membrane of ectoderm and endoderm that separates the hindgut from the proctodeum. After the cloacal membrane ruptures during the seventh week, the hindgut forms a continuous tube from mouth to anus.

The foregut develops into the *pharynx, esophagus, stomach*, and *part of the duodenum*. The midgut is transformed into the *remainder of the duodenum*, the *jejunum*, the *ileum*, and *portions of the large intestine* (cecum, appendix, ascending colon, and most of the transverse colon). The hindgut develops into the *remainder of the large intestine*, except for a portion of the anal canal that is derived from the proctodeum.

As development progresses, the endoderm at various places along the foregut develops into hollow buds that grow into the mesoderm. These buds will develop into the *salivary glands, liver, gallbladder*, and *pancreas*. Each of these organs retains a connection with the gastrointestinal tract through ducts.

◉ CHECKPOINT

38. What structures develop from the foregut, midgut, and hindgut?

AGING AND THE DIGESTIVE SYSTEM

◉ OBJECTIVE

• Describe the effects of aging on the digestive system.

Overall changes of the digestive system associated with aging include decreased secretory mechanisms, decreased motility of the digestive organs, loss of strength and tone of the muscular tissue and its supporting structures, changes in neurosensory feedback regarding enzyme and hormone release, and diminished response to pain and internal sensations. In the upper portion of the GI tract, common changes include reduced sensitivity to mouth irritations and sores, loss of taste, periodontal disease, difficulty in swallowing, hiatal hernia, gastritis, and peptic ulcer disease. Changes in the small intestine may include duodenal ulcers, appendicitis, malabsorption, and maldigestion. Other pathologies that increase in incidence with age are gallbladder problems, jaundice, cirrhosis, and acute pancreatitis. Large intestinal changes such as constipation, hemorrhoids, and diverticular disease may also occur. Cancer of the colon or rectum is quite common, as are bowel obstructions and impactions.

◉ CHECKPOINT

39. What overall changes in the digestive system appear with age?

KEY MEDICAL TERMS ASSOCIATED WITH THE DIGESTIVE SYSTEM

Achalasia (ak'-a-LĀ-zē-a; *a-*=without; *chalasis*=relaxation) A condition, caused by malfunction of the myenteric plexus, in which the lower esophageal sphincter fails to relax normally as food approaches. A whole meal may become lodged in the esophagus and enter the stomach very slowly. Distension of the esophagus results in chest pain that is often confused with pain originating from the heart.

Anal fissure (FISH-ur; *findo*=to cleave) A longitudinal tear in the mucosa of the anal canal frequently caused by injury from passing hard feces. Anal fissures may cause pain and bleeding, usually during or shortly after a bowel movement. Treatment consists of using stool softeners and lubricating suppositories and, in severe cases, surgery.

Anorexia nervosa A chronic disorder characterized by self-induced weight loss, negative perception of body image, and physiological changes that result from nutritional depletion. Patients have a fixation on weight control and often abuse laxatives, which worsens their fluid and electrolyte imbalances and nutrient deficiencies. Individuals may become emaciated and may ultimately die of starvation or one of its complications.

Barrett's esophagus A pathological change in the epithelium of the esophagus from nonkeratinized stratified squamous epithelium to columnar epithelium so that the lining resembles that of the stomach or small intestine. This abnormal development of tissue is due to long-term exposure of the esophagus to stomach acid and increases the risk of developing cancer of the esophagus.

Borborygmus (bor'-bō-RIG-mus) A rumbling noise caused by the propulsion of gas through the intestines.

Bulimia (*bu-*=ox; *limia*=hunger or **binge–purge syndrome**) A disorder that typically affects young, single, middle-class, white females, characterized by overeating at least twice a week followed by purging by self-induced vomiting, strict dieting or fasting, vigorous exercise, or use of laxatives or diuretics; it occurs in response to fears of being overweight or to stress, depression, and physiological disorders such as hypothalamic tumors.

Canker sore (KANG-ker) Painful ulcer on the mucous membrane of the mouth that affects females more often than males, usually between ages 10 and 40; may be an autoimmune reaction or a food allergy.

Celiac disease (SĒ-lē-ak; *koilia*=hollow) A common malabsorption disorder caused by sensitivity to gluten, a protein found in some grains such as wheat, rye, barley, and oats. The gluten damages intestinal villi and reduces the length of the microvilli, producing symptoms including foul-smelling diarrhea, bloating, weight loss, abdominal pain, and anemia.

Cholecystitis (kō'-lē-sis-TĪ-tis; *chole-*=bile; *cyst-*=bladder; *-itis*=inflammation of) An inflammation of the gallbladder, caused in some cases by an autoimmune reaction; other cases are caused by obstruction of the cystic duct by bile stones.

Cirrhosis Distorted or scarred liver as a result of chronic inflammation due to hepatitis, chemicals that destroy hepatocytes, parasites that infect the liver, or alcoholism; the hepatocytes are replaced by fibrous or adipose connective tissue. Symptoms include jaundice, edema in the legs, uncontrolled bleeding, and increased sensitivity to drugs.

Colitis (kō-LĪ-tis) Inflammation of the mucosa of the colon and rectum in which absorption of water and salts is reduced, producing watery, bloody feces and, in severe cases, dehydration and salt depletion. Spasms of the irritated muscularis produce cramps.

Colostomy (kō-LOS-tō-mē; *-stomy*=provide an opening) The diversion of feces through an opening in the colon, creating a surgical "stoma" (artificial opening) that is made in the exterior of the abdominal wall. This opening serves as a substitute anus through which feces are eliminated into a bag worn on the abdomen.

Dysphagia (dis-FĀ-jē-a; *dys-*=abnormal; *phagia*=to eat) Difficulty in swallowing that may be caused by inflammation, paralysis, obstruction, or trauma.

Flatus (FLĀ-tus) Air (gas) in the stomach or intestine, usually expelled through the anus. If the gas is expelled through the mouth, it is called **eructation** or **belching** (burping). Flatus may result from gas released during the breakdown of foods in the stomach or from swallowing air- or gas-containing substances such as carbonated drinks.

Food poisoning A sudden illness caused by ingesting food or drink contaminated by an infectious microbe (bacterium, virus, or protozoan) or a toxin (poison). The most common cause of food poisoning is the toxin produced by the bacterium *Staphylococcus aureus*. Most types of food poisoning cause diarrhea and/or vomiting, often associated with abdominal pain.

Gastroenteritis (gas'-trō-en-ter-Ī-tis; *gastro-*=stomach; *enteron* = intestine; *-itis*=inflammation) Inflammation of the lining of the stomach and intestine (especially the small intestine). It is usually caused by a viral or bacterial infection that may be acquired by contaminated food or water or by people in close contact. Symptoms include diarrhea, vomiting, fever, loss of appetite, cramps, and abdominal discomfort.

Gastroscopy (gas-TROS-kō-pē; *-scopy*=to view with a lighted instrument) Endoscopic examination of the stomach in which the examiner can view the interior of the stomach directly to evaluate an ulcer, tumor, inflammation, or source of bleeding.

Halitosis (hal'-i-TŌ-sis; *halitus-*=breath; *-osis*=condition) A foul odor from the mouth; also called **bad breath**.

Heartburn A burning sensation in a region near the heart due to irritation of the mucosa of the esophagus from hydrochloric acid in stomach contents. It is caused by failure of the lower esophageal sphincter to close properly, so that the stomach contents enter the inferior esophagus. It is not related to any cardiac problem.

Hemorrhoids (HEM-ō-royds; *hemi*=blood; *rhoia*=flow) Varicosed (enlarged and inflamed) superior rectal veins that develop when the veins are put under pressure and become engorged with blood. If the pressure continues, the wall of the vein stretches. Such a distended vessel oozes blood; bleeding or itching is usually the first sign that a hemorrhoid has developed. Stretching of a vein also favors clot formation, further aggravating swelling and pain. Hemorrhoids may be caused by constipation, which may be brought on by low-fiber diets. Also, repeated straining during defecation forces blood down into the rectal veins, increasing pressure in these veins and possibly causing hemorrhoids. Also called **piles**.

Hernia (HER-nē-a) Protrusion of all or part of an organ through a membrane or cavity wall, usually the abdominal cavity. *Hiatal hernia* is the protrusion of a part of the stomach into the thoracic cavity through the esophageal hiatus. *Inguinal hernia* is the protrusion of the hernial sac into the inguinal opening; it may contain a portion of the bowel in an advanced stage and may extend into the scrotal compartment in males, causing strangulation of the herniated part.

Indigestion (in-di-JES-chun) A nonspecific term used to describe many symptoms associated with abdominal distress, especially after eating. Symptoms include discomfort or a feeling of dullness in the upper abdomen, nausea, and a sensation of bloating, often relieved by belching. Among the factors related to indigestion are reactions to drugs, alcohol, and supplements; overproduction of stomach acid; infections; food sensitivity; stress; and psychological factors such as depression and anxiety. Also called **dyspepsia**.

Inflammatory bowel disease (in-FLAM-a-tō'-rē BOW-el) Inflammation of the gastrointestinal tract that exists in two forms. (1) **Crohn's disease** is an inflammation of any part of the gastrointestinal tract in which the inflammation extends from the mucosa through the submucosa, muscularis, and serosa. (2) **Ulcerative colitis** is an inflammation of the mucosa of the colon and rectum, usually accompanied by rectal bleeding.

Irritable bowel syndrome (IBS) Disease of the entire gastrointestinal tract in which a person reacts to stress by developing symptoms (such as cramping and abdominal pain) associated with alternating patterns of diarrhea and constipation. Excessive amounts of mucus may appear in feces; other symptoms include flatulence, nausea, and loss of appetite. The condition is also known as **irritable colon** or **spastic colitis**.

Laparoscopy (lap′-a-ROS-kō-pē; *laparo-*=flank, *-skopy*=to view) Examination of the contents of the peritoneum with a laparoscope (a type of endoscope) passed through the abdominal wall. Frequently used to examine female reproductive organs.

Malabsorption (mal-ab-SORP-shun; *mal-*=bad) A number of disorders in which nutrients from food are not absorbed properly. It may be due to disorders that result in the inadequate breakdown of food during digestion (due to inadequate digestive enzymes or juices), damage to the lining of the small intestine (surgery, infections, and drugs like neomycin and alcohol), and impairment of motility. Symptoms may include diarrhea, weight loss, weakness, vitamin deficiencies, and bone demineralization.

Malocclusion (mal′-ō-KLOO-zhun; *mal-*=bad; *occlusion*=to fit together) Condition in which the surfaces of the maxillary (upper) and mandibular (lower) teeth fit together poorly.

Nausea (NAW-sē-a; *nausia*=seasickness) Discomfort characterized by a loss of appetite and the sensation of impending vomiting. Its causes include local irritation of the gastrointestinal tract, a systemic disease, brain disease or injury, overexertion, or the effects of medication or drug overdosage.

Pruritus ani (proo-RĪ-tus; *pruire*=to itch) Anal itching that may be caused by pinworms (in children), infections, dermatitis, hemorrhoids, or dry, aging skin.

Traveler's diarrhea Infectious disease of the gastrointestinal tract that results in loose, urgent bowel movements, cramping, abdominal pain, malaise, nausea, and occasionally fever and dehydration. It is acquired through ingestion of food or water contaminated with fecal material typically containing bacteria (especially *Escherichia coli*); viruses or protozoan parasites are a less common cause.

Xerostomia (zē-rō-STŌ-mē-a; *xeros-*=dry; *-stoma*=mouth) Dryness of the mouth due to decreased or inhibited salivary secretion.

STUDY OUTLINE

Introduction (p. 781)
1. The breakdown of larger food molecules into smaller molecules is called digestion.
2. The passage of these smaller molecules into blood and lymph is termed absorption.

Overview of the Digestive System (p. 782)
1. The organs that collectively perform digestion and absorption constitute the digestive system and are usually divided into two main groups: the gastrointestinal (GI) tract and accessory digestive organs.
2. The GI tract is a continuous tube extending from the mouth to the anus.
3. The accessory digestive organs include the teeth, tongue, salivary glands, liver, gallbladder, and pancreas.
4. Digestion includes six basic processes: ingestion, secretion, mixing and propulsion, mechanical and chemical digestion, absorption, and defecation.
5. Mechanical digestion consists of mastication and movements of the gastrointestinal tract that aid chemical digestion.
6. Chemical digestion is a series of hydrolysis reactions that break down large carbohydrates, lipids, proteins, and nucleic acids in foods into smaller molecules that are usable by body cells.

Layers of the GI Tract (p. 782)
1. The basic arrangement of layers in most of the gastrointestinal tract, from deep to superficial, is the mucosa, submucosa, muscularis, and serosa.
2. Associated with the lamina propria of the mucosa are extensive patches of lymphatic tissue called mucosa-associated lymphoid tissue (MALT).

Peritoneum (p. 785)
1. The peritoneum is the largest serous membrane of the body; it lines the wall of the abdominal cavity and covers some abdominal organs.
2. Folds of the peritoneum include the mesentery, mesocolon, falciform ligament, lesser omentum, and greater omentum.

Mouth (p. 787)
1. The mouth is formed by the cheeks, hard and soft palates, lips, and tongue.
2. The vestibule is the space bounded externally by the cheeks and lips and internally by the teeth and gums.
3. The oral cavity proper extends from the vestibule to the fauces.
4. The tongue, together with its associated muscles, forms the floor of the oral cavity. It is composed of skeletal muscle covered with mucous membrane.
5. The upper surface and sides of the tongue are covered with papillae, some of which contain taste buds.
6. The major portion of saliva is secreted by the salivary glands, which lie outside the mouth and pour their contents into ducts that empty into the oral cavity.
7. There are three pairs of salivary glands: parotid, submandibular (submaxillary), and sublingual glands.
8. Saliva lubricates food and starts the chemical digestion of carbohydrates.
9. Salivation is controlled by the nervous system.
10. The teeth (dentes) project into the mouth and are adapted for mechanical digestion.
11. A typical tooth consists of three principal regions: crown, root, and neck.
12. Teeth are composed primarily of dentin and are covered by enamel, the hardest substance in the body.
13. There are two dentitions: deciduous and permanent.
14. Through mastication, food is mixed with saliva and shaped into a soft, flexible mass called a bolus.
15. Salivary amylase begins the digestion of starches, and lingual lipase acts on triglycerides.

Pharynx (p. 793)
1. Deglutition, or swallowing, moves a bolus from the mouth to the stomach.
2. Muscular contractions of the oropharynx and laryngopharynx propel a bolus into the esophagus.

Esophagus (p. 793)
1. The esophagus is a collapsible, muscular tube that connects the pharynx to the stomach.
2. It passes a bolus into the stomach by peristalsis.
3. It contains an upper and a lower esophageal sphincter.

Stomach (p. 794)
1. The stomach connects the esophagus to the duodenum.
2. The principal anatomic regions of the stomach are the cardia, fundus, body, and pylorus.
3. Adaptations of the stomach for digestion include rugae; glands that produce mucus, hydrochloric acid, pepsin, gastric lipase, and intrinsic factor; and a three-layered muscularis.
4. Mechanical digestion consists of mixing waves.
5. Chemical digestion consists mostly of the conversion of proteins into peptides by pepsin.
6. The stomach wall is impermeable to most substances.
7. Among the substances the stomach can absorb are water, certain ions, drugs, and alcohol.

Pancreas (p. 798)

1. The pancreas consists of a head, a body, and a tail and is connected to the duodenum via the pancreatic duct and accessory duct.
2. Endocrine pancreatic islets (islets of Langerhans) secrete hormones, and exocrine acini secrete pancreatic juice.
3. Pancreatic juice contains enzymes that digest starch (pancreatic amylase), proteins (trypsin, chymotrypsin, carboxypeptidase, and elastase), triglycerides (pancreatic lipase), and nucleic acids (ribonuclease and deoxyribonuclease).

Liver and Gallbladder (p. 800)

1. The liver has left and right lobes; the right lobe includes a quadrate and caudate lobe. The gallbladder is a sac located in a depression on the posterior surface of the liver that stores and concentrates bile.
2. The lobes of the liver are made up of lobules that contain hepatocytes (liver cells), sinusoids, stellate reticuloendothelial (Kupffer's) cells, a vein, and bile ducts.
3. Hepatocytes produce bile that is carried by a duct system to the gallbladder for concentration and temporary storage. Cholecystokinin (CCK) stimulates ejection of bile into the common bile duct.
4. Bile's contribution to digestion is the emulsification of dietary lipids.
5. The liver also functions in carbohydrate, lipid, and protein metabolism; processing of drugs and hormones; excretion of bilirubin; synthesis of bile salts; storage of vitamins and minerals; phagocytosis; and activation of vitamin D.
6. Bile secretion is regulated by neural and hormonal mechanisms.

Small Intestine (p. 805)

1. The small intestine extends from the pyloric sphincter to the ileocecal sphincter.
2. It is divided into three parts: duodenum, jejunum, and ileum.
3. Its glands secrete fluid and mucus, and the circular folds, villi, and microvilli of its wall provide a large surface area for digestion and absorption.
4. Brush-border enzymes digest α-dextrins, maltose, sucrose, lactose, peptides, and nucleotides at the surface of mucosal epithelial cells.
5. Pancreatic and intestinal brush-border enzymes break down carbohydrates, proteins, and nucleic acids.
6. Mechanical digestion in the small intestine involves segmentation and migrating motility complexes.
7. Absorption is the passage of digested nutrients from the gastrointestinal tract into blood or lymph.
8. The absorbed nutrients include monosaccharides, amino acids, fatty acids, monoglycerides, pentoses, and nitrogenous bases.

Large Intestine (p. 810)

1. The large intestine extends from the ileocecal sphincter to the anus.
2. Its regions include the cecum, colon, rectum, and anal canal.
3. The mucosa contains many goblet cells, and the muscularis consists of teniae coli and haustra.
4. Mechanical movements of the large intestine include haustral churning, peristalsis, and mass peristalsis.
5. The last stages of chemical digestion occur in the large intestine through bacterial action. Substances are further broken down, and some vitamins are synthesized.
6. The large intestine absorbs water, electrolytes, and vitamins.
7. Feces consist of water, inorganic salts, epithelial cells, bacteria, and undigested foods.
8. The elimination of feces from the rectum is called defecation.
9. Defecation is a reflex action aided by voluntary contractions of the diaphragm and abdominal muscles and relaxation of the external anal sphincter.

Development of the Digestive System (p. 814)

1. The endoderm of the primitive gut forms the epithelium and glands of most of the gastrointestinal tract.
2. The mesoderm of the primitive gut forms the smooth muscle and connective tissue of the gastrointestinal tract.

Aging and the Digestive System (p. 815)

1. General changes include decreased secretory mechanisms, decreased motility, and loss of tone.
2. Specific changes may include loss of taste, pyorrhea, hernias, peptic ulcer disease, constipation, hemorrhoids, and diverticular disease.

SELF-QUIZ QUESTIONS

Choose the one best answer to the following questions.

1. The cells of the gastric glands that produce secretions directly involved in chemical digestion are:
   a. mucous neck cells.
   b. parietal cells.
   c. chief cells.
   d. G cells.
   e. goblet cells.

2. Which anatomical region of the stomach is closest to the esophagus?
   a. body
   b. pylorus
   c. fundus
   d. cardia
   e. pyloric sphincter

3. *All* of the following are considered accessory organs of the digestive system *except* the:
   a. liver.
   b. stomach.
   c. pancreas.
   d. teeth.
   e. salivary glands.

4. Which of the following digestive juices contains enzymes that digest carbohydrates, lipids, and proteins?
   a. pancreatic juice
   b. bile
   c. saliva
   d. gastric juice
   e. none of the above (because no one digestive juice contains enzymes for digesting all three classes of foods)

5. The type of tissue that lines the stomach and intestines is:
   a. stratified squamous epithelium.
   b. pseudostratified ciliated columnar epithelium.
   c. simple columnar epithelium.
   d. simple cuboidal epithelium.
   e. endothelium.

6. The roots of a tooth are covered by a tissue that is harder and denser than bone and is called:
   a. gingivae.
   b. cementum.
   c. dentin.
   d. periodontal ligament.
   e. enamel.

7. The parotid glands are located:
   a. inferior and anterior to the ears.
   b. at the base of the tongue.
   c. inferior to the liver.
   d. in the gastric mucosa.
   e. in the folds of the peritoneum.

8. The region of the stomach that connects to the small intestine is the:
   a. fundus.
   b. body.
   c. cardia.
   d. rugae.
   e. pylorus.

Complete the following.

9. The largest peritoneal fold, which drapes over the transverse colon and small intestine, is the _____.

10. Place numbers in the blanks to indicate the correct order for the flow of bile. (a) bile canaliculi: _____; (b) common bile duct: _____; (c) common hepatic duct: _____; (d) right and left hepatic ducts: _____; (e) hepatopancreatic ampulla (ampulla of Vater) and duodenum: _____.

11. The scientific name for swallowing is _____.

12. The process by which contents of the gastrointestinal tract are propelled forward by coordinated contractions and relaxations of the circular and longitudinal muscles of the muscularis is called _____.

13. First and second premolars and first, second, and third molars are characteristic of the _____ dentition.

14. The _____ is an organ of the digestive system that assumes a role in phagocytosis, manufacture of plasma proteins, detoxification, and interconversions of nutrients.

15. The duct that conveys bile to and from the gallbladder is the _____ duct.

16. Parietal cells of the stomach secrete _____ and _____, and the chief cells secrete _____ and _____.

17. The two large lobes of the liver are the right and left lobes, and the two smaller lobes are the posterior _____ lobe and the inferior _____ lobe.

18. The three protein-digesting enzymes found in pancreatic juice are _____, _____, and _____.

Are the following statements true or false? If the statement is False, find and fix the mistake to make the statement True.

19. The submucosal plexus (plexus of Meissner) is important in controlling motility of the GI tract, while the myenteric plexus (plexus of Auerbach) controls secretions of the GI tract.

20. The gallbladder produces bile.

21. The largest lobe of the liver is the right lobe.

22. Acini are clusters of cells that make up the exocrine portions of the pancreas.

23. The sinusoids of the liver receive blood from two sources: oxygenated blood from branches of the hepatic artery, and deoxygenated blood from branches of the hepatic portal vein.

24. The duodenum is the most distal region of the small intestine.

Matching

25. Match the following structures and descriptions:

___ **(a)** secrete lysozyme in intestinal juice	**(1)** mesentery
___ **(b)** site of maltase, sucrase, and lactase in the small intestine	**(2)** parotid glands
	(3) Paneth cells
	(4) hepatocytes
___ **(c)** the pouches seen in the large intestine	**(5)** pancreas
___ **(d)** folds of the mucosa of the stomach and gallbladder	**(6)** brush border
	(7) rugae
___ **(e)** binds the small intestine to the posterior abdominal wall	**(8)** haustra
___ **(f)** produce bile	
___ **(g)** secrete saliva	
___ **(h)** secretes both enzymes and hormones	

CRITICAL THINKING QUESTIONS

1. If you could leave French fried potatoes in your mouth long enough after chewing them, they would start to taste sweeter and maybe even a little tart. Why?

2. When Zelda turned 50, her doctor said, "Well, it's time to schedule your first colonoscopy!" What is a colonoscopy? Describe the specific anatomical characteristics the doctor would be examining with the fiber optics.

3. The small intestine is the primary site of digestion and absorption in the gastrointestinal tract. What structural modifications are unique to the small intestine and what are their functions?

4. Krystal and her friend were giggling over their milk and fries at her birthday party. In mid-giggle, milk started coming out of Krystal's nose. How did this happen?

5. Four-year-old Billy was lying down with his head resting on his mother's abdomen. He started laughing and said, "Mommy, your tummy sure is making a lot of funny noises!" What (specifically) is Billy hearing?

25.1 Digestive enzymes are secreted by the salivary glands, tongue, stomach, pancreas, and small intestine.

25.2 The enteric plexuses help regulate secretions and motility of the gastrointestinal tract.

25.3 Mesentery binds the small intestine to the posterior abdominal wall.

25.4 The uvula helps prevent foods and liquids from entering the nasal cavity during swallowing.

25.5 The ducts of the submandibular glands empty on either side of the lingual frenulum.

25.6 The main component of teeth is a type of connective tissue called dentin.

25.7 The first, second, and third molars do not replace any deciduous teeth.

25.8 The esophageal mucosa and submucosa contain mucus-secreting glands that lubricate the gastrointestinal tract to allow food to pass through more easily.

25.9 Food is pushed along the gastrointestinal tract by contraction of smooth muscle behind the bolus and relaxation of smooth muscle in front of it.

25.10 After a large meal, the rugae stretch and disappear as the stomach fills; as the stomach empties, the rugae reappear.

25.11 In the gastric glands of the stomach, surface mucous cells and mucous neck cells secrete mucus; chief cells secrete pepsinogen and gastric lipase; parietal cells secrete HCl and intrinsic factor; and G cells secrete gastrin.

25.12 The pancreatic duct contains pancreatic juice (fluid and digestive enzymes); the common bile duct contains bile; the hepatopancreatic ampulla contains pancreatic juice and bile.

25.13 The epigastric region, which contains most of the liver, can be palpated in a clinical examination to check for enlargement.

25.14 The phagocytic cell in the liver is the stellate reticuloendothelial (Kupffer) cell.

25.15 The liver is often a site for metastasis of cancer that originates in the gastrointestinal tract because blood from the tract passes through the liver as part of the hepatic portal circulation.

25.16 The ileum is the longest part of the small intestine.

25.17 Nutrients being absorbed enter the blood via the capillaries or the lymph via the lacteals in the center of each villus.

25.18 The fluid secreted by duodenal glands—alkaline mucus—neutralizes gastric acid and protects the mucosal lining of the duodenum.

25.19 The ascending and descending portions of the colon are retroperitoneal.

25.20 The goblet cells of the large intestine secrete mucus to lubricate the colonic contents.

26 | THE URINARY SYSTEM

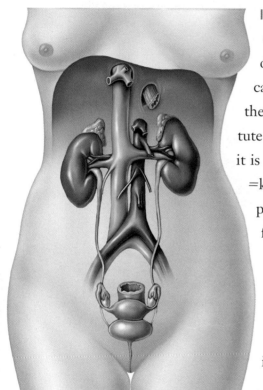

INTRODUCTION The **urinary system** consists of two kidneys, two ureters, one urinary bladder, and one urethra (Figure 26.1). Like the respiratory and digestive systems, the urinary system forms an extensive area of contact with the cardiovascular system. After the kidneys filter blood plasma, they return most of the water and solutes to the bloodstream. The remaining water and solutes constitute **urine**, which passes through the ureters and is stored in the urinary bladder until it is excreted from the body through the urethra. **Nephrology** (nef-ROL-ō-jē; *nephr-* =kidney;-*logy*=study of) is the scientific study of the anatomy, physiology, and pathology of the kidneys. The branch of medicine that deals with the male and female urinary systems and the male reproductive system is called **urology** (ū-ROL-ō-jē; *uro-* =urine). A physician who specializes in this branch of medicine is called a **urologist** (ū-ROL-ō-jist).

The kidneys do the major work of the urinary system. The other parts of the system are mainly passageways and storage areas. Functions of the kidneys include the following:

- **Regulation of blood ionic composition.** The kidneys help regulate the blood levels of several ions, most importantly sodium ions (Na^+), potassium ions (K^+), calcium ions (Ca^{2+}), chloride ions (Cl^-), and phosphate ions (HPO_4^{2-}).

- **Regulation of blood pH.** The kidneys excrete a variable amount of hydrogen ions (H^+) into the urine and conserve bicarbonate ions (HCO_3^-). Both of these activities help regulate blood pH.

- **Regulation of blood volume.** The kidneys adjust blood volume by conserving or eliminating water in the urine. An increase in blood volume increases blood pressure; a decrease in blood volume decreases blood pressure.

- **Enzymatic regulation of blood pressure.** The kidneys *also* help regulate blood pressure by secreting the enzyme renin, which causes an increase in blood pressure.
- **Maintenance of blood osmolarity.** By separately regulating loss of water and loss of solutes in the urine, the kidneys maintain a relatively constant blood osmolarity. The **osmolarity** of a solution is a measure

of the total number of dissolved particles per liter of solution.

- **Production of hormones.** The kidneys produce two hormones. *Calcitriol,* the active form of vitamin D, helps regulate calcium homeostasis, and *erythropoietin* stimulates the production of red blood cells.

Figure 26.1 Organs of the urinary system, shown in relation to the surrounding structures in a female.

Urine formed by the kidneys passes first into the ureters, then to the urinary bladder for storage, and finally through the urethra for elimination from the body.

(a) Anterior view

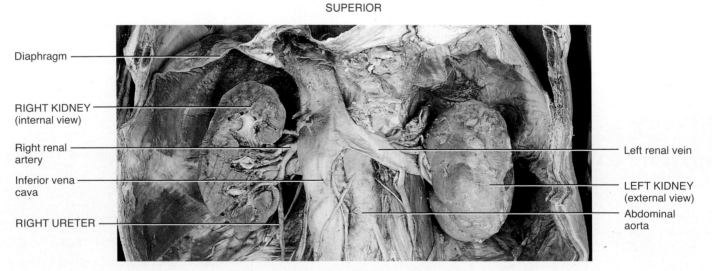

(b) Anterior view

- **Regulation of blood glucose level.** Like the liver, the kidneys can use the amino acid glutamine in *gluconeogenesis*, the synthesis of new glucose molecules. They can then release glucose into the blood to help maintain a normal blood glucose level.
- **Excretion of wastes and foreign substances.** By forming urine, the kidneys help excrete **wastes**—substances that have no useful function in the body. Some wastes excreted in urine result from metabolic reactions in the body. These include ammonia and urea from the deamination of amino acids; bilirubin from the catabolism of hemoglobin; creatinine from the breakdown of creatine phosphate in muscle fibers; and uric acid from the catabolism of nucleic acids. Other wastes excreted in urine are foreign substances from the diet, such as drugs and environmental toxins. •

SUPERIOR

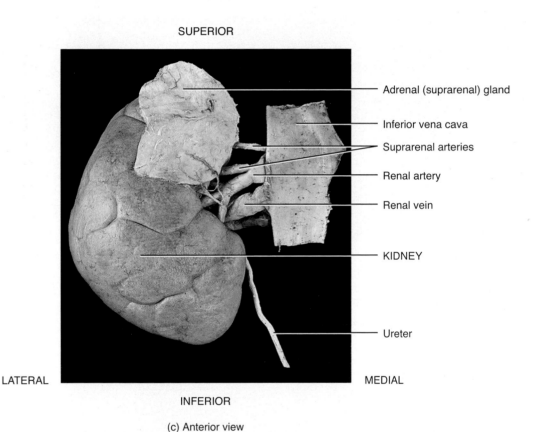

Adrenal (suprarenal) gland

Inferior vena cava

Suprarenal arteries

Renal artery

Renal vein

KIDNEY

Ureter

LATERAL

MEDIAL

INFERIOR

(c) Anterior view

Functions of the Urinary System

1. The kidneys regulate blood volume and composition, help regulate blood pressure, synthesize glucose, release erythropoietin, participate in vitamin D synthesis, and excrete wastes in the urine.
2. The ureters transport urine from the kidneys to the urinary bladder.
3. The urinary bladder stores urine.
4. The urethra discharges urine from the body.

? Which organs constitute the urinary system?

ANATOMY AND HISTOLOGY OF THE KIDNEYS

● OBJECTIVES
- Describe the external and internal gross anatomical features of the kidneys.
- Trace the path of blood flow through the kidneys.
- Describe the structure of renal corpuscles and renal tubules.

The paired **kidneys** are reddish, kidney-bean-shaped organs located just above the waist between the peritoneum and the posterior wall of the abdomen. Because their position is posterior to the peritoneum of the abdominal cavity, they are said to be **retroperitoneal** (re'-trō-per-i-tō-NĒ-al; *retro-*=behind) organs (Figure 26.2). The kidneys are located between the levels of the last thoracic and third lumbar vertebrae, a position where they are partially protected by the eleventh and twelfth pairs of ribs. The right kidney is slightly lower than the left (see Figure 26.1)

Figure 26.2 Position and coverings of the kidneys.

The kidneys are surrounded by a renal capsule, adipose capsule, and renal fascia.

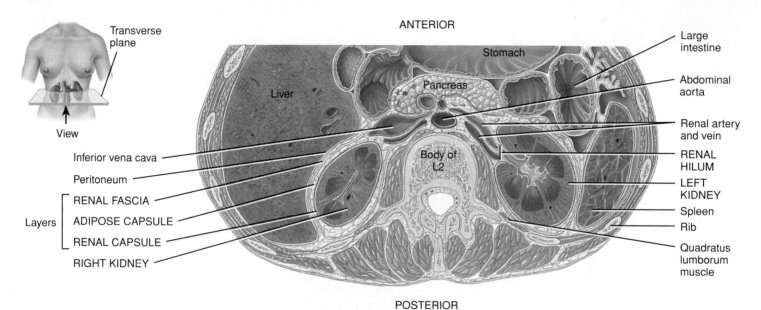

(a) Inferior view of transverse section of abdomen (L2)

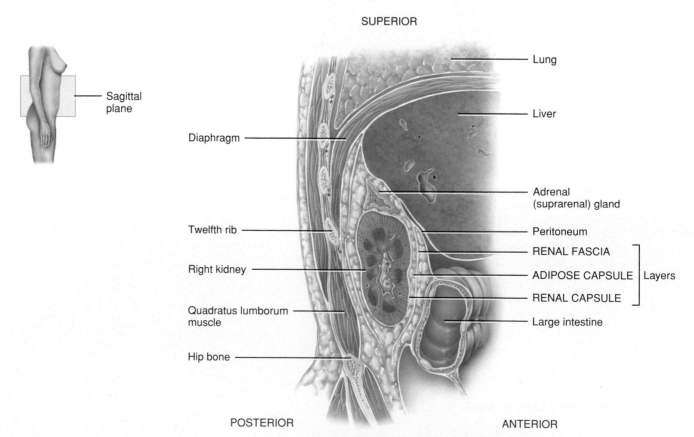

(b) Sagittal section through the right kidney

because the liver occupies considerable space on the right side superior to the kidney.

External Anatomy of the Kidneys

A typical kidney in an adult is 10–12 cm (4–5 in.) long, 5–7 cm (2–3 in.) wide, and 3 cm (1 in.) thick—about the size of a bar of bath soap—and has a mass of 125–170 g (4.5–5 oz). The concave medial border of each kidney faces the vertebral column (see Figure 26.1). Near the center of the concave border is an indentation called the **renal hilum** (RĒ-nal; *ren-*=kidney) (see Figure 26.3), through which the ureter emerges from the kidney along with blood vessels, lymphatic vessels, and nerves.

Three layers of tissue surround each kidney (see Figure 26.2). The deep layer, the **renal capsule**, is a smooth, transparent sheet of dense irregular connective tissue that is continuous with the outer coat of the ureter. It serves as a barrier against trauma and helps maintain the shape of the kidney. The middle layer, the **adipose capsule**, is a mass of fatty tissue surrounding the renal capsule. It also protects the kidney from trauma and holds it firmly in place within the abdominal cavity. The superficial layer, the **renal fascia**, is another thin layer of dense irregular connective tissue that anchors the kidney to the surrounding structures and to the abdominal wall. On the anterior surface of the kidneys, the renal fascia is deep to the peritoneum.

Internal Anatomy of the Kidneys

A frontal section through the kidney reveals two distinct regions: a superficial, light red area called the **renal cortex** (*cortex*=rind or bark) and a deep, darker red-brown region called the **renal medulla** (*medulla*=inner portion) (Figure 26.3). The renal medulla consists of several cone-shaped **renal pyramids**. The base (wider end) of each pyramid faces the renal cortex, and its apex (narrower end), called a **renal papilla**, points toward the renal hilum. The smooth-textured renal cortex extends from the renal capsule to the bases of the renal pyramids and into the spaces between them. It is divided into an outer *cortical zone* and an inner *juxtamedullary zone*. Those portions of the renal cortex that extend between renal pyramids are called **renal columns**. A **renal lobe** consists of a renal pyramid, its overlying area of renal cortex, and one-half of each adjacent renal column.

Together, the renal cortex and renal pyramids of the renal medulla constitute the **parenchyma** (functional portion) of the kidney. Within the parenchyma are the functional units of the kidney—about 1 million microscopic structures called **nephrons** (NEF-rons). Urine formed by the nephrons drains into large **papillary ducts**, which extend through the renal papillae of the pyramids. The papillary ducts drain into cuplike structures called **minor** and **major calyces** (KĀ-li-sēz=cups; singular is *calyx*). Each kidney has 8 to 18 minor calyces and 2 to 3 major calyces. A minor calyx receives urine from the papillary ducts of one renal papilla and delivers it to a major calyx. From the major calyces, urine drains into a single large cavity called the **renal pelvis** (*pelv-*=basin) and then out through the ureter to the urinary bladder.

The hilum expands into a cavity within the kidney called the **renal sinus**, which contains part of the renal pelvis, the calyces, and branches of the renal blood vessels and nerves. Adipose tissue helps stabilize the position of these structures in the renal sinus.

Blood and Nerve Supply of the Kidneys

Because the kidneys remove wastes from the blood and regulate its volume and ionic composition, it is not surprising that they are abundantly supplied with blood vessels. Although the kidneys constitute less than 0.5% of total body mass, they receive 20–25% of the resting cardiac output via the right and left **renal arteries** (Figure 26.4 on page 827). In adults, **renal blood flow**, the blood flow through the kidneys, is about 1200 mL per minute.

TEXT CONTINUES ON PAGE **828** ▶

ANTERIOR

Pancreas

Inferior vena cava

Liver

Parietal peritoneum

Renal hilum

RIGHT KIDNEY

Intervertebral disc between L2 and L3

Abdominal aorta

ADIPOSE CAPSULE ⎫ Layers

RENAL CAPSULE ⎭

LEFT KIDNEY

Body of L2

POSTERIOR

(c) Inferior view of transverse section of abdomen between L2 and L3

? Why are the kidneys said to be retroperitoneal?

Figure 26.3 Internal anatomy of the kidneys.

The two main regions of the kidney parenchyma are the renal cortex and the renal pyramids in the renal medulla.

Renal hilum

Nephron

Path of urine drainage:

Collecting duct
↓
Papillary duct in renal pyramid
↓
Minor calyx
↓
Major calyx
↓
Renal pelvis

Renal artery

Renal vein

Renal cortex

Renal medulla

Renal column

Renal pyramid in renal medulla

Renal sinus

Renal papilla

Fat in renal sinus

Renal capsule

Renal lobe

Ureter
↓
Urinary bladder

(a) Frontal section of right kidney

SUPERIOR

Renal capsule

Renal cortex

Minor calyx

Major calyx

Renal artery

Renal vein

Renal pelvis

Ureter

LATERAL

MEDIAL

(b) Anterior view of right kidney

? **What structures pass through the renal hilum?**

Figure 26.4 Blood supply of the kidneys. In the cast of the kidney shown in (c), the arteries are *red*, the veins are *blue*, and the urine-draining structures are *yellow*.

The renal arteries deliver 20–25% of the resting cardiac output to the kidneys.

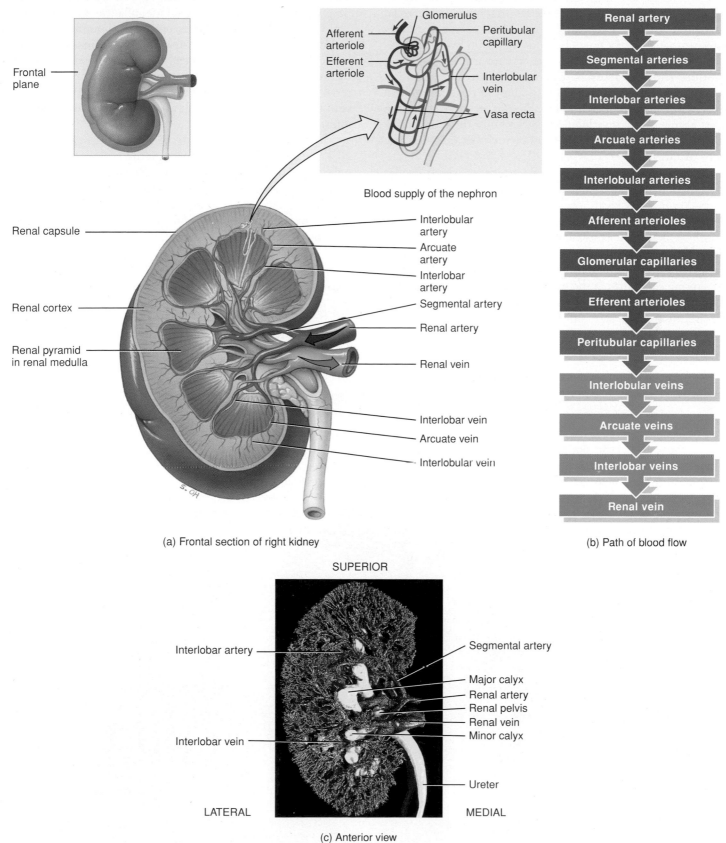

Glomerulus

Afferent arteriole

Efferent arteriole

Peritubular capillary

Interlobular vein

Vasa recta

Blood supply of the nephron

Frontal plane

Renal capsule

Renal cortex

Renal pyramid in renal medulla

Interlobular artery

Arcuate artery

Interlobar artery

Segmental artery

Renal artery

Renal vein

Interlobar vein

Arcuate vein

Interlobular vein

(a) Frontal section of right kidney

Renal artery → Segmental arteries → Interlobar arteries → Arcuate arteries → Interlobular arteries → Afferent arterioles → Glomerular capillaries → Efferent arterioles → Peritubular capillaries → Interlobular veins → Arcuate veins → Interlobar veins → Renal vein

(b) Path of blood flow

SUPERIOR

Interlobar artery

Interlobar vein

Segmental artery

Major calyx

Renal artery

Renal pelvis

Renal vein

Minor calyx

Ureter

LATERAL

MEDIAL

(c) Anterior view

? **What volume of blood enters the renal arteries per minute?**

Within the kidney, the renal artery divides into several **segmental arteries**, which supply different segments (areas) of the kidney. Each segmental artery gives off several branches that enter the parenchyma and pass through the renal columns between the lobes of the kidneys as the **interlobar arteries**. At the bases of the renal pyramids, the interlobar arteries arch between the renal medulla and cortex; here they are known as the **arcuate arteries** (AR-kū-āt=shaped like a bow), because they arch over the bases of the renal pyramids. Divisions of the arcuate arteries produce a series of **interlobular arteries**. These arteries are so named because they pass between lobules of the kidney (a **renal lobule** is a group of nephrons that open into branches of the same collecting duct; see Figure 26.5). Interlobular arteries enter the renal cortex and give off branches called **afferent arterioles** (*af-*=toward; *-ferrent*=to carry).

Each nephron receives one afferent arteriole, which divides into a tangled, ball-shaped capillary network called the **glomerulus** (glō-MER-ū-lus=little ball; plural is *glomeruli*). The glomerular capillaries then reunite to form an **efferent arteriole** (*ef-*=out), which has a smaller diameter than the afferent arteriole. The efferent arteriole carries blood out of the glomerulus toward a second capillary plexus. Glomerular capillaries are unique among capillaries in the body because they are positioned between two arterioles, rather than between an arteriole and a venule. Because of the size difference between the afferent and efferent capillaries, pressure is created in the thin-walled glomerulus. This is important in urine formation.

The efferent arterioles divide to form the **peritubular capillaries** (*peri-*=around), which surround tubular parts of the nephron in the renal cortex. Extending from some efferent arterioles are long loop-shaped capillaries called **vasa recta** (VĀ-sa REK-ta; *vasa*=vessels; *recta*=straight) that supply tubular portions of the nephron in the renal medulla (see Figure 26.5b).

The peritubular capillaries eventually reunite to form **peritubular venules** and then **interlobular veins**, which also receive blood from the vasa recta. Then the blood drains through the **arcuate veins** to the **interlobar veins** running between the renal pyramids. Blood leaves the kidney through a single **renal vein** that exits at the renal hilum and carries venous blood to the inferior vena cava.

Most renal nerves originate in the *celiac ganglion* (see Figure 20.4 on page 665) and pass through the *renal plexus* into the kidneys along with the renal arteries. Renal nerves are part of the sympathetic division of the autonomic nervous system. Most are vasomotor nerves that regulate the flow of blood through the kidney by causing vasodilation or vasoconstriction of renal arterioles.

The Nephron

Parts of a Nephron

Nephrons are the functional units of the kidneys. Each nephron (Figure 26.5) consists of two parts: a **renal corpuscle** (KOR-pus-

---

● CLINICAL CONNECTION **Kidney Transplant**

A **kidney transplant** is the transfer of a kidney from a living donor or a cadaver to a recipient whose kidney(s) no longer function. In the procedure, the donor kidney is placed in the pelvis of the recipient through an abdominal incision. The renal artery and vein of the transplanted kidney are attached to a nearby artery and vein in the pelvis and the ureter is then attached to the urinary bladder. During a kidney trans-

plant, the patient receives only one donor kidney, since only one kidney is needed to maintain sufficient renal function. The diseased kidneys are usually left in place. As with all organ transplants, kidney transplant patients must be ever vigilant for signs of infection or organ rejection. The transplant patient will take immunosuppressive drugs for the rest of his or her life to avoid rejection of the "foreign" organ. ●

Kidney transplant

Transplanted kidney

Donor: Functioning kidneys   Recipient: Functioning kidneys sul=tiny body), where blood plasma is filtered, and a **renal tubule** into which the filtered fluid passes. The two components of a renal corpuscle are the **glomerulus** (capillary network) and the **glomerular (Bowman's) capsule**, a dead-end tube that surrounds the glomerular capillaries in the form of a double-walled cup with a cavity between the two layers that receives filtrate from the capillaries. In the order that fluid (filtrate) passes through them, the renal tubule consists of a (1) **proximal convoluted tubule**, (2) **loop of Henle (nephron loop)**, and (3) **distal convoluted tubule**. *Proximal* denotes the part of the tubule attached to the glomerular capsule, and *distal* denotes the part that is farther away. *Convoluted* means the tubule is tightly coiled rather than straight. The renal corpuscle and both convoluted tubules lie within the renal cortex; some loops of Henle remain in the renal cortex and some extend into the renal medulla, make a hairpin turn, and then return to the renal cortex.

The distal convoluted tubules of several nephrons empty into a single **collecting duct**. Collecting ducts then unite and converge

Figure 26.5 **The structure of nephrons and associated blood vessels. Note that the collecting duct and papillary duct are not part of the nephron.** (a) A cortical nephron. (b) A juxtamedullary nephron.

Nephrons are the functional units of the kidneys.

FLOW OF FLUID THROUGH A CORTICAL NEPHRON

Glomerular (Bowman's) capsule
↓
Proximal convoluted tubule
↓
Descending limb of the loop of Henle
↓
Ascending limb of the loop of Henle
↓
Distal convoluted tubule (drains into collecting duct)

(a) Cortical nephron and vascular supply

FIGURE 26.5 CONTINUES ▶

until eventually there are only several hundred large **papillary ducts**, which drain into the minor calyces. The collecting ducts and papillary ducts extend from the renal cortex through the renal medulla to the renal pelvis. Although one kidney has about 1 million nephrons, it has a much smaller number of collecting ducts and even fewer papillary ducts.

In a nephron, the loop of Henle connects the proximal and distal convoluted tubules. The first part of the loop of Henle dips

into the renal medulla, where it is called the **descending limb of the loop of Henle** (Figure 26.5). It then makes that hairpin turn and returns to the renal cortex as the **ascending limb of the loop of Henle**. About 80–85% of the nephrons are **cortical nephrons**. Their renal corpuscles lie in the outer portion of the renal cortex, and they have *short* loops of Henle that lie mainly in the cortex and penetrate only into the outer region of the renal medulla (Figure 26.5a). The short loops of Henle receive their

○ **FIGURE 26.5 CONTINUED** ▶

FLOW OF FLUID THROUGH A
JUXTAMEDULLARY NEPHRON

Glomerular (Bowman's) capsule
↓
Proximal convoluted tubule
↓
Descending limb of the
loop of Henle
↓
Thin ascending limb of the
loop of Henle
↓
Thick ascending limb of the
loop of Henle
↓
Distal convoluted tubule
(drains into collecting duct)

Kidney labels: Renal cortex, Renal medulla, Renal papilla, Minor calyx

Main figure labels: Renal capsule, Distal convoluted tubule, Renal corpuscle: Glomerular (Bowman's) capsule, Glomerulus, Afferent arteriole, Efferent arteriole, Interlobular artery, Interlobular vein, Arcuate vein, Arcuate artery, Corticomedullary junction, Collecting duct, Loop of Henle: Descending limb, Thick ascending limb, Thin ascending limb, Papillary duct, Renal papilla, Minor calyx, Urine, Proximal convoluted tubule, Peritubular capillary, Renal cortex, Renal medulla

(b) Juxtamedullary nephron and vascular supply

❓ **What are the basic differences between cortical and juxtamedullary nephrons?**

blood supply from peritubular capillaries that arise from efferent arterioles. The other 15–20% of the nephrons are **juxtamedullary nephrons** (*juxta-*=near to). Their renal corpuscles lie deep in the cortex, close to the medulla, and they have *long* loops of Henle that extend into the deepest region of the medulla (Figure 26.5b). Long loops of Henle receive their blood supply from peritubular capillaries and from the vasa recta that arise from efferent arterioles. In addition, in juxtamedullary nephrons the ascending limb of the loop of Henle consists of two portions: a **thin ascending limb** followed by a **thick ascending limb** (Figure 26.5b). The lumen of the thin ascending limb is the same as in other areas of the renal tubule; it is only the epithelium that is thinner. Nephrons with long loops of Henle enable the kidneys to excrete very dilute or very concentrated urine.

Histology of the Nephron and Collecting Duct

A single layer of epithelial cells forms the entire wall of the glomerular capsule, renal tubule, and ducts. Each part, however, has distinctive histological features that reflect its particular functions. In the order that fluid flows through them, the parts are the glomerular capsule, the renal tubule, and the collecting duct.

GLOMERULAR CAPSULE The glomerular (Bowman's) capsule consists of visceral and parietal layers (Figure 26.6a). The visceral layer consists of modified simple squamous epithelial cells called **podocytes** (PŌ-dō-sīts; *podo-*=foot; *-cytes*=cells). The many footlike projections of these cells (pedicels) wrap around the single layer of endothelial cells of the glomerular capillaries and

Figure 26.6 Histology of a renal corpuscle.

A renal corpuscle consists of a glomerular (Bowman's) capsule and a glomerulus.

(a) Renal corpuscle (internal view)

(b) Renal corpuscle

LM 1380x

Is the photomicrograph in (b) from a section through the renal cortex or renal medulla? How can you tell?

form the inner wall of the capsule. The parietal layer of the glomerular capsule consists of simple squamous epithelium and forms the outer wall of the capsule. Fluid filtered from the glomerular capillaries enters the **capsular (Bowman's) space**, the space between the two layers of the glomerular capsule, which is the lumen of the urinary tube. Think of the relationship between the glomerulus and glomerular capsule in the following way. The glomerulus is a fist punched into a limp balloon (the glomerular capsule) until the fist is covered by two layers of balloon (visceral and parietal layers) with a space in between, the **capsular space**.

Renal Tubule and Collecting Duct Table 26.1 illustrates the histology of the cells that form the renal tubule and collecting duct. In the proximal convoluted tubule, the cells are simple cuboidal epithelial cells with a prominent brush border of microvilli on their apical surface (surface facing the lumen). These microvilli, like those of the small intestine, increase the surface area for reabsorption and secretion. The descending limb of the loop of Henle and the first part of the ascending limb of the loop of Henle (the thin ascending limb) are composed of simple squamous epithelium. (Recall that cortical or short-loop nephrons lack the thin ascending limb.) The thick ascending limb of the loop of Henle is composed of simple cuboidal to low columnar epithelium.

In each nephron, the final part of the ascending limb of the loop of Henle makes contact with the afferent arteriole serving that renal corpuscle (Figure 26.6a). Because the columnar tubule cells in this region are crowded together, they are known as the **macula densa** (*macula*=spot; *densa*=dense). Alongside the macula densa, the wall of the afferent arteriole (and sometimes the efferent arteriole) contains modified smooth muscle fibers called **juxtaglomerular (JG) cells**. Together with the macula densa, they constitute the **juxtaglomerular apparatus (JGA)**. The JGA helps regulate blood pressure within the kidneys. The distal convoluted tubule (DCT) begins a short distance past the macula densa. In the last part of the DCT and continuing into the collecting ducts, two different types of cells are present. Most are **principal cells**, which have receptors for both antidiuretic hormone (ADH) and aldosterone, two hormones that regulate their functions. A smaller number are **intercalated cells**, which play a role in the homeostasis of blood pH. The collecting ducts drain into large papillary ducts, which are lined by simple columnar epithelium.

The number of nephrons is constant from birth. Any increase in kidney size is due solely to the growth of individual nephrons. If nephrons are injured or become diseased, new ones do not form. Signs of kidney dysfunction usually do not become apparent until function declines to less than 25% of normal because the remaining functional nephrons adapt to handle a larger-than-

---

TABLE 26.1

Histological Features of the Renal Tubule and Collecting Duct

REGION AND HISTOLOGY		DESCRIPTION
Proximal convoluted tubule (PCT)		Simple cuboidal epithelial cells with prominent brush borders of microvilli.
Loop of Henle: descending limb and thin ascending limb		Simple squamous epithelial cells.
Loop of Henle: thick ascending limb		Simple cuboidal to low columnar epithelial cells.
Most of distal convoluted tubule (DCT)		Simple cuboidal epithelial cells.
Last part of DCT and all of collecting duct (CD)		Simple cuboidal epithelium consisting of principal cells and intercalated cells.

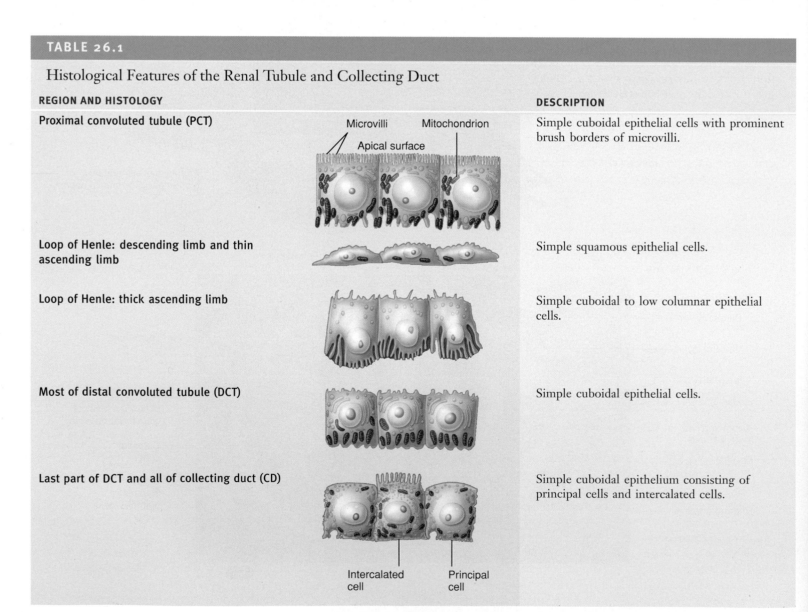

normal load. Surgical removal of one kidney, for example, stimulates hypertrophy (enlargement) of the remaining kidney, which eventually is able to filter blood at 80% of the rate of two normal kidneys.

◉ CHECKPOINT

1. Describe the location of the kidneys. Why are they said to be retroperitoneal?
2. Which branch of the autonomic nervous system innervates renal blood vessels?
3. How do cortical nephrons and juxtamedullary nephrons differ structurally?
4. Describe the histology of the various portions of a nephron and collecting duct.
5. Describe the structure of the juxtaglomerular apparatus (JGA).

FUNCTIONS OF NEPHRONS

◉ OBJECTIVES

- Identify the three basic tasks performed by nephrons and collecting ducts, and indicate where each task occurs.
- Describe the filtration membrane.

To produce urine, nephrons and collecting ducts perform three basic processes—glomerular filtration, tubular secretion, and tubular reabsorption (Figure 26.7):

❶ **Glomerular filtration.** In the first step of urine production, water and most solutes in blood plasma move across the wall of glomerular capillaries into the glomerular capsule and then into the renal tubule.

❷ **Tubular reabsorption.** As filtered fluid flows along the renal tubule and through the collecting duct, tubule cells reabsorb about 99% of the filtered water and many useful solutes. The water and solutes return to the blood as it flows through the peritubular capillaries and vasa recta. Note that the term *reabsorption* refers to the return of filtered water and solutes to

the bloodstream. The term *absorption*, by contrast, means entry of new substances into the body, as occurs in the gastrointestinal tract.

❸ **Tubular secretion.** As fluid flows along the tubule and through the collecting duct, the tubule and duct cells secrete other materials, such as wastes, drugs, and excess ions, into the fluid. Notice that tubular secretion *removes* a substance from the blood. In other instances of secretion—such as secretion of hormones—cells release substances into interstitial fluid and blood.

Solutes in the fluid that drains into the renal pelvis remain in the urine and are excreted. The rate of urinary excretion of any solute is equal to its rate of glomerular filtration, plus its rate of secretion, minus its rate of reabsorption.

By filtering, reabsorbing, and secreting, nephrons help maintain homeostasis of the blood's volume and composition. The situation is somewhat analogous to a recycling center: Garbage trucks dump refuse into an input hopper, where the smaller refuse passes onto a conveyor belt (glomerular filtration of plasma). As the conveyor belt carries the garbage along, workers remove useful items, such as aluminum cans, plastics, and glass containers (reabsorption). Other workers place additional garbage left at the center and larger items onto the conveyor belt (secretion). At the end of the belt, all remaining garbage falls into a truck for transport to the landfill (excretion of wastes in urine).

Glomerular Filtration

The fluid that enters the capsular space is called the **glomerular filtrate**. On average, the daily volume of glomerular filtrate in adults is 150 liters in females and 180 liters in males, a volume that represents about 65 times the entire blood plasma volume. More than 99% of the glomerular filtrate returns to the bloodstream via tubular reabsorption, however, so only 1–2 liters (about 1–2 qt) are excreted as urine.

Together, endothelial cells of glomerular capillaries and podocytes, which completely encircle the capillaries, form a leaky

Figure 26.7 **Relation of a nephron's structure to its three basic functions: glomerular filtration, tubular reabsorption, and tubular secretion.** Secreted substances remain in the urine and are subsequently excreted by the body.

🔑 Glomerular filtration occurs in the renal corpuscle; tubular reabsorption and tubular secretion occur all along the renal tubule and collecting duct.

❓ **When cells of the renal tubules secrete the drug penicillin, is the drug being added to or removed from the bloodstream?**

barrier referred to as the **filtration membrane** or **endothelial–capsular membrane**. This sandwichlike assembly permits filtration of water and small solutes but prevents filtration of most plasma proteins, blood cells, and platelets. Filtered substances move from the bloodstream through three barriers—a glomerular endothelial cell, the basal lamina, and a filtration slit formed by a podocyte (Figure 26.8):

❶ Glomerular endothelial cells are quite leaky because they have large **fenestrations** (pores) that are 0.07–0.1 μm in diameter. This size permits all solutes in blood plasma to exit glomerular capillaries but prevents filtration of blood cells and platelets.

Located among the glomerular capillaries and in the cleft between afferent and efferent arterioles are **mesangial cells** (*mes-*=in the middle; *-angi*=blood vessel), contractile cells that help regulate glomerular filtration (see Figure 26.6a).

❷ The **basal lamina**, a layer of material between the endothelium and the podocytes, consists of minute fibers in a glycoprotein matrix; it prevents filtration of larger plasma proteins.

❸ Extending from each podocyte are thousands of footlike processes termed **pedicels** (PED-i-sels=little feet) that wrap around glomerular capillaries. The spaces between pedicels are the **filtration slits**. A thin membrane, the **slit membrane**,

Figure 26.8 **The filtration (endothelial-capsular) membrane.** The size of the endothelial fenestrations and filtration slits in (a) have been exaggerated for emphasis. (Dr. Richard K. Kessel and Dr. Randy H. Kardon, *Tissues and Organs: A Text-Atlas of Scanning Electron Microscopy.* © 1979 by W. H. Freeman and Company. Reprinted by permission.)

🔑 During glomerular filtration, water and solutes pass from blood plasma into the capsular space.

Filtration slit
Pedicel

Podocyte of visceral layer of glomerular (Bowman's) capsule

❶ Fenestration (pore) of glomerular endothelial cell: prevents filtration of blood cells but allows all components of blood plasma to pass through

❷ Basal lamina of glomerulus: prevents filtration of larger proteins

❸ Slit membrane between pedicels: prevents filtration of medium-sized proteins

(a) Details of filtration membrane

Pedicel of podocyte   Filtration slit

Basal lamina

Lumen of glomerulus

TEM 78,000x

Fenestration (pore) of glomerular endothelial cell

(b) Filtration membrane

❓ **Which part of the filtration membrane prevents red blood cells from entering the capsular space?**

extends across each filtration slit; it permits the passage of molecules having a diameter smaller than 0.006–0.007 μm, including water, glucose, vitamins, amino acids, very small plasma proteins, ammonia, urea, and ions. Because the most plentiful plasma protein—albumin—has a diameter of 0.007 μm, less than 1% of it passes the slit membrane.

The principle of *filtration*—the use of pressure to force fluids and solutes through a membrane—is the same in glomerular capillaries as in capillaries elsewhere in the body. However, the volume of fluid filtered by the renal corpuscle is much larger than in other capillaries of the body for three reasons:

1. Glomerular capillaries present a large surface area for filtration because they are long and extensive. The mesangial cells regulate how much of this surface area is available for filtra-

tion. When mesangial cells are relaxed, surface area is maximal, and glomerular filtration is very high. Contraction of mesangial cells reduces the available surface area, and glomerular filtration decreases.

2. The filtration membrane is thin and porous. Despite having several layers, the thickness of the filtration membrane is only 0.1 μm. Glomerular capillaries also are about 50 times leakier than capillaries in most other tissues, mainly because of their large fenestrations.

3. Glomerular capillary blood pressure is high. Because the efferent arteriole is smaller in diameter than the afferent arteriole, resistance to the outflow of blood from the glomerulus is high. As a result, blood pressure in glomerular capillaries is considerably higher than in capillaries elsewhere in the body, and a higher pressure produces more filtrate.

• CLINICAL **CONNECTION** | Dialysis

If a person's kidneys are so impaired by disease or injury that they are unable to function adequately, then blood must be cleansed artificially by **dialysis**, which utilizes the same methods as kidney filtration: the separation of large solutes from smaller ones through use of a selectively permeable membrane. The leading cause of renal failure is diabetes. One method of dialysis is the artificial kidney machine, which performs **hemodialysis** (*hemo-*=blood) because it directly filters the patient's blood. As blood flows through tubing made of selectively permeable dialysis membrane, waste products diffuse from the blood into a dialysis solution surrounding the membrane. The dialysis solution is continuously replaced to maintain favorable concentration gradients for diffusion of solutes into and out of the blood. After passing through the

dialysis tubing, the cleansed blood flows back into the body. As a general rule, most affected people require 6–12 hours on dialysis each week (roughly every other day).

Continuous ambulatory peritoneal dialysis (CAPD) uses the peritoneal lining of the abdominal cavity as the dialysis membrane to filter the blood. The tip of a catheter is surgically placed in the patient's peritoneal cavity and connected to a sterile dialysis solution. The dialysis solution flows into the peritoneal cavity from a plastic container by gravity. The solution remains in the cavity until metabolic waste products, excess electrolytes, and extracellular fluid diffuse into the dialysis solution. The solution is then drained from the cavity by gravity into a sterile bag that is discarded. The procedure is repeated several times each day.   •

Patient undergoing hemodialysis

Tubular Reabsorption

The normal rate of glomerular filtration is so high that the volume of fluid entering the proximal convoluted tubules in half an hour is greater than the total volume of blood plasma. Reabsorption is the second function of the nephron and collecting duct. Epithelial cells all along the renal tubule and duct carry out reabsorption, but proximal convoluted tubule cells make the largest contribution. Solutes that are reabsorbed by both active and passive processes include glucose, amino acids, urea, and ions such as Na^+ (sodium), K^+ (potassium), Ca^{2+} (calcium), Cl^- (chloride), HCO_3^{2-} (bicarbonate), and HPO_4^{2-} (phosphate). Cells located more distally fine-tune the reabsorption processes to maintain the appropriate concentrations of water and selected ions. Most small proteins and peptides that pass through the filter also are reabsorbed, usually via bulk-phase endocytosis.

• CLINICAL **CONNECTION** Urinalysis

An analysis of the volume and physical, chemical, and microscopic properties of urine, called a **urinalysis** (ū-ri-NAL-i-sis), reveals much information about the health of the body. Of the 1–2 liters (about 1–2 quarts) of urine eliminated per day by a normal adult, about 95% is water. The remaining 5% consists of solutes, among them urea, sodium, potassium, phosphate, and sulfate ions; creatinine; and uric acid. In addition, much smaller amounts of calcium, magnesium, and bicarbonate ions are also found in urine. If disease alters body metabolism or kidney function, traces of substances not normally present may appear in the urine, or normal constituents may appear in abnormal amounts. •

Test stick previously placed in urine is compared to a chart to detect abnormal constituents in urine

Tubular Secretion

The third function of nephrons and collecting ducts is tubular secretion, the transfer of materials from the blood and tubule cells into tubular fluid. Secreted substances include H^+, K^+, ammonium ions (NH_4^+), creatinine, and certain drugs such as penicillin. Tubular secretion has two important outcomes: The secretion of H^+ helps control blood pH, and the secretion of other substances helps eliminate them from the body.

◉ CHECKPOINT

6. What are the functions of glomerular filtration, tubular reabsorption, and tubular secretion?
7. Describe the factors that allow a considerably greater filtration through glomerular capillaries than through capillaries elsewhere in the body.

URINE TRANSPORTATION, STORAGE, AND ELIMINATION

◉ OBJECTIVE

• Describe the anatomy, histology, and functions of the ureters, urinary bladder, and urethra.

Urine drains through papillary ducts into the minor calyces, which join to become major calyces that unite to form the renal pelvis (see Figure 26.3). From the renal pelvis, urine first drains into the ureters and then into the urinary bladder; urine is then discharged from the body through the single urethra (see Figure 26.1).

Ureters

Each of the two **ureters** (Ū-re-ters) transports urine from the renal pelvis of one kidney to the urinary bladder. Peristaltic contractions of the muscular walls of the ureters push urine toward the urinary bladder, but hydrostatic pressure and gravity also contribute. Peristaltic waves that pass from the renal pelvis to the urinary bladder vary in frequency from one to five per minute, depending on how fast urine is being formed.

The ureters are 25–30 cm (10–12 in.) long and are thick-walled, narrow tubes that vary in diameter from 1 mm to 10 mm along their course between the renal pelvis and the urinary bladder. Like the kidneys, the ureters are retroperitoneal. At the base of the urinary bladder the ureters curve medially and pass obliquely through the wall of the posterior aspect of the urinary bladder (Figure 26.9).

Even though there is no anatomical valve at the opening of each ureter into the urinary bladder, there is a physiological one that is quite effective. As the urinary bladder fills with urine, pressure within it compresses the oblique openings into the ureters and prevents the backflow of urine. When this physiological valve is not operating properly, it is possible for microbes to travel up the ureters from the urinary bladder to infect one or both kidneys.

Similar to other tubular structures in the body, the ureters are formed of three coats or layers (Figure 26.10 on page 838). The deepest coat, or **mucosa**, is a mucous membrane with **transitional epithelium** (see Table 3.1I on page 68) and an underlying **lamina propria** of areolar connective tissue with considerable collagen, elastic fibers, and lymphatic tissue. Transitional epithelium is able to stretch—a marked advantage for any organ that must accommodate a variable volume of fluid. Mucus secreted by the mucosa prevents the cells of the walls of the ureter from coming in contact with urine, which is important because the solute concentration and pH of urine may differ drastically from the cytosol of the cells of the walls of the ureter.

Throughout most of the length of the ureters, the intermediate coat, the **muscularis**, is composed of inner longitudinal and outer circular layers of smooth muscle fibers, an arrangement opposite that of the gastrointestinal tract, which contains inner circular and outer longitudinal layers; the muscularis of the distal third of the ureters also contains a third, outer layer of longitudinal muscle fibers. Peristalsis is the major function of the muscularis.

The superficial coat of the ureters is the **adventitia**, a layer of areolar connective tissue containing blood vessels, lymphatic

vessels, and nerves that serve the muscularis and mucosa. The adventitia blends in with surrounding connective tissue and anchors the ureters in place.

The arterial supply of the ureters is from the renal, testicular or ovarian, common iliac, and inferior vesical arteries (arising from the internal iliac artery, a trunk with the internal pudendal and superior gluteal arteries, or a branch of the internal pudendal artery). The veins terminate in the corresponding veins.

The ureters are innervated by the renal plexuses, which are supplied by sympathetic and parasympathetic fibers from the lesser and lowest splanchnic nerves.

Figure 26.9 Ureters, urinary bladder, and urethra in a female.

🔓 ⚙ Urine is stored in the urinary bladder before being expelled by micturition.

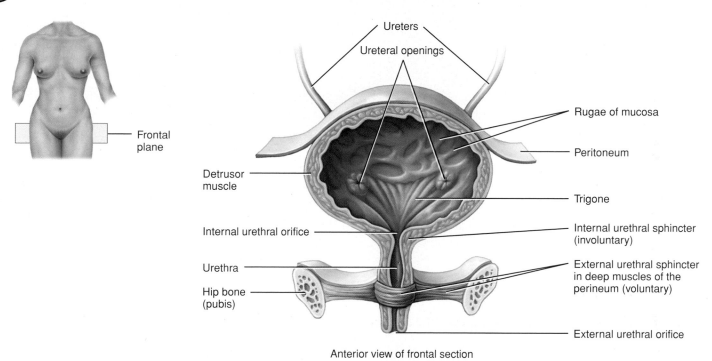

Anterior view of frontal section

(b) Midsagittal section

❓ **Why is transitional epithelium important in the ureters and urinary bladder?**

Figure 26.10 Histology of the ureter.

Three coats of tissue form the wall of the ureters: mucosa, muscularis, and adventitia.

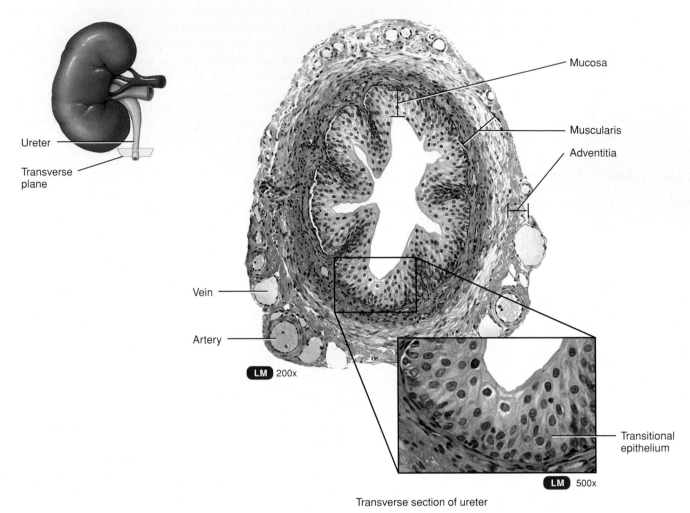

Transverse section of ureter

How does the muscularis of most of the ureters differ from that of the gastrointestinal tract?

Urinary Bladder

The **urinary bladder** is a hollow, distensible muscular organ situated in the pelvic cavity posterior to the pubic symphysis. In males, it is directly anterior to the rectum; in females it is anterior to the vagina and inferior to the uterus (see Figure 26.12 on page 840). It is held in position by folds of the peritoneum. The shape of the urinary bladder depends on how much urine it contains. Empty, it is collapsed; when slightly distended it becomes spherical; as urine volume increases it becomes pear-shaped and rises into the abdominal cavity. Urinary bladder capacity averages 700–800 mL; it is smaller in females because the uterus occupies the space just superior to the urinary bladder.

In the floor of the urinary bladder is a small triangular area called the **trigone** (TRĪ-gōn=triangle). The two posterior corners of the trigone contain the two ureteral openings; the opening into the urethra, the **internal urethral orifice**, lies in the anterior corner (see Figure 26.9). Because its mucosa is firmly bound to the muscularis, the trigone has a smooth appearance.

Three coats make up the wall of the urinary bladder (Figure 26.11). The deepest is the **mucosa**, a mucous membrane composed of **transitional epithelium** and an underlying **lamina propria** similar to that of the ureters. Rugae (the folds in the mucosa)

are also present. Surrounding the mucosa is the intermediate **muscularis**, also called the **detrusor muscle** (de-TROO-ser=to push down), which consists of three layers of smooth muscle fibers: the inner longitudinal, middle circular, and outer longitudinal layers. Around the opening to the urethra the circular fibers form an **internal urethral sphincter** (see Figure 26.9a); inferior to it is the **external urethral sphincter**, which is composed of skeletal muscle. The most superficial coat of the urinary bladder on the posterior and inferior surfaces is the **adventitia**, a layer of areolar connective tissue that is continuous with that of the ureters. Over the superior surface of the urinary bladder is the **serosa**, a layer of peritoneum.

Discharge of urine from the urinary bladder, called **micturition** (mik′-too-RISH-un; *mictur-*=urinate), is also known as *urination* or *voiding*. Micturition occurs via a combination of involuntary and voluntary muscle contractions. When the volume of urine in the urinary bladder exceeds 200–400 mL, pressure within the urinary bladder increases considerably, and stretch receptors in its wall transmit nerve impulses into the spinal cord. These impulses propagate to the **micturition center** in sacral spinal cord segments S2 and S3 and trigger a spinal reflex called the **micturition reflex**. In this reflex arc, parasympathetic impulses

Figure 26.11 Histology of the urinary bladder.

 Discharge of urine from the urinary bladder is a combination of voluntary and involuntary muscular contractions called micturition.

Lumen

Mucosa:
Transitional epithelium

Lamina propria

Muscularis:
Inner longitudinal

Middle circular

Outer longitudinal

Serosa

LM 40x

Transverse section of urinary bladder

? What is the trigone?

from the micturition center propagate to the urinary bladder wall and internal urethral sphincter. The nerve impulses cause *contraction* of the detrusor muscle and *relaxation* of the internal urethral sphincter muscle. Simultaneously, the micturition center inhibits somatic motor neurons that innervate skeletal muscle in the external urethral sphincter. Upon contraction of the urinary bladder wall and relaxation of the sphincters, urination takes place. Urinary bladder filling causes a sensation of fullness that initiates a conscious desire to urinate before the micturition reflex actually occurs. Although emptying of the urinary bladder is a reflex, in early childhood we learn to initiate it and stop it voluntarily. Through learned control of the external urethral sphincter muscle and certain muscles of the pelvic floor, the cerebral cortex can initiate micturition or delay its occurrence for a limited period of time.

The arteries of the urinary bladder are the superior vesical (arises from the umbilical artery), the middle vesical (arises from the umbilical artery or a branch of the superior vesical), and the inferior vesical (arises from the internal iliac artery, a trunk with the internal pudendal and superior gluteal arteries, or a branch of the internal pudendal artery). The veins from the urinary bladder pass to the internal iliac vein.

The nerves supplying the urinary bladder arise partly from the hypogastric sympathetic plexus and partly from the second and third sacral nerves (pelvic splanchnic nerve).

Urethra

The **urethra** is a small tube leading from the internal urethral orifice in the floor of the urinary bladder to the exterior of the body (see Figure 26.9a). In both males and females, the urethra is the terminal portion of the urinary system and the passageway for discharging urine from the body; in males it discharges semen as well.

In females, the urethra lies directly posterior to the pubic symphysis, is directed obliquely inferiorly and anteriorly, and has a length of 4 cm (1.5 in.) (Figure 26.12a). The opening of the urethra to the exterior, the **external urethral orifice**, is located between the clitoris and the vaginal opening. The wall of the female urethra consists of a deep **mucosa** and a superficial **muscularis**. The mucosa is a mucous membrane composed of **epithelium** and **lamina propria** (areolar connective tissue with elastic fibers and a plexus of veins). The muscularis consists of circularly arranged smooth muscle fibers and is continuous with that of the urinary bladder. Near the urinary bladder, the mucosa contains transitional epithelium that is continuous with that of the urinary bladder; near the external urethral orifice the epithelium is nonkeratinized stratified squamous epithelium. Between these areas, the mucosa contains stratified columnar or pseudostratified columnar epithelium.

In males, the urethra also extends from the internal urethral orifice to the exterior, but its length and passage through the body are considerably different than in females (Figure 26.12b). The male urethra first passes through the prostate, then through the deep perineal muscles, and finally through the penis, a distance of about 20 cm (8 in.).

The male urethra, which also consists of a deep **mucosa** and a superficial **muscularis**, is subdivided into three anatomical regions: (1) The **prostatic urethra** passes through the prostate; (2) the **membranous (intermediate) urethra**, the shortest portion, passes through the deep perineal muscles; and (3) the **spongy urethra**, the longest portion, passes through the penis. The mucosa of the prostatic urethra is continuous with that of the urinary bladder and consists of transitional epithelium that becomes stratified columnar or pseudostratified columnar epithelium more distally. The mucosa of the membranous urethra contains stratified columnar or pseudostratified columnar epithelium. The epithelium of the spongy urethra is stratified columnar or pseudostratified columnar epithelium, except near the external urethral orifice, which is nonkeratinized stratified squamous epithelium. The **lamina propria** of the male urethra is areolar connective tissue with elastic fibers and a plexus of veins.

The muscularis of the prostatic urethra is composed of wisps of mostly circular smooth muscle fibers superficial to the lamina propria; these circular fibers help form the internal urethral sphincter of the urinary bladder. The muscularis of the membranous urethra consists of circularly arranged skeletal muscle fibers of the urogenital diaphragm that help form the external urethral sphincter of the urinary bladder.

Several glands and other structures associated with reproduction deliver their contents into the male urethra (see Figure 27.9).

The prostatic urethra contains the openings of (1) ducts that transport secretions from the **prostate** and (2) the **seminal vesicles** and **ductus (vas) deferens**, which deliver sperm into the urethra and provide secretions that both neutralize the acidity of

Figure 26.12 Comparison between female and male urethras.

🔑 The urethra carries urine from the urinary bladder to the exterior.

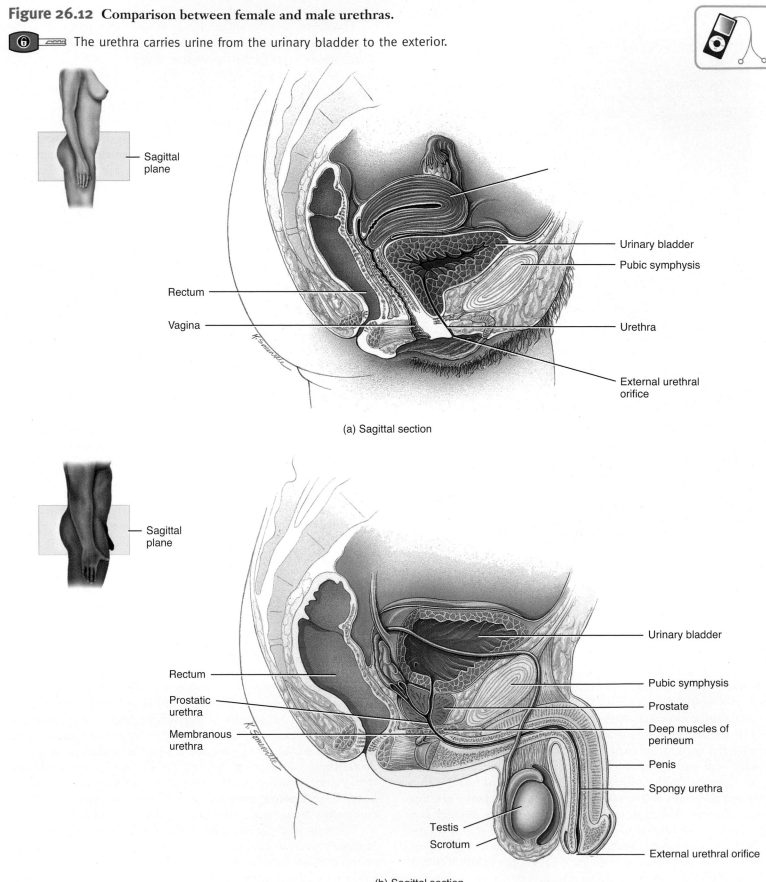

Sagittal plane

Rectum

Vagina

Urinary bladder

Pubic symphysis

Urethra

External urethral orifice

(a) Sagittal section

Sagittal plane

Rectum

Prostatic urethra

Membranous urethra

Urinary bladder

Pubic symphysis

Prostate

Deep muscles of perineum

Penis

Spongy urethra

Testis

Scrotum

External urethral orifice

(b) Sagittal section

❓ **Through which three main structures does the male urethra pass?**

the female reproductive tract and contribute to sperm motility and viability. The openings of the ducts of the **bulbourethral (Cowper's) glands** empty into the spongy urethra. They deliver an alkaline substance prior to ejaculation that neutralizes the acidity of the urethra. The glands also secrete mucus, which lubricates the end of the penis during sexual arousal. Throughout the urethra, but especially in the spongy urethra, the openings of the ducts of **urethral (Littré) glands** discharge mucus during sexual arousal and ejaculation.

⊙ CHECKPOINT

8. What forces help propel urine from the renal pelvis to the urinary bladder?
9. What is micturition? Describe the micturition reflex.
10. Compare the location, length, and histology of the urethra in males and females.

DEVELOPMENT OF THE URINARY SYSTEM

⊙ OBJECTIVE

• Describe the development of the urinary system.

Starting in the third week of fetal development, a portion of the mesoderm along the posterior aspect of the embryo, the **intermediate mesoderm**, differentiates into the kidneys. The intermediate mesoderm is located in paired elevations called **urogenital ridges**. Three pairs of kidneys form within the intermediate mesoderm in succession: the pronephros, the mesonephros, and the metanephros (Figure 26.13). Only the last pair remains as the functional kidneys of the newborn.

The first kidney to form, the **pronephros** (prō-NEF-rōs; *pro-* =before; *-nephros*=kidney), is the most superior of the three and

Figure 26.13 Development of the urinary system.

🔑 Three pairs of kidneys form within intermediate mesoderm in successive time periods: pronephros, mesonephros, and metanephros.

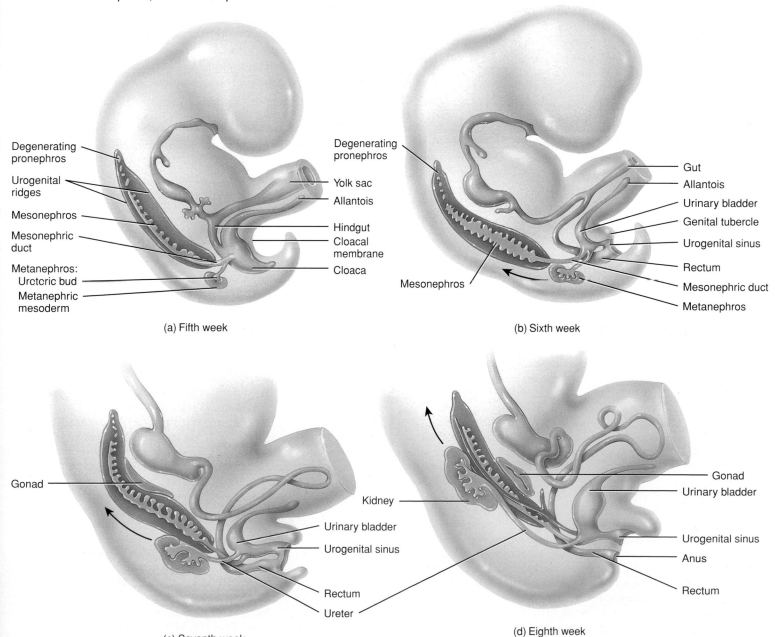

(a) Fifth week

(b) Sixth week

(c) Seventh week

(d) Eighth week

❓ **When do the kidneys begin to develop?**

has an associated **pronephric duct**. This duct empties into the **cloaca**, the expanded terminal part of the hindgut, which functions as a common outlet for the urinary, digestive, and reproductive ducts. The pronephros begins to degenerate during the fourth week and is completely gone by the sixth week.

The second kidney, the **mesonephros** (mez′-ō-NEF-rōs; *meso-*=middle), replaces the pronephros. The retained portion of the pronephric duct, which connects to the mesonephros, develops into the **mesonephric duct**. The mesonephros begins to degenerate by the sixth week and is almost gone by the eighth week.

At about the fifth week, a mesodermal outgrowth, called a **ureteric bud** (ū-rē-TER-ik), develops from the distal portion of the mesonephric duct near the cloaca. The **metanephros** (met-a-NEF-rōs; *meta-*=after), or ultimate kidney, develops from the ureteric bud and metanephric mesoderm. The ureteric bud forms the *collecting ducts, calyces, renal pelvis,* and *ureter.* The **metanephric mesoderm** forms the *nephrons* of the kidneys. By the third month, the fetal kidneys begin excreting urine into the surrounding amniotic fluid; indeed, fetal urine makes up most of the amniotic fluid.

During development, the cloaca divides into a **urogenital sinus**, into which urinary and genital ducts empty, and a *rectum* that discharges into the anal canal. The *urinary bladder* develops from the urogenital sinus. In females, the *urethra* develops as a result of lengthening of the short duct that extends from the urinary bladder to the urogenital sinus. In males, the urethra is considerably longer and more complicated, but it is also derived from the urogenital sinus.

Although the metanephric kidneys form in the pelvis, they ascend to their ultimate destination in the abdomen. As they do so, they receive renal blood vessels. Although the inferior blood vessels usually degenerate as superior ones appear, sometimes the inferior vessels do not degenerate. Consequently, some individuals (about 30%) develop multiple renal vessels.

In a condition called **unilateral renal agenesis** (ā-JEN-e-sis; *a-*=without; *genesis*=production; *unilateral*=one side) only one kidney develops (usually the right) due to the absence of a ureteric bud. The condition occurs once in every 1000 newborn infants and usually affects males more than females. Other kidney abnormalities that occur during development are **malrotated kidneys** (the hilum faces anteriorly, posteriorly, or laterally instead of medially); **ectopic kidney** (one or both kidneys may be in an abnormal position, usually inferior); and **horseshoe kidney** (the fusion of the two kidneys, usually inferiorly, into a single U-shaped kidney).

◉ **CHECKPOINT**

11. Which type of embryonic tissue develops into nephrons?
12. Which tissue gives rise to collecting ducts, calyces, renal pelves, and ureters?

AGING AND THE URINARY SYSTEM

◉ **OBJECTIVE**

• Describe the effects of aging on the urinary system.

With aging, the kidneys shrink in size, have a decreased blood flow, and filter less blood. The mass of the two kidneys decreases from an average of nearly 300 g in 20-year-olds to less than 200 g by age 80, a decrease of about one-third. Likewise, renal blood flow and filtration rates decline by 50% between ages 40 and 70. By age 80, about 40 percent of glomeruli are not functioning and thus filtration, reabsorption, and secretion decrease. Kidney diseases that become more common with age include acute and chronic kidney inflammations and renal calculi (kidney stones). Because the sensation of thirst diminishes with age, older individuals also are susceptible to dehydration. Urinary bladder changes that occur with aging include a reduction in size and capacity and weakening of the muscles. Urinary tract infections are more common among the elderly, as are polyuria (excessive urine production), nocturia (excessive urination at night), increased frequency of urination, dysuria (painful urination), urinary retention or incontinence, and hematuria (blood in the urine).

◉ **CHECKPOINT**

13. To what extent do kidney mass and filtration rate decrease with age?

KEY MEDICAL TERMS ASSOCIATED WITH THE URINARY SYSTEM

Azotemia (az-ō-TĒ-mē-a; *azot-*=nitrogen; *-emia*=condition of blood) Presence of urea or other nitrogen-containing substances in the blood.

Cystocele (SIS-tō-sēl; *cysto-*=bladder; *-cele*=hernia or rupture) Hernia of the urinary bladder.

Diabetic kidney disease A disorder caused by diabetes mellitus in which glomeruli are damaged. The result is the leakage of proteins into the urine and a reduction in the ability of the kidney to remove water and waste.

Dysuria (dis-Ū-rē-a; *dys*=painful; *uria*=urine) Painful urination.

Enuresis (en′-ū-RĒ-sis=to void urine) Involuntary voiding of urine after the age at which voluntary control has typically been attained.

Hydronephrosis (hī′-drō-ne-FRŌ-sis; *hydro-*=water; *nephros*=kidney; *-osis*=condition) Swelling of the kidney due to dilation of the renal pelvis and calyces as a result of an obstruction to the flow of urine. It may be due to a congenital abnormality, a narrowing of the ureter, a kidney stone, or an enlarged prostate.

Intravenous pyelogram (in′-tra-VĒ-nus PĪ-e-lō-gram′; *intra-*=within; *veno-*=vein; *pyelo-*=pelvis of kidney; *-gram*=record), or ***IVP*** Radiograph (x-ray) of the kidneys, ureters, and urinary bladder after venous injection of a radiopaque contrast medium.

Nephropathy (ne-FROP-a-thē; *neph-*=kidney; *-pathos*=suffering) Any disease of the kidneys. Types include *analgesic* (from long-term and excessive use of drugs such as ibuoprofen), *lead* (from ingestion of lead-based paint), and *solvent* (from carbon tetrachloride and other solvents).

Nocturnal enuresis (nok-TUR-nal en′-ū-RĒ-sis) Discharge of urine during sleep, resulting in bed-wetting; occurs in about 15% of 5-year-old children and generally resolves spontaneously, afflicting only about 1% of adults. It may have a genetic basis, as bed-wetting occurs more often in identical twins than in fraternal twins and more often in children whose parents or siblings were bed-wetters. Possible causes include smaller-than-normal urinary bladder capacity, failure to awaken in response to a full urinary bladder, and above-normal production of urine at night. Also referred to as **nocturia**.

Polyuria (pol′-ē-Ū-rē-a; *poly-*=too much) Excessive urine formation. It may occur in conditions such as diabetes mellitus and glomerulonephritis.

Stricture (STRIK-chur) Narrowing of the lumen of a canal or hollow organ, as may occur in the ureter, urethra, or any other tubular structure in the body.

Uremia (ū-RĒ-mē-a; *emia*=condition of blood) Toxic levels of urea in the blood resulting from severe malfunction of the kidneys.

Urinary retention A failure to completely or normally void urine; may be due to an obstruction in the urethra or neck of the urinary bladder, to nervous contraction of the urethra, or to lack of urge to urinate. In men, an enlarged prostate may constrict the urethra and cause urinary retention. If urinary retention is prolonged, a catheter (slender rubber drainage tube) must be placed into the urethra to drain the urine.

STUDY OUTLINE

Introduction (p. 821)

1. The organs of the urinary system are the kidneys, ureters, urinary bladder, and urethra.
2. After the kidneys filter blood and return most water and many solutes to the bloodstream, the remaining water and solutes constitute urine.
3. The kidneys regulate the ionic composition, osmolarity, volume, and pH of the blood, as well as blood pressure.
4. The kidneys also release calcitriol and erythropoietin, and excrete wastes and foreign substances.

Anatomy and Histology of the Kidneys (p. 824)

1. The kidneys are retroperitoneal organs attached to the posterior abdominal wall.
2. Three layers of tissue surround the kidneys: the renal capsule, adipose capsule, and renal fascia.
3. Internally, the kidneys consist of a renal cortex, a renal medulla, renal pyramids, renal papillae, renal columns, major and minor calyces, and a renal pelvis.
4. Blood flows into the kidney through the renal artery and successively into segmental, interlobar, arcuate, and interlobular arteries; afferent arterioles; glomerular capillaries; efferent arterioles; peritubular capillaries and vasa recta; and interlobular, arcuate, and interlobar veins before flowing out of the kidney through the renal vein.
5. Vasomotor nerves from the sympathetic division of the autonomic nervous system supply kidney blood vessels; they help regulate blood flow through the kidney.
6. The nephron is the functional unit of the kidneys. A nephron consists of a renal corpuscle (glomerulus and glomerular or Bowman's capsule) and a renal tubule.
7. A renal tubule consists of a proximal convoluted tubule, a loop of Henle, and a distal convoluted tubule, which drains into a collecting duct (shared by several nephrons). The loop of Henle consists of a descending limb and an ascending limb.
8. A cortical nephron has a short loop that dips only into the superficial region of the renal medulla; a juxtamedullary nephron has a long loop of Henle that stretches through the renal medulla almost to the renal papilla.
9. The wall of the entire glomerular capsule, renal tubule, and ducts consists of a single layer of epithelial cells. The epithelium has distinctive histological features in different parts of the tubule. Table 26.1 on page 832 summarizes the histological features of the renal tubule and collecting duct.
10. The juxtaglomerular apparatus (JGA) consists of the juxtaglomerular cells of an afferent arteriole and the macula densa of the final portion of the ascending limb of the loop of Henle.

Functions of Nephrons (p. 833)

1. Fluid that enters the capsular space is glomerular filtrate.

2. The filtration (endothelial-capsular) membrane consists of the glomerular endothelium, basal lamina, and filtration slits between pedicels of podocytes.
3. Most substances in plasma pass easily through the glomerular filter. However, blood cells and most proteins normally are not filtered.
4. Glomerular filtrate amounts to up to 180 liters of fluid per day. This large amount of fluid is filtered because the filter is porous and thin, the glomerular capillaries are long, and the capillary blood pressure is high.
5. Tubular reabsorption is a selective process that reclaims materials from tubular fluid and returns them to the bloodstream. Reabsorbed substances include water, glucose, amino acids, urea, and ions, such as sodium, chloride, potassium, bicarbonate, and phosphate.
6. Some substances not needed by the body are removed from the blood and discharged into the urine via tubular secretion. Included are ions (K^+, H^+, and NH_4^+), urea, creatinine, and certain drugs.

Urine Transportation, Storage, and Elimination (p. 836)

1. The ureters are retroperitoneal and consist of a mucosa, muscularis, and adventitia. They transport urine from the renal pelvis to the urinary bladder, primarily via peristalsis.
2. The urinary bladder is located in the pelvic cavity posterior to the pubic symphysis; its function is to store urine prior to micturition.
3. The urinary bladder consists of a mucosa with rugae, a muscularis (detrusor muscle), and an adventitia (serosa over the superior surface).
4. The micturition reflex discharges urine from the urinary bladder via parasympathetic impulses that cause contraction of the detrusor muscle and relaxation of the internal urethral sphincter muscle, and via inhibition of impulses in somatic motor neurons to the external urethral sphincter.
5. The urethra is a tube leading from the floor of the urinary bladder to the exterior. Its anatomy and histology differ in females and males. In both sexes the urethra functions to discharge urine from the body; in males it discharges semen as well.

Development of the Urinary System (p. 841)

1. The kidneys develop from intermediate mesoderm.
2. The kidneys develop in the following sequence: pronephros, mesonephros, and metanephros. Only the metanephros remains and develops into a functional kidney.

Aging and the Urinary System (p. 842)

1. With aging, the kidneys shrink in size, have a decreased blood flow, and filter less blood.
2. Common problems related to aging include urinary tract infections, increased frequency of urination, urinary retention or incontinence, and renal calculi.

SELF-QUIZ QUESTIONS

Choose the one best answer to the following questions.

1. Which of the following is *not* a function of the kidneys?
   a. participation in the formation of the active form of vitamin D
   b. regulation of the volume and composition of the blood
   c. removal of wastes from the blood in the form of urine
   d. production of white blood cells
   e. regulation of blood pressure by secretion of renin, which activates the renin–angiotensin pathway

2. Which of the following vessels split from the efferent arterioles?
   a. glomerular capillaries   b. peritubular capillaries
   c. afferent arterioles   d. vasa recta
   e. Both b and d are correct.

3. Urine leaving the distal convoluted tubule passes through various structures in which of the following sequences?
   a. collecting duct → renal hilus → calyx → ureter
   b. collecting duct → calyx → renal pelvis → ureter
   c. calyx → collecting duct → renal pelvis → ureter
   d. calyx → renal hilus → renal pelvis → ureter
   e. collecting duct → renal hilus → ureter → calyx

4. The trigone, a landmark in the urinary bladder, is a triangular area bounded by:
   a. the orifices of the ejaculatory ducts and the urethra.
   b. the internal urethral orifice and the inferior border of the detrusor muscle.
   c. the ureteral and the internal urethral orifices.
   d. the top of the fundus and the ureteral orifices.
   e. the major and minor calyces.

5. Fluid that filters into the glomerular capsule flows next into the:
   a. glomerulus.   b. efferent arteriole.
   c. proximal convoluted tubule.   d. distal convoluted tubule.
   e. loop of Henle.

6. The macula densa is located:
   a. between the glomerular capillaries and the glomerular capsule.
   b. between the renal papilla and the minor calyx.
   c. where the afferent arteriole branches from the interlobular artery.
   d. where the ascending limb of the loop of Henle makes contact with the afferent arteriole.
   e. at the bottom of the loop of Henle.

7. The detrusor muscle is:
   a. the middle layer of the wall of the urinary bladder.
   b. the muscle that controls the flow of urine from the renal pelvis into the ureter.
   c. another name for the muscularis of the ureter.
   d. a modification of the urogenital diaphragm muscle.
   e. Both a and d are correct.

Complete the following.

8. The apex of a renal pyramid, called the _____, points toward the interior of the kidney.

9. In the renal corpuscle, the _____ arteriole normally has a smaller diameter than the _____ arteriole.

10. The special simple squamous epithelial cells that have many projections and that are found in the visceral layer of the glomerular capsule are called _____.

11. The longest portion of the male urethra is the _____ urethra.

12. Most nephrons are cortical nephrons, but 15–20% are _____ nephrons.

13. The _____ carry urine from the kidney to the urinary bladder.

14. Place numbers in the blanks to arrange the following vessels in order. (a) arcuate arteries: _____; (b) interlobular arteries: _____; (c) renal arteries: _____; (d) peritubular capillaries and vasa recta: _____; (e) glomerular capillaries: _____; (f) efferent arteriole: _____; (g) afferent arteriole: _____; (h) interlobar arteries: _____; (i) segmental arteries: _____.

15. The process of emptying the urinary bladder is called _____.

Are the following statements true or false? If the statement is False, *find and fix the mistake to make the statement* True.

16. Signs of kidney dysfunction usually do not become apparent until more than 75% of normal kidney function is lost.

17. Most renal nerves arise from the celiac ganglion and are part of the sympathetic division of the autonomic nervous system.

18. The internal urethral sphincter is composed of voluntary skeletal muscle, and the external urethral sphincter is composed of involuntary smooth muscle.

19. The juxtaglomerular apparatus is part of the endothelial-capsular (filtration) membrane.

Matching

20. Match the anatomical feature(s) to its urinary organ. Answers may be used more than once.
   ____ (a) detrusor   (1) urinary bladder
   ____ (b) connects renal   (2) urethra
           pelvis to bladder   (3) kidney
   ____ (c) renal hilus   (4) ureter
   ____ (d) transitional
           epithelium
   ____ (e) prostatic, membranous,
           and spongy
   ____ (f) trigone
   ____ (g) major calyx
   ____ (h) peristalsis

CRITICAL THINKING QUESTIONS

1. Imagine that a new superbug has emerged from a nuclear waste dump. It produces a toxin that blocks renal tubule function but leaves the glomerulus unaffected. Predict the effects of this toxin.

2. Whenever 35-year-old Barbara jumps on the trampoline with her two young sons, she notices that she slightly "wets her pants," even though she just went to the bathroom before playing. Describe the muscles and nerves involved in the micturition reflex that may be causing Barbara's problem.

3. Although urinary catheters come in one length only, the number of centimeters that must be inserted to release the urine differs significantly in males and females. Why? Why is volume of urine released from a full bladder different in males and females?

4. As Juan focused his microscope on his own urine sample during his Human Anatomy lab, he was concerned to see many cells in the field of view; his urine had appeared to be fluid. There's no evidence of blood or infection. What are these cells and where do they come from?

5. Mr. Chase has suffered from *polycystic kidney disease* for his entire life, but it has only been in the past 10 years that the progression of the disease has destroyed a significant amount of kidney parenchyma (nephrons). In order to compensate for the destroyed nephron's functions, Mr. Chase undergoes *hemodialysis* four times per week. What are the functions of the nephron that are now being carried out by the dialysis machine?

? ANSWERS TO FIGURE QUESTIONS

26.1 The kidneys, ureters, urinary bladder, and urethra are the components of the urinary system.

26.2 The kidneys are retroperitoneal because they are posterior to the peritoneum of the abdominal cavity.

26.3 Blood vessels, lymphatic vessels, nerves, and a ureter pass through the renal hilum.

26.4 About 1200 mL of blood enters the renal arteries each minute.

26.5 Cortical nephrons have glomeruli in the superficial renal cortex, and their short loops of Henle penetrate only into the superficial renal medulla; juxtamedullary nephrons have glomeruli deep in the renal cortex, and their long loops of Henle extend through the renal medulla nearly to the renal papilla.

26.6 The section shown must be part of the renal cortex because there are no renal corpuscles in the renal medulla.

26.7 Penicillin secreted by the cells of the renal tubule is being removed from the bloodstream.

26.8 Endothelial fenestrations (pores) in glomerular capillaries prevent red blood cells from entering the capsular space because they are too small for red blood cells to pass through.

26.9 Transitional epithelium is important in the ureters and urinary bladder because it permits the organs to stretch to accommodate a variable amount of fluid.

26.10 The muscularis of most of the ureters consists of an inner longitudinal layer and an outer circular layer, an arrangement opposite that of the gastrointestinal tract.

26.11 The trigone is a triangular area in the urinary bladder formed by the ureteral openings (posterior corners) and the internal urethral orifice (anterior corner).

26.12 The male urethra passes through the prostate, urogenital diaphragm, and penis.

26.13 The kidneys start to develop during the third week of gestation.

27 | THE REPRODUCTIVE SYSTEMS

INTRODUCTION **Sexual reproduction** is the process by which organisms produce offspring through the union of germ cells called **gametes** (GAM-ēts=spouses). After the male gamete (sperm cell) unites with the female gamete (secondary oocyte)—an event called **fertilization**—the resulting cell contains one set of chromosomes from each parent. Males and females have anatomically distinct reproductive organs that are adapted for producing gametes, facilitating fertilization, and, in females, sustaining the growth of the embryo and fetus.

The male and female reproductive organs can be grouped by function. The **gonads**—testes in males and ovaries in females—produce gametes and secrete sex hormones. Various **ducts** then store and transport the gametes, and **accessory sex glands** produce substances that protect the gametes and facilitate their movement. Finally, **supporting structures**, such as the penis in males and the vagina in females, assist the delivery of gametes and, the uterus in females, assists in the growth of the embryo and fetus during pregnancy.

Gynecology (gī-ne-KOL-ō-jē; *gynec-*=woman; *-logy*=study of) is the specialized branch of medicine concerned with the diagnosis and treatment of diseases of the female reproductive system. As noted in Chapter 26, **urology** (ū-ROL-ō-jē) is the study of the urinary system. Urologists also diagnose and treat diseases and disorders of the male reproductive system. The branch of medicine that deals with male disorders, especially infertility and sexual dysfunction, is called **andrology** (an-DROL-ō-jē; *andro-*=masculine). •

MALE REPRODUCTIVE SYSTEM

● OBJECTIVES

- Describe the location, structure, and functions of the organs of the male reproductive system.
- Discuss the process of spermatogenesis in the testes.

The organs of the **male reproductive system** include the testes, a system of ducts (including the epididymis, ductus deferens, ejaculatory ducts, and urethra), accessory sex glands (seminal vesicles, prostate, and bulbourethral glands), and several supporting structures, including the scrotum and the penis (Figure 27.1). The testes (male gonads) produce sperm and secrete hormones. The

Figure 27.1 Male organs of reproduction and surrounding structures.

🔓 ⚏ Reproductive organs are adapted for producing new individuals and passing on genetic material from one generation to the next.

Functions of the Male Reproductive System

1. The testes produce sperm and the male sex hormone testosterone.
2. The ducts transport, store, and assist in maturation of sperm.
3. The accessory sex glands secrete most of the liquid portion of semen.
4. The penis contains the urethra, a passageway for ejaculation of semen and excretion of urine.

(a) Sagittal section

(b) Sagittal section

What are the groups of reproductive organs in males, and what are the functions of each group?

duct system transports and stores sperm, assists in their maturation, and conveys them to the exterior. Semen contains sperm plus the secretions provided by the accessory sex glands. The supporting structures have various functions. The penis delivers sperm into the female reproductive tract and the scrotum supports the testes.

Scrotum

The **scrotum** (SKRŌ-tum=bag), the supporting structure for the testes, consists of loose skin and an underlying subcutaneous layer that hangs from the root (attached portion) of the penis (Figure 27.1a). Externally, the scrotum looks like a single pouch of skin separated into lateral portions by a median ridge called the **raphe** (RĀ-fē =seam) (Figure 27.2). Internally, the **scrotal septum** divides the scrotum into two sacs, each containing a single testis. The septum is made up of a subcutaneous layer and muscle tissue called the **dartos muscle** (DAR-tōs=skinned), which is composed of bundles of smooth muscle fibers. The dartos muscle is also found in the subcutaneous layer of the scrotum. Associated with each testis in the scrotum is the **cremaster muscle** (krē-MAS-ter=suspender). The cremaster muscle consists of a series of small bands of skeletal muscle that descend, as an extension of the internal oblique muscle, through the spermatic cord to surround the testis.

The location of the scrotum and the contraction of its muscle fibers regulate the temperature of the testes. Normal sperm production requires a temperature about 2–3°C below core body temperature. This lowered temperature is maintained within the scrotum because it is outside the pelvic cavity. In response to cold temperatures, the cremaster and dartos muscles contract. Contraction of the cremaster muscles moves the testes closer to the body, where they can absorb body heat. Contraction of the dartos muscle causes the scrotum to become tight (wrinkled in appearance), which reduces heat loss. Exposure to warmth reverses these actions.

The blood supply of the scrotum arises from the internal pudendal branch of the internal iliac artery, the cremasteric branch of the inferior epigastric artery, and the external pudendal artery from the femoral artery. The scrotal veins follow the arteries.

The scrotal nerves arise from the pudendal nerve, posterior cutaneous nerve of the thigh, and ilioinguinal nerves.

Testes

The **testes** (TES-tēz), are paired oval glands in the scrotum measuring about 5 cm (2 in.) long and 2.5 cm (1 in.) in diameter (Figure 27.3 on page 849). Each **testis** (singular) weighs 10–15 grams. The testes develop near the kidneys, in the posterior portion of the abdomen, and they usually begin their descent into the scrotum through the inguinal canals (passageways in the anterior

Figure 27.2 The scrotum, the supporting structure for the testes.

🔑 The scrotum consists of loose skin and superficial fascia and supports the testes.

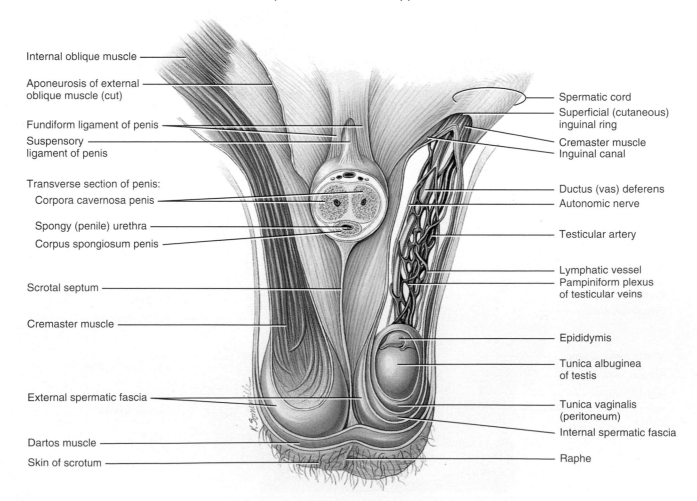

Anterior view of scrotum and testes and transverse section of penis

❓ **Which muscles help regulate the temperature of the testes?**

abdominal wall; see Figure 27.2) during the latter half of the seventh month of fetal development.

A serous membrane called the **tunica vaginalis** (*tunica*=sheath), which is derived from the peritoneum and forms during the descent of the testes, partially covers the testes. A collection of serous fluid in the tunica vaginalis is called a **hydrocele** (HĪ-drō-sēl; *hydro-*=water; *-kele*=hernia). It may be caused by injury to the testes or inflammation of the epididymis. Usually, no treatment is required. Internal to the tunica vaginalis is a white fibrous capsule composed of dense irregular connective tissue, the **tunica**

Figure 27.3 Internal and external anatomy of a testis.

The testes are the male gonads, which produce haploid sperm.

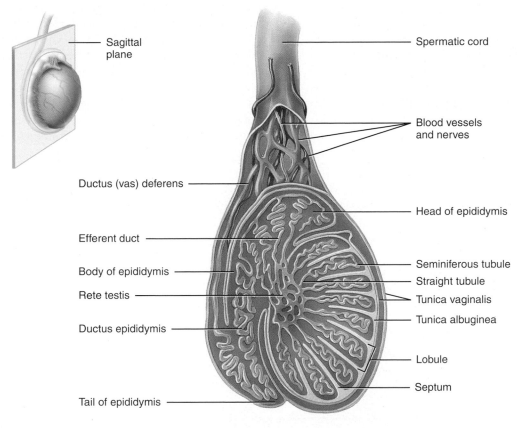

(a) Sagittal section of a testis showing seminiferous tubules

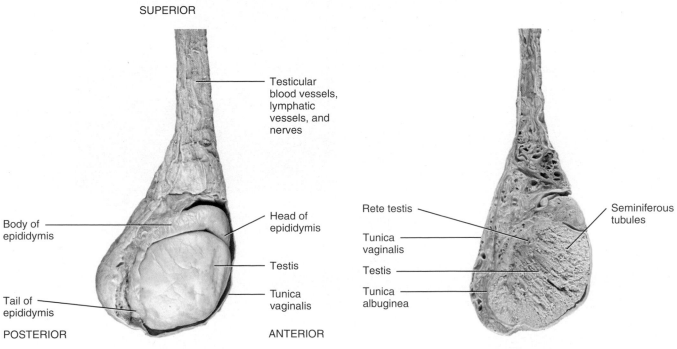

(b) Lateral view of testis and associated structures

(c) Sagittal section

? **What tissue layers cover and protect the testes?**

albuginea (al'-bū-JIN-ē-a; *albu-*=white); it extends inward, forming septa that divide each testis into a series of internal compartments called **lobules**. Each of the 200–300 lobules contains one to three tightly coiled tubules, the **seminiferous tubules** (*semin-*=seed; *fer-*=to carry), where sperm are produced (Figure 27.4). The process by which the seminiferous tubules of the testes produce sperm is called **spermatogenesis** (sper'-ma-tō-JEN-e-sis; *genesis*=beginning process or production).

The seminiferous tubules contain two types of cells: **spermatogenic cells**, the sperm-forming cells, and **Sertoli cells**, which have several functions in supporting spermatogenesis (Figure 27.4). Starting at puberty, sperm production begins at the periphery of the seminiferous tubules in stem cells called **spermatogonia** (sper'-ma-tō-GŌ-nē-a; *-gonia*=offspring; singular is *spermatogonium*). These cells develop from **primordial germ cells** (*primordi-*=primitive or early form) that arise from the yolk sac endoderm and enter the testes during the fifth week of development. In the embryonic testes, the primordial germ cells differentiate into spermatogonia, which remain dormant during childhood and become active at puberty. Toward the lumen of the tubule are layers of progressively more mature cells. In order of advancing maturity, these are primary spermatocytes, secondary spermatocytes, spermatids, and sperm. After a **sperm cell**, or **spermatozoon** (sper'-ma-tō-ZŌ-on; *-zoon*=life), has formed, it is released into the lumen of the seminiferous tubule. (The plural terms are *sperm* and *spermatozoa*.)

Embedded among the spermatogenic cells in the tubules are large **Sertoli cells**, or *sustentacular cells* (sus'-ten-TAK-ū-lar), which extend from the basement membrane to the lumen of the tubule. Internal to the basement membrane and spermatogonia, tight junctions join neighboring Sertoli cells to one another. These junctions form an obstruction known as the **blood–testis barrier** because substances must first pass through the Sertoli cells before they can reach the developing sperm. By isolating the developing gametes from the blood, the blood–testis barrier prevents an immune response against the spermatogenic cell's surface antigens, which are recognized as "foreign" by the immune system. The blood–testis barrier does not include spermatogonia.

Figure 27.4 **Microscopic anatomy of the seminiferous tubules and stages of sperm production (spermatogenesis).** Arrows in (b) indicate the progression of spermatogenic cells from least mature to most mature. The (*n*) and (*2n*) refer to haploid and diploid chromosome number, respectively.

Spermatogenesis occurs in the seminiferous tubules of the testes.

(a) Transverse section of several seminiferous tubules

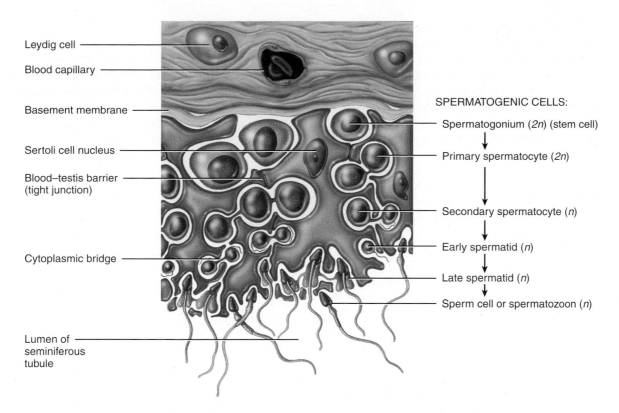

(b) Transverse section of a portion of a seminiferous tubule

? **Which cells produce testosterone?**

Figure 27.5 Events in spermatogenesis. Diploid cells (*2n*) have 46 chromosomes; haploid cells (*n*) have 23 chromosomes.

🔑 Spermiogenesis involves the maturation of spermatids into sperm.

? **What is "reduced" during meiosis I?**

Sertoli cells support and protect developing spermatogenic cells in several ways. They nourish spermatocytes, spermatids, and sperm; phagocytize excess spermatid cytoplasm as development proceeds; and control movements of spermatogenic cells and the release of sperm into the lumen of the seminiferous tubule. They also produce fluid for sperm transport, secrete the hormone inhibin, which decreases the rate of spermatogenesis, and regulate the effects of testosterone and FSH (follicle-stimulating hormone).

In the spaces between adjacent seminiferous tubules are clusters of cells called **Leydig (interstitial) cells** (Figure 27.4). These cells secrete testosterone, the most important androgen. An **androgen** (AN-drō-jen) is a hormone that promotes the development of masculine characteristics. Testosterone also promotes a man's *libido* (sex drive).

Before you read this section, please review the topic of reproductive cell division in Chapter 2 on page 48. Also, pay particular attention to Figures 2.20 and 2.21 on pages 49 and 50, respectively.

In humans, spermatogenesis takes about 65–75 days. It begins in the spermatogonia, which contain the diploid (*2n*) chromosome number (Figure 27.5). Spermatogonia are *stem cells* because when they undergo mitosis, some cells remain near the basement membrane of the seminiferous tubule in an undifferentiated state to serve as a reservoir of cells for future cell division and subsequent sperm production. The rest of the cells lose contact with the basement membrane, squeeze through the tight junctions of the blood–testis barrier, undergo developmental changes, and differentiate into **primary spermatocytes** (SPER-ma-tō-sītz′). Primary spermatocytes, like spermatogonia, are diploid (*2n*); that is, they have 46 chromosomes.

Shortly after it forms, each primary spermatocyte replicates its DNA and then meiosis begins (Figure 27.5). In meiosis I, homologous pairs of chromosomes line up at the metaphase plate, and crossing-over occurs. Then, the meiotic spindle pulls one (duplicated) chromosome of each pair to an opposite pole of the dividing cell. The two cells formed by meiosis I are called **secondary spermatocytes**. Each secondary spermatocyte has 23 chromosomes, the haploid number (*n*). Each chromosome within a secondary spermatocyte, however, is made up of two chromatids (two copies of the DNA) still attached by a centromere. No replication of DNA occurs in the secondary spermatocytes.

In meiosis II, no replication of DNA occurs. The chromosomes line up in single file along the metaphase plate, and the two chromatids of each chromosome separate. The four haploid cells resulting from meiosis II are called **spermatids**. A single primary spermatocyte therefore produces four spermatids through two rounds of cell division (meiosis I and meiosis II).

A unique process occurs during spermatogenesis. As the sperm cells proliferate following their production by spermatogonia, they fail to complete cytoplasmic separation (cytokinesis). The cells remain in contact by cytoplasmic bridges through their entire development (see Figures 27.4b and 27.5). This pattern of development most likely accounts for the synchronized production of sperm in any given area of a seminiferous tubule. It may have survival value in that half of the sperm contain an X chromosome and half contain a Y chromosome. The larger X chromosome may carry genes needed for spermatogenesis that are lacking on the smaller Y chromosome.

The final stage of spermatogenesis, **spermiogenesis** (sper′-mē-ō-JEN-e-sis), is the maturation of haploid spermatids into sperm. Because no cell division occurs in spermiogenesis, each spermatid develops into a single **sperm cell** (*sperma*=seed). During this process, spherical spermatids transform into elongated, slender sperm. An acrosome (described shortly) forms atop the condensing and elongating nucleus, a flagellum develops, and mitochondria multiply. Sertoli cells dispose of the excess cytoplasm that is sloughed off during this process. Finally, sperm are released from their connections to Sertoli cells, an event known as **spermiation**. Sperm then enter the lumen of the seminiferous tubule. Fluid secreted by Sertoli cells pushes sperm along their way, toward the ducts of the testes. At this point sperm are not yet able to swim.

Sperm

Spermatogenesis produces about 300 million sperm per day. A sperm is about 60 μm long and contains several structures that are highly adapted for reaching and penetrating a secondary oocyte (Figure 27.6). The major parts of a sperm are the head and the tail. The flattened, pointed **head** of the sperm is about 4–5 μm long. It contains a **nucleus** that has highly condensed haploid chromosomes (23). Covering the anterior two-thirds of the nucleus is the **acrosome** (*acro-*=atop; *-some*=body), a caplike vesicle filled with enzymes that help a sperm to penetrate a secondary oocyte to bring about fertilization. Among the enzymes are hyaluronidase and proteases. The **tail** of a sperm is subdivided into four parts: neck, middle piece, principal piece, and end piece. The **neck** is the constricted region just behind the head that contains centrioles. The centrioles form the microtubules that comprise the remainder of the tail. The **middle piece** contains mitochondria arranged in a spiral, which provide the energy (ATP) for locomotion of sperm to the site of fertilization and for sperm metabolism. The **principal piece** is the longest portion of the tail and the **end piece** is the terminal, tapering portion of the tail. Once ejaculated, most sperm do not survive more than 48 hours within the female reproductive tract.

■ CHECKPOINT
1. Describe the function of the scrotum in protecting the testes from temperature fluctuations.
2. Describe the internal structure of a testis. Where are sperm cells produced? What are the functions of Sertoli cells and Leydig cells?
3. Describe the principal events of spermatogenesis.
4. Identify the parts of a sperm cell, and list the functions of each.

Reproductive System Ducts in Males

Ducts of the Testis

Pressure generated by the fluid secreted by Sertoli cells pushes sperm and fluid along the lumen of seminiferous tubules and then into a series of very short ducts called **straight tubules**. The straight tubules lead to a network of ducts in the testis called the **rete testis** (RĒ-tē=network) (see Figure 27.3a). From the rete testis, sperm move into a series of coiled **efferent ducts** in the epididymis that empty into a single tube called the **ductus epididymis**.

Epididymis

The **epididymis** (ep′-i-DID-i-mis; *epi-*=above or over; *-didymis*=testis) is a comma-shaped organ about 4 cm (1.5 in.) long that lies along the posterior border of each testis (see Figure 27.3a). The plural is **epididymides** (ep′-i-did-ĪM-i-dēs). Each epididymis consists mostly of the tightly coiled **ductus epididymis**. The efferent ducts from the testis join the ductus epididymis at the larger, superior portion of the epididymis called the **head**. The **body** is the narrow midportion of the epididymis, and the **tail** is the smaller, inferior portion. At its distal end, the

Figure 27.6 Parts of sperm cell (spermatozoon).

About 300 million sperm mature each day.

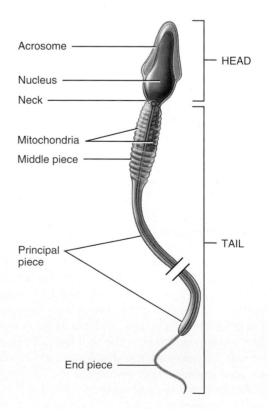

What are the functions of each part of a sperm cell?

tail of the epididymis continues as the ductus (vas) deferens (discussed shortly).

The ductus epididymis would measure about 6 m (20 ft) in length if it were straightened out. It is lined with pseudostratified columnar epithelium and encircled by a layer of smooth muscle (Figure 27.7). The free surfaces of the columnar cells contain **stereocilia**, long, branching microvilli that increase surface area

for the reabsorption of degenerated sperm. Connective tissue around the muscle layer attaches the loops of the ductus epididymis and carries blood vessels and nerves.

Functionally, the epididymis is the site of **sperm maturation**, the process by which sperm acquire motility and the ability to fertilize an ovum. This occurs over a period of about 14 days. The epididymis also stores sperm and helps propel them during sexual

Figure 27.7 **Histology of the ductus epididymis.**

Stereocilia increase the surface area for the reabsorption of degenerated sperm.

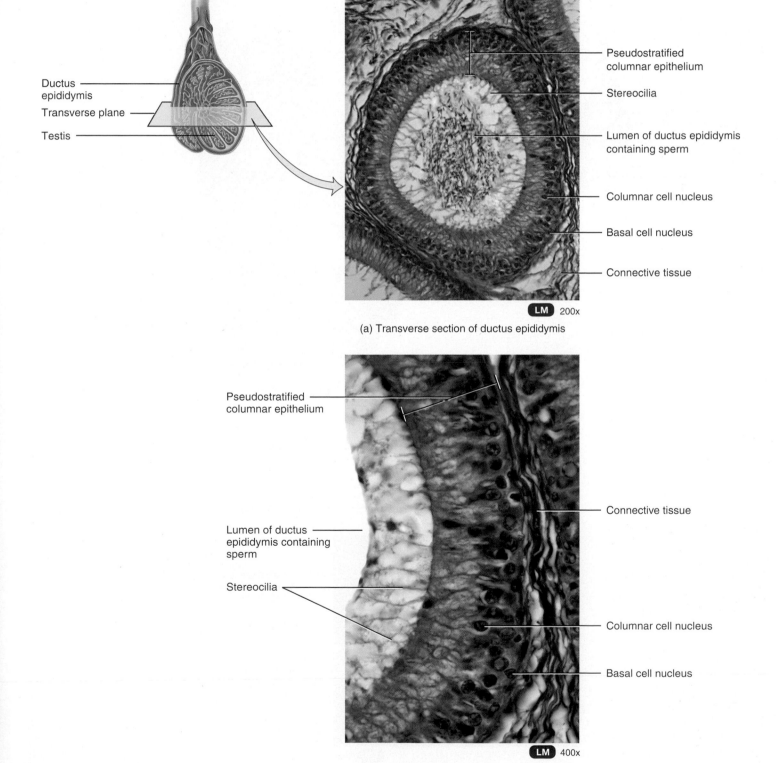

(a) Transverse section of ductus epididymis

(b) Details of the epithelium

? **What are the functions of the ductus epididymis?**

arousal by peristaltic contraction of its smooth muscle into the ductus (vas) deferens. Sperm may remain in storage in the epididymis for up to several months. Any stored sperm that are not ejaculated by that time are eventually reabsorbed.

Ductus Deferens

Within the tail of the epididymis, the ductus epididymis becomes less convoluted, and its diameter increases. Beyond this point, the duct is referred to as the **ductus deferens** or **vas deferens** (see Figure 27.3a). The ductus deferens, which is about 45 cm (18 in.) long, ascends along the posterior border of the epididymis, through the spermatic cord to the point in the lower abdominal wall where it passes through the inguinal canal (see Figure 27.2) to enter the pelvic cavity; there it loops over the ureter and passes over the side and down the posterior surface of the urinary bladder (see Figure 27.1a). The dilated terminal portion of the ductus deferens is known as the **ampulla** (am-POOL-la=little jar) (see Figure 27.9). The mucosa of the ductus deferens consists of

Figure 27.8 Histology of the ductus (vas) deferens.

🔑 The ductus (vas) deferens enters the pelvic cavity through the inguinal canal.

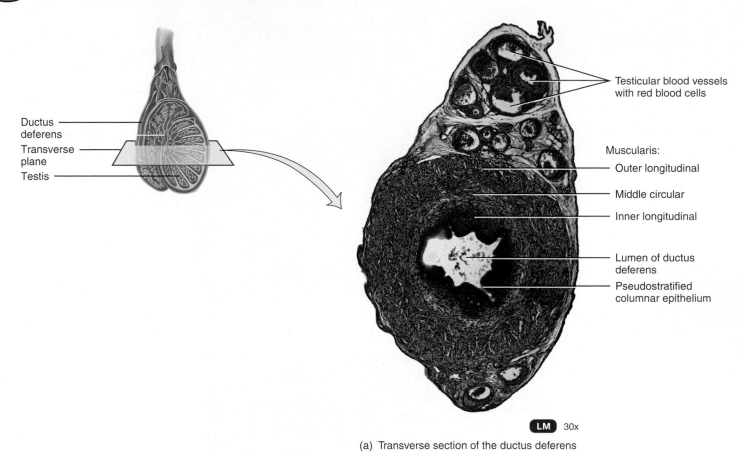

Ductus deferens
Transverse plane
Testis

Testicular blood vessels with red blood cells

Muscularis:
Outer longitudinal
Middle circular
Inner longitudinal

Lumen of ductus deferens
Pseudostratified columnar epithelium

LM 30x

(a) Transverse section of the ductus deferens

Lumen of ductus deferens

Pseudostratified columnar epithelium

Lamina propria

Inner longitudinal muscular layer

LM 400x

(b) Details of epithelium

❓ **What is the function of the ductus (vas) deferens?**

pseudostratified columnar epithelium and lamina propria (areolar connective tissue). The muscularis is composed of three layers; the inner and outer layers are longitudinal, and the middle layer is circular (Figure 27.8 on page 854).

Functionally, the ductus deferens conveys sperm during sexual arousal from the epididymis toward the urethra by peristaltic contractions of its muscular coat. Like the epididymis, the ductus deferens also can store sperm for several months. Any stored sperm that are not ejaculated by that time are eventually reabsorbed.

Ejaculatory Ducts

Each **ejaculatory duct** (e-JAK-ū-la-tō′-rē; *ejacul-*=to expel) is about 2 cm (1 in.) long and is formed by the union of the duct from the seminal vesicle and the ampulla of the ductus (vas) deferens (see Figure 27.9). The ejaculatory ducts form just superior to the base (superior portion) of the prostate and pass inferiorly and anteriorly through the prostate. They terminate in the prostatic urethra, where they eject sperm and seminal vesicle secretions just before the release of semen from the urethra to the exterior.

Figure 27.9 Locations of several accessory reproductive glands in males. The prostate, urethra, and penis have been sectioned to show internal details.

 The male urethra has three subdivisions: the prostatic, membranous (intermediate), and spongy (penile) urethra.

(a) Posterior view of male accessory organs of reproduction

Functions of Accessory Sex Gland Secretions

1. The seminal vesicles secrete alkaline, viscous fluid that helps neutralize acid in the female reproductive tract, provides fructose for ATP production by sperm, contributes to sperm motility and viability, and helps semen coagulate after ejaculation.

2. The prostate secretes a milky, slightly acidic fluid that helps semen coagulate after ejaculation and subsequently breaks down the clot.

3. The bulbourethral (Cowper's) glands secrete alkaline fluid that neutralizes the acidic environment of the urethra and mucus that lubricates the lining of the urethra and tip of the penis during sexual intercourse.

CONTINUES ▶

FIGURE **27.9** CONTINUED ▶

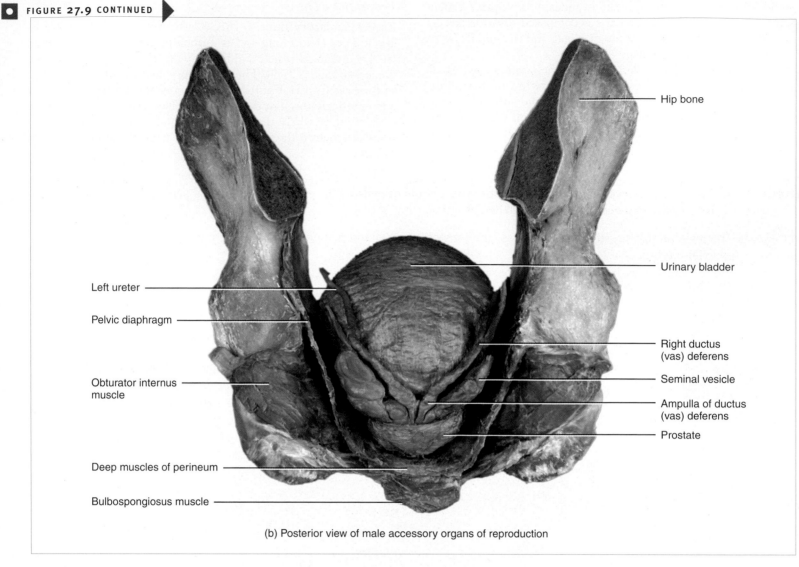

Hip bone

Urinary bladder

Left ureter

Pelvic diaphragm

Right ductus (vas) deferens

Seminal vesicle

Obturator internus muscle

Ampulla of ductus (vas) deferens

Prostate

Deep muscles of perineum

Bulbospongiosus muscle

(b) Posterior view of male accessory organs of reproduction

? **What accessory sex gland contributes the majority of the seminal fluid?**

Urethra

In males, the **urethra** is the shared terminal duct of the reproductive and urinary systems; it serves as a passageway for both semen and urine. About 20 cm (8 in.) long, it passes through the prostate, the deep muscles of the perineum, and the penis, and is subdivided into three parts (see Figures 27.1a and 27.9). The **prostatic urethra** is 2–3 cm (1 in.) long and passes through the prostate. As this duct continues inferiorly, it passes through the deep muscles of the perineum (see Figure 11.15 on page 352), where it is known as the **membranous (intermediate) urethra**. The membranous urethra is about 1 cm (0.5 in.) in length. As this duct passes through the corpus spongiosum of the penis, it is known as the **spongy (penile) urethra**, which is about 15–20 cm (6–8 in.) long. The spongy urethra ends at the **external urethral orifice**. The histology of the male urethra may be reviewed on page 839 of Chapter 26.

Spermatic Cord

The **spermatic cord** is a supporting structure of the male reproductive system that ascends out of the scrotum (see Figure 27.2). It consists of the ductus (vas) deferens as it ascends through the scrotum, the testicular artery, both somatic and autonomic nerves, veins that drain the testes and carry testosterone into the circulation (the *pampiniform plexus*), lymphatic vessels, and the

cremaster muscle. The term **varicocele** (VAR-i-kō-sēl; *varico-* =varicose; *-kele*=hernia) refers to a swelling in the scrotum due to varicosities in the veins that drain the testes. It usually disappears when the person lies down, and typically does not require treatment. The spermatic cord and ilioinguinal nerve pass through the **inguinal canal** (IN-gwin-al=groin), an oblique passageway in the anterior abdominal wall just superior and parallel to the medial half of the inguinal ligament. The canal, which is about 4–5 cm (about 2 in.) long, originates at the **deep (abdominal) inguinal ring**, a slitlike opening in the aponeurosis of the transversus abdominis muscle; the canal ends at the **superficial (subcutaneous) inguinal ring** (see Figure 27.2), a somewhat triangular opening in the aponeurosis of the external oblique muscle. In females, the round ligament of the uterus and ilioinguinal nerve pass through the inguinal canal.

◉ **CHECKPOINT**

5. Which ducts transport sperm within the testes?
6. Describe the location, structure, and functions of the ductus epididymis, ductus (vas) deferens, and ejaculatory duct.
7. Give the locations of the three subdivisions of the male urethra.
8. Trace the course of sperm through the system of ducts from the seminiferous tubules through the urethra.
9. List the structures within the spermatic cord.

Accessory Sex Glands in Males

The ducts of the male reproductive system store and transport sperm cells, but the **accessory sex glands** secrete most of the liquid portion of semen. The accessory sex glands are the seminal vesicles, the prostate, and the bulbourethral glands.

Seminal Vesicles

The paired **seminal vesicles** (VES-i-kuls) or **seminal glands** are convoluted pouchlike structures, about 5 cm (2 in.) in length, lying posterior to and at the base of the urinary bladder anterior to the rectum (Figure 27.9). They secrete through the seminal vesicle ducts (Figure 27.9) an alkaline, viscous fluid that contains fructose (a monosaccharide sugar), prostaglandins, and clotting proteins (discussed shortly) unlike those found in blood. The alkaline nature of the fluid helps to neutralize the acidic environment of the male urethra and female reproductive tract that otherwise would inactivate and kill sperm. The fructose is used for the production of ATP by sperm. Prostaglandins contribute to sperm motility and viability and may also stimulate muscular contractions within the female reproductive tract. The clotting proteins help semen coagulate after ejaculation. Fluid secreted by the seminal vesicles normally constitutes about 60% of the volume of semen.

Prostate

The **prostate** (PROS-tāt) is a single, doughnut-shaped gland about the size of a golf ball. It measures about 4 cm (1.6 in.) from side to side, about 3 cm (1.2 in.) from top to bottom, and about 2 cm (0.8 in.) from front to back. It is inferior to the urinary bladder and surrounds the prostatic urethra (Figure 27.9). The prostate slowly increases in size from birth to puberty, and then it expands rapidly. The size attained by age 30 typically remains stable until about age 45, when further enlargement may occur.

The prostate secretes a milky, slightly acidic fluid (pH about 6.5) that contains several substances. (1) *Citric acid* in prostatic fluid is used by sperm for ATP production. (2) Several *proteolytic enzymes*, such as *prostate-specific antigen (PSA)*, pepsinogen, lysozyme, amylase, and hyaluronidase, eventually break down the clotting proteins from the seminal vesicles. (3) The function of the *acid phosphatase* secreted by the prostate is unknown. (4) *Seminalplasmin* in prostatic fluid is an antibiotic that can destroy bacteria. Seminalplasmin may help decrease the number of naturally occurring bacteria in semen and in the lower female reproductive tract. Secretions of the prostate enter the prostatic urethra through many prostatic ducts. Prostatic secretions make up about 25% of the volume of semen and contribute to sperm motility and viability.

Bulbourethral Glands

The paired **bulbourethral glands** (bul′-bō-ū-RĒ-thral), or **Cowper's glands**, each about the size of a pea, lie inferior to the prostate on either side of the membranous urethra within the deep muscles of the perineum; their ducts open into the spongy urethra (Figure 27.9). During sexual arousal, the bulbourethral glands secrete an alkaline substance that protects the passing sperm by neutralizing acids from urine in the urethra. At the same time, they secrete mucus that lubricates the end of the penis and the lining of the urethra, thereby decreasing the number of sperm damaged during ejaculation.

• CLINICAL **CONNECTION** | Prostate Cancer

Prostate cancer is the leading cause of death from cancer in men in the United States, having surpassed lung cancer in 1991. Each year it is diagnosed in almost 200,000 U.S. men and causes nearly 40,000 deaths. The amount of PSA (prostate-specific antigen), which is produced only by prostate epithelial cells, increases with enlargement of the prostate (Figure B) and may indicate infection, benign enlargement, or prostate cancer. A blood test can measure the level of PSA in the blood. Males over the age of 40 should have an annual examination of the prostate gland. In a **digital rectal exam**, a physician palpates the gland through the rectum with the fingers (digits). Many physicians also recommend an annual PSA test for males over age 50. Treatment for prostate cancer may involve surgery, cryotherapy, radiation, hormonal therapy, and chemotherapy. Because many prostate cancers grow very slowly, some urologists recommend "watchful waiting" before treating small tumors in men over age 70. •

(A) Normal prostate

(B) Enlarged prostate

Semen

Semen (=seed) is a mixture of sperm and **seminal fluid**, a liquid that consists of the secretions of the seminiferous tubules, seminal vesicles, prostate, and bulbourethral glands. The volume of semen in a typical ejaculation is 2.5–5 mL, with a sperm count (concentration) of 50–150 million sperm/mL. A male whose sperm count falls below 20 million/mL is likely to be infertile. The very large number is required for successful fertilization because only a tiny fraction ever reach the secondary oocyte.

Despite the slight acidity of prostatic fluid, semen has a slightly alkaline pH of 7.2–7.7 due to the higher pH and larger volume of fluid from the seminal vesicles. The prostatic secretion gives semen a milky appearance, and fluids from the seminal vesicles and bulbourethral glands give it a sticky consistency. Seminal fluid provides sperm with a transportation medium, nutrients, and protection from the hostile acidic environment of the male's urethra and the female's vagina.

Once ejaculated, liquid semen coagulates within 5 minutes due to the presence of clotting proteins from the seminal vesicles. The functional role of semen coagulation is not known, but the proteins involved are different from those that cause blood coagulation. After about 10–20 minutes, semen reliquefies because prostate-specific antigen (PSA) and other proteolytic enzymes produced by the prostate break down the clot. Abnormal or delayed liquefaction of clotted semen may cause complete or partial immobilization of sperm, thereby inhibiting their movement through the cervix of the uterus. The presence of blood in semen is called **hemospermia** (hē-mō-SPER-mē-a; *hemo-*=blood; *-sperma*=seed). In most cases, it is caused by inflammation of the blood vessels lining the seminal vesicles; it is usually treated with antibiotics.

Penis

The **penis** contains the urethra and is a passageway for the ejaculation of semen and the excretion of urine (Figure 27.10). It is cylindrical in shape and consists of a body, glans penis, and a root. The **body of the penis** is composed of three cylindrical masses of tissue, each surrounded by fibrous tissue called the **tunica albuginea** (al-bū-JIN-ē-a; Figure 27.10). The two dorsolateral masses are called the **corpora cavernosa penis** (*corpora*=main bodies; *cavernosa*=hollow). The smaller midventral mass, the **corpus spongiosum penis**, contains the spongy urethra and keeps it open during ejaculation. Skin and a subcutaneous layer enclose all three masses, which consist of erectile tissue. *Erectile tissue* is composed of numerous blood sinuses (vascular spaces) lined by endothelial cells and surrounded by smooth muscle and elastic connective tissue.

The distal end of the corpus spongiosum penis is a slightly enlarged, acorn-shaped region called the **glans penis**; its margin is the **corona**. The distal urethra enlarges within the glans penis and forms a terminal slitlike opening, the **external urethral orifice**. Covering the glans in an uncircumcised penis is the loosely fitting **prepuce** (PRĒ-poos), or **foreskin**.

The **root of the penis** is the attached portion (proximal portion). It consists of the **bulb of the penis**, the expanded portion of the base of the corpus spongiosum penis, and the **crura of the penis** (singular is *crus*=resembling a leg), the two separated and tapered portions of the corpora cavernosa penis. The bulb of the penis is attached to the inferior surface of the deep muscles of the perineum and is enclosed by the bulbospongiosus muscle. Contraction of the bulbospongiosus muscle aids ejaculation. Each

crus of the penis bends laterally away from the bulb of the penis to attach to the ischial and inferior pubic rami and is surrounded by the ischiocavernosus muscle (see Figure 11.15 on page 352). The weight of the penis is supported by two ligaments that are continuous with the fascia of the penis. (1) The **fundiform ligament** arises from the inferior part of the linea alba. (2) The **suspensory ligament of the penis** arises from the pubic symphysis.

Upon sexual stimulation (visual, tactile, auditory, olfactory, or imagined), parasympathetic fibers from the sacral portion of the spinal cord initiate and maintain an **erection**, the enlargement and stiffening of the penis. The parasympathetic fibers release and cause local production of nitric oxide (NO). The NO causes smooth muscle in the walls of arterioles supplying erectile tissue to relax, which allows these blood vessels to dilate. This in turn causes large amounts of blood to enter the erectile tissue of the penis. NO also causes the smooth muscle within the erectile tissue to relax, resulting in widening of the blood sinuses. The combination of increased blood flow and widening of the blood sinuses results in an erection. Expansion of the blood sinuses also compresses the veins that drain the penis; the slowing of blood outflow helps to maintain the erection.

The term **priapism** (PRĪ-a-pizm) refers to a persistent and usually painful erection of the penis that does not involve sexual desire or excitement. The condition may last up to several hours and is accompanied by pain and tenderness. It results from abnormalities of blood vessels and nerves, usually in response to medication used to produce erections in males who otherwise cannot attain them. Other causes include a spinal cord disorder, leukemia, sickle-cell disease, or a pelvic tumor.

Ejaculation (ē-jak-ū-LĀ-shun; *ejectus*=to throw out), the powerful release of semen from the urethra to the exterior, is a sympathetic reflex coordinated by the lumbar portion of the spinal cord. As part of the reflex, the smooth muscle sphincter at the base of the urinary bladder closes. Therefore, urine is not expelled during ejaculation, and semen does not normally enter the urinary bladder. Even before ejaculation occurs, peristaltic contractions in the ampulla of the ductus deferens, seminal vesicles, ejaculatory ducts, and prostate propel semen into the penile portion of the urethra (spongy urethra). Typically, this leads to **emission** (ē-MISH-un), the discharge of a small volume of semen before ejaculation. Emission may also occur during sleep (nocturnal emission). The musculature of the penis (bulbospongiosus, ischiocavernosus, and superficial transverse perineus muscles), which is supplied by the pudendal nerve, also contracts at ejaculation (see Figure 11.15 on page 352).

Once sexual stimulation of the penis has ended, the arterioles supplying the erectile tissue of the penis constrict and the smooth muscle within erectile tissue contracts, making the blood sinuses smaller. This relieves pressure on the veins supplying the penis and allows the blood to drain through them. Consequently, the penis returns to its flaccid (relaxed) state.

The penis has a very rich blood supply from the internal pudendal artery and the femoral artery. The veins drain into corresponding vessels.

The sensory nerves to the penis are branches from the pudendal and ilioinguinal nerves. The corpora have a parasympathetic and a sympathetic supply. As noted previously, parasympathetic stimulation causes the blood vessels to dilate, increasing the flow of blood into the erectile tissue and trapping blood within the penis to maintain the erection. At ejaculation, sympathetic stimulation causes the smooth muscle located in the walls of the ducts and accessory sex glands of the reproductive tract to contract and propel the semen along its course.

Figure 27.10 Internal structure of the penis.
The inset in (b) shows details of the skin and fasciae. (See Tortora, *A Photographic Atlas of the Human Body*, 2e, Figure 14.6.)

The penis contains the urethra, a common pathway for semen and urine.

Internal urethral orifice
Prostatic urethra
Bulbourethral (Cowper's) gland
Deep muscles of perineum

Urinary bladder
Prostate
Orifice of ejaculatory duct
Membranous urethra
ROOT OF PENIS:
Bulb of penis
Crus of penis

BODY OF PENIS:
Corpora cavernosa penis
Corpus spongiosum penis

Spongy (penile) urethra

Frontal plane

Corona
GLANS PENIS
Prepuce (foreskin)
External urethral orifice

(a) Frontal section

Transverse plane

Deep dorsal vein
Dorsal artery

Skin
Superficial (subcutaneous) dorsal vein
Subcutaneous layer
Deep fascia

Dorsal

Corpora cavernosa penis
Tunica albuginea of corpora cavernosum
Deep artery of penis
Corpus spongiosum penis
Spongy (penile) urethra
Tunica albuginea of corpus spongiosum penis

Ventral

(b) Transverse section

Transverse plane
Skin
Subcutaneous layer

Superficial dorsal vein
Deep dorsal vein
Dorsal artery
Corpora cavernosa penis
Tunica albuginea of corpora cavernosa
Deep artery of penis
Spongy (penile) urethra
Corpus spongiosum penis
Tunica albuginea of corpus spongiosum penis

(c) Transverse section

Which tissue masses form the erectile tissue in the penis, and why do they become rigid during sexual arousal?

CHECKPOINT
10. Briefly explain the locations and functions of the seminal vesicles, the prostate, and the bulbourethral (Cowper's) glands.
11. What is semen? What is its function?
12. Explain the physiological processes involved in erection and ejaculation.

FEMALE REPRODUCTIVE SYSTEM

◉ **OBJECTIVES**

• Describe the location, structure, and functions of the organs of the female reproductive system.
• Discuss the process of oogenesis in the ovaries.

The organs of the **female reproductive system** (Figure 27.11) include the ovaries, which produce secondary oocytes and hormones, such as progesterone and estrogens (the female sex hormones), inhibin, and relaxin; the uterine (fallopian) tubes, or oviducts, which transport secondary oocytes and fertilized ova to the uterus; the uterus, in which embryonic and fetal development occur; the vagina; and external organs that constitute the vulva, or pudendum. The mammary glands also are considered part of the female reproductive system.

Ovaries

The **ovaries** (=egg receptacles) are paired glands that resemble unshelled almonds in size and shape; they are the female gonads and are homologous to the testes. (Here *homologous* means that two organs have the same embryonic origin.) The ovaries, one on either side of the uterus, descend to the brim of the superior portion of the pelvic cavity during the third month of development. A series of ligaments holds them in position (Figure 27.12 on page 862). The **broad ligament** of the uterus (see also Figure 27.11b), which is itself part of the parietal peritoneum, attaches to the ovaries by a double-layered fold of peritoneum called the **mesovarium**. The **ovarian ligament** anchors the ovaries to the uterus, and the **suspensory ligament** attaches them to the pelvic wall. Each ovary contains a **hilum**, the point of entrance and exit for blood vessels and nerves along which the mesovarium is attached (see Figure 27.13).

Figure 27.11 Female organs of reproduction and surrounding structures.

🔑 ▭ The organs of reproduction in females include the ovaries, uterine (fallopian) tubes, uterus, vagina, vulva, and mammary glands.

Functions of the Female Reproductive System

1. The ovaries produce secondary oocytes and hormones, including progesterone and estrogens (female sex hormones), inhibin, and relaxin.
2. The uterine tubes transport a secondary oocyte to the uterus and normally are the sites where fertilization occurs.
3. The uterus is the site of implantation of a fertilized ovum, development of the fetus during pregnancy, and labor.
4. The vagina receives the penis during sexual intercourse and is a passageway for childbirth.
5. The mammary glands synthesize, secrete, and eject milk for nourishment of the newborn.

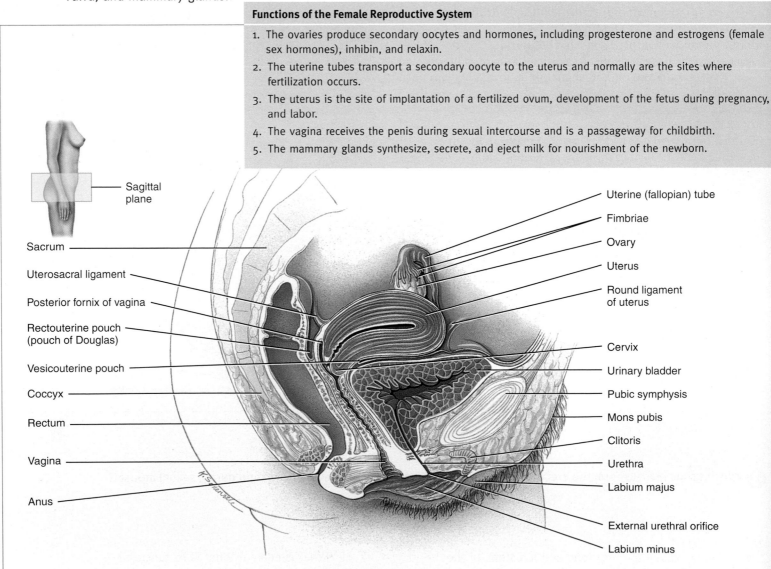

(a) Sagittal section

Histology of the Ovaries

Each ovary consists of the following parts (Figure 27.13 on page 863):

- The **germinal epithelium** (*germen*=sprout or bud) is a layer of simple epithelium (low cuboidal or squamous) that covers the surface of the ovary. It is continuous with the mesothelium that covers the mesovarium. We now know that the term germinal epithelium is inappropriate because it does not give rise to ova, but at one time people believed that it did. We have since learned that the cells that produce ova arise from the endoderm of the yolk sac and migrate to the ovaries during embryonic development.
- The **tunica albuginea** is a whitish capsule of dense, irregular connective tissue immediately deep to the germinal epithelium.
- The **ovarian cortex** is a region just deep to the tunica albuginea. It consists of ovarian follicles (described shortly) surrounded by dense irregular connective tissue that contains collagen fibers and fibroblast-like cells called stromal cells.
- The **ovarian medulla** is deep to the ovarian cortex. The border between the cortex and medulla is indistinct, but the medulla consists of more loosely arranged connective tissue and contains blood vessels, lymphatic vessels, and nerves.

- **Ovarian follicles** (*folliculus*=little bag) are located in the ovarian cortex and consist of **oocytes** in various stages of development, plus the cells surrounding them. When the surrounding cells form a single layer, they are called **follicular cells**; later in development, when they form several layers, they are referred to as **granulosa cells**. The surrounding cells nourish the developing oocyte and begin to secrete estrogens as the follicle grows larger.
- A **mature (graafian) follicle** is a large, fluid-filled follicle that is ready to rupture and expel its secondary oocyte, a process known as **ovulation**.
- A **corpus luteum** (=yellow body) contains the remnants of a mature follicle after ovulation. The corpus luteum produces progesterone, estrogens, relaxin, and inhibin until it degenerates into fibrous scar tissue called the **corpus albicans** (=white body).

The ovarian blood supply is furnished by the ovarian arteries, which anastomose with branches of the uterine arteries. The ovaries are drained by the ovarian veins. On the right side they drain into the inferior vena cava, and on the left side they drain into the renal vein.

Sympathetic and parasympathetic nerve fibers to the ovaries terminate on the blood vessels and enter the ovaries.

TEXT CONTINUES ON PAGE **864**

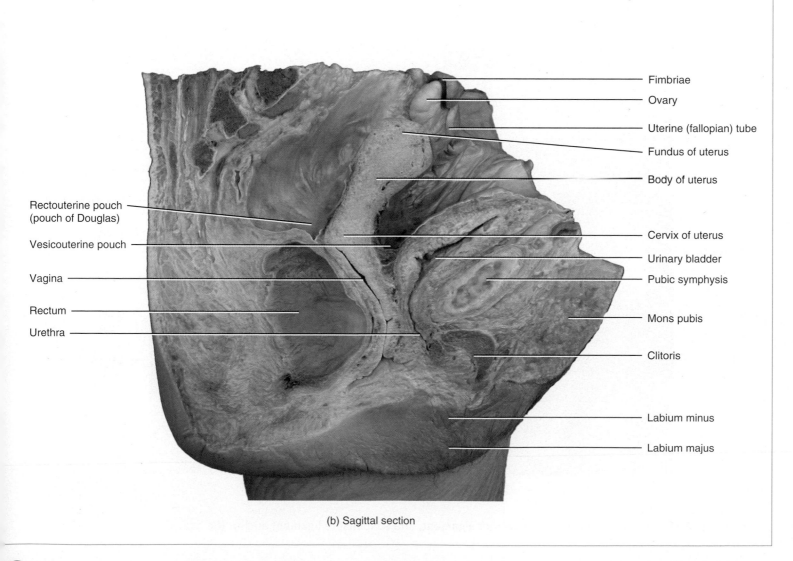

(b) Sagittal section

? **Which structures in males are homologous to the ovaries, the clitoris, the paraurethral glands, and the greater vestibular glands?**

Figure 27.12 Relative positions of the ovaries, the uterus, and the ligaments that support them.

Ligaments holding the ovaries in position include the mesovarium, the ovarian ligament, and the suspensory ligament.

ANTERIOR

Rectus abdominis muscle

Urinary bladder

Uterus

Round ligament

Ovarian ligament

Mesovarium

Cecum

Broad ligament

Vermiform appendix

Cardinal ligament

Uterosacral ligament

Common iliac artery

Uterine (fallopian) tube

Ovary

Rectouterine pouch (pouch of Douglas)

Suspensory ligament

Ureter

Ileum

Sigmoid colon

View

Transverse plane

(a) POSTERIOR

Superior view of transverse section

Rectus abdominis muscle

Uterus

Round ligament

Uterine (fallopian) tube

Ovarian ligament

Ovary

Fimbriae

Ileum

Cecum

Ascending colon

Sigmoid colon

Jejunum

(b) Anterosuperior view

To which structures do the mesovarium, ovarian ligament, and suspensory ligament anchor the ovary?

Figure 27.13 Histology of the ovary. The arrows in (a) indicate the sequence of developmental stages that occur as part of the maturation of an ovum during the ovarian cycle.

🔓 ⌕ The ovaries are the female gonads; they produce haploid oocytes.

(a) Frontal section

(b) Hemisection   LM 20x

(c) Ovulation of a secondary oocyte   LM 30x

❓ **What structures in the ovary contain endocrine tissue, and what hormones do they secrete?**

Oogenesis and Follicular Development

The formation of gametes in the ovaries is termed **oogenesis** (ō-ō-JEN-e-sis; *oo-*=egg). Unlike spermatogenesis, which begins in males at puberty, oogenesis begins in females before they are even born. Oogenesis occurs in essentially the same manner as spermatogenesis: Meiosis (see Chapter 2) takes place and the resulting germ cells undergo maturation.

During early fetal development, primordial (primitive) germ cells migrate from the yolk sac to the ovaries. There, germ cells differentiate within the ovaries into **oogonia** (ō′-ō-GŌ-nē-a; singular is

oogonium). Oogonia are diploid (*2n*) stem cells that divide mitotically to produce millions of germ cells. Even before birth, most of these germ cells degenerate in a process known as **atresia** (a-TRĒ-zē-a). A few, however, develop into larger cells called **primary oocytes** (Ō-ō-sītz) that enter prophase of meiosis I during fetal development but do not complete that phase until after puberty. During this arrested stage of development, each primary oocyte is surrounded by a single layer of flat follicular cells, and the entire structure is called a **primordial follicle** (Figure 27.14a).

Figure 27.14 Ovarian follicles.

🔓 ⚏ As an ovarian follicle enlarges, follicular fluid accumulates in a cavity called the antrum.

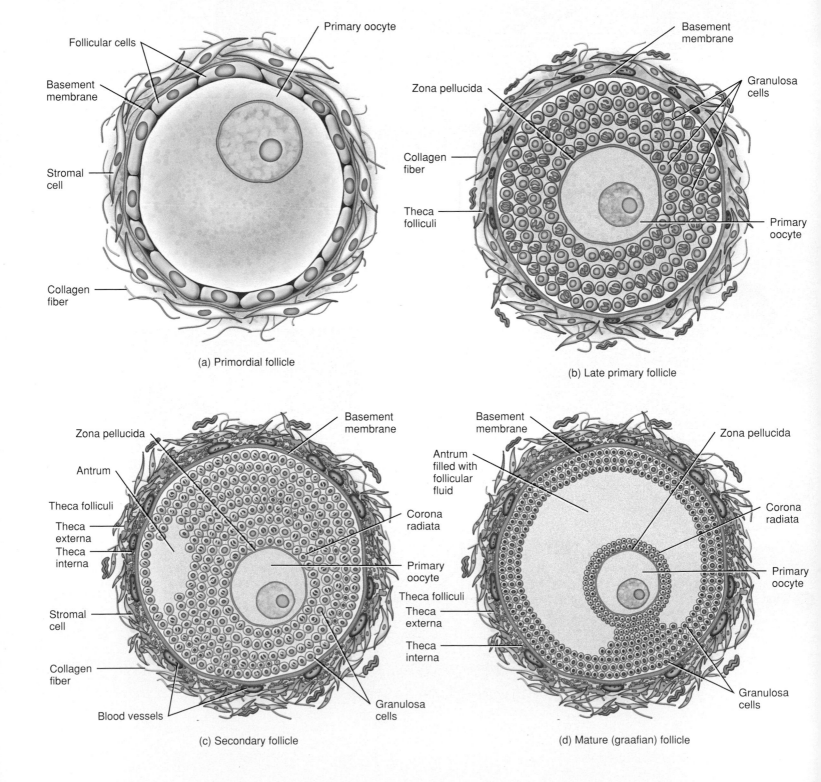

(a) Primordial follicle

(b) Late primary follicle

(c) Secondary follicle

(d) Mature (graafian) follicle

Germinal epithelium
Tunica albuginea
Primordial follicle
Ovarian cortex
Primary follicle granulosa cells
Theca folliculi
Zona pellucida
Primary oocyte
Secondary follicle granulosa cells
Secondary follicle
Corpus luteum

LM 30x

(e) Ovarian cortex

Theca folliculi

Antrum filled with follicular fluid

Secondary follicle granulosa cells
Corona radiata
Zona pellucida
Primary oocyte

LM 70x

(f) Secondary follicle

? **What happens to most ovarian follicles?**

The ovarian cortex surrounding the primordial follicles consists of collagen fibers and fibroblast-like **stromal cells**. At birth, approximately 200,000 to 2,000,000 primary oocytes remain in each ovary. Of these, about 40,000 are still present at puberty, and around 400 will mature and ovulate during a woman's reproductive lifetime. The remainder of the primary oocytes undergo atresia.

Each month after puberty until menopause, gonadotropins (FSH and LH) secreted by the anterior pituitary further stimulate the development of several primordial follicles, although only one will typically reach the maturity needed for ovulation. A few primordial follicles start to grow, developing into **primary follicles** (Figure 27.14b). Each primary follicle consists of a primary oocyte that is surrounded in a later stage of development by several layers of cuboidal and low-columnar cells called **granulosa cells**. The outermost granulosa cells rest on a basement membrane. As the primary follicle grows, it forms a clear glycoprotein layer called the **zona pellucida** (pe-LOO-si-da) between the primary oocyte and the granulosa cells. In addition, stromal cells surrounding the basement membrane begin to form an organized layer called the **theca folliculi** (THĒ-ka fo-LIK-ū-li).

With continuing maturation, a primary follicle develops into a secondary follicle (Figure 27.14c, f). In a **secondary follicle**, the theca differentiates into two layers: (1) the **theca interna**, a highly vascularized internal layer of cuboidal secretory cells that secrete estrogens and (2) the **theca externa**, an outer layer of stromal cells and collagen fibers. In addition, the granulosa cells begin to secrete follicular fluid, which builds up in a cavity called the **antrum** in the center of the secondary follicle. The innermost layer of granulosa cells becomes firmly attached to the zona pellucida and is now called the **corona radiata** (*corona*=crown; *radiata*=radiation) (Figure 27.14c, f).

The secondary follicle eventually becomes larger, turning into a **mature (graafian) follicle** (Figure 27.14d). While in this follicle, and just before ovulation, the diploid primary oocyte completes meiosis I, producing two haploid (*n*) cells of unequal size—each with 23 chromosomes (Figure 27.15). The smaller cell produced by meiosis I, called the **first polar body**, is essentially a packet of discarded nuclear material. The larger cell, known as the **secondary oocyte**, receives most of the cytoplasm. Once a secondary oocyte is formed, it begins meiosis II but then stops in metaphase. The mature (graafian) follicle soon ruptures and releases its secondary oocyte, a process known as **ovulation**.

At ovulation, the secondary oocyte is expelled into the pelvic cavity together with the first polar body and corona radiata. Normally these cells are swept into the uterine tube. If fertilization does not occur, the cells degenerate. If sperm are present in the uterine tube and one penetrates the secondary oocyte, however, meiosis II resumes. The secondary oocyte splits into two haploid cells, again of unequal size. The larger cell is the **ovum**, or mature egg; the smaller one is the **second polar body**. The nuclei of the sperm cell and the ovum then unite, forming a diploid **zygote**. If the first polar body undergoes another division to produce two polar bodies, then the primary oocyte ultimately gives rise to three haploid polar bodies, which all degenerate, and a single haploid ovum. Thus, one primary oocyte gives rise to a single gamete (an ovum). By contrast, recall that in males one primary spermatocyte produces four gametes (sperm).

Table 27.1 summarizes the events of oogenesis and follicular development.

Figure 27.15 Oogenesis. Diploid cells (*2n*) have 46 chromosomes; haploid cells (*n*) have 23 chromosomes.

In an oocyte, meiosis II is completed only if fertilization occurs.

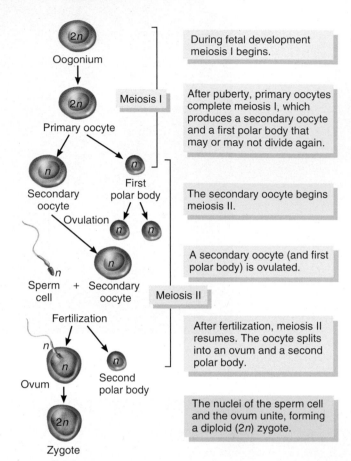

During fetal development meiosis I begins.

After puberty, primary oocytes complete meiosis I, which produces a secondary oocyte and a first polar body that may or may not divide again.

The secondary oocyte begins meiosis II.

A secondary oocyte (and first polar body) is ovulated.

After fertilization, meiosis II resumes. The oocyte splits into an ovum and a second polar body.

The nuclei of the sperm cell and the ovum unite, forming a diploid (*2n*) zygote.

? How does the age of a primary oocyte in a female compare with the age of a primary spermatocyte in a male?

◉ CHECKPOINT

13. How are the ovaries held in position in the pelvic cavity?
14. Describe the microscopic structure and functions of an ovary.
15. Describe the principal events of oogenesis.

Uterine Tubes

Females have two **uterine (fallopian) tubes**, or **oviducts**, that extend laterally from the uterus (Figure 27.16 on page 868). The tubes, which measure about 10 cm (4 in.) long and lie between the folds of the broad ligaments of the uterus, transport secondary oocytes and fertilized ova from the ovaries to the uterus. The funnel-shaped portion of each tube, called the **infundibulum**, is close to the ovary but is open to the peritoneal cavity. It ends in a fringe of fingerlike projections called **fimbriae** (FIM-brē-ē=fringe), one of which is attached to the lateral end of the ovary. From the infundibulum, the uterine tube extends medially and eventually inferiorly and attaches to the superior lateral angle of the uterus.

TABLE 27.1

Summary of Oogenesis and Follicular Development

Age	Oogenesis	Follicular development

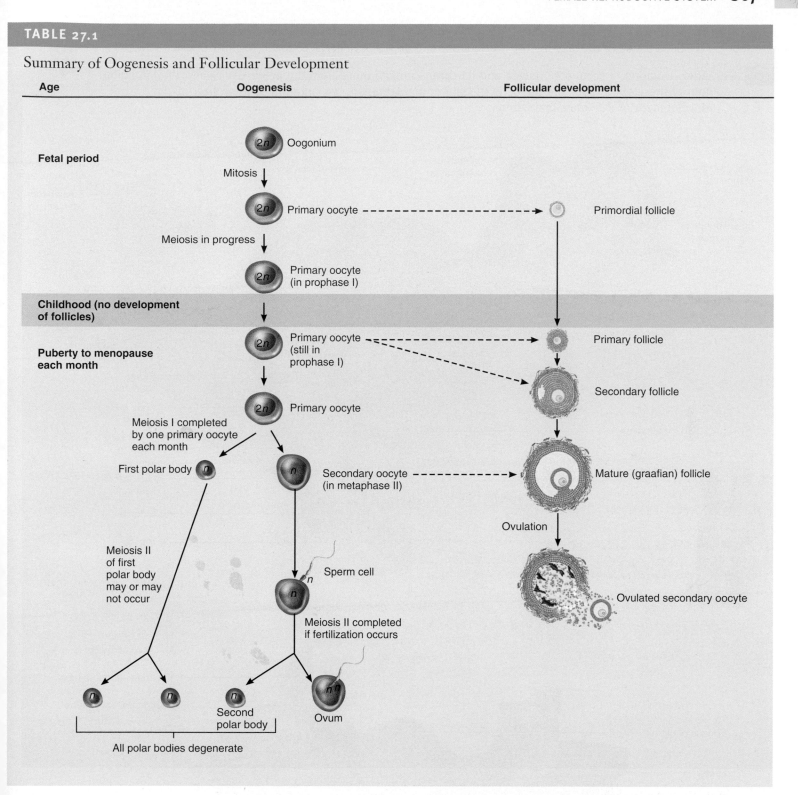

The **ampulla** of the uterine tube is the widest, longest portion, making up about the lateral two-thirds of its length. The **isthmus** of the uterine tube is the more medial, short, narrow, thick-walled portion that joins the uterus.

Histologically, the uterine tubes are composed of three layers: mucosa, muscularis, and serosa (Figure 27.17a on page 869). The mucosa consists of epithelium and lamina propria (areolar connective tissue). The epithelium contains ciliated simple columnar cells, which function as a ciliary "conveyor belt" to help move a fertilized ovum (or secondary oocyte) within the tube toward the uterus, and nonciliated (*peg*) cells, which have microvilli and

secrete a fluid that provides nutrition for the ovum (Figure 27.17b). The middle layer, the muscularis, is composed of an inner, thick, circular ring of smooth muscle and an outer, thin region of longitudinal smooth muscle. Peristaltic contractions of the muscularis and the ciliary action of the mucosa help move the oocyte or fertilized ovum toward the uterus. The outer layer of the uterine tubes is a serous membrane, the serosa.

Local currents produced by movements of the fimbriae, which surround the ovary during ovulation, sweep the ovulated secondary oocyte from the pelvic cavity into the uterine tube. A sperm cell usually encounters and fertilizes a secondary oocyte in

Figure 27.16 **Relationship of the uterine (fallopian) tubes to the ovaries, uterus, and associated structures.** In the left side of the drawing the uterine tube and uterus have been sectioned to show internal structures.

After ovulation, a secondary oocyte and its corona radiata move from the pelvic cavity into the infundibulum of the uterine tube. The uterus is the site of menstruation, implantation of a fertilized ovum, development of the fetus, and labor.

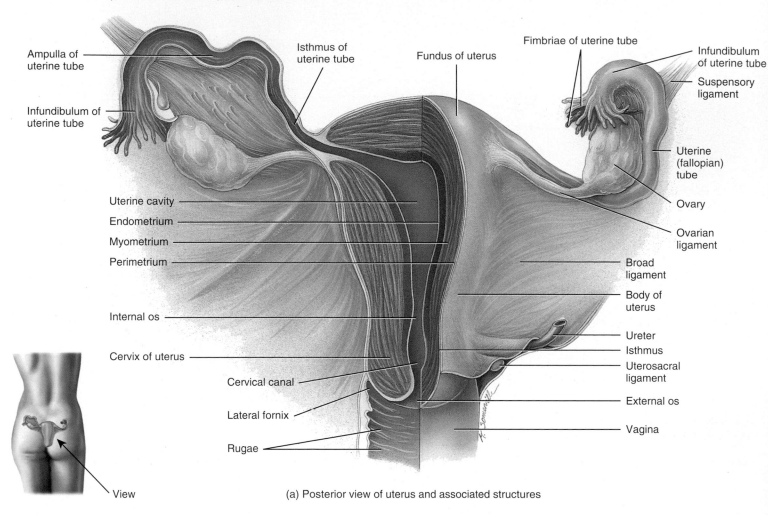

(a) Posterior view of uterus and associated structures

(b) Posterior view of uterus and associated structures

? **Where does fertilization usually occur?**

Figure 27.17 **Histology of the uterine (fallopian) tube.**

🔑 ⚏ Peristaltic contractions of the muscularis and ciliary action of the mucosa of the uterine tube help move the oocyte or fertilized ovum toward the uterus.

(a) Transverse section through uterine (fallopian) tube

(b) Details of epithelium

❓ **What types of cells line the uterine tubes?**

the ampulla of the uterine tube, although fertilization in the peritoneal cavity is not uncommon. Fertilization can occur at any time up to about 24 hours after ovulation. Some hours after fertilization, the nuclear materials of the haploid ovum and sperm unite. The diploid fertilized ovum is now called a **zygote** and begins to undergo cell divisions while moving toward the uterus. It arrives at the uterus 6 to 7 days after ovulation.

The uterine tubes are supplied by branches of the uterine (see Figure 27.19) and ovarian (see Figure 15.6b on page 494) arteries. Venous return is via the uterine veins.

The uterine tubes are supplied with sympathetic and parasympathetic nerve fibers from the hypogastric plexus and the pelvic splanchnic nerves. The fibers are distributed to the muscular coat of the tubes and their blood vessels.

Uterus

The **uterus** (womb) serves as part of the pathway for sperm deposited in the vagina to reach the uterine tubes (see Figure 27.16). It is also the site of implantation of a fertilized ovum, development of the fetus during pregnancy, and labor. During reproductive cycles when implantation does not occur, the uterus is the source of menstrual flow.

Situated between the urinary bladder and the rectum, the uterus is the size and shape of an inverted pear (see Figure 27.11).

In females who have never been pregnant, it is about 7.5 cm (3 in.) long, 5 cm (2 in.) wide, and 2.5 cm (1 in.) thick. The uterus is larger in females who have recently been pregnant, and smaller (atrophied) when sex hormone levels are low, as occurs after menopause.

Anatomical subdivisions of the uterus include the following (see Figure 27.16): (1) a dome-shaped portion superior to the uterine tubes called the **fundus**, (2) a tapering central portion called the **body**, and (3) an inferior narrow portion called the **cervix** that opens into the vagina. Between the body of the uterus and the cervix is the **isthmus** (IS-mus), a constricted region about 1 cm (0.5 in.) long. The interior of the body of the uterus is called the **uterine cavity**, and the interior of the narrow cervix is called the **cervical canal**. The cervical canal opens into the uterine cavity at the **internal os** (*os*=mouthlike opening) and into the vagina at the **external os**.

Normally, the body of the uterus projects anteriorly and superiorly over the urinary bladder in a position called **anteflexion**. The cervix projects inferiorly and posteriorly and enters the anterior wall of the vagina at nearly a right angle (see Figure 27.11). Several ligaments that are either extensions of the parietal peritoneum or fibromuscular cords maintain the position of the uterus (see Figure 27.12). The paired **broad ligaments** are double folds of peritoneum attaching the uterus to either side of the pelvic cavity. The paired **uterosacral ligaments**, also peritoneal extensions, lie on either side of the rectum and connect the uterus to the sacrum. The **cardinal (lateral cervical) ligaments** are located inferior to the bases of the broad ligaments and extend from the pelvic wall to the cervix and vagina. The **round ligaments** are bands of fibrous connective tissue between the layers of the broad ligament; they extend from a point on the uterus just inferior to the uterine tubes to a portion of the labia majora of the external genitalia. Although the ligaments normally maintain the anteflexed position of the uterus, they also allow the uterine body enough movement that the uterus may become malpositioned. A posterior tilting of the uterus, called **retroflexion** (*retro-*=backward or behind), is a harmless variation of the normal position of the uterus. There is often no cause for the condition, but it may occur after childbirth or because of an ovarian cyst.

Histologically, the uterus consists of three layers of tissue: the perimetrium, myometrium, and endometrium (Figure 27.18). The outer layer—the **perimetrium** (*peri-*=around; *-metrium*=uterus) or serosa—is part of the parietal peritoneum; it is composed of simple squamous epithelium and a thin layer of areolar connec-

tive tissue. Laterally, it becomes the broad ligament. Anteriorly, it covers the urinary bladder and forms a shallow pouch, the **vesicouterine pouch** (ves'-i-kō-Ū-ter-in; *vesico-*=bladder; see Figure 27.11) between the urinary bladder and the uterus. Posteriorly, it covers the rectum and forms a deep pouch between the uterus and the rectum, the **rectouterine pouch** (rek-tō-Ū-ter-in; *recto-*=rectum) or *pouch of Douglas*—the most inferior point in the peritoneal cavity.

The middle layer of the uterus, the **myometrium** (*myo-*=muscle), consists of three layers of smooth muscle fibers that are thickest in the fundus and thinnest in the cervix. The thicker middle layer is circular; the inner and outer layers are longitudinal or oblique. During labor and childbirth, coordinated contractions of the myometrium in response to stimulation by the hormone oxytocin from the posterior pituitary help expel the fetus from the uterus.

The inner layer of the uterus, the **endometrium** (*endo-*=within), is highly vascularized and has three components: (1) An innermost layer composed of simple columnar epithelium (ciliated and

Figure 27.18 Histology of the uterus.

🔓 The three layers of the uterus from superficial to deep are the perimetrium (serosa), the myometrium, and the endometrium.

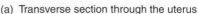

(a) Transverse section through the uterus

(b) Details of endometrium

❓ **What structural features of the endometrium and myometrium contribute to their functions?**

secretory cells) lines the lumen. (2) An underlying endometrial stroma is a very thick region of lamina propria (areolar connective tissue). (3) Endometrial (uterine) glands develop as invaginations of the luminal epithelium and extend almost to the myometrium. The endometrium is divided into two layers. The **stratum functionalis** (=functional layer) lines the uterine cavity and sloughs off during menstruation as a result of declining levels of progesterone from the ovaries. The deeper layer, the **stratum basalis** (=basal layer), is permanent and gives rise to a new stratum functionalis after each menstruation.

Branches of the internal iliac artery called **uterine arteries** (Figure 27.19) supply blood to the uterus. Uterine arteries give off branches called **arcuate arteries** (=shaped like a bow) that are arranged in a circular fashion in the myometrium. These arteries branch into **radial arteries** that penetrate deeply into the myometrium. Just before the branches enter the endometrium, they divide into two kinds of arterioles: **Straight arterioles** supply the stratum basalis with the materials needed to regenerate the stratum functionalis; **spiral arterioles** supply the stratum functionalis and change markedly during the menstrual cycle. Blood leaving the uterus is drained by the **uterine veins** into the internal iliac veins. The extensive blood supply of the uterus is essential to support regrowth of a new stratum functionalis after menstruation, implantation of a fertilized ovum, and development of the placenta.

The secretory cells of the mucosa of the cervix produce a secretion called **cervical mucus**, a mixture of water, glycoproteins, lipids, enzymes, and inorganic salts. During their reproductive years, females secrete 20–60 mL of cervical mucus per day. Cervical mucus is more hospitable to sperm at or near the time of ovulation because it is then less viscous and more alkaline (pH 8.5). At other times, a more viscous mucus forms a cervical plug that physically impedes sperm penetration. Cervical mucus supplements the energy needs of sperm, and both the cervix and cervical mucus protect sperm from phagocytes and the hostile environment of the vagina and uterus. They may also play a role in **capacitation**—a functional change that sperm undergo in the female reproductive tract before they are able to fertilize a secondary oocyte. Capacitation causes a sperm cell's tail to beat even more vigorously and it prepares the sperm cell's plasma membrane to fuse with the oocyte's plasma membrane.

◉ **CHECKPOINT**

16. Where are the uterine tubes located and what is their function?
17. What are the principal parts of the uterus? Where are they located in relation to one another?
18. Describe the arrangement of ligaments that hold the uterus in its normal position.
19. Describe the histology of the uterus.
20. Why is an abundant blood supply important to the uterus?

Vagina

The **vagina** (=sheath) is a tubular, 10-cm (4-in.)-long fibromuscular canal lined with mucous membrane that extends from the exterior of the body to the uterine cervix (see Figures 27.11 and 27.16). It is the receptacle for the penis during sexual intercourse, the outlet for menstrual flow, and the passageway for childbirth.

Figure 27.19 Blood supply of the uterus. The inset shows histological details of the blood vessels of the endometrium.

🔑 Straight arterioles supply the materials needed for regeneration of the stratum functionalis after menstruation.

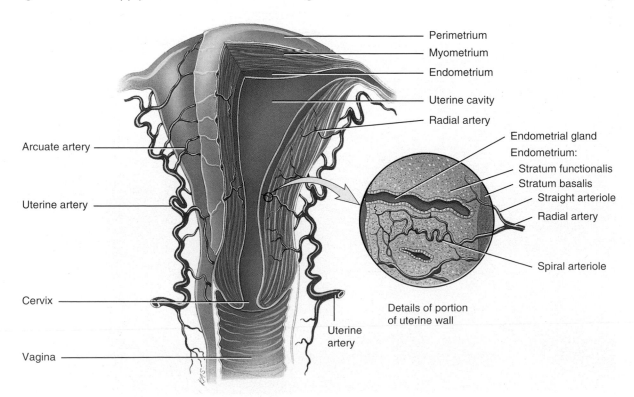

Details of portion of uterine wall

Anterior view with left side of uterus partially sectioned

❓ **What is the functional significance of the stratum basalis layer of the endometrium?**

Situated between the urinary bladder and the rectum, the vagina is directed superiorly and posteriorly, where it attaches to the uterus. A recess called the **fornix** (=arch or vault) surrounds the vaginal attachment to the cervix (see Figure 27.16a). When properly inserted, a contraceptive diaphragm rests in the fornix, where it is held in place as it covers the cervix.

Figure 27.20 Histology of the vagina.

The muscularis of the vagina can stretch considerably to accommodate the penis during sexual intercourse and an infant during birth.

(a) Transverse section through the vaginal wall

(b) Details of mucosa

? What are the functions of the vagina?

The mucosa of the vagina is continuous with that of the uterus. Histologically, it consists of nonkeratinized stratified squamous epithelium and lamina propria (areolar connective tissue) (Figure 27.20b on page 872) that lies in a series of transverse folds called **rugae** (ROO-gē; see Figure 27.16). Dendritic cells in the mucosa are antigen-presenting cells. Unfortunately, they also participate in the transmission of viruses—for example, HIV (the virus that causes AIDS)—to a female during intercourse with an infected male. The mucosa of the vagina contains large stores of glycogen, the decomposition of which produces organic acids. The resulting acidic environment retards microbial growth, but it is also harmful to sperm. Alkaline components of semen, mainly from the seminal vesicles, raise the pH of fluid in the vagina and increase viability of the sperm.

The muscularis is composed of an outer longitudinal layer and an inner circular layer of smooth muscle (Figure 27.20a) that can stretch considerably to accommodate the penis during sexual intercourse and an infant during birth.

The adventitia, the superficial layer of the vagina, consists of areolar connective tissue (Figure 27.20a). It anchors the vagina to adjacent organs such as the urethra and urinary bladder anteriorly, and the rectum and anal canal posteriorly.

A thin fold of vascularized mucous membrane, called the **hymen** (=membrane), forms a border around and partially closes the inferior end of the vaginal opening to the exterior, the **vaginal orifice** (see Figure 27.21). After its rupture, usually following the first sexual intercourse, only remnants of the hymen remain.

Sometimes the hymen completely covers the orifice, a condition called **imperforate hymen** (im-PER-fō-rāt). Surgery may be needed to open the orifice and permit the discharge of menstrual flow.

Vulva

The term **vulva** (VUL-va=to wrap around), or **pudendum** (pū-DEN-dum), refers to the external genitals of the female (Figure 27.21). The following are the components of the vulva:

- Anterior to the vaginal and urethral openings is the **mons pubis** (MONZ PŪ-bis; *mons*=mountain), an elevation of adipose tissue covered by skin and coarse pubic hair that cushions the pubic symphysis.
- From the mons pubis, two longitudinal folds of skin, the **labia majora** (LĀ-bē-a ma-JŌ-ra; *labia*=lips; *majora*=larger), extend inferiorly and posteriorly. The singular term is *labium majus*. The labia majora are covered by pubic hair and contain an abundance of adipose tissue, sebaceous (oil) glands, and apocrine sudoriferous (sweat) glands. They are homologous to the scrotum.
- Medial to the labia majora are two smaller folds of skin called the **labia minora** (mī-NŌ-ra; *minora*=smaller). The singular term is *labium minus*. Unlike the labia majora, the labia minora are devoid of pubic hair and fat and have few sudoriferous glands, but they do contain many sebaceous glands. The labia minora are homologous to the spongy (penile) urethra.

Figure 27.21 Components of the vulva (pudendum). (See Tortora, *A Photographic Atlas of the Human Body*, 2e, Figure 14.7.)

The vulva refers to the external genitals of the female.

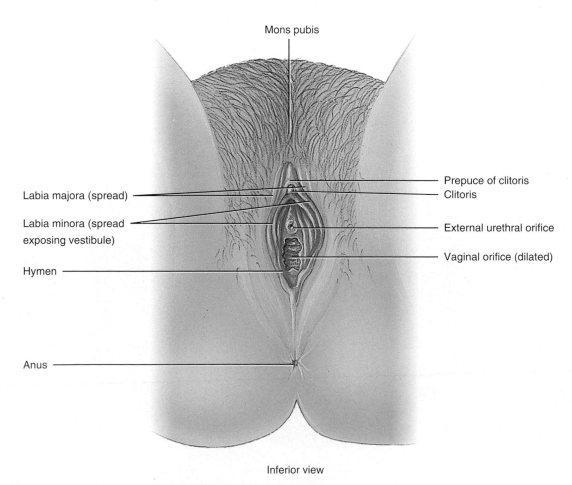

Mons pubis

Prepuce of clitoris
Clitoris

Labia majora (spread)

Labia minora (spread exposing vestibule)

External urethral orifice

Vaginal orifice (dilated)

Hymen

Anus

Inferior view

? What surface structures are anterior to the vaginal opening? Lateral to it?

- The **clitoris** (KLI-to-ris) is a small cylindrical mass of erectile tissue and nerves located at the anterior junction of the labia minora. A layer of skin called the **prepuce of the clitoris** is formed at the point where the labia minora unite and covers the body of the clitoris. The body of the clitoris has two bodies of erectile tissue, the corpora cavernosa. Like the similar bodies in the male penis, the corpora cavernosa bend posteriorly to attach to the rami of the pubis and ischium as the crura of the clitoris. The exposed portion of the clitoris is the **glans clitoris.** The glans clitoris is homologous to the glans penis in males. Like the male structure, the clitoris is capable of enlargement upon tactile stimulation and has a role in sexual excitement in the female.

- The region between the labia minora is the **vestibule**. Within the vestibule are the hymen (if still present), the vaginal orifice, the external urethral orifice, and the openings of the ducts of several glands. The vestibule is homologous to the membranous urethra of males. The **vaginal orifice**, the opening of the vagina to the exterior, occupies the greater portion of the vestibule and is bordered by the hymen. Anterior to the vaginal

Figure 27.22 Perineum of a female. (Figure 11.15 on page 352 shows the perineum of a male.)

The perineum is a diamond-shaped area medial to the thighs and buttocks that contains the external genitals and anus.

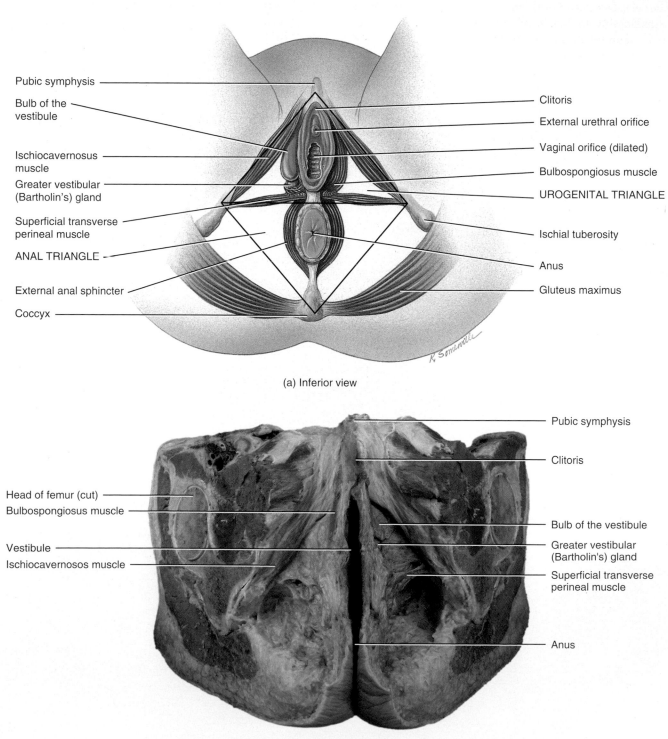

(a) Inferior view

(b) Inferior view

? Why is the anterior portion of the perineum called the urogenital triangle?

orifice and posterior to the clitoris is the **external urethral orifice**, the opening of the urethra to the exterior. On either side of the external urethral orifice are the openings of the ducts of the **paraurethral (Skene's) glands.** These mucus-secreting glands are embedded in the wall of the urethra. The paraurethral glands are homologous to the prostate. On either side of the vaginal orifice itself are the **greater vestibular (Bartholin's) glands** (see Figure 27.22), which open by ducts into a groove between the hymen and labia minora. They produce a small quantity of mucus during sexual arousal and intercourse that adds to cervical mucus and provides lubrication. The greater vestibular glands are homologous to the bulbourethral glands in males. Several **lesser vestibular glands** also open into the vestibule.

- The **bulb of the vestibule** (see Figure 27.22) consists of two elongated masses of erectile tissue just deep to the labia on either side of the vaginal orifice. The bulb of the vestibule becomes engorged with blood during sexual arousal, narrowing the vaginal orifice and placing pressure on the penis during intercourse. The bulb of the vestibule is homologous to the corpus spongiosum and bulb of the penis in males.

Table 27.2 summarizes the homologous structures of the female and male reproductive systems.

Perineum

The **perineum** (per'-i-NĒ-um) is the diamond-shaped area medial to the thighs and buttocks of both males and females (Figure 27.22 on page 874). It contains the external genitals and anus. The perineum is bounded anteriorly by the pubic symphysis, laterally by the ischial tuberosities, and posteriorly by the coccyx. A

TABLE 27.2	
Summary of Homologous Structures of the Female and Male Reproductive Systems	
FEMALE STRUCTURES	**MALE STRUCTURES**
Ovaries	Testes
Ovum	Sperm cell
Labia majora	Scrotum
Labia minora	Spongy (penile) urethra
Vestibule	Membranous urethra
Bulb of vestibule	Corpus spongiosum penis and bulb of penis
Clitoris	Glans penis and corpora cavernosa
Paraurethral glands	Prostate
Greater vestibular glands	Bulbourethral (Cowper's) glands

transverse line drawn between the ischial tuberosities divides the perineum into an anterior **urogenital triangle** (ū'-rō-JEN-i-tal) that contains the external genitalia and a posterior **anal triangle** that contains the anus.

Mammary Glands

Each **breast** is a hemispheric projection of variable size anterior to the pectoralis major and serratus anterior muscles and attached to them by a layer of fascia composed of dense irregular connective tissue (Figure 27.23).

Figure 27.23 Mammary glands.

The mammary glands function in the synthesis, secretion, and ejection of milk (lactation).

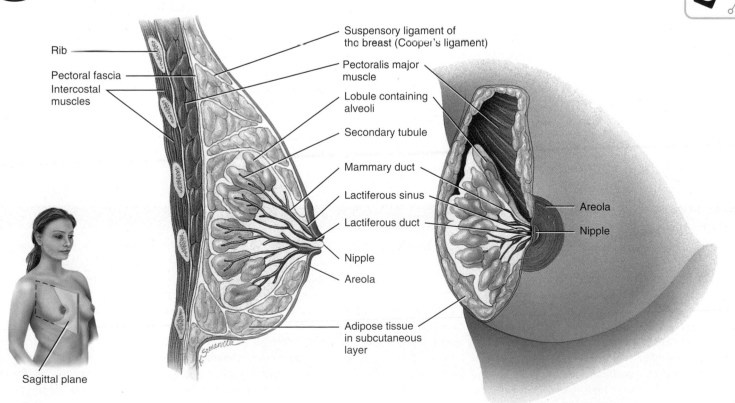

Rib

Pectoral fascia

Intercostal muscles

Suspensory ligament of the breast (Cooper's ligament)

Pectoralis major muscle

Lobule containing alveoli

Secondary tubule

Mammary duct

Lactiferous sinus

Lactiferous duct

Nipple

Areola

Adipose tissue in subcutaneous layer

Areola

Nipple

Sagittal plane

(a) Sagittal section

(b) Anterior view, partially sectioned

CONTINUES ▶

FIGURE 27.23 CONTINUED ▶

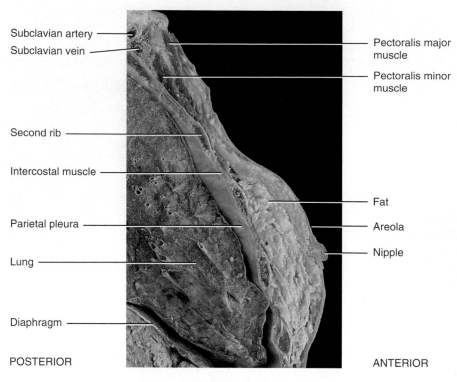

Subclavian artery

Subclavian vein

Second rib

Intercostal muscle

Parietal pleura

Lung

Diaphragm

POSTERIOR

Pectoralis major muscle

Pectoralis minor muscle

Fat

Areola

Nipple

ANTERIOR

(c) Sagittal section

? **What hormones regulate the synthesis and ejection of milk?**

Each breast has one pigmented projection, the **nipple**, that has a series of closely spaced openings of ducts called **lactiferous ducts**, where milk emerges. The circular pigmented area of skin surrounding the nipple is called the **areola** (a-RĒ-ō-la=small space); it appears rough because it contains modified sebaceous (oil) glands. Strands of connective tissue called the **suspensory ligaments of the breast (Cooper's ligaments)** run between the skin and fascia and support the breast. These ligaments become looser with age or with excessive strain, as occurs in long-term jogging or high-impact aerobics. Wearing a supportive bra can slow this process and help maintain the integrity of the suspensory ligaments.

Within each breast is a **mammary gland**, a modified sudoriferous (sweat) gland that produces milk (Figure 27.23). A mammary gland consists of 15 to 20 lobes, or compartments, separated by a variable amount of adipose tissue. In each lobe are several smaller compartments called **lobules**, composed of grapelike clusters of milk-secreting glands termed **alveoli** (=small cavities)

embedded in connective tissue (Figure 27.24). Contraction of **myoepithelial cells** surrounding the alveoli helps propel milk toward the nipples. When milk is being produced, it passes from the alveoli into a series of **secondary tubules** and then into the **mammary ducts**. Near the nipple, the mammary ducts expand to form sinuses called **lactiferous sinuses** (*lact-*=milk), where some milk may be stored before draining into a lactiferous duct. Each lactiferous duct typically carries milk from one of the lobes to the exterior.

The essential functions of the mammary glands are the synthesis, secretion, and ejection of milk; these functions, called **lactation**, are associated with pregnancy and childbirth. Milk production is stimulated largely by the hormone prolactin from the anterior pituitary, with contributions from progesterone and estrogens. The ejection of milk is stimulated by oxytocin, which is released from the posterior pituitary in response to the sucking of an infant on the mother's nipple (suckling).

•CLINICAL CONNECTION | Breast Cancer

One in eight women in the United States faces the prospect of **breast cancer**. After lung cancer, it is the second-leading cause of death from cancer in U.S. women. Breast cancer can occur in males but is rare. In females, breast cancer is seldom seen before age 30; its incidence rises rapidly after menopause. An estimated 5% of the 180,000 cases diagnosed each year in the United States, particularly those that arise in younger women, stem from inherited genetic mutations (changes in the DNA). Researchers have now identified two genes that increase susceptibility to breast cancer: *BRCA1 (breast cancer 1)* and *BRCA2*. Mutation of *BRCA1* also confers a high risk for ovarian cancer. In addition, mutations of the *p53* gene increase the risk of breast cancer in both

males and females, and mutations of the androgen receptor gene are associated with the occurrence of breast cancer in some males. Because breast cancer generally is not painful until it becomes quite advanced, any lump, no matter how small, should be reported to a physician at once. Early detection—by breast self-examination and mammograms—is the best way to increase the chance of survival.

The most effective technique for detecting tumors less than 1 cm (0.4 in.) in diameter is **mammography** (mam-OG-ra-fē; *-graphy*=to record), a type of radiography using very sensitive x-ray film. The image of the breast, called a **mammogram** (see also Table 1.3 on page 20), is best obtained by compressing the breasts, one at a time, using flat

Figure 27.24 **Histology of the mammary glands.**

Contraction of myoepithelial cells helps propel milk toward the nipples.

(a) Section of a nonlactating (inactive) mammary gland

LM 30x

Alveoli

Connective tissue

Lobule

Adipose tissue

(b) Section of a lactating (active) mammary gland

LM 50x

Alveoli

Lobule

Duct

Connective tissue

? **In which portion of the mammary gland are alveoli located?**

plates. A supplementary procedure for evaluating breast abnormalities is **ultrasound**. Although ultrasound cannot detect tumors smaller than 1 cm in diameter, it can be used to determine whether a lump is a benign, fluid-filled cyst or a solid (and therefore possibly malignant) tumor.

Among the factors that increase the risk of developing breast cancer are (1) a family history of breast cancer, especially in a mother or sister; (2) nulliparity (never having borne a child) or having a first child after age 35; (3) previous cancer in one breast; (4) exposure to ionizing radiation, such as x-rays; (5) excessive alcohol intake; and (6) cigarette smoking.

The American Cancer Society recommends the following steps to help diagnose breast cancer as early as possible:

- All women over 20 should develop the habit of monthly breast self-examination.
- A physician should examine the breasts every 3 years when a woman is between the ages of 20 and 40, and every year after age 40.
- A mammogram should be taken in women between the ages of 35 and 39, to be used later for comparison (baseline mammogram).
- Women with no symptoms should have a mammogram every year after age 40.
- Women of any age with a history of breast cancer, a strong family history of the disease, or other risk factors should consult a physician to determine a schedule for mammography.

(continues)

Treatment for breast cancer may involve hormone therapy, chemotherapy, radiation therapy, **lumpectomy** (removal of the tumor and the immediate surrounding tissue), a modified or radical mastectomy, or a combination of these approaches. A **radical mastectomy** (*mast-* =breast) involves removal of the affected breast along with the underlying pectoral muscles and the axillary lymph nodes. (Lymph nodes are removed because metastasis of cancerous cells usually occurs through lymphatic or blood vessels). Radiation treatment and chemotherapy may follow the surgery to ensure the destruction of any stray cancer cells. Several types of chemotherapeutic drugs are used to decrease the risk of relapse or disease progression. *Nolvadex®* (tamoxifen) is an antagonist to estrogens that binds to and blocks receptors for estrogens, thus decreasing the stimulating effect of estrogens on

Mammogram showing breast cancer (white spot)

breast cancer cells. Tamoxifen has been used for 20 years, and greatly reduces the risk of cancer recurrence. *Herceptin,®* a monoclonal antibody drug, targets an antigen on the surface of breast cancer cells. It is effective in causing regression of tumors and retarding progression of the disease. The early data from clinical trials of two new drugs, *Femara®* and *Amimidex,®* show relapse rates that are lower than those for tamoxifen. These drugs are inhibitors of aromatase, the enzyme needed for the final step in synthesis of estrogens. Finally, two drugs—tamoxifen and *Evista (raloxifene)*—are being marketed for breast cancer *prevention.* Interestingly, raloxifene blocks estrogen receptors in the breasts and uterus but activates estrogen receptors in bone. Thus, it can be used to treat osteoporosis without increasing a woman's risk of breast or endometrial (uterine) cancer. ●

● CHECKPOINT

21. What is the function of the vagina?
22. Describe the histology of the vagina.
23. List the parts of the vulva, and explain the functions of each.
24. Describe the structure of the mammary glands and explain how they are supported.
25. What is the route of milk from the alveoli of the mammary gland to the nipple?

FEMALE REPRODUCTIVE CYCLE

● OBJECTIVE

• Describe the major events of the female reproductive cycle.

During their reproductive years, nonpregnant females normally exhibit cyclical changes in the ovaries and uterus. Each cycle takes about a month and involves both oogenesis and preparation of the uterus to receive a fertilized ovum. Hormones secreted by the hypothalamus, anterior pituitary, and ovaries control the main events. The **ovarian cycle** is a series of events in the ovaries that occur during and after the maturation of an oocyte. Steroid hormones released by the ovaries control the **uterine (menstrual) cycle**, a concurrent series of changes in the endometrium of the uterus to prepare it for the arrival and development of a fertilized ovum. If fertilization does not occur, levels of ovarian hormones decrease, which causes the stratum functionalis of the endometrium to slough off. The general term **female reproductive cycle** encompasses the ovarian and uterine cycles, the hormonal changes that regulate them, and the related cyclical changes in the breasts and cervix.

Gonadotropin-releasing hormone (GnRH) secreted by the hypothalamus controls the events of the female reproductive cycle (Figure 27.25). GnRH stimulates the release of **follicle-stimulating hormone (FSH)** and **luteinizing hormone (LH)** from the anterior pituitary. FSH in turn initiates follicular growth and the secretion of estrogens by the growing ovarian follicles. LH stimulates the further development of follicles and their full secretion of estrogens. At midcycle, LH triggers ovulation and then promotes formation of the corpus luteum (the reason for the name luteinizing hormone). Stimulated by LH, the corpus luteum produces and secretes estrogens, progesterone, relaxin, and inhibin.

Figure 27.25 The female reproductive cycle. Events in the ovarian and uterine cycles and the release of anterior pituitary gland hormones are correlated with the sequence of the cycle's four phases. In the cycle shown, fertilization and implantation have not occurred.

The length of the female reproductive cycle typically is 24–36 days; the preovulatory phase is more variable in length than the other phases.

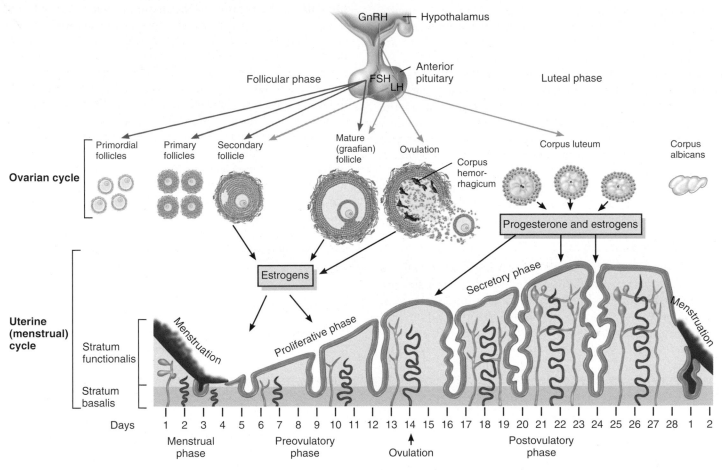

Hormonal regulation of changes in the ovary and uterus

? Which hormones are responsible for the proliferative phase of endometrial growth, for ovulation, for growth of the corpus luteum, and for the surge of LH at midcycle?

The duration of the female reproductive cycle typically ranges from 24 to 35 days. For this discussion, we assume a duration of 28 days and divide it into four phases: the menstrual phase, the preovulatory phase, ovulation, and the postovulatory phase (see Figure 27.25).

Menstrual Phase

The **menstrual phase** (MEN-stroo-al), also called **menstruation** (men′-stroo-Ā-shun) or **menses** (=month), lasts for roughly the first 5 days of the cycle. (By convention, the first day of menstruation is day one of a new cycle.)

Events in the Ovaries

Under the influence of FSH, several primordial follicles develop into primary follicles and then into secondary follicles. Follicular fluid, secreted by the granulosa cells and filtered from blood in the capillaries of the theca folliculi, accumulates in the enlarging antrum (space that forms within the follicle) while the oocyte remains near the edge of the follicle (see Figure 27.14b).

Events in the Uterus

Menstrual flow from the uterus consists of 50–150 mL of blood, tissue fluid, mucus, and epithelial cells shed from the endometrium. This discharge occurs because the declining level of ovarian progesterone and estrogens stimulates release of prostaglandins that cause the uterine spiral arterioles to constrict. As a result, the cells they supply become oxygen-deprived and start to die. Eventually, the entire stratum functionalis sloughs off. At this time the endometrium is very thin, about 2–5 mm, because only the stratum basalis remains. The menstrual flow passes from the uterine cavity through the cervix and vagina to the exterior.

Preovulatory Phase

The **preovulatory phase** is the time between the end of menstruation and ovulation. The preovulatory phase of the cycle is more variable in length than the other phases and accounts for most of the difference when cycles are shorter or longer than 28 days. It lasts from days 6 to 13 in a 28-day cycle.

Events in the Ovaries

Some of the secondary follicles in the ovaries begin to secrete estrogens and inhibin. By about day 6, a single secondary follicle in one of the two ovaries has outgrown all the others to become the **dominant follicle**. Estrogens and inhibin secreted by the dominant follicle decrease the secretion of FSH, which causes other, less well-developed follicles to stop growing and undergo atresia. Fraternal (nonidentical) twins or triplets result when two or three secondary follicles become codominant and later are ovulated and fertilized at about the same time.

Normally, the one dominant secondary follicle becomes the **mature (graafian) follicle**, which continues to enlarge until it is more than 20 mm in diameter and ready for ovulation (see Figure 27.13a). This follicle forms a blisterlike bulge due to the swelling antrum on the surface of the ovary. During the final maturation process, the mature follicle continues to increase its production of estrogens.

With reference to the ovarian cycle, the menstrual and pre-ovulatory phases together are termed the **follicular phase** (fō-LIK-ū-lar) because ovarian follicles are growing and developing.

Events in the Uterus

Estrogens liberated into the blood by growing ovarian follicles stimulate the repair of the endometrium; cells of the stratum basalis undergo mitosis and produce a new stratum functionalis. As the endometrium thickens, the short, straight endometrial glands develop, and the arterioles coil and lengthen as they penetrate the stratum functionalis. The thickness of the endometrium approximately doubles, to about 4–10 mm. The preovulatory phase is also referred to as the **proliferative phase** of the uterine cycle because the endometrium is proliferating.

Ovulation

Ovulation, the rupture of the mature (graafian) follicle and the release of the secondary oocyte into the pelvic cavity, usually occurs on day 14 in a 28-day cycle. During ovulation, the secondary oocyte remains surrounded by its zona pellucida and corona radiata. Development of a secondary follicle into a fully mature follicle generally takes a total of about 20 days (spanning the last 6 days of the previous cycle and the first 14 days of the current cycle). During this time the primary oocyte completes meiosis I to become a secondary oocyte; the secondary oocyte then begins meiosis II but halts in metaphase until it is fertilized. From time to time, an oocyte is lost into the pelvic cavity, where it later disintegrates. The small amount of blood that sometimes leaks into the peritoneal cavity from the ruptured follicle can cause pain, known as **mittelschmerz** (MIT-el-shmārts=pain in the middle), at the time of ovulation. An over-the-counter home test that detects a rising level of LH can be used to predict ovulation a day in advance.

Postovulatory Phase

The **postovulatory phase** of the female reproductive cycle is the time between ovulation and onset of the next menses. In duration,

---

• CLINICAL **CONNECTION** | Sexually Transmitted Diseases

A **sexually transmitted disease (STD)** is one that is spread by sexual contact. In most developed countries of the world, such as those of Western Europe, Japan, Australia, and New Zealand, the incidence of STDs has declined markedly during the past 25 years. In the United States, by contrast, STDs have been rising to near-epidemic proportions; they currently affect more than 65 million people. AIDS and hepatitis B, which are sexually transmitted diseases that also may be contracted in other ways, are discussed in Chapters 16 and 25, respectively.

Chlamydia

Chlamydia (kla-MID-ē-a) is a sexually transmitted disease caused by the bacterium *Chlamydia trachomatis* (*chlamy-*=cloak). (Figure A) This unusual bacterium cannot reproduce outside body cells; it "cloaks" it-self inside cells, where it divides. At present, chlamydia is the most prevalent sexually transmitted disease in the United States. In most cases, the initial infection is asymptomatic and thus difficult to recognize clinically. In males, urethritis is the principal result, causing a clear discharge, burning on urination, frequent urination, and painful urination. Without treatment, the epididymides may also become inflamed, leading to sterility. In 70% of females with chlamydia, symptoms are absent, but chlamydia is the leading cause of pelvic inflammatory disease. Moreover, the uterine tubes may also become inflamed, which increases the risk of ectopic pregnancy (implantation of a fertilized ovum outside the uterus) and infertility due to the formation of scar tissue in the tubes.

Gonorrhea

Gonorrhea (gon-ō-RĒ-a) (Figure B) or **"the clap"** is caused by the bacterium *Neisseria gonorrhoeae*. In the United States, 1–2 million new cases of gonorrhea appear each year, most among individuals aged 15–29 years. Discharges from infected mucous membranes are the source of transmission of the bacteria either during sexual contact or during the passage of a newborn through the birth canal. The infection site can be in the mouth and throat after oral–genital contact, in the vagina and penis after genital intercourse, or in the rectum after recto–genital contact.

Males usually experience urethritis with profuse pus drainage and painful urination. The prostate and epididymis may also become infected. In females, infection typically occurs in the vagina, often with a discharge of pus. Both infected males and females may harbor the disease without any symptoms, however, until it has progressed to a more advanced stage; about 5–10% of males and 50% of females are asymptomatic. In females, the infection and consequent inflammation can proceed from the vagina into the uterus, uterine tubes, and pelvic

Normal cell

Chlamydia-infected cell

LM

(A) Cervical smear showing normal and *Chlamydia*-infected cells it is the most constant part of the female reproductive cycle. It lasts for 14 days in a 28-day cycle, from day 15 to day 28 (see Figure 27.25).

Events in the Ovaries

After ovulation, the mature follicle collapses, and the basement membrane between the granulosa cells and theca interna breaks down. Once a blood clot forms from minor bleeding of the ruptured follicle, the follicle becomes the **corpus hemorrhagicum** (*hemo-*=blood; *rrhagic-*=bursting forth) (see Figure 27.13a). Theca interna cells mix with the granulosa cells as they all become transformed into **corpus luteum** cells under the influence of LH. Stimulated by LH, the corpus luteum secretes progesterone, estrogens, relaxin, and inhibin. The luteal cells also absorb the blood clot. This phase is also called the **luteal phase** of the ovarian cycle.

Later events in an ovary that has ovulated an oocyte depend on whether the oocyte is fertilized. If the oocyte is *not fertilized*, the corpus luteum has a life span of only 2 weeks. At the end of this time period, its secretory activity declines, and it degenerates into a corpus albicans (see Figure 27.13a). As the levels of progesterone, estrogens, and inhibin decrease, release of GnRH, FSH, and LH rise due to loss of negative feedback suppression by the ovarian hormones. Follicular growth resumes, and a new ovarian cycle begins.

If the secondary oocyte *is fertilized* and begins to divide, the corpus luteum persists past its normal 2-week life span. It is "rescued" from degeneration by **human chorionic gonadotropin**

(kō-rē-ON-ik) (**hCG**). This hormone is produced by the chorion of the embryo beginning about 8 days after fertilization. Like LH, hCG stimulates the secretory activity of the corpus luteum. The presence of hCG in maternal blood or urine is an indicator of pregnancy and is the hormone detected by home pregnancy tests.

Events in the Uterus

Progesterone and estrogens produced by the corpus luteum promote growth and coiling of the endometrial glands, vascularization of the superficial endometrium, and thickening of the endometrium to 12–18 mm (0.48–0.72 in.). Because of the secretory activity of the endometrial glands, which begin to secrete glycogen, this period is called the **secretory phase** of the uterine cycle. These preparatory changes peak about one week after ovulation, at the time a fertilized ovum might arrive in the uterus. If fertilization does not occur, the levels of progesterone and estrogens decline due to degeneration of the corpus luteum. Withdrawal of progesterone and estrogens causes menstruation.

◉ CHECKPOINT

26. Describe the function of each of the following hormones in the female reproductive cycle: GnRH, FSH, LH, estrogens, progesterone, and inhibin.
27. Briefly outline the major events of each phase of the female reproductive cycle, and correlate them with the events of the ovarian and uterine cycles.

LM 600x

(B) *Neisseria gonorrhoeae* bacteria (purple cells) in a vaginal smear

cavity. An estimated 50,000 to 80,000 women in the United States are made infertile by gonorrhea every year as a result of scar tissue formation that closes the uterine tubes. If bacteria in the birth canal are transmitted to the eyes of a newborn, blindness can result. Administration of a 1% silver nitrate solution in the infant's eyes prevents infection.

Syphilis

Syphilis, caused by the bacterium *Treponema pallidum* (Figure C), is transmitted through sexual contact or exchange of blood, or through the placenta to a fetus. The disease progresses through several stages.

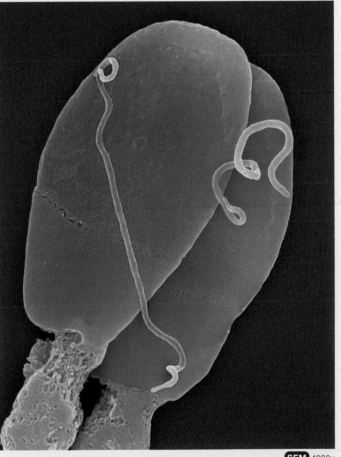

SEM 4000x

(C) *Treponema pallidum* bacteria (thread-like cells) on the heads of two sperm cells

(continues)

● CLINICAL **CONNECTION** | Sexually Transmitted Diseases (continued)

During the *primary stage,* the chief sign is a painless open sore, called a **chancre** (SHANG-ker), at the point of contact. The chancre heals within 1 to 5 weeks. From 6 to 24 weeks later, signs and symptoms such as a skin rash, fever, and aches in the joints and muscles usher in the *secondary stage,* which is systemic—the infection spreads to all major body systems. When signs of organ degeneration appear, the disease is said to be in the *tertiary stage.* If the nervous system is involved, the tertiary stage is called **neurosyphilis**. As motor areas become extensively damaged, victims may be unable to control urine and bowel movements. Eventually they may become bedridden and unable even to feed themselves. In addition, damage to the cerebral cortex produces memory loss and personality changes that range from irritability to hallucinations.

Genital Herpes

Genital herpes is an incurable STD. Type II herpes simplex virus (HSV-2) causes genital infections (Figure D), producing painful blisters on the prepuce, glans penis, and penile shaft in males and on the vulva or sometimes high up in the vagina in females. The blisters disappear and reappear in most patients, but the virus itself remains in the body. A related virus, type I herpes simplex virus (HSV-1), causes cold sores on the mouth and lips. Infected individuals typically experience recurrences of symptoms several times a year.

Genital Warts

Warts are an infectious disease caused by viruses. *Human papillomavirus (HPV)* causes **genital warts**, which is commonly transmitted sexually (Figure E). Nearly one million people a year develop genital warts in the United States. Patients with a history of genital warts may be at increased risk for cancers of the cervix, vagina, anus, vulva, and penis. There is no cure for genital warts. ●

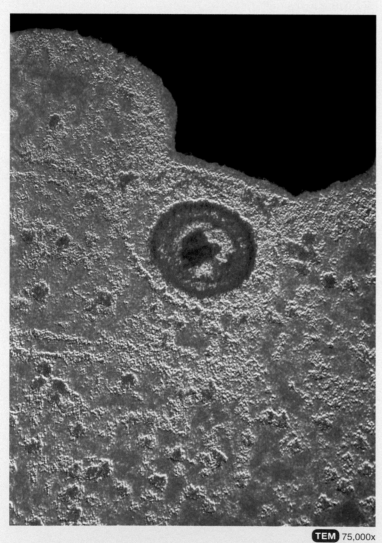

TEM 75,000x

(D) Type II herpes simplex virus (red-orange sphere) in the cytoplasm (blue) of an infected cell

TEM

(E) Human papillomaviruses (HPV), which cause genital warts

BIRTH CONTROL METHODS AND ABORTION

◉ OBJECTIVE

- Explain the differences among the various types of birth control methods and compare their effectiveness.

Birth Control Methods

Birth control refers to restricting the number of children by various methods designed to control fertility and prevent conception. No single, ideal method of birth control exists. The only method of preventing pregnancy that is 100% reliable is complete **abstinence**, the avoidance of sexual intercourse. Several other methods are available; each has its advantages and disadvantages. These include surgical sterilization, hormonal methods, intrauterine devices, spermicides, barrier methods, and periodic abstinence. Table 27.3 provides the failure rates for various methods of birth control. Although it is not a form of birth control, in this section we will also discuss abortion, the premature expulsion of the products of conception from the uterus.

TABLE 27.3

Failure Rates for Several Birth Control Methods

METHOD	FAILURE RATES* PERFECT USE†	TYPICAL USE
Complete abstinence	0%	0%
Surgical sterilization		
Vasectomy	0.10%	0.15%
Tubal ligation	0.5%	0.5%
Hormonal methods		
Oral contraceptives		
Combined pill	0.3%	1–2%
Seasonale®	0.3%	1–2%
Minipill	0.5%	2%
Non-oral contraceptives		
Contraceptive skin patch	0.1%	1–2%
Vaginal contraceptive ring	0.1%	1–2%
Emergency contraception	25%	25%
Hormone injections	0.3%	1–2%
Intrauterine devices (copper T380A®)	0.6%	0.8%
Spermicides (alone)	15%	29%
Barrier methods		
Male condom	2%	15%
Vaginal pouch	5%	21%
Diaphragm (with spermicide)	6%	16%
Cervical cap (with spermicide)	9%	16%
Periodic abstinence		
Rhythm	9%	25%
Sympto-thermal	2%	20%
No method	85%	85%

* Defined as percentage of women having an unintended pregnancy during the first year of use.

† Failure rate when the method is used correctly and consistently.

Surgical Sterilization

Sterilization is a procedure that renders an individual incapable of further reproduction. The principal method for sterilization of males is a **vasectomy** (vas-EK-tō-mē-; -*ectomy*=cut out), in which a portion of each ductus deferens is removed. An incision is made on either side of the scrotum, the ducts are located and cut, each is tied (ligated) in two places with stitches, and the portion between the ties is removed. Although sperm production continues in the testes, sperm can no longer reach the exterior. The sperm degenerate and are destroyed by phagocytosis. Because the blood vessels are not cut, testosterone levels in the blood remain normal, so vasectomy has no effect on sexual desire and performance. If done correctly, it is close to 100% effective. The procedure can be reversed, but the chance of regaining fertility is only 30–40%. Sterilization in females most often is achieved by performing a **tubal ligation** (lī-GĀ-shun) in which both uterine tubes are tied closed and then cut. This can be achieved in a few different ways. "Clips" or "clamps" can be placed on the uterine tubes, the tubes can be tied and/or cut, and sometimes they are cauterized. In any case the result is that the secondary oocyte cannot pass through the uterine tubes, and sperm cannot reach the oocyte.

Hormonal Methods

Aside from complete abstinence or surgical sterilization, hormonal methods are the most effective means of birth control. **Oral contraceptives** (the pill) contain hormones designed to prevent pregnancy. Some, called *combined oral contraceptives (COCs)*, contain both progestin (hormone with actions similar to progesterone) and estrogens. The primary action of COCs is to inhibit ovulation by suppressing the gonadotropins FSH and LH. The low levels of FSH and LH usually prevent the development of a dominant follicle in the ovary. As a result, levels of estrogens do not rise, the midcycle LH surge does not occur, and ovulation does not take place. Even if ovulation does occur, as it does in some cases, COCs may also block implantation in the uterus and inhibit the transport of ova and sperm in the uterine tubes. Progestins thicken cervical mucus and make it more difficult for sperm to enter the uterus. *Progestin-only pills* thicken cervical mucus and may block implantation in the uterus but they do not consistently inhibit ovulation.

Among the noncontraceptive benefits of oral contraceptives are regulation of the length of menstrual cycle and decreased menstrual flow (and therefore decreased risk of anemia). The pill also provides protection against endometrial and ovarian cancers and reduces the risk of endometriosis. However, oral contraceptives may not be advised for women with a history of blood clotting disorders, cerebral blood vessel damage, migraine headaches, hypertension, liver malfunction, or heart disease. Women who take the pill and smoke face far higher odds of having a heart attack or stroke than do nonsmoking pill users. Smokers should quit smoking or use an alternative method of birth control.

Following are several variations of *oral* hormonal methods of contraception:

- **Combined pill.** Contains both progestin and estrogens and is typically taken once a day for three weeks to prevent pregnancy and regulate the menstrual cycle. The pills taken during the fourth week are inactive (do not contain hormones) and permit menstruation to occur.
- **Seasonale.®** Contains both progestin and estrogens and is taken once a day in 3-month cycles of twelve weeks of hormone-containing pills followed by one week of inactive pills. Menstruation occurs during the thirteenth week.

- **Minipill.** Contains progestin only and is taken every day of the month.

Non-oral hormonal methods of contraception are also available. Among these are the following:

- **Contraceptive skin patch (Ortho Evra)®.** Contains both progestin and estrogens delivered in a skin patch placed on the skin (upper outer arm, back, lower abdomen, or buttocks) once a week for three weeks. After one week, the patch is removed from one location and then a new one is placed elsewhere. During the fourth week no patch is used.
- **Vaginal contraceptive ring (NuvaRing)®.** A flexible dough-nut-shaped ring about 5 cm (2 in.) in diameter that contains estrogens and progesterone and is inserted by the female herself into the vagina. It is left in the vagina for three weeks to prevent conception and then removed for one week to permit menstruation.
- **Emergency contraception (EC) (morning after pill).** Consists of progestin and estrogens or progestin alone to prevent pregnancy following unprotected sexual intercourse. The relatively high levels of progestin and estrogens in EC pills provide inhibition of FSH and LH secretion. Loss of the stimulating effects of these gonadtropic hormones causes the ovaries to cease secretion of their own estrogens and progesterone. In turn, declining levels of estrogens and progesterone induce shedding of the uterine lining, thereby blocking implantation. One pill is taken as soon as possible but within 72 hours of unprotected sexual intercourse. The second pill must be taken 12 hours after the first. The pills work in the same way as regular birth control pills.
- **Hormone injections (Depo-provera)®.** An injectable progestin given intramuscularly by a health-care practitioner once every three months.

Intrauterine Devices

An **intrauterine device (IUD)** is a small object made of plastic, copper, or stainless steel that is inserted by a health-care professional into the cavity of the uterus. IUDs prevent fertilization from taking place by blocking sperm from entering the uterine tubes. The IUD most commonly used in the United States today is the Copper T 380A®, which is approved for up to 10 years of use and has long-term effectiveness comparable to that of tubal ligation. Some women cannot use IUDs because of expulsion, bleeding, or discomfort.

Spermicides

Various foams, creams, jellies, suppositories, and douches that contain sperm-killing agents, or **spermicides**, make the vagina and cervix unfavorable for sperm survival and are available without prescription. They are placed in the vagina before sexual intercourse. The most widely used spermicide is *nonoxynol-9*, which kills sperm by disrupting their plasma membranes. A spermicide is more effective when used with a barrier method such as a male condom, vaginal pouch, diaphragm, or cervical cap.

Barrier Methods

Barrier methods use a physical barrier and are designed to prevent sperm from gaining access to the uterine cavity and uterine tubes. In addition to preventing pregnancy, certain barrier methods (male condom and vaginal pouch) may also provide some protection against sexually transmitted diseases (STDs) such as AIDS. In contrast, oral contraceptives and IUDs confer no such protection. Among the barrier methods are the male condom, vaginal pouch, diaphragm, and cervical cap.

A **male condom** is a nonporous, latex covering placed over the penis that prevents deposition of sperm in the female reproductive tract. A **vaginal pouch,** sometimes called a **female condom,** is designed to prevent sperm from entering the uterus. It is made of two flexible rings connected by a polyurethane sheath. One ring lies inside the sheath and is inserted to fit over the cervix; the other ring remains outside the vagina and covers the female external genitals. A **diaphragm** is a rubber, dome-shaped structure that fits over the cervix and is used in conjunction with a spermicide. It can be inserted by the female up to 6 hours before intercourse. The diaphragm stops most sperm from passing into the cervix and the spermicide kills most sperm that do get by. Although diaphragm use does decrease the risk of some STDs, it does not fully protect against HIV infection because the vagina is still exposed. A **cervical cap** resembles a diaphragm but is smaller and more rigid. It fits snugly over the cervix and must be fitted by a health-care professional. Spermicides should be used with the cervical cap.

Periodic Abstinence

A couple can use their knowledge of the physiological changes that occur during the female reproductive cycle to decide either to abstain from intercourse on those days when pregnancy is a likely result, or to plan intercourse on those days if they wish to conceive a child. In females with normal and regular menstrual cycles, these physiological events help to predict the day on which ovulation is likely to occur.

The first physiologically based method, developed in the 1930s, is known as the **rhythm method**. It involves abstaining from sexual activity on the days that ovulation is likely to occur in each reproductive cycle. During this time (3 days before ovulation, the day of ovulation, and 3 days after ovulation) the couple abstains from intercourse. The effectiveness of the rhythm method for birth control is poor in many women due to the irregularity of the female reproductive cycle.

Another system is the **sympto-thermal method**, in which couples are instructed to know and understand certain signs of fertility. The signs of ovulation include increased basal body temperature; the production of abundant clear, stretchy cervical mucus; and pain associated with ovulation (*mittelschmerz*). If a couple abstains from sexual intercourse when the signs of ovulation are present and for 3 days afterward, the chance of pregnancy is decreased. A big problem with this method is that fertilization is very likely if intercourse occurs one or two days *before* ovulation.

Abortion

Abortion refers to the premature expulsion of the products of conception from the uterus, usually before the twentieth week of pregnancy. An abortion may be **spontaneous** (naturally occurring; also called a *miscarriage*) or **induced** (intentionally performed).

There are several types of induced abortions. One involves **mifepristone**, called **miniprex** in the United States and **RU 486** in Europe. It is a hormone approved only for pregnancies 9 weeks or less when taken with a prostoglandin misopostol (a prostoglandin).

Mifepristone is an antiprogestin; it blocks the action of progesterone by binding to and blocking progesterone receptors.

Progesterone prepares the uterine endometrium for implantation and then maintains the uterine lining after implantation. If the level of progesterone falls during pregnancy or if the action of the hormone is blocked, menstruation occurs, and the embryo sloughs off along with the uterine lining. Within 12 hours after taking mifepristone, the endometrium starts to degenerate, and within 72 hours it begins to slough off. Misoprostol stimulates uterine contractions, and is given after mifepristone to aid in expulsion of the endometrium.

Another type of induced abortion is called **vacuum aspiration (suction)** and can be performed up to the sixteenth week of pregnancy. A small, flexible tube attached to a vacuum source is inserted into the uterus through the vagina. The embryo or fetus, placenta, and lining of the uterus are then removed by suction. For pregnancies between 13 and 16 weeks, a technique called **dilation and evacuation** is commonly used. After the cervix is dilated, suction and forceps are used to remove the fetus, placenta, and uterine lining. From the 16th to 24th week, a **late-stage abortion** may be induced using surgical methods similar to dilation and evacuation or nonsurgical methods utilizing a saline solution or medications. Labor may be induced by using vaginal suppositories, intravenous infusion, or injections into the amniotic fluid through the uterus.

◉ CHECKPOINT

28. How do oral contraceptives work to reduce the likelihood of pregnancy?
29. Why do some methods of birth control protect against sexually transmitted diseases, while others do not?

DEVELOPMENT OF THE REPRODUCTIVE SYSTEMS

◉ OBJECTIVE

• Describe the development of the male and female reproductive systems.

The *gonads* develop from the **intermediate mesoderm**. During the fifth week of development, they appear as bulges that protrude into the ventral body cavity (Figure 27.26). Adjacent to the gonads are the **mesonephric (Wolffian) ducts**, which eventually develop into structures of the reproductive system in males. A second pair of ducts, the **paramesonephric (Müllerian) ducts**, develop lateral to the mesonephric ducts and eventually form structures of the reproductive system in females. Both sets of ducts empty into the urogenital sinus. An early embryo has the potential to follow either the male or the female pattern of development because it contains both sets of ducts and primitive gonads that can differentiate into either testes or ovaries.

Cells of a male embryo have one X chromosome and one Y chromosome. The male pattern of development is initiated by a Y chromosome "master switch" gene named *SRY*, which stands for *Sex-determining Region* of the *Y* chromosome. When the *SRY* gene is expressed during development, its protein product causes the primitive Sertoli cells to begin to differentiate in the gonadal tissues during the seventh week. The developing Sertoli cells secrete a hormone called **Müllerian-inhibiting substance (MIS)**, which causes apoptosis of cells within the paramesonephric (Müllerian) ducts. As a result, those cells do not contribute any functional structures to the male reproductive system. Stimulated by human chorionic gonadotropin (hCG), primitive Leydig cells in the gonadal tissue begin to secrete the androgen

testosterone during the eighth week. Testosterone then stimulates development of the mesonephric duct on each side into the *epididymis, ductus (vas) deferens, ejaculatory duct*, and *seminal vesicle*. The *testes* connect to the mesonephric duct through a series of tubules that eventually become the *seminiferous tubules*. The *prostate* and *bulbourethral glands* are **endodermal** outgrowths of the urethra.

Cells of a female embryo have two X chromosomes and no Y chromosome. Because *SRY* is absent, the gonads develop into *ovaries*, and because MIS is not produced, the paramesonephric ducts flourish. The distal ends of the paramesonephric ducts fuse to form the *uterus* and *vagina*, and the unfused proximal portions become the *uterine (fallopian) tubes*. The mesonephric ducts degenerate without contributing any functional structures to the female reproductive system because testosterone is absent. The *greater* and *lesser vestibular glands* develop from **endodermal** outgrowths of the vestibule.

The *external genitals* of both male and female embryos (penis and scrotum in males and clitoris, labia, and vaginal orifice in females) also remain undifferentiated until about the eighth week. Before differentiation, all embryos have an elevated midline swelling called the **genital tubercle** (Figure 27.27 on page 887). The tubercle consists of the **urethral groove** (opening into the urogenital sinus), paired **urethral folds**, and paired **labioscrotal swellings**.

In male embryos, some testosterone is converted to a second androgen called **dihydrotestosterone (DHT)**. DHT stimulates development of the urethra, prostate, and external genitals (scrotum and penis). Part of the genital tubercle elongates and develops into a penis. Fusion of the urethral folds forms the *spongy (penile) urethra* and leaves an opening to the exterior only at the distal end of the penis, the *external urethral orifice*. The labioscrotal swellings develop into the *scrotum*. In the absence of DHT, the genital tubercle gives rise to the *clitoris* in female embryos. The urethral folds remain open as the *labia minora*, and the labioscrotal swellings become the *labia majora*. The urethral groove becomes the *vestibule*. After birth, androgen levels decline because hCG is no longer present to stimulate secretion of testosterone.

◉ CHECKPOINT

30. Describe the role of hormones in differentiation of the gonads, the mesonephric ducts, the paramesonephric ducts, and the external genitals.

AGING AND THE REPRODUCTIVE SYSTEMS

◉ OBJECTIVE

• Describe the effects of aging on the reproductive systems.

During the first decade of life, the reproductive system is in a juvenile state. At about age 10, hormone-directed changes start to occur in both sexes. **Puberty** (PŪ-ber-tē=a ripe age) is the period when secondary sexual characteristics begin to develop and the potential for sexual reproduction is reached. The onset of puberty is marked by pulses or bursts of LH and FSH secretion, each triggered by a pulse of GnRH. Most pulses occur during sleep. As puberty advances, the hormone pulses occur during the day as well as at night. The pulses increase in frequency during a three- to four-year period until the adult pattern is established. The stimuli that cause the GnRH pulses are still unclear, but a role for the hormone leptin is starting to unfold. Just before puberty, leptin levels rise in proportion to adipose tissue mass.

Figure 27.26 Development of the internal reproductive systems.

The gonads develop from intermediate mesoderm.

Which gene is responsible for the development of the gonads into testes?

Figure 27.27 **Development of the external genitals.**

🔑 The external genitals of male and female embryos remain undifferentiated until about the eighth week.

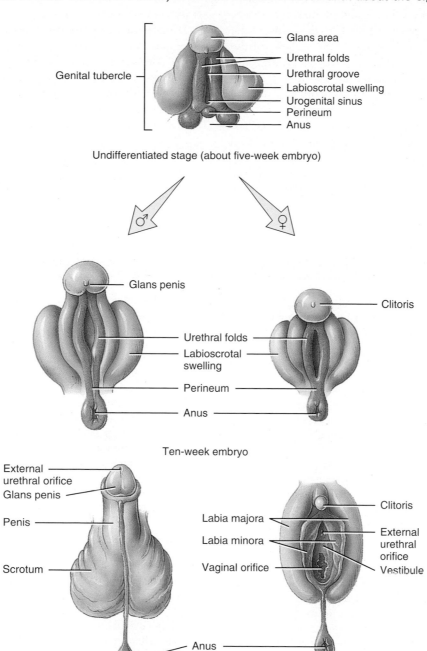

Undifferentiated stage (about five-week embryo)

Ten-week embryo

Near birth

MALE DEVELOPMENT   FEMALE DEVELOPMENT

❓ **Which hormone is responsible for the differentiation of the external genitals?**

Interestingly, leptin receptors are present in both the hypothalamus and anterior pituitary. Mice that lack a functional leptin gene from birth are sterile and remain in a prepubertal state. Giving leptin to such mice elicits secretion of gonadotropins, and they become fertile. Leptin may signal the hypothalamus that long-term energy stores (triglycerides in adipose tissue) are adequate for reproductive functions to begin.

In females, the reproductive cycle normally occurs once each month from **menarche** (me-NAR-kē), the first menses, to **menopause**, the permanent cessation of menses. Thus, the female reproductive system has a time-limited span of fertility between menarche and menopause. For the first 1 to 2 years after menarche, ovulation occurs in only about 10% of the cycles and the luteal phase is short. Gradually, the percentage of ovulatory cycles increases, and the luteal phase reaches its normal duration of 14 days. With age, fertility declines. Between the ages of 40 and 50 the pool of remaining ovarian follicles becomes exhausted. As a result, the ovaries become less responsive to hormonal stimulation. The production of estrogens declines, despite copious secretion of FSH and LH by the anterior pituitary. Many women experience hot flashes and heavy sweating, which coincide with bursts of GnRH release. Other symptoms of menopause are headache, hair loss, muscular pains, vaginal dryness, insomnia, depression, weight gain, and mood swings. Some atrophy of the ovaries, uterine tubes, uterus, vagina, external genitalia, and breasts occurs in postmenopausal women. Due to loss of estrogens, most women experience a decline in bone mineral density after menopause. Sexual desire (libido) does not show a parallel decline; it may be maintained by adrenal sex steroids. The risk of having uterine cancer peaks at about 65 years of age, but cervical cancer is more common in younger women.

In males, declining reproductive function is much more subtle than in females. Healthy men often retain reproductive capacity into their eighties or nineties. At about age 55 a decline in testos-terone synthesis leads to reduced muscle strength, fewer viable sperm, and decreased sexual desire. Although sperm production decreases 50–70% between ages 60 and 80, abundant sperm may still be present even in old age.

Enlargement of the prostate to two to four times its normal size occurs in approximately one-third of all males over age 60. This condition, called **benign prostatic hyperplasia (BPH)**, decreases the size of the prostatic urethra and is characterized by frequent urination, nocturia (bed-wetting), hesitancy in urination, decreased force of urinary stream, postvoiding dribbling, and a sensation of incomplete emptying.

◉ **CHECKPOINT**

31. What changes occur in males and females at puberty?
32. What do the terms menarche and menopause mean?

KEY MEDICAL TERMS ASSOCIATED WITH THE REPRODUCTIVE SYSTEMS

Castration (kas-TRĀ-shun=to prune) Removal, inactivation, or destruction of the gonads; commonly used in reference to removal of the testes only.

Colposcopy (kol-POS-kō-pē; *colpo-*=vagina; *-scopy*=to view) Visual inspection of the vagina and cervix of the uterus using a culposcope, an instrument that has a magnifying lens (between 5 and 50×) and a light.

Culdoscopy (kul-DOS-kō-pē; *cul-*=cul-de-sac; *-scopy*=to view) A procedure in which a culdoscope (endoscope) is inserted through the posterior wall of the vagina to view the rectouterine pouch in the pelvic cavity.

Dysmenorrhea (dis′-men-or-Ē-a; *dys-*=difficult or painful) Pain associated with menstruation; the term is usually reserved to describe menstrual symptoms that are severe enough to prevent a woman from functioning normally for one or more days each month. Some cases are caused by uterine tumors, ovarian cysts, pelvic inflammatory disease, or intrauterine devices.

Dyspareunia (dis-pa-ROO-nē-a; *dys-*=difficult; *para*=beside; *-enue*=bed) Pain during sexual intercourse. It may occur in the genital area or in the pelvic cavity, and may be due to inadequate lubrication, inflammation, infection, an improperly fitting diaphragm or cervical cap, endometriosis, pelvic inflammatory disease, pelvic tumors, or weakened uterine ligaments.

Endocervical curettage (kū-re-TAHZH; *curette*=scraper) A procedure in which the cervix is dilated and the endometrium of the uterus is scraped with a spoon-shaped instrument called a curette; commonly called a *D and C* (dilation and curettage).

Fibroids (FĪ-broyds; *fibro-*=fiber; *-eidos*=resemblance) Noncancerous tumors in the myometrium of the uterus composed of muscular and fibrous tissue. Their growth appears to be related to high levels of estrogens. They do not occur before puberty and usually stop growing after menopause. Symptoms include abnormal menstrual bleeding, and pain or pressure in the pelvic area.

Hermaphroditism (her-MAF-rō-dīt-izm) The presence of both ovarian and testicular tissue in one individual.

Hypospadias (hī′-pō-SPĀ-dē-as; *hypo-*=below) A common congenital abnormality in which the urethral opening is displaced. In males, the displaced opening may be on the underside of the penis, at the peno-scrotal junction, between the scrotal folds, or in the perineum; in females, the urethra opens into the vagina. The problem can be corrected surgically.

Leukorrhea (loo′-kō-RĒ-a; *leuko-*=white) A whitish (nonbloody) vaginal discharge containing mucus and pus cells that may occur at any age and affects most women at some time.

Menorrhagia (men-ō-RA-jē-a; *meno-*=menstruation; *-rhage*=to burst forth) Excessively prolonged or profuse menstrual period. May be due to a disturbance in hormonal regulation of the menstrual cycle, pelvic infection, medications (anticoagulants), fibroids (noncancerous uterine tumors composed of muscle and fibrous tissue), endometriosis, or intrauterine devices.

Oophorectomy (ō′-of-ō-REK-tō-mē; *oophor-*=bearing eggs) Removal of the ovaries.

Orchitis (or-KĪ-tis; *orchi-*=testes; *-itis*=inflammation) Inflammation of the testes, for example, as a result of the mumps virus or a bacterial infection.

Ovarian cyst The most common form of ovarian tumor, in which a fluid-filled follicle or corpus luteum persists and continues growing.

Papanicolaou test (pa-pa-ni-kō-LĀ-oo), or **PAP smear**. A test to detect uterine cancer in which a few cells from the cervix and the part of the vagina surrounding the cervix are removed with a swab and examined microscopically. Malignant cells have a characteristic appearance that allows diagnosis even before symptoms occur.

Pelvic inflammatory disease (PID) A collective term for any extensive bacterial infection of the pelvic organs, especially the uterus, uterine tubes, or ovaries, which is characterized by pelvic soreness, lower back pain, abdominal pain, and urethritis.

Polycystic ovary syndrome (pol-ē-SIS-tik) or **(PCOS)** A disorder that usually develops during puberty and is characterized by enlarged ovaries with many fluid-filled sacs (cysts) and a tendency to have high levels of male hormones (androgens). Symptoms include failure to menstruate, unpredictable menstrual periods, an unusual amount of facial or body hair, high blood sugar, obesity, and increased risk of cardiovascular disease.

Salpingectomy (sal′-pin-JEK-tō-mē; *salpingo*=tube) Removal of a uterine (fallopian) tube.

Smegma (SMEG-ma) The secretion, consisting principally of desquamated epithelial cells, found chiefly around the external genitalia and especially under the foreskin of the male.

Introduction (p. 846)

1. Reproduction is the process by which new individuals of a species are produced and the genetic material is passed from generation to generation.
2. The organs of reproduction are grouped as gonads (produce gametes), ducts (transport and store gametes), accessory sex glands (produce materials that support gametes), and supporting structures (have various roles in reproduction).

Male Reproductive System (p. 847)

1. The male structures of reproduction include the testes, ductus epididymis, ductus (vas) deferens, ejaculatory duct, urethra, seminal vesicles, prostate, bulbourethral (Cowper's) glands, and penis.
2. The scrotum is a sac that hangs from the root of the penis and consists of loose skin and superficial fascia; it supports the testes.
3. The temperature of the testes is regulated by contraction of the cremaster muscle and dartos muscle, which either elevates them and brings them closer to the pelvic cavity or relaxes and moves them farther from the pelvic cavity.
4. The testes are paired oval glands (gonads) in the scrotum containing seminiferous tubules, in which sperm cells are made; Sertoli cells (sustentacular cells), which nourish sperm cells and secrete inhibin; and Leydig (interstitial) cells, which produce the male sex hormone testosterone.
5. The testes descend into the scrotum through the inguinal canals during the seventh month of fetal development. Failure of the testes to descend is called cryptorchidism.
6. Secondary oocytes and sperm, both of which are called gametes, are produced in the gonads.
7. Spermatogenesis, which occurs in the testes, is the development of immature spermatogonia into mature sperm. The spermatogenesis sequence, which includes meiosis I, meiosis II, and spermiogenesis, results in the formation of four haploid sperm (spermatozoa) from each primary spermatocyte.
8. The principal parts of a mature sperm are a head and a tail. The function of sperm is to fertilize a secondary oocyte.
9. The duct system of the testes includes the seminiferous tubules, straight tubules, and rete testis. Sperm flow out of the testes through the efferent ducts.
10. The ductus epididymis is the site of sperm maturation and storage.
11. The ductus (vas) deferens stores sperm and propels them toward the urethra during ejaculation.
12. Each ejaculatory duct, formed by the union of the duct from the seminal vesicle and ductus (vas) deferens, is the passageway for ejection of sperm and secretions of the seminal vesicles into the first portion of the urethra, the prostatic urethra.
13. The urethra in males is subdivided into three portions: the prostatic, membranous, and spongy (penile) urethra.
14. The seminal vesicles secrete an alkaline, viscous fluid that constitutes about 60% of the volume of semen and contributes to sperm viability.
15. The prostate secretes a slightly acidic fluid that constitutes about 25% of the volume of semen and contributes to sperm motility.
16. The bulbourethral (Cowper's) glands secrete mucus for lubrication and an alkaline substance that neutralizes acid.
17. Semen is a mixture of sperm and seminal fluid; it provides the fluid in which sperm are transported, supplies nutrients, and neutralizes the acidity of the male urethra and the vagina.
18. The penis consists of a root, a body, and a glans penis.
19. Engorgement of the penile blood sinuses under the influence of sexual excitation is called erection.

Female Reproductive System (p. 860)

1. The female organs of reproduction include the ovaries (gonads), uterine (fallopian) tubes or oviducts, uterus, vagina, and vulva.
2. The mammary glands are considered part of the reproductive system in females.

3. The ovaries, the female gonads, are located in the superior portion of the pelvic cavity, lateral to the uterus.
4. Ovaries produce secondary oocytes, discharge secondary oocytes (ovulation), and secrete estrogens, progesterone, relaxin, and inhibin.
5. Oogenesis (production of haploid secondary oocytes) begins in the ovaries. The oogenesis sequence includes meiosis I and meiosis II, which goes to completion only after an ovulated secondary oocyte is fertilized by a sperm cell.
6. The uterine (fallopian) tubes transport secondary oocytes from the ovaries to the uterus and are the normal sites of fertilization. Ciliated cells and peristaltic contractions help move a secondary oocyte or fertilized ovum toward the uterus.
7. The uterus is an organ the size and shape of an inverted pear that functions in menstruation, implantation of a fertilized ovum, development of a fetus during pregnancy, and labor. It is also part of the pathway for sperm to reach the uterine tubes to fertilize a secondary oocyte. Normally, the uterus is held in position by a series of ligaments.
8. Histologically, the layers of the uterus are an outer perimetrium (serosa), a middle myometrium, and an inner endometrium.
9. The vagina is a passageway for sperm and the menstrual flow, the receptacle of the penis during sexual intercourse, and the inferior portion of the birth canal. It is capable of considerable distension.
10. The vulva, a collective term for the external genitals of the female, consists of the mons pubis, labia majora, labia minora, clitoris, vestibule, vaginal and urethral orifices, hymen, bulb of the vestibule, and the paraurethral (Skene's), greater vestibular (Bartholin's), and lesser vestibular glands.
11. The perineum is a diamond-shaped area at the inferior end of the trunk medial to the thighs and buttocks.
12. The mammary glands are modified sweat glands lying superficial to the pectoralis major muscles. Their function is to synthesize, secrete, and eject milk (lactation).
13. Mammary gland development depends on estrogens and progesterone.
14. Milk production is stimulated by prolactin, estrogens, and progesterone; milk ejection is stimulated by oxytocin.

Female Reproductive Cycle (p. 878)

1. The function of the ovarian cycle is to develop a secondary oocyte. The function of the uterine (menstrual) cycle is to prepare the endometrium each month to receive a fertilized egg. The female reproductive cycle includes both the ovarian and uterine cycles.
2. The female reproductive cycle is controlled by GnRH from the hypothalamus, which stimulates the release of FSH and LH by the anterior pituitary gland.
3. FSH stimulates development of secondary follicles and initiates secretion of estrogens by the follicles. LH stimulates further development of the follicles, secretion of estrogens by follicular cells, ovulation, formation of the corpus luteum, and the secretion of progesterone and estrogens by the corpus luteum.
4. During the menstrual phase, the stratum functionalis of the endometrium is shed, discharging blood, tissue fluid, mucus, and epithelial cells.
5. During the preovulatory phase, a group of follicles in the ovaries begins to undergo final maturation. One follicle outgrows the others and becomes dominant while the others degenerate. At the same time, endometrial repair occurs in the uterus. Estrogens are the dominant ovarian hormones during the preovulatory phase.
6. Ovulation is the rupture of the dominant mature (graafian) follicle and the release of a secondary oocyte into the pelvic cavity. It is brought about by a surge of LH. Signs and symptoms of ovulation include increased basal body temperature; clear, stretchy cervical mucus; changes in the uterine cervix; and ovarian pain.

7. During the postovulatory phase, both progesterone and estrogens are secreted in large quantity by the corpus luteum of the ovary, and the uterine endometrium thickens in readiness for implantation.

8. If fertilization and implantation do not occur, the corpus luteum degenerates, and the resulting low level of progesterone allows discharge of the endometrium followed by the initiation of another reproductive cycle.

9. If fertilization and implantation occur, the corpus luteum is maintained by placental hCG, and the corpus luteum and later the placenta secrete progesterone and estrogens to support pregnancy and breast development for lactation.

Birth Control Methods and Abortion (p. 883)

1. Methods include complete abstinence, surgical sterilization (vasectomy, tubal ligation), hormonal methods (combined pill, minipill, contraceptive skin patch, vaginal contraceptive ring, emergency contraception, hormonal injections), intrauterine devices, spermicides, barrier methods (male condom, vaginal pouch, diaphragm, cervical cap), and periodic abstinence (rhythm and sympto-thermal methods).

2. Contraceptive pills of the combination type contain progestin and estrogens in concentrations that decrease the secretion of FSH and LH and thereby inhibit development of ovarian follicles and ovulation, inhibit transport of ova and sperm in the uterine tubes, and block implantation in the uterus.

3. An abortion is the premature expulsion from the uterus of the products of conception; it may be spontaneous or induced.

Development of the Reproductive Systems (p. 885)

1. The gonads develop from intermediate mesoderm. In the presence of the *SRY* gene, the gonads begin to differentiate into testes during the seventh week. The gonads differentiate into ovaries when the *SRY* gene is absent.

2. In males, testosterone stimulates development of each mesonephric duct into an epididymis, ductus (vas) deferens, ejaculatory duct, and seminal vesicle, and Müllerian-inhibiting substance (MIS) causes the paramesonephric duct cells to die. In females, testosterone and MIS are absent; the paramesonephric ducts develop into the uterine tubes, uterus, and vagina and the mesonephric ducts degenerate.

3. The external genitals develop from the genital tubercle and are stimulated to develop into typical male structures by the hormone dihydrotestosterone (DHT). The external genitals develop into female structures when DHT is not produced, the normal situation in female embryos.

Aging and the Reproductive Systems (p. 885)

1. Puberty is the period when secondary sex characteristics begin to develop and the potential for sexual reproduction is reached.

2. Onset of puberty is marked by pulses or bursts of LH and FSH secretion, each triggered by a pulse of GnRH. The hormone leptin, released by adipose tissue, may signal the hypothalamus that long-term energy stores (triglycerides in adipose tissue) are adequate for reproductive functions to begin.

3. In females, the reproductive cycle normally occurs once each month from menarche, the first menses, to menopause, the permanent cessation of menses.

4. Between the ages of 40 and 50, the pool of remaining ovarian follicles becomes exhausted and levels of progesterone and estrogens decline. Most women experience a decline in bone mineral density after menopause, together with some atrophy of the ovaries, uterine tubes, uterus, vagina, external genitalia, and breasts. Uterine and breast cancer increase in incidence with age.

5. In older males, decreased levels of testosterone are associated with decreased muscle strength, waning sexual desire, and fewer viable sperm; prostate disorders are common.

SELF-QUIZ QUESTIONS

Choose the one best answer to the following questions.

1. The secondary sex gland(s) that contribute(s) the most volume to the semen is/are the:
   a. seminal vesicles.
   b. bulbourethral (Cowper's) glands.
   c. paraurethral (Skene's) glands.
   d. greater vestibular (Bartholin's) glands.
   e. prostate.

2. The dense irregular connective tissue that covers the testes and extends inward to subdivide the testes into lobules is the:
   a. tunica albuginea.   b. tunica vaginalis.   c. raphe.
   d. rete testis.   e. epididymis.

3. Sperm are produced in the _____, and testosterone is secreted by the _____.
   a. Leydig (interstitial) cells, seminiferous tubules
   b. seminal vesicles, Sertoli (sustentacular) cells
   c. prostate, primary spermatocytes
   d. seminiferous tubules, Leydig (interstitial) cells
   e. epididymis, spermatogonia

4. Which of the following is *not* a characteristic of the vagina?
   a. fornix
   b. alkaline
   c. nonkeratinized stratified squamous epithelium
   d. rugae
   e. low pH

5. The birth control method provided by the intrauterine device (IUD) causes changes to the _____, and thus prevents _____.
   a. serosa, gastrulation
   b. endometrium, implantation
   c. myometrium, cleavage
   d. perimetrium, ovulation
   e. endometrium, fertilization

6. In the postovulatory phase of the menstrual cycle, LH stimulates the corpus luteum to secrete hormones including:
   a. prolactin.   b. FSH.   c. hCG.
   d. progesterone.   e. Both c and d are correct.

Complete the following.

7. Place numbers in the blanks to arrange the following structures in the correct sequence for the pathway of sperm. (a) ejaculatory duct: _____; (b) testis: _____; (c) urethra: _____; (d) ductus (vas) deferens: _____; (e) epididymis: _____.

8. Branches of the radial arteries called _____ supply the stratum functionalis of the endometrium.

9. Fill in the blanks with the male homologues for each of the following. (a) labia majora: _____; (b) clitoris: _____; (c) paraurethral (Skene's) glands: _____; (d) greater vestibular (Bartholin's) glands: _____.

10. The layer of endometrium that is not shed during menstruation is the stratum _____.

11. The union of the ampulla of the ductus deferens and the duct from the seminal vesicle forms the _____.

12. If fertilization and implantation do not occur, the corpus luteum degenerates and becomes the _____.

13. The male accessory gland that is located inferior to the urinary bladder and that surrounds the urethra is the _____.

14. The ovary is attached to the uterus by the _____ ligament.

Are the following statements true or false? If the statement is False, *find and fix the mistake to make the statement* True.

15. The mass of tissue that surrounds the penile urethra is called the corpus spongiosum.

16. The female urethra is located posterior to the vagina.

17. Fertilization of a secondary oocyte by a sperm cell normally occurs in the uterus.

18. During the preovulatory phase of the menstrual cycle, levels of estrogens increase and the endometrium thickens.

19. The acrosome is the portion of a sperm that contains the genetic material.

20. The uterine tubes are open to the pelvic cavity.

Matching

21. Match the following terms to their descriptions:
____ **(a)** 200–300 of these per testis
____ **(b)** tightly coiled tubes (1–3 per lobule); composed of cells that develop into sperm
____ **(c)** cells located between developing sperm cells; form the blood–testis barrier and provide nourishment
____ **(d)** cells located between seminiferous tubules; secrete testosterone
____ **(e)** highly coiled duct in which sperm are stored for maturation
____ **(f)** ejects sperm and fluid into the urethra just before ejaculation
____ **(g)** enters the pelvic cavity through the inguinal canal

(1) Leydig (interstitial) cells
(2) ejaculatory duct
(3) Sertoli (sustentacular) cells
(4) seminiferous tubules
(5) ductus (vas) deferens
(6) ductus epididymis
(7) lobule

22. Match the descriptions to the major events of the female reproductive cycle (choices may be used more than once).
____ **(a)** Endometrial glands begin to secrete glycogen.
____ **(b)** The dominant follicle releases estrogens and inhibin, which in turn decreases the secretion of FSH.
____ **(c)** The entire stratum functionalis sloughs off.
____ **(d)** LH levels rise, causing release of the secondary oocyte into the pelvic cavity.
____ **(e)** Estrogens stimulate the repair and proliferation of the stratum basalis; endometrial glands develop.
____ **(f)** Theca interna cells mix and become transformed into corpus luteum cells under the influence of LH.
____ **(g)** Progesterone and estrogens produced by the corpus luteum promote vascularization of the superficial endometrium.
____ **(h)** Part of the "follicular phase" of the ovarian cycle.

(1) menstrual phase
(2) preovulatory phase
(3) ovulation
(4) postovulatory phase

CRITICAL THINKING QUESTIONS

1. Esther nearly died from peritonitis (inflammation of the peritoneum) that her doctor said had spread from an infection in her reproductive tract. How was this possible?

2. Occasionally, someone feels that he or she has been born the wrong gender and undergoes a "sex-change" or "gender-reassignment" process involving hormone treatment and surgery. However, a born male can never truly become a fully biological female or a born female become a fully biological male. Why not?

3. Thirty-nine-year-old Meg has been advised to have a hysterectomy due to medical problems. She is worried that the procedure will cause menopause. Is this a valid concern?

4. Darby has been doing peritoneal dialysis for kidney failure at home for several years. However, the weight of the dialysis fluid in his peritoneal cavity has caused him to develop inguinal hernias repeatedly. Darby is tired of this, and in consultation with his doctor, decides to have the inguinal canal surgically closed off. As a result, Darby will also have to have his testes removed. Why?

5. Phil has promised his wife that he will get a vasectomy after the birth of their next child. However, he's concerned about possible effects on his virility. How would you respond to Phil's concerns?

? ANSWERS TO FIGURE QUESTIONS

27.1 The *testes* produce gametes (sperm) and hormones; the *ducts* transport, store, and receive gametes; the *accessory sex glands* secrete materials that help transport and protect gametes; and *supporting structures*, such as the penis, convey semen to the exterior.

27.2 The cremaster and dartos muscles help regulate the temperature of the testes.

27.3 The tunica vaginalis and tunica albuginea are tissue layers that cover and protect the testes.

27.4 The Leydig (interstitial) cells of the testes secrete testosterone.

27.5 During meiosis I, the number of chromosomes in each cell is reduced by half.

27.6 The sperm head contains the nucleus with highly condensed haploid chromosomes and an acrosome that contains enzymes for penetration of a secondary oocyte; the neck contains centrioles that produce microtubules for the rest of the tail; the midpiece contains mitochondria for ATP production for locomotion and metabolism; the principal and end pieces of the tail provide motility.

27.7 The functions of the ductus epididymis are sperm maturation, sperm storage, and propulsion of sperm into the ductus (vas) deferens.

27.8 The ductus deferens stores sperm and conveys sperm toward the urethra.

27.9 Seminal vesicles are the accessory sex glands that contribute the largest volume to seminal fluid.

27.10 Two corpora cavernosa penis and one corpus spongiosum penis contain blood sinuses that fill with blood that cannot flow out of the penis as quickly as it flows in. The trapped blood engorges and stiffens the tissue, producing an erection. The corpus spongiosum penis keeps the spongy urethra open so that ejaculation can occur.

27.11 The testes are homologous to the ovaries; the glans penis is homologous to the clitoris; the prostate is homologous to the paraurethral glands; and the bulbourethral gland is homologous to the greater vestibular glands.

27.12 The mesovarium anchors the ovary to the broad ligament of the uterus and the uterine tube; the ovarian ligament anchors it to the uterus; the suspensory ligament anchors it to the pelvic wall.

27.13 Ovarian follicles secrete estrogens; the corpus luteum secretes progesterone, estrogens, relaxin, and inhibin.

27.14 Most ovarian follicles undergo atresia (degeneration).

27.15 Primary oocytes are present in the ovary at birth, so they are as old as the woman. In males, primary spermatocytes are continually being formed from stem cells (spermatogonia) and thus are only a few days old.

27.16 Fertilization most often occurs in the ampulla of the uterine tube.

27.17 Ciliated simple columnar epithelial cells and nonciliated (peg) cells with microvilli line the uterine tubes.

27.18 The endometrium is a highly vascular, secretory epithelium that provides the oxygen and nutrients needed to sustain a fertilized egg; the myometrium is a thick smooth muscle layer that supports the uterine wall during pregnancy and contracts to expel the fetus at birth.

27.19 The stratum basalis of the endometrium provides cells to replace those that shed (the stratum functionalis) during each menstruation.

27.20 The vagina receives the penis during sexual intercourse, serves as the outlet for menstrual flow, and is the passageway for childbirth.

27.21 Anterior to the vaginal opening are the mons pubis, clitoris, and prepuce. Lateral to the vaginal opening are the labia minora and labia majora.

27.22 The anterior portion of the perineum is called the urogenital triangle because its borders form a triangle that encloses the urethral (uro-) and vaginal (-genital) orifices.

27.23 Prolactin, estrogens, and progesterone are the hormones that regulate the synthesis of milk. Oxytocin regulates the ejection of milk.

27.24 Alveoli are located in lobules of the mammary glands.

27.25 The hormones responsible for the proliferative phase of endometrial growth are estrogens; for ovulation, LH; for growth of the corpus luteum, LH; and for the midcycle surge of LH, estrogens.

27.26 The *SRY* gene on the Y chromosome is responsible for the development of the gonads into testes.

27.27 The presence of dihydrotestosterone (DHT) stimulates differentiation of the external genitals in males; its absence allows differentiation of the external genitals in females.

MEASUREMENTS

U.S. Customary System

PARAMETER	UNIT	RELATION TO OTHER U.S. UNITS	SI (METRIC) EQUIVALENT
Length	inch	1/12 foot	2.54 centimeters
	foot	12 inches	0.305 meter
	yard	36 inches	0.914 meters
	mile	5,280 feet	1.609 kilometers
Mass	grain	1/1000 pound	64.799 milligrams
	dram	1/16 ounce	1.772 grams
	ounce	16 drams	28.350 grams
	pound	16 ounces	453.6 grams
	ton	2,000 pounds	907.18 kilograms
Volume (Liquid)	ounce	1/16 pint	29.574 milliliters
	pint	16 ounces	0.473 liter
	quart	2 pints	0.946 liter
	gallon	4 quarts	3.785 liters
Volume (Dry)	pint	1/2 quart	0.551 liter
	quart	2 pints	1.101 liters
	peck	8 quarts	8.810 liters
	bushel	4 pecks	35.239 liters

International System (SI)

BASE UNITS

UNIT	QUANTITY	SYMBOL
meter	length	m
kilogram	mass	kg
second	time	s
liter	volume	l
mole	amount of matter	mol

PREFIXES

PREFIX	MULTIPLIER	SYMBOL
tera-	$10^{12} = 1,000,000,000,000$	T
giga-	$10^{9} = 1,000,000,000$	G
mega-	$10^{6} = 1,000,000$	M
kilo-	$10^{3} = 1,000$	k
hecto-	$10^{2} = 100$	h
deca-	$10^{1} = 10$	da
deci-	$10^{-1} = 0.1$	d
centi-	$10^{-2} = 0.01$	c
milli-	$10^{-3} = 0.001$	m
micro-	$10^{-6} = 0.000,001$	μ
nano-	$10^{-9} = 0.000,000,001$	n
pico-	$10^{-12} = 0.000,000,000,001$	p

ANSWERS

Answers to Self-Quiz Questions

Chapter 1 1. (e) 2. (e) 3. (b) 4. (d) 5. (d) 6. (c) 7. epithelia, connective, muscular, nervous 8. peritoneum 9. epigastric 10. parasagittal 11. organs 12. ipsi 13. inferior, lateral (or superior, also) 14. F (The joints of the body are part of the *skeletal* system.) 15. F 16. F 17. F (The diaphragm divides the ventral cavity into *thoracic and abdominopelvic* portions.) 18. T 19. 1(d), 2(e), 3(g), 4(h), 5(a), 6(b), 7(c) 20. 1(c), 2(g), 3(a), 4(e), 5(d), 6(h), 7(f), 8(b)

Chapter 2 1. (a) 2. (a) 3. (c) 4. (e) 5. (d) 6. (e) 7. (c) 8. (b) 9. filtration 10. meiosis 11. receptor-mediated endocytosis 12. (phospho)lipids 13. ligands 14. interstitial fluid 15. osmosis 16. F (*Phospholipid* molecules make up the largest percentage of lipid molecules in the plasma membrane.) 17. F (*Transmembrane* proteins span the entire lipid bilayer.) 18. F (Facilitated diffusion is considered *a passive* form of membrane transport because it requires the assistance of a transporter protein.) 19. 1(j), 2(l), 3(c), 4(k), 5(e), 6(d), 7(h), 8(g), 9(a), 10(i), 11(b), 12(f) 20. 1(j), 2(a), 3(l), 4(h), 5(c), 6(g), 7(b), 8(f), 9(k), 10(e), 11(i), 12(d)

Chapter 3 1. (b) 2. (d) 3. (e) 4. (b) 5. (d) 6. (a) 7. (a) 8. (a) 9. cardiac 10. exocrine 11. gap junction 12. collagen, elastic, reticular 13. serous 14. lamina propria 15. T 16. T 17. F (A biopsy is the removal of a sample of *live* cells for microscopic examination.) 18. T 19. 1(e), 2(f), 3(h), 4(a), 5(c), 6(g), 7(b), 8(d) 20. 1(d), 2(e), 3(a), 4(h), 5(b), 6(g), 7(c), 8(f), 9(j), 10(i)

Chapter 4 1. (d) 2. (b) 3. (a) 4. (c) 5. (b) 6. (c) 7. syngamy 8. ectoderm, endoderm, mesoderm 9. gastrulation 10. amniotic fluid 11. angiogenesis 12. decidua 13. human chorionic gonadotropin (hCG) 14. F (Organ systems are almost completely developed by the end of the *second* trimester.) 15. F (The chorionic villi form the *fetal* portion of the placenta.) 16. F (Development of the placenta occurs during the *fourth* month of pregnancy.) 17. F (The primary germ layers, which give rise to all tissues and organs, are present at the time of *gastrulation*.) 18. F (The umbilical cord contains *one umbilical vein* that *carries* oxygenated blood to the fetus, and *two umbilical arteries* that *carry* fetal blood to the placenta.) 19. F (The most common site for an ectopic pregnancy is the *uterine tube*.) 20. (a)2, (b)1, (c)3, (d)3, (e)3, (f)1, (g)2 21. 1(d), 2(a), 3(f), 4(c), 5(g), 6(b), 7(e) 22. 1(d), 2(h), 3(b), 4(j), 5(a), 6(e), 7(c), 8(f), 9(i), 10(g)

Chapter 5 1. (c) 2. (b) 3. (a) 4. (d) 5. (a) 6. (d) 7. (c) 8. ceruminous 9. epidermis, dermis 10. sudoriferous 11. basal cell carcinomas, squamous cell carcinomas 12. 10, 25 13. F (Hair follicles and sebaceous glands are *absent* in areas covered by thick skin.) 14. F (The scientific name for the cuticle of a nail is the *eponychium*.) 15. F (The dermis receives its blood supply from two arterial networks or plexuses: the cutaneous plexus and the *papillary* plexus.) 16. T 17. T 18. F (Racial differences in skin color are mainly due to the *amount of melanin produced by melanocytes* in the epidermis.) 19. T 20. F (Increased incidence of heat stroke in the elderly probably results from *decreased* sweat production.) 21. 1(b), 2(d), 3(e), 4(a), 5(f), 6(c) 22. 1(c), 2(e), 3(b), 4(a), 5(d), 6(f) 23. 1(c), 2(b), 3(e), 4(a), 5(d)

Chapter 6 1. (e) 2. (a) 3. (b) 4. (e) 5. (d) 6. (d) 7. (d) 8. dense irregular fibrous connective tissue 9. endochondral 10. calcification 11. lacunae 12. yellow bone marrow 13. osteoclasts 14. appositional 15. Paget's 16. Osteogenic sarcoma 17. T 18. F (*Central canals* are microscopic canals running longitudinally through bone.) 19. F (Bones do have a nerve supply.) 20. F (A fracture hematoma is a *blood clot* that bridges the broken ends of the bones.) 21. T 22. F (*Osteoporosis* is a disease in which bone *resorption* outpaces bone *deposition*.) 23. F (*Osteomyelitis* is an infection of bone often caused by bacteria, usually *Staphylococcus aureus*.) 24. 1(a), 2(e), 3(c) 4(b), 5(d)

Chapter 7 1. (c) 2. (d) 3. (b) 4. (a) 5. (d) 6. (a) 7. (e) 8. (a) 9. (e) 10. (e) 11. atlas, axis 12. long, short, flat 13. Wormian 14. hyoid 15. T 16. T 17. T 18. F (The sacral promontory projects anteriorly from the *base* of the sacrum.) 19. F (The *head* of a rib articulates with demifacets on the bodies of adjacent vertebrae.) 20. 1(i), 2(j), 3(r), 4(o,k), 5(p), 6(n,a), 7(q), 8(d), 9(s), 10(g,m), 11(f,b)

Chapter 8 1. (c) 2. (e) 3. (d) 4. (e) 5. (c) 6. (b) 7. (a) 8. (b) 9. (e) 10. (b) 11. (d) 12. (b) 13.ischium 14. patella 15. posterior, femur 16. false 17. head, femur 18. proximal (or posterior) 19. trochlea, capitulum 20. F (The metatarsals are *distal* to the tarsals.) 21. T 22. T 23. F (The capitulum of the humerus articulates with the *head* of the radius.) 24. T 25. 1(j,f), 2(h,p), 3(a,k), 4(e,n), 5(b,c,j,m), 6(i,o), 7(c,q), 8(d), 9(g), 10(l) 26. 1(b), 2(e), 3(c), 4(a), 5(d), 6(f)

Chapter 9 1. (b) 2. (c) 3. (a) 4. (e) 5. (d) 6. (e) 7. (a) 8. (c) 9 (a) 10. (b) 11. cartilaginous joint (synchondrosis), hyaline cartilage 12. ligament 13. diarthrosis 14. hyaline cartilage (articular cartilage) 15. bursae 16. subluxation 17. tibial collateral ligament, anterior cruciate ligament, medial meniscus 18. F (Synovial fluid becomes *less* viscous as joint activity increases.) 19. T 20. F (Raising the arm to point straight ahead is an example of *flexion*.) 21. F 22. F (A shoulder dislocation is an injury to the *glenohumeral* joint.) 23. F (*Osteoarthritis* is a common *degenerative* disorder seen in most older persons.) 24.F (*Gout* refers to a disease *that often affects the joints of the foot*.) 25. 1(h), 2(d), 3(a), 4(e), 5(b), 6(i), 7(c), 8(f), 9(g) 26. 1(h), 2(e), 3(d), 4(b), 5(c), 6(a), 7(f), 8(g)

Chapter 10 1. (c) 2. (b) 3. (d) 4. (a) 5. (c) 6. (c) 7. (d) 8. superficial fascia 9. endomysium 10. Z lines 11. troponin, tropomyosin 12. S, U, U, S, S 13. smooth and cardiac 14. strain, sprain 15. F (Enlargement of existing muscle fibers is called *hypertrophy*.) 16. F (In adult life, growth of skeletal muscle is due to an increase in the *size of existing* cells rather than an increase in the *number of muscle cells*.) 17. F (Duchenne muscular dystrophy (DMD) typically strikes *boys between the ages of 2 and 5*.) 18. T 19. F (Thick filaments are made mostly of *myosin*.) 20. 1(b, e), 2 (c, g), 3 (a,d,f) 21. 1(e), 2(d), 3(b), 4(f), 5(a), 6(c)

Chapter 11 1. (d) 2. (b) 3. (d) 4. (e) 5. (b) 6. (e) 7. (d) 8.(e) 9. prime mover (agonist), synergists 10. origin, insertion 11. bones, joints, muscle contraction 12. semitendinosus, semimembranosus, biceps femoris 13. mastoid process 14. central 15. trapezius, rhomboid major, rhomboid minor 16. perineum 17. sphincter 18. flexion 19. anterior, four, tibia 20. gastrocnemius, soleus, plantaris 21. T 22. F (The action of the biceps brachii on the forearm at the elbow joint illustrates the structure and action of a *third-class* lever system.) 23. F (The muscles that plantar flex the foot at the ankle joint are located in the *posterior* compartment of the leg.) 24. F (The inferior free border of the *external oblique* aponeurosis forms the inguinal ligament, which runs from the anterior superior iliac spine to the pubic tubercle.) 25. (a)2 and 3, (b)3 and 6, (c)4 and 5, (d)5, (e)1 and 6, (f)3, (g)6 26. 1(f), 2(d), 3(b), 4(a), 5(c), 6(j), 7(e), 8(i), 9(g), 10(h) 27. (a)3, (b)6, (c)1, (d)11, (e)8, (f)2, (g)10, (h)5, (i)7, (j)9, (k)4

Chapter 12 1. (e) 2. (d) 3. (b) 4. (a) 5. (c) 6. (e) 7. (d) 8. Upper limb: axillary artery pulse, antecubital fossa, styloid process, pisiform; Lower limb: femoral artery pulse, popliteal fossa, malleolus, calcaneus 9. (a) inferior (b) inferior (c) distal (d) superior (e) anterior (f) lateral (g) inferior 10. latissimus dorsi, trapezius 11. xiphisternal 12. T 13. T 14. T 15. T 16. F (A vein of the upper limb often used for intravenous therapy is located in the *antecubital* fossa.) 17. 1(f), 2(e), 3(b), 4(a), 5(h), 6(g), 7(c), 8(d) 18. 1(b), 2(c), 3(f), 4(d), 5(a), 6(g), 7(h), 8(e)

Chapter 13 1. (b) 2. (d) 3. (a) 4. (d) 5. (a) 6. (c) 7. (a) 8. (b) 9. (a) 10. (b) 11. (e) 12. (b) 13. differential white blood cell 14. 45%, 55% 15. emigration 16. septicemia 17. reticulocyte 18. colony-stimulating factors, interleukins 19. iron 20. T 21. F (*White* blood cells have lobed nuclei and abundant mitochondria.) 22. T 23. F (Gamma globulin is a *solution of immunoglobulins*.) 24. (a)5, (b)3, (c)3, (d)1, (e)2, (f)4, (g)4 and 5, (h)2, (i)1, (j)4

Chapter 14 1. (d) 2. (b) 3. (a) 4. (d) 5. (e) 6. (a) 7. epicardium, visceral pericardium 8. right and left coronary arteries 9. atrioventricular 10. circumflex artery 11. open, closed 12. systole, diastole 13. left ventricle 14. inferior left 15. F (Trabeculae carneae are ridges located on the inner surface of the right and *left ventricles*.) 16. F (The *inferior* vena cava passes through the diaphragm.) 17. F (Blood passes into the aorta from the *left* ventricle.) 18. T 19. F (Most of the ventricular filling occurs during atrial *diastole*.) 20. T 21.

F (Chordae tendineae attach AV valve cusps to *papillary* muscles.) 22. F (The visceral layer of the serous pericardium is also called the *epicardium*.) 23. 1(d), 2(c), 3(a,f), 4(b,e)

Chapter 15   26. 1. (c) 2. (c) 3. (e) 4. (b) 5. (e) 6. (e) 7. (c) 8. (c) 9. (d) 10. (a) 11. (c) 12. common hepatic artery, left gastric artery, splenic artery 13. vasa vasorum 14. tunica intima 15. inferior vena cava 16. hepatic portal vein 17. anastomosis 18. media, vasoconstriction 19. primary, secondary 20. continuous, fenestrated, sinusoids 21. pulmonary 22. F (At rest, the largest portion of blood is located in the *veins*.) 23. F (The *right brachiocephalic vein joins* with the *left brachiocephalic vein* to form the superior vena cava.) 24. T 25. T 26. F (An increase in stimulation by sympathetic nerves typically stimulates *vasoconstriction* of blood vessels.) 27. F (Thrombi lodged in the ulnar artery *and the radial artery* would prevent any blood from reaching the hand.) 28. 1(f), 2(d), 3(h), 4(k), 5(g), 6(j) 7(b), 8(c), 9(i), 10(a), 11(e)

Chapter 16   1. (b) 2. (e) 3. (d) 4. (e) 5. (b) 6. (e) 7. (b) 8. (e) 9. lacteals 10. veins 11. red bone marrow, thymus 12. pharyngeal, palatine, lingual 13. pathogens 14. MALT (mucosa associated lymphatic tissue) 15. RNA, DNA 16. T 17. T 18. T 19. T 20. F (Because the T cell population *decreases* with age, elderly persons *are strongly* encouraged to get flu (influenza) shots every year.) 21. (a)4, (b)7, (c)1, (d)8, (e)2, (f)6, (g)5, (h)3

Chapter 17   1. (c) 2. (a) 3. (d) 4. (d) 5. (c) 6. (a) 7. (c) 8. (b) 9. ependymal cells 10. somatic, autonomic, enteric 11. tract 12. effectors 13. GABA and glycine 14. synaptic vesicles 15. oligodendrocytes 16. central, peripheral 17. T 18. T, 19. T 20. T 21. F (The nucleus of a neuron is located in the *soma/cell body*.) 22. F (*Smooth* muscles are effectors in the autonomic nervous system.) 23. (a)3, (b)6, (c)7, (d)8 (e)4, (f)5, (g)1, (h)2 24. (a)3, (b)1, (c)1, (d)2, (e)3, (f)2, (g)1

Chapter 18   1. (e) 2. (e) 3. (b) 4. (c) 5. (e) 6. (d) 7. (e) 8. (c) 9. medulla oblongata, second lumbar vertebra 10. denticulate ligament 11. skeletal muscles, cardiac muscle, smooth muscle, glands 12. anterior/ventral, C5-T1 13. anterior, lateral 14. dermatome 15. F (Cell bodies of motor neurons are located in the *anterior (ventral) horn*.) 16. F (The lateral spinothalamic tract is a *sensory* tract.) 17. T 18. T 19. T 20. F (Individual axons are protected by a connective tissue wrapping known as the *endoneurium*.) 21.F (Inability to feel touch around the umbilicus could indicate an injury to the *T10 spinal nerve*.) 22. T 23. (a)1, (b)2, (c)1, (d)5, (e)2, (f)4, (g)3 24. (a)7, (b)5, (c)6, (d)1, (e)2, (f)4, (g)3 25. 1(c,m), 2(g), 3(j,n), 4(b,e), 5(k), 6(h,l,o), 7(i), 8(f), 9(d), 10(a)

Chapter 19   1. (d) 2. (c) 3. (e) 4. (b) 5. (c) 6. (b) 7. (a) 8. (a) 9. (d) 10. choroid plexuses 11. third ventricle 12. apneustic, pneumotaxic 13. reticular activating system 14. association 15. cerebral cortex, gray, gyri, sulci 16. (a)w, (b)g, (c)g, (d)g, (e)w, (f)g 17. (a)9, (b)7, (c)5, (d)2, (e)4, (f)6, (g)3, (h)8, (i)1, (j)10 18. mamillary bodies 19. melatonin 20. seventh (facial) 21. F 22. T 23. F (The three parts of the brain stem are the medulla oblongata, pons, and *midbrain*.) 24. F (The *diencephalon contains* the supraoptic nucleus and the geniculate nuclei.) 25. F (The cranial nerve responsible for abducting your right eye as you read this sentence is *cranial nerve VI*.) 26. 1(f,j), 2(e,h), 3(c,l), 4(d,k), 5(g), 6(b), 7(a) 27. 1(g), 2(j), 3(a), 4(l), 5(c), 6(i), 7(d), 8(k), 9(f), 10(b), 11(h), 12(e)

Chapter 20   1. (b) 2. (b) 3. (a) 4. (a) 5. (d) 6. (e) 7. (e) 8. (c) 9. (d) 10. splanchnic 11. dual 12. MAO 13. alpha, beta 14. nicotinic, muscarinic 15. adrenal medulla 16. vagus 17. T 18. T 19. F (The ganglia that make up the sympathetic trunks are also called *paravertebral or vertebral chain ganglia*.) 20. F (Postganglionic neurons begin in the *autonomic ganglia* and extend to effectors.) 21. F (An event occurring during the fight-or-flight response is *decreased* urine production due to blood being shunted *away from the kidneys*.) 22. F (All sympathetic and parasympathetic preganglionic neurons release the neurotransmitter *acetylcholine*.) 23. (a)2, (b)5, (c)6, (d)4, (e)1, (f)3 24. (a)1, (b)2, (c)1, (d)2, (e)1, (f)1, (g)1, (h)1, (i)3, (j)1

Chapter 21   1. (c) 2. (e) 3. (a) 4. (d) 5. (d) 6. (c) 7. (d) 8. stimulus 9. exteroceptors, interoceptors (visceroceptors), proprioceptors 10. mechano- 11. brain stem, association 12. referred 13. spinocerebellar tracts 14. stereognosis 15. thalamus 16. lower motor neurons in the ventral horn 17. F (Stimulation of pain receptors in skeletal muscles, joints, and tendons results in *somatic pain*.) 18. T 19. F (The precentral gyrus of the cerebral cortex is the *primary motor area*.) 20. T 21. F (*First-order* neurons conduct impulses from somatic receptors to the central nervous system.) 22. F (Indirect (extrapyramidal) motor pathways arise from *brainstem gray matter nuclei*.) 23. (a)2, (b)3, (c)1, (d)1, (e)3, (f)2 24. (a)4, (b)1, (c)3, (d)2 25. 1(f), 2(h), 3(a), 4(e), 5(g), 6(d), 7(b), 8(c)

Chapter 22   1. (d) 2. (a) 3. (b) 4. (c) 5. (e) 6. (a) 7. (e) 8. (b) 9 (a) 10. stapedius, tensor tympani 11. papillae, fungiform 12. ciliary, scleral venous 13. basilar membrane, organ of Corti 14. conjunctiva 15. dynamic, static 16. (a)3, (b)1, (c)2, (d)4, (e)5 17. (a)1, (b)5, (c)3, (d)4, (e)6, (f)2 18. static equilibrium, utricle, saccule 19. F (*Axons* of olfactory receptors collectively make up the olfactory (I) nerves.) 20. T 21. F (The membranous labyrinth of the internal (inner) ear contains *endolymph*.) 22. F (The primary function of the *external auditory canal* is to conduct sound waves from the exterior to the tympanic membrane.) 23. F (The optic nerve is comprised of axons from retinal *ganglion cells*.) 24. (a)11, (b)6, (c)7, (d)1, (e)3, (f)10, (g)2, (h)9, (i)5, (j)4, (k)8, (l)12 25. (a)5, (b)4, (c)9, (d)3, (e)12, (f)6, (g)7, (h)8, (i)1, (j)2, (k)11, (l)10 26. 1(e,n), 2(a,f), 3(c,d,j,l), 4(b,k), 5(g,i) 6(c,h), 7(c,m)

Chapter 23   1. (a) 2. (c) 3. (a) 4. (d) 5. (d) 6. (e) 7. (a)3, (b)2, (c)1, (d)4 8. hypothalamus 9. tropic, target 10. alpha, increase 11. human growth hormone 12. external carotid artery 13. parathyroid hormone 14. sphenoid 15. F (*Decreasing* the number of receptors on a target cell decreases the cell's sensitivity to a hormone.) 16. T 17. F (Pituicytes are the *glial cells* of the posterior pituitary.) 18. F (Oxytocin stimulates *milk ejection*.) 19. F (*Pinealocytes* are the principal secreting cells of the pineal gland.) 20. T 21. F (The *head* of the pancreas is situated closer to the duodenum of the small intestine than to the spleen.) 22. 1(c), 2(f), 3(b), 4(g), 5(d), 6(a), 7(e) 23. (a)5, (b)6, (c)8, (d)3, (e)2, (f)9, (g)1, (h)10, (i)7, (j)4 24. 1(d), 2.(c), 3.(b), 4(e), 5(a)

Chapter 24   1. (c) 2. (a) 3. (c) 4. (e) 5. (d) 6. (e) 7. (d) 8. (a) 9 (d) 10. trachea 11. epiglottis 12. glottidis, vestibuli 13. lobule 14. (a)3, (b)2, (c)1 15. atelectasis 16. surfactant 17. inhalation 18. T 19. T 20. F (The trachea is *anterior* to the esophagus.) 21. F (The branches of the aorta that carry blood to the lungs are the *bronchial* arteries.) 22. T 23. (a)1, (b)2, (c)2, (d)1, (e)1 24. (a)1, (b)5, (c)2, (d)3, (e)4, (f)4, (g)2 and 3, (h)2 25. 1(c), 2(a), 3(g), 4(e), 5(b), 6(f), 7(d)

Chapter 25   1. (c) 2. (d) 3. (b) 4. (a) 5. (c) 6. (b) 7. (a) 8. (e) 9. greater omentum 10. (a)1, (b)4, (c)3, (d)2, (e)5 11. deglutition 12. peristalsis 13. permanent 14. liver 15. cystic 16. hydrochloric acid, intrinsic factor, pepsinogen, gastric lipase 17. caudate, quadrate 18. trypsin, chymotrypsin, carboxypeptidase 19. F (The submucosal plexus (plexus of Meissner) is important in controlling *secretions* of the GI tract, while the myenteric plexus (plexus of Auerbach) controls *motility* of the GI tract.) 20. F (The gallbladder *stores* bile.) 21. T 22. T 23. T 24. F (The duodenum is the most *proximal* region of the small intestine.) 25. 1(e), 2(g), 3(a), 4(f), 5(h), 6(b), 7(d), 8(c)

Chapter 26   1. (d) 2. (e) 3. (b) 4. (c) 5. (c) 6. (d) 7. (a) 8. renal papilla 9. efferent, afferent 10. podocytes 11. spongy 12. juxtaglomerular 13. ureters 14. (a)4, (b)5, (c)1, (d)9, (e)7, (f)8, (g)6, (h)3, (i)2 15. micturition 16. T 17. T 18. F (The *external* urethral sphincter is composed of voluntary skeletal muscle, and the *internal* urethral sphincter is composed of involuntary smooth muscle.) 19. F (The *endothelial cells of glomerular capillaries are* part of the endothelial-capsular (filtration) membrane.) 20. 1.(a,d,f), 2.(d,e), 3(c,g), 4(b,d,h)

Chapter 27   1. (a) 2. (a) 3. (d) 4. (b) 5. (b) 6. (d) 7. (a)4, (b)1, (c)5, (d)3, (e)2 8. spiral arterioles 9. scrotum, penis, prostate gland, bulbourethral (Cowper's) glands 10. basalis 11. ejaculatory duct 12. corpus albicans 13. prostate 14. ovarian 15. T 16. F (The female urethra is located *anterior* to the vagina.) 17. F (Fertilization of a secondary oocyte by a sperm cell normally occurs in the *uterine tube*.) 18. T 19. F (The acrosome is the portion of a sperm that contains *digestive enzymes*.) 20. T 21. (a)7, (b)4, (c)3, (d)1, (e) 6, (f)2, (g)5 22. 1(c,h), 2(b,e,h), 3(d), 4(a,f,g)

Answers to Critical Thinking Questions

Chapter 1

1. The kidneys are located behind the parietal peritoneum. To view the posterior surface of the kidney, a scope would penetrate the posterior body wall but would not enter the peritoneal cavity.

2. On the x-ray, the bone (humerus) of the arm (brachium) would be located most proximal. Two bones of the forearm (antebrachium) the ulna and radius, would be located distal to the humerus with the ulna situated medial to the radius. In the hand, 14 phalanges would comprise the distal fingers, while the 5 metacarpals would be found in the more proximal palm. Distal to the ulna and radius, and proximal to the metacarpals, the wrist would be comprised of 8 carpal bones.

3. The alien would have 2 tails, 4 arms, 2 legs, and a mouth where the navel is usually located.

4. Lung, diaphragm, stomach, large intestine, small intestine; possibly part of pancreas, ovary or uterine tube, kidney.

5. The peritoneum, the largest serous membrane in the body, covers most organs in the abdominal cavity. Therefore, an infection in this structure can spread to any or all organs in the cavity.

Chapter 2

1. An increase in neutrophils indicates an increased demand for phagocytosis, most likely due to bacterial infection.

2. Maternal inheritance is due to the DNA found in extranuclear organelles such as the mitochondria. The sperm contributes only nuclear DNA. Mitochondria are inherited only from the mother.

3. Water would move out of the red blood cells, where the concentration of water is higher, causing them to shrink.

4. Arsenic will poison the mitochondrial enzymes involved in cellular respiration. This will halt production of ATP by the mitochondria. Cells with a high metabolic rate such as muscle cells will be particularly affected.

5. Disruption of microtubules would halt cell division. Microtubules are also involved in cell motility, transportation, and movement of cilia, all of which would be affected.

6. The cell imports the virus via receptor-mediated endocytosis. A potential obstacle is having a healthy immune system react to this "foreign" virus and thus try to eliminate it along with the healthy gene.

Chapter 3

1. The drug will make it easier for the cilia on the epithelial cells to move the mucus, in which microbes are trapped, away from the lungs. Coughing up the thinned mucus should also be easier.

2. The cells are nonkeratinized stratified squamous epithelial cells which, when separated and spread over a glass microscope slide, would have the appearance of "tiles", i.e. thin and flat. In cervical dysplasia the cells appear altered in size, shape, and organization.

3. Janelle's kidneys may drop out of position from lack of supporting fat. There would also be less fat for padding in joints and buttocks, and eyes may appear sunken.

4. The tissue that Jonathan pierced is epithelium which constitutes the most superficial layer of skin. Epithelium is avascular, and therefore, there was no visible bleeding.

5. Bacteria would need a mechanism to pass through or between keratin-filled cells of the epidermis and to combat any phagocytic cells present. They would also need to survive any enzymes produced by cells the bacteria might encounter.

Chapter 4

1. Alchohol and other substances capable of producing birth defects are able to pass freely through the placenta.

2. All tissues and organs form from the embryo's three primary germ layers. The ectoderm, the outermost germ layer, develops into the nervous system and the epithelial layer of the skin. This early embryological connection may sometimes result in disorders showing signs in both tissues.

3. Josefina is in false labor. In true labor, the contractions are regular and the pain may localize in the back. The pain may be intensified by walking. True labor is indicated by the "show" of bloody mucus and cervical dilation.

4. Certain microorganisms can infect the placenta itself to infect the embryo/fetus (transplacental transmission). Risk is greatest during the first trimester, a crucial time in the differentiation of cells.

5. Pregnancy increases appetite to accommodate the increase in nutritional demands. Increasing pressure on the urinary bladder increases frequency of urination. As the uterus increases in size, abdominal contents are pushed upward, and the stomach and its contents may push up into the esophagus (heartburn). Elena may need to prop herself up higher at night to help prevent heartburn and make breathing easier.

Chapter 5

1. Chronic exposure to UV light causes damage to elastin and collagen, promoting wrinkles. The suspicious growth may be skin cancer caused by UV photodamage to skin cells.

2. Dilation of the blood vessels in the skin increases blood flow, which means more body heat will reach the surface of the skin and be radiated away. Therefore, although they will feel warmer superficially, body temperature will decrease. They should wear their coats!

3. "In part, this increase is due to depletion of the ozone layer, which absorbs some UV light high in the atmosphere. But the main reason for the increase is that more people are spending more time in the sun. Malignant melanomas metastasize rapidly and can kill a person within months of diagnosis."

4. The surface layer of the skin is keratinized stratified squamous epithelium. The keratinocytes are dead and filled with the protective protein keratin. Lipids from lamellar granules and sebum also function in waterproofing.

5. The higher the humidity, the less evaporation can occur, which means the less cooling can occur. Therefore people perceive the same temperature as hotter at a higher humidity.

Chapter 6

1. Lynda's diet lacks several nutrients necessary for bone health, including essential vitamins (such as A, D), minerals (such as calcium), and proteins. Her lack of exercise will weaken bone strength. Her age and smoking habit may result in a lack of estrogens, causing demineralization of bone.

2. In bone tissue, there is a balance between the actions of osteoblasts and osteoclasts. There also exists a balance between the inorganic (mineral) and organic (collagen) components of bone matrix. When these balances are altered, the consistency of the bone matrix will change accordingly. If osteoblast activity outpaces osteoclast activity, then too much new tissue will be formed, causing the bones to become abnormally thick and heavy. In this particular case, Scott noticed the spurs that were produced by too much mineral material being deposited in the bone. The surplus formed thick bumps, called spurs, on the bones that most likely interfered with movement at joints.

3. At Aunt Edith's age, the production of several hormones necessary for bone remodeling (such as estrogens and human growth hormone) would be decreased. She may have loss of bone mass, brittleness, and possibly osteoporosis. Increased susceptibility to fractures results in damage to the vertebrae and loss of height.

4. Exercise causes mechanical stress on bones but since there is no gravity in space, the pull of gravity on bones is missing. The lack of stress results in demineralization of bone and weakness.

5. Because babies have fontanels (soft spots), the cranial bones tend to be more movable than after the sutures close. As the baby passed through the birth canal (and possibly because of the use of forceps to assist the delivery), the bones could move into a more "cone-shaped" position. Likewise, laying the baby on its back all the time flattens the occipital bone. In addition, infant bones tend to be more flexible than older bones, which lose moisture and become more brittle over time.

Chapter 7

1. Fontanels, the soft spots between cranial bones, are fibrous connective tissue membrane–filled spaces. They allow the infant's head to be molded during its passage through the birth canal and allow for brain and skull growth in the infant.

2. The pituitary gland lies under the brain as it sits in the sella turcica of the sphenoid bone. Thus, the gland is heavily protected and difficult to reach from any direction. In addition, there are important nerves, such as the optic nerve, and blood vessels passing nearby.

3. Infants are born with a single concave curve. Adults have four curves in their vertebral column at the cervical, thoracic, lumbar, and sacral regions.

4. The occipital bone protects key areas of the cerebrum, but also the brain stem, which joins the spinal cord at the foramen magnum. Damage to the spinal cord and/or the vital centers in the medulla oblongata controlling breathing, heart activity, and blood pressure could be fatal.

5. The x-rays were taken to view the paranasal sinuses—the "fuzzy looking holes." John probably had sinusitis due to infected paranasal sinuses.

Chapter 8

1. There are 14 phalanges in each hand: 2 in the thumb and 3 in each of the other fingers. Farmer Ramsey has lost 5 phalanges on his left hand so he has 9 remaining on his left and 14 remaining on his right for a total of 23.

2. The talus receives all of the body's weight. Half that weight is normally transmitted to the calcaneus but the weight distribution will be shifted on the hallux by dancing. Problems may appear in the talocrural (ankle) joint and hallux. Rose also has a deviated hallux or bunion that may have been caused by tight ballet shoes. The dancing did strengthen the foot muscles and correct the flat feet.

3. Bone growth occurs in response to mechanical stress. Therefore, any activity that causes the deltoid muscle to pull on the deltoid tuberosity would

increase its growth. Such activity might include paddling the large kayaks frequently over time and/or against turbulent waters.

4. Grandmother Amelia has probably fractured the neck of her femur. This is a common fracture in the elderly and is often called a "broken hip." The sacrum does indeed articulate with the ilium, and Grandpa Jeremiah could have osteoarthritis in that joint, which could be quite painful. There are, however, other lower back issues (e.g., herniated discs, sciatica, etc.) that could be equally painful.

5. Derrick has patellofemoral syndrome (runner's knee). Running daily in the same direction on a banked track has stressed the downhill knee. The patella is tracking laterally to its proper position resulting in pain after exercise.

Chapter 9

1. Structurally the hip joint is a simple ball-and-socket synovial joint composed of the head of the femur (ball) and the acetabulum of the hip bone (socket). The knee is really three joints: patellofemoral, lateral tibiofemoral, and medial tibiofemoral. It is a combination gliding and hinge joint.

2. Flexion at knee, extension at hip, extension of vertebral column except for hyperextension at neck, abduction at shoulder, flexion of fingers, adduction of thumb, flexion/extension at elbow, and depression of mandible.

3. No more body surfing for Lars. He has a dislocated shoulder. The head of the humerus was displaced from the glenoid cavity causing tearing of the supporting ligaments and tendons (rotator cuff) of the shoulder joint.

4. Most likely the cartilage of the menisci were torn in the accident. Pieces of cartilage (and possibly bone) may be in the joint cavity. In addition, it is possible that the articular cartilage of the tibia (and possibly the femur) was damaged. Damage to the menisci and ligaments associated with the knee would contribute to stability problems.

5. There are many possibilities in Chuck's situation. The problems could be all unrelated, and the back pain could be completely related to vertebral column issues. On the opposite end of the spectrum, it is possible that a long-term ankle problem caused Chuck to walk differently in order to favor the damaged ankle, thus causing the bowing of the legs as the muscles pull the leg in a slightly different direction. From there, the change in gait could cause the muscles that attach to the pelvis to pull more forcefully on points that they normally would not, which could pull the pelvic bones slightly out of alignment in relation to the vertebral column, thus causing pain at the sacroiliac joint.

Chapter 10

1. The runner's leg muscles will contain a higher amount of slow fibers—rich in myoglobin, mitochondria, and blood. The weightlifter's arm muscles will contain abundant fast fibers–high in glycogen, with less myoglobin and blood capillaries than the slow fibers. The slow fibers appear red while the fast fibers are white in color.

2. Bill's muscles in the casted leg were not being used, so they decreased in size due to loss of myofibrils. Bill's leg shows disuse atrophy.

3. Acetylcholine (ACh) is the neurotransmitter used to "bridge the gap" at the neuromuscular junction. If ACh release is blocked, the neuron cannot send a signal to the muscle and the muscle will not contract.

4. Some cardiac muscle cells are autorhythmic and contraction is started intrinsically. Gap junctions connect the cardiac cells, spreading the impulse through the fiber network so it contracts as a unit. Cardiac muscle has a long refractory period so it will exhibit tetanus like skeletal muscle does.

5. First, it is the term "extensibility" (not "contractility") that refers to stretching of a muscle fiber. In addition, the contractile proteins never change length during muscle contraction or during stretching of a muscle fiber—they only change position relative to each other.

Chapter 11

1. The orbicularis oris protrudes the lips. The genioglossus protracts the tongue. The buccinator aids in sucking.

2. One likely possibility is the rhomboid major muscle.

3. Refer to Exhibit 11.3 "MUSCLES THAT MOVE THE MANDIBLE." You should have included temporalis (retract, elevate, i.e., closing the mouth); masseter (retract, elevate, i.e., closing the mouth); medial pterygoid (protract, elevate, i.e., closing the mouth); and lateral pterygoid (protract, depress, i.e., opening the mouth) Note that it is a suprahyoid group of muscles that aids the lateral pterygoid in depressing the mandible to opening the mouth (see Exhibit 11.5) but not with as much force as afforded by the temporalis and masseter muscles in closing the mouth; closing the mouth is a much more powerful movement.

4. The most likely injury is a rotator cuff injury. The rotator cuff includes tendons of the subscapularis, supraspinatus, infraspinatus, and teres minor muscles. The most common injury is to the supraspinatus muscle and/or tendon (impingement syndrome).

5. Wyman probably has an inguinal hernia. He should see a doctor, since constriction of the intestines pushing out through the opening can cause serious damage.

Chapter 12

1. The spine of C7 forms a prominence along the midline at the base of the neck. The spine of T1 forms a second prominence slightly inferior to C7.

2. Percussion is touching or examining using the hands. Palpitation is a forceful beating of the heart that can often be felt by the patient.

3. The projection (shape of an organ on the surface of the body) of the heart on the anterior surface of the chest is indicated by four points as follows: The *inferior left point* is the apex (inferior, pointed end) of the heart, which projects downward and to the left and can be palpated in the fifth intercostal space, about 9 cm (3.5 in.) to the left of the midline. The *inferior right point* is at the lower border of the costal cartilage of the right sixth rib, about 3 cm to the right of the midline. The *superior right point* is located at the superior border of the costal cartilage of the right third rib, about 3 cm to the right of the midline. The *superior left point* is located at the inferior border of the costal cartilage of the left second rib, about 3 cm to the left of the midline. If you connect the four points, you can determine the heart's location and size (about the size of a closed fist).

4. The *gluteus medius muscle* is the site for an intramuscular injection, not the gluteus maximus, which Randy seems to think. In order to give this injection, the buttock is divided into quadrants and the upper outer quadrant is used as an injection site, where there is less chance of injury to the sciatic nerve or major blood vessels.

5. The skin superior to the buttocks has a dimple or depression on each side of the spine, due to the attachment of the skin and fascia to bone.

Chapter 13

1. Blood doping is used by some athletes to improve performance but the practice strains the heart. A blood test would reveal polycythemia, an increased number of red blood cells.

2. Basophils appear as bluish-black granular cells in a stained blood smear. An elevated number of basophils may suggest an allergic response such as "hay fever."

3. As the kidney fails, it secretes less erythropoietin, which results in a lower rate of red blood cell production. Recombinant erythropoietin can be administered to boost red blood cell production.

4. Iron is an important element found in the heme portion of hemoglobin. If levels of iron are low, then hemoglobin production is low. Low levels of hemoglobin reduce the oxygen-carrying capacity of blood, and a person with low hemoglobin levels would demonstrate symptoms of anemia. Adding iron to the diet increases one of the raw materials needed for adequate hemoglobin production.

5. Raoul's doctor suggested autologous preoperative transfusion (predonation). If Raoul does need a transfusion during surgery it is safer to use his own blood. Possible problems with matching blood types and blood-borne disease are eliminated with predonation.

Chapter 14

1. The cusps of the valves "catch" the blood like the fabric of the parachute "catches" the air. The cusps are anchored to the papillary muscles of the ventricle by the chordae tendineae, just as the fabric of the parachute is anchored by the lines, so the cusps do not flip open backward.

2. The anterior ventricular branch is located in the anterior interventricular sulcus and supplies both ventricles. The circumflex branch is located in the coronary sulcus and supplies the left ventricle and the left atrium.

3. The pericardium surrounds the heart. The fibrous pericardium anchors the heart to the diaphragm, sternum, and major blood vessels. The intercalated discs contain desmosomes, which hold the cardiac muscle fibers together.

4. Some people develop an immune response to streptococcal infections; this can result in rheumatic fever, which can damage the valves of the heart, especially the mitral valve. It would be important to pinpoint the cause of the sore throat to ensure that antibiotic treatment is started if the infection is streptococcal to reduce the possibility of heart damage.

5. Aortic valve stenosis is a narrowing and/or stiffening of the valve, which makes the valve harder to open. Therefore, the heart must pump harder to open the valve sufficiently to pump out an amount of blood equal to that of a normal heart. If the heart is unable to pump strongly enough, more blood can remain in the left ventricle following ventricular systole than is normal, and the system backs up from there.

Chapter 15

1. The foramen ovale and ductus arteriosus close to establish the separate pulmonary and systemic circulations. The umbilical artery and veins close since the placenta is no longer functioning. The ductus venosus closes so that the liver is no longer bypassed.

2. In the skeletal system "sinuses" are cavities within certain cranial and facial bones near the nasal cavity. They are lined with mucous membranes that are continuous with the lining of the nasal cavity. Besides producing mucus, the paranasal sinuses serve as resonating chambers for sounds as we speak or sing. The four sinuses of the skull are the frontal, sphenoid, ethmoid, and maxillary. In the circulatory system, "sinuses" are enothelial-lined venous channels between layers of the cranial dura mater that circulate blood in the internal skull. The pathway of blood from the superior sagittal sinus goes to the right transverse sinus which communicates with the left transverse sinus, which was formed from the inferior sagittal sinus becoming the straight sinus. The caverous sinuses also flow into the transverse sinuses. The transverse sinuses flow into the sigmoid sinuses, which lead out of the skull through the jugular foramen. Outside of the skull, the sigmoid sinus is termed the internal jugular vein. Each internal jugular vein unites with its respective subclavian vein from the upper limb to form a brachiocephalic vein. The right brachiocephalic vein unites with the left brachicephalic vein to form the superior vena cava.

3. Varicose veins are caused by faulty valves. The weak valves allow pooling of blood in the veins. Valves are found only in veins and not in arteries.

4. The catheter would snake from the femoral artery to the external iliac artery to the common iliac artery to the abdominal aorta to the arch of the aorta to the left ventricle.

5. Gina's doctor will either say her blood pressure is too high or is "borderline high." The doctor should encourage Gina to stop smoking and to lose some weight, since both smoking and being overweight are known risk factors for high blood pressure. The doctor should also suggest regularly monitoring her blood pressure to make sure that this one reading was not unusual for Gina. If Gina's blood pressure stays high and she is making no progress with the weight loss and smoking cessation, she may require medication to control her blood pressure.

Chapter 16

1. The route is from lymph capillaries to lymphatic vessels to the popliteal nodes to the superficial inguinal nodes to the right lumbar trunk to the cisterna chyli to the thoracic duct to the junction of the left internal jugular and left subclavian veins.

2. The two palatine tonsils are in the lateral, posterior oral cavity. The two lingual tonsils are at the base of the tongue. The single pharyngeal tonsil is in the posterior nasopharynx. The five tonsils function in the immune response to inhaled or ingested foreign invaders.

3. The spleen is an important secondary lymphatic organ. By removing the organ, the doctor hopes to reduce the level of the immune response causing Kelly's symptoms. However, because the immune response would be reduced, taking antibiotics prior to invasive procedures, such as dental work, reduce the risk of serious infection.

4. Nan recognizes the difficulty breathing as a part of her anaphylactic allergic reaction to the eggplant. Release of histamine from mast cells causes her airways to constrict, thus reducing the flow of air. In very serious reactions, the airway can narrow so much that an affected individual cannot take in sufficient oxygen. In such cases, epinephrine should be administered; this will dilate the airways.

5. The left subclavian trunk drains lymph from the left upper limb. Blockage would cause a buildup of lymph and interstitial fluid in the upper limb, causing edema (swelling).

Chapter 17

1. The jingle that won't stop is similar to a reverberating circuit in that a signal will be repeated over and over again. Reverberating circuits function in breathing, memory, waking, and coordinated muscle activity.

2. The nervous system exhibits plasticity, "the capability to change based on experience." New pathways and synapses needed for foot manipulation replaced damaged pathways and synapses that were once needed for hand movements.

3. The endorphins are neuropeptides that function as the body's natural painkillers. They have been linked to feelings of pleasure or euphoria.

4. The somatic, afferent, peripheral division would detect sound and smell. The somatic, efferent, peripheral division sends messages to the skeletal muscles for stretching and yawning. Stomach rumbling and salivation are controlled by the autonomic, efferent, peripheral division. The enteric, efferent, peripheral division also controls GI tract organs.

5. The drug may not be able to pass the blood–brain barrier to reach the affected nervous tissue.

Chapter 18

1. The senior damaged the cervical region of his spinal cord. Because the upper limbs are paralyzed, the injury has to be superior to C5. This is a high cervical spinal cord injury. Recovery is unlikely.

2. The spinal cord is connected to the periphery by the spinal nerves. The nerves exit through the intervertebral foramina. Spaces are maintained between the vertebrae by the intervertebral discs. The dura mater of the spinal cord fuses with the epineurium. The nerves are connected to the cord by the posterior and anterior roots.

3. The spinal cord is anchored in place by the filum terminale and denticulate ligaments.

4. The spinal cord only extends to the level of the superior border of the second lumbar vertebra—well above the level of the coccygeal vertebrae.

5. The compression of the vertebrae may cause herniation of intervertebral discs, which then can affect spinal nerves. Nischal may have damage to one or more spinal nerves. In this case, damage to the sciatic nerve is a likely possibility.

Chapter 19

1. Movement of the right upper limb is controlled by the left hemisphere's primary motor area, located in the postcentral gyrus. Speech is controlled by Broca's area in the left hemisphere's frontal lobe just superior to the lateral cerebral sulcus.

2. While it is conceivable that he could still be drunk, the signs suggest Wolfgang may have suffered a subdural hemorrhage, which is very dangerous.

3. The amygdala is the center for fear, rage, and aggression. While it might seem useful to be "fearless" or to be nonaggressive, this could prove dangerous, if not fatal, in some circumstances in which fear might prevent someone from doing risky things or aggression might be an appropriate response.

4. There is such an enormous *loss of neurons* within the cerebrum that the gyri shrink and the sulci enlarge. While the lost neurons cannot be replaced, treatments for the disease center around increasing the availability of the neurotransmitter that is lost along with the neurons.

5. The dentist has injected anesthetic into the inferior alveolar nerve, a branch of the mandibular nerve that numbs the lower teeth and the lower lip. The tongue is numbed by blocking the lingual nerve. The upper teeth and lip are numbed by injecting the superior alveolar nerve, a branch of the maxillary branch.

Chapter 20

1. The exciting and potentially dangerous activities activate the sympathetic nervous system, resulting in a "fight-or-flight" reaction. The sympathetic nervous system stimulates the release of epinephrine (adrenalin) and norepinephrine from the adrenal medulla. These hormones prolong the fight-or-flight response.

2. The autonomic sensory neurons detect stretch in the stomach and send impulses to the brain. The hypothalamus sends impulses through the parasympathetic division, along the vagus nerve, to the heart, resulting in decreased heart rate and force of contraction.

3. Mando's fight-or-flight response is continually being stimulated, which means, among other effects, that her heart rate is often high, her blood glucose may become chronically high, and her adrenal glands are being continually stimulated. Such chronically altered physiological conditions can lead to other systemic problems. (See the endocrine system chapter (Chapter 23) for possibilities.)

4. The autonomic division controls the digestive system and many other organs, including the heart, lungs, and eyes. Most organs are controlled by dual innervation, with the sympathetic and parasympathetic divisions causing opposite effects. Stimulation of the sympathetic division results in widespread effects throughout the body, while the effects of parasympathetic stimulation are more localized. The enteric system controls only the digestive system and may function independently of the ANS.

5. The hypothalamus controls many aspects of behavior and physiology, including those regulated by the autonomic nervous system. Though it is a physically small area of the brain, its effects are wide-reaching.

Chapter 21

1. The warm (thermal) receptors in her hands were at first activated by the warmth of the cup, then they adapted to the stimulus as she continued to hold the cup. The high temperature of the hot cocoa stimulated the thermal receptors in her mouth as well as pain receptors.

2. Muscle spindles monitor the change in length of skeletal muscles, sending proprioceptive information (information about body position) to the brain. Without this input, Jenny would find it difficult to know where her body parts are without actually seeing them. If the spindle function does not return, Jenny may need to learn to move her arms, legs, facial muscles, etc., all over again in a different way.

3. Many brain areas are involved in controlling skeletal muscle activity, including the basal ganglia of the cerebrum, the premotor and primary motor areas of the cerebrum, the cerebellum, etc. In addition, sensory spinal and brain tracts are necessary for providing input and motor tracts for transmitting the "orders" for movement.

4. In an infant or very young child, the myelination of the CNS is not complete. The corticospinal tracts, which control fine, voluntary motor movement, are not fully myelinated until the child is about 2 years old. An infant would not be able to manipulate a knife or fork safely due to lack of complete motor control.

5. Jon's perception of feeling in his amputated foot is called phantom limb sensation. Impulses from the remaining proximal section of the sensory neuron are perceived by the brain as still coming from the amputated foot.

Chapter 22

1. The pressure from the tumor may be stimulating Suestia's olfactory areas in the temporal lobe, which would make her think she is smelling something that is not there.

2. Otitis media (an inner ear infection) would arise from the other regions of inflammation directly from the nasopharynx via the eustacian (auditory) tube. Left untreated, otitis media has the potential to spread into the mastoid air cells (mastoiditis) and even into the skull, producing inflammation of the meninges (meningitis).

3. The stick damaged the iris (the colored ring) which is part of the vascular tunic. The iris controls the diameter of the pupil but does not directly affect the focusing of light rays on the retina. If permanent damage was done to the cornea or ciliary muscles, then focusing may be affected.

4. The endolymph in the saccule, utricle, and membranous semicircular ducts moves in response to the up-and-down motion of the ship on the ocean. The vestibular apparatus is stimulated and sends impulses to the medulla, pons, and cerebellum. The eyes also send feedback to the cerebellum to help maintain balance. Excessive stimulation and conflicting messages (eyes say the ship is still, ears say it's moving) cause motion sickness.

5. Pigment in the iris and choroid are usually sufficient to absorb light entering the eye. A flash photograph captures the picture before the iris has time to constrict in response to the high intensity of light. The picture captures the reflection of the light back through the pupil (and iris if it is lightly pigmented). The eyes appear red due to the vascular tunic. Albinos have no pigment in their eyes so the "red eye" effect is more pronounced.

Chapter 23

1. Melatonin could be taken to help induce sleep during the evening at Tropicanaland. For the quickest adjustment, exposure to very bright light in the morning would help reset the body's clock. Sunlight would be better for this than a bright flashlight. The MSH may help with getting a tan but won't help reset the body's clock.

2. Amadu has a goiter, which is an increase in the size of the thyroid gland. In this case the thyroid gland has increased in size because Amadu's diet lacked sufficient iodine to produce normal levels of thyroid hormones. Because levels of the thyroid hormones were low, TSH levels increased in an attempt to stimulate the thyroid gland to produce more thyroid hormones, but without the iodine, this was not possible. Adding iodine back into the diet should provide the necessary raw materials, and the gland will not be stimulated to grow more.

3. Because blood glucose might be chronically elevated by such treatment, this could make the pancreas work harder to produce insulin to lower the levels, which could have an adverse effect on the pancreas.

4. The pituitary gland is located in the hypophyseal fossa in the base of the sphenoid bone. The sphenoid is posterior to the ethmoid bone, which makes up a major portion of the walls and roof of the nasal cavity. Due to its proximity to the nasal cavity, the pituitary gland could have been affected by the radiation.

5. All hormones, including insulin, need an adequate number of functioning receptors to perform their function. The problem in this patient is a lack of functioning receptors, not a problem with insulin level.

Chapter 24

1. The septum is composed of hyaline cartilage covered with mucous membrane. The lateral wall of the nose is skin and muscle, lined with mucous membrane. The upper ear is skin covering elastic cartilage.

2. Lower oxygen levels at the higher altitudes means lower oxygen levels in Suzanne's blood and cerebrospinal fluid. Central and peripheral chemoreceptors will detect changes in blood gas levels and proportions and signal the medullary rhythmicity center to increase the rate of breathing.

3. Once the endotracheal tube was introduced either nasally or orally, the progression of the tube did not enter the laryngopharynx to proceed to the larynx and trachea. Instead the tube remained in the pharynx and entered the esophagus, which eventually leads to the stomach. The sounds that the resident heard in the epigastric region were therefore coming from the stomach rather than the lungs.

4. Several factors decrease respiratory efficiency in smokers: (1) Nicotine constricts terminal bronchioles, which decreases airflow into and out of the lungs. (2) Carbon monoxide in smoke binds to hemoglobin and reduces its oxygen-carrying capability. (3) Irritants in smoke cause increased mucus secretion by the mucosa of the bronchial tree and swelling of the mucosal lining, both of which impede airflow into and out of the lungs. (4) Irritants in smoke also inhibit the movement of cilia and destroy cilia in the lining of the respiratory system. Thus, excess mucus and foreign debris are not easily removed, which further adds to the difficulty in breathing. (5) With time, smoking leads to destruction of elastic fibers in the lungs and is the prime cause of emphysema. These changes cause collapse of small bronchioles and trapping of air in alveoli at the end of exhalation. The result is less efficient gas exchange.

5. LaTonya's cerebral cortex can allow her to voluntarily hold her breath for a short time. Increasing levels of carbon dioxide and H^+ will stimulate the inspiratory area and normal breathing will resume despite LaTonya's desire to get her sister in trouble.

Chapter 25

1. Saliva contains salivary amylase, which will break down the starch of the potatoes into sugar (sweet), and lingual lipase, which will break down the fat from the frying into fatty acids (tart) and monoglycerides.

2. The doctor will be looking at the mucosa of the regions of the large intestine to check for smoothness (vs. polyp growth), normal-appearing simple columnar epithelium, normal-appearing haustra, etc.

3. Modifications to increase surface area include overall length, circular folds, villi, and microvilli (brush border). The circular folds also enhance absorption by imparting a spiraling motion to the chyme.

4. During swallowing, the soft palate and uvula close off the nasopharynx. If the nervous system sends conflicting signals (such as breathe, swallow, and giggle), the palate may be in the wrong position to block food from entering the nasal cavity.

5. Billy is hearing the normal bowel sounds produced as a result of the movement of intestinal contents via peristalsis, haustral churning, and mass peristalsis. In addition, gas is being produced as bacteria digest undigested materials, which contributes to the sounds as the bubbles move through the intestinal contents.

Chapter 26

1. The glomerulus would continue to filter the blood, producing filtrate containing water, glucose, ions, and other compounds. The toxin would block the normal reabsorption of 99% of the filtrate by the renal tubules. The infected person would rapidly become dehydrated from lost water and would also lose ions, glucose, and essential vitamins and nutrients. The infected person would most likely soon die.

2. The micturition reflex is primarily a parasympathetic reflex with input from the sympathetic system in "fight-flight-or fright" situations. Parasympathetic pathways traveling from the micturition center and through S2, 3, and 4 reach the inferior hypogastric (pelvic) plexus to be distributed to the smooth muscle of the urinary bladder and internal urethral sphincter. This will cause a contraction of the detrusor muscle and relaxation of the sphincter. Concurrently, relaxation of the external urethral sphincter is produced by inhibiting somatic efferent impulses to the pelvic floor (levator ani) musculature. The synergistic functions of contracting the bladder and relaxing the sphincters causes urination.

3. In females the urethra is about 4 cm long. In males, the urethra is about 15–20 cm long, including its passage through the penis, the urogenital diaphragm, and prostate. The female urinary bladder holds less urine than the male bladder due to the presence of the uterus in females.

4. Urine samples contain an abundance of epithelial cells that are shed from the mucosal lining of the urinary system.

5. Functions normally performed by the healthy nephrons include filtering the blood (glomerular filtration); *keeping* necessary substances for the body (tubular reabsorption); and *removing* unnecessary substances from the body (tubular secretion). Unnecessary substances (waste) is then excreted as a liquid (urine). Because of the large number of damaged nephrons resulting from the disease, these functions are now being carried out by a hemodialysis machine with the use of semipermeable membranes to separate the necessary from the unnecessary substances.

Chapter 27

1. The infection could pass through the vagina to the uterus through the uterine tubes, which are open to the pelvic cavity.

2. The presence or absence of the SRY gene on the Y chromosome determines the development of the urogenital structures before birth. The presence of SRY and secretion of testosterone by the fetal testes results in a male. Absence of SRY results in a female. Structures can be altered by surgery and hormones somewhat but the female will still lack receptors and organs necessary for sperm production and the male will lack receptors and organs necessary for the production of ova. The chromosome composition cannot be changed.

3. A hysterectomy will not cause menopause. A hysterectomy is the removal of the uterus, which does not produce hormones. The ovaries are left intact and continue to produce estrogens and progesterone.

4. The testicular artery and veins that drain the testes pass through the inguinal ring; if their blood supply is cut off by the closure of the inguinal canal, the testes would have no source of nourishment or waste removal.

5. A vasectomy cuts only the vas (ductus) deferens and leaves the testes untouched. Male secondary sex characteristics and libido (sex drive) are maintained by androgens including testosterone. The secretion of these hormones by the testes is not interrupted by the vasectomy.

GLOSSARY

Pronunciation Key

1. The most strongly accented syllable appears in capital letters, for example, bilateral (bī-LAT-er-al) and diagnosis (dī-ag-NŌ-sis).

2. If there is a secondary accent, it is noted by a prime ('), for example, constitution (kon'-sti-TOO-shun) and physiology (fiz'-ē-OL-ō-jē). Any additional secondary accents are also noted by a prime, for example, decarboxylation (dē'-kar-bok'-si-LĀ-shun).

3. Vowels marked by a line above the letter are pronounced with the long sound, as in the following common words:

    ā as in *māke*   ī as in *īvy*   ū as in *cūte*
    ē as in *bē*   ō as in *pōle*

4. Vowels not marked by a line above the letter are pronounced with the short sound, as in the following words:

    a as in *above* or *at*   i as in *sip*   u as in *bud*
    e as in *bet*   o as in *not*

5. Other vowel sounds are indicated as follows:

    oy as in *oil*
    oo as in *root*

6. Consonant sounds are pronounced as in the following words:

b as in *bat*	*k* as in *can*	*r* as in *rib*
ch as in *chair*	*ks* as in *tax*	*s* as in *so*
d as in *dog*	*kw* as in *quit*	*t* as in *tea*
f as in *father*	*l* as in *let*	*v* as in *very*
g as in *get*	*m* as in *mother*	*w* as in *welcome*
h as in *hat*	*n* as in *no*	*z* as in *zero*
j as in *jump*	*p* as in *pick*	*zh* as in *lesion*

A

Abdomen (ab-DŌ-men *or* AB-dō-men) The area between the diaphragm and pelvis.

Abdominal (ab-DOM-i-nal) **cavity** Superior portion of the abdominopelvic cavity that contains the stomach, spleen, liver, gallbladder, most of the small intestine, and part of the large intestine.

Abdominal thrust maneuver A first-aid procedure for choking. Employs a quick, upward thrust against the diaphragm that forces air out of the lungs with sufficient force to eject any lodged material.

Abdominopelvic (ab-dom'-i-nō-PEL-vic) **cavity** A body cavity that is subdivided into a superior abdominal cavity and an inferior pelvic cavity.

Abduction (ab-DUK-shun) Movement away from the midline of the body.

Abortion (a-BOR-shun) The premature loss (spontaneous) or removal (induced) of the embryo or nonviable fetus; miscarriage due to a failure in the normal process of developing or maturing.

Abscess (AB-ses) A localized collection of pus and liquefied tissue in a cavity.

Absorption (ab-SORP-shun) Intake of fluids or other substances by cells of the skin or mucous membranes; the passage of digested foods from the gastrointestinal tract into blood or lymph.

Accessory duct A duct of the pancreas that empties into the duodenum about 2.5 cm (1 in.) superior to the ampulla of Vater (hepatopancreatic ampulla). Also called the **duct of Santorini** (san'-tō-RĒ-nē).

Acetabulum (as'-e-TAB-ū-lum) The rounded cavity on the external surface of the hip bone that receives the head of the femur.

Acetylcholine (as'-e-til-KŌ-lēn) **(ACh)** A neurotransmitter liberated by many peripheral nervous system neurons and some central nervous system neurons. It is excitatory at neuromuscular junctions but inhibitory at some other synapses (for example, it slows heart rate).

Achalasia (ak'-a-LĀ-zē-a) A condition, caused by malfunction of the myenteric plexus, in which the lower esophageal sphincter fails to relax normally as food approaches. A whole meal may become lodged in the esophagus and enter the stomach very slowly. Distension of the esophagus results in chest pain that is often confused with pain originating from the heart.

Achilles tendon *See* **Calcaneal tendon.**

Acini (AS-i-nē) Groups of cells in the pancreas that secrete digestive enzymes. Also, a functional unit within the liver lobule.

Acoustic (a-KOOS-tik) Pertaining to sound or the sense of hearing.

Acquired immunodeficiency syndrome (AIDS) A disease caused by the human immunodeficiency virus (HIV). Characterized by a positive HIV-antibody test, low helper T cell count, and certain indicator diseases (for example Kaposi's sarcoma, pneumocystis carinii pneumonia, tuberculosis, fungal diseases). Other symptoms include fever or night sweats, coughing, sore throat, fatigue, body aches, weight loss, and enlarged lymph nodes.

Acrosome (AK-rō-sōm) A lysosomelike organelle in the head of a sperm cell containing enzymes that facilitate the penetration of a sperm cell into a secondary oocyte.

Actin (AK-tin) A contractile protein that is part of thin filaments in muscle fibers.

Action potential An electrical signal that propagates along the membrane of a neuron or muscle fiber (cell); a rapid change in membrane potential that involves a depolarization followed by a repolarization. Also called a **nerve action potential** or **nerve impulse** as it relates to a neuron, and a **muscle action potential** as it relates to a muscle fiber.

Active transport The movement of substances across cell membranes against a concentration gradient, requiring the expenditure of cellular energy (ATP).

Acute (a-KŪT) Having rapid onset, severe symptoms, and a short course; not chronic.

Adaptation (ad'-ap-TĀ-shun) The adjustment of the pupil of the eye to changes in light intensity. The property by which a sensory neuron relays a decreased frequency of action potentials from a receptor, even though the strength of the stimulus remains constant; the decrease in perception of a sensation over time while the stimulus is still present.

Adduction (ad-DUK-shun) Movement toward the midline of the body.

Adenoids (AD-e-noyds) The pharyngeal tonsils.

Adenosine triphosphate (a-DEN-ō-sēn trī-FOS-fāt) **(ATP)** The main energy currency in living cells; used to transfer the chemical energy needed for metabolic reactions. ATP consists of the purine base *adenine* and the five-carbon sugar *ribose*, to which are added, in linear array, three *phosphate* groups.

Adhesion (ad-HĒ-zhun) Abnormal joining of parts to each other.

Adipocyte (AD-i-pō-sīt) Fat cell, derived from a fibroblast.

Adipose (AD-i-pōz) **tissue** Tissue composed of adipocytes specialized for triglyceride storage and present in the form of soft pads between various organs for support, protection, and insulation.

Adrenal cortex (a-DRĒ-nal KOR-teks) The outer portion of an adrenal gland, divided into three zones; the zona glomerulosa secretes mineralocorticoids, the zona fasciculata secretes glucocorticoids, and the zona reticularis secretes androgens.

Adrenal glands Two glands located superior to each kidney. Also called the **suprarenal** (soo'-pra-RĒ-nal) **glands.**

Adrenal medulla (me-DUL-a) The inner part of an adrenal gland, consisting of cells that secrete epinephrine, norepinephrine, and a small amount of dopamine in response to stimulation by sympathetic preganglionic neurons.

Adrenergic (ad'-ren-ER-jik) **neuron** A neuron that releases epinephrine (adrenaline) or norepinephrine (noradrenaline) as its neurotransmitter.

Adrenocorticotropic (ad-rē′-nō-kor-ti-kō-TRŌP-ik) **hormone (ACTH)** A hormone produced by the anterior pituitary that influences the production and secretion of certain hormones of the adrenal cortex.

Adventitia (ad-ven-TISH-a) The outermost covering of a structure or organ.

Aerobic (air-Ō-bik) Requiring molecular oxygen.

Afferent arteriole (AF-er-ent ar-TĒ-rē-ōl) A blood vessel of a kidney that divides into the capillary network called a glomerulus; there is one afferent arteriole for each glomerulus.

Agglutination (a-gloo′-ti-NĀ-shun) Clumping of microorganisms or blood cells, typically due to an antigen–antibody reaction.

Aggregated lymphatic follicles Clusters of lymph nodules that are most numerous in the ileum. Also called **Peyer's** (PĪ-erz) **patches.**

Alar cartilages (Ā-lar) Cartilages that consist of the major (larger) and minor (smaller) alar cartilages. They form the inferior part of the lateral surface of the nose.

Albinism (AL-bin-izm) Abnormal, nonpathological, partial, or total absence of pigment in skin, hair, and eyes.

Aldosterone (al-DOS-ter-ōn) A mineralocorticoid produced by the adrenal cortex that promotes sodium and water reabsorption by the kidneys and potassium excretion in urine.

Allantois (a-LAN-tō-is) A small, vascularized outpouching of the yolk sac that serves as an early site for blood formation and development of the urinary bladder.

Alleles (a-LĒLZ) Alternate forms of a single gene that control the same inherited trait (such as type A blood) and are located at the same position on homologous chromosomes.

Allergen (AL-er-jen) An antigen that evokes a hypersensitivity reaction.

Alopecia (al′-ō-PĒ-shē-a) The partial or complete lack of hair as a result of factors such as genetics, aging, endocrine disorders, chemotherapy, and skin diseases.

Alpha (AL-fa) **cell** A type of cell in the pancreatic islets (islets of Langerhans) in the pancreas that secretes the hormone glucagon. Also termed an **A cell.**

Alpha receptor A type of receptor for norepinephrine and epinephrine; present on visceral effectors innervated by sympathetic postganglionic neurons.

Alveolar-capillary (al-VE-ō-lar) **membrane** Structure in the lungs consisting of the alveolar wall and basement membrane and a capillary endothelium and basement membrane through which the diffusion of respiratory gases occurs. Also called the **respiratory membrane.**

Alveolar duct Branch of a respiratory bronchiole around which alveoli and alveolar sacs are arranged.

Alveolar macrophage (MAK-rō-fāj) Highly phagocytic cell found in the alveolar walls of the lungs. Also called a **dust cell.**

Alveolar sac A cluster of alveoli that share a common opening.

Alveolus (al-VĒ-ō-lus) A small hollow or cavity; an air sac in the lungs; milk-secreting portion of a mammary gland. *Plural* is **alveoli** (al-VĒ-ol-ī).

Alzheimer (ALTZ-hī-mer) **disease (AD)** Disabling neurological disorder characterized by dysfunction and death of specific cerebral neurons, resulting in widespread intellectual impairment, personality changes, and fluctuations in alertness.

Amenorrhea (ā-men-ō-RĒ-a) Absence of menstruation.

Amnesia (am-NĒ-zē-a) A lack or loss of memory.

Amnion (AM-nē-on) A thin, protective fetal membrane that develops from the epiblast; holds the fetus suspended in amniotic fluid. Also called the "**bag of waters.**"

Amniotic (am′-nē-OT-ik) **fluid** Fluid in the amniotic cavity, the space between the developing embryo (or fetus) and amnion; the fluid is initially produced as a filtrate from maternal blood and later includes fetal urine. It functions as a shock absorber, helps regulate fetal body temperature, and helps prevent desiccation.

Amphiarthrosis (am′-fē-ar-THRŌ-sis) A slightly movable joint, in which the articulating bony surfaces are separated by fibrous connective tissue or fibrocartilage to which both are attached; types are syndesmosis and symphysis.

Ampulla (am-PUL-la) A saclike dilation of a canal or duct.

Ampulla of Vater *See* **Hepatopancreatic ampulla.**

Anabolism (a-NAB-ō-lizm) Synthetic, energy-requiring reactions whereby small molecules are built up into larger ones.

Anaerobic (an-ar-Ō-bik) Not requiring oxygen.

Anal (Ā-nal) **canal** The last 2 or 3 cm (1 in.) of the rectum; opens to the exterior through the anus.

Anal column A longitudinal fold in the mucous membrane of the anal canal that contains a network of arteries and veins.

Anal triangle The subdivision of the female or male perineum that contains the anus.

Analgesia (an-al-JĒ-zē-a) Pain relief; absence of the sensation of pain.

Anaphase (AN-a-fāz) The third stage of mitosis in which the chromatids that have separated at the centromeres move to opposite poles of the cell.

Anaphylaxis (an′-a-fi-LAK-sis) A hypersensitivity (allergic) reaction in which IgE antibodies attach to mast cells and basophils, causing them to produce mediators of anaphylaxis (histamine, leukotrienes, kinins, and prostaglandins) that bring about increased blood permeability, increased smooth muscle contraction, and increased mucus production. Examples are hay fever, hives, and anaphylactic shock.

Anastomosis (a-nas-tō-MŌ-sis) An end-to-end union or joining of blood vessels, lymphatic vessels, or nerves.

Anatomic dead space Spaces of the nose, pharynx, larynx, trachea, bronchi, and bronchioles totaling about 150 mL of the 500 mL in a quiet breath (tidal volume); air in the anatomic dead space does not reach the alveoli to participate in gas exchange.

Anatomical (an′-a-TOM-i-kal) **position** A position of the body universally used in anatomical descriptions in which the body is erect, the head is level, the eyes face forward, the upper limbs are at the sides, the palms face forward, and the feet are flat on the floor.

Anatomy (a-NAT-ō-mē) The structure or study of structure of the body and the relation of its parts to each other.

Androgens (AN-drō-jenz) Masculinizing sex hormones produced by the testes in males and the adrenal cortex in both sexes; responsible for libido (sexual desire); the two main androgens are testosterone and dihydrotestosterone.

Anemia (a-NĒ-mē-a) Condition of the blood in which the number of functional red blood cells or their hemoglobin content is below normal.

Anesthesia (an′-es-THĒ-zē-a) A total or partial loss of feeling or sensation; may be general or local.

Aneurysm (AN-ū-rizm) A saclike enlargement of a blood vessel caused by a weakening of its wall.

Angina pectoris (an-JI-na *or* AN-ji-na PEK-tō-ris) A pain in the chest related to reduced coronary circulation due to coronary artery disease (CAD) or spasms of vascular smooth muscle in coronary arteries.

Angiogenesis (an-jē-ō-JEN-e-sis) The formation of blood vessels in the extraembryonic mesoderm of the yolk sac, connecting stalk, and chorion at the beginning of the third week of development.

Ankylosis (ang′-ki-LŌ-sis) Severe or complete loss of movement at a joint as the result of a disease process.

Anomaly (a-NOM-a-lē) An abnormality that may be a developmental (congenital) defect; a variant from the usual standard.

Anoxia (an-OK-sē-a) Deficiency of oxygen.

Antagonist (an-TAG-ō-nist) A muscle that has an action opposite that of the prime mover (agonist) and yields to the movement of the prime mover.

Antagonistic (an-tag-ō-NIST-ik) **effect** A hormonal interaction in which the effect of one hormone on a target cell is opposed by another hormone. For example, calcitonin (CT) lowers blood calcium level, whereas parathyroid hormone (PTH) raises it.

Anterior (an-TER-ē-or) Nearer to or at the front of the body. Equivalent to **ventral** in bipeds.

Anterior pituitary Anterior lobe of the pituitary gland. Also called the **adenohypophysis** (ad′-e-nō-hī-POF-i-sis).

Anterior root The structure composed of axons of motor (efferent) neurons that emerges from the anterior aspect of the spinal cord and extends laterally to join a posterior root, forming a spinal nerve. Also called a **ventral root.**

Anterolateral (an′-ter-ō-LAT-er-al) **pathway** Sensory pathway that conveys information related to pain, temperature, crude touch, pressure, tickle, and itch.

Antibody (AN-ti-bod′-ē) A protein produced by plasma cells in response to a specific antigen; the antibody combines with that antigen to neutralize, inhibit, or destroy it. Also called an **immunoglobulin** (im-ū-nō-GLOB-ū-lin) or **Ig.**

Anticoagulant (an-tī-cō-AG-ū-lant) A substance that can delay, suppress, or prevent the clotting of blood.

Antidiuretic (an′-ti-dī-ū-RET-ik) Substance that inhibits urine formation.

Antidiuretic hormone (ADH) Hormone produced by neurosecretory cells in the paraventricular and supraoptic nuclei of the hypothalamus that stimulates water reabsorption from kidney tubule cells into the blood and vasoconstriction of arterioles. Also called **vasopressin** (vāz-ō-PRES-in).

Antigen (AN-ti-jen) A substance that has immunogenicity (the ability to provoke an immune response) and reactivity (the ability to react with the antibodies or cells that result from the immune response); contraction of *antibody generator.* Also termed a **complete antigen.**

Antrum (AN-trum) Any nearly closed cavity or chamber, especially one within a bone, such as a sinus.

Annulus fibrosus (AN-ū-lus fī-BRŌ-sus) A ring of fibrous tissue and fibrocartilage that encircles the pulpy substance (nucleus pulposus) of an intervertebral disc.

Anuria (an-Ū-rē-a) Absence of urine formation or daily urine output of less than 50 mL.

Anus (Ā-nus) The distal end and outlet of the rectum.

Aorta (ā-OR-ta) The main systemic trunk of the arterial system of the body that emerges from the left ventricle.

Aortic (ā-OR-tik) **body** Cluster of chemoreceptors on or near the arch of the aorta that respond to changes in blood levels of oxygen, carbon dioxide, and hydrogen ions (H^+).

Aortic reflex A reflex that helps maintain normal systemic blood pressure; initiated by baroreceptors in the wall of the ascending aorta and arch of the aorta. Nerve impulses from aortic baroreceptors reach the cardiovascular center via sensory axons of the vagus (X) nerves.

Aperture (AP-er-chur) An opening or orifice.

Apex (Ā-peks) The pointed end of a conical structure, such as the apex of the heart.

Aphasia (a-FĀ-zē-a) Loss of ability to express oneself properly through speech or loss of verbal comprehension.

Apnea (AP-nē-a) Temporary cessation of breathing.

Apneustic (ap-NOO-stik) **area** A part of the respiratory center in the pons that sends stimulatory nerve impulses to the inspiratory area that activate and prolong inhalation and inhibit exhalation.

Apocrine (AP-ō-krin) **gland** A type of gland in which the secretory products gather at the free end of the secreting cell and are pinched off, along with some of the cytoplasm, to become the secretion, as in mammary glands.

Aponeurosis (ap′-ō-noo-RŌ-sis) A sheetlike tendon joining one muscle with another or with bone.

Apoptosis (ap′-ō-TŌ-sis *or* ap′-ōp-TŌ-sis) Programmed cell death; a normal type of cell death that removes unneeded cells during embryological development, regulates the number of cells in tissues, and eliminates many potentially dangerous cells such as cancer cells. During apoptosis, the DNA fragments, the nucleus condenses, mitochondria cease to function, and the cytoplasm shrinks, but the plasma membrane remains intact. Phagocytes engulf and digest the apoptotic cells, and an inflammatory response does not occur.

Appositional (ap′-ō-ZISH-o-nal) **growth** Growth due to surface deposition of material, as in the growth in diameter of cartilage and bone. Also called **exogenous** (eks-OJ-e-nus) **growth.**

Aqueous humor (AK-wē-us HŪ-mer) The watery fluid, similar in composition to cerebrospinal fluid, that fills the anterior cavity of the eye.

Arachnoid (a-RAK-noyd) **mater** The middle of the three meninges (coverings) of the brain and spinal cord. Also termed the **arachnoid.**

Arachnoid villus (VIL-us) Berrylike tuft of the arachnoid mater that protrudes into the superior sagittal sinus and through which cerebrospinal fluid is reabsorbed into the bloodstream.

Arbor vitae (AR-bor VĪ-tē) The white matter tracts of the cerebellum, which have a treelike appearance when seen in midsagittal section.

Arch of the aorta The most superior portion of the aorta, lying between the ascending and descending segments of the aorta.

Areola (a-RĒ-ō-la) Any tiny space in a tissue. The pigmented ring around the nipple of the breast.

Areolar connective tissue (a-RĒ-ō-lar) A widely distributed connective tissue that contains randomly arranged collagen, elastic, and reticular fibers.

Arm The part of the free upper limb from the shoulder to the elbow.

Arousal (a-ROW-zal) Awakening from sleep, a response due to stimulation of the reticular activating system (RAS).

Arrector pili (a-REK-tor PI-lē) Smooth muscles attached to hairs; contraction pulls the hairs into a vertical position, resulting in "goose bumps."

Arrhythmia (a-RITH-mē-a) An irregular heart rhythm. Also called a **dysrhythmia.**

Arteriole (ar-TĒ-rē-ōl) A small, almost microscopic, artery that delivers blood to a capillary.

Arteriosclerosis (ar-tē-rē-ō-skle-RŌ-sis) Group of diseases characterized by thickening of the walls of arteries and loss of elasticity.

Artery (AR-ter-ē) A blood vessel that carries blood away from the heart.

Arthritis (ar-THRI-tis) Inflammation of a joint.

Arthrology (ar-THROL-ō-jē) The study or description of joints.

Arthroplasty (AR-thrō-plas′-tē) Surgical replacement of joints, for example, the hip and knee joints.

Arthroscopy (ar-THROS-co-pē) A procedure for examining the interior of a joint, usually the knee, by inserting an arthroscope into a small incision; used to determine extent of damage, remove torn cartilage, repair cruciate ligaments, and obtain samples for analysis.

Arthrosis (ar-THRŌ-sis) A joint or articulation.

Articular (ar-TIK-ū-lar) **capsule** Sleevelike structure around a synovial joint composed of a fibrous membrane and a synovial membrane.

Articular cartilage (KAR-ti-lij) Hyaline cartilage attached to articular bone surfaces.

Articular disc Fibrocartilage pad between articular surfaces of bones of some synovial joints. Also called a **meniscus** (men-IS-kus).

Articulation (ar-tik′-ū-LĀ-shun) A joint; a point of contact between bones, cartilage and bones, or teeth and bones.

Arytenoid (ar′-i-TĒ-noyd) **cartilages** A pair of small, pyramidal cartilages of the larynx that attach to the vocal folds and intrinsic pharyngeal muscles and can move the vocal folds.

Ascending colon (KŌ-lon) The part of the large intestine that passes superiorly from the cecum to the inferior border of the liver, where it bends at the right colic (hepatic) flexure to become the transverse colon.

Ascites (as-SĪ-tēz) Abnormal accumulation of serous fluid in the peritoneal cavity.

Aseptic (ā-SEP-tik) Free from any infectious or septic material.

Association areas Large cortical regions on the lateral surfaces of the occipital, parietal, and temporal lobes and on the frontal lobes anterior to the motor areas connected by many motor and sensory axons to other parts of the cortex. The association areas are concerned with motor patterns, memory, concepts of word-hearing and word-seeing, reasoning, will, judgment, and personality traits.

Asthma (AZ-ma) Usually allergic reaction characterized by smooth muscle spasms in bronchi resulting in wheezing and difficult breathing. Also called **bronchial asthma.**

Astigmatism (a-STIG-ma-tizm) An irregularity of the lens or cornea of the eye causing the image to be out of focus and producing faulty vision.

Astrocyte (AS-trō-sīt) A neuroglial cell having a star shape that participates in brain development and the metabolism of neurotransmitters, helps form the blood–brain barrier, helps maintain the proper balance of K^+ for generation of nerve impulses, and provides a link between neurons and blood vessels.

Ataxia (a-TAK-sē-a) A lack of muscular coordination, lack of precision.

Atherosclerotic (ath′-er-ō-skle-RO-tic) **plaque** (PLAK) A lesion that results from accumulated cholesterol and smooth muscle fibers (cells) of the tunica media of an artery; may become obstructive.

Atom Unit of matter that makes up a chemical element; consists of a nucleus (containing positively charged protons and uncharged neutrons) and negatively charged electrons that orbit the nucleus.

Atresia (a-TRĒ-zē-a) Degeneration and reabsorption of an ovarian follicle before it fully matures and ruptures; abnormal closure of a passage, or absence of a normal body opening.

Atrial fibrillation (Ā-trē-al fib-ri-LĀ-shun) Asynchronous contraction of cardiac muscle fibers in the atria that results in the cessation of atrial pumping.

Atrial natriuretic (na′-trē-ū-RET-ik) **peptide (ANP)** Peptide hormone, produced by the atria of the heart in response to stretching, that inhibits aldosterone production and thus lowers blood pressure; causes natriuresis, increased urinary excretion of sodium.

Atrioventricular (AV) (ā′-trē-ō-ven-TRIK-ū-lar) **bundle** The part of the conduction system of the heart that begins at the atrioventricular (AV) node, passes through the cardiac skeleton separating the atria and the ventricles, then extends a short distance down the interventricular septum before splitting into right and left bundle branches. Also called the **bundle of His** (HISS).

Atrioventricular (AV) node The part of the conduction system of the heart made up of a compact mass of conducting cells located in the septum between the two atria.

Atrioventricular (AV) valve A heart valve made up of membranous flaps or cusps that allows blood to flow in one direction only, from an atrium into a ventricle.

Atrium (Ā-trē-um) A superior chamber of the heart.

Atrophy (AT-rō-fē) Wasting away or decrease in size of a part, due to a failure, abnormality of nutrition, or lack of use.

Auditory ossicle (AW-di-tō-rē OS-si-kul) One of the three small bones of the middle ear called the **malleus, incus,** and **stapes.**

Auditory tube The tube that connects the middle ear with the nose and nasopharynx region of the throat. Also called the **eustachian** (ū-STĀ-shun *or* ū-STĀ-kē-an) **tube** or **pharyngotympanic tube.**

Auscultation (aws-kul-TĀ-shun) Examination by listening to sounds in the body.

Autoimmunity An immunological response against a person's own tissues.

Autolysis (aw-TOL-i-sis) Self-destruction of cells by their own lysosomal digestive enzymes after death or in a pathological process.

Autonomic ganglion (aw′-tō-NOM-ik GANG-lē-on) A cluster of cell bodies of sympathetic or parasympathetic neurons located outside the central nervous system.

Autonomic nervous system (ANS) Visceral sensory (afferent) and visceral motor (efferent) neurons. Autonomic motor neurons, both sympathetic and parasympathetic, conduct nerve impulses from the central nervous system to smooth muscle, cardiac muscle, and glands. So named because this part of the nervous system was thought to be self-governing or spontaneous.

Autonomic plexus (PLEK-sus) A network of sympathetic and parasympathetic axons; examples are the cardiac, celiac, and pelvic plexuses, which are located in the thorax, abdomen, and pelvis, respectively.

Autophagy (aw-TOF-a-jē) Process by which worn-out organelles are digested within lysosomes.

Autopsy (AW-top-sē) The examination of the body after death.

Autorhythmic cells Cardiac or smooth muscle fibers that are self-excitable (generate impulses without an external stimulus); act as the heart's pacemaker and conduct the pacing impulse through the conduction system of the heart; self-excitable neurons in the central nervous system, as in the inspiratory area of the brain stem.

Autosome (AW-tō-sōm) Any chromosome other than the X and Y chromosomes (sex chromosomes).

Axilla (ak-SIL-a) The small hollow beneath the arm where it joins the body at the shoulders. Also called the **armpit.**

Axon (AK-son) The usually single, long process of a nerve cell that propagates a nerve impulse toward the axon terminals.

Axon terminal Terminal branch of an axon where synaptic vesicles undergo exocytosis to release neurotransmitter molecules.

Azygos (AZ-ī-gos) An anatomical structure that is not paired; occurring singly.

B

B cell A lymphocyte that can develop into a clone of antibody-producing plasma cells or memory cells when properly stimulated by a specific antigen.

Babinski (ba-BIN-skē) **sign** Extension of the great toe, with or without fanning of the other toes, in response to stimulation of the outer margin of the sole; normal up to 18 months of age and indicative of damage to descending motor pathways such as the corticospinal tracts after that.

Back The posterior part of the body; the dorsum.

Ball-and-socket joint A synovial joint in which the rounded surface of one bone moves within a cup-shaped depression or socket of another bone, as in the shoulder or hip joint. Also called a **spheroid** (SFĒ-royd) **joint.**

Baroreceptor (bar′-ō-re-SEP-tor) Neuron capable of responding to changes in blood, air, or fluid pressure. Also called a **pressoreceptor.**

Basal ganglia (GANG-glē-a) Paired clusters of gray matter deep in each cerebral hemisphere including the globus pallidus, putamen, and caudate nucleus. Together, the caudate nucleus and putamen are known as the **corpus striatum.** Nearby structures that are functionally linked to the basal ganglia are the substantia nigra of the midbrain and the subthalamic nuclei of the diencephalon.

Basement membrane Thin, extracellular layer between epithelium and connective tissue consisting of a basal lamina and a reticular lamina.

Basilar (BĀS-i-lar) **membrane** A membrane in the cochlea of the internal ear that separates the cochlear duct from the scala tympani and on which the spiral organ (organ of Corti) rests.

Basophil (BĀ-sō-fil) A type of white blood cell characterized by a pale nucleus and large granules that stain blue-purple with basic dyes.

Belly The abdomen. The gaster or prominent, fleshy part of a skeletal muscle.

Beta (BĀ-ta) **cell** A type of cell in the pancreatic islets (islets of Langerhans) in the pancreas that secretes the hormone insulin.

Beta receptor A type of adrenergic receptor for epinephrine and norepinephrine; found on visceral effectors innervated by sympathetic postganglionic neurons.

Bicuspid (bī-KUS-pid) **valve** Atrioventricular (AV) valve on the left side of the heart. Also called the **mitral valve.**

Bilateral (bī-LAT-er-al) Pertaining to two sides of the body.

Bile (BĪL) A secretion of the liver consisting of water, bile salts, bile pigments, cholesterol, lecithin, and several ions; it emulsifies lipids prior to their digestion.

Bile canaliculi (kan′-a-LIK-ū-li) Small ducts between hepatocytes that collect bile produced by hepatocytes.

Bile duct Passageway that transports bile from the liver and gall bladder to the duodenum.

Bilirubin (bil-ē-ROO-bin) An orange pigment that is one of the end products of hemoglobin breakdown in the hepatocytes and is excreted as a waste material in bile.

Blastocyst (BLAS-tō-sist) In the development of an embryo, a hollow ball of cells that consists of a blastocele (the internal cavity), trophoblast (outer cells), and inner cell mass. Also called a **blastocele.**

Blastocyst cavity (BLAS-tō-sist) The fluid-filled cavity within the blastocyst.

Blastomere (BLAS-tō-mēr) One of the cells resulting from the cleavage of a fertilized ovum.

Blastula (BLAS-tyū-la) An early stage in the development of a zygote.

Blind spot Area in the retina at the end of the optic (II) nerve in which there are no photoreceptors.

Blood The fluid that circulates through the heart, arteries, capillaries, and veins and that constitutes the chief means of transport within the body.

Blood–brain barrier (BBB) A barrier consisting of specialized brain capillaries and astrocytes that prevents the passage of materials from the blood to the cerebrospinal fluid and brain.

Blood island Isolated mass of mesoderm derived from angioblasts and from which blood vessels develop.

Blood pressure (BP) Force exerted by blood against the walls of blood vessels due to contraction of the heart and influenced by the elasticity of the vessel walls; clinically, a measure of the pressure in arteries during ventricular systole and ventricular diastole. *See also* **mean arterial blood pressure.**

Blood reservoir (REZ-er-vwar) Systemic veins that contain large amounts of blood that can be moved quickly to parts of the body requiring the blood.

Blood–testis barrier (BTB) A barrier formed by Sertoli cells that prevents an immune response against antigens produced by spermatogenic cells by isolating the cells from the blood.

Body cavity A space within the body that contains various internal organs.

Bolus (BŌ-lus) A soft, rounded mass, usually food, that is swallowed.

Bony labyrinth (LAB-i-rinth) A series of cavities within the petrous portion of the temporal bone forming the vestibule, cochlea, and semicircular canals of the inner ear.

Bowman's capsule *See* **Glomerular capsule.**

Brachial plexus (BRĀ-kē-al PLEK-sus) A network of nerve axons of the ventral rami of spinal nerves C5, C6, C7, C8, and T1. The nerves that emerge from the brachial plexus supply the upper limb.

Bradycardia (brād′-i-KAR-dē-a) A slow resting heart or pulse rate (under 50 beats per minute).

Brain The part of the central nervous system contained within the cranial cavity.

Brain stem The portion of the brain immediately superior to the spinal cord, made up of the medulla oblongata, pons, and midbrain.

Brain waves Electrical signals that can be recorded from the skin of the head due to electrical activity of brain neurons.

Broad ligament A double fold of parietal peritoneum attaching the uterus to the side of the pelvic cavity.

Broca's (BRŌ-kaz) **area** Motor area of the brain in the frontal lobe that translates thoughts into speech. Also called the **motor speech area.**

Bronchi (BRONG-kē) Branches of the respiratory passageway including primary bronchi (the two divisions of the trachea), secondary or lobar bronchi (divisions of the primary bronchi that are distributed to the lobes of the lung), and tertiary or segmental bronchi (divisions of the secondary bronchi that are distributed to bronchopulmonary segments of the lung). *Singular is* **bronchus.**

Bronchial tree The trachea, bronchi, and their branching structures up to and including the terminal bronchioles.

Bronchiole (BRONG-kē-ōl) Branch of a tertiary bronchus further dividing into terminal bronchioles (distributed to lobules of the lung), which divide into respiratory bronchioles (distributed to alveolar sacs).

Bronchitis (brong-KĪ-tis) Inflammation of the mucous membrane of the bronchial tree; characterized by hypertrophy and hyperplasia of seromucous glands and goblet cells that line the bronchi which results in a productive cough.

Bronchopulmonary (brong′-kō-PUL-mō-ner-ē) **segment** One of the smaller divisions of a lobe of a lung supplied by its own branches of a bronchus.

Brunner's gland *See* **Duodenal gland.**

Buccal (BUK-al) Pertaining to the cheek or mouth.

Bulb of penis Expanded portion of the base of the corpus spongiosum penis.

Bulbourethral (bul′-bō-ū-RĒ-thral) **gland** One of a pair of glands located inferior to the prostate on either side of the urethra that secretes an alkaline fluid into the cavernous urethra. Also called a **Cowper's** (KOW-perz) **gland.**

Bulimia (boo-LIM-ē-a *or* boo-LĒ-mē-a) A disorder characterized by overeating at least twice a week followed by purging by self-induced vomiting, strict dieting or fasting, vigorous exercise, or use of laxatives or diuretics. Also called **binge–purge syndrome.**

Bulk-phase endocytosis A process by which most body cells can ingest membrane-surrounded droplets of interstitial fluid. Also called **pinocytosis.**

Bundle branch One of the two branches of the atrioventricular (AV) bundle made up of specialized muscle fibers (cells) that transmit electrical impulses to the ventricles.

Bundle of His *See* **Atrioventricular (AV) bundle.**

Bursa (BUR-sa) A sac or pouch of synovial fluid located at friction points, especially about joints.

Bursitis (bur-SĪ-tis) Inflammation of a bursa.

Buttocks (BUT-oks) The two fleshy masses on the posterior aspect of the inferior trunk, formed by the gluteal muscles.

C

Calcaneal (kal-KĀ-nē-al) **tendon** The tendon of the soleus, gastrocnemius, and plantaris muscles at the back of the heel. Also called the **Achilles** (a-KIL-ēz) **tendon.**

Calcification (kal-si-fi-KĀ-shun) Deposition of calcium salts, primarily hydroxyapatite, in a framework formed by collagen fibers in which the tissue hardens. Also called **mineralization** (min′-e-ral-i-ZĀ-shun).

Calcitonin (kal-si-TŌ-nin) **(CT)** A hormone produced by the parafollicular cells of the thyroid gland that can lower the amount of blood calcium and phosphates by inhibiting bone resorption (breakdown of bone matrix) and by accelerating uptake of calcium and phosphates into bone matrix.

Calculus (KAL-kū-lus) A stone, or insoluble mass of crystallized salts or other material, formed within the body, as in the gallbladder, kidney, or urinary bladder.

Callus (KAL-lus) A growth of new bone tissue in and around a fractured area, ultimately replaced by mature bone. An acquired, localized thickening.

Calyx (KĀL-iks) Any cuplike division of the kidney pelvis. *Plural is* **calyces** (KĀ-li-sēz).

Canal (ka-NAL) A narrow tube, channel, or passageway.

Canaliculus (kan′-a-LIK-ū-lus) A small channel or canal, as in bones, where they connect lacunae. *Plural is* **canaliculi** (kan′-a-LIK-u-li).

Canal of Schlemm *See* **Scleral venous sinus.**

Cancellous (KAN-sel-us) Having a reticular or latticework structure, as in spongy tissue of bone.

Capacitation (ka′-pas-i-TĀ-shun) The functional changes that sperm undergo in the female reproductive tract that allow them to fertilize a secondary oocyte.

Capillary (KAP-i-lar′-ē) A microscopic blood vessel located between an arteriole and venule through which materials are exchanged between blood and interstitial fluid.

Capillary bed A network of 10–100 capillaries that arises from a single metarteriole.

Carcinogen (kar-SIN-ō-jen) A chemical substance or radiation that causes cancer.

Cardiac (KAR-dē-ak) **arrest** Cessation of an effective heartbeat in which the heart is completely stopped or in ventricular fibrillation.

Cardiac cycle A complete heartbeat consisting of systole (contraction) and diastole (relaxation) of both atria plus systole and diastole of both ventricles.

Cardiac muscle Striated muscle fibers (cells) that form the wall of the heart; stimulated by an intrinsic conduction system and autonomic motor neurons.

Cardiac notch An angular notch in the anterior border of the left lung into which part of the heart fits.

Cardinal ligament A ligament of the uterus, extending laterally from the cervix and vagina as a continuation of the broad ligament.

Cardiogenic area (kar-dē-ō-JEN-ik) A group of mesodermal cells in the head end of an embryo that gives rise to the heart.

Cardiology (kar-dē-OL-ō-jē) The study of the heart and diseases associated with it.

Cardiovascular (kar-dē-ō-VAS-kū-lar) **center** Groups of neurons scattered within the medulla oblongata that regulate heart rate, force of contraction, and blood vessel diameter.

Carotene (KAR-o-tēn) Antioxidant precursor of vitamin A, which is needed for synthesis of photopigments; yellow-orange pigment present in the stratum corneum of the epidermis. Accounts for the yellowish coloration of skin. Also termed **beta-carotene.**

Carotid (ka-ROT-id) **body** Cluster of chemoreceptors on or near the carotid sinus that respond to changes in blood levels of oxygen, carbon dioxide, and hydrogen ions.

Carotid sinus A dilated region of the internal carotid artery just superior to where it branches from the common carotid artery; it contains baroreceptors that monitor blood pressure.

Carpal bones The eight bones of the wrist. Also called **carpals.**

Carpus (KAR-pus) A collective term for the eight bones of the wrist.

Cartilage (KAR-ti-lij) A type of connective tissue consisting of chondrocytes in lacunae embedded in a dense network of collagen and elastic fibers and a matrix of chondroitin sulfate.

Cartilaginous (kar′-ti-LAJ-i-nus) **joint** A joint without a synovial (joint) cavity where the articulating bones are held tightly together by cartilage, allowing little or no movement.

Cast A small mass of hardened material formed within a cavity in the body and then discharged from the body; can originate in different areas and can be composed of various materials.

Catabolism (ka-TAB-ō-lizm) Chemical reactions that break down complex organic compounds into simple ones, with the net release of energy.

Cataract (KAT-a-rakt) Loss of transparency of the lens of the eye or its capsule or both.

Cauda equina (KAW-da ē-KWĪ-na) A tail-like array of roots of spinal nerves at the inferior end of the spinal cord.

Caudal (KAW-dal) Pertaining to any tail-like structure; inferior in position.

Cecum (SĒ-kum) A blind pouch at the proximal end of the large intestine that attaches to the ileum.

Celiac plexus (PLEK-sus) A large mass of autonomic ganglia and axons located at the level of the superior part of the first lumbar vertebra. Also called the **solar plexus.**

Cell The basic structural and functional unit of all organisms; the smallest structure capable of performing all the activities vital to life.

Cell cycle Growth and division of a single cell into two identical cells; consists of interphase and cell division.

Cell division Process by which a cell reproduces itself that consists of a nuclear division (mitosis) and a cytoplasmic division (cytokinesis); types include somatic and reproductive cell division.

Cell junction Point of contact between plasma membrane of tissue cells.

Cementum (se-MEN-tum) Calcified tissue covering the root of a tooth.

Center of ossification (os′-i-fi-KĀ-shun) An area in the cartilage model of a future bone where the cartilage cells hypertrophy and then secrete enzymes that result in the calcification of their matrix, resulting in the death of the cartilage cells, followed by the invasion of the area by osteoblasts that then lay down bone.

Central canal A microscopic tube running the length of the spinal cord in the gray commissure. A circular channel running longitudinally in the center of an osteon (haversian system) of mature compact bone, containing blood and lymphatic vessels and nerves. Also called an **haversian** (ha-VER-shan) **canal.**

Central nervous system (CNS) That portion of the nervous system that consists of the brain and spinal cord.

Centrioles (SEN-trē-ōlz) Paired, cylindrical structures of a centrosome, each consisting of a ring of microtubules and arranged at right angles to each other.

Centromere (SEN-trō-mēr) The constricted portion of a chromosome where the two chromatids are joined; serves as the point of attachment for the microtubules that pull chromatids during anaphase of cell division.

Centrosome (SEN-trō-sōm) A dense network of small protein fibers near the nucleus of a cell, containing a pair of centrioles and pericentriolar material.

Cephalic (se-FAL-ik) Pertaining to the head; superior in position.

Cerebellar peduncle (ser-e-BEL-ar pe-DUNG-kul) A bundle of nerve axons connecting the cerebellum with the brain stem.

Cerebellum (ser-e-BEL-um) The part of the brain lying posterior to the medulla oblongata and pons; governs balance and coordinates skilled movements.

Cerebral aqueduct (SER-ē-bral AK-we-dukt) A channel through the midbrain connecting the third and fourth ventricles and containing cerebrospinal fluid. Also termed the **aqueduct of Sylvius.**

Cerebral arterial circle A ring of arteries forming an anastomosis at the base of the brain between the internal carotid and basilar arteries and arteries supplying the cerebral cortex. Also called the **circle of Willis.**

Cerebral cortex The surface of the cerebral hemispheres, 2–4 mm thick, consisting of gray matter; arranged in six layers of neuronal cell bodies in most areas.

Cerebral peduncle One of a pair of nerve axon bundles located on the anterior surface of the midbrain, conducting nerve impulses between the pons and the cerebral hemispheres.

Cerebrospinal (se-rē′-brō-SPĪ-nal) **fluid (CSF)** A fluid produced by ependymal cells that cover choroid plexuses in the ventricles of the brain; the fluid circulates in the ventricles, the central canal, and the subarachnoid space around the brain and spinal cord.

Cerebrovascular (se rē′-brō-VAS-kū-lar) **accident (CVA)** Destruction of brain tissue (infarction) resulting from obstruction or rupture of blood vessels that supply the brain. Also called a **stroke** or **brain attack.**

Cerebrum (SER-e-brum or se-RĒ-brum) The two hemispheres of the forebrain (derived from the telencephalon), making up the largest part of the brain.

Cerumen (se-ROO-men) Waxlike secretion produced by ceruminous glands in the external auditory meatus (ear canal). Also termed **ear wax.**

Ceruminous (se-ROO-mi-nus) **gland** A modified sudoriferous (sweat) gland in the external auditory meatus that secretes cerumen (ear wax).

Cervical ganglion (SER-vi-kul GANG-glē-on) A cluster of cell bodies of postganglionic sympathetic neurons located in the neck, near the vertebral column.

Cervical plexus (PLEK-sus) A network formed by nerve axons from the ventral rami of the first four cervical nerves and receiving gray rami communicantes from the superior cervical ganglion.

Cervix (SER-viks) Neck; any constricted portion of an organ, such as the inferior cylindrical part of the uterus.

Chemoreceptor (kē′-mō-rē-SEP-tor) Sensory receptor that detects the presence of a specific chemical.

Chiasm (KĪ-azm) A crossing; especially the crossing of axons in the optic (II) nerve.

Chief cell The secreting cell of a gastric gland that produces pepsinogen, the precursor of the enzyme pepsin, and the enzyme gastric lipase. Also called a **zymogenic** (zī′-mō-JEN-ik) **cell.** Cell in the parathyroid glands that secretes parathyroid hormone (PTH). Also called a **principal cell.**

Chiropractic (kī-rō-PRAK-tik) A system of treating disease by using one's hands to manipulate body parts, mostly the vertebral column.

Cholecystectomy (kō′-lē-sis-TEK-tō-mē) Surgical removal of the gallbladder.

Cholecystitis (kō′-lē-sis-TĪ-tis) Inflammation of the gallbladder.

Cholesterol (kō-LES-te-rol) Classified as a lipid, the most abundant steroid in animal tissues; located in cell membranes and used for the synthesis of steroid hormones and bile salts.

Cholinergic (kō′-lin-ER-jik) **neuron** A neuron that liberates acetylcholine as its neurotransmitter.

Chondrocyte (KON-drō-sīt) Cell of mature cartilage.

Chondroitin (kon-DROY-tin) **sulfate** An amorphous matrix material found outside connective tissue cells.

Chordae tendineae (KOR-dē TEN-di-nē-ē) Tendonlike, fibrous cords that connect atrioventricular valves of the heart with papillary muscles.

Chorion (KŌ-rē-on) The most superficial fetal membrane that becomes the principal embryonic portion of the placenta; serves a protective and nutritive function.

Chorionic villi (kō-rē-ON-ik VIL-lī) Fingerlike projections of the chorion that grow into the decidua basalis of the endometrium and contain fetal blood vessels.

Chorionic villi sampling (CVS) The removal of a sample of chorionic villus tissue by means of a catheter to analyze the tissue for prenatal genetic defects.

Choroid (KŌ-royd) One of the vascular coats of the eyeball.

Choroid plexus (PLEK-sus) A network of capillaries located in the roof of each of the four ventricles of the brain; ependymal cells around choroid plexuses produce cerebrospinal fluid.

Chromaffin (KRŌ-maf-in) **cell** Cell that has an affinity for chrome salts, due in part to the presence of the precursors of the neurotransmitter epinephrine; found, among other places, in the adrenal medulla.

Chromatid (KRŌ-ma-tid) One of a pair of identical connected nucleoprotein strands that are joined at the centromere and separate during cell division, each becoming a chromosome of one of the two daughter cells.

Chromatin (KRŌ-ma-tin) The threadlike mass of genetic material, consisting of DNA and histone proteins, that is present in the nucleus of a nondividing or interphase cell.

Chromatolysis (krō-ma-TOL-i-sis) The breakdown of Nissl bodies into finely granular masses in the cell body of a neuron whose axon has been damaged.

Chromosome (KRŌ-mō-sōm) One of the small, threadlike structures in the nucleus of a cell, normally 46 in a human diploid cell, that bears the genetic material; composed of DNA and proteins (histones) that form a delicate chromatin thread during interphase; becomes packaged into compact rodlike structures that are visible under the light microscope during cell division.

Chronic (KRON-ik) Long term or frequently recurring; applied to a disease that is not acute.

Chronic obstructive pulmonary disease (COPD) A disease, such as bronchitis or emphysema, in which there is some degree of obstruction of airways and consequent increase in airway resistance.

Chyle (KĪL) The milky-appearing fluid found in the lacteals of the small intestine after absorption of lipids in food.

Chyme (KĪM) The semifluid mixture of partly digested food and digestive secretions found in the stomach and small intestine during digestion of a meal.

Ciliary (SIL-ē-ar′-ē) **body** One of the three parts of the vascular tunic of the eyeball, the others being the choroid and the iris; includes the ciliary muscle and the ciliary processes.

Ciliary ganglion (GANG-glē-on) A very small parasympathetic ganglion whose preganglionic axons come from the oculomotor (III) nerve and whose postganglionic axons carry nerve impulses to the ciliary muscle and the sphincter muscle of the iris.

Cilium (SIL-ē-um) A hair or hairlike process projecting from a cell that may be used to move the entire cell or to move substances along the surface of the cell. *Plural is* **cilia.**

Circle of Willis *See* **Cerebral arterial circle.**

Circular folds Permanent, deep, transverse folds in the mucosa and submucosa of the small intestine that increase the surface area for absorption. Also called **plicae circulares** (PLĪ-kē SER-kū-lar-ēs).

Circumduction (ser′-kum-DUK-shun) A movement at a synovial joint in which the distal end of a bone moves in a circle while the proximal end remains relatively stable.

Circumferential lamellae (ser′-kum-fer-EN-chel) Thin concentric layers of bone that form the surface layers of compact bone.

Cirrhosis (si-RŌ-sis) A liver disorder in which the parenchymal cells are destroyed and replaced by connective tissue.

Cisterna chyli (sis-TER-na KĪ-lē) The origin of the thoracic duct.

Cleavage The rapid mitotic divisions following the fertilization of a secondary oocyte, resulting in an increased number of progressively smaller cells, called blastomeres.

Climacteric (klī-mak-TER-ik) Cessation of the reproductive function in the female or diminution of testicular activity in the male.

Climax The peak period or moments of greatest intensity during sexual excitement.

Clone (KLŌN) A population of identical cells.

Coarctation (kō′-ark-TĀ-shun) **of the aorta** A congenital heart defect in which a segment of the aorta is too narrow. As a result, the flow of oxygenated blood to the body is reduced, the left ventricle is forced to pump harder, and high blood pressure develops.

Coccyx (KOK-six) The fused bones at the inferior end of the vertebral column.

Cochlea (KŌK-lē-a) A winding, cone-shaped tube forming a portion of the inner ear and containing the spiral organ (organ of Corti).

Cochlear duct The membranous cochlea consisting of a spirally arranged tube enclosed in the bony cochlea and lying along its outer wall. Also called the **scala media** (SCA-la MĒ-dē-a).

Coitus (KŌ-i-tus) Sexual intercourse.

Collagen (KOL-a-jen) A protein that is the main organic constituent of connective tissue.

Collateral circulation The alternate route taken by blood through an anastomosis.

Colliculus (ko-LIK-ū-lus) A small elevation.

Colon The portion of the large intestine consisting of ascending, transverse, descending, and sigmoid portions.

Colony-stimulating factor (CSF) One of a group of molecules that stimulates development of white blood cells. Examples are macrophage CSF and granulocyte CSF.

Colostrum (kō-LOS-trum) A thin, cloudy fluid secreted by the mammary glands a few days prior to or after delivery before true milk is produced.

Column (KOL-um) Group of white matter tracts in the spinal cord.

Common bile duct A tube formed by the union of the common hepatic duct and the cystic duct that empties bile into the duodenum at the hepatopancreatic ampulla (ampulla of Vater).

Compact (dense) bone tissue Bone tissue that contains few spaces between osteons (haversian systems); forms the external portion of all bones and the bulk of the diaphysis (shaft) of long bones; is found immediately deep to the periosteum and external to spongy bone. Also called **cortical bone.**

Concentric contraction A contraction in which a muscle shortens as it produces a constant tension and overcomes the load it is moving.

Concentric lamellae (la-MEL-ē) Circular plates of mineralized extracellular matrix that form the rings of bony material in the osteon.

Concha (KONG-ka) A scroll-like bone found in the skull. *Plural is* **conchae** (KONG-kē).

Concussion (kon-KUSH-un) Traumatic injury to the brain that produces no visible bruising but may result in abrupt, temporary loss of consciousness.

Conduction system A group of autorhythmic cardiac muscle fibers that generates and distributes electrical impulses to stimulate coordinated contraction of the heart chambers; includes the sinoatrial (SA) node, the atrioventricular (AV) node, the atrioventricular (AV) bundle, the right and left bundle branches, and the Purkinje fibers.

Conductivity (kon′-duk-TIV-i-tē) The ability of a cell to propagate (conduct) action potentials along its plasma membrane; characteristic of neurons and muscle fibers (cells).

Condyloid (KON-di-loyd) **joint** A synovial joint structured so that an oval-shaped condyle of one bone fits into an elliptical cavity of another bone, permitting side-to-side and back-and-forth movements, such as the joint at the wrist between the radius and carpals. Also called an **ellipsoidal** (ē-lip-SOYD-al) **joint.**

Cone (KŌN) The type of photoreceptor in the retina that is specialized for highly acute color vision in bright light.

Congenital (kon-JEN-i-tal) Present at the time of birth.

Conjunctiva (kon′-junk-TĪ-va) The delicate membrane covering the eyeball and lining the eyes.

Connective tissue One of the most abundant of the four basic tissue types in the body, performing the functions of binding and supporting; consists of relatively few cells in a generous matrix (the ground substance and fibers between the cells).

Consciousness (KON-shus-nes) A state of wakefulness in which an individual is fully alert, aware, and oriented, partly as a result of feedback between the cerebral cortex and reticular activating system.

Continuous capillary Capillary in which the plasma membranes of endothelial cells form a continuous tube that is interrupted only by gaps between neighboring endothelial cells.

Contraception (kon′-tra-SEP-shun) The prevention of fertilization or impregnation without destroying fertility.

Contractility (kon′-trak-TIL-i-tē) The ability of cells or parts of cells to actively generate force to undergo shortening for movements. Muscle fibers (cells) exhibit a high degree of contractility.

Contralateral (kon′-tra-LAT-er-al) On the opposite side; affecting the opposite side of the body.

Conus medullaris (KŌ-nus med-ū-LAR-is) The tapered portion of the spinal cord inferior to the lumbar enlargement.

Convergence (con-VER-jens) A synaptic arrangement in which the synaptic end bulbs of several presynaptic neurons terminate on one postsynaptic neuron. The medial movement of the two eyeballs so that both are directed toward a near object being viewed in order to produce a single image.

Convulsion (con-VUL-shun) Violent, involuntary contractions or spasms of an entire group of muscles.

Cornea (KOR-nē-a) The nonvascular, transparent fibrous coat through which the iris of the eye can be seen.

Corneocyte (KOR-nē-ō-sīt) Thin, flat, plasma-membrane enclosed keratin cells form the stratum corneum of the epidermis. Also called **squame.**

Corona (kō-RŌ-na) Margin of the glans penis.

Corona radiata The innermost layer of granulosa cells that is firmly attached to the zona pellucida around a secondary oocyte.

Coronary artery disease (CAD) A condition such as atherosclerosis that causes narrowing of coronary arteries so that blood flow to the heart is reduced. The result is **coronary heart disease (CHD),** in which the heart muscle receives inadequate blood flow due to an interruption of its blood supply.

Coronary circulation The pathway followed by the blood from the ascending aorta through the blood vessels supplying the heart and returning to the right atrium. Also called **cardiac circulation.**

Coronary sinus (SĪ-nus) A wide venous channel on the posterior surface of the heart that collects the blood from the coronary circulation and returns it to the right atrium.

Corpus (KOR-pus) The principal part of any organ; any mass or body.

Corpus albicans (KOR-pus AL-bi-kanz) A white fibrous patch in the ovary that forms after the corpus luteum regresses.

Corpus callosum (kal-LŌ-sum) The great commissure of the brain between the cerebral hemispheres.

Corpuscle of touch The sensory receptor for the sensation of touch; found in the dermal papillae, especially in palms and soles. Also called a **Meissner** (MĪZ-ner) **corpuscle.**

Corpus luteum (LOO-tē-um) A yellowish body in the ovary formed when a follicle has discharged its secondary oocyte; secretes estrogens, progesterone, relaxin, and inhibin.

Corpus striatum (strī-Ā-tum) An area in the interior of each cerebral hemisphere composed of the caudate and putamen of the basal ganglia and white matter of the internal capsule, arranged in a striated manner.

Cortex (KOR-teks) An outer layer of an organ. The convoluted layer of gray matter covering each cerebral hemisphere.

Costal (KOS-tal) Pertaining to a rib.

Costal cartilage (KAR-ti-lij) Hyaline cartilage that attaches a rib to the sternum.

Cramp A spasmodic, usually painful contraction of a muscle.

Cranial (KRĀ-ne-al) **cavity** A subdivision of the dorsal body cavity formed by the cranial bones and containing the brain.

Cranial nerve One of 12 pairs of nerves that leave the brain; pass through foramina in the skull; and supply sensory and motor neurons to the head, neck, part of the trunk, and viscera of the thorax and abdomen. Each is designated by a Roman numeral and a name.

Craniosacral (krā-nē-ō-SĀ-kral) **outflow** The axons of parasympathetic preganglionic neurons, which have their cell bodies located in nuclei in the brain stem and in the lateral gray matter of the sacral portion of the spinal cord.

Cranium (KRĀ-nē-um) The skeleton of the skull that protects the brain and the organs of sight, hearing, and balance; includes the frontal, parietal, temporal, occipital, sphenoid, and ethmoid bones.

Crista (KRIS-ta) A crest or ridged structure. A small elevation in the ampulla of each semicircular duct that contains receptors for dynamic equilibrium.

Crossing-over The exchange of a portion of one chromatid with another during meiosis. It permits an exchange of genes among chromatids and is one factor that results in genetic variation of progeny.

Crus (KRUS) **of penis** Separated, tapered portion of the corpora cavernosa penis. *Plural is* **crura** (KROO-ra).

Crypt of Lieberkhn *See* **Intestinal gland.**

Cryptorchidism (krip-TOR-ki-dizm) The condition of undescended testes.

Cuneate (KŪ-nē-āt) **nucleus** A group of neurons in the inferior part of the medulla oblongata in which axons of the cuneate fasciculus terminate.

Cupula (KUP-ū-la) A mass of gelatinous material covering the hair cells of a crista; a sensory receptor in the ampulla of a semicircular canal stimulated when the head moves.

Cushing syndrome Condition caused by a hypersecretion of glucocorticoids characterized by spindly legs, "moon face," "buffalo hump," pendulous abdomen, flushed facial skin, poor wound healing, hyperglycemia, osteoporosis, hypertension, and increased susceptibility to disease.

Cutaneous (kū-TĀ-nē-us) Pertaining to the skin.

Cyanosis (si-a-NO-sis) A blue or dark purple discoloration, most easily seen in nail beds and mucous membranes, that results from an increased concentration of deoxygenated (reduced) hemoglobin (more than 5 gm/dL).

Cyst (SIST) A sac with a distinct connective tissue wall, containing a fluid or other material.

Cystic (SIS-tik) **duct** The duct that carries bile from the gallbladder to the common bile duct.

Cystitis (sis-TĪ-tis) Inflammation of the urinary bladder.

Cytokincsis (sī-tō-ki-NĒ-sis) Distribution of the cytoplasm into two separate cells during cell division; coordinated with nuclear division (mitosis).

Cytolysis (sī-TOL-i-sis) The rupture of living cells in which the contents leak out.

Cytoplasm (SĪ-tō-plazm) Cytosol plus all organelles except the nucleus.

Cytoskeleton Complex internal structure of cytoplasm consisting of microfilaments, microtubules, and intermediate filaments.

Cytosol (SĪ-tō-sol) Semifluid portion of cytoplasm in which organelles and inclusions are suspended and solutes are dissolved. Also called **intracellular fluid.**

D

Dartos (DAR-tōs) The contractile tissue deep to the skin of the scrotum.

Decidua (dē-SID-ū-a) That portion of the endometrium of the uterus (all but the deepest layer) that is modified during pregnancy and shed after childbirth.

Deciduous (dē-SID-ū-us) Falling off or being shed seasonally or at a particular stage of development. In the body, referring to the first set of teeth.

Decussation (dē-ku-SĀ-shun) A crossing-over to the opposite (contralateral) side; an example is the crossing of 90% of the axons in the large motor tracts to opposite sides in the medullary pyramids.

Deep Away from the surface of the body or an organ.

Deep inguinal (IN-gwi-nal) **ring** A slitlike opening in the aponeurosis of the transversus abdominis muscle that represents the origin of the inguinal canal.

Deep vein Vein that travels with arteries deep to the fascia and runs between the muscles of the body and within body cavities.

Deep-venous thrombosis (DVT) The presence of a thrombus in a vein, usually a deep vein of the lower limbs.

Defecation (def-e-KĀ-shun) The discharge of feces from the rectum.

Degeneration (dē-jen′-er-Ā-shun) A change from a higher to a lower state; a breakdown in structure.

Deglutition (dē-gloo-TISH-un) The act of swallowing.

Dehydration (dē-hī-DRĀ-shun) Excessive loss of water from the body or its parts.

Delta cell A cell in the pancreatic islets (islets of Langerhans) in the pancreas that secretes somatostatin. Also termed a **D cell.**

Demineralization (de-min′-er-al-i-ZĀ-shun) Loss of calcium and phosphorus from bones.

Dendrite (DEN-drīt) A neuronal process that carries electrical signals, usually graded potentials, toward the cell body.

Dendritic (den-DRIT-ik) **cell** One type of antigen-presenting cell with long branchlike projections that commonly is present in mucosal linings such as the vagina, in the skin (Langerhans cells in the epidermis), and in lymph nodes (follicular dendritic cells).

Dens (DENZ) Tooth.

Dental caries (KA-rēz) Gradual demineralization of the enamel and dentin of a tooth that may invade the pulp and alveolar bone. Also called **tooth decay.**

Denticulate (den-TIK-ū-lāt) Finely toothed or serrated; characterized by a series of small, pointed projections.

Dentin (DEN-tin) The bony tissues of a tooth enclosing the pulp cavity.

Dentition (den-TI-shun) The eruption of teeth. The number, shape, and arrangement of teeth.

Deoxyribonucleic (dē-ok′-sē-rī′-bō-noo-KLĒ-ik) **acid (DNA)** A nucleic acid constructed of nucleotides consisting of one of four bases (adenine, cytosine, guanine, or thymine), deoxyribose, and a phosphate group; encoded in the nucleotides is genetic information.

Depression (de-PRESH-un) Movement in which a part of the body moves inferiorly.

Dermal papilla (pa-PILL-a) Fingerlike projection of the papillary region of the dermis that may contain blood capillaries or corpuscles of touch (Meissner corpuscles).

Dermatology (der-ma-TOL-ō-jē) The medical specialty dealing with diseases of the skin.

Dermatome (DER-ma-tōm) The cutaneous area developed from one embryonic spinal cord segment and receiving most of its sensory innerva-

tion from one spinal nerve. An instrument for incising the skin or cutting thin transplants of skin.

Dermis (DER-mis) A layer of dense irregular connective tissue lying deep to the epidermis.

Descending colon (KŌ-lon) The part of the large intestine descending from the left colic (splenic) flexure to the level of the left iliac crest.

Detritus (de-TRĪ-tus) Particulate matter produced by or remaining after the wearing away or disintegration of a substance or tissue; scales, crusts, or loosened skin.

Detrusor (de-TROO-ser) **muscle** Smooth muscle that forms the wall of the urinary bladder.

Developmental biology The study of development from the fertilized egg to the adult form.

Diagnosis (dī-ag-NŌ-sis) Distinguishing one disease from another or determining the nature of a disease from signs and symptoms by inspection, palpation, laboratory tests, and other means.

Dialysis (dī-AL-i-sis) The removal of waste products from blood by diffusion through a selectively permeable membrane.

Diaphragm (DĪ-a-fram) Any partition that separates one area from another, especially the dome-shaped skeletal muscle between the thoracic and abdominal cavities. Also a dome-shaped device that is placed over the cervix, usually with a spermicide, to prevent conception.

Diaphysis (dī-AF-i-sis) The shaft of a long bone.

Diarrhea (dī-a-RE-a) Frequent defecation of liquid feces caused by increased motility of the intestines.

Diarthrosis (dī-ar-THRŌ-sis) Capsular joints with an outer connective tissue joint capsule surrounding a lubricated cavity. Types are gliding, hinge, pivot, condyloid, saddle, and ball-and-socket.

Diastole (dī-AS-tō-lē) In the cardiac cycle, the phase of relaxation or dilation of the heart muscle, especially of the ventricles.

Diastolic (dī-as-TOL-ik) **blood pressure** The force exerted by blood on arterial walls during ventricular relaxation; the lowest blood pressure measured in the large arteries, normally about 80 mmHg in a young adult.

Diencephalon (dī′-en-SEF-a-lon) A part of the brain consisting of the thalamus, hypothalamus, epithalamus, and subthalamus.

Diffusion (dif-Ū-zhun) A passive process in which there is a net or greater movement of molecules or ions from a region of high concentration to a region of low concentration until equilibrium is reached.

Digestion (dī-JES-chun) The mechanical and chemical breakdown of food to simple molecules that can be absorbed and used by body cells.

Dilate (DĪ-lāt) To expand or swell.

Diploe (DIP-lōu-ē) The spongy bone sandwiched between the outer plates of compact bone in flat bones.

Diploid (DIP-loyd) Having the number of chromosomes characteristically found in the somatic cells of an organism; having two haploid sets of chromosomes, one each from the mother and father. Symbolized 2*n*.

Direct motor pathways Collections of upper motor neurons with cell bodies in the motor cortex that project axons into the spinal cord, where they synapse with lower motor neurons or interneurons in the anterior horns. Also called the **pyramidal pathways.**

Disease Any change from a state of health.

Dislocation (dis-lō-KĀ-shun) Displacement of a bone from a joint with tearing of ligaments, tendons, and articular capsules. Also called **luxation** (luks-Ā-shun).

Dissect (di-SEKT) To separate tissues and parts of a cadaver or an organ for anatomical study.

Distal (DIS-tal) Farther from the attachment of a limb to the trunk; farther from the point of origin or attachment.

Diuretic (dī-ū-RET-ik) A chemical that increases urine volume by decreasing reabsorption of water, usually by inhibiting sodium reabsorption.

Divergence (dī-VER-jens) A synaptic arrangement in which the synaptic end bulbs of one presynaptic neuron terminate on several postsynaptic neurons.

Diverticulum (dī-ver-TIK-ū-lum) A sac or pouch in the wall of a canal or organ, especially in the colon.

Dorsal ramus (RĀ-mus) A branch of a spinal nerve containing motor and sensory axons supplying the muscles, skin, and bones of the posterior part of the head, neck, and trunk.

Dorsiflexion (dor′-si-FLEK-shun) Bending the foot in the direction of the dorsum (upper surface).

Down-regulation Phenomenon in which there is a decrease in the number of receptors in response to an excess of a hormone or neurotransmitter.

Duct of Santorini *See* **Accessory duct.**

Duct of Wirsung *See* **Pancreatic duct.**

Ductus arteriosus (DUK-tus ar-tē-rē-O-sus) A small vessel connect-ing the pulmonary trunk with the aorta; found only in the fetus.

Ductus (vas) deferens (DEF-er-ens) The duct that carries sperm from the epididymis to the ejaculatory duct. Also called the **seminal duct.**

Ductus epididymis (ep′-i-DID-i-mis) A tightly coiled tube inside the epididymis, distinguished into a head, body, and tail, in which sperm undergo maturation.

Ductus venosus (ve-NŌ-sus) A small vessel in the fetus that helps the circulation bypass the liver.

Duodenal (doo-ō-DĒ-nal) **gland** Gland in the submucosa of the duodenum that secretes an alkaline mucus to protect the lining of the small intestine from the action of enzymes and to help neutralize the acid in chyme. Also called **Brunner's** (BRUN-erz) **gland.**

Duodenal papilla (pa-PILL-a) An elevation on the duodenal mucosa that receives the hepatopancreatic ampulla (ampulla of Vater).

Duodenum (doo′-ō-DĒ-num *or* doo-OD-e-num) The first 25 cm (10 in.) of the small intestine, which connects the stomach and the ileum.

Dura mater (DOO-ra MĀ-ter) The outermost of the three meninges (coverings) of the brain and spinal cord.

Dynamic equilibrium (ē-kwi-LIB-rē-um) The maintenance of body position, mainly the head, in response to sudden movements such as rotation.

Dysfunction (dis-FUNK-shun) Absence of completely normal function.

Dysmenorrhea (dis′-men-ō-RĒ-a) Painful menstruation.

Dysplasia (dis-PLĀ-zē-a) Change in the size, shape, and organization of cells due to chronic irritation or inflammation; may either revert to normal if stress is removed or progress to neoplasia.

Dyspnea (DISP-nē-a) Shortness of breath.

Dystrophia (dis-TRO-fē-a) Progressive weakening of a muscle.

E

Ectoderm The primary germ layer that gives rise to the nervous system and the epidermis of skin and its derivatives.

Ectopic (ek-TOP-ik) Out of the normal location, as in ectopic pregnancy.

Eccentric contraction (ek-SEN-trik) When a muscle produces a constant tension but lengthens as it gives into the load it is moving.

Edema (e-DĒ-ma) An abnormal accumulation of interstitial fluid.

Effector (e-FEK-tor) An organ of the body, either a muscle or a gland, that is innervated by somatic or autonomic motor neurons.

Efferent arteriole (EF-er-ent ar-TĒ-rē-ōl) A vessel of the renal vascular system that carries blood from a glomerulus to a peritubular capillary.

Efferent (EF-er-ent) **ducts** A series of coiled tubes that transport sperm from the rete testis to the epididymis.

Ejaculation (e-jak-ū-LĀ-shun) The reflex ejection or expulsion of semen from the penis.

Ejaculatory (e-JAK-ū-la-tō-rē) **duct** A tube that transports sperm from the ductus (vas) deferens to the prostatic urethra.

Elastic fiber Thin, branched protein fiber that is strong and can be stretched up to 150% of its relaxed length.

Elasticity (e-las-TIS-i-tē) The ability of tissue to return to its original shape after contraction or extension.

Electrocardiogram (e-lek′-trō-KAR-dē-ō-gram) **(ECG or EKG)** A recording of the electrical changes that accompany the cardiac cycle that can be detected at the surface of the body; may be resting, stress, or ambulatory.

Elevation (el-e-VĀ-shun) Movement in which a part of the body moves superiorly.

Embolism (EM-bō-lizm) Obstruction or closure of a vessel by an embolus.

Embolus (EM-bō-lus) A blood clot, bubble of air or fat from broken bones, mass of bacteria, or other debris or foreign material transported by the blood.

Embryo (EM-brē-ō) The young of any organism in an early stage of development; in humans, the developing organism from fertilization to the end of the eighth week of development.

Embryoblast A region of cells of a blastocyst that differentiates into the three primary germ layers—ectoderm, mesoderm, and endoderm—from which all tissues and organs develop; also called an **inner cell mass.**

Embryology (em′-brē-OL-ō-jē) The study of development from the fertilized egg to the end of the eighth week of development.

Emesis (EM-e-sis) Vomiting.

Emigration (em′-e-GRA-shun) Process whereby white blood cells (WBCs) leave the bloodstream by rolling along the endothelium, sticking to it, and squeezing between the endothelial cells. Adhesion molecules help WBCs stick to the endothelium. Also known as **migration** or **extravasation.**

Emission (ē-MISH-un) Propulsion of sperm into the urethra due to peristaltic contractions of the ducts of the testes, epididymides, and ductus (vas) deferens as a result of sympathetic stimulation.

Emphysema (em-fi-SĒ-ma) A lung disorder in which alveolar walls disintegrate, producing abnormally large air spaces and loss of elasticity in the lungs; typically caused by exposure to cigarette smoke.

Emulsification (ē-mul′-si-fi-KĀ-shun) The dispersion of large lipid globules into smaller, uniformly distributed particles in the presence of bile.

Enamel (e-NAM-el) The hard, white substance covering the crown of a tooth.

Endocardium (en-dō-KAR-dē-um) The layer of the heart wall, composed of endothelium and smooth muscle, that lines the inside of the heart and covers the valves and tendons that hold the valves open.

Endochondral ossification (en′-dō-KON-dral os′-i-fi-KĀ-shun) The replacement of cartilage by bone. Also called **intracartilaginous** (in′-tra-kar′-ti-LAJ-i-nus) **ossification.**

Endocrine (EN-dō-krin) **gland** A gland that secretes hormones into interstitial fluid and then the blood; a ductless gland.

Endocrinology (en′-dō-kri-NOL-ō-jē) The science concerned with the structure and functions of endocrine glands and the diagnosis and treatment of disorders of the endocrine system.

Endocytosis (en′-dō-sī-TŌ-sis) The uptake into a cell of large molecules and particles in which a segment of plasma membrane surrounds the substance, encloses it, and brings it in; includes phagocytosis, bulk-phase endocytosis, and receptor-mediated endocytosis.

Endoderm (EN-dō-derm) A primary germ layer of the developing embryo; gives rise to the gastrointestinal tract, urinary bladder, urethra, and respiratory tract.

Endodontics (en′-dō-DON-tiks) The branch of dentistry concerned with the prevention, diagnosis, and treatment of diseases that affect the pulp, root, periodontal ligament, and alveolar bone.

Endogenous (en-DOJ-e-nus) Growing from or beginning within the organism.

Endolymph (EN-dō-limf′) The fluid within the membranous labyrinth of the internal ear.

Endometriosis (en′-dō-MĒ-trē-ō′-sis) The growth of endometrial tissue outside the uterus.

Endometrium (en′-dō-MĒ-trē-um) The mucous membrane lining the uterus.

Endomysium (en′-dō-MĪZ-ē-um) Invagination of the perimysium separating each individual muscle fiber (cell).

Endoneurium (en′-dō-NOO-rē-um) Connective tissue wrapping around individual nerve axons (cells).

Endoplasmic reticulum (en′-do-PLAZ-mik re-TIK-ū-lum) **(ER)** A network of channels running through the cytoplasm of a cell that serves in intracellular transportation, support, storage, synthesis, and packaging of molecules. Portions of ER where ribosomes are attached to the outer surface are called **rough ER;** portions that have no ribosomes are called **smooth ER.**

End organ of Ruffini *See* **Type II cutaneous mechanoreceptor.**

Endosteum (en-DOS-tē-um) The membrane that lines the medullary (marrow) cavity of bones, consisting of osteogenic cells and scattered osteoclasts.

Endothelial-capsular (en-dō-THĒ-lē-al) **membrane** A filtration membrane in a nephron of a kidney consisting of the endothelium and basement membrane of the glomerulus and the epithelium of the visceral layer of the glomerular (Bowman's) capsule.

Endothelium (en′-dō-THĒ-lē-um) The layer of simple squamous epithelium that lines the cavities of the heart, blood vessels, and lymphatic vessels.

Enteric (EN-ter-ik) **nervous system** The part of the nervous system that is embedded in the submucosa and muscularis of the gastrointestinal (GI) tract; governs motility and secretions of the GI tract.

Enteroendocrine (en-ter-ō-EN-dō-krin) **cell** A cell of the mucosa of the gastrointestinal tract that secretes a hormone that governs function of the GI tract; hormones secreted include gastrin, cholecystokinin, glucose-dependent insulinotropic peptide (GIP), and secretin.

Enzyme (EN-zīm) A substance that accelerates chemical reactions; an organic catalyst, usually a protein.

Eosinophil (ē′-ō-SIN-ō-fil) A type of white blood cell characterized by granules that stain red or pink with acid dyes.

Ependymal (e-PEN-de-mal) **cells** Neuroglial cells that cover choroid plexuses and produce cerebrospinal fluid (CSF); they also line the ventricles of the brain and probably assist in the circulation of CSF.

Epicardium (ep′-i-KAR-dē-um) The thin outer layer of the heart wall, composed of adipose tissue and mesothelium. The mesothelium alone is called the **visceral pericardium.**

Epidemiology (ep′-i-dē-mē-OL-ō-jē) Study of the occurrence and distribution of diseases and disorders in human populations.

Epidermis (ep-i-DERM-is) The superficial, thinner layer of skin, composed of keratinized stratified squamous epithelium.

Epididymis (ep′-i-DID-i-mis) A comma-shaped organ that lies along the posterior border of the testis and contains the ductus epididymis, in which sperm undergo maturation. *Plural is* **epididymides** (ep′-i-di-DIM-i-dēz).

Epidural (ep′-i-DOO-ral) **space** A space between the spinal dura mater and the vertebral canal, containing areolar connective tissue and a plexus of veins.

Epiglottis (ep′-i-GLOT-is) A large, leaf-shaped piece of cartilage lying on top of the larynx, attached to the thyroid cartilage; its unattached portion is free to move up and down to cover the glottis (vocal folds and rima glottidis) during swallowing.

Epimysium (ep′-i-MĪZ-ē-um) Fibrous connective tissue around muscles.

Epinephrine (ep-ē-NEF-rin) Hormone secreted by the adrenal medulla that produces actions similar to those that result from sympathetic stimulation. Also called **adrenaline** (a-DREN-a-lin).

Epineurium (ep′-i-NOO-rē-um) The superficial connective tissue covering around an entire nerve.

Epiphyseal (ep′-i-FIZ-ē-al) **line** The remnant of the epiphyseal plate in the metaphysis of a long bone.

Epiphyseal plate The hyaline cartilage plate in the metaphysis of a long bone; site of lengthwise growth of long bones. Also called the **growth plate.**

Epiphysis (ē-PIF-i-sis) The end of a long bone, usually larger in diameter than the shaft (diaphysis).

Epiphysis cerebri (se-RĒ-brē) Pineal gland.

Episiotomy (e-piz'-ē-OT-ō-mē) A cut made with surgical scissors to avoid tearing of the perineum at the end of the second stage of labor.

Epistaxis (ep'-i-STAK-sis) Loss of blood from the nose due to trauma, infection, allergy, neoplasm, and bleeding disorders. Also called **nosebleed.**

Epithalamus (ep'-i-THAL-a-mus) Part of the diencephalon superior and posterior to the thalamus, comprising the pineal gland and associated structures.

Epithelial (ep'-i-THĒ-lē-al) **tissue** The tissue that forms innermost and outermost surfaces of body structures and forms glands.

Eponychium (ep'-o-NIK-ē-um) Narrow band of stratum corneum at the proximal border of a nail that extends from the margin of the nail wall. Also called the **cuticle.**

Erectile dysfunction Failure to maintain an erection long enough for sexual intercourse. Also known as **impotence** (IM-pō-tens).

Erection (ē-REK-shun) The enlarged and stiff state of the penis or clitoris resulting from the engorgement of the spongy erectile tissue with blood.

Eructation (e-ruk'-TĀ-shun) The forceful expulsion of gas from the stomach. Also called **belching.**

Erythema (er'-i-THĒ-ma) Skin redness usually caused by dilation of the capillaries.

Erythrocyte (e-RITH-rō-sīt) A mature red blood cell.

Erythropoietin (e-rith'-rō-POY-e-tin) A hormone released by the juxtaglomerular cells of the kidneys that stimulates red blood cell production.

Esophagus (e-SOF-a-gus) The hollow muscular tube that connects the pharynx and the stomach.

Estrogens (ES-tro-jenz) Feminizing sex hormones produced by the ovaries; govern development of oocytes, maintenance of female reproductive structures, and appearance of secondary sex characteristics; also affect fluid and electrolyte balance, and protein anabolism. Examples are β-estradiol, estrone, and estriol.

Etiology (ē'-tē-OL-ō-jē) The study of the causes of disease, including theories of the origin and organisms (if any) involved.

Eupnea (ŪP-nē-a) Normal quiet breathing.

Eustachian tube *See* **Auditory tube.**

Eversion (ē-VER-zhun) The movement of the sole laterally at the ankle joint or of an atrioventricular valve into an atrium during ventricular contraction.

Exacerbation (eg-zas'-er-BĀ-shun) An increase in the severity of symptoms or of a disease.

Excitability (ek-sit'-a-BIL-ī-tē) The ability of muscle fibers to receive and respond to stimuli; the ability of neurons to respond to stimuli and generate nerve impulses.

Excrement (EKS-kre-ment) Material eliminated from the body as waste, especially fecal matter.

Excretion (eks-KRĒ-shun) The process of eliminating waste products from the body; also the products excreted.

Exocrine (EK-sō-krin) **gland** A gland that secretes its products into ducts that carry the secretions into body cavities, into the lumen of an organ, or to the outer surface of the body.

Exocytosis (ex'-ō-si-TŌ-sis) A process in which membrane-enclosed secretory vesicles form inside the cell, fuse with the plasma membrane, and release their contents into the interstitial fluid; achieves secretion of materials from a cell.

Exogenous (ex-SOJ-e-nus) Originating outside an organ or part.

Exhalation (eks-ha-LĀ-shun) Breathing out; expelling air from the lungs into the atmosphere. Also called **expiration.**

Extensibility (ek-sten'-si-BIL-i-tē) The ability of muscle tissue to stretch when it is pulled.

Extension (ek-STEN-shun) An increase in the angle between two bones; restoring a body part to its anatomical position after flexion.

External Located on or near the surface.

External auditory (AW-di-tōr-ē) **canal** or **meatus** (mē-Ā-tus) A curved tube in the temporal bone that leads to the middle ear.

External ear The outer ear, consisting of the pinna, external auditory canal, and tympanic membrane (eardrum).

External nares (NĀ-rez) The openings into the nasal cavity on the exterior of the body. Also called the **nostrils.**

External respiration The exchange of respiratory gases between the lungs and blood. Also called **pulmonary respiration.**

Exteroceptor (eks'-ter-ō-SEP-tor) A sensory receptor adapted for the reception of stimuli from outside the body.

Extracellular fluid (ECF) Fluid outside body cells, such as interstitial fluid and plasma.

Extracellular matrix (MĀ-triks) The ground substance and fibers between cells in a connective tissue.

Extravasation (eks-trav-a-SĀ-shun) The escape of fluid, especially blood, lymph, or serum, from a vessel into the tissues.

Extrinsic (eks-TRIN-sik) Of external origin.

Exudate (EKS-oo-dāt) Escaping fluid or semifluid material that oozes from a space and that may contain serum, pus, and cellular debris.

Eyebrow The hairy ridge superior to the eye.

F

F cell A cell in the pancreatic islets (islets of Langerhans) that secretes pancreatic polypeptide.

Face The anterior aspect of the head.

Falciform ligament (FAL-si-form LIG-a-ment) A sheet of parietal peritoneum between the two principal lobes of the liver. The ligamentum teres, or remnant of the umbilical vein, lies within its fold.

Falx cerebelli (FALKS ser'-e-BEL-lē) A small triangular process of the dura mater attached to the occipital bone in the posterior cranial fossa and projecting inward between the two cerebellar hemispheres.

Falx cerebri (FALKS SER-e-brē) A fold of the dura mater extending deep into the longitudinal fissure between the two cerebral hemispheres.

Fascia (FASH-ē-a) A fibrous membrane covering, supporting, and separating muscles.

Fascicle (FAS-i-kul) A small bundle or cluster, especially of nerve or muscle fibers (cells). Also called a **fasciculus** (fa-SIK-ū-lus). *Plural is* **fasciculi** (fa-SIK-yoo-lī).

Fasciculation (fa-sik'-ū-LĀ-shun) Abnormal, spontaneous twitch of all skeletal muscle fibers in one motor unit that is visible at the skin surface; not associated with movement of the affected muscle; present in progressive diseases of motor neurons, for example, poliomyelitis.

Fauces (FAW-sēz) The opening from the mouth into the pharynx.

Feces (FĒ-sēz) Material discharged from the rectum and made up of bacteria, excretions, and food residue. Also called **stool.**

Female reproductive cycle General term for the ovarian and uterine cycles, the hormonal changes that accompany them, and cyclic changes in the breasts and cervix; includes changes in the endometrium of a nonpregnant female that prepares the lining of the uterus to receive a fertilized ovum. Less correctly termed the **menstrual cycle.**

Fenestrated capillary (FEN-es-trāt'-ed) Capillary in which the plasma membranes of the endothelial cells have many small pores ranging from 70 to 100 nm in diameter.

Fertilization (fer'-ti-li-ZĀ-shun) Penetration of a secondary oocyte by a sperm cell, meiotic division of secondary oocyte to form an ovum, and subsequent union of the nuclei of the gametes.

Fetal circulation The cardiovascular system of the fetus, including the placenta and special blood vessels involved in the exchange of materials between fetus and mother.

Fetus (FĒ-tus) In humans, the developing organism *in utero* from the beginning of the third month to birth.

Fever An elevation in body temperature above the normal temperature of 37°C (98.6°F) due to a resetting of the hypothalamic thermostat.

Fibroblast (FĪ-brō-blast) A large, flat cell that secretes most of the extracellular matrix of areolar and dense connective tissues.

Fibrous (FĪ-brus) **joint** A synarthrosis in which the connective tissue core is dense collagenous connective tissue.

Fibrous tunic (TOO-nik) The superficial coat of the eyeball, made up of the posterior sclera and the anterior cornea.

Fight-or-flight response The effects produced upon stimulation of the sympathetic division of the autonomic nervous system.

Filiform papilla (FIL-i-form pa-PIL-a) One of the conical projections that are distributed in parallel rows over the anterior two-thirds of the tongue and lack taste buds.

Filtrate (fil-TRĀT) The fluid produced when blood is filtered by the endothelial-capsular membrane.

Filtration (fil-TRĀ-shun) The flow of a liquid through a filter (or membrane that acts like a filter) due to a hydrostatic pressure; occurs in capillaries due to blood pressure.

Filum terminale (FĪ-lum ter-mi-NAL-ē) Nonnervous fibrous tissue of the spinal cord that extends inferiorly from the conus medullaris to the coccyx.

Fimbriae (FIM-brē-ē) Fingerlike structures, especially the lateral ends of the uterine (Fallopian) tubes.

Fissure (FISH-ur) A groove, fold, or slit that may be normal or abnormal.

Fistula (FIS-tū-la) An abnormal passage between two organs or between an organ cavity and the outside.

Fixator A muscle that stabilizes the origin of the prime mover so that the prime mover can act more efficiently.

Fixed macrophage (MAK-rō-fāj) Stationary phagocytic cell found in the liver, lungs, brain, spleen, lymph nodes, subcutaneous tissue, and red bone marrow. Also called a **histiocyte** (HIS-tē-ō-sīt).

Flaccid (FLAS-sid) Relaxed, flabby, or soft; lacking muscle tone.

Flagellum (fla-JEL-um) A hairlike, motile process on the extremity of a bacterium, protozoan, or sperm cell. *Plural is* **flagella** (fla-JEL-a).

Flat bone Bone that is generally thin and composed of two nearly parallel plates of compact bone enclosing a layer of spongy bone.

Flatus (FLĀ-tus) Gas in the stomach or intestines; commonly used to denote expulsion of gas through the anus.

Flexion (FLEK-shun) Movement in which there is a decrease in the angle between two bones.

Follicle (FOL-i-kul) A small secretory sac or cavity; the group of cells that contains a developing oocyte in the ovaries.

Follicle-stimulating hormone (FSH) Hormone secreted by the anterior pituitary; it initiates development of ova and stimulates the ovaries to secrete estrogens in females, and initiates sperm production in males.

Fontanel (fon′-ta-NEL) A fibrous connective tissue membrane-filled space where bone formation is not yet complete, especially between the cranial bones of an infant's skull.

Foot The terminal part of the lower limb, from the ankle to the toes.

Foramen (fō-RĀ-men) A passage or opening; a communication between two cavities of an organ, or a hole in a bone for passage of vessels or nerves. *Plural is* **foramina** (fō-RAM-i-na).

Foramen ovale (fō-RĀ-men ō-VAL-ē) An opening in the fetal heart in the septum between the right and left atria. A hole in the greater wing of the sphenoid bone that transmits the mandibular branch of the trigeminal (V) nerve.

Forearm (FOR-arm) The part of the free upper limb between the elbow and the wrist.

Fornix (FOR-niks) An arch or fold; a tract in the brain made up of association fibers, connecting the hippocampus with the mammillary bodies; a recess around the cervix of the uterus where it protrudes into the vagina.

Fossa (FOS-a) A furrow or shallow depression.

Fourth ventricle (VEN-tri-kul) A cavity filled with cerebrospinal fluid within the brain lying between the cerebellum and the medulla oblongata and pons.

Fovea centralis (FŌ-vē-a Sen-TRAL-is) A depression in the center of the macula lutea of the retina, containing cones only and lacking blood vessels; the area of highest visual acuity (sharpness of vision).

Fracture (FRAK-choor) Any break in a bone.

Free lower limb The appendage attached at the pelvic (hip) girdle, consisting of the thigh, knee, leg, ankle, foot, and toes. Also called the **free lower extremity.**

Free upper limb The appendage attached at the shoulder girdle, consisting of the arm, forearm, wrist, hand, and fingers. Also called **free upper extremity.**

Frenulum (FREN-ū-lum) A small fold of mucous membrane that connects two parts and limits movement.

Frontal plane A plane at a right angle to a midsagittal plane that divides the body or organs into anterior and posterior portions. Also called a **coronal** (kō-RŌ-nal) **plane.**

Fundus (FUN-dus) The part of a hollow organ farthest from the opening.

Fungiform papilla (FUN-ji-form pa-PIL-a) A mushroomlike elevation on the upper surface of the tongue appearing as a red dot; most contain taste buds.

Furuncle (FYUR-ung-kul) A boil; painful nodule caused by bacterial infection and inflammation of a hair follicle or sebaceous (oil) gland.

G

Gallbladder A small pouch, located inferior to the liver, that stores bile and empties by means of the cystic duct.

Gallstone A solid mass, usually containing cholesterol, in the gallbladder or a bile-containing duct; formed anywhere between bile canaliculi in the liver and the hepatopancreatic ampulla (ampulla of Vater), where bile enters the duodenum. Also called a **biliary calculus.**

Gamete (GAM-ēt) A male or female reproductive cell; a sperm cell or secondary oocyte.

Ganglion (GANG-gle-on) Usually, a group of neuronal cell bodies lying outside the central nervous system (CNS). *Plural is* **ganglia** (GANG-glē-a).

Gastric (GAS-trik) **glands** Glands in the mucosa of the stomach composed of cells that empty their secretions into narrow channels called gastric pits. Types of cells are chief cells (secrete pepsinogen), parietal cells (secrete hydrochloric acid and intrinsic factor), surface mucous and mucous neck cells (secrete mucus), and G cells (secrete gastrin).

Gastroenterology (gas′-trō-en′-ter-OL-ō-jē) The medical specialty that deals with the structure, function, diagnosis, and treatment of diseases of the stomach and intestines.

Gastrointestinal (gas-trō-in-TES-ti-nal) **(GI) tract** A continuous tube running through the ventral body cavity extending from the mouth to the anus. Also called the **alimentary** (al′-i-MEN-tar-ē) **canal.**

Gastrulation (gas′-troo-LĀ-shun) The migration of groups of cells from the epiblast that transform a bilaminar embryonic disc into a trilaminar embryonic disc with three primary germ layers; transformation of the blastula into the gastrula.

Gene (JĒN) Biological unit of heredity; a segment of DNA located in a definite position on a particular chromosome; a sequence of DNA that codes for a particular mRNA, rRNA, or tRNA.

Genetic engineering The manufacture and manipulation of genetic material.

Genetics The study of genes and heredity.

Genitalia (jen′-i-TĀ-lē-a) Reproductive organs.

Genome (JĒ-nōm) The complete set of genes of an organism.

Genotype (JĒ-nō-tīp) The genetic makeup of an individual; the combination of alleles present at one or more chromosomal locations, as distinguished from the appearance, or phenotype, that results from those alleles.

Geriatrics (jer′-ē-AT-riks) The branch of medicine devoted to the medical problems and care of elderly persons.

Gestation (jes-TĀ-shun) The period of development from fertilization to birth.

Gingivae (jin-JI-vē) Gums. They cover the alveolar processes of the mandible and maxilla and extend slightly into each socket.

Gland Specialized epithelial cell or cells that secrete substances; may be exocrine or endocrine.

Glans clitoris (KLI-to-ris) An erectile organ of the female, located at the anterior junction of the labia minora, that is homologous to the male penis.

Glans penis (glanz PĒ-nis) The slightly enlarged region at the distal end of the penis.

Glaucoma (glaw-KŌ-ma) An eye disorder in which there is increased intraocular pressure due to an excess of aqueous humor.

Gliding joint A synovial joint having articulating surfaces that are usually flat, permitting only side-to-side and back-and-forth movements, as between carpal bones, tarsal bones, and the scapula and clavicle. Also called an **arthrodial** (ar-THRŌ-dē-al) **joint.**

Glomerular (glō-MER-ū-lar) **capsule** A double-walled globe at the proximal end of a nephron that encloses the glomerular capillaries. Also called **Bowman's** (BŌ-manz) **capsule.**

Glomerular filtrate (glō-MER-ū-lār FIL-trāt) The fluid produced when blood is filtered by the filtration membrane in the glomeruli of the kidneys.

Glomerular filtration The first step in urine formation in which substances in blood pass through the filtration membrane and the filtrate enters the proximal convoluted tubule of a nephron.

Glomerulus (glō-MER-ū-lus) A rounded mass of nerves or blood vessels, especially the microscopic tuft of capillaries that is surrounded by the glomerular (Bowman's) capsule of each kidney tubule. *Plural is* **glomeruli.**

Glottis (GLOT-is) The vocal folds (true vocal cords) in the larynx plus the space between them (rima glottidis).

Glucagon (GLOO-ka-gon) A hormone produced by the alpha cells of the pancreatic islets (islets of Langerhans) that increases blood glucose level.

Glucocorticoids (gloo-kō-KOR-ti-koyds) Hormones secreted by the cortex of the adrenal gland, especially cortisol, that influence glucose metabolism.

Glucose (GLOO-kōs) A hexose (six-carbon sugar), $C_6H_{12}O_6$, that is a major energy source for the production of ATP by body cells.

Glucosuria (gloo′-kō-SOO-rē-a) The presence of glucose in the urine; may be temporary or pathological. Also called **glycosuria.**

Glycogen (GLI-kō-jen) A highly branched polymer of glucose containing thousands of subunits; functions as a compact store of glucose molecules in liver and muscle fibers (cells).

Goblet cell A goblet-shaped unicellular gland that secretes mucus; present in epithelium of the airways and intestines.

Goiter (GOY-ter) An enlarged thyroid gland.

Golgi (GOL-jē) **complex** An organelle in the cytoplasm of cells consisting of four to six flattened sacs (cisternae), stacked on one another, with expanded areas at their ends;

functions in processing, sorting, packaging, and delivering proteins and lipids to the plasma membrane, lysosomes, and secretory vesicles.

Golgi tendon organ *See* **Tendon organ.**

Gomphosis (gom-FŌ-sis) A fibrous joint in which a cone-shaped peg fits into a socket.

Gonad (GŌ-nad) A gland that produces gametes and hormones; the ovary in the female and the testis in the male.

Gonadotropic hormone Anterior pituitary hormone that affects the gonads.

Gout (GOWT) Hereditary condition associated with excessive uric acid in the blood; the acid crystallizes and deposits in joints, kidneys, and soft tissue.

Graafian follicle *See* **Vesicular ovarian follicle.**

Gracile (GRAS-īl) **nucleus** A group of nerve cells in the inferior part of the medulla oblongata in which axons of the gracile fasciculus terminate.

Gray commissure (KOM-i-shur) A narrow strip of gray matter connecting the two lateral gray masses within the spinal cord.

Gray matter Areas in the central nervous system and ganglia containing neuronal cell bodies, dendrites, unmyelinated axons, axon terminals, and neuroglia; Nissl bodies impart a gray color and there is little or no myelin in gray matter.

Gray ramus communicans (RĀ-mus kō-MŪ-ni-kans) A short nerve containing axons of sympathetic postganglionic neurons; the cell bodies of the neurons are in a sympathetic chain ganglion, and the unmyelinated axons extend via the gray ramus to a spinal nerve and then to the periphery to supply smooth muscle in blood vessels, arrector pili muscles, and sweat glands. *Plural is* **rami communicantes** (RĀ-mē kō-mū-ni-KAN-tēz).

Greater omentum (ō-MEN-tum) A large fold in the serosa of the stomach that hangs down like an apron anterior to the intestines.

Greater vestibular (ves-TIB-ū-lar) **glands** A pair of glands on either side of the vaginal orifice that open by a duct into the space between the hymen and the labia minora. Also called **Bartholin's** (BAR-to-linz) **glands.**

Groin (GROYN) The depression between the thigh and the trunk; the inguinal region.

Gross anatomy The branch of anatomy that deals with structures that can be studied without using a microscope. Also called **macroscopic anatomy.**

Ground substance The fluid, semifluid, gelatinous, or calcified component of connective tissue between the cells and fibers. Also called the **interfibrillar extracellular matrix.**

Growth An increase in size due to an increase in (1) the number of cells, (2) the size of existing cells as internal components increase in size, or (3) the size of intercellular substances.

Gustatory (GUS-ta-tō′-rē) Pertaining to taste.

Gynecology (gī′-ne-KOL-ō-jē) The branch of medicine dealing with the study and treatment of disorders of the female reproductive system.

Gynecomastia (gīn′-e-kō-MAS-tē-a) Excessive growth (benign) of the male mammary glands due to secretion of estrogens by an adrenal gland tumor (feminizing adenoma).

Gyrus (JĪ-rus) One of the folds of the cerebral cortex of the brain. *Plural is* **gyri** (JĪ-rī). Also called a **convolution.**

H

Hair A threadlike structure produced by hair follicles that develops in the dermis. Also called a **pilus** (PĪ-lus).

Hair follicle (FOL-li-kul) Structure, composed of epithelium and surrounding the root of a hair, from which hair develops.

Hair root plexus (PLEK-sus) A network of dendrites arranged around the root of a hair as free or naked nerve endings that are stimulated when a hair shaft is moved.

Hand The terminal portion of a free upper limb, including the carpus, metacarpus, and phalanges.

Haploid (HAP-loyd) Having half the number of chromosomes characteristically found in the tic cells of an organism; characteristic of mature gametes. Symbolized *n.*

Hard palate (PAL-at) The anterior portion of the roof of the mouth, formed by the maxillae and palatine bones and lined by mucous membrane.

Haustra (HAWS-tra) A series of pouches that characterize the colon; caused by tonic contractions of the teniae coli. *Singular is* **haustrum.**

Haversian canal *See* **Central canal.**

Haversian system *See* **Osteon.**

Head The superior part of a human, cephalic to the neck. The superior or proximal part of a structure.

Heart A hollow muscular organ lying slightly to the left of the midline of the chest that pumps the blood through the cardiovascular system.

Heart block An arrhythmia (dysrhythmia) of the heart in which the atria and ventricles contract independently because of a blocking of electrical impulses through the heart at some point in the conduction system.

Heart murmur (MER-mer) An abnormal sound that consists of a flow noise that is heard before, between, or after the normal heart sounds, or that may mask normal heart sounds.

Hemangioblast (hē-MAN-jē-ō-blast) A precursor mesodermal cell that develops into blood and blood vessels.

Hematocrit (hē-MAT-ō-krit) **(Hct)** The percentage of blood made up of red blood cells. Usually measured by centrifuging a blood sample in a graduated tube and then reading the volume of red blood cells and dividing it by the total volume of blood in the sample.

Hematology (hē′-ma-TOL-ō-jē) The study of blood.

Hematoma (hē-ma-TO-ma) A tumor or swelling filled with blood.

Hemiplegia (hem-i-PLE-jē-a) Paralysis of the upper limb, trunk, and lower limb on one side of the body.

Hemoglobin (hē′-mō-GLŌ-bin) **(Hb)** A substance in red blood cells consisting of the protein globin and the iron-containing red pigment heme that transports most of the oxygen and some carbon dioxide in blood.

Hemolysis (hē-MOL-i-sis) The escape of hemoglobin from the interior of a red blood cell into the surrounding medium; results from disruption of the cell membrane by toxins or drugs, freezing or thawing, or hypotonic solutions.

Hemolytic disease of the newborn A hemolytic anemia of a newborn child that results from the destruction of the infant's erythrocytes (red blood cells) by antibodies produced by the mother; usually the antibodies are due to an Rh blood type incompatibility. Also called **erythroblastosis fetalis** (e-rith′-rō-blas-TŌ-sis fe-TAL-is).

Hemophilia (hē′-mō-FIL-ē-a) A hereditary blood disorder where there is a deficient production of certain factors involved in blood clotting, resulting in excessive bleeding into joints, deep tissues, and elsewhere.

Hemopoiesis (hē-mō-poy-Ē-sis) Blood cell production, which occurs in red bone marrow after birth. Also called **hematopoiesis** (hem′-a-to-poy-Ē-sis).

Hemorrhage (HEM-or-rij) Bleeding; the escape of blood from blood vessels, especially when the loss is profuse.

Hemorrhoids (HEM-ō-royds) Dilated or varicosed blood vessels (usually veins) in the anal region. Also called **piles.**

Hepatic (he-PAT-ik) Refers to the liver.

Hepatic acinus (AS-i-nus) The structural and functional unit of the liver. It is an oval mass of hepatocytes whose short axis is defined by the branches of the portal triad.

Hepatic duct A duct that receives bile from the bile capillaries. Small hepatic ducts merge to form the larger right and left hepatic ducts that unite to leave the liver as the common hepatic duct.

Hepatic laminae (LAM-i-nē) Plates of hepatocytes one cell thick bordered on either side by endothelial lined vascular spaces called hepatic sinusoids.

Hepatic portal circulation The flow of blood from the gastrointestinal organs to the liver before returning to the heart.

Hepatic sinusoid (SĪ-nū-soyd) Highly permeable blood capillary between rows of hepatocytes.

Hepatocyte (he-PAT-ō-cyte) A liver cell.

Hepatopancreatic (hep′-a-tō-pan′-krē-A-tik) **ampulla** A small, raised area in the duodenum where the combined common bile duct and main pancreatic duct empty into the duodenum. Also called the **ampulla of Vater** (VA-ter).

Hernia (HER-nē-a) The protrusion or projection of an organ or part of an organ through a membrane or cavity wall, usually the abdominal cavity.

Herniated (HER-nē-a′-ted) **disc** A rupture of an intervertebral disc so that the nucleus pulposus protrudes into the vertebral cavity. Also called a **slipped disc.**

Heterocrine (HET-er-ō-krin) **gland** A gland, such as the pancreas, that is both an exocrine and an endocrine gland.

Hiatus (hi-Ā-tus) An opening; a foramen.

Hilus (HĪ-lus) An area, depression, or pit where blood vessels and nerves enter or leave an organ. Also called a **hilum.**

Hinge joint A synovial joint in which a convex surface of one bone fits into a concave surface of another bone, such as the elbow, knee, ankle, and interphalangeal joints. Also called a **ginglymus** (JIN-gli-mus) **joint.**

Hirsutism (HER-soot-izm) An excessive growth of hair in females and children, with a distribution similar to that in adult males, due to the conversion of vellus hairs into large terminal hairs in response to higher-than-normal levels of androgens.

Histamine (HISS-ta-mēn) Substance found in many cells, especially mast cells, basophils, and platelets, that is released when the cells are injured; results in vasodilation, increased permeability of blood vessels, and constriction of bronchioles.

Histology (hiss-TOL-ō-jē) Microscopic study of the structure of tissues.

Holocrine (HŌL-ō-krin) **gland** A type of gland in which entire secretory cells, along with their accumulated secretions, make up the secretory product of the gland, as in the sebaceous (oil) glands.

Homeostasis (hō'-me-ō-STA-sis) The condition in which the body's internal environment remains relatively constant, within physiological limits.

Homologous (hō-MOL-ō-gus) Correspondence of two organs in structure, position, and origin.

Homologous chromosomes Two chromosomes that belong to a pair. Also called **homologs.**

Hormone (HOR-mōn) A secretion of endocrine cells that alters the physiological activity of target cells of the body.

Horn An area of gray matter (anterior, lateral, or posterior) in the spinal cord.

Human chorionic gonadotropin (kō-rē-ON-ik gō-nad-ō-TRŌ-pin) **(hCG)** A hormone produced by the developing placenta that maintains the corpus luteum.

Human chorionic somatomammotropin (sō-mat-ō-mam-ō-TRŌ-pin) **(hCS)** Hormone produced by the chorion of the placenta that stimulates breast tissue for lactation, enhances body growth, and regulates metabolism. Also called **human placental lactogen (hPL).**

Human growth hormone (hGH) Hormone secreted by the anterior pituitary that stimulates growth of body tissues, especially skeletal and muscular tissues. Also known as **somatotropin** and **somatotropic hormone (STH).**

Hyaluronic (hī'-a-loo-RON-ik) **acid** A viscous, amorphous extracellular material that binds cells together, lubricates joints, and maintains the shape of the eyeballs.

Hymen (HĪ-men) A thin fold of vascularized mucous membrane at the vaginal orifice.

Hyperextension (hī'-per-ek-STEN-shun) Continuation of extension beyond the normal range of motion.

Hyperplasia (hī'-per-PLĀ-zē-a) An abnormal increase in the number of normal cells in a tissue or organ, increasing its size.

Hypersecretion (hī'-per sc-KRĒ-shun) Overactivity of glands resulting in excessive secretion.

Hypersensitivity (hī'-per-sen-si-TI-vi-tē) Overreaction to an allergen that results in pathological changes in tissues. Also called **allergy.**

Hypertension (hī'-per-TEN-shun) High blood pressure.

Hyperthermia (hī'-per-THERM-ē-a) An elevated body temperature.

Hypertonia (hī'-per-TŌ-nē-a) Increased muscle tone that is expressed as spasticity or rigidity.

Hypertonic (hī'-per-TON-ik) Solution that causes cells to shrink due to loss of water by osmosis.

Hypertrophy (hī-PER-trō-fē) An excessive enlargement or overgrowth of tissue without cell division.

Hyperventilation (hī'-per-ven-ti-LĀ-shun) A rate of inhalation and exhalation higher than that required to maintain a normal partial pressure of carbon dioxide in the blood.

Hyponychium (hī'-pō-NIK-ē-um) Free edge of the fingernail.

Hypophyseal fossa (hī'-pō-FIZ-ē-al FOS-a) A depression on the superior surface of the sphenoid bone that houses the pituitary gland.

Hypophyseal (hī'-pō-FIZ-ē-al) **pouch** An outgrowth of ectoderm from the roof of the mouth from which the anterior pituitary develops.

Hypophysis (hī-POF-i-sis) Pituitary gland.

Hyposecretion (hī'-pō-se-KRĒ-shun) Underactivity of glands resulting in diminished secretion.

Hypothalamohypophyseal (hi-pō-tha-lam'-ō-hī-pō-FIZ-ē-al) **tract** A bundle of axons containing secretory vesicles filled with oxytocin or antidiuretic hormone that extend from the hypothalamus to the posterior pituitary.

Hypothalamus (hī'-pō-THAL-a-mus) A portion of the diencephalon, lying beneath the thalamus and forming the floor and part of the wall of the third ventricle.

Hypothermia (hi'-po-THER-mē-a) Lowering of body temperature below 35°C (95°F); in surgical procedures, it refers to deliberate cooling of the body to slow down metabolism and reduce oxygen needs of tissues.

Hypotonia (hi'-pō-TŌ-nē-a) Decreased or lost muscle tone in which muscles appear flaccid.

Hypotonic (hi'-pō-TON-ik) Solution that causes cells to swell and perhaps rupture due to gain of water by osmosis.

Hypoventilation (hi-pō-ven-ti-LĀ-shun) A rate of inhalation and exhalation lower than that required to maintain a normal partial pressure of carbon dioxide in plasma.

Hypoxia (hī-POKS-ē-a) Lack of adequate oxygen at the tissue level.

Hysterectomy (hiss-tē-REK-tō-mē) The surgical removal of the uterus.

I

Ileocecal (il-ē-ō-SĒ-kal) **sphincter** A fold of mucous membrane that guards the opening from the ileum into the large intestine. Also called the **ileocecal valve.**

Ileum (IL-ē-um) The terminal part of the small intestine.

Immunity (im-Ū-ni-tē) The state of being resistant to injury, particularly by poisons, foreign proteins, and invading pathogens.

Immunoglobulin (im-ū-nō-GLOB-ū-lin) **(Ig)** An antibody synthesized by plasma cells derived from B lymphocytes in response to the introduction of an antigen. Immunoglobulins are divided into five kinds (IgG, IgM, IgA, IgD, IgE).

Immunology (im'-ū-NOL-ō-jē) The study of the responses of the body when challenged by antigens.

Imperforate (im-PER-fō-rāt) Abnormally closed.

Implantation (im-plan-TĀ-shun) The insertion of a tissue or a part into the body. The attachment of the blastocyst to the stratum basalis of the endometrium about 6 days after fertilization.

Incontinence (in-KON-ti-nens) Inability to retain urine, semen, or feces through loss of sphincter control.

Indirect motor pathways Motor tracts that convey information from the brain down the spinal cord for automatic movements, coordination of body movements with visual stimuli, skeletal muscle tone and posture, and balance. Also known as **extrapyramidal pathways.**

Induction (in-DUK-shun) The process by which one tissue (inducting tissue) stimulates the development of an adjacent unspecialized tissue (responding tissue) into a specialized one.

Infarction (in-FARK-shun) A localized area of necrotic tissue, produced by inadequate oxygenation of the tissue.

Infection (in-FEK-shun) Invasion and multiplication of microorganisms in body tissues, which may be inapparent or characterized by cellular injury.

Inferior (in-FĒR-ē-or) Away from the head or toward the lower part of a structure. Also called **caudad** (KAW-dad).

Inferior vena cava (VĒ-na CĀ-va) **(IVC)** Large vein that collects blood from parts of the body inferior to the heart and returns it to the right atrium.

Infertility Inability to conceive or to cause conception. Also called **sterility.**

Inflammation (in'-fla-MĀ-shun) Localized, protective response to tissue injury designed to destroy, dilute, or wall off the infecting agent or injured tissue; characterized by redness, pain, heat, swelling, and sometimes loss of function.

Infundibulum (in'-fun-DIB-ū-lum) The stalklike structure that attaches the pituitary gland to the hypothalamus of the brain. The funnel-shaped, open, distal end of the uterine (Fallopian) tube.

Ingestion (in-JES-chun) The taking in of food, liquids, or drugs, by mouth.

Inguinal (IN-gwi-nal) Pertaining to the groin.

Inguinal canal An oblique passageway in the anterior abdominal wall just superior and parallel to the medial half of the inguinal ligament that transmits the spermatic cord and ilioinguinal nerve in the male and round ligament of the uterus and ilioinguinal nerve in the female.

Inhalation (in-ha-LĀ-shun) The act of drawing air into the lungs. Also termed **inspiration.**

Inheritance The acquisition of body traits by transmission of genetic information from parents to offspring.

Inhibin A hormone secreted by the gonads that inhibits release of follicle-stimulating hormone (FSH) by the anterior pituitary.

Inhibiting hormone Hormone secreted by the hypothalamus that can suppress secretion of hormones by the anterior pituitary.

Insertion (in-SER-shun) The attachment of a muscle tendon to a movable bone or the end opposite the origin.

Insula (IN-soo-la) A triangular area of the cerebral cortex that lies deep within the lateral cerebral fissure, under the parietal, frontal, and temporal lobes.

Insulin (IN-soo-lin) A hormone produced by the beta cells of a pancreatic islet (islet of Langerhans) that decreases the blood glucose level.

Integrins (IN-te-grinz) A family of transmembrane glycoproteins in plasma membranes that function in cell adhesion; they are present in hemidesmosomes, which anchor cells to a basement membrane, and they mediate adhesion of neutrophils to endothelial cells during emigration.

Integumentary (in-teg'-ū-MEN-tar-e) Relating to the skin.

Intercalated (in-TER-ka-lāt-ed) **disc** An irregular transverse thickening of sarcolemma that contains desmosomes, which hold cardiac muscle fibers (cells) together, and gap junctions, which aid in conduction of muscle action potentials from one fiber to the next.

Intercostal (in´-ter-KOS-tal) **nerve** A nerve supplying a muscle located between the ribs.

Intermediate Between two structures, one of which is medial and one of which is lateral.

Intermediate filament Protein filament, ranging from 8 to 12 nm in diameter, that may provide structural reinforcement, hold organelles in place, and give shape to a cell.

Internal Away from the surface of the body.

Internal capsule A large tract of projection fibers lateral to the thalamus that is the major connection between the cerebral cortex and the brain stem and spinal cord; contains axons of sensory neurons carrying auditory, visual, and somatic sensory signals to the cerebral cortex plus axons of motor neurons descending from the cerebral cortex to the thalamus, subthalamus, brain stem, and spinal cord.

Internal ear The inner ear or labyrinth, lying inside the temporal bone, containing the organs of hearing and balance.

Internal nares (NĀ-rez) The two openings posterior to the nasal cavities opening into the nasopharynx. Also called the **choanae** (ko-A-ne).

Internal respiration The exchange of respiratory gases between blood and body cells. Also called **tissue respiration.**

Interneurons (in´-ter-NOO-ronz) Neurons whose axons extend only for a short distance and contact nearby neurons in the brain, spinal cord, or a ganglion; they comprise the vast majority of neurons in the body.

Interoceptor (in´-ter-ō-SEP-tor) Sensory receptor located in blood vessels and viscera that provides information about the body's internal environment.

Interphase (IN-ter-fāz) The period of the cell cycle between cell divisions, consisting of the G1-(gap or growth) phase, when the cell is engaged in growth, metabolism, and production of substances required for division; S-(synthesis) phase, during which chromosomes are replicated; and G_2-phase.

Interstitial cell of Leydig *See* **Interstitial endocrinocyte.**

Interstitial (in´-ter-STISH-al) **endocrinocyte** A cell that is located in the connective tissue between seminiferous tubules in a mature testis that secretes testosterone. Also called an **interstitial cell of Leydig** (LĪ-dig).

Interstitial (in´-ter-STISH-al) **fluid** The portion of extracellular fluid that fills the microscopic spaces between the cells of tissues; the internal environment of the body. Also called **intercellular** or **tissue fluid.**

Interstitial growth Growth from within, as in the growth of cartilage. Also called **endogenous** (en-DOJ-e-nus) **growth.**

Interventricular (in´-ter-ven-TRIK-ū-lar) **foramen** A narrow, oval opening through which the lateral ventricles of the brain communicate with the third ventricle. Also called the **foramen of Monro.**

Intervertebral (in´-ter-VER-te-bral) **disc** A pad of fibrocartilage located between the bodies of two vertebrae.

Intervillous space Space within the placenta containing fetal capillaries that is bathed with maternal blood that leaves the uterine arteries.

Intestinal gland A gland that opens onto the surface of the intestinal mucosa and secretes digestive enzymes. Also called a **crypt of Lieberkühn** (LĒ-ber-kūn).

Intracellular (in´-tra-SEL-yū-lar) **fluid (ICF)** Fluid located within cells.

Intrafusal (in´-tra-FŪ-sal) **fibers** Three to ten specialized muscle fibers (cells), partially enclosed in a spindle-shaped connective tissue capsule, that make up a muscle spindle.

Intramembranous ossification (in´-tra-MEM-bra-nus os´-i-fi-KĀ-shun) The method of bone formation in which the bone is formed directly in membranous tissue.

Intraocular (in´-tra-OK-ū-lar) **pressure (IOP)** Pressure in the eyeball, produced mainly by aqueous humor.

Intrinsic (in-TRIN-sik) Of internal origin.

Intrinsic factor (IF) A glycoprotein, synthesized and secreted by the parietal cells of the gastric mucosa, that facilitates vitamin B_{12} absorption in the small intestine.

In utero (Ū-ter-ō) Within the uterus.

Invagination (in-vaj´-i-NĀ-shun) The pushing of the wall of a cavity into the cavity itself.

Inversion (in-VER-zhun) The movement of the sole medially at the ankle joint.

In vitro (VĒ-trō) Literally, in glass; outside the living body and in an artificial environment such as a laboratory test tube.

In vivo (VĒ-vō) In the living body.

Ipsilateral (ip´-si-LAT-er-al) On the same side, affecting the same side of the body.

Iris The colored portion of the vascular tunic of the eyeball seen through the cornea that contains circular and radial smooth muscle; the hole in the center of the iris is the pupil.

Irregular bone Bone that has a complex shape and cannot be placed into the other bone shape categories.

Irritable bowel syndrome (IBS) Disease of the entire gastrointestinal tract in which a person reacts to stress by developing symptoms (such as cramping and abdominal pain) associated with alternating patterns of diarrhea and constipation. Excessive amounts of mucus may appear in feces, and other symptoms include flatulence, nausea, and loss of appetite. Also known as **irritable colon** or **spastic colitis.**

Ischemia (is-KĒ-mē-a) A lack of sufficient blood to a body part due to obstruction or constriction of a blood vessel.

Islet of Langerhans *See* **Pancreatic islet.**

Isotonic (ī´-sō-TON-ik) Having equal tension or tone. A solution having the same concentration of impermeable solutes as cytosol.

Isthmus (IS-mus) A narrow strip of tissue or narrow passage connecting two larger parts.

J

Jaundice (JAWN-dis) A condition characterized by yellowness of the skin, the white of the eyes, mucous membranes, and body fluids because of a buildup of bilirubin.

Jejunum (je-JOO-num) The middle part of the small intestine.

Joint kinesthetic (kin´-es-THET-ik) **receptor** A proprioceptive receptor located in a joint, stimulated by joint movement.

Juxtaglomerular (juks-ta-glō-MER-ū-lar) **apparatus (JGA)** Consists of the macula densa (cells of the distal convoluted tubule adjacent to the afferent and efferent arteriole) and juxta-glomerular cells (modified cells of the afferent and sometimes efferent arteriole); secretes renin when blood pressure starts to fall.

K

Keratin (KER-a-tin) An insoluble protein found in the hair, nails, and other keratinized tissues of the epidermis.

Keratinocyte (ke-RAT-in´-ō-sīt) The most numerous of the epidermal cells; produces keratin.

Kidney (KID-nē) One of the paired reddish organs located in the lumbar region that regulates the composition, volume, and pressure of blood and produces urine.

Kidney stone A solid mass, usually consisting of calcium oxalate, uric acid, or calcium phosphate crystals, that may form in any portion of the urinary tract. Also called a **renal calculus.**

Kinesiology (ki-nē´-sē-OL-o-jē) The study of the movement of body parts.

Kinesthesia (kin-es-THĒ-zē-a) The perception of the extent and direction of movement of body parts; this sense is possible due to nerve impulses generated by proprioceptors.

Kinetochore (ki-NET-ō-kor) Protein complex attached to the outside of a centromere to which kinetochore microtubules attach.

Kupffer's cell *See* **Stellate reticuloendothelial cell.**

Kyphosis (kī-FŌ-sis) An exaggeration of the thoracic curve of the vertebral column, resulting in a "round-shouldered" appearance. Also called **hunchback.**

L

Labial frenulum (LĀ-bē-al FREN-ū-lum) A medial fold of mucous membrane between the inner surface of the lip and the gums.

Labia majora (LĀ-bē-a ma-JŌ-ra) Two longitudinal folds of skin extending downward and backward from the mons pubis of the female.

Labia minora (min-OR-a) Two small folds of mucous membrane lying medial to the labia majora of the female.

Labium (LĀ-bē-um) A lip. A liplike structure. *Plural is* **labia** (LA-bē-a).

Labor The process of giving birth in which a fetus is expelled from the uterus through the vagina.

Labrum (LĀ-brum) A fibrocartilaginous lip that extends from the edge of a ball-and-socket joint to deepen the socket.

Labyrinth (LAB-i-rinth) Intricate communicating passageway, especially in the internal ear.

Lacrimal canal A duct, one on each eyelid, beginning at the punctum at the medial margin of an eyelid and conveying tears medially into the nasolacrimal sac.

Lacrimal gland Secretory cells, located at the superior anterolateral portion of each orbit, that secrete tears into excretory ducts that open onto the surface of the conjunctiva.

Lacrimal sac The superior expanded portion of the nasolacrimal duct that receives the tears from a lacrimal canal.

Lactation (lak-TĀ-shun) The secretion and ejection of milk by the mammary glands.

Lacteal (LAK-tē-al) One of many lymphatic vessels in villi of the intes-tines that absorb triglycerides and other lipids from digested food.

Lacuna (la-KOO-na) A small, hollow space, such as that found in bones in which the osteocytes lie. *Plural is* **lacunae** (la-KOO-nē).

Lambdoid (lam-DOYD) **suture** The joint in the skull between the parietal bones and the occipital bone; sometimes contains sutural (Wormian) bones.

Lamellae (la-MEL-ē) Concentric rings of hard, calcified extracellular matrix found in compact bone.

Lamellar bone (la-MEL-er) The most common type of bone tissue. It has mineralized collagen fibers arranged in highly ordered plates.

Lamellated corpuscle Oval-shaped pressure receptor located in the dermis or subcutaneous tissue and consisting of concentric layers of connective tissue wrapped around the dendrites of a sensory neuron. Also called a **Pacinian** (pa-SIN-ē-an) **corpuscle.**

Lamina (LAM-i-na) A thin, flat layer or membrane, as the flattened part of either side of the arch of a vertebra. *Plural is* **laminae** (LAM-i-nē).

Lamina propria (PRŌ-prē-a) The connective tissue layer of a mucosa.

Langerhans (LANG-er-hans) **cell** Epidermal dendritic cell that functions as an antigen-presenting cell (APC) during an immune response. Also called an **intraepidermal macrophage.**

Lanugo (la-NOO-gō) Fine downy hairs that cover the fetus.

Large intestine The portion of the gastrointestinal tract extending from the ileum of the small intestine to the anus, divided structurally into the cecum, colon, rectum, and anal canal.

Laryngopharynx (la-rin′-gō-FAR-inks) The inferior portion of the pharynx, extending downward from the level of the hyoid bone that divides posteriorly into the esophagus and anteriorly into the larynx. Also called the **hypopharynx.**

Laryngotracheal (la-rin′-gō-TRĀ-ke-al) **bud** An outgrowth of endoderm of the foregut from which the respiratory system develops.

Larynx (LAR-inks) The voice box, a short passageway that connects the pharynx with the trachea.

Lateral (LAT-er-al) Farther from the midline of the body or a structure.

Lateral ventricle (VEN-tri-kul) A cavity within a cerebral hemisphere that communicates with the lateral ventricle in the other cerebral hemisphere and with the third ventricle by way of the interventricular foramen.

Leg The part of the free lower limb between the knee and the ankle.

Lens A transparent organ constructed of proteins (crystallins) lying posterior to the pupil and iris of the eyeball and anterior to the vitreous body.

Leptomeninges (lep-tō-me-NIN-jēz) The inner two meningeal membranes consisting of the pia mater and the arachnoid mater.

Lesion (LĒ-zhun) Any localized, abnormal change in a body tissue.

Lesser omentum (ō-MEN-tum) A fold of the peritoneum that extends from the liver to the lesser curvature of the stomach and the first part of the duodenum.

Lesser vestibular (ves-TIB-ū-lar) **gland** One of the paired mucus-secreting glands with ducts that open on either side of the urethral orifice in the vestibule of the female.

Leukemia (loo-KĒ-mē-a) A malignant disease of the blood-forming tissues characterized by either uncontrolled production and accumulation of immature leukocytes in which many cells fail to reach maturity (acute) or an accumulation of mature leukocytes in the blood because they do not die at the end of their normal life span (chronic).

Leukocyte (LOO-kō-sīt) A white blood cell.

Leydig (LĪ-dig) **cell** A type of cell that secretes testosterone; located in the connective tissue between seminiferous tubules in a mature testis. Also known as **interstitial cell of Leydig** or **interstitial endocrinocyte.**

Libido (li-BĒ-dō) Sexual desire.

Ligament (LIG-a-ment) Dense regular connective tissue that attaches bone to bone.

Ligand (LĪ-gand) A chemical substance that binds to a specific receptor.

Limbic system A part of the forebrain, sometimes termed the visceral brain, concerned with various aspects of emotion and behavior; includes the limbic lobe, dentate gyrus, amygdala, septal nuclei, mammillary bodies, anterior thalamic nucleus, olfactory bulbs, and bundles of myelinated axons.

Lingual frenulum (LIN-gwal FREN-ū-lum) A fold of mucous membrane that connects the tongue to the floor of the mouth.

Lipase An enzyme that splits fatty acids from triglycerides and phospholipids.

Lipid (LIP-id) An organic compound composed of carbon, hydrogen, and oxygen that is usually insoluble in water, but soluble in alcohol, ether, and chloroform; examples include triglycerides (fats and oils), phospholipids, steroids, and eicosanoids.

Lipid bilayer Arrangement of phospholipid, glycolipid, and cholesterol molecules in two parallel sheets in which the hydrophilic "heads" face outward and the hydrophobic "tails" face inward; found in cellular membranes.

Lipoprotein (lip′-ō-PRŌ-tēn) One of several types of particles containing lipids (cholesterol and triglycerides) and proteins that make it water soluble for transport in the blood; high levels of **low-density lipoproteins (LDLs)** are associated with increased risk of atherosclerosis, whereas high levels of **high-density lipoproteins (HDLs)** are associated with decreased risk of atherosclerosis.

Liver Large organ under the diaphragm that occupies most of the right hypochondriac region and part of the epigastric region. Functionally, it produces bile and synthesizes most plasma proteins; interconverts nutrients; detoxifies substances; stores glycogen, iron, and vitamins; carries on phagocytosis of worn-out blood cells and bacteria; and helps synthesize the active form of vitamin D.

Locus coeruleus (LŌ-kus sē-ROO-lē-us) A group of neurons in the brain stem where norepinephrine (NE) is concentrated.

Long bone Bone that has a greater length than width and consists of a shaft and a variable number of extremities (ends).

Long-term potentiation (LTP) Prolonged, enhanced synaptic transmission that occurs at certain synapses within the hippocampus of the brain; believed to underlie some aspects of memory.

Lordosis (lor-DŌ-sis) An exaggeration of the lumbar curve of the vertebral column. Also called **swayback.**

Lumbar (LUM-bar) Region of the back and side between the ribs and pelvis; loin.

Lumbar plexus (PLEK-sus) A network formed by the anterior (ventral) branches of spinal nerves L1 through L4.

Lumen (LOO-men) The space within an artery, vein, intestine, renal tubule, or other tubular structure.

Lungs Main organs of respiration that lie on either side of the heart in the thoracic cavity.

Lunula (LOO-noo-la) The moon-shaped white area at the base of a nail.

Luteinizing (LOO-tē-in′-iz-ing) **hormone (LH)** A hormone secreted by the anterior pituitary that stimulates ovulation, stimulates progesterone secretion by the corpus luteum, and readies the mammary glands for milk secretion in females; stimulates testosterone secretion by the testes in males.

Lymph (LIMF) Fluid confined in lymphatic vessels and flowing through the lymphatic system until it is returned to the blood.

Lymph node An oval or bean-shaped structure located along lymphatic vessels.

Lymphatic (lim-FAT-ik) **capillary** Closed-ended microscopic lymphatic vessel that begins in spaces between cells and converges with other lymphatic capillaries to form lymphatic vessels.

Lymphatic tissue A specialized form of reticular tissue that contains large numbers of lymphocytes.

Lymphatic vessel A large vessel that collects lymph from lymphatic capillaries and converges with other lymphatic vessels to form the thoracic and right lymphatic ducts.

Lymphocyte (LIM-fō-sīt) A type of white blood cell that helps carry out cell-mediated and antibody-mediated immune responses; found in blood and in lymphatic tissues.

Lysosome (LĪ-sō-sōm) An organelle in the cytoplasm of a cell, enclosed by a single membrane and containing powerful digestive enzymes.

Lysozyme (LĪ-sō-zīm) A bactericidal enzyme found in tears, saliva, and perspiration.

M

Macrophage (MAK-rō-fāj) Phagocytic cell derived from a monocyte; may be fixed or wandering.

Macula (MAK-ū-la) A discolored spot or a colored area. A small, thickened region on the wall of the utricle and saccule that contains receptors for static equilibrium.

Macula lutea (LOO-tē-a) The yellow spot in the center of the retina.

Major histocompatibility (MHC) antigens Surface proteins on white blood cells and other nucleated cells that are unique for each person (except for identical siblings); used to type tissues and help prevent rejection of transplanted tissues. Also known as **human leukocyte antigens (HLA).**

Malaise (ma-LĀYZ) Discomfort, uneasiness, and indisposition, often indicative of infection.

Malignant (ma-LIG-nant) Referring to diseases that tend to become worse and cause death, especially the invasion and spreading of cancer.

Mammary (MAM-ar-ē) **gland** Modified sudoriferous (sweat) gland of the female that produces milk for the nourishment of the young.

Mammillary (MAM-i-ler-ē) **bodies** Two small rounded bodies on the inferior aspect of the hypothalamus that are involved in reflexes related to the sense of smell.

Marrow (MAR-ō) Soft, spongelike material in the cavities of bone. Red bone marrow produces blood cells; yellow bone marrow contains adipose tissue that stores triglycerides.

Mast cell A cell found in areolar connective tissue that releases histamine, a dilator of small blood vessels, during inflammation.

Mastication (mas′-ti-KA-shun) Chewing.

Mature follicle A large, fluid-filled follicle containing a secondary oocyte and surrounding granulosa cells that secrete estrogens. Also called a **graafian** (GRAF-ē-an) **follicle.**

Meatus (mē-Ā-tus) A passage or opening, especially the external portion of a canal.

Mechanoreceptor (me-KAN-ō-rē-sep-tor) Sensory receptor that detects mechanical deformation of the receptor itself or adjacent cells; stimuli so detected include those related to touch, pressure, vibration, proprioception, hearing, equilibrium, and blood pressure.

Medial (MĒ-dē-al) Nearer the midline of the body or a structure.

Medial lemniscus (lem-NIS-kus) A white matter tract that originates in the gracile and cuneate nuclei of the medulla oblongata and extends to the thalamus on the same side; sensory axons in this tract conduct nerve impulses for the sensations of proprioception, fine touch, vibration, hearing, and equilibrium.

Median aperture (AP-er-choor) One of the three openings in the roof of the fourth ventricle through which cerebrospinal fluid enters the subarachnoid space of the brain and cord. Also called the **foramen of Magendie.**

Median plane A vertical plane dividing the body into right and left halves. Situated in the middle.

Mediastinum (mē′-dē-as-TĪ-num) The broad, median partition between the pleurae of the lungs, that extends from the sternum to the vertebral column in the thoracic cavity.

Medulla (me-DUL-la) An inner layer of an organ, such as the medulla of the kidneys.

Medulla oblongata (me-DUL-la ob′-long-GA-ta) The most inferior part of the brain stem. Also termed the **medulla.**

Medullary (MED-ū-lar′-ē) **cavity** The space within the diaphysis of a bone that contains yellow bone marrow. Also called the **marrow cavity.**

Medullary rhythmicity (rith-MIS-i-tē) **area** The neurons of the respiratory center in the medulla oblongata that control the basic rhythm of respiration.

Meibomian gland See **Tarsal gland.**

Meiosis (mī-Ō-sis) A type of cell division that occurs during production of gametes, involving two successive nuclear divisions that result in cells with the haploid *(n)* number of chromosomes.

Meissner's corpuscle See **Corpuscle of touch.**

Melanin (MEL-a-nin) A dark black, brown, or yellow pigment found in some parts of the body such as the skin, hair, and pigmented layer of the retina.

Melanocyte (MEL-a-nō-sīt′) A pigmented cell, located between or beneath cells of the deepest layer of the epidermis, that synthesizes melanin.

Melanocyte-stimulating hormone (MSH) A hormone secreted by the anterior pituitary that stimulates the dispersion of melanin granules in melanocytes in amphibians; continued administration produces darkening of skin in humans.

Melatonin (mel-a-TŌN-in) A hormone secreted by the pineal gland that helps set the timing of the body's biological clock.

Membrane A thin, flexible sheet of tissue composed of an epithelial layer and an underlying connective tissue layer, as in an epithelial membrane, or of areolar connective tissue only, as in a synovial membrane.

Membranous labyrinth (mem-BRA-nus LAB-i-rinth) The part of the labyrinth of the internal ear that is located inside the bony labyrinth and separated from it by the perilymph; made up of the semicircular ducts, the saccule and utricle, and the cochlear duct.

Memory The ability to recall thoughts; commonly classifed as shortterm (activated) and long-term.

Menarche (me-NAR-kē) The first menses (menstrual flow) and beginning of ovarian and uterine cycles.

Meninges (me-NIN-jēz) Three membranes covering the brain and spinal cord, called the dura mater, arachnoid mater, and pia mater. *Singular is* **meninx** (MEN-inks).

Menopause (MEN-ō-pawz) The termination of the menstrual cycles.

Menstrual (MEN-strū-al) **cycle** A series of changes in the endometrium of a nonpregnant female that prepares the lining of the uterus to receive a fertilized ovum.

Menstruation (men′-stroo-Ā-shun) Periodic discharge of blood, tissue fluid, mucus, and epithelial cells that usually lasts for 5 days; caused by a sudden reduction in estrogens and progesterone. Also called the **menstrual phase** or **menses.**

Merkel (MER-kel) **cell** Type of cell in the epidermis of hairless skin that makes contact with a tactile (Merkel) disc, which functions in touch. Also called a **tactile epithelial cell.**

Merocrine (MER-ō-krin) **gland** Gland made up of secretory cells that remain intact throughout the process of formation and discharge of the secretory product, as in the salivary and pancreatic glands.

Mesenchyme (MEZ-en-kīm) An embryonic connective tissue from which all other connective tissues arise.

Mesentery (MEZ-en-ter′-ē) A fold of peritoneum attaching the small intestine to the posterior abdominal wall.

Mesocolon (mez′-ō-KŌ-lon) A fold of peritoneum attaching the colon to the posterior abdominal wall.

Mesoderm The middle primary germ layer that gives rise to connective tissues, blood and blood vessels, and muscles.

Mesothelium (mez′-ō-THĒ-lē-um) The layer of simple squamous epithelium that lines serous membranes.

Mesovarium (mez′-ō-VAR-ē-um) A short fold of peritoneum that attaches an ovary to the broad ligament of the uterus.

Metabolism (me-TAB-ō-lizm) All the biochemical reactions that occur within an organism, including the synthetic (anabolic) reactions and decomposition (catabolic) reactions.

Metacarpus (met′-a-KAR-pus) A collective term for the five bones that make up the palm.

Metaphase (MET-a-phāz) The second stage of mitosis, in which chromatid pairs line up on the metaphase plate of the cell.

Metaphysis (me-TAF-i-sis) Region of a long bone between the diaphysis and epiphysis that contains the epiphyseal plate in a growing bone.

Metarteriole (met′-ar-TĒ-rē-ōl) A blood vessel that emerges from an arteriole. As it tapers toward the capillary junction, the distal-most muscular cell of a metarteriole forms a precapillary sphincter which monitors blood flow into a capillary.

Metastasis (me-TAS-ta-sis) The spread of cancer to surrounding tissues (local) or to other body sites (distant).

Metatarsus (met′-a-TAR-sus) A collective term for the five bones located in the foot between the tarsals and the phalanges.

Microcirculation The flow of blood from a metarteriole through capillaries and into a postcapillary venule.

Microfilament (mī-krō-FIL-a-ment) Rodlike protein filament about 6 nm in diameter; constitutes contractile units in muscle fibers (cells) and provides support, shape, and movement in nonmuscle cells.

Microglia (mi-krō-GLĒ-a) Neuroglial cells that carry on phagocytosis.

Microtubule (mī-krō-TOO-būl′) Cylindrical protein filament, from 18 to 30 nm in diameter, consisting of the protein tubulin; provides support, structure, and transportation.

Microvilli (mī′-krō-VIL-ē) Microscopic, finger-like projections of the plasma membranes of cells that increase surface area for absorption, especially in the small intestine and proximal convoluted tubules of the kidneys.

Micturition (mik′-choo-RISH-un) The act of expelling urine from the urinary bladder. Also called **urination** (ū-ri-NĀ-shun).

Midbrain The part of the brain between the pons and the diencephalon. Also called the **mesencephalon** (mes′-en-SEF-a-lon).

Middle ear A small, epithelial-lined cavity hollowed out of the temporal bone, separated from the external ear by the eardrum and from the internal ear by a thin bony partition containing the oval and round windows; extending across the middle ear are the three auditory ossicles. Also called the **tympanic** (tim-PAN-ik) **cavity.**

Midline An imaginary vertical line that divides the body into equal left and right sides.

Midsagittal plane A vertical plane through the midline of the body that divides the body or organs into *equal* right and left sides. Also called a **median plane.**

Mineralocorticoids (min′-er-al-ō-KOR-ti-koyds) A group of hormones of the adrenal cortex that help regulate sodium and potassium balance.

Mitochondrion (mī′-tō-KON-drē-on) A double-membraned organelle that plays a central role in the production of ATP; known as the "powerhouse" of the cell.

Mitosis (mī-TŌ-sis) The orderly division of the nucleus of a cell that ensures that each new nucleus has the same number and kind of chromosomes as the original nucleus. The process includes the replication of chromosomes and the distribution of the two sets of chromosomes into two separate and equal nuclei.

Mitotic spindle Collective term for a football-shaped assembly of microtubules (nonkinetochore, kinetochore, and aster) that is responsible for the movement of chromosomes during cell division.

Modality (mō-DAL-i-tē) Any of the specific sensory entities, such as vision, smell, taste, or touch.

Modiolus (mō-DĪ-ō′-lus) The central pillar or column of the cochlea.

Monocyte (MON-ō-sit′) The largest type of white blood cell, characterized by agranular cytoplasm.

Monounsaturated fat A fatty acid that contains one double covalent bond between its carbon atoms; it is not completely saturated with hydrogen atoms. Plentiful in triglycerides of olive and peanut oils.

Mons pubis (MONZ PŪ-bis) The rounded, fatty prominence over the pubic symphysis, covered by coarse pubic hair.

Morphology (mor-FOL-ō-jē) The study of the form and structure of things.

Morula (MOR-ū-la) A solid sphere of cells produced by successive cleavages of a fertilized ovum about four days after fertilization.

Motor area The region of the cerebral cortex that governs muscular movement, particularly the precentral gyrus of the frontal lobe.

Motor end plate Region of the sarcolemma of a muscle fiber (cell) that includes acetylcholine (ACh) receptors, which bind ACh released by synaptic end bulbs of somatic motor neurons.

Motor neurons (NOO-ronz) Neurons that conduct impulses from the brain toward the spinal cord or out of the brain and spinal cord into cranial or spinal nerves to effectors that may be either muscles or glands. Also called **efferent neurons.**

Motor unit A motor neuron together with the muscle fibers (cells) it stimulates.

Mucin (MŪ-sin) A protein found in mucus.

Mucosa-associated lymphatic tissue (MALT) Lymphatic nodules scattered throughout the lamina propria (connective tissue) of mucous membranes lining the gastrointestinal tract, respiratory airways, urinary tract, and reproductive tract.

Mucous (MŪ-kus) **cell** A unicellular gland that secretes mucus. Two types are mucous neck cells and surface mucous cells in the stomach.

Mucous connective tissue (MŪ-kus) This is an embryonic tissue found in the umbilical cord of the fetus. It is also called **Wharton's jelly.**

Mucous membrane A membrane that lines a body cavity that opens to the exterior. Also called the **mucosa** (mū-KŌ-sa).

Mucus The thick fluid secretion of goblet cells, mucous cells, mucous glands, and mucous membranes.

Muscarinic (mus′-ka-RIN-ik) **receptor** Receptor for the neurotransmitter acetylcholine found on all effectors innervated by parasympathetic postganglionic axons and on sweat glands innervated by cholinergic sympathetic postganglionic axons; so named because muscarine activates these receptors but does not activate nicotinic receptors for acetylcholine.

Muscle An organ composed of one of three types of muscle tissue (skeletal, cardiac, or smooth), specialized for contraction to produce voluntary or involuntary movement of parts of the body.

Muscle action potential A stimulating impulse that propagates along the sarcolemma and transverse tubules; in skeletal muscle, it is generated by acetylcholine, which increases the permeability of the sarcolemma to cations, especially sodium ions (Na^+).

Muscle fatigue (fa-TĒG) Inability of a muscle to maintain its strength of contraction or tension; may be related to insufficient oxygen, depletion of glycogen, and/or lactic acid buildup.

Muscle spindle An encapsulated proprioceptor in a skeletal muscle, consisting of specialized intrafusal muscle fibers and nerve endings; stimulated by changes in length or tension of muscle fibers.

Muscle tone A sustained, partial contraction of portions of a skeletal or smooth muscle in response to activation of stretch receptors or a baseline level of action potentials in the innervating motor neurons.

Muscular dystrophies (DIS-trō-fēz′) Inherited muscle-destroying diseases, characterized by degeneration of muscle fibers (cells), which causes progressive atrophy of the skeletal muscle.

Muscularis (MUS-ku-la′-ris) A muscular layer (coat or tunic) of an organ.

Muscularis mucosae (mū-KŌ-sē) A thin layer of smooth muscle fibers that underlie the lamina propria of the mucosa of the gastrointestinal tract.

Muscular tissue A tissue specialized to produce motion in response to muscle action potentials by its qualities of contractility, extensibility, elasticity, and excitability; types include skeletal, cardiac, and smooth.

Muscular venule Venule with thicker walls across which exchanges with the interstitial fluid cannot occur.

Mutation (mū-TĀ-shun) Any change in the sequence of bases in a DNA molecule resulting in a permanent alteration in some inheritable trait.

Myasthenia (mī-as-THĒ-nē-a) **gravis** Weakness and fatigue of skeletal muscles caused by antibodies directed against acetylcholine receptors.

Myelin (MĪ-e-lin) **sheath** Multilayered lipid and protein covering, formed by Schwann cells and oligodendrocytes, around axons of many peripheral and central nervous system neurons.

Myenteric plexus A network of autonomic axons and postganglionic cell bodies located in the muscularis of the gastrointestinal tract. Also called the **plexus of Auerbach** (OW-er-bak).

Myocardial infarction (mī′-ō-KAR-dē-al in-FARK-shun) **(MI)** Gross necrosis of myocardial tissue due to interrupted blood supply. Also called a **heart attack.**

Myocardium (mī′-ō-KAR-dē-um) The middle layer of the heart wall, made up of cardiac muscle tissue, lying between the epicardium and the endocardium and constituting the bulk of the heart.

Myoepitheliocyte (mī-ō-ep-i-THĒ-lē-ō-sīt) Muscle-like contractile epithelial cells that surround various glands.

Myofibril (mī-o-FĪ-bril) A threadlike structure, extending longitudinally through a muscle fiber (cell) consisting mainly of thick filaments (myosin) and thin filaments (actin, troponin, and tropomyosin).

Myoglobin (mī-ō-GLŌ-bin) The oxygen-binding, iron-containing protein present in the sarcoplasm of muscle fibers (cells); contributes the red color to muscle.

Myogram (MĪ-ō-gram) The record or tracing produced by a myograph, an apparatus that measures and records the force of muscular contractions.

Myology (mī-OL-ō-jē) The study of muscles.

Myometrium (mī′-ō-MĒ-trē-um) The smooth muscle layer of the uterus.

Myopathy (mī-OP-a-thē) Any abnormal condition or disease of muscle tissue.

Myopia (mī-Ō-pē-a) Defect in vision in which objects can be seen distinctly only when very close to the eyes; nearsightedness.

Myosin (MĪ-ō-sin) The contractile protein that makes up the thick filaments of muscle fibers.

Myotome (MĪ-ō-tōm) A group of muscles innervated by the motor neurons of a single spinal segment. In an embryo, the portion of a somite that develops into some skeletal muscles.

N

Nail A hard plate, composed largely of keratin, that develops from the epidermis of the skin to form a protective covering on the dorsal surface of the distal phalanges of the fingers and toes.

Nail matrix (MĀ-triks) The part of the nail beneath the body and root from which the nail is produced.

Nasal (NĀ-zal) **cavity** A mucosa-lined cavity on either side of the nasal septum that opens onto the face at the external nares and into the nasopharynx at the internal nares.

Nasal septum (SEP-tum) A vertical partition composed of bone (perpendicular plate of ethmoid and vomer) and cartilage, covered with a mucous membrane, separating the nasal cavity into left and right sides.

Nasolacrimal (nā′-zō-LAK-ri-mal) **duct** A canal that transports the lacrimal secretion (tears) from the nasolacrimal sac into the nose.

Nasopharynx (nā′-zō-FAR-inks) The superior portion of the pharynx, lying posterior to the nose and extending inferiorly to the soft palate.

Neck The part of the body connecting the head and the trunk. A constricted portion of an organ, such as the neck of the femur or uterus.

Necrosis (ne-KRŌ-sis) A pathological type of cell death that results from disease, injury, or lack of blood supply in which many adjacent cells swell, burst, and spill their contents into the interstitial fluid, triggering an inflammatory response.

Neonatal (nē-ō-NA-tal) Pertaining to the first four weeks after birth.

Neoplasm (NĒ-ō-plazm) A new growth that may be benign or malignant.

Nephron (NEF-ron) The functional unit of the kidney.

Nerve A cordlike bundle of neuronal axons and/or dendrites and associated connective tissue coursing together outside the central nervous system.

Nerve fiber General term for any process (axon or dendrite) projecting from the cell body of a neuron.

Nerve impulse A wave of depolarization and repolarization that self-propagates along the plasma membrane of a neuron; also called a **nerve action potential.**

Nervous tissue Tissue containing neurons that initiate and conduct nerve impulses to coordinate homeostasis, and neuroglia that provide support and nourishment to neurons.

Neuralgia (noo-RAL-jē-a) Attacks of pain along the entire course or branch of a peripheral sensory nerve.

Neural plate A thickening of ectoderm, induced by the notochord, that forms early in the third week of development and represents the beginning of the development of the nervous system.

Neural tube defect (NTD) A developmental abnormality in which the neural tube does not close properly. Examples are spina bifida and anencephaly.

Neuritis (noo-RĪ-tis) Inflammation of one or more nerves.

Neurofibral node *See* **Node of Ranvier.**

Neuroglia (noo-RŌG-lē-a) Cells of the nervous system that perform various supportive functions. The neuroglia of the central nervous system are the astrocytes, oligodendrocytes, microglia, and ependymal cells; neuroglia of the peripheral nervous system include Schwann cells and satellite cells. Also called **glial** (GLE-al) **cells.**

Neurohypophyseal (noo′-rō-hī′-pō-FIZ-ē-al) **bud** An outgrowth of ectoderm located on the floor of the hypothalamus that gives rise to the posterior pituitary.

Neurolemma (noo-rō-LEM-ma) The peripheral, nucleated cytoplasmic layer of the Schwann cell. Also called **sheath of Schwann** (SCHVON).

Neurology (noo-ROL-ō-jē) The study of the normal functioning and disorders of the nervous system.

Neuromuscular (noo-rō-MUS-kū-lar) **junction** A synapse between the axon terminals of a motor neuron and the sarcolemma of a muscle fiber (cell).

Neuron (NOO-ron) A nerve cell, consisting of a cell body, dendrites, and an axon.

Neurosecretory (noo-rō-SĒC-re-tō-rē) **cell** A neuron that secretes a hypothalamic releasing hormone or inhibiting hormone into blood capillaries of the hypothalmus; a neuron that secretes oxytocin or antidiuretic hormone into blood capillaries of the posterior pituitary.

Neurotransmitter One of a variety of molecules within axon terminals that are released into the synaptic cleft in response to a nerve impulse, and that change the membrane potential of the postsynaptic neuron.

Neurulation (noor-oo-LĀ-shun) The process by which the neural plate, neural folds, and neural tube develop.

Neutrophil (NOO-trō-fil) A type of white blood cell characterized by granules that stain pale lilac with a combination of acidic and basic dyes.

Nicotinic (nik′-ō-TIN-ik) **receptor** Receptor for the neurotransmitter acetylcholine found on both sympathetic and parasympathetic postganglionic neurons and on skeletal muscle in the motor end plate; so named because nicotine activates these receptors but does not activate muscarinic receptors for acetylcholine.

Nipple A pigmented, wrinkled projection on the surface of the breast that is the location of the openings of the lactiferous ducts for milk release.

Nociceptor (nō′-sē-SEP-tor) A free (naked) nerve ending that detects painful stimuli.

Node of Ranvier (ron-vē-Ā) A space, along a myelinated axon, between the individual Schwann cells that form the myelin sheath and the neurolemma. Also called a **neurofibral node.**

Norepinephrine (nor′-ep-ē-NEF-rin) **(NE)** A hormone secreted by the adrenal medulla that produces actions similar to those that result from sympathetic stimulation. Also called **noradrenaline** (nor-a-DREN-a-lin).

Notochord (NŌ-tō-cord) A flexible rod of mesodermal tissue that lies where the future vertebral column will develop and plays a role in induction.

Nuclear medicine The branch of medicine concerned with the use of radioisotopes in the diagnosis and therapy of disease.

Nucleic (noo-KLĒ-ic) **acid** An organic compound that is a long polymer of nucleotides, with each nucleotide containing a pentose sugar, a phosphate group, and one of four possible nitrogenous bases (adenine, cytosine, guanine, and thymine or uracil).

Nucleolus (noo-KLĒ-ō-lus) Spherical body within a cell nucleus composed of protein, DNA, and RNA that is the site of the assembly of small and large ribosomal subunits.

Nucleosome (NOO-klē-ō-sōm) Structural subunit of a chromosome consisting of histones and DNA.

Nucleus (NOO-klē-us) A spherical or oval organelle of a cell that contains the hereditary factors of the cell, called genes. A cluster of unmyelinated nerve cell bodies in the central nervous system. The central part of an atom made up of protons and neutrons.

Nucleus pulposus (pul-PŌ-sus) A soft, pulpy, highly elastic substance in the center of an intervertebral disc; a remnant of the notochord.

Nutrient A chemical substance in food that provides energy, forms new body components, or assists in various body functions.

O

Obesity (ō-BĒS-i-tē) Body weight more than 20% above a desirable standard due to excessive accumulation of fat.

Oblique (ō-BLĒK) **plane** A plane that passes through the body or an organ at an angle between the transverse plane and either the midsagittal, parasagittal, or frontal plane.

Obstetrics (ob-STET-riks) The specialized branch of medicine that deals with pregnancy, labor, and the period of time immediately after delivery (about 6 weeks).

Olfactory (ōl-FAK-tō-rē) Pertaining to smell.

Olfactory bulb A mass of gray matter containing cell bodies of neurons that form synapses with neurons of the olfactory (I) nerve, lying inferior to the frontal lobe of the cerebrum on either side of the crista galli of the ethmoid bone.

Olfactory receptor A bipolar neuron with its cell body lying between supporting cells located in the mucous membrane lining the superior portion of each nasal cavity; transduces odors into neural signals.

Olfactory tract A bundle of axons that extends from the olfactory bulb posteriorly to olfactory regions of the cerebral cortex.

Oligodendrocyte (ol′-i-gō-DEN-drō-sīt) A neuroglial cell that supports neurons and produces a myelin sheath around axons of neurons of the central nervous system.

Oligospermia (ol′-i-gō-SPER-mē-a) A deficiency of sperm cells in the semen.

Oliguria (ol′-i-GŪ-rē-a) Daily urinary output usually less than 250 ml.

Olive A prominent oval mass on each lateral surface of the superior part of the medulla oblongata.

Oncogenes (ONG-kō-jēnz) Cancer-causing genes; they derive from normal genes, termed proto-oncogenes, that encode proteins involved in cell growth or cell regulation but have the ability to transform a normal cell into a cancerous cell when they are mutated or inappropriately activated. One example is *p53*.

Oncology (ong-KOL-ō-jē) The study of tumors.

Oogenesis (ō′-ō-JEN-e-sis) Formation and development of female gametes (oocytes).

Oophorectomy (ō′-of-ō-REK-tō-me) Surgical removal of the ovaries.

Ophthalmic (of-THAL-mik) Pertaining to the eye.

Ophthalmologist (of′-thal-MOL-ō-jist) A physician who specializes in the diagnosis and treatment of eye disorders using drugs, surgery, and corrective lenses.

Ophthalmology (of′-thal-MOL-ō-je) The study of the structure, function, and diseases of the eye.

Optic (OP-tik) Refers to the eye, vision, or properties of light.

Optic chiasm (KĪ-azm) A crossing point of the two branches of the optic (II) nerve, anterior to the pituitary gland. Also called **optic chiasma.**

Optic disc A small area of the retina containing openings through which the axons of the ganglion cells emerge as the optic (II) nerve. Also called the **blind spot.**

Optician (op-TISH-an) A technician who fits, adjusts, and dispenses corrective lenses on prescription of an ophthalmologist or optometrist.

Optic tract A bundle of axons that carry nerve impulses from the retina of the eye between the optic chiasm and the thalamus.

Optometrist (op-TOM-e-trist) Specialist with a doctorate degree in optometry who is licensed to examine and test the eyes and treat visual defects by prescribing corrective lenses.

Ora serrata (Ō-ra ser-RĀ-ta) The irregular margin of the retina lying internal and slightly posterior to the junction of the choroid and ciliary body.

Orbit (OR-bit) The bony, pyramidal-shaped cavity of the skull that holds the eyeball.

Organ A structure composed of two or more different kinds of tissues with a specific function and usually a recognizable shape.

Organelle (or-gan-EL) A permanent structure within a cell with characteristic morphology that is specialized to serve a specific function in cellular activities.

Organism (OR-ga-nizm) A total living form; one individual.

Organogenesis (or′-ga-nō-JEN-e-sis) The formation of body organs and systems. By the end of the eighth week of development, all major body systems have begun to develop.

Orgasm (OR-gazm) Sensory and motor events involved in ejaculation for the male and involuntary contraction of the perineal muscles in the female at the climax of sexual intercourse.

Orifice (OR-i-fis) Any aperture or opening.

Origin (OR-i-jin) The attachment of a muscle tendon to a stationary bone or the end opposite the insertion.

Oropharynx (or'-ō-FAR-inks) The intermediate portion of the pharynx, lying posterior to the mouth and extending from the soft palate to the hyoid bone.

Orthopedics (or'-thō-PĒ-diks) The branch of medicine that deals with the preservation and restoration of the skeletal system, articulations, and associated structures.

Osmoreceptor (oz'-mō-re-CEP-tor) Receptor in the hypothalamus that is sensitive to changes in blood osmolarity and, in response to high osmolarity (low water concentration), stimulates synthesis and release of antidiuretic hormone (ADH).

Osmosis (os-MŌ-sis) The net movement of water molecules through a selectively permeable membrane from an area of higher water concentration to an area of lower water concentration until equilibrium is reached.

Osseous (OS-ē-us) Bony.

Ossicle (OS-si-kul) One of the small bones of the middle ear (malleus, incus, stapes).

Ossification (os'-i-fi-KĀ-shun) The production of a special mineralized extracellular matrix by bone cells.

Osteoblast (OS-tē-ō-blast) Cell formed from an osteogenic cell that participates in bone formation by secreting some organic components and inorganic salts.

Osteoclast (OS-tē-ō-clast') A large, multinuclear cell that resorbs (destroys) bone matrix.

Osteocyte (OS-tē-ō-sīt') A mature bone cell that maintains the daily activities of bone tissue.

Osteogenic (os'-tē-ō-JEN-ik) **cell** Stem cell derived from mesenchyme that has mitotic potential and the ability to differentiate into an osteoblast.

Osteogenic layer The inner layer of the periosteum that contains cells responsible for forming new bone during growth and repair.

Osteogenesis (os-tē-ō-JEN-i-sis) The formation of bone tissue.

Osteology (os'-tē-OL-ō-jē) The study of bones.

Osteon (OS-tē-on) The basic unit of structure in adult compact bone, consisting of a central (haversian) canal with its concentrically arranged lamellae, lacunae, osteocytes, and canaliculi. Also called a **haversian** (ha-VER-shan) **system.**

Osteoporosis (os'-tē-ō-pō-RŌ-sis) Age-related disorder characterized by decreased bone mass and increased susceptibility to fractures, often as a result of decreased levels of estrogens.

Otic (Ō-tik) Pertaining to the ear.

Otolith (Ō-tō-lith) A particle of calcium carbonate embedded in the otolithic membrane that functions in maintaining static equilibrium.

Otolithic (ō-tō-LITH-ik) **membrane** Thick, gelatinous, glycoprotein layer located directly over hair cells of the macula in the saccule and utricle of the internal ear.

Otorhinolaryngology (ō'-tō-rī-nō-lar'-in-GOL-ō-jē) The branch of medicine that deals with the diagnosis and treatment of diseases of the ears, nose, and throat.

Oval window A small, membrane-covered opening between the middle ear and inner ear into which the footplate of the stapes fits.

Ovarian (ē-VAR-ē-an) **cycle** A monthly series of events in the ovary associated with the maturation of a secondary oocyte.

Ovarian follicle (FOL-i-kul) A general name for oocytes (immature ova) in any stage of development, along with their surrounding epithelial cells.

Ovarian ligament (LIG-a-ment) A rounded cord of connective tissue that attaches the ovary to the uterus.

Ovary (Ō-var-ē) Female gonad that produces oocytes and the estrogens, progesterone, inhibin, and relaxin hormones.

Ovulation (ov-ū-LĀ-shun) The rupture of a mature ovarian (Graafian) follicle with discharge of a secondary oocyte into the pelvic cavity.

Ovum (Ō-vum) The female reproductive or germ cell; an egg cell; arises through completion of meiosis in a secondary oocyte after penetration by a sperm.

Oxyhemoglobin (ok'-sē-HĒ-mō-glō-bin) **(Hb–O₂)** Hemoglobin combined with oxygen.

Oxytocin (ok'-sē-TŌ-sin) **(OT)** A hormone secreted by neurosecretory cells in the paraventricular and supraoptic nuclei of the hypothalamus that stimulates contraction of smooth muscle in the pregnant uterus and myoepithelial cells around the ducts of mammary glands.

P

P wave The deflection wave of an electrocardiogram that signifies atrial depolarization.

Pacinian corpuscle See **Lamellated corpuscle.**

Palate (PAL-at) The horizontal structure separating the oral and the nasal cavities; the roof of the mouth.

Palpate (PAL-pāt) To examine by touch; to feel.

Pancreas (PAN-krē-as) A soft, oblong organ lying along the greater curvature of the stomach and connected by a duct to the duodenum. It is both an exocrine gland (secreting pancreatic juice) and an endocrine gland (secreting insulin, glucagon, somatostatin, and pancreatic polypeptide).

Pancreatic (pan'-krē-AT-ik) **duct** A single large tube that unites with the common bile duct from the liver and gallbladder and drains pancreatic juice into the duodenum at the hepatopancreatic ampulla (ampulla of Vater). Also called the **duct of Wirsung.**

Pancreatic islet A cluster of endocrine gland cells in the pancreas that secretes insulin, glucagon, somatostatin, and pancreatic polypeptide. Also called an **islet of Langerhans** (LANG-er-hanz).

Papanicolaou (pap'-a-NIK-ō-la-oo) **test** A cytological staining test for the detection and diagnosis of premalignant and malignant conditions of the female genital tract. Cells scraped from the epithelium of the cervix of the uterus are examined microscopically. Also called a **Pap test** or **Pap smear.**

Papilla (pa-PIL-a) A small nipple-shaped projection or elevation.

Paralysis (pa-RAL-a-sis) Loss or impairment of motor function due to a lesion of nervous or muscular origin.

Paranasal sinus (par'-a-NĀ-zal SĪ-nus) A mucus-lined air cavity in a skull bone that communicates with the nasal cavity. Paranasal sinuses are located in the frontal, maxillary, ethmoid, and sphenoid bones.

Paraplegia (par-a-PLĒ-jē-a) Paralysis of both lower limbs.

Parasagittal plane (par-a-SAJ-i-tal) A vertical plane that does not pass through the midline and that divides the body or organs into *unequal* left and right portions.

Parasympathetic (par'-a-sim-pa-THET-ik) **division** One of the two subdivisions of the autonomic nervous system, having cell bodies of preganglionic neurons in nuclei in the brain stem and in the lateral gray horn of the sacral portion of the spinal cord; primarily concerned with activities that conserve and restore body energy.

Parathyroid (par'-a-THĪ-royd) **gland** One of usually four small endocrine glands embedded in the posterior surfaces of the lateral lobes of the thyroid gland.

Parathyroid hormone (PTH) A hormone secreted by the chief (principal) cells of the parathyroid glands that increases blood calcium level and decreases blood phosphate level.

Paraurethral (par'-a-ū-RĒ-thral) **gland** Gland embedded in the wall of the urethra whose duct opens on either side of the urethral orifice and secretes mucus. Also called **Skene's** (SKĒNZ) **gland.**

Parenchyma (par-EN-ki-ma) The functional parts of any organ, as opposed to tissue that forms its stroma or framework.

Parietal (pa-RĪ-e-tal) Pertaining to or forming the outer wall of a body cavity.

Parietal cell A type of secretory cell in gastric glands that produces hydrochloric acid and intrinsic factor. Also called an **oxyntic cell.**

Parietal pleura (PLOO-ra) The outer layer of the serous pleural membrane that encloses and protects the lungs; the layer that is attached to the wall of the pleural cavity.

Parkinson's disease (PD) Progressive degeneration of the basal ganglia and substantia nigra of the cerebrum resulting in decreased production of dopamine (DA) that leads to tremor, slowing of voluntary movements, and muscle weakness.

Parotid (pa-ROT-id) **gland** One of the paired salivary glands located inferior and anterior to the ears and connected to the oral cavity via a duct (Stensen's) that opens into the inside of the cheek opposite the maxillary (upper) second molar tooth.

Pars intermedia A small avascular zone between the anterior and posterior pituitary glands.

Parturition (par'-too-RISH-un) Act of giving birth to young; childbirth, delivery.

Patent ductus arteriosus A congenital heart defect in which the ductus arteriosus remains open. As a result, aortic blood flows into the lower-pressure pulmonary trunk, increasing pulmonary trunk pressure and overworking both ventricles.

Pathogen (PATH-ō-jen) A disease-producing microbe.

Pathological (path'-ō-LOJ-i-kal) **anatomy** The study of structural changes caused by disease.

Pectinate (PEK-ti-nāt) **muscles** Projecting muscle bundles of the anterior atrial walls and the lining of the auricles.

Pectoral (PEK-tō-ral) Pertaining to the chest or breast.

Pediatrician (pē'-dē-a-TRISH-un) A physician who specializes in the care and treatment of children.

Pedicel (PED-i-sel) Footlike structure, as on podocytes of a glomerulus.

Pelvic (PEL-vik) **cavity** Inferior portion of the abdominopelvic cavity that contains the urinary bladder, sigmoid colon, rectum, and internal female and male reproductive structures.

Pelvic splanchnic (PEL-vik SPLANGK-nik) **nerves** Consist of preganglionic parasympathetic axons from the levels of S2, S3, and S4 that supply the urinary bladder, reproductive organs, and the descending and sigmoid colon and rectum.

Pelvis The basinlike structure formed by the two hip bones, the sacrum, and the coccyx. The expanded, proximal portion of the ureter, lying within the kidney and into which the major calyces open.

Penis (PĒ-nis) The organ of urination and copulation in males; used to deposit semen into the female vagina.

Pepsin Protein-digesting enzyme secreted by chief cells of the stomach in the inactive form pepsinogen, which is converted to active pepsin by hydrochloric acid.

Peptic ulcer An ulcer that develops in areas of the gastrointestinal tract exposed to hydrochloric acid; classified as a gastric ulcer if in the lesser curvature of the stomach and as a duodenal ulcer if in the first part of the duodenum.

Percussion (per-KUSH-un) The act of striking (percussing) an underlying part of the body with short, sharp blows as an aid in diagnosing the part by the quality of the sound produced.

Perforating canal A minute passageway by means of which blood vessels and nerves from the periosteum penetrate into compact bone. Also called **Volkmann's** (FŌLK-manz) **canal.**

Pericardial (per'-i-KAR-dē-al) **cavity** Small potential space between the visceral and parietal layers of the serous pericardium that contains pericardial fluid.

Pericardium (per'-i-KAR-dē-um) A loose-fitting membrane that encloses the heart, consisting of a superficial fibrous layer and a deep serous layer.

Perichondrium (per'-i-KON-drē-um) The membrane that covers cartilage.

Perilymph (PER-i-limf) The fluid contained between the bony and membranous labyrinths of the inner ear.

Perimetrium (per'-i-MĒ-trē-um) The serosa of the uterus.

Perimysium (per'-i-MĪZ-ē-um) Invagination of the epimysium that divides muscles into bundles.

Perineum (per'-i-NĒ-um) The portion of the pelvic floor beneath the pelvic diaphragm; the space between the anus and the scrotum in the male and between the anus and the vulva in the female.

Perineurium (per'-i-NOO-rē-um) Connective tissue wrapping around fascicles in a nerve.

Periodontal (per-ē-ō-DON-tal) **disease** A collective term for conditions characterized by degeneration of gingivae, alveolar bone, periodontal ligament, and cementum.

Periodontal ligament The periosteum lining the alveoli (sockets) for the teeth in the alveolar processes of the mandible and maxillae.

Periosteum (per'-ē-OS-tē-um) The covering of a bone that consists of connective tissue, osteogenic cells, and osteoblasts; is essential for bone growth, repair, and nutrition.

Peripheral (pe-RIF-er-al) Located on the outer part or a surface of the body.

Peripheral nervous system (PNS) The part of the nervous system that lies outside the central nervous system, consisting of nerves and ganglia.

Peristalsis (per'-i-STAL-sis) Successive muscular contractions along the wall of a hollow muscular structure.

Peritoneum (per'-i-to-NĒ-um) The largest serous membrane of the body that lines the abdominal cavity and covers the viscera.

Peritonitis (per'-i-tō-NĪ-tis) Inflammation of the peritoneum.

Peroxisome (per-OK-si-sōm) Organelle similar in structure to a lysosome that contains enzymes that use molecular oxygen to oxidize various organic compounds; such reactions produce hydrogen peroxide; abundant in liver cells.

Perspiration Sweat; produced by sudoriferous (sweat) glands and containing water, salts, urea, uric acid, amino acids, ammonia, sugar, lactic acid, and ascorbic acid. Helps maintain body temperature and eliminate wastes.

Peyer's patches See **Aggregated lymphatic follicles.**

pH A measure of the concentration of hydrogen ions (H^+) in a solution. The pH scale extends from 0 to 14, with a value of 7 expressing neutrality, values lower than 7 expressing increasing acidity, and values higher than 7 expressing increasing alkalinity.

Phagocytosis (fag'-ō-sī-TŌ-sis) The process by which phagocytes ingest and destroy microbes, cell debris, and other foreign matter.

Phalanx (FĀ-lanks) The bone of a finger or toe. *Plural is* **phalanges** (fa-LAN-jēz).

Pharmacology (far'-ma-KOL-ō-jē) The science of the effects and uses of drugs in the treatment of disease.

Pharynx (FAR-inks) The throat; a tube that starts at the internal nares and runs partway down the neck, where it opens into the esophagus posteriorly and the larynx anteriorly.

Phenotype (FĒ-nō-tīp) The observable expression of genotype; physical characteristics of an organism determined by genetic makeup and influenced by interaction between genes and internal and external environmental factors.

Phlebitis (fle-BĪ-tis) Inflammation of a vein, usually in a lower limb.

Photopigment A substance that can absorb light and undergo structural changes that can lead to the development of a receptor potential. An example is rhodopsin. In the eye, also called **visual pigment.**

Photoreceptor Receptor that detects light shining on the retina of the eye.

Physiology (fiz'-ē-OL-ō-jē) Science that deals with the functions of an organism or its parts.

Pia mater (PĪ-a-MĀ-ter *or* PĒ-a MA-ter) The innermost of the three meninges (coverings) of the brain and spinal cord.

Pineal (PĪN-ē-al) **gland** A cone-shaped gland located in the roof of the third ventricle that secretes melatonin. Also called the **epiphysis cerebri** (ē-PIF-i-sis se-RĒ-brē).

Pinealocyte (pin-ē-AL-ō-sīt) Secretory cell of the pineal gland that releases melatonin.

Pinna (PIN-na) The projecting part of the external ear composed of elastic cartilage and covered by skin and shaped like the flared end of a trumpet. Also called the **auricle** (OR-i-kul).

Pituicyte (pi-TOO-i-sīt) Supporting cell of the posterior pituitary.

Pituitary (pi-TOO-i-tār-ē) **gland** A small endocrine gland occupying the hypophyseal fossa of the sphenoid bone and attached to the hypothalamus by the infundibulum. Also called the **hypophysis** (hī-POF-i-sis).

Pivot joint A synovial joint in which a rounded, pointed, or conical surface of one bone articulates with a ring formed partly by another bone and partly by a ligament, as in the joint between the atlas and axis and between the proximal ends of the radius and ulna. Also called a **trochoid** (TRŌ-koyd) **joint.**

Placenta (pla-SEN-ta) The special structure through which the exchange of materials between fetal and maternal circulations occurs. Also called the **afterbirth.**

Plantar flexion (PLAN-tar FLEK-shun) Bending the foot in the direction of the plantar surface (sole).

Plaque (PLAK) A layer of dense proteins on the inside of a plasma membrane in adherens junctions and desmosomes. A mass of bacterial cells, dextran (polysaccharide), and other debris that adheres to teeth (dental plaque). *See also* **Atherosclerotic plaque.**

Plasma (PLAZ-ma) The extracellular fluid found in blood vessels; blood minus the formed elements.

Plasma cell Cell that develops from a B cell (lymphocyte) and produces antibodies.

Plasma (cell) membrane Outer, limiting membrane that separates the cell's internal parts from extracellular fluid or the external environment.

Platelet (PLĀT-let) A fragment of cytoplasm enclosed in a cell membrane and lacking a nucleus; found in the circulating blood; plays a role in hemostasis. Also called a **thrombocyte** (THROM-bō-sīt).

Platelet plug Aggregation of platelets (thrombocytes) at a site where a blood vessel is damaged that helps stop or slow blood loss.

Pleura (PLOOR-a) The serous membrane that covers the lungs and lines the walls of the chest and the diaphragm.

Pleural cavity Small potential space between the visceral and parietal pleurae.

Plexus (PLEK-sus) A network of nerves, veins, or lymphatic vessels.

Plexus of Auerbach See **Myenteric plexus.**

Plexus of Meissner See **Submucosal plexus.**

Pluripotent stem cell Immature stem cell in red bone marrow that gives rise to precursors of all the different mature blood cells.

Pneumotaxic (noo-mō-TAK-sik) **area** A part of the respiratory center in the pons that continually sends inhibitory nerve impulses to the inspiratory area, limiting inhalation and facilitating exhalation.

Podiatry (pō-DĪ-a-trē) The diagnosis and treatment of foot disorders.

Polar body The smaller cell resulting from the unequal division of primary and secondary oocytes during meiosis. The polar body has no function and degenerates.

Polycythemia (pol'-ē-sī-THĒ-mē-a) Disorder characterized by an above-normal hematocrit (above 55%) in which hypertension, thrombosis, and hemorrhage can occur.

Polyunsaturated fat A fatty acid that contains more than one double covalent bond between its carbon atoms; abundant in triglycerides of corn oil, safflower oil, and cottonseed oil.

Polyuria (pol′-ē-Ū-rē-a) An excessive production of urine.

Pons (PONZ) The part of the brain stem that forms a "bridge" between the medulla oblongata and the midbrain, anterior to the cerebellum.

Portal system The circulation of blood from one capillary network into another through a vein.

Portal triad (TRĪ-ad) A microscopic association of a bile duct, hepatic artery, and hepatic vein within the liver.

Postcapillary venule (post-KAP-i-lar-ē) Small venous vessel, fed by capillaries, with loosely organized intercellular junctions that are very porous. These vessels, like capillaries, are significant sites of exchange of nutrients and wastes.

Postcentral gyrus Gyrus of cerebral cortex located immediately posterior to the central sulcus; contains the primary somatosensory area.

Posterior (pos-TĒR-ē-or) Nearer to or at the back of the body. Equivalent to **dorsal** in bipeds.

Posterior column–medial lemniscus pathways Sensory pathways that carry information related to proprioception, fine touch, two-point discrimination, pressure, and vibration. First-order neurons project from the spinal cord to the ipsilateral medulla in the posterior columns (gracile fasciculus and cuneate fasciculus). Second-order neurons project from the medulla to the contralateral thalamus in the medial lemniscus. Third-order neurons project from the thalamus to the somatosensory cortex (postcentral gyrus) on the same side.

Posterior pituitary Posterior lobe of the pituitary gland. Also called the **neurohypophysis** (noo-rō-hī-POF-i-sis).

Posterior root The structure composed of sensory axons lying between a spinal nerve and the dorsolateral aspect of the spinal cord. Also called the **dorsal (sensory) root.**

Posterior root ganglion (GANG-glē-on) A group of cell bodies of sensory neurons and their supporting cells located along the posterior root of a spinal nerve. Also called a **dorsal (sensory) root ganglion.**

Postganglionic neuron (pōst′-gang-lē-ON-ik NOO-ron) The second autonomic motor neuron in an autonomic pathway, having its cell body and dendrites located in an autonomic ganglion and its unmyelinated axon ending at cardiac muscle, smooth muscle, or a gland.

Postsynaptic (pōst-sin-AP-tik) **neuron** The nerve cell that is activated by the release of a neurotransmitter from another neuron and carries nerve impulses away from the synapse.

Pouch of Douglas *See* **Rectouterine pouch.**

Precapillary sphincter (SFINGK-ter) A ring of smooth muscle fibers (cells) at the site of origin of capillaries that regulate blood flow into capillaries.

Precentral gyrus Gyrus of cerebral cortex located immediately anterior to the central sulcus; contains the primary motor area.

Preganglionic (pre′-gang-lē-ON-ik) **neuron** The first autonomic motor neuron in an autonomic pathway, with its cell body and dendrites in the brain or spinal cord and its myelinated axon ending at an autonomic ganglion, where it synapses with a postganglionic neuron.

Pregnancy Sequence of events that normally includes fertilization, implantation, embryonic growth, and fetal growth and terminates in birth.

Premenstrual syndrome (PMS) Severe physical and emotional stress occurring late in the postovulatory phase of the menstrual cycle and sometimes overlapping with menstruation.

Prepuce (PRĒ-poos) The loose-fitting skin covering the glans of the penis and clitoris. Also called the **foreskin.**

Presbyopia (prez-bē-Ō-pē-a) A loss of elasticity of the lens of the eye due to advancing age with resulting inability to focus clearly on near objects.

Presynaptic (prē-sin-AP-tik) **neuron** A neuron that propagates nerve impulses toward a synapse.

Prevertebral ganglion (prē-VER-te-bral GANG-glē-on) A cluster of cell bodies of postganglionic sympathetic neurons anterior to the spinal column and close to large abdominal arteries. Also called a **collateral ganglion.**

Primary germ layer One of three layers of embryonic tissue, called ectoderm, mesoderm, and endoderm, that give rise to all tissues and organs of the body.

Primary motor area A region of the cerebral cortex in the precentral gyrus of the frontal lobe of the cerebrum that controls specific muscles or groups of muscles.

Primary somatosensory area A region of the cerebral cortex posterior to the central sulcus in the postcentral gyrus of the parietal lobe of the cerebrum that localizes exactly the points of the body where somatic sensations originate.

Prime mover The muscle directly responsible for producing a desired motion. Also called an **agonist** (AG-ō-nist).

Primitive gut Embryonic structure formed from the dorsal part of the yolk sac that gives rise to most of the gastrointestinal tract.

Primordial (prī-MOR-dē-al) Existing first; especially primordial egg cells in the ovary.

Principal cell Cell type in the distal convoluted tubules and collecting ducts of the kidneys that is stimulated by aldosterone and antidiuretic hormone.

Proctology (prok-TOL-ō-jē) The branch of medicine concerned with the rectum and its disorders.

Progeny (PROJ-e-nē) Offspring or descendants.

Progesterone (prō-JES-te-rōn) A female sex hormone produced by the ovaries that helps prepare the endometrium of the uterus for implantation of a fertilized ovum and the mammary glands for milk secretion.

Prognosis (prog-NŌ-sis) A forecast of the probable results of a disorder; the outlook for recovery.

Prolactin (prō-LAK-tin) **(PRL)** A hormone secreted by the anterior pituitary that initiates and maintains milk secretion by the mammary glands.

Prolapse (PRŌ-laps) A dropping or falling down of an organ, especially the uterus or rectum.

Proliferation (prō-lif′-er-Ā-shun) Rapid and repeated reproduction of new parts, especially cells.

Pronation (prō-NĀ-shun) A movement of the forearm in which the palm is turned posteriorly.

Prophase (PRŌ-fāz) The first stage of mitosis during which chromatid pairs are formed and aggregate around the metaphase plate of the cell.

Proprioception (prō-prē-ō-SEP-shun) The perception of the position of body parts, especially the limbs, independent of vision; this sense is possible due to nerve impulses generated by proprioceptors.

Proprioceptor (prō′-prē-ō-SEP-tor) A receptor located in muscles, tendons, joints, or the internal ear (muscle spindles, tendon organs, joint kinesthetic receptors, and hair cells of the vestibular apparatus) that provides information about body position and movements.

Prostaglandin (pros′-ta-GLAN-din) **(PG)** A membrane-associated lipid; released in small quantities and acts as a local hormone.

Prostate (PROS-tāt) A doughnut-shaped gland inferior to the urinary bladder that surrounds the superior portion of the male urethra and secretes a slightly acidic solution that contributes to sperm motility and viability.

Proteasome (PRŌ-tē-a-sōm) Tiny barrel-shaped cellular organelle in cytosol and nucleus containing proteases that destroy unneeded, damaged, or faulty proteins.

Protein An organic compound consisting of carbon, hydrogen, oxygen, nitrogen, and sometimes sulfur and phosphorus; synthesized on ribosomes and made up of amino acids linked by peptide bonds.

Prothrombin (prō-THROM-bin) An inactive blood-clotting factor synthesized by the liver, released into the blood, and converted to active thrombin in the process of blood clotting by the activated enzyme prothrombinase.

Proto-oncogene (prō′-tō-ONG-kō-jēn) Gene responsible for some aspect of normal growth and development; it may transform into an oncogene, a gene capable of causing cancer.

Protraction (prō-TRAK-shun) The movement of the mandible or shoulder girdle forward on a plane parallel with the ground.

Proximal (PROK-si-mal) Nearer the attachment of a limb to the trunk; nearer to the point of origin or attachment.

Pseudopods (SOO-dō-pods) Temporary protrusions of the leading edge of a migrating cell; cellular projections that surround a particle undergoing phagocytosis.

Pterygopalatine ganglion (ter′-i-gō-PAL-a-tīn GANG-glē-on) A cluster of cell bodies of parasympathetic postganglionic neurons ending at the lacrimal and nasal glands.

Ptosis (TŌ-sis) Drooping, as of the eyelid or the kidney.

Puberty (PŪ-ber-tē) The time of life during which the secondary sex characteristics begin to appear and the capability for sexual reproduction is possible; usually occurs between the ages of 10 and 17.

Pubic symphysis A slightly movable cartilaginous joint between the anterior surfaces of the hip bones.

Puerperium (pū′-er-PER-ē-um) The period immediately after childbirth, usually 4–6 weeks.

Pulmonary (PUL-mo-ner′-ē) Concerning or affected by the lungs.

Pulmonary circulation The flow of deoxygenated blood from the right ventricle to the lungs and the return of oxygenated blood from the lungs to the left atrium.

Pulmonary edema (e-DĒ-ma) An abnormal accumulation of interstitial fluid in the tissue spaces and alveoli of the lungs due to increased pulmonary capillary permeability or increased pulmonary capillary pressure.

Pulmonary embolism (EM-bō-lizm) **(PE)** The presence of a blood clot or a foreign substance in a pulmonary arterial blood vessel that obstructs circulation to lung tissue.

Pulmonary ventilation The inflow (inhalation) and outflow (exhalation) of air between the atmosphere and the lungs. Also called **breathing**.

Pulp cavity A cavity within the crown and neck of a tooth, which is filled with pulp, a connective tissue containing blood vessels, nerves, and lymphatic vessels.

Pulse (PULS) The rhythmic expansion and elastic recoil of a systemic artery after each contraction of the left ventricle.

Pupil The hole in the center of the iris, the area through which light enters the posterior cavity of the eyeball.

Purkinje (pur-KIN-jē) **fiber** Muscle fiber (cell) in the ventricular tissue of the heart specialized for conducting an action potential to the myocardium; part of the conduction system of the heart.

Pus The liquid product of inflammation containing leukocytes or their remains and debris of dead cells.

Pyloric (pī-LOR-ik) **sphincter** A thickened ring of smooth muscle through which the pylorus of the stomach communicates with the duodenum. Also called the **pyloric valve.**

Pyogenesis (pī′-ō-JEN-e-sis) Formation of pus.

Pyorrhea (pī-ō-RĒ-a) A discharge or flow of pus, especially in the alveoli (sockets) and the tissues of the gums.

Pyramid (PIR-a-mid) A pointed or cone-shaped structure. One of two roughly triangular structures on the anterior aspect of the medulla oblongata composed of the largest motor tracts that run from the cerebral cortex to the spinal cord. A triangular structure in the renal medulla.

Pyramidal (pi-RAM-i-dal) **tracts (pathways).** *See* **Direct motor pathways.**

Pyuria (pi-Ū-rē-a) The presence of leukocytes and other components of pus in urine.

Q

QRS wave The deflection waves of an electrocardiogram that represent onset of ventricular depolarization.

Quadrant (KWOD-rant) One of four parts.

Quadriplegia (kwod′-ri-PLĒ-jē-a) Paralysis of four limbs: two upper and two lower.

R

Radiographic (rā′-dē-ō-GRAF-ic) **anatomy** Diagnostic branch of anatomy that includes the use of x-rays.

Rami communicantes (RĀ-mē kō-mū-ni-KAN-tēz) Branches of a spinal nerve. *Singular is* **ramus communicans** (RĀ-mus kō-MŪ-ni-kans).

Rathke's pouch *See* **Hypophyseal pouch.**

Receptor A specialized cell or a distal portion of a neuron that responds to a specific sensory modality, such as touch, pressure, cold, light, or sound, and converts it to an electrical signal (generator or receptor potential). A specific molecule or cluster of molecules that recognizes and binds a particular ligand.

Receptor-mediated endocytosis A highly selective process whereby cells take up specific ligands, which usually are large molecules or particles, by enveloping them within a sac of plasma membrane. Ligands are eventually broken down by enzymes in lysosomes.

Recombinant DNA Synthetic DNA, formed by joining a fragment of DNA from one source to a portion of DNA from another.

Rectouterine pouch A pocket formed by the parietal peritoneum as it moves posteriorly from the surface of the uterus and is reflected onto the rectum; the most inferior point in the pelvic cavity. Also called the **pouch** or **cul de sac of Douglas.**

Rectum (REK-tum) The last 20 cm (8 in.) of the gastrointestinal tract, from the sigmoid colon to the anus.

Recumbent (re-KUM-bent) Lying down.

Red bone marrow A highly vascularized connective tissue located in microscopic spaces between trabeculae of spongy bone tissue.

Red nucleus A cluster of cell bodies in the midbrain, occupying a large part of the tectum from which axons extend into the rubroreticular and rubrospinal tracts.

Red pulp That portion of the spleen that consists of venous sinuses filled with blood and thin plates of splenic tissue called splenic (Billroth's) cords.

Referred pain Pain that is felt at a site remote from the place of origin.

Reflex Fast response to a change (stimulus) in the internal or external environment that attempts to restore homeostasis.

Reflex arc The most basic conduction pathway through the nervous system, connecting a receptor and an effector and consisting of a receptor, a sensory neuron, an integrating center in the central nervous system, a motor neuron, and an effector.

Regional anatomy The division of anatomy dealing with a specific region of the body, such as the head, neck, chest, or abdomen.

Regurgitation (rē-gur′-ji-TĀ-shun) Return of solids or fluids to the mouth from the stomach; backward flow of blood through incompletely closed heart valves.

Relaxin (RLX) A female hormone produced by the ovaries and placenta that increases flexibility of the pubic symphysis and helps dilate the uterine cervix to ease delivery of a baby.

Releasing hormone Hormone secreted by the hypothalamus that can stimulate secretion of hormones of the anterior pituitary.

Remodeling Replacement of old bone by new bone tissue.

Renal (RĒ-nal) Pertaining to the kidneys.

Renal corpuscle (KOR-pus-l) A glomerular (Bowman's) capsule and its enclosed glomerulus.

Renal pelvis A cavity in the center of the kidney formed by the expanded, proximal portion of the ureter, lying within the kidney, and into which the major calyces open.

Renal pyramid A triangular structure in the renal medulla containing the straight segments of renal tubules and the vasa recta.

Reproduction (rē-prō-DUK-shun) The formation of new cells for growth, repair, or replacement; the production of a new individual.

Reproductive cell division Type of cell division in which gametes (sperm and oocytes) are produced; consists of meiosis and cytokinesis.

Respiration (res-pi-RĀ-shun) Overall exchange of gases between the atmosphere, blood, and body cells consisting of pulmonary ventilation, external respiration, and internal respiration.

Respiratory center Neurons in the pons and medulla oblongata of the brain stem that regulate the rate and depth of pulmonary ventilation.

Retention (rē-TEN-shun) A failure to void urine due to obstruction, nervous contraction of the urethra, or absence of sensation of desire to urinate.

Rete (RĒ-tē) **testis** The network of ducts in the testes.

Reticular (re-TIK-ū-lar) **activating system (RAS)** A portion of the reticular formation that has many ascending connections with the cerebral cortex; when this area of the brain stem is active, nerve impulses pass to the thalamus and widespread areas of the cerebral cortex, resulting in generalized alertness or arousal from sleep.

Reticular fiber (re-TIK-ū-lar FĪ-ber) Thin collagen fiber arranged in fine bundles with a coating of glycoprotein.

Reticular formation A network of small groups of neuronal cell bodies scattered among bundles of axons (mixed gray and white matter) beginning in the medulla oblongata and extending superiorly through the central part of the brain stem.

Reticulocyte (re-TIK-ū-lō-sīt) An immature red blood cell.

Reticulofibrous bone (re-TIK-ū-lō-fi-brus) Bone found at the initial sites of bone development in the fetus and at sites of bone repair in mature bones. It consists of a mixed array of fine and coarse mineralized collagen fibers in an irregular arrangement. Also called **primary or woven bone.**

Reticulum (re-TIK-ū-lum) A network.

Retina (RET-i-na) The deep coat of the posterior portion of the eyeball consisting of nervous tissue (where the process of vision begins) and a pigmented layer of epithelial cells that contact the choroid.

Retinaculum (ret-i-NAK-ū-lum) A thickening of deep fascia that holds structures in place, for example, the superior and inferior retinacula of the ankle.

Retraction (rē-TRAK-shun) The movement of a protracted part of the body posteriorly on a plane parallel to the ground, as in pulling the lower jaw back in line with the upper jaw.

Retrograde degeneration (RE-trō-grād dē-jen-er-Ā-shun) Changes that occur in the proximal portion of a damaged axon only as far as the first node of Ranvier; similar to changes that occur during Wallerian degeneration.

Retroperitoneal (re′-trō-pēr-i-tō-NĒ-al) External to the peritoneal lining of the abdominal cavity.

Rh factor An inherited antigen on the surface of red blood cells in Rh$^+$ individuals; not present in Rh$^-$ individuals.

Rhinology (rī-NOL-ō-jē) The study of the nose and its disorders.

Ribonucleic (rī-bō-noo-KLĒ-ik) **acid (RNA)** A single-stranded nucleic acid made up of nucleotides, each consisting of a nitrogenous base (adenine, cytosine, guanine, or uracil), ribose, and a phosphate group; three types are messenger RNA (mRNA), transfer RNA (tRNA), and ribosomal RNA (rRNA), each of which has a specific role during protein synthesis.

Ribosome (RĪ-bō-sōm) An organelle in the cytoplasm of cells, composed of a small subunit and a large subunit that contain ribosomal RNA and ribosomal proteins; the site of protein synthesis.

Right lymphatic (lim-FAT-ik) **duct** A vessel of the lymphatic system that drains lymph from the upper right side of the body and empties it into the right subclavian vein.

Rigidity (ri-JID-i-tē) Hypertonia characterized by increased muscle tone, but reflexes are not affected.

Rigor mortis State of partial contraction of muscles after death due to lack of ATP; myosin heads (cross bridges) remain attached to actin, thus preventing relaxation.

Rod One of two types of photoreceptor in the retina of the eye; specialized for vision in dim light.

Root canal A narrow extension of the pulp cavity lying within the root of a tooth.

Root of penis Attached portion of penis that consists of the bulb and crura.

Rotation (rō-TĀ-shun) Moving a bone around its own axis, with no other movement.

Round ligament (LIG-a-ment) A band of fibrous connective tissue enclosed between the folds of the broad ligament of the uterus, emerging from the uterus just inferior to the uterine tube, extending laterally along the pelvic wall and through the deep inguinal ring to end in the labia majora.

Round window A small opening between the middle and internal ear, directly inferior to the oval window, covered by the secondary tympanic membrane.

Rugae (ROO-gē) Large folds in the mucosa of an empty hollow organ, such as the stomach and vagina.

S

Saccule (SAK-ūl) The inferior and smaller of the two chambers in the membranous labyrinth inside the vestibule of the internal ear containing a receptor organ for static equilibrium.

Sacral hiatus (hī-Ā-tus) Inferior entrance to the vertebral canal formed when the laminae of the fifth sacral vertebra (and sometimes fourth) fail to meet.

Sacral plexus (SĀ-kral PLEK-sus) A network formed by the ventral branches of spinal nerves L4 through S3.

Sacral promontory (PROM-on-tor′-ē) The superior surface of the body of the first sacral vertebra that projects anteriorly into the pelvic cavity; a line from the sacral promontory to the superior border of the pubic symphysis divides the abdominal and pelvic cavities.

Saddle joint A synovial joint in which the articular surface of one bone is saddle-shaped and the articular surface of the other bone is shaped like the legs of the rider sitting in the saddle, as in the joint between the trapezium and the metacarpal of the thumb.

Sagittal (SAJ-i-tal) **plane** A plane that divides the body or organs into left and right portions. Such a plane may be **midsagittal (median),** in which the divisions are equal, or **parasagittal,** in which the divisions are unequal.

Saliva (sa-LĪ-va) A clear, alkaline, somewhat viscous secretion produced mostly by the three pairs of salivary glands; contains various salts, mucin, lysozyme, salivary amylase, and lingual lipase (produced by glands in the tongue).

Salivary amylase (SAL-i-ver-ē AM-i-lās) An enzyme in saliva that initiates the chemical breakdown of starch.

Salivary gland One of three pairs of glands that lie external to the mouth and pour their secretory product (saliva) into ducts that empty into the oral cavity; the parotid, submandibular, and sublingual glands.

Sarcolemma (sar′-kō-LEM-ma) The cell membrane of a muscle fiber (cell), especially of a skeletal muscle fiber.

Sarcomere (SAR-kō-mēr) A contractile unit in a striated muscle fiber (cell) extending from one Z disc to the next Z disc.

Sarcoplasm (SAR-kō-plazm) The cytoplasm of a muscle fiber (cell).

Sarcoplasmic reticulum (sar′-kō-PLAZ-mik re-TIK-ū-lum) A network of saccules and tubes surrounding myofibrils of a muscle fiber (cell), comparable to endoplasmic reticulum; functions to reabsorb calcium ions during relaxation and to release them to cause contraction.

Satellite cell (SAT-i-līt) Flat neuroglial cells that surround cell bodies of peripheral nervous system ganglia to provide structural support and regulate the exchange of material between a neuronal cell body and interstitial fluid.

Saturated fat A fatty acid that contains only single bonds (no double bonds) between its carbon atoms; all carbon atoms are bonded to the maximum number of hydrogen atoms; prevalent in triglycerides of animal products such as meat, milk, milk products, and eggs.

Scala tympani (SKA-la TIM-pan-ē) The inferior spiral-shaped channel of the bony cochlea, filled with perilymph.

Scala vestibuli (ves-TIB-ū-lē) The superior spiral-shaped channel of the bony cochlea, filled with perilymph.

Schwann (SCHVON) **cell** A neuroglial cell of the peripheral nervous system that forms the myelin sheath and neurolemma around a nerve axon by wrapping around the axon in a jelly-roll fashion.

Sciatica (sī-AT′-i-ka) Inflammation and pain along the sciatic nerve; felt along the posterior aspect of the thigh extending down the inside of the leg.

Sclera (SKLE-ra) The white coat of fibrous tissue that forms the superficial protective covering over the eyeball except in the most anterior portion; the posterior portion of the fibrous tunic.

Scleral venous sinus A circular venous sinus located at the junction of the sclera and the cornea through which aqueous humor drains from the anterior chamber of the eyeball into the blood. Also called the **canal of Schlemm** (SHLEM).

Sclerosis (skle-RŌ-sis) A hardening with loss of elasticity of tissues.

Scoliosis (skō′-lē-Ō-sis) An abnormal lateral curvature from the normal vertical line of the backbone.

Scrotum (SKRŌ-tum) A skin-covered pouch that contains the testes and their accessory structures.

Sebaceous (se-BĀ-shus) **gland** An exocrine gland in the dermis of the skin, almost always associated with a hair follicle, that secretes sebum. Also called an **oil gland.**

Sebum (SĒ-bum) Secretion of sebaceous (oil) glands.

Secondary sex characteristic A characteristic of the male or female body that develops at puberty under the influence of sex hormones but is not directly involved in sexual reproduction; examples are distribution of body hair, voice pitch, body shape, and muscle development.

Secretion (se-KRĒ-shun) Production and release from a cell or a gland of a physiologically active substance.

Selective permeability (per′-mē-a-BIL-i-tē) The property of a membrane by which it permits the passage of certain substances but restricts the passage of others.

Semen (SĒ-men) A fluid discharged at ejaculation by a male that consists of a mixture of sperm and the secretions of the seminiferous tubules, seminal vesicles, prostate, and bulbourethral (Cowper's) glands.

Semicircular canals Three bony channels (anterior, posterior, lateral), filled with perilymph, in which lie the membranous semicircular canals filled with endolymph. They contain receptors for equilibrium.

Semicircular ducts The membranous semicircular canals filled with endolymph and floating in the perilymph of the bony semicircular canals; they contain cristae that are concerned with dynamic equilibrium.

Semilunar (sem′-ē-LOO-nar) **valve** A valve between the aorta or the pulmonary trunk and a ventricle of the heart.

Seminal vesicle (SEM-i-nal VES-i-kul) One of a pair of convoluted, pouchlike structures, lying posterior and inferior to the urinary bladder and anterior to the rectum, that secrete a component of semen into the ejaculatory ducts. Also termed **seminal gland.**

Seminiferous tubule (sem′-i-NI-fer-us TOO-būl) A tightly coiled duct, located in the testis, where sperm are produced.

Senescence (se-NES-ens) The process of growing old.

Sensation A state of awareness of external or internal conditions of the body.

Sensory area A region of the cerebral cortex concerned with the interpretation of sensory impulses.

Sensory neurons (NOO-ronz) Neurons that carry sensory information from cranial and spinal nerves into the brain and spinal cord or from a lower to a higher level in the spinal cord and brain. Also called **afferent neurons.**

Septal defect An opening in the atrial septum (atrial septal defect) because the foramen ovale fails to close, or the ventricular septum (ventricular septal defect) due to incomplete development of the ventricular septum.

Septal nasal cartilage (SEP-tul) An unpaired cartilage that forms the superior part of the lateral sides of the nose as well as the cartilaginous septum that divides the external nose into right and left chambers.

Septum (SEP-tum) A wall dividing two cavities.

Serous (SĒR-us) **membrane** A membrane that lines a body cavity that does not open to the exterior. The external layer of an organ formed by a serous membrane. The membrane that lines the pleural, pericardial, and peritoneal cavities. Also called a **serosa** (se-RŌ-sa).

Sertoli (ser-TŌ-lē) **cell** A supporting cell in the seminiferous tubules that secretes fluid for supplying nutrients to sperm and the hormone inhibin, removes excess cytoplasm from spermatogenic cells, and mediates the effects of FSH and testosterone on spermatogenesis. Also called a **sustentacular** (sus´-ten-TAK-ū-lar) **cell.**

Serum Blood plasma minus its clotting proteins.

Sesamoid (SES-a-moyd) **bones** Small bones usually found in tendons.

Sex chromosomes The twenty-third pair of chromosomes, designated X and Y, which determine the genetic sex of an individual; in males, the pair is XY; in females, XX.

Sexual intercourse The insertion of the erect penis of a male into the vagina of a female. Also called **coitus** (KŌ-i-tus).

Sheath of Schwann *See* **Neurolemma.**

Shock Failure of the cardiovascular system to deliver adequate amounts of oxygen and nutrients to meet the metabolic needs of the body due to inadequate cardiac output. It is characterized by hypotension; clammy, cool, and pale skin; sweating; reduced urine formation; altered mental state; acidosis; tachycardia; weak, rapid pulse; and thirst. Types include hypovolemic, cardiogenic, vascular, and obstructive.

Short bone Bone that is somewhat cube-shaped and nearly equal in length and width.

Shoulder joint A synovial joint where the humerus articulates with the scapula.

Sigmoid colon (SIG-moyd KŌ-lon) The S-shaped part of the large intestine that begins at the level of the left iliac crest, projects medially, and terminates at the rectum at about the level of the third sacral vertebra.

Sign Any objective evidence of disease that can be observed or measured such as a lesion, swelling, or fever.

Sinoatrial (si-nō-Ā-trē-al) **(SA) node** A small mass of cardiac muscle fibers (cells) located in the right atrium inferior to the opening of the superior vena cava that spontaneously depolarize and generate a cardiac action potential about 100 times per minute. Also called the **pacemaker.**

Sinus (SĪ-nus) A hollow in a bone (paranasal sinus) or other tissue; a channel for blood (vascular sinus); any cavity having a narrow opening.

Sinusoid (SĪ-nū-soyd) A large, thin-walled, and leaky type of capillary, having large intercellular clefts that may allow proteins and blood cells to pass from a tissue into the bloodstream; present in the liver, spleen, anterior pituitary, parathyroid glands, and red bone marrow.

Skeletal muscle An organ specialized for contraction, composed of striated muscle fibers (cells), supported by connective tissue, attached to a bone by a tendon or an aponeurosis, and stimulated by somatic motor neurons.

Skene's gland *See* **Paraurethral gland.**

Skin The external covering of the body that consists of a superficial, thinner epidermis (epithelial tissue) and a deep, thicker dermis (connective tissue) that is anchored to the subcutaneous layer.

Skull The skeleton of the head consisting of the cranial and facial bones.

Sleep A state of partial unconsciousness from which a person can be aroused; associated with a low level of activity in the reticular activating system.

Sliding-filament mechanism The explanation of how thick and thin filaments slide relative to one another during striated muscle contraction to decrease sarcomere length.

Small intestine A long tube of the gastrointestinal tract that begins at the pyloric sphincter of the stomach, coils through the central and inferior part of the abdominal cavity, and ends at the large intestine; divided into three segments: duodenum, jejunum, and ileum.

Smooth muscle A tissue specialized for contraction, composed of smooth muscle fibers (cells), located in the walls of hollow internal organs, and innervated by autonomic motor neurons.

Sodium-potassium ATPase An active transport pump located in the plasma membrane that transports sodium ions out of the cell and potassium ions into the cell at the expense of cellular ATP. It functions to keep the ionic concentrations of these ions at physiological levels. Also called the **sodium-potassium pump.**

Soft palate (PAL-at) The posterior portion of the roof of the mouth, extending from the palatine bones to the uvula. It is a muscular partition lined with mucous membrane.

Somatic (sō-MAT-ik) **cell division** Type of cell division in which a single starting cell duplicates itself to produce two identical cells; consists of mitosis and cytokinesis.

Somatic nervous system (SNS) The portion of the peripheral nervous system consisting of somatic sensory (afferent) neurons and somatic motor (efferent) neurons.

Somite (SŌ-mīt) Block of mesodermal cells in a developing embryo that is distinguished into a myotome (which forms most of the skeletal muscles), dermatome (which forms connective tissues), and sclerotome (which forms the vertebrae).

Spasm (SPAZM) A sudden, involuntary contraction of large groups of muscles.

Spasticity (spas-TIS-i-tē) Hypertonia characterized by increased muscle tone, increased tendon reflexes, and pathological reflexes (Babinski sign).

Spermatic (sper-MAT-ik) **cord** A supporting structure of the male reproductive system, extending from a testis to the deep inguinal ring, that includes the ductus (vas) deferens, arteries, veins, lymphatic vessels, nerves, cremaster muscle, and connective tissue.

Spermatogenesis (sper´-ma-tō-JEN-e-sis) The formation and development of sperm in the seminiferous tubules of the testes.

Sperm cell A mature male gamete. Also termed **spermatozoon** (sper´-ma-tō-ZŌ-on).

Spermiogenesis (sper´-mē-ō-JEN-e-sis) The maturation of spermatids into sperm.

Sphincter (SFINGK-ter) A circular muscle that constricts an opening.

Sphincter of Oddi *See* **Sphincter of the hepatopancreatic ampulla.**

Sphincter of the hepatopancreatic ampulla A circular muscle at the opening of the common bile and main pancreatic ducts in the duodenum. Also called the **sphincter of Oddi** (OD-ē).

Spinal (SPĪ-nal) **cord** A mass of nerve tissue located in the vertebral canal from which 31 pairs of spinal nerves originate.

Spinal nerve One of the 31 pairs of nerves that originate on the spinal cord from posterior and anterior roots.

Spinal shock A period from several days to several weeks following transection of the spinal cord that is characterized by the abolition of all reflex activity.

Spinothalamic (spī-nō-tha-LAM-ik) **tracts** Sensory (ascending) tracts that convey information up the spinal cord to the thalamus for sensations of pain, temperature, crude touch, and deep pressure.

Spinous (SPĪ-nus) **process** A sharp or thornlike process or projection. Also called a **spine.** A sharp ridge running diagonally across the posterior surface of the scapula.

Spiral organ The organ of hearing, consisting of supporting cells and hair cells that rest on the basilar membrane and extend into the endolymph of the cochlear duct. Also called the **organ of Corti** (KOR-tē).

Splanchnic (SPLANK-nik) Pertaining to the viscera.

Spleen (SPLĒN) Large mass of lymphatic tissue between the fundus of the stomach and the diaphragm that functions in formation of blood cells during early fetal development, phagocytosis of ruptured blood cells, and proliferation of B cells during immune responses.

Spongy (trabeculae) bone tissue Bone tissue that consists of an irregular latticework of thin plates of bone called trabeculae; spaces between trabeculae of some bones are filled with red bone marrow; found inside short, flat, and irregular bones and in the epiphyses (ends) of long bones.

Sprain Forcible wrenching or twisting of a joint with partial rupture or other injury to its attachments without dislocation.

Squamous (SKWĀ-mus) Flat or scalelike.

Starvation (star-VĀ-shun) The loss of energy stores in the form of glycogen, triglycerides, and proteins due to inadequate intake of nutrients or inability to digest, absorb, or metabolize ingested nutrients.

Stasis (STĀ-sis) Stagnation or halt of normal flow of fluids, as blood or urine, or of the intestinal contents.

Static equilibrium (ē-kwi-LIB-rē-um) The maintenance of posture in response to changes in the orientation of the body, mainly the head, relative to the ground.

Stellate reticuloendothelial (STEL-āt re-tik´-ū-lō-en´-dō-THĒ-lē-al) **cell** Phagocytic cell bordering a sinusoid of the liver. Also called a **Kupffer** (KOOP-fer) **cell.**

Stem cell An unspecialized cell that has the ability to divide for indefinite periods and give rise to a specialized cell.

Stenosis (sten-Ō-sis) An abnormal narrowing or constriction of a duct or opening.

Stereocilia (ste´-rē-ō-SIL-ē-a) Groups of extremely long, slender, nonmotile microvilli projecting from epithelial cells lining the epididymis.

Sterile (STE-ril) Free from any living microorganisms. Unable to conceive or produce offspring.

Sterilization (ster´-i-li-ZĀ-shun) Elimination of all living microorganisms. Any procedure that renders an individual incapable of reproduction (for example, castration, vasectomy, hysterectomy, or oophorectomy).

Stimulus Any stress that changes a controlled condition; any change in the internal or external environment that excites a sensory receptor, a neuron, or a muscle fiber.

Stomach The J-shaped enlargement of the gastrointestinal tract directly inferior to the diaphragm in the epigastric, umbilical, and left hypochondriac regions of the abdomen, between the esophagus and small intestine.

Straight tubule (TOO-būl) A duct in a testis leading from a convoluted seminiferous tubule to the rete testis.

Stratum (STRĀ-tūm) A layer.

Stratum basalis (ba-SAL-is) The layer of the endometrium next to the myometrium that is maintained during menstruation and gestation and produces a new stratum functionalis following menstruation or parturition.

Stratum functionalis (funk'-shun-AL-is) The layer of the endometrium next to the uterine cavity that is shed during menstruation and that forms the maternal portion of the placenta during gestation.

Stretch receptor Receptor in the walls of blood vessels, airways, or organs that monitors the amount of stretching. Also termed **baroreceptor.**

Stroma (STRŌ-ma) The tissue that forms the ground substance, foundation, or framework of an organ, as opposed to its functional parts (parenchyma).

Subarachnoid (sub'-a-RAK-noyd) **space** A space between the arachnoid mater and the pia mater that surrounds the brain and spinal cord and through which cerebrospinal fluid circulates.

Subcutaneous (sub'-kū-TĀ-nē-us) Beneath the skin. Also called **hypodermic** (hi-pō-DER-mik).

Subcutaneous layer A continuous sheet of areolar connective tissue and adipose tissue between the dermis of the skin and the deep fascia of the muscles. Also called the **hypodermis.**

Subdural (sub-DOO-ral) **space** A space between the dura mater and the arachnoid mater of the brain and spinal cord that contains a small amount of fluid.

Sublingual (sub-LING-gwal) **gland** One of a pair of salivary glands situated in the floor of the mouth deep to the mucous membrane and to the side of the lingual frenulum, with a duct (Rivinus') that opens into the floor of the mouth.

Submandibular (sub'-man-DIB-ū-lar) **gland** One of a pair of salivary glands found inferior to the base of the tongue deep to the mucous membrane in the posterior part of the floor of the mouth, posterior to the sublingual glands, with a duct (Wharton's) situated to the side of the lingual frenulum. Also called the **submaxillary** (sub'-MAK-si-ler-ē) **gland.**

Submucosa (sub-mū-KŌ-sa) A layer of connective tissue located deep to a mucous membrane, as in the gastrointestinal tract or the urinary bladder; the submucosa connects the mucosa to the muscularis layer.

Submucosal plexus A network of autonomic nerve fibers located in the superficial part of the submucous layer of the small intestine. Also called the **plexus of Meissner** (MĪZ-ner).

Substrate A molecule upon which an enzyme acts.

Subthalamus (sub-THAL-a-mus) Part of the diencephalon inferior to the thalamus; the substantia nigra and red nucleus extend from the midbrain into the subthalamus.

Subthreshold stimulus A stimulus of such weak intensity that it cannot initiate an action potential (nerve impulse).

Sudoriferous (soo'-dor-IF-er-us) **gland** An apocrine or eccrine exocrine gland in the dermis or subcutaneous layer that produces perspiration. Also called a **sweat gland.**

Sulcus (SUL-kus) A groove or depression between parts, especially between the convolutions of the brain. *Plural is* **sulci** (SUL-sī).

Superficial (soo'-per-FISH-al) Located on or near the surface of the body or an organ.

Superficial inguinal (IN-gwi-nal) **ring** A triangular opening in the aponeurosis of the external oblique muscle that represents the termination of the inguinal canal.

Superficial vein Vein that courses through the subcutaneous layer unaccompanied by parallel arteries.

Superior (soo-PĒR-ē-or) Toward the head or upper part of a structure. Also called **cephalad** (SEF-a-lad) or **craniad.**

Superior vena cava (VĒ-na CĀ-va) **(SVC)** Large vein that collects blood from parts of the body superior to the heart and returns it to the right atrium.

Supination (soo-pi-NĀ-shun) A movement of the forearm in which the palm is turned anteriorly.

Surface anatomy The study of the structures that can be identified from the outside of the body.

Surfactant (sur-FAK-tant) Complex mixture of phospholipids and lipoproteins, produced by type II alveolar (septal) cells in the lungs, that decreases surface tension.

Susceptibility (sus-sep'-ti-BIL-i-tē) Lack of resistance to the damaging effects of an agent such as a pathogen.

Suspensory ligament (sus-PEN-so-rē LIG-a-ment) A fold of peritoneum extending laterally from the surface of the ovary to the pelvic wall.

Sutural (SOO-cher-al) **bone** A small bone located within a suture between certain cranial bones. Also called **Wormian** (WER-mē-an) **bone.**

Suture (SOO-cher) An immovable fibrous joint that joins skull bones.

Sympathetic (sim'-pa-THET-ik) **division** One of the two subdivisions of the autonomic nervous system, having cell bodies of preganglionic neurons in the lateral gray columns of the thoracic segment and the first two or three lumbar segments of the spinal cord; primarily concerned with processes involving the expenditure of energy.

Sympathetic trunk ganglion (GANG-glē-on) A cluster of cell bodies of sympathetic postganglionic neurons lateral to the vertebral column, close to the body of a vertebra. These ganglia extend inferiorly through the neck, thorax, and abdomen to the coccyx on both sides of the vertebral column and are connected to one another to form a chain on each side of the vertebral column. Also called **sympathetic chain** or **vertebral chain ganglia.**

Symphysis (SIM-fi-sis) A line of union. A slightly movable cartilaginous joint such as the pubic symphysis.

Symptom (SIMP-tum) A subjective change in body function not apparent to an observer, such as pain or nausea, that indicates the presence of a disease or disorder of the body.

Synapse (SYN-aps) The functional junction between two neurons or between a neuron and an effector, such as a muscle or gland; may be electrical or chemical.

Synapsis (sin-AP-sis) The pairing of homologous chromosomes during prophase I of meiosis.

Synaptic (sin-AP-tik) **cleft** The narrow gap at a chemical synapse that separates the axon terminal of one neuron from another neuron or muscle fiber (cell) and across which a neurotransmitter diffuses to affect the postsynaptic cell.

Synaptic end bulb Expanded distal end of an axon terminal that contains synaptic vesicles. Also called a **synaptic knob.**

Synaptic vesicle Membrane-enclosed sac in a synaptic end bulb that stores neurotransmitters.

Synarthrosis (sin'-ar-THRŌ-sis) Joint formed by a solid core of connective tissue between neighboring bones, such as a suture gomphosis, or syndesmosis.

Synchondrosis (sin'-kon-DRŌ-sis) A cartilaginous joint in which the connecting material is hyaline cartilage.

Syndesmosis (sin'-dez-MŌ-sis) A slightly movable joint in which articulating bones are united by fibrous connective tissue.

Syndrome (SIN-drōm) A group of signs and symptoms that occur together in a pattern that is characteristic of a particular disease or abnormal condition.

Synergist (SIN-er-jist) A muscle that assists the prime mover by reducing undesired action or unnecessary movement.

Synergistic (syn-er-JIS-tik) **effect** A hormonal interaction in which the effects of two or more hormones acting together is greater or more extensive than the sum of each hormone acting alone.

Synostosis (sin'-os-TŌ-sis) A joint in which the dense fibrous connective tissue that unites bones at a suture has been replaced by bone, resulting in a complete fusion across the suture line.

Synovial (si-NŌ-vē-al) **cavity** The space between the articulating bones of a synovial joint, filled with synovial fluid. Also called a **joint cavity.**

Synovial fluid Secretion of synovial membranes that lubricates joints and nourishes articular cartilage.

Synovial joint A fully movable or diarthrotic joint in which a synovial (joint) cavity is present between the two articulating bones.

Synovial membrane The deeper of the two layers of the articular capsule of a synovial joint, composed of areolar connective tissue that secretes synovial fluid into the synovial (joint) cavity.

System An association of organs that have a common function.

Systemic (sis-TEM-ik) Affecting the whole body; generalized.

Systemic anatomy The anatomic study of particular systems of the body, such as the skeletal, muscular, nervous, cardiovascular, or urinary systems.

Systemic circulation The routes through which oxygenated blood flows from the left ventricle through the aorta to all the organs of the body and deoxygenated blood returns to the right atrium.

Systole (SIS-tō-lē) In the cardiac cycle, the phase of contraction of the heart muscle, especially of the ventricles.

Systolic (sis-TOL-ik) **blood pressure** The force exerted by blood on arterial walls during ventricular contraction; the highest pressure measured in the large arteries, about 120 mmHg under normal conditions for a young adult.

T

T cell A lymphocyte that becomes immunocompetent in the thymus and can differentiate into a helper T cell or a cytotoxic T cell, both of which function in cell-mediated immunity.

T wave The deflection wave of an electrocardiogram that represents ventricular repolarization.

Tachycardia (tak′-i-KAR-dē-a) An abnormally rapid resting heartbeat or pulse rate (over 100 beats per minute).

Tactile (TAK-tīl) Pertaining to the sense of touch.

Tactile disc Modified epidermal cell in the stratum basale of hairless skin that functions as a cutaneous receptor for discriminative touch. Also called a **Merkel** (MER-kel) **disc.**

Target cell A cell whose activity is affected by a particular hormone.

Tarsal bones The seven bones of the ankle. Also called **tarsals.**

Tarsal gland Sebaceous (oil) gland that opens on the edge of each eyelid. Also called a **Meibomian** (mī-BŌ-mē-an) **gland.**

Tarsal plate A thin, elongated sheet of connective tissue, one in each eyelid, giving the eyelid form and support. The aponeurosis of the levator palpebrae superioris is attached to the tarsal plate of the superior eyelid.

Tarsus (TAR-sus) A collective term for the seven bones of the ankle.

Tectorial (tek-TŌ-rē-al) **membrane** A gelatinous membrane projecting over and in contact with the hair cells of the spiral organ (organ of Corti) in the cochlear duct.

Teeth (TETH) Accessory structures of digestion, composed of calcified connective tissue and embedded in bony sockets of the mandible and maxilla, that cut, shred, crush, and grind food. Also called **dentes** (DEN-tēz).

Telophase (TĒL-ō-fāz) The final stage of mitosis.

Tendon (TEN-don) A white fibrous cord of dense regular connective tissue that attaches muscle to bone.

Tendon organ A proprioceptive receptor, sensitive to changes in muscle tension and force of contraction, found chiefly near the junctions of tendons and muscles. Also called a **Golgi** (GOL-jē) **tendon organ.**

Tendon reflex A polysynaptic, ipsilateral reflex that protects tendons and their associated muscles from damage that might be brought about by excessive tension. The receptors involved are called tendon organs (Golgi tendon organs).

Tendon (synovial) sheath Tubelike bursa that wraps around a tendon to protect it from friction as it runs through a tunnel of connective tissue and bone.

Teniae coli (TĒ-nē-ē KO-lī) The three flat bands of thickened, longitudinal smooth muscle running the length of the large intestine, except in the rectum. *Singular is* **tenia coli.**

Tentorium cerebelli (ten-TŌ-rē-um ser′-e-BEL-ē) A transverse shelf of dura mater that forms a partition between the occipital lobe of the cerebral hemispheres and the cerebellum and that covers the cerebellum.

Teratogen (TER-a-tō-jen) Any agent or factor that causes physical defects in a developing embryo.

Terminal ganglion (TER-min-al GANG-glē-on) A cluster of cell bodies of parasympathetic postganglionic neurons either lying very close to the visceral effectors or located within the walls of the visceral effectors supplied by the postganglionic neurons.

Testis (TES-tis) Male gonad that produces sperm and the hormones testosterone and inhibin. Also called a **testicle.**

Testosterone (tes-TOS-te-rōn) A male sex hormone (androgen) secreted by interstitial endocrinocytes (Leydig cells) of a mature testis; needed for development of sperm; together with a second androgen termed **dihydrotestosterone (DHT),** controls the growth and development of male reproductive organs, secondary sex characteristics, and body growth.

Tetralogy of Fallot (tet-RAL-ō-jē of fal-Ō) A combination of four congenital heart defects: (1) constricted pulmonary semilunar valve, (2) interventricular septal opening, (3) emergence of the sorta from both ventricles instead of from the left only, and (4) enlarged right ventricle.

Thalamus (THAL-a-mus) A large, oval structure located bilaterally on either side of the third ventricle, consisting of two masses of gray matter organized into nuclei; main relay center for sensory impulses ascending to the cerebral cortex.

Thermoreceptor (THER-mō-rē-sep-tor) Sensory receptor that detects changes in temperature.

Thigh The portion of the free lower limb between the hip and the knee.

Third ventricle (VEN-tri-kul) A slitlike cavity between the right and left halves of the thalamus and between the lateral ventricles of the brain.

Thoracic (thō-RAS-ik) **cavity** Superior portion of the ventral body cavity that contains two pleural cavities, the mediastinum, and the pericardial cavity.

Thoracic duct A lymphatic vessel that begins as a dilation called the cisterna chyli, receives lymph from the left side of the head, neck, and chest, left arm, and the entire body below the ribs, and empties into the junction between the internal jugular and left subclavian veins. Also called the **left lymphatic** (lim-FAT-ik) **duct.**

Thoracolumbar (thō′-ra-kō-LUM-bar) **outflow** The axons of sympathetic preganglionic neurons, which have their cell bodies in the lateral gray columns of the thoracic segments and first two or three lumbar segments of the spinal cord.

Thorax (THŌ-raks) The chest.

Thoroughfare channel The distal end of a metarteriole that provides a direct route for blood from an arteriole to a venule, thus bypassing capillaries.

Thrombosis (throm-BŌ-sis) The formation of a clot in an unbroken blood vessel, usually a vein.

Thrombus A stationary clot formed in an unbroken blood vessel, usually a vein.

Thymus (THĪ-mus) A bilobed organ, located in the superior mediastinum posterior to the sternum and between the lungs, in which T cells develop immunocompetence.

Thyroid cartilage (THĪ-royd KAR-ti-lij) The largest single cartilage of the larynx, consisting of two fused plates that form the anterior wall of the larynx.

Thyroid follicle (FOL-i-kul) Spherical sac that forms the parenchyma of the thyroid gland and consists of follicular cells that produce thyroxine (T_4) and triiodothyronine (T_3).

Thyroid gland An endocrine gland with right and left lateral lobes on either side of the trachea connected by an isthmus; located anterior to the trachea just inferior to the cricoid cartilage; secretes thyroxine (T_4), triiodothyronine (T_3), and calcitonin.

Thyroid-stimulating hormone (TSH) A hormone secreted by the anterior pituitary that stimulates the synthesis and secretion of thyroxine (T_4) and triiodothyronine (T_3).

Thyroxine (thī-ROK-sēn) **(T4)** A hormone secreted by the thyroid gland that regulates metabolism, growth and development, and the activity of the nervous system.

Tic Spasmodic, involuntary twitching of muscles that are normally under voluntary control.

Tissue A group of similar cells and their intercellular substance joined together to perform a specific function.

Tissue rejection Phenomenon by which the body recognizes the protein (HLA antigens) in transplanted tissues or organs as foreign and produces antibodies against them.

Tongue A large skeletal muscle covered by a mucous membrane located on the floor of the oral cavity.

Tonsil (TON-sil) An aggregation of large lymphatic nodules embedded in the mucous membrane of the throat.

Topical (TOP-i-kal) Applied to the surface rather than ingested or injected.

Torn cartilage A tearing of an articular disc (meniscus) in the knee.

Trabecula (tra-BEK-ū-la) Irregular latticework of thin plates of spongy bone. Fibrous cord of connective tissue serving as supporting fiber by forming a septum extending into an organ from its wall or capsule. *Plural is* **trabeculae** (tra-BEK-ū-lē).

Trabeculae carneae (KAR-nē-ē) Ridges and folds of the myocardium in the ventricles.

Trachea (TRĀ-kē-a) Tubular air passageway extending from the larynx to the fifth thoracic vertebra. Also called the **windpipe.**

Tract A bundle of nerve axons in the central nervous system.

Transplantation (tranz-plan-TĀ-shun) The transfer of living cells, tissues, or organs from a donor to a recipient or from one part of the body to another in order to restore a lost function.

Transverse colon (trans-VERS KŌ-lon) The portion of the large intestine extending across the abdomen from the right colic (hepatic) flexure to the left colic (splenic) flexure.

Transverse fissure (FISH-er) The deep cleft that separates the cerebrum from the cerebellum.

Transverse plane A plane that divides the body or organs into superior and inferior portions. Also called a **horizontal plane.**

Transverse tubules (TOO-būls) **(T tubules)** Small, cylindrical invaginations of the sarcolemma of striated muscle fibers (cells) that conduct muscle action potentials toward the center of the muscle fiber.

Trauma (TRAW-ma) An injury, either a physical wound or psychic disorder, caused by an external agent or force, such as a physical blow or emotional shock; the agent or force that causes the injury.

Tremor (TREM-or) Rhythmic, involuntary, purposeless contraction of opposing muscle groups.

Triad (TRĪ-ad) A complex of three units in a muscle fiber composed of a transverse tubule and the sarcoplasmic reticulum terminal cisterns on both sides of it.

Tricuspid (trī-KUS-pid) **valve** Atrioventricular (AV) valve on the right side of the heart.

Triglyceride (trī-GLI-cer-īd) A lipid formed from one molecule of glycerol and three molecules of fatty acids that may be either solid (fats) or liquid (oils) at room temperature; the body's most highly concentrated source of chemical potential energy. Found mainly within adipocytes. Also called a **neutral fat** or a **triacylglycerol.**

Trigone (TRĪ-gon) A triangular region at the base of the urinary bladder.

Triiodothyronine (trī-ī-ō-dō-THĪ-rō-nēn) (**T₃**) A hormone produced by the thyroid gland that regulates metabolism, growth and development, and the activity of the nervous system.

Trophoblast (TRŌF-ō-blast) The superficial covering of cells of the blastocyst.

Tropic (TRŌ-pik) **hormone** A hormone whose target is another endocrine gland.

Trunk The part of the body to which the upper and lower limbs are attached.

Tubal ligation (lī-GĀ-shun) A sterilization procedure in which the uterine (Fallopian) tubes are tied and cut.

Tubular reabsorption The movement of filtrate from renal tubules back into blood in response to the body's specific needs.

Tubular secretion The movement of substances in blood into renal tubular fluid in response to the body's specific needs.

Tumor suppressor gene A gene coding for a protein that normally inhibits cell division; loss or alteration of a tumor suppressor gene called *p53* is the most common genetic change in a wide variety of cancer cells.

Tunica albuginea (TOO-ni-ka al′-bū-JIN-ē-a) A dense white fibrous capsule covering a testis or deep to the surface of an ovary.

Tunica externa (eks-TER-na) The superficial coat of an artery or vein, composed mostly of elastic and collagen fibers. Also called the **adventitia.**

Tunica interna (in-TER-na) The deep coat of an artery or vein, consisting of a lining of endothelium, basement membrane, and internal elastic lamina. Also called the **tunica intima** (IN-ti-ma).

Tunica media (ME-dē-a) The intermediate coat of an artery or vein, composed of smooth muscle and elastic fibers.

Tympanic (tim-PAN-ik) **membrane** A thin, semitransparent partition of fibrous connective tissue between the external auditory meatus and the middle ear. Also called the **eardrum.**

Tympanic antrum (tim-PAN-ik AN-trum) An air space in the middle ear that leads into the mastoid air cells or sinus.

Tympanic (tim-PAN-ik) **membrane** A thin, semitransparent partition of fibrous connective tissue between the external auditory meatus and the middle ear. Also called the **eardrum.**

Type II cutaneous mechanoreceptor A sensory receptor embedded deeply in the dermis and deeper tissues that detects stretching of skin. Also called a **Ruffini corpuscle.**

U

Umbilical cord The long, ropelike structure containing the umbilical arteries and vein that connect the fetus to the placenta.

Umbilicus (um-BIL-i-kus *or* um-bil-Ī-kus) A small scar on the abdomen that marks the former attachment of the umbilical cord to the fetus. Also called the **navel.**

Umbo (um-BŌ) The apex of the tympanic membrane that projects into the tympanic cavity.

Upper limb The appendage attached at the shoulder girdle, consisting of the arm, forearm, wrist, hand, and fingers. Also called the free **upper extremity.**

Uremia (ū-RĒ-mē-a) Accumulation of toxic levels of urea and other nitrogenous waste products in the blood, usually resulting from severe kidney malfunction.

Ureter (Ū-rē-ter) One of two tubes that connect the kidney with the urinary bladder.

Urethra (ū-RĒ-thra) The duct from the urinary bladder to the exterior of the body that conveys urine in females and urine and semen in males.

Urinalysis (ū-ri-NAL-i-sis) An analysis of the volume and physical, chemical, and microscopic properities of urine.

Urinary (Ū-ri-ner-ē) **bladder** A hollow, muscular organ situated in the pelvic cavity posterior to the pubic symphysis; receives urine via two ureters and stores urine until it is excreted through the urethra.

Urine The fluid produced by the kidneys that contains wastes and excess materials; excreted from the body through the urethra.

Urogenital (ū-′-rō-JEN-i-tal) **triangle** The region of the pelvic floor inferior to the pubic symphysis, bounded by the pubic symphysis and the ischial tuberosities, and containing the external genitalia.

Urology (ū--ROL-ō-jē) The specialized branch of medicine that deals with the structure, function, and diseases of the male and female urinary systems and the male reproductive system.

Uterine (Ū-ter-in) **tube** Duct that transports ova from the ovary to the uterus. Also called the **fallopian** (fal-LŌ-pē-an) **tube** or **oviduct.**

Uterosacral ligament (ū-′-ter-ō-SĀ-kral LIG-a-ment) A fibrous band of tissue extending from the cervix of the uterus laterally to the sacrum.

Uterovesical (yū′-ter-ō-VES-i-kal) **pouch** A shallow pouch formed by the reflection of the peritoneum from the anterior surface of the uterus, at the junction of the cervix and the body, to the posterior surface of the urinary bladder.

Uterus (Ū-te-rus) The hollow, muscular organ in females that is the site of menstruation, implantation, development of the fetus, and labor. Also called the **womb.**

Utricle (Ū-tri-kul) The larger of the two divisions of the membranous labyrinth located inside the vestibule of the inner ear, containing a receptor organ for static equilibrium.

Uvea (Ū-vē-a) The three structures that together make up the vascular tunic of the eye.

Uvula (Ū-vū-la) A soft, fleshy mass, especially the V-shaped pendant part, descending from the soft palate.

V

Vagina (va-JĪ-na) A muscular, tubular organ that leads from the uterus to the vestibule, situated between the urinary bladder and the rectum of the female.

Vallate papilla (VAL-at pa-PIL-a) One of the circular projections that is arranged in an inverted V-shaped row at the back of the tongue; the largest of the elevations on the upper surface of the tongue containing taste buds. Also called **circumvallate papilla.**

Varicocele (VAR-i-kō-sēl) A twisted vein; especially, the accumulation of blood in the veins of the spermatic cord.

Varicose (VAR-i-kōs) Pertaining to an unnatural swelling, as in the case of a varicose vein.

Vas A vessel or duct.

Vasa recta (VĀ-sa REK-ta) Extensions of the efferent arteriole of a juxtamedullary nephron that run alongside the loop of the nephron (Henle) in the medullary region of the kidney.

Vasa vasorum (va-SŌ-rum) Blood vessels that supply nutrients to the larger arteries and veins.

Vascular (VAS-kū-lar) Pertaining to or containing many blood vessels.

Vascular (venous) sinus A vein with a thin endothelial wall that lacks a tunica media and externa and is supported by surrounding tissue.

Vascular spasm Contraction of the smooth muscle in the wall of a damaged blood vessel to prevent blood loss.

Vascular tunic (TOO-nik) The middle layer of the eyeball, composed of the choroid, ciliary body, and iris. Also called the **uvea** (Ū-ve-a).

Vasectomy (va-SEK-tō-mē) A means of sterilization of males in which a portion of each ductus (vas) deferens is removed.

Vasoconstriction (vāz-ō-kon-STRIK-shun) A decrease in the size of the lumen of a blood vessel caused by contraction of the smooth muscle in the wall of the vessel.

Vasodilation (vāz′-ō-DĪ-lā-shun) An increase in the size of the lumen of a blood vessel caused by relaxation of the smooth muscle in the wall of the vessel.

Vein A blood vessel that conveys blood from tissues back to the heart.

Vena cava (VĒ-na KĀ-va) One of two large veins that open into the right atrium, returning to the heart all of the deoxygenated blood from the systemic circulation except from the coronary circulation.

Vena comitans (VĒ-na KOM-i-tans) Two to three veins accompanying medium-to-small sized muscular arteries.

Ventral (VEN-tral) Pertaining to the anterior or front side of the body; opposite of dorsal.

Ventral ramus (RĀ-mus) The anterior branch of a spinal nerve, containing sensory and motor fibers to the muscles and skin of the anterior surface of the head, neck, trunk, and the limbs.

Ventricle (VEN-tri-kul) A cavity in the brain filled with cerebrospinal fluid. An inferior chamber of the heart.

Ventricular fibrillation (ven-TRIK-ū-lar fib-ri-LĀ-shun) Asynchronous ventricular contractions; unless reversed by defibrillation, results in heart failure.

Venule (VEN-ūl) A small vein that collects blood from capillaries and delivers it to a vein.

Vermiform appendix (VER-mi-form a-PEN-diks) A twisted, coiled tube attached to the cecum.

Vermilion (ver-MIL-yon) The area of the mouth where the skin on the outside meets the mucous membrane on the inside.

Vermis (VER-mis) The central constricted area of the cerebellum that separates the two cerebellar hemispheres.

Vertebral (VER-te-bral) **canal** A cavity within the vertebral column formed by the vertebral foramina of all the vertebrae and containing the spinal cord. Also called the **spinal canal.**

Vertebral column The 26 vertebrae of an adult and 33 vertebrae of a child; encloses and protects the spinal cord and serves as a point of attachment for the ribs and back muscles. Also called the **backbone, spine,** or **spinal column.**

Vesicle (VES-i-kul) A small bladder or sac containing liquid.

Vesicouterine (ves′-ik-ō-Ū-ter-in) **pouch** A shallow pouch formed by the reflection of the peritoneum from the anterior surface of the uterus, at the junction of the cervix and the body, to the posterior surface of the urinary bladder.

Vestibular (ves-TIB-ū-lar) **apparatus** Collective term for the organs of equilibrium, which includes the saccule, utricle, and semicircular ducts.

Vestibular membrane The membrane that separates the cochlear duct from the scala vestibuli.

Vestibule (VES-ti-būl) A small space or cavity at the beginning of a canal, especially the inner ear, larynx, mouth, nose, and vagina.

Villus (VIL-lus) A projection of the intestinal mucosal cells containing connective tissue, blood vessels, and a lymphatic vessel; functions in the absorption of the end products of digestion. *Plural is* **villi** (VIL-ī).

Viscera (VIS-er-a) The organs inside the ventral body cavity. *Singular is* **viscus** (VIS-kus).

Visceral (VIS-er-al) Pertaining to the organs or to the covering of an organ.

Visceral effectors (e-FEK-torz) Organs of the ventral body cavity that respond to neural stimulation, including cardiac muscle, smooth muscle, and glands.

Vital signs Signs necessary to life that include temperature (T), pulse (P), respiratory rate (RR), and blood pressure (BP).

Vitamin An organic molecule necessary in trace amounts that acts as a catalyst in normal metabolic processes in the body.

Vitreous (VIT-rē-us) **body** A soft, jellylike substance that fills the vitreous chamber of the eyeball, lying between the lens and the retina.

Vocal folds Pair of mucous membrane folds below the ventricular folds that function in voice production. Also called **true vocal cords.**

Volkmann's canal *See* **Perforating canal.**

Vulva (VUL-va) Collective designation for the external genitalia of the female. Also called the **pudendum** (poo-DEN-dum).

W

Wallerian (wal-LE-rē-an) **degeneration** Degeneration of the portion of the axon and myelin sheath of a neuron distal to the site of injury.

Wandering macrophage (MAK-rō-fāj) Phagocytic cell that develops from a monocyte, leaves the blood, and migrates to infected tissues.

White matter Aggregations or bundles of myelinated and unmyelinated axons located in the brain and spinal cord.

White pulp The regions of the spleen composed of lymphatic tissue, mostly B lymphocytes.

White ramus communicans (RĀ-mus kō-MŪ-ni-kans) The portion of a preganglionic sympathetic axon that branches from the anterior ramus of a spinal nerve to enter the nearest sympathetic trunk ganglion.

X

Xiphoid (ZĪ-foyd). The inferior portion of the sternum is the **xiphoid process.**

Y

Yolk sac An extraembryonic membrane composed of the exocoelomic membrane and hypoblast. It transfers nutrients to the embryo, is a source of blood cells, contains primordial germ cells that migrate into the gonads to form primitive germ cells, forms part of the gut, and helps prevent desiccation of the embryo.

Z

Zona fasciculata (ZŌ-na fa-sik′-ū-LA-ta) The middle zone of the adrenal cortex consisting of cells arranged in long, straight cords that secrete glucocorticoid hormones, mainly cortisol.

Zona glomerulosa (glo-mer′-ū-LŌ-sa) The outer zone of the adrenal cortex, directly under the connective tissue covering, consisting of cells arranged in arched loops or round balls that secrete mineralocorticoid hormones, mainly aldosterone.

Zona pellucida (pe-LOO-si-da) Clear glycoprotein layer between a secondary oocyte and the surrounding granulosa cells of the corona radiata.

Zona reticularis (ret-ik′-ū-LAR-is) The inner zone of the adrenal cortex, consisting of cords of branching cells that secrete sex hormones, chiefly androgens.

Zygote (ZĪ-ḡot) The single cell resulting from the union of male and female gametes; the fertilized ovum.

CREDITS

Photo Credits

Chapter 1 Fig. 1.1: Rubberball Productions/Getty Images. Fig. 1.4a: Dissection Shawn Miller; Photograph Mark Nielsen. Fig. 1.4b: Dissection Shawn Miller; Photograph Mark Nielsen. Fig. 1.4c: Dissection Shawn Miller; Photograph Mark Nielsen. Fig. 1.6d: Dissection Shawn Miller; Photograph Mark Nielsen. Fig. 1.6c: Dissection Shawn Miller; Photograph Mark Nielsen. Fig. 1-7c: Dissection Shawn Miller; Photograph Mark Nielsen. Fig. 1.7d: Dissection Shawn Miller; Photograph Mark Nielsen. Fig. 1.7f: Dissection Shawn Miller; Photograph Mark Nielsen. Fig. 1.7e: Dissection Shawn Miller; Photograph Mark Nielsen. Fig. 1.8a: Andy Washnik. Table 1.3a Biophoto Associates/Photo Researchers, Inc. Table 1.3b Breast Cancer Unit, Kings College Hospital, London/Photo Researchers, Inc. Table 1.3c Zephyr/Photo Researchers, Inc. Table 1.3d Cardio-Thoracic Centre, Freeman Hospital, Newcastle-Upon-Tyne/Photo Researchers, Inc. Table 1.3e CNRI/Science Photo Library/Photo Researchers, Inc. Table 1.3f Science Photo Library/Photo Researchers, Inc. Table 1.3g Scott Camazine/Photo Researchers, Inc. Table 1.3g Simon Fraser/Photo Researchers, Inc. Table 1.3i Courtesy Andrew Joseph Tortora and Damaris Soler. Table 1.3j Howard Sochurek/ Medical Images, Inc. Table 1.3k SIU/Visuals Unlimited. Table 1.3l Dept. of Nuclear Medicine, Charing Cross Hospital/Photo Researchers, Inc. Table 1.3m Camal/Phototake. Page 4 (top): Scientifica/Visuals Unlimited. Page 4 (bottom left):_La/Bc.Aigo/Phototake. Page 4 (bottom center): Jose Luis Pelaez/Getty Images, Inc. Page 4 (bottom right): Malvina Mendil/Photo Researchers, Inc.

Chapter 2 Fig. 2.5b,c: Courtesy Abbott Laboratories. Fig. 2.7c: From McGill, M, Highfield, DP, Mohahan, TM. and Brinkley, BR, *J. Ultrastructure Research* 57, 42-42, 6, 1976, Academic Press. Reproduced with permission of Elsevier. Fig. 2.8b: P. Motta/Photo Researchers, Inc. Fig. 2.8c: David M. Phillips/Visuals Unlimited. Fig. 2.10b: D. W. Fawcett/Photo Researchers, Inc. Fig. 2.11b: Biophoto Associates/Photo Researchers, Inc. Fig. 2.13b: David M. Phillips/Visuals Unlimited. Fig. 2.14b: Don W. Fawcett/Visuals Unlimited. Fig. 2.15c: D.W. Fawcett/Photo Researchers, Inc. Fig. 2.19: Courtesy Michael Ross, University of Florida. Page 42 (right bottom): U.S. Department of Energy Genome Programs. Page 52 (left top): Steve Gschmeissner/Science Photo Library/Photo Researchers, Inc.

Chapter 3 Table 3.1a Biophoto Associates/ Photo Researchers, Inc. Table 3.1a Mark Nielsen. Table 3.1b Mark Nielsen. Table 3.1c Mark Nielsen. Table 3.1d Mark Nielsen. Table 3.1e Mark Nielsen. Table 3.1f Mark Nielsen. Table 3.1g Courtesy Michael Ross, University of Florida. Table 3.1h Courtesy Michael Ross, University of Florida. Table 3.1i Mark Nielsen. Table 3.2a Mark Nielsen. Table 3.2b Mark Nielsen. Table 3.3a Courtesy Michael Ross, University of Florida. Table 3.3b Mark Nielsen. Table 3.4a Mark Nielsen. Table 3.4b Mark Nielsen. Table 3.4b (inset) Mark Nielsen. Table 3.4c Mark Nielsen. Table 3.4d Mark Nielsen. Table 3.4e Mark Nielsen. Table 3.4f Mark Nielsen. Table 3.4g Mark Nielsen. Table 3.4g Mark Nielsen. Table 3.4h Mark Nielsen. Table 3.4i Mark Nielsen. Table 3.4j (Inset) Mark Nielsen. Table 3.4j Mark Nielsen. Table 3.4k Courtesy Michael Ross, University of Florida. Table 3.5a Courtesy Michael Ross, University of Florida. Table 3.5b Mark Nielsen. Table 3.5c Mark Nielsen. Table 3.5c Mark Nielsen. Table 3.6 Mark Nielsen. Page 59 (right): ALIX/PHANIE/Photo Researchers, Inc. Page 76 (right): Pascal Goetgheluck/Photo Researchers, Inc.

Chapter 4 Fig. 4.2b: David Phillips/Photo Researchers, Inc. Fig. 4.2c: Myriam Wharman/Phototake. Fig. 4.11b: Preparation and photography Mark Nielsen. Fig. 4.14a: Photo provided courtesy of Kohei Shiota, Congenital Anomaly Research Center, Kyoto University, Graduate School of Medicine. Fig. 4.14b: Courtesy National Museum of Health and Medicine, Armed Forces Institute of Pathology. Fig. 4.14c: Courtesy National Museum of Health and Medicine, Armed Forces Institute of Pathology. Fig. 4.14d: Courtesy National Museum of Health and Medicine, Armed Forces Institute of Pathology. Fig. 4.14e: Courtesy National Museum of Health and Medicine, Armed Forces Institute of Pathology. Fig. 4.14f: Photo by Lennart Nilsson/Albert Bonniers Fîrlag AB, A Child is Born, Dell Publishing Company. Reproduced with permission. Fig. 4.14g: Photo provided courtesy of Kohei Shiota, Congenital Anomaly Research Center, Kyoto University, Graduate School of Medicine. Fig. 4.14h: Photo provided courtesy of Kohei Shiota, Congenital Anomaly Research Center, Kyoto University, Graduate School of Medicine. Page 97 (top): Steve Gschmeissner/Photo Researchers, Inc. Page 98 (bottom left): Alamy Images.

Chapter 5 Fig. 5.1b: Courtesy Michael Ross, University of Florida. Fig. 5.1c: David Becker/Photo Researchers, Inc. Fig. 5.3b: Courtesy Michael Ross, University of Florida. Fig. 5.5b: VVG/Science Photo Library/Photo Researchers, Inc. Fig. 5.5e: 1Mark Nielsen. Fig. 5.6a: Mark Nielsen. Fig. 5.6b: Mark Nielsen. Fig. 5.6c: Mark Nielsen. Fig. 5.7c: Science Vu/Visuals Unlimited. Page 125 (left) Alain Dex/Photo Researchers, Inc. Page 125 (right) Biophoto Associates/Photo Researchers, Inc. Page 133 (left) Sheila Terry/Science Photo Library/ Photo Researchers, Inc. Page 133 (center) St. Stephen's Hospital/Science Photo Library/Photo Researchers, Inc. Page 133 (right) St. Stephen's Hospital/Science Photo Library/Photo Researchers, Inc.

Chapter 6 Fig. 6.2b: Mark Nielsen. Fig. 6.3: CNRI/Photo Researchers, Inc. Fig. 6.3: Dr. Richard Kessel and Randy Kardon/Tissues & Organs/Visuals Unlimited. Fig. 6.3: Dr. Richard Kessel and Randy Kardon/Tissues & Organs/Visuals Unlimited. Fig. 6.8a: The Bergman Collection. Fig. 6.8b: Mark Nielsen. Fig. 6.10a Courtesy Department of Medical Illustration, University of Wisconsin Medical School. Fig. 6.10b: Courtesy Department of Medical Illustration, University of Wisconsin Medical School. Fig. 6.10c: Courtesy Department of Medical Illustration, University of Wisconsin Medical School. Fig. 6.10d: Dr. Andrew Schmidt/The Bergman Collection. Fig. 6.10e: Courtesy Department of Medical Illustration, University of Wisconsin Medical School. Fig. 6.10f: Courtesy Department of Medical Illustration, University of Wisconsin Medical School. Page 148 (left): CNRI/Photo Researchers, Inc. Page 154 (left): P. Motta/Photo Researchers, Inc.

Chapter 7 Fig. 7.1c: Mark Nielsen. Fig. 7.1d: Mark Nielsen. Fig. 7.2b: Mark Nielsen. Fig. 7.3b: Mark Nielsen. Fig. 7.4b: Mark Nielsen. Fig. 7.5b: Mark Nielsen. Fig. 7.6b: Mark Nielsen. Fig. 7.6d: Mark Nielsen. Fig. 7.6f: Mark Nielsen. Fig. 7.7e: Mark Nielsen. Fig. 7.7g: Dissection Shawn Miller, Photograph Mark Nielsen. Fig. 7.8b: Mark Nielsen. Fig. 7.10b: Mark Nielsen. Fig. 7.11d: Dissection Shawn Miller, Photograph Mark Nielsen. Fig. 7.12b: Mark Nielsen. Fig. 7.14b: Mark Nielsen. Fig. 7.14c: Mark Nielsen. Fig. 7.12c: Mark Nielsen. Fig. 7.16c: Mark Nielsen. Fig. 7.16d: Mark Nielsen. Fig. 7.17b: Mark Nielsen. Fig. 7.17c: Mark Nielsen. Fig. 7.17d: Mark Nielsen. Fig. 7.17e: Mark Nielsen. Fig. 7.19b: Mark Nielsen. Fig. 7.19c: Mark Nielsen. Fig. 7.20a: Mark Nielsen. Fig. 7.20b: Mark Nielsen. Fig. 7.21b: Mark Nielsen. Fig. 7.21c: Mark Nielsen. Fig. 7.22d: Mark Nielsen. Fig. 7.22e: Mark Nielsen. Page 194 (left): Princess Margaret Rose Orthopaedic Hospital/Photo Researchers, Inc. Page 194 (center): Dr. P. Marazzi/Photo Researchers, Inc. Page 194 (right): Custom Medical Stock Photo, Inc.

Chapter 8 Fig. 8.2: Mark Nielsen. Fig. 8.3: Mark Nielsen. Fig. 8.4: Mark Nielsen. Fig. 8.5: Mark Nielsen. Fig. 8.6: Mark Nielsen. Fig. 8.7: Mark Nielsen. Fig. 8.8: Mark Nielsen. Fig. 8.10: Mark Nielsen. Fig. 8.11: Mark Nielsen. Fig. 8.12: Mark Nielsen. Page 216(right): _ISM/Phototake/Phototake. Page 232(right): Neil Borden/Photo Researchers, Inc.

Chapter 9 Fig. 9.1c: Dissection Richard Homer; Photograph Mark Nielsen. Fig. 9.3b: Mark Nielsen. Fig. 9.5: John Wilson White. Fig. 9.6a–g: John Wilson White. Fig. 9.7a–d: John Wilson White. Fig. 9.8a–b: John Wilson White. Fig. 9.9a–c: John Wilson White. Fig. 9.10a–h: John Wilson White. Fig. 9.10i: Andy Washnik. Fig. 9.11d: Dissection Shawn Miller, Photograph Mark Nielsen. Fig. 9.12d: Dissection Shawn Miller, Photograph Mark Nielsen. Fig. 9.14d: Dissection Shawn Miller, Photograph Mark Nielsen. Fig. 9.15e: Dissection Shawn Miller, Photograph Mark Nielsen. Fig. 9.15f: Dissection Shawn Miller, Photograph Mark Nielsen. Fig. 9.15g: Dissection Shawn Miller, Photograph Mark Nielsen. Fig. 9.15h: Dissection Shawn Miller, Photograph Mark Nielsen. Fig. 9.15i: Dissection Shawn Miller, Photograph Mark Nielsen. Page 265 (left): CNRI/Photo Researchers, Inc. Page 265 (right): Custom Medical Stock Photo. Page 283 (top left): SIU/Visuals Unlimited. Page 283 (top right): ISM/Phototake. Page 283 (bottom): Charles McRae/Visuals Unlimited.

Chapter 10 Fig. 10.1b: Dissection Richard Homer; Photograph Mark Nielsen. Fig. 10.4: Courtesy D.E. Kelley. Fig. 10.6d: Don Fawcett/Photo Researchers, Inc. Fig. 10.7a,b,c: Courtesy Hiroyouki Sasaki, Yale E. Goldman and Clara Franzini-Armstrong. Table 10.1 Biophoto Associates/Photo Researchers, Inc. Page 298 (right): Custom Medical Stock Photo, Inc.

Chapter 11 Fig. 11.4d: Dissection Shawn Miller, Photograph Mark Nielsen. Fig. 11.4e: Dissection Shawn Miller, Photograph Mark Nielsen. Fig. 11.4f: Dissection Shawn Miller, Photograph Mark Nielsen. Fig. 11.5c: Dissection Shawn Miller, Photograph Mark Nielsen. Fig. 11.6b: Dissection Shawn Miller, Photograph Mark Nielsen. Fig. 11.7b: Dissection Shawn Miller, Photograph Mark Nielsen. Fig. 11.8c: Dissection Shawn Miller, Photograph Mark Nielsen. Fig. 11.10c: Dissection Shawn Miller, Photograph Mark Nielsen. Fig. 11.11c: Dissection Shawn Miller, Photograph Mark Nielsen. Fig. 11.11d: Dissection Shawn Miller, Photograph Mark Nielsen. Fig. 11.12c: Dissection Nathan Mortensen and Shawn Miller; Photograph Mark Nielsen. Fig. 11.12d: Dissection Nathan Mortensen and Shawn Miller; Photograph Mark Nielsen. Fig. 11.12e: Dissection Nathan Mortensen and Shawn Miller; Photograph Mark Nielsen. Fig. 11.12f: Dissection Nathan Mortensen and Shawn Miller; Photograph Mark Nielsen. Fig. 11.13e: Dissection Shawn Miller, Photograph Mark Nielsen. Fig. 11.13f: Dissection Shawn Miller and Nathan Mortensen; Photograph Mark Nielsen. Fig. 11.15b: Dissection Nathan Mortensen; Photograph Mark Nielsen. Fig. 11.16e: Dissection Nathan Mortensen and Shawn Miller; Photograph Mark Nielsen. Fig. 11.16f: Dissection Shawn Miller, Photograph Mark Nielsen. Fig. 11.17f: Dissection Shawn Miller, Photograph Mark Nielsen. Fig. 11.17g: Dissection Nathan Mortensen and Shawn Miller; Photograph Mark Nielsen. Fig. 11.17h: Dissection Shawn Miller, Photograph Mark Nielsen. Fig. 11.17i: Dissection Shawn Miller, Photograph Mark Nielsen. Fig. 11.18h: Dissection Shawn Miller, Photograph Mark Nielsen. Fig. 11.18i: Dissection Shawn Miller, Photograph Mark Nielsen. Fig. 11.18j: Dissection Nathan Mortensen and Shawn Miller; Photograph Mark Nielsen. Fig. 11.18d: Dissection Shawn Miller, Photograph Mark Nielsen. Fig. 11.19f: Dissection Shawn Miller, Photograph Mark Nielsen. Fig. 11.19g: Dissection Shawn Miller, Photograph Mark Nielsen. Fig. 11.19h: Mark Nielsen and Mark Dissection Shawn Miller, Photograph Mark Nielsen. Fig. 11.19i: Dissection Shawn Miller, Photograph Mark Nielsen. Fig. 11.20h: Dissection Shawn Miller, Photograph Mark Nielsen. Fig. 11.20i: Dissection Shawn Miller, Photograph Mark Nielsen. Fig. 11.20g: Andy Washnik. Fig. 11.21d: Dissection Nathan Mortensen and Shawn Miller; Photograph Mark Nielsen. Fig. 11.21e: Dissection Nathan Mortensen and Shawn Miller; Photograph Mark Nielsen. Fig. 11.22h: Dissection Shawn Miller, Photograph Mark Nielsen. Fig. 11.22i: Dissection Shawn Miller, Photograph Mark Nielsen. Fig. 11.22j: Dissection Nathan Mortensen and Shawn Miller; Photograph Mark Nielsen.

Fig. 11.23b: Dissection Shawn Miller, Photograph Mark Nielsen. Fig. 11.24f: Dissection Shawn Miller, Photograph Mark Nielsen. Fig. 11.24g: Dissection Shawn Miller, Photograph Mark Nielsen. Fig. 11.24h: Dissection Shawn Miller, Photograph Mark Nielsen. Fig. 11.24i: Dissection Shawn Miller, Photograph Mark Nielsen. Fig. 11.25f: Dissection Nathan Mortensen; Photograph Mark Nielsen. Fig. 11.25g: Dissection Nathan Mortensen; Photograph Mark Nielsen. Fig. 11.25h: Dissection Nathan Mortensen; Photograph Mark Nielsen. Fig. 11.25d: Dissection Nathan Mortensen; Photograph Mark Nielsen.

Chapter 12 Fig. 12.1: John Wiley & Sons. Fig. 12.2a: Andy Washnik. Fig. 12.2b: John Wiley & Sons. Fig. 12.3: John Wiley & Sons. Fig. 12.4: John Wiley & Sons. Fig. 12.5: Courtesy Lynne Marie Barghesi. Fig. 12.6a: Andy Washnik. Fig. 12.6b: Andy Washnik. Fig. 12.7: John Wiley & Sons. Fig. 12.8: John Wiley & Sons. Fig. 12.9a: John Wiley & Sons. Fig. 12.9b: John Wiley & Sons. Fig. 12.9c: John Wiley & Sons. Fig. 12.10: Andy Washnik. Fig. 12.11b,c,d: John Wiley & Sons. Fig. 12.12a: John Wiley & Sons. Fig. 12.12b: John Wiley & Sons. Fig. 12.12c: John Wiley & Sons. Fig. 12.13a,b,c.d: John Wiley & Sons. Fig. 12.14: John Wiley & Sons. Fig. 12.15a,b,c: John Wiley & Sons. Fig. 12.16a,b,c,d: John Wiley & Sons.

Chapter 13 Fig. 13.2a: Juergen Berger/Photo Researchers, Inc. Fig. 13.2b: Mark Nielsen. Fig. 13.5: Courtesy Michael Ross, University of Florida. Fig. 13.6: Boehringer Ingelheim International. Photo by Lennart Nilsson/Bonnier Alba Publishers, THE INCREDIBLE MACHINE. Reproduced with permission. Page 445: Lewin/Royal Free Hospital/Photo Researchers, Inc.

Chapter 14 Fig. 14.1c: Dissection Shawn Miller, Photograph Mark Nielsen. Fig. 14.2b: Dissection Shawn Miller, Photograph Mark Nielsen. Fig. 14.1d: John Wiley & Sons. Fig. 14.3b: Dissection Shawn Miller, Photograph Mark Nielsen. Fig. 14.3d: Dissection Shawn Miller, Photograph Mark Nielsen. Fig. 14.4b: Dissection Shawn Miller, Photograph Mark Nielsen. Fig. 14.6c: Dissection Shawn Miller, Photograph Mark Nielsen. Fig. 14.6f: Dissection Shawn Miller, Photograph Mark Nielsen. Fig. 14.6g: Dissection Shawn Miller, Photograph Mark Nielsen. Fig. 14.8c: Dissection Shawn Miller, Photograph Mark Nielsen. Fig. 14.8d: Dissection Shawn Miller, Photograph Mark Nielsen. Page 466: CNRI/Photo Researchers, Inc. Page 474 (left): _Vu/ Cabisco/Visuals Unlimited. Page 474 (right): W. Ober/Visuals Unlimited. Page 476: _ISM/Phototake. Page 471: Scott Camazine/ Phototake.

Chapter 15 Fig. 15.1d: Dennis Strete. Fig. 15.1e: Courtesy Michael Ross, University of Florida. Fig. 15.4b: Dissection Shawn Miller, Photograph Mark Nielsen. Fig. 15.6c: Dissection Shawn Miller, Photograph Mark Nielsen. Fig. 15.7d: Dissection Shawn Miller, Photograph Mark Nielsen. Fig. 15.7e: Dissection Shawn Miller, Photograph Mark Nielsen. Fig. 15.8d: Dissection Shawn Miller, Photograph Mark Nielsen. Fig. 15.8e: Dissection Tory Meyer; Photograph Mark Nielsen. Fig. 15.8f: Dissection Tory Meyer; Photograph Mark Nielsen. Fig. 15.8g: Dissection Tory Meyer; Photograph Mark Nielsen. Fig. 15.9c: Dissection Shawn Miller, Photograph Mark Nielsen. Fig. 15.9d: Dissection Nathan Mortensen; Photograph Mark Nielsen. Fig. 15.10b: Dissection Shawn Miller, Photograph Mark Nielsen. Fig. 15.11b: Dissection Shawn Miller, Photograph Mark Nielsen. Fig. 15.11c: Dissection Shawn Miller, Photograph Mark Nielsen. Fig. 15.12d: Dissection Shawn Miller, Photograph Mark Nielsen. Fig. 15.13b: Dissection Shawn Miller, Photograph Mark Nielsen. Fig. 15.13c: Dissection Shawn Miller, Photograph Mark Nielsen. Fig. 15.14c: Dissection Shawn Miller, Photograph Mark Nielsen. Fig. 15.16c: Dissection Shawn Miller, Photograph Mark Nielsen. Page 488: Scott Camazine/Phototake.

Chapter 16 Fig. 16.5b: Courtesy Michael Ross, University of Florida. Fig. 16.5c: Courtesy Michael Ross, University of Florida. Fig. 16.6b: Mark Nielsen. Fig. 16.6c: Lester Bergman & Associates. Fig. 16.6d: Dissection Shawn Miller, Photograph Mark Nielsen. Fig. 16.7a: Dissection Shawn Miller, Photograph Mark Nielsen. Fig. 16.7d: Mark Nielsen. Fig. 16.10b: Dissection Shawn Miller, Photograph Mark Nielsen. Page 550: P. Marazzi/Photo Researchers, Inc.

Chapter 17 Fig. 17.3b: Mark Nielsen. Fig. 17.8c: Dennis Kunkel/Phototake. Fig. 17.8d: Martin Rotker/Phototake. Fig. 17.9c: Dissection Shawn Miller, Photograph Mark Nielsen. Fig. 17.9d: Dissection Shawn Miller, Photograph Mark Nielsen. Page 571: Image Source/Getty Images.

Chapter 18 Fig. 18.4b: Dissection Shawn Miller, Photograph Mark Nielsen. Fig. 18.1b: Dissection Shawn Miller, Photograph Mark Nielsen. Fig. 18.2b: Dissection Chris Roach; Photograph Mark Nielsen. Fig. 18.2c: Dissection Shawn Miller, Photograph Mark Nielsen. Fig. 18.3b: Courtesy Michael Ross, University of Florida. Fig. 18.5b: Dissection Shawn Miller, Photograph Mark Nielsen. Fig. 18.6c: Dissection Richard Homer; Photograph Mark Nielsen. Fig. 18.7c: Dissection Shawn Miller, Photograph Mark Nielsen. Fig. 18.8b: Dissection Shawn Miller, Photograph Mark Nielsen. Fig. 18.10: Dr. Richard Kessel and Randy Kardon/Tissues & Organs/Visuals Unlimited. Table 18.1 Dissection Shawn Miller, Photograph Mark Nielsen.

Chapter 19 Fig. 19.1b: Dissection Shawn Miller, Photograph Mark Nielsen. Fig. 19.2c: Dissection Shawn Miller, Photograph Mark Nielsen. Fig. 19.4a: Dissection Shawn Miller, Photograph Mark Nielsen. Fig. 19.5b: Dissection Shawn Miller, Photograph Mark Nielsen. Fig. 19.5c: Dissection Shawn Miller, Photograph Mark Nielsen. Fig. 19.7c: Dissection Shawn Miller, Photograph Mark Nielsen. Fig. 19.8d: Dissection Shawn Miller, Photograph Mark Nielsen. Fig. 19.11c: Dissection Shawn Miller, Photograph Mark Nielsen. Fig. 19.12: N. Gluhbegovic and T.H. Williams, *The Human Brain: A Photographic Guide*, Harper and Row, Publishers, Inc. Hagerstown, MD, 1980. Fig. 19.13c: Dissection Shawn Miller, Photograph Mark Nielsen. Table 19.2 Dissection Shawn Miller, Photograph Mark Nielsen. Table 19.4 Dissection Shawn Miller, Photograph Mark Nielsen. Page 619: Steve Allen/Photo Researchers, Inc.

Chapter 20 Fig. 20.4b: Dissection Shawn Miller, Photograph Mark Nielsen. Page 671: P. Marazzi/Photo Researchers, Inc.

Chapter 22 Fig. 22.1c: Courtesy Michael Ross, University of Florida. Fig. 22.2d: (inset) Mark Nielsen. Fig. 22.2d: Mark Nielsen. Fig. 22.6: Paul Parker/Photo Researchers, Inc. Fig. 22.9a: N. Gluhbegovic and T.H. Williams, *The Human Brain: A Photographic Guide*, Harper and Row, Publishers, Inc. Hagerstown, MD, 1980. Fig. 22.11b: Dissection Mark Nielsen; Photograph Mark Nielsen. Fig. 22.11c: Dissection Mark Nielsen; Photograph Mark Nielsen. Fig. 22.12: Dissection Mark Nielsen; Photograph Mark Nielsen. Fig. 22.12c: Dissection Mark Nielsen, Photograph Mark Nielsen. Fig. 22.13: John D. Cunningham/ Visuals Unlimited. Page 706: Science VU/NIH/Visuals Unlimited. Page 706: Science VU/NIH/Visuals Unlimited. Page 706: Science VU/NIH/Visuals Unlimited. Page 718: Dorling Kindersley Media Library.

Chapter 23 Fig. 23.2a: Dissection Shawn Miller, Photograph Mark Nielsen. Fig. 23.2c: Courtesy James Lowe, University of Nottingham, Nottingham, United Kingdom. Fig. 23.4b: Mark Nielsen. Fig. 23.4c: Dissection Shawn Miller, Photograph Mark Nielsen. Fig. 23.4d: Dissection Shawn Miller, Photograph Mark Nielsen. Fig. 23.5b: Mark Nielsen . Fig. 23.7d: Dissection Shawn Miller, Photograph Mark Nielsen. Fig. 23.5d: Dissection Shawn Miller, Photograph Mark Nielsen. Fig. 23.6c: Dissection Shawn Miller, Photograph Mark Nielsen. Fig. 23.6d: Mark Nielsen. Fig. 23.7c: Mark Nielsen. Page 744: CNRI/Phototake. Page 745 (left): Pulse Picture Library/CMP Images/Phototake. Page 745 (right): Ian Boddy/Photo Researchers, Inc

Chapter 24 Fig. 24.1b: Dissection Shawn Miller, Photograph Mark Nielsen. Fig. 24.2c: Dissection Shawn Miller, Photograph Mark Nielsen. Fig. 24.2d: Dissection Shawn Miller, Photograph Mark Nielsen. Fig. 24.3d,e: Dissection Shawn Miller, Photograph Mark Nielsen. Fig. 24.3g: Dissection Shawn Miller, Photograph Mark Nielsen. Fig. 24.3f: Dissection Shawn Miller, Photograph Mark Nielsen. Fig. 24.4c: CNRI/Photo Researchers, Inc. Fig. 24.5: Dissection Shawn Miller, Photograph Mark Nielsen. Fig. 24.6b: Dissection Shawn Miller, Photograph Mark Nielsen. Fig. 24.6c: Dissection Shawn Miller, Photograph Mark Nielsen. Fig. 24.6d: Dissection Shawn Miller, Photograph Mark Nielsen. Fig. 24.6e: Courtesy Michael Ross, University of Florida. Fig. 24.7: Dissection Shawn Miller, Photograph Mark Nielsen. Fig. 24.8f: Dissection Shawn Miller, Photograph Mark Nielsen. Fig. 24.8g: Dissection Shawn Miller, Photograph Mark Nielsen. Fig. 24.10b: Biophoto Associates/Photo Researchers, Inc. Fig. 24.10c Biophoto Associates/Photo Researchers, Inc. Page 771: Terry Williams/The Image Bank/ Getty Images. Page 775: Juergen Berger/Photo Researchers, Inc. Page 757: BSIP/Phototake.

Chapter 25 Fig. 25.1b: Dissection Shawn Miller, Photograph Mark Nielsen. Fig. 25.3e: Dissection Shawn Miller, Photograph Mark Nielsen. Fig. 25.3f: Dissection Shawn Miller, Photograph Mark Nielsen. Fig. 25.5b: Mark Nielsen. Fig. 25.5c: Dissection Shawn Miller, Photograph Mark Nielsen. Fig. 25.7c: Courtesy Dr. D. Gentry Steele. Fig. 25.7d: H. Hubatka/Maritus Gnbj/Phototake. Fig. 25.8b: Dissection Shawn Miller, Photograph Mark Nielsen. Fig. 25.8a: Mark Nielsen. Fig. 25.10b Dissection Shawn Miller, Photograph Mark Nielsen. Fig. 25.11c: Mark Nielsen. Fig. 25.12d: Dissection Shawn Miller, Photograph Mark Nielsen. Fig. 25.12e: Dissection Shawn Miller, Photograph Mark Nielsen. Fig. 25.13b: Dissection Shawn Miller, Photograph Mark Nielsen. Fig. 25.14c: Mark Nielsen. Fig. 25.16b: Dissection Shawn Miller, Photograph Mark Nielsen. Fig. 25.18a: Fred E. Hossler/Visuals Unlimited. Fig. 25.18b: G. W. Willis/Visuals Unlimited. Fig. 25.18c: Courtesy Michael Ross, University of Florida. Fig. 25.18d: Courtesy Michael Ross, University of Florida. Fig. 25.19c: Dissection Shawn Miller, Photograph Mark Nielsen. Fig. 25.20c: Courtesy Michael Ross, University of Florida. Fig. 25.20d: Courtesy Michael Ross, University of Florida. Page 814: David M. Martin/Photo Researchers, Inc.

Chapter 26 Fig. 26.2c: Dissection Shawn Miller, Photograph Mark Nielsen. Fig. 26.1b: Dissection Shawn Miller, Photograph Mark Nielsen. Fig. 26.1c: Dissection Shawn Miller, Photograph Mark Nielsen. Fig. 26.3b: Dissection Shawn Miller, Photograph Mark Nielsen. Fig. 26.4c: Mark Nielsen. Fig. 26.6b: Dennis Strete. 26.8b Courtesy Michael Ross, University of Florida. Fig. 26.9b: Dissection Shawn Miller, Photograph Mark Nielsen. Fig. 26.10: Mark Nielsen. Fig. 26.11: Mark Nielsen. Page 835: BSIP/Phototake. Page 836: Ian Hooton/Photo Researchers, Inc.

Chapter 27 Fig. 27.1b: Dissection Shawn Miller, Photograph Mark Nielsen. Fig. 27.3b: Dissection Shawn Miller, Photograph Mark Nielsen. Fig. 27.3c: Dissection Shawn Miller, Photograph Mark Nielsen. Fig. 27.4a: Mark Nielsen. Fig. 27.7a 853: Mark Nielsen. Fig. 27.7b: Mark Nielsen. Fig. 27.8a: Mark Nielsen. Fig. 27.8b: Mark Nielsen. Fig. 27.9b: Dissection Shawn Miller, Photograph Mark Nielsen. Fig. 27.10c: Dissection Shawn Miller, Photograph Mark Nielsen. Fig. 27.11b: Dissection Shawn Miller, Photograph Mark Nielsen. Fig. 27.12b: Dissection Shawn Miller, Photograph Mark Nielsen. Fig. 27.13b: Courtesy Michael Ross, University of Florida. Fig. 27.14e: Mark Nielsen. Fig. 27.14f: Mark Nielsen. Fig. 27.13c: Claude Edelmann/Photo Researchers, Inc. Fig. 27.16b: Dissection Shawn Miller, Photograph Mark Nielsen. Fig. 27.17a: Mark Nielsen. Fig. 27.17b: Mark Nielsen. Fig. 27.18a: Courtesy Michael Ross, University of Florida. Fig. 27.18b: Courtesy Michael Ross, University of Florida. Fig. 27.20a: Courtesy Michael Ross, University of Florida. Fig. 27.20b Mark Nielsen. Fig. 27.22b: Dissection Shawn Miller, Photograph Mark Nielsen. Fig. 27.23c: Dissection Shawn Miller, Photograph Mark Nielsen. Fig. 27.24a: Mark Nielsen. Fig. 27.24b: Mark Nielsen. Page 878: ISM/Phototake. Page 880: National Cancer Institute/Photo Researchers, Inc. Page 881 (left): A.M. Siegelman/Visuals Unlimited. Page 881 (right): Dennis Kunkel/Visuals Unlimited. Page 882 (left): Institut Pasteur/Photo Researchers, Inc. Page 882 (right): Science Vu/NCI/Visuals Unlimited.

INDEX

The letter *t* following a page number denotes a table; *f* denotes a figure.

EPONYMS USED IN THIS TEXT

An **eponym** is a term that includes reference to a person's name; for example, you may be more familiar with *Achilles tendon* than you are with the technical, but correct term *calcaneal tendon*. Because eponyms remain in frequent use, this glossary has been prepared to indicate which currect terms have been used to replace eponyms in this book. In the body of the text eponyms are cited in parentheses, immediately following the correct terms where they are used for the first time in a chapter or later in the book. In addition, although eponyms are included in the index, they have been cross-referenced to their currect terminology.

EPONYM	CURRENT TERMINOLOGY
Achilles tendon	calcaneal tendon
Adam's apple	thyroid cartilage
ampulla of Vater (VA-ter)	hepatopancreatic ampulla
Bartholin's (BAR-tō-linz) gland	greater vestibular gland
Billroth's (BIL-rōtz) cord	splenic cord
Bowman's (BŌ-manz) capsule	glomerular capsule
Bowman's (BŌ-manz) gland	olfactory gland
Broca's (BRŌ-kaz) area	motor speech area
Brunner's (BRUN-erz) gland	duodenal gland
bundle of His (HISS)	atrioventricular (AV) bundle
canal of Schlemm (SHLEM)	scleral venous sinus
circle of Willis (WIL-is)	cerebral arterial circle
Cooper's (KOO-perz) ligament	suspensory ligament of the breast
Cowper's (KOW-perz) gland	bulbourethral gland
crypt of Lieberkühn (LĒ-ber-kūn)	intestinal gland
duct of Rivinus (re-VĒ-nus)	lesser sublingual duct
duct of Santorini (san'-tō-RĒ-nē)	accessory duct
duct of Wirsung (VĒR-sung)	pancreatic duct
end organ of Ruffini (roo-FĒ-nē)	type II cutaneous mechanoreceptor
Eustachian (ū-STĀ-kē-an)	auditory tube
Fallopian (fal-LŌ-pē-an) tube	uterine tube
gland of Littré (LĒ-tra)	urethral gland
gland of Zeis (ZĪS)	sebaceous ciliary gland
Golgi (GOL-jē) tendon organ	tendon organ
Graafian (GRAF-ē-an) follicle	mature ovarian follicle
Hassall's (HAS-alz) corpuscle	thymic corpuscle
Haversian (ha-VĒR-shun) canal	central canal
Haversian (ha-VĒR-shun) system	osteon
Heimlich (HĪM-lik) maneuver	abdomial thrust maneuver
interstitial cell of Leydig (LĪ-dig)	interstitial endocrinocyte
islet of Langerhans (LANG-er-hanz)	pancreatic islet
Kupffer's (KOOP-ferz) cell	stellate reticuloendothelial cell

EPONYM	CURRENT TERMINOLOGY
loop of Henle (HEN-lē)	loop of the nephron
Luschka's (LUSH-kaz) aperture	lateral aperture
Magendie's (ma-JEN-dēz) aperture	median aperture
Malpighian (mal-PIG-ē-an) corpuscle	splenic nodule
Meibomian (mī-BŌ-mē-an) gland	tarsal gland
Meissner's (MĪS-nerz) corpuscle	corpuscle of touch
Merkel's (MER-kelz) disc	tactile disc
Müller's (MIL-erz) duct	paramesonephric duct
Nissl (NIS-l) bodies	chromatophilic substances
organ of Corti (KOR-tē)	spiral organ
Pacinian (pa-SIN-ē-an) corpuscle	lamellated corpuscle
Peyer's (PĪ-erz) patch	aggregated lymphatic follicle
plexus of Auerbach (OW-er-bak)	myenteric plexus
plexus of Meissner (MĪS-ner)	submucosal plexus
pouch of Douglas	rectouterine pouch
Rathke's (rath-KĒZ) pouch	hypophyseal pouch
Sertoli (ser-TŌ-lē) cell	sustentacular cell
Sharpey's (SHAR-pēz) fiber	perforating fiber
Sheath of Schwann (SCHVON)	neurolemma
Skene's (SKĒNZ) gland	paraurethral gland
sphincter of Oddi (OD-dē)	sphincter of the hepatopancreatic ampulla
Stensen's (STEN-senz) duct	parotid duct
Volkmann's (FŌLK-manz) canal	perforating canal
Wernicke's (VER-ni-kēz) area	auditory association area
Wharton's (HWAR-tunz) duct	submandibular duct
Wharton's (HWAR-tunz) jelly	mucous connective tissue
Wormian (WER-mē-an) bone	sutural bone